Micro-Econometrics

Second Edition

Myoung-jae Lee

Micro-Econometrics

Methods of Moments and Limited Dependent Variables

Second Edition

 Springer

Myoung-jae Lee
Department of Economics
Korea University
Anam-dong, Sungbuk-gu
Seoul, Korea 136-701

GAUSS is a trademarks of Aptech Systems, Inc.
STATA® and the STATA® logo are registered trademarks of StataCorp LP.

ISBN 978-0-387-95376-2   e-ISBN 978-0-387-68841-1
DOI 10.1007/b60971
Springer New York Dordrecht Heidelberg London

Library of Congress Control Number: 2009935059

Springer is part of Springer Science+Business Media (www.springer.com)

To my "Little Women"
Hyun-joo, Hyun and Young,
full of life, love and the cool

PREFACE

When I wrote the book *Methods of Moments and Semiparametric Econometrics for Limited Dependent Variable Models* published from Springer in 1996, my motivation was clear: there was no book available to convey the latest messages in micro-econometrics. The messages were that most econometric estimators can be viewed as method-of-moment estimators and that inferences for models with limited dependent variables (LDV) can be done without going fully parametric.

Time has passed and there are now several books available for the same purpose. These days, methods of moments are the mainstay in econometrics, not just in micro-, but also in macro-econometrics. Many papers have been published for semiparametric methods and LDV models. I, myself, learned much over the years since 1996, so much so that my own view on what should be taught, and how, has changed much. Particularly, my exposure to the "sample selection" and "treatment effect" literature has changed the way I look at econometrics now. When I set out to write the second edition of the 1996 book, these changes prompted me to re-title, reorganize, and re-focus the book.

This book, or the second edition of the book from Springer in 1996, differs greatly from the 1996 book in three aspects. First, I tried to write the book more as a textbook than as a monograph, so that the book can be used as a first year textbook in graduate econometrics courses. Second, differently from the 1996 book, many empirical examples have been added and estimators that work well in practice are given more coverage than the others. Third, the literature have been updated, or at least, the relevant new papers have been cited so that the reader can consult them if he/she desires so. These changes resulted in more than doubling the book length.

One may classify econometrics into two: micro-econometrics dealing with individual data, and macro-econometrics dealing with (aggregate) time-series data. Micro-econometrics may be further classified into "cross-section micro-econometrics" and "panel-data micro- econometrics"; an analogous classification can be done for macro-econometrics. In 2002, I published a book entitled *Panel Data Econometrics: Methods of Moments and Limited Dependent Variables* from Academic Press; for me, this leaves "*cross-section micro-econometrics*" to cover in micro-econometrics, which is what this book is mainly about, although panel data models are also examined occasionally.

One of the "buzz word" in micro-econometrics these days is "treatment effect." This topic has been studied extensively in epidemiology and medical science as well as in some social science disciplines. Treatment effect framework is, in fact, nothing but "switching regression" in micro-econometrics that was popular some time ago: the effect of a binary treatment is of interest, and if there is any treatment effect, we get to see two different (i.e., switching) regimes depending on the treatment. In 2005, I published a book entitled

Micro-Econometrics for Policy, Program, and Treatment Effects from Oxford
University Press. Hence, despite its prominence, treatment effect will be discussed at minimum, if at all, in this book.

Closely related to treatment effect is "sample selection" where the sample
at hand comes from only one regime while our interest is on both or on the
"averaged" regime. I am planning to write a book on sample selection in
the near future, and thus the coverage of sample selection in this book will
not be extensive. Sample selection is a fairly well-confined topic in micro-
econometrics, and the non-extensive coverage in this book would not distort
the overall picture of micro-econometrics.

The book consists of three parts in the main text, with each part having a number of chapters, and three appendices. The first part (Chapters
1 and 2) in the main text is for methods of moments for linear models, the
second part (Chapter 3–6) is for nonlinear models and parametric methods
for LDV models, and the third part (Chapter 7–9) is for semiparametric and
nonparametric methods. Appendix I contains one section on mathematical
and statistical backgrounds, and eight more sections of appendices for Chapters 2–9. Appendix II has further supporting materials. Both appendices are
technical, digressive or tentative, and Appendix II is more so than Appendix
I in this regard. Most things the reader may feel missing while reading the
main text can be found in the appendices, although what is available in the
appendix is not specifically mentioned in the main text. Some interesting
topics are put in the appendices to avoid lengthening the main text too much
and thus discouraging the reader prematurely.

Appendix III provides some GAUSS programs. I tried to select only
simple and numerically stable (i.e., reliable) programs. All programs use simulated data. Although I wrote this book so that the readers can write their
own programs, STATA commands are occasionally referred to, in case the
reader may think that the procedure under consideration is difficult to implement and not available in ready-made econometric packages.

As in my other books, small sample issues and matters of "second order
importance" will not be discussed much, because econometricians will be
making mistakes of large magnitude, if any. With this being the case, paying
attention to small sample improvement and low-order precision seems not so
meaningful. Of course, ideally, one should avoid mistakes of both large and
small magnitudes, but saying that would be ignoring econometricians' budget
and time constraints; politicians might feel comfortable saying that, but not
most economists.

Some glaring omissions in this book's coverage include weak instruments, factor analysis, stochastic frontiers, measurement errors (or errors
in variables), semiparametric efficiency, auction-related econometrics, spatial
dependence, demand system analysis, sampling, and missing data and imputation which are closely related to sample selection. Also it would be nicer
to have more detailed coverages of duration analysis, multinomial choices,
"bandwidth-dependent" semiparametric methods for LDV models, and so
on. All of these require much more time and efforts on my side, and cover-

ing them would mean this book not seeing the daylight for another several years—perhaps next time.

The target audience of this book are graduate students and researchers. The entire book may be covered in two to four semesters—one semester for each part plus the appendices—but covering essential topics selectively while omitting the optional starred topics (and some others) may be done in two semesters. Most estimators and tests have been tried with real or simulated data except some in the appendices. The reader will find intuitions for how estimators/tests work as well as various tips for hand-on experiences. About empirical examples in this book, it would be ideal to choose the "best" empirical examples for a given estimator/test. But, unfortunately, my time constraints prevented me from doing that; rather, most examples were chosen more or less "randomly"—i.e., I happened to run into, or just remember, the example when the topic was written about.

In this book, theoretically oriented readers will find an overview on micro-econometrics, and applied researchers will find helpful informations on how to apply micro-econometric techniques; there will be something for everybody—at least that is what I hope. The reader may also want to consult other good books with micro-econometric focus such as Wooldridge (2002), Cameron and Trivedi (2005, 2009), and Green (2007). Compared with these books, the theoretical coverage of this book is relatively at a higher level with a semi-(non) parametric bent.

I am grateful to the Springer Statistics Editor John Kimmel for his patience while this project was dragging on for eight-plus years after the initial talk. I am also grateful to the anonymous reviewers for their comments which led to substantial improvements and re-organizations of the book. Juaõ Santos-Silva provided valuable feedbacks on many occasions, and Jing-young Choi helped me much by proof-reading most parts of the book. Also Sang-hyeok Lee, Jong-hun Choi and Young-min Ju proof-read various chapters and gave me comments. I should admit, however, that I could not incorporate all the comments/feedbacks due to the book-length/time constraints, and also due to the fact that making too many changes near the final stage is a rather risky thing to do. Without implicating any reviewer or anybody for that matter, I will be solely responsible for any errors in the book.

REMARKS ON EXPRESSIONS AND NOTATIONS

Many acronyms will be used in lower/upper case letters: "rv" for random variable, "cdf" or "df" for (cumulative) distribution function, "rhs" for right-hand side, "lhs" for left-hand side, "dof" for degree of freedom, "wrt" for "with respect to", "cov" for covariance, "cor" for correlation, and so on.

For matrix A, "p.d." (n.d.) stands for "positive definite" (negative definite) and "p.s.d." (n.s.d.) stands for "positive semidefinite" (negative semidefinite). $tr(A)$ denotes its trace, and $|A|$ (or $det(A)$) denotes its determinant; $||A||$ will be then the absolute value of the determinant. But sometimes, $|A|$ or $||A||$ may mean the matrix norm $\{tr(A'A)\}^{1/2}$. For matrices $a_1, ..., a_M$, "$diag(a_1, ..., a_M)$" is the block diagonal matrix with $a_1, ..., a_M$ along the diagonal.

The notation "$\to^p$" means "convergence in probability", and "$\to^{ae}$" or "$\to^{as}$" means "convergence almost surely (a.s.)" or "convergence almost everywhere (a.e.)". The notation "$\rightsquigarrow$" denotes "convergence in distribution (or in law)", and "$\sim$" denotes the distribution of a rv; e.g., "$x \sim N(0,1)$" means that x follows the standard normal distribution. We will also use "$x \sim (\mu, \sigma^2)$" to mean that $E(x) = \mu$ and $V(x) = \sigma^2$ without its distribution specified. Frequently, ϕ and Φ will be used to denote the $N(0,1)$ density and distribution function, respectively. Uniform distribution on $[a, b]$ is denoted as $U[a, b]$, exponential distribution with parameter θ is denoted as $Expo(\theta)$, and Poission distribution with parameter λ is denoted as $Poi(\lambda)$. Other distributions are often denoted analogously; e.g., $Weibull(\alpha, \theta)$.

In many textbooks, an uppercase letter and its lowercase letter are used to denote, respectively, a rv and its realized value. In this case, for a rv Y with distribution function $F(y) \equiv P(Y \le y)$ and density $f(y)$, we have $E\{g(Y)\} = \int g(y)f(y)dy$ for a function $g(\cdot)$. But this distinction between Y and y will not be followed in most parts of this book, because upper case letters are frequently used to denote matrices in this book. A downside of not following the uppercase/lowercase convention can be seen in $E\{g(y)\} = \int g(y)f(y)dy$ where y in $E\{g(y)\}$ is a rv but y in $\int g(y)f(y)dy$ is just an integration dummy—$\int g(z)f(z)dz$ would mean just the same thing. In most cases, it will be clear from the context whether y is a rv or not. But if necessary, to avoid this kind of confusion, we may also write $\int g(y_o)f(y_o)dy_o$. Also from the given context, it will be clear whether $F(y)$ means $P(Y \le y)$ with Y random and y fixed, or $F(\cdot)$ taken on a rv y; if the meaning is not clear, we may write $F(y_o)$ for $P(y \le y_o)$ where y is a rv.

The conditional distribution function for $P(y \le y_o | x = x_o)$ is denoted as $F_{y|x}(y_o|x_o)$, $F_{y|x=x_o}(y_o)$, or $F_{y|x_o}(y_o)$. But if not interested in any particular values of y_o and x_o, we may just write $F_{y|x}(y|x)$, $F_{y|x}(y)$, or $F(y|x)$. The corresponding density function will be denoted as $f_{y|x}(y|x)$, $f_{y|x}(y)$, or $f(y|x)$, respectively. In these cases, y and x in parentheses are not rv's, but stand for some values that those rv's can take (just to indicate that F and f are for those rv's). $E_x(\cdot)$ and $E_{y|x}(\cdot)$ denote that the expected value is taken for x and $y|x$, respectively. $Med(y|x)$ and $Mode(y|x)$ denote the conditional median and mode, respectively. $Q_\alpha(y|x)$ (or $q_\alpha(y|x)$) denotes the conditional αth quantile. The independence between two random vectors x and y is denoted as $x \amalg y$, and the conditional independence between x and y given z is denoted as $x \amalg y|z$. The 'indicator function' $1[\cdot]$ is defined as $1[A] = 1$ if A holds and 0 otherwise. The 'sign function' $sgn(a) \equiv 2 \times 1[a \geq 0] - 1$ denotes the sign of a: $sgn(a) = 1$ if $a \geq 0$ and -1 if $a < 0$. Sometimes the sign function may be defined such that it becomes 0 or -1 when $a = 0$.

When $E(z)$ is used, it is implicitly assumed that $E(z) < \infty$; when $E^{-1}(z)$ is used (a shorthand for $\{E(z)\}^{-1}$), it is also assumed that $E(z)$ is invertible. Most vectors in this book are column vectors, and for a $m \times 1$ vector function $g(b)$ where b has dimension $k \times 1$, its first derivative matrix g_b has dimension $k \times m$; in comparison, $g_{b'} \equiv g'_b$ is a $m \times k$ matrix. R^k denotes the k-dimensional Euclidean space, and $R = R^1$ denotes the real space; $|\cdot|$ denotes the Euclidean norm in most cases. For a function $g(\cdot)$, "increasing" means "non-decreasing," and "strictly increasing" means increasing without the equality; the analogous usages hold "decreasing" and "strictly decreasing." $L(y|x)$ and $L_N(y|x)$ denote the linear projection $E(yx')E^{-1}(xx')x$ and its estimator, respectively; in comparison to $L(y|x)$, $E(y|x)$ may be called the (nonlinear) projection and $E_N(y|x)$ denotes an estimator for $E(y|x)$.

Since this book is mainly for cross-section micro-econometrics, unless otherwise noted, we will assume that data, say $z_1, z_2, ..., z_N$ from N subjects (individuals), are (independent and identically distributed) from a common distribution. The individuals will be indexed by $i = 1, ..., N$, and we will often drop the subscript i to write z_i just as z, if not interested in any particular subject. Hence, when $z_i = (z_{i1}, ..., z_{im})'$ is an "m-vector" (i.e., $m \times 1$ vector), its mth component z_{im} may be denoted as z_m with i omitted.

CONTENTS

CHAPTER 1
METHODS OF MOMENTS FOR SINGLE LINEAR EQUATION MODELS

Method-of-moment (MOM) estimator for single linear equation models is introduced here, whereas MOM for multiple linear equations will be examined in the next chapter. Least squares estimator (LSE) is reviewed to estimate the conditional mean (i.e., regression function) in a model with *exogenous* regressors. Not just conditional mean, but conditional variance also matters, and it is discussed under the headings "heteroskedasticity/ homoskedasticity" and generalized LSE (GLS). Instrumental variable estimator (IVE) and generalized method-of-moment (GMM) estimator allow *endogenous* regressors; IVE and GMM include LSE as a special case. Endogeneity matters greatly for policy variables, as the "ceteris paribus" effect of a policy is of interest but endogenous regressors lead to biased effect estimates. In addition to MOM estimation, testing linear hypotheses with "Wald test" is studied.

1 Least Squares Estimator (LSE)

This section introduces standard linear models with exogenous regressors, and then reviews least squares estimator (LSE) for regression functions, which is a "bread-and-butter" estimator in econometrics. Differently from the conventional approach, however, LSE will be viewed as a MOM. Also differently from the conventional approach, we will adopt a large sample framework and invoke only a few assumptions.

1.1 LSE as a Method of Moment (MOM)

1.1.1 Linear Model

Consider a linear model

$$y_i = x_i'\beta + u_i, \quad i = 1, ..., N$$

where x_i is a $k \times 1$ "regressor" vector with its first component being 1 (i.e., $x_i = (1, x_{i2}, ..., x_{ik})'$), $\beta \equiv (\beta_1, ..., \beta_k)'$ is a $k \times 1$ parameter vector reflecting effects of x_i on y_i, and u_i is an "error" term. In β, β_1 is called the "intercept" whereas $\beta_2, ..., \beta_k$ are called the "slopes." The left-hand side variable y_i is the "dependent" or "response" variable, whereas components of x_i are

Myoung-jae Lee, *Micro-Econometrics*, DOI 10.1007/b60971_1,
© Springer Science+Business Media, LLC 2010

"regressors," "explanatory variables," or "independent variables." Think of x_i as a collection of the observed variables affecting y_i through $x_i'\beta$, and u_i as a collection of the unobserved variables affecting y_i. Finding β with data (x_i', y_i), $i = 1, ..., N$, is the main goal in regression analysis. Assume that (x_i', y_i), $i = 1, ..., N$, are *independent and identically distributed (iid)* unless otherwise noted, which means that each (x_i', y_i) is an independent draw from a common probability distribution. We will often omit the subscript i indexing individuals.

The linear model is linear in β, but not necessarily linear in x_i, and it is more general than it looks. For instance, x_3 may be x_2^2, in which case $\beta_2 x_2 + \beta_3 x_2^2$ depicts a quadratic relationship between x_2 and y: the "effect" of x_2 on y is then $\beta_2 + 2\beta_3 x_2$—the first derivative of $\beta_2 x_2 + \beta_3 x_2^2$ with respect to (wrt) x_2. For instance, with y monthly salary and x_2 age, the effect of age on monthly salary may be quadratic: going up to a certain age and then declining after. Also x_4 may be $x_2 x_3$, in which case

$$\beta_2 x_2 + \beta_3 x_3 + \beta_4 x_2 x_3 = (\beta_2 + \beta_4 x_3) x_2 + \beta_3 x_3 :$$

the effect of x_2 on y is $\beta_2 + \beta_4 x_3$. For instance, x_3 can be education level: the effect of age on monthly salary is not the constant slope β_2, but $\beta_2 + \beta_4 x_3$ which varies depending on education level. The display can be written also as $\beta_2 x_2 + (\beta_3 + \beta_4 x_2) x_3$ to be interpreted analogously. The term $x_2 x_3$ is called the *interaction term* between x_2 and x_3, and its coefficient is the interaction effect. By estimating β with data (x_i', y_i), $i = 1, ..., N$, we can find these effects.

1.1.2   LSE and Moment Conditions

The *least squares estimator (LSE)* for β is obtained by minimizing

$$\frac{1}{N} \sum_i (y_i - x_i' b)^2$$

wrt b, where $y_i - x_i' b$ can be viewed as a "prediction error" in predicting y_i with the linear function $x_i' b$. LSE is also often called *ordinary LSE (OLS)*, relative to "generalized LSE" to appear later.

The first-order condition for the LSE b_{lse} is

$$\frac{1}{N} \sum_i x_i (y_i - x_i' b_{lse}) = 0 \iff \frac{1}{N} \sum_i x_i y_i = \frac{1}{N} \sum_i x_i x_i' \cdot b_{lse}.$$

Assuming that $N^{-1} \sum_i x_i x_i'$ is invertible, solve this for b_{lse} to get

$$b_{lse} = \left(\frac{1}{N} \sum_i x_i x_i' \right)^{-1} \cdot \frac{1}{N} \sum_i x_i y_i = \left(\sum_i x_i x_i' \right)^{-1} \cdot \sum_i x_i y_i.$$

The *residual* $\hat{u}_i \equiv y_i - x_i' b_{lse}$, which is an estimator for u_i, has zero sample mean and zero sample covariance with the regressors due to the first-order condition:

$$\frac{1}{N}\sum_i x_i\,(y_i - x_i'b_{lse}) = \left(\frac{1}{N}\sum_i \hat{u}_i, \ \frac{1}{N}\sum_i x_{i2}\hat{u}_i, \ ..., \ \frac{1}{N}\sum_i x_{ik}\hat{u}_i\right)' = 0.$$

Instead of minimizing $N^{-1}\sum_i(y_i - x_i'b)^2$, LSE can be motivated directly from a moment condition. Observe that the LSE first-order condition at $b = \beta$ is $N^{-1}\sum_i x_i u_i = 0$, and its population version is

$$E(xu) = 0 \iff \begin{bmatrix} E(u) \\ E(x_2 u) \\ \vdots \\ E(x_k u) \end{bmatrix} = \begin{bmatrix} 0 \\ 0 \\ \vdots \\ 0 \end{bmatrix}$$

$$\iff \quad E(u) = 0, \ COV(x_j, u) = 0 \ (\text{or } COR(x_j, u) = 0), \ j = 2, ..., k$$

as $COV(x_j, u) = E(x_j u) - E(x_j)E(u)$, where COV and COR stand for covariance and correlation, respectively.

Replacing u with $y - x'\beta$ yields

$$E\{x(y - x'\beta)\} = 0 \iff E(xy) = E(xx')\beta$$

which is a restriction on the joint distribution of (x', y). Assuming that $E(xx')$ is invertible, we get

$$\beta = \{E(xx')\}^{-1} \cdot E(xy).$$

LSE b_{lse} is just a *sample analog* of this expression of β, obtained by replacing $E(xx')$ and $E(xy)$ with their sample versions $N^{-1}\sum_i x_i x_i'$ and $N^{-1}\sum_i x_i y_i$. Instead of identifying β by minimizing the prediction error, here β is identified by the "information" (i.e., the assumption) that the observed x is "orthogonal" to the unobserved u.

For any $k \times 1$ constant vector γ,

$$\gamma' E(xx')\gamma = E(\gamma' xx'\gamma) = E\{(x'\gamma)'(x'\gamma)\} = E\{(x'\gamma)^2\} \geq 0.$$

Hence $E(xx')$ is positive semidefinite (p.s.d.). Assume that

$$E(xx') \text{ is of full rank.}$$

As $E(xx')$ is p.s.d., this full rank condition is equivalent to $E(xx')$ being positive definite (p.d.) and thus being invertible. Note that $E(xx')$ being p.d. is equivalent to $E^{-1}(xx')$ being p.d. where $E^{-1}(xx')$ means $\{E(xx')\}^{-1}$.

1.1.3   Zero Moments and Independence

The assumption $E(xu) = 0$ is the weakest for the LSE to be a valid estimator for β as can be seen in the next subsection. In econometrics, the following two assumptions have been used as well for LSE:

(i) $E(u|x) = 0$ { $\iff E(y|x) = x'\beta$ for the linear model}

(ii) u is independent of x and $E(u) = 0$.

Note that $E(u|x) = 0$ implies $E(u) = E\{E(u|x)\} = 0$. For the three assumptions, the following implications hold:

independence of u from x and $E(u) = 0 \implies E(u|x) = 0 \implies E(xu) = 0$;

the last implication holds because $E(xu) = E\{xE(u|x)\} = 0$.

We will use mainly $E(u|x) = 0$ from now on unless otherwise mentioned, because $E(xu) = 0$ would not take us much farther than LSE while the independence is too strong to be realistic. The regressor vector x is often said to be *exogenous* if any one of the three conditions holds. The function $E(y|x) = x'\beta$ is called the (*mean*) *regression function*, which is nothing but a location measure in the distribution of $y|x$. We can also think of other location measures, say quantiles, in the distribution of $y|x$, which then yield "quantile regression functions."

In β, the intercept β_1 shows the level of y, and each slope represents the effect of its regressor on $E(y|x)$ while controlling for (i.e., holding constant) the other regressors. This may be understood in

$$\frac{\partial E(y|x)}{\partial x_j} = \beta_j, \quad j = 1, ..., k.$$

The condition of "holding the other regressors constant"—reflected here with the partial differentiation symbol ∂—may be better understood when "partial regression" is explained later. The formal causal interpretation of regarding x_j as a cause and β_j as its effect on the response y requires a little deeper reasoning (see, e.g., Lee, 2005, and the references therein). This is because LSE is a MOM which depends only on the covariances of the variables involved, and the covariances per se do not designate any variable as a cause or the response.

1.2   Asymptotic Properties of LSE

As $N \to \infty$, the sample will be "close" to the population, and we would want b_{lse} to converge to β in some sense. This is necessary for b_{lse} to be a "valid" estimator for β. Going further, to be a "good" estimator for β, b_{lse} should converge fast to β. For instance, both N^{-1} and N^{-2} converge to 0, and they are valid "estimators" for 0, but N^{-2} is better than N^{-1} because N^{-2} converges to 0 faster. This subsection discusses these issues in the names "consistency" and "asymptotic distribution." The first-time readers may want to only browse this subsection instead of reading every detail, to come back later when better motivated theoretically. The upshot of this subsection is the display (*) showing the asymptotic distribution of b_{lse} (with its variance estimator in (*")) and its practical version (*') showing that b_{lse} will degenerate (i.e., converge) to β as $N \to \infty$. The main steps will also appear in the instrumental variable estimator (IVE) section.

1.2.1   LLN and LSE Consistency

A *law of large numbers (LLN)*, for an iid random variable (rv) sequence $z_1, ..., z_N$ with $E(z) < \infty$, holds that

$$\frac{1}{N} \sum_i z_i \to^p E(z) \text{ as } N \to \infty$$

where "$\to^p$" denotes convergence in probability:

$$P\left(|\frac{1}{N} \sum_i z_i - E(z)| < \varepsilon\right) \to 1 \text{ as } N \to \infty \text{ for any constant } \varepsilon > 0;$$

(the estimator) $\bar{z}_N \equiv N^{-1} \sum_i z_i$ is said to be "*consistent*" for (the parameter) $E(z)$.

If $\bar{z}_N$ is a matrix, the LLN applies to each component. This element-wise convergence in probability of $\bar{z}_N$ to $E(z)$ is equivalent to $|\bar{z}_N - E(z)| \to^p 0$ where $|A| \equiv \{tr(A'A)\}^{1/2}$ for a matrix A—the usual matrix norm—in the sense that the element-wise convergence implies the norm convergence and vice versa. As "$\bar{z}_N - E(z) \to^p 0$" means that the difference between $\bar{z}_N$ and $E(z)$ converges to 0 in probability, for two rv matrix sequences W_N and M_N, "$W_N - M_N \to^p 0$" (or $W_N \to^p M_N$) means that the difference between the two rv matrix sequences converges to zero in probability.

Substitute $y_i = x_i'\beta + u_i$ into b_{lse} to get

$$b_{lse} = \beta + \left(\frac{1}{N} \sum_i x_i x_i'\right)^{-1} \frac{1}{N} \sum_i x_i u_i.$$

Clearly, $b_{lse} \neq \beta$ due to the second term on the right-hand side (rhs) which shows that each $x_i u_i$ contributes to the deviation $b_{lse} - \beta$. Using the LLN, we have

$$\frac{1}{N} \sum_i x_i u_i \to^p E(xu) = 0 \text{ and } \frac{1}{N} \sum_i x_i x_i' \to^p E(xx').$$

Substituting these into the preceding display, we can get $b_{lse} \to^p \beta$, but we need to deal with the inverse: for a square random matrix W_N, when $W_N \to^p W$, will W_N^{-1} converge to W^{-1} in probability?

It is known that, for a rv matrix W_N and a constant matrix W_o,

$$f(W_N) \to^p f(W_o) \text{ if } W_N \to^p W_o \text{ and } f(\cdot) \text{ is continuous at } W_o.$$

The inverse $f(W) = W^{-1}$ of W, when it exists, is the adjoint of W divided by the determinant $\det(W)$. Because $\det(W)$ is a sum of products of elements of W and the adjoint consists of determinants, both $\det(W)$ and the adjoint are continuous in W, which implies that W^{-1} is continuous in W (see, e.g.,

Schott, 2005). Thus, W^{-1} is continuous at W_o so long as W_o^{-1} exists, and using the last display, we get $A_N^{-1} \to^p A^{-1}$ if $A_N \to^p A$ so long as A^{-1} exists; note that A_N^{-1} exists for a large enough N because $\det(A_N) \neq 0$ for a large enough N. Hence,

$$\left(\frac{1}{N} \sum_i x_i x_i' \right)^{-1} \to^p E^{-1}(xx') < \infty \quad \text{as } N \to \infty.$$

Therefore, b_{lse} is β plus a product of two terms, one consistent for a zero vector and the other consistent for a bounded matrix; thus the product is consistent for zero, and we have $b_{lse} \to^p \beta$: b_{lse} *is consistent for* β.

1.2.2   CLT and $\sqrt{N}$-Consistency

For the asymptotic distribution of the LSE, a *central limit theorem (CLT)* is needed that, for an iid random vector sequence $z_1, ..., z_N$ with finite second moments,

$$\frac{1}{\sqrt{N}} \sum_i \{z_i - E(z)\} \rightsquigarrow N\left(0,\, E\left[\{z - E(z)\}\{z - E(z)\}'\right]\right) \quad \text{as } N \to \infty$$

where "$\rightsquigarrow$" denotes *convergence in distribution*; i.e., letting $\Psi(\cdot)$ denote the df of $N(0, E[\{z - E(z)\}\{z - E(z)\}'])$,

$$\lim_{N \to \infty} P\left\{ \frac{1}{\sqrt{N}} \sum_i \{z_i - E(z)\} \leq t \right\} = \Psi(t) \quad \forall t.$$

When $w_N \to^p 0$, it is also denoted as $w_N = o_p(1)$; "$o_p(1)$" is the probabilistic analog for $o(1)$ where $o(1)$ is a sequence converging to 0. For $\bar{z}_N$, we thus have $\bar{z}_N - E(z) = o_p(1)$. In comparison to $w_N = o_p(1)$, "$w_N = O_p(1)$" means that $\{w_N\}$ is *bounded in probability (or stochastically bounded)*—i.e., "not explosive as $N \to \infty$" (even if it does not converge to anything) in the probabilistic sense. Note that $o_p(1)$ is also $O_p(1)$. Formally, $w_N = O_p(1)$ is that, for any constant $\varepsilon > 0$, there exists a constant δ_ε such that

$$\sup_N P\{|w_N| > \delta_\varepsilon\} < \varepsilon.$$

A single rv z always satisfies $P\{|z| > \delta_\varepsilon\} < \varepsilon$, because we can capture "all but ε" probability mass by choosing δ_ε large enough. The last display means that we can capture all but ε probability mass with δ_ε for any rv in the sequence $w_1, w_2, ...$ Any random sequence converging in distribution is $O_p(1)$, which implies $N^{-1/2} \sum_i \{z_i - E(z)\} = O_p(1)$.

To understand O_p better, consider N^{-1} and N^{-2}, both of which converge to 0. Observe $N^{-1}/N^{-1} = 1$, but $N^{-1}/N^{-1+\varepsilon} = 1/N^\varepsilon \to 0$ whereas $N^{-1}/N^{-1-\varepsilon} = N^\varepsilon \to \infty$ for any constant $\varepsilon > 0$. Thus the "(fastest) convergence rate" is N^{-1} which, when divided into N^{-1}, makes the resulting ratio bounded. Analogously, the convergence rate for N^{-2} is N^{-2}. Now consider $z_N \equiv z/\sqrt{N}$ where z is a rv. Then $\sqrt{N}z_N = z = O_p(1)$ (or $z_N = O_p(1/\sqrt{N})$) because we can choose δ_ε for any constant $\varepsilon > 0$ such that

$$\sup_N P\left(|\sqrt{N}z_N| > \delta_\varepsilon\right) = \sup_N P\left(|z| > \delta_\varepsilon\right) = P\left(|z| > \delta_\varepsilon\right) < \varepsilon.$$

For an estimator a_N for a parameter α, in most cases, we have $\sqrt{N}(a_N - \alpha) = O_p(1)$: a_N is "$\sqrt{N}$-consistent." This means that $a_N \to^p \alpha$, and that the convergence rate is $N^{-1/2}$ which, when divided into $a_N - \alpha$, makes the resulting product bounded in probability. For most cases in our discussion, it would be harmless to think of the $\sqrt{N}$-consistency of a_N as $\sqrt{N}(a_N - \alpha)$ converging to a normal distribution as $N \to \infty$.

Analogously to $o(1)O(1) = o(1)$—"a sequence converging to zero" times "a bounded sequence" converges to zero—it holds that $o_p(1)O_p(1) = o_p(1)$. Likewise, $o_p(1) + O_p(1) = O_p(1)$. Slutsky Lemma shows more: if $w_N \rightsquigarrow w$ (thus $w_N = O_p(1)$) and $m_N \to^p m_o$, then

(i) $m_N w_N \rightsquigarrow m_o w$

(ii) $m_N + w_N \rightsquigarrow m_o + w$.

Slutsky Lemma (i) states that, not just the product $m_N w_N$ is $O_p(1)$, its asymptotic distribution is that of w times the constant m_o. Slutsky Lemma (ii) can be understood analogously.

1.2.3   LSE Asymptotic Distribution

Observe

$$\sqrt{N}(b_{lse} - \beta) = \left(\frac{1}{N}\sum_i x_i x_i'\right)^{-1} \cdot \frac{1}{\sqrt{N}}\sum_i x_i u_i.$$

From the CLT, we have

$$\frac{1}{\sqrt{N}}\sum_i x_i u_i \rightsquigarrow N\{0, E(xx'u^2)\}.$$

Using Slutsky Lemma (i),

if $B_N \rightsquigarrow N(0, C)$ and $A_N \to^p A$, then $A_N B_N \rightsquigarrow N(0, ACA')$.

Apply this to

$$B_N = \frac{1}{\sqrt{N}}\sum_i x_i u_i \text{ and } A_N = \left(\frac{1}{N}\sum_i x_i x_i'\right)^{-1} \to^p E^{-1}(xx')$$

to get

$$\sqrt{N}(b_{lse} - \beta) \rightsquigarrow N(0, \Omega) \quad \text{where } \Omega \equiv E^{-1}(xx')E(xx'u^2)E^{-1}(xx'): \quad (*)$$

$\sqrt{N}(b_{lse} - \beta)$ is *asymptotically normal with mean 0 and variance* Ω. Often this convergence in distribution (or "in law") of $\sqrt{N}(b_{lse} - \beta)$ is informally stated as

$$b_{lse} \sim N\left\{\beta, \frac{1}{N}E^{-1}(xx')E(xx'u^2)E^{-1}(xx')\right\} \qquad (*')$$

The asymptotic variance Ω of $\sqrt{N}(b_{lse} - \beta)$ can be estimated consistently with (this point will be further discussed later)

$$\Omega_N \equiv \left(\frac{1}{N}\sum_i x_i x_i'\right)^{-1} \left(\frac{1}{N}\sum_i x_i x_i' \hat{u}_i^2\right) \left(\frac{1}{N}\sum_i x_i x_i'\right)^{-1}. \qquad (*'')$$

Alternatively (and informally), the asymptotic variance of b_{lse} is estimated consistently with

$$\frac{\Omega_N}{N} = \left(\sum_i x_i x_i'\right)^{-1} \left(\sum_i x_i x_i' \hat{u}_i^2\right) \left(\sum_i x_i x_i'\right)^{-1}.$$

Equipped with Ω_N and the asymptotic normality, we can test hypotheses involving β as to be seen later.

1.3   Matrices and Linear Projection

It is sometimes convenient (for computation) to express b_{lse} using matrices. Define $Y \equiv (y_1, ..., y_N)'$, $U \equiv (u_1, ..., u_N)'$, and $X \equiv (x_1, ..., x_N)'$ where $x_i = (x_{i1}, ..., x_{ik})'$ so that

$$\underset{N \times k}{X} \equiv \begin{bmatrix} x_1' \\ \vdots \\ x_N' \end{bmatrix} = \begin{bmatrix} x_{11}, \ x_{12}, \ \cdots \ , \ x_{1k} \\ \vdots \\ x_{N1}, \ x_{N2}, \ \cdots \ , \ x_{Nk} \end{bmatrix};$$

the numbers below X denote its dimension. In this matrix notation, the N linear equations $y_i = x_i'\beta + u_i$, $i = 1, ..., N$, become $Y = X\beta + U$, and

$$\frac{1}{N}\sum_i (y_i - x_i'\beta)^2 = \frac{1}{N}\sum_i u_i^2 = \frac{1}{N}U'U = \frac{1}{N}(Y - X\beta)'(Y - X\beta).$$

Differentiating this, the LSE first-order condition is $N^{-1}X'(Y - Xb_{lse}) = 0$, which is also a moment condition for MOM. This yields

$$b_{lse} = (X'X)^{-1}X'Y.$$

The parts $X'X$ and $X'Y$ are the same as $\sum_i x_i x_i'$ and $\sum_i x_i y_i$, respectively. For example, with $k = 2$ and $x_{i1} = 1 \ \forall i$,

$$X'X = \begin{bmatrix} 1 & \cdots & 1 \\ x_{12} & \cdots & x_{N2} \end{bmatrix} \begin{bmatrix} 1 & x_{12} \\ \vdots & \vdots \\ 1 & x_{N2} \end{bmatrix} = \begin{bmatrix} N & \sum_i x_{i2} \\ \sum_i x_{i2} & \sum_i x_{i2}^2 \end{bmatrix},$$

$$\sum_i x_i x_i' = \sum_i \begin{pmatrix} 1 \\ x_{i2} \end{pmatrix}(1, x_{i2}) = \sum_i \begin{bmatrix} 1 & x_{i2} \\ x_{i2} & x_{i2}^2 \end{bmatrix} = \begin{bmatrix} N & \sum_i x_{i2} \\ \sum_i x_{i2} & \sum_i x_{i2}^2 \end{bmatrix}.$$

Define the $N \times N$ "*(linear) projection matrix on* X"

$$P_X \equiv X(X'X)^{-1}X'$$

to get

$$\hat{Y} \equiv Xb_{lse} = X(X'X)^{-1}X'Y = P_X Y \quad (\text{"fitted value of Y"}),$$
$$\hat{U} \equiv Y - Xb_{lse} = Y - P_X Y = Q_X Y, \quad \text{where } Q_X \equiv I_N - P_X;$$

$\hat{U} = (\hat{u}_1, ..., \hat{u}_N)'$ is the $N \times 1$ residual vector. We may think of Y comprising X and the other components. Then P_X *extracts* the X part of Y, and Q_X *removes* the X part of Y (or Q_X extracts the non-X part of Y). The fitted value $\hat{Y} = Xb_{lse}$ is the part of Y explained by X, and the residual $\hat{U}$ is the part of Y unexplained by X as clear in the decomposition

$$Y = I_N Y = P_X Y + (I_N - P_X)Y = P_X Y + Q_X Y = Xb_{lse} + \hat{U}.$$

The LSE $(X'X)^{-1}X'Y$ is called the *sample (linear) projection coefficients of* Y *on* X. The population versions of the linear projection and linear projection coefficient are, respectively,

$$x'\beta \quad \text{and} \quad \beta \equiv E^{-1}(xx')E(xy).$$

The matrices P_X and Q_X are symmetric and idempotent:

$$P'_X = P_X, \quad P_X P_X = P_X \quad \text{and} \quad Q'_X = Q_X, \quad Q_X Q_X = Q_X.$$

Also note

$$P_X X = X \quad \text{and} \quad Q_X X = 0_N :$$

extracting the X part of X gives X itself, and removing the X part of X yields 0.

Suppose we use 1 as the only regressor. Defining 1_N as the $N \times 1$ vector of 1's and denoting Q_{1_N} just as Q_1,

$$
\begin{aligned}
Q_1 Y &= \left(I_N - 1_N \left(1'_N 1_N \right)^{-1} 1'_N \right) Y = \left(I_N - 1_N \frac{1}{N} 1'_N \right) Y \\[2mm]
&= \left(I_N - \frac{1}{N} 1_N 1'_N \right) Y \\[2mm]
&= \left\{
\begin{bmatrix}
1 & 0 & \cdots & 0 \\
0 & 1 & \cdots & 0 \\
\vdots & \vdots & \ddots & \vdots \\
0 & \cdots & 0 & 1
\end{bmatrix}
- \frac{1}{N}
\begin{bmatrix}
1 & 1 & \cdots & 1 \\
1 & 1 & \cdots & 1 \\
\vdots & \vdots & \ddots & \vdots \\
1 & \cdots & 1 & 1
\end{bmatrix}
\right\}
\cdot
\begin{bmatrix}
y_1 \\
y_2 \\
\vdots \\
y_N
\end{bmatrix} \\[2mm]
&=
\begin{bmatrix}
y_1 - \bar{y} \\
y_2 - \bar{y} \\
\vdots \\
y_N - \bar{y}
\end{bmatrix}.
\end{aligned}
$$

The part $(1_N'1_N)^{-1}1_N'Y = \bar{y}$ demonstrates that the LSE with 1 as the sole regressor is just the sample mean $\bar{y}$. Q_1 may be called the "mean-deviation" or "mean-subtracting" matrix.

1.4 R^2 and Two Examples

Before we present two examples of LSE, we introduce some terminologies frequently used in practice. Recall the LSE asymptotic variance estimator $\Omega_N/N \equiv [\omega_{N,hj}]$, $h, j = 1, ..., k$; i.e., the element of Ω_N/N in row h and column j is denoted as $\omega_{N,hj}$. The *t-values (t-ratios, or z-values)* are defined as

$$\frac{b_{lse,j}}{\sqrt{\omega_{N,jj}}}, \quad j = 1, ..., k, \quad \text{where } b_{lse} = (b_{lse,1}, ..., b_{lse,k})'.$$

Since the diagonal of Ω_N/N is the asymptotic variances of $b_{lse,j}$, $j = 1, ..., k$, the jth t-value asymptotically follows $N(0,1)$ under the $H_0 : \beta_j = 0$, and hence it is a test statistic for $H_0 : \beta_j = 0$. The off-diagonal terms of Ω_N/N are the asymptotic covariances for $b_{lse,j}$, $j = 1, ..., k$, and they are used for hypotheses involving multiple parameters.

The "*standard error* (of model)" s_N and "*R-squared*" R^2 are defined as

$$s_N \equiv \left(\frac{\sum_i \hat{u}_i^2}{N-k}\right)^{1/2} \to^p SD(u),$$

$$R^2 \equiv 1 - \frac{N^{-1}\sum_i \hat{u}_i^2}{N^{-1}\sum_i(y_i - \bar{y})^2} \to^p 1 - \frac{V(u)}{V(y)} = \frac{V(x'\beta)}{V(y)}, \quad \text{as}$$

$$V(y) = V(x'\beta + u) = V(x'\beta) + V(u) \quad \text{because } COV(x'\beta, u) = 0.$$

R^2 shows the proportion of $V(y)$ that is explained by $x'\beta$, and R^2 measures the "*model fitness*." In general, the higher the R^2 is the better, because the less is buried in the unobserved u. But this statement should be qualified, because R^2 keeps increasing by adding more regressors into the model. Using fewer regressors to explain y is desirable, which is analogous to using fewer shots to hit a target.

Recall $\hat{Y} = Xb_{lse}$, $Y = \hat{Y} + \hat{U}$, and the idempotent mean-subtracting matrix Q_1 to observe

$$Q_1\hat{U} = \hat{U} \text{ (because the sample mean of } \hat{U} \text{ is already zero)},$$

$$Y'Q_1Q_1Y \left\{ = \sum_i(y_i - \bar{y})^2 \right\} = Y'Q_1Y$$

$$= (\hat{Y} + \hat{U})'Q_1(\hat{Y} + \hat{U}) = \hat{Y}'Q_1\hat{Y} + \hat{U}'\hat{U} = \hat{Y}'Q_1\hat{Y} + \sum_i \hat{u}_i^2$$

$$\text{because } \hat{Y}'Q_1\hat{U} = b_{lse}'X'Q_1\hat{U} = b_{lse}'X'\hat{U} = b_{lse}'X'Q_XY = 0 \text{ for}$$
$$X'Q_X = (Q_XX)' = 0.$$

The last line also implies $\hat{Y}'Q_1Y = \hat{Y}'Q_1(\hat{Y} + \hat{U}) = \hat{Y}'Q_1\hat{Y}$. The key point of this display is the well-known decomposition

$$(Y'Q_1Y =)\ \underbrace{\sum_i (y_i - \bar{y})^2}_{\text{total variation in } y} = \underbrace{\hat{Y}'Q_1\hat{Y}}_{\text{explained (by } x) \text{ variation}} + \underbrace{\sum_i \hat{u}_i^2}_{\text{unexplained variation}} .$$

R^2 is defined as the ratio of the explained variation to the total variation:

$$R^2 \equiv \frac{\hat{Y}'Q_1\hat{Y}}{Y'Q_1Y} = \frac{\hat{Y}'Q_1Y \cdot \hat{Y}'Q_1\hat{Y}}{\hat{Y}'Q_1Y \cdot Y'Q_1Y} = \frac{\hat{Y}'Q_1Y \cdot \hat{Y}'Q_1Y}{\hat{Y}'Q_1\hat{Y} \cdot Y'Q_1Y}$$

$$= \frac{\left\{ \sum_i (\hat{y}_i - \bar{\hat{y}})(y_i - \bar{y}) \right\}^2}{\sum_i (\hat{y}_i - \bar{\hat{y}})^2 \cdot \sum_i (y_i - \bar{y})^2} = (\text{sample correlation of } Y \text{ and } \hat{Y})^2$$

R^2 falls in $[0, 1]$, being a squared correlation.

EXAMPLE: HOUSE SALE. A data set of size 467 was collected from the State College District in Pennsylvania for year 1991. State College is a small college town with the population of about 50,000. The houses sold during the year were sampled, and the sale prices and the durations until sale since the first listing in the market were recorded.

The dependent variable is the discount (DISC) percentage defined as 100 times the natural log of list price (LP) over sale price (SP) of a house:

$$100 \cdot \ln\left(\frac{LP}{SP}\right) = 100 \cdot \ln\left(1 + \frac{LP - SP}{SP}\right) \simeq 100\left(\frac{LP - SP}{SP}\right) = \text{discount } \%.$$

LP and SP are measured in $1000. Since LP is the initial list price, given LP, explaining DISC is equivalent to explaining SP. The following is the list of regressors—the measurement units should be kept in mind: the number of days on the market until sold (T), years built minus 1900 (YR), number of rooms (ROOM), number of bathrooms (BATH), dummy for heating by electricity (ELEC), property tax in $1,000 (TAX), dummy for spring listing (L1), summer listing (L2), and fall listing (L3), sale-month interest rate in % (RATE), dummy for sale by a big broker (BIGS), and number of houses on the market divided by 100 in the month when the house is listed (SUPPLY).

In Table 1, examine only the first three columns for a while. $\ln(T)$ appears before 1, because $\ln(T)$ is different from the other regressors—it is determined nearly simultaneously with DISC—and thus needs a special attention. Judging from the t-values in "tv-het," most regressors are statistically significant at 5% level, for their absolute t-values are greater than 1.96; "tv-ho" will be used in the next subsection where the qualifiers "het" and "ho" will be explained. A longer $\ln(T)$ implies the bigger DISC: with $\partial \ln T \simeq \partial T/T$, an increase of $\partial \ln T = 1$ (i.e., 100% increase in T) means 4.6% increase in DISC, which in turn means 1% increase in T leading to

Table 1: LSE for House-Sale Discount %

	b_{lse}	tv-het (tv-ho)	b_{lse} (T)	tv-het (tv-ho) (T)
$\ln(T)$	4.60	7.76 (12.2)	0.027	8.13 (13.9)
1	−2.46	−0.23 (−0.24)	8.73	0.82 (0.86)
BATH	0.11	0.18 (0.17)	0.31	0.51 (0.51)
ELEC	1.77	2.46 (2.60)	1.84	2.67 (2.80)
ROOM	−0.18	−0.67 (−0.71)	−0.26	−0.95 (−1.04)
TAX	−1.74	−1.28 (−1.65)	−1.92	−1.49 (−1.88)
YR	−0.15	−3.87 (−5.96)	−0.15	−4.11 (−6.17)
$\ln(LP)$	6.07	2.52 (3.73)	5.71	2.51 (3.63)
BIGS	−2.15	−2.56 (−3.10)	−1.82	−2.25 (−2.72)
RATE	−2.99	−3.10 (−3.25)	−2.12	−2.31 (−2.41)
SUPPLY	1.54	1.06 (1.02)	1.89	1.36 (1.30)
s_N, R^2		6.20, 0.34		5.99, 0.39

Variable	Mean	SD
DISC	7.16	7.64
L1	0.29	0.45
L2	0.31	0.46
L3	0.19	0.39
SP	115	57.7
T	188	150
BATH	2.02	0.67
ELEC	0.52	0.50
ROOM	7.09	1.70
TAX	1.38	0.65
YR	73.0	15.1
LP	124	64.9
BIGS	0.78	0.42
RATE	9.33	0.32
SUPPLY	0.62	0.19

0.046% increase in DISC. A newer house commands the less DISC: one year newer causes 0.15% less DISC, and thus 10 year newer causes 1.5% less DISC. A higher RATE means the lower DISC (1% increase in RATE causing 2.99% DISC drop); this finding seems, however, counter-intuitive, because a higher mortgage rate means the lower demand for houses. $R^2 = 0.34$ shows that 34% of the DISC variance is explained by $x'b_{lse}$, and $s_N = 6.20$ shows that about 95% of u_i's fall in the range $\pm 1.96 \times 6.20$ if u_i's follow $N(0, V(u))$.

As just noted, 1% increase in T causes 0.046% increase in DISC. Since this may not be easy to grasp, T is used instead of $\ln(T)$ for the LSE in the last two columns of the table. The estimate for T is significant with the magnitude 0.027, meaning that 100 day increase in T leads to 2.7% DISC increase, which seems reasonable. This kind of query—whether the popular logged variable $\ln(T)$, level T, or some other function of T should be used—will be addressed later when we deal with "transformation of variables" in nonlinear models.

EXAMPLE: INTEREST RATE. As another example of LSE, we use time-series data on three month US treasury bill rates monthly from $01/1982$ to $12/1999$ ($N = 216$). To see what extent the past interest rates can explain the current one, the LSE of y_i on 1, y_{i-1} and y_{i-2} was done (since two lags are used, the sample size becomes $N = 214$) with the following result:

$$
\begin{array}{ccccccc}
y_i & = & 0.216 & + & 1.298 \cdot y_{i-1} & - & 0.337 \cdot y_{i-2}, & s_N = 0.304, \\
\text{t-values:} & & (2.05) & & (10.29) & & (-2.61) & \\
\end{array}
$$

$$R^2 = 0.980.$$

Both y_{i-1} and y_{i-2} are statistically significant; i.e., $H_0 : \beta_2 = 0$ and $H_0 : \beta_3 = 0$ are rejected. The R^2 indicates that the two past rates predict very well the current rate. Since the unit of measurements are the same across the regressors and dependent variable, there is no complication in interpreting the estimates as in the house-sale example.

One curious point is that the estimate for y_{i-2} is significantly negative and differs too much from the estimate for y_{i-1}, casting some doubt over the linear model. One reason could be the "truncation bias": the other lagged regressors $(y_{i-3}, y_{i-4}, ...)$ were omitted from the regressors to become part of u_i, which means $COR(y_{i-1}, u_i) \neq 0$ and $COR(y_{i-2}, u_i) \neq 0$, violating the basic tenet of LSE. One counter argument, however, is $COR(\hat{u}_i, \hat{u}_{i-1}) = 0.104$ which means that $COR(u_i, u_{i-1})$ would not be far from zero. If omitting $y_{i-3}, y_{i-4}, \cdots$ really matters, then one would expect $COR(\hat{u}_i, \hat{u}_{i-1})$ to be higher than 0.104. Having $COR(\hat{u}_i, \hat{u}_{i-1}) = 0.104$ is also comforting for the iid assumption for u_i's. This data as well as the house sale data will be used again.

1.5   Partial Regression

Suppose x consists of two sets of regressors of dimension $k_f \times 1$ and $k_g \times 1$, respectively: $x = (x'_f, x'_g)'$ and $k = k_f + k_g$. Partition X and b_{lse} accordingly:

$$X = [X_f, X_g] \text{ and } b_{lse} = \begin{bmatrix} b_f \\ b_g \end{bmatrix} \implies X b_{lse} = X_f b_f + X_g b_g.$$

X_f can be written as

$$\underset{N \times k_f}{X_f} = \underset{N \times k}{X} \cdot \underset{k \times k_f}{S_f}$$

where S_f is a "selection matrix" consisting only of 1's and 0's to select the components of X for X_f; analogously we can get $X_g = X \cdot S_g$. For example, with $N = 3$, $k = 3$, and $k_f = 2$, the preceding display is

$$\begin{bmatrix} x_{11} & x_{12} \\ x_{21} & x_{22} \\ x_{31} & x_{32} \end{bmatrix} = \begin{bmatrix} x_{11} & x_{12} & x_{13} \\ x_{21} & x_{22} & x_{23} \\ x_{31} & x_{32} & x_{33} \end{bmatrix} \times \begin{bmatrix} 1 & 0 \\ 0 & 1 \\ 0 & 0 \end{bmatrix}.$$

Observe

$$
\begin{aligned}
P_{X_f} P_X &= P_{X_f} \quad \text{and} \quad Q_{X_f} Q_X = Q_X \quad \text{because} \\
P_{X_f} P_X &= X_f (X_f' X_f)^{-1} X_f' \cdot X (X'X)^{-1} X' \\
&= X_f (X_f' X_f)^{-1} S_f' X' \cdot X (X'X)^{-1} X' \\
&= X_f (X_f' X_f)^{-1} S_f' X' = X_f (X_f' X_f)^{-1} X_f' = P_{X_f}, \\
Q_{X_f} Q_X &= (I_N - P_{X_f})(I_N - P_X) = I_N - P_{X_f} - P_X + P_{X_f} = Q_X.
\end{aligned}
$$

In words, for $P_{X_f} P_X = P_{X_f}$, extracting first the X part (with P_X) and then its subset X_f part (with P_{X_f}) is the same as extracting only the X_f part. As for $Q_{X_f} Q_X = Q_X$, removing first the X part and then its subset X_f part is the same as removing the whole X part.

Multiply $Y = X_f b_f + X_g b_g + \hat{U}$ by Q_{X_f} to get

$$
\begin{aligned}
Q_{X_f} Y &= Q_{X_f} X_f b_f + Q_{X_f} X_g b_g + Q_{X_f} \hat{U} = Q_{X_f} X_g b_g + \hat{U}, \text{ because} \\
Q_{X_f} X_f &= O \quad \text{and} \quad Q_{X_f} \hat{U} = Q_{X_f} (Q_X Y) = Q_X Y = \hat{U}.
\end{aligned}
$$

Multiply both sides of $Q_{X_f} Y = Q_{X_f} X_g b_g + \hat{U}$ by $X_g' Q_{X_f}$ to get

$$X_g' Q_{X_f} Q_{X_f} Y = X_g' Q_{X_f} Q_{X_f} X_g \cdot b_g + X_g' Q_{X_f} \hat{U}.$$

Because

$$X_g' Q_{X_f} \hat{U} = X_g' Q_{X_f} Q_X Y = X_g' Q_X Y = S_g' X' Q_X Y = 0 \text{ for}$$
$$X' Q_X = 0_{k \times N},$$

the residual term disappears. Solving for b_g gives

$$b_g = (X_g' Q_{X_f} X_g)^{-1} X_g' Q_{X_f} Y.$$

This expression shows that the LSE b_g for β_g can be obtained in two stages. First, do the LSE of Y on X_f to get the *partial residual* $Q_{X_f} Y$, and then do the LSE of X_g on X_f to get the partial residual $Q_{X_f} X_g$. Second, do the LSE of $Q_{X_f} Y$ on $Q_{X_f} X_g$:

$$(X_g' Q_{X_f} Q_{X_f} X_g)^{-1} X_g' Q_{X_f} Q_{X_f} Y = (X_g' Q_{X_f} X_g)^{-1} X_g' Q_{X_f} Y.$$

This is the *partial regression* interpretation of b_g. The name "partial residual" is appropriate, for only the x_f part of x is used in the first regression. By using only the residuals in the second step, the presence of x_f is nullified, and thus b_g shows the effect of x_g on y with x_f controlled for. Put it differently, b_g shows the additional explanatory power of x_g for y, over and above what is already explained by x_f. When x_g is a scalar, it is informative to plot $Q_{X_f} Y$ (on the vertical axis) versus $Q_{X_f} X_g$ (on the horizontal axis) to isolate the effect of x_g on y. The correlation between the two residuals is called the *partial correlation* between y and x_g.

As a special case, suppose $x_f = 1$ and $x_g = (x_2, ..., x_k)'$. Denoting Q_{X_f} simply as Q_1, we already saw

$$Q_1 Y = (y_1 - \bar{y}, ..., y_N - \bar{y})' \quad \text{and} \quad \underset{N \times N}{Q_1} \underset{N \times (k-1)}{X_g} = (x_{1g} - \bar{x}_g, ..., x_{Ng} - \bar{x}_g)'.$$

Using the vector notations, the partial regression for the slopes b_g is nothing but the LSE with the mean-deviation variables: with $x_i = (1, \tilde{x}_i')'$ and $\bar{\tilde{x}} \equiv N^{-1} \sum_i \tilde{x}_i$,

$$b_g = \left\{ \sum_i (\tilde{x}_i - \bar{\tilde{x}})(\tilde{x}_i - \bar{\tilde{x}})' \right\}^{-1} \sum_i (\tilde{x}_i - \bar{\tilde{x}})(y_i - \bar{y}).$$

The role of "1" is to explain the level of ($\tilde{x}$ and) y.

1.6   Omitted Variable Bias

In the model $y = x_f' \beta_f + x_g' \beta_g + u$, what happens if x_g is not used in estimation? This is an important issue, as we may not have (or use) all relevant regressors in the data. With x_g not used, $x_g' \beta_g + u \equiv v$ becomes the new error term in the model, and the consequence of not using x_g depends on $COR(x_f, x_g)$. To simplify the discussion, assume that the model is written in mean-deviation form, i.e., $E(y) = E(x_f') \beta_f + E(x_g') \beta_g + E(u)$ is subtracted from the model to yield

$$y - E(y) = \{x_f - E(x_f)\}' \beta_f + \{x_g - E(x_g)\}' \beta_g + u - E(u)$$

and we redefine y as $y - E(y)$, x_f as $x_f - E(x_f)$ and so on. So long as we are interested in slopes in β_f, the mean deviation model is adequate.

If $COR(x_f, x_g) = 0$ (i.e., if $E(x_f x_g) = 0$), then β_f can still be estimated consistently by the LSE of y on x_f. The only downside is that, in general, $SD(v) > SD(u)$ as v has more terms than u, and thus R^2 will drop. If $COR(x_f, x_g) \neq 0$, however, then $COR(x_f, v) \neq 0$ makes x_f an *endogenous regressor* and the LSE becomes inconsistent. Specifically, the LSE of y on x_f is

$$b_f = \left(\frac{1}{N} \sum_i x_{if} x_{if}' \right)^{-1} \frac{1}{N} \sum_i x_{if} y_i$$

$$= \left(\frac{1}{N} \sum_i x_{if} x_{if}' \right)^{-1} \frac{1}{N} \sum_i x_{if} (x_{if}' \beta_f + v_i)$$

$$\begin{aligned}
&= \beta_f + \left(\frac{1}{N}\sum_i x_{if}x'_{if}\right)^{-1}\frac{1}{N}\sum_i x_{if}v_i \\
&= \beta_f + \left(\frac{1}{N}\sum_i x_{if}x'_{if}\right)^{-1}\frac{1}{N}\sum_i x_{if}\left(x'_{ig}\beta_g + u_i\right) \\
&= \beta_f + \left(\frac{1}{N}\sum_i x_{if}x'_{if}\right)^{-1}\frac{1}{N}\sum_i x_{if}x'_{ig}\cdot\beta_g \\
&\quad + \left(\frac{1}{N}\sum_i x_{if}x'_{if}\right)^{-1}\frac{1}{N}\sum_i x_{if}u_i
\end{aligned}$$

which is consistent for $\beta_f + E^{-1}\left(x_f x'_f\right)E\left(x_f x'_g\right)\cdot\beta_g$.

The term other than β_f is called the *omitted variable bias, which is 0 if either* $\beta_g = 0$ *(i.e., x_g is not omitted at all) or if* $E^{-1}(x_f x'_f)E(x_f x'_g) = 0$ which is the population linear projection coefficient of regressing x_g on x_f. In simple words, if $COR(x_f, x_g) = 0$, then there is no omitted variable bias. When LSE is run on some data and if resulting estimates do not make sense intuitively, in most cases, the omitted variable bias formula will provide a good guide on what might have gone wrong.

One question that might arise when $COR(x_f, x_g) \neq 0$ is what happens if a subvector x_{f2} of x_f is correlated to x_g while the other subvector x_{f1} of x_f is not where $x_f = (x'_{f1}, x'_{f2})'$. In this case, will x_{f1} still be subject to the omitted variable bias? The answer depends on $COR(x_{f1}, x_{f2})$ as can be seen in

$$\begin{aligned}
E^{-1}(x_f x'_f)E(x_f x'_g) &= \begin{bmatrix} E(x_{f1}x'_{f1}) & E(x_{f1}x'_{f2}) \\ E(x_{f2}x'_{f1}) & E(x_{f2}x'_{f2}) \end{bmatrix}^{-1}\begin{bmatrix} 0 \\ E(x_{f2}x'_g) \end{bmatrix} \\
&\quad \text{as } E(x_{f1}x'_g) = 0 \\
&= \begin{bmatrix} 0 \\ E^{-1}(x_{f2}x'_{f2})E(x_{f2}x'_g) \end{bmatrix} \text{ if } E(x_{f1}x'_{f2}) = 0.
\end{aligned}$$

Hence if $E(x_{f1}x'_{f2}) = 0$, then there is no omitted variable bias for x_{f1}. Otherwise, the bias due to $E(x_{f2}x'_g) \neq 0$ gets channeled to x_{f1} through $COR(x_{f1}, x_{f2})$.

In the case $COR(x_{f1}, x_{f2}) = 0$, $COR(x_{f1}, x_g) = 0$ but $COR(x_{f2}, x_g) \neq 0$, we can in fact use only x_{f1} as regressors—no omitted variable bias in this case. Nevertheless, using x_{f2} as regressors makes the model error term variance smaller, which leads to a higher R^2 and higher t-values for x_{f1}. In this case, we just have to be aware that the estimator for x_{f2} is biased.

As an example for omitted variable bias, imagine a state considering a mandatory seat belt law. Data is collected from N cities in the state, with y_i the yearly traffic fatality proportion per driver in city i, and x_{if} the proportion of drivers wearing seat belt in city i. LSE is run to find $b_f > 0$, which is counter-intuitive however. One possible scenario is that wearing the seat belt makes the driver go faster, which results in more accidents. That is, driving speed x_g in the error term is correlated with x_f, and the omitted variable bias dominates β_f so that the following sum becomes positive:

$$\underset{\text{negative}}{\beta_f} + \underset{\text{positive}}{E^{-1}(x_f x_f')E(x_f x_g')} \cdot \underset{\text{positive}}{\beta_g}$$

In this case, enacting the seat belt law will increase y, not because $\beta_f > 0$ but rather because it will cause x_g to increase.

What the state have in mind is the *ceteris paribus ("direct") effect* β_f with all the other variables held constant, but what is estimated is the total effect that is the sum of the direct effect β_f and the *indirect effect* of x_f on y through x_g. Both the direct and indirect effects can be estimated consistently using the LSE of y on x_f and x_g, but enacting only the seat belt law will not have the intended effect because the indirect effect will occur. A solution is enacting both the seat belt law and a speed limit law to assure $COR(x_f, x_g) = 0$ after the laws are passed.

In the example, omitted variable bias helped explaining an apparently nonsensical result. But it can also help negating an apparently plausible result. Suppose that there are two types of people, one cautious and the other reckless, with x_g denoting the proportion of the cautious people, and that the cautious people tend to wear seat belts more $(COR(x_f, x_g) > 0)$ and have fewer traffic accidents. Also suppose $\beta_f = 0$, i.e., no true effect of seat belt wearing. In this case, the LSE of y on x_f converges to a negative number

$$\underset{0}{\beta_f} + \underset{\text{positive}}{E^{-1}(x_f x_f')E(x_f x_g')} \cdot \underset{\text{negative}}{\beta_g}$$

and, due to omitting x_g, we may wrongly conclude that wearing seat belt will lower y to enact the seat belt law. Here the endogeneity problem of x_f leads to an ineffective policy as the seat belt law will have no true effect on y. Note that, differently from the $x_g = speed$ example, there is no indirect effect of forcing seat belt wearing because seat belt wearing will not change the people's type.

2   Heteroskedasticity and Homoskedasticity

The assumption $E(u|x) = 0$ for LSE is a restriction on the conditional first moment of $u|x$. We do not need restrictions on higher moments of $u|x$ to estimate β, but whether $E(u^2|x)$ varies or not as x changes matters in the LSE asymptotic inference, which is the topic of this section. $E(u^2|x)$ will also appear prominently later for generalized LSE (GLS).

Observe that $E(u^2|x) = V(u|x)$ because $E(u|x) = 0$, and also that $V(u|x) = V(y|x)$ because $y|x$ is a $x'\beta$-shifted version of $u|x$. If $V(u|x)$ is a non-constant function of x, then u is "heteroskedastic" (or there is "heteroskedasticity"). If $V(u|x)$ is a constant, say σ^2, then u is "homoskedastic" (or there is "*homoskedasticity*"). Although we assume that (u_i, x_i') are iid across i, $u_i | x_i$ are not iid across i under heteroskedasticity.

2.1   Heteroskedasticity Sources

2.1.1   Forms of Heteroskedasticity

A well-known source for heteroskedasticity is *random coefficients*. Suppose the coefficient vector is β_i that is random around a constant β:

$$y_i \;=\; x_i'\beta_i + u_i, \quad \beta_i = \beta + v_i, \; E(v) = 0, \; E(vv') \equiv \Lambda,$$
$$v \text{ is independent of } x \text{ and } u.$$

Substituting the β_i equation yields a constant coefficient model:

$$y_i = x_i'\beta + (x_i'v_i + u_i), \quad E(x'v + u|x) = 0, \; V(x'v + u|x) = x'\Lambda x + E(u^2|x).$$

Even if $E(u^2|x) = \sigma^2$, still the error term $\varepsilon \equiv x'v + u$ is heteroskedastic. Here the functional form of $V(\varepsilon|x) = x'\Lambda x + \sigma^2$ is known (up to Λ and σ^2) due to the random coefficients and the homoskedasticity of u.

Heteroskedasticity does not necessarily have to be motivated by random coefficients. If x is income and y is consumption per month, we can simply imagine the variation of $y|x$ increasing as x increases. In this case, we may postulate, say,

$$y_i \;=\; x_i'\beta + u_i, \quad V(u|x_i) = \exp(x_i'\theta),$$
$$\text{where } \theta \text{ is an unknown parameter vector;}$$

again, this is heteroskedasticity *of known form* as in the random coefficient model.

The linear model assumption $E(y|x) = x'\beta$ is restrictive because $E(y|x)$ may not be a linear function. Assuming $V(y|x) = \exp(x'\theta)$ additionally is even more restrictive, in which we would have even less confidence than in $E(y|x) = x'\beta$. If we just allow $V(u|x) = V(y|x)$ to be an unknown function of x instead of specifying the functional form of $V(u|x)$, then we allow for a heteroskedasticity *of unknown form*. The consistency and asymptotic distribution results of the LSE hold under heteroskedasticity of unknown form, because we did not impose any assumption on $V(y|x)$ in their derivations. If $V(u|x) = \sigma^2$, then because

$$E(xx'u^2) = E\{xx'E(u^2|x)\} = \sigma^2 E(xx'),$$

the asymptotic variance of $\sqrt{N}(b_{lse} - \beta)$ is

$$E^{-1}(xx')E(xx'u^2)E^{-1}(xx') = \sigma^2 \cdot E^{-1}(xx')$$

which appears often in introductory econometrics textbooks. The right-hand side (rhs) is valid only under homoskedasticity; the left-hand side (lhs) is called a "*heteroskedasticity-robust (or -consistent) variance*," which is valid with or without the homoskedasticity assumption.

2.1.2   Heteroskedasticity due to Aggregation

When "averaging with different numbers of observations" takes place, heteroskedasticity can arise without x getting involved. Suppose a model holds at individual level, but we have only a city-level aggregate data:

$$y_{j_i} = x'_{j_i}\beta + u_{j_i}, \quad j_i = 1, ..., n_i \text{ where } j_i \text{ denotes individual}$$
$$j \text{ in city } i = 1, ..., N$$
$$\Longrightarrow \quad y_i = x'_i\beta + u_i \quad \text{where} \quad y_i \equiv \frac{1}{n_i}\sum_{j_i} y_{j_i}, \ x_i \equiv \frac{1}{n_i}\sum_{j_i} x_{j_i},$$
$$u_i \equiv \frac{1}{n_i}\sum_{j_i} u_{j_i}.$$

That is, what is available is a random sample on cities with (n_i, x'_i, y_i), $i = 1, ..., N$, where n_i is the total number of people in city i and N is the number of the sampled cities. Suppose that u_{j_i} is independent of x_{j_i}, and that u_{j_i}'s are iid with zero mean and variance σ^2 (i.e., $u_{j_i} \sim (0, \sigma^2)$). Then $u_1, ..., u_N$ are independent, and $u_i|(x_i, n_i) \sim (0, \sigma^2/n_i)$: the error terms in the city-level model are heteroskedastic wrt n_i, but not wrt x_i. Note that all of n_i, x_i, and y_i are random as we do not know which city gets drawn.

This type of heteroskedasticity can be dealt with by minimizing $\sum_i (y_i - x'_i b)^2 n_i$, which is equivalent to applying LSE to the transformed equation

$$y_i^* = x_i^{*'}\beta + u_i^*, \quad \text{where } y_i^* \equiv y_i\sqrt{n_i}, \ x_i^* \equiv x_i\sqrt{n_i} \text{ and } u_i^* \equiv u_i\sqrt{n_i}.$$

In the transformed equation, as (x_i, n_i) is "finer" than x_i^*,

$$E(u_i^*|x_i^*) = E\{ E(u_i\sqrt{n_i}|x_i, n_i)|x_i^*\} = 0 \quad \text{as } E(u_i\sqrt{n_i}|x_i, n_i) = 0,$$
$$V(u_i^*|x_i^*) = E\{ V(u_i\sqrt{n_i}|x_i, n_i)|x_i^*\} = \sigma^2 \quad \text{as } V(u_i\sqrt{n_i}|x_i, n_i) = \sigma^2.$$

Hence $u_1^*, ..., u_N^*$ are iid $(0, \sigma^2)$. This LSE motivates "weighted LSE" to appear later.

Two remarks on the city-level data example. First, there is no unity in the transformed regressors because 1 is replaced with $\sqrt{n_i}$. This requires a different definition of R^2 which was defined using $Q_1\hat{U} = \hat{U}$. R^2 for the transformed model can be defined as $\{\text{sample } COR(y, \hat{y})\}^2$, not as $\{\text{sample } COR(y^*, \hat{y}^*)\}^2$, where $\hat{y}_i = x'_i b^*_{lse}$, $\hat{y}_i^* = x_i^{*'}b^*_{lse}$, and b^*_{lse} is the LSE for the transformed model. This definition of R^2 can also be used for "weighted LSE" below. Second, we assumed above that sampling is done at city level and what is available is the averaged variables y_i and x_i along with n_i. If, instead, all cities are included but n_i individuals get sampled in city i where n_i is a pre-determined (i.e., fixed) constant ahead of sampling, then n_i is not random (but still may vary across i); in contrast, (x'_i, y_i) is still random because it depends on the sampled individuals. In this case, u_i's are *independent but non-identically distributed (inid)* due to $V(u_i) = \sigma^2/n_i$ where $V(u_i)$ is the marginal variance of u_i. Clearly, how sampling is done matters greatly.

2.1.3   Variance Decomposition

Observe

$$
\begin{aligned}
V(u) &= E(u^2) - \{E(u)\}^2 = E[E(u^2|x)] - [E\{E(u|x)\}]^2 \\
&= E[V(u|x) + \{E(u|x)\}^2] - [E\{E(u|x)\}]^2 \\
&= E[V(u|x)] + E[g(x)^2] - [E\{g(x)\}]^2 \text{ (defining } g(x) \equiv E(u|x)) \\
&= E\{V(u|x)\} + V\{g(x)\} \quad \text{(in general)} \\
\{ &= E\{V(u|x)\} \quad \text{as } g(x) = E(u|x) = 0\}.
\end{aligned}
$$

Under homoskedasticity, $V(u|x) = \sigma^2 \ \forall x$, and thus $V(u) = E(\sigma^2) = \sigma^2$.

For a rv y, this display gives the *variance decomposition* of $V(y)$

$$
V(y) = E\{V(y|x)\} + V\{E(y|x)\}
$$

which can help understand the sources of $V(y)$. Suppose that x is a rv taking on 1, 2, or 3. Decompose the population with x into 3 groups (i.e., subpopulations):

Group	$P(x=1)=1/2$	$P(x=2)=1/4$	$P(x=3)=1/4$
Group mean (level)	$E(y|x=1)$	$E(y|x=2)$	$E(y|x=3)$
(Within-) Group Variance	$V(y|x=1)$	$V(y|x=2)$	$V(y|x=3)$

Each group has its conditional variance, and we may be tempted to think that $E\{V(y|x)\}$ which is the weighted average of $V(y|x)$ with $P(y|x)$ as the weight yields the marginal variance $V(y)$. But the variance decomposition formula demonstrates $V(y) \neq E\{V(y|x)\}$ unless $E(y|x) = 0 \ \forall x$, although $E(y) = E\{E(y|x)\}$ always. That is, the source of the variance is not just the *"within-group variance"* $V(y|x)$, but also the *"between-group variance"* $V\{E(y|x)\}$ of the group mean $E(y|x)$.

If the variance decomposition is done with an observable variable x, then we may dig deeper by estimating $E(y|x)$ and $V(y|x)$. But a decomposition with an unobservable variable u can be also thought of, as we can choose any variable we want in the variance decomposition: $V(y) = E\{V(y|u)\} + V\{E(y|u)\}$. In this case, the decomposition can help us imagine the sources depending on u. For instance, if y is income and u is ability (whereas x is education), then the income variance is the weighted average of ability-group variances plus the variance between the average group-incomes.

Two polar cases are of interest. Suppose that y is income and x is education group: 1 for "below high school graduation," 2 for "high school graduation" to "below college graduation," and 3 for college graduation or above. One extreme case is the same mean income for all education groups:

$$
E(y|x) = E(y) \ \forall x \implies V\{E(y|x)\} = 0 \implies V(y) = E\{V(y|x)\}.
$$

The other extreme case is the same variance in each education group:

$$V(y|x) = \sigma^2 \ \forall x \implies E\{V(y|x)\} = \sigma^2 \implies V(y) = \sigma^2 + V\{E(y|x)\};$$

if $\sigma^2 = 0$, then $V(y) = V\{E(y|x)\}$: the variance comes solely from the differences of $E(y|x)$ across the groups.

2.1.4 Analysis of Variance (ANOVA)*

The variance decomposition formula is the basis for *Analysis of Variance (ANOVA)*, where x stands for treatment categories (with one category being no treatment). In ANOVA, the interest is on the "mean treatment effect," i.e., whether $E(y|x)$ changes across the treatment groups/categories or not. The classical approach—*one-way ANOVA*—assumes normality for y and homoskedasticity across the groups $(V(y|x) = \sigma^2 \ \forall x)$ so that $V(y) = \sigma^2 + V\{E(y|x)\}$. One-way ANOVA decomposes the sample variance into sample versions of σ^2 and $V\{E(y|x)\}$ which are two independent χ^2 rv's. "$H_0 : E(y|x)$ is a constant $\forall x$" is tested using the ratio of the two sample versions, and the ratio follows a F-distribution.

Let $x^{(j)}$ denote the x-value for group j, and define the group-j mean $\mu_j \equiv E(y|x = x^{(j)})$. In one-way ANOVA, y gets indexed as in y_{ij}, $i = 1, ..., N_j$, $j = 1, ..., J$ where j denotes the jth group (category) and i denotes the ith observation in the jth group; there are N_j observations in group j. The model is

$$y_{ij} = \mu_j + u_{ij}, \quad u_{ij} \sim iid \ N(0, \sigma^2) \text{ across } i \text{ and } j.$$

Define the total sample size, group-j average and the "grand average" as, respectively,

$$N \equiv \sum_{j=1}^{J} N_j, \quad \bar{y}_j \equiv \frac{1}{N_j} \sum_{i=1}^{N_j} y_{ij} \quad \bar{y} \equiv \frac{1}{N} \sum_{j=1}^{J} \sum_{i=1}^{N_j} y_{ij}.$$

Then the decomposition $y_{ij} - \bar{y} = (y_{ij} - \bar{y}_j) + (\bar{y}_j - \bar{y})$ is used in one-way ANOVA where $\bar{y}_j - \bar{y}$ is for $V\{E(y|x)\}$.

Specifically, take $\sum_{j=1}^{J} \sum_{i=1}^{N_j}$ on $(y_{ij} - \bar{y})^2 = \{(y_{ij} - \bar{y}_j) + (\bar{y}_j - \bar{y})\}^2$ to see that the cross-product term is zero because

$$\sum_{j=1}^{J} \sum_{i=1}^{N_j} (y_{ij} - \bar{y}_j)(\bar{y}_j - \bar{y}) = \sum_{j=1}^{J} (\bar{y}_j - \bar{y}) \sum_{i=1}^{N_j} (y_{ij} - \bar{y}_j)$$

$$= \sum_{j=1}^{J} (\bar{y}_j - \bar{y})(N_j \bar{y}_j - N_j \bar{y}_j) = 0.$$

Thus we get

$$\underbrace{\sum_{j=1}^{J} \sum_{i=1}^{N_j} (y_{ij} - \bar{y})^2}_{\text{total variation}} = \underbrace{\sum_{j=1}^{J} \sum_{i=1}^{N_j} (y_{ij} - \bar{y}_j)^2}_{\text{unexplained variation}} + \underbrace{\sum_{j=1}^{J} N_j (\bar{y}_j - \bar{y})^2}_{\text{explained variation}}$$

where the two terms on the rhs are for $\sigma^2 + V\{E(y|x)\}$ when divided by N. The aforementioned test statistic for mean equality is

$$\frac{(J-1)^{-1} \sum_{j=1}^{J} N_j (\bar{y}_j - \bar{y})^2}{(N-J)^{-1} \sum_{j=1}^{J} \sum_{i=1}^{N_j} (y_{ij} - \bar{y}_j)^2} \sim F(J-1, N-J).$$

To understand the dof's, note that there are J-many "observations" ($\bar{y}_j$'s) in the numerator, and 1 is subtracted in the dof because the grand mean gets estimated by $\bar{y}$. In the denominator, there are N-many observations y_{ij}'s, and J is subtracted in the dof because the group means get estimated by $\bar{y}_j$'s. Under the H_0, the test statistic is close to zero as the numerator is so because of $V\{E(y|x)\} = 0$.

The model $y_{ij} = \mu_j + u_{ij}$ can be rewritten as a familiar linear model. Define $J-1$ dummy variables, say $x_{i2}, ..., x_{iJ}$, where $x_{ij} = 1$ if observation i belongs to group j and $x_{ij} = 0$ otherwise. Then

$$y_i = x_i'\beta + u_i, \quad \text{where} \ \underset{J \times 1}{x} = (1, x_{i2}, ..., x_{iJ})' \ \text{and}$$

$$\beta = (\mu_1, \mu_2 - \mu_1, ..., \mu_J - \mu_1)'.$$

Here the intercept is for μ_1 and the slopes are for the deviations from μ_1; group 1 is typically the "control (i.e., no-treatment) group" whereas the other groups are the "treatment groups." For instance, if observation i belongs to treatment group 2, then

$$x_i'\beta = (1, 1, 0, ..., 0)'\beta = \mu_1 + (\mu_2 - \mu_1) = \mu_2.$$

Instead of the above F-test, we can test for $H_0 : \mu_1 =, ..., = \mu_J$ with "Wald test" to appear later without assuming normality; the Wald test checks out whether all slopes are zero or not.

"*Two-way ANOVA*" generalizes one-way ANOVA. There are two "factors" now, and we get y_{ijk} where j and k index the group (j, k), $j = 1, ..., J$ and $k = 1, ..., K$; group (j, k) has N_{jk} observations. The model is

$$y_{ijk} = \alpha_j + \beta_k + \gamma_{jk} + u_{ijk}, \quad u_{ijk} \sim N(0, \sigma^2) \ iid \ \text{across all indices}$$

where α_j is the factor-1 effect, β_k is the factor-2 effect, and γ_{jk} is the interaction effect between the two factors. The relevant decomposition is

$$y_{ijk} - \bar{y} = (y_{ijk} - \bar{y}_{j.} - \bar{y}_{.k} + \bar{y}) + (\bar{y}_{j.} - \bar{y}) + (\bar{y}_{.k} - \bar{y})$$

where $\bar{y}$ is the grand mean, $\bar{y}_{j.}$ is the average of all observations with j fixed (i.e., $\bar{y}_{j.} \equiv \sum_{k=1}^{K} \sum_{i=1}^{N_{jk}} y_{ijk} / \sum_{k=1}^{K} N_{jk}$), and $\bar{y}_{.k}$ is analogously defined. Various F-test statistics can be devised by squaring and summing up this display, but the two-way ANOVA model can be also written as a familiar linear model, to which "Wald tests" can be applied.

2.2   Weighted LSE (WLS)

Suppose $E(u^2|x) = (= V(u|x)) = m'\theta$ where m consists of elements of x and functions of those, and suppose that we know this functional form; e.g., with $k = 4$,

$$m_i'\theta = \theta_1 + \theta_2 x_{i2} + \theta_3 x_{i3} + \theta_4 x_{i2}^2 + \theta_5 x_{i2} x_{i3}.$$

Then we can do "*Weighted LSE (WLS)*":

- First, apply LSE to $y_i = x_i'\beta + u_i$ to get the residuals $\hat{u}_i$.

- Second, estimate θ by the LSE of $\hat{u}_i^2$ on m_i to get the LSE $\hat{\theta}$ for θ; this is motivated by $E(u^2|x) = m'\theta$.

- Third, assuming $m_i'\hat{\theta} > 0$ for all m_i, estimate β again by minimizing the weighted minimand $N^{-1}\sum_i(y_i - x_i'b)^2/(m_i'\hat{\theta})$ wrt b.

In Chapter 3.3.3, it will be shown that replacing θ with $\hat{\theta}$ is innocuous, and WLS is asymptotically equivalent to applying LSE to (with $SD(u|x_i) = (m_i'\theta)^{1/2}$)

$$\frac{y_i}{SD(u|x_i)} = \frac{x_i'}{SD(u|x_i)}\beta + \frac{u_i}{SD(u|x_i)}, \quad \text{where}$$

$$V\{\frac{u}{SD(u|x)}|x\} = \frac{V(u|x)}{SD(u|x)^2} = 1.$$

As in the above averaged data case, we can define $y_i^* \equiv y_i/SD(u|x_i)$ and $x_i^* \equiv x_i/SD(u|x_i)$. The error term in the transformed equation is homoskedastic with known variance 1. Inserting 1 and $x_i/SD(u|x_i)$, respectively, into σ^2 and x in $\sigma^2 E^{-1}(xx')$, we get

$$\sqrt{N}(b_{wls} - \beta) \rightsquigarrow N(0, \; E^{-1}\{\frac{xx'}{V(u|x)}\}).$$

The assumption $m_i'\hat{\theta} > 0$ for all m_i can be avoided if $V(u|x) = \exp(m'\theta)$ and if θ is estimated with "nonlinear LSE" that will appear later. The assumption $m_i'\hat{\theta} > 0$ for all m_i is simply to illustrate WLS using LSE in the first step.

An easy practical alternative to guarantee positive estimated weights is adopting a log-linear model $\ln u_i^2 = m_i'\zeta + v_i$ with v_i being an error term. The log-linear model is equivalent to

$$u_i^2 = e^{m_i'\zeta}e^{v_i} = (e^{m_i'\zeta/2}\nu_i)^2 \quad \text{where } \nu_i \equiv e^{v_i/2}$$

and $e^{m_i'\zeta/2}$ may be taken as the scale factor $SD(u|x_i)$ for ν_i (but $e^{m_i'\zeta/2}\nu_i > 0$ and thus the error u_i cannot be $e^{m_i'\zeta/2}\nu_i$ although $u_i^2 = (e^{m_i'\zeta/2}\nu_i)^2$). This suggests using $SD(u|x_i) \simeq e^{m_i'\hat{\zeta}/2}$ for WLS weighting where $\hat{\zeta}$ is the LSE for ζ. Strictly speaking, this "suggestion" is not valid because, for $SD(u|x_i) = e^{m_i'\zeta/2}$ to hold, we need

$$\ln E(u^2|x_i) = m_i'\zeta \iff E(u^2|x_i) = \exp(m_i'\zeta)$$

but $\ln u_i^2 = m_i'\zeta + v_i$ postulates instead $E(\ln u^2|x_i) = m_i'\zeta$. Since $\ln E(u^2|x_i) \neq E(\ln u^2|x_i)$, $\ln u_i^2 = m_i'\zeta + v_i$ is not compatible with $SD(u|x_i) = e^{m_i'\zeta/2}$. Despite this, however, defining $\hat{u}_i^* \equiv y_i^* - x_i^{*\prime} b_{wls}$ where b_{wls} is the WLS with weight $\exp(m_i'\hat{\zeta}/2)$, so long as the LSE of $\hat{u}_i^{*2}$ on m_i returns insignificant slopes, we can still say that the weight $\exp(m_i'\hat{\zeta}/2)$ is adequate because the heteroskedasticity has been removed by the weight, no matter how it was obtained.

In short, each one of the following has different implications on how we go about LSE.

- heteroskedasticity of unknown form: LSE to use $E^{-1}(xx')$
                                         $E(xx'u^2)E^{-1}(xx')$
- homoskedasticity: LSE to use $\sigma^2 E^{-1}(xx')$
- heteroskedasticity of known form: WLS to use
                                         $E^{-1}\{xx'/V(u|x)\}$.

Under homoskedasticity, all three variance matrices agree; otherwise, they differ in general.

Later, we will see that, under the known form of heteroskedasticity, WLS is more efficient than LSE; i.e.,

$$E^{-1}\left(xx'\right) E\left(xx'u^2\right) E^{-1}\left(xx'\right) \geq E^{-1}\left\{\frac{xx'}{V(u|x)}\right\}$$

in the matrix sense (for two matrices A and B, $A \geq B$ means that $A - B$ is p.s.d). For instance, if u_i is specified as

$$u_i = w_i \exp(x_i'\theta/2), \quad \text{where } w_i \text{ is independent of } x_i \text{ with } V(w) = 1,$$

then $V(u|x) = \exp(x'\theta)$, and we can do WLS with this. This is also convenient in viewing y_i: y_i is obtained by generating x_i and w_i first and then summing up $x_i'\beta$ and $w_i \exp(x_i'\theta)$. But if the specified form of heteroskedasticity $\exp(x'\theta)$ is wrong, then the asymptotic variance of the WLS is no longer $E^{-1}\{xx'/V(u|x)\}$. So, it is safer to use LSE with heteroskedasticity-robust variance. From now on, we will not invoke homoskedasticity assumption, unless it gives helpful insights for the problem at hand, which does happen from time to time.

2.3   Heteroskedasticity Examples

EXAMPLE: HOUSE SALE (continued). In the preceding section, the t-values under heteroskedasticity of unknown form (tv-het) were shown along with the t-values under homoskedasticity (tv-ho). Comparing the two sets of t-values, the differences are small other than for $\ln(T)/T$ and YR, and tv-ho tends to be greater than tv-het. This indicates that the extent of heteroskedasticity would be minor, if any. Three courses of action are conceivable from this observation:

- First, test for the H_0 of homoskedasticity using a test, say, in White (1980). This test does the LSE of $\hat{u}_i^2$ on 1 and some polynomial functions of regressors to see if the slopes are all zero or not; all zero slopes mean homoskedasticity. $N \cdot R^2 \rightsquigarrow \chi^2_{\#slopes}$ can be used as an asymptotic test statistic where R^2 is the R^2 for the $\hat{u}_i^2$ equation LSE. If the null is not rejected, then tv-ho may be used.

- Second, if the null is rejected, then model the form of heteroskedasticity using the aforementioned forms (or some others) to do WLS, where the weighted error term $u/SD(u|x)$ should have variance one regardless of x (this can be checked out using the method in the first step).

- Third, instead of testing for the H_0 or modelling heteroskedasticity, simply use tv-het. This is the simplest and most robust procedure. Also, the gain in the above two procedures tends to be small in micro-data; see, e.g., Deaton (1995).

EXAMPLE: INTEREST RATE (continued). Recall the interest rate example:

$$
\begin{array}{cccccc}
y_i & = & 0.216 & + & 1.298 & \cdot y_{i-1} - & 0.337 & \cdot y_{i-2}. \\
t-vlaues: & & 2.05\,(3.42) & & 10.29\,(21.4) & & -2.61\,(-5.66) &
\end{array}
$$

We list both tv-het and tv-ho; the latter is in $(\cdot)$ and was computed with $b_{lse,j}/\sqrt{v_{N,jj}}$, $j = 1, ..., k$, where $V_N \equiv [v_{N,hj}]$, $h, j = 1, ..., k$, is defined as $s_N^2(\sum_i x_i x_i')^{-1}$. The large differences between the two types of t-values indicate that the homoskedasticity assumption would not be valid for this model. In this time-series data, if the form of heteroskedasticity is correctly modeled, the gain in significance (i.e., the gain in the precision of the estimators) would be substantial. Indeed, such modeling is often done in financial time-series.

3   Testing Linear Hypotheses

3.1   Wald Test

Suppose we have an estimator b_N with

$$\sqrt{N}(b_N - \beta) \rightsquigarrow N(0, C);$$

b_N is said to be a "$\sqrt{N}$-consistent asymptotically normal estimator with asymptotic variance C." LSE and WLS are two examples of b_N and more examples will appear later. Given b_N, often we want to test linear null hypotheses such as

$$H_0 : R'\beta = c, \quad \text{where } rank(R) = g,$$

R is a $k \times g$ ($g \leq k$) known constant matrix and c is a $g \times 1$ known constant vector. Since $b_N \to^p \beta$, we have $R'b_N \to^p R'\beta$, because $R'b$ is a continuous function of b. If $R'\beta = c$ is true, $R'b_N$ should be close to c. Hence testing for $R'\beta = c$ can be based on the difference $\sqrt{N}(R'b_N - c)$.

As an example of $R'\beta = c$, suppose $k = 4$ and $H_0 : \beta_2 = 0, \beta_3 = 2$. For this, set

$$R' = \begin{bmatrix} 0 & 1 & 0 & 0 \\ 0 & 0 & 1 & 0 \end{bmatrix}, \quad c = \begin{bmatrix} 0 \\ 2 \end{bmatrix}$$

to get $R'\beta = c$ equivalent to $\beta_2 = 0$ and $\beta_3 = 2$. If we want to add another hypothesis, say $\beta_1 - \beta_4 = 0$, then set

$$R' = \begin{bmatrix} 0 & 1 & 0 & 0 \\ 0 & 0 & 1 & 0 \\ 1 & 0 & 0 & -1 \end{bmatrix}, \quad c = \begin{bmatrix} 0 \\ 2 \\ 0 \end{bmatrix}.$$

Typically, we test for some chosen elements of β being zero jointly. In that case, R consists of the column vectors picking up the chosen elements of β (each column consists of $k - 1$ zeros and 1) and c is a zero vector.

Given the above C and R, define H and Λ such that

$$R'CR = H\Lambda H'.$$

H is a matrix whose g columns are orthonormal eigenvectors of $R'CR$ and Λ is the diagonal matrix of the eigenvalues; $H\Lambda H'$ exists because $R'CR$ is real and symmetric. By construction, $H'H = I_g$. Also, pre-multiplying $H'H = I_g$ by H to get $(HH')H = H$, we obtain $HH' = I_g$ because H is of full rank. Observe now

$$\begin{aligned} S &\equiv H\Lambda^{-0.5}H' \implies S'S = (R'CR)^{-1} \quad \text{because} \\ S'S &= H\Lambda^{-0.5}H'H\Lambda^{-0.5}H' = H\Lambda^{-1}H' \quad \text{and} \\ S'S \cdot R'CR &= H\Lambda^{-1}H' \cdot H\Lambda H' = I_g. \end{aligned}$$

Further observe

$$\begin{aligned} \sqrt{N} \cdot R'(b_N - \beta) &\rightsquigarrow N(0, R'CR) \\ \left\{ \text{from } \sqrt{N}(b_N - \beta) \right. &\rightsquigarrow N(0, C) \text{ ``times } R''' \left.\right\}, \\ \sqrt{N}SR'(b_N - \beta) &\rightsquigarrow N(0, I_g), \quad \text{since} \\ S \cdot R'CR \cdot S' &= H\Lambda^{-0.5}H' \cdot H\Lambda H' \cdot H\Lambda^{-0.5}H' = I_g, \end{aligned}$$

$$\begin{aligned} N(R'b_N - R'\beta)'S'S(R'b_N - R'\beta) \\ = N(R'b_N - R'\beta)'(R'CR)^{-1}(R'b - R'\beta) &\rightsquigarrow \chi_g^2, \end{aligned}$$

because $N(R'b_N - R'\beta)'S'S(R'b_N - R'\beta)$ is a sum of g-many squared, asymptotically uncorrelated $N(0,1)$ random variables (rv). Replacing $R'\beta$ with c under $H_0 : R'\beta = c$, we get a *Wald test* statistic

$$N(R'b_N - c)'(R'C_NR)^{-1}(R'b_N - c) \rightsquigarrow \chi_g^2 \quad \text{where } C_N \rightarrow^p C.$$

The matrix $(R'C_NR)^{-1}$ in the middle standardizes the vector $\sqrt{N}(R'b_N - c)$.

3.2   Remarks

When b_N is the LSE of y on x, we get

$$C \equiv E^{-1}(xx')E(xx'u^2)E^{-1}(xx'),$$

$$C_N = (\frac{1}{N}\sum_i x_i x_i')^{-1} \cdot \frac{1}{N}\sum_i x_i x_i' \hat{u}_i^2 \cdot (\frac{1}{N}\sum_i x_i x_i')^{-1}$$

$$[\ = N(X'X)^{-1}X'DX(X'X)^{-1}, \quad \text{in matrices where}$$
$$D \equiv diag(\hat{u}_1^2, ..., \hat{u}_N^2)]$$

If homoskedasticity holds, then instead of C and C_N, we can use C_o and C_{oN} where

$$C_{oN} \equiv s_N^2 \left(\frac{1}{N}\sum_i x_i x_i'\right)^{-1} = s_N^2 \left(\frac{X'X}{N}\right)^{-1} \to^p C_o \equiv \sigma^2 E^{-1}(xx').$$

To show $C_N \to^p C$, since $(N^{-1}\sum_i x_i x_i')^{-1} \to^p E^{-1}(xx')$ was noted already, we have to show

$$\frac{1}{N}\sum_i x_i x_i' \hat{u}_i^2 - E(xx'u^2) = o_p(1).$$

Here, we take the "working proposition" that, for the expected value $E(h(x,y,\beta))$ where $h(x,y,\beta)$ is a (matrix-valued) function of x, y, and β, it holds in general that

$$\frac{1}{N}\sum_i h(x_i, y_i, b_N) - E(h(x, y, \beta)) = o_p(1), \quad \text{if } b_N \to^p \beta.$$

Then, setting $h(x,y,b) = xx'(y - x'b)^2$ establishes $C_N \to^p C$. For the preceding display to hold, $h(\cdot,\cdot,b)$ should not be too variable as a function of b so that the LLN holds uniformly over b. In almost all cases we encounter, the preceding display holds.

Instead of C_N, MacKinnon and White (1985) suggested to use, for a better small sample performance,

$$\tilde{C}_N \equiv (N-1)(X'X)^{-1}\left(X'\tilde{D}X - \frac{X'\tilde{r}\tilde{r}'X}{N}\right)(X'X)^{-1}, \quad \text{where}$$

$$\tilde{D} \equiv diag(\tilde{r}_1^2, ..., \tilde{r}_N^2), \quad \tilde{r}_i \equiv \frac{y_i - x_i' b_{lse}}{1 - d_{ii}}, \quad \tilde{r} \equiv (\tilde{r}_1, ..., \tilde{r}_N)', \quad \text{and}$$

$$d_{ii} \text{ is the } i\text{th diagonal element of the matrix } X(X'X)^{-1}X'.$$

$\tilde{C}_N$ and C_N are asymptotically equivalent as the term $X'\tilde{r}\tilde{r}'X/N$ in $\tilde{C}_N$ is of smaller order than $X'\tilde{D}X$.

Although the two variance estimators C_N and C_{No} numerically differ in finite samples, we have $C_N - C_{No} = o_p(1)$ under homoskedasticity. As already noted, too much difference between C_N and C_{No} would indicate the presence of heteroskedasticity, which is the basis for *White (1980) test for heteroskedasticity*. We will not, however, test for heteroskedasticity; instead, we will just allow it by using the heteroskedasticity-robust variance estimator C_N. There have been criticisms on the heteroskedasticity-robust variance estimator. For instance, Kauermann and Carroll (2001) showed that, when homoskedasticity holds, the heteroskedasticity-robust variance estimator has the higher variance than the variance estimator under homoskedasticity, and that confidence intervals based on the former have the coverage probability lower than the nominal value.

Suppose

$$y_i = x_i'\beta + d_i\beta_d + d_i w_i'\beta_{dw} + u_i$$

where d_i is a dummy variable of interest (e.g., a key policy variable on $(d = 1)$ or off $(d = 0)$), and w_i consists of elements of x_i interacting with d_i. Here, the effect of d_i on y_i is $\beta_d + w_i'\beta_{dw}$ which varies across i; i.e., we get N different individual effects. A way to summarize the N-many effects is using $\beta_d + E(w')\beta_{dw}$ (the effect evaluated at the "mean person") or $\beta_d + Med(w')\beta_{dw}$ (the effect evaluated at the "median person"). Observe

$$E(\beta_d + w_i'\beta_{dw}) = \beta_d + E(w')\beta_{dw} \quad \text{but}$$
$$Med(\beta_d + w_i'\beta_{dw}) \neq \beta_d + Med(w')\beta_{dw};$$

$Med(z_1 + z_2) \neq Med(z_1) + Med(z_2)$ in general for two rv's z_1 and z_2. The former is that the mean effect is also the effect at the mean person, whereas the latter is that the median effect is not the effect at the median person.

If we want to Wald-test "$H_0 : \beta_d + E(w')\beta_{dw} = 0$," then replace $E(w)$ with $\bar{w}$ to set $c = 0$ and $R' = (0_{k_x}', 1, \bar{w}')$ where 0_{k_x} is the $k_x \times 1$ vector of zero's. In this test, we may worry about the difference $\bar{w} - E(w)$ in replacing the unknown $(0_{k_x}', 1, E(w'))$ with its estimator $(0_{k_x}', 1, \bar{w}')$. But $\bar{w} - E(w)$ can be ignored, as we can just declare that we want to evaluate the effect at the sample mean $\bar{w}$. "$H_0 : \beta_d + Med(w')\beta_{dw} = 0$" can be tested in the analogous way, replacing $Med(w)$ with the sample median.

3.3   Empirical Examples

EXAMPLE: HOUSE SALE (continued). The two variables BATH and ROOM looked insignificant. Since BATH and ROOM tend to be highly correlated, using the two individual t-values for BATH and ROOM for the two separate hypotheses $H_0 : \beta_{bath} = 0$ and $H_0 : \beta_{room} = 0$ may be different from testing the joint null hypothesis $H_0 : \beta_{bath} = \beta_{room} = 0$ with Wald test, because the latter involves the asymptotic covariance between $b_{N,bath}$ and $b_{N,room}$ that is not used for the two t-values. It does happen in practice that, when two regressors are highly correlated (*"multicollinearity"*), the two separate null hypotheses may not be rejected while the single joint null hypothesis is rejected. This is because either one of them has explanatory power, but adding the other to the model when one is already included does not add any new explanatory power. With $k = 11$, $g = 2$, $c = (0,0)'$, and

$$
\underset{2 \times 11}{R'} = \begin{bmatrix} 0 & 0 & 1 & 0 & 0 & 0 & ,..., & 0 \\ 0 & 0 & 0 & 0 & 1 & 0 & ,..., & 0 \end{bmatrix}
$$

$$
\underset{11 \times 1}{\beta} = (\beta_{\ln(T)}, \beta_1, \beta_{bath}, \beta_{elec}, \beta_{room}, ..., \beta_{supply})',
$$

the Wald test statistic is 0.456 with the p-value $0.796 = P(\chi_2^2 > 0.456)$ for the model with $\ln(T)$ and C_N: the joint null hypothesis is not rejected. When C_{oN} is used instead of C_N, the Wald test statistic is 0.501 with the p-value 0.779—hardly any change. Although BATH and ROOM are important variables for house prices, they do not explain the discount % DISC. The t-values with C_N and $\tilde{C}_N$ shown below for the 11 regressors are little different (tv-C_N was shown already) because $N = 467$ is not too small for the number of regressors:

x:	$\ln(T)$	1	BATH	ELEC	RM	TAX	YR	$\ln(LP)$	BIGS	RATE	SUPPLY
tv-C_N:	7.76	−0.23	0.18	2.46	−0.67	−1.28	−3.87	2.52	−2.56	−3.10	1.06
tv-$\tilde{C}_N$:	7.29	−0.22	0.17	2.38	−0.65	−1.22	−3.66	2.40	−2.45	−2.98	1.03

EXAMPLE: TRANSLOG PRODUCTION FUNCTION. Consider a *"translog production function"*:

$$
\ln y = \beta_0 + \sum_{p=1}^{m} \beta_p \ln x_p + \sum_{p=1}^{m} \sum_{q=1}^{m} \beta_{pq} \frac{1}{2} \ln x_p \ln x_q + u \quad \text{where } \beta_{pq} = \beta_{qp}.
$$

This becomes a Cobb-Douglas production function if $\beta_{pq} = 0 \ \forall p, q$. To see why the restriction $\beta_{pq} = \beta_{qp}$ appears, observe

$$
\beta_{pq} \frac{1}{2} \ln x_p \ln x_q + \beta_{qp} \frac{1}{2} \ln x_q \ln x_p = \frac{\beta_{pq} + \beta_{qp}}{2} \ln x_p \ln x_q = \beta_{pq} \ln x_p \ln x_q :
$$

we can only identify the average of β_{pq} and β_{qp}, and $\beta_{pq} = \beta_{qp}$ essentially redefines the average as β_{pq}.

If we take the translog function as a second-order approximation to an underlying smooth function, say, $y = \exp\{f(x)\}$, then $\beta_{pq} = \beta_{qp}$ is a natural restriction from the symmetry of the second-order matrix. Specifically, observe

$$
\ln y = f(x) \implies \ln y = f\{\exp(\ln x)\} = \tilde{f}(\ln x) \quad \text{where } \tilde{f}(t) \equiv f\{\exp(t)\}.
$$

Now $\tilde{f}(\ln x)$ can be linearized around $x = 1$ (i.e., around $\ln x = 0$) with its second-order approximation where the β-parameters depend on the approximation point $x = 1$.

For a production function $y = f(x) + u$, it is "homogeneous of degree h" if $t^h y = f(tx) + u\ \forall t$. To test for the h-homogeneity, apply $t^h y = f(tx) + u$ to the translog production function to get

$$h \ln t + \ln y = \beta_0 + \sum_{p=1}^{m} \beta_p (\ln t + \ln x_p)$$

$$+ \sum_{p=1}^{m} \sum_{q=1}^{m} \beta_{pq} \frac{1}{2} (\ln t + \ln x_p)(\ln t + \ln x_q) + u$$

$$\implies\ h \ln t + \ln y = \beta_0 + \ln t \sum_{p=1}^{m} \beta_p + \sum_{p=1}^{m} \beta_p \ln x_p + \frac{(\ln t)^2}{2} \sum_{p=1}^{m} \sum_{q=1}^{m} \beta_{pq}$$

$$+ \frac{\ln t}{2} \left\{ \sum_{p=1}^{m} (\ln x_p \sum_{q=1}^{m} \beta_{pq}) + \sum_{q=1}^{m} (\ln x_q \sum_{p=1}^{m} \beta_{pq}) \right\}$$

$$+ \sum_{p=1}^{m} \sum_{q=1}^{m} \beta_{pq} \frac{1}{2} \ln x_p \ln x_q + u.$$

For both sides to be equal for all t, it should hold that

$$\sum_{p=1}^{m} \beta_p = h \quad \text{and} \quad \sum_{q=1}^{m} \beta_{pq} = 0\ \forall p$$

$$\left(\iff \sum_{q=1}^{m} \beta_{qp} = 0\ \forall p \iff \sum_{p=1}^{m} \beta_{pq} = 0\ \forall q \right).$$

To be specific, for $m = 2$ and $h = 1$, there are six parameters (β_0, β_1, β_2, β_{11}, β_{22}, β_{12}) to estimate in

$$\ln y = \beta_0 + \beta_1 \ln x_1 + \beta_2 \ln x_2 + \beta_{11} \frac{(\ln x_1)^2}{2} + \beta_{22} \frac{(\ln x_2)^2}{2} + \beta_{12} \ln x_1 \ln x_2 + u.$$

Bear in mind $\beta_{12} = \beta_{21}$, and we use only β_{12} with the first subscript smaller than the second. The 1-homogeneity (i.e., constant returns to scale) restrictions are

$$H_0 : \beta_1 + \beta_2 = 1, \quad \beta_{11} + \beta_{12} = 0 \text{ and } \beta_{12} + \beta_{22} = 0 \text{ (from } \beta_{21} + \beta_{22} = 0).$$

Clearly, we can estimate the model with LSE to test for this linear H_0.

If H_0 is accepted, then one may want to impose the H_0 on the model using its equivalent form

$$\beta_2 = 1 - \beta_1, \quad \beta_{11} = -\beta_{12}, \quad \beta_{22} = -\beta_{12}.$$

That is, the H_0-imposed model is

$$\ln y - \ln x_2 = \beta_0 + \beta_1 (\ln x_1 - \ln x_2) + \beta_{12} \left\{ -\frac{(\ln x_1)^2}{2} - \frac{(\ln x_2)^2}{2} \right.$$

$$\left. + \ln x_1 \ln x_2 \right\} + u.$$

This can be estimated by the LSE of $\ln y - \ln x_2$ on the rhs regressors.

If $m = 3$ and $h = 1$, then there will be 10 parameters (β_0, β_1, β_2, β_3, $\beta_{11}, \beta_{22}, \beta_{33}$, $\beta_{12}, \beta_{13}, \beta_{23}$), and the 1-homogeneity restrictions are

$$\beta_1 + \beta_2 + \beta_3 = 1, \quad \beta_{11} + \beta_{12} + \beta_{13} = 0,$$
$$\beta_{12} + \beta_{22} + \beta_{23} = 0 \quad \text{(from } \beta_{21} + \beta_{22} + \beta_{23} = 0\text{) and}$$
$$\beta_{13} + \beta_{23} + \beta_{33} = 0 \quad \text{(from } \beta_{31} + \beta_{32} + \beta_{33} = 0\text{)}.$$

4   Instrumental Variable Estimator (IVE)

When $E(xu) \neq 0$, LSE becomes inconsistent. A solution is dropping (i.e., substituting out) the endogenous components of x from the model, but the ensuing LSE does not deliver what is desired: the "other-things-being-equal" effect. Another solution is to extract only the exogenous part of the endogenous regressors, which is the topic of this and the following sections.

4.1   IVE Basics

4.1.1   IVE in Narrow Sense

For the linear model $y = x'\beta + u$, suppose we have a $k \times 1$ moment condition

$$E(zu) = E(z(y - x'\beta)) = 0,$$

instead of $E(xu) = 0$, where z is a $k \times 1$ random vector such that $E(zx')$ is invertible. Solve the equation for β to get

$$\beta = E^{-1}(zx') \cdot E(zy).$$

The sample analog of this is the *instrumental variable estimator (IVE)*

$$b_{ive} = \left(\frac{1}{N} \sum_i z_i x_i' \right)^{-1} \frac{1}{N} \sum_i z_i y_i \quad \{= (Z'X)^{-1} Z'Y \quad \text{in matrices}\}.$$

While IVE in its broad sense includes any estimator using instruments, here we define IVE in its narrow sense as the one taking this particular form. IVE includes LSE as a special case when $z = x$ (or $Z = X$ in matrices).

Substitute $y_i = x_i'\beta + u_i$ into the b_{ive} formula to get

$$b_{ive} = \left(\frac{1}{N} \sum_i z_i x_i' \right)^{-1} \frac{1}{N} \sum_i z_i (x_i'\beta + u_i) = \beta + \left(\frac{1}{N} \sum_i z_i x_i' \right)^{-1}$$
$$\times \frac{1}{N} \sum_i z_i u_i.$$

The consistency of the IVE follows simply by applying the LLN to the terms other than β in the last equation. As for the asymptotic distribution, observe

$$\sqrt{N}\,(b_{ive} - \beta) = \left(\frac{1}{N}\sum_i z_i x_i'\right)^{-1} \frac{1}{\sqrt{N}}\sum_i z_i u_i.$$

Applying the LLN to $N^{-1}\sum_i z_i x_i'$ and the CLT to $N^{-1/2}\sum_i z_i u_i$, it holds that

$$\sqrt{N}\,(b_{ive} - \beta) \rightsquigarrow N\left\{0,\ E^{-1}\left(zx'\right) E\left(zz'u^2\right) E^{-1}\left(xz'\right)\right\}.$$

This is informally stated as

$$b_{ive} \sim N\left\{\beta,\ \frac{1}{N}E^{-1}(zx')E(zz'u^2)E^{-1}(xz')\right\}$$

the variance of which can be estimated with (defining $r_i \equiv y_i - x_i' b_{ive}$)

$$\left(\sum_i z_i x_i'\right)^{-1}\left(\sum_i z_i z_i' r_i^2\right)\left(\sum_i x_i z_i'\right)^{-1} = (Z'X)^{-1} Z'DZ\,(X'Z)^{-1},$$

in matrices,

where $D \equiv diag(r_1^2, ..., r_N^2)$ and $r_i = y_i - x_i' b_{ive}$, *not* $y_i - z_i' b_{ive}$.

4.1.2   Instrumental Variable (IV) qualifications

IVE is useful when LSE is not applicable because some regressors are endogenous in the sense $E(xu) \neq 0$. For instance, suppose $x_i = (1, x_{i2}, x_{i3}, x_{i4})'$ (thus $y_i = \beta_1 + \beta_2 x_{i2} + \beta_3 x_{i3} + \beta_4 x_{i4} + u_i$) and

$$E(u) = E(x_2 u) = E(x_3 u) = 0,\ \text{but } E(x_4 u) \neq 0;$$

x_2 and x_3 are *exogenous regressors* in the y-equation whereas x_4 is an *endogenous regressor*. If there is a rv w such that

(i) $COR(w, u) = 0$ ($\Longleftrightarrow$ $E(wu) = 0$)

(ii) $0 \neq COR(w, x_4)$ ("inclusion restriction")

(iii) w does not appear in the y equation   ("exclusion restriction")

then w is a valid *instrumental variable (IV)*—or just *instrument*—for x_4, and we can use $z_i = (1, x_{i2}, x_{i3}, w_i)'$. The reason why (ii) is called "inclusion restriction" is that w should be in the x_4 equation for (ii) to hold. Conditions (ii) and (iii) together are simply called "*inclusion/exclusion restrictions*."

As an example, suppose that y is blood pressure, x_2 is age, x_3 is gender, x_4 is exercise, u includes health concern, and w is a randomized education dummy variable on health benefits of exercise (i.e., a coin is flipped to give person i the education if head comes up). Those who are health-conscious may exercise more, which means $COR(x_4, u) \neq 0$. Checking out $(i-iii)$ for w, first, w satisfies (i) because w is randomized. Second, those who received the health education are likely to exercise more, thus implying (ii). Third, receiving the education alone cannot affect blood pressure, and hence (iii) holds. (iii) does not mean that w should not influence y at all: (iii) is that w can affect y only indirectly through x_4.

Condition (i) is natural in view of $E(zu) = 0$. Condition (ii) is necessary as w is used as a "proxy" for x_4; if $COR(w, x_4) = 0$, then w cannot represent x_4—a rv from a coin toss is independent of x_4 and fails (ii) despite satisfying (i) and (iii). Condition (iii) is necessary to make $E(zx')$ invertible; an exogenous regressor x_2 (or x_3) already in the y-equation cannot be used as an instrument for x_4 despite it satisfies (i) and possibly (ii), because $E(zx')$ is not invertible if $z = (1, x_2, x_3, x_2)'$.

Recalling partial regression, only the part of x_4 not explained by the other regressors $(1, x_2, x_3)$ in the y equation contributes to explaining y. Among the part of x_4, w picks only the part uncorrelated with u, because w is uncorrelated with u by condition (i). *The instrument w is said to extract the "exogenous variation" in x_4.* In view of this, to be more precise, (ii) should be replaced with

$(ii)'$ $\qquad$ $0 \neq COR[w, \ \{\text{part of } x_4 \text{unexplained by the other}$
$$\text{regressors}(1, x_2, x_3)\}]$$
$$\Longleftrightarrow 0 \neq COR[w, \ \{\text{residual of the linear projection of}$$
$$x_4 \text{ on } (1, x_2, x_3)\}].$$

Condition $(ii)'$ can be (and should be) verified by the LSE of x_4 on w and the other regressors: the slope coefficient of w should be non-zero in this LSE for w to be a valid instrument. But conditions (i) and (iii) cannot be checked out; they can be only "argued for." In short, *an instrument should be excluded from the response equation and included in the endogenous regressor equation with zero correlation with the error term.*

There are a number of sources for the endogeneity of x_4:

- First, a *simultaneous relation* when x_4 is affected by y (as well as affecting y). For example, $x_{i4} = q_i'\gamma + \alpha y_i + v_i$ where q_i are regressors and v_i is an error term. This implies that u is correlated with x_4 through y: $u \rightarrow y \rightarrow x_4$. If y is the work hours of a spouse and x_4 is the work hours of the other spouse in the same family, then the simultaneous relation may occur.

- Second, a *recursive relation with correlated errors*. For example, $x_{i4} = q_i'\gamma + v_i$ and $COR(v, u) \neq 0$ holds (no simultaneity). Here x_4 is correlated with u through v: $x_4 \longleftarrow v \longrightarrow u$. In the preceding family work hour case, if x_4 is for the "leader" (i.e., the dominant spouse), y is for the "follower," and local labor-market-condition variables influencing both spouses are omitted, then these variables will lurk in u and v, leading to $COR(u, v) \neq 0$.

- Third, *errors-in-variables*. Here, x_4 is not observed, but its error-ridden version $x_{i4}^e = x_{i4} + e_i$ is. In this case, we can rewrite the y equation as

$$y_i = \ldots, +\beta_4 x_{i4} + u_i = \ldots, +\beta_4(x_{i4}^e - e_i) + u_i = \ldots, +\beta_4 x_{i4}^e + (u_i - \beta_4 e_i)$$

and use x_4^e as a regressor. But the new error $u - \beta_4 e$ is correlated with x_4^e through e.

4.1.3 Further Remarks

What if there is no variable available for instruments? In this case, it is tempting to use functions of exogenous regressors. Recall the above example:

$$y_i = \beta_1 + \beta_2 x_{i2} + \beta_3 x_{i3} + \beta_4 x_{i4} + u_i, \quad E(x_2 u) = E(x_3 u) = 0 \text{ but } E(x_4 u) \neq 0.$$

Functions of x_2 and x_3 (such as x_2^2 or $x_2 x_3$) qualify as instruments for x_4, if we know a priori that those functions are excluded from the y equation. But "smooth" functions such as x_2^2 are typically not convincing instruments, because x_2^2 may very well be included in the y equation if x_2 is so. Instead of smooth functions, non-smooth functions of exogenous regressors may be used as instruments if there are due justifications that they appear in the x_4 equation, but not in the y equation. Such examples can be seen in relation to "*regression discontinuity design*" in the treatment-effect literature; see Lee (2005a) and the references there. Those discontinuous functions then serve as "local instruments" around the discontinuity points.

In case of no instrument, the endogenous regressors may be dropped and LSE may be applied. But this leads to an omitted variable bias as examined already. For instance, suppose $x_{i4} = \gamma_1 + \gamma_2 x_{i3} + v_i$. Substitute this into the y_i equation to get

$$\begin{aligned} y_i &= \beta_1 + \beta_2 x_{i2} + \beta_3 x_{i3} + \beta_4(\gamma_1 + \gamma_2 x_{i3} + v_i) + u_i \\ &= (\beta_1 + \beta_4 \gamma_1) + \beta_2 x_{i2} + (\beta_3 + \beta_4 \gamma_2)x_{i3} + (u_i + \beta_4 v_i). \end{aligned}$$

When this is estimated by LSE, the slope estimator for x_3 is consistent for $\beta_3 + \beta_4 \gamma_2$, where $\beta_4 \gamma_2$ is nothing but the bias due to omitting x_4 in the LSE. The slope parameter $\beta_3 + \beta_4 \gamma_2$ of x_3 consists of two parts: the "direct effect" of x_3 on y, and the "indirect part" of x_3 on y through x_4. If x_3 affects x_4 but not the other way around, then the indirect part can be interpreted as the "indirect effect" of x_3 on y through x_4. So long as we are interested in the total effect $\beta_3 + \beta_4 \gamma_2$, the LSE is all right. But usually in economics, the desired effect is the "ceteris paribus" effect of changing x_3 while holding all the other variables (including x_4) constant.

The IVE can also be cast into a minimization problem. The sample analog of $E(zu)$ is $N^{-1} \sum_i z_i u_i$. Since u_i is unobservable, replace u_i by $y_i - x_i' b$

to get $N^{-1} \sum_i z_i (y_i - x_i' b)$. We can get the IVE by minimizing the deviation of $N^{-1} \sum_i z_i (y_i - x_i' b)$ from 0. Since $N^{-1} \sum_i z_i (y_i - x_i' b)$ is a $k \times 1$ vector, we need to choose how to measure the distance from 0. Adopting the squared Euclidean norm as usual and ignoring N^{-1}, we get

$$\left\{ \sum_i z_i (y_i - x_i' b) \right\}' \cdot \sum_i z_i (y_i - x_i' b) = \{ Z'(Y - X'b) \}' \cdot Z'(Y - X'b)$$

$$= (Y - Xb)' ZZ'(Y - Xb) = Y'ZZ'Y - 2b'X'ZZ'Y + b'X'ZZ'Xb.$$

The first-order condition of minimization is

$$0 = -2X'ZZ'Y + 2X'ZZ'Xb \implies b_{ive} = (Z'X)^{-1} Z'Y.$$

Although IVE can be cast into a minimization problem, it minimizes the distance of $N^{-1} \sum_i z_i (y_i - x_i' b)$ from 0. For LSE, we would be minimizing the distance of $N^{-1} \sum_i x_i (y_i - x_i' b)$ from 0, which is different from minimizing the scalar $N^{-1} \sum_i (y_i - x_i' b)^2$. This scalar minimand shows that LSE is a "prediction-error minimizing estimator" where y_i is the target and $x_i' b_{lse}$ is the predictor for the target. In minimizing $N^{-1} \sum_i (y_i - x_i' b)^2$, there is no concern for endogeneity: regardless of $E(xu) = 0$ holding or not, we can always minimize $N^{-1} \sum_i (y_i - x_i' b)^2$. The resulting estimator is, however, consistent for β in $y_i = x_i' \beta + u_i$ only if $E(xu) = 0$. The usual model fitness and R^2 are irrelevant for IVE, because, if they were, we would be using LSE, not IVE. Nevertheless, there is a pseudo R^2 to appear later that may be used for model selection with IVE, as the usual R^2 is used for the same purpose with LSE.

4.2   IVE Examples

Here we provide three empirical examples for IVE, using the same four regressor model as above with x_4 being endogenous. The reader will see that some instruments are more convincing than others. Among the three examples, the instruments in the first example will be the most convincing, followed by those in the second which are in turn more plausible than those in the third. More examples of instruments can be found in Angrist and Krueger (2001) and the references therein. It is not clear, however, who invented IVE. See Stock and Trebbi (2003) for some "detective work" on the origin of IVE.

EXAMPLE: FERTILITY EFFECT ON WORK. Understanding the relationship between fertility and female labor supply matters greatly in view of increasing labor market participation of women and declining fertility rates in many countries; the latter is also a long-term concern for pension systems. But finding a causal effect for either direction has proven difficult, as females are likely to decide on fertility and labor supply jointly, leading to a simultaneity problem. Angrist and Evans (1998) examined the effect of the number of children (x_4) on labor market outcomes. Specifically, their x_4 is a dummy variable for more than two children.

One instrument for x_4 is the dummy for the same sex children in a household: having only girls or boys in the first two births may result in more children than the couple planned otherwise. The random event (gender) for the first two children gives an exogenous variation to x_4, and the dummy for the same sex children is to take advantage of this variation. Another instrument is the dummy for twin second birth: having a twin second birth means an exogenous increase to the third child. Part of Table 2 (using a 1990 data set in US for women aged 21–35 with two or more children) in Angrist and Evans (1998) shows descriptive statistics (SD is in ($\cdot$)):

	#Children ever	More than two	First birth boy	Same sex	Twin second birth	First birth age
All women	2.50	0.375	0.512	0.505	0.012	21.8
	(0.76)	(0.48)	(0.50)	(0.50)	(0.108)	(3.5)
Wives	2.48	0.367	0.514	0.503	0.011	22.4
	(0.74)	(0.48)	(0.50)	(0.50)	(0.105)	(3.5)

The table shows that there is not much difference across all women data and wives only data, that the probability of boy is slightly higher than the probability of girl, and that the probability of twin birth is about 1%.

Part of Table 8 for the all women data in Angrist and Evans (1998) is:

y	Worked or not	Weeks worked	Hours per week	Labor income
LSE (SD)	−0.155 (0.002)	−8.71 (0.08)	−6.80	−3984
			(0.07)	(44.2)
IVE (SD)	−0.092 (0.024)	−5.66 (1.16)	−4.08	−2100
			(0.98)	(664.0)

For instance, having the third child decreases weeks worked by 6–9% and hours worked by 4–7 hours per week. Overall, IVE magnitudes are about 50–100% smaller than the LSE magnitudes.

EXAMPLE: POLICE IMPACTS ON CRIME. Whether the number of policemen (x_4) lowers crime rates (y) has been an important question in criminology. The main difficulty in assessing the effects has been the simultaneity problem between y and x_4. Suppose that x_4 decreases y, and y increases x_4 (a higher crime rates leads to the more policemen):

$$y_i \;=\; \beta_1 + \beta_2 x_{i2} + \beta_3 x_{i3} + \beta_4 x_{i4} + u_i, \quad x_{i4} = q_i'\gamma + \alpha y_i + v_i,$$
$$\text{(with } \beta_4 < 0 \text{ and } \alpha > 0)$$
$$\Longrightarrow \quad y_i = \beta_1 + \beta_2 x_{i2} + \beta_3 x_{i3} + \beta_4(q_i'\gamma + \alpha y_i + v_i) + u_i,$$
$$\text{(substituting the } x_4 \text{ equation)}$$
$$\Longrightarrow \quad y_i = \frac{1}{1 - \beta_4 \alpha}(\beta_1 + \beta_2 x_{i2} + \beta_3 x_{i3} + \beta_4 q_i'\gamma + \beta_4 v_i + u_i)$$

which is the y "*reduced form (RF).*" Substituting the y RF into $x_{i4} = q_i'\gamma + \alpha y_i + v_i$, we also get the x_4 RF:

$$x_{i4} = q_i'\gamma + \frac{\alpha}{1 - \beta_4 \alpha}(\beta_1 + \beta_2 x_{i2} + \beta_3 x_{i3} + \beta_4 q_i'\gamma + \beta_4 v_i + u_i) + v_i.$$

Judging from the u's slope $\alpha(1 - \beta_4 \alpha)^{-1} > 0$, we get $COR(x_4, u) > 0$. Suppose that LSE is run for the y equation ignoring the simultaneity. Then with $x_i = (1, x_{i2}, x_{i3}, x_{i4})'$, the LSE of y on x will be inconsistent by the magnitude

$$E^{-1}(xx')E(xu) = E^{-1}(xx')\{0, 0, 0, E(x_4 u)\}' :$$

the LSE for β_4 is upward biased and hence the LSE for β_4 can even be positive. Recalling the discussion on omitted variable bias, we can see that the bias is not restricted to β_4 if x_4 is correlated with x_2 or x_3, because the last column of $E^{-1}(xx')$ can "spread" $E(x_4 u) \neq 0$ to all components of the LSE.

One way to overcome the simultaneity problem is to use data for short periods. For instance, if y is a monthly crime number for city i and x_4 is the number of policemen in the same month, then it is unlikely that y affects x_4 as it takes time to adjust x_4, whereas x_4 can affect y almost instantly. Another way is to find an instrument. Levitt (1997) noted that the change in x_4 takes place almost always in election years, mayoral or gubernatorial. Thus he sets up a "panel (or longitudinal) data" model where y_{it} is a change in crime numbers for city i and year t, $x_{it,4}$ is a change in policemen, and $w_{it} = 1$ if year t is an election year at city i and 0 otherwise, because w_{it} is unlikely to be correlated with the error term in the crime number change equation. Levitt (1997) concluded that the police force size reduces (violent) crimes.

As McCrary (2002) noted, however, there was a small error in Levitt (1997). Levitt (2002) thus proposed the number of firefighters per capita as a new instrument for the number of policemen per capita. The panel data model used is

$$\Delta \ln(y_{it}) = \beta_p \ln(police_{i,t-1}) + x_{it}'\beta_x + \delta_i + \lambda_t + u_{it}$$

where i indexes large US cities with $N = 122$, t indexes years 1975–1995, $police_{i,t-1}$ instead of $police_{it}$ is used to mitigate the endogeneity problem, and x_{it} are the regressors other than police; δ_i is for the "city effect" (estimated by city dummies) and λ_t is for the "year effect" (estimated year dummies).

Part of Table 3 in Levitt (2002) for police effect is shown below with SD in ($\cdot$), where "LSE without city dummies" means the LSE without city dummies but with year dummies. By not using city dummies, the parameters are identified mainly with cross-city variation because cross-city variation is much greater than over-time variation, and this LSE is thus similar to cross-section LSE pooling all panel data.

y	Violent crimes per capita	Property crimes per capita
LSE without city dummies	0.562 (0.056)	0.113 (0.038)
LSE with city/year dummies	−0.076 (0.061)	−0.218 (0.052)
IVE with city/year dummies	−0.435 (0.231)	−0.501 (0.235)

This table shows that the LSE's are upward biased as analyzed above although the bias is smaller when the city dummies are used, and that police force expansion indeed reduces the number of crimes. The number of firefighters is an attractive instrument, but somewhat less convincing than the instruments in the fertility example.

EXAMPLE: ECONOMIC IMPACTS ON CRIME. In the preceding examples for IVE, the justification of the instruments was strong. Here is an IVE example with a weaker justification—this kind of cases are more common in practice.

Gould et al. (2002) analyzed the effect of local labor market conditions on crime rates in the US for 1979–1997. They set up a panel data model

$$y_{it} = x'_{it}\beta + \delta_i + u_{it}, \quad i = 1, ..., N, \ t = 1, ..., T$$

where y_{it} is the number of various offenses per 100,000 people in county i at year t, x_{it} includes the mean log weekly wage of non-college educated men ($wage_{i_t}$), unemployment rate of non-college educated men (ur_{it}), and the mean log household income (inc_{it}), and time dummies, δ_i is a time-constant error and u_{it} is a time-variant error. Our presentation in the following is a rough simplification of their longer models.

Since δ_i represents each county's unobserved long-standing culture and practice such as how extensively crimes are reported and so on, δ_i is likely to be correlated with x_{it}. They take the difference between 1979 and 1989 to remove δ_i and get (removing δ_i by differencing is a "standard" procedure in panel data)

$$\Delta y_i = \Delta x'_i \beta + \Delta u_i,$$

where $\Delta y_i \equiv y_{i,1989} - y_{i,1979}$, Δx_i and Δu_i are analogously defined, and $N = 564$. Their estimation results are as follows with SD in ($\cdot$):

LSE: $\quad b_{wage} = -1.13\,(0.38), \ b_{ur} = 2.35\,(0.62), \ b_{inc} = 0.71\,(0.35)$
$\qquad R^2 = 0.094,$

IVE: $\quad b_{wage} = -1.06\,(0.59), \ b_{ur} = 2.71\,(0.97), \ b_{inc} = 0.093\,(0.55);$

the instruments will be explained below. All three estimates in LSE are significant and show that low wage, high unemployment rate, and high household income increase the crime rate. The IVE is close to the LSE in $wage_{it}$ and ur_{it}, but much smaller for inc_{it} and insignificant. See Freeman (1999) for a survey on crime and economics.

It is possible that crime rates influence local labor market conditions, because firms may move out in response to high crime rates or firms may offer higher wages to compensate for high crime rates. This means that a simultaneous relation problem may occur between crime rates and labor market conditions. To avoid this problem, Gould et al. (2002) constructed a number of instruments. One of the instruments is

$$\sum_j (\text{employment share of industry } j \text{ in county } i \text{ in 1979})$$

$$\cdot(\text{national growth rate of industry } j \text{ for 1979–1989}).$$

The two conditions to check are $COR(\Delta u, z) = 0$ and $COR(\Delta x, z) \neq 0$. For $COR(\Delta u, z) = 0$, the primary reason to worry for endogeneity was the influence of the crime rate on the labor market conditions. But it is unlikely that a county's crime rates over 1979–1989 influenced the national industry growth rates over 1979–1989. Also the employment shares had been taken in 1979 before the crime rates were measured. These points support $COR(\Delta u, z) = 0$. For $COR(\Delta x, z) \neq 0$, consider a county in Michigan: if the national auto industry shrank during 1979–1989 and if the share of auto industry was large in the county in 1979, then the local labor market condition would have deteriorated.

4.3   IVE with More than Enough Instruments

4.3.1   IVE in Wide Sense

If a random variable w is independent u, then we get not just COR $(w, u) = 0$, but also $COR(w^2, u) = 0$. This means that if w is an instrument for an endogenous regressor x_4, then we may use both w and w^2 as instruments for x_4. In this case, $z_i = (1, x_{i2}, x_{i3}, w_i, w_i^2)'$, the dimension of which is bigger than the dimension of x_i: $E(zx')$ is not a square matrix as in the preceding subsection, and hence not invertible. There arises the question of selecting or combining more than enough instruments (i.e., more than enough moment conditions) for only k-many parameters. While a complete answer will be provided later, here we just provide one simple answer which also turns out to be optimal under homoskedasticity.

Suppose $E(zu) = 0$, where z is $p \times 1$ with $p \geq k$, the rank of $E(xz') = k$, and $E^{-1}(zz')$ exists. Observe

$$E\{z(y - x'\beta)\} = 0 \iff E(zy) = E(zx')\beta$$
$$\implies E(xz')E^{-1}(zz') \cdot E(zy) = E(xz')E^{-1}(zz') \cdot E(zx')\beta$$
$$\implies \beta = \{E(xz')E^{-1}(zz')E(zx')\}^{-1} \cdot E(xz')E^{-1}(zz')E(zy).$$

For the product AB of two matrices A and B where B^{-1} exists, $rank\,(AB) = rank(A)$; i.e., multiplication by a non-singular matrix B does not alter the rank of A. This fact implies that $E(xz')E^{-1}(zz')\ E(zx')$ has rank k and thus is invertible. If $E(zx')$ is invertible, then β in the last display becomes $E^{-1}(zx') \cdot E(zy)$ and the resulting b_{ive} is the IVE when the number of instruments is the same as the number of parameters.

The sample analog for β is the following *instrumental variable estimator*

$$b_{ive} = \left\{ \sum_i x_i z_i' \left(\sum_i z_i z_i' \right)^{-1} \sum_i z_i x_i' \right\}^{-1} \cdot \sum_i x_i z_i' \left(\sum_i z_i z_i' \right)^{-1} \sum_i z_i y_i$$

where many N^{-1}'s are ignored that cancel one another out. The consistency is obvious, and the asymptotic distribution of $\sqrt{N}(b_{ive} - \beta)$ is

$$N(0,\ G \cdot E(zz'u^2) \cdot G'), \quad \text{where } G \equiv \{E(xz')E^{-1}(zz')E(zx')\}^{-1}E(xz')E^{-1}(zz').$$

A consistent estimator for the asymptotic variance is

$$G_N \cdot \frac{1}{N} \sum_i z_i z_i' r_i^2 \cdot G_N',$$

where $r_i \equiv y_i - x_i' b_{ive}$, and

$$G_N \equiv \left\{ \frac{1}{N} \sum_i x_i z_i' \left(\frac{1}{N} \sum_i z_i z_i' \right)^{-1} \frac{1}{N} \sum_i z_i x_i' \right\}^{-1} \cdot \frac{1}{N} \sum_i x_i z_i' \left(\frac{1}{N} \sum_i z_i z_i' \right)^{-1}.$$

4.3.2   Various Interpretations of IVE

It is informative to see the IVE in matrices:

$$
\begin{aligned}
b_{ive} &= \left\{ X'Z(Z'Z)^{-1}Z'X \right\}^{-1} X'Z(Z'Z)^{-1}Z'Y \\
&= \left[\{Z(Z'Z)^{-1}Z'X\}'X \right]^{-1} \{Z(Z'Z)^{-1}Z'X\}'Y \\
&= (\hat{X}'X)^{-1}\hat{X}'Y, \quad \text{where } \hat{X} \equiv Z(Z'Z)^{-1}Z'X;
\end{aligned}
$$

$\hat{X}$ that has dimension $N \times k$ is "x fitted by z," or the part of x explained by z; $(Z'Z)^{-1}Z'X$ is the LSE of z on x. $\hat{X}$ combines more than k instruments into just k many.

Using $P_Z \equiv Z(Z'Z)^{-1}Z'$, b_{ive} can also be written as (recall that P_Z is idempotent)

$$b_{ive} = \{(P_Z X)'P_Z X\}^{-1}(P_Z X)'P_Z Y = (\hat{X}'\hat{X})^{-1}\hat{X}'Y$$

as if b_{ive} were the LSE for the equation $Y = \hat{X}\beta + error$ where the "error" is $Y - \hat{X}\beta$. This rewriting accords an interesting interpretation of b_{ive}. As X is endogenous, we can decompose X as $\hat{X} + (X - \hat{X})$ where the first component is exogenous, "sifted" from X using P_Z, and the second component is the remaining endogenous component. Then

$$Y = X\beta + U = \{\hat{X} + (X - \hat{X})\}\beta + U = \hat{X}\beta + \{(X - \hat{X})\beta + U\}.$$

Hence, β can be estimated by the LSE of Y on $\hat{X}$ so long as $\hat{X}$ is "asymptotically uncorrelated" with the error term $(X - \hat{X})\beta + U$, which is shown in the following.

Recall

"$A_N = o_p(1)$" means $A_N \to^p 0$ (and $B_N = C_N + o_p(1)$ means $B_N - C_N \to^p 0$).

The error vector $(X - \hat{X})\beta + U$ satisfies $N^{-1}\hat{X}'\{(X - \hat{X})\beta + U\} = o_p(1)$ because

$$
\begin{aligned}
\frac{1}{N}\hat{X}'\{(X - \hat{X})\beta + U\} &= \frac{1}{N}\left(\hat{X}'X\beta - \hat{X}'\hat{X}\beta + \hat{X}'U\right) = \frac{1}{N}\hat{X}'U, \\
&\quad \text{for } \hat{X}'X = \hat{X}'\hat{X} \\
&= \frac{1}{N}\{Z(Z'Z)^{-1}Z'X\}'U \\
&= \frac{1}{N}X'Z \cdot \left(\frac{1}{N}Z'Z\right)^{-1} \cdot \frac{1}{N}ZU = o_p(1).
\end{aligned}
$$

The expression $(\hat{X}'\hat{X})^{-1}\hat{X}'Y$ also demonstrates that the so-called "two-stage LSE (2SLSE)" for simultaneous equations is nothing but IVE. For simplification, consider two simultaneous equations with two endogenous variables y_1 and y_2:

$$
\begin{aligned}
y_1 &= \alpha_1 y_2 + x_1'\beta_1 + u_1, \quad y_2 = \alpha_2 y_1 + x_2'\beta_2 + u_2, \\
COR(x_j, u_{j'}) &= 0, \; j, j' = 0, 1 \quad \text{and} \quad x_1 \neq x_2.
\end{aligned}
$$

Let z denote the system exogenous regressors (i.e., the collection of the elements in x_1 and x_2). Denoting the regressors for the y_1 equation as $x \equiv (y_2, x_1')'$, the first step of 2SLSE for (α_1, β_1) is the LSE of y_2 on z to obtain the fitted value $\hat{y}_2$ of y_2, and the second step is the LSE of y_1 on $(\hat{y}_2, x_1)$. This 2SLSE is nothing but the IVE where the first step is $P_Z X$ to obtain the LSE fitted value of x on z—the LSE fitted value of x_1 on z is simply x_1—and the second step is the LSE of y_1 on $P_Z X$.

4.3.3   Further Remarks

We already mentioned that the usual R^2 is irrelevant for IVE. Despite this, sometimes

$$1 - \frac{\sum_i (y_i - x_i' b_{ive})^2}{\sum_i (y_i - \bar{y})^2}$$

is reported in practice as a measure of model fitness. Pesaran and Smith (1994) showed, however, that this should not be used as a model selection criterion. Instead, they propose the following *pseudo R^2 for IVE*:

$$R_{ive}^2 = 1 - \frac{\sum_i (y_i - \hat{x}_i' b_{ive})^2}{\sum_i (y_i - \bar{y})^2}$$

where $\hat{x}_i$ is the ith row of $\hat{X}$. R_{ive}^2 satisfies $0 \leq R_{ive}^2 \leq 1$ and takes 1 if $y = \hat{x}_i' b_{ive}$ and 0 if all slope components of b_{ive} are zero. The intuition for R_{ive}^2 was given ahead already: $Y = \hat{X}\beta + error$ with the error term asymptotically orthogonal to $\hat{X}$.

Observe

$$
\begin{aligned}
E(u^2|z) &= \sigma^2 \text{ (homoskedasticity)} \text{ implies } G \cdot E(zz'u^2) \cdot G' \\
&= \sigma^2 \{ E(xz')E^{-1}(zz')E(zx') \}^{-1};
\end{aligned}
$$

here, homoskedasticity is wrt z, not x. To compare this to the LSE asymptotic variance $\sigma^2 E^{-1}(xx')$ under homoskedasticity, observe

$$
\begin{aligned}
E(xx') &= E(xz')E^{-1}(zz')E(zx') + E\{(x - \gamma'z)(x - \gamma'z)'\} \quad \text{where} \\
\gamma &\equiv E^{-1}(zz')E(zx').
\end{aligned}
$$

This is a decomposition of $E(xx')$ into two parts, one explained by z and the other unexplained by z; $x - \gamma'z$ is the "residual" (compared with the linear projection, $E(x|z)$ is often called the *projection of x on z*).

From the decomposition, we get

$$
\begin{aligned}
E(xx') \geq E(xz')E^{-1}(zz')E(zx') &\iff E^{-1}(xx') \\
&\leq \{ E(xz')E^{-1}(zz')E(zx') \}^{-1};
\end{aligned}
$$

the former is called *generalized Cauchy-Schwarz inequality*. This shows that the "explained variation" $E(xz')E^{-1}(zz')E(zx')$ is not greater than the "total variation" $E(xx')$. Hence, under homoskedasticity, LSE is more efficient than IVE; under homoskedasticity, there is no reason to use IVE unless $E(xu) \neq 0$. Under heteroskedasticity, however, the asymptotic variances of LSE and IVE are difficult to compare, because the comparison depends on the functional forms of $V(u|x)$ and $V(u|z)$.

5   Generalized Method-of-Moment Estimator (GMM)

In IVE, we saw an answer to the question of how to combine more than enough moment conditions. There, the idea was to multiply the more than enough p-many equations $E(zy) = E(zx')\beta$ from $E(zu) = 0$ with a $k \times p$ matrix. But, there are many candidate $k \times p$ matrices, with $E(xz')E^{-1}(zz')$

for IVE being just one of them. If we use $E(xz')W^{-1}$ where W is a $p \times p$ p.d. matrix, we will get

$$E(xz')W^{-1}E(zy) = E(xz')W^{-1}E(zx')\beta$$
$$\implies \quad \beta = \{E(xz')W^{-1}E(zx')\}^{-1}E(xz')W^{-1}E(zy).$$

As it turns out, $W = E(zz'u^2)$ is optimal for iid samples, which is the theme of this section.

5.1   GMM Basics

Suppose there are $p \, (\geq k)$ population moment conditions

$$E\psi(y, x, z, \beta) = 0$$

which may be nonlinear in β; we will often write $\psi(y, x, z, \beta)$ simply as $\psi(\beta)$. The *generalized method-of-moment (GMM)* estimator is a class of estimators indexed by W that is obtained by minimizing the following wrt b:

$$\frac{1}{\sqrt{N}} \sum_i \psi(b)' \cdot W^{-1} \cdot \frac{1}{\sqrt{N}} \sum_i \psi(b).$$

The question in GMM is which W to use. Hansen (1982) showed that the W yielding the smallest variance for the class of GMM estimator is

$$V \left\{ \frac{1}{\sqrt{N}} \sum_i \psi(\beta) \right\} \quad [= E\{\psi(\beta)\psi(\beta)'\} \text{ for iid samples}];$$

this becomes $E(zz'u^2)$ when $\psi(\beta) = z(y - x'\beta)$.

The intuition for $W = V\{N^{-1/2} \sum_i \psi(\beta)\}$ is that, in the minimization, it is better to standardize $N^{-1} \sum_i \psi(b)$; otherwise one component with a high variance can unduly dominate the minimand. The optimal GMM is simply called (the) GMM. The GMM with $W = I_p$ is sometimes called the "un-weighted (or equally weighted) GMM"; the name "equally weighted GMM," however, can be misleading, for the optimal GMM has this interpretation. It may seem that we may be able to do better than GMM by using a dis-tance other than the quadratic distance. But Chamberlain (1987) showed that the GMM is the efficient estimator under the given moment condition $E\psi(\beta) = 0$. In statistics, $\psi(y, x, z, \beta) = 0$ is called "estimating functions" (Godambe, 1960) and $E\psi(y, x, z, \beta) = 0$ "*estimating equations*"; see Owen (2001) and the references therein.

While GMM with nonlinear models will be examined in another chapter in detail, for the linear model, we have

$$E\psi(\beta) = E\{z(y - x'\beta)\} = E(zu) = 0.$$

In matrices, the GMM minimand with W is

$$\begin{aligned}
&\{Z'(Y - Xb)\}'W^{-1}\{Z'(Y - Xb)\} \\
&= (Y'ZW^{-1} - b'X'ZW^{-1}) \cdot (Z'Y - Z'Xb) \\
&= Y'ZW^{-1}Z'Y - 2b'X'ZW^{-1}Z'Y + b'X'ZW^{-1}Z'Xb.
\end{aligned}$$

From the first-order condition of minimization, we get $X'ZW^{-1}Z'Y = X'ZW^{-1}Z'Xb$. Solve this to obtain

$$
\begin{aligned}
b_W &= (X'ZW^{-1}Z'X)^{-1} \cdot (X'ZW^{-1}Z'Y) \\
&= \left(\sum_i x_i z_i' W^{-1} \sum_i z_i x_i' \right)^{-1} \cdot \sum_i x_i z_i' W^{-1} \sum_i z_i y_i, \quad \text{in vectors.}
\end{aligned}
$$

Clearly b_W is consistent for β, and its asymptotic distribution is

$$
\sqrt{N}(b_W - \beta) \rightsquigarrow N(0, C_W), \quad \text{where}
$$
$$
\begin{aligned}
C_W \equiv\ & \{E(xz')W^{-1}E(zx')\}^{-1}E(xz')\ W^{-1}E(zz'u^2)W^{-1}\ E(zx') \\
& \{E(xz')W^{-1}E(zx')\}^{-1}.
\end{aligned}
$$

With $W = E(zz'u^2)$, this matrix becomes $\{E(xz')E^{-1}(zz'u^2)E(zx')\}^{-1}$, and we get the GMM with

$$
\sqrt{N}(b_{gmm} - \beta) \rightsquigarrow N(0,\ \{E(xz')E^{-1}(zz'u^2)E(zx')\}^{-1}).
$$

Since $W = E(zz'u^2)$ can be estimated consistently with

$$
\frac{1}{N}\sum_i z_i z_i' r_i^2 = \frac{1}{N}Z'DZ,
$$

where $r_i = y_i - x_i' b_{ive}$ and $D = diag(r_1^2, ..., r_N^2)$, we get

$$
\begin{aligned}
b_{gmm} &= \left\{ \sum_i x_i z_i' \left(\sum_i z_i z_i' r_i^2 \right)^{-1} \sum_i z_i x_i' \right\}^{-1} \cdot \\
& \quad \sum_i x_i z_i' \left(\sum_i z_i z_i' r_i^2 \right)^{-1} \sum_i z_i y_i \\
&= (X'Z(Z'DZ)^{-1}Z'X)^{-1}(X'Z(Z'DZ)^{-1}Z'Y) \quad \text{in matrices.}
\end{aligned}
$$

Differently from IVE, $Z(ZDZ')^{-1}Z'$ is no longer the linear projection matrix of Z. A consistent estimator for the GMM asymptotic variance $\{E(xz')E^{-1}(zz'u^2)E(zx')\}^{-1}$ is easily obtained: it is simply the first part $(X'Z(Z'DZ)^{-1}Z'X)^{-1}$ of b_{gmm} times N.

5.2   GMM Remarks

A nice feature of GMM is that it also provides a specification test, called "GMM over-identification test": with $u_{Ni} \equiv y_i - x_i' b_{gmm}$,

$$
\frac{1}{\sqrt{N}}\sum_i z_i' u_{Ni} \cdot \left(\frac{1}{N}\sum_i z_i z_i' u_{Ni}^2 \right)^{-1} \cdot \frac{1}{\sqrt{N}}\sum_i z_i u_{Ni} \rightsquigarrow \chi_{p-k}^2.
$$

Too big a value, greater than an upper quantile of χ^2_{p-k}, indicates that some moment conditions do not hold (or some other assumptions of the model may be violated). The reader may wonder how we can test for the very moment conditions that were used to get the GMM. If there are only k moment conditions, this concern is valid. But when there are more than k moment conditions (p-many), essentially only k of them get to be used in obtaining the GMM. The GMM over-identification test checks if the remaining $p - k$ moment conditions are satisfied by the GMM, as can be seen in the degrees of freedom ("dof") of the test.

The test statistics may be viewed as

$$\sum_i \left[\left\{ \frac{u_{Ni} z_i'}{\sqrt{N}} \left(\sum_i \frac{z_i u_{Ni}}{\sqrt{N}} \frac{z_i' u_{Ni}}{\sqrt{N}} \right)^{-1} \sum_i \frac{z_i u_{Ni}}{\sqrt{N}} 1 \right\}' \cdot \right.$$
$$\left. \left\{ \frac{u_{Ni} z_i'}{\sqrt{N}} \left(\sum_i \frac{z_i u_{Ni}}{\sqrt{N}} \frac{z_i' u_{Ni}}{\sqrt{N}} \right)^{-1} \sum_i \frac{z_i u_{Ni}}{\sqrt{N}} 1 \right\} \right].$$

Defining the matrix version for $z_i u_{Ni}/\sqrt{N}$ as G—i.e., the ith row of G is $z_i' u_{Ni}/\sqrt{N}$—this display can be written as

$$\{G(G'G)^{-1} G' 1_N\}' \{G(G'G)^{-1} G' 1_N\} = 1_N' G (G'G)^{-1} G' 1_N$$

The inner-product form shows that the test statistic is non-negative at least.

Using the GMM over-identification test, a natural thing to do is to use only those moment conditions that are not rejected by the test. This can be done in practice by doing GMM on various subsets of the moment conditions, which would be ad hoc, however. Andrews (1999) and Hall and Peixe (2003) provided a formal discussion on this issue of selecting valid moment conditions, although how popular these suggestions will be in practice remains to be seen.

Under the homoskedasticity $E(u^2|z) = \sigma^2$, $W = \sigma^2 E(zz') = \sigma^2 Z'Z/N + o_p(1)$. But any multiplicative scalar in W is irrelevant for the minimization. Hence setting $W = Z'Z$ is enough, and b_{gmm} becomes b_{ive} under homoskedasticity; the aforementioned optimality of b_{ive} comes from the GMM optimality under homoskedasticity. Under homoskedasticity, we do not need an initial estimator to get the residuals r_i's. But when we do not know whether homoskedasticity holds or not, GMM is obtained in two stages: first apply IVE to get the r_i's, then use $\sum_i z_i z_i' r_i^2$ to get the GMM. For this reason, the GMM is sometimes called a "two-stage IVE."

We can summarize our analysis for the linear model under $p \times 1$ moment condition $E(zu) = 0$ and heteroskedasticity of unknown form as follows. First, the efficient estimator when $p \geq k$ is

$$b_{gmm} = \{X'Z(Z'DZ)^{-1} Z'X\}^{-1} X'Z(Z'DZ)^{-1} Z'Y.$$

If homoskedasticity prevails,

$$b_{ive} = \{X'Z(Z'Z)^{-1} Z'X\}^{-1} X'Z(Z'Z)^{-1} Z'Y$$

is the efficient estimator to which b_{gmm} becomes asymptotically equivalent. If $p = k$ and $(N^{-1} \sum_i z_i x_i')^{-1}$ exists, then

$$b_{gmm} = \{X'Z(Z'DZ)^{-1}Z'X\}^{-1}X'Z(Z'DZ)^{-1}Z'Y = (Z'X)^{-1}(Z'Y);$$

i.e., $b_{gmm} = b_{ive}$. Furthermore, if $z = x$, then $b_{gmm} = b_{ive} = b_{lse}$. Since GMM is efficient under $E(zu) = 0$, IVE is also efficient under homoskedasticity; if $(\sum_i z_i x_i')^{-1}$ exists, IVE inherits the efficiency from GMM because $b_{gmm} = b_{ive}$, homoskedasticity or not; furthermore, LSE is efficient when $z = x$, because $b_{gmm} = b_{lse}$. This way of characterizing the LSE efficiency is more relevant to economic data than using the conventional Gauss–Markov theorem in many econometric textbooks that requires non-random regressors.

There have been some further developments in linear-model IVE/ GMM. The main issues there are weak instruments (i.e., small correlations between instruments and endogenous variables), small sample performance of IVE/GMM (e.g., small sample bias), small sample distribution (i.e., non-normal contrary to the asymptotic normality), and estimation of the variance (e.g., under-estimation of the variance). Just to name a few studies for readers interested in these topics, Donald and Newey (2001) showed how to choose the number of instruments by minimizing a mean-squared error criterion for IVE and other estimators, Stock et al. (2002) provided a survey on the literature, and Windmeijer (2005) suggested a correction to avoid the variance under-estimation problem. See also Hall (2005) for an extensive review on GMM.

5.3   GMM Examples

As an example of GMM moment conditions, consider a "rational expectation" model:

$$\begin{aligned} y_t &= \rho \cdot E(y_{t+1}|I_t) + x_t'\beta + \varepsilon_t, \quad t = 1, ..., T, \\ E(\varepsilon_t x_{t-j}) &= E(\varepsilon_t y_{t-j}) = 0 \ \forall j = 1, ..., t \end{aligned}$$

where I_t is the information available up to period t including x_t, y_{t-1}, x_{t-1}, ... Here ρ captures the effect of the expectation of y_{t+1} on y_t. One way to estimate ρ and β is to replace $E(y_{t+1}|I_t)$ by y_{t+1}:

$$\begin{aligned} y_t &= \rho y_{t+1} + x_t'\beta + \varepsilon_t + \rho\{E(y_{t+1}|I_t) - y_{t+1}\} \\ &\equiv \rho y_{t+1} + x_t'\beta + u_t, \quad t = 1, ..., T-1, \quad \text{where } u_t \\ &\equiv \varepsilon_t + \rho\{E(y_{t+1}|I_t) - y_{t+1}\}. \end{aligned}$$

Then y_{t-1}, x_{t-1}, y_{t-2}, x_{t-2}, ... are all valid instruments because the error term $E(y_{t+1}|I_t) - y_{t+1}$ is uncorrelated with all available information up to t. If $E(x_t \varepsilon_t) = 0$, then x_t is also a good instrument.

EXAMPLE: HOUSE SALE (continued). In estimating the DISC equation, ln(T) is a possibly endogenous variable as already noted. The endogeneity can be dealt with IVE and GMM. We will use L1, L2, and L3 as instruments. An argument for the instruments would be that, while the market conditions and the characteristics of the house and realtor may influence DISC directly, it is unlikely that DISC is affected directly by when to list the house in the market. The reader may object to this argument, in which case the following should be taken just as an illustration.

Although we cannot test for the exclusion restriction, we can at least check whether the three variables have explanatory power for the potentially endogenous regressor $\ln(T)$. For this, the LSE of $\ln(T)$ on the instruments and the exogenous regressors was done to yield (heteroskedasticity-robust variance used):

$$\ln(T_i) = -1.352 +, ..., -\underset{(-2.77)}{0.294 \cdot L_1} - \underset{(-2.21)}{0.269 \cdot L_2} - \underset{(-1.22)}{0.169 \cdot L_3}, \ R^2 = 0.098,$$
$$\underset{(t-value)}{} \ \ \ \underset{(-0.73)}{}$$

which shows that indeed L1, L2, and L3 have explanatory power for $\ln(T)$. Table 2 shows the LSE, IVE, and GMM results (LSE is provided here again for the sake of comparison). The pseudo R^2 for the IVE is 0.144; compare this to the $R^2 = 0.34$ of the LSE. The GMM over-identification test statistic value and its p-value are, respectively, 2.548 and 0.280, not rejecting the moment conditions.

Table 2: LSE, IVE, and GMM for House Sale Discount %

	LSE			IVE			GMM	
	b_{lse}	tv-ho	tv-het	b_{ive}	tv-ho	tv-het	b_{gmm}	tv
Ln(T)	4.60	12.23	7.76	10.57	2.66	2.84	10.35	2.78
1	−2.46	−0.24	−0.23	3.51	0.26	0.21	1.65	0.10
BATH	0.11	0.17	0.18	−0.15	−0.19	−0.19	−0.43	−0.54
ELEC	1.77	2.60	2.46	0.75	0.69	0.68	0.87	0.80
RM	−0.18	−0.71	−0.67	0.08	0.22	0.20	0.14	0.36
TAX	−1.74	−1.65	−1.28	−1.27	−0.94	−0.86	−1.05	−0.71
YR	−0.15	−5.96	−3.87	−0.15	−4.93	−3.71	−0.15	−3.75
Ln(LP)	6.07	3.73	2.52	3.15	1.12	0.97	2.79	0.87
BIGS	−2.15	−3.10	−2.56	−1.57	−1.66	−1.56	−1.47	−1.46
RATE	−2.99	−3.25	−3.10	−5.52	−2.73	−2.51	−5.05	−2.39
SUPPLY	1.54	1.02	1.06	1.96	1.03	1.14	2.11	1.23

In the IVE, the tv–ho's are little different from the tv–het's other than for YR. The difference between the tv for GMM and the tv–het for IVE is also negligible. The LSE have far more significant variables than the IVE and GMM which are close to each other. Ln(T)'s estimate is about 50% smaller in LSE than in IVE and GMM. ELEC and ln(LP) lose its significance in the IVE and GMM and the estimate sizes are also halved. YR has almost the same estimates and t-values across the three estimators. BIGS have similar estimates across the three estimators, but not significant in the IVE and GMM. RATE is significant for all three estimators, and its value changes from -3 in LSE to -5 in the IVE and GMM. Overall, the signs of the significant estimates are the same across all estimators, and most earlier remarks made for LSE apply to IVE and GMM.

6   Generalized Least Squares Estimator (GLS)

In WLS, we assumed that $u_1, ..., u_N$ are independent and that $E(u_i^2|x_i) = \omega(x_i, \theta)$ is a parametric function of x_i with some unknown parameter vector θ. In this section, we generalize WLS further by allowing $u_1, ..., u_N$ to be correlated and $u_i u_j$ to be heteroskedastic, which leads to "Generalized LSE (GLS)." Although GLS is not essential for our main theme of using "simple" moment conditions, understanding GLS is helpful in understanding GMM and its efficiency issue. GLS will appear later in other contexts as well.

6.1   GLS Basics

Suppose

$$E(u_i u_j | x_1, ..., x_N) = \omega(x_1, ..., x_N, \theta) \quad \forall i, j$$

for some parametric function ω. The product $u_i u_j$ may depend only on x_i and x_j, but for more generality, we put all $x_1, ..., x_N$ in the conditioning set. For example, if the data come from a small town, then $u_i u_j$ may depend on all $x_1, ..., x_N$. If we set $E(u_i u_j | x_1, ..., x_N) = \sigma$ which is a non-zero constant for all $i \neq j$, then we are allowing for dependence between u_i and u_j while ruling out heteroskedasticity. Recall that the consistency of LSE requires only $E(xu) = 0$; there is no restriction on the dependence among $u_1, ..., u_N$ nor on the form of heteroskedasticity. Hence, correlations among $u_1, ..., u_N$ or unknown forms of heteroskedasticity do not make LSE inconsistent; they may make either the LSE asymptotic variance matrix formula invalid or the LSE inefficient.

Writing the assumption succinctly using matrix notations, we have

$$E(UU'|X) = \Omega(X; \theta);$$

denote $\Omega(X; \theta)$ just as Ω to simply notation, and pretend that θ is known for a while. As we transformed the original equation in WLS so that the resulting error term variance matrix becomes homoskedastic with unit variance, multiply $Y = X\beta + U$ by $\Omega^{-1/2} = H\Lambda^{-0.5}H'$ where $\Omega = H\Lambda H'$, Λ is the diagonal matrix of the eigenvalue of Ω, and H is a matrix whose columns are orthonormal eigenvectors) to get

$$\Omega^{-1/2}Y = \Omega^{-1/2}X\beta + U^* \quad \text{where } U^* \equiv \Omega^{-1/2}U$$

$$\implies \quad E(U^*U^{*\prime}|X) = E(\Omega^{-1/2}UU'\Omega^{-1/2}|X) = \Omega^{-1/2}\Omega\Omega^{-1/2}$$

$$= \quad H\Lambda^{-0.5}H'H\Lambda H'H\Lambda^{-0.5}H' = I_N.$$

Define

$$X^* \equiv \Omega^{-1/2}X \quad \text{and} \quad Y^* \equiv \Omega^{-1/2}Y,$$

and apply LSE to get the *Generalized LSE (GLS)*

$$b_{gls} = (X^{*\prime}X^*)^{-1}(X^{*\prime}Y^*)$$

$$= (X'\Omega^{-1}X)^{-1}X'\Omega^{-1}Y = \beta + (X'\Omega^{-1}X)^{-1} \cdot X'\Omega^{-1}U.$$

As in WLS, we need to replace θ with a first-stage $\sqrt{N}$-consistent estimator, say $\hat{\theta}$; call the GLS with $\hat{\theta}$ the *"feasible GLS"* and the GLS with θ the *"infeasible GLS."* Whether the feasible GLS is consistent with the same asymptotic distribution as the infeasible GLS follows depends on the form of $\Omega(X;\theta)$, but in all cases we will consider GLS for, this will be the case as to be shown in a later chapter. In the transformed equation, the error terms are iid and homoskedastic with unit variance. Thus we get

$$\sqrt{N}(b_{GLS} - \beta) \rightsquigarrow N(0, E^{-1}(x^* x^{*\prime})).$$

The variance matrix $E(x^* x^{*\prime})$ can be estimated consistently with

$$\frac{1}{N} \sum_i x_i^* x_i^{*\prime} = \frac{1}{N} X^{*\prime} X^* = \frac{1}{N} X' \Omega(X;\hat{\theta})^{-1} X.$$

6.2   GLS Remarks

If we define $Z \equiv \Omega^{-1} X$, then $b_{gls} = (Z'X)^{-1} Z'Y$, which is reminiscent of IVE. But, differently from that IVE was motivated to avoid inconsistency of LSE, the main motivation for GLS is a more efficient estimation than GMM. In GMM, the functional form $E(UU'|X)$ is not specified; rather, GMM just allows $E(UU'|X)$ to be an arbitrary unknown function of X. In contrast, GLS specifies fully the functional form $E(UU'|X)$. Hence, GLS makes use of more assumptions than GMM, and as a consequence, GLS is more efficient than GMM—more on this later. But the obvious disadvantage of GLS is that the functional form assumption on $E(UU'|X)$ can be wrong, which then nullifies the advantage and makes the GLS asymptotic variance formula invalid.

Recall that, when Ω is diagonal, we have two ways to proceed: one is doing LSE with an asymptotic variance estimator allowing for an unknown form of heteroskedasticity, and the other is specifying the form of Ω to do WLS; the latter is more efficient if the specified form is correct. When Ω is not diagonal, an example of which is provided below, again we can think of two ways to proceed, one of which is specifying the form of Ω to do GLS. The other way would be doing LSE with an asymptotic variance estimator allowing for an unknown form of Ω. When nonlinear GMM is discussed later, we will see asymptotic variance matrix estimators allowing for an unknown form of heteroskedasticity and correlations among $u_1, ..., u_N$.

To see an example of GLS with a specified non-diagonal Ω, consider the following model with dependent error terms (the so-called *"auto-regressive errors of order one"*):

$$y_t = x_t'\beta + u_t, \quad u_t = \rho u_{t-1} + v_t, \quad |\rho| < 1,$$
$$u_0 = 0, \, t = 1, ..., T.$$
$$\{v_t\} \text{ are iid with } E(v) = 0 \text{ and } E(v^2) \equiv \sigma_v^2 < \infty, \text{ and independent}$$
$$\text{of } x_1, ..., x_T.$$

By substituting $u_{t-1}, u_{t-2}, \ldots$ successively, we get

$$u_t = \rho u_{t-1} + v_t = \rho^2 u_{t-2} + v_t + \rho v_{t-1} = \rho^3 u_{t-3} + v_t + \rho v_{t-1} + \rho^2 v_{t-2} = \ldots,$$

from which $E(u_t^2) \to \sigma_u^2 \equiv \sigma_v^2/(1-\rho^2)$ as $t \to \infty$. Also observe

$$
\begin{aligned}
E(u_t u_{t-1}) &= E\{(\rho u_{t-1} + v_t)u_{t-1}\} = \rho E(u_{t-1}^2) \simeq \rho \sigma_u^2, \\
E(u_t u_{t-2}) &= E\{(\rho^2 u_{t-2} + v_t + \rho v_{t-1})u_{t-2}\} = \rho^2 E(u_{t-1}^2) \simeq \rho^2 \sigma_u^2.
\end{aligned}
$$

As $\{v_t\}$ are independent of $x_1, \ldots, x_T$ and u_t consists of $\{v_t\}$, u_t is independent of $x_1, \ldots, x_T$. Hence, $E(UU'|X) = E(UU')$ and

$$
\begin{aligned}
\Omega &= \begin{bmatrix} E(u_1 u_1) & \cdots & E(u_1 u_N) \\ \vdots & & \vdots \\ E(u_N u_1) & \cdots & E(u_N u_N) \end{bmatrix} \\[2ex]
&\simeq \sigma_u^2 \begin{bmatrix} 1 & \rho & \rho^2 & \rho^3 & \cdots & \rho^{N-1} \\ \rho & 1 & \rho & \rho^2 & \cdots & \rho^{N-2} \\ \vdots & \vdots & \vdots & \vdots & & \vdots \\ \rho^{N-1} & \rho^{N-2} & \rho^{N-3} & \cdots & \rho & 1 \end{bmatrix}.
\end{aligned}
$$

To implement the GLS, first do the LSE of y_t on x_t to get the residual $\hat{u}_t$. Second, replace ρ with the LSE estimator $\hat{\rho}$ of $\hat{u}_t$ on $\hat{u}_{t-1}$; σ_u^2 can be replaced by 1 because any scale factor in Ω is canceled in the GLS formula. Third, transform the equation with $\hat{\Omega}^{-1/2}$ and carry out the final LSE on the transformed equation.

6.3   Efficiency of LSE, GLS, and GMM

One may ask why we use LSE instead of some other estimators. For instance, minimizing $N^{-1}\sum_i |y_i - x_i'b|$ may be more natural than LSE. The usual answer found in many econometric textbooks is that LSE has the smallest variance among the "unbiased linear estimators" where a linear estimator a_N should be written as $A \cdot Y$ for some $N \times N$ constant matrix A, and a_N is said to be unbiased for β if $E(a_N) = \beta$. However, this answer is not satisfactory, for unbiasedness is hard to establish for nonlinear estimators. Also, focusing on the linear estimators is too narrow. In the following, we provide a modern answer which shows an optimality of LSE and efficiency comparison of LSE and GLS. The optimality of LSE is implied by the optimality of GMM.

Chamberlain (1987) showed the smallest possible variance (or, the "efficiency bound") under a general moment condition. His results are valid for nonlinear as well as linear models. For the linear model under the iid assumption on observations, suppose we have a moment condition

$$E(zu) = E\{z(y - x'\beta)\} = 0$$

where $y = x'\beta + u$ and z has at least k components; z may include x and u may be a vector. Using the moment condition only, the smallest possible variance for ("regular") estimators for β is

$$\{E(xz') \cdot E^{-1}(zuuz') \cdot E(zx')\}^{-1}.$$

This is the asymptotic variance of GMM, which means that GMM is efficient under $E(zu) = 0$, $y = x'\beta + u$, and the iid assumption. When $z = x$ and u is a scalar, the efficiency bound becomes

$$E^{-1}(xx') \cdot E(xx'u^2) \cdot E^{-1}(xx')$$

which is the asymptotic variance of LSE. Thus LSE is the most efficient under the moment condition $E(xu) = 0$, the linear model, and the iid assumption.

Chamberlain (1987) also showed that if

$$E(u|z) = E(y - x'\beta|z) = 0$$

then the smallest possible variance (or the efficiency bound) is

$$E_z^{-1} \left\{ E\left(\frac{\partial \left(y - x'\beta \right)}{\partial \beta} | z \right) \cdot E^{-1}(uu'|z) \cdot E\left(\frac{\partial \left(y - x'\beta \right)}{\partial \beta'} | z \right) \right\}.$$

If z includes x, then this becomes

$$E_z^{-1} \left\{ -x \cdot E^{-1}(uu'|z) \cdot (-x') \right\}.$$

If $z = x$ and u is a scalar, then the bound becomes the asymptotic variance of GLS

$$E^{-1} \left\{ \frac{xx'}{V(u|x)} \right\}.$$

Interestingly, if the error term is homoskedastic, then the two bounds under $E(xu) = 0$ and $E(u|x) = 0$ agree:

$$\sigma^2 E^{-1}(xx').$$

This observation might be, however, misleading, because the homoskedasticity condition is an extra information which could change the efficiency bound.

Observe $E\{xx'/V(u|x)\} = E[\{x/SD(u|x)\}\{x'/SD(u|x)\}]$ which is the "variation" of $x/SD(u|x)$. Also observe

$$E(xx')E^{-1}(xx'u^2)E(xx') = E(xx')E_x^{-1} \left\{ xx' E_{u|x}(u^2) \right\} E(xx')$$

$$= E \left\{ \frac{x}{SD(u|x)} x'SD(u|x) \right\} E^{-1} \left[\{x \ SD(u|x)\} \ \{x'SD(u|x)\} \right] E \left\{ xSD(u|x) \frac{x'}{SD(u|x)} \right\}$$

which is the variation of the projection of $x/SD(u|x)$ on $x \cdot SD(u|x)$. This shows that

$$E\left\{\frac{xx'}{V(u|x)}\right\} \geq E(xx')E^{-1}(xx'u^2)E(xx')$$

$$\Longleftrightarrow E^{-1}\left\{\frac{xx'}{V(u|x)}\right\} \leq E^{-1}(xx')E(xx'u^2)E^{-1}(xx') :$$

GLS is more efficient than LSE.

The condition $E(u|z) = 0$ is stronger than $E(zu) = 0$, because $E(u|z) = 0$ implies $E\{g(z)u\} = E\{g(z)E(u|z)\} = 0$ for any square-integrable function $g(z)$ (i.e., $E\{g(z)^2\} < \infty$). Under the stronger moment condition, the efficiency bound becomes smaller, which is attained by GLS. But this comes at the price that GLS should specify correctly the form of heteroskedasticity. Also, the known parametric form of heteroskedasticity is an extra information which may change the efficiency bound.

CHAPTER 2
METHODS OF MOMENTS FOR MULTIPLE LINEAR EQUATION SYSTEMS

Going further from single linear equation models in the preceding chapter, *multiple linear equation systems* are examined. As in single linear equation models, there are LSE, IVE, and GMM available for multiple linear equation systems. An extra issue that arises now is whether there is any benefit in doing the system estimation relative to estimating each equation separately one by one; the answer is a qualified yes. Simultaneous equations that impose further structures on multiple linear equation systems are studied, which explore how variables can be related and how their "relation parameters" are identified and then estimated. Multiple linear equation systems include panel data models as special cases, and thus linear panel data models are briefly studied as well.

1 System LSE, IVE, and GMM

1.1 System LSE

1.1.1 Multiple Linear Equations

Consider H-many linear equations:

$$y_{hi} = x'_{hi}\beta_h + u_{hi} \ (\Longleftrightarrow y_{hi} - x'_{hi}\beta_h = u_{hi}), \quad E(x_h u_h) = 0, \quad h = 1, ..., H.$$

Let x_i denote the collection of all components of $x_{1i}, ..., x_{Hi}$; x_{hi} has the dimension $k_h \times 1$ with $E(x_h x'_h)$ having the full rank, and x_i has the dimension $K \times 1$ with $E(xx')$ having the full rank. Define

$$\gamma \equiv \begin{bmatrix} \beta_1 \\ \vdots \\ \beta_H \end{bmatrix} \quad \text{and its row dimension} \quad k \equiv k_1 +, ..., +k_H.$$

For instance, $H = 3$, $k_1 = 2$, $k_2 = 3$, $k_3 = 4$ ($\Longrightarrow k = 9$), $K = 4$, and

$$x_{1i} = (1, q_{i2})', \ x_{2i} = (1, q_{i3}, q_{i4})',$$
$$x_{3i} = (1, q_{i2}, q_{i3}, q_{i4})', \quad x_i = (1, q_{i2}, q_{i3}, q_{i4})'.$$

When there is no endogenous regressor, the equation system is known as "seemingly unrelated regressions (SUR)," as the equations look unrelated to one another other than through the overlapping elements in the regressors; they can be, however, related through the error term relations.

In this subsection, we will apply LSE simultaneously to the system of equations, which turns out to be the same as applying LSE to each equation separately one by one. Nevertheless, verifying this would be beneficial. The gain of doing a "system estimation" (e.g., "*system GMM*") over "single equation estimation" (e.g., "*separate GMM*") will be realized through the error term correlations when GMM is applied. Also, system estimation is convenient in testing for H_0 involving coefficients across different equations. For instance, there may be two periods (or regions) for $h = 1, 2$, and we may desire to test for the equality of model parameters for the two periods (or regions). For example, are the consumption patterns before and after a financial crisis the same? If no, then there was a structural change/break.

Define

$$y_i \equiv \begin{bmatrix} y_{1i} \\ \vdots \\ y_{Hi} \end{bmatrix}, \quad u_i \equiv \begin{bmatrix} u_{1i} \\ \vdots \\ u_{Hi} \end{bmatrix}, \quad \text{and} \quad w_i = diag(x_{1i}, , ..., x_{Hi}).$$

To simplify exposition, set $H = 3$ in the rest of this subsection; the general case can be easily inferred from this special case. For $H = 3$, we have

$$w \equiv \begin{bmatrix} x_1 & 0 & 0 \\ 0 & x_2 & 0 \\ 0 & 0 & x_3 \end{bmatrix} \implies wu = \begin{bmatrix} x_1 & 0 & 0 \\ 0 & x_2 & 0 \\ 0 & 0 & x_3 \end{bmatrix} \begin{bmatrix} u_1 \\ u_2 \\ u_3 \end{bmatrix} = \begin{bmatrix} x_1 u_1 \\ x_2 u_2 \\ x_3 u_3 \end{bmatrix},$$

and the three equations can be written as

$$y_i = w_i' \gamma + u_i.$$

1.1.2   System LSE and Motivation

The moment conditions $E(x_h u_h) = 0, h = 1, 2, 3$, can be written as

$$\begin{aligned} E(wu) &= 0 \iff E\{w(y - w'\gamma)\} = 0 \\ \iff E(wy) &= E(ww')\gamma \implies \gamma = E^{-1}(ww') \cdot E(wy). \end{aligned}$$

Hence, the system LSE g_{lse} for γ is

$$g_{lse} = \left(\frac{1}{N} \sum_i w_i w_i' \right)^{-1} \left(\frac{1}{N} \sum_i w_i y_i \right) = \left(\sum_i w_i w_i' \right)^{-1} \left(\sum_i w_i y_i \right);$$

the consistency of g_{lse} for γ is easy to see.

Observe

$$E(ww') = \begin{bmatrix} E(x_1x_1') & 0 & 0 \\ 0 & E(x_2x_2') & 0 \\ 0 & 0 & E(x_3x_3') \end{bmatrix} \quad \text{and}$$

$$E(wy) = \begin{bmatrix} E(x_1y_1) \\ E(x_2y_2) \\ E(x_3y_3) \end{bmatrix}.$$

From this, we get

$$E^{-1}(ww')E(wy) = \begin{bmatrix} E^{-1}(x_1x_1')E(x_1y_1) \\ E^{-1}(x_2x_2')E(x_2y_2) \\ E^{-1}(x_3x_3')E(x_3y_3) \end{bmatrix}.$$

This shows that g_{lse} is the stacked version of the three LSE's applied to each equation separately.

While the "separate LSE" is obtained by applying LSE three times separately to the three equations, the "system LSE" is obtained simultaneously. The disadvantage of the system LSE is that it is cumbersome to construct w_i and y_i, but the advantage is that the system LSE gives asymptotic covariances between $g_{h,lse}$ and $g_{j,lse}$ for $h,j = 1,2,3$ as shown in the following where

$$g_{lse} = \begin{bmatrix} g_{1,lse} \\ g_{2,lse} \\ g_{3,lse} \end{bmatrix}.$$

For example, when $k_1 = k_2 = k_3 \equiv k'$, $H_0 : \beta_1 = \beta_2 = \beta_3$ (all three equations share the same parameters) can be tested with Wald test using

$$H_0 : R'\gamma = 0 \quad \text{where} \quad R' = \begin{bmatrix} I_{k'} & -I_{k'} & 0 \\ 0 & I_{k'} & -I_{k'} \end{bmatrix}.$$

This H_0 is relevant for panel data models where h indexes time periods. If $H = 2$, then $R' = (I_{k'}, -I_{k'})$ for $H_0 : \beta_1 = \beta_2$.

1.1.3   Asymptotic Variance

Substitute $y_i = w_i'\gamma + u_i$ to rewrite the system LSE as

$$g_{lse} = \gamma + \left(\sum_i w_i w_i' \right)^{-1} \sum_i w_i u_i$$

$$\implies \sqrt{N}(g_{lse} - \gamma) \rightsquigarrow N\{0, \ E^{-1}(ww')E(wuu'w')E^{-1}(ww')\}.$$

The two outside matrices in the asymptotic variance are block-diagonal, and for the middle matrix, observe

$$
wuu'w' = \begin{bmatrix} x_1u_1 \\ x_2u_2 \\ x_3u_3 \end{bmatrix} \begin{bmatrix} u_1x_1' & u_2x_2' & u_3x_3' \end{bmatrix}
$$

$$
\implies E(wuu'w') = \begin{bmatrix} E(x_1x_1'u_1^2) & E(x_1x_2'u_1u_2) & E(x_1x_3'u_1u_3) \\ E(x_2x_1'u_2u_1) & E(x_2x_2'u_2^2) & E(x_2x_3'u_2u_3) \\ E(x_3x_1'u_3u_1) & E(x_3x_2'u_3u_2) & E(x_3x_3'u_3^2) \end{bmatrix}.
$$

This sandwich form asymptotic variance has two types of entries. First, the diagonal terms are the variance of $b_{h,lse}$, $h = 1, 2, 3$:

$$
E^{-1}(x_hx_h')E(x_hx_h'u_h^2)E^{-1}(x_hx_h'), \quad h = 1, 2, 3.
$$

Second, the off-diagonal matrices are the asymptotic covariance matrices, which take the form

$$
E^{-1}(x_hx_h')E(x_hx_j'u_hu_j)E^{-1}(x_jx_j'), \quad \forall h, j;
$$

these are needed, for instance, to test hypotheses involving both β_h and β_j. If $E(x_hx_j'u_hu_j) = 0$, then the asymptotic covariance between $b_{h,lse}$ and $b_{j,lse}$ is zero. The condition holds, for instance, if $E(u_hu_j|x) = 0$.

Sometimes, when we have $E(u_hx_h) = 0$ $\forall h$, we may also have $E(u_hx_j) = 0$ $\forall h \neq j$. In this case, for equation h, $E(u_hx_h) = 0$ gets augmented to yield

$$
E(xu_h) = 0.
$$

One may think of applying IVE instead of LSE to equation h to take advantage of the more-than-enough moment conditions. But as mentioned already, IVE to equation h under $E(xu_h) = 0$ is just the LSE under $E(u_hx_h) = 0$, because x includes x_h and thus projecting x_h on x renders only x_h. Taking advantage of the augmented moment conditions makes sense only when some elements of x_h are endogenous and thus are not included in x. In this case, we would consider the "system IVE" and "system GMM" in the following subsections.

1.2   System IVE and Rank Condition

1.2.1   Moment Conditions

Consider H-many linear equations:

$$
y_{hi} = x_{hi}'\beta_h + u_{hi} (\iff y_{hi} - x_{hi}'\beta_h = u_{hi}), \quad E(xu_h) = 0, \quad h = 1, ..., H.
$$

Differently from the preceding subsection for LSE, x does not include all components of $x_1, ..., x_H$; i.e., some endogenous regressors in $x_1, ..., x_H$ are excluded from x. But we will still denote the dimension of x as $K \times 1$ to save notations. IVE (and GMM) can be applied to this equation system. To simplify exposition, set again $H = 3$ in this subsection. This and the following subsections draw on Lee (2008).

For a $m \times n$ matrix $A = [a_{ij}]$ and a $p \times q$ matrix B, the Kronecker product $A \otimes B$ is

$$\underset{mp \times nq}{A \otimes B} \equiv \left[\begin{array}{cccc} a_{11}B & a_{12}B & \cdots & a_{1n}B \\ \vdots & \vdots & \vdots & \vdots \\ a_{m1}B & a_{m2}B & \cdots & a_{mn}B \end{array} \right].$$

It can be verified that

$$\begin{array}{rcl} (A \otimes B)(C \otimes D) & = & (AC \otimes BD) \quad \text{if } AC \text{ and } BD \text{ exist} \\ (A \otimes B)' & = & (A' \otimes B') \\ (A \otimes B)^{-1} & = & (A^{-1} \otimes B^{-1}) \quad \text{if } A^{-1} \text{ and } B^{-1} \text{ exist} \\ A \otimes (B + C) & = & (A \otimes B) + (A \otimes C). \end{array}$$

Rewrite the moment conditions $E(xu_h) = 0, h = 1, 2, 3$, compactly as (recall $u \equiv (u_1, u_2, u_3)'$)

$$\underset{(3K) \times 1}{E(u \otimes x)} = \left[\begin{array}{c} E(u_1 x) \\ E(u_2 x) \\ E(u_3 x) \end{array} \right] = 0.$$

Define

$$\underset{(3K) \times 3}{z \equiv I_3 \otimes x} \implies \underset{(3K) \times 3}{zu = (I_3 \otimes x)} \cdot \underset{3 \times 1}{u} = \left[\begin{array}{ccc} x & 0 & 0 \\ 0 & x & 0 \\ 0 & 0 & x \end{array} \right] \left[\begin{array}{c} u_1 \\ u_2 \\ u_3 \end{array} \right] = \left[\begin{array}{c} xu_1 \\ xu_2 \\ xu_3 \end{array} \right].$$

With this, we can see that the moment condition $E(u \otimes x) = 0$ is equivalent to

$$E(zu) \ [= E\{(instruments) \times error\}] = 0$$

which will lead to "system IVE."

1.2.2   System IVE and Separate IVE

Recall $w = diag(x_1, x_2, x_3)$ and observe

$$\begin{array}{rcl} E(zu) & = & E\{z \cdot (y - w'\gamma)\} = 0 \iff E(zy) = E(zw')\gamma \implies \\ \gamma & = & \{E(wz')E^{-1}(zz')E(zw')\}^{-1} \cdot E(wz')E^{-1}(zz')E(zy); \end{array}$$

conditions are needed for the matrix inversion, which will be discussed shortly. Hence, the system IVE g_{ive} for γ is

$$g_{ive} = \left\{ \sum_i w_i z_i' \left(\sum_i z_i z_i' \right)^{-1} \sum_i z_i w_i' \right\}^{-1} \cdot$$

$$\sum_i w_i z_i' \left(\sum_i z_i z_i' \right)^{-1} \sum_i z_i y_i;$$

the consistency is easy to see.

Substitute $y_i = w_i' \gamma + u_i$ into g_{ive} to rewrite g_{ive} as

$$g_{ive} = \gamma + \left\{ \sum_i w_i z_i' \left(\sum_i z_i z_i' \right)^{-1} \sum_i z_i w_i' \right\}^{-1} \cdot$$

$$\sum_i w_i z_i' \left(\sum_i z_i z_i' \right)^{-1} \sum_i z_i u_i.$$

From this, we get

$$\sqrt{N}(g_{ive} - \gamma) \;\rightsquigarrow\; N(0, \, A \cdot E(zuu'z') \cdot A') \quad \text{where}$$
$$A \;\equiv\; \{E(wz')E^{-1}(zz')E(zw')\}^{-1} \cdot E(wz')E^{-1}(zz').$$

For the population moments in g_{ive}, observe

$$E(zy) \;=\; \begin{bmatrix} E(xy_1) \\ E(xy_2) \\ E(xy_3) \end{bmatrix} \quad \text{and} \quad E(wz')E^{-1}(zz')E(zy)$$

$$=\; \begin{bmatrix} E(x_1 x')E^{-1}(xx')E(xy_1) \\ E(x_2 x')E^{-1}(xx')E(xy_2) \\ E(x_3 x')E^{-1}(xx')E(xy_3) \end{bmatrix}.$$

Combining this with the block-diagonal $E(wz')E^{-1}(zz')E(zw')$, we can see that

$$g_{ive} = \begin{bmatrix} \{\sum_i x_{1i}x_i'(\sum_i x_i x_i')^{-1}\sum_i x_i x_{1i}'\}^{-1} \cdot \sum_i x_{1i}x_i'(\sum_i x_i x_i')^{-1}\sum_i x_i y_{1i} \\ \{\sum_i x_{2i}x_i'(\sum_i x_i x_i')^{-1}\sum_i x_i x_{2i}'\}^{-1} \cdot \sum_i x_{2i}x_i'(\sum_i x_i x_i')^{-1}\sum_i x_i y_{2i} \\ \{\sum_i x_{3i}x_i'(\sum_i x_i x_i')^{-1}\sum_i x_i x_{3i}'\}^{-1} \cdot \sum_i x_{3i}x_i'(\sum_i x_i x_i')^{-1}\sum_i x_i y_{3i} \end{bmatrix}.$$

That is, the system IVE is nothing but the stacked version of the three separate IVE's

$$b_{h,ive} \;=\; \left\{ \sum_i x_{hi}x_i' \left(\sum_i x_i x_i' \right)^{-1} \sum_i x_i x_{hi}' \right\}^{-1} \cdot$$

$$\sum_i x_{hi}x_i' \left(\sum_i x_i x_i' \right)^{-1} \sum_i x_i y_{hi}, \; h = 1, 2, 3.$$

The system IVE differs from the separate IVE's only in that the former yields asymptotic covariances between $b_{h,ive}$ and $b_{j,ive}$ $\forall h \neq j$. This situation is analogous to the system LSE and the separate LSE in the preceding subsection.

1.2.3   Identification Conditions

Observe

$$
w_i z_i' = \begin{bmatrix} x_{1i} & 0 & 0 \\ 0 & x_{2i} & 0 \\ 0 & 0 & x_{3i} \end{bmatrix} \cdot \begin{bmatrix} x_i' & 0 & 0 \\ 0 & x_i' & 0 \\ 0 & 0 & x_i' \end{bmatrix} = \begin{bmatrix} x_{1i} x_i' & 0 & 0 \\ 0 & x_{2i} x_i' & 0 \\ 0 & 0 & x_{3i} x_i' \end{bmatrix} ;
$$

$w_i z_i'$ is of dimension $(k \times 3) \cdot (3 \times 3K) = k \times 3K$, which is block-diagonal. $E(zz')$ is also block diagonal. Thus $E(wz') E^{-1}(zz') E(zw')$ is block-diagonal:

$$
\begin{aligned}
diag\{ & E(x_1 x') E^{-1}(xx') E(xx_1'), \; E(x_2 x') E^{-1}(xx') E(xx_2'), \\
& E(x_3 x') E^{-1}(xx') E(xx_3') \}.
\end{aligned}
$$

For this to be invertible, each diagonal matrix should be invertible, for which it is necessary and sufficient to have

$$
rank\{E(x_h x')\} = k_h \text{ (requiring } k_h \leq K) \quad \forall h.
$$

This is the "*rank condition of identification*," whereas "$k_h \leq K$" is called the "*order condition of identification*"; the order condition is only a necessary condition.

If the rank condition does not hold for some h, the hth equation should be dropped for g_{ive}; alternatively, from a "model-building perspective," more IV's may be added to the system so that the rank condition holds for the hth equation. The following terminologies are often used:

$k_h < K$:   equation h is "*over-identified*" (more instruments than necessary)

$k_h = K$:   equation h is "*just-identified*" (just enough instruments)

$k_h > K$:   equation h is "*under-identified*" (not enough instruments).

In solving $E(zy) = E(zw')\gamma$ for γ, we multiplied the equation by $E(wz') E^{-1}(zz')$. In fact, we could have used another matrix of the same dimension, say L, so long as $L \cdot E(zw')$ is invertible. It might look as if the above identification result holds only for the particular choice $L = E(wz') E^{-1}(zz')$. That this is not the case can be shown using the following fact for a system of equations. Consider a system of HK equations with k unknowns:

$$
\underset{(HK) \times k}{A} \cdot \underset{k \times 1}{\gamma} = \underset{HK \times 1}{c}.
$$

For these equations to have any solution, c has to be in the column space of A, as $A\gamma$ is a linear combination of the columns of A. Then the equation system is said to be "*consistent*". Given that the equation system is consistent, a necessary and sufficient condition for the solution to be unique is $rank(A) = k$; see, e.g., Searle (1982, p. 233) or Schott (2005, p. 227). For $E(zw')\gamma = E(zy)$ with $A = E(zw')$ and $c = E(zy)$, $rank(A) = k$ is equivalent to $rank\{E(x_h x')\} = k_h \; \forall h$.

Our framework includes linear simultaneous equation systems. For instance, consider three simultaneous equations with three endogenous variables y_1, y_2, and y_3:

$$
\begin{bmatrix}
y_1 - \alpha_{12}y_2 - \alpha_{13}y_3 - m_1'\eta_1 \\
y_2 - \alpha_{21}y_1 - \alpha_{23}y_3 - m_2'\eta_2 \\
y_3 - \alpha_{31}y_1 - \alpha_{32}y_2 - m_3'\eta_3
\end{bmatrix}
=
\begin{bmatrix}
u_1 \\
u_2 \\
u_3
\end{bmatrix}
\quad \text{where } E(m_j u_h) = 0 \; \forall j, h.
$$

Define

$$
\begin{aligned}
x_1 &= (y_2, y_3, m_1')', \quad x_2 = (y_1, y_3, m_2')', \quad x_3 = (y_1, y_2, m_3')', \\
\gamma &\equiv (\alpha_{12}, \alpha_{13}, \eta_1', \; \alpha_{21}, \alpha_{23}, \eta_2', \; \alpha_{31}, \alpha_{32}, \eta_3')',
\end{aligned}
$$

and let x consist of the elements in m_1, m_2, and m_3; x is the exogenous regressor vector of the system. Under $E(x u_h) = 0 \; \forall h$, γ can be estimated by the system IVE. The order condition $k_h \le K$ for equation h is that the *number of the regressors in equation h should be smaller than or equal to the number of the elements in x*. The discussion of the preceding paragraph on rank and order conditions for simultaneous equation identification is much simpler than the typical econometric textbook discussion on rank and order conditions that will be seen later.

Notice that

k_h: (# exo. regressors in eq. h) + (# endo. regressors in eq. h)

K: (# exo. regressors in eq. h) + (# exo. regressors in all eq.'s but h).

Removing the common first term in k_h and K, the order condition $k_h \le K$ can be written as

(# endo. regressors in eq. h) $\le$ (# exo. regressors in all eq.'s but h).

Since the right-hand side variables are instruments for the left-hand side variables, this explains why the over-, just-, and under-identified cases are described above with the number of instruments.

1.3   System GMM and Link to Panel Data

1.3.1   System GMM

The "system GMM" corresponding to the system IVE is obtained by taking one step from g_{ive}. Define the residuals $\hat{u}_i \equiv y_i - w_i' g_{ive}$ to get

$$g_{gmm} = \left\{ \sum_i w_i z_i' \left(\sum_i z_i \hat{u}_i \hat{u}_i' z_i' \right)^{-1} \sum_i z_i w_i' \right\}^{-1} \cdot$$

$$\sum_i w_i z_i' \left(\sum_i z_i \hat{u}_i \hat{u}_i' z_i' \right)^{-1} \sum_i z_i y_i;$$

$$\sqrt{N}(g_{gmm} - \gamma) \rightsquigarrow N(0, \{E(wz')E^{-1}(zuu'z')E(zw')\}^{-1});$$

$$\frac{1}{\sqrt{N}} \sum_i \tilde{u}_i' z_i' \cdot \left(\frac{1}{N} \sum_i z_i \tilde{u}_i \tilde{u}_i' z_i' \right)^{-1} \cdot \frac{1}{\sqrt{N}} \sum_i z_i \tilde{u}_i \rightsquigarrow \chi^2_{3K-k},$$

$$\tilde{u}_i \equiv y_i - w_i' g_{gmm}.$$

The inverse of the asymptotic variance matrix is

$$\begin{bmatrix} E(x_1 x') & 0 & 0 \\ 0 & E(x_2 x') & 0 \\ 0 & 0 & E(x_3 x') \end{bmatrix} \cdot$$

$$\begin{bmatrix} E(xx' u_1^2) & E(xx' u_1 u_2) & E(xx' u_1 u_3) \\ E(xx' u_1 u_2) & E(xx' u_2^2) & E(xx' u_2 u_3) \\ E(xx' u_1 u_3) & E(xx' u_2 u_3) & E(xx' u_3^2) \end{bmatrix}^{-1} \cdot$$

$$\begin{bmatrix} E(xx_1') & 0 & 0 \\ 0 & E(xx_2') & 0 \\ 0 & 0 & E(xx_3') \end{bmatrix} \cdot$$

If

$$E(xx' u_h u_j) = 0 \ \forall h \neq j,$$

then, all off-diagonal terms in $E(zuu'z')$ are zero, and the system GMM asymptotic variance becomes diagonal with the diagonal terms

$$\{E(x_h x')E^{-1}(xx' u_h^2)E(xx_h')\}^{-1}, \quad h = 1, ..., H.$$

This is the same as the asymptotic variance of the "separate GMM" (i.e., the GMM applied to each equation separately). If the off-diagonal terms are not zero, then differently from the system LSE and IVE, the system GMM's diagonal terms are not in general equal to the separate GMM's asymptotic variance. Having $\sum_i z_i \hat{u}_i \hat{u}_i' z_i'$ in the system GMM that is not block-diagonal, instead of the block-diagonal $\sum_i z_i z_i'$ in the system IVE, makes the difference.

Lee (2004a) showed that, for the efficiency gain for equation h of the system GMM over the separate GMM, it is necessary to have

(i) the rank condition holds for each equation,

(ii) $E(xx'u_h u_j) \neq 0$ for some $j \neq h$,

(iii) There is at least one equation j $(\neq h)$ that is over-identified

In a two-equation system with one equation just identified, (iii) implies that the other equation will get no efficiency gain with the system GMM.

EXAMPLE: HOUSE SALE (continued). Recall the house sale example in the previous chapter. After a model specification search using "RESET test" with $\hat{y}^2$ as an "artificial regressor" (i.e., second-order terms are added into the model until $\hat{y}^2$ becomes insignificant; RESET test will appear later), BATH^2, TAX^2, and BIGSYR (= BIGS × YR) were added as can be seen in Table 1. The endogenous variables are DISC and $\ln T$, and all the other variables are regarded as exogenous. Two equations are constructed with exclusion restrictions obtained following the approach in Lee and Chang (2007), which will be explained later in a separate section under the heading "avoiding arbitrary exclusion restrictions." The first two columns show the system GMM result in Lee and Chang (2007), and the last two columns show the separate GMM result. Although some t-values are lower in the system GMM, overall more variables have higher t-values in the system GMM. The efficiency gain for the system GMM is high in BATH and BATH^2, but otherwise seems low or non-existent in this empirical example. The system IVE and the separate IVE are not presented in the table as they are exactly the same, although they were used to get the GMM's.

1.3.2   System GMM and Panel Data

So far we considered multiple equations for cross-section data where the same x can be used as instruments for all equations. If we have panel data with H "waves" (i.e., each subject is observed for H-many periods), then there will be H equations—one for each period. In this case, typically, different instruments are used for different equations to result in moment conditions

$$E(u_h z_h) = 0 \quad h = 1, ..., H.$$

This case can be dealt with simply by redefining z as

$$z = \begin{bmatrix} z_1 & 0 & 0 \\ 0 & z_2 & 0 \\ 0 & 0 & z_3 \end{bmatrix} \implies E(zu) = \begin{bmatrix} E(z_1 u_1) \\ E(z_2 u_2) \\ E(z_3 u_3) \end{bmatrix}.$$

With this redefined z, the rest of IVE and GMM is the same as those for $z = I_3 \otimes x$. Clearly, the new z in the display includes the old $z = I_3 \otimes x$ as a special case when $z_1 = z_2 = z_3 = x$.

For example, suppose we have a panel data model

$$y_{it} = x'_{it}\beta_t + u_{it}, \quad i = 1, ..., N, \ t = 1, 2, 3,$$

Table 1: System GMM vs. Separate GMM with House Sale Data (tv in (·))

	System GMM		Separate GMM	
	SF1 for DISC	SF2 for lnT	SF1 for DISC	SF2 for lnT
ln T	12.4 (6.13)		12.8 (6.01)	
DISC		0.034 (1.60)		0.027 (1.20)
1	18.1 (1.20)	−1.07 (−0.86)	19.9 (1.27)	−1.56 (−0.92)
BATH	−1.81 (−1.08)		−1.37 (−0.35)	
BATH^2	0.53 (1.22)		0.31 (0.30)	
ELEC		0.120 (1.65)		0.116 (1.27)
ROOM		−0.017 (−1.15)		−0.025 (−0.82)
TAX		−0.244 (−1.09)		−0.411 (−1.25)
TAX^2		0.039 (0.90)		0.080 (1.30)
YR	−0.23 (−3.26)	0.0052 (1.57)	−0.25 (−3.26)	0.0047 (1.32)
L1		−0.162 (−1.87)		−0.226 (−2.10)
L2		−0.190 (−1.81)		−0.163 (−1.26)
L3		−0.109 (−1.64)		−0.089 (−0.73)
lnLP		0.346 (1.46)		0.404 (1.37)
BIGS	−9.42 (−1.69)		−11.5 (−1.84)	
BIGSYR	0.11 (1.58)		0.14 (1.73)	
RATE	−5.85 (−3.57)	0.441 (4.01)	−6.05 (−3.64)	0.496 (3.71)
SUPPLY	0.019 (1.01)	0.0011 (0.51)	0.018 (0.93)	0.0013 (0.53)
$SD(u_j)$	8.62	0.68	8.80	0.68
$COR(u_1, u_2)$	−0.89		−0.86	

where y_{it} is work-hour$_{it}$, $x_{it} = (1, wage_{it})'$, and the index $t = 1, ..., T$ is used now instead of $h = 1, ..., H$ to make it clear that different equations refer to different time periods. Note a minor point that the subscript positions have been switched from "hi" to "it" for panel data, because of the perception that each equation h forms a separate "block" in "hi" whereas a block is each individual i in panel data.

In a given period, wage may be simultaneously determined with work hour, which implies $COR(x_{it}, u_{it}) \neq 0$. But we may have $COR(x_{is}, u_{it}) = 0 \ \forall s < t$ but $\neq 0 \ \forall s \geq t$, which happens, e.g., if x_{is} is adjusted after observing the current or past u_{it}'s. Then we get the following moment conditions: omitting $E(u_t \cdot 1) = 0$ and the corresponding 1 in $z_t \ \forall t$,

$$E(u_2 x_1) = 0, \ E(u_3 x_1) = 0, \ E(u_3 x_2) = 0$$
$$\implies \quad z_1 \text{ is empty}, \ z_2 = wage_1, \ z_3 = (wage_1, wage_2)';$$

only the second and third period equations are jointly estimable. Since the second period is just identified, the efficiency gain of the joint estimation is restricted only to the second equation. If $COR(x_{is}, u_{it}) = 0 \ \forall s \neq t$ that are stronger than $COR(x_{is}, u_{it}) \ \forall s < t$, then the moment conditions are

$$E(u_1 x_2) = 0, \ E(u_1 x_3) = 0, \ E(u_2 x_1) = 0, \ E(u_2 x_3) = 0,$$
$$E(u_3 x_1) = 0, \ E(u_3 x_2) = 0$$
$$\implies \quad z_1 = (wage_2, wage_3)', \ z_2 = (wage_1, wage_3)',$$
$$z_3 = (wage_1, wage_2)'.$$

Now all three equations are jointly estimable, with efficiency gain possible for all equations.

The panel data model $y_{it} = x_{it}'\beta_t + u_{it}$ is in fact more general than the restricted panel models with $\beta_t = \beta \ \forall t$. As already noted in the section for system LSE, this restriction can be tested with Wald test. If not rejected, imposing the restriction can be done using, not $w_i = diag(x_{i1}, x_{i2}, x_{i3})$, but $w_i = (x_{i1}, x_{i2}, x_{i3})$ so that

$$\begin{bmatrix} y_{i1} \\ y_{i2} \\ y_{i3} \end{bmatrix} = \begin{bmatrix} x_{i1}' \\ x_{i2}' \\ x_{i3}' \end{bmatrix} \beta + \begin{bmatrix} u_{i1} \\ u_{i2} \\ u_{i3} \end{bmatrix} \iff y_i = w_i'\beta + u_i.$$

The GMM b_{gmm} to this is the same as the above g_{gmm} except that w_i is not a block-diagonal matrix.

A panel data model popular in practice is neither $y_{it} = x_{it}'\beta_t + u_{it}$ nor $y_{it} = x_{it}'\beta + u_{it}$, but the one that allows only the intercept to change over time. This falls in between the two models, and can be accommodated with

$$w_i' = \begin{bmatrix} 1 & 0 & 0 & \tilde{x}_{i1}' \\ 0 & 1 & 0 & \tilde{x}_{i2}' \\ 0 & 0 & 1 & \tilde{x}_{i3}' \end{bmatrix} \quad \text{and} \quad \beta = \begin{bmatrix} \tau_1 \\ \tau_2 \\ \tau_3 \\ \tilde{\beta} \end{bmatrix}$$

where $\tilde{x}_{it}$ is the regressors other than 1, $\tilde{\beta}$ is the coefficient for $\tilde{x}_{it}$, and τ_1, τ_2, τ_3 are the time-varying intercepts.

A panel data model perhaps even more popular than $y_{it} = x'_{it}\beta + u_{it}$ is

$$y_{it} = x'_{it}\beta + \delta_i + u_{it}$$

where δ_i is a time-constant error possibly correlated with components of x_{it}; δ_i is also called "unit-specific effect" or "individual-specific effect." To get rid of the possible endogeneity, often the model is transformed into

$$\Delta y_{it} = \Delta x'_{it}\beta + \Delta u_{it} \quad \text{where} \quad \Delta y_{it} \equiv y_{it} - y_{i,t-1}$$

renaming the time periods as $0, 1, ..., T-1$. Note that the time-constant elements of x_{it} get removed in Δx_{it}, and in this case, β here is not necessarily the same as β for x_{it}. Applying GMM to this panel data model is essentially the same as the above GMM: find a vector of instruments for each differenced equation and apply system GMM.

For instance, with $T = 3$, we get only two equations after first-differencing:

$$\begin{bmatrix} \Delta y_{i1} \\ \Delta y_{i2} \end{bmatrix} = \begin{bmatrix} \Delta x'_{i1} \\ \Delta x'_{i2} \end{bmatrix} \beta + \begin{bmatrix} \Delta u_{i1} \\ \Delta u_{i2} \end{bmatrix}.$$

Let z_{it} be the instrument vector of dimension $s_j \times 1$ for the Δy_{it} equation, and let β be $k \times 1$. We have

$$b_{ive} = \left\{ \sum_i w_i z'_i \left(\sum_i z_i z'_i \right)^{-1} \sum_i z_i w'_i \right\}^{-1}$$

$$\sum_i w_i z'_i \left(\sum_i z_i z'_i \right)^{-1} \sum_i z_i \zeta_i,$$

where $\underset{k \times 2}{w_i} \equiv [\Delta x_{i1}, \Delta x_{i2}]$, $\underset{(s_1+s_2) \times 2}{z_i} = \begin{bmatrix} z_{i1} & 0 \\ 0 & z_{i2} \end{bmatrix}$, $\underset{2 \times 1}{\zeta_i} \equiv \begin{bmatrix} \Delta y_{i1} \\ \Delta y_{i2} \end{bmatrix}$.

With $v_i \equiv \zeta_i - w'_i\beta$ and $\hat{v}_i \equiv \zeta_i - w'_i b_{ive}$, we obtain

$$b_{gmm} = \left\{ \sum_i w_i z'_i \left(\sum_i z_i \hat{v}_i \hat{v}'_i z'_i \right)^{-1} \sum_i z_i w'_i \right\}^{-1}$$

$$\cdot \sum_i w_i z'_i \left(\sum_i z_i \hat{v}_i \hat{v}'_i z'_i \right)^{-1} \sum_i z_i \zeta_i,$$

$$\sqrt{N}(b_{gmm} - \beta) \rightsquigarrow N(0, \{E(wz')E^{-1}(zvv'z')E(zw')\}^{-1}).$$

Although straightforward conceptually, panel IVE and GMM have some special features and "pitfalls" that deserve a closer look. We will do this later in a separate section.

2   Simultaneous Equations and Identification

This section studies simultaneous equations in detail which appeared sporadically up to this point. Rank and order conditions are presented in a manner that is coherent with the usual textbook presentation, yet easier to understand and more readily applicable to simultaneous equations in limited dependent variables to appear in a later chapter.

2.1   Relationship Between Endogenous Variables

Suppose we have two response variables y_1 and y_2 which may be related in a couple of different ways. One is SUR: with x being the collection of the elements in x_1 and x_2,

$$y_h = x_h'\beta_h + u_h, \ h = 1, 2, \quad E(u_h x) = 0;$$

the two equations are related through $COR(u_1, u_2) \neq 0$. To be precise, what matters is not $COR(u_1, u_2)$, but $COR(u_1, u_2|x)$ as can be seen in the system GMM asymptotic variance where terms like $E(xx'u_1u_2) = E\{xx'E(u_1u_2'|x)\}$ appeared. We will, however, use just $COR(u_1, u_2)$ in this subsection as this does not hinder conveying the main idea; also, equations systems have been traditionally discussed with errors independent of x, in which case $COR(u_1, u_2) = COR(u_1, u_2|x)$. Both β_1 and β_2 in SUR can be consistently estimated with LSE. Note that, differently from our notations in some of the preceding subsections where x included both endogenous and exogenous regressors, here we are denoting endogenous regressors with y and exogenous regressors with x.

A relation between y_1 and y_2 "closer" than in SUR is seen in the following *recursive system*

$$y_1 = x_1'\beta_1 + u_1, \ y_2 = \alpha_{21}y_1 + x_2'\beta_2 + u_2, \ \alpha_{21} \neq 0, \ E(u_h x) = 0, \ h = 1, 2,$$

where y_1 influences y_2 but y_2 does not influence y_1. Differently from SUR, β_2 cannot be consistently estimated by LSE if $COR(u_1, u_2) \neq 0$, because y_1 and u_2 are related (through u_1) in the y_2 equation. Certainly, α_{21} and β_2 can be estimated by IVE if x_1 provides an instrument for y_1. The equivalent 2SLSE view is that β_1 is estimated by the LSE of y_1 on x_1, and α_{21} and β_2 are estimated by the LSE of y_2 on $x_1'\tilde{\beta}_1$ and x_2 where $\tilde{\beta}_1$ is the first LSE for the y_1 equation. If the y_1 equation is not specified, then the first stage of the 2SLSE is the LSE of y_1 on x.

An "even closer" relation than the recursive relation is a *simultaneous relation* where

$$y_1 \ = \ \alpha_{12}y_2 + x_1'\beta_1 + u_1, \ y_2 = \alpha_{21}y_1 + x_2'\beta_2 + u_2,$$
$$\alpha_{12}, \alpha_{21} \neq 0, \ E(u_h x) = 0, \ h = 1, 2.$$

In the y_1 equation, $COR(y_2, u_1) \neq 0$ regardless of $COR(u_1, u_2)$, because u_1 affects y_1 which in turn affect y_2; i.e., u_1 and y_2 are related through y_1. Thus, the LSE for the y_1 equation is inconsistent, and analogously, the LSE for the second equation is inconsistent as well. This equation system as well as the preceding one can be estimated with IVE and GMM.

An example of simultaneous system analogous to the police/crime nexus is doctors/disease-incidence. Suppose y_1 is disease incidence and y_2 is the number of doctors. It is possible that the more doctors the lesser the disease incidence ($\alpha_{12} < 0$). However, it is also possible that the higher disease incidence results in the more doctors ($\alpha_{21} > 0$). The relationship between disease incidence and doctors can be simultaneous, or recursive if only one of α_{12} and α_{21} is zero; if both are zero, we get SUR. For policy purpose, the hypothesis of interest would be $\alpha_{12} = 0$: no deterring effect of doctors on disease incidence.

Yet another relation "weaker" than simultaneous relation can be seen in an "expectation-based relation" as in

$$y_1 = \alpha_{12}y_2 + x_1'\beta_1 + u_1, \ y_2 = \alpha_{21}E(y_1|x) + x_2'\beta_2 + u_2,$$
$$\alpha_{12}, \alpha_{21} \neq 0, \ E(u_h x) = 0, \ h = 1, 2$$

where the expected value of y_1, not y_1 per se, affects y_2. In the doctor/disease example, more doctors may be deployed in advance if a higher disease incidence is expected. The model can be transformed into simultaneous relation by rewriting the second equation as

$$y_2 = \alpha_{21}y_1 + x_2'\beta_2 + u_2 - \alpha_{21}(y_1 - E(y_1|x))$$

where the error term is $u_2 - \alpha_{21}(y_1 - E(y_1|x))$. This can be estimated by IVE with x as instruments for y_1 because $E[x\{u_2 - \alpha_{21}(y_1 - E(y_1|x))\}] = 0$. If desired, we can certainly entertain the possibility that $E(y_2|x)$ instead of y_2 appears in the y_1 equation. Since simultaneous relation is the most general, we will examine simultaneous relation in detail in the the next subsection.

A long-winded, but more informative way to look at $E(y_1|x)$ is invoking "rational expectation." To find $E(y_1|x)$, substitute the y_2 "structural form" (SF) into the y_1 SF to get

$$y_1 = \alpha_{12}\{\alpha_{21}E(y_1|x) + x_2'\beta_2 + u_2\} + x_1'\beta_1 + u_1$$
$$= \alpha_{12}\alpha_{21}E(y_1|x) + x_1'\beta_1 + x_2'\beta_2\alpha_{12} + (u_1 + \alpha_{12}u_2);$$

The "original" y_1 and y_2 equations are called SF's, relative to their RF's; the y_2 RF is derived below. Take $E(\cdot|x)$ to obtain, assuming $\alpha_{12}\alpha_{21} \neq 1$,

$$E(y_1|x) = \alpha_{12}\alpha_{21}E(y_1|x) + x_1'\beta_1 + x_2'\beta_2\alpha_{12}$$
$$\implies E(y_1|x) = \frac{x_1'\beta_1 + x_2'\beta_2\alpha_{12}}{1 - \alpha_{12}\alpha_{21}}.$$

The fact that we used both SF's to obtain $E(y_1|x)$ means that we appealed to the *rational expectation* principle: the economic agent knows the entire economic system in forming their expectations.

Substitute $E(y_1|x)$ back into the y_2 SF to obtain

$$
\begin{aligned}
y_2 &= \frac{\alpha_{21}}{1 - \alpha_{12}\alpha_{21}}(x_1'\beta_1 + x_2'\beta_2\alpha_{12}) + x_2'\beta_2 + u_2 \\
&= x_1'\beta_1\frac{\alpha_{21}}{1 - \alpha_{12}\alpha_{21}} + x_2'\beta_2(1 + \frac{\alpha_{12}\alpha_{21}}{1 - \alpha_{12}\alpha_{21}}) + u_2 \\
&= x_1'\beta_1\frac{\alpha_{21}}{1 - \alpha_{12}\alpha_{21}} + x_2'\beta_2\frac{1}{1 - \alpha_{12}\alpha_{21}} + u_2.
\end{aligned}
$$

This is the y_2 RF, as neither y_1 nor $E(y_1|x)$ appears on the rhs. In the y_2 RF, the RF parameters are functions of the SF parameters, and x_1 appears which is not in the y_2 SF.

2.2   Conventional Approach to Rank Condition

We already saw two and three simultaneous equations. More generally, suppose we have H-many SF equations:

$$
\underset{H\times H}{\Gamma}\,\underset{H\times 1}{y_i} - \underset{H\times K}{B}\,\underset{K\times 1}{x_i} = \underset{H\times 1}{u_i},
$$

where $E(u\otimes x) = 0$, Γ is invertible, $E(xx')$ is of full rank, y_i is the "endogenous variables," x_i is the "exogenous variables," Γ and B are the "*SF parameters*," and u_i is the SF error possibly correlated to one another. Sometimes, we call Γ the "*endogenous SF parameters*", and B the "*exogenous SF parameters*." In simultaneous equations, the aim is to learn about the SF parameters. Solving the SF equations for y_i, we get the RF equations

$$
y_i = \Gamma^{-1}Bx_i + \Gamma^{-1}u_i = \Pi x_i + v_i,
$$

where $\Pi \equiv \Gamma^{-1}B$, $v_i \equiv \Gamma^{-1}u_i$, Π is the RF parameters, and v_i is the RF errors. Even if the components of u_i are uncorrelated to one another, those of v_i are correlated in general.

For example, consider two simultaneous equations:

$$
\begin{aligned}
y_1 &= \alpha_{12}y_2 + \beta_{11} + \beta_{12}x_2 + \beta_{13}x_3 + \beta_{14}x_4 + u_1, \\
y_2 &= \alpha_{21}y_1 + \beta_{21} + \beta_{22}x_2 + \beta_{25}x_5 + u_2, \\
\beta_{13} &\neq 0,\ \beta_{14} \neq 0,\ \text{and}\ \beta_{25} \neq 0,
\end{aligned}
$$

where

$$
\begin{aligned}
y_1 &= \text{wife work hours,} \quad y_2 = \text{husband work hours,} \\
x_2 &= \text{household wealth,} \quad x_3 = \text{number of preschool kids,} \\
x_4 &= \text{number of primary school kids,} \\
x_5 &= \text{husband schooling years.}
\end{aligned}
$$

Presumably, y_1 and y_2 are chosen jointly by the couple, maximizing the household utility; the y_1 equation (y_2 equation) may be thought of as a linear approximation to the first-order condition wrt y_1 (y_2) for the maximization problem.

Getting the RF equation for y_1 with $\gamma \equiv (1 - \alpha_{12}\alpha_{21})^{-1}$, we get

$$
\begin{aligned}
y_1 &= \alpha_{12} \cdot (\alpha_{21}y_1 + \beta_{21} + \beta_{22}x_2 + \beta_{25}x_5 + u_2) + \beta_{11} + \beta_{12}x_2 \\
&\quad + \beta_{13}x_3 + \beta_{14}x_4 + u_1 \\
&= \gamma \cdot \{(\alpha_{12}\beta_{21} + \beta_{11}) + (\alpha_{12}\beta_{22} + \beta_{12})x_2 + \beta_{13}x_3 + \beta_{14}x_4 \\
&\quad + \alpha_{12}\beta_{25}x_5 + u_1 + \alpha_{12}u_2\}.
\end{aligned}
$$

Doing analogously, we can get the RF for y_2. Since both RF's are linear functions of $x \equiv (1, x_2, x_3, x_4, x_5)'$ and $(u_1, u_2)'$, they can be written as

$$
\begin{aligned}
y_1 &= \pi_{11} + \pi_{12}x_2 + \pi_{13}x_3 + \pi_{14}x_4 + \pi_{15}x_5 + v_1, \\
y_2 &= \pi_{21} + \pi_{22}x_2 + \pi_{23}x_3 + \pi_{24}x_4 + \pi_{25}x_5 + v_2.
\end{aligned}
$$

The RF parameters and error terms depend on the SF parameters and errors. For example, comparing π_{11} and v_1 to the intercept and the error in the y_1 SF with γ in, we have

$$
\pi_{11} = \gamma\alpha_{12}\beta_{21} + \gamma\beta_{11} \quad \text{and} \quad v_1 = \gamma u_1 + \gamma\alpha_{12}u_2.
$$

If our only goal is predicting the (equilibrium) value of y_1 and y_2, the RF's would be enough, which cannot, however, answer causal questions such as how the number of policeman affects the crime rate. The RF's are enough for the query "which values of y_1 and y_2 are to be attained in the system," but for the query on the *effects of y_2-intervention on y_1*, we need the SF's.

In MOM, we estimate the SF parameters directly in a single step, either with a system IVE/GMM or with their separate versions for each equation. But, they can be estimated also in two steps using the relation between the SF and RF parameters embodied in

$$
\Pi \equiv \Gamma^{-1}B.
$$

Specifically, first estimate Π with, say, the LSE $\tilde{\Pi}$; then a two-stage SF estimator $(\tilde{\Gamma}, \tilde{B})$ can be obtained if $\tilde{\Pi} \equiv \tilde{\Gamma}^{-1}\tilde{B}$ can be solved for $\tilde{\Gamma}$ and $\tilde{B}$. Thus, whether $\Pi \equiv \Gamma^{-1}B$ can be solved for the SF parameters Γ and B is called the *identification (ID) issue in simultaneous equations*. The key conditions for the ID are the order and rank conditions, which have been shown already to some extent. In the following, we take a different approach drawing on Lee and Kimhi (2005). This will turn out to be helpful later for simultaneous systems with limited dependent variables.

2.3   Simpler Approach to Rank Condition

SF equation h can be written such that x_i appears in the regression function; let x_i be a $K \times 1$ vector. Define the $K \times k_h$ selection matrix S_h such that

$$
\underset{1 \times k_h}{x'_{hi}} = \underset{1 \times K}{x'_i} \cdot \underset{K \times k_h}{S_h} ;
$$

S_h is a known matrix consisting of ones and zeros. With both x_i and x_{hi} having 1 as their first element, the first row of S_h is $(1, 0, ..., 0)$ $\forall h$. For example, if $K = 3$, $x_i = (1, p_i, q_i)'$, $k_h = 2$ and $x_{hi} = (1, q_i)'$, then

$$S_h = \begin{bmatrix} 1 & 0 \\ 0 & 0 \\ 0 & 1 \end{bmatrix} \quad \text{and } x'_{hi} = x'_i S_h \text{ is } [1, q_i] = [1, p_i, q_i] \begin{bmatrix} 1 & 0 \\ 0 & 0 \\ 0 & 1 \end{bmatrix}.$$

Using S_h, SF h can be written as

$$y_{hi} = \sum_{m=1, m \neq h}^{H} \alpha_{hm} y_{mi} + x'_i S_h \beta_h + u_{hi},$$

whereas RF h can be written as

$$y_{hi} = x'_i \eta_h + v_{hi}.$$

Now insert all RF's into SF h to get

$$x'_i \eta_h + v_{hi} = \sum_{m=1, m \neq h}^{H} \alpha_{hm} (x'_i \eta_m + v_{mi}) + x'_i S_h \beta_h + u_{hi}$$

$$= x'_i \left(\sum_{m=1, m \neq h}^{H} \alpha_{hm} \eta_m + S_h \beta_h \right) + \left(\sum_{m=1, m \neq h}^{H} \alpha_{hm} v_{mi} + u_{hi} \right).$$

Pre-multiply both sides by x_i and take expectation to get rid of the error terms. Since $E(x_i x'_i)$ is of full rank, the resulting equation is equivalent to

$$\eta_h = \sum_{m=1, m \neq h}^{H} \alpha_{hm} \eta_m + S_h \beta_h, \qquad h = 1, ..., H.$$

This equation links the RF parameters η_h's to the SF parameters α_{hm}'s and β_h's.

Since the RF parameters are easily identified and S_h is a known matrix, imagine the LSE of

$$\eta_h \quad \text{on} \quad D_h \equiv (\eta_1, ..., \eta_{h-1}, \eta_{h+1}, ..., \eta_H, S_h)$$

to estimate $\alpha_{h1}, ..., \alpha_{h,h-1}, \alpha_{h,h+1}, ..., \alpha_{hH}$ and β_h. Since the dimension of D_h is $K \times (H - 1 + k_h)$, we need $D'_h D_h$ to be of full column rank for this LSE. That is, the *rank condition* should hold:

$$rank(D_h) = H - 1 + k_h.$$

The order condition, which is a necessary condition for the rank condition, is that the column dimension of D_h should be equal to or smaller than the row dimension:

$$H - 1 + k_h \leq K.$$

To better appreciate the order condition, rewrite it as

$$H - 1 \leq K - k_h.$$

$H - 1$ is the *number of the included endogenous variables* in SF h, and $K - k_h$ is the *number of the excluded exogenous variables* from SF h. The order condition requires that the former be less than or equal to the latter. Essentially, the latter are used as instruments for the former, and the order condition is nothing but the requirement that there be enough instruments.

The rank condition is not easy to check: even if $rank(D_h) < H - 1 - k_h$, $rank(\hat{D}_h) = H - 1 - k_h$ is possible for a consistent estimator $\hat{D}_h$ for D_h as $\hat{D}_h$ is not exactly equal to D_h. But the order condition is easy to check and should be checked out for each SF equation. For instance, consider SF j ($\neq H$) with $\alpha_{jH} = 0$ to have only $H - 2$ included endogenous variables. Then the order condition for SF j is $H - 2 \leq K - k_j$.

2.4   Avoiding Arbitrary Exclusion Restrictions*

In simultaneous equations, exclusion restrictions are crucial, as obvious from the preceding subsections. Typically, the researcher imposes them on an intuitive ground and tries to convince the reader why the assumptions are plausible. Is there any way to derive exclusion restrictions from the data at hand? The answer is a qualified yes, and here we explain how it can be done, drawing on Lee and Chang (2007) who present an empirical illustration using the house sale data.

2.4.1   Grouping and Assigning

Consider two SF equations:

$$
\begin{aligned}
\text{SF1: } y_1 &= \alpha_1 y_2 + x_1' \beta_1 + x_c' \gamma_1 + u_1, \quad \alpha_1 \neq 0 \\
\text{SF2: } y_2 &= \alpha_2 y_1 + x_2' \beta_2 + x_c' \gamma_2 + u_2, \quad \alpha_2 \neq 0
\end{aligned}
$$

where x_j is a $k_j \times 1$ exogenous regressor vector, $j = 1, 2$, such that x_1 and x_2 share no common elements, x_c is a $k_c \times 1$ exogenous regressor vector common to the two SF's. If $\alpha_1 \alpha_2 = 1$, then the system is singular; otherwise the system can be solved for the RF's. There are *two types of decisions involved in the exclusion restrictions* (hence, two types of arbitrariness):

1. *Grouping*: which variables to remove jointly (two groups of variables are needed);

2. *Assigning*: which group to remove from which equation.

The main idea to avoid arbitrary exclusion restrictions is examining the ratios of the two RF estimates. For this, we will say that two exogenous regressors belong to the same *group* if their RF ratios are the same, and call an exogenous regressor a *"loner"* if it does not belong to any group. As will be shown below, the RF ratios satisfy the following:

- First, if both SF's are over-identified, then there are two groups and a loner/loners in the RF ratios.

- Second, if one SF is over-identified and the other SF is just-identified, then there are only one group and a loner/loners.

- Third, if both SF's are just-identified, then there are only loners.

Hence, examining the RF ratios provide valuable information for the grouping part of the exclusion restrictions. As for the assigning part, there is yet no clear-cut rule to apply. But the empirical example in Lee and Chang (2007) showed that some practical answers, if not the solutions, are available.

Define

$$\theta \equiv 1 - \alpha_1 \alpha_2.$$

Assuming $\alpha_1 \alpha_2 \neq 1$, solve the SF's to get their RF's:

$$y_1 = \frac{1}{\theta} \{ x_1' \beta_1 + x_c'(\gamma_1 + \alpha_1 \gamma_2) + x_2' \alpha_1 \beta_2 + u_1 + \alpha_1 u_2 \}$$

$$y_2 = \frac{1}{\theta} \{ x_2' \beta_2 + x_c'(\gamma_2 + \alpha_2 \gamma_1) + x_1' \alpha_2 \beta_1 + u_2 + \alpha_2 u_1 \}$$

$$\implies \text{RF1: } y_1 = w' \eta_1 + v_1 \quad \text{and} \quad \text{RF2: } y_2 = w' \eta_2 + v_2$$

defining w, η_1, η_2, v_1, and v_2 appropriately such that w is the system exogenous regressors, η_1 and η_2 are the RF parameters, and v_j's are the RF error terms.

From the RF's involving θ, we can see that the ratios of the coefficients of RF1 and RF2 for x_1 is α_2^{-1} because the x_1 coefficient in RF1 is β_1/θ and the x_1 coefficient in RF2 is $\alpha_2 \beta_1/\theta$; analogously, the ratios for x_2 is α_1—the ratios can be anything for x_c. Hence we get the following structure on the RF ratios:

$$x_1 : \alpha_2^{-1}, \qquad x_2 : \alpha_1, \qquad x_c : \text{possibly all different.}$$

Thus, examining the ratios of RF1 and RF2 is informative for learning about the SF exclusion restrictions. In the following, we examine various possibilities of the RF ratio structures, assuming that there are six exogenous regressors $w_1, ..., w_6$.

2.4.2   Patterns in Reduced-Form Ratios

Two Groups and Some Loners

Suppose that the ratios of the RF coefficients for $w_1, ..., w_6$ are

$$0.5, 0.5, \quad 3, 3, \quad 10, 20;$$

these ratios are "pure" numbers as they are unit-free. The first group (w_1, w_2) must be in one SF, the second group (w_3, w_4) in the other SF, and the loners w_5 and w_6 in both SF's. Turning to assigning, if we have a prior knowledge that (w_1, w_2) should be excluded from SF1, then the following SF's are obtained from two linear combinations of the RF's:

$$y_1 - 0.5y_2 = \dots \text{ no } (w_1, w_2) \text{ and } y_1 - 3y_2 = \dots \text{ no } (w_3, w_4)$$

$$\implies \quad y_1 = 0.5y_2 +, \dots \text{ no } (w_1, w_2) \text{ and } y_2 = \frac{1}{3}y_1 +, \dots \text{ no } (w_3, w_4).$$

Hence, $\alpha_1 = 0.5$ and $\alpha_2 = 1/3$ in the SF's. This display shows that it is wrong to start with a SF excluding, for instance, w_1 and w_3 together. Excluding (w_1, w_2) from SF1 is equivalent to solving the $y_1 - 0.5y_2$ equation for y_1, not for y_2; if we exclude (w_1, w_2) from SF2, we would solve the $y_1 - 0.5y_2$ equation for y_2.

One Group and Some Loners

Suppose we have the following RF ratios:

$$0.5,\ 0.5,\quad 3,\ 10,\ 20,\ 30.$$

In this case, (w_1, w_2) should be removed together from a SF that is over-identified, and one variable out of $(w_3, \dots, w_6)$ should be removed from the other SF that is just-identified. Differently from the two-groups-and-some-loner case, however, now it is not clear at all which one to choose among $w_3, \dots, w_6$ to exclude from the just-identified SF. In this sense, *just-identified SF's are more arbitrary than over-identified SF's*, although the former may look less restrictive excluding only a single variable than the latter. Depending on the arbitrary choice of a single variable to exclude, the just-identified SF parameter will be different. Suppose we exclude (w_1, w_2) from SF1 and w_6 from SF2. Then we get the following SF's:

$$y_1 - 0.5y_2 = \dots \text{ no } (w_1, w_2) \text{ and } y_1 - 30y_2 = \dots \text{ no } w_6$$

$$\implies \quad y_1 = 0.5y_2 +, \dots \text{ no } (w_1, w_2) \text{ and } y_2 = \frac{1}{30}y_1 +, \dots \text{ no } w_6.$$

All Loners

Now suppose we have no group structure whatsoever in the RF ratios; for instance,

$$0.5,\ 1,\ 3,\ 10,\ 20,\ 30.$$

In this case, we can obtain two just-identified SF's by removing any two variables. Depending on which ones are removed, the SF coefficients will be different. Suppose we remove w_1 and w_2 to get

$$y_1 - 0.5y_2 = \dots \text{ no } w_1 \qquad \text{and} \qquad y_1 - y_2 = \dots \text{ no } w_2$$

$$\implies \quad y_1 = 0.5y_2 +, \dots \text{ no } w_1 \qquad \text{and} \qquad y_1 = 1 \cdot y_2 +, \dots \text{ no } w_2.$$

Imagine that w_1 is a policy variable which can be controlled to attain some
target level of y_1 or y_2. The $y_1 - 0.5y_2$ equation simply shows that we cannot
hit the two targets (y_1 and y_2) freely with one "tool" w_1: no matter how we
choose w_1, still y_1 and y_2 will maintain the relationship $y_1 - 0.5y_2$ because
w_1 is absent in the $y_1 - 0.5y_2$ equation. This seems rather trivial. It would be
more interesting to see the relationship $y_1 - 0.5y_2$ between y_1 and y_2 being
undisturbed even if we control multiple tools, which will be the case if the
tools are removable together. That is, *viewed from the RF's, over-identified
SF's are meaningful while just-identified SF's are not.* Put it differently, removing one regressor can be always done with a linear combination of the y_1
and y_2 RF as in

$$y_1 + \lambda y_2 = w'(\eta_1 + \lambda \eta_2) + v_1 + \lambda v_2,$$

which is thus trivial, whereas removing multiple regressors with a single linear combination requires a group structure in the RF ratios, which is thus
nontrivial.

One Group Without any Loner

Suppose we have

$$0.5, \quad 0.5, \quad 0.5, \quad 0.5, \quad 0.5, \quad 0.5 :$$

only one group without any loners. In this case, the only possible linear
combination of y_1 and y_2 is $y_1 - 0.5y_2$ to yield "no exogenous-variable relation"

$$y_1 - 0.5y_2 = v_1 - 0.5v_2.$$

If we try to get two SF's out of this, the only way is solving this single
equation twice for y_1 and y_2 respectively, which then yields the following
singular system with $\alpha_1 \alpha_2 = 1$:

$$y_1 = 0.5y_2 + v_1 - 0.5v_2 \quad \text{and} \quad y_2 = 2y_1 + (v_2 - 2v_1).$$

Note that the correlation coefficient between the two SF error terms is -1.
This singular system is interesting in that no matter how we control all
exogenous variables, still we cannot change the structural relationship $y_1 -$
$0.5y_2$ in the singular system. Equation $y_1 - 0.5y_2 = v_1 - 0.5v_2$ can be viewed
as an "equilibrium (or stable)" relationship between y_1 and y_2: regardless of
the controllable variables, y_1 and y_2 will maintain the relation $y_1 - 0.5y_2 =$
$v_1 - 0.5v_2$.

2.4.3   Meaning of Singular Systems

The concept of singular relationship can be applied also to the above one-
group-and-some-loner case. Suppose we are not sure of the assigning decisions

in that case. Then the only sensible thing to do is to remove w_1 and w_2 and present the result as

$$y_1 - 0.5y_2 = \ldots \text{ no } (w_1, w_2), \ldots + v_1 - 0.5v_2$$

without trying to solve this for y_1 nor for y_2. This display depicts a "conditionally (on $w_3, \ldots, w_6$) stable relationship" between y_1 and y_2, showing that, even if we control the two exogenous variables w_1 and w_2, we cannot hit the two targets y_1 and y_2 freely. This is because the two control variables affect y_1 and y_2 in the (proportionally) same way. If we solve this display twice first for y_1 and then for y_2, we will get two singular SF's. Here the point is that singular SF's are not useless; we just have to take "one half" of them and interpret *a singular system as a stable relation between the response variables that is invariant to certain regressor changes*. This statement may answer the query of how to interpret a singular simultaneous equation system.

Although formally extending above RF-ratio-based approach to three or more SF's is not available yet, it is still possible to use the above approach when there are more than two SF's. Suppose y_1 has $y_2, y_3, \ldots, y_J$ as the endogenous regressors with coefficients $\alpha_{12}, \alpha_{13}, \ldots, \alpha_{1J}$. Then $(y_3, \ldots, y_J)$ can be substituted out with their RF's to leave only y_2 (and y_1), with which α_{12} can be estimated by our method. Analogously, $(y_2, y_4, \ldots, y_J)$ can be substituted out to leave only y_3 (and y_1), with which α_{13} can be estimated, and so forth. This is somewhat long-winded, but it shows that the approach is viable for more than two SF's as well.

In short, there are two ways to get SF's. One is building and estimating SF's first, say with IVE for each equation, and then obtaining their RF's—call this *"top down."* The other is building and estimating the RF's first and then deriving the SF's—call this *"bottom-up."* Since we would have more confidence in the bottom-up RF's, if these RF's contradict the top-down RF's, then we would conclude an error in the top-down SF's. The procedure in this subsection in essence suggests to go "bottom-up" in getting SF's rather than "top-down" as usually done. There will be then no risk of spending much time on building SF's first, only to see them negated by a disagreement between their derived top-down RF's and the directly built bottom-up RF's.

3   Methods of Moments for Panel Data

Panel data models such as

$$y_{it} = x'_{it}\beta + \delta_i + u_{it}, \quad i = 1, \ldots, N, \; t = 1, \ldots, T,$$

δ_i is a time-constant error have appeared already. Typically, if δ_i is assumed to be uncorrelated with x_{it}, the model is called a *"random effect model"*; otherwise, it is called a *"fixed effect model."* But these names are misnomers, as they do not mean what they are supposed to mean. "Unrelated effect models" and "related effect models" are better names. In the former, the main issue is how to estimate the model more efficiently to take advantage of the error term structure $\delta_i + u_{it}$. In the latter, the main issue is how to overcome the endogeneity problem. As the latter seems far more important than the former, we will discuss mostly the latter in this section while touching upon the former on a couple of occasions.

Other than having more observations than in a cross-section, the main attraction of panel data is its ability to handle endogenous regressors. Suppose that the only source of endogeneity is some components, say m_{it}, of x_{it} being related to δ_i. Then the first-differenced model

$$\Delta y_{it} = \Delta x_{it}'\beta + \Delta u_{it}$$

can be estimated with LSE. A problem with this *model-differencing (or "error-differencing")* approach, however, is that all time-constant regressors are removed along with δ_i and consequently their effect cannot be assessed. Also the temporal variations of x_{it} and y_{it} tend to be small relative to the interpersonal variation and, consequently, none of the estimation results in the differenced model may come out significant.

If $m_{it} = g_i(\delta_i) + \lambda_{it}$ for a function $g_i(\cdot)$ where $COR(\lambda_{it}, \delta_i) = 0$, then there is another approach available to deal with $COR(m_{it}, \delta_i) \neq 0$: use $\Delta m_{it} = m_{it} - m_{i,t-1} = \lambda_{it} - \lambda_{i,t-1}$ as an instrument for the original model in levels, not for the differenced model. This *"regressor-differencing"* approach can estimate the effects of time-constant regressors. Of course, there are other sources for endogeneity—relations between some of $x_{i1}, ..., x_{iT}$ to some of $u_{i1}, ..., u_{iT}$—but the two approaches to deal with δ_i provide a convenient forum to discuss panel data estimation on, which is the topic of this section.

In panel data, each period cross-section data is called a *"wave,"* and to simplify exposition, we will often set $T = 3$ ("three-wave panel"). Assume that $\{(x_{it}', y_{it})', \ t = 1, ..., T\}$ *are iid across i while allowing for arbitrary dependence and heteroskedasticity across t within a given i.* Assume that T is small but N is large so that the asymptotics is applied as $N \rightarrow \infty$. We will first set up our panel data model and then introduce panel IVE and GMM which are "regressor-differencing." This will then be followed by "error-differencing" approaches as above. After these approaches are examined, we will turn to "random effect" models. This section draws partly on Lee (2002).

3.1   Panel Linear Model

3.1.1   Typical Panel Data Layout

Although the model $y_{it} = x_{it}'\beta + \delta_i + u_{it}$ is convenient for exposition, the actual implementation of panel data estimation requires a more detailed modeling. For this, suppose

$$y_{it} = \underset{1\times 1}{\tau_t} + \underset{1\times k_{\tilde{c}}}{\tilde{c}_i'\ \tilde{\alpha}} + \underset{1\times k_x}{x_{it}'\ \beta} + \delta_i + u_{it}$$

where τ_t, $\tilde{\alpha}$, and β are the parameters to estimate, $\tilde{c}_i$ is a time-constant regressor vector, x_{it} is a time-variant regressor vector, δ_i is a time-constant error, and u_{it} is a time-varying error. The error term decomposition $\delta_i + u_{it}$ does not necessarily mean $COR(\delta_i, u_{it}) = 0$. For example, if $u_{it} = \delta_i \varepsilon_{it}$ for some time varying ε_{it}, then $COR(\delta_i, u_{it}) \neq 0$ in general. An example for the linear model is

y_{it}: hourly ln(wage) of men aged 40–60 at $t = 1$

τ_t: effect of the economy on y_{it} common to all i

$\tilde{c}_i$: race, schooling years

x_{it}: work hours, local unemployment rate, self-employment dummy

δ_i: genes and innate ability/IQ

u_{it}: unobserved time variants such as detailed residential location information

Define further

$$\tilde{k} \equiv k_{\tilde{c}} + k_x, \quad \tilde{\gamma} \equiv \begin{bmatrix} \tilde{\alpha} \\ \beta \end{bmatrix}, \quad \tilde{w}_{it} \equiv \begin{bmatrix} \tilde{c}_i \\ x_{it} \end{bmatrix}, \quad v_{it} \equiv \delta_i + u_{it},$$

to compactly rewrite the model as

$$y_{it} = \tau_t + \underset{1 \times \tilde{k}}{\tilde{w}'_{it} \tilde{\gamma}} + v_{it}.$$

Stack the equations for unit i across $t = 1, ..., T$:

$$\begin{bmatrix} y_{i1} \\ \vdots \\ y_{iT} \end{bmatrix} = I_T \begin{bmatrix} \tau_1 \\ \vdots \\ \tau_T \end{bmatrix} + \begin{bmatrix} \tilde{c}'_i \\ \vdots \\ \tilde{c}'_i \end{bmatrix} \tilde{\alpha} + \begin{bmatrix} x'_{i1} \\ \vdots \\ x'_{iT} \end{bmatrix} \beta + \begin{bmatrix} \delta_i \\ \vdots \\ \delta_i \end{bmatrix} + \begin{bmatrix} u_{i1} \\ \vdots \\ u_{iT} \end{bmatrix}$$

where I_T is the time dummy variables to estimate τ_t's which may be called "*period-specific intercepts.*" In view of the stacked variables, define

$$\underset{T \times 1}{y_i} \equiv \begin{bmatrix} y_{i1} \\ \vdots \\ y_{iT} \end{bmatrix}, \quad \underset{T \times k_x}{x'_i} \equiv \begin{bmatrix} x'_{i1} \\ \vdots \\ x'_{iT} \end{bmatrix}, \quad \text{and} \quad \underset{T \times 1}{u_i} \equiv \begin{bmatrix} u_{i1} \\ \vdots \\ u_{iT} \end{bmatrix}.$$

A typical panel data set is made up of y_i, $\tilde{c}_i$ and x'_i along with time dummies. For the wage example, the first six lines of a three-wave panel data for individual 1 and 2 look like Table 2. In the table, the wage column is for y_i; "td2" and "td3" are the time dummies for period 2 and 3, respectively; "unity" is the usual unity regressor in cross-section regression; race and education columns are for $\tilde{c}_i$; work-hr (work hour), ur (local unemployment rate), and self (self-employment or not) are for x'_i. Having td2, td3 and unity are equivalent to having td1 (time dummy for period 1), td2, and td3 as will be shown shortly.

Table 2: An Example of Three-Wave Panel Data

i	wage ($/hr)	td2	td3	unity	race	education (yr)	work-hr/ week	ur (%)	self
1	8.8	0	0	1	1	14	42	5.4	1
1	8.2	1	0	1	1	14	45	4.6	0
1	3.2	0	1	1	1	14	27	4.5	1
2	17.4	0	0	1	0	13	32	7.8	0
2	15.7	1	0	1	0	13	33	7.2	0
2	17.7	0	1	1	0	13	34	6.7	0

3.1.2   Panel Model with a Cross-Section Look

Our desire is to express the panel data model in such a way that it looks like the usual cross-section data model, as we have done in preceding sections. This requires more notations as follows.

Define

$$
\underset{T\times 1}{\tau} \equiv \begin{bmatrix} \tau_1 \\ \vdots \\ \tau_T \end{bmatrix}, \quad \underset{(T-1)\times 1}{\Delta\tau} \equiv \begin{bmatrix} \tau_2 - \tau_1 \\ \vdots \\ \tau_T - \tau_1 \end{bmatrix},
$$

$$
\underset{T\times 1}{\tau^*} \equiv \begin{bmatrix} \Delta\tau \\ \tau_1 \end{bmatrix} = \begin{bmatrix} \tau_2 - \tau_1 \\ \vdots \\ \tau_T - \tau_1 \\ \tau_1 \end{bmatrix}
$$

$$
\underset{T\times(T-1)}{m_T} \equiv \text{ the second to the last columns of } I_T.
$$

Observe the following equivalent parameterization for τ when $T = 3$:

$$
\begin{bmatrix} 1 & 0 & 0 \\ 0 & 1 & 0 \\ 0 & 0 & 1 \end{bmatrix} \begin{bmatrix} \tau_1 \\ \tau_2 \\ \tau_3 \end{bmatrix} = I_3\tau = \begin{bmatrix} 0 & 0 & 1 \\ 1 & 0 & 1 \\ 0 & 1 & 1 \end{bmatrix} \begin{bmatrix} \tau_2 - \tau_1 \\ \tau_3 - \tau_1 \\ \tau_1 \end{bmatrix}
$$

$$
= (m_3, 1_3)\tau^*;
$$

for a generic T, $I_T\tau = (m_T, 1_T)\tau^*$. We can estimate τ using I_T as the regressors, or equivalently, estimate τ^* using $(m_T, 1_T)$ as the regressors.

Define

$$
w_{it} \equiv \begin{bmatrix} 1 \\ \tilde{w}_{it} \end{bmatrix} = \begin{bmatrix} 1 \\ \tilde{c}_i \\ x_{it} \end{bmatrix} = \begin{bmatrix} c_i \\ x_{it} \end{bmatrix}, \quad \underset{k_c\times 1}{c_i} = \begin{bmatrix} 1 \\ \tilde{c}_{it} \end{bmatrix}, \text{ and } k \equiv k_c + k_x
$$

With the $k_c \times T$ matrix

$$
1_T' \otimes c_i = [c_i, ..., c_i],
$$

the model can be further written as

$$
\underset{T\times 1}{y_i} = \underset{T\times(T-1)}{m_T} \cdot \underset{(T-1)\times 1}{\Delta\tau} + \underset{T\times k_c}{(1_T' \otimes c_i)'\alpha} + \underset{T\times k_x}{x_i'\,\beta} + 1_T\delta_i + \underset{T\times 1}{u_i}\,,
$$

$$
\text{where } \underset{k_c \times 1}{\alpha} \equiv \begin{bmatrix} \tau_1 \\ \tilde{\alpha} \end{bmatrix}
$$

$$
= \underset{T\times k}{m_T\Delta\tau + w_i'\,\gamma + v_i}, \quad \text{where } w_i' \equiv ((1_T' \otimes c_i)', x_i'),
$$

$$
\underset{k\times 1}{\gamma} \equiv \begin{bmatrix} \alpha \\ \beta \end{bmatrix}, \quad \underset{T\times 1}{v_i} \equiv 1_T\delta_i + u_i.
$$

$$
= \underset{T\times(k+T-1)}{q_i'}\,\eta + v_i, \quad \text{where } q_i' \equiv (m_T, w_i'), \quad \underset{(k+T-1)\times 1}{\eta} \equiv \begin{bmatrix} \Delta\tau \\ \gamma \end{bmatrix}.
$$

In this final form, (y_i, q_i') corresponds to the above table entries. If $T = 1$, then $\Delta\tau = 0$ and $(1_1' \otimes c_i)' = c_i'$: we get the usual cross-section model $y_i = w_i'\gamma + v_i$.

3.1.3   Remarks*

The above panel model allows endogenous regressors including lagged response variables $y_{i,t-1}$, $y_{i,t-2}$, ... But using these as regressors requires a justification; models with lagged response variables in the regressors are said to be "dynamic." Sometimes, y_{it} may truly depend on $y_{i,t-1}$. One example is that the current wage y_{it} may be determined by the previous wage $y_{i,t-1}$ plus some adjustment term. Another example is that the current consumption amount of a good depends on the previous consumption amount. Other than for these true "state dependence" cases, $y_{i,t-1}$ are sometimes used to control for unobserved variables (both time-constant or time-variant) when w_{it} is not adequate to explain y_{it}. Dynamic models need "care" as can be seen in the following.

As a regressor, $y_{i,t-1}$ is necessarily endogenous regardless of any assumption on δ_i. For instance, consider a simple *dynamic model*

$$
y_{it} = \alpha y_{i,t-1} + \delta_i + u_{it}, \quad \text{where } |\alpha| < 1
$$
$$
\{ = \alpha(\alpha y_{i,t-2} + \delta_i + u_{i,t-1}) + \delta_i + u_{it}
$$
$$
= \delta_i + \alpha\delta_i + u_{it} + \alpha u_{i,t-1} + \alpha^2 y_{i,t-2} \}.
$$

By repeatedly substituting away the lagged responses, y_{it} can be written as a sum of all past errors and δ_i. Because $y_{i,t-1} = \alpha y_{i,t-2} + \delta_i + u_{i,t-1}$ includes δ_i which is also in $\delta_i + u_{it}$, $y_{i,t-1}$ is an endogenous regressor in the dynamic model. Here a regressor-differencing approach is applicable:

$\Delta y_{i,t-1} \equiv y_{i,t-1} - y_{i,t-2} = u_{i,t-1} - u_{i,t-2}$ can be used as an instrument for $y_{i,t-1}$. But the validity of $\Delta y_{i,t-1}$ as an instrument depends on the assumption for the relations among $\delta_i, u_{i1}, ..., u_{iT}$. If $u_{i1}, ..., u_{iT}$ are iid and independent of δ_i, then $\Delta y_{i,t-1} = u_{i,t-1} - u_{i,t-2}$ is independent of $\delta_i + u_{it}$ and thus $\Delta y_{i,t-1}$ is a valid instrument. If $COR(u_{it}, u_{i,t-1}) \neq 0$ but $COR(u_{it}, u_{i,t-s}) = 0$ for $\forall s \geq 2$ while $u_{i1}, ..., u_{iT}$ are still independent of δ_i, then $\Delta y_{i,t-2} = u_{i,t-2} - u_{i,t-3}$ is a valid instrument whereas $\Delta y_{i,t-1}$ is not.

One may wonder if using all time dummies instead of using only unity as in cross-section data is indeed necessary. To answer this query, assume that $E_t(u_{it}) \equiv \int c f_t(c) dc$ may be a function of t where f_t is the density of u_t at time t. Rewrite the model $y_{it} = \tau_t + \tilde{c}_i'\tilde{\alpha} + x_{it}'\beta + \delta_i + u_{it}$ as

$$y_{it} = \{\tau_t + E(\delta_i) + E_t(u_{it})\} + \tilde{c}_i'\tilde{\alpha} + x_{it}'\beta + \{\delta_i + u_{it} - E(\delta_i) - E_t(u_{it})\}$$

where the period-specific intercept is now $\tau_t + E(\delta_i) + E_t(u_{it})$ and the new error term has mean zero by construction. With 1 used as an instrument for $v_{it} = \delta_i + u_{it} \; \forall t$, which is equivalent to using the time dummy matrix I_T as a regressor for the y_i equation, we get to estimate $\tau_t + E(\delta_i) + E_t(u_{it})$, not τ_t (if all time dummies are not used, then it may happen that $E_t(v_{it}) \neq 0$ for some t). Redefining δ_i as $\delta_i - E(\delta_i)$, u_{it} as $u_{it} - E_t(u_{it})$, and τ_t as $\tau_t + E(\delta_i) + E_t(u_{it})$, we get back to the model $y_{it} = \tau_t + \tilde{c}_i'\tilde{\alpha} + x_{it}'\beta + \delta_i + u_{it}$ where the error term has zero mean for all periods. We will stick to this model with all time dummies and all zero-mean errors.

It is helpful to think of how a panel data set gets generated for two period case. Denote the density/probability of $\lambda | (\mu = \mu_o)$ as "$L(\lambda|\mu_o)$" for a while. In the first period, $(\tilde{c}_i, \delta_i)$ gets drawn from $L(\tilde{c}, \delta)$ to be fixed throughout all periods; then (x_{i1}, u_{i1}) gets drawn from $L(x_1, u_1|\tilde{c}_i, \delta_i)$; and adding τ_1 to $\tilde{c}_i'\tilde{\alpha} + x_{i1}'\beta + \delta_i + u_{i1}$ yields y_{i1}. It goes without saying that if $(x_1, u_1) \amalg (\tilde{c}, \delta)$, then (x_{i1}, u_{i1}) gets drawn from $L(x_1, u_1)$. Also note that τ_1 may be drawn from a distribution, but this is inconsequential, as τ_1 will be estimated. Drawing from $L(x_1, u_1|\tilde{c}_i, \delta_i)$ can be further decomposed into drawing first from $L(x_1|\tilde{c}_i, \delta_i)$ and then from $L(u_1|x_{i1}, \tilde{c}_i, \delta_i)$.

In the second period, (x_{i2}, u_{i2}) gets drawn from $L(x_2, u_2|\tilde{c}_i, \delta_i)$ which may differ from $L(x_1, u_1|\tilde{c}_i, \delta_i)$, and adding τ_2 to $\tilde{c}_i'\tilde{\alpha} + x_{i2}'\beta + \delta_i + u_{i2}$ yields y_{i2}. More generally, (x_{i2}, u_{i2}) may get drawn from $L(x_2, u_2|x_{i1}, u_{i1}, \tilde{c}_i, \delta_i)$. In this case, $L(x_2, u_2|x_{i1}, u_{i1}, \tilde{c}_i, \delta_i)$ differs from $L(x_1, u_1|\tilde{c}_i, \delta_i)$ in general. Hence the joint density/probability for each person is

$$L(x_2, u_2, x_1, u_1, \tilde{c}, \delta) = L(x_2, u_2|x_1, u_1, \tilde{c}, \delta) \cdot L(x_1, u_1|\tilde{c}, \delta) \cdot L(\tilde{c}, \delta).$$

From this, $L(u_2)$ and $L(u_1)$ can be found, with which $E_2(u_2)$ and $E_1(u_1)$ can be obtained. Also from this display, we get the joint density/probability for the observed variables $L(y_2, x_2, y_1, x_1, \tilde{c})$; the observations are iid draws from this.

3.2   Panel GMM and Constructing Instruments

3.2.1   Panel IVE and GMM

With some moments conditions, we can estimate η in $y_i = q_i'\eta + v_i$. Suppose we have $E(z_t v_t) = 0$, $t = 1, ..., T$, where z_t is the period-t instruments. The reader should specify (and justify) each z_t essentially in the same way an instrument vector is specified in the usual cross-section IVE. This could be the most difficult step, but suppose this has been done—we will discuss finding z_t shortly. Let

$$
\underset{K \times T}{z_i} \equiv \begin{bmatrix} z_{i1} & 0 & 0 \\ 0 & \ddots & 0 \\ 0 & 0 & z_{iT} \end{bmatrix} \implies E(zv) = \begin{bmatrix} E(z_1 v_1) \\ \vdots \\ E(z_T v_T) \end{bmatrix} = 0.
$$

Assume $K \geq k + T - 1$: there are at least as many moment conditions as the number of parameters.

With the moment conditions, the panel IVE and GMM for $y_i = q_i'\eta + v_i$ is

$$
h_{ive} = \left\{ \sum_i q_i z_i' \left(\sum_i z_i z_i' \right)^{-1} \sum_i z_i q_i' \right\}^{-1} \cdot
$$

$$
\sum_i q_i z_i' \left(\sum_i z_i z_i' \right)^{-1} \sum_i z_i y_i
$$

$$
h_{gmm} = \left(\sum_i q_i z_i' C_N^{-1} \sum_i z_i q_i' \right)^{-1} \sum_i q_i z_i' C_N^{-1} \sum_i z_i y_i
$$

$$
\text{where} \quad C_N \equiv \frac{1}{N} \sum_i z_i \hat{v}_i \hat{v}_i' z_i' \quad \text{and} \quad \hat{v}_i \equiv y_i - q_i' h_{ive};
$$

$$
\sqrt{N}(h_{gmm} - \eta) \rightsquigarrow N(0, C_{gmm}) \quad \text{where}
$$

$$
\left(\frac{1}{N} \sum_i q_i z_i' \cdot C_N^{-1} \cdot \frac{1}{N} \sum_i q_i z_i' \right)^{-1} \rightarrow^p C_{gmm}.
$$

With $\tilde{v}_i \equiv y_i - q_i' h_{gmm}$, the GMM over-id test statistic for $H_0 : E(z_i v_i) = 0$ is

$$
\left(\frac{1}{\sqrt{N}} \sum_i z_i \tilde{v}_i \right)' \left(\frac{1}{N} \sum_i z_i \tilde{v}_i \tilde{v}_i' z_i' \right)^{-1} \frac{1}{\sqrt{N}} \sum_i z_i \tilde{v}_i \rightsquigarrow \chi^2_{K-(k+T-1)}
$$

Note that $N^{-1} \sum_i \hat{v}_i \hat{v}_i'$ shows the auto-correlation pattern of $v_{i1}, ..., v_{iT}$. Due to δ_i present in all of $v_{i1}, ..., v_{iT}$, $v_{i1}, ..., v_{iT}$ are auto-correlated even if $u_{i1}, ..., u_{iT}$ are not.

3.2.2   Instrument Construction

Turning to the specifics of constructing z, suppose x_t has *strictly exogeneity (EXO)* wrt v_t in the sense that x_t is uncorrelated with v_t at all leads and lags. Then all of (x_1', x_2', x_3') get included in all of z_1, z_2, and z_3. For instance, for $T = 3$ and c having EXO wrt v_t as well, set

$$
\begin{aligned}
z &= diag\{(x_1', x_2', x_3', c')', (x_1', x_2', x_3', c')', (x_1', x_2', x_3', c')'\} \\
&= I_3 \otimes \underset{\{3\cdot(k_x\cdot 3 + k_c)\}\times 3}{(x_1', x_2', x_3', c')'}.
\end{aligned}
$$

But the dimension of z can be too high if T is large, which means too long a time to implement IVE and GMM. Also EXO assumption might be too strong.

An often-used assumption weaker than EXO is *predeterminedness (PRE)*: $COR(x_s, v_t) = 0 \ \forall s \leq t$. This allows the economic agents to modify its future regressors after observing the current or past error terms. For instance, observing soil quality v_t, farmers may adjust their future fertilizer levels x_t to affect the output y_t. In rational expectation models, $E(v_t | x_1, ..., x_t) = 0$ is assumed, which allows PRE; compare this to $E(v_t | x_1, ..., x_T) = 0$ which rules out PRE as $COR(v_t, x_T) \neq 0$ is not allowed. To illustrate moment conditions under PRE, suppose c is EXO. Then set

$$
z = diag\{w_1, (x_1', w_2')', (x_1', x_2', w_3')'\};
$$

compare this to the above z for the EXO case.

A further modification may be needed in practice. For instance, if simultaneity is a concern, then further weakening PRE, we may adopt only $COR(x_s, v_t) = 0 \ \forall s < t$, which allows a contemporaneous relation between x_t and v_t. An often-encountered problem in practice of panel IVE and GMM is near singularity due to some time variants having little variation over time. For example, if the time span for the panel is short, $marriage_t$ may change very little over time. When $T = 3$, using all of $marriage_1$, $marriage_2$, and $marriage_3$ as three separate instruments may result in $N^{-1} \sum_i z_i z_i'$ near singular. If this problem occurs, then treat $marriage_t$ as time invariant when used as instruments; i.e., use only $marriage_1$ as an instrument in c. Another variable that requires care is age which varies across i and t, but only in a deterministic way for t; treating age as time variant can cause a trouble as will be shown later. In the following, we provide a more specific example using the wage case with $T = 3$, where the subscript i will not be omitted to prevent confusion.

3.2.3   Specific Examples of Instruments

Suppose δ_i is ability, IQ, or productivity, and y_{it} is hourly wage to be explained by time-constant c_i and time variants x_{it} where

c_i: 1, RC (race), ED (education in years),

x_{it}: WH (working hours), UR (local unemployment rate), SF (1 if self-employed and 0 otherwise),

u_{it}: Residential location.

In this example, δ_i may be observable to the employer of individual i but not to econometricians. Comparing $v_{it} = \delta_i + u_{it}$ with c_i and x_{it}, suppose the following holds:

(a) RC_i is not correlated with v_{it}.

(b) ED_i is correlated with v_{it} only through δ_i.

(c) UR_{it} is not correlated with v_{it}.

(d) SF_{it} is correlated with v_{it} only through δ_i.

(e) WH_{it} is correlated with v_{it} only through u_{it}.

The best way to understand these would be pointing out when these assumption might not hold. In (a), $COR(RC_i, \delta_i) = 0$ and $COR(RC_i, u_{it}) = 0$, but the former will not hold if certain race is innately less (or more) productive, and the latter will not hold if residential location is informative for race. In (c), $COR(UR_{it}, v_{it}) = 0$ will not hold if residential location is correlated with UR_{it}. In (d), $COR(SF_{it}, \delta_i) \neq 0$ is allowed (self-employed people may be more able because they can survive on their own, or less able if nobody wants to hire them), but (d) would not hold if residential location is informative for SF_{it}. In (e), $COR(WH_{it}, \delta_i) = 0$, but this may not hold if able or productive people tend to work more (or less).

Suppose we take a simple-minded approach of not using any variable correlated with v_{it} in any fashion. Then only 1, RC_i and UR_{it} can be used as instruments. Because UR_{it} is time-variant, we can impose various moment conditions on UR_{it}. Suppose we use PRE type conditions:

$$E(v_{it}) = 0 \quad \forall t, \quad E(RC_i v_{it}) = 0 \quad \forall t, \quad E(UR_{is} v_{it}) = 0 \quad \forall s \le t.$$

Then,

$$z_i' = \left[\begin{array}{ccc} (1, RC_i, UR_{i1}) & & 0 \\ 0 & (1, RC_i, UR_{i1}, UR_{i2}) & 0 \\ 0 & 0 & (1, RC_i, UR_{i1}, UR_{i2}, UR_{i3}) \end{array} \right]$$

which is block-diagonal with each block as each period instruments. For EXO type conditions,

$$E(v_{it}) = 0 \quad \forall t, \quad E(RC_i v_{it}) = 0 \quad \forall t, \quad E(UR_{is} v_{it}) = 0 \quad \forall s, t$$
$$\implies z_i' = I_3 \otimes (1, RC_i, UR_{i1}, UR_{i2}, UR_{i3}).$$

So far, we might have been too conservative using only 1, RC_i, and UR_{it} as instruments that are not related with v_{it} at all. Under some assumptions, however, we may use WH_{it} and SF_{it} as well, for they are correlated only with a part of v_{it}. For WH_{it}, assume

$$E(WH_{is}v_{it}) = 0 \quad \forall s < t$$

which is PRE without the contemporaneous orthogonality. As for SF_{it}, suppose

$$SF_{it} = g_i(\delta_i) + \lambda_{it} \quad \text{where } COR(\lambda_{it}, \delta_i) = 0.$$

Any transformation of SF_{it} that removes $g_i(\delta_i)$ can be used as instruments: e.g.,

$$SF_{it} - SF_{i,t-1} = \lambda_{it} - \lambda_{i,t-1} \quad \text{and} \quad SF_{it} - \frac{1}{T}\sum_t SF_{it} = \lambda_{it} - \frac{1}{T}\sum_t \lambda_{it}.$$

One caution is that, although these differences can be valid instruments in theory, in practice, the temporal variation of SF_{it} may be too small, or just a pure "error" independent of endogenous regressors in the model. That is, what is left in SF_{it} after removing the δ_i-related part of SF_{it} may not be much.

Define $\Delta SF_{it} \equiv SF_{it} - SF_{i,t-1}$. Under the PRE type conditions

v_{i1} is orthogonal to $z_{i1} = (1, RC_i, UR_{i1})'$

v_{i2} is orthogonal to $z_{i2} = (1, RC_i, UR_{i1}, UR_{i2}, WH_{i1}, \Delta SF_{i2})'$

v_{i3} is orthogonal to $z_{i3} = (1, RC_i, UR_{i1}, UR_{i2}, UR_{i3}, WH_{i1}, WH_{i2}, \Delta SF_{i2}, \Delta SF_{i3})'.$

Then $z_i' = diag(z_{i1}', z_{i2}', z_{i3}')$. If we use EXO type conditions for UR_{it}, then

$$
\begin{aligned}
z_{i1} &= (1, RC_i, UR_{i1}, UR_{i2}, UR_{i3})' \\
z_{i2} &= (1, RC_i, UR_{i1}, UR_{i2}, UR_{i3}, WH_{i1}, \Delta SF_{i2})' \\
z_{i3} &= (1, RC_i, UR_{i1}, UR_{i2}, UR_{i3}, WH_{i1}, WH_{i2}, \Delta SF_{i2}, \Delta SF_{i2})'.
\end{aligned}
$$

3.3   Within-Group and Between-Group Estimators

3.3.1   Within Group Estimator (WIT)

Define an idempotent and symmetric mean-differencing matrix

$$\underset{T \times T}{Q_T} \equiv I_T - \frac{1}{T}1_T 1_T'.$$

Q_T is a mean-subtracting matrix because

$$Q_T y_i = \begin{bmatrix} y_{i1} - \bar{y}_{i.} \\ \vdots \\ y_{iT} - \bar{y}_{i.} \end{bmatrix}, \quad \text{where } \bar{y}_{i.} \equiv \frac{1}{T}\sum_t y_{it}.$$

Multiply $y_i = m_T \Delta\tau + (1'_T \otimes c_i)'\alpha + x'_i\beta + 1_T\delta_i + u_i$ by Q_T:

$$Q_T y_i = Q_T m_T \cdot \Delta\tau + Q_T(1'_T \otimes c_i)'\alpha + Q_T x'_i\beta + Q_T 1_T\delta_i + Q_T u_i.$$

The mean differencing eliminates the second and fourth terms on the right-hand side (i.e., all time-constants are removed, observed or not) to leave

$$Q_T y_i \;=\; Q_T m_T \cdot \Delta\tau + Q_T x'_i\beta + Q_T u_i = Q_T w_i^{*\prime}\gamma^* + Q_T u_i,$$

where

$$\underset{T\times(T-1+k_x)}{w_i^{*\prime}} \;\equiv\; (m_T, x'_i), \quad \gamma^* \equiv (\Delta\tau', \beta')'.$$

Apply LSE to this model under $E(w_i^* Q_T u_i) = 0$ to get *"within-group estimator (WIT)"*—a "group" refers to each individual:

$$g_{wit} = \left(\sum_i w_i^* Q_T w_i^{*\prime}\right)^{-1} \cdot \sum_i w_i^* Q_T y_i,$$

$$\sqrt{N}(g_{wit} - \gamma^*) \rightsquigarrow N(0, C_{wit}),$$

where $\quad \Psi_N^{-1} \cdot \dfrac{1}{N}\sum_i (w_i^* Q_T \hat{u}_i \hat{u}'_i Q_T w_i^{*\prime}) \cdot \Psi_N^{-1} \to^p C_{wit},$

$$\Psi_N \equiv \frac{1}{N}\sum_i w_i^* Q_T w_i^{*\prime} \quad \text{and} \quad Q_T \hat{u}_i = Q_T y_i - Q_T w_i^{*\prime} g_{wit}.$$

In WIT, the coefficients of time-constants are not identified, which is the main weakness of WIT: typically, in practice, there are more time-constant regressors than time-variant ones. But the advantage of WIT, in addition to removing δ_i, can be seen in the following example. In finding the effect of age on salary with cross-section data, essentially we compare people with different ages to infer how one person's salary would change when his/her own age increases. WIT avoids this type of "less-than-ideal" interpersonal "level" comparisons.

In WIT, the regressors and the error become, respectively, $x_{it} - \bar{x}_{i.}$ and $u_{it} - \bar{u}_{i.}$, where $\bar{x}_{i.} \equiv T^{-1}\sum_t x_{it}$ and $\bar{u}_{i.} \equiv T^{-1}\sum_t u_{it}$. For these to be orthogonal, we need EXO. Hence, although panel IVE can be applied to the model $Q_T y_i = Q_T w_i^{*\prime}\gamma^* + Q_T u_i$, it is not easy to think of instruments for the mean-differenced model. Instead, consider the first-differenced model

$$\Delta y_{it} = \Delta x'_{it}\beta + \Delta u_{it}.$$

For this, PRE $E(x_{is}u_{it}) = 0 \,\forall s \leq t$ implies $E(\Delta x_{is}\Delta u_{it}) = 0 \,\forall s \leq t-1$. That is, if removing δ_i is not enough to remove all endogeneity problems, then it is better to first-difference the model and then apply panel IVE/GMM to the differenced model.

Removing δ_i requires care in practice, because some time-variants such as age_{it} vary over time in a deterministic way: when demeaning or first-differencing is applied to such variables, demeaned or first-differenced versions become linearly dependent on demeaned or first-differenced m_T, respectively. To see this, let $T = 3$ and compare m_3 and age_{it} before and after mean differencing:

$$
\begin{bmatrix}
0 & 0 & \text{age}_{i1} \\
1 & 0 & \text{age}_{i2} \\
0 & 1 & \text{age}_{i3}
\end{bmatrix}
\Rightarrow
\begin{bmatrix}
-1/3 & -1/3 & -1 \\
2/3 & -1/3 & 0 \\
-1/3 & 2/3 & 1
\end{bmatrix}
\Rightarrow
\begin{bmatrix}
2/3 & -1/3 & 0 \\
-1/3 & 2/3 & 1
\end{bmatrix}
$$

where the second arrow follows from losing one row in mean differencing (this is essentially the same as losing one wave in first-differencing). In the last 2×3 matrix, one column is linearly dependent on the other two, and any one column should be removed to prevent the singularity of the inverted matrix in g_{wit}. This also holds more generally for $T > 3$: remove either one column in m_T or age_{it} from x_{it}.

3.3.2   Between Group Estimator (BET) and Panel LSE and GLS

If we take temporal average of all variables for each individual, then the time dimension is gone. The LSE to this cross-section data is called "*between-group estimator (BET)*" as only between-group variations are used for estimation. Formally, with $\bar{q}_{i.} \equiv T^{-1} \sum_t q_{it}$ and $\bar{v}_{i.} \equiv T^{-1} \sum_t v_{it}$, BET is the LSE applied to

$$
\bar{y}_{i.} = \bar{q}'_{i.}\eta + \bar{v}_{i.}, \quad i = 1, ..., N.
$$

For *unbalanced panels*—not everybody observed for all periods—BET may be the easiest estimator to use.

In practice, BET is little different from *panel LSE*

$$
h_{lse} \equiv \left(\sum_i q_i q'_i \right)^{-1} \cdot \sum_i q_i y_i; \quad \text{with } \hat{v}_i \equiv y_i - q'_i h_{lse},
$$

$$
\sqrt{N}(h_{lse} - \eta) \rightsquigarrow N(0, C_{lse}),
$$

$$
\left(\frac{1}{N} \sum_i q_i q'_i \right)^{-1} \left(\frac{1}{N} \sum_i q_i \hat{v}_i \hat{v}'_i q'_i \right) \left(\frac{1}{N} \sum_i q_i q'_i \right)^{-1} \to^p C_{lse}.
$$

BET does averaging across time and then across individuals. In Panel LSE, $q_i q'_i$ includes summing across time, and there is summing across individuals as well in $\sum_i q_i q'_i$. In this regard, both BET and panel LSE use the *pooled panel data*, i.e., the data treating each individual's time-series observations as if they come from different independent individuals.

Although BET and panel LSE do not take advantage of panel data structure—they do not deal with δ_i, not to mention removing δ_i—they remove measurement errors due to the temporal averaging. In practice, one may want to weigh the biases due to $COR(\delta_i, w_{it}) \neq 0$ and measurement error in w_{it}. If the latter problem is severe, then BET and panel LSE may be preferred to WIT and IVE.

Sometimes, under the homoskedasticity assumption

$$E(v_i v_i' | w_{i1}, ..., w_{iT}) = \Omega, \quad \text{a constant matrix,}$$

panel GLS is done using

$$\Omega_N \equiv \frac{1}{N} \sum_i \hat{v}_i \hat{v}_i'.$$

The *panel GLS* is

$$h_{gls} = \left\{ \sum_i q_i \Omega_N^{-1} q_i' \right\}^{-1} \cdot \sum_i q_i \Omega_N^{-1} y_i;$$

$$\sqrt{N}(h_{gls} - \eta) \rightsquigarrow N(0, C_{lse}) \quad \text{where} \quad \left(\frac{1}{N} \sum_i q_i \Omega_N^{-1} q_i' \right)^{-1} \rightarrow^p C_{gls}.$$

3.3.3   WIT as Fixed-Effect Estimator*

In the literature, WIT is also called *"fixed-effect estimator."* To see why, suppose we construct the dummy variables for all individuals to estimate $\delta_1, ..., \delta_N$. Since all δ_i's are treated as parameters in this approach, δ_i's are called fixed effects. Using the partial regression removing the presence of the individual dummies from the other variables in the model, we will show that the partial regression is nothing but WIT in the following. That is, as far as estimating the WIT parameters γ^* goes, estimating γ^* along with $\delta_1, ..., \delta_N$ using q_i and the individual dummies is exactly the same as estimating γ^* with WIT.

To show the partial regression claim, it helps to write the model for all individuals together. For this, define

$$\underset{NT \times 1}{Y} \equiv \begin{bmatrix} y_1 \\ \vdots \\ y_N \end{bmatrix}, \quad \underset{NT \times k_{\tilde{c}}}{\tilde{C}} \equiv \begin{bmatrix} (1_T \otimes \tilde{c}_1') \\ \vdots \\ (1_T \otimes \tilde{c}_N') \end{bmatrix}, \quad \underset{NT \times k_x}{X} \equiv \begin{bmatrix} x_1' \\ \vdots \\ x_N' \end{bmatrix},$$

$$\underset{N \times 1}{\delta} \equiv \begin{bmatrix} \delta_1 \\ \vdots \\ \delta_N \end{bmatrix}, \quad \underset{NT \times 1}{U} \equiv \begin{bmatrix} u_1 \\ \vdots \\ u_N \end{bmatrix}.$$

For the time effect $\Delta \tau$, the regressor matrix is

$$\underset{NT \times (T-1)}{1_N \otimes m_T} = \begin{bmatrix} m_T \\ \vdots \\ m_T \end{bmatrix} \implies \underset{NT \times (T-1)}{(1_N \otimes m_T)} \underset{(T-1) \times 1}{\Delta \tau} = \begin{bmatrix} m_T \Delta \tau \\ \vdots \\ m_T \Delta \tau \end{bmatrix}.$$

For individual effect δ_i, the regressor matrix is

$$D \equiv \underset{NT \times N}{I_N \otimes 1_T} = \begin{bmatrix} 1_T & 0 & \cdots & 0 \\ 0 & 1_T & 0 & \vdots \\ \vdots & 0 & \ddots & 0 \\ 0 & \cdots & 0 & 1_T \end{bmatrix} \implies D \begin{bmatrix} \delta_1 \\ \vdots \\ \delta_N \end{bmatrix} = \begin{bmatrix} 1_T \delta_1 \\ \vdots \\ 1_T \delta_N \end{bmatrix}.$$

Since the column sum of $I_N \otimes 1_T$ is 1_{NT}, the unity regressor in c_i should be dropped and $\tilde{c}_i$ should be used instead. Hence the model with all individuals together becomes

$$Y = (1_N \otimes m_T)\Delta\tau + \tilde{C}\tilde{\alpha} + X\beta + D\delta + U.$$

With

$$P_D \equiv D(D'D)^{-1}D' \quad \text{and} \quad Q_D = I_{NT} - D(D'D)^{-1}D',$$

observe

$$\begin{aligned} P_D &= (I_N \otimes 1_T)\{(I_N \otimes 1_T)'(I_N \otimes 1_T)\}^{-1}(I_N \otimes 1_T)' \\ &= (I_N \otimes 1_T)(I_N \otimes 1_T'1_T)^{-1}(I_N \otimes 1_T') \\ &= (I_N \otimes 1_T)(I_N \otimes T^{-1})(I_N \otimes 1_T') = I_N \otimes T^{-1}1_T1_T' \end{aligned}$$

$$= \begin{bmatrix} T^{-1}1_T1_T' & 0 & \cdots & 0 \\ 0 & T^{-1}1_T1_T' & 0 & \vdots \\ \vdots & 0 & \ddots & 0 \\ 0 & \cdots & 0 & T^{-1}1_T1_T' \end{bmatrix}$$

$$\text{where } T^{-1}1_T1_T' = \frac{1}{T}\begin{bmatrix} 1 & \cdots & 1 \\ \vdots & \ddots & \vdots \\ 1 & \cdots & 1 \end{bmatrix}.$$

Each diagonal block is to compute the temporal mean for each individual. Multiply this by Y to get

$$\underset{NT \times 1}{P_D Y} = \begin{bmatrix} \underset{T \times 1}{1_T \otimes \bar{y}_{1.}} \\ \vdots \\ \underset{T \times 1}{1_T \otimes \bar{y}_{N.}} \end{bmatrix} \implies \underset{NT \times 1}{Q_D Y} = \begin{bmatrix} y_1 - 1_T \otimes \bar{y}_{1.} \\ \vdots \\ y_N - 1_T \otimes \bar{y}_{N.} \end{bmatrix}:$$

$Q_D Y$ is Y with individual temporal mean subtracted from each element of Y. A similar statement can be made for $Q_D X$. Hence the partial regression eliminating D is nothing but WIT.

As for the time dummy part, it is instructive to observe

$$P_D(1_N \otimes m_T) = (I_N \otimes T^{-1}1_T1_T')(1_N \otimes m_T)$$
$$= 1_N \otimes T^{-1}1_T1_T'm_T$$
$$= 1_N \otimes T^{-1}1_T1_{T-1}'$$
$$(\text{as } 1_T'm_T = 1_{T-1}'; \text{ the sum of each time dummy vector is 1})$$
$$\Longrightarrow Q_D(1_N \otimes m_T) = (1_N \otimes m_T) - (1_N \otimes T^{-1}1_T1_{T-1}')$$
$$= 1_N \otimes (m_T - T^{-1}1_T1_{T-1}').$$

For instance, when $T = 3$,

$$m_T - T^{-1}1_T1_{T-1}' = \begin{bmatrix} 0 & 0 \\ 1 & 0 \\ 0 & 1 \end{bmatrix} - \frac{1}{3}\begin{bmatrix} 1 & 1 \\ 1 & 1 \\ 1 & 1 \end{bmatrix} = \begin{bmatrix} -1/3 & -1/3 \\ 2/3 & -1/3 \\ -1/3 & 2/3 \end{bmatrix}.$$

This is m_T minus its temporal mean which is $1/3$ for each column of m_T.

CHAPTER 3
M-ESTIMATOR AND MAXIMUM
LIKELIHOOD ESTIMATOR (MLE)

Least square estimator (LSE) minimizes an objective function, and the estimator itself is obtained in a closed form. There are many other estimators maximizing/minimizing some objective functions, but most of them are not written in closed forms; those estimators, called "M-estimators", are reviewed here. Typically, the first-order conditions of M-estimators are moment conditions, and this links M-estimator to MOM estimator/test. A well-known example of M-estimator is maximum likelihood estimator (MLE), which is studied in this chapter along with three tests associated with MLE. MOM tests, some of which are derived from M-estimators, are also examined in this chapter.

1 M-Estimator

LSE, IVE, and GMM are rare cases where the estimators are written in closed forms. Often, an estimator is defined implicitly by

$$b_N \equiv \text{argmax}_{b \in B} Q_N(b),$$

where B is a (compact) parameter space in R^k, "argmax" means the b maximizing $Q(b)$, and

$$Q_N(b) \equiv \frac{1}{N} \sum_i q(x_i, y_i, b) = \frac{1}{N} \sum_i q(z_i, b), \quad \text{where } z_i \equiv (x_i', y_i)';$$

This kind of estimators, defined implicitly as a maximand (or minimand) of a function, is called a "*M-estimator*" (or "extremum estimator"). The prefix "M" comes from "M" in MLE, meaning "MLE-like estimator."

1.1 Four Issues and Main Points

M-estimators abound. For instance, $q(z, b) = -(y - x'b)^2$ yields LSE. If the regression function is nonlinear in β, say $\rho(x, \beta)$, then $q(z, b) = -\{y - \rho(x, b)\}^2$, which leads to a nonlinear LSE. If $q(z, b) = -|y - x'b|$ where the function $x'b$ used to predict y is linear but the "outside" function is not, then we get a "least absolute deviation (LAD)" estimator, which seems more natural than LSE but less popular due to its relative computational and analytical difficulties compared with LSE. There are many more examples of M-estimators yet to appear later, and for this, we will discuss M-estimators in general terms. To simplify notations, sometimes z_i will be omitted in $q(z_i, b)$, or $q(z_i, b)$ may be denoted just as q_i.

For an M-estimator b_N, four questions arise:

1. Identification (ID): for which population parameter is b_N designed?

2. Consistency: does b_N converge to the parameter, say β?

3. Asymptotic distribution of $\sqrt{N}(b_N - \beta)$; if this is normal (as it almost always is), what is the asymptotic variance?

4. Consistent estimation of the asymptotic variance which involves β and other unknown components: how do we estimate the variance then?

In a nutshell, the answers to these questions are as follows: defining $Q(b) = Eq(z, b)$,

1. ID: b_N is designed for $\beta \equiv \text{argmax}_{b \in B} Q(b)$.

2. Consistency: $b_N \to^p \beta$.

3. Asymptotic distribution is

$$\sqrt{N}(b_N - \beta) \rightsquigarrow N(0, \, E^{-1}\{q_{bb'}(\beta)\} E\{q_b(\beta)q_{b'}(\beta)\} E^{-1}\{q_{bb'}(\beta)\})$$

where $E^{-1}(\cdot) = \{E(\cdot)\}^{-1}$, $q_{b'} = q'_b$, and q_b and $q_{bb'}$ are the first and second derivative matrices, respectively. The asymptotic variance takes a "sandwich" form; i.e., $E\{q_b(\beta)q_{b'}(\beta)\}$ is sandwiched by $E^{-1}\{q_{bb'}(\beta)\}$.

4. For estimating the asymptotic variance consistently, one can use

$$\left\{ \frac{1}{N} \sum_i q_{bb'}(b_N) \right\}^{-1} \left\{ \frac{1}{N} \sum_i q_b(b_N)q_{b'}(b_N) \right\} \left\{ \frac{1}{N} \sum_i q_{bb'}(b_N) \right\}^{-1}.$$

1.2   Remarks for Asymptotic Distribution

Here, three remarks are made for M-estimator asymptotic distribution. First, assuming that $q(z, b)$ is twice continuously differentiable (i.e., the first- and second-order derivatives exist and are continuous) with respect to (wrt) b on some open convex set B_β including β as an interior point for all z, we show the key equations for the asymptotic distribution of M-estimator. Second, "δ-method" and "continuous mapping theorem" which are convenient in deriving the asymptotic distribution of functions of (M-) estimators are introduced. Third, a theorem for the asymptotic distribution of M-estimator using weak assumptions are presented.

First, the asymptotic distribution for $\sqrt{N}(b_N - \beta)$ is derived from the first-order condition for b_N:

$$\frac{1}{\sqrt{N}} \sum_i q_b(b_N) = 0, \quad \text{where } q_b(b_N) \equiv \frac{\partial q(b)}{\partial b}\Big|_{b=b_N} .$$

Apply the mean-value theorem to b_N around β to get

$$0 = \frac{1}{\sqrt{N}} \sum_i q_b(\beta) + \frac{1}{N} \sum_i q_{bb'}(b_N^*)\sqrt{N}(b_N - \beta) \quad \text{where } b_N^* \in (b_N, \beta)$$

Note that, as there are k components in $q_b(b_N)$, the mean-value theorem requires a different b_N^* for each element, say b_{Nj}^*, $j = 1, ..., k$, although we just use a single b_N^* in the display. Invert the second-order matrix to solve for $\sqrt{N}(b_N - \beta)$:

$$\sqrt{N}(b_N - \beta) = \left\{ -\frac{1}{N} \sum_i q_{bb'}(b_N^*) \right\}^{-1} \cdot \frac{1}{\sqrt{N}} \sum_i q_b(\beta).$$

For a function $m(z, b)$, in almost all cases, it holds that

$$(i): \frac{1}{N} \sum_i m(z_i, b_N) - \frac{1}{N} \sum_i m(z_i, \beta) = o_p(1) \quad \text{which implies}$$

$$\frac{1}{N} \sum_i m(z_i, b_N) \to^p 0 \quad \text{because}$$

$$\frac{1}{N} \sum_i m(z_i, \beta) \to^p E\{m(z, \beta)\} = 0, \quad \text{although}$$

$$(ii) \quad : \quad \frac{1}{\sqrt{N}} \sum_i m(z_i, b_N) - \frac{1}{\sqrt{N}} \sum_i m(z_i, \beta) \neq o_p(1);$$

note the difference in the norming factors (N versus $\sqrt{N}$). Applying this to the preceding equation for $\sqrt{N}(b_N - \beta)$, as $b_N^* \to^p \beta$ because b_N^* falls between b_N and β, we get

$$\sqrt{N}(b_N - \beta) = -E^{-1}\{q_{bb'}(\beta)\} \frac{1}{\sqrt{N}} \sum_i q_b(\beta) + o_p(1).$$

From this, the asymptotic normality and the sandwich form variance follow. Display (i) also justifies the estimator for the asymptotic variance matrix.

Second, the so-called "δ-method" is convenient in deriving the asymptotic distribution of a function of an estimator whose asymptotic distribution is known already. Suppose we have a $g \times 1$ nonlinear and continuously differentiable function $h(b)$ of b on some open convex set including β as an interior point with $rank(h_{b'}(\beta)) = g$, where $h_{b'}(\beta) \equiv \partial h(\beta)/\partial b'$ is the first derivative matrix wrt b of dimension $g \times k$ evaluated at $b = \beta$. Also suppose that

$$\sqrt{N}(b_N - \beta) \rightsquigarrow N(0, V) \quad \text{and} \quad V_N \to^p V.$$

In this case, the asymptotic distribution of $h(b_N)$ can be obtained from that of b_N. Taylor-expand $\sqrt{N}h(b_N)$ as

$$\sqrt{N}h(b_N) - \sqrt{N}h(\beta) = \sqrt{N}h_{b'}(b_N^*)(b_N - \beta),$$

where $b_N^* \in (b_N, \beta)$. Since $b_N \to^p \beta$, we have $b_N^* \to^p \beta$, and owing to the continuity of $h_{b'}$, $h_{b'}(b_N^*) \to^p h_{b'}(\beta)$. Adding and subtracting $h_{b'}(\beta)\sqrt{N}(b_N - \beta)$ gives

$$\sqrt{N}\{h(b_N) - h(\beta)\} \simeq h_{b'}(\beta)\sqrt{N}(b_N - \beta) \rightsquigarrow N\{0, h_{b'}(\beta) \cdot V \cdot h_b(\beta)\}.$$

As usual, the asymptotic variance can be consistently estimated by $h_{b'}(b_N)$ $V_N h_b(b_N)$. Lehmann and Romano (2005, p. 436) showed that δ-method still holds under the differentiability of $h(b)$ at $b = \beta$ instead of the continuous differentiability.

As an example of δ-method, suppose $k = 1$, $\beta \neq 0$, and $h(b) = b^2$. Then, $h_b'(\beta) = 2\beta$, and thus

$$\sqrt{N}\{b_N^2 - \beta^2\} = 2\beta\sqrt{N}(b_N - \beta) + o_p(1) \rightsquigarrow N(0, 4\beta^2 V)$$
$$\text{where } 4b_N^2 V_N \longrightarrow^p 4\beta^2 V.$$

This result can be in fact also obtained using Slutsky lemma:

$$\sqrt{N}\{b_N^2 - \beta^2\} = \sqrt{N}(b_N - \beta)(b_N + \beta) \rightsquigarrow N(0, V) \cdot 2\beta.$$

As another example of δ-method, suppose $g = 2$, $k = 3$, $\beta_1 \neq 0$, and

$$h(b) = \begin{bmatrix} b_2^{-1} \\ b_2 + b_3 \end{bmatrix} \implies h_{b'}(\beta) = \begin{bmatrix} 0 & -\beta_2^{-2} & 0 \\ 0 & 1 & 1 \end{bmatrix}$$

$$\implies \begin{bmatrix} \sqrt{N}(b_2^{-1} - \beta_2^{-1}) \\ \sqrt{N}\{b_2 + b_3 - (\beta_2 + \beta_3)\} \end{bmatrix} \rightsquigarrow N\{0, h_{b'}(\beta) \cdot V \cdot h_b(\beta)\}.$$

In relation to δ method, we introduce *continuous mapping theorem*: for a function $f(w)$ of a random vector w that is continuous at each point of a set A such that $P(w \in A) = 1$,

(i): $f(w_N) \rightsquigarrow f(w)$ if $w_N \rightsquigarrow w$

(ii): $f(w_N) \to^p f(w)$ if $w_N \to^p w$.

(ii) was in fact used in proving the LSE consistency where $f(w_N) = w_N^{-1}$ and w is a constant matrix. As an example of (i), for the $k = 1$ case,

$$f\left\{\sqrt{N}(b_N - \beta)/\sqrt{V}\right\} \equiv \left\{\sqrt{N}(b_N - \beta)/\sqrt{V}\right\}^2 \rightsquigarrow \{N(0,1)\}^2 = \chi_1^2.$$

As another example with $f(w_{N1}, w_{N2}) = w_{N1}/w_{N2}$, suppose $w_N = (w_{N1}, w_{N2}) \rightsquigarrow (w_1, w_2)$ that follows $N(0, I_2)$. Let $A = \{(a_1, a_2)\}$ be the two-dimensional real space except the points with $a_2 = 0$. Then the function a_1/a_2 is continuous on A, and

$$P\{(w_1, w_2) \in A\} = 1 \quad \text{because } P\{N(0,1) = 0\} = 0.$$

Hence

$$f(w_{N1}, w_{N2}) \rightsquigarrow \frac{N(0,1)}{N(0,1)} \text{ which follows the Cauchy distribution.}$$

Third, there are many theorems available for M-estimators' asymptotic normality. Here we present one in Van der Vaart (1998, p. 53). We show the assumptions of the theorem changing notations and simplifying the contents a little. The assumptions weakens the twice continuous differentiability of $q(z, b)$ wrt b on B_β for all z that was invoked above.

- $q(z, b)$ is differentiable at $b = \beta$ for all z with derivative $q_b(z, \beta)$, which is weaker than $q_b(z, b)$ being continuously differentiable at $b = \beta$.

- For every b_1 and b_2 in a neighborhood of β, there exists a function $\tilde{q}(z)$ with $E\tilde{q}(z)^2 < \infty$ such that

$$|q(z, b_1) - q(z, b_2)| \le \tilde{q}(z)|b_1 - b_2|.$$

- $Eq(z, b)$ admits a second-order Taylor expansion at $b = \beta$ with a non-singular second-order derivative matrix $Eq_{bb'}(z, \beta)$.

1.3   Computation

The reader may wonder how to compute the M-estimator b_N in practice. Luckily, one of the equations we saw already shows how: recall

$$\sqrt{N}(b_N - \beta) \simeq -\left\{\frac{1}{N}\sum_i q_{bb'}(b_N^*)\right\}^{-1} \cdot \frac{1}{\sqrt{N}}\sum_i q_b(\beta)$$

and replace b_N with $b^{(2)}$, and b_N^* and β with $b^{(1)}$ to get

$$b^{(2)} = b^{(1)} - \left\{\sum_i q_{bb'}(b^{(1)})\right\}^{-1}\sum_i q_b(b^{(1)}).$$

A numerical algorithm using this goes as follows (we will examine numerical maximization in detail later after MLE is discussed):

- Step 1. Choose $b^{(1)}$, say 0_k or the LSE of y on x.

- Step 2. Compute $b^{(2)}$ using the equation; numerical derivatives may be used instead of analytic derivatives, where the numerical first derivative consists of

$$q_{bj}(b^{(1)}) = \frac{q(b^{(1)} + \varepsilon e_j) - q(b^{(1)} - \varepsilon e_j)}{2\varepsilon}, \quad j = 1, ..., k,$$

e_j is the $k \times 1$ vector with 1 in its jth row and 0 elsewhere, and ε is a small positive constant, say 10^{-7}. For instance, with $k = 2$ and $\beta = (\beta_1, \beta_2)'$, the numerical first derivatives are

$$\frac{q\left\{b^{(1)} + \varepsilon \binom{1}{0}\right\} - q\left\{b^{(1)} - \varepsilon \binom{1}{0}\right\}}{2\varepsilon} \quad \text{and} \quad \frac{q\left\{b^{(1)} + \varepsilon \binom{0}{1}\right\} - q\left\{b^{(1)} - \varepsilon \binom{0}{1}\right\}}{2\varepsilon}.$$

The Hessian matrix can also be obtained numerically doing analogously to $q_{bj}(b^{(1)})$, but it is often replaced by $-N^{-1} \sum_i q_b(b^{(1)}) \, q_{b'}(b^{(1)})$ which is at least n.s.d.

- Step 3. If $b^{(1)} \simeq b^{(2)}$ or $Q_N(b^{(1)}) \simeq Q_N(b^{(2)})$, then stop to take either $b^{(1)}$ or $b^{(2)}$ as b_N; otherwise, replace $b^{(1)}$ with $b^{(2)}$ and go to Step 1. Unless computing $Q_N(b)$ is time-consuming, using the stopping criterion $Q_N(b^{(1)}) \simeq Q_N(b^{(2)})$ is preferred to using $b^{(1)} \simeq b^{(2)}$, because the scales of components of b are different in general.

Showing more details on second-order derivatives including Hessian, consider $q(a, b)$ where both a and b are 2×1, and we want to obtain the 2×2 cross-derivative matrix $\partial^2 q(b, a)/\partial b \partial a'$ at β and α. The term at row 1 and column 2 is the cross-derivative wrt b_1 first and then wrt a_2, which can be computed with

$$\frac{1}{2\varepsilon} \left[q_{b1} \left\{ \beta, \ \alpha + \varepsilon \binom{0}{1} \right\} - q_{b1} \left\{ \beta, \ \alpha - \varepsilon \binom{0}{1} \right\} \right]$$

$$= \frac{1}{2\varepsilon} \left[\frac{q\left\{\beta + \varepsilon \binom{1}{0}, \ \alpha + \varepsilon \binom{0}{1}\right\} - q\left\{\beta - \varepsilon \binom{1}{0}, \ \alpha + \varepsilon \binom{0}{1}\right\}}{2\varepsilon} \right.$$

$$\left. - \frac{q\left\{\beta + \varepsilon \binom{1}{0}, \ \alpha - \varepsilon \binom{0}{1}\right\} - q\left\{\beta - \varepsilon \binom{1}{0}, \ \alpha - \varepsilon \binom{0}{1}\right\}}{2\varepsilon} \right].$$

2   Maximum Likelihood Estimator (MLE)

Although MOM has gained popularity in recent years in econometrics as researchers try to avoid imposing strong assumptions on the model at hand, as an estimation principle, "maximum likelihood estimation (MLE)" has been dominant for a long time. This section studies MLE, which is a special case of M-estimator as already noted.

2.1 MLE Basics

Let $z_i = (x_i', y_i)$, $i = 1, ..., N$ be an iid sample drawn from a known form of distribution $F(z_i, \beta)$ up to β where β is a $k \times 1$ vector of unknown parameters. Let $f_{y|x}(y, \beta)$ denote the *likelihood function* of $y|x$, which is the density function of $y|x$ if $y|x$ is continuously distributed or the probability of $y|x$ if $y|x$ is discrete. Define $f_x(x)$ analogously, which is not a function of β. The *maximum likelihood estimator* maximizes the likelihood of the sample at hand:

$$P\{(x_1, y_1), ..., (x_N, y_N)\} = \prod_{i=1}^{N} P(x_i, y_i, b) = \prod_{i=1}^{N} f_{y|x_i}(y_i, b) \cdot f_x(x_i)$$

wrt b.

Equivalently, taking ln, MLE maximizes the log-likelihood function

$$\sum_i \ln\{f_{y|x_i}(y_i, b) \cdot f_x(x_i)\} = \sum_i [\ln\{f_{y|x_i}(y_i, b)\} + \ln\{f_x(x_i)\}].$$

Dropping $f_x(x)$ that is not a function of b, MLE maximizes

$$\sum_i \ln\{f_{y|x_i}(y_i, b)\}$$

which still depends on x_i as well as on y_i and b.

If we observe only $\{y_i\}$, then the conditional likelihood cannot be obtained. But if the marginal likelihood $f_y(y_i)$ of y is a function of β, β may be estimable by maximizing $\sum_i \ln\{f_y(y_i)\}$. This shows that there are different likelihood functions: joint, conditional, and marginal; there are in fact more ("partial likelihood", "profile likelihood", and so on). Which likelihood to use in practice will depend on data availability and the parameter we want to know. Unless otherwise mentioned, we will always refer to the joint likelihood function for $z = (x', y)'$, and with the marginal likelihood for x being free of β, maximizing the joint likelihood is equivalent to maximizing the conditional likelihood of $y|x$.

For MLE b_{mle} maximizing the log-likelihood function $\sum_i \ln f(z_i, b)$, it will be shown shortly that $b_{mle} \to^p \beta$ and

$$\sqrt{N}(b_{mle} - \beta) \rightsquigarrow N[0, E^{-1}\{s(z, \beta)s(z, \beta)'\}], \text{ where}$$
$$s(z, \beta) \equiv \frac{\partial \ln f(z, b)}{\partial b} \Big|_{b=\beta};$$

$s(z, \beta)$ is called the *score function*.

A simplest example of MLE for regression analysis is obtained when we assume

$$y_i = x_i'\beta + u_i, \quad u_i \sim N(0, \sigma^2) \text{ independently of } x_i \text{ where } \sigma \text{ is an}$$
unknown parameter,

which implies $y_i | x_i \sim N(x_i'\beta, \sigma^2)$. The conditional likelihood to maximize for (b', s), which is for (β', σ), is

$$\max_{b,s} \prod_{i=1}^{N} \frac{1}{\sqrt{2\pi s^2}} \exp\left\{ -\frac{1}{2} \left(\frac{y_i - x_i'b}{s} \right)^2 \right\} \implies$$

$$\max_{b,s} \sum_i \left\{ -\frac{1}{2} \ln(2\pi s^2) - \frac{1}{2} \left(\frac{y_i - x_i'b}{s} \right)^2 \right\}.$$

It is interesting to see that, regardless of s, maximizing this for b is the same as LSE. Of course, this is only a simple example, and we can easily think of more complicated ones as in the following.

If u has a known form of heteroskedasticity, say $u_i | x_i \sim N(0, e^{2x_i'\gamma})$ for an unknown parameter vector γ, then we get

$$\sum_i \left\{ -\frac{1}{2} \ln \left(2\pi e^{2x_i'g} \right) - \frac{1}{2} \left(\frac{y_i - x_i'b}{e^{x_i'g}} \right)^2 \right\},$$

which is to be maximized wrt b and g. This includes the homoskedastic case when $x = (1, x_2, ..., x_k)'$ and $\gamma = (\ln \sigma, 0, ..., 0)'$ because $\exp(2x_i'\gamma) = \exp(2\ln(\sigma)) = \sigma^2$.

Yet another example of MLE is

u_i follows "double exponential with scale parameter θ" independently of x_i

$$\iff \quad f(y_i | x_i) = \frac{1}{2\theta} \exp\left(-\frac{|y_i - x_i'\beta|}{\theta} \right),$$

then the MLE maximizes, wrt q and b,

$$\sum_i \left\{ -\ln(2q) - \frac{|y_i - x_i'b|}{q} \right\}.$$

Regardless of q, maximizing this for b is the same as minimizing $\sum_i |y_i - x_i'b|$ wrt b. Thus, in the double exponential distribution, which is also called the "Laplace" distribution, MLE becomes the least absolute deviation (LAD) estimator.

As LSE is consistent for β if $E(y|x) = x'\beta$ even when u is not normal, LAD estimator is consistent if the median of $y|x$ is $x'\beta$ even when u is not double exponential. Also both LSE and LAD do not have to be motivated by MLE, because they can be motivated as minimizing the prediction error $y_i - x_i'b$ using the squared and absolute value loss functions, respectively. What this illustrates is that MLE sometimes yields interesting estimators that are consistent even when the underlying distributional assumption fails. When we maximize the MLE maximand in this case, we can derive its asymptotic distribution following the general steps for M-estimators while allowing for the underlying distribution to differ from the specified one. In this case, the asymptotic variance would take the usual sandwich form instead of the simpler $E^{-1}(ss')$ shown above. The estimation procedure in this case is sometimes called *"quasi-MLE" or "pseudo-MLE."*

The disadvantage of MLE is clear: we need to specify the distribution of $y|x$; if heteroskedasticity is present, its form should be spelled out as shown above, which was not required for MOM. The advantage of MLE is twofold: one is its applicability to a variety of problems which are hard to approach without taking advantage of the specified likelihood function, and the other is its asymptotic efficiency—having the smallest asymptotic variance—among a wide class of estimators. Due to these advantages, MLE is popular in applied works and MLE can serve as a benchmark when we compare different estimators.

As shown already, MOM can be cast into an optimization framework and MLE is an M-estimator. In fact, all estimating principles may be regarded as minimizing a distance between two entities, thus yielding a general estimating principle "minimum distance." Although only MLE is examined in this section, interested readers may refer to Bera and Bilias (2002) for the history of various estimation principles and the relevant references including those for MLE.

2.2   MLE Identification

As already noted, MLE is an M-estimator and, as such, its asymptotic properties can be derived from those of M-estimators. Among the four issues of M-estimators, three of them, other than identification (ID), are easy to see. In this subsection, we examine the ID for MLE.

Let $F(z,b)$ denote the probability distribution of z when the parameter is $b \in B$. If $F(z,\gamma) = F(z,\beta)$ for all z, then we cannot distinguish β from γ, since observations on z cannot tell beyond $F(z,\cdot)$. In this case, γ and β are said to be "observationally equivalent." Thus β is identifiable if there are no other observationally equivalent elements in B.

Suppose we specify the form of the distribution $F(z,\cdot)$, or equivalently the form of the likelihood function $f(z,\cdot)$. For the true parameter β to be identified, it is necessary to have

$$P_\beta(z \in Z_\beta) \equiv E_\beta(1[z \in Z_\beta]) > 0$$
$$\text{where } Z_\beta \equiv \{z_o : f(z_o,\beta) \neq f(z_o,b), \text{ for any } b \neq \beta \text{ and } b \in B\}$$

and $E_\beta(\cdot)$ is the expected value when z follows $F(z,\beta)$. The subscript β is used here to avoid confusion; typically, the subscript is omitted.

Assume $f(z,b) > 0$ for all z and $b \in B$, and define the *Kullback-Leibler information criterion (KLIC)*

$$H(\beta,b) \equiv E_\beta\left\{\ln\frac{f(z,\beta)}{f(z,b)}\right\}.$$

Observe $H(\beta, \beta) = E_\beta(\ln 1) = 0$, and for a continuously distributed z,

$$
\begin{aligned}
-H(\beta, b) &= E_\beta \left\{ \ln \frac{f(z, b)}{f(z, \beta)} \right\} \\
&< \ln E_\beta \left\{ \frac{f(z, b)}{f(z, \beta)} \right\} = \ln \int f(z, b) dz = 0, \quad \forall b \neq \beta;
\end{aligned}
$$

for the inequality, $P_\beta(z \in Z_\beta) > 0$ is invoked, which means that Jensen's inequality holds with "$<$," instead of "$=$." For a discretely distributed z, the expression $\ln \int f(z, b) dz = 0$ should be replaced with $\ln \sum f(z, b)$, which is zero as the sum of the probabilities is one.

The above display means that the ID of β can be viewed as an maximization problem of $-H(\beta, b)$ wrt $b \in B$ where $-H(\beta, \beta) = 0$ is the maximum value. The sample version of $-H(\beta, b)$ is

$$
\frac{1}{N} \sum_i \ln f(z, b) - \frac{1}{N} \sum_i \ln f(z, \beta).
$$

Maximizing the first term wrt b renders MLE. Since $H(\beta, b) = 0$ iff $b = \beta$, β is identified in MLE. This way of viewing ID in parametric models with the Kullback-Leibler information criterion appears, e.g., in Bowden (1973).

2.3   Asymptotic Variance Relative to M-estimator

The asymptotic variance of MLE is straightforward to obtain as a special case of M-estimator asymptotic variance, and there exists an interesting interpretation of the asymptotic variance of M-estimator in relation to that of MLE, which is presented here. To simplify exposition, suppose that z is continuously distributed.

Consider differentiating the first-order condition $E\{q_b(\beta)\} = 0$ of an M-estimator maximizing the sample version of $E\{q(b)\}$. It holds that $\partial E\{q_b(\beta)\}/\partial \beta = 0$, and bringing $\partial(\cdot)/\partial$ inside of $E\{q_b(\beta)\} = \int q_b(z, \beta) f(z, \beta) dz$, we get

$$
\begin{aligned}
\int \left[\frac{\partial \{q_b(z, \beta)\, f(z, \beta)\}}{\partial \beta'} \right] dz &= \int \left\{ \frac{\partial q_b(z, \beta)}{\partial \beta'} f(z, \beta) \right\} dz \\
&\quad + \int \left\{ q_b(z, \beta) \frac{\partial f(z, \beta)}{\partial \beta'} \right\} dz \\
&= \int q_{bb'}(z, \beta) f(z, \beta) dz + \int q_b(z, \beta) \frac{\partial f(z, \beta)/\partial \beta'}{f(z, \beta)} f(z, \beta) dz.
\end{aligned}
$$

That is,

$$
0 = \frac{\partial E\{q_b(\beta)\}}{\partial \beta} = E(q_{bb'}) + E(q_b s') \iff E(q_b s') = -E(q_{bb'})
$$

where $s(z, \beta) \equiv f_b(z, \beta)/f(z, \beta)$ is the score function. Note that we need a regularity condition to interchange the order of $\partial(\cdot)/\partial \beta$ and $\int$; e.g., for all b

in a neighborhood of β,

$$|q_{bb'}(z,b)f(z,b)| + |q_b(z,b)\frac{\partial f(z,b)}{\partial b'}| \le g(z), \quad \text{where} \int g(z)dz < \infty.$$

With $q(z,\beta) = \ln\{f(z,\beta)\}$ for MLE, $E(q_b s') = -E(q_{bb'})$ becomes

$$E(ss') = -E\left\{\frac{\partial^2 \ln f(z,\beta)}{\partial b \partial b'}\right\}.$$

Both terms are called the "*information matrix*," and this equality is called the "*information equality*" whereas the more general version $E(q_b s') = -E(q_{bb'})$ is called the "*generalized information equality*." With the information inequality, the asymptotic variance of MLE can be obtained from the M-estimator asymptotic variance:

$$E^{-1}\left\{\frac{\partial^2 \ln f(z,\beta)}{\partial b \partial b'}\right\} \cdot E(ss') \cdot E^{-1}\left\{\frac{\partial^2 \ln f(z,\beta)}{\partial b \partial b'}\right\} = E^{-1}(ss').$$

Using $E(q_b s') = -E(q_{bb'})$, the sandwich form asymptotic variance matrix of M-estimator can be written as

$$E^{-1}(sq_b') \cdot E(q_b q_b') \cdot E^{-1}(q_b s')$$
$$= E^{-1}[E(sq_b')E^{-1}(q_b q_b')q_b \cdot q_b'E^{-1}(q_b q_b')E(q_b s')]$$

which is the inverse of the "square" of the projection of s on q_b. Since

$$E(sq_b') \cdot E^{-1}(q_b q_b') \cdot E(q_b s') \le E(ss'),$$

M-estimator asymptotic variance is greater than the MLE asymptotic variance, and thus M-estimator (or MOM) is less efficient than MLE. This fact as well as the equation $E(q_{bb'}) = -E(q_b s')$ appear in Godambe (1960).

In fact, it is known that MLE is the most efficient (i.e., MLE has the smallest asymptotic variance) among the $\sqrt{N}$-consistent ("regular") asymptotically normal estimators; see, e.g., Lehmann (1983, p. 406, Theorem 1.1) and the references therein. But the obvious advantage of an M-estimator (or MOM) is that there is no need to specify the likelihood function. An intuitive understanding for the MLE efficiency may be gained from the following special case.

Suppose we have an $\sqrt{N}$-consistent estimator $b_N = b_N(z_1, ..., z_N)$ that is also *unbiased* for β (namely, $E(b_N) = \beta$). Observe

$$\begin{aligned}
I_k &= \frac{\partial E(b_N)}{\partial \beta'} = \frac{\partial \int b_N \cdot \prod_i f(z_i,\beta)dz_1 \cdots dz_N}{\partial \beta'} \\
&= \int b_N \cdot \frac{\partial \prod_i f(z_i,\beta)}{\partial \beta'} dz_1 \cdots dz_N \\
&= \int b_N \frac{\partial \exp\left\{\sum_i \ln f(z_i,\beta)\right\}}{\partial \beta'} dz_1 \cdots dz_N \\
&\quad \left(\text{as} \prod_i f(z_i,\beta) = \exp\left\{\sum_i \ln f(z_i,\beta)\right\}\right)
\end{aligned}$$

$$= \int b_N \sum_i \frac{\partial \ln f(z_i, \beta)}{\partial \beta'} \cdot \exp\left\{ \sum_i \ln f(z_i, \beta) \right\} dz_1 \cdots dz_N$$

$$= \int \left(b_N \sum_i s_i' \right) \cdot \prod_i f(z_i, \beta) dz_1 \cdots dz_N$$

$$= COV\left(b_N, \sum_i s_i' \right) \leq V(b_N) \cdot N \cdot E(ss').$$

From the first and last expressions, we get $I_k \leq V(b_N) \cdot N \cdot E(ss')$, and hence,

$$V(b_N) \geq \frac{E^{-1}(ss')}{N} \iff V(\sqrt{N}(b_N - \beta)) \geq E^{-1}(ss').$$

3　M-Estimator with Nuisance Parameters

3.1　Two-Stage M-Estimator Basics

One important generalization of M-estimator is a *two stage M-estimator*:

$$b_N \equiv \text{argmax}_{b \in B} Q_N(b, a_N) = \frac{1}{N} \sum_i q(z_i, b, a_N)$$

where a_N is a first-stage estimator for a *nuisance parameter* α which is not of interest per se but should be estimated nonetheless before β. For instance, the variance matrix of the error terms is not interesting per se in Generalized LSE (GLS), but for GLS, we need to estimate the variance matrix first. If q is a log-likelihood function, then we get a two-stage MLE. Strictly speaking, a two-stage MLE is not a "real" MLE which would maximize $Q_N(b, a)$ jointly with (b, a) in one step. Nevertheless, two-stage MLE does appear in practice, because the one-step MLE may be difficult to implement.

Omitting z_i in $q(z_i, b, a_N)$, let the two-stage M-estimator b_N be

$$b_N = \text{argmax}_{b \in B} \frac{1}{N} \sum_i q(b, a_N)$$

where a_N is a first-stage $\sqrt{N}$-consistent estimator for α. As done just now, often we will omit z in $q(z, b, a)$ and in the derivatives $q_b(z, b, a)$, $q_{bb'}(z, b, a)$, and $q_{ba'}(z, b, a)$. Furthermore, we may also omit the arguments b and a in $q(b, a)$, particularly when $b = \beta$ and $a = \alpha$.

For the two-stage M-estimator, having a_N in place of α does not affect the consistency of b_N for β, which can be established by doing analogously to what was done for M-estimator, but the asymptotic distribution of two-estimator M-estimator may be affected by the first-stage error $a_N - \alpha$. Finding when the first stage affects the second stage, and if it does then in which way, are interesting questions.

The estimator b_N satisfies the first-order condition

$$\frac{1}{\sqrt{N}} \sum_i q_b(b_N, a_N) = 0.$$

Apply the mean value theorem to b_N around β to get

$$\frac{1}{\sqrt{N}} \sum_i q_b(\beta, a_N) + \frac{1}{N} \sum_i q_{bb'}(b_N^*, a_N) \cdot \sqrt{N}(b_N - \beta) = 0$$

$$\Longrightarrow \sqrt{N}(b_N - \beta) = \left\{ -\frac{1}{N} \sum_i q_{bb'}(b_N^*, a_N) \right\}^{-1} \cdot \frac{1}{\sqrt{N}} \sum_i q_b(\beta, a_N).$$

The asymptotic distribution of the M-estimator without the nuisance parameter α was derived already (using this display with a_N replaced by α). Due
to a_N, we should go one step further applying the mean value theorem to a_N
as follows.

Expand $N^{-1/2} \sum_i q_b(\beta, a_N)$ around α to get

$$\frac{1}{\sqrt{N}} \sum_i q_b(\beta, a_N) = \frac{1}{\sqrt{N}} \sum_i q_b(\beta, \alpha) + \frac{1}{N} \sum_i q_{ba'}(\beta, a_N^*) \cdot \sqrt{N}(a_N - \alpha).$$

Substitute this into the preceding display and replace $N^{-1} \sum_i q_{bb'}(b_N^*, a_N)$
and $N^{-1} \sum_i q_{ba'}(\beta, a_N^*)$, respectively, with $E\{q_{bb'}(\beta, \alpha)\}$ and $E\{q_{ba'}(\beta, \alpha)\}$
(invoking the uniform LLN) to obtain

$$\sqrt{N}(b_N - \beta) = -E^{-1}(q_{bb'}) \cdot \left\{ \frac{1}{\sqrt{N}} \sum_i q_b + E(q_{ba'})\sqrt{N}(a_N - \alpha) \right\} + o_p(1).$$

The asymptotic distribution of $\sqrt{N}(b_N - \beta)$ depends on the covariance of
the two terms on the rhs. We will take a detailed look at this in the next
subsection.

3.2   Influence Function and Correction Term

Suppose

$$\sqrt{N}(a_N - \alpha) = \frac{1}{\sqrt{N}} \sum_i \eta_i + o_p(1), \quad \{\eta_i\} \text{ are iid, } E(\eta) = 0,$$

$$\text{and } E(\eta\eta') < \infty \text{ is p.d.;}$$

this implies $\sqrt{N}(a_N - \alpha) \rightsquigarrow N(0, E(\eta\eta'))$. For instance, if a_N is the LSE for
$z_i = w_i'\alpha + \varepsilon_i$, then η_i is $E^{-1}(ww') \cdot w_i\varepsilon_i$ so that the asymptotic variance
matrix becomes the LSE asymptotic variance:

$$E(\eta\eta') = E\{E^{-1}(ww')w\varepsilon \cdot \varepsilon w' E^{-1}(ww')\}$$
$$= E^{-1}(ww')E(ww'\varepsilon^2)E^{-1}(ww').$$

The idea here is that if $\sqrt{N}(a_N - \alpha)$ has an asymptotic variance Ω, then we can think of a random vector η such that $E(\eta\eta') = \Omega$ and $\sqrt{N}(a_N - \alpha)$ is a sum of iid "error" η_i's; η_i is called the "*influence function.*"

In the display with $\sqrt{N}(b_N - \beta)$, replace $\sqrt{N}(a_N - \alpha)$ with $N^{-1/2}\sum_i \eta_i + o_p(1)$ to get

$$\sqrt{N}(b_N - \beta) = -E^{-1}(q_{bb'}) \cdot \frac{1}{\sqrt{N}} \sum_i \{q_b(z_i) + E(q_{ba'})\eta_i\} + o_p(1).$$

Hence

$$\sqrt{N}(b_N - \beta) \rightsquigarrow N\{0,\ E^{-1}(q_{bb'})\ C_a\ E^{-1}(q_{bb'})\} \qquad \text{where}$$

$$C_a \equiv E[\ \{q_b(z_i) + E(q_{ba'})\eta_i\} \cdot \{q_b(z_i) + E(q_{ba'})\eta_i\}'\ ].$$

Although C_a may look complicated, it can be consistently estimated with $N^{-1}\sum_i \delta_i\delta_i'$, where

$$\delta_i \equiv q_b(b_N, a_N) + \left\{\frac{1}{N}\sum_i q_{ba'}(b_N, a_N)\right\} \cdot \eta_i(a_N)$$

and $\eta_i(a_N)$ is a consistent estimator for $\eta_i = \eta_i(\alpha)$; δ_i is an influence function for b_N. For instance, if a_N is the LSE for $z_i = w_i'\alpha + \varepsilon_i$, then $\eta_i(\alpha) = E^{-1}(ww') \cdot w_i(z_i - w_i'\alpha)$, and thus

$$\eta_i(a_N) = \left(\frac{1}{N}\sum_i w_iw_i'\right)^{-1} \cdot w_i(z_i - w_i'a_N).$$

In $E(q_{ba'})\eta_i$ of C_α, η_i is the first-stage error, and $E(q_{ba'})$ may be called the "*link matrix*" for the first and the second stage. The part $E(q_{ba'})\eta_i$ is sometimes called the (first-stage) "*correction term.*" This shows that the first-stage estimation error η_i is channeled through $E(q_{ba'})$ because α in $q_b(z, \beta, \alpha)$ gets estimated. Because $E\{q_b(z, \beta, \alpha)\} = 0$, we have

$$0 = \frac{\partial E\{q_b(z, \beta, \alpha)\}}{\partial\alpha'} = \frac{\partial \int q_b(z, \beta, \alpha)f(z, \beta, \alpha)dz}{\partial\alpha'}$$

$$= \int \frac{\partial q_b(z, \beta, \alpha)}{\partial\alpha'} f(z, \beta, \alpha)dz + \int q_b(z, \beta, \alpha)\frac{\partial f(z, \beta, \alpha)}{\partial\alpha'}dz$$

$$= \int \frac{\partial q_b(z, \beta, \alpha)}{\partial\alpha'} f(z, \beta, \alpha)dz + \int q_b(z, \beta, \alpha)\frac{\partial f(z, \beta, \alpha)/\partial\alpha'}{f(z, \beta, \alpha)}$$

$$f(z, \beta, \alpha)dz$$

$$\implies E(q_{ba'}) = -E(q_b s^{(a)'}) \quad \text{where } s^{(a)}(z, \beta, \alpha) \equiv \frac{\partial f(z, \beta, \alpha)/\partial\alpha}{f(z, \beta, \alpha)}.$$

That is, the correction term can be written as $-E(q_b s^{(a)'})\eta_i$: the first-stage estimation error η_i is channeled through $-E(q_b s^{(a)'})$ as if α in $s^{(a)}(z, \beta, \alpha)$ were estimated.

3.3   Various Forms of Asymptotic Variances

Let
$$C_\alpha \equiv E(q_b q_{b'})$$

which is the middle outer-product matrix of the asymptotic variance $E^{-1}(q_{bb'})$ $E(q_b q_{b'}) E^{-1}(q_{bb'})$ of the infeasible M-estimator with α known. In comparing C_a (α estimated) to C_α (α known), three special cases are worth mention-ing. First, if $E(q_{ba'}) = 0$, then $C_a = C_\alpha$ and there is no first-stage effect on the second: the two-stage M-estimator is as good as the infeasible M-estimator; some examples will be provided shortly for this case. Second, if $E(q_b \eta') = 0$, then C_a can be decomposed into two p.s.d. matrices $E(q_b q_b')$ and $E(q_{ba'}) E(\eta \eta') E(q_{ab'})$, which makes $C_a \geq C_\alpha$. This case occurs if a_N is obtained by another sample not used for b_N (e.g., the sample is split into two, and the first is used for a_N while the second is used for b_N), or a_N is a function of terms "orthogonal" to $N^{-1/2} \sum_i q_b(z_i)$. An example for this will appear later in Chapter 5.5 when "sample selection" is discussed. Third, as C_a consists of four terms:

$$C_a = E(q_b q_{b'}) + E(q_b \eta) E(q_{ab'}) + E(q_{ba'}) E(\eta q_{b'}) + E(q_{ba'}) E(\eta \eta') E(q_{ab'}),$$

if

$$E(q_{ba'}) E(\eta q_{b'}) = -E(q_{ba'}) E(\eta \eta') E(q_{ab'}),$$

then

$$C_a = E(q_b q_{b'}) - E(q_{ba'}) E(\eta \eta') E(q_{ab'}) \leq C_\alpha.$$

The last display that the two-stage estimator estimating the unknown α can be better than the infeasible estimator with α known is counter-intuitive, but it has been known at least since Pierce (1982). See Hitomi et al. (2008) and the references therein; they examine a more difficult case of a nonpara-metric α. To illustrate when this can happen, let a_N be the MLE. Then

$$\eta_i = E^{-1}(s^{(a)} s^{(a)'}) s_i^{(a)} \quad \text{which implies}$$

$$E(q_{ba'}) E(\eta q_{b'}) = E(q_{ba'}) E^{-1}(s^{(a)} s^{(a)'}) E(s^{(a)} q_{b'})$$

$$= -E(q_{ba'}) E^{-1}(s^{(a)} s^{(a)'}) E(q_{ab'}), \quad \text{as } E(s^{(a)} q_{b'}) = -E(q_{ab'})$$

$$= -E(q_{ba'}) E(\eta \eta') E(q_{ab'}), \quad \text{as } E(\eta \eta') = E^{-1}(s^{(a)} s^{(a)'});$$

this is the above case $C_a \leq C_\alpha$. Also observe

$$\frac{1}{\sqrt{N}} \sum_i \{q_b(z_i) + E(q_{ba'}) \eta_i\}$$

$$= \frac{1}{\sqrt{N}} \sum_i \left\{ q_b(z_i) - E(q_b s^{(a)'}) E^{-1}(s^{(a)} s^{(a)'}) s_i^{(a)} \right\} \left\{ = \frac{1}{\sqrt{N}} \sum_i \delta_i \right\}.$$

Since the summand δ_i is the residual of projecting $q_b(z_i)$ on $s_i^{(a)}$, its asymptotic variance is smaller than $E(q_b q_{b'})$.

The last display shows the main point of Pierce (1982). If a_N is an efficient estimator for α, then the influence function δ_i for b_N and the influence function η_i for a_N should be uncorrelated. Otherwise, the residual of the linear projection of η_i on δ_i would yield a better estimator for α than a_N is, with its variance reduced due to $COR(\delta, \eta)$. Since this is a contradiction for MLE, we should have $COR(\delta, \eta) = 0$. In the last display, this holds because δ is the residual of projecting q_b on $s^{(a)}$ and η consists of $s^{(a)}$. This "efficiency-orthogonality nexus" appears in various forms in econometrics and statistics.

3.4   Examples of Two-Stage M-Estimators

3.4.1   No First-Stage Effect

One example for $Eq_{ba'} = 0$ is the weighted LSE (WLS) or the feasible GLS for the linear model $y = x'\beta + u$ minimizing

$$\frac{1}{N} \sum_i \left(\frac{y_i - x_i' b}{s_i} \right)^2 \quad \text{where} \quad E(u|x) = 0 \text{ and } s_i^2 \to^p \sigma_i^2 \equiv E(u_i^2|x_i).$$

Suppose

$$\sigma_i^2 = (\alpha_1 + \alpha_2 x_{ik})^2, \quad s_i^2 = (a_1 + a_2 x_{ik})^2, \quad a_N \equiv (a_1, a_2)'.$$

Differentiate $N^{-1} \sum_i \{(y_i - x_i' b)/s_i\}^2$ wrt b to get

$$\frac{1}{N} \sum_i (-2) \, x_i (y_i - x_i' b) \, s_i^{-2}.$$

Differentiate this wrt a_N to get

$$\frac{1}{N} \sum_i 4 \, x_i (y_i - x_i' b) \, s_i^{-3} \frac{\partial s_i}{\partial a_N}.$$

Evaluating this at β and α, this "link matrix" becomes $N^{-1} \sum_i u_i g(x_i)$ for a function $g(x_i)$, because s_i and $\partial s_i / \partial a_N$ are functions of x_i. But

$$\frac{1}{N} \sum_i u_i g(x_i) \to^p E\{ug(x)\} = E\{g(x) \cdot E(u|x)\} = 0;$$

i.e., $E(q_{ba'}) = 0$. This explains why the feasible generalized LSE has the same asymptotic distribution as the (infeasible) generalized LSE.

As another example of $E(q_{ba'}) = 0$ (the reader may skip the rest of this subsection, simply taking the point that *replacing an instrument with a consistent estimator is innocuous while replacing a regressor with a consistent estimator alters the asymptotic variance*), consider

$$y = x_1' \beta_1 + x_2' \beta_2 + u \equiv x'\beta + u \quad \text{and} \quad E(u|x_2) = 0$$

for two-stage LSE (2SLSE), where x_j is a $k_j \times 1$ vector, $j = 1, 2$, and $E(x_1 u) \neq 0$. This equation can be regarded as the first equation of a simultaneous equation system, and x_1 is the endogenous regressors in the first equation.

Let a $s \times 1$ ($s \geq k_1 + k_2$) vector z denote the exogenous variables in the system ($E(zu) = 0$); z is an instrument for x_1 (and x_2). Then a_N is the LSE of x_1 on z (since z is a $s \times 1$ vector, a_N is a $s \times k_1$ matrix, not a vector), and α is $E^{-1}(zz')E(zx_1')$—the $s \times k_1$ matrix of the projection coefficient of x_1 on z. The moment condition for the 2SLSE is

$$\frac{1}{N} \sum_i (y_i - x_i' b_N) \cdot \{(z_i' a_N), x_{i2}'\}' = 0.$$

The dimension of $(z_i' a_N)'$ is the same as that of x_1 ($k_1 \times 1$), and the dimension of $((z_i' a_N), x_{i2}')'$ is $(k_1 + k_2) \times 1$. In this display, the instrument $z_i' \alpha$ for x_{i1} is estimated by $z_i' a_N$.

Stack the $s \times k_1$ matrix α as a $(s \cdot k_1) \times 1$ vector α^*; see the next paragraph for an example. Differentiate the preceding display wrt a_N^*, the version of a_N stacked analogously to α^*. Then we get a $(k_1 + k_2) \times (s \cdot k_1)$ matrix $N^{-1} \sum_i q_{ba'}$ whose typical element is either 0 or

$$\frac{1}{N} \sum_i (y_i - x_i' b_N) z_{ij} = 0, \quad j = 1, ..., s.$$

Since this is consistent for $E\{(y - x'\beta)z_j\} = E(uz_j) = 0$, $j = 1, ..., s$, $E(q_{ba'})$ is zero in the 2SLSE, implying no first-stage estimation effect on the second stage. This result illustrates that estimating instruments does not affect the second stage in general.

To be specific about $N^{-1} \sum_i (y_i - x_i' b_N) z_{ij} = 0$, let $s = 3$ and $k_1 = 2$. Then α is a 3×2 matrix and α^* can be set as

$$\alpha^* = (\alpha_{11}, \alpha_{21}, \alpha_{31}, \alpha_{12}, \alpha_{22}, \alpha_{32})'.$$

Differentiating $((z_i' a_N), x_{i2}')'$ wrt a_N^*, $\partial(z_i' a_N)/\partial a_N^*$ is nonzero, while $\partial x_2 / \partial a_N^* = 0$. Observe that (omitting N in a_N)

$$z_i' a = (z_{i1} a_{11} + z_{i2} a_{21} + z_{i3} a_{31}, \ z_{i1} a_{12} + z_{i2} a_{22} + z_{i3} a_{32}).$$

Differentiate this wrt $a^* = (a_{11}, a_{21}, a_{31}, a_{12}, a_{22}, a_{32})'$ to get

$$\begin{bmatrix} z_{i1} & z_{i2} & z_{i3} & 0 & 0 & 0 \\ 0 & 0 & 0 & z_{i1} & z_{i2} & z_{i3} \end{bmatrix}.$$

Attaching a $k_2 \times 6$ zero matrix at the bottom for $\partial x_2 / \partial a_N^* = 0$, we get the desired $(k_1 + k_2) \times (s \cdot k_1) = (2 + k_2) \times 6$ matrix. Taking $N^{-1} \sum_i (\cdot)$ on "this matrix times u_i" yields a null link matrix.

3.4.2   First-Stage Effect

As an example of $E(q_b\eta') \neq 0$ and $E(q_{ba'}) \neq 0$, consider

$$y = x'\beta + u,$$

where the kth variable x_k is not observable. Suppose $x_k = E(w|z) = z'\alpha$ where z is a $g \times 1$ vector with $E(zu) = 0$ and $E(xz') \neq 0$. Then x_{ik} can be consistently estimated with

$$\hat{x}_{ik} \equiv z_i'a_N,$$

where a_N is the LSE of w on z. Let

$$\hat{x} \equiv (x_1, ..., x_{k-1}, \hat{x}_k)'.$$

The issue here is the effect of using the "generated regressor" $\hat{x}$ instead of x in the LSE b_N of y on $\hat{x}$ to estimate β. It is easy to prove $b_N \to^p \beta$; the first-stage error $a_N - \alpha$ matters only for the asymptotic variance of b_N.

The first-order condition of the LSE is

$$\frac{1}{N}\sum_i \hat{x}_i(y_i - \hat{x}_i'b_N) = \frac{1}{N}\sum_i \left[\begin{pmatrix} x_1 \\ \vdots \\ x_{k-1} \\ z_i'a_N \end{pmatrix} \{y_i - (x_1, ..., x_{k-1}, z_i'a_N)b_N\} \right]$$

$$= 0.$$

Differentiate this wrt a_N to get $N^{-1}\sum_i q_{ba'}$:

$$\frac{1}{N}\sum_i q_{ba'} = \left[\begin{array}{c} 0_{(k-1)\times g} \\ N^{-1}\sum_i(y_i - \hat{x}_i'b_N)z_i' \end{array} \right] - b_{Nk}\frac{1}{N}\sum_i \hat{x}_i z_i'$$

$$\to^p - \beta_k E(xz') \neq 0.$$

Hence the first-stage error is felt in the second stage. This illustrates that *estimating explanatory variables affects the second-stage variance (while estimating instruments does not* as in 2SLSE).

One caution is that estimating explanatory variables is not the same as the so-called "errors-in-variable" problem where the parameters cannot even be consistently estimated. In the errors-in-variable problem, x_k is observed as $x_k + \varepsilon$ where ε does not converges to 0, while $\hat{x}_k$ above can be written as $x_k + v$ with $v = o_p(1)$. See Pagan and Ullah (1988) for the same point made for the typical erroneous practice of using a risk term as a regressor in the macro-finance literature.

4   Method-of-Moment Tests (MMT)

4.1   Basics

Suppose that we set up a model $y = x'\alpha + u$ with a suspicion that w may be wrongly omitted in the model. One way to test for the possible omission is to see if $E\{(y - x'\alpha)w\} = 0$. If w is indeed omitted, $y - x'\alpha$ should be correlated with w because y includes w, resulting in $E\{(y - x'\alpha)w\} \neq 0$. More generally, for a rv z, suppose that a parameter α satisfies a moment condition $E\{m(z,\alpha)\} = 0$ which is *implied by the model specification but not used* in getting an estimator a_N for α. Then we can test the validity of the model specification by checking if

$$\frac{1}{\sqrt{N}} \sum_i m(z_i, a_N) \quad \text{is centered at zero}$$

because this will have $E\{m(z,\alpha)\} = 0$ as its expected value. Testing model specifications using moment conditions is called a *method-of-moment test (MMT)* as in Newey (1985), Tauchen (1985), and Pagan and Vella (1989).

As method-of-moment estimators include many known estimators as special cases, MMT includes many known tests as special cases. In this section, we examine MMT where a_N is a first-stage estimator for a nuisance parameter α. In deriving the asymptotic distribution of MMT, since this is a simpler special case of two-stage M-estimator, the technique of the previous section can be applied with a minor modification. Let a_N have an influence function η_i:

$$\sqrt{N}(a_N - \alpha) = \frac{1}{\sqrt{N}} \sum_i \eta_i + o_p(1) \quad \text{with } E(\eta) = 0 \text{ and } E(\eta\eta') < \infty,$$

Observe

$$\frac{1}{\sqrt{N}} \sum_i m(z_i, a_N) = \frac{1}{\sqrt{N}} \sum_i \{m(z_i, \alpha) + E(m_{a'}) \eta_i\} + o_p(1)$$

$$\rightsquigarrow N(0, C), \quad \text{where } m_{a'} \equiv \frac{\partial m(\alpha)}{\partial a'} \text{ and}$$

$$C = E[\{m(z_i, \alpha) + E(m_{a'})\eta_i\}\{m(z_i, \alpha) + E(m_{a'})\eta_i\}'].$$

The asymptotic variance C can be estimated consistently with

$$C_N \equiv \frac{1}{N} \sum_i \delta_{Ni}\delta'_{Ni}, \quad \delta_{Ni} \equiv m(z_i, a_N) + \left\{ \frac{1}{N} \sum_i m_{a'}(z_i, a_N) \right\} \eta_i(a_N)$$

where $\eta_i(a_N)$ is a consistent estimator for $\eta_i(\alpha) = \eta_i$. In the rest of this section, we may omit either argument z or a in $m(z,a)$ and $m_{a'}(z,a)$.

4.2   Examples

Recall the MMT for $H_0 : w$ is not omitted in $y = x'\alpha + u$. Let x be a $p \times 1$ vector and w be a $k \times 1$ vector. Assume that a_N is the LSE. A test statistic is

$$\frac{1}{\sqrt{N}} \sum_i w_i(y_i - x'_i a_N) = \frac{1}{\sqrt{N}} \sum_i m(z_i, a_N),$$

where $z \equiv (y, x', w')'$. Then $m_{a'} = -wx'$ which is a $k \times p$ matrix. Since a_N is the LSE of y on x, we have

$$\sqrt{N}(a_N - \alpha) = \frac{1}{\sqrt{N}} \sum_i E^{-1}(xx')x_i u_i + o_p(1).$$

Thus

$$\frac{1}{\sqrt{N}} \sum_i \{m(z_i, \alpha) + E(m_{a'})\eta_i\} = \frac{1}{\sqrt{N}} \sum_i \{w_i u_i - E(wx')\, E^{-1}(xx')x_i u_i\}.$$

In estimating the variance matrix, u_i can be replaced by $y_i - x_i' a_N$ under the H_0, and $E(wx')$ and $E(xx')$ can be replaced by their sample versions. Dividing the test statistic by the SD, we get an asymptotic $N(0,1)$ test statistic.

Although it is not a MMT, here we note another (easier) way for omitted variable test. With $H_0 : y = x'\alpha + u$, suppose there is a reason to believe that w may be a relevant variable for y. Then we may consider an alternative $y = x'\alpha + w'\gamma + u$. More generally, we may set up

$$H_0 : y = x'\alpha + \eta \cdot g(w) + u,$$

which nests H_0 with $\eta = 0$, where $g(w)$ is a known function of w. Here, α and η can be easily estimated and tested with the LSE of y on x and $g(w)$. By employing a sufficiently general $g(w)$, we can detect departures from H_0 into various directions. If η is significantly different from 0, then the model $y = x'\alpha + u$ must be misspecified. As a matter of fact, we can try almost anything in the place of $g(w)$. In this sense, $g(w)$ is an "artificial regressor" and the model is an artificial regression: we do not necessarily think that the model in H_a is true, but so long as $g(w)$ can detect a misspecification, using $g(w)$ is justified.

As another example of MMT, consider the linear model $y = x'\alpha + u$ where u has density function f_u. Suppose that we assumed the symmetry of f_u but estimated α by LSE a_N, which does not use the symmetry assumption. A symmetry test can be done for $H_0 : E(u^3) = 0$, with

$$\frac{1}{\sqrt{N}} \sum_i (y_i - x_i' a_N)^3 \equiv \frac{1}{\sqrt{N}} \sum_i r_i^3, \text{ where } r_i \equiv y_i - x_i' a_N$$

since the symmetry implies $E(u^3) = 0$. Note that we cannot test $E(u) = 0$ that is also implied by the symmetry, because we used $E(u) = 0$ in getting the LSE. Rejecting $E(u^3) = 0$ negates symmetry, but accepting $E(u^3) = 0$ does not necessarily imply symmetry. Observe

$$m_{a'}(z_i, a_N) = -3r_i^2 x_i' \Rightarrow m_{a'}(z_i, \alpha) = -3u_i^2 x_i'.$$

Thus

$$\frac{1}{\sqrt{N}} \sum_i \{m(z_i, \alpha) + E(m_{a'})\eta_i\} = \frac{1}{\sqrt{N}} \sum_i \{u_i^3 - 3E(u^2 x')\, E^{-1}(xx')x_i u_i\}.$$

The asymptotic variance should not be estimated with $N^{-1} \sum_i r_i^6$, because

$$\frac{1}{N} \sum_i r_i^6 = \frac{1}{N} \sum_i u_i^6 + o_p(1) \to^P E(u^6)$$

which ignores the correction term $3E(u^2 x') \cdot E^{-1}(xx')x_i u_i$.

EXAMPLE: HOUSE SATE (continued). Recall the house sale data in the preceding chapter. Let discount % be y and let x consist of 1, BATH, ELEC, RM, TAX, YR, L1, L2, L3, ln(LP), BIGS, RATE, and SUPPLY. Applying LSE to get the residuals and defining

$$\delta_{Ni} \equiv r_i^3 - 3 \sum_i r_i^2 x_i' \left(\sum_i x_i x_i' \right)^{-1} x_i r_i,$$

we obtained

$$\frac{1}{\sqrt{N}} \sum_i r_i^3 = 15776, \quad \left(\frac{1}{N} \sum_i \delta_{Ni}^2 \right)^{1/2} = 6174$$

$$\Longrightarrow \text{ Right Test Statistic } 2.555,$$

$$\left(\frac{1}{N} \sum_i r_i^6 \right)^{1/2} = 7694 \Longrightarrow \text{ Wrong Test Statistic } 2.050.$$

Although the null hypothesis $E(u^3) = 0$ is rejected with both test statistics, using the right SD gives the more powerful test in this example. One may wonder why the correct SD with α estimated is smaller than the wrong SD with "α known." This can be understood recalling that a two-stage estimator may have a smaller variance than the infeasible one-stage estimator with α known if a_N is an efficient estimator for α. For instance, if the first-stage linear model of y has a normal error, then the LSE is an efficient estimator for α, which may be happening in the house sale example.

Besides the above omitted variable and symmetry tests, there are other examples which can be easily thought of. For instance, if we suspect that the error terms may be auto-correlated (in time series data), we may test if $E(u_i u_{i-1}) = 0$. The appropriate test statistic is $N^{-1/2} \sum_{i=2}^N r_i r_{i-1}$, although it is in fact simpler to use r_{i-1} as an artificial regressor in the original model to see if the slope of r_{i-1} is significant or not. If we want heteroskedasticity test, then we may examine if

$$E\{x(u^2 - \sigma^2)\} = E[\, E\{x(u^2 - \sigma^2)|x\}\,] = E[x\{E(u^2|x) - \sigma^2\}] = 0$$
$$\text{if } E(u^2|x) = \sigma^2.$$

A test statistic for this is $N^{-1/2} \sum_i x_i(r_i^2 - s_N^2)$ where $s_N^2 = N^{-1} \sum_i r_i^2$.

4.3   Conditional Moment Tests

More generally than $E(vz) = 0$, a conditional moment condition $E(v|z) = 0$ implies $E\{v \cdot g(z)\} = 0$ for any square-integrable function $g(z)$ (i.e., $E\{g(z)^2\} < \infty$). This can be tested with $N^{-1/2} \sum_i \hat{v}_i g(z_i)$, where $\hat{v}_i$ is an estimator for v_i. The test includes the aforementioned homoskedasticity test as a special case with $v_i = u_i^2 - \sigma^2$ and $g(z_i) = x_i$. When the moment condition in a MMT is derived from a conditional moment condition, the MMT may be called a *conditional moment test* (Newey, 1985).

For a conditional moment test with $E(v|z) = 0$, one can use many different functions for $g(z)$ to test $E\{v \cdot g(z)\} = 0$. In principle, if we use sufficiently many functions, say $g_1(z), ..., g_m(z)$, for $g(z)$ such that any function of z can be well approximated by $g_j(z)$, $j = 1, ..., m$, then a test for $E\{v \cdot g_j(z)\} = 0$, $j = 1, ..., m$, may be as good as the (infeasible) test for $E(v|z) = 0$. The test statistic in this case is

$$G'_N C_N^{-1} G_N \rightsquigarrow \chi_m^2 \text{ where } G_N \equiv \left\{ \frac{1}{\sqrt{N}} \sum_i \hat{v}_i g_1(x_i), ..., \frac{1}{\sqrt{N}} \sum_i \hat{v}_i g_m(x_i) \right\}'$$

and $C_N \rightarrow^p C$ that is the asymptotic variance for G_N. In general, $\hat{v}$ includes a nuisance parameter, and its estimation will affect C, which should be thus accounted for.

De Jong and Bierens (1994) showed that if the number m of functions goes to infinity and if the sequence of functions spans the space of square-integrable functions, then MMT using the infinite moment conditions can detect any kind of violation of H_0, that is, the test is "consistent." The test statistic they propose is

$$\sqrt{2m}\{G'_N C_N^{-1} G_N - m\} \rightsquigarrow N(0, 1), \quad \text{as } m \to \infty$$

which results from CLT applied to m-many χ_1^2 random variables. In practice, one can use low-order polynomial functions of x for g_N. Unless m is very large, however, it is likely that χ_m^2 provides a better approximation for the asymptotic distribution of $G'_N C_N^{-1} G_N$ than this display does. In practice, since we will be using only a finite m, there will be a set D for z such that $E(v|z) \neq 0$ for $z \in D$ which fails to be detected by tests with a finite m. Better conditional moment tests will be seen later in Chapters 8 and 9.

5   Tests Comparing Two Estimators

We saw moment-based tests, which is an important idea of testing. Another idea, probably as important, is testing by comparing two estimators, which is the topic of this section; Wald tests are in fact based on this idea as shown below. Although the topic is not quite relevant to M-estimation, it is discussed here for two reasons. One is to contrast the testing principle to MMT, and the other is that influence functions that appeared for

M-estimators with nuisance parameters come handy for tests based on the difference of two estimators.

5.1   Two Estimators for the Same Parameter

The Wald test for the linear hypothesis $R'\beta = c$ looks at the difference between the two estimators $R'b_N$ and c for the same parameter $R'\beta$. $R'b_N$ is consistent for $R'\beta$ under both H_0 and H_a, whereas c is consistent for $R'\beta$ only under H_0. Thus if H_0 holds, $R'b_N - c \to^p 0$; otherwise $R'b_N - c$ is not consistent for 0. Extending this idea in the Wald test, we can test for the model assumptions by comparing two estimators which are supposed to be close if the assumptions are right.

Consider two estimators a_N and b_N for β, where b_N is $\sqrt{N}$-consistent under both H_0 and H_a and a_N is $\sqrt{N}$-consistent only under H_0. Assume that

$$\sqrt{N}(a_N - \beta) \rightsquigarrow N(0, A) \quad \text{and} \quad \sqrt{N}(b_N - \beta) \rightsquigarrow N(0, B).$$

Then, for a variance matrix C, we might get

$$\sqrt{N}(a_N - b_N) = \sqrt{N}(a_N - \beta) - \sqrt{N}(b_N - \beta) \rightsquigarrow N(0, C)$$

and with $C_N \to^p C$,

$$\sqrt{N}(a_N - b_N)'C_N^{-1}\sqrt{N}(a_N - b_N) = N(a_N - b_N)'C_N^{-1}(a_N - b_N) \rightsquigarrow \chi^2_{rank(C)}.$$

Specifically, suppose a_N and b_N have influence functions:

$$\sqrt{N}(a_N - \beta) = \frac{1}{\sqrt{N}}\sum_i v_i + o_p(1) \rightsquigarrow N(0, E(vv')),$$

$$\sqrt{N}(b_N - \beta) = \frac{1}{\sqrt{N}}\sum_i w_i + o_p(1) \rightsquigarrow N(0, E(ww')).$$

Then

$$\sqrt{N}(a_N - b_N) = \sqrt{N}(a_N - \beta) - \sqrt{N}(b_N - \beta) = \frac{1}{\sqrt{N}}\sum_i (v_i - w_i) + o_p(1).$$

This shows that

$$C = E\{(v_i - w_i)(v_i - w_i)'\} \quad \text{and} \quad C_N = \frac{1}{N}\sum_i (\hat{v}_i - \hat{w}_i)(\hat{v}_i - \hat{w}_i)',$$

where $\hat{v}_i \to^p v_i$ and $\hat{w}_i \to^p w_i$.

In addition to the assumptions on a_N and b_N, further assume that a_N is efficient under H_0 while b_N is not. Then, for the χ^2 test statistic, C_N can be replaced with a consistent estimator, say $B_N - A_N$ for $B - A$:

$$N \cdot (a_N - b_N)'(B_N - A_N)^{-1}(a_N - b_N) \rightsquigarrow \chi^2_{rank(B-A)},$$

which is called a *Hausman test* (Hausman, 1978) in econometrics. In Hausman test, H_0 and H_a are not specific: H_0 is the set of model assumptions which make a_N efficient and consistent and b_N inefficient and consistent, while H_a is the set of model assumptions which make a_N inconsistent and b_N consistent. We will show why $B - A$ is p.s.d. below. But before that, two examples of Hausman tests are given next.

First, consider two moment conditions

$$E\{ \underset{p_1 \times 1}{z_1} \, (y - x'\beta)\} \; = \; 0 \;\; \text{and} \;\; E\{ \underset{p_2 \times 1}{z_2} \, (y - x'\beta)\} = 0$$

$$\text{where } p_1, p_2 \; \geq \; k \, (= \dim(\beta)).$$

Suppose $H_0 : E\{z_2(y - x'\beta)\} = 0$. The GMM using both moment conditions is consistent and efficient under H_0 but inconsistent under H_a. The GMM using only the first moment condition is consistent under both H_0 and H_a, but inefficient under H_0.

Second, for a panel data model

$$y_{it} = x_{it}'\beta + v_{it}, \quad v_{it} = \delta_i + u_{it}, \quad i = 1, ..., N, \; t = 1, ..., T,$$

suppose that $\delta_i \sim N(0, \sigma_\delta^2)$, $u_{it} \sim N(0, \sigma_u^2)$ and that, with $u_i \equiv (u_{i1}, ..., u_{iT})'$, (δ_i, u_i') is independent of one another and is independent of $(x_{i1}, ..., x_{iT})$. This implies that $v_i \equiv (v_{i1}, ..., v_{iT})' = 1_T \otimes \delta_i + u_i$ is jointly normal with 0 mean and variance

$$E(vv') = E\{(1_T \otimes \delta_i + u_i)(1_T \otimes \delta_i + u_i)'\}$$
$$= \; E(1_T 1_T' \otimes \delta_i^2) + E(u_i u_i') = 1_T 1_T' \otimes \sigma_\delta^2 + I_T \sigma_u^2.$$

With this, MLE can be done which is efficient under "H_0: δ_i is independent of $(x_{i1}, ..., x_{iT})$" but inconsistent if H_0 is violated. Since this MLE standardizes all errors terms and then does LSE, the MLE is equivalent to panel GLS using $N^{-1} \sum_i \hat{v}_i \hat{v}_i'$ as the weighing matrix. The within group estimator WIT is consistent under both H_0 and H_a but inefficient under H_0.

In practice, the estimator $B_N - A_N$ in the Hausman test statistic may not be invertible, nor p.s.d. even if invertible. If the sample size is small, then a poor estimate for the covariance matrix may be the reason for this problem. However, if the sample size is large, the problem should be taken as rejecting H_0, since $B - A$ being p.s.d. is valid only under H_0. One way to avoid this problem is to use the above C_N instead of $B_N - A_N$.

Turning to the question on why the asymptotic variance of $\sqrt{N}(b_N - a_N)$ is the difference $B - A$ of two individual variances in Hausman test, if a_N is the efficient estimator, then under certain regularity conditions, it holds that

$$\sqrt{N}(b_N - \beta) = \frac{1}{\sqrt{N}} \sum_i w_i = \frac{1}{\sqrt{N}} \sum_i (v_i + \eta_i)$$

where $E(v\eta') = 0$. That is, an inefficient asymptotically normal $\sqrt{N}$-consistent estimator is a sum of $N^{-1/2} \sum_i v_i$ (from the efficient estimator) and $N^{-1/2}\sum_i \eta_i$ orthogonal to $N^{-1/2}\sum_i v_i$. This has been known since 1950s; see, e.g. Bickel et al. (1993). Intuitively, if $E(v\eta') \neq 0$, then we can get the following estimator more efficient than a_N by linearly projecting v on η:

$$\frac{1}{\sqrt{N}}\sum_i \{v_i - E(v\eta')E^{-1}(\eta\eta')\eta_i\}.$$

This contradicts the efficiency of a_N, and so $E(v\eta')$ must be zero. Therefore we have

$$\sqrt{N}(b_N - a_N) = \frac{1}{\sqrt{N}}\sum_i \eta_i \rightsquigarrow N(0, E(\eta\eta')), \text{ and } A_N - B_N \to^p E(\eta\eta').$$

The idea of comparing two estimators is old, and $C = B - A$ under the efficiency of a_N was known well before Hausman (1978). Also in practice, frequently one has to use the influence-function-based C_N, because $B_N - A_N$ is not p.s.d.. In view of these observations, the term "Hausman test" is somewhat over-used in econometrics. Hausman test is an example of the "efficiency-orthogonality nexus", which was noted in relation to two-stage M-estimator being more efficient than its infeasible version with the known first-stage parameter.

5.2   Two Estimators for the Same Variance

Although somewhat different from Wald-type tests, it is possible to test model specifications using variance-matrix estimators. Consider two estimators Ω_{N1} and Ω_{N2} for Ω where $\sqrt{N}(b_N - \beta) \rightsquigarrow N(0,\Omega)$. Suppose that both Ω_{N1} and Ω_{N2} are consistent for Ω under H_0, whereas only Ω_{N1} is consistent for Ω under H_a. In the following, we introduce two tests based on this idea.

White's Heteroskedasticity Test (White, 1980) for the linear model $y = x'\beta + u$ uses a heteroskedasticity-robust variance estimator and a variance estimator under homoskedasticity. The test looks at the difference

$$E^{-1}(xx') \, E(xx'u^2) \, E^{-1}(xx') \, - \sigma^2 E^{-1}(xx').$$

Stacking some elements of the sample version of this, we get a vector which follows a normal distribution with mean 0 when normed by $\sqrt{N}$; the matrix in the preceding display is symmetric and it has $k(k+1)/2$ distinct elements at maximum where x is a $k \times 1$ vector. Then a Wald test statistic can be formed. As such, however, the test is somewhat cumbersome to implement; instead, one may use an artificial regression version of the White test: do LSE of r_i^2 on 1, x_i, and quadratic terms of elements of x_i where $r_i \equiv y_i - x_i'b_{lse}$. If any slope coefficient is significant, then homoskedasticity is rejected.

Another specification test based on covariance matrix comparison is *information matrix test* in White (1982); the rest of this subsection requires some knowledge on MLE, and thus may be read after more on MLE is covered. In MLE with the likelihood function f, the second-order matrix times minus one should equal the outer-product of the score function. That is, if the MLE specification is correct, the following k by k matrix moment condition holds:

$$E\left(\frac{\partial \ln f}{\partial b}\frac{\partial \ln f}{\partial b'} + \frac{\partial^2 \ln f}{\partial b \partial b'}\right) = 0.$$

To see how to implement the information matrix test, let

$$E\left(\frac{\partial \ln f(z,\beta)}{\partial b}\frac{\partial \ln f(z,\beta)}{\partial b'}\right) = [g_{jl}(z,\beta)] \quad \text{and}$$

$$E\left(\frac{\partial^2 \ln f(z,\beta)}{\partial b \partial b'}\right) = [h_{jl}(z,\beta)].$$

Due to the symmetry of these matrices, we can compare only the upper- (or lower) triangular components. Let $\tau(z,\beta)$ denote the stacked version of those elements; τ has $k' \equiv k(k+1)/2$ elements. For instance, with $k=2$, $\tau(z,\beta)$ is a 3×1 vector such that $\tau = (g_{11}+h_{11}, g_{12}+h_{12}, g_{22}+h_{22})'$. Observe now, with $\tau_{b'}(z,\beta) \equiv \partial \tau(z,\beta)/\partial b'$ and s denoting the score vector,

$$\frac{1}{\sqrt{N}}\sum_i \underset{k' \times 1}{\tau(z_i, b_N)} = \frac{1}{\sqrt{N}}\sum_i \tau(z_i, \beta)$$

$$+\frac{1}{N}\sum_i \underset{k' \times k}{\tau_{b'}(z_i, \beta)}\sqrt{N}(b_N - \beta) + o_p(1)$$

$$= \frac{1}{\sqrt{N}}\sum_i \eta(z_i, \beta) + o_p(1), \quad \text{where}$$

$$\eta(z_i, \beta) \equiv \tau(z_i, \beta) + E\{\tau_{b'}(z,\beta)\}E^{-1}\{s(z,\beta)s(z,\beta)'\}\, s(z_i, \beta).$$

From this, with the population means replaced by their sample versions, the test statistic is

$$\left\{\frac{1}{\sqrt{N}}\sum_i \tau(z_i, b_N)\right\}' \left\{\frac{1}{N}\sum_i \eta(z_i, b_N)\eta(z_i, b_N)'\right\}^{-1}$$

$$\left\{\frac{1}{\sqrt{N}}\sum_i \tau(z_i, b_N)\right\} \rightsquigarrow \chi^2_{k'}.$$

The information matrix test (and the heteroskedasticity test) also suffers from the same drawback as Hausman test: the rejection of H_0 does not indicate exactly which assumption in the null model is violated. Also deriving the second derivatives (h_{jl}) and the cross-derivatives ($\tau_{b'}$) of the log-likelihood is difficult; in practice, it would be safer to use numerical derivatives for both. One way to avoid $\tau_{b'}$ involving third-order derivatives comes from noting

$$\frac{\partial E\{\tau(z,\beta)\}}{\partial \beta'} = 0 \quad \text{under } H_0 : E\{\tau(z,\beta)\} = 0$$

$$\implies \quad 0 = \frac{\partial \int \tau(z,\beta) f(z,\beta) dz}{\partial \beta'} = \int \frac{\partial \tau(z,\beta)}{\partial \beta'} f(z,\beta) dz$$
$$+ \int \tau(z,\beta) \frac{\partial f(z,\beta)}{\partial \beta'} dz$$

$$= \int \frac{\partial \tau(z,\beta)}{\partial b'} f(z,\beta) dz + \int \tau(z,\beta) s(z,\beta)' f(z,\beta) dz$$
$$\text{where } s(z,\beta) = \frac{\partial f(z,\beta)/\partial \beta}{f(z,\beta)}$$
$$\implies \quad E\{\tau_{b'}(z,\beta)\} = -E\{\tau(z,\beta) s(z,\beta)\}.$$

Substituting the last display into $\eta(z,\beta)$ yields

$$\eta(z_i,\beta) = \tau(z,\beta) - E\{\tau(z,\beta) s(z,\beta)\} E^{-1}\{s(z,\beta) s(z,\beta)'\} \cdot s(z,\beta)$$

which is the linear projection residual of τ on s. This form of information matrix test is called the "outer-product of gradient (OPG)" form and appeared in Lancaster (1984). As once noted already, by plugging in the MLE b_N for β, the asymptotic variance of $N^{-1/2} \sum_i \tau(z_i, b_N)$ is smaller than that of its infeasible version $N^{-1/2} \sum_i \tau(z_i, \beta)$.

The information matrix test and its modifications as the OPG form tend to over-reject H_0; see, e.g., Orme (1990), Davidson and MacKinnon (1992) and the references therein. One possible reason for the problem is second-order derivatives. When $x'\beta$ is in the likelihood, the second-order derivatives have x_j^2, $j = 1, ..., k$, and the asymptotic variance estimator involves x_j^4: a big value (an "outlier") in x_j will be four-fold magnified in x_j^4. This can make the asymptotic variance estimator unstable. For real data, the over-rejection problem can also occur because the first and second derivatives are "fine details" of f: even if f provides a good approximation to the true likelihood, say g, f' and f'' may not be close to g' and g''. The over-rejection problem may be avoided using "parametric bootstrap" as suggested by Horowitz (1994). Overall, a caution is warranted in using information matrix test in practice.

6   Three Tests for MLE

So far we have seen only Wald and method-of-moment tests mostly for linear hypotheses. In this section, more tests for MLE are examined. First, we quickly review Wald test for linear hypotheses, and then examine nonlinear hypotheses. Second, "likelihood ratio (LR) test" for MLE is presented. Third, "LM test" or "score test" is studied. The three tests (Wald, LR, LM) together are sometimes called the "trilogy" in MLE tests.

6.1   Wald Test and Nonlinear Hypotheses

Suppose we want to test $H_0 : R'\beta = c$, where R is a $k \times g$ known matrix with rank $g \leq k$, and c is a $g \times 1$ known vector; both R and c do not involve β. For an estimator b_N for β such that $\sqrt{N}(b_N - \beta) \rightsquigarrow N(0, V)$, a Wald test statistic is

$$N(R'b_N - c)'(R'V_N R)^{-1}(R'b_N - c) \rightsquigarrow \chi_g^2 \quad \text{where } V_N \to^p V.$$

Using δ-method, we can construct a Wald test for $g \times 1$ nonlinear hypotheses

$$H_0 : h(\beta) = c \text{ where } rank(h_b(\beta)) = g \text{ and } h_b(\beta) \equiv \frac{\partial h(b)}{\partial b}|_{b=\beta};$$

this includes $R'\beta = c$ as a special case.

Observe

$$h(b_N) - c = h(b_N) - h(\beta) \quad \text{under } H_0$$
$$\implies \quad \sqrt{N}\{h(b_N) - c\} \rightsquigarrow N\{0, \, h_{b'}(\beta) \, V \, h_b(\beta)\}.$$

Assuming (see Andrews (1987b))

$$P\{rank(h_b(b_N)) = rank(h_b(\beta))\} \to 1 \quad \text{as} \quad N \to \infty,$$

the Wald test statistic for the nonlinear hypothesis is

$$N\{h(b_N) - c\}' \, \{h_{b'}(b_N) \, V_N \, h_b(b_N)\}^{-1}\{h(b_N) - c\} \rightsquigarrow \chi_g^2,$$
$$\text{where } V_N \to^p V.$$

As an example, suppose we have

$$H_0 : \beta_2\beta_3 = 1 \text{ and } \frac{\beta_4}{\beta_5} = \frac{\beta_6}{\beta_7}, \qquad \text{where } g = 2, k = 7.$$

Rewrite H_0 as

$$\begin{bmatrix} \beta_2\beta_3 \\ \beta_4\beta_7 - \beta_5\beta_6 \end{bmatrix} = \begin{bmatrix} 1 \\ 0 \end{bmatrix}$$

$$\implies \underset{2 \times 7}{h_{b'}(\beta)} = \begin{bmatrix} 0 & \beta_3 & \beta_2 & 0 & 0 & 0 & 0 \\ 0 & 0 & 0 & \beta_7 & -\beta_6 & -\beta_5 & \beta_4 \end{bmatrix}$$

where $c = (1, 0)'$ and $h(b_N) = (b_2 b_3, b_4 b_7 - b_5 b_6)'$ with $b_N = (b_1, ..., b_k)'$. Substitute c, $h_b(b_N)$, and $h(b_N)$ into the Wald test statistic to implement the test.

There is a problem in Wald tests with nonlinear hypotheses. For example, in the H_0 above, $\beta_4/\beta_5 = \beta_6/\beta_7$ can be reformulated in many algebraically equivalent ways. Instead of $\beta_4\beta_7 - \beta_5\beta_6$, if the original form $\beta_4\beta_5^{-1} - \beta_6\beta_7^{-1} = 0$ is used, then the second row of $h_{b'}(\beta)$ becomes

$$0, 0, 0, \beta_5^{-1}, -\beta_4\beta_5^{-2}, -\beta_7^{-1}, \beta_6\beta_7^{-2}.$$

Using this renders a different value for the Wald test in general. A more drastic example is testing $H_0: \beta_2 = 1$, which can be rewritten as nonlinear hypotheses $\beta_2^2 = 1$, $\beta_2^3 = 1$, ..., $\beta_2^{1000} = 1$ (Lafontaine and White, 1986).

The following two points may help choosing a nonlinear hypothesis in practice (Gregory and Veall (1985) and Phillips and Park (1988)). First, there may be a hypothesis more natural than others; in the preceding example, $\beta_2 = 1$ is a more natural choice than $\beta_2^{1000} = 1$, because we do not actually think that β_2 is exactly one (β_2^{1000} will be either almost 0 or almost ∞ depending on whether $|\beta_2| < 1$ or $|\beta_2| > 1$). Second, a nonlinear hypothesis in a multiplicative form seems better than that in a ratio form in the sense that the linear approximation of the nonlinear function holds better, which is why we used $\beta_4 \beta_7 - \beta_5 \beta_6 = 0$ rather than $\beta_4 \beta_5^{-1} - \beta_6 \beta_7^{-1} = 0$.

6.2   Likelihood Ratio (LR) Test

Consider a twice continuously differentiable likelihood function $f(z, b)$ wrt b for all z, which will be often denoted just as $f(b)$. Suppose we have a null hypothesis $H_0: R'\beta = c$. The unrestricted MLE b_N maximizes $\sum_i \ln f(b)$ and thus satisfies $\sum_i s(b_N) = 0$ where $s(b)$ is the score function for $f(b)$, whereas the restricted MLE b_{Nr} maximizes $\sum_i \ln f(b)$ subject to $R'b = c$. Namely, b_{Nr} maximizes

$$\sum_i \ln f(b) + N\lambda'(R'b - c)$$

where λ is a $g \times 1$ Lagrangian multiplier. Likelihood ratio test compares two maximized log-likelihood functions at b_N and b_{Nr}. But before we proceed further, it helps to look at a linear model case first.

6.2.1   Restricted LSE

Suppose $y = x'\beta + u$ and we want to have a restricted LSE b_{Nr} maximizing

$$-\frac{1}{2} \sum_i (y_i - x_i' b_{Nr})^2 + N\lambda'(R'b - c)$$

$$\implies \quad \text{first-order conditions for } (b_{Nr}', \lambda_N')' :$$

$$\sum_i x_i(y_i - x_i' b_{Nr}) + NR\lambda_N = 0 \quad \text{and} \quad R'b_{Nr} = c.$$

From the first equation,

$$\sum_i x_i y_i - \sum_i x_i x_i' \cdot b_{Nr} + NR\lambda_N = 0$$

$$\implies \quad b_{Nr} = b_N + \left(\frac{1}{N}\sum_i x_i x_i'\right)^{-1} R\lambda_N \quad \text{as}$$

$$b_N = \left(\frac{1}{N} \sum_i x_i x_i' \right)^{-1} \frac{1}{N} \sum_i x_i y_i.$$

Substitute this into $R' b_{Nr} = c$ to get

$$R' b_N + R' \left(\frac{1}{N} \sum_i x_i x_i' \right)^{-1} R \lambda_N = c$$

$$\implies \lambda_N = \left\{ R' \left(\frac{1}{N} \sum_i x_i x_i' \right)^{-1} R \right\}^{-1} (c - R' b_N).$$

Substitute this back into the b_{Nr} equation to get

$$
\begin{aligned}
b_{Nr} &= b_N + \left(\frac{1}{N} \sum_i x_i x_i' \right)^{-1} \cdot R \left\{ R' \left(\frac{1}{N} \sum_i x_i x_i' \right)^{-1} R \right\}^{-1} (c - R' b_N) \\
&= \left[I_k - \left(\frac{1}{N} \sum_i x_i x_i' \right)^{-1} R \left\{ R' \left(\frac{1}{N} \sum_i x_i x_i' \right)^{-1} R \right\}^{-1} R' \right] \cdot b_N \\
&\quad + \left(\frac{1}{N} \sum_i x_i x_i' \right)^{-1} \cdot R \left\{ R' \left(\frac{1}{N} \sum_i x_i x_i' \right)^{-1} R \right\}^{-1} c.
\end{aligned}
$$

Multiplying the first and last expressions in this display by R', we can see $R' b_{Nr} = c$ holds.

Suppose $c = 0$. If we replace $N^{-1} \sum_i x_i x_i'$ with I_k in b_{Nr}, then we get a different (inefficient in general, but easier to interpret) restricted estimator

$$b_{Nr}^* \equiv [I_k - R(R'R)^{-1} R'] \cdot b_N \ (\implies R' b_{Nr}^* = 0):$$
$$b_{Nr}^* \text{ is the projection "residual" of } b_N \text{ on } R.$$

Suppose $k > 3$ and H_0: the last 3 components of β are zeros. Then $c = 0$ and R' is $0_{3 \times k}$ with its last 3×3 matrix replaced by I_3:

$$
R' = \begin{bmatrix} 0 & \cdots & 0 & 0 & 1 & 0 & 0 \\ \vdots & \cdots & \vdots & \vdots & 0 & 1 & 0 \\ 0 & \cdots & 0 & 0 & 0 & 0 & 1 \end{bmatrix}.
$$

In this case, $I_k - R(R'R)^{-1} R'$ is I_k with its lower right I_3 matrix replaced by $0_{3 \times 3}$; i.e., b_{Nr}^* is b_N with its last three elements replaced by zeros.

It is not difficult to see $b_{Nr}^* \neq b_{Nr}$. For instance, denoting $x = (x_f', x_g')'$ with x_g for the last three elements of x, suppose $COR(x_f, x_g)$ is very high. Then this will make b_N poor (multicollinearity problem), but b_{Nr}^* just takes over the poor estimates from b_N while b_{Nr} takes this $COR(x_f, x_g)$ into account by using $N^{-1} \sum_i x_i x_i'$ in transforming b_N to b_{Nr}. The interesting question is then, will the LSE of y on x_f be the same as b_{Nr}? This LSE also takes advantage of $COR(x_f, x_g)$ as x_g is dropped to remove the multicollinearity problem. The answer can be shown to be yes, as it is intuitively plausible.

6.2.2   Restricted MLE and LR Test

Turning back to MLE, apply Taylor expansion of second order to $\sum_i \ln f(b_{Nr})$ around b_N to get, for some $b_N^* \in (b_{Nr}, b_N)$,

$$
\sum_i \ln f(b_{Nr}) = \sum_i \ln f(b_N) + \sum_i s(b_N)'(b_{Nr} - b_N)
$$
$$
+ \frac{1}{2}(b_{Nr} - b_N)' \sum_i s_{b'}(b_N^*)(b_{Nr} - b_N)
$$

where $s_{b'} \equiv \partial s/\partial b'$ is a $k \times k$ matrix. Since $\sum_i s(b_N) = 0$ by the construction of b_N, under the H_0, this display can be rewritten as

$$
2\left\{ \sum_i \ln f(b_N) - \sum_i \ln f(b_{Nr}) \right\}
$$
$$
= \sqrt{N}(b_{Nr} - b_N)' \left\{ -\frac{1}{N} \sum_i s_{b'}(b_N^*) \right\} \sqrt{N}(b_{Nr} - b_N)
$$
$$
= \sqrt{N}(b_{Nr} - b_N)' I_f \sqrt{N}(b_{Nr} - b_N) + o_p(1)
$$

for $b_N \to^p \beta$ and $b_{Nr} \to^p \beta$, where $I_f = -E\{s_b(\beta)\} = E(s(\beta)s(\beta)')$ is the information matrix. We need to know the relation between b_{Nr} and b_N to deal with this equation, which is done in the following. The reader may skip the derivation and go directly to "$2\{\sum_i \ln f(b_N) - \sum_i \ln f(b_{Nr})\} \rightsquigarrow \chi_g^2$" near the end of this subsection.

Expand the first-order condition $\sum_i s(b_{Nr}) + NR\lambda = 0$ for b_{Nr} around b_N: for some $b_{Nr}^* \in (b_{Nr}, b_N)$, because $\sum_i s(b_N) = 0$,

$$
0 = \sum_i \{s(b_N) + s_{b'}(b_{Nr}^*)(b_{Nr} - b_N)\} + NR\lambda
$$
$$
= \sum_i s_{b'}(b_{Nr}^*)(b_{Nr} - b_N) + NR\lambda
$$
$$
\Longrightarrow 0 = \frac{1}{N} \sum_i s_{b'}(b_{Nr}^*)\sqrt{N}(b_{Nr} - b_N) + \sqrt{N}R\lambda
$$
$$
= -I_f \sqrt{N}(b_{Nr} - b_N) + R\sqrt{N}\lambda + o_p(1).
$$

Multiply both sides by $R'I_f^{-1}$ to get

$$
0 = -R'\sqrt{N}(b_{Nr} - b_N) + R'I_f^{-1}R \cdot \sqrt{N}\lambda + o_p(1)
$$
$$
\Longrightarrow \sqrt{N}\lambda = -(R'I_f^{-1}R)^{-1}\sqrt{N}(R'b_N - c) + o_p(1), \quad \text{for } R'b_{Nr} = c.
$$

Substitute this back into $0 = -I_f \sqrt{N}(b_{Nr} - b_N) + R\sqrt{N}\lambda + o_p(1)$ to get

$$
0 = -I_f \sqrt{N}(b_{Nr} - b_N) - R(R'I_f^{-1}R)^{-1}\sqrt{N}(R'b_N - c) + o_p(1)
$$
$$
\Longrightarrow \sqrt{N}(b_N - b_{Nr}) = I_f^{-1}R(R'I_f^{-1}R)^{-1}\sqrt{N}(R'b_N - c).
$$

Substitute this into the above $2\left\{\sum_i \ln f(b_N) - \sum_i \ln f(b_{Nr})\right\}$ equation to get

$$2\left\{\sum_i \ln f(b_N) - \sum_i \ln f(b_{Nr})\right\} = N(Rb_N - c)'(R'I_f^{-1}R)^{-1}(Rb_N - c)$$
$$+ o_p(1).$$

The right-hand side is the Wald test statistic (with I_f replaced with an estimator), and we get

$$2\left\{\sum_i \ln f(b_N) - \sum_i \ln f(b_{Nr})\right\} \rightsquigarrow \chi_g^2$$

which is called the *likelihood ratio (LR) test* in MLE.

The above equation $b_N - b_{Nr} = I_f^{-1}R(R'I_f^{-1}R)^{-1}(R'b_N - c)$ can be solved for b_{Nr} to show that b_{Nr} can be obtained from b_N:

$$b_{Nr} = \{I_k - I_f^{-1}R(R'I_f^{-1}R)^{-1}R'\}b_N + I_f^{-1}R(R'I_f^{-1}R)^{-1}c$$

where I_f should be replaced with $N^{-1}\sum_i s(b_N)s(b_N)'$. When $c = 0$, b_{Nr} is reminiscent of the projection residual of b_N on the columns of R. This formula comes handy when imposing $Rb_{Nr} = c$ directly on the estimation procedure is difficult. If this is easy—e.g., H_0 specifies some coefficients to be zero—then getting b_{Nr} directly is easier: just drop the corresponding regressors from the model.

LR is also good for nonlinear hypotheses so long as b_{Nr} satisfies the hypothesis. Compared with the Wald test, the LR test has the disadvantage of requiring both b_N and b_{Nr}, but the LR test has the advantage of good "invariance properties" such as invariance to re-parametrizations of the model. See Dagenais and Dufour (1991) and the references therein for more on this; an example for this point will appear in the chapter for limited dependent variable models. LR test also appears in the celebrated "Neyman-Pearson Lemma" for optimality in testing simple (and one-sided) hypotheses.

6.3   Score (LM) Test and Effective Score Test

When we discussed MMT, the idea was to test for (zero) moment conditions implied by the model which were, however, not used in obtaining the estimator. Applying the idea to MLE, we can devise *score test* or *Lagrangian multiplier test (LM)*. When we get b_{Nr} under H_0, we do not use all the first-order conditions used in getting b_N. Thus if H_0 is correct, then b_{Nr} should satisfy the unused first-order conditions, which is the key idea.

Denote the score vector for b_N evaluated at b_{Nr} as $s_i(b_{Nr}) = \partial \ln\{f(b_{Nr})\}/\partial b$. Then the score test statistic is

$$\sum_i s_i(b_{Nr})' \left\{\sum_i s_i(b_{Nr})s_i(b_{Nr})'\right\}^{-1} \sum_i s_i(b_{Nr}) \rightsquigarrow \chi_g^2$$

which requires only b_{Nr} (no need to compute b_N). LM test is the most useful when getting b_N is difficult while getting b_{Nr} is easy. As noted already, this is the case if the H_0 is linear stating that some components of β are zero. There exist other forms of LM tests, and the above one is called the "outer-product of gradient (OPG)" form. In the score test, the dimension of $s_i(b_{Nr})$ is $k \times 1$, while the degree of freedom in χ^2 is still g, which may be a little puzzling; we show why in the following.

Recall that b_{Nr} satisfies $\sum_i s_i(b_{Nr}) = -NR\lambda$, and thus the score test statistic can be written as

$$\sqrt{N}\lambda' \cdot R' \left\{ \frac{1}{N} \sum_i s_i(b_{Nr}) s_i(b_{Nr})' \right\}^{-1} R \cdot \sqrt{N}\lambda$$

which gives justice to the name LM test. The middle matrix is consistent for $R'I_f^{-1}R$ under the H_0, which is the inverse of the asymptotic variance of $\sqrt{N}\lambda$, because, as we already saw in the preceding subsection,

$$\sqrt{N}\lambda = (R'I_f^{-1}R)^{-1}\sqrt{N}(R'b_N - c) + o_p(1)$$

$$\rightsquigarrow N\{0, (R'I_f^{-1}R)^{-1}R'I_fR(R'I_f^{-1}R)^{-1}\} = N(0, (R'I_f^{-1}R)^{-1}).$$

Hence, the LM test statistic is a quadratic form of an asymptotically normal $g \times 1$ vector inversely weighted by its asymptotic variance matrix, which follows χ_g^2.

As an example of LM test, consider H_0: the first k_1 components of β are zero. Define

$$\beta \equiv (\beta_1', \beta_2')', \quad s_i(b) \equiv (s_{1i}(b)', s_{2i}(b)')'$$

where the dimension of β_j and s_j is $k_j \times 1$, $j = 1, 2$. Under the H_0, the first k_1 elements of x is not used. That is, only the condition $N^{-1}\sum_i s_{2i}(b_{Nr}) = 0$ is used to get b_{Nr}, and the LM test examines if b_{Nr} satisfies $N^{-1}\sum_i s_{1i}(b_{Nr}) = 0$. Since estimating b_{Nr} is certainly easier than b_N in this example, the LM test has a practical advantage over the Wald and LR tests. This advantage, however, may not hold for a complicated H_0.

In the LM example, only a part $s_{1i}(b)$ of the score vector $s_i(b)$ is effectively used for the LM test. If we follow the idea of MMT, we should test, not $N^{-1}\sum_i s_i(b_{Nr}) \to^p 0$, but only $N^{-1}\sum_i s_{1i}(b_{Nr}) \to^p 0$. The asymptotic distribution of the test statistic $N^{-1/2}\sum_i s_{1i}(b_{Nr})$ can be found by doing analogously to deriving the asymptotic distribution of the MMT's, but the asymptotic distribution would be more complicated than that of the LM test, because b_{Nr} becomes essentially a nuisance parameter in $N^{-1/2}\sum_i s_{1i}(b_{Nr})$, and this fact influences the asymptotic distribution of the "second-stage" test. By using $N^{-1}\sum_i s_i(b_{Nr})$ in LM, this complication is avoided.

Suppose we want to test $N^{-1}\sum_i s_{1i}(b_{Nr}) \to^p 0$. As just mentioned, the asymptotic variance of $\sqrt{N}(b_{Nr} - \beta)$ will appear in the asymptotic distribution of $N^{-1/2} \sum_i s_{1i}(b_{Nr})$. However, there is a way to avoid this problem. In the following paragraph, we will show that

$$\frac{1}{\sqrt{N}} \sum_i [s_{1i}(b_2) - I_{12}I_{22}^{-1}s_{2i}(b_2)] \rightsquigarrow N(0, I_{11} - I_{12}I_{22}^{-1}I_{21})$$

where b_2 is an any $\sqrt{N}$-consistent estimator for β_2, $I_{12} \equiv E\{s_1(\beta)s_2(\beta)'\}$, and I_{11}, I_{22}, and I_{21} are analogously defined. From this display, we then obtain

$$\sum_i S_i' \left(\sum_i S_i S_i' \right)^{-1} \sum_i S_i \rightsquigarrow \chi_g^2, \quad \text{where}$$

$$S_i \equiv s_{1i}(b_2) - \sum_i s_{1i}(b_2)s_{2i}(b_2)' \left\{ \sum_i s_{2i}(b_2)s_{2i}(b_2)' \right\}^{-1} s_{2i}(b_2).$$

This is convenient, for the asymptotic variance matrix does not depend on the asymptotic variance of $\sqrt{N}(b_2 - \beta_2)$. Instead of using s_1, the test uses the part of s_1 not explained by s_2. The test using the *effective score* (s_1 not explained by s_2) is called the *Neyman's "C(α) test"* or *"effective score test."* According to Bera and Premaratne (2001) who provided a brief survey on hypothesis testing, the "C" in C(α) refers to Cramér (the author of a classical statistics book) and α refers to the usual level of significance.

Apply Taylor expansion to $s_{1i}(b_2) - I_{12}I_{22}^{-1}s_{2i}(b_2)$ around β_2 to get

$$\frac{1}{\sqrt{N}} \sum_i \{s_{1i}(\beta_2) - I_{12}I_{22}^{-1}s_{2i}(\beta_2)\}$$

$$+ \left[\frac{1}{N} \sum_i \frac{\partial s_{1i}(b_2^*)}{\partial b_2'} - I_{12}I_{22}^{-1} \frac{1}{N} \sum_i \frac{\partial s_{2i}(b_2^*)}{\partial b_2'} \right] \sqrt{N}(b_2 - \beta_2).$$

Recall $E(s(\beta)s(\beta)') = -E(\partial s/\partial b')$ and observe

$$\frac{1}{N} \sum_i \frac{\partial s_{1i}(b_2^*)}{\partial b_2'} \rightarrow_p E\left\{ \frac{\partial s_1(\beta_2)}{\partial b_2'} \right\} = -I_{12},$$

$$\frac{1}{N} \sum_i \frac{\partial s_{2i}(b_2^*)}{\partial b_2'} \rightarrow_p E\left\{ \frac{\partial s_2(\beta_2)}{\partial b_2'} \right\} = -I_{22}.$$

Substituting these into $[\cdots]$ in the Taylor expansion makes the term in $[\cdots]$ to be $o_p(1)$. Hence only the first term in the Taylor expansion remains, establishing the desired asymptotic normality that is the same for any $\sqrt{N}$-consistent b_2.

6.4   Further Remarks and an Empirical Example

Wald, LR, and score tests are called the three classical tests. All three follow χ_g^2 asymptotically under H_0, but they differ in terms of computational ease, performance in small samples, and invariance properties. The three tests are based on *different kinds of distances* which are small under H_0 and large otherwise (Engle, 1984):

$$\text{Wald : the estimators } |b_N - b_{Nr}|$$

$$\text{LR : the maximands } |Q_N(b_N) - Q_N(b_{Nr})|$$

$$\text{Score (LM) : the slopes } \left| \frac{\partial Q_N(b_{Nr})}{\partial b} - \frac{\partial Q_N(b_N)}{\partial b} \right| = \left| \frac{\partial Q_N(b_{Nr})}{\partial b} - 0 \right|.$$

Asymptotically, the three tests are equivalent in having the same significance level under $H_0 : R\beta = c$ and the same power against a local alternative $H_a : R\beta = c + \delta/\sqrt{N}$; under this H_a, they follow the same non-central χ^2 with the non-centrality parameter (NCP) $\delta'(RI_f^{-1}R')^{-1}\delta$. See Bera and Bilias (2001) for historical perspectives and references.

To illustrate the three MLE tests, we use an US cross-section data for males with $N = 545$ for 1987, which is in fact part of an eight-year panel data set used in Lee (2002) who in turn drew the data from Vella and Verbeek (1998). The response variable is whether the male is in a labor union ($y = 1$) or not (the mean is 0.26), and the regressors are 1 and (in the following, the numbers in $(\cdot)$ are mean and SD; for dummies, only the mean is shown)

edu: years of schooling (11.77, 1.75)

exr: job experience computed as $age - 6 - edu$ (10.01, 1.65)

exredu: interaction between exr and edu (116.17, 16.71)

blc: race dummy for black (0.12)

hisp: race dummy for Hispanic (0.16)

mar: dummy for married (0.61)

rur: dummy for living in a rural area (0.06)

sou: dummy for living in south (0.36)

The other regressors used are industry dummies (mean in $(\cdot)$): agr (agriculture; 0.02), bus (business and repair service; 0.10), cst (construction; 0.08), fin (finance; 0.04), man (manufacturing; 0.30), pro (professional; 0.07), pub (public administration; 0.22), trad (trade; 0.20), and tran (transportation; 0.08).

Under $y_i = 1[x_i'\beta + u_i > 0]$ and $u_i \sim N(0, \sigma^2)$ independently of x_i, we have the log-likelihood function to maximize for a (the details of this MLE—*probit*—for binary response will appear in a separate chapter)

$$\sum_i \{y_i \ln \Phi(x_i'a) + (1 - y_i)(1 - \Phi(x_i'a))\} \quad \text{where } a \text{ is for}$$

$$\alpha \equiv \frac{\beta}{\sigma} = (\frac{\beta_1}{\sigma}, ..., \frac{\beta_k}{\sigma})'.$$

Table 1: Probit for Joining Labor Union or Not

x	Unrestricted (tv)	Restricted (tv)	x	Unrestricted (tv)	Restricted (tv)
one	6.438 (2.42)	6.488 (2.62)	agr	1.090 (1.91)	0
edu	−0.724 (−3.12)	−0.693 (−3.19)	bus	−0.148 (−0.32)	0
exr	−0.721 (−3.09)	−0.695 (−3.12)	cst	−0.256 (−0.45)	0
exredu	0.068 (3.23)	0.066 (3.28)	fin	0.614 (1.41)	0
blc	0.763 (4.02)	0.766 (4.08)	man	0.575 (1.42)	0
hisp	0.175 (0.96)	0.165 (0.99)	pro	0.815 (1.80)	0
mar	0.131 (0.94)	00.210 (1.65)	pub	−0.083 (−0.51)	0
rur	1.294 (2.91)	0.824 (3.67)	trad	0.226 (0.55)	0
sou	−0.019 (−0.14)	0.015 (0.12)	tran	0.934 (2.14)	0
log-like.		−275.649 (Unrestricted),		−289.311 (Restricted)	

The column "Unrestricted" in the probit table for joining labor union or not shows the estimates and the t-values (Table 1). Since most of the nine industry dummies look insignificant, we test the null hypothesis

$$H_0 : \beta_{10} =, ..., = \beta_{18} = 0 \quad (9 \text{ restrictions}).$$

With this H_0 imposed, we obtained the column "Restricted" in the table. The three test statistics (p-value) are

$$Wald : 24.77 \ (0.003), \quad LR : 27.33 \ (0.001), \quad LM : 31.33 \ (0.000);$$

All three tests reject the H_0.

7   Numerical Optimization and One-Step Efficient Estimation

For LSE, estimators are written in closed forms. However, M-estimators are defined implicitly by maximands which are functions of b. Hence, obtaining an M-estimator in practice requires some type of numerical searching process, which has been shown already briefly. We substitute a number for b in the sample maximand and evaluate the maximand. Then we decide whether the number is the maximizer or other numbers for b could increase the maximand. If the latter is the case, we choose another number for b and repeat the process. Searching for an estimator in this trial and error fashion is called *numerical maximization*, which is the topic of this section. First, we introduce the popular Newton–Raphson algorithm. Second, some remarks are made for Newton–Raphson algorithm and then other numerical optimization methods are briefly mentioned. Third, we show that only one iteration is enough asymptotically if we start from a $\sqrt{N}$-consistent estimator. The reader may want to read this section when he/she actually has to obtain an M-estimator, say, a MLE.

7.1   Newton–Raphson Algorithm

Intuitively, numerical searching is like being deserted in a foggy mountainous area and trying to reach the highest point from a given spot. If the sky were clear, it would be easier to visually locate the peak. But with the foggy sky, the vision is impaired, and we have to decide on two things from a given spot: *which direction to move and how far to move in that chosen direction*. If we go too far in the wrong direction, then it will be difficult to come back. But if we don't go far enough, we will never know what lies in the far area.

Let $Q(b)$ be a (approximately) quadratic concave maximand where b is a scalar. Then b attains the maximum if $Q'(b) = 0$. If $Q'(b)$ is positive (negative), we should increase (decrease) b, for we are to the left (right) of the peak. The direction to move is decided by the first derivative, and the magnitude of the move depends on $Q''(b)$. In the following, we formalize this idea to obtain "Newton–Raphson algorithm." See, for example, Press et al. (1986) for the numerical maximization methods in this section.

Let b_0 and b_1, respectively, denote the initial and the next estimates. Choose b_1 by maximizing the following wrt b:

$$Q(b) \simeq Q(b_0) + G(b_0)'(b - b_0) + \frac{1}{2}(b - b_0)' \cdot H(b_0) \cdot (b - b_0)$$

where G is the first-derivative vector (gradient) and H is the second-derivative matrix (Hessian). Differentiating the right-hand side (rhs) wrt b gives

$$b_1 = b_0 - H(b_0)^{-1}G(b_0).$$

Repeating this while updating b_0 with b_1 each time until certain stopping criterion is met is the *Newton–Raphson Algorithm*. This iteration formula is valid even if $Q(b)$ is a minimand, because $-Q(b)$ is then a maximand and the minus sign gets canceled in $H(b_0)^{-1}G(b_0)$.

Two popular stopping criteria are: with $b_1 = (b_{11}, ..., b_{1k})'$, $b_0 = (b_{01}, ..., b_{0k})'$, and $\varepsilon = 10^{-7}$,

$$(i) \; \sum_{j=1}^{k} \frac{|b_{1j} - b_{0j}|}{|b_{0j}|} < \varepsilon \quad \text{and} \quad (ii) \; |Q(b_1) - Q(b_0)| < \varepsilon$$

where the choice of ε is arbitrary and division by $|b_{0j}|$ is to remove the scale differences across different variables. The advantage of (i) over (ii) is that using (i) provides an uniform stopping criterion to compare different numerical algorithms (for possibly different maximands) for the same parameter β. The advantage of (ii) is that it is free from different scale or parametrization problems.

Although using analytical derivatives makes the iteration scheme run faster, getting them is often difficult. Thus, using numerical derivatives is recommended in most cases at the cost of slower iteration, unless one has access to a software that can give analytical derivatives. When $k = 2$, the first numerical derivatives for the maximand $Q_N(b)$ at $b = \beta$ where $b = (b_1, b_2)'$ and $\beta = (\beta_1, \beta_2)'$ are

$$\frac{\partial Q_N(\beta)}{\partial b_1} \equiv \frac{Q_N\left(\beta + \varepsilon\binom{1}{0}\right) - Q_N\left(\beta - \varepsilon\binom{1}{0}\right)}{2\varepsilon} \quad \text{and}$$

$$\frac{\partial Q_N(\beta)}{\partial b_2} \equiv \frac{Q_N\left(\beta + \varepsilon\binom{0}{1}\right) - Q_N\left(\beta - \varepsilon\binom{0}{1}\right)}{2\varepsilon}.$$

There are $4 = 2 \times 2$ second derivatives, with the detail shown only for (i):

(i) :

$$\frac{\partial^2 Q_N(\beta)}{\partial b_1 \partial b_1} \equiv \frac{1}{2\varepsilon} \left\{ \frac{\partial Q_N \left(\beta + \varepsilon \binom{1}{0} \right)}{\partial b_1} - \frac{\partial Q_N \left(\beta - \varepsilon \binom{1}{0} \right)}{\partial b_1} \right\}$$

$$= \frac{1}{2\varepsilon} \left\{ \frac{Q_N \left(\beta + \varepsilon \binom{1}{0} + \varepsilon \binom{1}{0} \right) - Q_N \left(\beta + \varepsilon \binom{1}{0} - \varepsilon \binom{1}{0} \right)}{2\varepsilon} \right.$$

$$\left. - \frac{Q_N \left(\beta - \varepsilon \binom{1}{0} + \varepsilon \binom{1}{0} \right) - Q_N \left(\beta - \varepsilon \binom{1}{0} - \varepsilon \binom{1}{0} \right)}{2\varepsilon} \right\},$$

$$= \frac{1}{4\varepsilon} \left\{ Q_N \left(\beta + 2\varepsilon \binom{1}{0} \right) - 2 Q_N(\beta) + Q_N \left(\beta - 2\varepsilon \binom{1}{0} \right) \right\}$$

(ii) :

$$\frac{\partial^2 Q_N(b)}{\partial b_1 \partial b_2} \left\{ = \frac{\partial^2 Q_N(\beta)}{\partial b_2 \partial b_1} \right\}$$

$$\equiv \frac{1}{2\varepsilon} \left\{ \frac{\partial Q_N \left(\beta + \varepsilon \binom{0}{1} \right)}{\partial b_1} - \frac{\partial Q_N (\beta - \varepsilon \binom{0}{1})}{\partial b_1} \right\},$$

(iii) :

$$\frac{\partial^2 Q_N(b)}{\partial b_2 \partial b_2} \equiv \frac{1}{2\varepsilon} \left\{ \frac{\partial Q_N \left(\beta + \varepsilon \binom{0}{1} \right)}{\partial b_2} - \frac{\partial Q_N \left(\beta - \varepsilon \binom{0}{1} \right)}{\partial b_2} \right\}.$$

7.2   Newton–Raphson Variants and Other Methods

For LSE, $H(b_0) = -N^{-1} \sum_i x_i x_i'$, which is at least n.s.d. In general, however, there is no guarantee that $H(b_0)$ is n.d.; if not n.d., the Newton–Raphson method fails. One way to avoid this problem is to use $(1 - \lambda) H(b_0) + \lambda M$ instead of $H(b_0)$, where M is a chosen n.d. matrix; the scalar weight λ should be chosen too. Depending on λ and M, many variations of Newton–Raphson are possible.

Sometimes $H(b)$ does not give a good "magnitude of the movement" (or *step size*) along the direction given by $G(b)$. With the step size too small, it will take a long time to reach the peak. With the step size too big, we may overstep, going from one side of the peak to the other side resulting in an oscillation around the peak. The latter is the more serious problem. One way to avoid this is to modify the Newton–Raphson algorithm as

$$b_1 = b_0 - \eta \cdot H(b_0)^{-1} G(b_0)$$

where η is a positive constant. The smaller η is, the smaller is the step size. The choice of η is arbitrary.

Often $H(b)$ is complicated. One way to simplify $H(b)$ is to use only the terms in $H(b)$ that do not disappear in $E\{H(\beta)\}$. This is called the *method of scoring*. Owing to this approximation of $H(b)$, the method of scoring may be slower in areas far away from β. But near β, the ignored term is almost zero so that the method of scoring should perform comparably to the Newton-Raphson. In MLE, $-E\{H(\beta)\}$ is the same as the expected outer product of the score function. So often we use

$$-\frac{1}{N}\sum_i \left\{ \frac{\partial \ln f(z;b)}{\partial b} \frac{\partial \ln f(z;b)}{\partial b'} \right\}$$

for $H(b)$, which saves the burden of deriving $H(b)$ analytically or numerically. One advantage of this outer-product is that it is always at least n.s.d. If numerical derivatives are employed throughout, then there is no need to bother even with deriving the analytic first derivative. Numerical derivatives, however, increase the computation time.

Suppose that $Q(b)$ is differentiable only once, or that $Q''(b)$ is too complicated to obtain analytically (or too time-consuming to derive numerically). In this case, we have only the gradient available. In this case Newton–Raphson type iteration is infeasible. In the following, we present a simple algorithm employing solely the gradient.

Let b_0 denote the current estimate and $b_1 = b_0 + \eta\delta$ denote the next candidate with its vector direction δ and scalar step size η. Then

$$Q(b_0 + \eta\delta) - Q(b_0) \simeq \eta \cdot G(b_0)'\delta,$$

where G denotes the gradient. For the right-hand side to be positive, δ should be chosen such that $G(b_0)'\delta$ is always positive. One obvious choice is $\delta = G(b_0)$. Hence the direction of improvement is determined. Since η is a positive scalar, it is not too difficult to find the optimal step size for the direction δ. "Grid search" is a possibility. Better yet, "line search by bracketing" (see, e.g., Press et al., 1986) provides a good way to find the optimal η.

Newton–Raphson type algorithms are simpler to implement, but not necessarily superior to algorithms combining the gradient and a line search method. If $Q(b)$ is not approximately quadratic, the Newton–Raphson can be misleading. If $Q(b)$ is shaped like a normal density, the Newton–Raphson may continue to search forever without finding the optimal b because the normal density is not concave in the tails. In this case, success or failure depends on the starting point of the algorithm. Hence employing an algorithm using only the gradient or no gradient at all such as "downhill simplex" could be more robust. Downhill simplex is not to be confused with the simplex method in linear programming. STATA provides an option technique(nm) in its optimization algorithm choice to implement downhill simplex where "nm" stands for Nelder and Mead who invented the algorithm.

Although we showed that β attains a unique global maximum in MLE, there may be multiple global maxima or local maxima in M-estimators in general. Since numerical search procedures may stop at any local maximum, they may stop the iteration prematurely. The only way to avoid this pitfall is to try a number of different starting values and obtain the local maximum for each starting value. If the starting values are scattered enough to be "dense" in the entire parameter space, then one of the local maxima is likely to be a global maximum. Thus by choosing the estimate that yields the maximum among the local maxima, we will get a better chance of finding a global maximizer. Except a few known cases, usually the maximands in M-estimators are not globally concave and thus have multiple local maxima.

7.3   One-Step Efficient Estimation

Sometimes, in trying to implement MLE, we may be able to obtain an initial $\sqrt{N}$-consistent (but inefficient) estimator easily, and start the iteration from the estimator to get the MLE. In this case, surprisingly, doing the iteration only once is asymptotically as good as doing it many times. We will show this for a M-estimator maximizing $N^{-1}\sum_i q(z_i, b)$ for more generality.

Let q_b and $q_{bb'}$ denote the first- and second-order matrices for q. Take one step from a $\sqrt{N}$-consistent estimator b_0:

$$b_N = b_0 - \left\{\frac{1}{N}\sum_i q_{bb'}(b_0)\right\}^{-1} \cdot \frac{1}{N}\sum_i q_b(b_0)$$

$$\Longrightarrow \sqrt{N}(b_N - \beta) = \sqrt{N}(b_0 - \beta) - \left\{\frac{1}{N}\sum_i q_{bb'}(b_0)\right\}^{-1} \cdot \frac{1}{\sqrt{N}}\sum_i q_b(b_0).$$

Apply the mean value theorem to $q_b(b_0)$ around β to get, for some $b^* \in (b_0, \beta)$,

$$\sqrt{N}(b_N - \beta) = \sqrt{N}(b_0 - \beta) - E^{-1}q_{bb'}(\beta)\frac{1}{\sqrt{N}}\sum_i \{q_b(\beta)$$

$$+ q_{bb'}(b^*)(b_0 - \beta)\} + o_p(1)$$

$$= -E^{-1}q_{bb'}(\beta)\frac{1}{\sqrt{N}}\sum_i q_b(\beta)$$

$$+ \left\{I_k - E^{-1}q_{bb'}(\beta)\frac{1}{N}\sum_i q_{bb'}(b^*)\right\}\sqrt{N}(b_0 - \beta) + o_p(1).$$

But the term in $\{\cdot\}$ is $I_k - I_k + o_p(1) = o_p(1)$. Therefore, we get

$$\sqrt{N}(b_N - \beta) = -E^{-1}q_{bb'}(\beta)\frac{1}{\sqrt{N}}\sum_i q_b(\beta) + o_p(1)$$

which follows the same asymptotic distribution as the desired M-estimator. The iteration may be repeated, but there is no asymptotic gain in doing so. This is "one-step efficient estimation."

Suppose we have two estimators b_1 and b_2 for β, and

$$\sqrt{N}(b_j - \beta) = \frac{1}{\sqrt{N}}\sum_i \lambda_{ji} + \frac{1}{N}\sum_i \delta_{ji} + o_p(1), \quad j = 1, 2.$$

Then the asymptotic variances of $\sqrt{N}(b_1 - \beta)$ and $\sqrt{N}(b_2 - \beta)$ depend only on $V(\lambda_1)$ and $V(\lambda_2)$, respectively. If $V(\lambda_1) = V(\lambda_2)$, then b_1 and b_2 have the same *first-order efficiency*. The terms δ_1 and δ_2 will determine the *second-order efficiency*. More generally, if there are more terms with $N^{-3/2}$, $N^{-2}, \ldots$ attached, then we may consider higher-order efficiencies. The above result that taking one step from a $\sqrt{N}$-consistent estimator is enough for M-estimator is also based on the first-order efficiency. Repeating the iteration many times may yield different higher-order efficiencies. Unless otherwise noted, however, we will stick to the first-order efficiency in this book.

CHAPTER 4
NONLINEAR MODELS AND ESTIMATORS

Nonlinearity can be added to LSE at least in two ways. One is using a nonlinear regression function, and the other is using a loss function other than the squared one. Of particular interest is the absolute deviation loss function, which leads to "least absolute deviation (LAD) estimator" and median regression. Its generalization then yields quantile regression. The first-order conditions of these nonlinear minimization problems lead to IVE/GMM for nonlinear moment conditions. There are estimation principles other than MOM and M-estimation, and one of them—"minimum distance estimator" (and its special case "minimum χ^2 estimator")—is examined as well.

1 Nonlinear Least Squares Estimator (NLS)

Consider a nonlinear regression model

$$y_i = r(x_i, \beta) + u_i, \quad E(u|x) = 0,$$

where β is a $k \times 1$ parameter vector and the functional form of $r(\cdot)$ is known up to β; e.g. $r(x, \beta) = \exp(x'\beta)$. In contrast to the linear model, the dimension of x is not necessarily the same as that of β. Depending on cases, we may omit either x or β in $r(x, \beta)$. Since $y = r(x, \beta) + u$ includes the linear model $y = x'\beta + u$ as a special case when $r(x, \beta) = x'\beta$, we can estimate β by minimizing

$$\frac{1}{N} \sum_i \{y_i - r(x_i, b)\}^2$$

with respect to (wrt) b. The estimator is *nonlinear least squares estimator (NLS)*, which is a topic of this section.

A model more general than $y = r(x, \beta) + u$ is

$$\rho(y, x, \beta) = u, \quad E(u|x) = 0$$

which includes the above model as a special case when $\rho(y, x, \beta) = y - r(x, \beta)$. The model $\rho(y, x, \beta) = u$ allows the y part to depend on an unknown parameter, say α, to have $y^\alpha - r(x'\beta) = u$ for instance. But NLS is not applicable to this kind of models. To see why, consider minimizing $E\{h(y, a) - x'b\}^2$ wrt a and b where $h(y, a) = y^a$. The first-order condition at α and β is

$$2 \cdot E\left\{ u \frac{\partial h(y, \alpha)}{\partial a} \right\} = 0, -2 \cdot E(ux) = 0 \quad \text{where} \quad \frac{\partial h(y, \alpha)}{\partial a} \equiv \frac{\partial h(y, a)}{\partial a} \, | \, a = \alpha.$$

Myoung-jae Lee, *Micro-Econometrics*, DOI 10.1007/b60971_4,
© Springer Science+Business Media, LLC 2010

Since $\partial h(y,\alpha)/\partial a$ is a function of y and y includes u, the first equation is unlikely to hold. But nonlinear IVE/GMM with x or functions of x as instruments are still applicable.

1.1   Various Nonlinear Models

There are different ways to allow nonlinearity of $r(x,\beta)$ while still retaining linearity to some extent. Well-known models are "index models," "transformation-of-variable models," and "additive models", which are reviewed in this section. Those models will reappear in Chapters 8 and 9 in semiparametric (i.e., less parametric) contexts.

1.1.1   Index Models

In a (multiple) *index model*, x affects $E(y|x)$ through J number of linear indices $x'_{(j)}\beta_{(j)}$, $j = 1, ..., J$:

$$y = r\{x'_{(1)}\beta_{(1)}, ..., x'_{(J)}\beta_{(J)}\} + u$$

where $x_{(j)}$ is a subset of x (overlaps in $x_{(j)}$ and $x_{(m)}$ are allowed). Special cases of this are

$$y = \sum_{j=1}^{J} r_j\{x'_{(j)}\beta_{(j)}\} + u \quad \text{and} \quad y = \sum_{j=1}^{J} r\{x'_{(j)}\beta_{(j)}\} + u.$$

The simplest index model is *single index model*:

$$y = s(x'\beta) + u.$$

Here, the effect of x_k on y is gauged by

$$\frac{\partial s(x'\beta)}{\partial x_k} = \beta_k \cdot s'(x'\beta), \quad \text{where } s'(x'\beta) \equiv \frac{ds(x'\beta)}{d(x'\beta)}.$$

Although $\beta_k s'(x'\beta)$ depends on $x'\beta$, the relative effect

$$\frac{\partial s(x'\beta)/\partial x_j}{\partial s(x'\beta)/\partial x_k} = \frac{\beta_j}{\beta_k}$$

does not. Sometimes, models with only monotonic $s(\cdot)$ are called single index models.

An example of single index model is $s(x'\beta) = \exp(x'\beta)$, which is often used when y takes only non-negative values so that $E(y|x)$ should be positive. A more "elaborate" example occurs in a binary response model with a known error term distribution. Let F be a known df (e.g., "logistic": $F(a) = e^a/(1 + e^a)$), and

$$y_i = 1[x'_i\beta + u_i > 0], \quad \text{where } \frac{u_i}{\sigma} \sim F \text{ independently of } x_i$$

with $\sigma > 0$ unknown.

Then

$$E(y|x) = P(y = 1|x) = P(u > -x'\beta|x)$$

$$= P(\frac{u}{\sigma} > -x'\frac{\beta}{\sigma}|x) = 1 - F(-x'\frac{\beta}{\sigma});$$

here, $s(x'\beta) = 1 - F(-x'\beta/\sigma)$. Define $\alpha \equiv \beta/\sigma = (\beta_1/\sigma, ..., \beta_k/\sigma)'$ to get $E(y|x) = \tilde{s}(x'\alpha)$ where $\tilde{s}(x'\alpha) = 1 - F(-x'\alpha)$. Then the form of $\tilde{s}(\cdot)$ is fully known and we can estimate α (but not β).

1.1.2   Transformation-of-Variable Models

A *transformation-of-variable model* is

$$g_0(y) = g_1(x_1) \cdot \beta_1 + \cdots + g_k(x_k) \cdot \beta_k + u$$

where $g_j(\cdot)$ is a transformation, $j = 0, 1, ..., k$, which may be indexed by an unknown parameter α_j, $j = 0, 1, ..., k$. Often we restrict the transformation to be (strictly) monotonic; e.g., $g_0(y) = y^{1/2}$ or $\ln y$ where $y > 0$. One example is a "CES production function" $y^\alpha = \sum_{j=1}^{k} \beta_j x_j^\alpha + u$ where "power transformation" indexed by a single common parameter α is on y and x_j's.

A well-known transformation is *Box-Cox transformation* (Box and Cox, 1964): for $y > 0$,

$$g_0(y) = \frac{y^\alpha - 1}{\alpha} \text{ if } \alpha \neq 0 \quad \text{and} \quad g_0(y) = \ln(y) \text{ if } \alpha = 0.$$

Using the L'Hospital's rule, $(y^\alpha - 1)/\alpha \rightarrow \ln(y)$ as $\alpha \rightarrow 0$, because

$$\frac{d(y^\alpha - 1)/d\alpha}{d\alpha/d\alpha} = \frac{y^\alpha \ln y}{1} \rightarrow \ln y \text{ as } \alpha \rightarrow 0;$$

i.e., the transformation is continuous at $\alpha = 0$. This display holds in fact for $(y^\alpha - c)/\alpha$ with any constant c, but $c = 1$ makes $(y^\alpha - 1)/\alpha = 0 = \ln y$ when $y = 1 \; \forall\alpha$. If the transformation is applied to all variables $y, x_1, ..., x_k$ with the parameters $\alpha_0, ..., \alpha_k$, then all α_j's as well as β have to be estimated.

Often $(y^\alpha - 1)/\alpha = x'\beta + u$ is adopted with the assumption $u \sim N(0, \sigma^2)$ independently of x to apply MLE. But the normality assumption is not tenable unless $\alpha = 0$. To see this, let $\beta = 0$ for simplification, under which $(y^\alpha - 1)/\alpha$ should follow $N(0, \sigma^2)$. Suppose $\alpha > 0$. Then

$$P\left(\frac{y^\alpha - 1}{\alpha} \leq t\right) = P(y^\alpha \leq 1 + \alpha t).$$

For this probability to be positive because $P(N(0, \sigma^2) \leq t) > 0 \; \forall t$, we need $1 + \alpha t > 0 \Longleftrightarrow t > -1/\alpha$ because $y^\alpha > 0$: $(y^\alpha - 1)/\alpha$ is bounded from below by $-1/\alpha$. Analogously, if $\alpha < 0$, then

$$P\left(\frac{y^\alpha - 1}{\alpha} \geq t\right) = P(y^\alpha \leq 1 + \alpha t) \implies 1 + \alpha t > 0 \implies t < -\frac{1}{\alpha}:$$

$(y^\alpha - 1)/\alpha$ is bounded from above by $-1/\alpha$.

To relax the restriction $y > 0$, Bickel and Doksum (1981) suggested the following transformation: for $\alpha > 0$,

$$g_0(y) = \frac{|y|^\alpha \cdot sign(y) - 1}{\alpha}.$$

Also available is "shifted power-transformation":

$$g_0(y) = (y - \mu)^\alpha.$$

See Caroll and Ruppert (1988) for more on transformation, and Breiman and Friedman (1985) and Tibshirani (1988) on "optimal transformations." Also see MacKinnon and Magee (1990) for another alternative to the Box-Cox transformation.

1.1.3   Mean, Median, and More Nonlinear Models

One problem with transformed variables (and many nonlinear models) is that it becomes complicated to assess the effect of x on y. To see this, observe that $(y^\alpha - 1)/\alpha = x'\beta + u$ implies

$$y = (\alpha x'\beta + \alpha u + 1)^{1/\alpha} = (x'\alpha\beta + \alpha u)^{1/\alpha},$$

absorbing 1 into the intercept in $x'(\alpha\beta)$.

Although the response variable no longer has the parameter α with it, this model is not in the form $y = r(x, \alpha, \beta) + u$, and hence NLS is still not applicable. One may try to "force" the form $y = r(x, \alpha, \beta) + u$ by writing $y = E(y|x) + v$ with $v \equiv y - E(y|x)$ so that $E(v|x) = 0$, but getting

$$E(y|x) = \int \{x'(\alpha\beta) + \alpha u\}^{1/\alpha} f_{u|x}(u)du$$

$$\left(= \int \{x'(\alpha\beta) + \alpha u\}^{1/\alpha} f_u(u)du \quad \text{if } u \amalg x \right)$$

requires specifying $f_{u|x}$ and then doing the integration.

The condition $u \amalg x$ makes things a little easier though, because $E(y|x = x_o)$ can be estimated easily with

$$\frac{1}{N} \sum_i (x_o'\hat\alpha\hat\beta + \hat\alpha\hat u_i)^{1/\hat\alpha} \quad \text{where } \hat\alpha, \hat\beta, \hat u_i \text{ are estimators for } \alpha, \beta, u_i.$$

This estimator does not require specifying f_u, and $(\hat\alpha, \hat\beta', \hat u_i)$ can be obtained using nonlinear GMM and its residual; nonlinear GMM will be examined later. From this display, we can get the "mean-effect" at x_o

$$\frac{\partial E(y|x_o)}{\partial x}\Big|_{x=x_o} \simeq \frac{1}{N} \sum_i \frac{1}{\hat\alpha}(x_o'\hat\alpha\hat\beta + \hat\alpha\hat u_i)^{1/\hat\alpha\,-1} \cdot \hat\alpha\hat\beta$$

$$= \hat\beta \cdot \frac{1}{N} \sum_i (x_o'\hat\alpha\hat\beta + \hat\alpha\hat u_i)^{1/\hat\alpha\,-1}.$$

See Abrevaya (2002), Ai and Norton (2008) and the references therein if further interested in this literature.

Instead of focusing on mean, it is much simpler to use $Med(y|x)$ where $Med(y|x)$ stands for the median of $y|x$, because of the well-known fact

$$Med\{\tau(\cdot)\} = \tau\{Med(\cdot)\} \quad \forall \text{ increasing transformation } \tau(\cdot):$$
$$\tau(m_1) \leq \tau(m_2) \text{ iff } m_1 \leq m_2.$$

It may be easier to understand this with sample median as follows. Suppose that data on y is ordered such that $y_1 \leq, ..., \leq y_N$ with an even N. Let m_N be the sample median: $m_N = y_{N/2}$. Now consider $\tau(y) = \ln(y)$ for which $\ln(y_1) \leq, ... \leq \ln(y_N)$ holds. The sample median of the log-transformed data is then $\ln(y_{N/2}) = \ln(m_N)$. That is, the sample median of $\tau(y)$ equals τ(sample median) as in the last display which also holds for a "crude" transformation such as $\tau(y) = 1[y \geq 0]$.

The assumption $Med(u|x) = 0$ and $\alpha > 0$ gives $Med(y|x)$ and the "median effect" at x_o:

$$Med(y|x) = \{x'\alpha\beta + Med(\alpha u|x)\}^{1/\alpha} = (x'\alpha\beta)^{1/\alpha}$$
$$\implies \quad \frac{\partial Med(y|x)}{\partial x}|_{x=x_o} = \beta(x'_o\alpha\beta)^{1/\alpha \, -1} \simeq \hat{\beta} \cdot (x'_o\hat{\alpha}\hat{\beta})^{1/\hat{\alpha} \, -1}.$$

This is reminiscent of the above mean-effect estimator under $u \amalg x$.

Both the mean-effect and median-effect estimators become the constant $\hat{\beta}$ regardless of x_o when $\hat{\alpha} = 1$ for the usual linear model. Differently from the mean case, however, the median effect does not require $u \amalg x$. Also we can rewrite the model as $y = (x'\alpha\beta)^{1/\alpha} + v$ with $v \equiv y - Med(y|x)$ so that $Med(v|x) = 0$, to which "nonlinear least absolute deviation estimation" can be applied minimizing $N^{-1}\sum_i |y_i - (x'_i\alpha\beta)^{1/\alpha}|$. More generally than median, we may look at quantiles; this topic as well as the details of estimating $Med(y|x)$ will be studied later.

To overcome the problem with mean effect under Box-Cox transformation of y, Wooldridge (1992a) specified

$$E(y|x) = (1 + \alpha x'\beta)^{1/\alpha} \text{ if } \alpha \neq 0 \quad \text{and} \quad E(y|x) = \exp(x'\beta) \quad \text{if } \alpha = 0.$$

This may be taken as "the inverse Box-Cox transformation applied to $x'\beta$," and the transformation is continuous at $\alpha = 0$, because

$$\lim_{\alpha \to 0} (1 + \alpha x'\beta)^{1/\alpha} = \lim_{\alpha \to 0} (1 + \frac{x'\beta}{1/\alpha})^{1/\alpha} = \exp(x'\beta).$$

Here α and β can be estimated by NLS as the model can be written in the form $y = E(y|x) + v$ with $v \equiv y - E(y|x)$. Getting $\partial E(y|x)/\partial x$ now is only as difficult as getting $\partial Med(y|x)/\partial x$ in the preceding paragraph. Also observe the mean-effect at x:

$$\frac{\partial E(y|x)}{\partial x} = \beta \frac{(1 + x'\alpha\beta)^{1/\alpha}}{1 + x'\alpha\beta} \quad \text{if } \alpha \neq 0 \quad \text{and}$$

$$\frac{\partial E(y|x)}{\partial x} = \beta \exp(x'\beta) \quad \text{if } \alpha = 0.$$

The two expressions coincide as $\alpha \to 0$, using the preceding display.

Since the set of all polynomial functions on $[a_1, a_2]$ with rational coefficients can approximate any continuous function on $[a_1, a_2]$ arbitrarily well, if $s(\cdot)$ is unknown in a single index model, then we may consider a polynomial in $x'\beta$ or in $g(x)'\beta$ where $g(x) \equiv (g_1(x_1), ..., g_k (x_k))'$: for some M,

$$y = \sum_{m=1}^{M} \gamma_m (x'\beta)^m + u \quad \text{or} \quad y = \sum_{m=1}^{M} \gamma_m \{g(x)'\beta\}^m + u.$$

It is possible to further classify nonlinear models. For instance, $r(x) = \sum_{j=1}^{k} r_j(x_j)$ is an "additive model"; $r(x) = r_1(x_1, ..., x_j) + r_2(x_{j+1}, ..., x_k)$ is a "partially additive model." Since it is cumbersome to treat these models one by one, we will discuss only the general models $y = r(x, \beta) + u$ or $\rho(y, x, \beta) = u$ in the remainder of this chapter unless otherwise necessary.

1.2   NLS and Its Asymptotic Properties

The first-order condition of NLS is

$$\frac{1}{N} \sum_i -2\{y_i - r(x_i, b)\} r_b(x_i, b) = 0 \quad \text{where} \quad r_b(x, b) \equiv \frac{\partial r(x, b)}{\partial b}.$$

With this, NLS may be viewed as a method-of-moment estimator (MOM) with the population moment condition

$$E[\{(y - r(x, \beta)\} r_b(x, \beta)] = E\{u \cdot r_b(x, \beta)\} = 0.$$

Differently from LSE, however, *there could be many solutions to the moment condition even when $E\{(y - r(x, b))^2\}$ has a unique minimizer*; viewing everything as MOM has its peril. In practice, we may end up with many estimates from MOM. In this case, the one minimizing $E[\{y - r(x, b)\}^2]$ should be chosen. That is, the M-estimator framework provides a guidance on which one to choose. For a pure MOM which is not the first-order condition of a M-estimator, no such guidance is available.

Let $Q(b)$ denote the population version $E\{y - r(x, b)\}^2$ for $N^{-1} \sum_i \{y_i - r(x_i, b)\}^2$. Assume that r is twice continuously differentiable wrt b. The first two derivatives of $Q(b)$ are

$$Q_b(b) \equiv -2E[\{y - r(b)\} r_b(b)];$$

$$Q_{bb'}(b) \equiv 2E\{r_b(b) \cdot r_{b'}(b)\} - 2E[\{y - r(b)\} r_{bb'}(b)] \quad \text{where}$$

$$r_{bb'}(b) \equiv \frac{\partial r(b)}{\partial b \partial b'}.$$

The following three conditions together are sufficient for the identification of β in NLS: for a parameter space B,

$$Q_b(\beta) = 0, \quad Q_{bb'}(\beta) \text{ is p.d. and } Q_{bb'}(b) \text{ is p.s.d. for any } b \in B.$$

The first two conditions together make β a local minimum, and the last condition assures that β is an unique global minimum.

Observe that

$$Q_b(\beta) = -2E\{u \cdot r_b(\beta)\} = 0 \text{ and } Q_{bb'}(\beta)$$
$$= 2E\{r_b(\beta)r_{b'}(\beta)\} \text{ is p.s.d always.}$$

Assuming that $E\{r_b(\beta)r_b(\beta)'\}$ is of full rank analogously to the assumption that $E(xx')$ is of full rank in LSE, $Q_{bb'}(\beta)$ is p.d. Hence the two conditions, $Q_b(\beta) = 0$ and $Q_{bb'}(\beta)$ being p.d, are easily satisfied, and β is a local minimum. The third condition—$Q_{bb'}(b)$ being p.s.d. for any $b \in B$—is impossible to check unless $r(b)$ is specified.

The asymptotic distribution of NLS is straightforward using the asymptotics of M-estimator:

$$\sqrt{N}(b_{nls} - \beta) \rightsquigarrow N\{0, E^{-1}(r_b r_b') \, E(r_b r_b' u^2) \, E^{-1}(r_b r_b')\}$$
$$\implies \sqrt{N}(b_{nls} - \beta) \rightsquigarrow N(0, \sigma^2 E^{-1}(r_b r_b')) \quad \text{if } E(u^2|x) = \sigma^2$$

(homoskedasticity).

With $r_b = x$, the asymptotic variance of NLS becomes that of LSE.

Let Φ be the $N(0,1)$ df. In the binary response model

$$y_i = 1[x_i'\beta + u_i > 0], \text{ and } \frac{u_i}{\sigma} \sim N(0,1) \text{ independently of } x_i \text{ where}$$
$$\sigma > 0 \text{ is unknown,}$$

it holds that $E(y|x) = P(y = 1|x) = P(u/\sigma > x'\beta/\sigma) = \Phi(x'\alpha)$. Hence

$$y = \Phi(x'\alpha) + v, \quad \text{where } \alpha \equiv \frac{\beta}{\sigma} \text{ and } v \equiv y - \Phi(x'\alpha);$$

$E(v|x) = 0$ by construction. As for the variance of $v|x$,

$$
\begin{aligned}
V(v|x) &= E[\{y - E(y|x)\}^2 \, |x] = E[\{y - \Phi(x'\alpha)\}^2 \, |x] \\
&= E\{y^2 - 2y\Phi(x'\alpha) + \Phi(x'\alpha)^2 \, |x\} \\
&= E(y|x) - 2E(y|x)\Phi(x'\alpha) + \Phi(x'\alpha)^2 \\
&= \Phi(x'\alpha)\{1 - \Phi(x'\alpha)\}, \quad \text{as } y^2 = y.
\end{aligned}
$$

Thus v *has heteroskedasticity of a known form.*

Suppose we apply NLS to $y = \Phi(x'\alpha) + v$ and get the NLS estimator a_{N0}. Then we can estimate $V(v_i|x_i)$ with $w_i^2 \equiv \Phi(x_i'a_{N0})\{1 - \Phi(x_i'a_{N0})\}$. Transform $y_i = \Phi(x_i'\alpha) + v_i$ into

$$\frac{y_i}{w_i} = \frac{\Phi(x_i'\alpha)}{w_i} + \frac{v_i}{w_i}.$$

Applying NLS, we get a more efficient estimator. This two-stage procedure is a *weighted NLS (WNLS)* where the first-stage estimation error $a_{N0} - \alpha$ does not affect the second stage as in GLS. The asymptotic variance matrix of this WNLS is

$$E^{-1}\left[\frac{\phi(x'\alpha)\{y - \Phi(x_i\alpha)\}^2 x}{\Phi(x_i'\alpha)\{1 - \Phi(x_i'\alpha)\}}\right].$$

As can be seen later, this is the asymptotic variance of the corresponding MLE for the binary response model, illustrating that WNLS can be as efficient as MLE. Sometimes a simple weighing does "wonders."

EXAMPLE: PERFORMANCE EFFECT ON PAY. Schaefer (1998) examined pay-performance relation for CEOs. Using an "agency-model for optimal contract", Schaefer (1998) derived a nonlinear model linking CEO compensation to the firm value: for a panel data with a small T,

$$y_{it} = \alpha_t + \frac{V_{it}}{1 + \beta S_{i,t-1}^\gamma} + u_{it}, \quad i = 1, ..., N, \, t = 1, ..., T$$

where y_{it} is a compensation measure (salary plus bonus, or pay-related CEO wealth), S_{it} is a size of the firm (market value or assets) and V_{it} is the value of the firm (shareholder wealth), and α_t, β, γ are parameters to estimate. The data are for large American firms between 1991 and 1995; NT is about 3000 to 4000, depending on the variables in use.

Not just the nonlinear model, other models were used also in Schaefer (1998) including the linear one $y_{it} = \alpha_t + \beta_l V_{it}$ where the slope β_l of V_{it} does not depend on the firm size S_{it}, whereas the slope of V_{it} does depends on S_{it} inversely in the nonlinear model. In the Schaefer's model, β is in fact a product of three structural form parameters: the CEO's coefficient of absolute risk aversion, the second derivative of the CEO's cost of effort function, and the variance of the firm's market value; also, γ is a difference of two structural form parameters.

When the nonlinear model was estimated with industry dummies used as well, the result in Table 1 with SD in $(\cdot)$ was obtained (in fact, Schaefer (1998) took the first-difference and estimated the differenced model with NLS, but this aspect is irrelevant for our discussion here).

Other than for Pay-Related Wealth with Market Value as Firm Size, both β and γ are significantly different from 0 rejecting the linear model, and γ seems to be about 0.4. $R^2 \equiv 1 - \sum_i \hat{u}_i^2 / \sum_i (y_i - \bar{y})^2$ where $\hat{u}_i \equiv y_i - r(b_N)$ hovers around 0.1, which is rather low.

Table 1: CEO Compensation to Firm Value

y_{it}	Market Value as Firm Size		Assets as Firm Size	
	Salary+Bonus	Pay-Related Wealth	Salary+Bonus	Pay-Related Wealth
β	878 (432)	38.0 (41)	448 (200)	1.92 (0.58)
γ	0.382 (0.054)	0.0760 (0.124)	0.462 (0.054)	0.362 (0.032)
R^2	0.108	0.083	0.115	0.120

1.3   Three Tests for NLS

Testing a linear hypothesis $H_0 : R'\beta = c$ where $rank(R) = g \leq k$ with NLS is similar to that with MLE: there are three kinds of tests corresponding to Wald, Lagrangian Multiplier (LM), and Likelihood Ratio (LR) tests. It goes without saying that there are also three analogous tests available for linear models, as linear models are special cases of nonlinear models.

Let b_N be a NLS with $\sqrt{N}(b_N - \beta) \rightsquigarrow N(0, C)$. Defining $\hat{u}_i \equiv y_i - r(b_N)$, C can be estimated consistently with

$$C_N \equiv \left\{ \frac{1}{N} \sum_i r_b(b_N) r_{b'}(b_N) \right\}^{-1} \cdot \left\{ \frac{1}{N} \sum_i \hat{u}_i^2 r_b(b_N) r_{b'}(b_N) \right\}$$
$$\cdot \left\{ \frac{1}{N} \sum_i r_b(b_N) r_{b'}(b_N) \right\}^{-1}.$$

The Wald test statistic is

$$Wald_{nls} \equiv N(R'b_N - c)'(RC_N R')^{-1}(R'b_N - c) \rightsquigarrow \chi_g^2.$$

If $E(u^2|x) = \sigma^2$, then C becomes $\sigma^2 E^{-1}\{r_b(\beta) r_b(\beta)'\}$ and

$$C_N = \frac{1}{N} \sum_i \hat{u}_i^2 \cdot \left\{ \frac{1}{N} \sum_i r_b(b_N) r_{b'}(b_N) \right\}^{-1}.$$

In the following, we present LM and LR type test statistics for this homoskedastic case. The derivations are analogous to LM and LR test derivations for MLE.

The LR-type test statistic for NLS under homoskedasticity is

$$LR_{nls} \equiv \frac{\sum_i \{y_i - r(x_i, b_{Nr})\}^2 - \sum_i \{y_i - r(x_i, b_N)\}^2}{N^{-1} \sum_i \{y_i - r(x_i, b_N)\}^2} \rightsquigarrow \chi_g^2 \quad \text{where}$$

$$b_{Nr} = \{I_k - H^{-1}R(R'H^{-1}R)^{-1}R'\}b_N + H^{-1}R(R'H^{-1}R)^{-1}c \quad \text{and}$$

$$H \equiv E\{r_b(\beta) r_b(\beta)'\};$$

recall that the b_{Nr} for MLE takes the same form with H replaced by I_f^{-1}. Also recall

$$LR_{mle} \equiv 2\{L(\text{unrestricted MLE}) - L(\text{restricted MLE})\} \quad \text{where}$$

$$L \text{ is the log-likelihood.}$$

Compared with LR_{mle}, LR_{nls} does not have the number 2 but has the denominator $\hat{\sigma}^2 \equiv N^{-1} \sum_i \{y_i - r(x_i, b_N)\}^2$ which is not present in LR_{mle}. First, the reason why LR_{mle} has the number 2 is that the second-order Taylor's expansion of L yields

$$\{L(\text{unrestricted MLE}) - L(\text{restricted MLE})\} \simeq \frac{1}{2} Wald_{mle};$$

multiplying both sides by 2 yields LR_{mle}. But in LR_{nls}, the exponent 2 in NLS minimand cancels $1/2$ in the second-order expansion, which is why there is no number 2 in LR_{nls}. Second, while the second-order expansion of L yields the second-order matrix which is the inverse of the MLE asymptotic variance, the second-order expansion of the NLS minimand gives $E\{r_b(\beta)r_{b'}(\beta)\}$ which becomes the inverse of the NLS asymptotic variance only when divided by σ^2 under homoskedasticity. The denominator in LR_{nls} is for σ^2.

In the linear model with $u \sim N(0, \sigma^2)$ and x being non-random, there is an exact (not asymptotic) F test:

$$\frac{1}{g} \frac{\sum_i \{y_i - r(x_i, b_{Nr})\}^2 - \sum_i \{y_i - r(x_i, b_N)\}^2}{(N-k)^{-1} \sum_i \{y_i - r(x_i, b_N)\}^2} \sim F(g, N-k).$$

It is well known that $g \cdot F(g, \infty) \sim \chi_g^2$; i.e., g times a rv following $F(g, \infty)$ follows χ_g^2. Hence, $g \cdot F(g, N-k) \rightsquigarrow \chi_g^2$ as $N \to \infty$. Using this fact and multiplying the F test statistic by g, we get also LR_{nls}. That is, under homoskedasticity, the exact F-test is asymptotically equivalent to LR_{nls}; in large samples, CLT replaces the normality assumption. LR_{nls} appears in many textbooks as, with "SS" standing for "sum of squares"

$$\frac{(\text{restricted error SS}) - (\text{unrestricted error SS})}{\hat{\sigma}^2} \rightsquigarrow \chi_g^2 \quad \text{where } \hat{\sigma}^2$$

$$\equiv \frac{1}{N} \sum_i \{y_i - r(x_i, b_N)\}^2.$$

Turning to the LM-type test, we have

$$LM_{nls} \equiv \sum_i \tilde{u}_i r_{b'}(b_{Nr}) \left\{ \sum_i \tilde{u}_i^2 r_b(b_{Nr}) r_{b'}(b_{Nr}) \right\}^{-1} \sum_i \tilde{u}_i r_b(b_{Nr}) \rightsquigarrow \chi_g^2$$

which is reminiscent of the LM test in MLE. For a linear model with $y = m'\beta_m + w'\beta_w + u$ and $H_0 : \beta_w = 0$, we get, defining $x \equiv (m', w')'$,

$$b_N = \left(\frac{1}{N} \sum_i x_i x_i' \right)^{-1} \left(\frac{1}{N} \sum_i x_i y_i \right)^{-1},$$

$$b_{Nr} = \left(\frac{1}{N} \sum_i m_i m_i' \right)^{-1} \left(\frac{1}{N} \sum_i m_i y_i \right)^{-1}$$

$$\tilde{u}_i = y_i - m_i' b_{Nr}, \quad r_b(b_{Nr}) = x_i, \quad \text{because } r_b(b) = x_i \text{ for any } b.$$

1.4   Gauss–Newton Algorithm

One well-known way to implement NLS is the *Gauss–Newton algorithm*. Taylor expand $r(b)$ around b_0, an initial estimator:

$$r(b) \simeq r(b_0) + r_{b'}(b_0)(b - b_0).$$

Substitute this into $r(b)$ and minimize the following wrt b:

$$\frac{1}{2N} \sum_i \{y_i - r(b_0) - r_{b'}(b_0)(b - b_0)\}^2.$$

The first-order condition is

$$-\frac{1}{N} \sum_i r_b(b_0)\{y_i - r(b_0) - r_{b'}(b_0)(b - b_0)\} = 0.$$

Solve this for b and denote the solution by b_1 to get an iteration scheme:

$$b_1 = b_0 - \left\{ \frac{1}{N} \sum_i r_b(b_0)r_{b'}(b_0) \right\}^{-1} \cdot \frac{1}{N} \sum_i -r_b(b_0)\{y_i - r(b_0)\}.$$

Repeat this until a stopping criterion is met.

The Gauss–Newton method may be viewed somewhat differently. Replace b by β in $r(b) \simeq r(b_0)+r_{b'}(b_0)(b-b_0)$ to get $r(\beta) \simeq r(b_0)+r_{b'}(b_0)(\beta-b_0)$, which is then substituted into $y = r(\beta) + u$ to yield

$$y - r(b_0) + r_{b'}(b_0)b_0 = r_{b'}(b_0)\beta + u.$$

Treat the left-hand side as a new dependent variable and $r_b(b_0)$ as the regressor. Applying LSE to this model, we get

$$b_1 = \left\{ \frac{1}{N} \sum_i r_b(b_0)r_{b'}(b_0) \right\}^{-1} \cdot \frac{1}{N} \sum_i r_b(b_0)\{y - r(b_0) + r_{b'}(b_0)b_0\}$$

$$= b_0 + \left\{ \frac{1}{N} \sum_i r_b(b_0)r_{b'}(b_0) \right\}^{-1} \cdot \frac{1}{N} \sum_i r_b(b_0)\{y - r(b_0)\}$$

which is nothing but the Gauss–Newton algorithm. That is, the Gauss–Newton algorithm is equivalent to applying LSE repeatedly to the linearized version of the nonlinear model.

An analog of Gauss–Newton algorithm was also used in MLE. Recall the Newton–Raphson algorithm with the second-order matrix approximated by the outer-product of the score function times minus one:

$$\text{new estimate} = \text{old estimate} - (-\text{outerproduct})^{-1} \cdot (\text{gradient}).$$

This version has essentially the same form as the above NLS iteration scheme except in two aspects. One is that the second-order matrix is n.d. here, whereas it is p.d. in the NLS iteration scheme. The other is that the inverted matrix in Gauss–Newton is not exactly the outer-product of the first-order derivative vector.

1.5   NLS-LM Test for Linear Models*

In this subsection, a specification test for linear regression functions is introduced. The test is motivated by the LM test for NLS with a single-index model and transformed regressors, and generalizes the popular RESET test of Ramsey (1969). Rejection of the specification test would suggest using power transformations of regressors or adding high order terms of regressors.

Define

$$x_i(\alpha)'\beta \equiv x_{i1}(\alpha_1)\beta_1 + x_{i2}(\alpha_2)\beta_2 +, \cdots, +x_{ik}(\alpha_k)\beta_k,$$
$$\text{where } x_{i1}(\alpha_1) = 1 \text{ always,}$$

$$x_{ij}(\alpha_j) \equiv \frac{x_{ij}^{\alpha_j} - 1}{\alpha_j} \text{ if } \alpha_j \neq 0 \quad \text{and} \quad \ln(x_{ij}) \text{ if } \alpha_j = 0, \ j = 2, ..., k;$$

if x_{ij} can be non-positive, set $x_{ij}(\alpha_j) = x_{ij}$. Consider a polynomial single-index regression function: for some M,

$$y = \sum_{m=1}^{M} \gamma_m \{x(\alpha)'\beta\}^m + u.$$

This general alternative will be used below when we test the null linear model with the parameter values

$$\alpha = (1, ..., 1)', \quad \gamma_1 = 1, \ \gamma_2 = 0, \ ..., \ \gamma_m = 0.$$

Imagine estimating $\alpha = (\alpha_2, ..., \alpha_k)'$, β, and $\gamma = (\gamma_2, ..., \gamma_M)'$ with NLS by maximizing ($a_1 = 1$ and $g_1 = 1$ always)

$$Q_N \equiv -\frac{1}{2N} \sum_{i=1}^{N} \left[y_i - \sum_{m=1}^{M} g_m \{x_i(a)'b\}^m \right]^2$$

wrt a, b, and g. The first derivatives evaluated at the true values are (assume that all regressors are transformed to simplify notations):

$$\frac{\partial Q_N}{\partial g_m} = \frac{1}{N} \sum_i u_i \{x_i(\alpha)'\beta\}^m, \quad m = 2, ..., M$$

$$\frac{\partial Q_N}{\partial a_j} = \frac{1}{N} \sum_i \left[u_i \beta_j \left\{ \frac{x_{ij}^{\alpha_j} \ln(x_{ij})}{\alpha_j} - \frac{x_{ij}^{\alpha_j} - 1}{\alpha_j^2} \right\} \sum_{m=1}^{M} \gamma_m m \{x_i(\alpha)'\beta\}^{m-1} \right],$$
$$j = 2, ..., k$$

$$\frac{\partial Q_N}{\partial b_j} = \frac{1}{N} \sum_i \left[u_i x_{ij}(\alpha_j) \sum_{m=1}^{M} \gamma_m m (x_i(\alpha)'\beta)^{m-1} \right], \quad j = 1, 2, ..., k.$$

Substitute the null linear model parameter values into the gradient to
set the gradient equal to 0. This yields the moment conditions

(i) $N^{-1}\sum_i u_i (x_i'\beta)^m = 0, \quad m = 2, ..., M$

(ii) $N^{-1}\sum_i u_i \beta_j \{x_{ij}\ln(x_{ij}) - x_{ij} - 1\} = 0, \quad j = 2, ..., k$

(iii) $N^{-1}\sum_i u_i x_i = 0.$

A LM-type linear regression function specification test is possible, because
LSE is obtained only with (iii). The test examines if the LSE also satisfies (i)
and (ii). Substituting (iii) into (ii), (ii) becomes $N^{-1}\sum_i u_i \beta_j x_{ij}\ln(x_{ij}) = 0$.
Here β_j is irrelevant because using $x_{ij}\ln(x_{ij})$ as an extra regressor is the same
as using $\beta_j\{x_{ij}\ln(x_{ij})\}$—the "scale change" β_j does not alter the t-value for
$x_{ij}\ln(x_{ij})$.

Specifically, the LM-type test procedure is the following.

- Step 1: Estimate the linear model with LSE.

- Step 2: Set up the following artificial regression model

$$y = x'\beta + \sum_{m=2}^{M}\delta_m (x'b_N)^m + \sum_{j=2}^{k}\theta_j\{x_j\ln(x_j)\} + \varepsilon,$$

  to estimate β, δ_m's, and θ_j's with LSE.

- Step 3: Test "$H_0 : \delta_m = 0$ for $m = 2, ..., M$, and $\theta_j = 0$ for $j = 2, ..., k$"
  in the artificial model using the heteroskedasticity-robust covariance
  matrix for

$$E^{-1}(zz') \cdot E(\varepsilon^2 zz') \cdot E^{-1}(zz')$$

  where z denotes the regressors in Step 2.

Typically $M = 2$ or 3 with a number of $x_j\ln(x_j)$ will be enough. Without
the terms in $\sum_{j=2}^{k}(\cdot)$, the test becomes the RESET test of Ramsey (1969).

2   Quantile and Mode Regression

In the preceding section, nonlinearity came from the regression function.
But, as we have seen already in M-estimator, nonlinearity can come also from
the "external" part of the optimand even when the "internal" regression function itself is linear. This section examines such nonlinear estimators; they are
also M-estimators just as most estimators are. We start with "median regression" which is "one-degree less smooth" than the usual mean regression.
Median regression is generalized to quantile regression. Then "mode regression" which is "one-degree less smooth" than median regression is introduced.
We will also touch on "treatment effect" and "sample selection" issues in this
section.

2.1   Median Regression

Consider predicting y with a function $r(x)$ of x. Under the quadratic loss function of mis-prediction, we seek to find $r(x)$ minimizing

$$E\{y - r(x)\}^2 = E_x E_{y|x}\{y - r(x)\}^2.$$

Differentiate $E_{y|x}\{y - r(x)\}^2$ wrt $r(x)$ and set the derivative to zero to get

$$-2 \cdot E_{y|x}\{y - r(x)\} = 0 \iff E(y|x) = r(x).$$

Since this holds for any x, $E\{y - r(x)\}^2$ is minimized when $r(x) = E(y|x)$. Since the second derivative is positive (i.e., 2), the minimizer is unique.

In the quadratic loss function, a misprediction is penalized by the squared distance. More generally, we can think of minimizing $E|y - q(x)|^p$, $p > 0$, where the predictor is denoted as $q(x)$ now. There is no a priori reason to set $p = 2$ except for analytic convenience. Perhaps it is more intuitive to set $p = 1$ and minimize

$$E|y - q(x)| = E_x E_{y|x}|y - q(x)|$$

$$= E_x \left[\int_{-\infty}^{q(x)} \{q(x) - y\}f(y|x)dy + \int_{q(x)}^{\infty} \{y - q(x)\}f(y|x)dy \right]$$

where $y|x$ is assumed to be continuously distributed and $f(y|x)$ is the density.

Note the Leibniz's rule:

$$\frac{d\int_{a(t)}^{b(t)} g(x,t)dx}{dt} = \int_{a(t)}^{b(t)} \frac{\partial g(x,t)}{\partial t}dx + \frac{db(t)}{dt}g\{b(t),t\} - \frac{da(t)}{dt}g\{a(t),t\}.$$

Differentiate $E|y - q(x)|$ wrt $q(x)$ using Leibniz's rule to get the first-order condition

$$E\left\{ \int_{-\infty}^{q(x)} f(y|x)dy - \int_{q(x)}^{\infty} f(y|x)dy \right\} = 0.$$

This first-order condition is satisfied by $q(x) = Med(y|x)$, where Med stands for median. The second derivative is $2 \cdot E\{f(q(x)|x)\}$. Thus assuming that $E\{f(q(x)|x)\} > 0$, the minimizer is unique.

With $q(x) = x'\beta$, the estimator obtained by minimizing the sample analog of $E|y - q(x)|$

$$\frac{1}{N}\sum_i |y_i - x_i'b|$$

is the least absolute deviation estimator (LAD) as appeared once already. LAD has as long a history as LSE has. In LAD, we get to estimate the conditional *median regression function*. In this regard, we use the word "regression" for any location measure in the conditional distribution of $y|x$, not just for $E(y|x)$. As in NLS, we may use a nonlinear function of b, instead of $x'b$, for LAD. The LAD minimand is piecewise linear and continuous as a function of $x'_i b$, which is "one-degree less smooth" than the minimand of LSE that is quadratic.

If the distribution of $y|x$ is asymmetric as in income distribution, then the mean and median can differ much. In this case, estimating the median would be as much of interest as estimating the mean. In fact, compared with mean, median has the well-known advantage of being robust to outliers. To see this, imagine observations (y_i's) on a horizontal axis, 50% to the left of the median m and the other 50% to the right. Suppose we take some observations on the lhs to $-\infty$ and some on the rhs to ∞. This, however, does not affect the median m, and median can resist almost up to 50% data contamination. This fact for the marginal distribution of y applies to the median regression: median regression is robust to outliers in y (but not in x).

For regression analysis, certainly robustness matters a great deal. But the real attraction of median regression comes from its generalization to "quantile regression" in the following subsection. Namely, we can estimate not just the center (median) of the $y|x$ distribution, but also the lower and upper quantiles. It is possible that some components of x affect only a tail area, but not necessarily the center. Viewed differently, suppose that the effect of a regressor x_k depends on the level of y—e.g., the effect is greater for low y. Dependence of x_k-effect on, say, x_{k-1} can be easily taken into account by adding the interaction term $x_{k-1}x_k$ into $x'\beta$. But the dependence on y cannot. One may think of quantile regression as a way to take the dependence on y-level into account since using $x_k y$ as a regressor would not make sense: the effect for low (high) y level can be seen by low (high) quantiles in the $y|x$ distribution. More on this will be seen in the following.

2.2 Quantile Regression

2.2.1 Asymmetric Absolute Loss and Quantile Function

Generalizing median regression, suppose we use an *asymmetric absolute loss function* which penalizes positive and negative prediction errors differently. With the predictor denoted as $q_\alpha(y|x)$ for some α with $0 < \alpha < 1$, the loss function is

$$\alpha|y - q_\alpha(y|x)| \text{ if } y - q_\alpha(y|x) > 0 \text{ and } (1-\alpha)|y - q_\alpha(y|x)| \text{ if } y - q_\alpha(y|x) < 0.$$

This includes the symmetric absolute loss function for median regression as a special case when $\alpha = 0.5$. The expected loss becomes

$$E\{\alpha(y - q_\alpha(y|x)) \, 1[y > q_\alpha(y|x)] + (1-\alpha)(q_\alpha(y|x) - y)1[y < q_\alpha(y|x)]\}$$

$$= E\left[\alpha \int_{q_\alpha(y|x)}^{\infty} \{y - q_\alpha(y|x)\}f(y|x)dy + (1-\alpha)\int_{-\infty}^{q_\alpha(y|x)} \{q_\alpha(y|x) - y\}f(y|x)dy\right].$$

Before we show that the αth quantile of $y|x$ minimizes this, we define quantile function formally.

For a df $F(t) \equiv P(y \leq t)$ of a rv y, define its *quantile function* $F^{-1}(\cdot)$ as

$$F^{-1}(\alpha) \equiv \min\{t : F(t) \geq \alpha\}, \quad 0 < \alpha < 1.$$

The domain of $F^{-1}(\alpha)$ is $(0,1)$; otherwise the minimum does not exist (i.e., $\min\{t : F(t) \geq 0\}$ is not bounded from below). It follows that

$$F^{-1}(\alpha) \leq t_o \iff \alpha \leq F(t_o);$$

in words, if we know $t_\alpha \equiv F^{-1}(\alpha) \leq t_o$, then $F(t_o)$ must be at least as large as α because t_α is the smallest value to accumulate the probability mass α, and if we know $\alpha \leq F(t_o)$ for some t_o, t_o must be at least as large as t_α. This holds for any df including discontinuous ones. It can be shown that $F\{F^{-1}(\alpha)\} = \alpha$ for $0 < \alpha < 1$ iff F is continuous, and $F^{-1}\{F(t)\} = t$ $\forall t \in R$ iff F is strictly increasing; hence, F^{-1} *is the usual (one-to-one and onto) inverse of F iff F is continuous and strictly increasing* (see, e.g., Van der Vaart 1998).

The last display implies that, for any $u \sim U(0,1)$,

$$P\{F^{-1}(u) \leq t\} = P\{u \leq F(t)\} = F(t) :$$

the *quantile transformation* $F^{-1}(u)$ is a rv with df F. Now assume that F is continuous. Then $F(y)$ follows $U[0,1]$ because

$$F(t) \equiv P(y \leq t) = P\{F(y) \leq F(t)\}.$$

Although the second equality does not require F to be continuous, if F is not continuous with a jump at y_o, then $F(y)$ is not continuously distributed. Rather, $F(y)$ has a probability mass at $F(y_o)$ and thus $F(y)$ cannot be $U[0,1]$. $F(y)$ is called a *probability integral transform*.

Turning back to the expected loss function, differentiate it wrt $q_\alpha(y|x)$ to get the first-order condition

$$E[-\alpha \int_{q_\alpha(y|x)}^{\infty} f(y|x)dy + (1 - \alpha) \int_{-\infty}^{q_\alpha(y|x)} f(y|x)dy] = 0.$$

Suppose that $y|x$ is continuously distributed. If $q_\alpha(y|x) = F_{y|x}^{-1}(\alpha)$ where $F_{y|x}^{-1}(\alpha)$ denotes the quantile function of $y|x$, then $\int_{-\infty}^{q_\alpha(y|x)} f(y|x)\,dy = \alpha$ and the first-order condition is satisfied because $-\alpha(1 - \alpha) + (1 - \alpha)\alpha = 0$. The second derivative is $E_x\{f(q_\alpha(y|x)|x)\}$: the minimizer is unique, assuming $E_x\{f(q_\alpha(y|x)|x)\} > 0$. Therefore, under the asymmetric loss function, the αth quantile minimizes the expected loss. Newey and Powell (1987) explored asymmetric quadratic loss functions, calling the minimizing location measure "expectile"; see also Efron (1991).

In view of the notation $q_\alpha(y|x)$, $q(x)$ for median regression could have been denoted $q_{0.5}(y|x)$. Bear in mind that $q_\alpha(y|x)$ is not a function of y, although it is a function of α and x. As mentioned above, if $F_{y|x}(t)$ is continuous and strictly increasing in t, then the αth quantile can be defined simply as $q_\alpha(y|x) = F_{y|x}^{-1}(\alpha)$ where $F_{y|x}^{-1}(\cdot)$ is the usual inverse of $F_{y|x}(\cdot)$. If $F_{y|x}$ is continuous but not strictly increasing, then $F_{y|x}$ has a flat portion which makes $F_{y|x}$ many-to-one, not one-to-one; if $F_{y|x}$ is discontinuous, then $F_{y|x}$ is not "onto." In these cases, we need the above "generalized inverse" $\min\{t : F_{y|x}(t) \geq \alpha\}$. Since the continuous and strictly increasing case is easier to understand, unless otherwise mentioned, we will assume from now on that the df under consideration is continuous and strictly increasing whenever quantiles are examined.

For the linear model $y = x'\beta + u$, we get

$$q_\alpha(y|x) = x'\beta + q_\alpha(u|x) = \beta_1 + \tilde{x}'\tilde{\beta} + q_\alpha(u|x) \quad \text{where } x = (1, \tilde{x}')';$$

i.e., β_1 is the intercept and $\tilde{\beta}$ is the slope vector. If $q_\alpha(u|x) = 0$, then $q_\alpha(y|x) = x'\beta$; otherwise, $q_\alpha(y|x)$ consists of $x'\beta$ and $q_\alpha(u|x)$. If u is independent of x, then

$$q_\alpha(y|x) = \beta_1 + \tilde{x}'\tilde{\beta} + q_\alpha(u) = \{\beta_1 + q_\alpha(u)\} + \tilde{x}'\tilde{\beta} :$$

the quantiles are parallel to one another, being intercept-shifted versions of $\tilde{x}'\tilde{\beta}$ as α varies.

If u is dependent on x, however, then the quantiles are no longer parallel. For instance, suppose $u = (x'\theta) \cdot v$ where v is an error term independent of x—a heteroskedasticity model with $SD(u|x) = |x'\theta|$. Then

$$q_\alpha(y|x) = \quad x'\beta + q_\alpha(u|x) = x'\beta + x'\theta \cdot q_\alpha(v) = x'\beta_\alpha, \quad \text{where}$$

$$\beta_\alpha \;\equiv\; \beta + \theta \cdot q_\alpha(v), \quad \text{i.e.,} \quad \beta_{\alpha j} = \beta_j + \theta_j q_\alpha(v), \; j = 1, ..., k.$$

If $q_{0.5}(v) = 0$, then $\beta_{0.5} = \beta$. If $E(v) = 0$ instead of $q_{0.5}(v) = 0$, then $E(y|x) = x'\beta$ as $E(u|x) = (x'\theta)E(v) = 0$. Note that, as $y|x$ is $x'\beta$-shifted version of $u|x$, $f_{u|x}(0) = f_{y|x}(x'\beta)$, and more generally, $f_{u|x}\{q_\alpha(u|x)\} = f_{y|x}\{x'\beta + q_\alpha(u|x)\}$.

We mentioned earlier that quantile regression may be viewed as a way of allowing the effect of x to depend on y-level, i.e., x interacting with y. The last display shows better what it means: the effect of x depends on the quantile level of u and x interacts with itself in the scale parameter. Suppose $q_{0.5}(v) = 0$ and $\theta_j > 0$. This implies

$$\beta_{\alpha j} = \beta_j + \theta_j q_\alpha(v) < \beta_j \; \forall \alpha < 0.5 \quad \text{because } \theta_j > 0 \text{ and } q_\alpha(v) < 0;$$

$$\beta_{\alpha j} = \beta_j + \theta_j q_\alpha(v) > \beta_j \; \forall \alpha > 0.5 \quad \text{because } \theta_j > 0 \text{ and } q_\alpha(v) > 0.$$

Hence, it is certainly possible that the median effect is zero because $\beta_j = 0$ while the other quantile effects are negative or positive. When u is independent of x, with 1 included in x, all θ_j's are zero except θ_1; this is the above case of parallel quantiles. Note that, if we assume just $q_{\alpha_o}(u|x) = 0$ for some α_o, not the more structured $u = (x'\beta)v$ and $v \amalg x$, then we can state only

$$q_\alpha(y|x) = x'\beta + q_\alpha(u|x) \ \{\simeq x'(\beta + \delta_\alpha) \quad \text{if } q_\alpha(u|x) \simeq x'\delta_\alpha\}.$$

2.2.2   Quantile Regression Estimator

For a given α, to simplify notation, denote β_α just as β:

$$y = x'\beta + u \quad \text{where } q_\alpha(u|x) = 0 \text{ and } u|x \text{ has density } f_{u|x}.$$

Then the sample version to minimize for the αth *quantile regression* is

$$\frac{1}{N}\sum_i \{\alpha(y_i - x_i'b)\,1[y_i > x_i'b] + (1-\alpha)(x_i'b - y_i)\,1[y_i < x_i'b]\}$$

$$= \frac{1}{N}\sum_i (y_i - x_i'b)\cdot(\alpha - 1[y_i - x_i'b < 0]).$$

The function $t\cdot(\alpha - 1[t<0])$ is called a "*check function*." Differently from LSE, this cannot be solved for b. But the minimand is convex in $x'b$, and thus a Gauss–Newton-type optimization with numerical (two-sided) gradients works fairly well in practice. See Koenker and Bassett (1978), Koenker and Hallock (2001), and Koenker (2005) for more on quantile regression, which is getting popular these days.

As it turns out, the quantile regression estimator satisfies the "asymptotic first-order condition" obtained by differentiating the minimand while treating $1[y_i - x_i'b < 0]$ as a constant:

$$\frac{1}{\sqrt{N}}\sum_i -(\alpha - 1[y_i - x_i'b < 0])x_i \simeq 0$$

$$\Longrightarrow \quad \frac{1}{\sqrt{N}}\sum_i -\frac{1}{2}sgn(y_i - x_i'b)x_i \simeq 0 \text{ when } \alpha = 0.5.$$

As intuitively explained in the following paragraph, the asymptotic variance is

$$E^{-1}\{f_{u|x}(0)\ xx'\}\cdot\alpha(1-\alpha)E(xx')\cdot E^{-1}\{f_{u|x}(0)\ xx'\}$$

$$= \frac{\alpha(1-\alpha)}{f_u(0)^2}\cdot E^{-1}(xx') \quad \text{(under the independence of } u \text{ from } x\text{)}$$

$$= \frac{1}{4f_u(0)^2}\cdot E^{-1}(xx') \quad \text{(further under } \alpha = 0.5\text{)}.$$

The presence of $f_{u|x}(0)$ in the second-order matrix can be understood in view of $E\{f_{y|x}(q_\alpha(y|x))\}$ $(= E\{f_{y|x}(x'\beta)\}$ when $q_\alpha(y|x) = x'\beta)$ that appeared in the second-order condition for the population minimand of quantile regression.

Another way of understanding the presence of $f_{u|x}(0)$ is "differentiating" the asymptotic first-order condition for the case $\alpha = 0.5$. Although $sgn(u)$ is not differentiable in $-N^{-1/2}\sum_i 0.5sgn(y_i - x_i'b)x_i$, it changes by 2 (from -1 to 1) at $u = 0$. Hence, take $2\delta_0(u)$ as the derivative of $sgn(u)$ where $\delta_0(u) = 1$ if $u = 0$ and 0 otherwise. The population version of the second-order derivative matrix is then

$$-E\{-\frac{1}{2}2\delta_0(u)\cdot xx'\} = E\{\delta_0(u)\cdot xx'\} \simeq E\{f_{u|x}(0)\cdot xx'\}.$$

As for $\alpha(1-\alpha)$, it comes from the outer-product in the asymptotic first-order condition at $b = \beta$: since $P(u < 0|x) = \alpha$ in $y = x'\beta + u$ with $q_\alpha(y|x) = x'\beta$,

$$\frac{1}{N}\sum_i (\alpha - 1[y_i - x_i'\beta < 0])^2 x_i x_i' \to^p E\{(\alpha - 1[u < 0])^2 xx'\}$$

$$= \quad E\{ E(\alpha^2 - 2\alpha 1[u < 0] + 1[u < 0]|x)\cdot xx' \}$$

$$= \quad E\{(\alpha^2 - 2\alpha^2 + \alpha)xx'\} = \alpha(1-\alpha)E(xx').$$

Getting the residual $r_i \equiv y_i - x_i b_N$ where b_N is the αth quantile linear regression estimator, an estimator for $E\{f_{u|x}(0)xx'\}$ is

$$\frac{1}{N}\sum_i \frac{1[-h < r_i < h]}{2h}x_i x_i' \quad \text{where } h \to 0 \text{ as } N \to \infty.$$

The part $1[\cdot]/(2h)$ is to estimate $f_{u|x}(0)$, and the role of h is that of the grouping interval in a histogram construction. In practice, h should be a small positive number, e.g. $SD(r)\cdot N^{-1/5}$, such that there be a non-trivial proportion of residuals with $-h < r_i < h$.

A smooth alternative to the last display is

$$\frac{1}{N}\sum_i \frac{1}{h}\phi(\frac{r_i}{h})\cdot x_i x_i'$$

where ϕ is the $N(0,1)$ density. In practice, using this for

$$h = 0.5\cdot SD(r)\cdot N^{-1/5}, \quad h = SD(r)\cdot N^{-1/5} \quad \text{and} \quad h = 2\cdot SD(r)\cdot N^{-1/5}$$

and reporting all three results may be adequate as the choice of h is more or less arbitrary. As a way to avoid choosing h, a "nonparametric bootstrap" resampling from the original sample can be applied to construct confidence intervals. Due to $f_{u|x}$, the asymptotic variance and estimation of quantile regression will be further discussed after nonparametrics is covered.

2.2.3   Empirical Examples

EXAMPLE: SMOKING EFFECT ON BIRTH WEIGHT. Abrevaya (2001) applied quantile regression to find the effects of maternal characteristics on birth weight measured in grams. Using the Natality Data Set from the National Center for Health Statistics for 1992 June births, he compiled a data set of $N = 199,108$ with singleton births in the US for white or black women aged 18–45. A part of his Table 2 is Table 2 (SD in $(\cdot)$).

Table 2: Effects on Birth Weight (in grams)

	10%	50%	90%	LSE
black	−253 (8.1)	−199 (4.2)	−182 (5.5)	−220 (3.9)
no smoke	171 (13)	159 (6.5)	147 (9.2)	161 (5.7)
college	82.9 (8.6)	37.5 (5.2)	−3.2 (7.9)	37.2 (4.6)
no prenatal visit	−389 (55)	−145 (14)	−102 (21)	−194 (16)

In the table, the effect of college is relative to less-than-highschool education, and no-prenatal-visit is relative to visit in the first trimester. The different quantile effects across the tails (10% and 90%) and the center (50%) are clearly visible. In no-smoke and college, LSE is almost the same as the median regression, but in black and no prenatal visit, LSE differs nontrivially from the median regression. Black and no-prenatal-visit have large negative effects while no-smoke has a large positive effect. College has a large effect only at 10% and a nearly zero effect at 90%. Of course, the effects of black and college are unlikely to be genuine; rather, they reflect the care/behavior that the mothers give to the unborns. Whether or not the effect of no-smoke is real is also debatable, as it can also reflect the mother's care/behavior. If it is real, quitting smoking will have effects as in the table; otherwise, quitting smoking will have no effect unless the mother changes her care/behavior.

The smoking variable demonstrates why economists pay attention to endogeneity problems. The key point is that ignoring endogeneity and applying LSE (or median regression) leads to estimating

$$\beta_{no-smoke} + \beta_{good-care} \frac{COV(\text{no-smoke, good-care})}{V(\text{no-smoke})}$$

using the well-known LSE omitted variable bias formula. The above positive effect of no smoking might have been obtained because

$$\beta_{no-smoke} = 0, \quad \beta_{good-care} > 0 \quad \text{and} \quad COR(\text{no-smoke, good-care}) > 0.$$

That is, there may be no genuine effect of no smoking. Taking no-smoking as a policy variable, if one only quits smoking, then there will be no effect on birth weight; it is only when the mother changes her "good-care" that the birth weight increases. Taking care of the endogeneity problem will lead to estimating $\beta_{no-smoke}$ correctly, and if the policy maker sees $\beta_{no-smoke} = 0$, then the no smoking policy will not be adopted.

Koenker and Geling (2001) applied quantile regression to duration data on medflies with $N > 1,000,000$. They found that the mortality rate declines at advanced ages. They also found that males have lower motality rates than females up to the 95-percentile age, and then have higher mortality after that. Compared with human mortality, these findings are quite surprising. Koenker and Geling also examined the effects of other variables such as the initial density in medfly cages and proportion of males and so on. Overall, they presented findings that would not have been seen, had quantile regression not been employed and had N not been so large to permit investigation into the extreme right tail area. Chernozhukov (2005) examined "extremal quantile regression" where $\alpha \to 0$ and αN either diverge or converge to a constant.

In the above birth-weight study, some regressors are possibly endogenous. There exist two-stage versions of quantile regression estimator that can handle endogenous regressors. See Kim and Muller (2004) and the references therein. An application can be seen in Kan and Tsai (2004). Since the moment condition from the first-order condition cannot be solved for the error term, IVE versions are infeasible, and this problem holds also for M-estimator in general. Finally, Angrist et al. (2006, p. 542) accorded an weighted-LSE-approximation to linear quantile regression functions, and presented (in p. 547) an omitted-variable bias formula for quantile regression.

2.3   Mode Regression

Suppose that the loss function is $1[|y - m(x)| > \delta]$ where δ is a known positive constant and the predictor is denoted now as $m(x)$. That is, if the prediction falls within $\pm\delta$ of y, then there is no loss; otherwise, the loss is one. The expected loss is

$$E\{1[|y - m(x)| > \delta]\} = 1 - E\{1[|y - m(x)| \le \delta]\}$$
$$= 1 - E_x[F_{y|x}\{m(x) + \delta\} - F_{y|x}\{m(x) - \delta\}].$$

This is minimized by choosing $m(x)$ such that the interval $[m(x) - \delta, m(x) + \delta]$ captures the most probability mass under $f_{y|x}$. Manski (1991) calls it "δ-mode." If $f_{y|x}$ is unimodal and δ is small, then $m(x)$ is approximately equal to the mode of $f_{y|x}$.

Lee's (1989) *mode regression* estimator maximizes the following sample analog:

$$\frac{1}{N} \sum_i 1\left[|y_i - x_i' b| \le \delta\right].$$

Following Kim and Pollard (1990), it can be shown that this estimator is $N^{1/3}$-consistent with no practical asymptotic distribution. Compared with the optimands of LSE and LAD, the optimand here (as a function of $x_i' b$) consists of flat lines and is discontinuous, thus "one-degree less smooth" than the piecewise linear and continuous optimand of LAD. Since each observation is given only the weight 0 or 1, the influence of any outlier can be no more than its share $1/N$. Mode regression estimator is robust to outliers in x as well as in y, differently from LAD or LSE.

If the reader thinks that distinction between various measures of central tendency in a distribution is trivial, consider the following litigation in Freedman (1985). In early 1980s, a number of lawsuits were filed by railroad companies against several states' taxing authorities in USA. The companies argued that their property tax rate should be equalized to the median of the rates for the other property tax rates, while the state authorities argued that the mean is more appropriate. The problem was that the probability distribution of the property tax rate has a long right tail. As a result, the median was smaller than the mean, and the difference had an implication of millions of dollars. Eventually, the states won the case, not because mean is a better measure than median, but because the Courts concluded that the word "average" in the law (so-called "4-R Act") meant mean, not median.

2.4   Treatment Effects

Having seen many different regressions such as mean, quantile, and mode regressions, here we digress a little to discuss "treatment effects," as there are various treatment effects such as mean and quantile effects. Detailed discussion on treatment effects can be found in Lee (2005a).

Often we want to know the effect(s) of a "treatment" or "cause" d_i on a response (or outcome) variable of interest y_i; the effects are called "treatment effects" or "causal effects." The following is examples of treatments and responses:

Treatment:	job training	college education	drug	exercise
Response:	wage	lifetime earnings	cholesterol	blood pressure

Let y_i^j, $j = 0, 1$, denote the "potential outcome" when individual i receives treatment j exogenously (i.e., when treatment j is forced in ($j = 1$) or out ($j = 0$), in comparison to treatment j self-selected by the individual). For the drug/cholesterol example,

y_i^1: cholesterol level with drug "forced in";

y_i^0: cholesterol level with drug "forced out".

A person may choose to take the drug (*self-selection*), but we can imagine the drug getting injected regardless of the person's will (*intervention*). The interest is on the effect of a treatment intervened, not self-selected.

Among the two potential outcomes corresponding to the two potential treatments, only one outcome is observed while the other (called "counterfactual") is not. For example, in the effect of college education on life time earnings, only one outcome (earnings with college education or earnings without) is available per person. One may argue that for some other cases, say drug on cholesterol, both y_i^1 and y_i^0 can be observed sequentially. But strictly speaking, if two treatments are administered one by one sequentially, we cannot say that we observe both y_i^1 and y_i^0, for the subject changes over time, however little the change may be. The *observed response* y_i is

$$y_i = (1 - d_i) \cdot y_i^0 + d_i \cdot y_i^1.$$

Imagine that person i is endowed with (y_i^0, y_i^1) but shows either y_i^0 and y_i^1 depending on $d_i = 0$ or 1. In a given data set, the group with $d_i = 1$ revealing y_i^1 is the *treatment group*, whereas the group with $d_i = 0$ revealing y_i^0 is the *control group*.

Define

the treatment effect for individual $i : y_i^1 - y_i^0$

which is, however, not identified, because only one of y_i^1 and y_i^0 can be observed, never both together. *Causal relation is different from associative relation* such as correlation or covariance: we need (d_i, y_i^0, y_i^1) in the former to get $y_i^1 - y_i^0$, while we need only (d_i, y_i) in the latter. Of course, an associative relation suggests a causal relation.

Although the individual treatment effect is not identified, its mean version $E(y_i^1 - y_i^0)$ is identified under one assumption. Suppose that y^0 and y^1 are mean-independent of d: $E(y^j|d) = E(y^j)$, $j = 0, 1$. *Under the mean-independence, the mean treatment effect is identified with the group-mean difference*:

$$E(y|d = 1) - E(y|d = 0) = E(y^1|d = 1) - E(y^0|d = 0)$$
$$= E(y^1) - E(y^0) = E(y^1 - y^0).$$

Randomized experiments are often used in clinical trials (e.g., cancer patients are randomly assigned to treatment and control groups to see whether or not a drug is effective for the cancer), and the mean-independence condition holds for such randomized experiments where d is assigned with "coin flips." Other than for randomized experiments, the condition may hold if d is forced on people by a law or regulation for reasons unrelated to y^0 and y^1 ("quasi experiment") or by nature such as weather and geography ("natural experiments").

Not just mean effect, we can also think of other effects, such as median effect $Med(y^1 - y^0)$ or more generally quantile effect $q_\alpha(y^1 - y^0)$. Sometimes different effects agree. For instance, suppose

$$y_i^0 \;=\; \beta_1 + u_i \text{ and } y_i^1 = \beta_1 + \beta_d + u_i$$
$$\{ \;\implies\; y_i = (1 - d_i) \cdot y_i^0 + dy_i^1 = \beta_1 + \beta_d d_i + u_i \}.$$

In this case, the treatment effect for individual i is $y_i^1 - y_i^0 = \beta_d$, which is a constant and thus the same for all i. Hence all mean, median, and mode effects agree, as the distribution of $y_i^1 - y_i^0$ is degenerate at β_d. Even if $y_i^1 - y_i^0 = \beta_{di}$ that varies across i, if β_{d_i} is symmetric around β_d, then at least the mean and median agree.

Suppose the last display holds and we have

$$M(y|d = 0) = \beta_1 + M(u|d = 0) \text{ and } M(y|d = 1) = \beta_1 + \beta_d + M(u|d = 1)$$

where $M(y|d)$ can be any of mean, median (or αth quantile), or mode. From this, the identified "M treatment effect" is

$$M(y|d = 1) - M(y|d = 0) = \beta_d + M(u|d = 0) - M(u|d = 1)$$

whereas the true effect is β_d. Suppose $M(y|d) = q_\alpha(y|d)$ and the df for $y|d = 0$ and $y|d = 1$ are denoted, respectively, as F_0 and F_1. Then we get

$$M(y|d = 1) - M(y|d = 0) = F_1^{-1}(\alpha) - F_0^{-1}(\alpha)$$

which is a vertical difference between F_1^{-1} and F_0^{-1}; see Lee (2000) if further interested in quantile treatment effect.

The identified effect $M(y|d = 1) - M(y|d = 0)$ equals the true effect β_d only when

$$M(u|d = 0) - M(u|d = 1) = 0$$

which holds if u is independent of d. $M(u|d = 0) \neq M(u|d = 1)$ is an endogeneity problem because u differs systematically across the control and treatment groups; i.e., u is related to d in $y = \beta_1 + \beta_d d + u$. This kind of endogeneity problem due to the relation between u and a binary group indicator d is often called a "*sample selection problem*" (or "selected-sample problem") as will be discussed further in a later chapter.

When d is self-selected, not intervened, often this problem occurs because people different in u (genes or abilities) make different choices. If the two groups differ only in some observed variables x, then this observed difference may be controlled by using $y = \beta_1 + \beta_d d + \beta_x x + u$; in contrast, the difference in the unobserved u is hard to handle. When u is related to d, there is a scope that certain location measures may be less affected (i.e., biased) than others by this dependence. For instance, it may happen that $E(u|d = 0) \neq E(u|d = 1)$, but $q_\alpha(u|d = 0) = q_\alpha(u|d = 1)$ for some α. This sort of perspective prompted various semiparametric methods exploiting different location measures as will be seen in a later chapter.

3   GMM for Nonlinear Models

GMM has appeared already many times. In this section, GMM for nonlinear moment conditions will be studied. As IVE appeared for GMM with linear models, nonlinear IVE will appear in this section as well. The first-order conditions for M-estimator and NLS may be taken as moment conditions, to which nonlinear GMM may apply. But it may not be possible to go the other way around. For instance, a nonlinear moment condition for GMM may not be convertible to NLS if the moment condition is not solvable for y.

3.1 GMM for Single Nonlinear Equation

Let

$$\underset{s\times 1}{E\,\psi\,(z;\,\underset{k\times 1}{\beta\,})} = 0$$

be a $s \times 1$ moment condition where β is a $k \times 1$ vector. For instance, in NLS for $y = r(x,\beta) + u$,

$$E\psi(z,\beta) = 0 \iff E[r_b(x,\beta)\{y - r(x,\beta)\}] = 0, \quad \text{where } r_b \equiv \frac{\partial r}{\partial b}.$$

The sample version is

$$\frac{1}{N}\sum_i \psi(z_i,\beta) = 0 \iff \frac{1}{N}\sum_i r_b(x_i,\beta)\{y_i - r(x_i,\beta)\} = 0.$$

Regardless of $\psi(\beta)$ being linear or not, if $\{z_i\}$ is iid, then the most efficient way to combine s moment restrictions $E\psi(z,\beta) = 0$ is to minimize the quadratic form

$$Q_N(b) \equiv \frac{1}{N}\sum_i \psi(b)' \left\{ \frac{1}{N}\sum_i \psi(b_0)\psi(b_0)' \right\}^{-1} \frac{1}{N}\sum_i \psi(b)$$

wrt b, where b_0 is an initial consistent estimator such that

$$W_N \equiv \frac{1}{N}\sum_i \psi(b_0)\psi(b_0)' \to^p W \equiv V\left[\frac{1}{\sqrt{N}}\sum_i \psi(\beta)\right].$$

Intuitively, whatever metric we may use on R^s, only its quadratic approximation matters for the asymptotic variance. Hence GMM indexed by the weighting matrix W_N is a large enough class for the moment condition. For b_0, NLS or some IVE may be used; the first-stage estimation error $b_0 - \beta$ does not affect the second-stage GMM, because the error appears only in the weighting.

Earlier we deferred dealing with a general nonlinear model $\rho(y,x,\beta) = u$. Suppose β is estimated by minimizing $N^{-1}\sum_i \rho(y_i,x_i,b)^2$ as in NLS. The first-order condition is

$$\frac{1}{N}\sum_i \left\{ \rho(y_i,x_i,b)\frac{\partial\rho(y_i,x_i,b)}{\partial b} \right\} = 0.$$

Setting $\psi(y,x,b) = \rho(y,x,b)\cdot\partial\rho(y,x,b)/\partial b$ leads to GMM. But one caution should be noted before we proceed further: even when the minimizer of $N^{-1}\sum_i \rho(y_i,x_i,b)^2$ is unique, there may be multiple solutions to the first-order condition. Thus, although GMM provides a convenient unifying framework, if there are multiple solutions, then we may have to revert back to the original M-estimator setup.

Turning to the asymptotic distribution of GMM, it cannot be derived as a special case of the M-estimator asymptotic distribution, because the minimand is not of the form $N^{-1}\sum_i q(z_i, b)$. Still, Taylor expansion of the first-order condition yields the asymptotic distribution. For a general weighting matrix W_N, not just the optimal one, the first-order condition is

$$\frac{1}{N}\sum_i \psi_b(b_N)W_N^{-1}\frac{1}{\sqrt{N}}\sum_i \psi(b_N) = 0, \quad \text{where } \psi_b \equiv \frac{\partial\psi}{\partial b}.$$

Taylor-expand $N^{-1/2}\sum_i \psi(b_N)$ around β to get, for some $b_N^* \in (b_N, \beta)$,

$$\frac{1}{\sqrt{N}}\sum_i \psi(\beta) + \frac{1}{N}\sum_i \psi_{b'}(b_N^*)\sqrt{N}(b_N - \beta).$$

Substitute this into the first-order condition to get

$$
\begin{aligned}
0 &= \frac{1}{N}\sum_i \psi_b(b_N)W_N^{-1}\left\{\frac{1}{\sqrt{N}}\sum_i \psi(\beta) + \frac{1}{N}\sum_i \psi_{b'}(b_N^*)\sqrt{N}(b_N - \beta)\right\}\\
&= E\psi_b(\beta)\cdot W^{-1}\frac{1}{\sqrt{N}}\sum_i \psi(\beta) + E\psi_b(\beta)\cdot W^{-1}E\psi_{b'}(\beta)\\
&\quad \cdot\sqrt{N}(b_N - \beta) + o_p(1).
\end{aligned}
$$

Hence,

$$\sqrt{N}(b_N - \beta) = -(E\psi_b W^{-1}E\psi_{b'})^{-1}E\psi_b W^{-1}\cdot\frac{1}{\sqrt{N}}\sum_i \psi(\beta) + o_p(1)$$

$$
\begin{aligned}
&\rightsquigarrow N\{0, (E\psi_b W^{-1}E\psi_{b'})^{-1}\cdot E\psi_b W^{-1}E\psi\psi' W^{-1}E\psi_{b'}\\
&\quad \cdot(E\psi_b W^{-1}E\psi_{b'})^{-1}\}.
\end{aligned}
$$

Choosing $W = E\psi\psi'$ simplifies the asymptotic variance matrix, yielding the most efficient one for the GMM class:

$$\sqrt{N}(b_{gmm} - \beta) \rightsquigarrow N(0, \{E\psi_b\cdot E^{-1}(\psi\psi')\cdot E\psi_{b'}\}^{-1}).$$

If $E\psi_b$ is invertible with $rank(E\psi_b) = s = k$, then the GMM asymptotic variance takes the "sandwich form" $E^{-1}(\psi_b)E(\psi\psi')E^{-1}(\psi_{b'})$ regardless of W; the asymptotic distribution is the same as the one that would be obtained by applying the mean value theorem to $N^{-1/2}\sum_i \psi(b_{gmm}) = 0$. Hence, in this case, we can just use the "unweighting" $W = I_s$. This result is analogous to that GMM becomes LSE when $s = k$ in the linear model; i.e., if $s = k$, then there is nothing to gain by weighting the moment conditions. For the nonlinear model $y = r(\beta)+u$ with the $k\times 1$ moment condition $E\{r_b(\beta)u\} = 0$, NLS becomes GMM, and the asymptotic variance of $\sqrt{N}(b_{nls} - \beta)$ can be obtained by substituting $\psi = r_b(\beta)\{y - r(\beta)\}$ and $E\psi_b = E\{r_b(\beta)r_{b'}(\beta)\}$

into $E\psi_b E^{-1}(\psi\psi')E\psi_{b'}$. Hence, the relationship between GMM and LSE in the linear model holds between GMM and NLS in the nonlinear model.

When $s > k$, if the moment conditions are correct, then GMM should not just minimize $Q_N(b)$ but also make $Q_N(b_{gmm}) \to^p 0$. Hansen (1982) suggested a *GMM over-identification ("over-id") test* examining whether the moment conditions more than k can be satisfied by only k many parameters. The test statistic is

$$\frac{1}{\sqrt{N}}\sum_i \psi(b_{gmm})' \left\{ \frac{1}{N}\sum_i \psi(b_0)\psi(b_0)' \right\}^{-1} \frac{1}{\sqrt{N}}\sum_i \psi(b_{gmm}) \rightsquigarrow \chi^2_{s-k}.$$

Any consistent estimator can be used for the middle inverted matrix, and using b_0 means that the over-id test statistic can be obtained simply from the GMM minimand. In practice, however, to assure that the test statistic is p.s.d., it is better to use

$$J_N \equiv \frac{1}{\sqrt{N}}\sum_i \psi(b_{gmm})' \left\{ \frac{1}{N}\sum_i \psi(b_{gmm})\psi(b_{gmm})' \right\}^{-1}$$
$$\frac{1}{\sqrt{N}}\sum_i \psi(b_{gmm}) \rightsquigarrow \chi^2_{s-k}.$$

The dof of the over-id test can be intuitively understood as follows. First, observe

$$\frac{1}{\sqrt{N}}\sum_i \psi(\beta)' \left\{ \frac{1}{N}\sum_i \psi(b_0)\psi(b_0)' \right\}^{-1} \frac{1}{\sqrt{N}}\sum_i \psi(\beta) \rightsquigarrow \chi^2_s.$$

Second, minimizing $Q_N(b)$ wrt b to get b_{gmm}, k-many moment conditions are used up (in the first-order condition). Third, plugging b_{gmm} back into its minimand leaves only $s - k$ dof.

GMM (more broadly IVE) is not valid for errors-in-variable models that are nonlinear in x, whereas IVE is applicable to errors-in-variable models linear both in x and in β. To see this, consider $y_i = r(x_i^*, \beta) + u_i$ where $r(x^*, \beta)$ is nonlinear in x^*, $x_i = x_i^* + v_i$, and (x', y) is observed. Then, for some $x_i^{o*} \in (x_i^*, x_i)$, denoting the gradient of $r(x^*, \beta)$ wrt x^* as $r_x(x^*, \beta)$,

$$y_i = r(x_i, \beta) + r_x(x_i^{o*})(x_i^* - x_i) + u_i = r(x_i, \beta) + \{u_i - r_x(x_i^{o*})v_i\}.$$

Even if there is an instrument z_i that is correlated with x_i and uncorrelated with u_i and v_i, the new error term $u_i - r_x(x_i^{o*})v_i$ is correlated with z_i through $r_x(x_i^{o*})$ as x_i^{o*} is correlated with x_i.

3.2 Implementation and Examples

GMM can be implemented by taking one step from an initial $\sqrt{N}$-consistent estimator; one may also iterate further if desired. The one-step formula comes from the above expression for $\sqrt{N}(b_N - \beta)$: replace β with b_0, put b_0 on the right-hand side and use an optimal weighting to get

$$b_N \;=\; b_0 - \left[\sum_i \psi_b(b_0) \left\{ \sum_i \psi(b_0)\psi(b_0)' \right\}^{-1} \sum_i \psi_{b'}(b_0) \right]^{-1}$$

$$\cdot \sum_i \psi_b(b_0) \left\{ \sum_i \psi(b_0)\psi(b_0) \right\}^{-1} \sum_i \psi(b_0).$$

If there is no such initial consistent estimator, then GMM estimation can be implemented iteratively by starting with an arbitrary b_0 and updating b_0 until convergence.

It is also possible to minimize

$$\frac{1}{N}\sum_i \psi(b)' \left\{ \frac{1}{N}\sum_i \psi(b)\psi(b)' \right\}^{-1} \frac{1}{N}\sum_i \psi(b)$$

wrt b as in Hansen et al. (1996); note that b appears also in the weighting matrix. This procedure, called "continuously updated GMM," is computationally more difficult, although it seems to have a better small-sample performance in connection with (generalized) "empirical likelihood method"; we will not further examine the continuously updated GMM. In the following, we present specific nonlinear moment condition examples, and then an empirical example that is a shortened version of the nonlinear GMM example in Lee (2002).

Suppose $y \geq 0$ always, and thus we posit $E(y|x) = \exp(x'\beta)$ to assure $E(y|x) > 0$. In this case, we may use

$$E\psi(x, y, b) = E[x\{y\exp(-x'b) - 1\}] = 0, \quad \text{because}$$

$$= E[\, x \cdot E\{y\exp(-x'b) - 1|x\} \,]$$

$$= E[\, x\{\exp(x'\beta - x'b) - 1\} \,] = 0 \text{ if } \beta = b;$$

i.e., β satisfies the moment condition. Here, $y\exp(-x'b) - 1$ plays the role of an error term orthogonal to x. In applying GMM, we will need $\psi_b(\beta) = -xx'y\exp(-x'\beta)$. Instead of the moment condition in the last display, we may also use

$$E[x\{y - \exp(x'b)\}] = 0$$

where $y - \exp(x'b) = y - E(y|x)$ is an error term; this will appear later for count responses. The preceding moment condition is a weighted version of this moment condition because

$$x\{y\exp(-x'b) - 1\} = x\frac{y - \exp(x'b)}{\exp(x'b)}.$$

This demonstrates that there are different moment conditions for the same parameters. Also note that, instead of "x times an error," "functions of x" times an error can be used for the moment conditions.

Another example of GMM moment conditions comes from the Box-Cox transformation model $(y^\alpha - 1)/\alpha = x'\beta + u$. As noted already, MLE under $u \sim N(0, \sigma^2)$ is not tenable for this model. But GMM can be applied to estimate α and β jointly. Under $E(u|x) = 0$, we can use

$$E(ux_j) = 0 = E(ux_j^2)$$

$$\Longleftrightarrow \; E\left\{ (\frac{y^\alpha - 1}{\alpha} - x'\beta)x_j \right\} = 0 = E\left\{ (\frac{y^\alpha - 1}{\alpha} - x'\beta)x_j^2 \right\} \; \forall j.$$

This GMM may not converge well due to estimating α. One simple solution is to fix α to get $\hat\beta(\alpha)$ in the resulting linear model with GMM. Repeating this for all possible values of α and then comparing the GMM minimand values, we can find $\hat\alpha$ and $\hat\beta(\hat\alpha)$. Whenever possible, it is advantageous to turn a nonlinear problem into a linear one.

EXAMPLE: INTEREST RATE (continued). Consider a "stochastic differential equation" for an interest rate $y(t)$:

$$dy(t) = \{\beta_1 - \beta_2 y(t)\} \cdot dt \; + \; \sigma y(t)^\eta \cdot dw(t)$$

where β_1, β_2, σ, and η are positive parameters, and $w(t)$ for $0 \le t \le T$ is a "*Brownian motion*" or "*Wiener process*": $w(t)$ has continuous sample path that starts from 0 (i.e., $w(0) = 0$) and for any finite number of points $t_1, ..., t_m$,

$$\{w(t_1), ..., w(t_m)\} \text{ is Gaussian (i.e., jointly normal) with}$$
$$\text{0-mean and } E\{w(t_j)w(t_k)\} = \min(t_j, t_k).$$

Setting $t_j = t_k = t$, this display implies $w(t) \sim N(0, t)$ because $V\{w(t)\}$ $= E\{w(t)^2\} = t$. From this display, it follows that

$$w(t) - w(s) \sim N(0, \, t - s), \quad \forall \, s < t \quad \text{because}$$

$$E\{w(t) - w(s)\}^2 = E\{w(t)^2\} - 2E\{w(t)w(s)\} + E\{w(s)^2\}$$

$$= t - 2s + s = t - s.$$

Also $w(t)$ has independent increments:

$$w(t_4) - w(t_3) \text{ is independent of } w(t_2) - w(t_1), \quad \forall \, t_1 < t_2 \le t_3 < t_4$$

because $E[\{w(t_4) - w(t_3)\}\{w(t_2) - w(t_1)\}] = t_2 - t_2 - t_1 + t_1 = 0.$

Approximate the differential equation in discrete time with

$$y_t \; \simeq \; y_{t-1} + (\beta_1 - \beta_2 y_{t-1})\Delta t + \sigma y_{t-1}^\eta (\Delta t)^{1/2}\varepsilon_t, \quad [\{\varepsilon_t\} \text{ iid } N(0,1)]$$

$$= \; y_{t-1} + \beta_2(\beta_1/\beta_2 - y_{t-1})\Delta t + \sigma y_{t-1}^\eta (\Delta t)^{1/2}\varepsilon_t.$$

The stationary level for y_t (with $\varepsilon_t = 0$) is β_1/β_2. The second term on the right-hand side makes y_t revert to β_1/β_2, and β_2 is the "speed of adjustment." With $\Delta t = 1$, the discrete version becomes

$$y_t = y_{t-1} + \beta_1 - \beta_2 y_{t-1} + \sigma y_{t-1}^{\eta} \varepsilon_t.$$

We will estimate the four parameters β_1, β_2, σ, and η with GMM.
   Define

$$u_t \equiv \sigma y_{t-1}^{\eta} \varepsilon_t = y_t - y_{t-1} - (\beta_1 - \beta_2 y_{t-1}), \quad v_t \equiv u_t^2 - \sigma^2 y_{t-1}^{2\eta}$$

$$\Longrightarrow \quad E(u_t | y_1, ..., y_{t-1}) = 0 \quad \text{and} \quad E(v_t | y_1, ..., y_{t-1}) = 0.$$

This gives many moment conditions; we will use only the following six unconditional moment conditions for illustration:

$$E(u_t) \;=\; 0, \quad E(u_t y_{t-1}) = 0, \quad E(u_t y_{t-2}) = 0,$$

$$E(v_t) \;=\; 0, \quad E(v_t y_{t-1}) = 0, \quad E(v_t y_{t-2}) = 0.$$

The main source of difficulty in this GMM is η, the identification of which becomes almost impossible when σ is close to zero.
   Using the US 3-month treasury bill rate (in percentages) data monthly from January 1982 to December 1999 ($N = 216$), the nonlinear GMM estimation result (with t-values in $(\cdot)$) is

β_1	β_2	σ	η
0.162 (2.13)	0.029 (2.18)	0.024 (2.58)	1.24 (6.47)

The over-id test statistic and p-value from $\chi_{6-4}^2 = \chi_2^2$ are, respectively, 5.858 and 0.053, indicating a possible misspecification. The steady state rate is $0.162/0.029 = 5.33\%$, which looks reasonable. As noted already, the estimation is sensitive to η—a contentious parameter in the literature. Although GMM provides a nice unifying theme for econometrics, nonlinear GMM often does not behave well computationally.
   Recall that, in a preceding chapter, we estimated the linear model of the form $y_t = \beta_0 + \beta_1 y_{t-1} + \beta_2 y_{t-2} + v_t$ with the same data and noted the potential problem of heteroskedasticity. Because η is significant, the heteroskedasticity factor σy_{t-1}^{η} indeed matters. This indicates that the conditional variance given y_{t-1} is large if y_{t-1} is so, which seems plausible, because some people expect y_t to drop toward the steady-state level while some other people expect y_t to go up even higher. Specification and estimation of short-term interest-rate stochastic differential equations is a contentious issue in finance; see, e.g., Lee and Li (2005) and the references therein. Jagannathan et al. (2002) provided a survey on GMM applications to finance.

3.3   Three Tests in GMM

There are GMM analogs for Wald, LR, and LM tests. Consider moment conditions and a null hypothesis

$$E\{ \underset{s\times 1}{\psi} (z, \underset{k\times 1}{\beta})\} = 0, \quad H_0 : \underset{g\times 1}{h} (\beta) = 0 \text{ with } rank \left(\frac{\partial h(\beta)}{\partial b'} \right) = g.$$

If H_0: $h(\beta) = c$, redefine $h(\beta)$ as $h(\beta) - c$ to get $H_0 : h(\beta) = 0$. Clearly $h(\beta)$ includes the linear version $R'\beta$ as a special case. Define the efficient GMM as b_N, and its restricted version as b_{Nr} minimizing

$$Q_N(b) \equiv 0.5 \cdot \frac{1}{N} \sum_i \psi(b)' \, W_N^{-1} \, \frac{1}{N} \sum_i \psi(b) \; + \lambda'h(b)$$

where λ is a Lagrangean vector.

Define a $k \times k$ matrix estimator for the inverse of the asymptotic variance of $\sqrt{N}(b_N - \beta)$:

$$\Omega_N(b_N) \quad \equiv \quad \frac{1}{N} \sum_i \psi_b(b_N) \, W_N^{-1} \, \frac{1}{N} \sum_i \psi_{b'}(b_N)$$

$$\rightarrow \quad ^P \; H \equiv E\psi_b(\beta) \; W^{-1} \; E\psi_{b'}(\beta).$$

By doing analogously to what was done for the restricted estimators in MLE and NLS, it holds that

$$b_{Nr} = \left[I_k - \Omega_N^{-1} \frac{\partial h(b_N)}{\partial b} \left\{ \frac{\partial h(b_N)}{\partial b'} \, \Omega_N^{-1}(b_N) \, \frac{\partial h(b_N)}{\partial b} \right\}^{-1} \frac{\partial h(b_N)}{\partial b'} \right] \cdot b_N.$$

The following χ_g^2 test statistics hold for GMM (see Newey and McFadden (1994) for more related test statistics):

$$Wald_{gmm} \quad = \quad N \cdot h(b_N)' \left\{ \frac{\partial h(b_N)}{\partial b'} \, \Omega_N^{-1}(b_N) \, \frac{\partial h(b_N)}{\partial b} \right\}^{-1} h(b_N),$$

$$LR_{gmm} \quad = \quad N \left\{ \frac{1}{N} \sum_i \psi(b_{Nr})' W_N^{-1} \frac{1}{N} \sum_i \psi(b_{Nr}) \right.$$

$$\left. -\frac{1}{N} \sum_i \psi(b_N)' W_N^{-1} \frac{1}{N} \sum_i \psi(b_N) \right\},$$

$$LM_{gmm} \quad = \quad \left\{ \frac{1}{N} \sum_i \psi_b(b_{Nr}) W_N^{-1} \frac{1}{\sqrt{N}} \sum_i \psi(b_{Nr}) \right\}' \Omega_N(b_{Nr})^{-1}$$

$$\left\{ \frac{1}{\sqrt{N}} \sum_i \psi_b(b_{Nr}) W_N^{-1} \frac{1}{N} \sum_i \psi(b_{Nr}) \right\}$$

The steps to derive LR_{gmm} and LM_{gmm} are analogous to those for LR_{nls} and LM_{nls}. In practice, it is advisable to use the same estimator W_N for the two terms in LR_{gmm}; otherwise LR_{gmm} may take a negative number. The observations made for Wald, LR, and LM tests in relation to MLE and NLS also hold for the GMM tests.

3.4   Efficiency of GMM

Earlier we showed that, for an M-estimator b_N maximizing $N^{-1}\sum_i$ $q(z_i, b)$, the asymptotic variance of b_N can be written as

$$[E(sq_{b'})\ E^{-1}(q_b q_{b'})\ E(q_b s')]^{-1}$$

where s is the score function for β and $q_b \equiv \partial q/\partial b$. Doing analogously, the GMM asymptotic variance can be rewritten as

$$[E(s\psi')\ E^{-1}(\psi\psi')\ E(\psi s')]^{-1}.$$

This is the inverse of the square of the part of $E(ss')$ explained by ψ. Since $E(ss')$ is larger than the part explained by ψ, GMM is less efficient than MLE. If ψ is s, then the GMM is MLE; the closer ψ is to s, the more efficient GMM becomes.

Let $z = (x', y)'$. If the moment condition is $E\{\psi(y, x)|x\} = 0$, not $E\psi(y, x) = 0$, then the GMM under $E\psi(y, x) = 0$ is not efficient. The reason is that $E\{\psi(y, x)|x\}$ is much stronger than $E\psi(y, x) = 0$: $E\{\psi(y, x)|x\} = 0$ implies $E\{g(x)\cdot\psi(y, x)\} = 0$ for any square-integrable function $g(x)$, rendering infinitely many unconditional moment conditions. The asymptotic variance of the efficient estimator (or the "efficiency bound") under $E\{\psi(y, x)|x\} = 0$ is (Chamberlain, 1987)

$$E_x^{-1}\{E(\psi_b|x)\ E^{-1}(\psi\psi'|x)\ E(\psi_{b'}|x)\}.$$

If $g(x)$ is continuous, it can be well approximated by polynomial functions of x. Then $E(g\cdot\psi) = 0$ for an arbitrary $g(x)$ is equivalent to $E(\zeta_j(x)\cdot\psi) = 0$, $j = 1, ..., J$, where $\zeta_j(x)$ are polynomial functions of x. The GMM with these unconditional moment conditions attains the preceding bound as $J \to \infty$ (Chamberlain, 1987). Donald et al. (2003) provided a related "empirical-likelihood" approach for the efficient estimation. Kitamura et al. (2004) showed that the efficient estimation can also be done using a "localized version of empirical likelihood" approach, which is a combination of empirical-likelihood and "nonparametric smoothing." All these approaches require choosing a "smoothing parameter" or J typical in nonparametric methods, which is a disadvantage. This theme of turning a conditional moment to unconditional moments will be picked up later.

As an example of the efficiency bound under $E\{\psi(y, x)|x\} = 0$, suppose $y = x'\beta + u$ with $E(\psi|x) = 0$ where $\psi(z, \beta) = y - x'\beta = u$. Then the bound becomes

$$E_x^{-1}\{x\ E^{-1}(u^2|x)\ x'\} = E^{-1}\left\{\frac{xx'}{V(u|x)}\right\}$$

which is the asymptotic variance of the GLS. If $E(u^2|x) = \sigma^2$, then the efficiency bound is $\sigma^2 E^{-1}(xx')$; we already know that LSE is efficient under

$E(ux) = 0$. Therefore, for linear model,

(i) if $E(u|x) = 0$ and $E(u^2|x) = \sigma^2$, then LSE is efficient;

(ii) if $E(u|x) = 0$ and $E(u^2|x) = h(x)$ whose form is known, then GLS attains the efficiency bound;

(iii) if $E(u|x) = 0$, whether homoskedasticity holds or not, there exists a GMM which is efficient attaining the efficiency bound.

One relevant question for (ii) is: with $\psi = (y - x'\beta, (y - x'\beta)^2 - h(x))$, can we get an estimator more efficient than the GLS? The answer is yes if $E(u^3|x) \neq 0$ or β enters the $h(x)$; see Newey (1993, p. 427).

3.5   Weighting Matrices for Dependent Data

Although we deal mostly with cross-section data, here we digress to consider dependent data, which will demonstrate the "versatility" of GMM. If we have dependent data, then the optimal weighting matrix in GMM requires some adjustment, although estimation of $E\psi_b(z, \beta)$ can be done in the same way as with iid data by $N^{-1} \sum_i \psi_b(z_i, b_N)$. Recall that the optimal weighting matrix is the inverse of $W = V\{N^{-1/2} \sum_i \psi(z_i, \beta)\}$, which is

$$W = E\left[\frac{1}{N} \sum_i \sum_j \psi(z_i, \beta)\psi(z_j, \beta)' \right] = \frac{1}{N} \sum_i \sum_j E\{\psi(z_i, \beta)\psi(z_j, \beta)'\}.$$

If $\{z_i\}$ are iid, then all cross products disappear and W can be estimated consistently with $N^{-1} \sum_i \psi(z_i, b_N)\psi(z_i, b_N)'$.

It is important to realize that W cannot be estimated consistently with (the "wrong estimator")

$$\frac{1}{N} \sum_i \sum_j \psi(z_i, b_N)\psi(z_j, b_N)' = \frac{1}{\sqrt{N}} \sum_i \psi(z_i, b_N) \cdot \frac{1}{\sqrt{N}} \sum_j \psi(z_j, b_N)'$$

because $N^{-1/2} \sum_i \psi(z_i, b_N) = 0$ by the first-order condition. White and Domowitz (1984) suggested an estimator for W: omitting b_N,

$$\hat{W}_N \equiv \frac{1}{N} \sum_{j=1}^{N} \psi(z_j)\psi(z_j)' + \frac{1}{N} \sum_{i=1}^{m} \sum_{j=i+1}^{N} \{\psi(z_j)\psi(z_{j-i})' + \psi(z_{j-i})\psi(z_j)'\},$$

where $m < N - 1$. The two terms in $\{\cdot\}$ guarantee the symmetry of the estimator; e.g., with $\psi(z_j) = (w_j, w_{j-1})'$,

$$\psi(z_j)\psi(z_{j-i})' = \begin{bmatrix} w_j \\ w_{j-1} \end{bmatrix} \begin{bmatrix} w_{j-i} & w_{j-i-1} \end{bmatrix}$$

$$= \begin{bmatrix} w_j w_{j-i} & w_j w_{j-i-1} \\ w_{j-1} w_{j-i} & w_{j-1} w_{j-i-1} \end{bmatrix}.$$

Clearly, this matrix is not symmetric, and a symmetrization is done in $\{\cdot\}$.

If $m = N - 1$, then $\hat{W}_N$ is the same as the wrong estimator. Hence by removing some terms in the wrong estimator, we get a valid estimator $\hat{W}_N$. Essentially $\hat{W}_N$ makes the wrong estimator non-zero by limiting the dependence over time:

$$E\{\psi(z_j)\psi(z_{j-i})' + \psi(z_{j-i})\psi(z_j)'\} = 0 \quad \text{for all } i > m.$$

There is, however, no good practical guideline on how to select the truncation number m. See White and Domowitz (1984, pp.153–154) for more.

Newey and West (1987) proposed another estimator which is guaranteed to be p.s.d. for a given N. Their estimator is

$$\tilde{W}_N = \frac{1}{N}\sum_j \psi(z_j)\psi(z_j)' + \frac{1}{N}\sum_{i=1}^{m}\left(1 - \frac{i}{m+1}\right)$$

$$\sum_{j=i+1}^{N}\{\psi(z_j)\psi(z_{j-i})' + \psi(z_{j-i})\psi(z_j)'\}.$$

For instance, with $m = 2$,

$$\tilde{W}_N = \frac{3}{3}\cdot\frac{1}{N}\sum_{j=1}^{N}\psi(z_j)\psi(z_j)' + \frac{2}{3}\cdot\frac{1}{N}\sum_{j=2}^{N}\{\psi(z_j)\psi(z_{j-1})' + \psi(z_{j-1})\psi(z_j)'\}$$

$$+ \frac{1}{3}\cdot\frac{1}{N}\sum_{j=3}^{N}\{\psi(z_j)\psi(z_{j-2})' + \psi(z_{j-2})\psi(z_j)'\};$$

smaller weights are given to the terms with more lags. If $m = 1$, use the first two of the preceding display with the weights being $2/2$ and $1/2$, respectively.

More estimators for the weighting matrix have appeared under the name "heteroskedasticity- and autocorrelation-consistent (HAC) variance"; see, e.g., Andrews and Monahan (1992), Newey and West (1994), West (1997), Smith (2005), and the references therein. Hall (2000) suggested using the centered version $\psi(z_i) - N^{-1}\sum\psi(z_i)$ instead of $\psi(z_i)$ in the weighting matrix to raise the power of the over-id test.

3.6   GMM for Multiple Nonlinear Equations*

Suppose we have a nonlinear equation system

$$\underset{s\times 1}{\rho}\,(\,\underset{s\times 1}{y}\,,\,\underset{K\times 1}{x}\,,\,\underset{k\times 1}{\gamma}\,) = \underset{s\times 1}{u}$$

where y is the endogenous variable and x is the exogenous variable of the system. Assume

$$E(\underset{(sK)\times 1}{x\otimes u}) = 0 \iff E\left\{\underset{(sK)\times s}{(I_s\otimes x)}\cdot\underset{s\times 1}{u}\right\} = 0,$$

which is the $s \times K$ moment conditions of the system where the error term in each equation has zero covariance with x. We thus get

$$E(z \cdot u) = 0 \iff E\{z \cdot \rho(y, x, \gamma)\} = 0 \quad \text{where} \quad \underset{sK \times s}{z_i} \equiv I_s \otimes x_i.$$

This type of moment conditions appeared when linear equation systems were discussed.

Consider a *nonlinear IVE (NIV)* minimizing

$$\left\{ \frac{1}{\sqrt{N}} \sum_i z_i \rho(y_i, x_i, g) \right\}' \left(\frac{1}{N} \sum_i z_i z_i' \right)^{-1} \left\{ \frac{1}{\sqrt{N}} \sum_i z_i \rho(y_i, x_i, g) \right\}$$

wrt g. This was first suggested by Amemiya (1974) under a different name. The form of NIV is similar to IVE in the linear model. Since ρ is nonlinear, NIV is obtained with an iterative method. Observe that

$$\rho(g_1) \simeq \rho(g_0) + \rho_{g'}(g_0)(g_1 - g_0), \quad \underset{s \times k}{\rho_{g'}} \equiv \frac{\partial \rho}{\partial g'}.$$

Substitute this into the minimand and solve the first-order condition wrt g_1 to obtain the following iterative scheme for NIV:

$$g_1 = g_0 - \left\{ \sum_i \rho_g(g_0) z_i' \left(\sum_i z_i z_i' \right)^{-1} \sum_i z_i \rho_{g'}(g_0) \right\}^{-1} \sum_i \rho_g(g_0) z_i'$$

$$\left(\sum_i z_i z_i' \right)^{-1} \sum_i z_i \rho(g_0).$$

Once NIV is obtained, the corresponding GMM is obtained minimizing

$$\frac{1}{N} \sum_i \{z_i \rho(y_i, x_i, g)\}' \; W_N^{-1} \; \frac{1}{N} \sum_i z_i \rho(y_i, x_i, g)$$

where W_N is a consistent estimator for the $sK \times sK$ variance matrix of $N^{-1/2} \sum_i z_i u_i$:

$$W = E(z u u' z') = E\{(I_s \otimes x) u u' (I_s \otimes x')\}.$$

With $\hat{u}_i \equiv \rho(y_i, x_i, g_{niv})$, GMM is obtained as, with $g_0 = g_{niv}$,

$$g_{gmm} = g_0 - \left\{ \sum_i \rho_g(g_0) z_i' \left(\sum_i z_i \hat{u}_i \hat{u}_i' z_i' \right)^{-1} \sum_i z_i \rho_{g'}(g_0) \right\}^{-1}$$

$$\sum_i \rho_g(g_0) z_i' \left(\sum_i z_i \hat{u}_i \hat{u}_i' z_i' \right)^{-1} \sum_i z_i \rho(g_0).$$

The asymptotic variance matrix for the GMM is

$$\underset{k \times (sK)}{[E(\rho_g z')\ E^{-1}(zuu'z')\ E(z\rho_{g'})]^{-1}.}$$

Replacing ρ_g with an instrument matrix, this includes the GMM for multiple linear equations as a special case.

Suppose we have the conditional moment condition $E(u|x) = 0$, not just $E(x \otimes u) = 0$. Then the preceding GMM is not efficient, but we can get an efficient estimator by augmenting the moment conditions as shown already when conditional moment conditions were discussed. Newey (1988, 1993) suggested another way to get an efficient estimator for the model $\rho(y, x, \gamma) = u$. Consider a $r \times s$ $(r \geq k)$ known instrument matrix $A(x)$ and the moment condition

$$E\{A(x)\rho(y, x, \gamma)\} = 0.$$

Applying GMM to this, the asymptotic variance is

$$[E\{\rho_g A(x)'\}\ E^{-1}\{A(x)uu'A(x)'\}\ E\{A(x)\rho_{g'}\}]^{-1}.$$

If we set $r = k$ and

$$A(x) = E(\rho_g|x) \cdot E^{-1}(uu'|x),$$

then the preceding asymptotic variance becomes

$$E_x^{-1}\{E(\rho_g|x)\ E^{-1}(uu'|x)\ E(\rho_{g'}|x)\}$$

which is nothing but the efficiency bound $E_x^{-1}\{E(\psi_b|x)E^{-1}(\psi\psi'|x)\ E(\psi_{b'}|x)\}$; i.e., if we can get $A(x)$, we can attain the efficiency bound with GMM.

To get $A(x)$, we need $E(\rho_g|x)$ and $E(uu'|x)$. If homoskedasticity is assumed, then $E(uu'|x)$ is a constant and thus can be estimated easily with the residuals from an initial estimator. Getting $E(\rho_g|x)$ is more problematic, because we need the conditional density $f_{y|x}$ and the integration $\int \rho_g(y, x, \gamma) f_{y|x}(y)dy$ to get $E(\rho_g|x)$. But using "nonparametric techniques" to appear, it is possible to obtain $A(x)$ without specifying the conditional density; see Robinson (1991), Newey (1993), and Newey (1990b) who treated the homoskedastic case only.

4   Minimum Distance Estimation (MDE)

Consider a parameter β, and a function $\psi(\beta)$ which is consistently estimable with an estimator θ_N without knowing β. *Minimum distance estimation (MDE)*, in a wide sense, estimates β by minimizing a distance between θ_N and $\psi(\beta)$. That is, for a parameter space B,

$$b_{mde} = \mathrm{argmin}_{b \in B} \|\theta_N - \psi(b)\|$$

where $\|\cdot\|$ is a norm. For instance, $\|\cdot\|$ can be the Euclidean norm or the absolute deviation norm. In a narrow sense, MDE uses a quadratic norm and minimizes, for instance, $\{\theta_N - \psi(b)\}' W_N^{-1} \{\theta_N - \psi(b)\}$ for a weighting matrix W_N. In this case, MDE is also called "minimum χ^2 estimation." We will use the term MDE in the narrow sense unless otherwise noted.

If θ_N is a sample moment, say $\theta_N = N^{-1} \sum_i \mu(z_i)$ for a known function $\mu(\cdot)$, then

$$\theta_N - \psi(b) = \frac{1}{N} \sum_i \{\mu(z_i) - \psi(b)\}.$$

This is a moment condition, and with this MDE becomes MOM; indeed the reader will see much similarity between MDE and GMM. Thus, MDE is more interesting when θ_N is not sample moments, and such MDE is useful in estimating simultaneous equations [see Lee (1995) and Lee and Kimhi (2005), and the references therein] and panel data models [see Chamberlain (1982) and Lee (2002)]. MDE has in fact a long history; see Malinvaud (1970) and Chamberlain (1982) among many others, and the references therein.

MDE is a way of "indirect inference" as β gets estimated indirectly through θ_N, but the expression "indirect inference" is typically used for the special case of MDE combined with simulated data; see, e.g., Gourieroux et al. (1993). This approach tends to be fully parametric involving likelihoods, which presents different ways to measure the difference between θ and $\psi(\beta)$: Wald-based, LR-based, and score-based differences. The Wald-based difference is a special case of the above quadratic difference.

In the following, we will start with a motivational subsection first for MDE, generalize it further, and present an empirical example that is a shortened version of an example in Lee (2002).

4.1   MDE Basics

Suppose there is a panel data set with two waves ($T = 2$) for a model $y_{it} = x_{it}'\beta + v_{it}$. Also suppose that we can consistently estimate β with each wave. If we estimate β with wave 1 and wave 2 separately, we will get two estimators, say b_1 and b_2, for the same parameter β. Then the question arises: how can we combine the two estimators to come up with a single estimator for β? This question in fact came up because we assumed the same parameter for both waves to begin with, and it would be more general to posit $y_{it} = x_{it}'\beta_t + v_{it}$ and then test for $H_o: \beta_1 = \beta_2$; if this H_o is accepted, then we would proceed to combining b_1 and b_2. As it turns out, the answer to the question of combining b_1 and b_2 also provides a test statistic for $H_o: \beta_1 = \beta_2$.

The first-step in MDE is expressing the constraint on the parameters into equations. In the panel data example, the constraint is $\beta_1 = \beta_2$ ($\equiv \beta$). With β being of dimension $k \times 1$, the constraint can be written as

$$\begin{bmatrix} \beta_1 \\ \beta_2 \end{bmatrix} = \begin{bmatrix} I_k \\ I_k \end{bmatrix} \cdot \beta \implies \begin{bmatrix} b_1 \\ b_2 \end{bmatrix} \simeq \begin{bmatrix} I_k \\ I_k \end{bmatrix} \cdot \beta.$$

In the second equation, the left-hand side $(b_1', b_2')'$ is θ_N, and the right-hand side is $\psi(\beta) = \psi\beta$ where $\psi \equiv [I_k, I_k]'$. Our goal is to estimate β using the equation, and there is an easy way of doing that: regard $\theta_N \equiv (b_1', b_2')'$ as a response variable and ψ as a regressor to do the LSE of θ_N on ψ. Because $\psi'\psi = I_k + I_k = 2I_k$, and $\psi'\theta_N = b_1 + b_2$, the LSE is nothing but the simple average of b_1 and b_2:

$$(\psi'\psi)^{-1}\psi'\theta_N = \frac{b_1 + b_2}{2}.$$

If we recall GLS, an weighted average will be better than this simple average, which leads to the aforementioned MDE using the quadratic norm with the weighting matrix W_N. With the basic idea of MDE fixed now, we turn to a more general setup.

The main question that MDE answers is *how to optimally impose overidentifying restrictions*. Consider a $k \times 1$ parameter vector β and K $(K \geq k)$ restrictions $\theta = \psi(\beta)$ where $\psi(\beta)$ has the $K \times k$ continuous first derivative matrix $\psi_{b'}$. Suppose θ can be estimated with an estimator θ_N. Then $\theta = \psi(\beta)$ can be written as

$$\theta_N - \theta = \theta_N - \psi(\beta).$$

Now choose an estimator b_N for β such that $\theta_N - \psi(b_N)$ is as small as possible. Since $\theta_N - \psi(b_N)$ is a $K \times 1$ vector, we turn it into a scalar using a quadratic norm. Often, we call θ a "reduced form (or "shallow") parameter" and β a "structural form (or "deep") parameter."

An (inefficient) MDE for β is obtained by minimizing

$$N\{\theta_N - \psi(b)\}' \cdot W^{-1} \cdot \{\theta_N - \psi(b)\}$$

wrt b, where W is a $K \times K$ symmetric p.d. matrix (for instance, $W = I_K$). The efficient MDE is obtained by setting

$$W = V_N \to^p V \quad \text{where} \quad \sqrt{N}(\theta_N - \theta) \rightsquigarrow N(0, V).$$

Often the efficient MDE is simply called (the) MDE.

Denoting the MDE as b_N, it will be shown shortly that

$$\sqrt{N}(b_N - \beta) \rightsquigarrow N[0, \{\psi_b(\beta)V^{-1}\psi_{b'}(\beta)\}^{-1}]$$

$$N\{\theta_N - \psi(b_N)\}'V_N^{-1}\{\theta_N - \psi(b_N)\} \rightsquigarrow \chi^2_{K-k} \quad \text{where } V_N \to^p V.$$

The latter is the MDE over-id test for the restriction $\theta = \psi(\beta)$, analogous to the GMM over-id test. Simply minimizing the quadratic form does not necessarily mean that the restriction $\theta = \psi(\beta)$ holds; if it does, then we should have $\theta_N \simeq \psi(b_N)$, which is verified with the over-id test. In the following, we derive the asymptotic distribution result.

To see the minimization process, rewrite the minimand (with b and W replaced by b_N and W_N, respectively) as N times

$$\theta_N'W_N^{-1}\theta_N + \psi(b_N)'W_N^{-1}\psi(b_N) - 2\psi(b_N)'W_N^{-1}\theta_N.$$

Differentiate this wrt b_N to get the first-order condition

$$2 \cdot \psi_b(b_N) W_N^{-1} \psi(b_N) - 2 \cdot \psi_b(b_N) W_N^{-1} \theta_N = 2\psi_b(b_N) W_N^{-1}$$
$$\cdot \{\psi(b_N) - \theta_N\} = 0$$

$$\iff \psi_b(b_N) W_N^{-1} \{\psi(\beta) + \psi_{b'}(b_N^*)(b_N - \beta) - \theta_N\} = 0$$

where $b_N^* \in (b_N, \beta)$. Solve this for b_N to get

$$b_N = \beta + \{\psi_b(b_N) W_N^{-1} \psi_{b'}(b_N^*)\}^{-1} \psi_b(b_N) W_N^{-1} \{\theta_N - \psi(\beta)\}$$

$$\implies \sqrt{N}(b_N - \beta) = \{\psi_b(\beta) W^{-1} \psi_{b'}(\beta)\}^{-1} \psi_b(\beta) W^{-1} \sqrt{N}(\theta_N - \theta) + o_p(1).$$

This yields the asymptotic distribution for MDE with the weighting matrix W, and $W = V$ renders the efficient MDE with the asymptotic variance $\{\psi_b(\beta) V^{-1} \psi_{b'}(\beta)\}^{-1}$.

The preceding display with $b_N = \beta + \dots$ also yields an iterative computational algorithm: replace b_N and β with b_1 and b_0, respectively, to get

$$b_1 = b_0 + \{\psi_b(b_0) W_N^{-1} \psi_{b'}(b_0)\}^{-1} \cdot \psi_b(b_0) W_N^{-1} \{\theta_N - \psi(b_0)\};$$

update b_0 and repeat this until convergence.

There is an interesting point to be noted for MDE and MLE. If θ_N is a MLE and $\theta = \psi(\beta)$ where the dimension of θ is at least as large as β, then there are two ways to estimate β. The first is MDE: obtain θ_N by maximizing $N^{-1} \sum_i \ln f(\theta)$ and then use MDE to find β. The second is the direct MLE maximizing $N^{-1} \sum_i \ln f\{\psi(b)\}$ wrt b. Now we will show that the two methods are equivalent. Denote the score vector for θ as s_θ and define $I_\theta \equiv E(s_\theta s_\theta')$. The MDE b_N has

$$\sqrt{N}(b_N - \beta) \rightsquigarrow N[0, \{\psi_b(\beta) I_\theta \psi_{b'}(\beta)\}^{-1}].$$

If we maximize $N^{-1} \sum_i \ln f\{\psi(b)\}$ directly wrt b, then the score function for b is (by the chain rule) $\psi_b(\beta) s_\theta$. Thus the information matrix is $E\{\psi_b(\beta) s_\theta s_\theta' \psi_b (\beta)\} = \psi_b(\beta) I_\theta \psi_{b'}(\beta)$: the direct MLE has the same asymptotic variance as the MDE has. In short, *if θ_N is a MLE, then MDE is the efficient way to impose over-identified parameter restrictions.*

4.2   Various MDE Cases

The simplest MDE occurs when the restriction $\theta = \psi(\beta)$ is a known linear function. Suppose

$$\theta = \Psi\beta, \quad \Psi \text{ is a known } K \times k \text{ matrix}, \quad K \geq k, \quad \text{rank}(\Psi) = k.$$

First consider $K = k$. Then $\beta = \Psi^{-1}\theta$, and the natural estimator for β is

$$b_N = \Psi^{-1}\theta_N.$$

Since $\theta_N \to^p \theta$, we get $b_N \to^p \Psi^{-1}\theta = \beta$. Also the asymptotic distribution of $\sqrt{N}(b_N - \beta)$ is straightforward to obtain: recalling $\sqrt{N}(\theta_N - \theta) \rightsquigarrow N(0, V)$,

$$\sqrt{N}(b_N - \beta) \rightsquigarrow N(0, \Psi^{-1}V\Psi^{-1\prime}).$$

Now consider $K > k$. For a $K \times K$ p.d. matrix W, $\Psi\beta = \theta$ implies

$$\Psi'W^{-1}\Psi \cdot \beta = \Psi'W^{-1} \cdot \theta.$$

Solve this for β and replace θ with θ_N to get

$$b_N = (\Psi'W^{-1}\Psi)^{-1}\Psi'W^{-1} \cdot \theta_N.$$

This b_N also minimizes $N(\theta_N - \Psi b)'W^{-1}(\theta_N - \Psi b)$ and the asymptotic distribution is

$$\sqrt{N}(b_N - \beta) \rightsquigarrow N\{0, \ (\Psi'W^{-1}\Psi)^{-1}\Psi'W^{-1}VW^{-1}\Psi(\Psi'W^{-1}\Psi)\};$$

with $W = I_K$, the MDE is the "unweighted MDE." The covariance matrix attains its minimum when $W = V$. Thus, the efficient MDE is

$$b_N = (\Psi'V_N^{-1}\Psi)^{-1}\Psi'V_N^{-1} \cdot \theta_N \implies \sqrt{N}(b_N - \beta) \rightsquigarrow N\{0, \ (\Psi'V^{-1}\Psi)^{-1}\}.$$

Returning to the case $K = k$, we can use W as in the case $K > k$. But since there is only one b_N satisfying $\Psi b_N = \theta_N$, b_N is the same regardless of W when $K = k$. The variance matrix $(\Psi'V^{-1}\Psi)^{-1}$ when $K > k$ becomes $\Psi^{-1}V\Psi'^{-1}$ when $K = k$, for Ψ is invertible under $K = k$.

Consider a more complicated situation where the linear restriction Ψ is a function of θ:

$$\theta = \Psi(\theta)\beta, \quad \Psi(\theta) \text{ is } K \times k, \quad K \geq k, \quad rank\{\Psi(\theta)\} = k;$$

the above nonlinear case does not include this as a special case, because if we linearize $\theta = \psi(b)$ around $b = 0$, we would get $\theta = \psi_{b'}(b^*)b$ that has no θ on the right-hand side. Rewrite $\Psi(\theta)\beta - \theta = 0$ as

$$\theta_N = \Psi(\theta_N)\beta + [\theta_N - \theta - \{\Psi(\theta_N) - \Psi(\theta)\}\beta].$$

Define an $K \times 1$ error vector ε as

$$\varepsilon \ \equiv \ \theta_N - \theta - \{\Psi(\theta_N) - \Psi(\theta)\}\beta = A \cdot (\theta_N - \theta) + o_p(1) \quad \text{where}$$

$$\underset{K \times k}{\Psi} \ = \ \begin{bmatrix} \underset{1 \times k}{\Psi_1} \\ \vdots \\ \underset{1 \times k}{\Psi_K} \end{bmatrix} \quad \text{and} \quad A \equiv \left(I_K - \begin{bmatrix} \underset{k \times K}{\beta' \cdot \partial\Psi_1/\partial\theta'} \\ \vdots \\ \underset{k \times K}{\beta' \cdot \partial\Psi_K/\partial\theta'} \end{bmatrix} \right)$$

and denote $\Psi(\theta_N)$ as Ψ_N to rewrite the preceding display as

$$\theta_N = \Psi_N\beta + \varepsilon.$$

This equation looks like the usual linear model with θ_N as the response variable and Ψ_N as the regressors. We can estimate β $\sqrt{N}$-consistently with the LSE b_c:

$$b_c = (\Psi_N' \Psi_N)^{-1} \Psi_N' \theta_N.$$

Since the asymptotic variance of $\sqrt{N}\varepsilon$ is, in general, not diagonal, we can do better with generalized LSE (GLS). Note that

$$\sqrt{N}\varepsilon \rightsquigarrow N(0, \ AVA')$$

Since A depends on β and θ, let A_N be the version with b_c and θ_N plugged into β and θ, respectively, in A. Then the GLS is

$$b_N = \{\Psi_N'(A_N V_N A_N')^{-1} \Psi_N\}^{-1} \Psi_N'(A_N V_N A_N')^{-1}\theta_N$$

$$\sqrt{N}(b_N - \beta) \rightsquigarrow N[0, \ \{\Psi(\theta)'(AVA')^{-1}\Psi(\theta)\}^{-1}].$$

The reader may wonder how we can apply LSE or GLS with only K many "observations." The answer is that the error term ε is not the usual error term; ε has a degenerate distribution converging to 0, because it is proportional to $\theta_N - \theta$. Thus $\theta_N = \Psi_N\beta + \varepsilon$ behaves like K linear deterministic equations. Examples for $\theta = \Psi(\theta)\beta$ arise when MDE is applied to simultaneous equations with LDV's; see Lee (1995) and Lee and Kimhi (2005).

Further generalizing MDE, consider (L.F. Lee, 1992)

$$\underset{q\times 1}{f} (\underset{K\times 1}{\theta}, \underset{k\times 1}{\beta}) = 0$$

where f is twice continuously differentiable, $\sqrt{N}(\theta_N - \theta) \rightsquigarrow N(0, V)$, and $k \leq q \leq K$. By δ-method, Taylor-expanding $f(\theta_N, \beta)$ around $\theta_N = \theta$, we get

$$\sqrt{N}\{f(\theta_N, \beta) - f(\theta, \beta)\} \rightsquigarrow N(0, \Omega) \quad \text{where } \Omega \equiv \frac{\partial f(\theta, \beta)}{\partial \theta'} V \frac{\partial f(\theta, \beta)}{\partial \theta}.$$

Consider a (generalized) MDE b_N for β by minimizing wrt b

$$f(\theta_N, b)' \cdot \Omega_N^{-1} \cdot f(\theta_N, b) \quad \text{where } \Omega_N \to^p \Omega.$$

Then

$$\sqrt{N}(b_N - \beta) \rightsquigarrow N \left(0, \ \left\{\frac{\partial f(\theta, \beta)}{\partial \beta}\Omega^{-1}\frac{\partial f(\theta, \beta)}{\partial \beta'}\right\}^{-1}\right) \quad \text{and}$$

$$N \cdot f(\theta_N, b_N)'\Omega_N^{-1}f(\theta_N, b_N) \rightsquigarrow \chi^2_{q-k};$$

the latter is the MDE over-id test for $f(\theta, \beta) = 0$. Equating $f(\theta, \beta)$ to $\theta - \psi(\beta)$, $\theta - \Psi\beta$, or $\theta - \Psi(\theta)\beta$, this result includes all the preceding cases with $q = K$.

Estimating Ω with Ω_N requires a $\sqrt{N}$-consistent estimator for β, say b_{N0}, so that

$$\Omega_N \equiv \frac{\partial f(\theta_N, b_{N0})}{\partial \theta'} V_N \frac{\partial f(\theta_N, b_{N0})}{\partial \theta} \to^p \Omega.$$

Such an estimator can be obtained by minimizing $f(\theta_N, b)' f(\theta_N, b)$, and it is in general less efficient than b_N. In practice, a quadratic approximation to $f(\theta_N, b)' \Omega_N^{-1} f(\theta_N, b)$ at $b = b_{N0}$ may be used to obtain b_N. This would yield an iterative formula. Theoretically, however, taking one step from b_{N0} is enough to get b_N.

4.3   An Empirical Example from Panel Data

Consider a panel data with three waves $(T = 3)$ and a panel model

$$y_{it} = \tau_t + \tilde{x}_{it}' \mu_t + v_{it} \quad \text{where } x_{it} = (1, \tilde{x}_{it}')' \text{ and } \beta_t \equiv (\tau_t, \mu_t')'.$$

Suppose we want to impose the restriction that the slope coefficients μ_t in β_t, $t = 1, 2, 3$, are the same, whereas the intercepts τ_t's may be different. Denoting the $(k-1) \times 1$ common slope parameter vector as μ, these restrictions (i.e., the $\theta = \psi(\beta)$ equation) are

$$
\begin{bmatrix} \tau_1 \\ \mu_1 \\ \tau_2 \\ \mu_2 \\ \tau_3 \\ \mu_3 \end{bmatrix}
=
\begin{bmatrix}
1 & 0 & 0 & 0 \\
0 & 0 & 0 & I_{k-1} \\
0 & 1 & 0 & 0 \\
0 & 0 & 0 & I_{k-1} \\
0 & 0 & 1 & 0 \\
0 & 0 & 0 & I_{k-1}
\end{bmatrix}
\cdot
\begin{bmatrix} \tau_1 \\ \tau_2 \\ \tau_3 \\ \mu \end{bmatrix}.
$$

Denote the middle matrix with zeros and I_{k-1} as ψ; some of the zeros in ψ are zero vectors defined conformably for the identity matrices. Let b_t be the estimator for β_t in wave t. Also define

$$b_N \equiv (b_1', b_2', b_3')', \quad \beta \equiv (\beta_1', \beta_2', \beta_3')', \quad \beta_o \equiv (\tau_1, \tau_2, \tau_3, \mu')'.$$

For the MDE, the main difficulty in practice is in getting W_N: because b_1, b_2, and b_3 are obtained separately in the first stage, their covariance matrices are not readily available in the first stage.

To get the asymptotic variance matrix estimator $V_N \to^p V$ where $\sqrt{N}$ $(b_N - \beta) \rightsquigarrow N(0, V)$, let the influence function for b_t be η_{it} and denote its estimator as $\hat{\eta}_{it}$. Define

$$\hat{\eta}_i \equiv (\hat{\eta}_{i1}', \hat{\eta}_{i2}', \hat{\eta}_{i3}')' \quad \text{and} \quad V_N = N^{-1} \sum_i \hat{\eta}_i \hat{\eta}_i'.$$

Then the MDE is

$$b_{mde} = (\psi' V_N^{-1} \psi)^{-1} \cdot \psi' V_N^{-1} b_N$$

$$\sqrt{N}(b_{mde} - \beta_o) \rightsquigarrow N(0, (\psi' V^{-1} \psi)^{-1})$$

$$N \cdot (b_N - \psi \cdot b_{mde})' V_N^{-1} (b_N - \psi \cdot b_{mde}) \rightsquigarrow \chi^2_{2(k-1)}$$

Table 3: GMM and MDE for lnWage (t-values in (·))

	Wave-1-GMM	Wave-2-GMM	Wave-3-GMM	MDE
τ_1	7.64 (0.54)			−0.15 (−0.10)
τ_2		−0.72 (−0.22)		−0.09 (−0.06)
τ_3			0.32 (0.12)	−0.12 (−0.08)
age	−0.07 (−0.23)	0.17 (0.34)	0.08 (0.82)	0.11 (1.65)
$age^2/100$	0.11 (0.30)	−0.20 (−0.34)	−0.09 (−0.68)	−0.12 (−1.36)
edu	0.07 (0.51)	−0.07 (−0.15)	0.04 (0.75)	0.03 (0.70)
kids	0.39 (0.51)	−0.55 (−0.45)	−0.12 (−0.27)	−0.17 (−0.68)
ln(hr)	−0.30 (−0.33)	0.21 (0.32)	−0.12 (−0.55)	0.02 (0.13)
mar	−3.85 (−0.79)	0.11 (0.01)	0.93 (0.52)	0.06 (0.07)
sal	0.43 (0.56)	0.32 (0.60)	0.44 (2.22)	0.29 (2.58)
self	−0.18 (−0.25)	−0.24 (−0.37)	−0.39 (−1.41)	−0.37 (−2.71)
ur	0.00 (0.02)	−0.03 (−0.26)	−0.01 (−0.26)	−0.02 (−0.91)
p-value (over-id):	0.94	0.023	0.61	
p-value for slope constancy (over-id)				0.000

which is a test statistic for the over-identifying restriction. The dimension of the leftmost vector in $\theta = \psi(\beta)$ is $3k$, whereas that of the rightmost vector is $(k-1)+3$, which yields the dof $3k - \{(k-1)+3\} = 2(k-1)$. If the restriction is rejected, then this would mean that some slope coefficients are time-variant.

For $y_{it} = \ln(\text{hourly-wage}_{it})$ of married men of age 40–60, the following regressors are used: 1, age, age^2, edu (schooling years), kids (# children), ln(yearly work hours), mar (1 if married), sal (1 if salaried worker), self (1 if self-employed), and ur (local unemployment %). The data size N is 334, and the data were drawn from the Panel Study of Income Dynamics (PSID). In the regressors, only sal, self, and ur are time-varying (age is time-varying but only in the deterministic fashion). For each wave, there are 10 parameters (the intercept plus nine slopes as shown in Table 3), and the following 12 instruments were used: with i omitted in the subscripts,

$$1,\ \text{age}_1,\ \text{age}_1^2,\ \text{sal}_1,\ \text{sal}_2,\ \text{sal}_3,\ \text{self}_1,\ \text{self}_2,\ \text{self}_3,\ \text{ur}_1,\ \text{ur}_2,\ \text{ur}_3.$$

Table 3 shows the GMM for each wave and MDE with the above time-constant slope restriction.

Maybe due to weak instruments, all but one (sal for wave 3) estimates are insignificant for waves 1–3. When they are combined with MDE, however, sal and self become significant, and age and age^2 gain much in significance. This is natural as MDE uses all three waves in two stages while each GMM uses a single wave. In comparison to MDE, a panel data estimator would use all three waves in a single stage. MDE is mostly an average of the three corresponding estimates in waves 1–3. Somewhat surprisingly, the GMM over-id test for wave 2 is rejecting, whereas those for waves 1 and 3 are not. The MDE over-id test also rejects the restriction (constancy of the slope parameters) with ease.

CHAPTER 5
PARAMETRIC METHODS FOR SINGLE EQUATION LDV MODELS

Single equation models with limited dependent variables (LDV) are reviewed here whereas multiple equations with LDV will be reviewed in the next chapter. LDV models include discrete responses: binary responses taking 0 or 1, ordered discrete responses taking R-many ordered categories 0, 1,...,$R-1$, and count responses taking integers 0, 1, 2,... Also discussed are non-discrete response LDV models such as censored or truncated responses. Many economic empirical examples are provided throughout this chapter.

1   Binary Response

1.1   Basics

Binary response is the starting point for LDV. As already shown, the "threshold-crossing" binary response model is

$y_i^* = x_i'\beta + u_i$ is a "latent" continuous response

$x_i = (1, x_{i2}, ..., x_{ik})'$, $\beta = (\beta_1, ..., \beta_k)'$

$y_i = 1[y_i^* \geq 0] = 1[x_i'\beta + u_i \geq 0]$

$u_i \amalg x_i$ and u has df G with $E(u) = 0$ and $V(u) = \sigma^2$ (σ unknown)

(x_i', y_i), $i = 1, ..., N$, are iid and observed.

Further assume that G is continuous and that the distribution is symmetric about 0 to get

$$E(y|x) = 0 \cdot P(y = 0|x) + 1 \cdot P(y = 1|x) = P(y = 1|x)$$

$$= P(u \geq -x'\beta|x) = 1 - G(-x'\beta) = G(x'\beta).$$

If u depends on x, then we would have $G(x'\beta|x)$ instead of $G(x'\beta)$, in which case x affects $E(y|x)$ through two separate routes—a complicating scenario.

Since y_i is discrete, we cannot just assume the linear model $y = x'\beta + u$ (unless u is defined as $y - x'\beta$—more on this below) because both x and u can be continuous. Instead, $E(y|x)$ that is a fractional number is modeled as

Myoung-jae Lee, *Micro-Econometrics*, DOI 10.1007/b60971_5,
© Springer Science+Business Media, LLC 2010

a nonlinear function of x. While u in y^* may be called the "structural form" error, we can define the "reduced form" error v as

$$v \equiv y - E(y|x) = y - G(x'\beta).$$

Then v satisfies $E(v|x) = 0$ and we have a nonlinear regression model $y = G(x'\beta) + v$.

The following is examples of the binary response model:

(i) Loan approval: $y = 1$ if a loan application is approved (and 0 otherwise), and x is a list of the characteristics of the applicant and the loan. Here y^* is "loan-worthiness."

(ii) Accepting an offer: $y = 1$ if an offer is accepted, and x is a list of the characteristics of the offer and the decision maker.

(iii) Surviving a situation: $y = 1$ if survival, and x is a list of the characteristics of the subject and the situation. Here y^* may be construed as the difference between the hardship of the situation and the durability of the subject (measured in the same unit).

As in the usual linear model, almost always $E(u)$ is a non-zero constant, and $E(u) = 0$ is forced on the model by rewriting

$$x_i'\beta + u_i = \{\beta_1 + E(u)\} + \beta_2 x_{i2} +, ..., + \beta_k x_{ik} + \{u_i - E(u)\};$$

redefine β_1 as $\beta_1 + E(u)$ and u_i as $u_i - E(u)$. Furthermore, if the threshold is a non-zero constant γ_1 as in $y_i = 1[x_i'\beta + u_i > \gamma_1]$, then we can absorb the constant γ_1 into the intercept β_1 to have $y_i = 1[\beta_1 - \gamma_1 + \beta_2 x_{i2} +, ..., + \beta_k x_{ik} + u_i > 0]$ where $\beta_1 - \gamma_1$ is the intercept. This shows that, as in the linear model, interpreting the intercept is difficult in binary response models. For the above loan approval example, the bankers may have a formula giving a weight to each regressor to make a "loan-worthiness score' $x'\beta$. For example, if $x'\beta > 50$ points, then the application may be approved. In the binary response model, we cannot identify the threshold 50, as this gets absorbed into the intercept.

If the threshold varies across i and is not observed as in $y_i = 1[x_i'\beta + u_i > \gamma_{1i}]$, then γ_{1i} should be absorbed into u_i to render $y_i = 1[x_i'\beta + u_i^* > 0]$ where $u_i^* \equiv u_i - \gamma_{1i}$. In this case, the independence of u from x may not hold if γ_{1i} is related to x_i. For instance, in the above example (iii), suppose $x_i'\beta + u_i$ is the hardship, γ_{1i} is the subject's durability, and both x_i and γ_{1i} are related to some (physical) training. If subjects with more training tend to put themselves in more severe hardship, then x_i and u_i^* get related.

To apply MLE to the binary response model, we should specify G. Although $E(u) = 0$ and $V(u) = \sigma^2$ are assumed, σ is still unknown. Divide $x'\beta + u$ by σ to get $x'(\beta/\sigma) + (u/\sigma)$ and $V(u/\sigma) = 1$. If G is indexed by the mean and variance (as in normal distributions), then u/σ has the standardized distribution function Φ of G. Thus we have

$$P(y = 1|x) = P\left(\frac{u}{\sigma} \geq -x'\frac{\beta}{\sigma}|x\right) = \Phi\left(x'\frac{\beta}{\sigma}\right) = \Phi\left(x'\alpha\right),$$

$$\alpha \equiv \frac{\beta}{\sigma} = \left(\frac{\beta_1}{\sigma}, ..., \frac{\beta_k}{\sigma}\right)'.$$

With this, we can specify the likelihood to do MLE, but what is estimated by the MLE is α not β. Still the sign of β_j can be obtained from the sign of α_j as $\sigma > 0$. Also ratios β_j/β_h can be obtained from the ratios α_j/α_h as σ gets canceled: $\alpha_j/\alpha_h = (\beta_j/\sigma)/(\beta_h/\sigma) = \beta_j/\beta_h$.

If u is heteroskedastic, say $V(u|x) = \sigma(x)^2$ where $\sigma(x)$ is a function of x, then we need to divide $x'\beta + u$ by $\sigma(x)$ to get $\{x'\beta/\sigma(x)\} + \{u/\sigma(x)\}$ and the (x-conditionally) standardized error term $u/\sigma(x)$. Suppose $x'\beta/\sigma(x) \simeq x'\delta$ for some parameter δ. If we ignore heteroskedasticity, then we will be estimating δ, not β, with the MLE. The parameter δ is a mixture of the mean and variance function parameters, and it will be impossible to find β from the estimates for δ unless $\sigma(x)$ is known. For instance, $u = \nu \exp(x'\gamma)$ where $\nu \sim N(0,1)$ independently of x. Here $\sigma(x) = \exp(x'\gamma)$, and

$$P(y = 1|x) = P\left\{\frac{u}{\sigma(x)} \geq -\frac{x'\beta}{\sigma(x)}\right\} \simeq P(\nu \geq -x'\delta) = \Phi(x'\delta)$$

and we will be estimating δ, not α.

Although y is binary, sometimes in practice, linear model is adopted. Consider the *linear probability model*:

$y = x'\beta + \varepsilon$,   where $\varepsilon \equiv y - x'\beta$ (ε is binary given x, taking on $-x'\beta$ and $1 - x'\beta$)

$E(\varepsilon|x) = (0 - x'\beta)\{1 - G(x'\beta)\} + (1 - x'\beta)G(x'\beta) = -x'\beta + G(x'\beta),$

$V(\varepsilon|x) = V(y|x) = P(y = 1|x)P(y = 0|x) = G(x'\beta)\{1 - G(x'\beta)\},$

for y is binary.

In this model, $E(\varepsilon|x) \neq 0$ and the error term has a heteroskedasticity of a known form. Hence, as long as $x'\beta \neq G(x'\beta)$—$x'\beta$ can go out of $[0,1]$ while $G(x'\beta)$ falls in $[0,1]$—LSE is not valid. Despite this, however, the LSE with heteroskedasticity-robust variance or WLS are sometimes used for binary responses, and their results tend to be similar to those of the following MLE's.

1.2   Logit and Probit

One choice of the df $G(\cdot)$ of u is the logistic df:

$$G(u, \delta) = \frac{e^{u/\delta}}{1 + e^{u/\delta}} = \frac{1}{1 + e^{-u/\delta}}$$

which has mean 0 and variance $\delta^2 \pi^2/3$, and G is symmetric about 0. Choosing $\delta = \sqrt{3}/\pi$ renders the standardized logistic distribution. But usually $\delta = 1$ is used (i.e., $V(u) = \pi^2/3$ and $SD(u) \simeq 1.8$), and for this logistic distribution, the MLE is called *logit*. Another well-known choice for G is $N(0, \sigma^2)$. In this case, the standardized error-term distribution function Φ is for $N(0,1)$, and the MLE is called *probit*. Since logit is similar to probit, from now on we focus on probit. Denote the $N(0,1)$ density function as ϕ. Typically, since the form of the logit density is not much different from that of ϕ, the signs of logit estimates are almost the same as those of the probit estimates. The logit estimates, however, tend to be bigger than the probit estimates by about 80%, because logit has $SD(u) = 1.8$. An intuitive way to see this point is multiplying $y^*/\sigma = x'\alpha + u/\sigma$ by 1.8 to get $x'(1.8\alpha) + (u/\sigma)1.8$: the new error term has SD 1.8 and the new regression function parameters are 1.8 times the old parameters.

The probit likelihood function is

$$L(a) = \prod_{i=1}^{N} \Phi(x_i'a)^{y_i} \left\{1 - \Phi(x_i'a)\right\}^{1-y_i}$$

$$\{= \Phi(x_i'a) \text{ if } y_i = 1 \text{ and } 1 - \Phi(x_i'a) \text{ if } y_i = 0\}$$

$$\implies \text{log-likelihood function is}$$

$$Q_N(a) = \sum_i \{y_i \ln \Phi(x_i'a) + (1 - y_i) \ln(1 - \Phi(x_i'a))\}.$$

Denote $\Phi(x_i'a)$ and $\phi(x_i'a)$ as Φ_i and ϕ_i, respectively, to get

$$\frac{\partial Q_N(a)}{\partial a} = \sum_i s_i(a), \quad \text{where}$$

$$s_i(a) \equiv \frac{(y_i - \Phi_i)\phi_i x_i}{\Phi_i(1 - \Phi_i)} \text{ is the score function.}$$

We have $\sqrt{N}(a_{mle} - \alpha) \rightsquigarrow N(0, I_f^{-1})$, where

$$I_f \equiv E\{s(\alpha)s(\alpha)'\} = E\left[\frac{\phi(x'\alpha)^2 \cdot xx'\{y - \Phi(x'\alpha)\}^2}{\{\Phi(x'\alpha)(1 - \Phi(x'\alpha))\}^2}\right]$$

$$= E\left[\frac{\phi(x'\alpha)^2 \cdot xx'}{\Phi(x'\alpha)\{1 - \Phi(x'\alpha)\}}\right] \quad \text{because} \quad E[\{y - \Phi(x'\alpha)\}^2|x]$$

$$= \Phi(x'\alpha)(1 - \Phi(x'\alpha)).$$

The equation $N^{-1}\partial Q_N(a)/\partial a$ can be viewed as a moment condition $N^{-1} \sum_i v_i w_i = 0$, where $v_i = y_i - \Phi_i$ and $w_i \equiv x_i \phi_i/\{\Phi_i(1 - \Phi_i)\}$ is an instrument. Alternatively, regard $x_i \phi_i/\{\Phi_i(1 - \Phi_i)\}^{1/2}$ as the instrument and $v_i/\{\Phi_i(1 - \Phi_i)\}^{1/2}$ as the weighted or standardized residual, because $V(v|x_i) = \Phi_i(1 - \Phi_i)$.

Once a binary response model is estimated, one would like to have a goodness-of-fit measure as R^2 for linear models. There are a number of pseudo R^2's, but

$$\frac{a'_{mle} N^{-1} \sum_i (x_i - \bar{x})(x_i - \bar{x})' a_{mle}}{a'_{mle} N^{-1} \sum_i (x_i - \bar{x})(x_i - \bar{x})' a_{mle} + 1},$$

suggested by Mckelvey and Zavoina (1975) seems to be the best for the following reason.

Define $\tilde{a}_{mle}$ as the slope components of a_{mle}, $x_{is} \equiv (x_{i2}, ..., x_{ik})'$, and $\bar{x}_s \equiv N^{-1} \sum_i x_{is}$. Since the element of $x_i - \bar{x}$ corresponding to $x_{i1} = 1$ is zero, the pseudo R^2 can be written as

$$R_S^2 \equiv \frac{\tilde{a}'_{mle} N^{-1} \sum_i (x_{is} - \bar{x}_s)(x_{is} - \bar{x}_s)' \tilde{a}_{mle}}{\tilde{a}'_{mle} N^{-1} \sum_i (x_{is} - \bar{x}_s)(x_{is} - \bar{x}_s)' \tilde{a}_{mle} + 1}$$

$$= \frac{\tilde{b}'_{mle} N^{-1} \sum_i (x_{is} - \bar{x}_s)(x_{is} - \bar{x}_s)' \tilde{b}_{mle}}{\tilde{b}'_{mle} N^{-1} \sum_i (x_{is} - \bar{x}_s)(x_{is} - \bar{x}_s)' \tilde{b}_{mle} + \sigma^2}$$

where the second term is obtained by multiplying all terms by σ^2 and $\tilde{b}_{mle}$ is defined as $\tilde{a}_{mle}\sigma$. In the last expression, the numerator is the explained variation and the denominator is the total variation in the usual linear-model R^2. This is the reason why R_S^2 is recommended. For logit, an analogous pseudo R^2 is R_S^2 with the number 1 in the denominator replaced by the logit error term variance $\pi^2/3$. See Laitila (1993) and Veall and Zimmermann (1996) for more on pseudo R^2's for LDV models.

Recall the probit example for union membership that appeared already in Chapter 3.3.3 for three MLE tests. Table 1 shows probit and logit results (t-values are in $(\cdot)$). As noted already, logit is greater in magnitude than probit by a proportion somewhere around 80%; for two variables bus (business) and sou (living in South), however, the proportion is much greater. The signs of probit and logit are exactly the same and the t-values are almost the same except for the two variables bus and sou. This is understandable if we regard logit as 1.8 times probit; such multiplication does not change the t-value.

From the table, for instance, $sgn(\alpha_{edu})$ $\{= sgn(\beta_{edu})\}$ is negative, showing the less educated tend to join labor union more. Other than signs, we can also learn about

$$\frac{\alpha_{blc}}{\alpha_{rur}} \left(= \frac{\beta_{blc}}{\beta_{rur}} \right) = \frac{0.763}{1.294} = 0.59 :$$

the blc's (black) influence on y^* is only 59% of the rur's (living in rural area) influence. Other than signs and ratios of β, we can also interpret $\alpha_j = \beta_j/\sigma$ directly as the "magnitude of β_j relative to σ." For instance, $\alpha_j = 2$ means $\beta_j = 2\sigma$, which is a very large number: if x_j changes from 0 to 1, then $\alpha_j x_j$ increases by 2, which can push almost any negative y^*/σ to the positive side (i.e., y changes from 0 to 1) because the probability of y^*/σ

Table 1: Binary Response: Joining Union or Not

Regressor	probit (tv)	logit (tv)	logit/probit
one	6.438 (2.42)	11.550 (2.38)	1.796
edu	−0.724 (−3.12)	−1.283 (−3.03)	1.773
exr	−0.721 (−3.09)	−1.296 (−2.98)	1.799
exredu	0.068 (3.23)	0.122 (3.12)	1.795
blc	0.763 (4.02)	1.287 (4.05)	1.686
hisp	0.175 (0.96)	0.327 (1.07)	1.869
agr	1.090 (1.91)	1.862 (1.91)	1.709
bus	−0.148 (−0.32)	−0.362 (−0.41)	2.430
cst	0.614 (1.41)	1.046 (1.35)	1.706
fin	−0.256 (−0.45)	−0.469 (−0.45)	1.830
man	0.575 (1.42)	0.975 (1.33)	1.699
pro	0.815 (1.80)	1.402 (1.75)	1.721
pub	−0.083 (−0.51)	−0.146 (−0.53)	1.769
trad	0.226 (0.55)	0.382 (0.51)	1.691
tran	0.934 (2.14)	1.556 (2.00)	1.667
mar	0.131 (0.94)	0.231 (0.98)	1.769
rur	1.294 (2.91)	2.159 (2.74)	1.669
sou	−0.019 (−0.14)	−0.055 (−0.23)	2.902
log-like	−275.649	−275.705	

smaller than -2 is less than 2.5% "on the $N(0,1)$ scale." Indeed, in the above union membership example, all slope estimates are smaller than 2 in absolute magnitude.

Earlier we mentioned the possibility of u being heteroskedastic with $V(u|x) = \sigma(x)^2$. Generalizing this slightly, suppose that u depends on w, and with m consisting of all elements in x and w,

$$u = e^{w'\zeta}v = e^{\zeta_1 + w_2\zeta_2 + , \dots , + w_\nu\zeta_\nu}v \text{ where } v \sim N(0,1)$$
$$\text{independently of } m \implies \frac{u}{e^{w'\zeta}} = v.$$

We may have $w = x$, but here we allow the possibility that different variables are used for the regression parameter for y^* and its variance function: $E(y^*|m) = x'\beta$ and $V(y^*|m) = \exp(2w'\zeta)$. For this model, we get

$$P(y = 1|m) = \Phi\left(\frac{x'\beta}{e^{w'\zeta}}\right) = \Phi\left(\frac{\beta_1 + \beta_2 x_2 + , \dots , + \beta_k x_k}{e^{\zeta_1 + w_2\zeta_2 + , \dots , + w_\nu\zeta_\nu}}\right).$$

"Heteroskedastic probit" can be done with this in the probit likelihood function, subject to the following modification to deal with the unknown scale problem.

By dividing the numerator and denominator by $\theta \equiv e^{\zeta_1}$, rewrite the last $\Phi(\cdot)$ expression as

$$\Phi\left\{\frac{\beta_1/\theta \ + (\beta_2/\theta)x_2+, \ \ldots, \ +(\beta_k/\theta)x_k}{e^{w_2\zeta_2+,\ldots,+w_\nu\zeta_\nu}}\right\}.$$

The heteroskedastic probit estimates

$$\alpha \equiv \left(\frac{\beta_1}{\theta}, \frac{\beta_2}{\theta}, \ldots, \frac{\beta_k}{\theta}, \ \zeta_2, \ldots, \zeta_\nu\right)' = \left(\frac{\beta'}{\theta}, \ \zeta_2, \ldots, \zeta_\nu\right)'.$$

If $H_0 : \zeta_2 =,\ldots,= \zeta_\nu = 0$ is accepted, then $\alpha = \beta/\theta$, which is nothing but the usual probit under $\theta = \sigma$ and the independence of u from x.

In micro data, heteroskedasticity is often present, and this way of allowing for heteroskedasticity may be adopted also for other LDV models. For LDV models to appear later where y^* is observed at least for part of the data, the scale of y^* is observed, and thus the scale normalization as in (heteroskedastic) probit is not necessary and β is fully identified.

1.3   Marginal Effects

In the linear model $y_i = x_i'\beta + u_i$, β_k is taken as the effect of x_{ik} on $E(y|x_i) \ \forall i$ because $\beta_k = \partial E(y|x_i)/\partial x_{ik} \ \forall \ i$. But in probit, we have

$$\frac{\partial E(y|x_i)}{\partial x_{ik}} = \frac{\partial \Phi(x_i'\alpha)}{\partial x_{ik}} = \phi(x_i'\alpha) \cdot \alpha_k.$$

Since this varies across i, we have N-many individual effects $\phi(x_i'a_{mle})a_{mle,k}$. Thus we may get a summary measure out of the N effects. For instance, we can think of a number of "marginal effects":

(i) $N^{-1}\sum_i \phi(x_i'a_{mle})a_{mle,k}$,

(ii) $\phi(\bar{x}'a_{mle})a_{mle,k}$,

(iii) Sample median of $\phi(x_i'a_{mle})a_{mle,k}$, $i = 1, \ldots, N$.

The word "marginal" is used relative to "conditional" when $\phi(x_i'\alpha)\alpha_k$ is viewed as the effect of x_k on $E(y|x)$ with the value of x set at x_i (or "conditional on $x = x_i$"). Marginal effect (i) is the sample average of conditional effects, whereas (ii) is the conditional effect evaluated at the "average person"—but the average person may be fictitious (think of x_i being a gender dummy). Since any of (i) to (iii) being zero is equivalent to $a_{mle,k} = 0$, testing for a zero marginal effect is equivalent to testing for $\alpha_k = 0$.

The preceding marginal effect is appropriate when x_k is continuously distributed. Suppose now that x_k is a dummy variable. For the linear function $E(y|x) = x'\beta$, we get

$$\frac{\partial E(y|x)}{\partial x_k} = \beta_k = E(y|x_1, \ldots, x_k = 1) - E(y|x_1, \ldots, x_k = 0) :$$

the derivative with an infinitesimal change in x_k agrees with the difference when $\Delta x_k = 1$. But, for nonlinear models in general, this equality does not hold and the derivative is only a linear approximation for the change in $E(y|x)$ when x_k changes from 0 to 1. Thus,

effect of x_k changing from 0 to 1 :

$$E(y|x_1, ..., x_k = 1) - E(y|x_1, ..., x_k = 0)$$
$$= A(x, \alpha) \equiv \Phi(\alpha_1 + \alpha_2 x_2 +, ..., +\alpha_{k-1} x_{k-1} + \alpha_k)$$
$$- \Phi(\alpha_1 + \alpha_2 x_2 +, ..., +\alpha_{k-1} x_{k-1})$$

is preferred to $\partial E(y|x)/\partial x_k$. As in (i) to (iii) above, we can think of the following "average" versions for $A(x_i, \alpha)$, $i = 1, ..., N$:

(iv) $N^{-1} \sum_i A(x_i, a_{mle})$,

(v) $A(\bar{x}, a_{mle})$,

(vi) Sample median of $A(x_i, \alpha)$, $i = 1, ..., N$.

Turning back to non-dummy x_k, if x_k^2 or interaction terms with x_k appear in the regression function, then the above marginal effects need modifications. Suppose that $\alpha_{k-1,k} x_{k-1} x_k + \alpha_{k2} x_k^2$ is added to the regression function. Then, for instance, the above (i) should be replaced by

$$\frac{1}{N} \sum_i \phi(x_i'\alpha)(\alpha_k + \alpha_{k-1,k} x_{i,k-1} + 2\alpha_{k2} x_{ik})$$

as $\dfrac{\partial \Phi(x_i'\alpha)}{\partial x_k} = \phi(x'\alpha)(\alpha_k + \alpha_{k-1,k} x_{k-1} + 2\alpha_{k2} x_k)$

where α is used instead of a_{mle} to simplify presentation.

Getting the marginal effects using the same labor union data, the marginal effect of rur (dummy variable for living in a rural area) is

0.369 for (i), 0.393 for (ii), 0.405 for (iii);

0.430 for (iv), 0.476 for (v), 0.455 (vi).

Depending on which marginal effect is looked at, the effect ranges from 0.369 to 0.476. For confidence intervals (CI) of marginal effects, one may use the following "*(nonparametric) bootstrap*": (1) sample N times with replacement from the original sample to construct a "pseudo-sample" of size N, (2) apply probit to the pseudo sample to get a pseudo estimate $a_N^{(b)}$, and (3) repeat steps (1) and (2), say $B = 200$ times, to get $a_N^{(1)}, ..., a_N^{(B)}$. Then use the empirical distribution of $a_N^{(1)}, ..., a_N^{(B)}$ for a CI. For instance, the lower 2.5 percentile and the upper 2.5 percentile in $a_{Nj}^{(1)}, ..., a_{Nj}^{(B)}$ yields an asymptotic 95% CI for α_j, $j = 1, ..., k$.

The standard errors can also be obtained using the two-stage M-estimator framework—the reader may skip the following. For more generality, instead of (i) above, consider its $k \times 1$ vector version centered at $E\{\phi(x'\alpha)\alpha\}$ and normalized by $\sqrt{N}$. Expand it around $a_{mle} = \alpha$ to get

$$\frac{1}{\sqrt{N}} \sum_i [\phi(x_i'a_{mle})a_{mle} - E\{\phi(x'\alpha)\alpha\}] = \frac{1}{\sqrt{N}} \sum_i [\phi(x_i'\alpha)\alpha - E\{\phi(x'\alpha)\alpha\}]$$

$$+ \frac{1}{N} \sum_i \left[\underset{k \times k}{\phi(x_i'\alpha) \, I_k} + \underset{k \times k}{\phi'(x_i'\alpha)\alpha x_i'} \right] \cdot \sqrt{N}(a_{mle} - \alpha) + o_p(1)$$

$$= \frac{1}{\sqrt{N}} \sum_i \xi_i, \quad \text{where } \xi_i \equiv [\phi(x_i'\alpha)\alpha - E\{\phi(x'\alpha)\alpha\}] + L \cdot E^{-1}(ss') \cdot s_i,$$

$L \equiv E[\phi(x'\alpha)I_k + \phi'(x'\alpha)\alpha x']$ and s_i is the score function for a_{mle}; note $\phi'(t) = -t\phi(t)$. Hence

$$\frac{1}{\sqrt{N}} \sum_i [\phi(x_i'a_{mle})a_{mle} - E\{\phi(x'\alpha)\alpha\}] \rightsquigarrow N(0, C) \text{ and } C_N \to^p C$$

where

$$C_N \equiv \frac{1}{N} \sum_i \hat{\xi}_i \hat{\xi}_i' \quad \text{and } \hat{\xi}_i \text{ is } \xi_i \text{ with } \alpha \text{ replaced by } a_{mle} \text{ and } E(\cdot)$$

by its sample version.

The t-value for the rur average-effect (i) using this is 2.722.

1.4   Willingness to Pay and Treatment Effect

1.4.1   Willingness to Pay (WTP)

Sometimes we are interested in something other than marginal effects. We examine one interesting case in this subsection: "*willingness to pay (WTP).*" This case will also demonstrate the importance of regression-functional form assumption, transformation-of-variable issue, and error-distributional assumption. We will draw on Haab and McConnell (2002, pp. 36–39 and p. 97), which the reader can refer to for extensive LDV model applications to environmental economics in general.

Suppose there is a project to improve water quality. A survey is conducted where each respondent is shown an amount t_i that he/she should pay for the project in terms of higher monthly water bills. Then the respondent is asked whether he/she would vote for the project ($y = 1$) or not ($y = 0$). The data consist of (y_i, t_i, x_i', m_i) where x_i is individual characteristics and m_i is income.

Let the "indirect utility functions" with and without the project done be

$$x_i'\beta_1 + \delta_1 \ln(m_i - t_i) + u_{1i} \quad \text{and} \quad x_i'\beta_0 + \delta_0 \ln m_i + u_{0i}.$$

The marginal utility of income takes the form δ_j/m, which is decreasing in m. We get

$$y_i = 1 \quad \text{iff} \quad x_i'\beta + \delta_1 \ln(m_i - t_i) - \delta_0 \ln m_i + u_i > 0 \quad \text{where}$$
$$\beta \equiv \beta_1 - \beta_0, u_i \equiv u_{1i} - u_{0i}.$$

Under $u \sim N(0,\sigma^2)$, we can estimate $(\beta'/\sigma, \delta_1/\sigma, \delta_0/\sigma)$ with probit where $(x', \ln(m-t), -\ln m)$ is the regressor vector.

To see WTP—i.e., to what extent people can be pushed to pay—replace t_i with w_i and set the utility difference to zero to solve it for w_i:

$$x_i'\beta + \delta_1 \ln(m_i - w_i) - \delta_0 \ln m_i + u_i = 0$$
$$\Longrightarrow \ln(m_i - w_i) = \frac{\delta_0 \ln m_i - x_i'\beta - u_i}{\delta_1}$$
$$\Longrightarrow m_i - w_i = \exp\left(\frac{\delta_0 \ln m_i - x_i'\beta - u_i}{\delta_1}\right)$$
$$\Longrightarrow w_i = m_i - \exp\left(\frac{\delta_0}{\delta_1}\ln m_i - x_i'\frac{\beta}{\delta_1} - \frac{u_i}{\delta_1}\right)$$
$$\Longrightarrow w_i = m_i - \exp\left(\frac{\delta_0/\sigma}{\delta_1/\sigma}\ln m_i - x_i'\frac{\beta/\sigma}{\delta_1/\sigma} - \frac{u_i/\sigma}{\delta_1/\sigma}\right).$$

If $\beta_1 = \beta_0$, then $\beta = 0$ and x will drop out from this display. If $\delta_1 = \delta_0$ (but not necessarily $\beta_1 = \beta_0$), then

$$w = m - m\exp\left(-x'\frac{\beta/\sigma}{\delta_1/\sigma} - \frac{u/\sigma}{\delta_1/\sigma}\right).$$

Both $\beta_1 = \beta_0$ and $\delta_1 = \delta_0$ are regression-functional form issues (i.e., whether the regression function changes or not as the "regime" changes).

To remove u in WTP, suppose we get $E(w|x,m)$: using $E[\exp\{N(\mu, c^2)\}] = \exp(\mu + c^2/2)$,

$$E(w|x,m) = m - \exp\left(\frac{\delta_0/\sigma}{\delta_1/\sigma}\ln m - x'\frac{\beta/\sigma}{\delta_1/\sigma}\right) \cdot E\left\{\exp\left(-\frac{u/\sigma}{\delta_1/\sigma}\right)\right\}$$
$$= m - \exp\left\{\frac{\delta_0/\sigma}{\delta_1/\sigma}\ln m - x'\frac{\beta/\sigma}{\delta_1/\sigma} + \frac{1}{2}\left(\frac{\sigma}{\delta_1}\right)^2\right\} \quad \text{as}$$
$$-\frac{u/\sigma}{\delta_1/\sigma} \sim N\left\{0, \left(\frac{\sigma}{\delta_1}\right)^2\right\}.$$

If u is not normal, then $E(w|x,m)$ would differ, which is an error-term distributional issue.

As an alternative to $E(w|x,m)$, because median is equivariant to increasing transformations, consider

$$Med(w|x,m) = m - \exp\left(\frac{\delta_0/\sigma}{\delta_1/\sigma}\ln m - x'\frac{\beta/\sigma}{\delta_1/\sigma}\right) \cdot Med\left\{\exp\left(-\frac{u/\sigma}{\delta/\sigma}\right)\right\}$$

$$= m - \exp\left(\frac{\delta_0/\sigma}{\delta_1/\sigma}\ln m - x'\frac{\beta/\sigma}{\delta_1/\sigma}\right).$$

This median WTP would be the same for any median-zero u including logistic error. This means that $Med(w|x,m)$ is robust to error term distribution assumption violations (so long as the error term median is zero).

Instead of taking $E(w|x,m)$ (or $Med(w|x,m)$) on w, yet another way of getting rid of u is taking $E(\cdot|x,m)$ on the utility difference first. This then would yield yet another WTP estimator. There appears to be no single best answer to the question of which WTP estimator to use by removing u at which stage and how. The above WTP estimators are affected by the probit error for $(\beta'/\sigma, \delta_1/\sigma, \delta_0/\sigma)$, which can be dealt with the two-stage procedure techniques in Chapter 3. Alternatively, nonparametric bootstrap can be used to construct a CI.

1.4.2   Remarks for WTP Estimation

Suppose m, not $\ln m$, appears, which is a transformation-of-variable issue. Then (compare to the above equation with "$y = 1$ iff..."):

$$y_i = 1 \quad \text{iff} \quad x_i'\beta + (\delta_1 - \delta_0)m_i - \delta_1 t_i + u_i > 0$$

$$\Longleftrightarrow -\frac{u_i}{\sigma} < x_i'\frac{\beta}{\sigma} + \frac{\delta_1 - \delta_0}{\sigma}m_i + \frac{\delta_1}{\sigma}(-t_i).$$

Applying probit with $(x_i', m_i, -t_i)$ as the regressors,

$$\left(\frac{\beta'}{\sigma}, \frac{\delta_1 - \delta_0}{\sigma}, \frac{\delta_1}{\sigma}\right) \text{ that is equivalent to } \left(\frac{\beta'}{\sigma}, \frac{\delta_1}{\sigma}, \frac{\delta_0}{\sigma}\right) \text{ can be estimated.}$$

To get WTP, rewrite $\delta_1 m_i - \delta_0 m_i$ as $\delta_1(m - w) - \delta_0 m$ and set the utility difference to 0 to solve it for w:

$$x'\beta + \delta_1(m - w) - \delta_0 m + u = 0 \implies w = x'\frac{\beta}{\delta_1} + (1 - \frac{\delta_0}{\delta_1})m + \frac{u}{\delta_1}$$

$$E(w|x,m) = x'\frac{\beta/\sigma}{\delta_1/\sigma} + \left(1 - \frac{\delta_0/\sigma}{\delta_1/\sigma}\right)m \quad \text{for any } u \text{ with } E(u|x,m) = 0$$

which differs much from the preceding $E(w|x,m)$ expression involving exp. In terms of the identified probit parameters,

$$E(w|x,m) = x'\frac{\text{slope } x}{\text{slope } -t} + \{1 - \frac{(\text{slope } -t) - (\text{slope } m)}{\text{slope } -t}\}m$$

$$= x'\frac{\text{slope } x}{\text{slope } -t} + \frac{\text{slope } m}{\text{slope } -t}m.$$

Because w_i makes the utility difference zero, we get

$$t_i < w_i \iff \text{utility difference is positive} \iff y_i = 1.$$

Suppose we *model* w_i *directly* as $w_i = x_i'\beta_w + \delta_w m_i + v_i$ with $v_i \sim N(0, \sigma_v^2)$. Then

$$P(y = 1 | x, m, t) = P(w > t | x, m, t) = P(v > t - x'\beta_w - \delta_w m | x, m, t)$$

$$= \Phi\left\{ x'\frac{\beta_w}{\sigma_v} + \frac{\delta_w}{\sigma_v} m + \frac{1}{\sigma_v}(-t) \right\}$$

where $(\beta_w'/\sigma_v, \delta_w/\sigma_v, 1/\sigma_v)$ is estimated by probit; the monetary scale, not utility scale, factor σ_v is identified. We get, just as above,

$$E(w|x, m) = x'\beta_w + \delta_w m = x\frac{\beta_w/\sigma_v}{1/\sigma_v} + \frac{\delta_w/\sigma_v}{1/\sigma_v} m$$

$$= x'\frac{\text{slope } x}{\text{slope } -t} + \frac{\text{slope } m}{\text{slope } -t} m.$$

WTP w should fall in $[0, m]$, i.e., $0 \le m/w \le 1$ should hold. In the above model with $\ln m$, the upper bound holds because $w = m - \exp(\cdot)$, but not necessarily the lower bound. In the model with m not with $\ln m$, however, neither bound holds, which may result in negative numbers. Hence, to impose the bounds, suppose we model w/m as a fraction: $w/m = F(x'\beta + u)$ where $0 \le F(\cdot) \le 1$. Specifically, suppose $F(x'\beta + u) = [1 + \exp\{-(x'\beta + u)\}]^{-1}$ to get

$$w = \frac{m}{1 + \exp\{-(x'\beta + u)\}} > t \iff u > -x'\beta - \ln\frac{m - t}{t}.$$

Because $y = 1 \iff w > t$, this leads to

$$P(y = 1 | x, m) = \Phi\left(x\frac{\beta}{\sigma} + \frac{1}{\sigma} \ln\frac{m - t}{t} \right) \quad \text{and}$$

$$Med(w|x, m) = \frac{m}{1 + \exp(-x'\beta)}.$$

Because β is identified as the slope of x divided by the slope of $\ln\{(m - t)/t\}$ in the probit, this median WTP is also identified. Clearly, this median WTP respects the bound $[0, m]$. One shortcoming in this approach is, however, that $w/m = F(x'\beta + u)$ with a smooth F rules out w/m taking the boundary values $0, 1$. There are better ways to model proportions or "fractional responses" such as w/m, and they will be seen later in the section for censored responses.

1.4.3   Comparison to Treatment Effect

It is instructive to view WTP from the treatment effect framework that appeared in the preceding chapter. Let the water quality improvement project be the treatment $d = 0, 1$. If there is a data on (d_i, z_i, x_i', m_i) where z is a response variable measuring the benefit of the project to person i, say blood pressure, then $E(z^1 - z^0)$ is the mean effect of the treatment on the response z. But the consequences of the treatment can be felt on many other things, not just on blood pressure: tooth decay level, cholesterol, bottled water expense etc. How do we combine them all? The answer is the ubiquitous "*utility*" (or satisfaction) as done above with the indirect utility function.

Using utility as the "summary response" in the treatment effect framework, however, poses a couple of problems. First of all, utility is not observed: we cannot get, for instance, the group mean difference $E(\text{utility}|d = 1) - E(\text{utility}|d = 0)$. Second, the treatment might never have been implemented, which means no available data. The above framework of asking people whether they would vote for the treatment or not is a way of overcoming these two problems. By including income and the treatment cost in the indirect utility function and then getting WTP, we can see the monetary benefit *WTP that essentially combines all consequences, good or bad, of the treatment.*

Finding the indirect utility function parameters by asking directly whether the respondent would vote for the treatment or not requires some caution: the respondents should be well informed of all possible consequences of the treatment including the direct cost t_i. Otherwise, their welfare can be quite different when they get better informed (or realize unexpected consequences) later. Even when the treatment has been implemented and thus people's preference have been observed, some caution is still needed for applying the above method to non-market goods. For instance, suppose that the treatment is a national park access, and that we have data on (y_i, t_i, x_i', m_i) where $y_i = 1$ means visiting a national park, which is the realized version of "voting yes for the national park" if the park had not been available. In this case, the cost t_i should include not just the park entrance fee, but also the travel cost to the park including gas, lodging, and time opportunity cost.

Relatively speaking, the treatment effect framework is a reduced form (RF) approach whereas the above way of finding WTP is a structural form (SF) approach. For the former, we can imagine a "black box" which a treatment goes into and an outcome comes out of; what is going on inside the box is left unexplained. In the latter SF approach, we try to find parameters governing our behavior, and WTP gets obtained with those parameters. *When the treatment has never been tried and consequently no outcome data is available, doing the SF approach which calls for more assumptions is unavoidable;* more structures and assumptions are substitutes for data.

2   Ordered Discrete Response

2.1   Basics

Suppose that y_i^* $(= x_i'\beta + u_i)$ is continuously distributed with $SD(u) = \sigma$, but the observed response y_i is an ordered discrete response (ODR) taking $0, 1, ..., R - 1$ (R ordered categories) determined by fixed thresholds γ_r's:

$$y_i = r \quad \text{if } \gamma_r \le y_i^* < \gamma_{r+1}, \quad r = 0, ..., R-1, \ \gamma_0 = -\infty, \ \gamma_R = \infty$$

$$\implies y_i = \sum_{r=1}^{R-1} 1[y_i^* \ge \gamma_r] = \sum_{r=1}^{R-1} 1[u_i \ge \gamma_r - x_i'\beta];$$

see Figure 1. That is, omitting the subscript i,

$$
\begin{aligned}
y \ &= 0 && \text{if } x'\beta + u < \gamma_1, \\
&= 1 && \text{if } \gamma_1 \le x'\beta + u < \gamma_2, \\
&\ \vdots && \quad \vdots \\
&= R-1 && \text{if } \gamma_{R-1} \le x'\beta + u.
\end{aligned}
$$

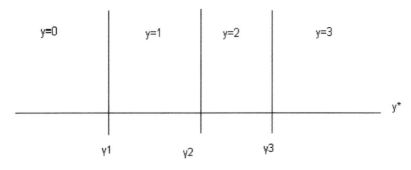

Figure 1: Observed Response (y) and Latent Response (y*)

ODR includes binary response as a special case when $R = 2$: $y = 1[x'\beta + u \ge \gamma_1]$ and γ_1 gets absorbed into the intercept in $x'\beta$ as shown in the binary response section. More specifically, our ODR model is

$$x_i = (1, x_{i2}, ..., x_{ik})', \ \ \beta = (\beta_1, ..., \beta_k)', \ \ y_i = \sum_{r=1}^{R-1} 1\left[y_i^* \ge \gamma_r\right],$$

$$y_i^* = x_i'\beta + u_i$$

$$u_i \amalg x_i \quad \text{and} \quad u \text{ has df } G \text{ with } E(u) = 0 \text{ and } V(u) = \sigma^2 \ (\sigma \text{ unknown})$$

$$(x_i', y_i), \ i = 1, ..., N, \text{ are iid and observed}$$

One example of ODR is income data in which individual income is not recorded, but the bracket to which the income belongs is known. Another example is the number of durable goods purchased (car or TV). Yet another example is the answer to a question such as "Are you satisfied with life?"; the answer can be very negative (0), negative (1), neutral (2), positive (3), or very positive (4). Depending on restrictions placed on the γ_r's, various specifications are possible: γ_r's may be known, or unknown but limited in its range, or completely unknown except for their ordering $\gamma_1 < \gamma_2 <, ..., < \gamma_{R-1}$.

Suppose the γ_r's are unknown. As in binary response, parameters are not fully identified; specifically, γ_1 is absorbed into the intercept β_1 and β has to be divided by a scale factor such as $SD(u)$. To see this, subtract γ_1 from $\gamma_r \leq x'\beta + u < \gamma_{r+1}$ and divide the inequalities by σ to get

$$\frac{\gamma_r - \gamma_1}{\sigma} \leq \frac{x'\beta - \gamma_1}{\sigma} + \frac{u}{\sigma} < \frac{\gamma_{r+1} - \gamma_1}{\sigma}, \quad r = 0, ..., R - 1.$$

Here γ_1 is absorbed into β_1, and what is identified is

$$\alpha \equiv \left(\frac{\beta_1 - \gamma_1}{\sigma}, \frac{\beta_2}{\sigma}, ..., \frac{\beta_k}{\sigma} \right)', \quad \tau_r \equiv \frac{\gamma_r - \gamma_1}{\sigma}, \quad r = 2, ..., R - 1.$$

For instance, with three categories $(R = 3)$, α and one normalized threshold difference $\tau_2 = (\gamma_2 - \gamma_1)/\sigma$ are identified; compare this to binary response where no threshold parameter is identified. With four categories, α and two normalized threshold differences $\tau_2 = (\gamma_2 - \gamma_1)/\sigma$ and $\tau_3 = (\gamma_3 - \gamma_1)/\sigma$ are identified. Each additional category adds one more identified threshold parameter.

Although γ_r's are not identified, τ_r's are still useful in finding whether the threshold differences are the same or not. For instance,

$$\tau_{r+2} - \tau_{r+1} = \frac{\gamma_{r+2} - \gamma_{r+1}}{\sigma} \quad \text{and} \quad \tau_{r+1} - \tau_r = \frac{\gamma_{r+1} - \gamma_r}{\sigma} :$$

$\tau_{r+2} - \tau_{r+1} \neq \tau_{r+1} - \tau_r$ is equivalent to $\gamma_{r+2} - \gamma_{r+1} \neq \gamma_{r+1} - \gamma_r$.

2.2   Digression on Re-parametrization in MLE

Pinning down the identified parameters in ODR as done above is not the only possibility. An alternative is to subtract the intercept β_1, not γ_1, from $\gamma_r \leq x'\beta + u < \gamma_{r+1}$ and then divide through by σ to get

$$\frac{\gamma_r - \beta_1}{\sigma} \leq \frac{x'\beta - \beta_1}{\sigma} + \frac{u}{\sigma} < \frac{\gamma_{r+1} - \beta_1}{\sigma}, \quad r = 0, ..., R - 1.$$

Here the intercept is zero, and what is identified is

$$\tilde{\alpha} \equiv \left(\frac{\beta_2}{\sigma}, ..., \frac{\beta_k}{\sigma} \right)', \quad \tilde{\tau}_r \equiv \frac{\gamma_r - \beta_1}{\sigma}, \quad r = 1, ..., R - 1.$$

But this re-parametrization is equivalent to the initial parametrization in the sense that α and τ_r's are one-to-one to $\tilde{\alpha}$ and $\tilde{\tau}_r$'s and vice versa. For instance, with $k = 2$ and $R = 4$, the re-parametrization can be written as an one-to-one transformation of the initial parametrization and vice versa (set $\sigma = 1$ for simplification that would otherwise appear in all terms):

$$\begin{bmatrix} \beta_2 \\ \gamma_1 - \beta_1 \\ \gamma_2 - \beta_1 \\ \gamma_3 - \beta_1 \end{bmatrix} = \begin{bmatrix} 0 & 1 & 0 & 0 \\ -1 & 0 & 0 & 0 \\ -1 & 0 & 1 & 0 \\ -1 & 0 & 0 & 1 \end{bmatrix} \begin{bmatrix} \beta_1 - \gamma_1 \\ \beta_2 \\ \gamma_2 - \gamma_1 \\ \gamma_3 - \gamma_1 \end{bmatrix}.$$

Defining the 4×4 matrix as T, $rank(T) = 4$: this transformation and its inverse are one-to-one. Of course, if desired, we can do the "location normalization" with something other than γ_1 and β_1 (say, with γ_2). The reason we used γ_1 is that it is coherent with binary response models with threshold 0.

More generally, consider MLE with the likelihood function $L(\theta)$ with its parameter θ. Let θ_N be the MLE which uniquely maximizes $L(\theta)$. Now consider a reparametrization $\tilde{\theta} \equiv T\theta$ (as in the last display) with T^{-1} existing. Then we maximize $L(\theta) = L(T^{-1}T\theta) = L(T^{-1}\tilde{\theta}) \equiv L^*(\tilde{\theta})$ wrt $\tilde{\theta}$ to get $\tilde{\theta}_N$. Suppose $\tilde{\theta}_N = T\theta_N + c$ for some c. Then

$$L^*(\tilde{\theta}_N) = L\{T^{-1}(T\theta_N + c)\} = L(\theta_N + T^{-1}c)$$

which is smaller than $L(\theta_N)$ unless $T^{-1}c = 0 \iff c = 0$ as θ_N is the unique maximizer of $L(\theta)$. Thus it should be that $c = 0$ and $\tilde{\theta}_N = T\theta_N$: *the MLE for $T\theta$ is just the T-transformation of the initial MLE.*

Consider a linear hypothesis

$$H_0 : R\theta = r \iff RT^{-1}T\theta = r \iff RT^{-1}\tilde{\theta} = r$$
$$\iff \tilde{R}\tilde{\theta} = r \quad \text{where } \tilde{R} \equiv RT^{-1}.$$

If we conduct a likelihood-ratio (LR) test, then the re-parametrization does not affect the test result, because the maximized likelihood does not change. This is the main attraction of LR test, despite its disadvantage of requiring two estimators, with and without the H_0 restriction. Although we used a linear transformation $T\theta$, our discussion involving (L, L^*) and T also holds for nonlinear one-to-one and onto transformations $T(\theta)$.

2.3   Ordered Probit

Suppose u follows $N(0, \sigma^2)$ independently of x as in probit for binary response. Denote the $N(0, 1)$ distribution function and density as Φ and ϕ, respectively. Observe

$$y = r \iff \gamma_r - x'\beta \leq u < \gamma_{r+1} - x'\beta$$
$$\iff \frac{\gamma_r - \gamma_1}{\sigma} + \frac{\gamma_1 - x'\beta}{\sigma} \leq \frac{u}{\sigma} < \frac{\gamma_{r+1} - \gamma_1}{\sigma} + \frac{\gamma_1 - x'\beta}{\sigma}$$
$$\iff \tau_r - x'\alpha \leq \frac{u}{\sigma} < \tau_{r+1} - x'\alpha.$$

Hence, we get

$$P(y = r|x) = P(\tau_r - x'\alpha \leq \frac{u}{\sigma} < \tau_{r+1} - x'\alpha|x)$$
$$= \Phi(\tau_{r+1} - x'\alpha) - \Phi(\tau_r - x'\alpha).$$

Define, for all i,

$$y_{ir} = 1 \text{ if } y_i = r, \text{ and } y_{ir} = 0 \text{ otherwise}, \quad \text{for } r = 0, ..., R - 1.$$

Assuming that the γ_r's are unknown, *ordered probit* maximizes

$$Q_N(a,t) \equiv \sum_{i=1}^{N} \sum_{r=0}^{R-1} y_{ir} \ln\{\Phi(t_{r+1} - x_i'a) - \Phi(t_r - x_i'a)\}$$

wrt a and t_r's. The first derivatives are ($r = 2, ..., R - 1$ for t_r)

$$\frac{\partial Q_N}{\partial a} = \sum_{i=1}^{N} \sum_{r=0}^{R-1} y_{ir} \frac{\phi(t_{r+1} - x_i'a) - \phi(t_r - x_i'a)}{\Phi(t_{r+1} - x_i'a) - \Phi(t_r - x_i'a)} (-x_i);$$

$$\frac{\partial Q_N}{\partial t_r} = \sum_{i=1}^{N} \phi(t_r - x_i'a) \left\{ \frac{y_{i,r-1}}{\Phi(t_r - x_i'a) - \Phi(t_{r-1} - x_i'a)} \right.$$

$$\left. - \frac{y_{ir}}{\Phi(t_{r+1} - x_i'a) - \Phi(t_r - x_i'a)} \right\}.$$

With these, ordered probit can be easily implemented. The Newton–Raphson algorithm converges straightforwardly for ordered probit as the maximand is concave; see Pratt (1981). The same pseudo R^2 as in binary response can be used. If we use the logistic distribution for u, then we would get *ordered logit*.

Applying ODR models with unknown thresholds, one often confronts the question of threshold constancy: given that the regression function depends on x, would it not be likely that the thresholds are functions of x as well? Terza (1985) used a bond-rating ODR variable: the bond rating companies may not be applying the same standard (the thresholds) to all companies, and indeed this was found to be the case. As another example, suppose that y^* is promotability and y is the observed rank. If thresholds depend on race (or sex), then this is an evidence for discrimination in promotion. Winter-Ebmer and Zweimuller (1997) and Pudney and Shields (2000) applied an ODR model with varying thresholds to promotion processes. When the thresholds depend on x, there is an identification issue to deal with, because the first threshold is subtracted from the regression function and the other thresholds. Lee and Kimhi (2005) dealt with the issue in detail, although we do not pursue regressor-dependent thresholds any further here.

Lee and Kimhi (2005) also proposed a test for threshold constancy using the difference between probit and ordered probit under constant thresholds ("COPRO")—note that ODR can be collapsed into binary responses in $R - 1$ ways. The test works because probit does not require threshold constancy other than for the normalizing threshold. If the thresholds are constant, then both probit and COPRO regression function estimators are consistent; otherwise, while probit is still consistent, COPRO is not. Hence a Wald test based on the difference between probit and COPRO can be devised.

Marginal effects of x on each choice probabilities analogous to those for probit can be derived:

$$P(y = r|x) = \Phi(\tau_{r+1} - x'\alpha) - \Phi(\tau_r - x'\alpha)$$

$$\implies \frac{\partial P(y = r|x)}{\partial x} = \{\phi(\tau_{r+1} - x'\alpha) - \phi(\tau_r - x'\alpha)\}(-\alpha).$$

A consistent estimator for the marginal effect $E\{\partial P(y = r|x)/\partial x\}$ is

$$A_{rN}a_N, \quad \text{where} \quad A_{rN} \equiv \frac{1}{N}\sum_{i=1}^{N}\{\phi(t_{Nr} - x'a_N) - \phi(t_{N,r+1} - x'a_N)\}.$$

In interpreting these effects, a caution is warranted because

$$\sum_{r=0}^{R}P(y = r|x) = 1 \implies \sum_{r=0}^{R}\frac{\partial P(y = r|x)}{\partial x} = 0:$$

an increase in some choice probability necessarily entails a decrease in some other choice probabilities.

If the categories are not just ordinal but cardinal, then we may use $E\{\partial E(y|x)/\partial x\}$ as a marginal effect where

$$\frac{\partial E(y|x)}{\partial x} = \sum_{r=1}^{R-1}r\{\phi(\tau_{r+1} - x'\alpha) - \phi(\tau_r - x'\alpha)\}(-\alpha).$$

A consistent estimator for this

$$\left[\frac{1}{N}\sum_{i=1}^{N}\sum_{r=1}^{R-1}r\{\phi(t_{Nr} - x'a_N) - \phi(t_{N,r+1} - x'a_N)\}\right]a_N.$$

This is convenient because we do not have to look at $\partial P(y = r|x)/\partial x$ $\forall r$. Even if the categories are not cardinal, sometimes $E(y|x)$ is used in practice (particularly if the category number is high). For instance, letter grades are ordinal, but often numbers are assigned to them and grade point average (GPA) is computed.

2.4   An Empirical Example: Contingent Valuation

Contingent valuation methods (CVM) are frequently used in environmental and tourism economics to value non-market goods. In the early versions of CVM, the respondents were asked how much they are willing to pay to improve a public good (e.g., national park or air quality) or to avoid deterioration. Since this often led to implausible numbers, the later versions of CVM became somewhat conservative and ask whether the respondents are willing to pay a specified amount. The answer is recorded as 0/1, and this way of finding WTP has been examined in the preceding section already.

Going one step further from binary response, if the answer to the first question is yes, the amount is increased and the question is asked again; if the answer to the first question is no, the amount is lowered and the question is asked again. In this case, we get four ordered categories with known thresholds:

$$\text{no/no: } y = 0, \text{ no/yes: } y = 1, \text{ yes/no: } y = 2, \text{ yes/yes: } y = 3.$$

The asked amounts are typically drawn from several preset values, and one randomly selected value among them is presented to each individual; i.e., the thresholds are known and typically vary across i.

In ordered probit, $y^*|x$ is symmetric with its support $(-\infty, \infty)$. For WTP in CVM, the range of $y^*|x$ should be $[0, \infty)$ and it may be more likely for y^* to take small values with high probabilities and large values with low probabilities. One distribution for these features is "Weibull distribution": using the notation T instead of y^* for a while, if T^α follows the exponential distribution with parameter $\theta > 0$; i.e., if

$$P(T^\alpha \leq t) = 1 - \exp(-\theta t)$$

then T follows Weibull distribution with parameters θ and α ($\theta, \alpha > 0$). Weibull includes exponential as a special case with $\alpha = 1$. The Weibull df is $1 - \exp(-\theta t^\alpha)$ and its density is $f(t) = \theta \alpha t^{\alpha-1} \exp(-\theta t^\alpha)$. When α is small, the distribution puts most probability mass around 0, and as α increases, the Weibull density takes an up-and-down shape and becomes more symmetric about its mode. Weibull will reappear later for duration analysis, and $f(t)$ will be understood better there. See Figure 2.

In using Weibull distribution, usually regressors are introduced in nonlinear fashion with $\theta = \exp(-x'\beta)$. Here, $\exp(\cdot)$ is used to assure $\theta > 0$ for all x, and the minus sign is attached because the mean of Weibull distribution is inversely related to θ: with the minus sign, a higher $x'\beta$ would mean the higher WTP. The θ parametrization yields

$$F(t; \alpha, x'\beta) \equiv P(T \leq t|x) = P(T^\alpha \leq t^\alpha|x) = 1 - \exp(-e^{-x'\beta}t^\alpha);$$

notations α and β are different from those of ordered probit. Denoting the observed thresholds as $\gamma_{i1}, \gamma_{i2}, \gamma_{i3}$ for person i, the Weibull ODR log-likelihood to maximize for α and β is

$$\sum_{i=1}^{N} [\, y_{i0} \ln F(\gamma_{i1}; \alpha, x_i'\beta) + y_{i1} \ln\{F(\gamma_{i2}; \alpha, x_i'\beta) - F(\gamma_{i1}; \alpha, x_i'\beta)\}$$
$$+ y_{i2} \ln\{F(\gamma_{i3}; \alpha, x_i'\beta) - F(\gamma_{i2}; \alpha, x_i'\beta)\} + y_{i3} \ln\{1 - F(\gamma_{i3}; \alpha, x_i'\beta)\} \,].$$

Whereas the distribution of the error term is parametrized in ordered probit, here the distribution of T (i.e., y^*) is parametrized. Since T carries a known monetary unit, there is no unknown scale parameter σ in this Weibull ODR,

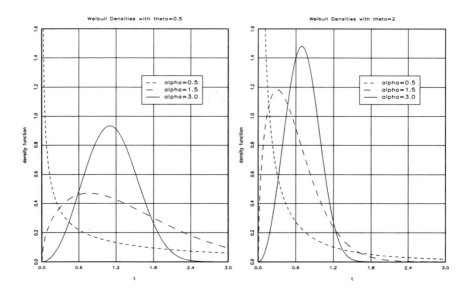

Figure 2: Weibull Densities

as this was the case in the binary response model in the previous section where WTP amount was directly modeled.

Werner (1999) applied this model to see how much people are willing to pay to preserve Kakadu Conservation Zone in Australia from mining in 1990. With $N = 1827$, the Table 2 shows some estimates in Table 3 of Werner (1999). The findings are self-explanatory: people with more environmental concern, younger age, and higher income are more willing to pay for the preservation. With the parameter estimates, Werner (1999) computed the household median WTP t_m by solving

$$0.5 = \frac{1}{N} \sum_i F\left(t_m; a_N, x_i' b_N\right)$$

Table 2: Contingent Valuation

Regressor	Estimate (SD)
jobs are important	-0.377 (0.0657)
financial benefits are important	-0.291 (0.0679)
mining hurts the value of the park	0.251 (0.0652)
there should be more parks	0.157 (0.0668)
environmentally conscious	0.341 (0.147)
age	-0.019 (0.0045)
income in \$1000	0.0223 (0.0058)
α	0.714 (0.0431)

Table 3: Inequality Effect on Crime

Regressor	Violent: b_N (SD)	Property: b_N (SD)
Population density per mile ^{2}	0.0582 (0.0168)	0.0362 (0.0122)
Income Gini coefficient	1.33 (0.151)	−0.154 (0.117)
% female-head family	1.58 (0.0942)	0.792 (0.0709)
% nonwhite	0.0568 (0.0391)	−0.0305 (0.0251)
% unemployed male	−0.0489 (0.0775)	0.0158 (0.0599)
% below poverty line	−0.216 (0.0666)	0.345 (0.0483)
% lived elsewhere 5 years ago	1.604 (0.123)	1.338 (0.0879)
% age 16–24	−0.906 (0.102)	−1.036 (0.0786)
% college-educated	−0.297 (0.0659)	0.184 (0.0475)
Police expenditure per capita 1987	−0.0209 (0.0387)	−0.105 (0.0341)

which turns out to be \$85.71; for individual i, solving $0.5 = F(t_m; a_N, x_i' b_N)$ would yield the median WTP. When t_m is multiplied by the number of the Australian households in 1990, this gives the final desired willingness-to-pay estimate of the Australian households in 1990. Compared with the WTP analysis with binary response, here we have more information by bounding WTP with ODR.

Werner (1999) in fact examined more general models than the Weibull ODR to allow for people who is not willing to pay any positive amount. Such models are worth considering, because there is a worry that numbers derived from CVM for WTP might be overblown. Clinch and Murphy (2001) went one step further allowing for negative WTP. This can happen, because some public goods are costs to some people. For instance, environmentalists may like to re-introduce wolves into an area, which can, however, be a cost to the local farmers. A more interesting example of CVM involving the infamous 1989 Exxon Valdez oil spill along the Alaskan coast can be found in Carson et al. (2003). The State of Alaska and the US government settled the case with Exxon for \$1 billion. Additionally, Exxon spent more than \$2 billion for restoration.

One curious feature in the above four-category ODR is that the respondent tends to behave differently on the follow-up (i.e., second) question. For instance, the proportion of the yes responses to \$50 price tag differs depending on whether it was asked in the first question or in the second—this can be verified in a given data set. In general, regardless of the price tag, those respondents who said yes to the first question are more likely to say no to the second question. This would lead to a lower WTP estimate in the four-category ODR than otherwise. One solution would be to merge the top two categories (yes/no and yes/yes) into one and do three-category ODR. If the opposite problem appears for no/no and no/yes—i.e., more likely to say "yes" following an initial "no"—then the bottom two categories may be merged into one. See Carson and Hanemann (2005) and the references therein if further interested in CVM.

3   Count Response

3.1   Basics and Poisson MLE

Suppose y takes integers $0, 1, 2...$, which are cardinal, not just ordinal. Then y is a count response. For example, y can be a number of accidents, crimes, visits to certain places, successes, etc. over some time period. The preceding ODR methods are applicable to count responses, but there are other approaches that make use of the fact that count responses are cardinal; that is, the differences in category numbers such as $3 - 2$ and $1 - 0$ are comparable. This section reviews "Poisson-like" MLE and NLS for count responses.

In *Poisson MLE*, we assume that $y|x$ follows Poisson distribution with parameter $\lambda(x) > 0$, denoted as $Poi\{\lambda(x)\}$:

$$P(y = r|x) = \frac{\lambda(x)^r}{r!} e^{-\lambda(x)}, \quad r = 0, 1, 2, ...$$

For $Poi\{\lambda(x)\}$, it holds that $E(y|x) = V(y|x) = \lambda(x)$. Focus on the first term $\lambda(x)^r/r!$. Since $r!$ increases much faster than $\lambda(x)^r$ as $r \to \infty$, the probability of y taking a large integer decreases rapidly. The second term $e^{-\lambda(x)}$ is a normalizing factor for the sum of the first term over r: $\sum_{r=0}^{\infty} \lambda(x)^r/r! = e^{\lambda(x)}$.

To assure $\lambda(x) > 0$ while keeping the linear model, the usual specification for $\lambda(x)$ is

$$\lambda(x) = \exp(x'\beta).$$

This yields the log-likelihood function

$$Q_N(b) = \sum_i \{y_i(x_i'b) - \exp(x_i'b) - \ln(y_i!)\} \quad \Longrightarrow$$

$$\text{score function } s_i = \{y_i - \exp(x_i'b)\}x_i.$$

Formally, the Poisson MLE model is

$$x_i = (1, x_{i2}, ..., x_{ik})', \quad \beta = (\beta_1, ..., \beta_k)'$$

$$P(y_i = r|x_i) = \frac{\{\exp(x_i'\beta)\}^r}{r!} e^{-\exp(x_i'\beta)}, \quad r = 0, 1, 2, ...$$

$$(x_i', y_i), \ i = 1, ..., N, \text{ are iid and observed.}$$

The main objective in ordered discrete response was to link a discrete y to possibly continuous $x'\beta$. In ordered probit, $x'\beta$ affects y through $\Phi(\cdot)$, whereas $x'\beta$ affects y through $\lambda(\cdot)$ in Poisson MLE. Both $\Phi(\cdot)$ and $\lambda(\cdot)$ can take continuous values.

Differentiate the log-likelihood function wrt b to get

$$\frac{\partial Q_N(b)}{\partial b} = \sum_i \{y_i - \exp(x_i'b)\} x_i, \quad \frac{\partial^2 Q_N(b)}{\partial b \partial b'} = \sum_i \{-\exp(x_i'b)\}x_i x_i'$$

which is n.d. for all b if $\sum_i x_i x_i'$ is p.d. Hence, for Poisson MLE, a Newton–Raphson-type algorithm converges straightforwardly. Defining $v_i \equiv y - \exp(x_i'\beta)$, v is heteroskedastic by construction because $V(v|x) = V(y|x) = \exp(x'\beta)$.

A pseudo R^2 for Poisson MLE is $1 - (l_{ur}/l_r)$, where l_{ur} is the maximized log-likelihood at b_{mle} and l_r is the maximized log-likelihood using only 1 as the regressor—both l_r and l_{ur} exclude the constant $\sum_i \ln(y_i!)$. For instance, if $l_{ur} = -203$ and $l_r = -286$, then $1 - (l_{ur}/l_r) = 1 - (203/286) = 0.29$. This pseudo R^2, which is applicable to other MLE's as well, is motivated by the fact that the usual linear model R^2 is $1 - \sum_i r_i^2 / \sum_i (y_i - \bar{y})^2$ where $\sum_i r_i^2$ is the residual when LSE is used and $\sum_i (y_i - \bar{y})^2$ is the residual when only 1 is used as the regressor. If the pseudo R^2 does not work well perhaps because the MLE assumption is false, then one may use $\{COR(y, \exp(x'b_N))\}^2$ that is a prediction-based pseudo R^2; this may be preferred to $1 - (l_{ur}/l_r)$.

If one follows the Poisson model "faithfully," defining $z_i \equiv 1[y_i > 0]$, the model for the binary response z_i becomes

$$P(z_i = 1|x_i) = 1 - P(z_i = 0|x_i) = 1 - e^{-\exp(x_i'\beta)}.$$

A MLE for binary response, which is called "gompit," can be done with this. Certainly, this differs from probit and logit.

As for marginal effects, observe

$$\frac{\partial E(y|x)}{\partial x} = \beta \exp(x'\beta) \implies E\left\{\frac{\partial E(y|x)}{\partial x}\right\} = \beta E\left\{\exp(x'\beta)\right\}$$

which can be estimated consistently with $b_{mle} \cdot N^{-1} \sum_i \exp(x_i'b_{mle})$. But a better marginal effect might be

$$\frac{\partial E(y|x)/\partial x}{E(y|x)} = \frac{\partial E(y|x)/E(y|x)}{\partial x} = \beta;$$

that is, β shows the proportional change in $E(y|x)$ as x changes by one unit, and this marginal effect is the same for all x.

If $\ln x_k$ is used as a regressor instead of x_k when $x_k > 0$, then β_k can be interpreted as an elasticity. To see this, observe

$$E(y|x) = \exp(\beta_1 x_1 +, ..., +\beta_k \ln x_k) \implies \ln E(y|x)$$
$$= \beta_1 x_1 +, ..., +\beta_k \ln x_k$$
$$\implies \frac{\partial E(y|x)/\partial x_k}{E(y|x)} = \frac{1}{x_k}\beta_k \implies \frac{\partial E(y|x)/E(y|x)}{\partial x_k/x_k} = \beta_k.$$

This elasticity, however, differs from the elasticity in the usual "double-log linear model" $\ln y = \beta_1 x_1 +, ..., +\beta_k \ln x_k + u$ where

$$\frac{\partial y/\partial x_k}{y} = \frac{1}{x_k}\beta_k \implies \frac{\partial y/y}{\partial x_k/x_k} = \beta_k.$$

3.2   Poisson Over-dispersion Problem and Other Estimators

One problem of Poisson MLE is the restriction $E(y|x) = V(y|x)$, which is unlikely to be satisfied in practice; usually we have $V(y|x) > E(y|x)$, an "*over-dispersion problem*" which invalidates Poisson MLE. To see why this occurs, consider an omitted variable ε independent of x, and suppose that $y|(x,\varepsilon)$, not $y|x$, follows $Poi\{\exp(x'\beta + \varepsilon)\}$:

$$E(y|x,\varepsilon) = e^{x'\beta+\varepsilon} = e^{x'\beta}e^{\varepsilon} = e^{x'\beta}E(e^{\varepsilon})\frac{e^{\varepsilon}}{E(e^{\varepsilon})} = e^{x'\beta+\ln E(e^{\varepsilon})}w,$$

$$\text{where } w \equiv \frac{e^{\varepsilon}}{E(e^{\varepsilon})};$$

note that $E(w) = 1$. With $x'\beta = \beta_1 + \beta_2 x_2 +, \cdots, +\beta_k x_k$, redefine β_1 as $\beta_1 + \ln E(e^{\varepsilon})$ to rewrite $e^{x'\beta+\ln E(e^{\varepsilon})}w$ as $E(y|x,w) = e^{x'\beta}w$. From this,

$$E(y|x) = e^{x'\beta}E(w) = e^{x'\beta}.$$

Although the presence of ε does not alter $E(y|x)$ in essence, ε does alter $V(y|x)$ as follows—that is, $y|x$ does not follow Poisson anymore. Keep in mind that $y|(x,w)$ follows $Poi(e^{x'\beta}w)$.

Observe

$$E(y^2|x) = E_{w|x}\{E(y^2|x,w)\} = E_w\{E(y^2|x,w)\}$$
$$\text{(because } w \text{ is independent of } x)$$
$$= E_w\{V(y|x,w) + E^2(y|x,w)\} = E_w(e^{x'\beta}w + e^{2x'\beta}w^2)$$
$$= e^{x'\beta} + e^{2x'\beta}E(w^2).$$

With $E^2(y|x) = e^{2x'\beta}$,

$$V(y|x) = E(y^2|x) - E^2(y|x) = e^{x'\beta} + e^{2x'\beta}E(w^2) - e^{2x'\beta}$$
$$= e^{x'\beta} + e^{2x'\beta}\{E(w^2) - 1\}$$
$$= e^{x'\beta} + e^{2x'\beta}\{E(w^2) - E^2(w)\} > e^{x'\beta} = E(y|x).$$

Over-dispersion due to omitted variables is analogous to the following fact in the usual linear model: if $y = x_1'\beta_1 + x_2'\beta_2 + u$ holds where x_1 and x_2 are independent, then ignoring x_2 does not cause inconsistency in estimating β_1 with x_1 alone, but the error term variance increases since $x_2'\beta_2 + u$ becomes the error term. In the Poisson MLE, ε plays the role of $x_2'\beta_2$. But differently from the linear model, omitting ε causes the Poisson MLE to be inconsistent despite that ε is independent of x_2, because $E(y|x) \neq V(y|x)$.

3.2.1   Negative Binomial (NB) MLE

Cameron and Trivedi (1998) and Winkelmann (2003) showed various parametric approaches generalizing Poisson MLE while relaxing the restriction $E(y|x) = V(y|x)$. One well-known parametric method allowing for over-dispersion is *"negative binomial MLE."* Define

$$\lambda_i \equiv \exp(x_i'\beta) \quad \text{and} \quad \psi_i \equiv \alpha^{-1}\lambda_i^\kappa$$

where β, α (> 0) and κ are parameters. The negative binomial (NB) probability for $y_i|x_i$ is (with $\Gamma(s) \equiv \int_0^\infty z^{s-1}e^{-z}dz$ for $s > 0$)

$$P(y_i|x_i) = \frac{\Gamma(y_i + \psi_i)}{\Gamma(\psi_i)\Gamma(y_i + 1)} \cdot \left(\frac{\psi_i}{\lambda_i + \psi_i}\right)^{\psi_i} \left(\frac{\lambda_i}{\lambda_i + \psi_i}\right)^{y_i}.$$

In NB, it is known that

$$E(y|x_i) = \lambda_i \quad \text{and} \quad V(y|x_i) = \lambda_i + \alpha\lambda_i^{2-\kappa} > \lambda_i$$

allowing for over-dispersion. Poisson is the limiting case when $\alpha \to 0^+$. $\Gamma(s)$, $s > 0$, is called the *"gamma function"*; gamma function satisfies $\Gamma(s+1) = s\Gamma(s)$, and hence $\Gamma(s) = (s-1)!$ for a positive integer s.

Although all parameters β, α, κ can be estimated, often κ is set at 0 to yield $\psi_i \equiv \alpha^{-1}$ and $V(y_i|x_i) = \lambda_i + \alpha\lambda_i^2$, which is quadratic in λ_i. One reason for $\kappa = 0$ is to avoid a possible identification problem: κ is not identified well if $\alpha \simeq 0$. Also, when $\kappa = 0$, the information matrix of the NB MLE becomes block-diagonal for β and α, meaning that the estimation of β would not be affected by the estimation of α.

A rv z following *Gamma distribution with two parameters* $\lambda, s > 0$ has density

$$f(z_o) = \frac{\lambda}{\Gamma(s)}(\lambda z_o)^{s-1}e^{-\lambda z_o}, \quad z_o \in (0, \infty) \quad \text{with} \quad E(z) = \frac{s}{\lambda}$$
$$\text{and} \quad V(z) = \frac{s}{\lambda^2};$$

$\Gamma(s)$ is the normalizing constant for $\int_0^\infty f(z_o)dz_o = 1$. The sum of s-many iid rv's from $Expo(\lambda)$ follows $Gamma(\lambda, s)$, which thus gives the interpretation of the *waiting time until the sth occurrence of the Poisson event* as the waiting time between two Poisson events follows $Expo(\lambda)$. The above $\kappa = 0$ case in the preceding paragraph is also obtained when the omitted "unobserved heterogeneity" e^ε in $E(y|x, \varepsilon) = e^{x'\beta}e^\varepsilon$ follows Gamma distribution with parameter $(\alpha^{-1}, \alpha^{-1})$. The restriction $\lambda = s = \alpha^{-1}$ yields $E(e^\varepsilon) = 1$ and $V(e^\varepsilon) = \alpha$. For this Gamma-distributed e^ε case, we get $E(y|x) = e^{x'\beta}$ as $E(e^\varepsilon) = 1$, and

$$V(y|x) = E[V(y|x, \varepsilon)|x] + V[E(y|x, \varepsilon)|x] = E(e^{x'\beta}e^\varepsilon|x) + V[e^{x'\beta}e^\varepsilon|x]$$
$$= e^{x'\beta} + \alpha e^{2x'\beta}$$

verifying the above display $V(y|x) = \lambda + \alpha\lambda^{2-\kappa}$ with $\lambda = e^{x'\beta}$ and $\kappa = 0$.

3.2.2   Zero-Inflated Count Responses

Another way to deal with the over-dispersion problem is *"zero-inflated"* *count response*. Here, we introduce the idea only briefly, with the details deferred to the sample-selection part in the next chapter because the idea is based on a multivariate LDV model.

Imagine y^* being the number of fish caught by angling, where $y^* > 0$ only for those who "participate" in angling. The observed count response is

$$y = 1[x'\alpha + v > 0] \cdot y^* \text{ where } 1[x'\alpha + v > 0] \text{ denotes participation, } v \amalg y^*|x$$

and v has a symmetric df $F(\cdot)$. Call y^* "performance" (following the participation). Observe (keeping in mind $y \neq y^*$), as $v \amalg y^*|x$,

$$E(y|x) = E(1[x'\alpha + v > 0]|x) \ E(y^*|x) = F(x'\alpha)E(y^*|x) \quad \text{and}$$
$$E(y^2|x) = E(1[x'\alpha + v > 0]|x) \ E(y^{*2}|x) = F(x'\alpha)E(y^{*2}|x)$$
$$\implies V(y|x) = F(x'\alpha)E(y^{*2}|x) - F(x'\alpha)^2 E^2(y^*|x).$$

If $V(y^*|x) = E(y^*|x)$ as in Poisson, then $E(y^{*2}|x) - E^2(y^*|x) = E(y^*|x)$, from which $E(y^{*2}|x) = E^2(y^*|x) + E(y^*|x)$. Substitute this into the last display for $V(y|x)$ to get

$$V(y|x) = F(x'\alpha)\{E^2(y^*|x) + E(y^*|x)\} - F(x'\alpha)^2 E^2(y^*|x)$$
$$= \{F(x'\alpha) - F(x'\alpha)^2\}E^2(y^*|x) + F(x'\alpha)E(y^*|x)$$
$$> F(x'\alpha)E(y^*|x) = E(y|x):$$

over-dispersion in y occurs despite $V(y^*|x) = E(y^*|x)$. MLE can be implemented with logistic $F(\cdot)$ and Poisson $y^*|x$ as done in Lambert (1992), which is a "Zero-Inflated Poisson (ZIP)." Clearly, zero-inflated NB can be used as well. See Lam et al. (2006) for more development and references.

3.2.3   Methods of Moments

Yet another, less parametric, way to avoid the over-dispersion problem is to view $E(y|x) = e^{x'\beta}$ just as a nonlinear regression function without $V(y|x)$ specified. Then we get the NLS minimand $N^{-1}\sum_i\{y_i - \exp(x_i'b)\}^2$, which yields the first-order condition

$$\frac{1}{N}\sum_i\{y_i - \exp(x_i'b)\}x_i \exp(x_i'b) = 0.$$

This differs from the Poisson MLE first-order condition by the factor $\exp(x_i'b)$ next to x_i. Following the distribution theory of M-estimator, the asymptotic variance is

$$E^{-1}\{xx' \exp(2x'\beta)\} \cdot E[\{y - \exp(x'\beta)\}^2 \exp(2x_i'\beta)xx'] \cdot$$
$$E^{-1}\{xx' \exp(2x'\beta)\}.$$

In Poisson MLE, the first-order condition can be taken as a moment condition $N^{-1} \sum_i v_i x_i = 0$ where $v_i \equiv y_i - \exp(x_i'b)$. Since $E(v|x) = 0$ implies $E\{v \cdot g(x)\} = 0$ for any (square integrable) function of x, Poisson MLE specification can be tested by a method-of-moment test. Another advantage of viewing the first-order condition as a moment condition is that we can forget the Poisson MLE origin and derive the asymptotic distribution of the moment estimator. In this case, the resulting estimator—"quasi Poisson MLE"—is asymptotically normal with variance

$$E^{-1}\{xx' \exp(x'\beta)\} \cdot E[\{y - \exp(x'\beta)\}^2 xx'] \cdot E^{-1}\{xx' \exp(x'\beta)\}.$$

Deriving this formula is essentially the same as deriving the preceding nonlinear LSE asymptotic variance, except that the multiplicative factor $\exp(x'\beta)$ in the nonlinear LSE first-order condition does not appear in $N^{-1} \sum_i v_i x_i$ and hence neither in the asymptotic variance. This display can be used as a "overdispersion-robust variance" for Poisson MLE, which otherwise tends to return too big t-values.

The above MOM approach for count responses seems to be a good method to use in practice. Recalling the Poisson MLE gradient and Hessian, the last display can be also viewed as

$$(\text{Poisson Hessian})^{-1} \times (\text{Poisson gradient outer-product})$$
$$\times (\text{Poisson Hessian})^{-1}.$$

Imposing the information equality reduces this sandwich form into just a single matrix $E^{-1}[\{y - \exp(x'\beta)\}^2 xx']$ that is the Poisson MLE asymptotic variance. A good way to understand the two asymptotic variances is recalling the LSE moment condition $E(ux) = 0$. The LSE asymptotic variance takes the usual sandwich form. But if we assume $u \sim N(0, \sigma^2)$ independently of x, then we would be doing MLE whose first-order condition is also $E(ux) = 0$ but its asymptotic variance is just a single matrix, the inverse of the information matrix.

3.3   An Empirical Example: Inequality Effect on Crime

As an example of Poisson MLE, Kelly (2000) analyzed effects of inequality on the number of property and violent crimes. The crime data are from the 1991 FBI Uniform Crime Reports for year 1990, and the unit of observations is a US metropolitan county ($N = 829$). The response variables are the number of violent crimes and property crimes. Other than crime numbers examined here, count responses are also popular in insurance where the number of accidents/claims is of keen interest; see Gourieroux and Jasiak (2007) for applications of count responses (and other LDV models in general) for insurance, banking, and finance.

Part of Table 3 in Kelly (2000) omitting the intercept and population size is shown in the Table 3; the SD's there were obtained using a "quasi-likelihood" approach, which allows an over-dispersion parameter. See Kelly

(2000) for the detailed description of the regressors which are all in logarithm (taking ln on % variables is unwarranted, however, in view of our discussion on marginal effects).

High population density increases crime probably by increasing the number of potential victims and reducing the chance of getting caught. Income inequality measured by Gini coefficient increases violent crime but is insignificant for property crime. The percentages of female-head family and those who moved show breakdowns of family structure and instability of the community, which increase both types of crimes. One caution is that the percentages of female-head family may be an outcome variable from inequality just as crimes are, in which case using it as a regressor would be inappropriate. Percentage non-white and unemployed male are insignificant. Poverty increases property crime but decreases violent crime. Surprisingly, educated population increases property crime. Finally, police activity is insignificant for violent crime but reduces property crime; note that using 1987 variable is to avoid the simultaneity problem.

3.4   IVE for Count or Positive Responses

Suppose the error term u in the linear model $y = x'\beta + u$ consists of two error terms ε and v. In this case, we can rewrite the model in two equivalent ways: $y = (x'\beta + \varepsilon) + v = x'\beta + (\varepsilon + v)$. But the equivalence no longer holds for nonlinear models: for a nonlinear model $y = \exp(x'\beta) + u$,

$$y = \exp(x'\beta + \varepsilon) + v \neq \exp(x'\beta) + \varepsilon + v.$$

Suppose $y = \exp(x'\beta + \varepsilon) + v$ is true, and some components of x are endogenous only because of their relationship with ε. One example is errors-in-variables: $y = \exp(\tilde{x}'\beta) + v$ where $x = \tilde{x} + \tilde{\varepsilon}$ and x (not $\tilde{x}$) is observed. In this case we get $y = \exp(x'\beta + \varepsilon) + v$ with $\varepsilon = -\tilde{\varepsilon}'\beta$; x is clearly related to ε. In the following, we show an innovative IVE for this kind of model, drawing on Mullahy (1997). In the literature, a term like ε is often called "unobserved heterogeneity."

Suppose

$$E(y|x,\varepsilon) = \exp(x'\beta + \varepsilon) = \exp(x'\beta) \cdot e^\varepsilon$$

where some components of x are related to ε.

Recall that an instrument w for an endogenous regressor x_k is a regressor for the x_k-equation (inclusion restriction), but the instrument does not appear as a regressor for the y-equation (exclusion restriction). Let z be an instrument vector for x; if x_k is the only endogenous regressor with the instrument w, then $z = (x_1, ..., x_{k-1}, w)'$. The exclusion restriction of w from the y equation can be expressed as

$$E(y|x,z,\varepsilon) = E(y|x,\varepsilon)$$

because x is in the conditioning set: so long as x (i.e., x_k) is fixed, z (i.e., w) cannot influence y due to the exclusion restriction. Further suppose, analogously to $E(\varepsilon|z) = 0$,

$$E(e^\varepsilon|z) = \tau, \text{ an unknown constant.}$$

Define

$$v \equiv y - E(y|x, z, \varepsilon) = y - E(y|x, \varepsilon) = y - \exp(x'\beta) \cdot e^\varepsilon$$
$$\implies y = \exp(x'\beta) \cdot e^\varepsilon + v \quad \text{and } E(v|x, z, \varepsilon) = 0$$
$$\implies y = \exp(x'\beta) \cdot \frac{e^\varepsilon}{\tau} + v, \quad \text{redefining } \beta_1 \text{ as } \beta_1 + \ln \tau.$$

Multiply $y = \exp(x'\beta)e^\varepsilon/\tau + v$ by $\exp(-x'\beta)$ and subtract 1 to get

$$y\exp(-x'\beta) - 1 = \frac{e^\varepsilon}{\tau} - 1 + v\exp(-x'\beta)$$
$$\implies E\{y\exp(-x'\beta) - 1|z\} = E\left(\frac{e^\varepsilon}{\tau} - 1\Big|z\right) + E\{v\exp(-x'\beta)|z\}.$$

The first term on the right-hand side is zero, and the second term $E[E\{v \exp(-x'\beta)|x, z, \varepsilon\}|z]$ is also zero because $E(v|x, z, \varepsilon) = 0$. Therefore, we get a conditional moment condition

$$E\{y\exp(-x'\beta) - 1|z\} = 0,$$

expressed in observed variables and the parameter β. With this moment condition, GMM can be done for β. Compare this moment condition to $E[x\{y\exp(-x'\beta) - 1\}] = 0$ that appeared for GMM examples.

The IVE here takes advantage of the exponential regression function, and cannot be easily extended to other LDV's, which is why we have not discussed endogenous regressors up to now for LDV's in general. This will be done later when multiple equations with LDV's are examined.

Instead of the multiplicative endogeneity, Windmeijer and Santos-Silva (1997) considered an additive endogeneity model

$$y = \exp(x'\beta) + u, \quad \text{where } E(u|x) \neq 0 \text{ but } E(u|z) = 0.$$

Under this, the moment condition to use is

$$E[z\{y - \exp(x'\beta)\}] = 0.$$

Cameron and Trivedi (2009, p. 596) showed in detail how to implement this with STATA.

If the additive model holds, the above conditional moment condition becomes

$$E\{y\exp(-x'\beta) - 1|z\} = E[\{\exp(x'\beta) + u\}\exp(-x'\beta) - 1|z]$$
$$= E\{u\exp(-x'\beta)|z\}:$$

this cannot be 0 when $E(u|z) = 0$. Hence the multiplicative and additive models are incompatible, and which model holds in a given problem is an empirical matter. In their empirical application, Windmeijer and Santos-Silva (1997) found the additive model more plausible for their chosen instrument. More applications of the above IVE's, which are easy to apply using some GMM software, can be seen in the literature, e.g., in Schellhorn (2001).

4   Censored Response and Related LDV Models

4.1   Censored Models

So far, we have been dealing with discrete responses where the latent continuous variable is either not considered or not observed at all even if considered. There are cases where the latent continuous response is observed only when y_i^* is less (or greater) than a "censoring point" c_i; the model is called a "*censored response (regression) model.*"

Suppose

$$y_i^* = x_i'\beta + u_i, \quad d_i = 1[y_i^* \le c_i], \quad y_i = \min(y_i^*, c_i)$$
$$c_i \text{ and } y_i^* \text{ are independent given } x_i,$$
$$(x_i', d_i, y_i), \ i = 1, ..., N, \text{ are iid and observed.}$$

That is, whereas x_i is observed always, $y_i = y_i^*$ only when $y_i^* \le c_i$; otherwise $y_i = c_i$. A well-known example is that y_i^* is a duration, x_i is individual characteristics, and c_i is determined by when person i enters the data and when the data collection (or follow-up) ends; see the table below where the duration is in months.

If the actual duration is shorter than the censoring point, then the actual duration is observed; otherwise, the censoring point is observed along with whether the actual duration is shorter than the censoring point or not; i.e., only $y_i = \min(y_i^*, c_i)$ and d_i are observed along with x_i. The duration of a state continues until the state changes, and the change is the event of interest in a given duration problem. For example, the duration of unemployed state continues until the person gets employed, and the event of interest is finding a job.

	Data-entering (y*-starting) Date	Study-Ending Date	y^*	c	y	d
person 1	01/01/2006	31/12/2006	6	12	6	1
person 2	01/03/2006	31/12/2006	12	10	10	0
person 3	01/09/2006	31/12/2006	5	4	4	0

There are a number of distinctions to be made for c_i when y_i is a duration.

1. If all subjects are followed up for, say, 1 year then $c_i = 365$ days for all i; c_i is not random and the same for all i.

2. If the follow-up ends on a given date, say $31/12/2006$, then $c = 365$ for a person who entered the data on $1/1/2006$ and $c = 183$ for a person who entered the data on $1/6/2006$. In this case, c_i is random (so long as the data-entering date is so) and observed for everybody. For instance, suppose a person enters the data on $1/6/2006$ with the duration ending on $30/6/2006$, which means $y_i = y_i^* = 30$ and $c_i = 183$; c_i is observed although $y_i^* < c_i$.

3. If the data collection ends because some event occurs that is not the event of interest (e.g., accidental death or moving overseas while the interest is on a diseased death), then c_i is random and $\min(y_i^*, c_i)$ is observed. When $y_i^* < c_i$, the subject is no longer followed and c_i is not observed.

Cases 1 and 2 are similar in that c_i is observed $\forall i$. What is observed in these cases is then (x_i', c_i, y_i); compare this to (x_i', d_i, y_i) above. Cases 2 and 3 are similar in that c_i is random.

In the literature, somewhat confusing terminologies have been used as follows. Case 1 is called "type I censoring"; Case 2 is called either type I censoring or "generalized type I censoring"; Case 3 is "competing risks censoring" or "random censoring." We will call Cases 1 and 2 *type-I censoring* and Case 3 *random censoring* although c_i can be random in Case 2 as well. The above model and table are for random censoring, which is the weakest in terms of the data requirement as c_i does not have to be observed $\forall i$. Note, however, that if variation in c is required in estimation, then random censoring assumption is not necessarily weaker than Type-I censoring with a constant c. When c is random, c is assumed to be either independent of (x', y^*) (i.e., of (x', u)), or more weakly, independent of y^* given x. In the latter, c may be related to y^* but only through x. In *type-II censoring*, which is rarely used in econometrics, the follow-up continues until N_0 units' durations of interest end where N_0 is predetermined, i.e., until the smallest N_0 durations among $y_1^*, ..., y_N^*$ are observed.

In type I censoring, upper censoring can be turned into lower censoring. To see this,

$$y_i = \min(y_i^*, c_i) = \min(x_i'\beta + u_i, c_i) = -\max(-x_i'\beta - u_i, -c_i).$$

Multiply $y_i = -\max(-x_i'\beta - u_i, -c_i)$ with -1 and add c_i to get

$$c_i - y_i = \max(-x_i'\beta + c_i - u_i, 0).$$

Redefine y_i as $c_i - y_i$ and absorb c_i into $x_i'\beta$: if c_i is constant, then it is absorbed into the intercept in $x_i'\beta$; otherwise, c_i becomes a regressor with known coefficient 1 (not to be estimated). Thus we get, redefining $-x_i$ as x_i,

$$y_i = \max(x_i'\beta + u_i, 0).$$

The fixed "lower censoring" model, called "*tobit*" owing to Tobin (1958), has been used for female labor supply, expenditure on durable goods, etc. where a nontrivial portion of y is 0. It is not clear, however, whether these are appropriate uses of the censored model. In the duration example, there is y_i^* actually occurring although not observed when $y_i^* > c_i$; y is an observed version of y^* that exists. In the labor supply and durable expenditure examples, in contrast, there is no negative labor supply nor negative expenditure when $y_i^* < 0$. Negative something would be a "ghost" observation and it is difficult to think of y as an observed version of y^* that may not exist. These cases are called "*corner solution*" cases. A better approach might be a bivariate model as in ZIP with two equations: one equation for "participation" d and the other for "performance" y^* that is observed only when $d = 1$. Tobit is a special case of this bivariate "sample-selection model" with $d = 1[y^* > 0]$.

4.2   Censored-Model MLE

Recall the random upper-censoring linear model with $c \amalg y^*|x$ (i.e., $c \amalg u|x$). Assume further that $u \sim N(0, \sigma^2)$ and $u \amalg x$. Then the log-likelihood function to maximize wrt b (for β) and s (for σ) is

$$Q_N(b, s) = \sum_i \left[(1 - d_i) \ln \left\{ 1 - \Phi \left(\frac{y_i - x_i'b}{s} \right) \right\} + d_i \ln \left\{ \frac{1}{s} \phi \left(\frac{y_i - x_i'b}{s} \right) \right\} \right].$$

To understand this log-likelihood function, observe that the likelihood function is

$$\{ f_{y^*}(y|x) S_c(y|x) f_x(x) \}^d \{ f_c(y|x) S_{y^*}(y|x) f_x(x) \}^{1-d}$$

where $f_{y^*}(\cdot|x)$ is the $y^*|x$ density, $S_c(\cdot|x)$ is the $c|x$ survival function, f_x is the x density, $f_c(\cdot|x)$ is the $c|x$ density, and $S_{y^*}(\cdot|x)$ is the $y^*|x$ survival function. In the $d = 1$ part, the observed $y = \min(y^*, c)$ is y^*, which is why $f_{y^*}(y|x)$ appears; $S_c(y|x)$ is there for $P(c > y^*|x)$ when $d = 1$. The $d = 0$ part can be understood analogously. Taking $\ln$ on the likelihood yields

$$d \ln \{ f_{y^*}(y|x) S_c(y|x) f_x(x) \} + (1 - d) \ln \{ f_c(y|x) S_{y^*}(y|x) f_x(x) \}$$
$$\implies d \ln f_{y^*}(y|x) + (1 - d) \ln S_{y^*}(y|x)$$

because $\ln S_c(y|x)$, $\ln f_c(y|x)$, and $\ln f_x(x)$ do not depend on (β', σ) and thus they drop out. Under normality, $f_{y^*}(y|x)$ and $S_{y^*}(y|x)$ take the forms with

ϕ and Φ in the above log-likelihood function. Note that $c \amalg y^*|x$ is necessary; otherwise we would not get the product expressions $f_{y^*}(y|x)S_c(y|x)$ and $f_c(y|x)S_{y^*}(y|x)$.

If $y_i = \max(x_i'\beta + u_i, 0)$—"left-censoring at 0"—then

$$d_i = 0 \iff x_i'\beta + u_i < 0 \iff \frac{u_i}{\sigma} < \frac{-x_i'\beta}{\sigma}.$$

Since the censoring point is not random, but a constant, the likelihood function is the same as the above one with S_c and f_c replaced by one. Hence the log-likelihood function becomes

$$\sum_i \left[(1 - d_i) \ln \Phi \left(\frac{-x_i'b}{s} \right) + d_i \ln \left\{ \frac{1}{s} \phi(\frac{y_i - x_i'b}{s}) \right\} \right].$$

As for pseudo R^2 for censored (and truncated) response models, we can use

$$\frac{\tilde{b}_N' N^{-1} \sum_i (x_{is} - \bar{x}_s)(x_{is} - \bar{x}_s)' \tilde{b}_N}{\tilde{b}_N' N^{-1} \sum_i (x_{is} - \bar{x}_s)(x_{is} - \bar{x}_s)' \tilde{b}_N + s_N^2}$$

where $\tilde{b}_N$ is the slope estimators and x_s are the regressors other than unity. Since β is identified, b_{mle} can be taken as a measure of x-effect on y^*.

Define the score functions s_{bi}, s_{si}, and $s_i \equiv (s_{bi}', s_{si}')'$ such that

$$\frac{\partial Q_N}{\partial b} = \sum_i s_{bi} \quad \text{and} \quad \frac{\partial Q_N}{\partial s} = \sum_i s_{si}, \quad \text{where}$$

$$s_{bi} = \left[d_i \frac{(y_i - x_i'b)}{s^2} - (1 - d_i) \frac{\phi(x_i'b/s)}{s\{1 - \Phi(x_i'b/s)\}} \right] \cdot x_i$$

$$s_{si} = d_i \left\{ \frac{(y_i - x_i'b)^2}{s^3} - \frac{1}{s} \right\} + (1 - d_i) \frac{\phi(x_i'b/s) \cdot x_i'b}{s^2\{1 - \Phi(x_i'b/s)\}}.$$

One might think that the asymptotic variance of $\sqrt{N}(b_N - \beta)$ is $E^{-1}(s_b s_b')$, which is, however, false as can be seen in the following.

Define $\gamma \equiv (\beta', \sigma')'$ and $g_N = (b_N', s_N')'$ to observe

$$\text{asymptotic variance of } \sqrt{N}(g_N - \gamma) = I_f^{-1} = \begin{bmatrix} E(s_b s_b') & E(s_b s_s') \\ E(s_s s_b') & E(s_s s_s') \end{bmatrix}^{-1}.$$

Using the formula for the inverse of a partitioned matrix, the upper left $k \times k$ submatrix of I_f^{-1} is

$$[E(s_b s_b') - E(s_b s_s') \cdot E^{-1}(s_s s_s') \cdot E(s_s s_b')]^{-1}$$
$$= E^{-1}\{(s_b - \mu' s_s)(s_b - \mu' s_s)'\} \equiv E^{-1}(s_b^* s_b^{*'})$$

where $\mu \equiv E^{-1}(s_s s_s') \cdot E(s_s s_b')$, and s_b^* is called the *effective score function* for β. Hence,

$$\sqrt{N}(b_N - \beta) \rightsquigarrow N\{0, \ E^{-1}(s_b^* s_b^{*'})\};$$

the variance becomes $E^{-1}(s_b s_b')$ only when $E(s_b s_s') = 0$. If $E(s_b s_s') = 0$, then β is said to be "*adaptively estimable*" despite the nuisance parameter σ.

The parameter μ is the population regression coefficient of s_b on s_s, and s_b^* is the "residual," or the part of s_b not explained by s_s. Since $\mu \neq 0$ makes $E^{-1}(s_b^* s_b^{*\prime})$ larger than for the case $\mu = 0$, the estimation of β is hampered by the correlation between s_b and s_s. For a future reference, note that

$$\sqrt{N}(g_N - \gamma) = \frac{1}{\sqrt{N}} \sum_i E^{-1}(ss') \cdot s_i \left(= \frac{1}{\sqrt{N}} \sum_i I_f^{-1} s_i \right) + o_p(1),$$

$$\sqrt{N}(b_N - \beta) = \frac{1}{\sqrt{N}} \sum_i E^{-1}(s_b^* s_b^{*\prime}) \cdot s_{bi}^* + o_p(1),$$

$$\sqrt{N}(s_N - \sigma) = \frac{1}{\sqrt{N}} \sum_i E^{-1}(s_s^* s_s^{*\prime}) \cdot s_{si}^* + o_p(1),$$

where s_s^* is defined analogously to s_b^*.

4.3   Truncated Regression and Fractional Response

For the censored response model, x_i is observed always regardless of d_i. That is, $(x_i', d_i, d_i y_i^*)$ is observed. If x_i is observed also only when $d_i = 1$, that is, if we observe only

$$d_i x_i', \, d_i y_i^*$$

then the model is called a *truncated response (or truncated regression) model*. For instance, with y being income and x being individual characteristics, we may have data only for the people with income greater than $1,000 per month; x_i and y_i are not observed at all for those with $y_i^* \leq 1,000$.

In duration analysis, if the duration origin precedes the study origin, then the truncation occurs, because the durations which had ended before the study began are not observed all. Since the truncation is from the "left" side— short durations are not observed—this is an example of "*left-truncation*" whereas the above censoring $\min(y_i^*, c_i)$ is from the "right" side, and thus called *right-censoring*. The combination of these two, called "*LTRC*," occurs sometimes. For instance, in the income example, the income may be top-coded at $20,000; i.e., any income higher than $20,000 per month is recorded just as $20,000. In censored response models, x can be used to gauge what y^* would be when y^* is not observed, which is, however, impossible for truncated response models because x is not available when $d = 0$. With less information, estimating β is more difficult in truncated response models than in censored response models.

Specifically, in the tobit model $y = \max(y^*, 0)$, if the observations with $y = 0$ are removed from the data set, we get a truncated response model with left-truncation at 0:

$$y_i^* = x_i'\beta + u_i, \quad d_i = 1[y_i^* > 0],$$

(x_i', y_i^*), $i = 1, ..., N$, are iid and observed only when $y_i^* > 0$.

For this model, under normality, Hausman and Wise (1977) proposed to maximize wrt b and s

$$\sum_i \left[\ln \left\{ \frac{1}{s} \phi(\frac{y_i - x_i'b}{s}) \right\} - \ln \Phi \left(\frac{x_i'b}{s} \right) \right]$$

where the second term is the normalizing factor for the truncation of u/σ from below at $-x'\beta/\sigma$. Truncation is, in fact, a special case of "endogenous sampling" where sampling is not random but involves the response variable.

Generalizing one-sided censoring, both lower (or left) and upper (or right) censoring may occur together. For example, the share of riskless assets in financial wealth in Hochguertel (2003) is bounded by $[0,1]$. This is sometimes called a *"fractional response."* If y does not take the extreme values, then the usual linear model may be used for y; alternatively, one may posit $E[\ln\{y/(1-y)\}|x] = x'\beta$. If $P(y = 0) > 0$ and $P(y = 1) > 0$, it may be more appropriate to use an *interval-censored model*

$$y = \max\{0, \ \min(x'\beta + u, \ 1)\}.$$

A log-likelihood function for this is analogous to that for tobit.

For fractional responses, Papke and Wooldridge (1996) assumed $E(y|x) = G(x'\beta)$ for a known df G, e.g. $G(t) = \exp(t)/\{1+\exp(t)\}$, to do a quasi MLE maximizing wrt b

$$\sum_i [y_i \ln G(x_i'b) + (1 - y_i) \ln \{1 - G(x_i'b)\}]$$

which looks like a log-likelihood for binary y. This works due to the fact that $s \ln t + (1-s) \ln(1-t)$ is maximized when $s = t$. Papke and Wooldridge (1996) applied this estimator to data on employee participation proportion in 401(k) pension plans. The asymptotic variance can be found using the usual M-estimator sandwich-form asymptotic variance formula.

4.4   Marginal Effects for Censored/Selection Models

So far we introduced a number of marginal effects for various LDV models. In tobit which consists of probit and the truncated model, there are different versions of marginal effects if we treat the probit part as "participation-or-not decision" and the truncation model as "performance following participation" as in sample-selection models. Those marginal effects are examined here. Although tobit is used below, the following discussion can be done under any distributional assumption. The two-part view with participation and performance appeared already for zero-inflated Poisson (ZIP). As already noted, a critical difference between ZIP and tobit is that there is only one equation in tobit with zero occurring as a "corner solution," whereas two equations appear in ZIP with "non-participation zero" being structurally different from "participation zero."

Observe

$$E(y|x) = E(y|x, d = 1)P(d = 1|x) + E(0|x, d = 0)P(d = 0|x)$$
$$= E(y|x, d = 1)P(d = 1|x)$$

$$E(y|x, d = 1) = x'\beta + E\left(u \Big| x, \frac{u}{\sigma} > -x'\frac{\beta}{\sigma}\right)$$

$$= x'\beta + \sigma E\left(\frac{u}{\sigma} \Big| x, \frac{u}{\sigma} > -x'\frac{\beta}{\sigma}\right) = x'\beta + \sigma\frac{\phi(x'\beta/\sigma)}{\Phi(x'\beta/\sigma)};$$

the last expression will be derived in the next chapter when sample selection is discussed. Regard $d = 1$ as a "participation," and $y|d = 1$ as a "performance" given participation; y^* is then the latent performance. For example, y^* is the latent demand for a product, $d = 1$ is participating in the buying activity, and $y|d = 1$ is the number of units to buy given participation. Another example is that y^* is the latent market work hours, $d = 1$ is working in the labor market, and $y|d = 1$ is the market work hours given $d = 1$. These concepts provide interesting interpretations as follows.

Noting $y = dy = dy^*$ and $\phi'(t) = -t\phi(t)$, various (mean) effects can be thought of:

$$E(y^*|x) = x'\beta \implies \frac{\partial E(y^*|x)}{\partial x} = \beta \quad \text{(effect on the latent } y^*\text{)};$$

$$E(y|x, d = 1) = x'\beta + \sigma\frac{\phi(x'\beta/\sigma)}{\Phi(x'\beta/\sigma)}$$

$$\implies \frac{\partial E(y|x, d = 1)}{\partial x} = \beta + \frac{\phi'(x'\beta/\sigma)}{\Phi(x'\beta/\sigma)}\beta - \frac{\phi(x'\beta/\sigma)^2}{\Phi(x'\beta/\sigma)^2}\beta$$

(effect on "$y|$participation")

$$= \beta - \frac{\phi(x'\beta/\sigma)x'\beta/\sigma}{\Phi(x'\beta/\sigma)}\beta - \frac{\phi(x'\beta/\sigma)^2}{\Phi(x'\beta/\sigma)^2}\beta$$

$$= \beta\left\{I_k - \frac{\phi(x'\beta/\sigma)}{\Phi(x'\beta/\sigma)}x'\frac{\beta}{\sigma} - \frac{\phi(x'\beta/\sigma)^2}{\Phi(x'\beta/\sigma)^2}\right\};$$

$$E(dy|x) = E(y|x, d = 1)P(d = 1|x) = x'\beta \cdot \Phi\left(x'\frac{\beta}{\sigma}\right) + \sigma\phi\left(x'\frac{\beta}{\sigma}\right)$$

$$\implies \frac{\partial E(dy|x)}{\partial x} = \frac{\partial E(y|x, d = 1)}{\partial x}P(d = 1|x)$$

$$+ E(y|x, d = 1)\frac{\partial P(d = 1|x)}{\partial x} \quad \text{(effect on observed } y\text{)}$$

$$= \beta\Phi\left(x'\frac{\beta}{\sigma}\right) + x'\beta\phi\left(x'\frac{\beta}{\sigma}\right)\frac{\beta}{\sigma} + \phi'\left(x'\frac{\beta}{\sigma}\right)\beta$$

$$= \beta\left\{\Phi\left(x'\frac{\beta}{\sigma}\right) + \phi\left(x'\frac{\beta}{\sigma}\right)x'\frac{\beta}{\sigma} - \phi\left(x'\frac{\beta}{\sigma}\right)x'\frac{\beta}{\sigma}\right\} = \beta\Phi\left(x'\frac{\beta}{\sigma}\right).$$

When x changes, (i) β is its "potential effect" (β can increase the desire to work although some people do not work), (ii) $\partial E(y|x, d = 1)/\partial x$ is the "participant effect" showing the effect on the $d = 1$ group, and (iii) $\partial E(dy|x)/\partial x$ is the "observed effect" consisting of $\partial E(y|x, d = 1)/\partial x$ and the effect due to new entries/exits reflected in $\partial P(d = 1|x)/\partial x$.

It should be kept in mind that the above specific results such as $\partial E(dy|x)/\partial x = \beta \Phi(x'\beta/\sigma)$ are based on the normality assumption in tobit and the assumption that the participation and performance are both governed by a single equation. When these assumptions are relaxed, the effect formulas will change much. Interestingly though, Greene (1999) showed that when the error term is independent of x,

$$\partial E(dy|x)/\partial x = \beta \cdot P(y^* > 0|x)$$

regardless of the u distribution in the model $y = \max(y^*, 0)$. That is, the observed effect is the latent effect β times the participation proportion $P(y^* > 0|x)$. If x is a policy variable, an intriguing question is which one between β and $\beta \cdot P(y^* > 0|x)$ should be taken as the policy effect. Note that the participant effect may not be a good effect measure, because x can change the participation status.

4.5   Empirical Examples

EXAMPLE: JOB-TRAINING EFFECT ON UNEMPLOYMENT DURATION FOR RIGHT CENSORING. As an example of type 1 right-censored response $y_i = \min(y_i^*, c_i)$ with c_i varying across i, Lee and Lee (2005) analyzed job-training effects on Korean women using a data set, where $N_1 = 5031$ unemployed women received a job training ("treatment group") and $N_0 = 47060$ unemployed women did not receive any job training ("control group")—the control group received unemployment insurance benefit instead—during January 1999–March 2000. In the data, the right-censoring was done on the same calendar date for everybody, but since the subjects entered the study on different dates, their censoring time c_i varies across i. Table 4 shows the summary statistics for some variables.

Table 4: Descriptive Statistics for Job Training Data

	Treatment Group		Control Group	
	Mean	SD	Mean	SD
dummy for being censored	0.71	0.002	0.72	0.006
age (years)	27.79	5.57	34.92	10.82
education (years)	13.07	1.76	12.13	2.50
unemp. duration (days) before enrollment	65.1	53.2	33.6	32.5

The table shows that the treatment group is younger, more-educated, and searched longer before enrolling in the job training program.

Applying the censored response MLE with ln(*unemployment duration*) as the response variable—the training duration is included in the unemployment duration—Table 5 shows part of the results; dummy variables for ex-job's industry, ex-job's type, job-training type, reason for leaving the previous job, and interactions between these dummies and the job-training dummy (d) were used but not shown in the table. Relative to high school graduation which is the base education case, finishing only primary school increases the duration whereas junior college or higher education decreases the duration. The previous unemployment duration increases the duration. Unfortunately, the job-training dummy d increases the unemployment duration. Lee and Lee (2005) attributed this failure of job-training to that the job-training takes about 4 months while it reduces the subsequent unemployment duration by about 2 months on average. The effect of d looks too big, but this is because the interaction terms between d and many regressors are omitted, almost all of which carry negative estimates. That is, taking the interaction terms into account, d still increases the unemployment duration but by a much smaller magnitude than shown in the table—in some groups, d does decrease the duration. In fact, an analogous caveat applies to the other regressors as well due to their interaction terms.

Table 5: Censored Regression for Job Training Data

	Estimate	SD	t-Value
primary school	0.457	0.187	2.44
middle school	0.036	0.089	0.41
junior college	−0.396	0.041	−9.74
college	−0.419	0.073	−5.71
graduate school	−0.645	0.169	−3.81
previous unemployment duration	0.004	0.0003	13.15
d	1.829	0.438	4.17

EXAMPLE: WTP FOR BEACH VISITS FOR LEFT CENSORING. As an empirical example for $y = \max(0, x'\beta + u)$, Haab and McConnell (2002, pp. 157–163) examined a model where y is the number of yearly visits to a beach to find out the "access value" to the beach. In their data with $N = 499$, x includes the travel cost to the beach (say, x_k) and the travel cost to the nearest alternative beach. Applying tobit, they found $b_k = -5.48$ ($SD = 1.25$). Rewriting $x_i'\beta + u_i$ as $\beta_{0i} + \beta_k x_{ik}$ where β_{0i} is the part other than $\beta_k x_{ik}$ in $x_i'\beta + u_i$, the person-i "choke price" x_{ic} for zero demand (i.e., $y = 0$)—imagine charging an entrance fee until person i stops using the beach—is obtained from

$$0 = \beta_{0i} + \beta_k x_{ic} \implies x_{ic} \equiv -\frac{\beta_{0i}}{\beta_k}.$$

The WTP (in addition to the travel cost paid already) can be approximated with the consumer surplus, which is obtained by integrating the beach-visit demand equation wrt x_k over $[x_{ik}, x_{ic}]$. Note that the estimated equation is a demand equation because the "supply" is virtually unlimited unless over-crowding occurs in the beach. As in the WTP analysis in binary response, we are trying to find the maximum amount that can be "squeezed out" of the beach-goers, which is the travel cost plus the consumer surplus.

The integral is the person-i consumer surplus

$$\int_{x_{ik}}^{x_{ic}} (\beta_{0i} + \beta_k m) dm = \beta_{0i}(x_{ic} - x_{ik}) + \frac{\beta_k}{2}(x_{ic}^2 - x_{ik}^2)$$

$$= \beta_{0i}\left(-\frac{\beta_{0i}}{\beta_k} - \frac{y_i - \beta_{0i}}{\beta_k}\right) + \frac{\beta_k}{2}\left\{\frac{\beta_{0i}^2}{\beta_k^2} - \left(\frac{y_i - \beta_{0i}}{\beta_k}\right)^2\right\}$$

$$\text{(substituting } x_{ic} \equiv -\frac{\beta_{0i}}{\beta_k} \text{ and } x_{ik} = \frac{y_i - \beta_{0i}}{\beta_k})$$

$$= \beta_{0i}\frac{-y_i}{\beta_k} + \frac{1}{2}\left(\frac{-y_i^2 + 2\beta_{0i}y_i}{\beta_k}\right) = \frac{y_i^2}{-2\beta_k} = \frac{y_i^2}{2|\beta_k|} \quad \text{as } \beta_k < 0.$$

Hence the estimated mean consumer surplus is

$$\frac{1}{N}\sum_i \frac{y_i^2}{2(5.48)} = \$10.77.$$

We can also think of the median consumer surplus: because $y^2/(2|\beta_k|)$ is increasing with $y \geq 0$,

$$Med\left(\frac{y^2}{2|\beta_k|}\right) = \frac{\{Med(y)\}^2}{2|\beta_k|}.$$

Haab and McConnell (2002) do not show the sample median. As the sample median is likely to be smaller than the sample mean (some regular beach-goers would have rather high values for y, making $\bar{y}$ too big), using $\bar{y} = 3.84$ as an upper bound for the median, we get

$$Med\left(\frac{y^2}{2|\beta_k|}\right) \leq \frac{\bar{y}^2}{2(5.48)} = \$1.35.$$

Charging the median consumer surplus for the beach use, about 50% of the beach users will be willing to pay the amount. More generally, charging the αth quantile consumer surplus, about $(1 - \alpha)100\%$ of the beach users will be willing to pay the amount as their WTP is higher than the amount. Charging the mean consumer surplus, probably only the regular users will be willing to pay the amount, because the difference between the mean and median consumer surpluses is large as shown above.

The number of visits to a beach is a count response, and thus it would be better to apply a count response estimator than tobit. In this case, the demand equation will be no more linear, but exponential. Other than this, the remaining steps are analogous: find the person-i choke price x_{ic}, integrate the demand equation over $[x_{ik}, x_{ic}]$, and then come up with a sample "average" estimator of one kind or another.

5   Parametric Estimators for Duration

5.1   Basics

Suppose we want to explain duration y of a certain state with explanatory variables x. The state can be being unemployed or being alive (with some disease). In the former, the duration (or survival) is bad, whereas it is good in the later. Typically a duration gets right-censored as we cannot observe the subjects forever; i.e., we get to observe only the minimum of the actual duration and the censoring time. The actual duration comes from the *event of interest* (e.g., getting employed or dying of the disease), whereas the censoring duration comes from other events (e.g., lost in the follow-up or dying of an accident). Assuming that x_i *is not time-varying*, we can use cross-section data $(x_i', y_i), i = 1, ..., N$, to explain the duration. Duration was briefly examined for censored regression, and here we take a detailed look.

Duration is by nature a continuous variable over $[0, \infty)$. If the duration is observed in ordinal grouped intervals (e.g., survival up to 1 month, 2–3 months, or 4–12 months), then we have an ODR. In the Weibull distribution application of the contingent valuation example for ODR, we in fact showed how to deal with grouped durations. In this section, we will assume that y is continuously distributed. Discrete durations or grouped durations will reappear in the chapter for semiparametrics.

In the preceding section, we saw a linear model for ln(duration) in censored regression. There is another more "classical" approach for duration analysis, where the duration is assumed to follow some distribution (exponential, Weibull,...) and x enters the distribution. This classical approach is the focus of this section. *Duration analysis* is also called *"survival analysis,"* *"failure-time analysis"* or *"reliability analysis."* See Lancaster (1992) and Van den Berg (2001) for the econometric literature, and Hougaard (2000), Therneau and Grambsch (2000), Klein and Moeschberger (2003), and the references therein for the statistical literature. We will use unemployment duration as the main example throughout.

5.1.1   Survival and Hazard Functions

Ignore x for a while, and let $F(\cdot)$ denote the df of y:

$$F(t) = P(y \leq t); \; F \text{ has density } f.$$

Bear in mind that $F(t)$ is a function describing where y accumulates its probability, and $F(t)$ itself is not a rv. Define the *"survival function"* $S(t) \equiv 1 - F(t) = P(y > t)$. A key concept for duration models is *hazard function*, or *hazard rate* $\lambda(t)$, which is defined by

$$\lambda(t) \equiv \frac{f(t)}{S(t^-)}, \quad \text{where } S(t^-) = P(y \geq t);$$

$S(t) = S(t^-)$ because $S(t)$ is continuous, but here we use $S(t^-)$ just to be coherent with our future study on discrete durations. The relation between $\lambda(t)$ and $S(t)$ is a "two-way street": we can specify $S(t)$ which determines $\lambda(t)$, or we can specify $\lambda(t)$ which then determine $S(t)$. Certain functional forms are more easily handled in one of $\lambda(t)$ and $S(t)$ than in the other.

If $f(t)$ stays the same over a small interval $[t, t + dt)$, then $f(t)dt$ can be interpreted as $P(t \leq y < t + dt)$. Analogously,

$$\lambda(t)dt \simeq \frac{f(t)dt}{S(t)} = \frac{P(t \leq y < t + dt)}{P(y \geq t)} = \frac{P(t \leq y < t + dt, \ y \geq t)}{P(y \geq t)}$$

$$= P(t \leq y < t + dt | y \geq t) = P(\text{leaving the state in } [t, t + dt)|$$
$$\text{survived up to } t).$$

Observe, due to the division by $S(t)$ in $\lambda(t)$,

$$\lambda(t)dt \simeq P(t \leq y \leq t + dt | y \geq t) > P(t \leq y \leq t + dt) \simeq f(t)dt.$$

For instance, let $t = 100$ and $dt = 1$. The probability of dying between age 100 and 101 for a person who survived up to age 100 is much greater than the probability of a person dying between age 100 and 101, because most people die before age 100. If we plot $S(t)$ and $\lambda(t)$, $S(t)$ will be near zero for $t \geq 90$, but $\lambda(t)$ will be well above 0 for $t \geq 90$ and increasing as t further goes up. For a patient diagnosed with a cancer, $S(t)$ shows the probability of survival beyond t and $\lambda(t)$ shows the death probability over $[t, t+1)$ if he/she survives somehow up to t; both are interesting.

Owing to $\lambda(t) = f(t)/S(t)$ and $S(0) = 1$, we get

$$\lambda(t) = \frac{-d\ln S(t)}{dt} \ \left\{ \Longleftrightarrow \ \int_0^t \lambda(v)dv = -\ln S(t) + \ln S(0) = -\ln S(t) \right\}$$

$$\Longleftrightarrow \ S(t) = \exp\{-\Lambda(t)\}, \quad \text{where } \Lambda(t) \equiv \int_0^t \lambda(v)dv;$$

$\Lambda(t)$ is the "*integrated (or cumulative) hazard.*" For instance, if $\lambda(t) = \lambda_o$, a constant, then $\Lambda(t) = \lambda_o t$. Splitting the interval $[0, t]$ into n-many small intervals of length t/n,

$$S(t) = \exp(-\lambda_o t) = \lim_{n \to \infty} \left(1 - \frac{\lambda_o t}{n}\right)^n \simeq \prod_{i=1}^{n} P(\text{surviving interval } i)$$

as $1 - (\lambda_o t/n)$ is the probability of surviving each small interval of length t/n. This accords a "product-limit" view of the survival function: survival up to t is the limit of the product of survivals over infinitesimal intervals. If the hazard rate changes over time, then $S(t)$ takes the general form $\exp\{-\Lambda(t)\}$ instead of $\exp(-\lambda_o t)$.

5.1.2   Log-Likelihood Functions

Suppose we have randomly censored duration data (d_i, y_i, x_i'), $i = 1, ...,$ N—some y_i's are the upper-censoring points and d_i is the non-censoring indicator. Assume, as before, that c is independent of the true duration y^* given x. Let $f(\cdot|x; \theta)$ denote density for the true duration that is parametrized by θ. Then the log-likelihood function to maximize for θ is

$$\sum_i [d_i \ln f(y_i|x_i; \theta) + (1 - d_i) \ln S(y_i|x_i; \theta)]$$

where $d_i = 1$ if observation i is uncensored and 0 otherwise. Using $f = \lambda \cdot S$ and $\ln S = -\Lambda$, this can be rewritten in terms of λ and Λ:

$$\sum_i [d_i \ln\{\lambda(y_i|x_i; \theta) \cdot S(y_i|x_i; \theta)\} + (1 - d_i) \ln S(y_i|x_i; \theta)]$$

$$= \sum_i [d_i \ln \lambda(y_i|x_i; \theta) + \ln S(y_i|x_i; \theta)] = \sum_i [d_i \ln \lambda(y_i|x_i; \theta) - \Lambda(y_i|x_i; \theta)].$$

When left-truncation occurs as well at a point, say t_i (i.e., $y_i > t_i \ \forall i$ in the data although $y_i \leq t_i$ can occur in the population) so that we have *LTRC (left-truncation and right-censoring)*, then the above log-likelihood with f and S can be modified to

$$\sum_i \left[d_i \ln \frac{f(y_i|x_i; \theta)}{S(t_i|x_i, \theta)} + (1 - d_i) \ln \frac{S(y_i|x_i; \theta)}{S(t_i|x_i, \theta)} \right]$$

where $S(t_i|x_i, \theta) = P(y_i^* > t_i|x_i, \theta)$ is the normalizing factor for the truncation so that the truncated density still integrates to one. The log-likelihood can be rewritten with (cumulated) hazards only:

$$\sum_i \left[d_i \ln \frac{f(y_i|x_i; \theta)}{S(y_i|x_i; \theta)} \frac{S(y_i|x_i; \theta)}{S(t_i|x_i, \theta)} + (1 - d_i) \ln \frac{S(y_i|x_i; \theta)}{S(t_i|x_i, \theta)} \right]$$

$$= \sum_i \left[d_i \ln \lambda(y_i|x_i, \theta) + \ln \frac{S(y_i|x_i; \theta)}{S(t_i|x_i, \theta)} \right]$$

$$= \sum_i [d_i \ln \lambda(y_i|x_i, \theta) - \Lambda(y_i|x_i, \theta) + \Lambda(t_i|x_i, \theta)].$$

To see how t_i's are obtained in reality, assume no right-censoring for a while. Suppose that the *population of interest is those who are unemployed at any time during the calendar time period* $[0, \tau]$ but sampling is done only over $[\tau - \varepsilon, \tau]$ for some constant $0 < \varepsilon < \tau$. This results in missing those who become unemployed in $[0, \tau - \varepsilon)$ to end the duration before $\tau - \varepsilon$—assume that the probability of multiple unemployment spells is zero for the period. Suppose that (x', y, q) is observed where q is the calendar time of getting unemployed and that $(x, y) \amalg q$. If $q_i \in [\tau - \varepsilon, \tau]$, then y_i is observed always; if

$q_i \in [0, \tau - \varepsilon)$, then y_i is observed only when $q_i + y_i \geq \tau - \varepsilon \iff y_i \geq \tau - \varepsilon - q_i$. Thus the conditional likelihood for $y_i | (x_i, q_i)$ is

$$\frac{f(y_i | x_i; \theta)}{S(t_i | x_i; \theta)} \text{ where } t_i = 0 \text{ if } q_i \in [\tau - \varepsilon, \tau] \text{ and } t_i \equiv \tau - \varepsilon - q_i \text{ if } q_i \in [0, \tau - \varepsilon).$$

Sometimes we may be interested in *mean residual life function*:

$$\mu(t) \equiv E(y - t | y > t) = \frac{E\{(y - t)1[y > t]\}}{S(t)} = \frac{\int_t^\infty (\tau - t) f(\tau) d\tau}{S(t)}$$

$$= \frac{\int_t^\infty S(\tau) d\tau}{S(t)}$$

$$\text{as } \int_t^\infty (\tau - t) f(\tau) d\tau = -(\tau - t) S(\tau)|_t^\infty + \int_t^\infty S(\tau) d\tau =$$

$$\int_t^\infty S(\tau) d\tau.$$

This shows how long more a subject will survive on average, given that the subject has survived up to t. Setting $t = 0$, this display also shows $E(y) = \int_0^\infty S(\tau) d\tau$, which holds for all nonnegative rv's with $E(y) < \infty$.

If y is discrete taking $0, 1, 2, 3, \ldots$ with the probabilities $p_0, p_1, p_2, \ldots$, then $E(y) = \sum_j j p_j$ can be written as, because each p_j appears j times in $\sum_j j p_j$,

$$(p_1 + p_2 + p_3, \ldots) + (p_2 + p_3+, \ldots) + (p_3+, \ldots) +, \ldots$$

This is a discrete analog for $\int_0^\infty S(\tau) d\tau$. More succinctly, write y as

$$\sum_{t=0}^\infty 1[y > t] = 1[y > 0] + 1[y > 1] + 1[y > 2], +, \ldots$$

to get $E(y) = \sum_{t=0}^\infty E(1[y > t]) = \sum_{t=0}^\infty S(t)$, which is analogous to $\int_0^\infty S(\tau) d\tau$.

5.2   Exponential Distribution for Duration

Although normal distribution is a basic building block for MLE, it is not good for duration analysis, as duration should be positive; also duration is often asymmetric with a long right tail. The basic distribution in duration analysis is *exponential distribution* indexed by one parameter $\theta > 0$. With y following $Expo(\theta)$, we get the following facts:

(i) $f(t) = \theta \cdot \exp(-\theta t)$;

(ii) $S(t) = \exp(-\theta t), \quad F(t) = 1 - \exp(-\theta t)$;

(iii) $\lambda(t) = \theta, \quad \Lambda(t) = \theta t$;

(iv) $E(y) = 1/\theta, \quad V(y) = 1/\theta^2$;

(v) $E\{\ln(y)\} \simeq -\ln \theta - 0.577, \quad V\{\ln(y)\} \simeq 1.645$.

$E(y) = \theta^{-1}$ follows from

$$E(y) = \int_0^\infty t \cdot \theta e^{-\theta t} dt = -t e^{-\theta t}\Big|_0^\infty + \int_0^\infty e^{-\theta t} dt = 0 - \frac{1}{\theta} e^{-\theta t}\Big|_0^\infty = \frac{1}{\theta}.$$

The most notable about $Expo(\theta)$ is the constant hazard θ that is not a function of t. This implies implausible "memoryless property" of exponential distribution:

$$P(y > T + t | y > t) = \frac{P(y > T + t, \ y > t)}{P(y > t)} = \frac{P(y > T + t)}{P(y > t)}$$

$$= \frac{\exp\{-\theta(T + t)\}}{\exp(-\theta t)} = \exp(-\theta T) = P(y > T);$$

the probability of surviving T more given the survival up to t is just the same as the probability of survival up to T from the initial time point as if there is no "wear and tear." Also notable is that $\Lambda(t) = \theta t$ is a *linear* function of t.

To account for x, usually we specify

$$\theta(x) = \exp(x'\beta)$$

as in Poisson regression ($\exp(\cdot)$ guarantees $\theta(x) > 0$) to assume that $y|x$ is $Expo\{\theta(x)\}$. Then, using (v), we may assume

$$E(\ln y | x) \simeq -\ln \theta(x) - 0.577, \quad V(\ln y | x) \simeq 1.645$$
$$\implies \ln y = (-x)'\beta + u, \quad \text{where } u \equiv \ln y + x'\beta, \quad E(u|x) \simeq -0.577,$$
$$V(u|x) \simeq 1.645.$$

This linear model can be estimated with LSE (barring censoring problems).

The major problem with the LSE is, however, that $V(\ln y | x)$ should be the known constant 1.645; the non-zero $E(u|x)$ can be absorbed into the intercept in $-x'\beta$ to make $E(u|x) = 0$. If we estimate β with the LSE of $\ln y$ on $-x$ to estimate the error term variance with $N^{-1}\sum_i \hat{u}_i^2$ where $\hat{u}$ is the residual, then it is unlikely to get $N^{-1}\sum_i \hat{u}_i^2 \simeq 1.645$. Another problem is that the hazard function $\exp(x'\beta)$ is not a function of time, not allowing the hazard rate to change across time. For instance, as unemployment duration goes up, the unemployed may be more willing to accept a job offer. Then the hazard rate will go up as time goes. Weibull distribution to appear shortly overcomes these problems in exponential distribution.

If one is to do MLE with exponential duration allowing for right-censoring, then the log-likelihood function to maximize for b is (recall $\Lambda(t) = t\theta$)

$$\sum_i [d_i \ln \lambda(y_i | x_i; \theta) - \Lambda(y_i | x_i; \theta)] = \sum_i \{d_i x_i' b - y_i \exp(x_i' b)\}.$$

This log-likelihood function is reminiscent of the Poisson MLE log-likelihood function $\sum_i \{y_i x_i' b - \exp(x_i' b)\}$ where y_i is a count.

5.3   Weibull Distribution for Duration

Weibull distribution with two parameters $\theta > 0$ and $\alpha > 0$ is

(i) $f(t) = \theta \alpha t^{\alpha-1} \exp(-\theta t^\alpha)$;

(ii) $S(t) = \exp(-\theta t^\alpha)$, $F(t) = 1 - \exp(-\theta t^\alpha)$;

(iii) $\lambda(t) = \theta \alpha t^{\alpha-1}$, $\Lambda(t) = \theta t^\alpha$;

(iv) $E(y^r) = \theta^{-r/\alpha} \Gamma(1 + r\alpha^{-1})$,   where $\Gamma(w) = \int_0^\infty z^{w-1} e^{-z} dz$
     for $w > 0 \Rightarrow E(y) = \theta^{-1/\alpha} \Gamma(1 + \alpha^{-1})$ and
     $V(y) = \theta^{-2/\alpha} \{\Gamma(1 + 2\alpha^{-1}) - \Gamma^2(1 + \alpha^{-1})\}$;

(v) $E \ln(y) \simeq \alpha^{-1}(-\ln\theta - 0.577)$ and $V\{\ln(y)\} \simeq \alpha^{-2} 1.645$.

For Weibull distribution, $\lambda(t)$ is increasing in t if $\alpha > 1$ and decreasing if $\alpha < 1$; see Figure 3. "$d\lambda(t)/dt < 0$" is called negative *duration dependence* ($d\lambda(t)/dt > 0$ is *positive duration dependence*): as time progresses, the duration becomes less likely to end. Weibull distribution includes exponential distribution as a special case when $\alpha = 1$. Viewed differently, as clear in $S(t)$, Weibull distribution becomes exponential distribution by redefining t^α as t. If $\alpha > 1$, then time accelerates, which is equivalent to $\lambda(t)$ increasing over time. Also, a notable feature is that $\ln \Lambda(t) = \ln \theta + \alpha \ln t$ is a linear function of $\ln t$. This feature can be used to check out the plausibility of the Weibull distribution assumption when $\Lambda(t)$ is estimated nonparametrically as will be seen later in the nonparametrics chapter.

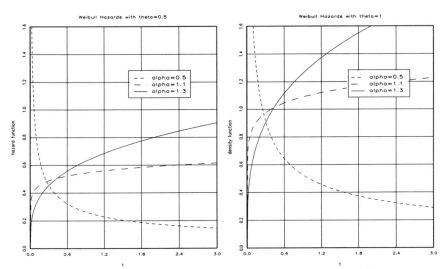

Figure 3: Weibull Hazards

Owing to $S(\infty) = 0$, eventually everybody will leave unemployment. This, however, may be too restrictive, because there can be "super-survivors" surviving forever. In this case, there are two sub-populations—"split population"—one group subject to failure and the other never. This case can be accommodated with

$$S(t) = pS^*(t) + (1 - p)$$

where $1 - p > 0$ is the proportion for the super-survivors and $S^*(t)$ is a latent survival function of those destined to fail; observe $S(\infty) = 1 - p > 0$. Regressors can be introduced into p and $S^*(t)$ to result in $p(z)$ and $S^*(t|x)$ with z and x possibly overlapping, and $p(z)$ may be specified as logit and $S^*(t|x)$ as Weibull. See, e.g., Zhang and Peng (2007) and the references therein.

Suppose now that $y|x$ follows Weibull with parameters $\theta(x) = \exp(x'\beta)$ and α. Then

$$\ln y = (-x)' \left(\frac{\beta}{\alpha}\right) + u, \quad E(u|x) = \frac{-0.577}{\alpha}, \quad V(u|x) = \frac{1.645}{\alpha^2}.$$

Unlike exponential distribution, there is an unknown constant α in $V(u|x)$. Doing the LSE of $\ln(y)$ on $-x$, we can estimate $\gamma \equiv \beta/\alpha$ consistently with g_N, and

$$s_N^2 \equiv \frac{1}{N} \sum_i \{\ln(y_i) + x_i'g_N\}^2 \to^p \frac{1.645}{\alpha^2}.$$

From this, an estimator a_N for α is $(1.645)^{0.5}/s_N$. An estimate b_N for β is then obtained by $b_N = g_N a_N$. The intercept is still off the target due to $E(u|x) = -0.577/\alpha$, but this is not of the main concern. If censoring is present, this LSE is inconsistent. Instead, we can do MLE in the following.

Using the Weibull hazard specification with $\theta = \exp(x'\beta)$, the Weibull log-likelihood function and the gradient allowing for right-censoring become

$$Q(a, b) = \sum_i [d_i\{\ln(a) + (a - 1)\ln(y_i) + x_i'b\} - y_i^a \exp(x_i'b)];$$

$$\frac{\partial Q(a, b)}{\partial a} = \sum_i \left[\frac{d_i}{a} + \{d_i - y_i^a \exp(x_i'b)\}\ln(y_i)\right],$$

$$\frac{\partial Q(a, b)}{\partial b} = \sum_i \{d_i - y_i^a \exp(x_i'b)\}x_i.$$

With this, the MLE can be implemented; $Q(a, b)$ includes the exponential duration log-likelihood function as a special case when $a = 1$.

The Weibull hazard with $\theta = \exp(x'\beta)$ is

$$\alpha t^{\alpha-1} \cdot \exp(x'\beta)$$

where the part $\alpha t^{\alpha-1}$ is called the "baseline hazard"—"baseline" as it does not depend on the individual characteristics x. The part t^α reflects the commonalities behind the durations (e.g., a common pattern of recovery from a disease). If all individuals' durations start at the same calendar time, then t^α

may reflect the macro (weather, economy, wide spread diseases, etc.) effects common to everybody. This kind of hazard with a baseline hazard $\lambda_o(t)$ and a function $g(x)$ of x multiplied together $(\lambda_o(t) \cdot g(x))$ as in the last display is called a "*proportional hazard*."

The above log-linear model, now written as $\ln y = -x'\gamma + u$, may still hold without the Weibull assumption, in which case the LSE is still valid and there will be no need to bother with α. But the interpretation with α is buried in γ $(= \beta/\alpha)$ if indeed Weibull holds: if $\alpha > 1$, then γ gets smaller in magnitude—time "accelerates." In the log-linear model, the predicted duration of an individual with x_i can be easily done with $\hat{y}_i = \exp(x_i'\hat{\gamma})$, whereas such a prediction is cumbersome if $E_N(y|x) \equiv \int_0^\infty \hat{S}(t|x)dt$ is used after the Weibull MLE where $\hat{S}(t|x) \equiv \exp(-\exp(x'\hat{\beta}) \cdot t^{\hat{\alpha}})$.

5.4   Unobserved Heterogeneity and Other Parametric Hazards

In the usual linear regression model, one way to view the error term is that it is a combination of omitted variables uncorrelated with the regressors. Suppose we include an unobserved term v_i in the Weibull parameter θ to take into account omitted variables uncorrelated with x_i as already done once for count responses:

$$\theta(x_i, v_i) = \exp(x_i'\beta + v_i) \implies \ln(y_i) = (-x)_i'\frac{\beta}{\alpha} - \frac{v_i}{\alpha} + u_i$$

where $-v_i/\alpha + u_i$ is the error term of this linear model. This can be estimated with LSE if there is no censoring. The unobserved term v is often called an "*unobserved heterogeneity*," whereas u is called just an error term.

If we want to apply MLE either for the efficiency or to handle a censoring problem, the presence of v poses a difficulty. Since v is not observed—we only observe (x', y)—we need to specify a distribution for v to integrate it out of the likelihood function. If we allow v to depend on x, the precise form of the dependence should be spelled out (i.e., how x enters the distribution function of $v|x$). Consequently the estimation of α and β depends critically on the assumed distribution of $v|x$, which is not needed in the above LSE. Not allowing for v in the MLE will cause a downward bias in duration dependence estimation as will be shown shortly. Thus, if the censoring percentage is low, using LSE for the linear model rather than MLE may be a good idea.

Ignore x and imagine $\lambda(v) = v\theta$; that is, the hazard rate depends only on v and a constant θ. Assume v takes 1 and 2 with the equal probability. Then one half of the population have hazard θ (Group 1) and the other half has hazard 2θ (Group 2). Initially there are the equal proportions of Group 1 and Group 2 subjects in the population, for $P(v = 1) = P(v = 2) = 0.5$. As time progresses, however, subjects in Group 2 with the higher hazard rate will leave the state, and the remaining population will have more and more Group-1 subjects. This scenario is indistinguishable from the situation

$\lambda'(t) < 0$. Thus even when we have $\lambda'(t) = 0$ for all t for each subject, if we estimate $\lambda(t)$ ignoring v, we will end up with $\lambda'(t) < 0$, which is a downward bias.

Unlike Exponential or Weibull, some distributions exhibit an "up-and-down" or "down-and up" duration dependence. For example,

$$Log\text{-}logistic\ integrated\ hazard : \Lambda(t) = \ln(1 + \theta t^{\alpha}), \quad \alpha, \theta > 0$$

$$\implies \lambda(t) = \frac{\theta \alpha t^{\alpha-1}}{1 + \theta t^{\alpha}} \quad \text{and} \quad S(t) = \frac{1}{1 + \theta t^{\alpha}}.$$

The numerator of $\lambda(t)$ is the same as the Weibull hazard. When $\alpha \leq 1$, the hazard declines monotonically; otherwise, it goes up and down. See Figure 4. In contrast, *Exponential power hazard* with $\alpha < 1$ shows an U-shaped (down-and-up) hazard with two parameters θ and α: with $\theta, \alpha > 0$,

$$\lambda(t) = \theta \alpha t^{\alpha-1} \exp(\theta t^{\alpha}) \quad \text{and} \quad S(t) = \exp\{1 - \exp(\theta t^{\alpha})\}.$$

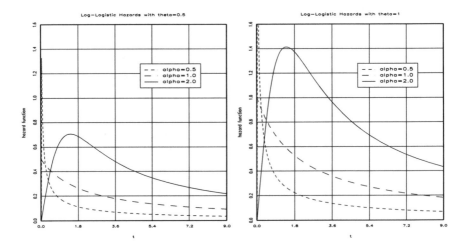

Figure 4: Log-Logistic Hazards

Klein and Moeschberger (2003, pp. 38 and 44) listed many other parametric hazard functions, and one fairly comprehensive distribution is *generalized gamma* with three parameters:

$$\lambda(t) = \frac{\theta^{\beta} \alpha t^{\alpha\beta-1} \exp(-\theta t^{\alpha})}{\int_{\theta t^{\alpha}}^{\infty} z^{\beta-1} e^{-z} dz} = \frac{\theta^{\beta} \alpha t^{\alpha\beta-1} \exp(-\theta t^{\alpha})/\Gamma(\beta)}{\int_{\theta t^{\alpha}}^{\infty} z^{\beta-1} e^{-z} dz/\Gamma(\beta)} \quad \left(= \frac{f(t)}{S(t)}\right);$$

the name comes from *gamma distribution* when $\alpha = 1$. Generalized gamma includes Weibull when $\beta = 1$ and Exponential when $\alpha = \beta = 1$; it also includes *log-normal hazard* as a limiting case when $\beta \to \infty$ which can also have an up-and-down hazard similar to log-logistic hazard. But generalized gamma hazard does not allow for U-shaped hazard.

Saha and Hilton (1997) generalized the above exponential power hazard to a three-parameter family:

$$\lambda(t) = \gamma \alpha t^{\alpha-1} \exp(\theta t^{\alpha}) \quad \text{and} \quad S(t) = \exp\left[\frac{\gamma}{\theta}\{1 - \exp(\theta t^{\alpha})\}\right], \quad \text{where}$$

$$\alpha, \gamma > 0.$$

This includes the above exponential power hazard as a special case when $\gamma = \theta$, and Weibull hazard as a limiting case as $\theta \to 0$. With $\gamma = \exp(x'\beta)$ (or $\theta = \exp(x'\beta)$), x can be introduced into the model. This hazard is quite flexible, allowing for constant, increasing/decreasing, U-shaped, and inverse U-shaped hazards; for (inverse) U-shaped hazard, the turning point is $\{(1-\alpha)/\alpha\theta\}^{1/\alpha}$. In the empirical comparison study of Seetharaman and Chintagunta (2003), this "generalized exponential power hazard" performed better than other parametric hazards. Of course, it is possible that the true hazard (e.g., double humps) is not covered by a specified parametric hazard in use. This motivates "semiparametric estimators" which do not require specifying the distribution of $y|x$ fully.

5.5   Invariances and Extreme Value Distributions*

With $\Lambda(t)$ strictly increasing in t,

$$\exp(-\Lambda(t)) = S(t) = P(y > t) = P\{\Lambda(y) > \Lambda(t)\}.$$

From the first and last terms, defining $a \equiv \Lambda(t) > 0$,

$$P(\Lambda(y) > a) = \exp(-a)$$

which is the survival function for $Expo(1)$. That is,

$$\Lambda(y) \; follows \; Expo(1)$$

regardless of the distribution of y. This invariance is analogous to $F(y) \sim U[0,1]$ where $F(\cdot)$ is the df of a continuously distributed rv y.

Observe

$$1 - \exp(-a) = P(\Lambda(y) \le a) = P(-\ln \Lambda(y) \ge -\ln(a)).$$

Define $w = -\ln \Lambda(y)$ and $b = -\ln(a)\,(\Longleftrightarrow a = e^{-b})$ to get

$$1 - \exp(-e^{-b}) = P(w \ge b).$$

The left-hand side is the survival function for "type-I extreme value distribution" with location parameter 0 and scale parameter 1, which is explained further below. That is,

$$-\ln \Lambda(y) \; follows \; the \; \text{``standard''} \; type\text{-}I \; extreme \; value \; distribution$$

regardless of the distribution of y.

When a rv z follows *Type I extreme value distribution* (or *Gumbel distribution*) with parameter μ for location and ψ for scale, it holds that

(a): $P(z \leq z_o) = F(z_o) = \exp(-e^{-(z_o-\mu)/\psi})$, $\quad -\infty < z < \infty$,
$-\infty < \mu < \infty$, $\psi > 0$

(b): $E(z) = \mu + \psi \cdot \gamma$ where γ is the "Euler's constant" $\simeq 0.577$

(c): $V(z) = \psi^2 \pi^2/6$.

Figure 5 shows three densities with $\mu = 0$ and $\psi = 0.5, 1.0, 1.5$.

When $\mu = 0$ and $\psi = 1$ (the "standard" case), differentiate the type I extreme value distribution function $\exp(-e^{-t})$ wrt t to get the density

$$\exp(-e^{-t}) \cdot \exp(-t) = \exp(-e^{-t} - t)$$

which is asymmetric around 0 and unimodal at 0; this can be seen by differentiating the density, as well as by looking at Figure 5.

From the invariant distribution of $-\ln \Lambda(y)$, when y follows $Expo(1)$, $-\ln \Lambda(y) = -\ln y$ follows the "standard" type-1 extreme value distribution. More generally, when y follows $Expo(\theta)$, $-\ln \Lambda(y) = -\ln(\theta y) = -\ln y - \ln \theta$ follows the "standard" type-I extreme value distribution; equivalently, $-\ln y$ follows type-I extreme distribution with $\mu = \ln \theta$ and $\psi = 1$. Hence, using (b) and (c) above, we get

$$E(-\ln y) \simeq \ln \theta + 0.577 \; (\implies \; E(\ln y) \simeq -\ln \theta - 0.577),$$
$$V\{-\ln(y)\} \simeq \pi^2/6 \simeq 1.645$$

which is one of the properties mentioned for exponential distribution.

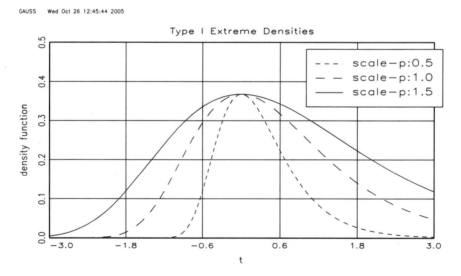

Figure 5: Type-I Extreme Value Distribution Densities

Other than Type-1 extreme value distributions, there are two more types of extreme value distributions (Johnson et al., 1995): for some parameters μ, ψ, and α such that $\psi, \alpha > 0$,

$$\text{Type } 2 : P(z \leq z_o) = 0 \quad \text{when } z_o < \mu,$$
$$= \exp\{-(\frac{z_o - \mu}{\psi})^{-\alpha}\} \quad \text{when } z_o \geq \mu$$

$$\text{Type } 3 : P(z \leq z_o) = \exp\{-(\frac{\mu - z_o}{\psi})^{\alpha}\} \quad \text{when } z_o \leq \mu$$
$$= 1 \quad \text{when } z_o > \mu.$$

In each type, sometimes the distribution of $-z$ is also called an extreme value distribution.

In Type 3, replace z with $-z$, z_o with $-t$, and μ with 0 to get, when $t \geq 0$,

$$P(-z \leq -t) = P(z \geq t) = \exp\{-(\frac{t}{\psi})^{\alpha}\} = \exp(-\theta t^{\alpha}) \quad \text{where } \theta \equiv \psi^{-\alpha}.$$

This is nothing but the Weibull distribution with parameters θ and α; Type 3 is sometimes called Weibull. Whereas type 1 has an exponential function of t in $\exp(\cdot)$, type 3 has a polynomial function of t in $\exp(\cdot)$. The reason for the qualifier "extreme" is that all three types of distributions are obtained as the limiting distributions of the maximum among N iid rv's.

CHAPTER 6
PARAMETRIC METHODS FOR MULTIPLE EQUATION LDV MODELS

Going further from single equation models with limited dependent variables (LDV) in the preceding chapter, multiple equations with LDV's are examined. Multiple equations with LDV include multinomial (or multiple) choice, sample-selection (or selected-sample) models, LDV models with endogenous regressors, simultaneous equations with LDV's, and panel data binary-response models. Also, duration analysis for competing risks is introduced.

1 Multinomial Choice Models

This section deals with multinomial choice which is a multiple-equation LDV model. With many equations in LDV's, the limited observability in each equation becomes "aggravated or compounded" and the estimation is more difficult than for single equation LDV models or linear multiple equations. While ordered response models extend the binary response model "horizontally" by allowing more ordered categories, multinomial (or multiple) choice models extend the binary response model "vertically" by considering multiple response equations jointly.

Suppose we have N individuals with each one having J many alternatives to choose from. Each person will choose one alternative that yields the highest utility or satisfaction, depending on his/her attributes and the characteristics of the alternatives. One example is a transportation mode choice problem where the alternatives are bus, car, and train. Let i index the individuals and j index the alternatives. Here, the explanatory variables are of three types: the first type varies across i and j (cost and time for each mode), the second type varies only across i (attributes of people such as income, sex, and race), and the third type varies only across j (some characteristics of mode common to all i such as whether the transportation mode has a dining facility). Another example is industry choice for job where the alternatives are manufacturing, service, government, and so on. Recent reviews on multinomial choice econometrics can be found in McFadden (2001), Train (2003), and Hensher et al. (2005). Part of this section draws on Lee and Kim (2007).

Myoung-jae Lee, *Micro-Econometrics*, DOI 10.1007/b60971_6,
© Springer Science+Business Media, LLC 2010

1.1   Basics

Person i gets "utility" or "satisfaction" s_{ij} from alternative j. In linking satisfaction to observed and unobserved variables, consider two models

$$(i) : s_{ij} = x'_{ij}\delta + u_{ij}, \quad (ii) : s_{ij} = z'_i\eta_j + u_{ij}$$

where x_{ij} and z_i are $k_x \times 1$ and $k_z \times 1$ regressor vectors, respectively, and u_{ij} is an error term. Let the first component of z_i be 1 for all i so that the first component of η_j is the "alternative-specific intercept," which includes alternative-specific characteristics common to all i. A model combining (i) and (ii) is

$$(iii) : s_{ij} = x'_{ij}\delta + z'_i\eta_j + u_{ij}.$$

The term $z'_i\eta_j$ in (ii) and (iii) needs some justification before it is used. Suppose we have $z'_i\eta$ in (iii) instead of $z'_i\eta_j$. Then the effect of z_i on all s_{ij}, $j = 1, ...J$, is the same. This means that $z'_i\eta$ plays no role in the choice and z_i drops out of the picture. If we use $z'_i\eta_j$ to include z_i in our choice analysis, we need to justify why z_i has different coefficients for the alternatives. For instance, if z_i is income, then we should question why an unit increase in income changes the utilities of different transportation modes in different ways. When we use $z'_i\eta_j$, we can also think of using $x_{ij}\delta_j$ instead of $x_{ij}\delta$, which leads to a model more general than (iii):

$$(iv) : s_{ij} = x'_{ij}\delta_j + z'_i\eta_j + u_{ij}.$$

In (iv), x_{ij} and z_i are treated equally in the sense that both are allowed to have alternative-variant slopes. Model (iii) is a special case of (iv) with the restriction $\delta_1 = \delta_2 =, ..., = \delta_J$. We will use (iv) from now on.

Clearly, z_i is the "individual-variant but alternative-constant" regressors and x_{ij} is the "individual- and alternative-variant" regressors. But x_{ij} may also include interaction terms such as $z_i m_j$ where m_j is an alternative-variant variable; e.g., z_i is income and $m_j = 1$ if transportaion mode j has a first-class section and 0 otherwise. Here, the effect of m_j on the utility depends on person i's income level. It is also possible to have more "elaborate" interaction terms in x_{ij} such as the original x_{ij} times z_i or m_j although they may be rarely used in practice; e.g., x_{ij} is the cost of alternative j for person i and z_i is income.

Among the J-many alternatives, alternative j will be chosen if it gives the maximum satisfaction, that is, if (ignoring ties)

$$s_{ij} > s_{im} \,\forall m \neq j \iff s_{ij} - s_{i1} > s_{im} - s_{i1} \,\forall m \neq j;$$

this is a "location normalization" using the first alternative as the base. From this, we use not $s_{ij} = x'_{ij}\delta_j + z'_i\eta_j + u_{ij}$, but

$$s_{ij} - s_{i1} = x'_{ij}\delta_j - x'_{i1}\delta_1 + z'_i(\eta_j - \eta_1) + u_{ij} - u_{i1}$$

with parameters

$$\delta_1, ..., \delta_J, \ \eta_2 - \eta_1, ..., \eta_J - \eta_1.$$

Although (iv) is more general than (i) and (ii), in practice, we transform our model into one that looks like (i). Suppose $J = 3$ and consider the following differences of the regression functions:

2 and 1: $x'_{i2}\delta_2 + z'_i\eta_2 - (x'_{i1}\delta_1 + z'_i\eta_1) = x'_{i2}\delta_2 - x_{i1}{}'\delta_1 + z'_i(\eta_2 - \eta_1)$;

3 and 1: $x'_{i3}\delta_3 + z'_i\eta_3 - (x'_{i1}\delta_1 + z'_i\eta_1) = x'_{i3}\delta_3 - x_{i1}{}'\delta_1 + z'_i(\eta_3 - \eta_1)$.

Define w_{i2}, w_{i3}, and β as

$$w_{i2} \equiv (-x'_{i1}, x'_{i2}, 0'_{k_x}, z'_i, 0'_{k_z})', \quad w_{i3} \equiv (-x'_{i1}, 0'_{k_x}, x'_{i3}, 0'_{k_z}, z'_i)',$$

$$\beta \equiv (\delta'_1, \delta'_2, \delta'_3, \eta'_2 - \eta'_1, \eta'_3 - \eta'_1)',$$

where 0_k is the $k \times 1$ zero vector. Then the above *regression function differences can be written simply as* $w'_{i2}\beta$ *and* $w'_{i3}\beta$, respectively, which yields the same form as (i).

If (iii) is adopted under $\delta_1 = \delta_2 = \delta_3 \equiv \delta$, then

2 and 1: $x'_{i2}\delta + z'_i\eta_2 - (x'_{i1}\delta + z'_i\eta_1) = (x'_{i2} - x_{i1})'\delta + z'_i(\eta_2 - \eta_1)$;

3 and 1: $x'_{i3}\delta + z'_i\eta_3 - (x'_{i1}\delta + z'_i\eta_1) = (x'_{i3} - x_{i1})'\delta + z'_i(\eta_3 - \eta_1)$.

In this case, to get the same form as (i), w_{i2}, w_{i3}, and β should be defined as

$$w_{i2} \equiv (x'_{i2} - x'_{i1}, z'_i, 0'_{k_z})', \quad w_{i3} \equiv (x'_{i3} - x'_{i1}, 0'_{k_Z}, z'_i)' \quad \text{and}$$

$$\beta \equiv (\delta', \eta'_2 - \eta'_1, \eta'_3 - \eta'_1)'.$$

Define w_i as the collection of all components of w_{ij}'s, $j = 1, ..., J$. Also define

$$y_{ij} = 1 \text{ if person } i \text{ chooses } j, \text{ and } 0 \text{ otherwise.}$$

Then we get

$$\sum_{j=1}^{J} y_{ij} = 1, \text{ and } \sum_{j=1}^{J} P(y_{ij} = 1|w_i) = 1 \quad \text{for all } i.$$

These identities will come handy later.

1.2   Multinomial Probit (MNP)

1.2.1   Choice Probabilities and Identified Parameters

Setting $J = 3$ and omitting i for simplicity, we get

$$
\begin{aligned}
P(y_1 &= 1|w) = P(s_1 > s_2, \ s_1 > s_3 \ |w) \\
&= P(s_2 - s_1 < 0, \ s_3 - s_1 < 0 \ |w) \\
&= P(w'_2\beta + u_2 - u_1 < 0, \ w'_3\beta + u_3 - u_1 < 0|w) \\
&= P\{u_2 - u_1 < -w'_2\beta, \ u_3 - u_1 < -w'_3\beta|w\}
\end{aligned}
$$

$$
\begin{aligned}
P(y_2 &= 1|w) = P(s_1 < s_2,\ s_2 > s_3\ |w) \\
&= P(s_2 - s_1 > 0,\ s_2 - s_1 - (s_3 - s_1) > 0\ |w) \\
&= P\{u_2 - u_1 > -w_2'\beta,\ (u_2 - u_1) - (u_3 - u_1) > \\
&\quad -(w_2 - w_3)'\beta\ |w\} \\
P(y_3 &= 1|w) = P(s_3 > s_1,\ s_3 > s_2\ |w) \\
&= P(s_3 - s_1 > 0,\ s_3 - s_1 - (s_2 - s_1) > 0\ |w) \\
&= P\{u_3 - u_1 > -w_3'\beta,\ (u_3 - u_1) - (u_2 - u_1) > \\
&\quad -(w_3 - w_2)'\beta\ |w\}.
\end{aligned}
$$

Here all choice probabilities are written in terms of "the difference from the first alternative."

Define

$$
v_2 \equiv u_2 - u_1, \quad v_3 \equiv u_3 - u_1, \quad \sigma^2 \equiv V(v_2)
$$

and rewrite $P(y_j = 1|w)$, $j = 1, 2, 3$, as

$$
\begin{aligned}
P(y_1 &= 1|w) = P\{v_2 < -w_2'\beta,\ v_3 < -w_3'\beta\ |w\} \\
&= P\left\{ \frac{v_2}{\sigma} < -w_2'\frac{\beta}{\sigma},\ \frac{v_3}{\sigma} < -w_3'\frac{\beta}{\sigma}\ |w \right\}, \\
P(y_2 &= 1|w) = P\{v_2 > -w_2'\beta,\ v_2 - v_3 > -(w_2 - w_3)'\beta\ |w\} \\
&= P\left\{ \frac{v_2}{\sigma} > -w_2'\frac{\beta}{\sigma},\ \frac{v_2 - v_3}{\sigma} > -(w_2 - w_3)'\frac{\beta}{\sigma}\ |w \right\}, \\
P(y_3 &= 1|w) = P\{v_3 > -w_3'\beta,\ v_3 - v_2 > -(w_3 - w_2)'\beta\ |w\} \\
&= P\left\{ \frac{v_3}{\sigma} > -w_3'\frac{\beta}{\sigma},\ \frac{v_3 - v_2}{\sigma} > -(w_3 - w_2)'\frac{\beta}{\sigma}\ |w \right\}.
\end{aligned}
$$

In view of this, the case with J can be written as

$$
\begin{aligned}
P(y_1 &= 1|w) = P\left\{ \frac{v_2}{\sigma} < -w_2'\frac{\beta}{\sigma},\ ...,\ \frac{v_J}{\sigma} < -w_J'\frac{\beta}{\sigma}\ |w \right\}, \\
P(y_2 &= 1|w) = P\left\{ \frac{v_2}{\sigma} > -w_2'\frac{\beta}{\sigma},\ \frac{v_2 - v_3}{\sigma} > -(w_2 - w_3)'\frac{\beta}{\sigma}, ..., \right. \\
&\quad \left. \frac{v_2 - v_J}{\sigma} > -(w_2 - w_J)'\frac{\beta}{\sigma}\ |w \right\}, \\
&\quad\vdots \\
P(y_J &= 1|w) = P\left\{ \frac{v_J}{\sigma} > -w_J'\frac{\beta}{\sigma},\ \frac{v_J - v_2}{\sigma} > -(w_J - w_2)'\frac{\beta}{\sigma}, ..., \right. \\
&\quad \left. \frac{v_J - v_{J-1}}{\sigma} > -(w_J - w_{J-1})'\frac{\beta}{\sigma}\ |w \right\}
\end{aligned}
$$

where v_j and w_j for $j \geq 4$ are defined analogously to v_3 and w_3. The choice probabilities depend on the joint distribution of $(v_2/\sigma, ..., v_J/\sigma)$. The parameters to estimate are β/σ and the variance matrix of $(v_2/\sigma, ..., v_J/\sigma)$.

To see better what is identified in the variance matrix, consider the case $J = 3$. In the variance matrix of $(v_2/\sigma, v_3/\sigma)$, only

$$\sigma_{23} \equiv COV\left(\frac{v_2}{\sigma}, \frac{v_3}{\sigma}\right) \quad \text{and} \quad \sigma_3^2 \equiv V\left(\frac{v_3}{\sigma}\right)$$

are unknown ($V(v_2/\sigma) = 1$ by construction). Overall, the parameters to estimate when $J = 3$ is

$$\gamma \equiv \left(\frac{\beta'}{\sigma}, \sigma_{23}, \sigma_3^2\right)' \Longleftrightarrow \left(\frac{\beta'}{\sigma}, \rho_{23}, \sigma_3^2\right)' \text{where } \rho_{23} \equiv COR\left(\frac{v_2}{\sigma}, \frac{v_3}{\sigma}\right) = \frac{\sigma_{23}}{\sigma_3}$$

where β is identified only up to the scale σ as in the binary model. If $J = 4$, then γ becomes

$$\gamma \equiv \left(\frac{\beta'}{\sigma}, \rho_{23}, \rho_{24}, \sigma_3^2, \rho_{34}, \sigma_4^2\right)'.$$

In the general case with J,

$$\gamma = \left(\frac{\beta'}{\sigma}, \rho_{23}, ..., \rho_{2J}, \sigma_3^2, \rho_{34}, ..., \rho_{3J}, \sigma_4^2, ..., ..., ..., \sigma_{J-1}^2, \rho_{J-1,J}, \sigma_J^2\right)'.$$

1.2.2   Log-Likelihood Function and MOM

Defining $P_{ij}(\gamma)$ as

$$P_{ij}(\gamma) = P(i \text{ chooses } j|w_i),$$

the log-likelihood function to be maximized wrt g is

$$\sum_{i=1}^{N}\sum_{j=1}^{J} y_{ij} \ln P_{ij}(g) \implies \text{the first derivatives are } \sum_{i=1}^{N}\sum_{j=1}^{J} y_{ij}\frac{\partial \ln P_{ij}(g)}{\partial g}.$$

The score function is $\sum_{j=1}^{J} y_{ij}\partial \ln P_{ij}(g)/\partial g$, and $\sqrt{N}(g_N - \gamma)$ follows $N(0, I_f^{-1})$ where I_f can be estimated consistently with the outer-product of the score function as usual:

$$\frac{1}{N}\sum_{i=1}^{N}\left\{\sum_{j=1}^{J} y_{ij}\frac{\partial \ln P_{ij}(g_N)}{\partial g}\right\}\left\{\sum_{j} y_{ij}\frac{\partial \ln P_{ij}(g_N)}{\partial g'}\right\} \to^p I_f.$$

Differentiate the identity $\sum_{j=1}^{J} P(y_{ij} = 1|w_i) = 1$, which holds for any given g, to get

$$\sum_{j=1}^{J} P_{ij}(g)\frac{\partial P_{ij}(g)/\partial g}{P_{ij}(g)} = \sum_{j=1}^{J} P_{ij}(g)\frac{\partial \ln P_{ij}(g)}{\partial g} = 0.$$

Subtract this from the first-order condition of the MLE to get

$$\sum_{i=1}^{N}\sum_{j=1}^{J}\{y_{ij}-P_{ij}(g)\}\frac{\partial \ln P_{ij}(g)}{\partial g}=0.$$

This can be viewed as a moment condition where $y_{ij}-P_{ij}(g)$ is the error term and $\partial \ln P_{ij}(g)/\partial g$ is the instrument.

If we assume that u_1 to u_J are jointly normally distributed, then v_2 to v_J are also jointly normally distributed. In this case, we get *multinomial probit (MNP)* where $P_{ij}(g)$ is obtained by integrating a $J-1$ dimensional normal density function; see Hausman and Wise (1978) who also considered random coefficients. Numerical integration in high dimensions (say, greater than 4) is computationally burdensome and not reliable. Although "method of simulated moments" to be examined later can solve this integration problem, the real problem with MNP is the difficulty in estimating the variance matrix parameters. Even with $J=3$, often it is difficult to estimate both σ_{23} and σ_3. Put it differently, the log-likelihood function often hardly changes as σ_{23} or σ_3 changes. STATA has the command `mprobit` to implement MNP.

The identification of parameters in MNP is "fragile" when there are no exclusion restrictions that some explanatory variables in s_{ij} are excluded from s_{im} for all $m \neq j$ (i.e., the variables have zero coefficients in s_{im}), as noted by Keane (1992). Keane (1992, p. 194) also cited a working paper by Bunch and Kitamura stating that nearly half the existing applications of MNP have used non-identified models. Train (2003) remarked also on non-identified MNP models in the literature. Some specific examples for using non-identified MNP models in political science can be found in Lee and Kang (2009).

1.2.3 Implementation

Suppose there is a program to implement MNP and the program provides a routine to compute

$$\Psi(c_1, c_2, \rho) \equiv \int_{-\infty}^{c_1}\int_{-\infty}^{c_2} \psi(\varepsilon_1, \varepsilon_2, \rho)d\varepsilon_1 d\varepsilon_2$$

where ψ is the joint density for two $N(0,1)$ rv's with correlation ρ:

$$\psi(\varepsilon_1, \varepsilon_2, \rho) = \frac{1}{2\pi\sqrt{1-\rho^2}}\exp\left\{-\frac{\varepsilon_1^2 - 2\rho\varepsilon_1\varepsilon_2 + \varepsilon_2^2}{2(1-\rho^2)}\right\}.$$

To make use of this when $J=3$, we need to express the choice probabilities in terms of two error terms; the obvious choice is v_2 and v_3. Setting $\sigma = 1$ to simplify notations, define $COR(v_2, v_3) = \rho$. Then,

$$P(\text{choose } 1|w) = P(v_2 < -w_2'\beta, \frac{v_3}{\sigma_3} < -w_3'\frac{\beta}{\sigma_3}|w)$$

$$= \Psi(-w_2'\beta, -w_3'\frac{\beta}{\sigma_3}, \rho),$$

$$P(\text{choose } 2|w) = P(v_2 > -w_2'\beta, v_3 - v_2 < -w_3'\beta + w_2'\beta \,|w)$$

$$= P(-v_2 < w_2'\beta, v_3 - v_2 < -w_3'\beta + w_2'\beta \,|w),$$

which needs to be rewritten to make use of Ψ as follows; once this is done, $P(\text{choose } 3|w)$ is $1 - P(\text{choose } 1|w) - P(\text{choose } 2|w)$. Observe

$$V(v_3 - v_2) = \sigma_3^2 + 1 - 2\rho\sigma_3 = (\sigma_3 - \rho)^2 + 1 - \rho^2 > 0$$

$$\implies \sigma^* \equiv SD(v_3 - v_2) = (\sigma_3^2 + 1 - 2\rho\sigma_3)^{1/2},$$

$$COR(-v_2, v_3 - v_2) = \frac{E(v_2(v_2 - v_3))}{\sigma^*} = \frac{1 - \rho\sigma_3}{\sigma^*}.$$

Hence $P(\text{choose } 2|w)$ is

$$\Psi\left\{w_2'\beta, \frac{w_2'\beta - w_3'\beta}{(\sigma_3^2 + 1 - 2\rho\sigma_3)^{1/2}}, \frac{1 - \rho\sigma_3}{(\sigma_3^2 + 1 - 2\rho\sigma_3)^{1/2}}\right\},$$

showing that $(\beta', \rho, \sigma_3)'$ is to be estimated.

With the choice probabilities expressed in terms of the identified parameters, the rest of MNP iteration can be done with numerical derivatives. Because $v_2 = u_2 - u_1$ and $v_3 = u_3 - u_1$, if u_1, u_2, u_3 are iid, then $COR(v_2, v_3) = 0.5$ and $\sigma_3 = 1$. That is, $\rho = 0.5$ and $\sigma_3 = 1$ may serve as a good starting value for the numerical maximization.

1.3   Multinomial Logit (MNL)

1.3.1   Choice Probabilities and Implications

A rv u following (the standard) type-1 extreme value distribution has the df $\exp(-e^{-u})$ and density $\exp(-u - e^{-u})$. If $u_{i1}, .., u_{iJ}$ are iid with type I extreme value distribution, then we get the following *multinomial logit (MNL)* specification as McFadden (1974) showed:

$$P_{ij}(\beta) = \frac{\exp(x_{ij}'\delta_j + z_i'\eta_j)}{\sum_{j=1}^{J} \exp(x_{ij}'\delta_j + z_i'\eta_j)}.$$

Although the scale factor is not explicit in β, β should be viewed as divided by an unknown positive scale factor as this was the case for binary logit. The assumption on $u_{i1}, ..., u_{iJ}$ is restrictive, because the error terms should have the same variance and be independent of one another. This essentially amounts to throwing away the covariance matrix estimation problem in MNP.

It is convenient to define q_{ij} and α_j as follows:

$$q_{ij} \equiv (x'_{ij}, z'_i)' \quad \text{and} \quad \alpha_j \equiv (\delta'_j, \eta'_j)'$$
$$\implies \quad s_{ij} = x'_{ij}\delta_j + z'_i\eta_j + u_{ij} = q'_{ij}\alpha_j + u_{ij}.$$

The MNL specification is then $P_{ij}(\beta) = \exp(q'_{ij}\alpha_j) / \sum_j \exp(q'_{ij}\alpha_j)$, and thus

$$\frac{\exp(q'_{ij}\alpha_j)}{\sum_{j \neq m} \exp(q'_{ij}\alpha_j)} = \frac{\exp(q'_{ij}\alpha_j)}{\sum_{j=1}^J \exp(q'_{ij}\alpha_j)} \frac{\sum_{j=1}^J \exp(q'_{ij}\alpha_j)}{\sum_{j \neq m} \exp(q'_{ij}\alpha_j)}$$

$$= P_{ij}(\beta) \cdot \frac{\sum_{j=1}^J \exp(q'_{ij}\alpha_j)}{\sum_{j \neq m} \exp(q'_{ij}\alpha_j)} :$$

if alternative m is dropped from the choice set, then the new choice probability in the remaining alternatives is the old choice probability $P_{ij}(\beta)$ times a factor common to all alternatives.

Also observe that the ratio of the probabilities P_{ij} and P_{il}, $j \neq l$, is

$$\frac{P_{ij}(\beta)}{P_{il}(\beta)} = \exp(x'_{ij}\delta_j + z'_i\eta_j - x'_{il}\delta_l - z'_i\eta_l) = \exp\{(w_{ij} - w_{il})'\beta\} :$$

the availability of the other alternatives does not play any role in this ratio because the attributes of the other alternatives are not in $w_{ij} - w_{il}$. This feature, known as *independence of irrelevant alternatives (IIA)*, is unlikely to hold in real life when some other alternatives are similar to either j or l in their attributes. A well-known example against IIA is three commuting alternatives, a blue bus, a red bus, and a car, where the probability ratio of red bus to car should depend on the blue bus availability because blue bus will take away one half of the red bus commuters.

Consider MNL with J alternatives and "J-but-m" alternatives. In both cases, the ratio of the choice probabilities $P_{ij}(\beta)/P_{il}(\beta)$ where $j, l \neq m$ is the same. Thus one way to test for IIA is applying MNL twice (first with J alternatives, and then with J minus alternative m) to see if choice-probability ratios change or not. Another way is to do "nested logit" explained later which relaxes IIA and includes MNL as a special case. See McFadden (1987) and Fry and Harris (1996) for more on specification tests for MNL.

Divide both the numerator and denominator of $P_{ij}(\beta)$ with exp $(x'_{i1}\delta_1 + z'_i\eta_1)$ to get the *normalized* choice probabilities

$$P_{i1} = \frac{1}{\left\{1 + \sum_{j=2}^J \exp(w'_{ij}\beta)\right\}}, \quad P_{ij} = \frac{\exp(w'_{ij}\beta)}{\left\{1 + \sum_{j=2}^J \exp(w'_{ij}\beta)\right\}},$$

$$j = 2, ..., J;$$

recall $\beta = (\delta'_1, ..., \delta'_J, \eta'_2 - \eta'_1, ..., \eta'_J - \eta'_1)'$. Once this is substituted into the log-likelihood $\sum_{i=1}^N \sum_{j=1}^J y_{ij} \ln P_{ij}(\beta)$, MNL can be implemented. The likelihood function is well behaving and multinomial logit is computationally attractive despite its rigid specification; it may be called the "LSE" of multinomial choice.

1.3.2   Further Remarks

MNL renders a simple moment condition of the form $E\left\{\sum_{j=2}^{J} w_{ij}\right.$
$\left.(y_{ij} - E(y_{ij}|w_i))\right\} = 0$. To see this, observe

$$\frac{\partial \ln P_{i1}(b)}{\partial b} = -\sum_{j=2}^{J} w_{ij} P_{ij}, \quad \frac{\partial \ln P_{ij}(b)}{\partial b} = w_{ij} - \sum_{j=2}^{J} w_{ij} P_{ij}, \; j = 2, ..., J.$$

Substitute this into the first-order condition $\sum_{i=1}^{N} \sum_{j=1}^{J} y_{ij} \partial \ln P_{ij}(b)/$
$\partial b = 0$ to get

$$\sum_{i=1}^{N} \left\{ -y_{i1} \sum_{j=2}^{J} w_{ij} P_{ij} + \sum_{j=2}^{J} y_{ij} w_{ij} - \sum_{j=2}^{J} y_{ij} \sum_{j=2}^{J} w_{ij} P_{ij} \right\}.$$

Substitute $\sum_{j=2}^{J} y_{ij} = 1 - y_{i1}$ to obtain

$$= \sum_{i=1}^{N} \left\{ \sum_{j=2}^{J} y_{ij} w_{ij} - \sum_{j=2}^{J} w_{ij} P_{ij} \right\} = \sum_{i=1}^{N} \sum_{j=2}^{J} w_{ij}(y_{ij} - P_{ij}) = 0.$$

McFadden and Train's (2000) "mixed logit" generalizes MNL by allowing for random coefficients in MNL. In mixed logit, β in $w'_{ij}\beta$ of MNL follows, say, a joint normal distribution with density $\varphi(\cdot|\beta_o, \Gamma_o)$ where β_o is the mean and Γ_o is the variance matrix; conditional on the realized β, we get MNL and $\varphi(\cdot|\beta_o, \Gamma_o)$ is a "mixing density." Let the dimension of w_{ij} be $\tau \times 1$. The mixed logit log-likelihood function to maximize for b_o and G_o is

$$\sum_{i=1}^{N} \sum_{j=1}^{J} y_{ij} \ln \left\{ \int \frac{\exp(w'_{ij}b)}{\sum_{j=1}^{J} \exp(w'_{ij}b)} \varphi(b|b_o, G_o) db \right\}$$

$$= \sum_{i=1}^{N} \sum_{j=1}^{J} y_{ij} \ln \left\{ \int \frac{\exp(w'_{ij}b)}{\sum_{j=1}^{J} \exp(w'_{ij}b)} \frac{\varphi(b|b_o, G_o)}{\varphi(b|0_\tau, I_\tau)} \varphi(b|0_\tau, I_\tau) db \right\}$$

$$\simeq \sum_{i=1}^{N} \sum_{j=1}^{J} y_{ij} \ln \left\{ \frac{1}{S} \sum_{s=1}^{S} \frac{\exp(w'_{ij}b_s)}{\sum_{j=1}^{J} \exp(w'_{ij}b_s)} \frac{\varphi(b_s|b_o, G_o)}{\varphi(b_s|0_\tau, I_\tau)} \right\}.$$

where b_s is drawn from the τ-variate iid $N(0,1)$ and S is the random draw number; the last step is a "simulated MLE" to be explained further later. One caution for the last equality is that, although $\varphi(b_s|0_\tau, I_\tau)$ is used there because zero-mean normal random numbers are readily available, if b_o differs much from 0, then $\varphi(b_s|b_o, G_o)/ \varphi(b_s|0_\tau, I_\tau)$ may become almost infinity for all s. For instance, when $b_o = 100$ and $b_s = 120$, we might get $\varphi(120|100, G_o)/\varphi(120|0_\tau, I_\tau) \simeq \infty$ as $\varphi(120|0_\tau, I_\tau) \simeq 0$.

Another formulation of mixed logit is an "error component model," say $u_{ij} = \tilde{u}_{ij} + \tilde{\varepsilon}_{ij}$, where $\tilde{u}_{i1}, ..., \tilde{u}_{iJ}$ are iid with type I extreme-value distribution (for MNL feature) and $\tilde{\varepsilon}_{i1}, ..., \tilde{\varepsilon}_{iJ}$ are allowed to be related to one another and follow a specified distribution up to some unknown parameters; $\tilde{\varepsilon}_{i1}, ..., \tilde{\varepsilon}_{iJ}$ then give flexible substitution patterns among the alternatives. Implementing mixed logit, however, seems to require much care despite its attractive theoretical properties as can be seen in Hensher and Greene (2003). For instance, without no restriction on Γ_o, mixed logit would run into the same kind of problem as MNP has. A simplification is using a diagonal Γ_o or allowing random coefficients only for some regressors. Deciding on these matters, specifying the distribution for $\tilde{\varepsilon}_{ij}$ and interpreting the estimation results (such as finding marginal effects) would be challenging, to say the least. An empirical example of mixed logit for mobile phone choice can be found in Ida and Kuroda (2009).

1.3.3   Marginal Effects

As in binary response and ODR, it is interesting to see marginal effects of the regressors. As in ODR, we use the marginal effects defined as $E(\partial P_{ij}/\partial x_{im})$ and $E(\partial P_{ij}/\partial z_i)$. Even if a variable appears only in w_{ij}, it influences all choice probabilities, because w_{ij} appears in the denominator $1 + \sum_{j=2}^{J} \exp(w_{ij}'\beta)$. The following marginal effects apply only when there are no functionally dependent regressors. For example, if $x_{i,j+1} = x_{i,j}^2$, then the marginal effect wrt x_{ij} involves the derivatives wrt x_{ij} and $x_{i,j+1}$, and this will require some modifications.

Define

$$S_i \equiv 1 + \sum_{m=2}^{J} \exp(w_{im}'\beta)$$

and observe

$$
\begin{aligned}
\frac{\partial P_{ij}}{\partial x_{i1}} &= S_i^{-2}\left\{ S_i \exp(w_{ij}'\beta)(-\delta_1) - \exp(w_{ij}'\beta)\sum_{m=2}^{J} \exp(w_{im}'\beta)(-\delta_1)\right\} \\
&= (-\delta_1)P_{ij} - P_{ij}\sum_{m=2}^{J}(-\delta_1)P_{im} \\
&= (-\delta_1)P_{ij} - (-\delta_1)P_{ij}(1 - P_{i1}) = -\delta_1 P_{ij} P_{i1} \ \forall j \neq 1;
\end{aligned}
$$

$$
\begin{aligned}
\frac{\partial P_{ij}}{\partial x_{ij}} &= S_i^{-2}[S_i \exp(w_{ij}'\beta)\delta_j - \exp(w_{ij}'\beta)\exp(w_{ij}'\beta)\delta_j] \\
&= P_{ij}\delta_j - P_{ij}^2\delta_j = \delta_j P_{ij}(1 - P_{ij}) \ \forall j;
\end{aligned}
$$

$$
\frac{\partial P_{ij}}{\partial x_{im}} = S_i^{-2}[-\exp(w_{ij}'\beta)\exp(w_{im}'\beta)\cdot\delta_m] = -\delta_m P_{ij} P_{im} \ \forall m \neq 1, j;
$$

∂P_{i1} is omitted, because a derivative for P_{i1} can be found from those for $P_{i2}, ..., P_{i3}$ using the restriction $\sum_j P_{ij} = 1 \Longrightarrow \sum_j \partial P_{ij} = 0$.

Combining the three cases in the preceding display, we get

$$E\left(\frac{\partial P_{ij}}{\partial x_{i1}}\right) = -\delta_1 E(P_{ij}P_{i1}); E\left(\frac{\partial P_{ij}}{\partial x_{ij}}\right) = \delta_j E\{P_{ij}(1 - P_{ij})\} \text{ for } j \neq 1;$$

$$E\left(\frac{\partial P_{ij}}{\partial x_{im}}\right) = -\delta_m E(P_{ij}P_{im}) \ \forall m \neq 1, j, \quad j = 2, ..., J.$$

If $\delta_1, ..., \delta_J > 0$, then the effect on its own probability is always positive, whereas the effects on the other probabilities are always negative. Also,

$$E\left\{\frac{\partial P_{ij}/P_{ij}}{\partial x_{im}/x_{im}}\right\} = E\left\{\frac{\partial P_{ij}}{\partial x_{im}}\frac{x_{im}}{P_{ij}}\right\} = -\delta_m E(x_{im}P_{im}) \ \ \forall j \neq m :$$

the cross-elasticities of P_j wrt x_m are the same across j, which is highly restrictive.

As for the effect of z_i, define

$$\bar{\eta}_j \equiv \eta_j - \eta_1, \quad j = 2, ..., J$$

and observe, for $j \neq 1$,

$$\frac{\partial P_{ij}}{\partial z_i} = S_i^{-2}\left\{S_i \exp(w'_{ij}\beta)\bar{\eta}_j - \exp(w'_{ij}\beta)\sum_{m=2}^{J} \exp(w'_{im}\beta)\bar{\eta}_m\right\}$$

$$= P_{ij}\bar{\eta}_j - P_{ij}\sum_{m=2}^{J} P_{im}\bar{\eta}_m = P_{ij}(1 - P_{ij})\bar{\eta}_j - P_{ij}\sum_{m\neq 1,j} P_{im}\bar{\eta}_m$$

$$\Longrightarrow E\left(\frac{\partial P_{ij}}{\partial z_i}\right) = E\{P_{ij}(1 - P_{ij})\}\bar{\eta}_j - E\left\{P_{ij}\sum_{m\neq 1,j} P_{im}\bar{\eta}_m\right\}.$$

Finally, if desired, then $1 - (l_{ur}/l_r)$ may be used as a pseudo R^2 where l_{ur} is the maximized log-likelihood and l_r is the maximized log-likelihood using only 1 as the regressor for each alternative.

1.3.4   An Empirical Example: Presidential Election

Turning to an empirical example, we show part of the results in Lee and Kang (2009) who analyzed the 1992 US presidential election with three major candidates (Clinton, Bush, Perot); Perot is the normalizing alternative.

Among the regressors used, only age and age^2/100 are cardinal, whereas all the others are either dummy or ordinal variables. Two dummy variables

appear in the table below (more dummies were used in Lee and Kang (2009)),
with "R" denoting the respondent:

fin.-better:	1 if R feels his/her personal finance is better;
female:	1 if R is female.

Also, the following ordinal variables appear in the table below (again, more
ordinal variables were used in Lee and Kang (2009)):

ideology-P:	1 to 7 on R's placement of Perot ideology for liberal to conservative;
ideology-C:	1 to 7 on R's placement of Clinton ideology for liberal to conservative;
ideology-B:	1 to 7 on R's placement of Bush ideology for liberal to conservative;
ideology-R:	1 to 3 for R being liberal to conservative;
education:	1 to 7 for grade 8 or less to graduate study;
income:	1 to 24 for below \$3,000 to above \$105,000 per year;
party:	1 to 3 for Democrat to Republican;
abortion:	1 to 3 for anti-abortion to pro-abortion;
health:	1 to 7 on health insurance opinion for private insurance to government insurance;
welfare:	1 to 7 on welfare program opinion for removal to increase.

These ordinal variables pose a problem: unless they are cardinal, differences
do not make sense. For instance, the difference $2 - 1$ does not necessarily
mean the same magnitude as the difference $7 - 6$ in the variables taking 1 to
7. In principle, those 7-category ordinal variables should be used to generate
6 dummy variables. But, given the data size ($N = 894$) and the number
of the ordinal variables, this would lead to excessively many parameters to
estimate. Hence, the ordinal variables were regarded as cardinal. Among the
regressors, only ideology-P, -C, -B are alternative-variant.

In Table 1 where the numbers in ($\cdot$) are t-values, one category increment
(out of seven) in Clinton ideology increases the Clinton choice probability
by 0.052 and decreases the Bush choice probability by 0.027; "moving to the
center" of Clinton seems to have worked. The two intercepts are significantly
negative with substantial magnitude. If most regressors have little explanatory power for Perot, then the Perot's intercept being much larger than both
Clinton's and Bush's might be attributed to "voters angry with politics as
usual." In the following, we will interpret the column $\eta_{clinton} - \eta_{bush}$ that is
the difference of the two columns on its left:

$$\eta_{clinton} - \eta_{bush} \ = \ \eta_{clinton} - \eta_{perot} - (\eta_{bush} - \eta_{perot})$$

which is free of non-identified η_{perot}.

Table 1: MNL and Average Effects

	Clinton	Bush	$\eta_{clinton} - \eta_{bush}$	Clinton-effect	Bush-effect
ideology-Perot	0.008 (0.1)			-0.001	-0.000
ideology-Clinton	0.349 (3.7)			0.052	-0.027
ideology-Bush		0.115 (1.2)		-0.009	0.015
1	-7.566(-3.7)	-5.807(-2.5)	-1.759	-0.679	-0.179
age	0.119 (2.2)	-0.034(-0.6)	0.153	0.021	-0.014
age^2/100	-0.095(-1.7)	0.068 (1.1)	-0.163	-0.020	0.017
education	0.085 (0.9)	0.384 (3.4)	-0.299	-0.018	0.045
fin.-better	-0.092(-0.3)	0.756 (2.1)	-0.848	-0.073	0.108
ideology-R	-0.542(-3.2)	0.461 (2.3)	-1.003	-0.118	0.104
income	-0.055(-2.0)	-0.031(-1.0)	-0.024	-0.006	0.000
party	-0.207(-0.9)	1.145 (4.0)	-1.352	-0.121	0.169
female	0.370 (1.4)	0.639 (2.2)	-0.269	0.005	0.056
abortion	-0.072(-0.3)	-0.732(-3.1)	0.660	0.047	-0.092
health	-0.108(-1.4)	-0.239(-2.9)	0.131	0.003	-0.023
welfare	0.531 (3.1)	0.192 (0.9)	0.339	0.065	-0.016

Along with this, the two marginal effect columns will be examined, as these are easier to interpret.

The age function slopes upward for Clinton (relative to Bush) and then downward against Clinton, showing that Clinton is more popular than Bush among relatively younger voters. The age "turning point" may be seen by solving $0.153 - 0.163(2 \cdot age/100) = 0$ where the two fractional numbers are from the $\eta_{clinton} - \eta_{bush}$ column, and $2 \cdot age/100$ is the derivative of $age^2/100$. The solution is about 47.

Education works strongly for Bush (relative to Clinton); one category increment out of seven increases the Bush choice probability by 0.045. If one feels financially better off, then that increases the Bush choice probability by 0.108. The respondent ideology also has strong effects, -0.118 and 0.104, for Clinton and Bush, respectively; almost all of Bush's gain is Clinton's loss. Low income group prefer Clinton; the rather small effect magnitude on Clinton -0.006 for income is because income has 24 categories. Party affiliation has a substantial effect of 0.169 on the Bush choice probability, and the pattern is similar to that of ideology-R.

Females prefer Bush relative to Clinton with the effect on Bush being 0.056, which may be a surprise given the perception that Bush is unpopular among women while Clinton is. Abortion has a big negative impact (-0.092) on Bush—Bush's abortion stance may be the actual reason why Bush looked unpopular among women. One category increase out of seven in health insurance opinion lowers the Bush choice probability by 0.023, whereas one category increase out of seven in welfare opinion raises the Clinton choice probability by 0.065.

1.4   Nested Logit (NES)

Sometimes, alternatives are grouped or nested. For example, a town is chosen first, and then a house in the chosen town. In this case, we can imagine a "choice tree with branches," and the alternatives within a given branch are "nested." Nested logit (NES) is suitable for this kind of nested alternatives, and relaxes the IIA assumption of MNL as the nested alternatives are more related to one another than to "out-of-branch" alternatives. In econometrics, NES is attributed often to McFadden (1978, 1981), but Ortúzar (2001) pointed out that the credit should go to Ben-Akiva (1974), Williams (1977), and Daly and Zachary (1978) as well as McFadden (1978). See Carrasco and Ortúzar (2002) for the history of NES and partial review on various issues involving NES.

As type I extreme distribution leads to MNL, "*generalized extreme-value distribution*" leads to NES: the df of $u_1, ..., u_J$ is

$$\exp\{-h(e^{-u_1}, e^{-u_2}, ..., e^{-u_J})\}, \quad \text{for some known function } h(\cdot)$$

which becomes the df for J-many iid type I extreme distribution rv's if $h(a_1, a_2, ..., a_J) = \Sigma_{j=1}^{J} a_j$. NES for $J = 3$ with alternatives 2 and 3 nested postulates

$$h(a_1, a_2, a_3) = a_1 + (a_2^{1/\sigma} + a_3^{1/\sigma})^{\sigma}, \text{ for an unknown constant } 0 < \sigma \leq 1.$$

$(a_2^{1/\sigma} + a_3^{1/\sigma})^{\sigma}$ can be taken as "averaging" a_2 and a_3, to be compared with a_1. More generally, with $J = 6$, if we want to nest alternatives 2 to 3 and 4 to 6 because 2 and 3 are similar and 4 to 6 are similar, then set

$$h(a_1, ..., a_6) = a_1 + (a_2^{1/\sigma_1} + a_3^{1/\sigma_1})^{\sigma_1} + (a_4^{1/\sigma_2} + a_5^{1/\sigma_2} + a_6^{1/\sigma_2})^{\sigma_2}.$$

When $\sigma_1 = \sigma_2 = 1$, this yields the df of 6-many iid type-I extreme distribution rv's. For "nest j", $1 - \sigma_j$ may be thought of as a degree of the positive relationship among the nested alternatives.

For three alternative case, denoting the regression function part of the utility from alternative j as λ_j, the choice probability can be shown to be

$$
\begin{aligned}
P(\text{choose } 1|w) &= \frac{\exp(\lambda_1)}{\exp(\lambda_1) + \{\exp(\lambda_2/\sigma) + \exp(\lambda_3/\sigma)\}^{\sigma}} \\
&= \frac{\exp(\lambda_1)}{\exp(\lambda_1) + \exp[\sigma \ln\{\exp(\lambda_2/\sigma) + \exp(\lambda_3/\sigma)\}]}, \\
P(\text{choose } 2|w) &= P(\text{choose } 2|w, \text{ not choosing } 1) \cdot P(\text{not choosing } 1|w) \\
&= \frac{\exp(\lambda_2/\sigma)}{\exp(\lambda_2/\sigma) + \exp(\lambda_3/\sigma)}\{1 - P(\text{choose } 1|w)\}, \\
P(\text{choose } 3|w) &= P(\text{choose } 3|w, \text{ not choosing } 1) \cdot P(\text{not choosing } 1|w) \\
&= \frac{\exp(\lambda_3/\sigma)}{\exp(\lambda_2/\sigma) + \exp(\lambda_3/\sigma)}\{1 - P(\text{choose } 1|w)\}
\end{aligned}
$$

With this, the MLE for NES can be implemented using MNL and $\sigma = 1$ as the initial values, but the actual implementation is somewhat complicated due to location normalization for each "nest." Alternatively, two-stage procedures are available as well where the first stage is a "within-nest" estimation and the second stage is a "between-nest" estimation. See, e.g., Brownstone and Small (1989) for more on the computational aspects of NES. STATA has the command `nlogit` to implement NES.

A couple of remarks are in order. First, if $\sigma = 1$, then the choice probabilities become those for MNL. Second, $\ln(\exp(\lambda_2/\sigma) + \exp(\lambda_3/\sigma))$ is called the "*inclusive value*" for the nested alternatives 2 and 3. In the two-stage implementation of NES, λ_j/σ is estimated for each nest in the first stage, and then in the second stage, a normalized version of inclusive value can be used as a regressor to estimate σ, with which we can test for MNL: $H_0 : \sigma = 1$. Third, the ratio $P(\text{choose } 1|w)$ to $P(\text{choose } 2|w)$ depends on the regression function of alternative 3 when $\sigma \neq 1$; IIA in MNL is thus relaxed. A disadvantage with NES is that nesting may not be obvious. For instance, in the presidential election example, Perot may be nested with Clinton or with Bush. If J is large, there are so many different ways to form the sequential choice tree.

As an empirical example of NES, Hoffman and Duncan (1988) analyzed marrying and welfare-use decisions of divorced or separated women. There are three choices: marry, remain unmarried without welfare receipt, and remain unmarried with welfare receipt; the latter two are nested. The data is drawn from the Panel Study of Income Dynamics (PSID) for 1983 with 460 white women and 307 black women. Part of Table 2 in Hoffman and Duncan (1988) is shown in Table 2 (SD in ($\cdot$)).

Table 2: Nested Logit for Marriage and Welfare

	White Women		Black Women	
	marry or not	welfare or not	marry or not	welfare or not
spouse income	−0.010 (0.021)		0.051 (0.032)	
welfare income		0.329 (0.092)		0.288 (0.068)
norm.inc.value	0.114 (0.129)		0.079 (0.203)	

Spouse income is insignificant for whites women while nearly significant for black women. Welfare income is significant for welfare decision. The normalized inclusive value (norm.inc.value) coefficients are σ whose estimates are close to zero. $H_0 : \sigma = 1$ (not $H_0 : \sigma = 0$) is easily rejected, meaning that MNL would be the wrong model.

2   Methods of Simulated Moments (MSM)

In this section, we introduce method of simulated moments which solves multidimensional integration problems as in multinomial probit. Although we introduce *method of simulated moments* (MSM) here for multinomial probit (MNP) as in McFadden (1989), it is certainly applicable whenever a high dimensional integration is required. MSM recasts MNP in a method-of-moment framework and estimates the choice probabilities using simulated rv's drawn from the same distribution as that of u_{ij}'s.

2.1   Basic Idea with Frequency Simulator

Recall the simple linear utility (or satisfaction) model for multinomial choice:

$$(i)\ s_{ij} = x'_{ij}\beta + u_{ij}, \quad i = 1, ..., N,\ j = 1, ..., J.$$

A slightly different specification, a *random coefficient model*, is

$$(v)\ s_{ij} = x'_{ij}\alpha_i = x'_{ij}\beta + x'_{ij}\Gamma e_i, \quad \text{where } \alpha_i \equiv \beta + \Gamma e_i,$$

e_i is a $J \times 1$ error vector from a known distribution and Γ is an unknown constant matrix defined conformably. Defining $x'_{ij}\Gamma e_i$ as u_{ij}, the random coefficient model becomes $s_{ij} = x'_{ij}\beta + u_{ij}$ except that now u_{ij} has a known form of heteroskedasticity: e.g., with $e_i \sim N(0, I_J)$, we get $u_{ij}|x_{ij} \sim N(0, x'_{ij}\Gamma\Gamma'x_{ij})$. More generally, we may consider, for some error term ζ_{ij},

$$(vi) \quad s_{ij} = x'_{ij}\alpha_i + \zeta_{ij} = x'_{ij}\beta + (x'_{ij}\Gamma e_i + \zeta_{ij})$$

which includes both (i) and (v); McFadden (1989) used (v) for MNP. We will accommodate both models. For (i), β and the error-term variance matrix are to be estimated; for (v), β and Γ.

Denote the identified parameters in MNP as a $k \times 1$ vector γ. Recall the method-of-moments interpretation of MNP:

$$\frac{1}{N}\sum_{i=1}^{N}\sum_{j=1}^{J}\frac{\partial \ln P_{ij}(g)}{\partial g}\{y_{ij} - P_{ij}(g)\} = 0.$$

Regard the true parameter as the one satisfying the population version of this moment condition; there may be multiple such parameters, which is an inherent problem whenever we turn a M-estimator into a MOM. Define the instrument vector

$$\underset{k\times 1}{z_{ij}} \equiv \frac{\partial \ln P_{ij}(g)}{\partial g} = \frac{\partial P_{ij}(g)/\partial g}{P_{ij}(g)}$$

which depends on the unknown parameters, differently from the usual case of instruments being observed variables.

Further define y_i, P_i, and z_i as (the dimension is shown under each variable)

$$\underset{J\times 1}{y_i} \equiv \begin{bmatrix} y_{i1} \\ \vdots \\ y_{iJ} \end{bmatrix}, \quad \underset{J\times 1}{P_i} \equiv \begin{bmatrix} P_{i1} \\ \vdots \\ P_{iJ} \end{bmatrix}, \quad \underset{J\times 1}{u_i} \equiv \begin{bmatrix} u_{i1} \\ \vdots \\ u_{iJ} \end{bmatrix}, \quad \underset{k\times J}{z_i} \equiv [z_{i1}, ..., z_{iJ}]$$

to rewrite the sample moment as $N^{-1}\sum_i(\text{score function}_i)$ which is

$$\frac{1}{N}\sum_{i=1}^{N}\underset{k\times J}{z_i}\underset{J\times 1}{(y_i - P_i)} = 0.$$

The key step in the MSM is simulating $P_i(g)$. Let n be the simulated (or generated) sample size. Then the MSM can be done as follows. Denoting the dimension of x_{ij} as $k_x \times 1$,

1. Generate $J \times 1$ vectors $\{\varepsilon_i(t)\}_{t=1}^{n}$ for each $i = 1, ..., N$ such that $\varepsilon_i(t)$, $t = 1, ..., n$, are iid following $N(0, I_J)$. $\{\varepsilon_i(t)\}_{t=1}^{n}$ is independent of $\{\varepsilon_{i'}(t)\}_{t=1}^{n}$ for all $i' \neq i$. Overall, $N \cdot J \cdot n$ rv's are generated.

2. For the random coefficient model (v), fix a matrix Γ to obtain $\eta_{ij}(\Gamma, t)$— a "pseudo u_{ij}"—with

$$\underset{1\times 1}{\eta_{ij}(\Gamma, t)} \equiv \underset{1\times k_x}{x'_{ij}}\underset{k_x\times J}{\Gamma}\underset{J\times 1}{\varepsilon_i(t)}, \quad j = 1, ..., J \text{ and } t = 1, ..., n.$$

For (i), let $\Omega \equiv V(u_i)$ and fix C such that $CC' = \Omega$—choose C to be lower-triangular ("*Cholesky decomposition*" of Ω). Then

$$\{\eta_{i1}(C,t), ..., \eta_{iJ}(C,t)\}' \equiv \underset{J \times J}{C} \underset{J \times 1}{\varepsilon_i(t)}, \quad \text{for } t = 1, ..., n.$$

3. Fix b and generate

$$\begin{aligned} \text{for } (v) \quad &: \quad s_{ij}(\Gamma, b, t) \equiv x'_{ij}b + \eta_{ij}(\Gamma, t) \quad \text{for } t = 1, ..., n; \\ \text{for } (i) \quad &: \quad s_{ij}(C, b, t) \equiv x'_{ij}b + \eta_{ij}(C, t) \quad \text{for } t = 1, ..., n. \end{aligned}$$

4. Find the relative frequency that person i chooses alternative j: for (i),

$$f_{ij}(g) \equiv \frac{1}{n} \sum_{t=1}^{n} 1[s_{ij}(C, b, t) > s_{im}(C, b, t), \ m \neq j]$$

which is a "simulated estimator" for $P_{ij}(g)$; recall that g consists of b and C. Stack $f_{ij}(g)$'s to get the $J \times 1$ simulated estimator vector $f_i(g)$, which is for $P_i(g)$. The steps for (v) are analogous.

5. To get $f_i(g)$ for different values of g, repeat Step 2 through Step 4 with the same $\varepsilon_i(t)$'s obtained in Step 1.

6. With $f_i(g)$ replacing $P_i(g)$, iterate until convergence using

$$g_1 = g_0 \ + \ \left\{ \sum_i z_i(g_0)\{y_i - P_i(g_0)\}\{y_i - P_i(g_0)\}'z_i(g_0)' \right\}^{-1} \\ \sum_i z_i(g_0)\{y_i - P_i(g_0)\}.$$

which follows the usual iteration scheme for MLE with the inverted matrix of the outer-product of the score function $z_i(y_i - P_i)$.

The iteration formula follows from rewriting $P_i(g_1)$ as $P_i(g_0) + \{\partial P_i(g_0)/\partial g'\}(g_1 - g_0)$ and solving the moment condition for g_1 (and then using the information equality). Differently from the other cases, the instrument $z_i(g)$ also depends on g. But if $g_0 \simeq \gamma$, this dependence can be ignored; i.e., there is no need to update $z_i(g)$ at every iteration. Recall our discussion in relation to two-stage M-estimators: estimated instruments are as good as the "true" instruments.

In order to get z_{ij}, we need $\partial P_{ij}/\partial g$, which can be obtained analytically as in McFadden (1989) and Hajivassilious and Ruud (1994). But the derivative is somewhat complicated. Instead, we may use a numerical derivative

$$\frac{\partial P_{ij}(g)}{\partial g_m} \simeq \frac{P_{ij}(g + he_m) - P_{ij}(g - he_m)}{2h}, \quad m = 1, ..., k,$$

where e_m is the $k \times 1$ basis vector with 1 in its mth row and 0's elsewhere, and h is a small positive scalar close to 0. Since z_i should be orthogonal to $y_i - P_i$, the generated ε_{it}'s used for $y_i - P_i$ are not good for simulating z_i. Instead, generate new error vectors and use them to simulate z_i through a procedure analogous to the above steps 1 to 5. Since the frequency simulator $f_{ij}(g)$ has indicator functions, the numerical derivatives would be difficult to get; better simulators replacing the indicator function with a smooth function will be shown later.

As for the asymptotic distribution of the MSM, it can be shown that

$$\sqrt{N}(g_{msm} - \gamma) \rightsquigarrow N\left\{0, \; \left(1 + \frac{1}{n}\right) I_f^{-1}\right\},$$

$$\text{where} \quad I_f \;=\; E[z(\gamma)\{y - P(\gamma)\}\{y - P(\gamma)\}'z(\gamma)'].$$

which is the information matrix of MNP; as usual, it holds that

$$\frac{1}{N}\sum_i [z_i(g_{msm})\{y_i - P_i(g_{msm})\}\{y_i - P_i(g_{msm})\}'z_i(g_{msm})'] \rightarrow^p I_f.$$

The simulation error increases the asymptotic variance by I_f^{-1}/n, which can be ignored if n is large.

2.2   GHK Smooth Simulator

The frequency simulator $f_{ij}(g)$ has a number of disadvantages. First, $f_{ij}(g)$ is not a smooth function of g due to the indicator function. Second, f_{ij} can take 0—a trouble for $\ln P_{ij}$ due to $\ln(0)$. Third, it may take too many draws (too large a n) to estimate a small P_{ij}. Replacing the indicator function with a smooth function taking a value in $(0, 1)$ can solve these problems. This subsection introduces "GHK (Geweke, Hajivassiliou, Kean)" simulator for the model $s_{ij} = x'_{ij}\beta + u_{ij}$. See Hajivassiliou et al. (1996) and the references therein for more on various simulators; Hajivassiliou et al. (1996) recommended GHK simulator.

Observe that the events $s_{ij} > s_{im}, \forall m \neq j$, can be written as

$$\underset{(J-1)\times 1}{a_{ij}} \quad < \quad \underset{(J-1)\times(J-1)}{A_j} \quad \frac{v_i}{\sigma}, \quad j = 2, ..., J$$

for a matrix A_j, a vector $a_{ij} \equiv (a_{ij1}, ..., a_{ij(J-1)})'$, and $v_i \equiv (v_{i2}, ..., v_{iJ})'$; recall the definition of $v_{ij} \equiv u_{ij} - u_{i1}$ for $j \geq 2$ and $\sigma \equiv SD(v_{i2})$. For instance, when $J = 3$, for $s_{i2} > s_{im} \; \forall m \neq 2$, $a_{i2} < A_2 v_i/\sigma$ becomes

$$a_{i2} \;=\; \left[\begin{array}{c} -w'_{i2}\beta/\sigma \\ -(w_{i2} - w_{i3})'\beta/\sigma \end{array}\right], \quad A_2 = \left[\begin{array}{cc} 1 & 0 \\ 1 & -1 \end{array}\right], \quad v_i = \left[\begin{array}{c} v_{i2}/\sigma \\ v_{i3}/\sigma \end{array}\right]$$

$$\implies \left[\begin{array}{c} 0 < w_{i2}'\beta/\sigma + v_{i2}/\sigma \\ w_{i3}'\beta/\sigma + v_{i3}/\sigma < w_{i2}'\beta/\sigma + v_{i2}/\sigma \end{array} \right].$$

For $s_{i3} > s_{im} \; \forall m \neq 2$, $a_{i3} < A_3 v_i/\sigma$ becomes

$$a_{i3} = \left[\begin{array}{c} -w_{i3}'\beta/\sigma \\ -(w_{i3}-w_{i2})'\beta/\sigma \end{array} \right], \quad A_3 = \left[\begin{array}{cc} 0 & 1 \\ -1 & 1 \end{array} \right], \quad v_i = \left[\begin{array}{c} v_{i2}/\sigma \\ v_{i3}/\sigma \end{array} \right]$$

$$\implies \left[\begin{array}{c} 0 < w_{i3}'\beta/\sigma + v_{i3}/\sigma \\ w_{i2}'\beta/\sigma + v_{i2}/\sigma < w_{i3}'\beta/\sigma + v_{i3}/\sigma \end{array} \right].$$

Let $\Sigma \equiv V(v_i/\sigma)$; when $J = 3$,

$$\Sigma = \left[\begin{array}{cc} 1 & \sigma_{23} \\ \sigma_{23} & \sigma_3^2 \end{array} \right] = \left[\begin{array}{cc} 1 & \rho\sigma_3 \\ \rho\sigma_3 & \sigma_3^2 \end{array} \right].$$

Recall that we estimate β/σ and Σ. Then

$$V \left(\underset{(J-1)\times 1}{A_j \frac{v_i}{\sigma}} \right) = \underset{(J-1)\times(J-1)}{A_j \Sigma A_j'}, \quad j = 2,...,J.$$

To simplify explaining GHK simulator, suppose $J = 3$. Set

$$H_j H_j' = A_j \Sigma A_j', \quad j = 2, 3$$

where H_j is the lower-triangular Cholesky decomposition (or "square-root") of $A_j \Sigma A_j'$. Then

$A_j \dfrac{v_i}{\sigma}$ follows the same distribution as $\left[\begin{array}{cc} h_{j11} & 0 \\ h_{j21} & h_{j22} \end{array} \right] \left[\begin{array}{c} e_1 \\ e_2 \end{array} \right]$ does

where $h_{jll'}$, $l, l' = 1, 2$, denotes an element in H_j, and e_1, e_2 are iid $N(0,1)$.
With $a_{ij} = (a_{ij1}, a_{ij2})'$,

$$\begin{aligned} P_{ij} &\equiv P(i \text{ chooses } j) = P\left(a_{ij} < A_j \frac{v_i}{\sigma} \right) \\ &= P(a_{ij1} < h_{j11}e_1, \; a_{ij2} < h_{j21}e_1 + h_{j22}e_2) \\ &= P\left(\frac{a_{ij1}}{h_{j11}} < e_1, \frac{a_{ij2}-h_{j21}e_1}{h_{j22}} < e_2 \right) \\ &= P\left(\frac{a_{ij1}}{h_{j11}} < e_1 \right) \cdot P\left(\frac{a_{ij2}-h_{j21}e_1}{h_{j22}} < e_2 \; \bigg| \; \frac{a_{ij1}}{h_{j11}} < e_1 \right) \\ &\simeq \left\{ 1 - \Phi\left(\frac{a_{ij1}}{h_{j11}} \right) \right\} \cdot \frac{1}{n} \sum_{t=1}^{n} \left\{ 1 - \Phi\left(\frac{a_{ij2}-h_{j21}e_{1t}^*}{h_{j22}} \right) \right\} \end{aligned}$$

where $\{e_{1t}^*\}_{t=1}^n$ are iid, drawn from the $N(0,1)$ truncated from below at a_{ij1}/h_{j11}. This simulator is unbiased for the desired choice probability, and is differentiable wrt the parameters.

If the truncation threshold a_{ij1}/h_{j11} is not too big (say, smaller than 3), then e_{1t}^* can be generated from $N(0,1)$ by keeping only those greater than a_{ij1}/h_{j11}. Otherwise, e_{1t}^* should be drawn directly from the $N(0,1)$ truncated at a_{ij1}/h_{j11}, using, e.g., *"acceptance/rejection method"* as follows. On the two-dimensional plane, draw the truncated density $\tilde{\phi}(\cdot) \equiv \phi(\cdot)/\{1 - \Phi(a_{ij1}/h_{j11})\}$ over $(a_{ij1}/h_{j11}, \infty)$, and then generate two independent uniform numbers (v_1, v_2) to place (v_1, v_2) on the plane where, for some big constant $M \simeq \infty$,

$$v_1 \sim U[a_{ij1}/h_{j11}, M] \quad \text{and} \quad v_2 \sim U[0, \tilde{\phi}(0)].$$

If (v_1, v_2) falls within the density (i.e., if $v_2 < \tilde{\phi}(v_1)$), then take v_1 as a desired random number; otherwise, reject (i.e., throw away) (v_1, v_2) and draw two uniform numbers anew. This procedure preserves the "relative frequency" of $\tilde{\phi}$. For instance, if $\tilde{\phi}(\tau_2) = 2\tilde{\phi}(\tau_1)$, then τ_2 is twice more likely to get realized (i.e., retained) than τ_1.

If $J = 4$, then only one more term is needed in P_{ij}: with $H_j = [h_{jll'}]$, $l, l' = 1, 2, 3$,

$$P_{ij} = P\left(a_{ij} < A_j \frac{v_i}{\sigma}\right)$$

$$= \ P(a_{ij1} < h_{j11}e_1, \ a_{ij2} < h_{j21}e_1 + h_{j22}e_2, \ a_{ij3} < h_{j31}e_1$$

$$+ h_{j32}e_2 + h_{j33}e_3)$$

$$= \ P\left(\frac{a_{ij1}}{h_{j11}} < e_1, \ \frac{a_{ij2} - h_{j21}e_1}{h_{j22}} < e_2, \ \frac{a_{ij3} - h_{j31}e_1 - h_{j32}e_2}{h_{j33}} < e_3\right)$$

$$= \ P\left(\frac{a_{ij1}}{h_{j11}} < e_1\right) \cdot P\left(\frac{a_{ij2} - h_{j21}e_1}{h_{j22}} < e_2 \ \bigg| \ \frac{a_{ij1}}{h_{j11}} < e_1\right)$$

$$\cdot P\left(\frac{a_{ij3} - h_{j31}e_1 - h_{j32}e_2}{h_{j33}} < e_3 \ \bigg| \ \frac{a_{ij1}}{h_{j11}} < e_1, \ \frac{a_{ij2} - h_{j21}e_1}{h_{j22}} < e_2\right)$$

$$\simeq \ \left\{1 - \Phi\left(\frac{a_{ij1}}{h_{j11}}\right)\right\} \cdot \frac{1}{n}\sum_{t=1}^{n}\left\{1 - \Phi\left(\frac{a_{ij2} - h_{j21}e_{1t}^*}{h_{j22}}\right)\right\}$$

$$\cdot \left\{1 - \Phi\left(\frac{a_{ij3} - h_{j31}e_{1t}^* - h_{j32}e_{2t}^*}{h_{j33}}\right)\right\}$$

where e_{1t}^* is drawn from $N(0,1)$ truncated from below at a_{ij1}/h_{j11} as in the $J = 3$ case, and e_{2t}^* is drawn from the $N(0,1)$ truncated from below at $(a_{ij2} - h_{j21}e_{1t}^*)/h_{j22}$—generate e_{1t}^* first and then many $N(0,1)$ rv's to retain them only if they are greater than $(a_{ij2} - h_{j21}e_{1t}^*)/h_{j22}$.

2.3   Methods of Simulated Likelihood (MSL)

To introduce methods of simulated likelihood (MSL), consider a duration problem where $\{(y_i, x_i)\}_{i=1}^N$ are observed and y_i is the duration. Suppose $y_i|(x_i, \varepsilon_i)$ follows a Weibull distribution, where ε is an unobserved heterogeneity term following $N(0, \sigma^2)$. As in the usual linear model, ε_i can be regarded as a collection of unobserved variables affecting y_i. The density and survival functions for $y|(x, \varepsilon)$ are

$$f(y, x, \varepsilon; \alpha) \;=\; \alpha y^{\alpha-1}\theta(x, \varepsilon) \cdot \exp\{-y^\alpha \theta(x, \varepsilon)\} \;\; \text{and}$$

$$S(y, x, \varepsilon; \alpha) \;=\; \exp\{-y^\alpha \theta(x, \varepsilon)\} \;\; \text{where } \theta(x, \varepsilon) = \exp(x'\beta + \varepsilon).$$

Since ε is not observed, we need to obtain the density and survival functions for $y|x$ with ε integrated out.

Denoting $N(0, 1)$ density as ϕ and defining $v \equiv \varepsilon/\sigma$, the density for $y|x$ with ε integrated out is

$$f(y, x; \alpha, \beta, \sigma) = \int \alpha y^{\alpha-1} \exp\left(x'\beta + \sigma\frac{\varepsilon}{\sigma}\right)$$

$$\cdot \exp\left\{-y^\alpha \exp\left(x'\beta + \sigma\frac{\varepsilon}{\sigma}\right)\right\} \phi\left(\frac{\varepsilon}{\sigma}\right) d\left(\frac{\varepsilon}{\sigma}\right)$$

$$= \int \alpha y^{\alpha-1} \exp(x'\beta + \sigma v) \cdot \exp\{-y^\alpha \exp(x'\beta + \sigma v)\} \phi(v) dv$$

$$\implies S(y, x; \alpha, \beta, \sigma) = \int \exp\{-y^\alpha \exp(x'\beta + \sigma v)\} \phi(v) dv.$$

Then the log-likelihood function for the MLE is ($d_i = 1$ if uncensored and 0 otherwise)

$$\sum_i d_i \ln f(y_i, x_i; a, b, s) + \sum_i (1 - d_i) \ln S(y_i, x_i; a, b, s)$$

to be maximized for a, b, and s. Note that, since the hazard and the cumulative hazard are not likelihoods, it is erroneous to specify them with ε in and then integrate ε out. If we do that, we will be using, with $E_\varepsilon(\cdot)$ denoting the integral wrt ε,

$$E_\varepsilon \lambda(y, x, \varepsilon) = E_\varepsilon \frac{f(y, x, \varepsilon)}{S(y, x, \varepsilon)} \neq \frac{E_\varepsilon f(y, x, \varepsilon)}{E_\varepsilon S(y, x, \varepsilon)}$$

whereas the last term is the right hazard to use.

The integration can be done numerically. But we can use a simulator as well: generate n-many $N(0, 1)$ rv's $\{\varepsilon_i(t)\}$ for each (y_i, x_i') and (a, b', s) to get

$$f_n(y_i, x_i; a, b, s) \equiv \frac{1}{n} \sum_{t=1}^{n} a y_i^{a-1} \exp\{x_i'b + s\varepsilon_i(t)\}\cdot$$

$$\exp[-y_i^a \exp\{x_i'b + s\varepsilon_i(t)\}]$$

$$S_n(y_i, x_i; a, b, s) \equiv \frac{1}{n} \sum_{t=1}^{n} \exp[-y_i^a \exp\{x_i'b + s\varepsilon_i(t)\}].$$

Then we can maximize the log-likelihood with f and S replaced by the simulators f_n and S_n, respectively, which are smooth in the parameters.

More generally, consider a log-likelihood function $\sum_{i=1}^{N} \ln f(z_i, b)$ for $\{z_i\}_{i=1}^{N}$ where $z_i = (x_i', y_i)'$. The estimation can be done iteratively with (omit z in f for a while)

$$b_1 = b_0 + \left\{ \sum_i \frac{\nabla f(b_0)}{f(b_0)} \frac{\nabla f(b_0)'}{f(b_0)} \right\}^{-1} \sum \frac{\nabla f(b_0)}{f(b_0)},$$

$$\text{where } \nabla f(b_0) \equiv \frac{\partial f(b)}{\partial b} \big|_{b=b_0}.$$

Suppose $\ln f(b)$ is an integrated entity as in the above duration example, and we want to apply MSL. Then we need to simulate $\nabla f(b)$ as well as $f(b)$. In MSM, we showed that the asymptotic variance is $(1 + n^{-1})I_f^{-1}$; i.e., the simulation error resulted in one additional term I_f^{-1}/n. In MSL, however, it is difficult to get the variance with a finite n as shown in the following.

The MSL estimator b_{msl} should satisfy

$$\sqrt{N}(b_{msl} - \beta) = \left\{ \frac{1}{N} \sum_i \frac{\nabla f_n(\beta)}{f_n(\beta)} \frac{\nabla f_n(\beta)'}{f_n(\beta)} \right\}^{-1} \frac{1}{\sqrt{N}} \sum_i \frac{\nabla f_n(\beta)}{f_n(\beta)} + o_p(1)$$

where f_n denotes a simulator for f as seen above. It can be shown that the inverted matrix converges to the information matrix I_f^{-1}. If the simulator satisfies

$$E_\varepsilon \left\{ \frac{\nabla f_n(\beta)}{f_n(\beta)} \right\} = \frac{\nabla f(\beta)}{f(\beta)}$$

where $E_\varepsilon(\cdot)$ is the expected value wrt the simulated ε, then

$$\frac{1}{\sqrt{N}} \sum_i \frac{\nabla f_n(\beta)}{f_n(\beta)} = \frac{1}{\sqrt{N}} \sum_i \frac{\nabla f(\beta)}{f(\beta)} + \frac{1}{\sqrt{N}} \sum_i \left[\frac{\nabla f_n(\beta)}{f_n(\beta)} - E_\varepsilon \left\{ \frac{\nabla f_n(\beta)}{f_n(\beta)} \right\} \right].$$

The first term on the rhs yields the asymptotic variance of the MLE, while a CLT can be applied to the second term which reflects the pure simulation error. Thus if the condition $E\{\nabla f_n(\beta)/f_n(\beta)\} = \nabla f(\beta)/f(\beta)$ holds, then the asymptotic variance of $\sqrt{N}(b_{msl} - \beta)$ will be a sum of two terms analogously to the asymptotic variance for MSM. However, the condition does not hold in general because $\nabla f_n(\beta)/f_n(\beta)$ is a ratio of two rv's: for two rv's λ_1 and

λ_2, $E(\lambda_1/\lambda_2) \neq E(\lambda_1)/E(\lambda_2)$. It can be shown that, if $n/\sqrt{N} \to \infty$, then the second term in the preceding display is negligible (i.e., MSL becomes as efficient as the MLE) under certain conditions; e.g., see Proposition 4 and 5 in Hajivassiliou and Ruud (1994). In short, although we can simulate unbiased estimates for f and ∇f separately, it is not clear how to simulate an unbiased estimate for the score function.

The problem $E(\lambda_1/\lambda_2) \neq E(\lambda_1)/E(\lambda_2)$ due to simulating $\nabla f(z, b)$ and $f(z, b)$ separately in MSL may look avoidable, if we simulate $\ln f(z, b)$ instead of $f(z, b)$; differentiating this simulator wrt b would yield an unbiased simulator for the score function. But this idea does not work. To see this, recall the above duration example: we need $\ln f(z, b) = \ln E_\varepsilon f(z, \varepsilon, b)$, but we will get only $E_\varepsilon \ln f(z, \varepsilon, b)$ if we try to simulate $\ln f(z, \varepsilon, b)$. Due to $E_\varepsilon \ln(\cdot) < \ln E_\varepsilon(\cdot)$, we get $E_\varepsilon \ln f(z, b) < \ln E_\varepsilon f(z, \varepsilon, b)$. By using the moment condition (i.e., the first-order condition), MSM avoids this problem associated with MSL; MSM integrates out ε in the first-order moment condition, rather than ε in the likelihood function. In MSM, the simulation error appears linearly (recall $y_i - P_i$)—simulating the "instrument" $\partial \ln P_{ij}(\gamma)/\partial g$ for the moment condition is innocuous there—rather than log-linearly as in MSL.

Nonetheless, one should not take MSM as a panacea; as mentioned already, due to the potential multiple solutions in the moment condition for MSM, it is not clear how to establish identification in MSM, which is automatic in MSL under the MLE principle. In practice, surprisingly, MSL seems to work well even with a very small n, say 10. Readers with further interest in simulation-based methods can refer to Gourieroux and Monfort (1996), Hajivassiliou et al. (1996), Stern (1997), Hajivassiliou and McFadden (1998), and the references therein.

3   Sample-Selection Models

In surveys, the respondent may choose not to answer certain questions. For example, in income surveys, the respondent may not answer if his or her income is too big (or too small). If the respondent does not answer only some questions in the survey, then this is called *item nonresponse*; if the respondent does not answer at all, then this is called *unit nonresponse*. Obviously, the former poses less of a problem, for we may use the respondent's answered items to fill in the nonresponses.

An easy solution to the nonresponse problem is to use only the respondents with complete answers, but this can cause a bias. For example, if we are interested in the population mean income, the sample mean without the high income group will be biased downward. Sampling from a special subpopulation and consequently incurring a bias in estimating a population parameter is known as a *"sample-selection problem."* (Sample) selection problem is specific to the parameter of interest. In the income example, the sample is biased for the population mean if the sample-selection problem is ignored, but it may not be biased for the population mode. In the real world, the sample-selection problem seems to be the rule rather than the exception. The very fact that the subject ever entered the data voluntarily may be a source for selection bias. L.F. Lee (2001) provided a survey on sample selection issues.

Sample selection is a big issue deserving an extensive treatment in a separate book. This section will examine some often-used sample-selection models. We will start with a basic model where y is continuously distributed but observed only when $d = 1$; this model will be given an extensive coverage. Then we will examine cases where y is an LDV. Also multivariate generalizations of selection models called "hurdle models" will be introduced.

3.1 Various Selection Models

The best-known *basic (sample-) selection model* is, with $i = 1, ..., N$,

$$
\begin{aligned}
d_i^* &= w_i'\alpha + \varepsilon_i, \quad d_i = 1[d_i^* > 0] \\
y_i &= x_i'\beta + u_i, \quad w_i = (x_i', c_i')', \quad \varepsilon \text{ and } u \text{ are zero-mean errors} \\
&\quad \alpha_c, \text{ the coefficient of } c_i \text{ in } w_i'\alpha, \text{ is not a zero vector} \\
&\quad (d_i, w_i', d_i y_i) \text{ is observed}
\end{aligned}
$$

where w_i is a regressor vector with its first component 1, ε_i and u_i are error terms, and α and β are conformable parameter vectors. The d-equation is the "*selection equation*" determining the selection (decision), and the y-equation is the "*outcome equation*" observed only when $d_i = 1$. The model includes an inclusion/exclusion restriction that c is included in the selection equation while excluded from the outcome equation.

An example is: y is wage, $d = 1$ if working and 0 otherwise, x is a vector of explanatory variables relevant for wage and work decision, and c is an explanatory variable relevant for the work decision but not for the wage (e.g., c is the parents' education levels); wage is observed only for those who choose to work. Another example is that y is export volume, $d = 1$ if deciding to export, x is a vector of regressors for the export decision and volume, and c is an explanatory variable relevant for the export decision but not for the export volume (e.g., an export barrier not related to export volume).

The above main selection model includes the "fixed-threshold censored model" $y_i = \max(y_i^*, 0) = 1[y_i^* > 0]y_i^*$ as a special case when $d_i^* = y_i$. This may be a good way to view the censored model in judging whether the censored model is appropriate or not. For instance, in the export example, if the decision to export is driven by the export volume equation, then the censored model is appropriate. Ideally, instead of imposing the prior restriction $\alpha = \beta$ (setting $c_i = 0$), one should estimate the selection model and test for $\alpha = \beta$, whose acceptance leads to the simpler censored model. But as will be shown later, dealing with the selection model is not always straightforward; e.g., the selection model includes an exclusion restriction which is often not easy to justify.

In the sample-selection model, w is fully observed regardless of d; i.e., w, not dw, is observed. If we have

$$(d_i, d_i w'_i, d_i y_i) \text{ is observed}, \ i = 1, ..., N,$$

then the model would be called a "*truncated selection model*," which is harder to deal with than the main model above, because the observability is more restricted here. In the non-response problem, the main model corresponds to item (y) nonresponse because only y has missings, whereas the truncated selection model corresponds to unit nonresponse because all variables (w and y) have missings. This truncated selection model includes the "fixed-threshold truncated response model" as a special case when $d_i^* = y_i$. While the MLE for truncated regression model is fairly well-behaving computationally, the truncated selection model does not seem to be so.

If we have

$$(d_i^* d_i, w'_i, d_i y_i) \text{ observed}$$

then the model would be called a "*censored selection model*" or "*tobit-type selection model*," which is easier to deal with than the main model, because the observability is less restricted as $d_i^* d_i = \max(d_i^*, 0)$ is a censored variable, not just binary. One example for tobit-type selection is: d_i^* is the latent work hours and y_i is wage; both d_i^* and y_i are observed only when $d_i = 1$. Tobit-type selection is much rarer compared with the main selection model. See Lee and Vella (2006) and the references therein.

In treatment effect framework with a binary treatment $d = 0, 1$, we assumed two potential responses y^0 and y^1, depending on $d = 0$ or $d = 1$, respectively. Suppose

$$y_i^j = x'_i \beta_j + u_i^j, \ COR(u^j, \varepsilon) \neq 0, \ j = 0, 1, \quad \text{and}$$
$$(d_i, w'_i, \ d_i y_i^1 + (1 - d_i) y_i^0) \text{ is observed}.$$

Note

$$
\begin{aligned}
y_i &= d_i y_i^1 + (1 - d_i) y_i^0 = d_i(x'_i \beta_1 + u_i^1) + (1 - d_i)(x'_i \beta_0 + u_i^0) \\
&= d_i x_i \beta_1 + (1 - d_i) x_i \beta_0 + u_i, \quad \text{where } u_i \equiv d_i u_i^1 + (1 - d_i) u_i^0.
\end{aligned}
$$

This is a "multi-regime" generalization of the selection model where only one regime for $d = 1$ is observed. In the above selection models, we omitted the superscript/subscript j because there is no other observed regime.

Th treatment effect model is also called "*switching regression model*" with endogenous switching because d is related to (u^0, u^1). If $d \amalg (u^0, u^1)$, then the model is called "switching regression with exogenous switching," which is, however, not a sample-selection model as there is no selection problem. In the time-series econometric literature, switching regression with an unobserved regime indicator (d) has been also considered. It is certainly possible to have a partially observed d, or d observed with error.

3.2 Selection Addition, Bias, and Correction Terms

Turning back to the "best-known" selection model, the LSE of dy on dx is inconsistent for β in general. To see this, observe the moment condition

$$
\begin{aligned}
E(dxu) &= E\{E(dxu|w)\} = E\{xE(du|w)\} \\
&= E\{xE(u|w, d = 1)P(d = 1|w)\} \\
&= E\{xE(u|w, \varepsilon > -w'\alpha)P(d = 1|w)\} \neq 0
\end{aligned}
$$

for $E(u|w, \varepsilon > -w'\alpha) \neq 0$ in general.

Put it differently, suppose we select the observations with $d = 1$ and do LSE of y on x. Then

$$
E(y|w, d = 1) = x'\beta + E(u|w, d = 1) = x'\beta + E(u|w, \varepsilon > -w'\alpha) \neq x'\beta.
$$

$E(u|w, \varepsilon > -w'\alpha)$ is the troublesome term, which is zero if ε and u are independent given w:

$$
E(u|w, \varepsilon > -w'\alpha) = E(u|w) = 0.
$$

Because the LSE ignores $E(u|w, \varepsilon > -w'\alpha)$, this term which is often called "selection bias" becomes part of the error term and causes an omitted variable bias

$$
E^{-1}(xx') \cdot E\{x\, E(u|w, \varepsilon > -w'\alpha)\}
$$

for the LSE, because $E(u|w, \varepsilon > -w'\alpha)$, a function of w, is correlated with x in general. This shows also that the selection bias can be easily mistaken for a regression function misspecification (or the other way around). It seems more appropriate to call the LSE omitted variable bias in estimating β, not $E(u|w, \varepsilon > -w'\alpha)$ per se, the *selection bias*. It is better to call $E(u|w, \varepsilon > -w'\alpha)$ the *selection addition term* as it is added due to the selection equation. We will follow these terminologies in the following.

To see the form of the selection addition term better, observe as a prerequisite that, when $v \sim N(0, 1)$,

$$
\begin{aligned}
\frac{d\phi(s)}{ds} &= \frac{d(2\pi)^{-1/2}\exp(-0.5s^2)}{ds} = -s(2\pi)^{-1/2}\exp(-0.5s^2) \\
&= -s\phi(s);
\end{aligned}
$$

$$
\begin{aligned}
E(v|v > -t) &= \frac{\int_{-t}^{\infty} s\phi(s)ds}{\int_{-t}^{\infty} \phi(s)ds} = \frac{-\int_{-t}^{\infty} \phi'(s)ds}{\Phi(t)} = \frac{-\phi(\infty) + \phi(-t)}{\Phi(t)} \\
&= \frac{\phi(t)}{\Phi(t)}.
\end{aligned}
$$

Denote the joint density for $(\varepsilon = a, u = b)|w$ as $f(a, b|w) = f_{\varepsilon, u|w}(a, b)$ and the marginal density for $(\varepsilon = a)|w$ as $f(a|w) = f_{\varepsilon|w}(a)$. Assume that $\varepsilon|w$ has unbounded support for almost every (a.e.) w. Then the selection addition term is

$$
\begin{aligned}
E(u|w, \varepsilon > -w'\alpha) \quad &= \frac{\int_{-\infty}^{\infty} \int_{-w'\alpha}^{\infty} u \cdot f(\varepsilon, u|w) d\varepsilon du}{\int_{-\infty}^{\infty} \int_{-w'\alpha}^{\infty} f(\varepsilon, u|w) d\varepsilon du} \\[2mm]
&= \frac{\int_{-\infty}^{\infty} \int_{-w'\alpha}^{\infty} u \cdot f(u|\varepsilon, w) \cdot f(\varepsilon|w) d\varepsilon du}{\int_{-w'\alpha}^{\infty} f(\varepsilon|w) d\varepsilon} \\[2mm]
&= \frac{\int_{-w'\alpha}^{\infty} \int_{-\infty}^{\infty} u \cdot f(u|\varepsilon, w) du \cdot f(\varepsilon|w) d\varepsilon}{\int_{-w'\alpha}^{\infty} f(\varepsilon|w) d\varepsilon} \\[2mm]
&= \frac{\int_{-w'\alpha}^{\infty} E(u|\varepsilon, w) \cdot f(\varepsilon|w) d\varepsilon}{\int_{-w'\alpha}^{\infty} f(\varepsilon|w) d\varepsilon} .
\end{aligned}
$$

Furthermore, with $\sigma_{\varepsilon u} \equiv COV(\varepsilon, u)$, $\sigma_\varepsilon^2 \equiv V(\varepsilon)$, $\sigma_u^2 \equiv V(u)$, and $\rho \equiv COR(\varepsilon, u)$, if

$$
E(u|\varepsilon, w) = \frac{\sigma_{\varepsilon u}}{\sigma_\varepsilon^2} \varepsilon,
$$

then the selection addition term becomes

$$
\begin{aligned}
\frac{\sigma_{\varepsilon u}}{\sigma_\varepsilon^2} \frac{\int_{-w'\alpha}^{\infty} \varepsilon f(\varepsilon|w) d\varepsilon}{\int_{-w'\alpha}^{\infty} f(\varepsilon|w) d\varepsilon} &= \frac{\sigma_{\varepsilon u}}{\sigma_\varepsilon^2} E(\varepsilon|w, \varepsilon > -w'\alpha) \\[2mm]
&= \frac{\sigma_{\varepsilon u}}{\sigma_\varepsilon} E\left(\frac{\varepsilon}{\sigma_\varepsilon} \middle| w, \frac{\varepsilon}{\sigma_\varepsilon} > -w'\frac{\alpha}{\sigma_\varepsilon}\right) \\[2mm]
&= \rho\sigma_u \cdot E\left(\frac{\varepsilon}{\sigma_\varepsilon} \middle| w, \frac{\varepsilon}{\sigma_\varepsilon} > -w'\frac{\alpha}{\sigma_\varepsilon}\right) \quad \text{as } \rho\sigma_u = \frac{\sigma_{\varepsilon u}}{\sigma_\varepsilon}.
\end{aligned}
$$

In addition, if $\varepsilon \sim N(0, \sigma_\varepsilon^2)$ independently of w, then $\varepsilon/\sigma_\varepsilon \sim N(0, 1)$ and the selection addition term becomes

$$
\rho\sigma_u \cdot \frac{\phi(w'\alpha/\sigma_\varepsilon)}{\Phi(w'\alpha/\sigma_\varepsilon)}.
$$

If $\rho = 0$ ($\Longleftrightarrow \sigma_{\varepsilon u} = 0$), then the selection addition term is zero. The selection addition term was obtained only under

$$
E(u|\varepsilon, w) = \frac{\sigma_{\varepsilon u}}{\sigma_\varepsilon^2} \varepsilon, \quad \text{the marginal normality of } \varepsilon, \quad \text{and} \quad \varepsilon \amalg x.
$$

That is, the joint normality of (ε, u) is not necessary. As will be shown shortly, $\phi(w'\alpha/\sigma_\varepsilon)/\Phi(w'\alpha/\sigma_\varepsilon)$ can be used as an extra regressor, in which case we will call it the "*selection correction term,*" as it is used to correct for the selection problem.

3.3   MLE

Suppose that (ε, u) is bivariate normal and independent of w. The log-likelihood function for the basic selection model MLE for $(\alpha'/\sigma_\varepsilon, \beta', \sigma_u, \rho)'$ is

$$\ln L\left(\frac{\alpha}{\sigma_\varepsilon}, \beta, \sigma_u, \rho\right) = \sum_{i=1}^{N}\left[(1-d_i)\cdot\ln\Phi\left(-w_i'\frac{\alpha}{\sigma_\varepsilon}\right)\right.$$
$$+ d_i\cdot\left[\ln\Phi\left\{\left(w_i'\frac{\alpha}{\sigma_\varepsilon} + \rho\frac{y_i - x_i'\beta}{\sigma_u}\right)(1-\rho^2)^{-1/2}\right\}\right.$$
$$\left.\left. - \ln\sigma_u + \ln\phi\left(\frac{y_i - x_i'\beta}{\sigma_u}\right)\right]\right].$$

For $d_i = 0$, the contribution of observation i to the log-likelihood is only $\Phi(-w_i'\alpha/\sigma_\varepsilon) = P(d = 0|w_i)$. For $d_i = 1$, there are two terms in the log-likelihood function which may be thought of as

$$\ln\{P(d = 1|u, w)\cdot P(u|w)\} = \ln P(d = 1|u, w) + \ln P(u|w).$$

The first term with $\Phi\{(w_i'...\}$ can be understood in the light of standardizing $\varepsilon|u$:

$$E(\varepsilon|u) = \frac{\sigma_{\varepsilon u}}{\sigma_u^2}u = \frac{\rho\sigma_\varepsilon}{\sigma_u}u, \quad V(\varepsilon|u) = \sigma_\varepsilon^2 - \frac{\sigma_{\varepsilon u}^2}{\sigma_u^2} = \sigma_\varepsilon^2(1-\rho^2)$$
$$\Longrightarrow \quad P(d = 1|u, w) = P(\varepsilon > -w'\alpha|u, w)$$
$$= P\left[\left(\varepsilon - \frac{\rho\sigma_\varepsilon}{\sigma_u}u\right)/\{\sigma_\varepsilon(1-\rho^2)^{1/2}\}\right.$$
$$> \left(-w'\alpha - \frac{\rho\sigma_\varepsilon}{\sigma_u}u\right)/\{\sigma_\varepsilon(1-\rho^2)^{1/2}\} \,|u, w]$$
$$= \Phi\left\{\left(w'\frac{\alpha}{\sigma_\varepsilon} + \rho\frac{u}{\sigma_u}\right)\cdot(1-\rho^2)^{-1/2}\right\}.$$

The second term, $-\ln(\sigma_u) + \ln\phi\{(y_i - x_i'\beta)/\sigma_u\}$, comes from the logged density for $u|w$. Note that only $\alpha/\sigma_\varepsilon = (\alpha_1/\sigma_\varepsilon, ..., \alpha_{k_w}/\sigma_\varepsilon)'$, not α per se, is identified as in probit because d_i is binary with no scale information for d_i^* available.

If $\rho = 0$, then the log-likelihood becomes

$$\sum_{i=1}^{N}\left[(1-d_i)\ln\Phi\left(-w_i'\frac{\alpha}{\sigma_\varepsilon}\right) + d_i\ln\Phi\left(w_i'\frac{\alpha}{\sigma_\varepsilon}\right)\right.$$
$$\left. + d_i\left\{-\ln(\sigma_u) + \ln\phi(\frac{y_i - x_i'\beta}{\sigma_u})\right\}\right].$$

This has two parts: the first part (the first two terms) is the probit log-likelihood function with $\alpha/\sigma_\varepsilon$ being the only parameter, and the second (the last term) is the log-likelihood function involving only σ_u and β; maximizing this wrt β is then equivalent to doing LSE using only the $d = 1$ observations. That is, if $\rho = 0$, then the LSE for β with the subsample $d = 1$ is the MLE.

Estimating ρ poses the main difficulty in practice due to the bound $(-1, 1)$; the estimate for ρ may go out of the bound ± 1. It is thus recommended to estimate first

$$\zeta(\rho) \equiv \left(\frac{\alpha'}{\sigma_\varepsilon}, \beta', \sigma_u\right)'$$

for each fixed grid point of ρ on $(-1, 1)$ (Nawata and Nagase, 1996), and then choose ρ maximizing the log-likelihood function. Let $m_i(\zeta_o, \rho_o)$ denote the score function for ζ for the i-th observation evaluated at ζ_o and ρ_o. Then one can use a Newton-Raphson type iteration scheme

$$\zeta_1(\rho) = \zeta_o(\rho) + \left\{\frac{1}{N}\sum_i m_i\left(\zeta_o, \rho\right) \cdot m_i(\zeta_o, \rho)'\right\}^{-1} \cdot \frac{1}{N}\sum_i m_i(\zeta_o, \rho)$$

until convergence; the MLE for ζ under $\rho = 0$ can be used as the initial value.

When $\zeta(\rho)$ is plugged into the likelihood function, we get the "*profile likelihood*" function for ρ which has only ρ as its parameter. The optimal ρ (and thus the optimal $\zeta(\rho)$) can be chosen with the profile likelihood. One caveat on profile likelihood is that usually the name is used in the context where ρ is the parameter of interest while the other parameter ζ is a "nuisance parameter." In the example with ζ and ρ, these roles are reversed.

3.4   Two-Stage Estimator

For Heckman two-stage estimator (TSE) in Heckman (1979), assume

(a) $\varepsilon \amalg w$, and $\varepsilon \sim N(0, \sigma_\varepsilon^2)$;
(b) linear regression of u on ε: $E(u|w, \varepsilon) = (\sigma_{\varepsilon u}/\sigma_\varepsilon^2)\varepsilon$.

The selection addition term $\rho\sigma_u\phi(w'\alpha/\sigma_\varepsilon)/\Phi(w'\alpha/\sigma_\varepsilon)$ holds under these assumptions; the joint normality of (ε, u) is not necessary. Heckman TSE, which accounts for the selection addition term explicitly, proceeds as follows.

The first step is applying probit to the selection equation to estimate $\alpha/\sigma_\varepsilon$ and replacing the selection correction term

$$\lambda(w_i'\frac{\alpha}{\sigma_\varepsilon}) \equiv \frac{\phi(w_i'\alpha/\sigma_\varepsilon)}{\Phi(w_i'\alpha/\sigma_\varepsilon)}$$

with its feasible version $\lambda_{Ni} \equiv \lambda(w_i'a_N)$ where a_N is the probit for $\alpha/\sigma_\varepsilon$. Defining

$$z_{Ni} \equiv (x_i', \lambda_{Ni})' \quad \text{and} \quad z_i \equiv \left(x_i', \lambda\left(w_i'\frac{\alpha}{\sigma_\varepsilon}\right)\right)',$$

the second step is getting the LSE g_N of $d_i y_i$ on $d_i z_{Ni}$ for the parameter

$$\gamma \equiv (\beta', \rho\sigma_u)';$$

for this LSE, certainly $E^{-1}(dzz')$ should exist, much as the outer-product of the score functions should exist for the preceding MLE. Despite the selection problem, β is estimated consistently due to the presence of λ_{Ni} in the second stage LSE. STATA has the command `heckman` to implement this estimator.

For the asymptotic distribution of Heckman TSE, define the probit score function evaluated at the true value as $S_i = S_i(\alpha)$ and its "influence function" as η_i so that

$$\sqrt{N}\left(a_N - \frac{\alpha}{\sigma_\varepsilon}\right) = \frac{1}{\sqrt{N}}\sum_i \eta_i + o_p(1),$$

$$\text{where } \phi_i \equiv \phi\left(w_i'\frac{\alpha}{\sigma_\varepsilon}\right), \ \Phi_i \equiv \Phi\left(w_i'\frac{\alpha}{\sigma_\varepsilon}\right),$$

$$\eta_i \equiv E^{-1}(SS')S_i = E^{-1}(SS')\frac{w_i\phi_i(d_i - \Phi_i)}{\Phi_i(1 - \Phi_i)};$$

$$\eta_i\text{'s are iid with } E(\eta) = 0 \text{ and } E(\eta\eta') < \infty.$$

Further define a "*link matrix*"

$$L \equiv \rho\sigma_u \cdot E\{d \cdot \lambda'(w'\frac{\alpha}{\sigma_\varepsilon}) \cdot zw'\}, \quad \text{where } \lambda'(a) \equiv \frac{\partial\lambda(a)}{\partial a} = -a\lambda(a) - \lambda(a)^2$$

and the "reduced form" error and its residual as, respectively,

$$v_i \equiv y_i - z_i'\gamma \quad \text{and} \quad v_{Ni} \equiv y_i - z_{Ni}'g_N.$$

Then,

$$\sqrt{N}(g_N - \gamma) \rightsquigarrow N(0, \ E^{-1}(dzz') \cdot E\{(dvz - L\eta)(dvz - L\eta)'\} \cdot E^{-1}(dzz')).$$

The asymptotic variance of $\sqrt{N}(g_N - \gamma)$ can be estimated consistently by replacing $E(\cdot)$ with $N^{-1}\sum_i(\cdot)$ and the other unknowns by consistent estimates: specifically,

$$\left(\frac{1}{N}\sum_i d_i z_{Ni} z_{Ni}'\right)^{-1} \frac{1}{N}\sum_i (d_i v_{Ni} z_{Ni} - L_N\eta_{Ni})(d_i v_{Ni} z_{Ni} - L_N\eta_{Ni})'$$

$$\left(\frac{1}{N}\sum_i d_i z_{Ni} z_{Ni}'\right)^{-1},$$

where

$$\eta_{Ni} \equiv \left\{\frac{1}{N}\sum_i S_i(a_N)S_i(a_N)'\right\}^{-1} S_i(a_N) \quad \text{and}$$

$$L_N \equiv g_{\rho\sigma_u}\sum_i d_i\lambda'(w_i'a_N)z_{Ni}w_i',$$

and $g_{\rho\sigma_u}$ is the last component of g_N—the estimator for $\rho\sigma_u$.

The part $L\eta$ in the asymptotic variance is the first-stage correction term: the error $a_N - \alpha/\sigma_\varepsilon$ affects the second-stage asymptotic variance. The term a_N appears in many places in the second-stage LSE, but the estimation error matters only when it appears in the residual v_{Ni}. Specifically, $\sqrt{N}(g_N - \gamma)$ can be shown to be $o_p(1)$-equal to

$$E^{-1}(dzz') \frac{1}{\sqrt{N}} \sum_i d_i z_i \{ y_i - x_i'\beta - \rho\sigma_u \lambda(w_i'a_N) \}$$

$$= E^{-1}(dzz') \frac{1}{\sqrt{N}} \sum_i d_i z_i \left[y_i - x_i'\beta - \rho\sigma_u \left\{ \lambda \left(w_i' \frac{\alpha}{\sigma_\varepsilon} \right) \right. \right.$$
$$\left. \left. + \lambda' \left(w_i' \frac{\alpha}{\sigma_\varepsilon} \right) w_i' \left(a_N - \frac{\alpha}{\sigma_\varepsilon} \right) \right\} \right] + o_p(1)$$

$$= E^{-1}(dzz') \frac{1}{\sqrt{N}} \sum_i \left[d_i z_i v_i \ - \rho\sigma_u d_i \lambda' \left(w_i' \frac{\alpha}{\sigma_\varepsilon} \right) z_i w_i' \left(a_N - \frac{\alpha}{\sigma_\varepsilon} \right) \right]$$
$$+ o_p(1)$$

$$= E^{-1}(dzz') \cdot \left[\frac{1}{\sqrt{N}} \sum_i d_i z_i v_i \ - \rho\sigma_u \frac{1}{N} \sum_i d_i \lambda' \left(w_i' \frac{\alpha}{\sigma_\varepsilon} \right) z_i w_i' \right.$$
$$\left. \cdot \sqrt{N} \left(a_N - \frac{\alpha}{\sigma_\varepsilon} \right) \right] + o_p(1)$$

$$= E^{-1}(dzz') \cdot \left[\frac{1}{\sqrt{N}} \sum_i d_i z_i v_i \ - L \cdot \frac{1}{\sqrt{N}} \sum_i \eta_i \right] + o_p(1)$$

$$= E^{-1}(dzz') \cdot \frac{1}{\sqrt{N}} \sum_i (d_i z_i v_i - L\eta_i) + o_p(1)$$

where the mean-value theorem is invoked for the first equality in probability. In the last equation, the second term explains the first-stage correction term.

The selection bias can be tested with $H_0 : \rho\sigma_u = 0$ in the second-stage LSE. Under the H_0, one may set $L\eta = 0$ in the asymptotic variance because $L = 0$ if $\rho = 0$, which means that there is no need to bother with the first-stage correction as far as testing the H_0 goes. If $\rho = 0$, as mentioned ahead, the MLE is numerically equal to the LSE of dy on dx. Thus, if $\rho = 0$, the only difference between the MLE and the TSE is that the selection correction term is added as another regressor in the LSE stage of the TSE.

In principle, Heckman TSE does not need an exclusion restriction. If the regression function is misspecified, however, then one may easily reach a false conclusion, because $\lambda(\cdot)$ may pick up the misspecified regression function; e.g., x_j^2 may be omitted in the outcome equation regression function where x_j is the j-th component of x, and λ may come out significant picking up x_j^2 due to the nonlinearity of λ despite $\rho = 0$. More generally, because λ can be approximated arbitrarily well by a polynomial function of $x'\alpha/\sigma_\varepsilon$, one cannot tell the true selection bias term from the regression function including the polynomial function. Exclusion restrictions avoid this problem.

Even when we are sure of the outcome equation regression functional form, still Heckman TSE has a major practical problem if an exclusion restriction is not used (i.e., if $w = x$), because $\lambda(\cdot)$ is almost linear: $\lambda(t)$ decreases monotonically toward zero and is almost linear for, say, $t \leq -1$, and then approaches zero becoming slightly nonlinear; $\lambda(t)$ is almost zero, say, for $t \geq 3$. Thus λ tends to be highly collinear with x, and quite often the standard error of the estimate for $\rho\sigma_u$ is too large, leading to a failure to reject no selection problem. It is advisable to check the range of $w'a_N$: if there are not many $w_i'a_N$'s in $[-1,3]$, Heckman TSE may work poorly. In this regard, the MLE may be preferable to Heckman TSE, which Nawata and Nagase (1996) also pointed out. The multicollinearity problem of λ and x can be checked with the R^2 in the LSE of $\lambda(w_i'a_N)$ on x_i.

Although an exclusion restriction helps alleviate the identification and multicollinearity problems in the two preceding paragraphs, there always is a chance that the exclusion restriction is false. Lee (2003) analyzed the bias resulting from a false exclusion restriction in the sample selection; the form of bias differs across different estimators. Interestingly, Lee (2003) further showed that the outcome equation parameters for *regressors with zero coefficients in the selection equation are immune to exclusion bias if only one regressor is excluded.*

One way to make Heckman TSE less sensitive to its assumptions is as follows. Assume that $\varepsilon \amalg w$ still holds with ε following $N(0,\sigma_\varepsilon^2)$. But, instead of $E(u|w,\varepsilon) = (\sigma_{\varepsilon u}/\sigma_\varepsilon^2)\varepsilon$, assume, for some parameters γ_1 and γ_2,

$$E(y|w,d=1) = x'\beta + E(u|w'\alpha + \varepsilon > 0) = x'\beta + \gamma_1\left(w'\frac{\alpha}{\sigma_\varepsilon}\right) + \gamma_2\left(w'\frac{\alpha}{\sigma_\varepsilon}\right)^2.$$

Here $E(u\,|w,\,w'\alpha + \varepsilon > 0)$ depends on w only through the linear index $w'\alpha/\sigma_\varepsilon$, and that $E(u|w'\alpha + \varepsilon > 0)$ can be approximated by a quadratic function of $w'\alpha/\sigma_\varepsilon$; extension to higher-order polynomials is straightforward. To show the asymptotic variance of this robustified version of Heckman TSE, define

$$z_i \equiv \left(x_i', \left(w_i'\frac{\alpha}{\sigma_\varepsilon}\right), \left(w_i'\frac{\alpha}{\sigma_\varepsilon}\right)^2\right)', \quad z_{Ni} \equiv (x_i', (w_i'a_N), (w_i'a_N)^2)',$$

$$\theta \equiv (\beta', \gamma_1, \gamma_2)', \quad \theta_N \equiv (b_N', g_{N1}, g_{N2})',$$

$$\varepsilon_i \equiv y_i - x_i'\beta - \gamma_1\left(w_i'\frac{\alpha}{\sigma_\varepsilon}\right) - \gamma_2\left(w_i'\frac{\alpha}{\sigma_\varepsilon}\right)^2.$$

Then it can be shown that θ is estimated consistently by the second-stage LSE θ_N with

$$\sqrt{N}(\theta_N - \theta) \rightsquigarrow N(0, \Omega_q),$$

$$\Omega_q \equiv E^{-1}(dzz') \cdot E\{(d\varepsilon z - A_q\eta)(d\varepsilon z - A_q\eta)'\} \cdot E^{-1}(dzz'),$$

$$A_q \equiv E[d\{\gamma_1 + 2\gamma_2(w'\frac{\alpha}{\sigma_\varepsilon})\}zw'].$$

3.5   Selection Models for Some LDV's

So far we dealt with selection models for continuously distributed y. This section considers cases where y is a LDV. First, binary y is studied, and then count response y will be examined. ODR y can be also examined in this section, but to save space, we will leave this task out. An ODR can be collapsed into binary responses in multiple ways, and each binary response model can be estimated; the binary response estimation results can be then combined using minimum distance estimation (MDE). Censored y will be studied in the following subsection when "double hurdle" models are introduced. Certainly more involving cases are possible; e.g., Lee (2004b) considered ODR d and count response y, which included binary d and count response y in Terza (1998) as a special case; Terza's (1998) approach will be examined below. A review on sample-selection estimators when the selection equation is multinomial (not binary) can be found in Bourguignon et al. (2007).

3.5.1   Binary-Response Selection MLE

Our "binary-response selection model" is the same as the basic selection model with the only difference being that $y_i = 1[y_i^* > 0]$ is observed where y_i^* is a latent continuous response:

$$d_i = 1[d_i^* > 0] \text{ where } d_i^* = w_i'\alpha + \varepsilon_i \quad \text{and} \quad y_i = 1[y_i^* > 0]$$
$$\text{where } y_i^* = x_i'\beta + u_i(d_i, w_i', d_i y_i)' \text{ is observed.}$$

Since no scale information is available for both d_i^* and y_i^*, neither σ_ε nor σ_u are identified; set $\sigma_\varepsilon = 1 = \sigma_u$ to simplify notations. Although the exclusion restriction is not explicit here, it will be hard to estimate the model without it—this in fact holds for all selection models including the ones to be introduced in the remainder of this section.

The binary-response selection case is different from "*bivariate binary response* models" where both d and y are observed separately. In the latter, there are four "cells" from 2×2 possibilities, whereas there are three "cells" $d = 0$, $(d = 1, y = 0)$, and $(d = 1, y = 1)$ in the former. See Lee (1999) and the references therein for more on bivariate binary response models. Another distinction to be made is between binary-response selection and the case of only (w', dy) being observed where the observability is further restricted to only two cells $dy = 0$ and $dy = 1$; see Meng and Schmidt (1985) and the references therein. This difficult case occurs if the observed binary response (dy) is one when two conditions $d = 1$ and $y = 1$ are met; e.g., two political parties should agree for a bill to pass. Here one would like to know what determines each condition d and y, but only the product dy is observed. It would be very difficult, if not impossible, to identify separately the parameters for the d-equation and the y-equation.

Adding the joint normality of (ε, u) and its independence from w, the parameters α, β, ρ can be estimated by MLE. For the MLE, define $\psi(u_1, u_2; \rho)$ as the standard joint normal density function with correlation ρ, and

$$\Psi(e_1, e_2, \rho) \equiv \int_{-\infty}^{e_2} \int_{-\infty}^{e_1} \psi(u_1, u_2; \rho) du_1 du_2.$$

The log-likelihood function for α, β, ρ consists of three terms corresponding to three cases $d = 0$, $(d = 1, y = 1)$, and $(d = 1, y = 0)$:

$$\sum_{i=1}^{N} [\, (1 - d_i) \cdot \ln \Phi(-w_i'\alpha) + d_i y_i \cdot \ln P(\varepsilon_i > -w_i'\alpha, \; u_i > -x_i'\beta)$$
$$+ \, d_i(1 - y_i) \cdot \ln P(\varepsilon_i > -w_i'\alpha, \; u_i < -x_i'\beta) \,]$$

where

$$P(\varepsilon_i > -w_i'\alpha, u_i > -x_i'\beta) = 1 - P(\varepsilon_i < -w_i'\alpha) - P(u_i < -x_i'\beta)$$
$$+ P(\varepsilon_i < -w_i'\alpha, u_i < -x_i'\beta)$$
$$= 1 - \Phi(-w_i'\alpha) - \Phi(-x_i'\beta) + \Psi(-w_i'\alpha, -x_i'\beta, \rho),$$
$$P(\varepsilon_i > -w_i'\alpha, \; u_i < -x_i'\beta) = 1 - \Psi(-w_i'\alpha, -x_i'\beta, \rho) - \Phi(x_i'\beta).$$

These can be understood by drawing the region $\varepsilon_i > -w_i'\alpha$ and $u_i > -x_i'\beta$ on a two-dimensional plane with ε_i and u_i as the two axes. As in other multivariate MLE's, estimating ρ is troublesome and it is recommended to do grid-search over ρ while doing a Newton–Raphson type iteration for α and β.

Dubin and Rivers (1989) provided an empirical example where $d = 1$ if votes and $y = 1$ if votes for Reagan using the 1984 US National Election Study data with $N = 2237$ but only 1347 people voting. Part of their Table 1 and 2 is reproduced in Table 3 with SD in parentheses, where all variables are dummies, "new resident" means living at the current address for less than a year, college is 1 if ever attended a college, and TV/news is 1 if watches TV news or reads a newspaper on a daily basis.

Several remarks are in order. First, there is a significant selection problem: ρ is -0.41 with the absolute t-value higher than 3; this means that non-voters are likely to prefer Reagan, and the Democratic loss in 1984 is not due to low turnout. Second, the selection equation is virtually identical across PRO and MLE despite ρ being -0.41, which is the source of efficiency of the MLE over PRO. Third, comparing the outcome equations, the biggest difference is in "under 30," which is however insignificant; despite the significant selection problem, the resulting bias for α and β seems little. Fourth, in the variables excluded from the outcome equation, college is not a plausible variable for it can easily influence y; Dubin and Rivers (1989) in fact exclude marriage dummy as well, which is also problematic. Fifth, the signs of the estimates make sense mostly other than for "under 30" and "over 55" in the outcome equation.

Table 3: Voting for Reagan: Probit v. Sample-Selection MLE

	PRO		Sample-Selection MLE		$\bar{w}$ (full sample)
	Selection	Outcome	Selection	Outcome	
black	−0.27 (0.09)	−1.37 (0.18)	−0.27 (0.09)	−1.22 (0.20)	0.11
female	0.14 (0.06)	−0.09 (0.07)	0.14 (0.06)	−0.11 (0.07)	0.56
union	0.20 (0.07)	−0.51 (0.09)	0.20 (0.07)	−0.55 (0.08)	0.21
under 30	−0.22 (0.07)	0.03 (0.10)	−0.22 (0.07)	0.14 (0.10)	0.28
over 55	0.18 (0.07)	−0.19 (0.09)	0.19 (0.07)	−0.24 (0.08)	0.29
new resident	−0.53 (0.07)		−0.53 (0.07)		0.21
college	0.62 (0.07)		0.62 (0.06)		0.41
TV/news	0.32 (0.06)		0.31 (0.06)		0.62
ρ			−0.41 (0.14)		

3.5.2   Count-Response Zero-Inflated MLE

Turning to count responses, we will consider two cases separately: y^* takes $0, 1, 2...$ and y^* takes $1, 2, ...$ An example for the former is angling $(d = 1)$: one may try but catch no fish in a given time interval unless the fish "cooperate," which means $P(y^* = 0) > 0$. An example for the latter is donation $(d = 1)$ and its frequency where $y^* = 0$ is unthinkable given $d = 1$. In this case, we may use a "zero-truncated density" for y^*; e.g., the zero-truncated Poisson distribution with the support points $1, 2, ...$

For the former case $P(y^* = 0) > 0$, suppose (i) $P(d = 1|x)$ is logistic, (ii) $y^*|w$ follows Poisson, and (iii) $\varepsilon \amalg y^*|w$. This gives a zero-inflated Poisson (ZIP) model where $P(d = 1|x)$ can be modeled as probit if desired:

$$y_i = d_i y_i^*, \quad y^*|w \sim Poi\{\exp(x'\beta)\}, \quad (d_i, w_i', y_i) \text{ observed}$$
$$P(y = 0|w) = P(d = 0|w) + P(d = 1, y^* = 0|w)$$
$$= P(d = 0|w) + P(d = 1|w)P(y^* = 0|w)$$
$$= \frac{1}{1 + \exp(w'\alpha)} + \frac{\exp(w'\alpha)}{1 + \exp(w'\alpha)} \exp(-e^{x'\beta}),$$
$$P(y = j|w) = P(d = 1, y^* = j|w)$$
$$= \frac{\exp(w'\alpha)}{1 + \exp(w'\alpha)} \frac{\{\exp(x'\beta)\}^j}{j!} \exp(-e^{x'\beta}), \quad \forall j = 1, 2, ...$$

Certainly, other distributions such as negative binomial (NB) can be used for $y^*|x$ to result in "ZINB." STATA provides the commands ZIP and ZINB to implement ZIP and the zero-inflated NB. The assumption $\varepsilon \amalg y^*|w$ in ZIP, which is equivalent to $d \amalg y^*|w$, rules out selection problem as the selection and outcome equations are independent given w. That is, ZIP is not a sample-selection model. Nevertheless, ZIP is examined here, for ZIP is a multivariate LDV model motivating a count-response selection MOM to appear shortly.

For the latter case $P(y^* = 0) = 0$ with the zero-truncated Poisson,

$$P(y = 0|w) = P(d = 0|w) = \frac{1}{1 + \exp(w'\alpha)}$$
$$P(y = j|w) = P(d = 1, y^* = j|x) = P(d = 1|w)P(y^* = j|x)$$
$$= \frac{\exp(w'\alpha)}{1 + \exp(w'\alpha)} \cdot \frac{1}{1 - \exp(-e^{x'\beta})} \frac{\{\exp(x'\beta)\}^j}{j!} \exp(-e^{x'\beta}),$$
$$j = 1, 2, ...$$

The term $\{1 - \exp(-e^{x'\beta})\}^{-1}$ is the normalizing factor $\{1 - P(y^* = 0|w)\}^{-1}$ for the truncation at zero. The individual likelihood function is

$$\left\{ \frac{1}{1 + \exp(w'\alpha)} \right\}^{1-d} \cdot \left\{ \frac{\exp(w'\alpha)}{1 + \exp(w'\alpha)} \right\}^d \cdot$$
$$\left[\frac{1}{1 - \exp(-e^{x'\beta})} \frac{\{\exp(x'\beta)\}^j}{j!} \exp(-e^{x'\beta}) \right]^d .$$

When ln is taken on this, this gets split into the logit log-likelihood and the truncated-Poisson log-likelihood. Hence, the MLE $\hat{\alpha}$ and $\hat{\beta}$ is nothing but, respectively, the logit of d on w and the truncated Poisson MLE with the subsample $d_i = 1$. Such a split does not occur in the preceding case $P(y^* = 0) > 0$, because $P(y = 0|w)$ is a sum of two terms involving zero Poisson probability.

Suppose that ZIP has been implemented for a case with $P(y^* = 0) > 0$, but its $\hat{\alpha}$ differs much from logit. Because logit tends to be similar to probit except that logit estimates are about 1.8 times greater than the probit estimates, and because not much can be done other than logit/probit for binary responses, we should retain the logit specification to question the other two aspects of ZIP. That is, we might doubt the Poisson specification or question the assumption $\varepsilon \amalg y^*|w$. For the former, we may try zero-inflated NB, for instance. But the latter is hard to address, because the joint distribution for (ε, y^*) is not easy to specify—y^* is discrete while ε is not. This awkward situation motivates the next method-of-moment approach based partly on normality.

3.5.3   Count-Response Selection MOM

Terza (1998) assumed, in essence,

$$E(y^*|w, \varepsilon) = \exp(x'\beta_x + \beta_\varepsilon \varepsilon) \text{ and } \varepsilon \text{ in } d = 1[w'\alpha + \varepsilon > 0] \text{ follows } N(0,1).$$

Under this, $E(y|w, d = 1)$ can be derived as a function of β_x, β_ε, and α:

$$E(y|w, \varepsilon > -w'\alpha) = \int_{-w'\alpha}^{\infty} \exp(x'\beta_x + \beta_\varepsilon t) \frac{1}{\sqrt{2\pi}} \exp\left(-\frac{1}{2}t^2\right) dt \frac{1}{\Phi(w'\alpha)}$$

$$= \exp(x'\beta_x) \int_{-w'\alpha}^{\infty} \frac{1}{\sqrt{2\pi}} \exp\left(\beta_\varepsilon t - \frac{1}{2}t^2 + \frac{1}{2}\beta_\varepsilon^2 - \frac{1}{2}\beta_\varepsilon^2\right) dt \cdot \frac{1}{\Phi(w'\alpha)}$$

$$= \exp\left(x'\beta_x + \frac{1}{2}\beta_\varepsilon^2\right) \int_{-w'\alpha}^{\infty} \frac{1}{\sqrt{2\pi}} \exp\left\{-\frac{1}{2}(t - \beta_\varepsilon)^2\right\} dt \cdot \frac{1}{\Phi(w'\alpha)}$$

$$= \exp\left(x'\beta_x + \frac{1}{2}\beta_\varepsilon^2\right) \int_{-w'\alpha-\beta_\varepsilon}^{\infty} \frac{1}{\sqrt{2\pi}} \exp\left(-\frac{1}{2}\tau^2\right) d\tau \frac{1}{\Phi(w'\alpha)}$$

$$= \exp(x'\beta_x + \frac{1}{2}\beta_\varepsilon^2) \frac{\Phi(w'\alpha + \beta_\varepsilon)}{\Phi(w'\alpha)}.$$

With the subsample $d = 1$, we can do a two-stage MOM estimation using

$$E(y|w, d = 1) = \exp(x'\beta_x + \frac{1}{2}\beta_\varepsilon^2) \frac{\Phi(w'\alpha + \beta_\varepsilon)}{\Phi(w'\alpha)}.$$

The first-stage is probit for α, and the second stage is GMM for β_x and β_ε with α replaced by the probit $\hat{\alpha}$. Test for $H_0 : \beta_\varepsilon = 0$ is a test for the selection problem.

Instead of entering ε directly into the y^* equation, we may specify

$$E(y^*|w,v) \;\; = \;\; \exp(x'\beta_x + \beta_v v) \text{ where } (v,\varepsilon) \text{ is jointly normal with}$$
$$COR(v,\varepsilon) \neq 0.$$

But this makes no real difference from the above approach. To see this, decompose v as $\gamma_\varepsilon \varepsilon + \gamma_e e$ with $\varepsilon \amalg e$ and assume

$$E(y^*|w,\varepsilon,e) = \exp(x'\beta_x + \beta_v \gamma_\varepsilon \varepsilon + \beta_v \gamma_e e)$$

$$\Longrightarrow E(y|w,d=1) = \int\!\!\int_{-w'\alpha} \exp(x'\beta_x + \beta_v \gamma_\varepsilon t + \beta_v \gamma_e \tau)\frac{1}{\sqrt{2\pi}}$$

$$\cdot \exp\left(-\frac{1}{2}t^2\right) dt\, \frac{\phi(\tau/\sigma_e)}{\sigma_e} d\tau$$

$$= \int \exp(\beta_v \gamma_e \tau)\frac{\phi(\tau/\sigma_e)}{\sigma_e} d\tau \cdot \int_{-w'\alpha} \exp(x'\beta_x + \beta_v \gamma_\varepsilon t)\frac{1}{\sqrt{2\pi}} \exp(-\frac{1}{2}t^2) dt.$$

The first term is a constant to be absorbed into the intercept in $x'\beta_x$, and in the second term, only $\beta_v \gamma_\varepsilon$ is identified which can be taken as β_ε above.

The two-stage MOM is a selection correction method where the selection addition term takes the particular multiplicative form $\exp(0.5\beta_\varepsilon^2)\Phi(w'\alpha + \beta_\varepsilon)/\Phi(w'\alpha)$. The generalization of Lee (2004b) for ordered d and a sensitivity analysis there send a warning that this specification may be too tight. Combined with the "explosive" nature of the exponential function, the two-stage MOM might be sensitive to violations of the model assumptions. One way to robustify the procedure is, as done for the Heckman two-stage estimator in the linear selection model, to replace the selection addition term with an exponential function with polynomials of $w'\alpha$ in; e.g.

$$\exp\{1 + \gamma_1(w'\alpha) + \gamma_2(w'\alpha)^2\}.$$

"$H_0 : \gamma_1 = 0$ and $\gamma_2 = 0$" means no selection problem.

3.6   Double and Multiple Hurdle Models

Suppose that y is a response variable censored from below at 0. Combining the censoring problem with a selection problem, consider a model:

$$d_i^* \;\; = \;\; w_i'\alpha + \varepsilon_i, \quad d_i = 1[d_i^* > 0]$$
$$y_i^* \;\; = \;\; x_i'\beta + u_i, \quad y_i = y_i^* 1[y_i^* > 0] = \max(y_i^*, 0)$$
$$(d_i, w_i', d_i y_i) \;\; = \;\; (d_i, w_i', d_i y_i^* 1[y_i^* > 0]) \text{ is observed}$$

Here, both d_i^* and y_i^* are latent variables, and y_i^* is observed when two conditions (or hurdles), $d_i = 1$ and $y_i^* > 0$, are met. This is an example of a *double hurdle model* originally due to Cragg (1971).

There are variations on this kind of double hurdle models, depending on the reason why y_i^* is not observed. Suppose y_i^* is expenditure on tobacco. For y_i^* to be positive, a couple of conditions may be needed. First, consumption decision should be made by the consumer ($f_i = 1$); nonsmokers will have $y_i^* = 0$. Second, during the period, purchase should be made at least once ($g_i = 1$); some people purchase a large amount only occasionally, while some others may purchase small amounts frequently. Third, the consumer should overcome the social hurdle of anti-smoking sentiment ($h_i = 1$). Putting these three conditions together, we get $f_i g_i h_i y_i^*$, which may be dubbed a "multiple hurdle model," special cases of which are

$g_i y_i^*$: a necessary good ($f_i = h_i = 1$ for all i),

$f_i h_i y_i^*$: a perishable good, bought frequently ($g_i = 1$ for all i),

$f_i g_i y_i^*$: a "no stigma" good ($h_i = 1$ for all i).

The above model can be viewed also as a special case where the two conditions $d_i = 1$ and $1[y_i^* > 0]$ are two of $f_i = 1$, $g_i = 1$, and $h_i = 1$.

In the multiple hurdle case, we have to consider one equation for each hurdle, which means multiple correlated error terms. Doing MLE for this is troublesome because several correlation coefficients and σ_u has to be estimated, which typically results in non-convergence in the MLE. In practice, often zero correlation assumptions are imposed. For instance, the error term in the purchase frequency equation is assumed to be independent of the other error terms. But this may be false, because, for instance, lazy people may purchase less frequently and being lazy may also influence y.

For the above double hurdle model, if ε is independent of u, then the log-likelihood function becomes

$$\sum_{i=1}^{N} \left[(1 - d_i) \ln \left\{ 1 - \Phi(w_i'\alpha)\Phi\left(x_i'\frac{\beta}{\sigma_u}\right) \right\} \right.$$
$$\left. + d_i \ln \left\{ \Phi(w_i'\alpha) \cdot \frac{1}{\sigma_u}\phi\left(\frac{y_i - x_i'\beta}{\sigma_u}\right) \right\} \right]$$

where $1 - \Phi(w_i'\alpha)\Phi(x_i'\beta/\sigma_u) = 1 - P(d_i = 1, y_i^* > 0|w_i)$ and $\Phi(w_i'\alpha) \cdot \phi\{(y_i - x_i'\beta)/\sigma_u\}/\sigma_u$ comes from

$$P(d_i = 1|w_i) \cdot P(y_i^* > 0|w_i)\frac{\sigma_u^{-1}\phi\{(y_i - x_i'\beta)/\sigma_u\}}{P(y_i^* > 0|w_i)}.$$

Certainly the likelihood function for the general case $COR(\varepsilon, u) \neq 0$ can be derived, but it often fails to work as mentioned already. See Smith (2002) for the literature on double hurdle models.

4 LDV's with Endogenous Regressors

When we consider multivariate equations with LDV's, we can think of three types of systems. First, "RF-like" multivariate equations where no LDV or its latent version appears on the right-hand side; no endogeneity issue comes up here. Second, recursive equations where some LDV's or their latent versions appear recursively on the right-hand side. Third, simultaneous equations where all LDV's or their latent versions appear on the right-hand side non-recursively.

For instance, consider

$$y_1 = f_1(\alpha_1 y_2 + x_1' \beta_1 + u_1) \quad \text{and} \quad y_2 = f_2(\alpha_2 y_1 + x_2' \beta_2 + u_2),$$

$$u_1 \text{ and } u_2 \text{ are related to each other, and } (u_1, u_2)$$

$$\text{are independent of } (x_1, x_2)$$

where f_1 and f_2 are known functions such as $f_1(\cdot) = 1[\cdot > 0]$ for binary responses and $f_2(\cdot) = \max(\cdot, 0)$ for responses censored from below at 0. If $\alpha_1 = \alpha_2 = 0$, then the system is of RF-type; if only one of α_1 and α_2 is zero, then the system is recursive; if $\alpha_1 \neq 0$ and $\alpha_2 \neq 0$, then the system is simultaneous. For recursive or simultaneous systems, we can also think of the versions with the latent continuous variables y_1^* and y_2^*, instead of the observed LDV's y_1 and y_2, appearing on the right-hand side.

Since simultaneous systems include recursive systems as special cases, which in turn include RF-like systems as special cases, we will devote most space to simultaneous systems, whereas recursive or RF-like systems will be discussed briefly. See Heckman (1978), Maddala (1983), Amemiya (1985), Blundell and Smith (1993), L.F. Lee (1993), and the references therein for more details and the literature on multivariate systems with LDV's.

When endogenous regressors appear in a LDV model, there are a number of ways to deal with them:

1. IVE with an orthogonality moment condition.

2. Replacing the regressors x with $E(x|z)$ where z is an instrument; for the overlapping (i.e., exogenous) components of x and z, the replacement is the same as using the original variable.

3. Decompose the error term u to explicitly account for (or remove) the part of u related to x.

4. Specify the equations determining the endogenous regressors and then deal with the multiple equations simultaneously.

5. Add the instrument z as an artificial regressor with its slope ψ; fixing the endogenous parameter α initially, estimate $\psi(\alpha)$ and $\beta(\alpha)$ with $\hat{\psi}(\alpha)$ and $\hat{\beta}(\alpha)$; and get $\hat{\alpha}$ such that $\hat{\psi}(\hat{\alpha}) = 0$, and then obtain $\hat{\beta}(\hat{\alpha})$.

In the following subsections, first, we discuss these five approaches drawing on Lee (2008, semiparametric estimates for limited dependent variable
(LDV) models with endogenous regressors, unpublished paper). Second, we
examine recursive systems, which are relatively easy to deal with. Third, we
study simultaneous equations in LDV's, not in their latent versions, to discuss the "coherecy conditions" and show the problems in this type of models.
Fourth, simultaneous equations in the latent variables are examined in detail
for the fourth approach above. Throughout this section, bivariate binary responses will be used frequently for illustration. Kang and Lee (2009) applied
the above approaches 2-5 to censored models to recommend 2-4 for practitioners; Kang and Lee (2009) also showed how to estimate the asymptotic
variances using numerical derivatives.

4.1   Five Ways to Deal with Endogenous LDV's

First, if LDV's appear as endogenous regressors for a continuous response
variable, then the endogeneity can be dealt easily with IVE if instruments are
available. If LDV's appear as endogenous regressors in a LDV model, however,
IVE is difficult to do. To appreciate the difficulty, consider a linear model
$y = x'\beta + u$ with an instrument vector z for x such that $E(zu) = 0$. To take
advantage of this moment (orthogonality) condition, we solve the equation for
u to obtain $E\{z(y - x'\beta)\}$ that is expressed in terms of observed variables and
the parameter β. For LDV models, to do an IVE for endogenous regressors,
we need the same thing: an orthogonality condition in terms of observed
variables and an error term. For a binary response model $y = 1[x'\beta + u > 0]$
with z as instrument for x, however, it is impossible to solve the equation for
u, differently from the linear model.

Earlier, we saw an IVE for count response models where a new error term
was devised instead of the original error u to come up with an orthogonality
moment condition; this shows that IVE is not impossible—IVE's for LDV
models requiring nonparametric methods will be examined in the last chapter.
Nonetheless, the IVE for count response takes advantage of the exponential
regression function in a special way—i.e., $\exp(a + b) = \exp(a)\exp(b)$—and
cannot be easily generalized for other LDV models. An ill-advised idea is
applying IVE with the RF error $v = y - \Phi(x'\beta/\sigma)$ under $u \sim N(0, \sigma^2)$
because the RF has a smooth regression function; this fails, however, even if
$E(uz) = 0$ because $E(vz) \neq 0$ in general.

Second, consider $y_1 = f(\alpha y_2 + x'\beta + u)$ where f is a known function and
u is related to y_2 but not to x. For instance $f(\cdot) = 1[\cdot > 0]$ for binary y_1 and
$f(\cdot) = \max(\cdot, 0)$ for y_1 censored from below at 0. With an instrument vector
z for y_2 with $E(u|z) = 0$, rewrite the model as

$$y_1 = f\{\alpha E(y_2|x, z) + x'\beta + \alpha(y_2 - E(y_2|x, z)) + u\}$$
$$= f\{\alpha E(y_2|x, z) + x'\beta + v\}$$

where $v \equiv \alpha(y_2 - E(y_2|x,z)) + u$ is the new error term satisfying at least $E(v|x,z) = 0$ by construction. Replacing $E(y_2|x,z)$ with an estimator E_N $(y_2|x,z)$, one may estimate α and β. For instance, if y_2 is binary, then probit is run for y_2 on $(x',z')'$ in the first stage, and the fitted value, say $\Phi\{(x',z')g_N\}$, is used for $E_N(y_2|x,z)$. In this approach, the SF parameters α and β are estimated directly—relative to another approach below where the SF parameters are estimated indirectly through the RF estimators.

But within the parametric framework, strictly speaking, this approach is not tenable despite that it is sometimes used in practice, because the distribution of $v|(x,z)$ is not known even if the distribution of $u|(x,z)$ is so; the idea would be, however, subject to lesser criticisms if one uses semiparametric estimation methods to appear in a later chapter. *Replacing endogenous regressors with their projections is an analog of 2SLSE* for linear models where the linear projection of y_2 on (x,z) is used instead of $E(y_2|x,z)$. Despite the theoretical shortcoming, however, this method as well as the following two methods seem to work fairly well in practice.

Third, suppose y_1 is censored from below at 0, y_2 is continuous, and

$$y_1 = \max(0, \alpha y_2 + x_1'\beta_1 + u_1), \quad y_2 = x_2'\beta_2 + u_2,$$
$$(u_1, u_2) \sim N(0, \Omega) \text{ independently of } (x_1, x_2).$$

Here, the endogeneity of y_2 is due to $COR(u_1, u_2) \equiv \rho \neq 0$. Decompose u_1 as

$$u_1 = \frac{COV(u_1, u_2)}{V(u_2)} u_2 + \varepsilon = \frac{\rho \sigma_1}{\sigma_2} u_2 + \varepsilon,$$
$$\text{where } \varepsilon \sim N(0, \sigma_\varepsilon^2) \text{ independently of } u_2;$$

this can be done always thanks to the joint normality of u_1 and u_2. Then

$$y_1 = \max\left(0, \alpha y_2 + x_1'\beta_1 + \frac{\rho \sigma_1}{\sigma_2} u_2 + \varepsilon\right).$$

Replacing u_2 with the residual $\hat{u}_2$ from the LSE of y_2 on x_2, this equation can be estimated with tobit where y_2, x_1, and $\hat{u}_2$ are the regressors.

Although using $\hat{u}_2$ instead of u_2 affects the asymptotic distribution of the tobit, the test for no endogeneity $H_0 : \rho = 0$ can be done with the t-value for $\hat{u}_2$ ignoring the correction term; this situation is similar to that in Heckman two-stage for sample selection models. This way of explicitly accounting for the part of the error term causing endogeneity is called *"control function approach"*; $\hat{u}_2$ is a control function for the endogeneity of y_2. Heckman two-stage method for sample selection is also a control function approach.

Control function approach is advantageous when functions of an endogenous regressor appear as regressors. For instance, suppose m is an endogenous regressor and m, m^2, and m^3 are used together as regressors. Taking care of this problem with IVE might need three instruments. But a single control function for m removes the endogeneity of all three terms. A disadvantage of control function approach is that the form of the control function could be difficult to find. As in the preceding approach, the SF parameters α and β_1 are estimated directly in the control function approach, rather than indirectly through the RF parameter estimators.

Fourth, other than the preceding two-stage estimation methods, we can also think of MLE for simultaneous systems. Consider a simultaneous system

$$y_1 = f_1(y_1^*) = f_1(\alpha_1 y_2^* + x_1'\beta_1 + u_1) \text{ where } y_1^* = \alpha_1 y_2^* + x_1'\beta_1 + u_1,$$

$$y_2 = f_2(y_2^*) = f_2(\alpha_2 y_1^* + x_2'\beta_2 + u_2) \text{ where } y_2^* = \alpha_2 y_1^* + x_2'\beta_2 + u_2,$$

where f_1 and f_2 are known functions linking the latent variables to the observed LDV's. The *key feature of this system is that a linear system holds in terms of y_1^* and y_2^**. Then the linear system can be solved for their RF's, say,

$$y_1^* = x'\gamma_1 + v_1, \quad y_2^* = x'\gamma_2 + v_2$$

$$\implies y_1 = f_1(x'\gamma_1 + v_1), \quad y_2 = f_2(x'\gamma_2 + v_2)$$

where x consists of the elements in x_1 and x_2.

In this approach, the SF parameters are estimated through their relationship to the RF parameters using minimum distance estimation (MDE). Differently from the two preceding approaches, here the SF parameters are estimated indirectly through the RF parameter estimators. There are in fact two varieties in this approach: MLE is applied to each RF equation separately or to the two equations jointly. The former is simpler because there is no need to estimate the variance matrix of the RF errors, which entails, however, efficiency loss. The latter is complicated due to estimating the variance matrix of the RF errors, but efficient—as efficient as estimating the SF parameters directly using MLE.

Fifth, to fix the idea, consider the censored model for y_1 with an instrument z augmented artificially:

$$y_1 = \max(0, \alpha_1 y_2 + x_1'\beta_1 + \psi z + u_1)$$

$$\iff y_{1i} - \alpha_1 y_{2i} = \max(-\alpha_1 y_{2i}, x_{1i}'\beta_1 + \psi z_i + u_{1i}).$$

Take this as a censored model with the response variable $y_{1i} - \alpha_1 y_{2i}$ and the known censoring point $-\alpha_1 y_{2i}$ that varies across i. Let the tobit estimator for β and ψ with α_1 fixed be $\hat{\beta}_1(\alpha_1)$ and $\hat{\psi}(\alpha_1)$. Repeat this for all possible values of α_1. Then, because we know $\psi = 0$, choose $\hat{\alpha}_1$ as the one that makes $\hat{\psi}(\hat{\alpha}_1) = 0$ as closely as possible, which also gives $\hat{\beta}_1(\hat{\alpha}_1)$. A reasonable range for α_1 may be obtained by initially ignoring the endogeneity of y_2. This approach was adopted by Chernozhukov and Hansen (2006, 2008); also see Sakata (2007). The idea seems to work well in practice, although it is computationally quite burdensome as the maximization has to be repeated for each possible value of α_1. Chernozhukov and Hansen (2006, 2008) listed several applied studies using the approach.

4.2   A Recursive System

Before we get to recursive system, in passing, we briefly discuss RF-like multivariate equations with LDV's, which has in fact appeared already. For instance, consider

$$y_1 = f_1(x_1'\beta_1 + u_1), \ y_2 = f_2(x_2'\beta_2 + u_2),$$
$$(u_1, u_2) \text{ is jointly normal and is independent of } (x_1, x_2),$$

where f_1 and f_2 are known functions such as $f_1(\cdot) = 1[\cdot > 0]$ for binary responses and $f_2(\cdot) = \max(\cdot, 0)$ for responses censored from below at 0. Each equation can be estimated consistently on its own using univariate estimation methods, but accounting for the relationship between u_1 and u_2 will enhance the efficiency of the estimators. For instance, if y_1 and y_2 are binary (see Lee (1999) and the references therein) with $COR(u_1, u_2) = \rho$, then $(\beta_1'/\sigma_1, \beta_2'/\sigma_2, \rho)$ can be estimated simultaneously using the standard bivariate normal distribution function with correlation ρ where $\sigma_1 = SD(u_1)$ and $\sigma_2 = SD(u_2)$; the exact likelihood is a special case of that for a recursive case to be shown below. Suppose u_1 and u_2 share a common "dominant" component, say c. Then $\rho > 0$ ($\rho < 0$) suggests that c affects y_1 and y_2 in the same (opposite) direction. Other than this interpretation, the system estimation is simply to estimate β_1/σ_1 and β_2/σ_2 more efficiently. In multivariate models with LDV's, estimating the covariance matrix of the error terms could be troublesome.

Turning to recursive systems in LDV's, differently from linear recursive systems, there are many varieties depending on the form of the LDV's. We will show MLE only for a bivariate binary response model with a binary response appearing as an endogenous regressor. A simpler recursive system than this is

$$y_1 = x'\beta_1 + u_1 \quad \text{and} \quad y_2 = 1[\alpha_2 y_1 + x'\beta_2 + u_2 > 0]$$
$$(u_1, u_2) \sim N(0, \Omega) \text{ independently of } x$$

where y_1 is a continuously distributed, (possibly) endogenous regressor in the y_2 equation. STATA provides the command $\texttt{ivprobit}$ to estimate the y_2 equation.

Consider a recursive system with both endogenous variables being binary:

$$y_1 = 1[x'\beta_1 + u_1 > 0] \quad \text{and} \quad y_2 = 1[\alpha_2 y_1 + x'\beta_2 + u_2 > 0]$$
$$(u_1, u_2) \sim N(0, \Omega) \text{ independently of } x.$$

The same x is used for both equations to simplify notations; there is no loss of generality, because $x'_1\beta_1$ can be written always in the form $x'\beta_1$ by including 0's in β_1. For the likelihood, we need the following four probabilities, which sum to one:

$$
\begin{aligned}
P(y_1 &= 1, y_2 = 1|x) = P(u_1 > -x'\beta_1, \ u_2 > -\alpha_2 - x'\beta_2 \,|x), \\
P(y_1 &= 1, y_2 = 0|x) = P(u_1 > -x'\beta_1, \ u_2 < -\alpha_2 - x'\beta_2 \,|x), \\
P(y_1 &= 0, y_2 = 1|x) = P(u_1 < -x'\beta_1, \ u_2 > -x'\beta_2 \,|x), \\
P(y_1 &= 0, y_2 = 0|x) = P(u_1 < -x'\beta_1, \ u_2 < -x'\beta_2|x).
\end{aligned}
$$

One example for this is $y_1 = 1$ for attending a Catholic high school and $y_2 = 1$ for high school graduation (or college entrance) as in Evans and Schwab (1995). When $COR(u_1, u_2) \neq 0$, y_1 is an endogenous regressor in the y_2 equation.

Examine the first probability, which is

$$
P\left(-\frac{u_1}{\sigma_1} < x'\frac{\beta_1}{\sigma_1}, \ -\frac{u_2}{\sigma_2} < \frac{\alpha_2}{\sigma_2} + x'\frac{\beta_2}{\sigma_2}\,|x\right) = \Psi\left(x'\frac{\beta_1}{\sigma_1}, \ \frac{\alpha_2}{\sigma_2} + x'\frac{\beta_2}{\sigma_2}; \rho\right)
$$

where we use $COR(-u_1, -u_2) = COR(u_1, u_2)$ and $\Psi(v_1, v_2, \rho)$ is the integral of the standard bivariate normal density with correlation ρ over $(-\infty, v_1)$ and $(-\infty, v_2)$. Analogously, the second and third probabilities are, respectively,

$$
P\left(-\frac{u_1}{\sigma_1} < x'\frac{\beta_1}{\sigma_1}, \ \frac{u_2}{\sigma_2} < -\frac{\alpha_2}{\sigma_2} - x'\frac{\beta_2}{\sigma_2}\,|x\right) = \Psi\left(x'\frac{\beta_1}{\sigma_1}, \ -\frac{\alpha_2}{\sigma_2} - x'\frac{\beta_2}{\sigma_2}; -\rho\right),
$$

$$
P\left(\frac{u_1}{\sigma_1} < -x'\frac{\beta_1}{\sigma_1}, \ -\frac{u_2}{\sigma_2} < x'\frac{\beta_2}{\sigma_2}\,|x\right) = \Psi\left(-x'\frac{\beta_1}{\sigma_1}, \ x'\frac{\beta_2}{\sigma_2}; -\rho\right),
$$

because $COR(-u_1, u_2) = -COR(u_1, u_2)$. Proceeding this way, the four probabilities are obtained as functions of $\beta_1/\sigma_1, \alpha_2/\sigma_2, \beta_2/\sigma_2, \rho$, and the ensuing MLE can be done straightforwardly by maximizing

$$
\sum_i \left[y_{1i}y_{2i}\ln\Psi\left(x'\frac{\beta_1}{\sigma_1}, \frac{\alpha_2}{\sigma_2} + x'\frac{\beta_2}{\sigma_2}; \rho\right) + y_{1i}(1 - y_{2i})\ln\Psi\left(x'\frac{\beta_1}{\sigma_1}, -\frac{\alpha_2}{\sigma_2} - x'\frac{\beta_2}{\sigma_2}; -\rho\right) \right.
$$

$$
\left. + (1 - y_{1i})y_{2i}\ln\Psi\left(-x'\frac{\beta_1}{\sigma_1}, \ x'\frac{\beta_2}{\sigma_2}; -\rho\right) + (1 - y_{1i})(1 - y_{2i})\ln\Psi\left(-x'\frac{\beta_1}{\sigma_1}, -x'\frac{\beta_2}{\sigma_2}; \rho\right) \right].
$$

If we set $\alpha_2 = 0$, then the likelihood is for the bivariate binary-response RF-like system.

4.3   Simultaneous Systems in LDV's and Coherency Conditions

4.3.1   Incoherent System in Binary Responses

Suppose y_1 and y_2 are both binary and we have a simultaneous system with LDV's, not the latent continuous variables, on the rhs:

$$y_1 = 1[\alpha_1 y_2 + x'\beta_1 + u_1 > 0] \quad \text{and} \quad y_2 = 1[\alpha_2 y_1 + x'\beta_2 + u_2 > 0].$$

Consider the four probabilities:

$$
\begin{aligned}
P(y_1 &= 1, y_2 = 1|x) = P(u_1 > -\alpha_1 - x'\beta_1,\ u_2 > -\alpha_2 - x'\beta_2\,|x), \\
P(y_1 &= 1, y_2 = 0|x) = P(u_1 > -x'\beta_1,\ u_2 < -\alpha_2 - x'\beta_2\,|x), \\
P(y_1 &= 0, y_2 = 1|x) = P(u_1 < -\alpha_1 - x'\beta_1\ u_2 > -x'\beta_2\,|x), \\
P(y_1 &= 0, y_2 = 0|x) = P(u_1 < -x'\beta_1,\ u_2 < -x'\beta_2|x).
\end{aligned}
$$

The sum of these four probabilities becomes one only if

$$\alpha_1 \alpha_2 = 0 \iff \text{either one of } \alpha_1 \text{ and } \alpha_2 \text{ is } 0.$$

That is, the simultaneous system is not allowed, although a recursive system is.

The condition $\alpha_1 \alpha_2 = 0$ is called a *"coherency condition."* Despite their obvious mathematical necessity, this coherency condition is hard to take. For instance, if y_1 and y_2 are binary work-or-not decisions of a couple in a household, it is not clear why the influence should be unidirectional. Tamer (2003) defined "coherency" as the solvability of a SF equation system for the RF's, and "completeness" as the solution being unique. Lewbel (2007) derived necessary and sufficient conditions for two-equation SF systems with a binary y_1 to be coherent and complete. The main finding of Lewbel (2007) is that the system should be triangular (i.e., no simultaneity allowed), but the direction of causality can vary across individuals. For example, y_1 SF has $d_i y_{2i}$ on the right-hand side and y_2 SF has $(1 - d_i)y_{1i}$ on the right-hand side. Then, depending on d_i, the direction of the causality changes. Alternatively, Lewbel (2007) suggests to "nest" the two SF's in one larger behavior model that determines both; e.g., if both y_1 and y_2 are binary, then nest them in one multinomial choice as can be seen in Lewbel (2007, 1388–1389).

4.3.2   Coherent System in Censored Responses

Consider a simultaneous system in censored responses:

$$
\begin{aligned}
y_1^* &= \alpha_1 y_2 + x'\beta_1 + u_1, \quad y_2^* = \alpha_2 y_1 + x'\beta_2 + u_2, \\
y_1 &= \max(y_1^*, 0),\ y_2 = \max(y_2^*, 0), \quad (x', y_1, y_2) \text{ is observed.}
\end{aligned}
$$

To construct the likelihood function, we need to express (u_1, u_2) as an one-to-one function of (y_1, y_2). In the linear model with $y_1 = y_1^*$ and $y_2 = y_2^*$, we would have

$$\Gamma \cdot \begin{bmatrix} y_1 \\ y_2 \end{bmatrix} - \begin{bmatrix} \beta_1' \\ \beta_2' \end{bmatrix} x = \begin{bmatrix} u_1 \\ u_2 \end{bmatrix}$$

and this is solvable for (y_1, y_2) if Γ^{-1} exists.

In the simultaneous system in censored responses, the equation like this holds, but the form of Γ depends on the values of (y_1, y_2):

$$y_1 > 0, \ y_2 > 0 : \Gamma_{++} \equiv \begin{bmatrix} 1 & -\alpha_1 \\ -\alpha_2 & 1 \end{bmatrix},$$

$$y_1 > 0, \ y_2 = 0 : \Gamma_{+-} \equiv \begin{bmatrix} 1 & 0 \\ -\alpha_2 & 1 \end{bmatrix},$$

$$y_1 = 0, \ y_2 > 0 : \Gamma_{-+} \equiv \begin{bmatrix} 1 & -\alpha_1 \\ 0 & 1 \end{bmatrix},$$

$$y_1 = 0, \ y_2 = 0 : \Gamma_{--} \equiv \begin{bmatrix} 1 & 0 \\ 0 & 1 \end{bmatrix}.$$

Thus we get

$$\{\Gamma_{++}1[y_1 > 0, \ y_2 > 0] \ + \Gamma_{+-}1[y_1 > 0, \ y_2 = 0] \ + \Gamma_{-+}1[y_1 = 0,$$
$$y_2 > 0] + \Gamma_{--}1[y_1 = 0, \ y_2 = 0]\} \cdot \begin{bmatrix} y_1 \\ y_2 \end{bmatrix} - \begin{bmatrix} \beta_1' \\ \beta_2' \end{bmatrix} x = \begin{bmatrix} u_1 \\ u_2 \end{bmatrix}.$$

Gourieroux et al. (1980) showed that, for the invertibility of the term $\{\cdot\}$, it is necessary and sufficient (under some conditions) to have the same sign for all determinants $|\Gamma_{++}| = 1 - \alpha_1\alpha_2$, $|\Gamma_{+-}| = 1$, $|\Gamma_{-+}| = 1$, and $|\Gamma_{--}| = 1$. That is, the coherency condition for the simultaneous system in censored responses is

$$1 - \alpha_1\alpha_2 > 0.$$

This allows for simultaneous systems, differently from the bivariate binary SF's above. It is not clear, however, how to interpret the condition. One may hope MLE ignoring the coherency condition to be all right. But Van Soest et al. (1993) showed a counter example where MLE is inconsistent when the coherency condition is ignored. Nevertheless, if the resulting MLE meets the coherency condition, then ignoring it in the first place seems harmless.

Sometimes a coherency condition is satisfied with ease. For instance, consider (we omit exogenous regressors to simplify the presentation)

$c_i = u_{ci}$, where c_i is crop in period i,

$p_i = \gamma_d d_i + u_{pi}$, inverse demand function with p_i being price, $\gamma_d < 0$,

$h_i = \beta_c c_i + \beta_p p_i + u_{hi}$, where h_i is the desired harvest in period i, $\beta_p > 0$,

$s_i = \min(c_i, h_i)$ actual harvest (supply).

With p_i establishing the equilibrium $d_i = s_i$, we get

$$c_i = u_{ci}, \quad h_i = \beta_c c_i + \beta_p p_i + u_{hi}, \quad p_i = \gamma_d \min(c_i, h_i) + u_{pi},$$

with three endogenous variable c_i, h_i, p_i. There are two regimes (compared with four in the simultaneous system with censored variables): $c_i < h_i$ and $c_i \geq h_i$. We thus get

$$
\left(\begin{bmatrix} 1 & 0 & 0 \\ -\beta_c & 1 & -\beta_p \\ -\gamma_d & 0 & 1 \end{bmatrix} 1[c_i < h_i] + \begin{bmatrix} 1 & 0 & 0 \\ -\beta_c & 1 & -\beta_p \\ 0 & -\gamma_d & 1 \end{bmatrix} 1[c_i \geq h_i] \right)
$$

$$
\times \begin{bmatrix} c_i \\ h_i \\ p_i \end{bmatrix} = \begin{bmatrix} u_{ci} \\ u_{hi} \\ u_{pi} \end{bmatrix}.
$$

The determinants of the first and second matrices are 1 and $1 - \beta_p \gamma_d$, respectively. Hence, $1 - \beta_p \gamma_d > 0$ is the coherency condition for the model, which is satisfied because $\beta_p > 0$ and $\gamma_d < 0$.

4.3.3 Control Function Approach with a Censored Response

Whereas estimating simultaneous equations in latent continuous variables is fairly straightforward as will be shown shortly, estimating simultaneous equations in LDV's is not; we already noted the difficulty associated with coherency conditions. In the following, we review a control function approach in Blundell and Smith (1993) for a relatively simple model with two variables y_1 and y_2 where y_1 is censored from below at 0, y_2 is continuously distributed, and they are related in the following way:

$$
SF1 : y_1 = \max(0, \ \alpha_1 y_2 + x_1' \beta_1 + u_1), \qquad SF2 : y_2 = \alpha_2 y_1 + x_2' \beta_2 + u_2.
$$

The main difficulty is that y_1, not y_1^*, appears in SF2: substitute SF2 into SF1 to get

$$
y_1 = \max(0, \ \alpha_1 (\alpha_2 y_1 + x_2' \beta_2 + u_2) + x_1' \beta_1 + u_1)
$$

which cannot be solved for y_1 to get RF1. The coherency condition can be shown to be (again) $1 - \alpha_1 \alpha_2 > 0$. Had we had $y_2 = \alpha_2 y_1^* + x_2' \beta_2 + u_2$ as SF2, then we would have

$$
y_1^* = \alpha_1 (\alpha_2 y_1^* + x_2' \beta_2 + u_2) + x_1' \beta_1 + u_1,
$$

which could be easily solved for y_1^* so long as $\alpha_1 \alpha_2 \neq 1$ ($\Longleftarrow 1 - \alpha_1 \alpha_2 > 0$); this would in turn yield RF1 for $y_1 = \max(0, y_1^*)$.

Suppose $(u_1, u_2)'$ follows $N(0, \Omega)$ independently of (x_1, x_2). The main idea of the control function approach is in observing that, defining $y_3 \equiv y_2 - \alpha_2 y_1 \Leftrightarrow y_2 = y_3 + \alpha_2 y_1$,

$$
y_1 = \max(0, \ \alpha_1 y_3 + \alpha_1 \alpha_2 y_1 + x_1' \beta_1 + u_1), \quad y_3 = x_2' \beta_2 + u_2,
$$

and $u_1 = \lambda u_2 + \varepsilon$, where ε is independent of u_2 and λ is an unknown constant;

the latter can be arranged always thanks to the joint normality. Pretend that α_2 and u_2 are observed for a while. We then get

$$y_1 > 0 \implies y_1 = \frac{1}{1 - \alpha_1 \alpha_2}(\alpha_1 y_3 + x_1' \beta_1 + \rho u_2 + \varepsilon),$$

$$y_1 = 0 \implies \frac{1}{1 - \alpha_1 \alpha_2}(\alpha_1 y_3 + x_1' \beta_1 + \rho u_2 + \varepsilon) < 0,$$

where the coherency condition $1 - \alpha_1 \alpha_2 > 0$ is used. This can be written succinctly as

$$y_1 = \max(0, \ \alpha_1^* y_3 + x_1' \beta_1^* + \rho^* u_2 + \varepsilon^*), \quad \text{where}$$

$$\alpha_1^* \equiv \frac{\alpha_1}{1 - \alpha_1 \alpha_2}, \ \beta_1^* \equiv \frac{\beta_1}{1 - \alpha_1 \alpha_2}, \ \rho^* \equiv \frac{\rho}{1 - \alpha_1 \alpha_2}, \ \varepsilon^* \equiv \frac{\varepsilon}{1 - \alpha_1 \alpha_2}.$$

With u_2 as a regressor, y_3 is independent of ε, and the censored model with regressors y_3, x_1 and u_2 can be estimated for α_1^*, β_1^*, and ρ^* using tobit.

Turning to the question of getting α_2 and u_2, note that SF2 $y_2 = \alpha_2 y_1 + x_2' \beta_2 + u_2$ is a linear model, which can be estimated with IVE. Denoting the IVE for α_2 and β_2 as a_{N2} and b_{N2}, respectively, use

$$y_2 - a_{N2} y_1 \text{ for } y_3 \quad \text{and} \quad \hat{u}_2 \equiv y_2 - a_{N2} y_1 - x_2' b_{N2} \text{ for } u_2.$$

Assessing the effect of the estimation errors $a_{N2} - \alpha_2$ and $\hat{u}_2 - u_2$ on the second stage can be done following the approach for M-estimators with nuisance parameters where the M-estimator is the tobit. Alternatively, bootstrap may be used.

Blundell and Smith (1993) presented an empirical illustration ($N = 2539$) for married women in UK in 1981 where y_1 is work hours and y_2 is the household income other than the woman's. The main results are (SD in $(\cdot)$)

$(i) \ \alpha_1 = -0.093 \ (0.022), \ \alpha_2 = -0.466 \ (0.180);$

$(ii) \ \alpha_1 = -0.121 \ (0.011) \quad$ when tobit is applied ignoring the
$\qquad\qquad\qquad\qquad\qquad\qquad y_2$ endogeneity;

$(iii) \ \alpha_1 = -0.048 \ (0.029) \quad$ when y_1^* appears in SF2.

First, (i) shows that the coherency condition is easily satisfied. Second, (i) and (ii) show that ignoring endogeneity does not make much difference in this example. Third, (i) and (iii) show that it matters substantially whether the latent continuous variable or the observed LDV appears in the SF.

4.4   Simultaneous Systems in Latent Continuous Variables

4.4.1   Motivations and Justifications

Differently from the preceding methods, if the simultaneous system in LDV's are in terms of the latent continuous variables, then the estimation methods for linear simultaneous equations can be combined with single equation LDV-model estimation methods. The idea goes as follows. First, the SF's in latent variables y^*'s are solved for RF's. Second, each RF is a LDV model and thus the RF parameters can be estimated with ease. Third, the SF parameters are estimated using the restrictions involving the RF and SF parameters in the framework of minimum distance estimation (MDE).

The procedure has advantage that what is identified in the SF's can be presented lucidly. The procedure also has two disadvantages. One is ineffi-ciency resulting from the fact that the covariance matrix of the RF errors is not estimated; this disadvantage, however, disappears if the multivariate RF-like equations are estimated jointly. This disadvantage is not really a dis-advantage, because covariance matrices are not well estimated in practice. The other disadvantage is that we should be able to justify why the simulta-neous relations hold in the latent, not the actual observed limited, variables. This latent/actual variable issue depends on whether the LDV at hand was the actual choice variable the economic agents faced or only a limitation in observability to econometricians.

For instance, suppose y_1^* and y_2^* are the incomes of two spouses in a household, which can be simultaneously related. If y_1^* and y_2^* are observed as ordered categories (y_1 and y_2) for confidentiality, then the simultaneous system in y_1^* and y_2^* is the right model, because the choice variables for the economic agents are y_1^* and y_2^*, not y_1 and y_2 that are restrictions on the observability of y_1^* and y_2^* to econometricians. In this case, the effect of y_2^* on y_1^* is not fully identified; i.e., α_1 in $y_1^* = \alpha_1 y_2^* + x'\beta_1 + u_1$ is not fully identified, because only the ODR's y_1 and y_2 are observed.

Now, suppose y_1^* and y_2^* are "propensity" to work, but the labor law allows only one of no work, 50% part-time, or full-time work. In this case, we can think of three-category ODR's y_1 and y_2, and a simultaneous system in y_1 and y_2 would be more appropriate. But, as simultaneous relation between two binary responses is not allowed due to the coherency condition, simultaneous relations between ODR variables are not allowed either. Hence, despite that simultaneous relations in the actual LDV's are better suited in this case, still the simultaneous system in latent continuous variables is often used.

A little different from the above multi-step procedure using MDE is the following MLE. Recall the two equation recursive system

$$y_1 = x'\beta_1 + u_1 \quad \text{and} \quad y_2 = 1[\alpha_2 y_1 + x'\beta_2 + u_2 > 0]$$
$$(u_1, u_2) \sim N(0, \Omega) \text{ independently of } x.$$

One can proceed at least in two different ways for this model. One is the above procedure with MDE where the estimator uses the RF errors (v_1, v_2) marginal densities (not efficient) or joint density (efficient), and the other is the MLE where the likelihood funtion uses the conditional density of $u_2|u_1$ along with the marginal density of u_1. For the conditional density of $u_2|(x, u_1)$, $i.e.$ $u_2|u_1$, y_1's endogeneity does not matter any more as (x, u_1) gets fixed. STATA has the command `ivprobit` for the second approach; when y_2 is of tobit type, `ivtobit` can be used.

In the following, we examine a simultaneous system with three response variables (censored, binary, and ODR with R-categories):

$$\max(y_{1i}^*,0),\ \ 1[y_{2i}^* > 0],\ \ \sum_{r=1}^{R-1} 1[y_{3i}^* \geq \gamma_r],$$

drawing on Lee (1995) and Kimhi and Lee (1996). Although we do not have any particular real example in mind for this case, dealing with this will illustrate well how to handle simultaneous systems in latent continuous variables. For more generality, we will proceed with general H-many equations and resort to the case $H = 3$ when the specific details should be shown.

4.4.2   Individual RF-Based Approach with MDE

Consider simultaneous equations in y_j^*'s:

$$y_{hi}^* = \sum_{m=1,m\neq h}^{H} \alpha_{hm} y_{mi}^* + x_{hi}' \beta_h + u_{hi},\ \ h = 1,...,H,$$

$$(u_{1i},...,u_{Hi})' \sim N(0,\Omega) \text{ independently of } x_i,$$

where x_h has dimension $k_h \times 1$ and x_i of dimension $K \times 1$ consists of all elements in x_{hi}'s. Let the $K \times k_h$ selection matrix be S_h consisting of 0's and 1's such that $x_{hi}' = x_i' S_h$. Using S_h, rewrite the SF's as

$$y_{hi}^* = \sum_{m=1,m\neq h}^{H} \alpha_{hm} y_{mi}^* + x_i' S_h \beta_h + u_{hi} \quad \forall h.$$

Denoting the hth RF as $y_{hi}^* = x_i' \eta_h + v_{hi},\ h = 1,...,H$, insert them into the hth SF to get

$$x_i' \eta_h + v_{hi} = x_i' \left(\sum_{m=1,m\neq h}^{H} \alpha_{hm} \eta_m + S_h \beta_h \right) + \left(\sum_{m=1,m\neq h}^{H} \alpha_{hm} v_{mi} + u_{hi} \right) \quad \forall h.$$

Pre-multiply both sides by x_i and take expectation to get rid of the error terms. Assuming that $E(x_i x_i')$ is of full rank, the resulting equation is equivalent to

$$\eta_h = \sum_{m=1,m\neq h}^{H} \alpha_{hm} \eta_m + S_h \beta_h, \quad \forall h.$$

This equation links the RF parameter η_h's to the SF parameter α_{hm}'s and β_h's.

Recall the $H = 3$ case with y_1, y_2, y_3 being censored, binary, and ODR, respectively. Whereas η_1 is fully identified, as for η_2 and η_3, only η_2/σ_2 and

η_3/σ_3 are identified where $\sigma_2 \equiv SD(v_2)$ and $\sigma_3 \equiv SD(v_3)$. Rewrite the preceding display for η_1, η_2/σ_2 and η_3/σ_3:

$$\eta_1 = \alpha_{12}\sigma_2\frac{\eta_2}{\sigma_2} + \alpha_{13}\sigma_3\frac{\eta_3}{\sigma_3} + S_1\beta_1,$$

$$\frac{\eta_2}{\sigma_2} = \frac{\alpha_{21}}{\sigma_2}\eta_1 + \alpha_{23}\frac{\sigma_3}{\sigma_2}\frac{\eta_3}{\sigma_3} + S_2\frac{\beta_2}{\sigma_2},$$

$$\frac{\eta_3}{\sigma_3} = \frac{\alpha_{31}}{\sigma_3}\eta_1 + \alpha_{32}\frac{\sigma_2}{\sigma_3}\frac{\eta_2}{\sigma_2} + S_3\frac{\beta_3}{\sigma_3}.$$

Define

$$\delta_1 \equiv \eta_1, \; \delta_2 \equiv \frac{\eta_2}{\sigma_2}, \; \delta_3 \equiv \frac{\eta_3}{\sigma_3} \quad \text{(identified RF parameters)}$$

$$\mu_{12} \equiv \alpha_{12}\sigma_2, \; \mu_{13} \equiv \alpha_{13}\sigma_3 \quad \text{(identified endogenous SF parameters}$$
$$\text{for SF 1)}$$

$$\mu_{21} \equiv \frac{\alpha_{21}}{\sigma_2}, \; \mu_{23} \equiv \alpha_{23}\frac{\sigma_3}{\sigma_2}, \quad \text{(identified endogenous SF parameters}$$
$$\text{for SF 2)}$$

$$\mu_{31} \equiv \frac{\alpha_{31}}{\sigma_3}, \; \mu_{32} \equiv \alpha_{32}\frac{\sigma_2}{\sigma_3}, \quad \text{(identified endogenous SF parameters}$$
$$\text{for SF 3)}$$

$$\bar\beta_1 \equiv \beta_1, \; \bar\beta_2 \equiv \frac{\beta_2}{\sigma_2}, \; \bar\beta_3 \equiv \frac{\beta_3}{\sigma_3}, \quad \text{(identified exogenous SF parameters)}$$

to get

$$\delta_h = \sum_{m=1, m\neq h}^{H} \mu_{hm}\delta_m + S_h\bar\beta_h, \quad \forall h.$$

One thing we ignored is that the first threshold is absorbed into the intercept in binary response and ODR; consequently, the intercepts in β_1, β_2, and β_3 are not identified in the second and third equations. This usually does not matter, because intercepts are not of interest per se. But there are cases it matters, including cases of regressor-dependent thresholds; see Lee and Kimhi (2005) for more on this.

Consider the LSE of

$$\underset{K\times 1}{\delta_h} \quad \text{on} \quad D_h \equiv \left(\underset{K\times 1}{\delta_1}, ..., \delta_{h-1}, \delta_{h+1}, ..., \delta_H, \underset{K\times k_h}{S_h} \right)$$

to estimate $\mu_{h1}, ..., \mu_{h,h-1}, \mu_{h,h+1}, ..., \mu_{hH}$ and $\bar\beta_h$. For the LSE, $D_h' D_h$ should be of full rank, and since the dimension of D_h is $K \times (H - 1 + k_h)$, the *rank condition* is

$$rank(D_h) = H - 1 + k_h.$$

The *order condition*, which is a necessary condition for the rank condition, is that the column dimension of D_h should be equal to or smaller than the row dimension:

$$H - 1 + k_h \leq K \iff H - 1 \leq K - k_h$$

$$\iff \quad \text{\# included endogenous variables}$$

$$\leq \text{\# excluded exogenous variables.}$$

This settles the identification issue.

With the identification issue settled, we turn to estimation with MDE. Let d_h denote an estimator for δ_h; for the example above, d_1 is tobit, d_2 is probit, and d_3 is ordered probit. Then $\delta_h = \sum_{m=1,m\neq h}^{H} \mu_{hm}\delta_m + S_h\bar{\beta}_h$, $h = 1,2,3$, can be written as

$$d_1 = (d_2, d_3, S_1)(\mu_{12}, \mu_{13}, \bar{\beta}_1')' + w_1 = D_1\gamma_1 + w_1,$$

$$d_2 = (d_1, d_3, S_2)(\mu_{21}, \mu_{23}, \bar{\beta}_2')' + w_2 = D_2\gamma_2 + w_2,$$

$$d_3 = (d_1, d_2, S_3)(\mu_{31}, \mu_{32}, \bar{\beta}_3')' + w_3 = D_3\gamma_3 + w_3,$$

where

$$w_1 \equiv (d_1 - \delta_1) - \mu_{12}(d_2 - \delta_2) - \mu_{13}(d_3 - \delta_3), \quad \gamma_1 \equiv (\mu_{12}, \mu_{13}, \bar{\beta}_1')',$$

$$w_2 \equiv (d_2 - \delta_2) - \mu_{21}(d_1 - \delta_1) - \mu_{23}(d_3 - \delta_3), \quad \gamma_2 \equiv (\mu_{21}, \mu_{23}, \bar{\beta}_2')',$$

$$w_3 \equiv (d_3 - \delta_3) - \mu_{31}(d_1 - \delta_1) - \mu_{32}(d_2 - \delta_2), \quad \gamma_3 \equiv (\mu_{31}, \mu_{32}, \bar{\beta}_3')'.$$

Stack these equations to get

$$d = D\gamma + w \quad \text{where}$$

$$d \equiv \begin{bmatrix} d_1 \\ d_2 \\ d_3 \end{bmatrix}, \quad D \equiv \begin{bmatrix} D_1 & 0 & 0 \\ 0 & D_2 & 0 \\ 0 & 0 & D_3 \end{bmatrix}, \quad \gamma \equiv \begin{bmatrix} \gamma_1 \\ \gamma_2 \\ \gamma_3 \end{bmatrix},$$

$$w \equiv \begin{bmatrix} w_1 \\ w_2 \\ w_3 \end{bmatrix}.$$

If the asymptotic variance for w is C with a consistent estimator C_N, then the MDE for γ is

$$g_N \equiv (D'C_N^{-1}D)^{-1}D'C_N^{-1}d \quad \text{and} \quad \sqrt{N}(g_N - \gamma) \rightsquigarrow N(0, (D'C^{-1}D)^{-1});$$

the MDE is reminiscent of GLS. In the following, we show how to get C_N.

A simpler alternative to $(D'C_N^{-1}D)^{-1}D'C_N^{-1}d$ is its LSE variety $(D'D)^{-1}D'd$, which is the same as the stacked version of each equation estimators. The asymptotic variance of $(D'D)^{-1}D'd$ is more complicated than $(D'C^{-1}D)^{-1}$, but one may use nonparametric bootstrap for asymptotic inferences involving $(D'D)^{-1}D'd$.

To do the "GLS," first do LSE for each $d_h = D_h\gamma_h + w_h$ separately, which gives a $\sqrt{N}$-consistent estimator, say $\hat{g}_h$. From $\hat{g}_h$'s, pull out the estimators $\hat{\mu}_{hm}$ for μ_{hm}. Let $\hat{q}_{hi}$ be a consistent estimator for the influence function q_{hi} of

$\sqrt{N}(d_h - \delta_h)$. For $\sqrt{N}(d_2 - \delta_2)$, we have $q_{2i} = E^{-1}(s_{2i}s'_{2i})s_{2i}$, where s_{2i} is the probit score function. But for $\sqrt{N}(d_1 - \delta_1)$, since there is a nuisance parameter σ_1 while δ_1 is the regression coefficients of RF1, we need an effective influence function for δ_1. Let the tobit score function be $(s'_{\delta i}, s'_{\sigma i})'$ where $s_{\delta i}$ is for the regression coefficients and $s_{\sigma i}$ is for σ_1. Then the effective score function and the desired influence function for $\sqrt{N}(d_1 - \delta_1)$ are, respectively,

$$s_{1i} = s_{\delta i} - E(s_{\delta i}s'_{\sigma i})E^{-1}(s_{\sigma i}s'_{\sigma i})s_{\sigma i} \quad \text{and} \quad q_{1i} = E^{-1}(s_{1i}s'_{1i})s_{1i}.$$

Doing analogously, let the ordered probit score function be $s_{oi} = (s'_{\zeta i}, s'_{\tau i})'$ where $s_{\zeta i}$ is for the regression coefficients and $s_{\tau i}$ is for the thresholds. Then the effective score function and the desired influence function for $\sqrt{N}(d_3 - \delta_3)$ are, respectively,

$$s_{3i} = s_{\zeta i} - E(s_{\zeta i}s'_{\tau i})E^{-1}(s_{\tau i}s'_{\tau i})s_{\tau i} \quad \text{and} \quad q_{3i} = E^{-1}(s_{3i}s'_{3i})s_{3i}.$$

Now, we have

$$\sqrt{N}w_1 = \frac{1}{\sqrt{N}}\sum_i Q_{1i} + o_p(1), \quad \text{where } Q_{1i} \equiv q_{1i} - \hat{\mu}_{12}q_{2i} - \hat{\mu}_{13}q_{3i},$$

$$\sqrt{N}w_2 = \frac{1}{\sqrt{N}}\sum_i Q_{2i} + o_p(1), \quad \text{where } Q_{2i} \equiv q_{2i} - \hat{\mu}_{21}q_{1i} - \hat{\mu}_{23}q_{3i},$$

$$\sqrt{N}w_3 = \frac{1}{\sqrt{N}}\sum_i Q_{3i} + o_p(1), \quad \text{where } Q_{3i} \equiv q_{3i} - \hat{\mu}_{31}q_{1i} - \hat{\mu}_{32}q_{2i}.$$

Finally, we get

$$C_N = \frac{1}{N}\sum_i Q_i Q'_i \quad \text{where } Q_i \equiv (Q'_{1i}, Q'_{2i}, Q'_{3i})'.$$

4.4.3   An Empirical Example

As an empirical example, Lee and Kimhi (2005) analyzed 1995 farm household data from Israel ($N = 1337$). The data pertains to the joint time allocation decisions of farm couples, including four endogenous ODR's in four categories. To simplify the presentation however, we show the simultaneous system in four binary responses (work or not) instead of ODR's:

	y_1: male farm work	y_2: male market work	y_3: female farm	y_4: female market
$y_j = 1$:	83%	48%	48%	58%

The regressors are as follows (Lee and Kimhi (2005) in fact used more regressors than listed here): with sample average in $(\cdot)$, age: husband age divided by 10 (5.08)

age2: square of husband age divided by 10 (27.03)

cap: ln(farm capital stock+1) (4.43)

catt: dummy for raising cattle or other livestock (0.08)

ed1h: dummy for male's high school completion (0.61)

ed1c: dummy for male's education more than high school (0.13)

ed2h: dummy for female's high school completion (0.60)

ed2c: dummy for female's education more than high school (0.14)

land: ln(landholdings+1) [in dunams (0.23 acre)] (3.26)

nkid: number of children (age under 15) in the household (1.55)

nadu: number of adults (age above 21) in the household (3.17)

is1: dummy for male born in Israel (0.52)

is2: dummy for female born in Israel (0.56)

In estimating the SF's with MDE, the order condition requires at least three excluded exogenous variables from each SF. There are two justifications for variable exclusions. First, farm variables (cap, catt, and land) are unlikely to affect the market variables y_2 and y_4 directly. Second, education and ethnicity are unlikely to be directly relevant for the spouse's labor supply. These considerations suggest the following list of excluded variables:

y_1 :   ed2h, ed2c, is2 y_2 :   ed2h, ed2c, is2, cap, catt, land
y_3 :   ed1h, ed1c, is1 y_4 :   ed1h, ed1c, is1, cap, catt, land

This shows that the order conditions for y_1 and y_3 are just enough, while those for y_2 and y_4 are relatively plentiful. Table 4 shows the result for the y_2 and y_4 (market work) SF's.

Examining the endogenous SF coefficients, there is a significant negative effect of male's farm work on the male's market work. Note that -0.494 is for $\mu_{21} = \alpha_{21}(\sigma_1/\sigma_2)$, and as such it is hard to interpret. But if $\sigma_1 = \sigma_2$, then $-0.494 = \alpha_{21}$. Also, somewhat surprisingly, the female's market work increases the male's market work, and the effect is almost significant with the magnitude 0.647 which is for $\alpha_{24}(\sigma_4/\sigma_2)$. As for the female, there is significant negative effect of the female's farm work on the female's market work with the magnitude -1.681 that is for $\mu_{43} = \alpha_{43}(\sigma_3/\sigma_4)$. If both $\sigma_1 = \sigma_2$ and $\sigma_3 = \sigma_4$ hold, then μ_{21} and μ_{43} are comparable, and the magnitude is about three times greater for the females. Using ODR instead of binary responses, Lee and

Table 4: Simultaneous System for Binary Responses

	y_2 SF est.(male mart.)	t-value	y_4 SF est.(female mart.)	t-value
y_1	−0.494	−3.417	−0.004	−0.007
y_2			−0.402	−0.391
y_3	0.135	0.242	−1.681	−2.033
y_4	0.647	1.811		
1	1.004	0.768	−3.639	−1.988
age	−0.113	−0.213	1.600	1.786
age2	0.002	0.042	−0.163	−1.699
ed1h	0.035	0.365		
ed1c	0.267	1.522		
ed2h			0.125	0.412
ed2c			0.200	0.509
nkid	0.001	0.024	0.020	0.425
nadu	0.021	0.511	−0.083	−1.354
is1	−0.171	−1.525		
is2			0.234	1.545

Kimhi (2005) showed ways to identify α_{jk}'s from the threshold information in the ODR's.

Turning to the exogenous SF coefficients, recall that what is identified in the y_2-SF (y_4-SF) is β_2/σ_2 (β_4/σ_4). Differently from the endogenous SF parameters μ_{jm}'s, there is no possibility to get rid of the unknown scale factors for the exogenous SF coefficients. Age matters much for female market work, whereas it hardly does for males. College education seems to be advantageous only for the male market work, and the number of adults seems to decrease the female market work. Being born in Israel may work in the opposite directions for the male and female.

5 Panel-Data Binary-Response Models

Many panel data models have LDV's. With each individual having T many periods (i.e., equations), they are also multivariate LDV models. Drawing on Lee (2002), this section introduces some binary-response panel data models, as binary response is basic for LDV's. For more on panel data LDV models, see Lee (2002), Hsiao (2003), Baltagi (2005), and the references therein.

5.1 Panel Conditional Logit

Recall the linear panel data model with a time-constant regressor vector $\tilde{c}_i$, a time-variant regressor vector x_{it}, a time-constant error δ_i, a time-varying error u_{it}, and the continuous response denoted now as y_{it}^*:

$$
\begin{aligned}
y_{it}^* &= \tau_t + \tilde{c}_i'\tilde{\alpha} + x_{it}'\beta + \delta_i + u_{it} \quad (\tau_t \text{ is time-varying intercept}) \\
&= \tau_t + \tilde{w}_{it}'\tilde{\gamma} + v_{it}, \quad \text{where } \tilde{w}_{it} = (\tilde{c}_i', x_{it}')', \quad \tilde{\gamma} = (\tilde{\alpha}', \beta')', \\
v_{it} &= \delta_i + u_{it}.
\end{aligned}
$$

Also recall

$$
c_i \equiv \begin{bmatrix} 1 \\ \tilde{c}_i \end{bmatrix}, \quad w_{it} \equiv \begin{bmatrix} 1 \\ \tilde{c}_i \\ x_{it} \end{bmatrix}, \quad \alpha \equiv \begin{bmatrix} \tau_1 \\ \tilde{\alpha} \end{bmatrix}, \quad \text{and} \quad \gamma \equiv \begin{bmatrix} \alpha \\ \beta \end{bmatrix}.
$$

What is identified in the linear model is

$$
\tau_1, \tau_2 - \tau_1, ..., \tau_T - \tau_1, \gamma \iff \tau_1, ..., \tau_T, \gamma.
$$

Here, only the intercept is allowed to vary over time whereas the slopes are fixed. In panel binary responses, instead of y_{it}^*, we observe $y_{it} = 1[y_{it}^* > 0]$ and assume

$$
\{(y_{it}, \tilde{w}_{it}')'\}_{t=1}^T, \quad i = 1, ..., N, \quad \text{are iid across } i.
$$

5.1.1 Two Periods with Time-Varying Intercept

To simplify exposition, set $T = 2$ for a while. Assume

u_{it}/σ is logistic independently of $(\delta_i, \tilde{c}_i, x_{i1}, x_{i2})$, *iid across i and t*

$$
\implies P(y_{it} = 1 | \delta_i, \tilde{c}_i, x_{i1}, x_{i2})
$$

$$
= P(u_{it} > -\tau_t - \tilde{c}_i'\tilde{\alpha} - x_{it}'\beta - \delta_i | \delta_i, \tilde{c}_i, x_{i1}, x_{i2})
$$

$$
= \frac{\exp(\tau_t/\sigma + \tilde{c}_i'\tilde{\alpha}/\sigma + x_{it}'\beta/\sigma + \delta_i/\sigma)}{1 + \exp(\tau_t/\sigma + \tilde{c}_i'\tilde{\alpha}/\sigma + x_{it}'\beta/\sigma + \delta_i/\sigma)}
$$

for an unknown positive constant σ that is not a function of t. With this understood, set $\sigma = 1$ from now on to simplify notations.

Observe, omitting i and "$|\delta_i, \tilde{c}_i, x_{i1}, x_{i2}$,"

$$
P(y_1 = 0, y_2 = 1 | y_1 + y_2 = 1)
$$

$$
= P\{y_1 = 0, y_2 = 1 \mid (y_1 = 0, y_2 = 1) \text{ or } (y_1 = 1, y_2 = 0)\}
$$

$$
= \frac{P(y_1 = 0)P(y_2 = 1)}{P(y_1 = 0)P(y_2 = 1) + P(y_1 = 1)P(y_2 = 0)}
$$

$$
= \frac{\exp(\tau_2 + \tilde{c}'\tilde{\alpha} + x_2'\beta + \delta)}{\exp(\tau_2 + \tilde{c}'\tilde{\alpha} + x_2'\beta + \delta) + \exp(\tau_1 + \tilde{c}'\tilde{\alpha} + x_1'\beta + \delta)}
$$

because all three products of probabilities share the same denominator

$$\{1 + \exp(\tau_1 + \tilde{c}_i'\tilde{\alpha} + x_{i1}'\beta + \delta_i)\} \cdot \{1 + \exp(\tau_2 + \tilde{c}_i'\tilde{\alpha} + x_{i2}'\beta + \delta_i)\}.$$

Divide through by $\exp(\tau_1 + \tilde{c}'\tilde{\alpha} + x_1'\beta + \delta)$ to obtain

$$P(y_1 \;=\; 0, y_2 = 1 | y_1 + y_2 = 1) = \frac{\exp(\Delta\tau + \Delta x'\beta)}{1 + \exp(\Delta\tau + \Delta x'\beta)} \quad \text{where}$$

$$\Delta\tau \;\equiv\; \tau_2 - \tau_1 \quad \text{and} \quad \Delta x_i \equiv x_{i2} - x_{i1}.$$

Subtract this from 1 to get also

$$P(y_1 = 1, y_2 = 0 | y_1 + y_2 = 1) = \frac{1}{1 + \exp(\Delta\tau + \Delta x'\beta)}.$$

The iid assumption across t is essential for this derivation. Because $\tilde{c}_i$ is removed along with δ_i, redefine δ_i as $\tilde{c}_i'\tilde{\alpha} + \delta_i$ to ignore $\tilde{c}_i'\tilde{\alpha}$ from now on unless otherwise necessary.

The probabilities conditional on $y_1 + y_2 = 1$ are free of δ_i, and we can estimate $\Delta\tau$ and β by maximizing the likelihood function conditional on $y_1 + y_2 = 1$. Define

$$d_i = 1 \quad \text{if } y_{i1} + y_{i2} = 1, \quad \text{and 0 otherwise.}$$

The panel conditional-logit log-likelihood function to maximize wrt $\Delta\tau$ and β corresponding to the regressors 1 and Δx, respectively, is

$$\sum_i d_i \left[y_{i1} \ln \left\{ \frac{1}{1 + \exp(\Delta\tau + \Delta x_i'\beta)} \right\} + y_{i2} \ln \left\{ \frac{\exp(\Delta\tau + \Delta x_i'\beta)}{1 + \exp(\Delta\tau + \Delta x_i'\beta)} \right\} \right].$$

The log-likelihood function is almost the same as that for the cross-section logit.

With $\Delta y_i \equiv y_{i2} - y_{i1}$,

$$d_i y_{i1} \;=\; 1 \iff y_{i1} = 1 \text{ and } y_{i2} = 0 \iff \Delta y_i = -1$$
$$d_i y_{i2} \;=\; 1 \iff y_{i1} = 0 \text{ and } y_{i2} = 1 \iff \Delta y_i = 1.$$

This shows that the log-likelihood function depends on $(x_{i1}, x_{i2}, y_{i1}, y_{i2})$ only through the first-differences Δx_i and Δy_i; conditional logit is an "error-differencing" (or "model-differencing") type estimator, analogous to the first-differencing estimator for linear panel data models. The idea is that $\sum_t y_{it}$ is a "sufficient statistic" for δ_i: given x_{i1} and x_{i2}, the likelihood of (y_{i1}, y_{i2}) does not depend on δ_i when conditioned on $\sum_t y_{it}$—more on sufficient statistic later. This idea of conditioning on a sufficient statistic appears in Anderson (1970) and Chamberlain (1980). The log-likelihood can be maximized and the estimator's asymptotic variance can be estimated in the usual MLE way, although no efficiency claim can be made for the MLE.

5.1.2   Three or More Periods

Suppose $T = 3$. Assume $\tau_t = \tau$ for all t to simplify exposition—it will be shown below how to relax this restriction. Define

$$d_{i1} \; = \; 1 \quad \text{if } \sum_t y_{it} = 1, \quad \text{and 0 otherwise,}$$

$$d_{i2} \; = \; 1 \quad \text{if } \sum_t y_{it} = 2, \quad \text{and 0 otherwise,}$$

$$x_i \; \equiv \; (x_{i1}, x_{i2}, x_{i3}) \underset{3 \times 1}{\Longrightarrow} x_i'\beta = (x_{i1}'\beta, x_{i2}'\beta, x_{i3}'\beta)'.$$

Given $d_{i1} = 1$, there are three possibilities for $y_i' \equiv (y_{i1}, y_{i2}, y_{i3})$:

$$(1,0,0), \quad (0,1,0), \quad (0,0,1).$$

Doing analogously to the derivation of $P(y_1 = 0, y_2 = 1 | y_1 + y_2 = 1)$, the probability of observing a particular y_i given x_i, δ_i, and $d_{i1} = 1$ is

$$\frac{\exp(y_i' \cdot x_i'\beta)}{\exp\{(1,0,0)x_i'\beta\} + \exp\{(0,1,0)x_i'\beta\} + \exp\{(0,0,1)x_i'\beta\}}$$

$$= \frac{\exp(y_i' \cdot x_i'\beta)}{\exp(x_{i1}'\beta) + \exp(x_{i2}'\beta) + \exp(x_{i3}'\beta)}.$$

Doing analogously, the conditional probability of observing a particular y_i given x_i, δ_i, and $d_{i2} = 1$ is

$$\frac{\exp(y_i' \cdot x_i'\beta)}{\exp\{(1,1,0) \cdot x_i'\beta\} + \exp\{(1,0,1) \cdot x_i'\beta\} + \exp\{(0,1,1) \cdot x_i'\beta\}}$$

$$= \frac{\exp(y_i' \cdot x_i'\beta)}{\exp((x_{i1}+x_{i2})'\beta) + \exp((x_{i1}+x_{i3})'\beta) + \exp((x_{i2}+x_{i3})'\beta)}.$$

Hence, the three-period conditional-logit log-likelihood function to maximize for β is

$$\sum_i \left[d_{i1} \ln \left\{ \frac{\exp(y_i' \cdot x_i'\beta)}{\exp(x_{i1}'\beta) + \exp(x_{i2}'\beta) + \exp(x_{i3}'\beta)} \right\} \right.$$

$$\left. + d_{i2} \ln \left\{ \frac{\exp(y_i' \cdot x_i'\beta)}{\exp((x_{i1}+x_{i2})'\beta) + \exp((x_{i1}+x_{i3})'\beta) + \exp((x_{i2}+x_{i3})'\beta)} \right\} \right].$$

For a general $T \geq 3$, the panel conditional-logit log-likelihood function is

$$\sum_i \ln \left\{ \frac{\exp(y_i' \cdot x_i'\beta)}{\sum_{\lambda \in G_i} \exp(\lambda' \cdot x_i'\beta)} \right\} \quad \text{where}$$

$$G_i \equiv \left\{ \lambda \equiv (\lambda_1, ..., \lambda_T)' | \lambda_t = 0, 1 \text{ and } \sum_t \lambda_t = \sum_t y_{it} \right\}.$$

Here each i is classified depending on $\sum_t y_{it}$. For example, if $\sum_t y_{it} = 1$, then we can think of all possible sequences of $\{\lambda_t\}$ such that $\sum_t \lambda_t = 1$. The denominator $\sum_{\lambda \in G_i} \exp(\lambda' \cdot x_i'\beta)$ is nothing but the sum of the possibilities corresponding to all such $\{\lambda_t\}$ sequences. In STATA, this estimator can be implemented by the command xtlogit with the option fe.

Conditional logit has the main advantage of allowing δ_i to be related to x_{it} in an arbitrary fashion because δ_i is removed. But the dynamics allowed by conditional logit is restricted in a couple of ways. First, $u_{i1}, ..., u_{iT}$ are iid, and thus $v_{it} = \delta_i + u_{it}, t = 1, ..., T$, are allowed to be related only through δ_i; the serial correlation of v_{it} does not change at all over time. Second, u_{it} is independent of $(\delta_i, x_{i1}, ..., x_{iT})$, not just of (δ_i, x_{it}), nor just of $(\delta_i, x_{i1}, ..., x_{it})$; this rules out economic agents who adjust the future x_{it} in view of the past u_{is}, $s < t$. Third, the lagged response $y_{i,t-1}$ is not allowed in x_{it}: if $y_{i,t-1}$ is in x_{it}, then u_{it} cannot be independent of $x_{i1}, ..., x_{iT}$, because y_{it} including u_{it} appears in $x_{i,t+1}$.

Regarding time-varying parameters, compare three cases of the $T \times 1$ vector $x_i'\beta$: with τ and $\tilde{\beta}$ denoting the intercept and slope, respectively,

$$
\begin{bmatrix} x_{i1}'\beta \\ \vdots \\ x_{iT}'\beta \end{bmatrix} = \begin{bmatrix} x_{i1}'(\tau, \tilde{\beta}') \\ \vdots \\ x_{iT}'(\tau, \tilde{\beta}') \end{bmatrix}, \quad \begin{bmatrix} x_{i1}'\beta_1 \\ \vdots \\ x_{iT}'\beta_T \end{bmatrix} = \begin{bmatrix} x_{i1}'(\tau_1, \tilde{\beta}') \\ \vdots \\ x_{iT}'(\tau_T, \tilde{\beta}') \end{bmatrix},
$$

$$
\begin{bmatrix} x_{i1}'\beta_1 \\ \vdots \\ x_{iT}'\beta_T \end{bmatrix} = \begin{bmatrix} x_{i1}'(\tau_1, \tilde{\beta}_1') \\ \vdots \\ x_{iT}'(\tau_T, \tilde{\beta}_T') \end{bmatrix}.
$$

These correspond to, respectively, (i) time-constant β, (ii) time-varying intercept τ_t with time-constant slope $\tilde{\beta}$, and (iii) fully time-varying β_t. Let $x_{it} = (1, \tilde{x}_{it}')'$ to rewrite $x_{it}'\beta_t - x_{i1}'\beta_1$ for the three cases as, respectively,

$$(\tilde{x}_{it} - \tilde{x}_{i1})'\tilde{\beta}, \quad \tau_t - \tau_1 + (\tilde{x}_{it} - \tilde{x}_{i1})'\tilde{\beta}, \quad \text{and} \quad \tau_t - \tau_1 + \tilde{x}_{it}'\tilde{\beta}_t - \tilde{x}_{i1}'\tilde{\beta}_1.$$

Hence, only $\tilde{\beta}$ is identified in (i); $\tau_t - \tau_1 \ \forall t = 2, ..., T$ and $\tilde{\beta}$ are identified in (ii); and $\tau_t - \tau_1 \ \forall t = 2, ..., T$ and $\tilde{\beta}_t \ \forall t = 1, ..., T$ are identified in (iii).

5.1.3   Digression on Sufficiency

Given an iid data $z_1, ..., z_N$ from a density $f(z; \theta)$, a statistic (or a vector of statistics) $\tau_N \equiv \tau(z_1, ..., z_N)$ is said to be a *sufficient statistic* for θ if the distribution of $(z_1, ..., z_N)|\tau_N$ does not depend on θ. In panel conditional logit, recall that the likelihood of (y_{i1}, y_{i2}) does not depend on δ_i when conditioned on $\sum_t y_{it}$; i.e., $\sum_t y_{it}$ is a sufficient statistic for δ_i.

The condition that $(z_1, ..., z_N)|\tau_N$ does not depend on θ is known to be equivalent to "*factorization theorem*": for some non-negative functions g and h

$$\prod_i f(z_i; \theta) = g\{\tau(z_1, ..., z_N), \theta\} \cdot h(z_1, ..., z_N)$$

where g depends on $(z_1, ..., z_N)$ only through τ_N and h does not depend on θ. For instance, if $z_i \sim N(\theta, 1)$, then

$$
\begin{aligned}
\prod_i f(z_i; \theta) &= \left(\frac{1}{\sqrt{2\pi}}\right)^N \exp\left\{-\frac{1}{2}\sum_i (z_i - \theta)^2\right\} \\
&= \left(\frac{1}{\sqrt{2\pi}}\right)^N \exp\left\{-\frac{1}{2}\sum_i (z_i^2 - 2z_i\theta + \theta^2)\right\} \\
&= \exp\left(\theta\sum_i z_i - \frac{N}{2}\theta^2\right) \cdot \left(\frac{1}{\sqrt{2\pi}}\right)^N \exp\left(-\frac{1}{2}\sum_i z_i^2\right):
\end{aligned}
$$

the first term depends on $(z_1, ..., z_N)$ only through $\sum_i z_i$, and thus $\sum_i z_i$ (i.e., $\bar{z} = N^{-1}\sum_i z_i$) is sufficient for θ.

The factorization theorem shows that, as far as θ goes, the information provided by $(z_1, ..., z_N)$ is fully contained in τ_N. That is, once we know the value of τ_N, nothing more can be learned about θ using the data $(z_1, ..., z_N)$. Put it differently, we can "summarize" the data into τ_N for θ without losing anything. Certainly then, we would desire a minimal set of sufficient statistics to summarize the data: τ_N is *minimal sufficient*, if τ_N can be written as a function of $\zeta(z_1, ..., z_N)$ for any other sufficient statistic $\zeta(z_1, ..., z_N)$. Since the function can be many-to-one, τ_N is "coarser" than $\zeta(z_1, ..., z_N)$, and τ_N is minimal in this sense; τ_N is equivalent to $\zeta(z_1, ..., z_N)$ if the function is one-to-one and onto.

Suppose that τ_N is minimal-sufficient for θ and that τ_N is also a good estimator for θ. Consider a function $\alpha(\theta)$ of θ (e.g., $\alpha(\theta) = \theta^2$), and our interest is on $\alpha(\theta)$, not necessarily on θ. One immediate way to proceed is getting the "plug-in" estimator $\alpha(\tau_N)$. But is this a good (or an optimal) estimator for $\alpha(\theta)$? In the following, we show that the plug-in estimator is not necessarily "optimal," although it is still likely to be a function of τ_N because τ_N contains all the information about θ.

The "*Rao-Blackwell theorem*" states that, given an estimator a_N for a function $\alpha = \alpha(\theta)$ and a sufficient statistic τ_N for θ,

$$a_N^* \equiv E(a_N | \tau_N)$$

is better than a_N in the sense $E(a_N^* - \alpha)^2 \leq E(a_N - \alpha)^2$. This follows from

$$
\begin{aligned}
E(a_N - \alpha)^2 &= E(a_N - a_N^* + a_N^* - \alpha)^2 \\
&= E\{(a_N - a_N^*)^2 + 2(a_N - a_N^*)(a_N^* - \alpha) + (a_N^* - \alpha)^2\}
\end{aligned}
$$

$$= E(a_N - a_N^*)^2 + E(a_N^* - \alpha)^2$$

for $\quad E\{(a_N - a_N^*)(a_N^* - \alpha)\} = 0$ as shown below.

Furthermore, if a_N is unbiased for α (i.e., $E(a_N) = \alpha$), then a_N^* is also unbiased for α, because $E(a_N^*) = E\{E(a_N|\tau_N)\} = E(a_N) = \alpha$.

To show that the cross-product term is 0, observe

$$E\{(a_N - a_N^*)(a_N^* - \alpha)\} = E[\ E\{(a_N - a_N^*)(a_N^* - \alpha)|\tau_N\}\ ]$$
$$= E\{(a_N^* - \alpha) \cdot E(a_N - a_N^*|\tau_N)\} \text{ (for } a_N^* - \alpha \text{ is a function of } \tau_N)$$
$$= 0 \text{ (for } E(a_N - a_N^*|\tau_N) = E(a_N|\tau_N) - a_N^* = 0).$$

The theorem suggests finding a crude (unbiased) estimator a_N first, and then "optimizing" it by conditioning a_N on a minimal-sufficient statistic. This theme will reappear later when U-statistic is discussed.

5.2   Unrelated-Effect Panel Probit

Although conditional logit allows the unobserved time-constant hetero- geneity term δ_i to be related to be x_{it}, it cannot estimate the coefficients of time-constant regressors. In this and the following subsections, we explore "random-effect" approaches to avoid the problem. Following the terminology in Lee (2002), we call the model "unrelated-effect" panel probit. To ease ex- position, this subsection explores a "static" model where no lagged response appears as a regressor; the next section will examine a dynamic model.

Suppose that x_{it} includes both time-varying and time-constant regres- sors; for unrelated-effect estimators, it is unlikely that this causes confusion. Assume

$$\delta_i \ \sim \ N(0, \sigma_\delta^2) \text{ independently of } (x_{i1}, ..., x_{iT},\ u_{i1}, ..., u_{iT})$$
$$u_{i1}, ..., u_{iT} \text{ are iid } N(0, \sigma_u^2) \text{ and independent of } (x_{i1}, ..., x_{iT}).$$

Then, for the model

$$y_{it} = 1[\tau_t + x_{it}'\beta + \delta_i + u_{it} > 0],$$

the unrelated-effect panel probit likelihood function for individual i is

$$P(y_{i1}, ..., y_{iT},\ x_{i1}, ..., x_{iT},\ \delta_i)$$
$$= P(y_{i1}, ..., y_{iT}|x_{i1}, ..., x_{iT}, \delta_i)f(x_{i1}, ..., x_{iT}, \delta_i)$$
$$= f(x_{i1}, ..., x_{iT}, \delta_i) \prod_t P(y_{it}|x_{i1}, ..., x_{iT}, \delta_i)$$

(for y_{it}'s are independent given $x_{i1}, ..., x_{iT}, \delta_i$)

$$= f(x_{i1}, ..., x_{iT}, \delta_i) \prod_t P(y_{it}|x_{it}, \delta_i) \text{ (for } y_{it} \text{ depends only on } x_{it}, \delta_i)$$

$$= f(x_{i1}, ..., x_{iT})f(\delta_i) \prod_t P(y_{it} = 1|x_{it}, \delta_i)^{y_{it}} P(y_{it} = 0|x_{it}, \delta_i)^{1-y_{it}}$$

(as $(x_{i1}, ..., x_{iT})$ Ⅱ δ_i)

$$= \quad f(x_{i1}, ..., x_{iT}) \prod_t \Phi \left(\frac{\tau_t}{\sigma_u} + x'_{it}\frac{\beta}{\sigma_u} + \frac{\delta}{\sigma_u} \right)^{y_{it}}$$

$$\left\{ 1 - \Phi \left(\frac{\tau_t}{\sigma_u} + x'_{it}\frac{\beta}{\sigma_u} + \frac{\delta}{\sigma_u} \right) \right\}^{1-y_{it}} f(\delta_i).$$

Integrating out δ in the last display and dropping $f(x_{i1}, ..., x_{iT})$ that is not a function of the model parameters of interest, we get

$$\int \prod_t \Phi \left(\frac{\tau_t}{\sigma_u} + x'_{it}\frac{\beta}{\sigma_u} + \frac{\delta}{\sigma_u} \right)^{y_{it}}$$

$$\left\{ 1 - \Phi \left(\frac{\tau_t}{\sigma_u} + x'_{it}\frac{\beta}{\sigma_u} + \frac{\delta}{\sigma_u} \right) \right\}^{1-y_{it}} \phi \left(\frac{\delta}{\sigma_\delta} \right) \frac{1}{\sigma_\delta} d\delta$$

$$= \quad \int \prod_t \Phi \left(\frac{\tau_t}{\sigma_u} + x'_{it}\frac{\beta}{\sigma_u} + \frac{\sigma_\delta}{\sigma_u}\frac{\delta}{\sigma_\delta} \right)^{y_{it}}$$

$$\left\{ 1 - \Phi \left(\frac{\tau_t}{\sigma_u} + x'_{it}\frac{\beta}{\sigma_u} + \frac{\sigma_\delta}{\sigma_u}\frac{\delta}{\sigma_\delta} \right) \right\}^{1-y_{it}} \phi \left(\frac{\delta}{\sigma_\delta} \right) \frac{1}{\sigma_\delta} d\delta$$

$$= \quad \int \prod_t \Phi \left(\frac{\tau_t}{\sigma_u} + x'_{it}\frac{\beta}{\sigma_u} + \frac{\sigma_\delta}{\sigma_u}\zeta \right)^{y_{it}}$$

$$\left\{ 1 - \Phi \left(\frac{\tau_t}{\sigma_u} + x'_{it}\frac{\beta}{\sigma_u} + \frac{\sigma_\delta}{\sigma_u}\zeta \right) \right\}^{1-y_{it}} \phi(\zeta)d\zeta.$$

The sample log-likelihood function to maximize wrt

$$\frac{\tau_1}{\sigma_u}, ... \frac{\tau_T}{\sigma_u}, \frac{\beta}{\sigma_u}, \frac{\sigma_\delta}{\sigma_u}$$

can be written simply as

$$\sum_i \ln \int \prod_t \Phi \left\{ \left(\frac{\tau_t}{\sigma_u} + x'_{it}\frac{\beta}{\sigma_u} + \frac{\sigma_\delta}{\sigma_u}\zeta \right) \cdot (2y_{it} - 1) \right\} \phi(\zeta)d\zeta.$$

The integration can be done with Monte Carlo simulation. STATA has the command xtprobit with the option RE to implement this estimator. If desired, the unrelated-effect panel logit can be done with xtlogit and the option RE.

If δ_i is related to x_{it}, then we may as well assume that

$$\delta_i = x'_{i1}\xi_1 +, ..., +x'_{iT}\xi_T + \tilde{\delta}_i \quad \text{where} \quad \tilde{\delta}_i \text{ Ⅱ } (x_{i1}, ..., x_{iT})$$

and $\xi_1, ..., \xi_T$ are unknown parameters. The logic is that *if δ_i is related to x_{it}, then δ_i is likely to be related to all of* $x_{i1}, ..., x_{iT}$ because δ_i is time-constant. In this case, merging $x'_{i1}\xi_1 +, ..., +x'_{iT}\xi_T$ into the regression function, the model time-constant error term becomes $\tilde{\delta}$ that is independent of $(x_{i1}, ..., x_{iT})$. This approach is based on Chamberlain (1982), and may be called "Chamberlain's all-period approach."

The "all-period approach," however, will make the model too long. Also, if $x_{i1}, ..., x_{iT}$ are included in period-t equation, then

$$x_{it}'\beta + x_{i1}'\xi_1 +, ..., +x_{iT}'\xi_T = x_{i1}'\xi_1 +, ..., +x_{it}'(\xi_t + \beta) +, ..., +x_{iT}'\xi_T :$$

the coefficient of x_{it} becomes $\xi_t + \beta$, complicating the identification of β somewhat. A reasonable practical alternative might be including the temporal average $x_{i.} \equiv T^{-1} \sum_t x_{it}$ or some lagged regressors $x_{i,t-1}, x_{i,t-2}$ in x_{it} instead of including all $x_{i1}, ..., x_{iT}$.

5.3   Dynamic Panel Probit

In panel data, typically there is a "momentum" (i.e., persistence) in the data. There are (at least) three ways to capture the momentum. The first is using δ_i that appears in all period equations. The second is serial correlations in the error terms $u_{i1}, ..., u_{iT}$. The third is using lagged responses $y_{i,t-1}, y_{i,t-2}, ...$ as regressors. Using lagged regressors is similar to this, but can be accommodated by redefining x_{it} as x_{it} and its lagged versions. Conditional logit and unrelated-effect panel probit capture the data persistence using only δ_i. This subsection allows for $y_{i,t-1}$ to get "dynamic panel probit," drawing on Lee and Tae (2005). Of course, if desired, one may consider all sources of persistence simultaneously, but this will be cumbersome; see Keane and Sauer (2009) and the reference therein.

Suppose

$$y_{it}^* = \beta_y y_{i,t-1} + x_{it}'\beta + v_{it}, \quad v_{it} = \delta_i + u_{it}.$$

Differently from the "static" model, this dynamic model brings up the issue of how to model the first period response y_{i1} equation. For this, we list three approaches: for some parameters $\alpha_1, ..., \alpha_T$ and α_δ,

(i) y_{i1} is not random to treat y_{i1} only as a fixed regressor in the y_{i2} equation

(ii) $y_{i1} = 1[x_{i1}'\alpha_1 +, ..., +x_{iT}'\alpha_T + v_{i1} > 0]$, $COR(v_{i1}, v_{it}) \equiv \rho_v$ $\forall t = 2, ..., T$

(iii) $y_{i1} = 1[x_{i1}'\alpha_1 +, ..., +x_{iT}'\alpha_T + \alpha_\delta \delta_i + u_{i1} > 0]$, $COR(u_{i1}, u_{it}) = 0$ $\forall t = 2, ..., T$

The first is the simplest but unrealistic. The second is general, but difficult to implement, requiring a high-dimensional integration. The third falls in between (i) and (ii) in terms of its strength of assumptions, which we will adopt; $V(u_{i1}) \equiv \sigma_1^2$ in (iii) is allowed to differ from $V(u_{it}) \equiv \sigma_u^2$ $\forall t = 2, ..., T$. In (ii) and (iii), $\alpha_1, ..., \alpha_T$ may get further restricted in practice; e.g., $\alpha_t = 0$ $\forall t \neq 1$; we will also adopt this.

Assume

δ_i is independent of $u_{i1}, ..., u_{iT}$, and $(\delta, u_{i1}, ..., u_{iT})$ is independent of $(x_{i1}, ..., x_{iT})$;

$u_{i2}, ..., u_{iT}$ are iid $N(0, \sigma_u^2)$ and independent of u_{i1} that follows $N(0, \sigma_1^2)$;

δ_i follows $N(0, \sigma_\delta^2)$.

Define
$$\sigma_v \equiv SD(v_{it}) = SD(\delta_i + u_{it}) \quad \forall t = 2, ..., T.$$

Dividing the period-1 latent equation by σ_1 and the period-t equation by σ_u, we get the log-likelihood function

$$\sum_i \ln \left[\int \Phi \left\{ \left(x_{i1}' \frac{\alpha}{\sigma_1} + \frac{\delta\alpha_\delta}{\sigma_1} \right) (2y_{i1} - 1) \right\} \right.$$
$$\left. \prod_{t=2}^{T} \Phi \left\{ \left(y_{i,t-1} \frac{\beta_y}{\sigma_u} + x_{it}' \frac{\beta}{\sigma_u} + \frac{\delta}{\sigma_u} \right) (2y_{it} - 1) \right\} \phi \left(\frac{\delta}{\sigma_\delta} \right) \frac{1}{\sigma_\delta} d\delta \right]$$

where only x_1 is used in the y_{i1} equation as noted above. Further rewrite this as

$$\sum_i \ln \left[\int \Phi \left\{ \left(x_{i1}' \frac{\alpha}{\sigma_1} + \frac{\delta}{\sigma_\delta} \frac{\alpha_\delta \sigma_\delta}{\sigma_1} \right) (2y_{i1} - 1) \right\} \right.$$
$$\left. \cdot \prod_{t=2}^{T} \Phi \left\{ \left(y_{i,t-1} \frac{\beta_y}{\sigma_u} + x_{it}' \frac{\beta}{\sigma_u} + \frac{\delta}{\sigma_\delta} \frac{\sigma_\delta}{\sigma_u} \right) (2y_{it} - 1) \right\} \phi \left(\frac{\delta}{\sigma_\delta} \right) \frac{1}{\sigma_\delta} d\delta \right]$$
$$= \sum_i \ln \left[\int \Phi \left\{ \left(x_{i1}' \frac{\alpha}{\sigma_1} + \zeta \frac{\alpha_\delta \sigma_\delta}{\sigma_1} \right) (2y_{i1} - 1) \right\} \right.$$
$$\left. \cdot \prod_{t=2}^{T} \Phi \left\{ \left(y_{i,t-1} \frac{\beta_y}{\sigma_u} + x_{it}' \frac{\beta}{\sigma_u} + \zeta \frac{\sigma_\delta}{\sigma_u} \right) (2y_{it} - 1) \right\} \phi(\zeta) d\zeta \right].$$

The identified parameters here are

$$\frac{\alpha}{\sigma_1}, \quad \frac{\alpha_\delta \sigma_\delta}{\sigma_1}, \quad \frac{\beta_y}{\sigma_u}, \quad \frac{\beta}{\sigma_u}, \quad \frac{\sigma_\delta}{\sigma_u}.$$

The last term σ_δ/σ_u shows how important δ_i is relative to u_{it}. If we treat y_{i1} as fixed, then we just have to drop the first period likelihood component. But as shown in Lee and Tae's (2005) empirical example, treating y_{i1} as fixed does not look sensible.

Going further, to allow for relationship between δ_i and x_{it}, we may assume

$$\delta_i = x_{i1}'\mu_1+, ..., +x_{iT}'\mu_T + \eta_i, \text{ or a simpler version } \delta_i = \bar{x}_i'\mu + \eta_i,$$

$$\bar{x}_i \equiv T^{-1} \sum_{t=1}^{T} x_{it}.$$

Since the former is computationally too demanding, suppose we adopt the simpler version that includes the restriction $\mu_1 =, ..., = \mu_T \equiv \mu_0$:

$$\delta_i = \left(\sum_t x'_{it} \right) \mu_0 + \eta_i = \bar{x}'_i (\mu_0 T) + \eta_i = \bar{x}'_i \mu + \eta_i, \quad \text{where } \mu \equiv \mu_0 T.$$

Writing x_{it} as $x_{it} - \bar{x}_i + \bar{x}_i$, the effect of x_{it} can be decomposed into two: the temporary (or transitory) effect from $x_{it} - \bar{x}_i$ and the permanent effect from $\bar{x}_i$. But only the permanent effect remains in

$$\sum_t x_{it} = \sum_t (x_{it} - \bar{x}_i + \bar{x}_i) = \sum_t (x_{it} - \bar{x}_i) + \sum_t \bar{x}_i = \bar{x}_i T.$$

Equation $\mu = \mu_0 T$ states that the permanent effect is the sum of "one-shot time-invariant" effects over T periods. The permanent effect is a level change, for which expressions such as "tendency" or "propensity" are often used.

Substitute $\delta_i = \bar{x}'_i \mu + \eta_i$ into the above likelihood function before δ/σ_δ gets replaced by ζ to obtain

$$\sum_i \ln \left[\int \Phi \left\{ \left(x'_{i1} \frac{\alpha}{\sigma_1} + (\bar{x}'_i \mu + \eta_i) \frac{\alpha_\delta}{\sigma_1} \right) (2y_{i1} - 1) \right\} \cdot \right.$$

$$\prod_{t=2}^{T} \Phi \left\{ \left(y_{i,t-1} \frac{\beta_y}{\sigma_u} + x'_{it} \frac{\beta}{\sigma_u} + (\bar{x}'_i \mu + \eta_i) \frac{1}{\sigma_u} \right) (2y_{it} - 1) \right\}$$

$$\left. \phi \left(\frac{\eta}{\sigma_\eta} \right) \frac{1}{\sigma_\eta} d\eta \right].$$

With $\zeta = \eta/\sigma_\eta$, this can be rewritten as

$$\sum_i \ln \left[\int \Phi \left\{ \left(x'_{i1} \frac{\alpha}{\sigma_1} + \bar{x}'_i \frac{\mu \alpha_\delta}{\sigma_1} + \zeta \frac{\alpha_\delta \sigma_\eta}{\sigma_1} \right) (2y_{i1} - 1) \right\} \right.$$

$$\left. \cdot \prod_{t=2}^{T} \Phi \left\{ \left(y_{i,t-1} \frac{\beta_y}{\sigma_u} + x'_{it} \frac{\beta}{\sigma_u} + \bar{x}'_i \frac{\mu}{\sigma_u} + \zeta \frac{\sigma_\eta}{\sigma_u} \right) (2y_{it} - 1) \right\} \phi(\zeta) d\zeta \right].$$

Now, the identified parameters are

$$\frac{\alpha}{\sigma_1}, \quad \frac{\mu \alpha_\delta}{\sigma_1}, \quad \boxed{\frac{\alpha_\delta \sigma_\eta}{\sigma_1}}, \quad \frac{\beta_y}{\sigma_u}, \quad \frac{\beta}{\sigma_u}, \quad \frac{\mu}{\sigma_u}, \quad \boxed{\frac{\sigma_\eta}{\sigma_u}};$$

the two underlined terms did not appear previously and the two boxed terms appeared with σ_η replaced by σ_δ. Although x_{it} includes both time-constant and time-variant variables, in practice, $\bar{x}_i$ should consist only of time-variants.

Otherwise the time-constant variables get to be used twice as regressors in x_{it} and $\bar{x}_i$ for the same equation.

6 Competing Risks*

Suppose there are different reasons for a duration to end; e.g., different causes for one's life to end. Let y_j^* denote the latent cause-j duration, $j = 1, ..., J$. The person-i duration y_i^* without censoring and the observed duration y_i with censoring are, respectively,

$$y_i^* = \min(y_{i1}^*, ..., y_{iJ}^*) \quad \text{and} \quad y_i = \min(y_i^*, c_i).$$

The observed data is (x_i', y_i, r_i, d_i) where $r_i = j$ means that the duration ended with reason j and d_i is the non-censoring indicator. The reasons/causes "compete to get" individual i, and this explains the name "competing risks." We examine this topic in this section drawing heavily on Crowder (2001); often we will set $J = 2$ for illustrations.

6.1 Observed Causes and Durations

Define the cause-j "sub-distribution function" $F(j,t)$, cause-j "sub-survival function" $S(j,t)$, and cause-j "sub-density function" $f(j,t)$:

$$F(j,t) \equiv P(r = j, y^* \leq t), S(j,t) \equiv P(r = j, y^* > t), \ f(j,t) \equiv -\frac{dS(j,t)}{dt}$$

$$\implies F(j,t) + S(j,t) = P(r = j) \equiv p_j = F(j,\infty) = S(j,0) < 1.$$

Both $F(j,t)$ and $S(j,t)$ are bounded from above by p_j, not by 1, and thus they are not proper distribution/survival function. The conditional version $P(r = j|x)$ of p_j can be used to predict the eventual cause for ending the duration for a person with trait x.

With the superscript "m" standing for "minimum," the *(minimum) survival function* $S^m(t)$ and the *(minimum) density* $f^m(t)$ are, respectively,

$$S^m(t) \equiv P(y^* > t) = \sum_{j=1}^{J} P(r = j, y^* > t) = \sum_{j=1}^{J} S(j,t)$$

$$f^m(t) \equiv -\frac{dS^m(t)}{dt} = -\frac{d\left\{\sum_{j=1}^{J} S(j,t)\right\}}{dt} = -\sum_{j=1}^{J} \frac{dS(j,t)}{dt} = \sum_{j=1}^{J} f(j,t).$$

Observe (and think about the meaning of)

$$\frac{S(j,t)}{p_j} = P(y^* > t|r = j) \quad \text{and} \quad \frac{S(j,t)}{S^m(t)} = P(r = j|y^* > t).$$

Define the cause-j "sub-hazard function" $\lambda(j,t)$ and compare $\lambda(j,t)$ to $\lambda^m(t)$:

$$\lambda(j,t) \equiv \frac{f(j,t)}{S^m(t)} : \text{failing with cause } j \text{ at } t \text{ given survival up to } t$$

$$\lambda^m(t) \equiv \frac{f^m(t)}{S^m(t)} = \frac{\sum_{j=1}^J f(j,t)}{S^m(t)} = \sum_j \lambda(j,t):$$

failing at t for any cause given survival up to t.

Pay attention to that the denominator has $S^m(t)$, not $S(j,t)$, as one has to overcome all competing causes to survive up to t. As in the single duration case,

$$\lambda^m(t) = -\frac{d\ln S^m(t)}{dt} \iff S^m(t) = \exp\left\{-\int_0^t \lambda^m(s)ds\right\}.$$

For instance, suppose that a Weibull sub-hazard holds for each $\lambda(j,t)$:

$$\lambda(j,t) = \theta_j \alpha_j t^{\alpha_j - 1} \implies \lambda^m(t) = \sum_j \theta_j \alpha_j t^{\alpha_j - 1}$$

$$\implies \Lambda^m(t) \equiv \int_0^t \lambda^m(s)ds = \sum_j \theta_j t^{\alpha_j}$$

$$S^m(t) = \exp\{-\Lambda^m(t)\} = \exp\left(-\sum_j \theta_j t^{\alpha_j}\right) \quad \left\{= \prod_j \exp(-\theta_j t^{\alpha_j})\right\}$$

$$f(j,t) = \lambda(j,t)S^m(t) = \theta_j \alpha_j t^{\alpha_j - 1} \cdot \exp\left(-\sum_j \theta_j t^{\alpha_j}\right).$$

With $\theta_j = \theta_j(x, \beta_j) = \exp(x'\beta_j)$ as usual, defining

$$\delta_{ij} \equiv 1[r_i = j] \quad \text{and} \quad \gamma_j \equiv (\alpha_j, \beta_j')'$$

the log-likelihood function for $\gamma_1, ..., \gamma_J$ is

$$\sum_i \{d_i \ln f(r_i, y_i) + (1 - d_i) \ln S^m(y_i)\}$$

$$= \sum_i [d_i \ln \{\lambda(r_i, y_i)S^m(y_i)\} + (1 - d_i) \ln S^m(y_i)]$$

$$= \sum_i [d_i \ln \lambda(r_i, y_i) + \ln S^m(y_i)] = \sum_i [d_i \ln \lambda(r_i, y_i) - \Lambda^m(y_i)]$$

$$= \sum_i \left[d_i \sum_j \delta_{ij} \ln\{\theta_j(x_i, \beta_j)\alpha_j y_i^{\alpha_j - 1}\} - \sum_j \theta_j(x_i, \beta_j) y_i^{\alpha_j} \right]$$

$$= \sum_i \left[d_i \sum_j \delta_{ij}\{x_i'\beta_j + \ln \alpha_j + (\alpha_j - 1) \ln y_i\} - \sum_j \exp(x_i'\beta_j) y_i^{\alpha_j} \right].$$

6.2   Latent Causes and Durations

The above MLE is close to estimating a reduced form (RF) rather than a structural form (SF) equation, because we would be more interested in the parameters governing the latent durations. That is, not just the parameters for a given cause and duration as in $\lambda(j,t) = \exp(x'\beta_j) \cdot \alpha_j t^{\alpha_j - 1}$, we would like to know the parameters for the joint distribution of latent durations. We will examine joint latent durations in this subsection. It will be useful if we can see, for instance, whether increasing one latent duration increases (or decreases) some other latent durations.

Consider a *joint survival function* $S^{1 \cdots J}(t_1, ..., t_J)$:

$$S^{1 \cdots J}(t_1, ..., t_J) \equiv P(y_1^* > t_1, ..., y_J^* > t_J)$$

$$\{= S^m(t) \text{ when } t_1 =, ..., = t_J = t\}.$$

The *marginal survival function* $S^1(t_1)$ for y_1^* is obtained by setting all the other durations at zero:

$$S^1(t_1) \equiv P(y_1^* > t_1) = S^{1 \cdots J}(t_1, 0, ..., 0);$$

$S^j(t_j)$ are defined analogously. *Each marginal $S^j(t_j)$ defines its own $f^j(t_j)$ and hazard $\lambda^j(t)$.* It is important to be aware of the differences between $S^{1 \cdots J}(t_1, ..., t_J)$, $S^m(t)$, and $S^j(t_j)$ as in the following.

For instance, with $J = 2$, a "joint exponential survival function" is

$$S^{12}(t_1, t_2) = \exp(-\theta_1 t_1 - \theta_2 t_2 - \nu t_1 t_2) \quad \text{where } \theta_1, \theta_2, \nu > 0$$

$$\implies S^1(t_1) = S^{12}(t_1, 0) = \exp(-\theta_1 t_1) \quad \text{and}$$

$$S^m(t) = \exp\{-(\theta_1 + \theta_2)t - \nu t^2\}.$$

Although each $S^j(t_j)$ is $Expo(\theta_j)$, $S^{12}(t_1, t_2)$ and $S^m(t)$ are not. If $\nu = 0$, i.e., if the interaction term $t_1 t_2$ drops out, then

$$S^{12}(t_1, t_2) = \exp(-\theta_1 t_1 - \theta_2 t_2) = \exp(-\theta_1 t_1) \cdot \exp(-\theta_2 t_1)$$
$$= S^1(t_1) \cdot S^2(t_2):$$

y_1^* and y_2^* are independent when $\nu = 0$.

Observe

$$f(1, t) = \lim_{q \to 0} \frac{P(y_j^* > y_1^* \ \forall j \neq 1 \ \text{and} \ t < y_1^* \leq t + q)}{q}.$$

In the event $t < y_1^* \leq t + q$, y_1^* is bounded by $(t, t + q]$. This implies, as $y_j^* > t + q \implies y_j^* > y_1^* \implies y_j^* > t$ in this case,

$$\lim_{q \to 0} \frac{P(y_j^* > t + q \ \forall j \neq 1, \ t < y_1^* \leq t + q)}{q} \leq f(1, t)$$

$$\leq \lim_{q \to 0} \frac{P(y_j^* > t \ \forall j \neq 1, \ t < y_1^* \leq t + q)}{q}.$$

The lower and the upper bounds are, respectively,

$$\lim_{q \to 0} \frac{S^{1 \cdots J}(t, t+q, ..., t+q) - S^{1 \cdots J}(t+q, t+q, ..., t+q)}{q}$$

$$= \frac{-\partial S^{1 \cdots J}(t_1, ..., t_J)}{\partial t_1} \bigg|_{t_1 = \cdots = t_J = t}$$

$$\lim_{q \to 0} \frac{S^{1 \cdots J}(t, t, ..., t) - S^{1 \cdots J}(t+q, t, ..., t)}{q}$$

$$= \frac{-\partial S^{1 \cdots J}(t_1, ..., t_J)}{\partial t_1} \bigg|_{t_1 = \cdots = t_J = t}.$$

Therefore, we get an equation relating $S^{1 \cdots J}(t_1, ..., t_J)$ for the latent durations to $f(1, t)$ for the observed duration and its cause (the above derivation is due to Tsiatis 1975):

$$f(1, t) = \frac{-\partial S^{1 \cdots J}(t_1, ..., t_J)}{\partial t_1} \bigg|_{t_1 = \cdots = t_J = t}$$

$$\Longrightarrow \lambda(1, t) = \frac{f(1, t)}{S^m(t)} = \frac{-\partial \ln S^{1 \cdots J}(t_1, ..., t_J)}{\partial t_1} \bigg|_{t_1 = \cdots = t_J = t};$$

$f(j, t)$ and $\lambda(j, t)$, $j = 2, ..., J$, satisfy analogous equations.

Using the last display,

$$S^{12}(t_1, t_2) = \exp(-\theta_1 t_1 - \theta_2 t_2 - \nu t_1 t_2)$$

$$\Longrightarrow \quad \lambda(1, t) = \theta_1 + \nu t, \lambda(2, t) = \theta_2 + \nu t.$$

Compare these to the earlier finding

$$S^1(t_1) = \exp(-\theta_1 t_1), \ S^2(t_2) = \exp(-\theta_2 t_2) \Longrightarrow \lambda^1(t) = \theta_1, \ \lambda^2(t) = \theta_2.$$

Clearly, $\lambda(j, t) \neq \lambda^j(t)$ unless $\nu = 0$ so that y_1^* and y_2^* are independent.

It is instructive to further compare the *cause-1 marginal hazard* $\lambda^1(t_1)$ *and the cause-1 sub-hazard* $\lambda(1, t)$. $S^1(t_1)$ was obtained by setting $t_2 = 0$ in $S^{12}(t_1, t_2)$ to eliminate y_2^* from consideration. Thus $\lambda^1(t_1)$ *is the cause-1 hazard when cause 2 is not operating*; it is a SF ceteris paribus hazard for cause 1 while holding the other cause constant. The interest on $\lambda^1(t_1)$ stems from the usual policy intervention scenario. For instance, if we can intervene on a treatment for an illness (i.e., cause 1) without affecting the other cause, then $\lambda^1(t_1)$ would be the right parameter of interest where the treatment appears as a regressor. In contrast, $\lambda(1, t)$ *is more of a RF parameter; it is the hazard while the other cause is still in operation* so that the two causes can interact. The two causes may exchange influences, and $\lambda(1, t)$ reflects the final outcome when the exchanges get settled; in the above example

$\lambda(1,t) = \theta_1 + \nu t$, the interaction parameter ν appears in $\lambda(1,t)$. Although we desire to know $\lambda^1(t_1)$, it is hard to find this—more on this below.

The "*dependence ratio*" is defined as

$$\frac{S^{12}(t_1,t_2)}{S^1(t_1)S^2(t_2)} = \frac{P(y_1^* > t_1, y_2^* > t_2)}{P(y_1^* > t_1) \cdot P(y_2^* > t_2)}.$$

Suppose that this is less than one, and multiply the inequality by $P(y_2^* > t_2)$ to get

$$\frac{P(y_1^* > t_1, y_2^* > t_2)}{P(y_1^* > t_1)} < P(y_2^* > t_2)$$

$$\iff \quad P(y_2^* > t_2 | y_1^* > t_1) < P(y_2^* > t_2) : \text{"\textit{negative dependence}"}$$
between y_1^* and y_2^*.

That is, given that one latent duration gets longer $(y_1^* > t_1)$, the other latent duration tends to get shorter. In the above exponential case, the dependence ratio is $\exp(-\nu t_1 t_2) < 1$, and a negative dependence holds.

6.3   Dependent Latent Durations and Identification

Recall the joint exponential survival function with $\nu > 0$ (i.e., the latent durations are not independent) and the ensuing derivations:

$$\begin{aligned}
S^m(t) &= \exp\{-(\theta_1 + \theta_2)t - \nu t^2\}, \quad \Lambda^m(t) = (\theta_1 + \theta_2)t + \nu t^2, \\
&\quad \lambda^m(t) = \theta_1 + \theta_2 + 2\nu t \\
\lambda(j,t) &= \theta_j + \nu t, \quad \lambda^j(t) = \theta_j.
\end{aligned}$$

But, surprisingly, $S^m(t)$ can be written as the product of two marginal survival functions as if the latent durations are independent. Specifically, consider

$$S^m(t) = \exp\left(-\theta_1 t - \frac{\nu t^2}{2}\right) \cdot \exp\left(-\theta_2 t - \frac{\nu t^2}{2}\right) \equiv \tilde{S}^1(t)\tilde{S}^2(t),$$

$$\tilde{S}^j(t_j) \equiv \exp\left(-\theta_j t_j - \frac{\nu t_j^2}{2}\right)$$

$$\left\{ \implies \tilde{\lambda}^j(t_j) = \frac{-\partial \ln \tilde{S}^j(t_j)}{\partial t_j} = \theta_j + \nu t_j \right\}.$$

Define further

$$\begin{aligned}
\tilde{S}^{12}(t_1,t_2) &\equiv \tilde{S}^1(t_1)\tilde{S}^2(t_2) = \exp\left\{-\theta_1 t_1 - \theta_2 t_2 - \frac{\nu}{2}(t_1^2 + t_2^2)\right\} \\
&\neq S^{12}(t_1,t_2) = \exp(-\theta_1 t_1 - \theta_2 t_2 - \nu t_1 t_2).
\end{aligned}$$

Observe

$$\tilde{\lambda}(j,t) \equiv \frac{-\partial \ln \tilde{S}^{12}(t_1, t_2)}{\partial t_j} \Big|_{t_1=t_2=t} = \theta_j + \nu t = \lambda(j,t) :$$

although the joint survival functions differ (i.e., $\tilde{S}^{12}(t_1, t_2) \neq S^{12}(t_1, t_2)$), the independence case yields exactly the same sub-hazards as the dependent case does. Hence, because $\lambda(j,s)$ determines $S^m(t) = \exp\{-\sum_j \int_0^t \lambda(j,s)ds\}$, we get

$$\left[\exp\left\{ -\sum_j \int_0^t \tilde{\lambda}(j,s)ds \right\} = \right] \tilde{S}^m(t) = S^m(t).$$

Since only $\lambda(j,t)$ and $S^m(t)$ appear in the likelihood function, the independence model is not distinguishable from the dependence model with the given data on (x_i', y_i, r_i, d_i). Note that $\tilde{\lambda}^j(t) = \tilde{\lambda}(j,t)$ in the independence model whereas $\lambda^j(t) \neq \lambda(j,t)$ in the dependent model.

In short, *for each dependent latent durations with given sub-hazards $\lambda(j,t)$ (and the survival function $S^m(t)$), there exist independent latent durations with the same sub-hazards (and thus the same survival function),* although $\lambda^j(t)$ in the dependent model differs from $\tilde{\lambda}^j(t) = \tilde{\lambda}(j,t)$ in the independent model. That is, there is no problem identifying $\lambda(j,t)$, but $\lambda^j(t)$ in the dependent model cannot be identified. This is the identification problem in competing risks shown by Tsiatis (1975). Despite this "grim" picture, however, still there are positive findings for identifications as in Heckman and Honoré (1989), Abbring and Van den Berg (2003), and S.B. Lee (2006) when regressors are added into the model. Fermanian (2003) implemented a "kernel nonparametric estimation" of the Heckman and Honoré idea. Honoré and Lleras-Muney (2006) proposed a "set-identification" (or "bounding") approach.

Finally in this section, we make a few remarks for competing risks and univariate duration models. First, if we look at the survival for all causes perhaps to avoid the dependence/identification issue, i.e., if we take y^* as a single duration, then the usual univariate duration analyses apply to $y = \min(y^*, c)$. This, however, needs some care; e.g., even if each of y_1^* and y_2^* follows Weibull, $\min(y_1^*, y_2^*)$ does not. Second, if we focus on one cause only, say cause 1, to take any other causes as censoring variables, then univariate duration analyses apply to the observed duration $\min(y_1^*, c)$ where c is redefined as $\min(y_2^*, ..., y_J^*,$ "old c"). This would, however, require the independent latent duration assumption; otherwise, the new censoring variable becomes dependent on y_1^* given x. Third, suppose that $\lambda(1,t)$ is decreasing while $\lambda(2,t)$ is increasing. As $\lambda^m(t) = \sum_j \lambda(j,t)$, $\lambda^m(t)$ may exhibit an up-down or down-up shape. That is, a highly variable univariate hazard may be the outcome of competing risks. For instance, an apparently down-up hazard of a single disease may be in fact a sum of two monotonic hazards with different effect signs from two separate causes related to the single disease.

CHAPTER 7
KERNEL NONPARAMETRIC ESTIMATION

Some regression models are fully *parametric* in that both the regression function and the error term distribution are parametrically specified, whereas some are *semiparametric* in the sense that only the regression function is parametrically specified—LSE is semiparametric in this sense. Going further, it is possible to go *nonparametric* in that the regression function $E(y|x)$ gets estimated with neither its functional form nor the error term distribution specified. Since $E(y|x) = \int y f(y|x) dy = \int y \{ f(y,x)/f(x) \} dy$, if the densities $f(y,x)$ and $f(x)$ can be estimated nonparametrically, then $E(y|x)$ can be as well. Kernel nonparametric estimation for density $f(x)$ is introduced first, and then for $E(y|x)$. Also, nonparametric hazard function estimation is studied.

1 Kernel Density Estimator

Assume that x of dimension $k \times 1$ has a continuous density function $f(x)$. In this section, first, "kernel density estimators" are introduced; to simplify exposition, we start out with the $k = 1$ case and then allow an arbitrary k later. Second, kernel estimators for density derivatives (and integrals) are examined. Third, further remarks are provided. If x is discretely distributed, then we can estimate $P(x = x_o)$ either by the sample average of observations with $x_i = x_o$ (i.e., $N^{-1} \sum_i 1[x_i = x_o]$) or by nonparametric methods explained in this section. There are many nonparametric estimators available other than kernel estimators, but they are not reviewed here. See Prakasa Rao (1983), Silverman (1986), Bierens (1987), Müller (1988), Härdle (1990), Izenman (1991), Rosenblatt (1991), Scott (1992), Härdle and Linton (1994), Wand and Jones (1995), and Wasserman (2006) among many others.

1.1 Density Estimators

Suppose x_i is a scalar ($k = 1$) and $x_1, ..., x_N$ are observed. If our interest is in $P(x \leq x_o) \equiv F(x_o)$, then $P(x \leq x_o)$ can be estimated by the *empirical distribution function*

$$F_N(x_o) \equiv \frac{1}{N} \sum_{i=1}^{N} 1[x_i \leq x_o].$$

Myoung-jae Lee, *Micro-Econometrics*, DOI 10.1007/b60971_7,

Although this converges to $F(x_o)$ in various senses, $F_N(x_o)$ is not differentiable while $F(x_o)$ is so. Hence we cannot estimate $f(x_o)$ by differentiating the empirical distribution function. It is conceivable, however, to estimate $f(x_o)$ by approximating $dF(x_o)$ and dx in view of $f(x_o) = dF(x_o)/dx$. Bear in mind that, whereas x_i, $i = 1, ..., N$, are observations, x_o denotes a fixed "evaluation point" of interest; x_o may or may not be equal to any of $x_1, ..., x_N$.

Let h be a small positive number. Set $dx \simeq h$ and observe

$$dF(x_o) \simeq P(x_o < x < x_o + h)$$

$$\simeq \frac{1}{N} \sum_{i=1}^{N} 1[x_o < x_i < x_o + h] \simeq \frac{1}{N} \sum_i \frac{1[x_o - h < x_i < x_o + h]}{2}.$$

Hence, a nonparametric density function estimator approximating $dF(x_o)/dx$ is

$$\frac{1}{Nh} \sum_i \frac{1[x_o - h < x_i < x_o + h]}{2} = \frac{1}{Nh} \sum_i \frac{1[-h < x_i - x_o < h]}{2}.$$

For this approximation to work, h should be small. If h is too small, however, there may be no observation satisfying $-h < x_i - x_o < h$. Thus, we can let $h \to 0^+$ only as $N \to \infty$.

Viewing the role of the indicator function as a weighting function giving the weight 1 if x_i is within h-distance from x_o and 0 otherwise, we can generalize the above estimator with a smooth weighting function K (Rosenblatt, 1956):

$$f_N(x_o) \equiv \frac{1}{Nh} \sum_i K\left(\frac{x_i - x_o}{h}\right)$$

where K is called a *kernel* and $f_N(x_o)$ is a (nonparametric) *kernel density estimator*. This kernel estimator includes the preceding one as a special case when $K(z) = 1[-1 < z < 1]/2$ because

$$1\left[-1 < \frac{x_i - x_o}{h} < 1\right]/2 = \frac{1[-h < x_i - x_o < h]}{2}.$$

For instance, with the $N(0, 1)$ density ϕ as K, we get

$$f_N(x_o) = \frac{1}{Nh} \sum_i \frac{1}{\sqrt{2\pi}} \exp\left\{-\frac{1}{2}\left(\frac{x_i - x_o}{h}\right)^2\right\}.$$

Getting $f_N(x_o)$ over $x_o = -1.5, -1.4, ..., 0, ..., 1.4, 1.5$ and then connecting them, we can see the shape of $f(x)$ over $[-1.5, 1.5]$. As x_o varies over a chosen range, $f_N(x_o)$ traces a curve. In this case, we may as well write $f_N(x)$ instead of $f(x_o)$, for our interest is in the entire range, not in any particular evaluation point such as x_o.

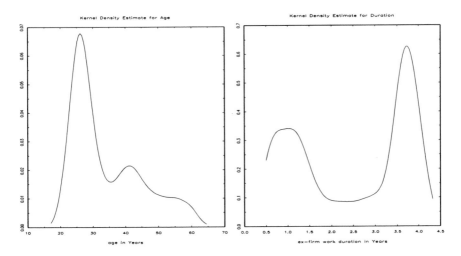

Figure 1: Two Marginal Density Estimates

Figure 1 presents two kernel density estimates using some Korean female data with $N = 9312$. The Korean women became unemployed in 1999 to receive unemployment insurance benefit. The kernel is the $N(0,1)$ density ϕ, and $h = SD(x) \cdot N^{-1/5}$. The first figure for age shows that the global mode of age is about 27 with another possible (local) mode around 42. The second figure is for the work duration in the previous workplace where bimodality is more visible with a local mode at around 1 year and the global mode at around 3.7. The first mode may be due to one year probationary employment period, and the second may be due to many young women quitting after marriage.

Now suppose $x_i = (x_{i1}, ..., x_{ik})'$ is a $k \times 1$ vector and let $x_o = (x_{o1}, ..., x_{ok})'$ accordingly. Then $dx \simeq h^k$ and $dF(x_o)$ takes the same form as the $dF(x_o)$ above to yield an estimator

$$\frac{1}{Nh^k} \sum_i \frac{1[-h < x_i - x_o < h]}{2}$$

$$= \frac{1}{Nh^k} \sum_i \frac{1[-h < x_{i1} - x_{o1} < h] \cdots 1[-h < x_{ik} - x_{ok} < h]}{2}.$$

With K having k arguments now, the kernel estimator becomes

$$f_N(x_o) = \frac{1}{Nh^k} \sum_i K\left(\frac{x_i - x_o}{h}\right)$$

$$= \frac{1}{Nh^k} \sum_i K\left(\frac{x_{i1} - x_{o1}}{h}, ..., \frac{x_{ik} - x_{ok}}{h}\right).$$

One example of K with $k = 2$ is a product of ϕ: $K(z_1, z_2) = \phi(z_1)\phi(z_2)$. Another example is the standard bivariate normal density with a specified correlation, say $\rho = 0.5$:

$$K(z_1, z_2) = \frac{1}{2\pi\sqrt{1-\rho^2}} \exp\left\{ -\frac{z_1^2 - 2\rho z_1 z_2 + z_2^2}{2(1-\rho^2)} \right\}.$$

More generally, for $k \geq 2$, the product $K(z) = \phi(z_1) \cdots \phi(z_k)$ may be used where $z = (z_1, ..., z_k)'$. Also a multivariate zero-mean normal density with a specified variance matrix Ω can be used as well:

$$K(z) = \frac{1}{(2\pi)^{k/2}\{\det(\Omega)\}^{1/2}} \cdot \exp\left(-\frac{z'\Omega^{-1}z}{2} \right).$$

The form of kernel will be discussed further later.

Instead of using the same h for all components of x, we may use different bandwidths for different components: h_j for x_j, $j = 1, ..., k$, to get

$$f_N(x_o) = \frac{1}{N \cdot h_1 \cdots h_k} \sum_i K\left(\frac{x_{i1} - x_{o1}}{h_1}, ..., \frac{x_{ik} - x_{ok}}{h_k} \right).$$

But choosing k-many bandwidths is troublesome. Instead, set

$$h_j = SD(x_j) \cdot h_o$$

where h_o is a "base" bandwidth and choose only the "base bandwidth" h_o. In this case,

$$f_N(x_o) = \frac{1}{N \cdot h_o^k \cdot SD(x_1) \cdots SD(x_k)} \sum_i K\left(\frac{x_{i1} - x_{o1}}{SD(x_1)h_o}, ..., \frac{x_{ik} - x_{ok}}{SD(x_k)h_o} \right).$$

Typically $h_o \simeq c \cdot N^{-1/(k+4)}$ where c ranges over, say $[0.5, 2.5]$; recall that $h_o = N^{-1/5}$ was used for the univariate density figures. *This display with* $K(z_1, ..., z_k) = \Pi_j \phi(z_j)$ *and* $h_o = N^{-1/(k+4)}$ *would be a "practical first estimator" that one can easily try.* Because using different bandwidths for different variables is notationally cumbersome, we will use h for all variables unless otherwise necessary.

Figure 2 presents a kernel estimate for age and (ex-firm work) duration using the same data as in the preceding figure. The kernel is $K(z_1, z_2) = \phi(z_1)\phi(z_2)$ and the bandwidth is the same as in the marginal densities: $SD(age) \cdot N^{-1/5}$ for age and $SD(duration) \cdot N^{-1/5}$ for duration, although we might use $N^{-1/6}$ as well instead of $N^{-1/5}$. The age bimodality, or possible trimodality, can be seen for duration around 4 years, whereas the age density seems nearly unimodal at around 30 for durations lower than 3. The duration bimodality is highly visible for women of age 20–40, but the size of the peaks is much lower for higher ages. Bivariate density reveals features that could not be seen with the two marginal density estimates.

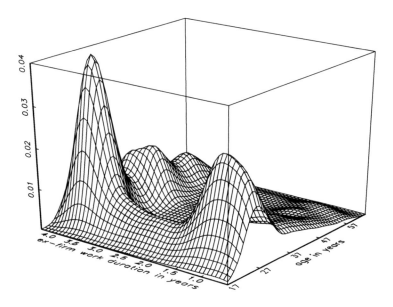

Figure 2: Bivariate Density Estimate

1.2   Density-Derivative Estimators

Suppose we are interested in the first derivative $f'(x_o)$. Again, we start out with the $k = 1$ case and then allow an arbitrary k later. Suppose K is differentiable; all kernels in use from now and onward will be assumed to be differentiable.

Let $k = 1$. One simple way to estimate $f'(x_o)$ is using a numerical derivative: for a small constant ε, say $\varepsilon = 10^{-5}$,

$$\tilde{f}'_N(x_o) \equiv \frac{f_N(x_o + \varepsilon) - f_N(x_o - \varepsilon)}{2\varepsilon}$$

$$= \frac{1}{Nh2\varepsilon} \sum_i \left\{ K\left(\frac{x_i - x_o - \varepsilon}{h}\right) - K\left(\frac{x_i - x_o + \varepsilon}{h}\right) \right\}$$

which is an approximation to the analytic derivative

$$f'_N(x_o) = -\frac{1}{Nh^2} \sum_i K'\left(\frac{x_i - x_o}{h}\right).$$

For example, with $K(z) = \phi(z)$, since

$$\phi'(z) = -z \frac{1}{\sqrt{2\pi}} \exp(-\frac{1}{2}z^2) = -z\phi(z),$$

$$f'_N(x_o) = \frac{1}{Nh^2} \sum_i \left(\frac{x_i - x_o}{h}\right) \frac{1}{\sqrt{2\pi}} \exp\left\{-\frac{1}{2}\left(\frac{x_i - x_o}{h}\right)^2\right\}.$$

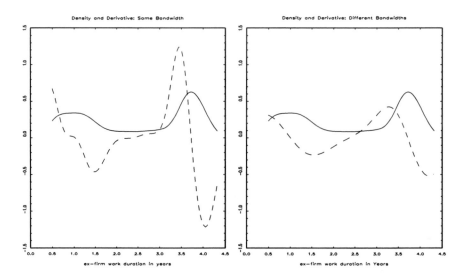

Figure 3: Density and First Derivative

Since f' shows a "finer" aspect of f, f' is more difficult to estimate than f. That is, f' requires more data than f; in a given data set, this translates into a larger bandwidth, which gives more data locally around x_o. Figure 3 shows the duration density estimate (solid line) and its first derivative (dashed line) for the same ex-firm work-duration data used ahead. The numerical derivative was employed with $\varepsilon = 10^{-5}$. The left figure uses the same bandwidth for both density and derivative, whereas the right figure uses twice greater bandwidth for the derivative. The derivative in the left figure looks under-smoothed compared with that in the right.

It is important to pay attention to the scales of figures. For example, the earlier univariate $f_N(x)$ figure looks more bimodal than the current figure despite that both figures are exactly the same. It is only that the scales are different: the vertical axis of the current diagram is longer running from -1.5 to 1.5, and this makes the current figure "weaker." At the extreme, if the vertical axis gets blown off to, say -150 to 150, then $f_N(x)$ will look like a flat line with $f_N(x) \simeq 0 \;\forall x$. At the other extreme, if the scale gets too small, $f_N(x)$ will look varying too much. As elementary as this caution may sound, mistakes of this sort are frequently made in practice.

Doing analogously to the first derivative, we can also estimate the second derivative $f_N''(x_o)$ if necessary. A numerical second derivative is

$$\frac{1}{2\varepsilon} \left\{ \tilde{f}_N'(x_o + \varepsilon) - \tilde{f}_N'(x_o - \varepsilon) \right\}$$

$$= \frac{1}{2\varepsilon} \left\{ \frac{f_N(x_o + \varepsilon + \varepsilon) - f_N(x_o + \varepsilon - \varepsilon)}{2\varepsilon} \right.$$

$$\left. -\frac{f_N(x_o - \varepsilon + \varepsilon) - f_N(x_o - \varepsilon - \varepsilon)}{2\varepsilon} \right\}$$
$$= \frac{f_N(x_o + 2\varepsilon) + f_N(x_o - 2\varepsilon) - 2f_N(x_o)}{4\varepsilon^2}.$$

As in the first derivative, the second derivative can be estimated so long as f can be estimated. The analytic second derivative is

$$f_N''(x_o) = \frac{1}{Nh^3} \sum_i K'' \left(\frac{x_i - x_o}{h} \right).$$

As f_N' needs a bandwidth larger than that for f, f_N'' needs a bandwidth ever larger than that for f_N'.

Not just derivatives, if necessary, a *functional* $G(f)$ of f such as the integral $\int f(x)^2 dx$ of f^2 can be estimated with $G(f_N) = \int f_N(x)^2 dx$. This is a *plug-in estimator*, for f_N is plugged into $G(\cdot)$. Later, $\int f''(x)^2 dx$ will appear when the choice of h is discussed. It goes without saying that $\int f''(x)^2 dx$ can be estimated by $\int f_N''(x)^2 dx$.

Suppose now that x is a $k \times 1$ vector and we want to estimate $\partial f(x_o)/\partial x_j$ where x_j is the jth component in x. As above, we can use a numerical derivative: defining S_j as the $k \times 1$ null vector with its jth zero replaced by 1, a numerical derivative for $\partial f(x_o)/\partial x_j$ is

$$\frac{f_N(x_o + S_j\varepsilon) - f_N(x_o - S_j\varepsilon)}{2\varepsilon}$$
$$= \frac{1}{Nh^k \cdot 2\varepsilon} \sum_i \left\{ K \left(\frac{x_i - x_o - S_j\varepsilon}{h} \right) - K \left(\frac{x_i - x_o + S_j\varepsilon}{h} \right) \right\}.$$

Analytically, we can differentiate $f_N(x_o)$ with respect to (wrt) x_{oj}:

$$\frac{\partial f_N(x_o)}{\partial x_{oj}} \equiv -\frac{1}{Nh^{k+1}} \sum_i \partial K \left(\frac{x_i - x_o}{h} \right) /\partial x_{oj}.$$

Again, either a numerical derivative or the analytic one can be used. Doing analogously, we can also estimate the second derivative $\partial^2 f_N(x_o)/\partial x_j^2$ if desired. In practice, to avoid differentiation mistakes, it will be safer to use numerical derivatives, although they tend to take more time than the analytic ones.

1.3   Further Remarks

Choosing a kernel is up to the researcher, but usually functions with the following properties are used:

(i) $K(z)$ is symmetric around zero and continuous.

(ii) $\int K(z)dz = 1$, $\int K(z)zdz = 0_k$, and $\int |K(z)|dz < \infty$.

(iii) (a) $K(z) = 0$ if $|z| > z_o$ for some z_o or   (b) $|zK(z)| \to 0$ as $|z| \to \infty$.

(iv) $K(z) = \prod_{j=1}^{k} L(z_j)$, where L satisfies (i) to (iii) for $k = 1$.

Obviously, the condition (iii)(a) implies (iii)(b). In addition to these conditions, often we require $K(z) \geq 0$; this condition combined with $\int K(z)dz = 1$ in (ii) makes various density functions good candidates for K. But so-called "high order kernels" to be seen later take on negative as well as positive values. We may impose further restrictions on K, but so long as we can find a kernel satisfying the restrictions, imposing them should not matter. Almost all kernels used in practice satisfy (i)–(iii).

Examples of K satisfying the preceding conditions for $k = 1$ are

(i) $1[|z| < 1]/2$: uniform kernel

(ii) $N(0,1)$ density $\phi(z)$; "normal" or "Gaussian" kernel

(iii) $(3/4) \cdot (1 - z^2) \cdot 1[|z| < 1]$: "(trimmed) quadratic" kernel

(iv) $(15/16) \cdot (1 - z^2)^2 \cdot 1[|z| < 1]$: "quartic" or "biweight" kernel.

The uniform kernel is not smooth and so rarely used (it renders a histogram), while the other three are frequently used. The normal kernel has the unbounded support and is continuously differentiable up to any order. The trimmed quadratic kernel has a bounded support and continuously differentiable up to the second order over $(-1, 1)$ with non-zero derivatives; it is not smooth at ± 1. The quartic kernel has a bounded support and continuously differentiable once; it is continuously differentiable up to the fourth order over $(-1, 1)$ with non-zero derivatives.

It seems widely agreed that the *choice of kernel makes little difference.* For instance, hardly any difference will be noticeable using (ii), (iii) or (iv) in practice. Also, when $k > 1$, despite some arguments favoring multivariate kernels, it appears that product kernels are simpler to use and do just as well; see, e.g., Kondo and Lee (2003). One point worth mentioning though is bounded versus unbounded supports for K. For example, the trimmed quadratic kernel has a bounded support, whereas ϕ has the unbounded support. A kernel estimator at x_o does a "local weighted averaging" around x_o. A bounded-support kernel gives zero weight to observations far away from x_o, whereas ϕ gives non-zero positive weights to all observations. For the sake of robustness, this aspect favors bounded-support kernels, for they constraint influence from outlying observations (but if one believes that the tail areas of f are informative, then unbounded-support kernels might be better). When x_o is near a boundary point of the x-support, all observations for the local estimation at x_o come only from one side of x_o, which is not desirable.

This point also tends to favor bounded-support kernels, because their weight-ing is more local than unbounded-support kernels. The boundary problem, however, can be avoided simply by restricting x_o to a subset of the support of x.

Differently from choice of kernel, choice of h is crucial. Later we will examine how to choose h in detail; h is called a *bandwidth, smoothing param-eter*, or *window size*. From the above kernel estimators, one can see that, if h is small, then only a few observations are used in calculating $f_N(x_o)$, making $f_N(x_o)$ too jagged as x_o varies. If h is too large, then $f_N(x_o)$ hardly changes as x_o does; at the extreme, $f_N(x_o)$ may become a constant, not changing at all as x_o changes. Thus, a "good" h should be found between these two ex-tremes. Figure 4 illustrates three density estimates: one under-smoothed (the jagged line) with too small a h, one just about right-smoothed (the thick solid line), and one over-smoothed (the relatively flat line) with too big a h. The over-smoothed line dampens the curvature and under-estimates the peak of the density. Since it is easier to smooth with eyes than "un-smooth," if anything, it is recommended to present an estimate slightly under-smoothed rather than over-smoothed.

As will be shown in the following section, the asymptotic distribution of $f_N(x_o)$ is relatively complicated, compared with that of the empirical df $F_N(x_o) = N^{-1}\sum_{i=1}^{N} 1[x_i \le x_o]$. Subtracting $F(x_o)$ and then multiplying by $\sqrt{N}$, we get

$$\sqrt{N}\{F_N(x_o) - F(x_o)\} = \frac{1}{\sqrt{N}}\sum_i \{1[x_i \le x_o] - F(x_o)\}$$
$$\rightsquigarrow N[0, \ F(x_o)\{1 - F(x_o)\}];$$

the asymptotic variance follows from

$$E\{1[x_i \le x_o] - F(x_o)\}^2 = E\{1[x_i \le x_o] - 2 \cdot 1[x_i \le x_o]F(x_o) + F(x_o)^2\}$$
$$= F(x_o) - 2F(x_o)^2 + F(x_o)^2 = F(x_o)\{1 - F(x_o)\}.$$

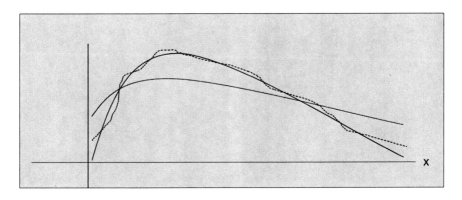

Figure 4: Three Density Estimates (under-, right-, over-smoothed)

$\sqrt{N}\{F_N(x_o) - F(x_o)\}$ is nothing but the centered and scale-normalized version of the binomial rv $\sum_i 1[x_i \leq x_o]$. Note that $F_N(x_o)$ is $\sqrt{N}$-consistent (while $f_N(x_o)$ is not, as will be seen shortly).

1.4 Adaptive Kernel Estimator

One weakness of kernel estimator is that the same h (i.e., the same size neighborhood) is used for all evaluation points. This means that only a few observations get used when x_o falls near the boundary of the x-support. "Adaptive kernel estimator" can overcome the weakness as follows; set $k = 1$ to simplify exposition.

Adaptive kernel estimation is a two step procedure. First, get an initial estimate $f_N(x_o)$ as x_o varies. Second, define a local smoothing parameter λ_i for each x_i such that

$$\lambda_i = \left\{ \frac{g}{f_N(x_i)} \right\}^{\alpha} \quad \text{where } 0 \leq \alpha \leq 1 \text{ (e.g., } \alpha = 0.5) \text{ and}$$

$$g \equiv \prod_i f_N(x_i)^{1/N} \iff \ln(g) = \frac{1}{N} \sum_i \ln f_N(x_i);$$

g is the geometric average of $f_N(x_i)$'s. Then an *adaptive kernel estimator* is

$$\frac{1}{N} \sum_i \frac{1}{h\lambda_i} K \left(\frac{x_i - x_o}{h\lambda_i} \right).$$

The bandwidth $h\lambda_i$ is stretched when f is small, i.e., when only a few observations are available near x_o. Setting $\alpha = 0$ gives the usual kernel estimator. As α goes up, the bandwidth becomes more flexible (large if $f_N(x_o)$ is small, and small if $f_N(x_o)$ is large). Note that λ_i requires getting f_N at each x_i, $i = 1, ..., N$; i.e., each observation becomes an evaluation point.

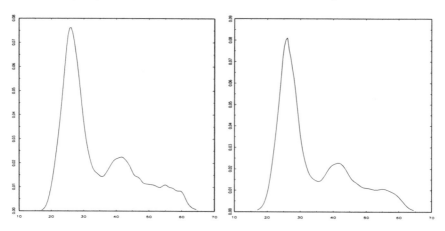

Figure 5: Ordinary and Adaptive Kernel Estimates

Recall the Korean women data. Estimating the age density, Figure 5 presents the usual kernel estimate on the left panel and an adaptive kernel estimate with $\alpha = 0.5$ on the right panel, where the $N(0,1)$ kernel was used with $h = 0.5 \cdot SD(x) \cdot N^{-1/5}$. The usual kernel estimate looks a little under-smoothed on the right tail with this bandwidth, and this problem gets attenuated in the adaptive kernel estimate. But increasing h twice to $SD(x) \cdot N^{-1/5}$ also solves this problem for the usual kernel estimator as can be seen in the earlier age density figure. Thus, although adaptive kernel estimator may be better, its two-stage feature adds more arbitrariness and complications; adaptive kernel estimator will not be further discussed.

2   Consistency and Bandwidth Choice

2.1   Bias and Order of Kernel

Before we examine the bias of f_N, we need the following "change of variables." For an integral $\int \int_A Q(w_1, w_2) dw_1 dw_2$ over a set A for (w_1, w_2), suppose

$$z_j = g_j(w_1, w_2), \ j = 1, 2 \quad \text{with the inverses} \quad w_j = m_j(z_1, z_2), \ j = 1, 2$$

where $\{g_1(w_1, w_2), g_2(w_1, w_2)\}$ is one-to-one, and $m_j, \ j = 1, 2$, are assumed to be continuously differentiable. With the "Jacobian $|J|$ of transformation" being the determinant of the matrix $J \equiv [\partial m_j / \partial z'_{j'}, \ j, j' = 1, 2]$, it holds that

$$\int \int_A Q(w_1, w_2) dw_1 dw_2 = \int \int_B Q\{m_1(z_1, z_2), m_2(z_1, z_2)\} |J| \cdot dz_1 dz_2$$

where B is the set of $\{g_1(w_1, w_2), g_2(w_1, w_2)\}$ as (w_1, w_2) ranges over A and $|J|$ is assumed to be non-zero on B. The formula for more than two variables can be easily inferred from this display. This display also shows that, *if $Q(w_1, w_2)$ is the density of (w_1, w_2), then the density of (z_1, z_2) is $Q\{m_1(z_1, z_2), m_2(z_1, z_2)\}|J|$*. The condition of (g_1, g_2) being one-to-one is critical. If not, split the domain of (g_1, g_2) such that it becomes one-to-one on each segment; then the display is good on each segment.

Turning back to the bias of f_N, for a continuously distributed x, there is no observation that exactly equals the evaluation point x_o unless x_o is chosen such that it is equal to some x_i. One thus has to "borrow" neighboring observations to estimate $f(x_o)$, and this entails the *bias* $Ef_N(x_o) - f(x_o)$. Since $f_N(x_o)$ is a sample average of $h^{-k}K((x_i - x_o)/h), \ i = 1, ..., N$, we get

$$Ef_N(x_o) = \int_{-\infty}^{\infty} \frac{1}{h^k} K\left(\frac{x - x_o}{h}\right) f(x) dx$$

$$= \int_{-\infty}^{\infty} K(z) f(x_o + hz) dz \text{ setting } \underset{k \times 1}{z} = \frac{x - x_o}{h};$$

h^{-k} disappears due to the Jacobian of the transformation:

$$x = (x_{o1} + hz_1, ..., x_{ok} + hz_k)' \implies \underset{k \times k}{\frac{\partial x}{\partial z}} = diag(h, ..., h)$$

$$\implies \det\left(\frac{\partial x}{\partial z}\right) = h^k.$$

Assuming that f has a $k \times 1$ bounded continuous first derivative vector ∇f, for some $x^* \in (x_o, x_o + hz)$, we get

$$Ef_N(x_o) = \int K(z)\{f(x_o) + hz'\nabla f(x^*)\}dz$$

$$= \int K(z)f(x_o)dz + \int hK(z)z'\nabla f(x^*)dz$$

$$= f(x_o) + h \cdot \int K(z)z'\nabla f(x^*)dz.$$

Hence the bias is $O(h)$.

If f has a $k \times k$ bounded continuous second derivative matrix $\nabla^2 f$, then with $\int K(z)zdz = 0$,

$$Ef_N(x_o) = \int K(z)\left\{f(x_o) + hz'\nabla f(x_o) + \frac{h^2}{2}z'\nabla^2 f(x^*)z\right\}dz$$

$$= f(x_o) + O(h^2).$$

If we further assume that f has continuous partial derivatives up to an order $m \geq 3$, then we get ($\sum_{j=1}^{J}(\cdot) = 0$ when $J < j$)

$$Ef_N(x_o) = f(x_o) + \frac{h^2}{2}\int z'\nabla^2 f(x_o)zK(z)dz$$

$$+ \sum_{q=3}^{m-1}\frac{h^q}{q!}\int \sum_{i_1=1}^{k}\cdots\sum_{i_q=1}^{k}\left\{\frac{\partial^q f(x_o)}{\partial x_{i_1}\cdots\partial x_{i_q}}\prod_{j=1}^{q}z_{i_j}\right\}K(z)dz$$

$$+ \frac{h^m}{m!}\int \sum_{i_1=1}^{k}\cdots\sum_{i_m=1}^{k}\left\{\frac{\partial^m f(x^*)}{\partial x_{i_1}\cdots\partial x_{i_m}}\prod_{j=1}^{m}z_{i_j}\right\}K(z)dz.$$

In general, the second-order term does not disappear and thus the bias is still $O(h^2)$. To better understand the second term, observe that, when $m = 4$ and $k = 2$, it becomes $h^3/3!$ times

$$\int \sum_{i_1=1}^{2}\sum_{i_2=1}^{2}\sum_{i_3=1}^{2}\frac{\partial^3 f(x_o)}{\partial x_{i_1}\cdots\partial x_{i_3}}z_{i_1}z_{i_2}z_{i_3}K(z)dz = \int\left\{\frac{\partial^3 f(x_o)}{\partial x_1^3}z_1z_1z_1\right.$$

$$+ \frac{\partial^3 f(x_o)}{\partial x_1^2\partial x_2}z_1z_1z_2 + \frac{\partial^3 f(x_o)}{\partial x_1^2\partial x_2}z_1z_2z_1 + \frac{\partial^3 f(x_o)}{\partial x_1\partial x_2^2}z_1z_2z_2$$

$$+, ..., + \frac{\partial^3 f(x_o)}{\partial x_2^3} z_2 z_2 z_2 \Big\} \cdot K(z) dz.$$

From now on, unless otherwise noted, we will always assume that $\nabla^2 f$ is continuous and bounded over the range of x on which f is estimated.

If we choose $K(z)$ such that the terms of higher order than the first disappear, then we can make the bias smaller than $O(h^2)$. Such a kernel is called a *high order kernel*, which, however, has the following problem. For a scalar x, the second-order term becomes $(h^2/2)f''(x_o) \int K(z)z^2 dz$. To have $\int K(z)z^2 dz = 0$, $K(z)$ should be negative for some values of z, which backs away from the notion of weighting by $K(z)$. For instance, with ϕ being the $N(0,1)$ density, consider

$$K(z) = a_0 \phi(z) + a_1 z^2 \phi(z)$$

where a_0 and a_1 are chosen such that

$$\int K(z) dz = 1 \implies a_0 + a_1 = 1$$

$$\left(\text{note} : \int z K(z) dz = 0 \text{ for any } a_0 \text{ and } a_1 \right),$$

$$\int z^2 K(z) dz = 0 \implies a_0 + a_1 \int z^4 \phi(z) dz = 0$$

$$\implies a_0 + a_1 3 = 0 \text{ because } \int z^4 \phi(z) dz = 3.$$

Solving these two equations, we get $a_0 = 3/2$ and $a_1 = -1/2$: the ϕ-based kernel with $\int z^2 K(z) dz = 0$ is

$$\frac{3}{2} \phi(z) - \frac{1}{2} z^2 \phi(z).$$

In general, if $K(\cdot)$ satisfies $\int z^j K(z) dz = 0 \ \forall j = 1, ..., \kappa - 1$, then $K(\cdot)$ is said to be an order-κ (or κth order) kernel.

Going further, if desired, the ϕ-based kernel with $\int z^2 K(z) dz = 0$ and $\int z^4 K(z) dz = 0$ is

$$\frac{3}{2} \phi(z) - \frac{1}{2} z^2 \phi(z) + \frac{1}{35} z^4 \phi(z).$$

Because $\int z^5 K(z) dz = 0$, this is an order-6 kernel, which yields an $O(h^6)$ or smaller bias. See Bierens (1987, p. 112) for multivariate versions of these. A polynomial kernel with $\int z^2 K(z) dz = 0$ is

$$\frac{15}{32} (3 - 10z^2 + 7z^4) \cdot 1[|z| < 1]$$

which is an extension of the quadratic kernel. As $\int z^3 K(z) dz = 0$, this is an order-4 kernel. A polynomial kernel with $\int z^4 K(z) dz = 0$ is

$$\frac{315}{2048} (15 - 140z^2 + 378z^4 - 396z^6 + 143z^8) 1[|z| < 1].$$

As $\int z^5 K(z)dz = 0$, this is an order-6 kernel. A table in Müller (1988, p. 68) lists univariate polynomial kernels indexed by three parameters ν, k, and μ; set $\nu = 0$ in the table and choose k and μ which stand for, respectively, $\int z^j K(z)dz = 0$ for $0 \leq j < k$ and $\mu - 1$ times continuous differentiability of $K(z)$ on the real line R.

A k-variate kernel $K(z)$ is an order-m kernel if

$$\int z_1^{j_1} \cdots z_k^{j_k} K(z)dz = 0 \quad \text{for all } 0 < j_1 + \cdots + j_k \leq m - 1.$$

Using an order-m kernel makes the bias at most $O(h^m)$ in view of the above Taylor's expansion of order m. In a small sample, high-order kernels may make $f_N(x_o)$ negative. Unless otherwise noted, we will always use kernels with $K(z) \geq 0$.

2.2   Variance and Consistency

The $O(h^2)$ bias term disappears if $h \to 0$ as $N \to \infty$. If $V\{f_N(x_o)\} \to 0$ as well, then we get $f_N(x_o) \to^p f(x_o)$. In this subsection, we show this first and then prove $V f_N(x_o) \to 0$.

Invoking the triangle inequality,

$$|f_N(x_o) - f(x_o)| \leq |f_N(x_o) - Ef_N(x_o)| + |Ef_N(x_o) - f(x_o)|.$$

The event $|f_N(x_o) - f(x_o)| > 2\varepsilon$ for a constant $\varepsilon > 0$ implies that the right-hand side (rhs) is greater than 2ε. This implication yields

$$P(|f_N(x_o) - f(x_o)| \geq 2\varepsilon)$$
$$\leq \ P(|f_N(x_o) - Ef_N(x_o)| + |Ef_N(x_o) - f(x_o)| \geq 2\varepsilon)$$

because "A implying B" means $A \subseteq B \iff P(A) \leq P(B)$.

Since $\lim_{N \to \infty} Ef_N(x_o) = f(x_o)$ in the last term, there exists $N_o = N_o(\varepsilon)$ such that $|Ef_N(x_o) - f(x_o)| \leq \varepsilon \ \forall N \geq N_o$, which implies that the neighboring term $|f_N(x_o) - Ef_N(x_o)|$ should be greater than or equal to $\varepsilon \ \forall N \geq N_o$; otherwise the sum is less than 2ε. Hence for all $N \geq N_o$, the rhs should be less than $P(|f_N(x_o) - Ef_N(x_o)| \geq \varepsilon)$. By Chebyshev's inequality,

$$P(|f_N(x_o) - Ef_N(x_o)| \geq \varepsilon) \leq \frac{V f_N(x_o)}{\varepsilon^2}.$$

Therefore, because $V f_N(x_o) \to 0$ as $N \to \infty$, $P(|f_N(x_o) - Ef_N(x_o)| \geq \varepsilon) \to 0$, and we get

$$P(|f_N(x_o) - f(x_o)| \geq 2\varepsilon) \to 0 \iff f_N(x_o) \to^p f(x_o).$$

To show $V f_N(x_o) \to 0$, observe

$$V f_N(x_o) = \frac{\text{each term's variance}}{N}$$

$$= \frac{1}{N} \left(E\left\{ \frac{1}{h^k} K\left(\frac{x - x_o}{h}\right)\right\}^2 - \left[E\left\{ \frac{1}{h^k} K\left(\frac{x - x_o}{h}\right)\right\}\right]^2 \right)$$

$$= \frac{1}{N} \left(\frac{1}{h^{2k}} \int K\left(\frac{x - x_o}{h}\right)^2 f(x) dx - \{f(x_o) + O(h^2)\}^2 \right)$$

$$= \frac{1}{N} \left(\frac{1}{h^k} \int K(z)^2 f(x_o + hz) dz + O(1) \right)$$

$$= \frac{1}{Nh^k} f(x_o) \int K(z)^2 dz + o\left(\frac{1}{Nh^k}\right).$$

Therefore, if $Nh^k \to \infty$, then we get $V f_N(x_o) \to 0$ as $N \to \infty$. Adding the condition $h \to 0$ as $N \to \infty$ for the bias, we get the desired result:

$$f_N(x_o) \to^p f(x_o) \quad \text{if } h \to 0 \text{ and } Nh^k \to \infty \text{ as } N \to \infty.$$

The *bandwidth h should be small for the bias, but not too small for the variance.* If h is too small, the small number of data points around x_o will result in a high variance for $f_N(x_o)$. If h is too large, the large number of data far away from x_o will result in a high bias for $f_N(x_o)$. This way of selecting h by balancing the bias and variance will be formalized later.

Although it is good to have $f_N(x_o) \to^p f(x_o)$ at a given point x_o (*pointwise consistency*), it would be better to have *uniform consistency*

$$\sup_{x \in X} |f_N(x) - f(x)| = o_p(1)$$

where X is a chosen range of x. Uniform consistency means that the maximum deviation of $f_N(x_o)$ from $f(x_o)$ converges to zero in probability. In pointwise consistency, for any small constants $\varepsilon, \varepsilon' > 0$, there exists $N(\varepsilon, \varepsilon', x_o)$ such that

$$P(|f_N(x_o) - f(x_o)| > \varepsilon\} < \varepsilon' \quad \text{for all } N \geq N(\varepsilon, \varepsilon', x_o).$$

In uniform consistency, there exists $N(\varepsilon, \varepsilon')$ not depending on x_o such that

$$P(|f_N(x_o) - f(x_o)| > \varepsilon\} < \varepsilon' \quad \text{for all } N \geq N(\varepsilon, \varepsilon').$$

The latter assures that $f_N(x_o)$ is arbitrarily close to $f(x_o)$ *regardless of x_o* so long as N is greater than some threshold, whereas no such assurance is available in the former because the threshold varies across x_o.

Put it differently, what we desire is the "graph" $f(x)$, $x \in X$. But what $\lim_{N \to \infty} f_N(x)$, $x \in X$, offers is only the "graph of pointwise limits"—imagine connecting $f_N(x^{(1)}), ..., f_N(x^{(M)})$. Under pointwise consistency, there is no guarantee that the graph of pointwise limits is close to $f(x)$, $x \in X$. Uniform consistency assures that the graph of pointwise limits is close to the "limit of graphs," because the maximum difference between the estimated graph and the true graph goes to zero in probability. In most cases, whenever pointwise consistency holds, the uniform consistency holds as well under some extra regularity conditions, the most notable of which is

$$h \to 0 \text{ and } \frac{Nh^k}{\ln(N)} \to \infty \text{ as } N \to \infty.$$

The h in $Nh^k / \ln(N) \to \infty$ should be greater than the h in $Nh^k \to \infty$. Further information (e.g., uniform convergence rate) can be found in Giné et al. (2004), Einmahl and Mason (2005), and the references therein.

2.3   Choosing Bandwidth with MSE

In this and the following subsections, we discuss how to choose the band-width h. First, in this subsection, we explore choosing h by minimizing the mean squared error $E\{f_N(x_o) - f(x_o)\}^2$ or its integrated version. Then in the following subsection, a "cross-validation" method is introduced, which is a data-driven automatic method of choosing h.

As well known, *mean squared error (MSE)* is Variance + Bias^2. For $f_N(x_o)$,

$$MSE\{f_N(x_o),\ f(x_o)\} = E\{f_N(x_o) - f(x_o)\}^2$$

$$\simeq \frac{1}{Nh^k} f(x_o) \int K(z)^2 dz + \frac{h^4}{4} \left\{ \int z' \nabla^2 f(x_o) z K(z) dz \right\}^2.$$

To simplify exposition, consider a product kernel $K(z) = \prod_{j=1}^{k} L(z_j)$ to get

$$\int zz'K(z)dz = \kappa I_k \quad \text{where} \quad \kappa \equiv \int z_j^2 L(z_j)dz_j;$$

e.g., if $L = \phi$, then $\kappa = 1$ and $\int zz'K(z)dz$ is the covariance matrix of k-many independent $N(0, 1)$ rv's. With the product kernel, the term inside $\{\cdot\}$ in the preceding display is

$$\int trace\{z'\nabla^2 f(x_o)z\}K(z)dz = \int trace\{zz'\nabla^2 f(x_o)\}K(z)dz$$

$$= \kappa \sum_{j=1}^{k} \frac{\partial^2 f(x_o)}{\partial x_{oj}^2}.$$

Therefore

$$MSE\{f_N(x_o), f(x_o)\} \simeq \frac{1}{Nh^k} f(x_o) \int K(z)^2 dz + \frac{h^4}{4} \{\kappa \sum_{j=1}^{k} \frac{\partial^2 f(x_o)}{\partial x_{oj}^2}\}^2.$$

MSE measures the "local error" of estimation around x_o. The "global error" can be measured by the *integrated mean squared error*, which removes x_o by integration:

$$IMSE\{f_N(x_o), f(x_o)\} \equiv \int E\{f_N(x_o) - f(x_o)\}^2 dx_o$$

$$\simeq \frac{1}{Nh^k} \int K(z)^2 dz + \frac{h^4}{4} \kappa^2 \int \{\sum_{j=1}^{k} \frac{\partial^2 f(x_o)}{\partial x_{oj}^2}\}^2 dx_o$$

$$\equiv \frac{1}{Nh^k} A + \frac{h^4}{4} B, \left(A \equiv \int K(z)^2 dz, \ B \equiv \kappa^2 \int \left\{ \sum_{j=1}^{k} \frac{\partial^2 f(x_o)}{\partial x_{oj}^2} \right\}^2 dx_o \right)$$

$$= \frac{A}{N} h^{-k} + \frac{B}{4} h^4.$$

This can be controlled by the choice of h and K. Choice of K is much more difficult to address than that of h, for K is a function. But it is known in the literature that the choice of K is not crucial, with $f_N(x_o)$ varying not much as K varies. Hence we first choose h by minimizing this display. Then we discuss how to choose K in the simple case $k = 1$. Interchanging $\int$ and E, integrated mean squared error is usually called *mean integrated squared error (MISE)*, and we will use this terminology from now on.

Differentiating MISE wrt h, the optimal h, say h_0, is

$$h_0 = \left(\frac{kA}{B} \right)^{1/(k+4)} \cdot N^{-1/(k+4)} \quad \text{and} \quad N \cdot h_0^k = \left(\frac{kA}{B} \right)^{k/(k+4)} \cdot N^{4/(k+4)}.$$

The result is more illuminating for $k = 1$:

$$MISE\{f_N(x_o), f(x_o)\} = \frac{1}{Nh} \int K(z)^2 dz + \frac{h^4}{4} \kappa^2 \int f''(x_o)^2 dx_o,$$

$$h_0 = \left(\frac{\int K(z)^2 dz}{\kappa^2 \int f''(x_o)^2 dx_o} \right)^{1/5} N^{-1/5}.$$

Here $\int f''(x_o)^2 dx_o$ measures a variation in f. Hence, if $f(x)$ is highly variable, then h_0 is small. If $K = \phi$ and f is also a normal density, then $h_0 \simeq N^{-1/5} SD(x)$ as can be seen in Silverman (1986, p. 45). This simple h_0 often works well in practice, even if K and f are not normal. Extending the *rule of thumb* to $k > 1$ yields (see Scott (1992, p. 152))

$$h_j = N^{-1/(k+4)} SD(x_j), \quad j = 1, ..., k.$$

This rule-of-thumb bandwidth can be used as an initial bandwidth in practice, and it works well.

With $h = O(N^{-1/(k+4)})$, we get

$$bias^2 = O(h^4) = O(N^{-4/(k+4)}) \text{ and}$$

$$variance = O\{(Nh^k)^{-1}\} = O(N^{-4/(k+4)}).$$

That is, both *variance and bias^2 in MISE converge to zero at the same rate if $h = h_0$*. Decreasing one faster than the other increases MISE. Recall that for a sample mean $\bar{x}_N = N^{-1}\sum_i x_i$ where $V(x) = \sigma^2$, its MSE goes to zero at $O(N^{-1})$, for $MSE(\bar{x}_N) = V(\bar{x}_N) = \sigma^2/N$. In kernel density estimation, MISE with h_0 converges to zero at $O(N^{-4/(k+4)})$—a bit slower than $O(N^{-1})$. MISE converges to zero more slowly as k increases.

Suppose we want to assure a constant MISE as we change k. This requires increasing N accordingly. To see how much higher N is needed, take ln on $MISE = N^{-4/(k+4)}$ (times a constant) to get $\ln(MISE) = -\{4/(k+4)\}\ln N$ (plus a constant). To make $\ln(MISE)$ constant, as k increases linearly, N should increase exponentially. Viewed differently, as k goes up, the "volume" of a local neighborhood increases exponentially, and N should increase accordingly to fill the volume proportionally. This problem in nonparametric local averaging is called *"the curse of dimensionality,"* and it in fact applies to all nonparametric methods one way or another.

Although h_0 shows what is involved in choosing h, h_0 is not useful as such, because h_0 depends on the unknown f''. Plugging an estimator f''_N for f'' into f'', we can estimate h_0, which is a "plug-in" estimator. But estimating f'' to estimate f is unattractive, for f'' is more difficult to find than f [see, e.g., Scott (1992, pp. 131–132)]. We need an easier bandwidth choice method that is automatic once a data set is given, which is explored in the following.

2.4   Choosing Bandwidth with Cross-Validation

There are many automatic data-driven methods to select h. One popular method is the *least squares cross validation* minimizing an estimator for the *integrated squared error (ISE)*:

$$\int \{f_N(x_o) - f(x_o)\}^2 dx_o = \int f_N(x_o)^2 dx_o - 2\int f_N(x_o)f(x_o)dx_o$$
$$+ \int f(x_o)^2 dx_o$$

wrt h. Drop the last term that is independent of h. To minimize ISE in a data-driven way, we should rewrite the first and second terms into a function of h and the data. It is shown in the following two paragraphs that the first two terms of the ISE can be approximated with

$$\frac{1}{N^2 h^k}\sum_i\sum_j K^{(2)}\left(\frac{x_i - x_j}{h}\right) - \frac{2}{N(N-1)h^k}\sum_i\sum_{j,i\neq j} K\left(\frac{x_i - x_j}{h}\right)$$

where

$$K^{(2)}(a) \equiv \int K(a - z)K(z)dz.$$

Minimizing this wrt h yields an optimal bandwidth, which works fairly well in practice.

The first term $\int f_N(x_o)^2 dx_o$ in the ISE is

$$\int \frac{1}{N^2 h^{2k}} \sum_i \sum_j K\left(\frac{x_i - x}{h}\right) K\left(\frac{x_j - x}{h}\right) dx$$

$$= \frac{1}{N^2 h^{2k}} \sum_i \sum_j \int K\left(\frac{x_i - x}{h}\right) K\left(\frac{x_j - x}{h}\right) dx.$$

Let $-z = (x_j - x)/h \iff x/h = x_j/h + z$ to get

$$\int K\left(\frac{x_i}{h} - \frac{x}{h}\right) K\left(\frac{x_j - x}{h}\right) dx = h^k \int K\left(\frac{x_i}{h} - \frac{x_j}{h} - z\right) K(z) dz$$

$$= h^k \int K\left(\frac{x_i - x_j}{h} - z\right) K(z) dz = h^k K^{(2)}\left(\frac{x_i - x_j}{h}\right).$$

Hence

$$\int f_N(x_o)^2 dx_o = \frac{1}{N^2 h^k} \sum_i \sum_j K^{(2)}\left(\frac{x_i - x_j}{h}\right).$$

As for the second term $-2 \int f_N(x_o) f(x_o) dx_o$ in the ISE, observe

$$\int f_N(x) f(x) dx \simeq \frac{1}{N} \sum_{j=1}^N f_N(x_j) \quad \left(\text{replacing } E(\cdot) \text{ with } \frac{1}{N} \sum_j (\cdot)\right)$$

$$\simeq \frac{1}{N} \sum_j f_{N-1}(x_j) \left(\text{with } f_{N-1}(x_j) \equiv \frac{1}{(N-1)h^k}\right.$$

$$\left. \sum_{i,i\neq j} K\left(\frac{x_i - x_j}{h}\right)\right)$$

$$= \frac{1}{N(N-1)h^k} \sum_i \sum_{j,i\neq j} K\left(\frac{x_i - x_j}{h}\right).$$

Replacing $f_N(x_j)$ with $f_{N-1}(x_j)$ means that x_j is not used in estimating $f(x_j)$. This "leave-one-out" scheme yields the name "cross-validation." The qualifier "least squares" comes from the squared loss function in ISE.

Least squares cross-validation has been initiated by Rudemo (1982) and Bowman (1984), and there are slight variations in approximating ISE; e.g., using $N^{-1} \sum_j \int f_{N-1}(x)^2 dx$ for the first term of ISE, and using N^2 instead of $N(N-1)$ for the second term. There are other cross-validation schemes, e.g., "likelihood cross-validation" maximizing $\Pi_j f_{N-1}(x_j)$, which is, however, known to be susceptible to tails of f. The likelihood cross-validation shows the motivation to leave-one-out: if we maximize $\Pi_j f_N(x_j)$ wrt h without leave-one-out, then we will get $h \simeq 0$ because $h \simeq 0$ yields (for a given N)

$$\frac{1}{Nh^k} K\left(\frac{x_j - x_j}{h}\right) \simeq \infty \text{ as } K(.) \text{ is bounded.}$$

Consider two iid rv's z_1 and z_2 drawn from density K. Then

$$P(z_1 + z_2 \leq a) = \int P(z_2 \leq a - z_1 | z_1) K(z_1) dz_1$$

$$= \int P(z_2 \leq a - z_1) K(z_1) dz_1 \quad \text{due to the independence of } z_1 \text{ and } z_2.$$

Differentiating this wrt a, we get $\int K(a - z_1) K(z_1) dz_1$, the $z_1 + z_2$-density evaluated at a. This shows that $K^{(2)}(a)$ is the density (evaluated at a) of the sum of two iid rv's from K. For instance, if $K(z) = \phi(z)$, then $K^{(2)}(z)$ is simply $(1/\sqrt{2})\phi(z/\sqrt{2})$, the density of $N(0,2)$, because $z_1 + z_2$ follows $N(0,2)$ when z_1 and z_2 are iid $N(0,1)$. More generally, if z_1 is from density K, z_2 is from density M, and z_1 and z_2 are independent, then the density (at a) of the sum is $\int M(a - z_1) K(z_1) dz_1$, which is called the *convolution* of K and M; $K^{(2)}$ is thus the "convolution" of K with itself. When K is not ϕ, $K^{(2)}(a)$ can be obtained with "Monte Carlo Integration" as will be seen shortly.

One drawback of the least-squares cross-validation idea is that the minimand tends to have several local minima particularly in h-small areas (Marron, 1988, p. 196); i.e., there is a good chance for under-smoothing. Also since the minimand is ISE, not MISE, the optimal choice is only good for the particular data set, not for any sample of the same size; another data set from the same population may require a rather different h. Nevertheless, Hall (1983) and Stone (1984) proved that the least squares cross validation choice h_{cv} minimizing the above ISE is optimal in the sense that, as $N \to \infty$, the ISE minimand evaluated at h_{cv} converges a.s. to

$$\min_h \int \{f_N(x, h) - f(x)\}^2 dx - \int f(x)^2 dx$$

where we write $f_N(x, h)$ instead of $f_N(x)$ to make the dependence on h explicit. See Marron (1988) and Park and Marron (1990) for more on choosing h and data-driven methods. Despite some new developments in bandwidth choice methods as in Jones et al. (1996), cross validation (CV) bandwidth seems to be a reliable method in practice as Loader (1999b) also advocates. But if k is small, say 1 or 2, then nothing beats choosing bandwidth by drawing $f_N(x)$ over a range of x—called "eye-balling" or "trial and error" method.

3   Asymptotic Distribution

Since
$$V f_N(x_o) = \frac{1}{Nh^k} f(x_o) \int K(z)^2 dz + o\left(\frac{1}{Nh^k}\right)$$
multiplying $f_N(x_o) - f(x_o)$ by $\sqrt{Nh^k}$ will give a non-degenerate asymptotic variance with the asymptotic distribution being normal. That is, the main

result to be established in this subsection is

$$\sqrt{Nh^k}\{f_N(x_o) - f(x_o)\} \rightsquigarrow N\left(0,\ f(x_o)\int K(z)^2 dz\right).$$

First, we introduce the Lindeberg CLT. Second, this asymptotic distribution is proved using the CLT, and confidence intervals for $f(x_o)$ are constructed. Third, instead of the single point x_o, we may be interested in multiple evaluation points $x^{(1)}, ..., x^{(m)}$, for which "confidence bands" are introduced.

3.1   Lindeberg CLT

Observe that

$$\sqrt{Nh^k}f_N(x_o) = \frac{1}{\sqrt{Nh^k}}\sum_i K\left(\frac{x_i - x_o}{h}\right) = \sum_i \frac{1}{\sqrt{Nh^k}}K\left(\frac{x_i - x_o}{h}\right).$$

We will be applying the Lindeberg CLT for "triangular arrays" or "double arrays," because the summands in the sum are changing with N. In an ordinary CLT, we deal with independent arrays: for $N = 10$, we draw 10 observations, and for $N = 11$, we throw away those 10 and redraw 11 new observations to have each array (the observations with a given N) independent of the other arrays. In the above sum, h_N (now we use h_N to make the dependence of h on N explicit) gets smaller as $N \to \infty$; for instance, $h_{11} < h_{10}$. Thus an "observation" $K((x_i - x_o)/h_{11})$ with $N = 11$ tends to be smaller than an observation $K((x_i - x_o)/h_{10})$ with $N = 10$. Hence an array with a given N is not independent of another array with a different N. A CLT good for a triangular array allows dependence across arrays. The expression "triangular" is due to that putting each array horizontally and stacking the arrays vertically yields a triangular shape; with N being the row number, each row has one more term than the preceding row. The following is the Lindeberg CLT for triangular arrays.

Lindeberg CLT for Triangular Arrays: For triangular arrays $\{z_{Ni}\}$ with independence holding within each array, let

$$S_N \equiv \sum_{i=1}^{N} z_{Ni}, \quad E z_{Ni} = 0, \quad \sigma_N^2 \equiv \sum_{i=1}^{N} E z_{Ni}^2.$$

Then

$$\frac{S_N}{\sigma_N} \rightsquigarrow N(0,1) \quad \text{if} \quad \sum_{i=1}^{N} E\left\{\left(\frac{z_{Ni}}{\sigma_N}\right)^2 \cdot 1\left[\left|\frac{z_{Ni}}{\sigma_N}\right| \geq \varepsilon\right]\right\} \to 0 \quad \text{as } N \to 0$$

for any constant $\varepsilon > 0$; this condition is called the "Lindeberg condition."

An example of the Lindeberg CLT can be seen in $S_N = (1/\sqrt{N})$ $\sum_i w_i = \sum_i (1/\sqrt{N}) w_i$ where w_i's are iid with $E(w_i) = 0$ and $V(w_i) = 1$. Set $z_{Ni} = (1/\sqrt{N}) w_i$. Then $\sigma_N^2 = 1$ and the Lindeberg condition is

$$\sum_i E\left\{ \frac{w_i^2}{N} \cdot 1\left[\left|\frac{w_i}{\sqrt{N}}\right| \geq \varepsilon\right]\right\} = \frac{1}{N}\sum_i E(w_i^2 1\left[|w_i| \geq \varepsilon\sqrt{N}\right])$$

$$= E\left(w^2 1\left[|w| \geq \varepsilon\sqrt{N}\right]\right).$$

The term $E(w^2 1[\cdot])$ is dominated by $E(w^2) = 1$. Since $1[|w| \geq \varepsilon\sqrt{N}] \to^p 0$, $E(w^2 1[|w| \geq \varepsilon\sqrt{N}]) \to 0$ due to the "dominated convergence theorem."

3.2   Confidence Intervals

In applying the Lindeberg CLT, instead of $(Nh^k)^{-1/2} \sum_i K((x_i - x_o)/h)$, consider its centered version to have $Ez_{Ni} = 0$:

$$\sqrt{Nh^k}\{f_N(x_o) - Ef_N(x)\} = \sum_i \frac{1}{\sqrt{Nh^k}}\left\{K\left(\frac{x_i - x_o}{h}\right) - EK\left(\frac{x_i - x_o}{h}\right)\right\}.$$

Let $z_{Ni} = (Nh^k)^{-0.5}\{K((x_i - x_o)/h) - EK((x_i - x_o)/h)\}$ and note

$$\sigma_N^2 = V[\sqrt{Nh^k}\{f_N(x_o) - Ef_N(x)\}] = Nh^k \cdot Vf_N(x_o)$$

$$= f(x_o)\int K(z)^2 dz + o(1).$$

The Lindeberg condition holds by doing analogously to what was done for the example $S_N = (1/\sqrt{N}) \sum_i w_i$ above. Therefore we have

$$\sqrt{Nh^k}\{f_N(x_o) - Ef_N(x_o)\} \rightsquigarrow N\left(0, \ f(x_o)\int K(z)^2 dz\right).$$

Observe now

$$\sqrt{Nh^k}\{f_N(x_o) - f(x_o)\} = \sqrt{Nh^k}\{f_N(x_o) - Ef_N(x_o)\}$$
$$+ \sqrt{Nh^k}\{Ef_N(x_o) - f(x_o)\}.$$

To get the asymptotic distribution for $\sqrt{Nh^k}\{f_N(x_o) - f(x_o)\}$, we need the second (bias) term to disappear:

$$\sqrt{Nh^k}\{Ef_N(x_o) - f(x_o)\} = o(1).$$

Since we know $Ef_N(x_o) - f(x_o) = O(h^2)$, this display holds if

$$O\{(Nh^k)^{0.5} \cdot h^2\} = O\{(Nh^{k+4})^{0.5}\} = o(1).$$

Hence, by choosing h such that $Nh^{k+4} \to 0$ (while $Nh^k \to \infty$ for the asymptotic variance to converge to zero), we get the desired asymptotic distribution result: as $N \to \infty$,

$$\sqrt{Nh^k}\{f_N(x_o) - f(x_o)\} \rightsquigarrow N\left(0,\ f(x_o)\int K(z)^2 dz\right),\ \text{if}$$

$$Nh^{k+4} \to 0 \text{ and } Nh^k \to \infty.$$

Here, $f_N(x_o)$ is "$\sqrt{Nh^k}$-consistent" for $f(x_o)$. From this display, an asymptotic 95% "point-wise" confidence interval (CI) for $f(x_o)$ is

$$f_N(x_o) \pm 1.96 \left\{\frac{f_N(x_o)\int K(z)^2 dz}{Nh^k}\right\}^{0.5}.$$

In the CI, $\int K(t)^2 dt$ can be evaluated analytically, but it can also be done numerically as follows. Observe, with ϕ being the $N(0,1)$ density,

$$\int K(t)^2 dt = \int \frac{K(t)^2}{\phi(t)}\phi(t)dt = E\left\{\frac{K(z)^2}{\phi(z)}\right\} \quad (\text{where } z \sim N(0,1))$$

$$\simeq \frac{1}{T}\sum_{j=1}^{T} \frac{K(z_j)^2}{\phi(z_j)} \quad \text{where } z_1, ..., z_T \text{ are iid } N(0,1).$$

That is, $\int K(t)^2 dt$ can be numerically obtained by generating T-many $N(0,1)$ variables $z_1, ..., z_T$ and then taking the sample average of $K(z_j)^2/\phi(z_j)$, $j = 1, ..., T$. This way of approximating an integral with a pseudo sample is called *Monte Carlo integration*—the idea already employed in methods of simulated moments. One caution is that, if the support of K deviates too much from, say, $[-3.3,\ 3.3]$ (the "99.9% support" of ϕ), then the Monte Carlo integration may not work well in practice. In this case, a density with its support close to that of K should be used. As an extreme example, if K's support is $[-5,5]$ with large masses near ± 5, then use the density of $U[-5,5]$ instead of ϕ because $z = \pm 5$ from $N(0,1)$ may never be realized unless $T \simeq \infty$.

One problem with the above rate $Nh^{k+4} \to 0$ is that the optimal h, h_0, in the preceding subsection does not satisfy $Nh_0^{k+4} \to 0$: h_0 is such that Nh_0^{k+4} converges to a non-zero constant, because h_0 is proportional to $N^{-1/(k+4)}$ which implies $Nh_0^{k+4} = NN^{-1} = 1$. Hence, $\sqrt{Nh_0^k}\{Ef_N(x_o) - f(x_o)\}$ converges to a non-zero constant, which is called the *asymptotic bias*. The optimal bandwidth decreases both variance and bias^2 in MISE at the same rate. For the above asymptotic distribution, the bias gets reduced faster than the variance with $Nh^{k+4} \to 0$, which is a deliberate "under-smoothing."

If desired, one can estimate and remove the asymptotic bias instead of doing under-smoothing. Recall that, when $K(z) = \prod_{j=1}^{k} L(z_j)$ is used with $\kappa \equiv \int z_j^2 L(z_j)dz_j$, we get

$$\sqrt{Nh^k}\{Ef_N(x_o) - f(x_o)\} \simeq \sqrt{Nh^k}\frac{h^2}{2}\int z'\nabla^2 f(x_o)zK(z)dz$$

$$= (Nh^{k+4})^{0.5} \frac{\kappa}{2} \sum_{j=1}^{k} \frac{\partial^2 f(x_o)}{\partial x_{oj}^2} \equiv B_N$$

$$\implies \sqrt{Nh^k}\{f_N(x_o) - f(x_o)\} \rightsquigarrow N\left(B_N, \ f(x_o) \int K(z)^2 dz\right).$$

Consistently estimating B_N with $\hat{B}_N$, which requires in turn estimating the second derivatives, we would get

$$\text{95\% CI for } f(x_o): f_N(x_o) - \frac{\hat{B}_N}{\sqrt{Nh^k}} \pm 1.96 \frac{\{f_N(x_o) \int K(z)^2 dz\}^{0.5}}{\sqrt{Nh^k}}.$$

Although this may be theoretically pleasing, this is not necessarily a good idea, because estimating B_N is harder than estimating $f(x_o)$.

3.3   Confidence Bands

It can be shown that the asymptotic covariance between $(Nh^k)^{1/2}$ $\{f_N(x_a) - f(x_a)\}$ and $(Nh^k)^{1/2}\{f_N(x_b) - f(x_b)\}$ for $x_a \neq x_b$ is zero, so that the multivariate normal asymptotic distribution holds straightforwardly with a diagonal covariance matrix. In essence, this is because the averaging at x_a and x_b is local, and as $N \to \infty$, the number of observations used for both $f_N(x_a)$ and $f_N(x_b)$ becomes negligible. But $(1-\alpha)100\%$ *pointwise CI's* at $x^{(1)}, ..., x^{(m)}$ does not give a $(1-\alpha)100\%$ *joint CI's*, for the coverage probability of the former becomes $(1-\alpha)^m$. That is, if we desire a $(1-\alpha)100\%$ joint CI's for m evaluation points, then under the asymptotic independence, we should set

$$(1-\alpha)^m = 0.95 \iff m\ln(1-\alpha) = \ln 0.95 \iff \alpha = 1 - \exp\left(\frac{\ln 0.95}{m}\right).$$

Viewed more simply using $\ln(1+z) \simeq z$ when $z \simeq 0$, the middle equation in the display becomes

$$m \cdot \alpha = 0.05 \iff \alpha = \frac{0.05}{m}.$$

For instance, with $m = 20$, we get $\alpha = 0.05/20 = 0.0025$: it takes 99.75% pointwise CI's to get a 95% joint CI's across 20 evaluation points. With $m = 50$, $\alpha = 0.05/50 = 0.001$, it takes 99.9% pointwise CI's (the critical values ± 3.3) to get a 95% joint CI's across 50 evaluation points.

The above procedure can be too conservative, which can be seen from the *Bonferroni's inequality*: for possibly dependent events $A_1, ..., A_m$,

$$P(\cap_{j=1}^m A_j) \geq \sum_{j=1}^{m} P(A_j) - (m-1).$$

Suppose $P(A_j) = 1 - \alpha/m \ \forall j$. Then

$$P(\cap_{j=1}^m A_j) \geq (1 - \frac{\alpha}{m})m - (m-1) = m - \alpha - (m-1) = 1 - \alpha:$$

having a pointwise confidence level $1 - \alpha/m$, the joint CI's over m evaluation points has the coverage probability *at least* $1-\alpha$. As m gets larger and thus as x_a gets closer to x_b, there would be dependence between $f_N(x_a)$ and $f_N(x_b)$ in a finite sample, which might yield a higher coverage probability than that under the asymptotic independence.

Suppose now we connect the m-many joint CI's to get an "*artificial confidence band.*" Looking at the artificial confidence band instead of looking at the m-many joint CI's, we may get an overly optimistic picture, because the correct band should be based on infinite number of evaluation points while only a finite m is used in the artificial confidence band. As m increases, the critical value at each pointwise CI increases as well. The "limit" of this may be taken as an 'uniform confidence band'.

Instead of the artificial confidence band connecting joint CI's, Bickel and Rosenblatt (1973) presented an (*uniform*) *confidence band* when $k = 1$ and the range of x is $[0,1]$. For an unbounded-support kernel, their uniform confidence band is the expression in $\{\cdot\}$ in the following:

$$\lim_{N \to \infty} P \left[f(x) \in \left\{ f_N(x) \pm \frac{\{f_N(x) \int K(t)^2 dt\}^{0.5}}{\sqrt{Nh}} \left(d_N + \frac{\lambda}{\sqrt{2\delta \ln N}} \right) \right\} \forall x \right]$$

$$= \exp(-2e^{-\lambda})$$

where $d_N \equiv \sqrt{2\delta \ln N} + \dfrac{\ln\{(\hat{K}/2)^{1/2}\pi^{-1}\}}{\sqrt{2\delta \ln N}}$,

$$h = N^{-\delta}, \quad \frac{1}{5} < \delta < \frac{1}{2}, \quad \hat{K} \equiv \frac{\int \{K'(t)\}^2 dt}{2 \int K(t)^2 dt};$$

$\hat{K}$ can be obtained numerically, $\sqrt{2\delta \ln N} = \sqrt{2 \ln h^{-1}}$ for $h = N^{-\delta}$, and λ should be chosen such that $\exp(-2e^{-\lambda})$ equals the desired coverage probability. For instance, with $\exp(-2e^{-\lambda}) = 0.95$, we get

$$-2e^{-\lambda} = \ln 0.95 \implies e^{-\lambda} = \frac{-\ln 0.95}{2} \simeq \frac{0.05}{2} \implies \lambda = -\ln 0.025 = 3.69.$$

In the uniform band, since $1/5 < \delta$, under-smoothing is done to avoid the asymptotic bias. The range restriction $[0,1]$ for x is not really a restriction, because x can be re-centered and re-scaled so that its range falls in $[0,1]$. In the Bonferroni-type band (i.e., joint CI's), the multiplicative factor $d_N + \lambda/\sqrt{2\delta \ln N}$ gets replaced by the fixed critical value for the pointwise $1 - \alpha/m$ confidence level.

3.4   An Empirical Example of Confidence Bands

Now we illustrate the preceding confidence intervals/bands with a Dutch data set that was used in Lee and Melenberg (1998), where $N = 1815$ and x is the logarithm of 1981 Dutch family total expenditure in "Guilders"— the Dutch monetary unit at that time. We will show a points-connecting band and an uniform band.

To get $0 \le x_i \le 1 \; \forall i$, we transformed the data with

$$\frac{x_i - \min_j x_j}{\max_j x_j - \min_j x_j}.$$

Then $f(x)$ was estimated at 40 equally spaced points over $[0.1, 0.9]$. We set $\alpha = 0.05$ so that the coverage probability is 0.95. The $N(0,1)$ kernel was used.

In getting the points-connecting band, at each point, the confidence level should be $0.05/40 = 0.00125$. The critical value is then 3.227 because $\Phi(3.227) = 0.00125/2$. As for the bandwidth, $h = 0.5 \times SD(x)N^{-1/5}$ was chosen with "eyeballing." Also, $\int \phi(t)^2 dt$ was found by simulation:

$$\int \phi(t)^2 dt = \int \phi(t) \cdot \phi(t) dt \simeq \frac{1}{T} \sum_{j=1}^{T} \phi(z_j) \simeq 0.283,$$

where $z_1, ..., z_T$ are iid $N(0,1)$.

The estimate $f_N(x)$ is the middle curve on the left panel in Figure 6, and the two lines around $f_N(x)$ form the artificial confidence band. There is a hint of bimodality, but one could easily fit an unimodal asymmetric density within the artificial confidence band.

In getting the uniform band, we found δ in $h = N^{-\delta}$ by equating $N^{-\delta} = 0.5N^{-1/5}$:

$$\delta = \frac{1}{5} - \frac{\ln 0.5}{\ln N} = 0.292;$$

this step is not really necessary unless one desires to know the rate for h, because we can just use $\sqrt{2 \ln h^{-1}}$ instead of $\sqrt{2\delta \ln N}$ in constructing the

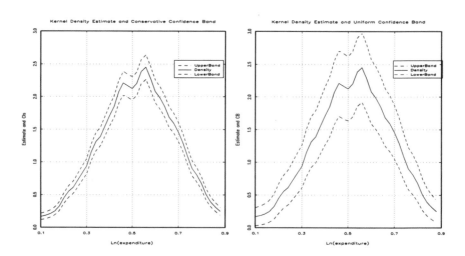

Figure 6: Points-Connecting Band and Uniform Band

uniform band. Also for $\int \{K'(t)\}^2 dt$, since $\phi'(t) = -t\phi(t)$, we get

$$\int \{\phi'(t)\}^2 dt = \int t^2 \phi(t)^2 dt \simeq \frac{1}{T} \sum_{j=1}^{T} z_j^2 \phi(z_j) \simeq 0.141.$$

This gave $\hat{K} = 0.141/(2 \times 0.283)$. The uniform band on the right panel is much wider than the points-connecting band. One could easily fit an unimodal symmetric density within the confidence band.

4 Finding Modes*

One of the interesting features in a density is multi-modality, as multi-modality reflects multiple groups in the population. Many tests for modality have appeared in the literature, and as an application of kernel density estimation, we present a modality test in Silverman (1981) and its application in this section. The test has been applied to world income distribution data in Bianchi (1997) and Kang and Lee (2005), and to US income data in Zhu (2005); the exposition here draws on Kang and Lee (2005).

4.1 Graphical Detection

Before conducting a formal test as in Silverman (1981), it is sensible to estimate the density first and try to see the number of modes graphically; around each (local) mode, the density estimate goes up and down. Figure 7 shows four boxes of kernel density estimates with the $N(0,1)$ ($= \phi$) kernel for the year 1970 and 1989 world income distributions. The left two are for GDP and the right two are for $LGDP \equiv \ln GDP$.

In the top two boxes, the bandwidth is a rule of thumb bandwidth $h_o = 0.9 \cdot SD \cdot N^{-1/5}$ for the given year. In the bottom two boxes, the bandwidth is the least squares CV bandwidth h_{cv}. For GDP, the two density estimates for 1970 look unimodal with "shoulders" (the flat portion near the upper tail; formally y_o is a shoulder point for the density f if $f'(y_o) = f''(y_o) = 0$ but $f'''(y_o) \neq 0$), while the two density estimates for 1989 look "relatively" bimodal. The two bandwidths h_o and h_{cv} are close to each other for GDP, and thus both yield similar results. For LGDP, however, h_o turned out to be much smaller than h_{cv}, and consequently h_o renders single mode for 1970 and double modes for 1989 whereas h_{cv} renders only single mode for both years.

As the figure demonstrates, bandwidth choice is critical in finding the number of modes: too small a bandwidth can easily yield multiple modes particularly in the tail areas, whereas too big a bandwidth can render an unimodal density even when the true density has multiple modes. Jones et al. (1996) compared several bandwidth choice methods in univariate kernel density estimation to recommend the one in Sheather and Jones (1991).

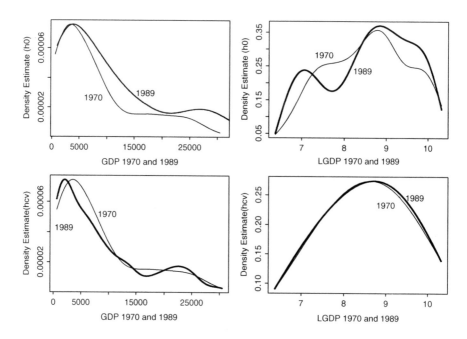

Figure 7: Bandwidth and Modality

Jones et al. (1996) referred to Park and Turlach (1992) for implementation of the bandwidth choice methods examined in their study. One thing
particular about Park and Turlach (1992) is that they compare bandwidth
choices not only in the traditional criterion of mean integrated squared (or
absolute) error but also in detecting the number of modes and their loca-
tions. Overall, Park and Turlach (1992) recommended the least squares CV
method.

4.2   A Multimodality Test

Turning to the Silverman's (1981) multimodality test, consider a sample
$y_1, ..., y_N$ from a density f. In testing for

$$H_o : k \text{ modes} \quad \text{versus} \quad H_a : \text{more than } k \text{ modes}$$

define the "k-critical bandwidth"

$$h_{crit} \equiv \min\{h : \hat{f}(y_o, h) \text{ has at most } k \text{ modes}\},$$
$$\text{where } \hat{f}(y_o, h) \equiv \frac{1}{Nh} \sum_i \phi\left(\frac{y_i - y_o}{h}\right).$$

As Silverman (1981) proved, using ϕ implies that

$$\hat{f}(y_o, h) \text{ has more than } k \text{ modes iff } h < h_{crit}.$$

The main idea is that if h_{crit} is "large," then the H_o is rejected, because it takes a "deliberate" over-smoothing with the large h_{crit} to get only k modes.

The p-value for this test can be obtained by a "smooth bootstrap": resample $w_1^*, ..., w_N^*$ from $y_1, ..., y_N$ and get

$$y_i^* = \bar{w}^* + \frac{w_i^* - \bar{w}^* + h_{crit}\varepsilon_i}{\{1 + (h_{crit}/s_y)^2\}^{1/2}}, \quad \text{where } \varepsilon_i\text{'s iid } N(0,1) \text{ and } s_y^2$$

$$= \frac{1}{N-1}\sum_i (y_i - \bar{y})^2$$

to see whether

$$\hat{f}^*(y_o, h_{crit}) \equiv \frac{1}{N \cdot h_{crit}} \sum_i \phi\left(\frac{y_i^* - y_o}{h_{crit}}\right)$$

has more than k modes or not. Repeating this, say B times, the bootstrap approximation to the p-value is $B^{-1}\sum_{j=1}^{B} 1[h_{crit,j}^* > h_{crit}]$ where $h_{crit,j}^*$ is the k-critical bandwidth for the jth pseudo sample. Equivalently, the p-value is

$$\frac{1}{B}\sum_{j=1}^{B} 1[\hat{f}^*(y_o, h_{crit}) \text{ has more than } k \text{ modes in the } j\text{th pseudo sample}].$$

In the above figures, the mode was found from the whole density function estimate, which necessarily leads to the sensitivity problem due to bandwidth choice. The above modality test with h_{crit} is not the same as this less sophisticated procedure, but an analogous problem exists in the modality test. Efron and Tibshirani (1993, p. 233) showed an example where the modality test accepts 2 modes and then 7 modes again. This problem has been avoided in the literature by proceeding sequentially: start with k versus more than k modes, $k = 1, 2, ...$, to proceed to $k + 1$ only if k modes are rejected; otherwise stop to conclude k modes. This sequential procedure is adopted in the empirical example below.

Whereas one may call the bandwidth-sensitivity problem a "practical" problem, the Silverman's modality test has a theoretical problem as well, as the "bootstrap consistency" was not proved in Silverman (1981). Hall and York (2001) showed that Silverman's test systematically under-rejects and suggest two ways to minor-modify the test. Between the two, the simpler is replacing $B^{-1}\sum_{j=1}^{B} 1[h_{crit,j}^* > h_{crit}]$ with $B^{-1}\sum_{j=1}^{B} 1[h_{crit,j}^* > h_{crit}1.13]$ for nominal level 5% test, which makes the p-value smaller (and thus the test rejects more easily). The number 1.13 comes from Equation (4.1) of Hall and York (2001); this modification is used in the following example.

4.3   An Empirical Example: World Income Distribution

Convergence of world income distribution is an issue that has been hotly debated in the economic growth literature. If equity is a virtue, then one would like to see the income distribution converge over time. But "convergence" can be defined in different ways: (i) "β-convergence" is that a low initial income level causes a high growth rate; (ii) "σ-convergence" is that the standard deviation of the income distribution declines over time; (iii) "modal convergence" is that the number of modes declines over time; (iv) "Q-convergence" is that the interquartile range declines over time. See Kang and Lee (2005) for a brief review and the references on income convergence. Here we show the modal convergence part in Kang and Lee (2005).

The reason why modal convergence was analyzed in the literature is that each mode reflects an income class. Thus unimodality is construed as a single income class—no division of the rich and the poor. But, just as the other convergence concepts, modal convergence also has its weaknesses. For instance, unimodality does not necessarily mean "no rich and no poor," as one can always take the upper and lower tails as the rich and poor, respectively. Also dealing with modes is more difficult than dealing with SD or interquartiles. With these caveat in mind, now we present the empirical results.

Table 1: Modal Convergence Test

	$k = 1$			$k = 2$			$k = 3$		
	h_{crit}	$p(1)$	$p(1.13)$	h_{crit}	$p(1)$	$p(1.13)$	h_{crit}	$p(1)$	$p(1.13)$
	GDP								
1970	3358	0.33	0.13	2060	0.32	0.15	1250	0.70	0.45
1980	3929	0.42	0.27	2112	0.57	0.40	1848	0.35	0.18
1989	4808	0.12	0.03	1891	0.78	0.55	1426	0.86	0.68
	LGDP								
1970	0.3404	0.48	0.29	0.3282	0.10	0.02	0.1625	0.82	0.62
1980	0.3754	0.53	0.33	0.2901	0.20	0.09	0.2351	0.14	0.03
1989	0.5115	0.20	0.08	0.3620	0.11	0.03	0.1406	0.92	0.78

Table 1 reports the critical bandwidth h_{crit} and the p-values using the world income distribution in years 1970, 1980, and 1989; $p(1)$ is the p-value with the original Silverman's test and $p(1.13)$ is the p-value for the modification of Hall and York (2001). Our interpretation will be based on $p(1.13)$. With GDP, unimodality is not rejected in 1970 and 1980 but rejected in 1989 in favor of bimodality, which means two groups (i.e., "income divergence" in 1989 or sometime before). With LGDP, unimodality is not rejected in all three years. Although we are using 5% significance level, LGDP for 1989 has 8%—too close to 5%—and if we proceed further raising the significance level above 8%, then we would accept three modes for LGDP 1989.

Not just the bandwidth choice issue, this example also raises the issue of which transformation of variable to use, although using GDP seems more natural than using LGDP. See Zhu (2005) for an "adaptive kernel"-method version of the Silverman's test and more references on modality tests.

5   Survival and Hazard Under Random Right-Censoring*

Suppose $y_1, ..., y_N$ are durations until an event of interest with df F and density f. If the durations are observed fully, then the survival function $S(t)$ and hazard function $\lambda(t)$ can be easily estimated with

$$S_N(t) \equiv \frac{1}{N} \sum_i 1[y_i > t] \quad \text{and} \quad \lambda_N(t) \equiv \frac{f_N(t)}{S_N(t)}$$

where $f_N(t)$ is a kernel estimator for $f(t)$. But in duration data, typically there is a right-censoring problem and these estimators need modifications. This section examines nonparametric survival and hazard function estimation under random right-censoring. Section 5.1 introduces a nonparametric estimator for cumulative hazard. Section 5.2 reviews two nonparametric estimators for survival function. Section 5.3 presents nonparametric estimators for density and hazard.

5.1   Nelson–Aalen Cumulative-Hazard Estimator

Suppose $(y_1, d_1), ..., (y_N, d_N)$ are observed with $y_i = \min(y_i^*, c_i)$ where y_i^* is an *event duration* of interest, c_i is a *censoring duration* with $y^* \amalg c$, y_i is the observed (or "recorded") duration, and d_i is a non-censoring indicator; i.e., $d_i = 1$ means that y_i is the event duration, and $d_i = 0$ means that y_i is the censoring duration.
Define

$$R_i(t) \equiv 1[y_i \geq t] \quad \text{and} \quad N_i(t) \equiv 1[y_i \leq t, d_i = 1], \quad t \in (0, T];$$

$R_i(t)$ is the indicator function for whether subject i is in the *risk set* or not at time t (i.e., whether subject i is available for the event or not). As t increases from 0, $R_i(t)$ changes from 1 to 0 only once at $t = y_i$, and $N_i(t)$ changes from 0 to 1 only once at $t = y_i$ *if subject i is not censored*. Also define

$$R(t) \equiv \sum_{i=1}^{N} R_i(t) \quad \text{and} \quad N(t) \equiv \sum_{i=1}^{N} N_i(t), \quad t \in (0, T] :$$

$R(t)$ is the size of the risk set at t, and $N(t)$ counts the number of event durations up to t.

The *Nelson–Aalen (nonparametric) estimator* for cumulative hazard is

$$\Lambda_N(t) = \sum_{i:y_i \leq t} \frac{\Delta N(y_i)}{R(y_i)} \quad \text{for } t \text{ with } R(t) > 0$$

(i.e., at least one subject remaining).

For instance, suppose $N = 5$ and $(y_1, d_1),...,(y_5, d_5)$ are

$$(2,1),\ (3,0),\ (5,1),\ (5,0),\ (8,1) = \{2,\ 3^+,\ 5,\ 5^+,\ 8\}$$

where "+" indicates the censored observations, meaning that their event duration is greater than the number next to "+." Thus it makes sense to put 5 before 5^+ in ranking y_i's, because the censored observation's event duration should be greater than 5. In this data, the event durations are 2, 5, 8. $\Lambda_N(t)$ is increasing and step-shaped with

$$\Lambda_N(t) = 0 \quad \text{for } 0 \leq t < 2$$

$$= \frac{1}{5} = 0.200 \quad \text{for } 2 \leq t < 5$$

$$= \frac{1}{5} + \frac{1}{3} = 0.533 \quad \text{for } 5 \leq t < 8$$

$$= \frac{1}{5} + \frac{1}{3} + 1 = 1.533 \quad \text{for } 8 = t$$

$\Lambda_N(t)$ jumps only at the event times 2, 5, and 8 (no jump at the censoring time 3), and the jump magnitude is the hazard rate at the point. For the jump magnitude, the censored observations do not count; e.g., 5^+ is ignored at $t = 5$.

Some remarks are in order. First, there is no basis to estimate $\Lambda(t)$ beyond the last observed event time. Second, $\Lambda_N(t)$ jumps at the event times while $\Lambda(t)$ accumulates the smooth hazard $\lambda(t)$ over time continuously; imagine a geyser ($\Lambda_N(t)$) which sends up hot water and steam occasionally while the pressure ($\Lambda(t)$) builds up continuously underneath with the underlying "intensity to jump" over dt being $d\Lambda(t) = \lambda(t)dt$. Third, in continuous time, no two subjects can experience the event at the same time; i.e., only one subject from a given risk set can end the duration at a given time. But in real life, often t is discrete (grouped duration), and multiple subjects may experience the event together. We will deal with grouped durations in a later chapter.

The asymptotic distribution of $\sqrt{N}\{\Lambda_N(t) - \Lambda(t)\}$ with continuous t is not easy to derive, as it requires understanding "counting processes" and "stochastic integrals." In the following, we provides some intuition and estimators for the asymptotic variance, after recalling the following defining characteristics of Poisson distribution. For $Poi(\lambda)$, there is a small interval of length dt such that

$$P(\text{no occurrence}) \simeq 1 - \lambda dt \quad \text{and} \quad P(\text{one occurrence}) \simeq \lambda dt$$

which implies $P($more than one occurrence$)$ is almost zero on the interval. $Poi\,(\lambda m)$ is obtained by connecting m-many such intervals; the occurrences in non-overlapping intervals of the same size are iid.

On a small interval of $[t + dt)$, suppose $\Delta N_i(t) = 1$ with "intensity" $R_i(t)\lambda(t)dt$ $(= 0$ if $R_i(t) = 0$, and $\lambda(t)dt$ otherwise): $\Delta N_i(t) \sim Poi\{R_i(t)\lambda(t)\}$. Then its sum $\Delta N(t)$ across i follows $Poi\{R(t)\lambda(t)\}$ because a sum of independent $Poi(\mu_i)$ rv's follow $Poi(\sum_i \mu_i)$. Given that one event occurred over $[t + dt)$, the "probability" that the event occurred to subject i with $R_i(t) = 1$ is

$$\frac{\text{intensity of } i}{\text{the sum of intensities}} = \frac{R_i(t)\lambda(t)}{R(t)\lambda(t)} = \frac{1}{R(t)}.$$

$\Lambda_N(t)$ is a weighted sum of the Poisson rv's with the weight being this "probability" $R(t)^{-1}$.

The increments of $\Delta N(t)$ at different times are uncorrelated with one another, and hence the asymptotic variance of $\Lambda_N(t) - \Lambda(t)$ is the sum of the variances at the event times. The asymptotic variance can be estimated with

$$\sum_{i:y_i \le t} \frac{\Delta N(y_i)}{R(y_i)^2}$$

because the expected value of $\Delta N(y_i)$ is the intensity which is also the variance as $\Delta N(t) \sim Poi\{R_i(t)\lambda(t)\}$, and $R(y_i)^2$ is the multiplicative factor getting squared for the variance. Another estimator for the asymptotic variance, called "Greenwood formula," is

$$\sum_{i:y_i \le t} \frac{\Delta N(y_i)}{R(y_i)\{R(y_i) - \Delta N(y_i)\}}.$$

Ignoring the fact that N is small, we apply the two variance estimators to the above numerical example to get (the Greenwood formula is in $(\cdot)$)

$$V\{\Lambda_N(t)\} \simeq \frac{1}{5^2} = 0.04 \ (\frac{1}{5 \cdot (5-1)} = 0.05) \text{ for } 2 \le t < 5$$

$$\simeq \frac{1}{5^2} + \frac{1}{3^2} = 0.151 \ (\frac{1}{5 \cdot (5-1)} + \frac{1}{3 \cdot 2} = 0.217) \text{ for } 5 \le t < 8$$

$$= \frac{1}{5^2} + \frac{1}{3^2} + \frac{1}{1^2} = 1.151 \text{ (not defined) for } 8 = t.$$

The two variance estimators may look much different, but this is only because of the extremely small N or the evaluation points close to the end point. Typically, the two estimators differ little.

With $\Lambda_N(t)$, one can check out some parametric assumptions, because they imply functional form restrictions on $\Lambda(t)$. For instance,

$$Weibull(\theta, \alpha) : \ln \Lambda(t) = \ln \theta + \alpha \ln t \implies \text{plot } \ln \Lambda(t) \text{ vs. } \ln t$$
$$Log-Logistic(\theta, \alpha) \quad : \quad \ln[\exp\{\Lambda(t)\} - 1] = \ln \theta + \alpha \ln t$$

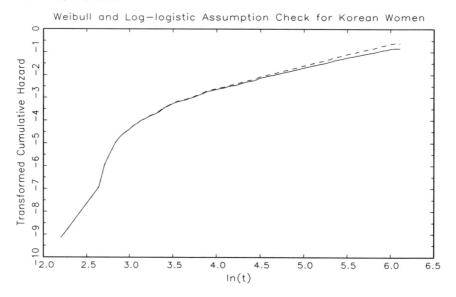

Figure 8: Check on Weibull and Log-Logistic Assumptions

$$\Longrightarrow \quad \text{plot } \ln[\exp\{\Lambda(t)\} - 1] \text{ vs. } \ln t.$$

In both plots, a linear function of $\ln t$ should be seen if the assumed distribution is correct. For Weibull, if the slope α is one, then we get $Expo(\theta)$.
Although we are looking at the marginal, not conditional, distribution of y,
still the parametric fit for the y-distribution might be suggestive of which
distribution to use for $y|x$.

Using the Korean women data that appeared in the density estimation
section, Figure 8 shows two curves from plotting $\ln \Lambda(t)$ for Weibull (solid
line) and $\ln[\exp\{\Lambda(t)\} - 1]$ for Log-logistic (dashed line). The curves reveal
that both distributions are inappropriate, as the lines are nonlinear.

5.2   Survival-Function Estimators

There are two easy-to-use, well-known estimators for $S(t)$. One is using
$\Lambda_N(t)$ and the equation $S(t) = \exp\{-\Lambda(t)\}$, and the other is the so-called
"Kaplan-Meier product limit estimator." We examine the former first, an
approximation of which then leads to the latter.

5.2.1   Cumulative-Hazard-Based Estimator

Once $\Lambda_N(t)$ is obtained, $S(t)$ can be estimated with

$$S_{NA}(t) \equiv \exp\{-\Lambda_N(t)\}$$

where the subscript "NA" stands for Nelson–Aalen. For the above numerical example, $S_{NA}(t)$ is decreasing and step-shaped with

$$
\begin{aligned}
S_{NA}(t) &= \exp(0) = 1 && \text{for } 0 \le t < 2 \\
&= \exp(-0.200) = 0.819 && \text{for } 2 \le t < 5 \\
&= \exp(-0.533) = 0.587 && \text{for } 5 \le t < 8 \\
&= \exp(-1.533) = 0.216 && \text{for } 8 = t
\end{aligned}
$$

As $\Lambda(t)$ cannot be estimated beyond the last event time, $S(t)$ cannot be estimated beyond the last event time either. Having $S_{NA}(t)$ reaching short of 0 as in this example is not pleasing, and this shortcoming may be overcome by the second estimator below.

As for estimating the asymptotic variance of $S_{NA}(t)$, which is denoted simply as $V\{S_{NA}(t)\}$, we can use δ-method to get

$$
\begin{aligned}
V\{S_{NA}(t)\} &\simeq [\exp\{-\Lambda_N(t)\}]^2 \cdot V\{\Lambda_N(t)\} = S_{NA}(t)^2 \cdot V\{\Lambda_N(t)\} \\
&\implies SD\{S_{NA}(t)\} \simeq S_{NA}(t) \cdot SD\{\Lambda_N(t)\}.
\end{aligned}
$$

For CI's, we list a couple of different ways among many others in the next paragraph. Let $c_{1-\alpha/2}$ denote the $(1 - \alpha/2)$th quantile of the $N(0,1)$ distribution.

For an asymptotic $100(1-\alpha)\%$ CI, first, we can use $S_{NA}(t) \pm c_{1-\alpha/2}SD\{S_{NA}(t)\}$ which is

$$
(i): [S_{NA}(t) - c_{1-\alpha/2}S_{NA}(t) \cdot SD\{\Lambda_N(t)\}, \quad S_{NA}(t) + c_{1-\alpha/2}S_{NA}(t) \cdot SD\{\Lambda_N(t)\}].
$$

But (i) may yield a CI going out of the bound $[0,1]$ for $S(t)$. Second, "going inside" of $\exp(-\Lambda_N(t))$ and using the CI for $\Lambda(t)$, we get

$$
(ii): [\exp\{-\Lambda_N(t) - c_{1-\alpha/2}SD(\Lambda_N(t))\}, \quad \exp\{-\Lambda_N(t) + c_{1-\alpha/2}SD(\Lambda_N(t))\}]
$$

which at least respects the lower bound 0. Third, as the upper bound of (ii) can easily go over 1 due to the exp function, we may use an asymmetric CI combining the upper bound of (i) and the lower bound of (ii):

$$
(iii): [\exp\{-\Lambda_N(t) - c_{1-\alpha/2}SD(\Lambda_N(t))\}, \quad S_{NA}(t) + c_{1-\alpha/2}S_{NA}(t) \cdot SD\{\Lambda_N(t)\}].
$$

For the above numerical example with $V\{\Lambda_N(t)\}$ estimated by $\sum_{i:y_i \le t} R(y_i)^{-2}\Delta N(y_i)$, the 95% CI's based on (iii) are, pretending again that N is large,

$$
\begin{aligned}
t = 2 &: \{\exp(-0.200 - 1.96 \times 0.200) = 0.553, \quad 0.819 + 1.96 \times 0.819 \\
&\quad \times 0.200 = 1.139\} \\
t = 5 &: \{\exp(-0.533 - 1.96 \times \\
&\quad 0.389) = 0.274, \quad 0.587 + 1.96 \times 0.587 \times 0.389 = 1.035\} \\
t = 8 &: \{\exp(-1.533 - 1.96 \\
&\quad \times 1.073) = 0.026, \quad 0.216 + 1.96 \times 0.216 \times 1.073 = 0.670\}.
\end{aligned}
$$

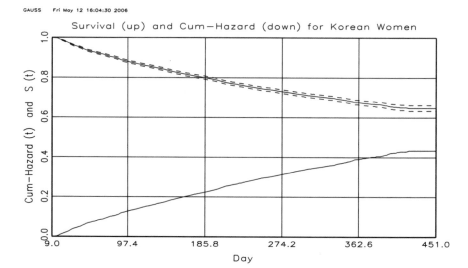

Figure 9: Survival Function and Cumulative Hazard Function

Figure 9 presents $S_{NA}(t)$ (upper curve) and $\Lambda_N(t)$ (lower curve) using the Korean women data. The 95% point-wise CI for $S_{NA}(t)$ is very tight because N is large. In the figure, one curve can be almost obtained by "flipping over" the other curve. Since the censoring % is high, $S_{NA}(t)$ ends at about 0.65. Judging from $S_{NA}(t)$, it may never hit the horizontal axis: there may be "super-surviviors" who never find a job.

5.2.2   Kaplan–Meier Product Limit Estimator

Using the approximation $e^{-a} \simeq 1 - a$ when a is small, observe

$$S_{NA}(t) = \exp\{- \sum_{i:y_i \leq t} \frac{\Delta N(y_i)}{R(y_i)}\} = \prod_{i:y_i \leq t} \exp(-\frac{\Delta N(y_i)}{R(y_i)})$$

$$\simeq \prod_{i:y_i \leq t} \{1 - \frac{\Delta N(y_i)}{R(y_i)}\} \equiv S_{KM}(t)$$

where KM in $S_{KM}(t)$ stands for "Kaplan–Meier" (Kaplan and Meier, 1958); $S_{KM}(t)$ is the *Kaplan–Meier product limit estimator*. The fact that the survival function can be written as a product of one minus small-interval hazards provides a direct motivation for $S_{KM}(t)$: being alive today means having survived each small interval in the past. The asymptotic distribution of $S_{KM}(t)$ is the same as that of $S_{NA}(t)$, as evident from the display. Note that, since $e^{-a} \geq 1 - a$ always, it holds that $S_{NA}(t) \geq S_{KM}(t)$, although the difference between the two is typically negligible.

For the above numerical example $\{2, 3^+, 5, 5^+, 8\}$, we get

$$S_{KM}(t) = 1 \quad \text{for } 0 \le t < 2$$

$$= 1(1 - \frac{1}{5}) = \frac{4}{5} \quad \text{for } 2 \le t < 5$$

$$= \frac{4}{5}(1 - \frac{1}{3}) = \frac{8}{15} \quad \text{for } 5 \le t < 8$$

$$= \frac{4}{5}(1 - \frac{1}{3})(1 - 1) = 0 \quad \text{for } 8 \le t$$

Unlike $S_{NA}(t)$, $S_{KM}(t)$ reaches 0, which is why we have "$8 \le t$", not "$8 = t$" in the last line. This happens only when there is no censored duration at the last duration.

For $S_{KM}(t)$ (and $S_{NA}(t)$), it is curious why *using the censored observations only for the risk sets* delivers consistent estimators. Recall the numerical example again: 2, 3^+, 5, 5^+, 8. Suppose *each censored observation redistributes its probability mass equally to all remaining observations* as in Table 2. The redistribution makes sense, because we do not know when the censored durations actually ended.

Table 2: Censored-Data Mass Redistribution

t	Initial Mass	First Adjustment	Second Adjustment	Final Mass
2	1/5	1/5	1/5	1/5
3^+	1/5	0	0	0
5	1/5	1/5 +(1/3)(1/5)	1/5 +(1/3)(1/5)	4/15
5^+	1/5	1/5 +(1/3)(1/5)	0	0
8	1/5	1/5 +(1/3)(1/5)	1/5 +(1/3)(1/5) +1/5 +(1/3)(1/5)	8/15

In the table, the initial probability mass is $1/5$ for all observations. In column "First Adjustment", the mass $1/5$ for 3^+ gets equally redistributed to the three remaining observations including the censored one. In column "Second Adjustment", the mass $1/5 +(1/3)(1/5)$ for 5^+ gets equally redistributed to the remaining observations (only one left though). In column "Final Mass," the probability masses for event times are tallied up.

Estimating $S(t)$ with the final column of the table, we get $S_{KM}(t)$ again, but in a different way:

$$S_{KM}(t) = 1 \quad \text{for } 0 \le t < 2$$

$$= 1 - \frac{1}{5} = \frac{4}{5} \text{ for } 2 \le t < 5 \quad \left(1 \text{ from above}, \frac{1}{5} \text{ from the table}\right)$$

$$= \frac{4}{5} - \frac{4}{15} \left(= \frac{4}{5} - \frac{4}{5}\frac{1}{3}\right) = \frac{8}{15} \text{ for } 5 \le t < 8 \quad \left(\frac{4}{5} \text{ from above},\right.$$

$$\frac{4}{15} \text{ from the table} \Bigg)$$

$$= \frac{8}{15} - \frac{8}{15} \left(= \frac{8}{15} - \frac{8}{15}\frac{1}{1}\right) = 0 \text{ for } 8 \leq t \quad \left(\frac{8}{15} \text{ from above,}\right.$$

$$\left.\frac{8}{15} \text{ from the table}\right).$$

Pay attention to the row for $5 \leq t < 8$ where two subjects remain:

$$\frac{4}{5} - \frac{4}{5}\frac{1}{3} = \frac{4}{5}\left(1 - \frac{1}{3}\right) = (\text{prob. mass just before } t = 5)$$

$$\times \text{ (share of the remaining subjects)};$$

from 4/5, one subject's share is subtracted and the remaining two subjects' share get carried over to the future.

5.3   Density and Hazard Estimators

5.3.1   Kernel Density Estimator

Recall the usual kernel density estimator

$$f_N(t) \equiv \frac{1}{N}\sum_i \frac{1}{h}K\left(\frac{y_i - t}{h}\right) = \int \frac{1}{h}K\left(\frac{y - t}{h}\right)\partial F_N(y),$$

where $F_N(y) \equiv \frac{1}{N}\sum_i 1[y_i \leq y]$. Since F_N is step-shaped and jumps at $y_1, ..., y_N$ with the jump magnitude N^{-1}, the integral should be understood as attaching N^{-1} to $h^{-1}K((y-t)/h)$ at $y = y_i$ and summing across i, which is exactly what the definition of $f_N(t)$ states. Although this definition fails under-right censoring because $F_N(t)$ is not consistent for $F(t)$, the expression with $\partial F_N(y)$ still works once ∂F_N is replaced by $\partial(1 - S_{KM})$; see Mielniczuk (1986) and the references therein. The estimator $1 - S_{KM}$ is step-shaped but its jump magnitude is not N^{-1} due to the mass-redistribution of the censored observations seen above.

Since we have to stick to continuous time in discussing density/ hazard in this subsection, assume no ties in the observations and modify the data $\{2, 3^+, 5, 5^+, 8\}$ to $\{2, 3^+, 5, 6^+, 8\}$ to rule out tied observations at $t = 5$. For the new data, $S_{KM}(t)$ is still the same as before:

$$S_{KM}(t) = 1 \quad \text{for } 0 \leq t < 2$$

$$= 1 - \frac{1}{5} = \frac{4}{5} \quad \text{for } 2 \leq t < 5 \text{ (1 from above, } \frac{1}{5} \text{ from the table)}$$

$$= \frac{4}{5} - \frac{4}{15}\left(= \frac{4}{5} - \frac{4}{5}\frac{1}{3}\right) = \frac{8}{15} \quad \text{for } 5 \leq t < 8$$

$$\left(\frac{4}{5} \text{ from above, } \frac{4}{15} \text{ from the table}\right)$$

$$= \frac{8}{15} - \frac{8}{15}\left(= \frac{8}{15} - \frac{8}{15}\frac{1}{1}\right) = 0 \quad \text{for } 8 \le t$$

$(\frac{8}{15}$ from above, $\frac{8}{15}$ from the table$)$.

Recalling $y_i = \min(y_i^*, c_i)$, as we can estimate $S(t)$ with $S_{KM}(t)$, we can also estimate the survival function $\bar{G}_{KM}(t)$ of c by reversing the role of the event and censoring durations—the reason for the notation $\bar{G}_{KM}(t)$ will become clear soon. That is, suppose we have $\{2^+, 3, 5^+, 6, 8^+\}$ and obtain its KM estimator:

$$\bar{G}_{KM}(t) = 1 \quad \text{for } 0 \le t < 3$$

$$= 1\left(1 - \frac{1}{4}\right) = \frac{3}{4} \quad \text{for } 3 \le t < 6$$

$$= \frac{3}{4}\left(1 - \frac{1}{2}\right) = \frac{3}{8} \quad \text{for } 6 \le t \le 8;$$

$t = 8$ is included in $6 \le t \le 8$ because the last observation at $t = 8$ is censored. What is interesting is that $\{N\bar{G}_{KM}(t)\}^{-1}$ is the drop magnitude of $S_{KM}(t)$ at the event times:

$$\text{at } t = 2: \frac{1}{N\bar{G}_{KM}(t)} = \frac{1}{5 \cdot 1} = \frac{1}{5} \left\{= S_{KM}(2^-) - S_{KM}(2) = 1 - \frac{4}{5}\right\}$$

$$\text{at } t = 5: \frac{1}{N\bar{G}_{KM}(t)} = \frac{1}{5 \cdot (3/4)} = \frac{4}{15}$$

$$\left\{= S_{KM}(5^-) - S_{KM}(5) = \frac{4}{5} - \frac{8}{15}\right\}$$

$$\text{at } t = 8: \frac{1}{N\bar{G}_{KM}(t)} = \frac{1}{5 \cdot (3/8)} = \frac{8}{15}$$

$$\left\{= S_{KM}(8^-) - S_{KM}(8) = \frac{8}{15} - 0\right\}.$$

Denote the distribution function of c as G and the distribution function of y as H; also define $\bar{G}(t) \equiv 1 - G(t)$. Then, as $y^* \, \text{II} \, c$,

$$P(y > t) = H(t) = P(y^* > t)P(c > t) = S(t)\bar{G}(t).$$

As Susarla et al. (1984) stated (and as illustrated just above), the jump magnitude of $1 - S_{KM}(t)$ is $d_i\{N\bar{G}_{KM}(t)\}^{-1}$; d_i is attached because the jumps take place at the event times only. Define now a kernel density estimator under random right-censoring

$$f_{KM}(t) \equiv \int \frac{1}{h} K\left(\frac{y - t}{h}\right) \partial\{1 - S_{KM}(y)\}$$

$$= \frac{1}{N} \sum_i \frac{1}{h} K\left(\frac{y_i - t}{h}\right) \frac{d_i}{\bar{G}_{KM}(y_i)}.$$

The sum in $f_{KM}(t)$ converges at a rate slower than $N^{-1/2}$, but $\bar{G}_{KM}$ converges to $\bar{G}$ at $N^{-1/2}$, which implies that the asymptotic properties of this estimator is the same as the following one with $\bar{G}_{KM}$ replaced by $\bar{G}$ (e.g., Marron and Padgett, 1987):

$$\tilde{f}_{KM}(t) \equiv \frac{1}{N} \sum_i \frac{1}{h} K\left(\frac{y_i - t}{h}\right) \frac{d_i}{\bar{G}(y_i)}.$$

Compared with $f_N(t)$, the only difference is that the N^{-1} weighting in $f_N(t)$ is replaced with $d_i/\{N\bar{G}(y_i)\}$; since $d_i = 0$ for the censored observations, the mass N^{-1} is blown up by $\bar{G}(y_i)$ for the non-censored observations.

Examine a typical sum in $E\tilde{f}_{KM}(t)$ where $E(\cdot)$ is taken wrt y and c with d_i replaced by $1[y_i < c_i]$: using $P(y < c|y) = \bar{G}(y)$,

$$\int \frac{1}{h} K\left(\frac{y_i - t}{h}\right) \frac{1[y_i < c_i]}{\bar{G}(y_i)} \partial G(c_i) \partial F(y_i) = \int \frac{1}{h} K\left(\frac{y_i - t}{h}\right) \frac{\bar{G}(y_i)}{\bar{G}(y_i)} \partial F(y_i)$$

$$= \int \frac{1}{h} K\left(\frac{y_i - t}{h}\right) \partial F(y_i) = \int K(z) f(t + zh) dz$$

$$\simeq f(t) + \frac{h^2}{2} f''(t) \int z^2 K(z) dz.$$

Hence the asymptotic bias takes the usual form. As for the variance, examine the second moment:

$$\int \frac{1}{h^2} K\left(\frac{y_i - t}{h}\right)^2 \frac{1[y_i < c_i]}{\bar{G}(y_i)^2} \partial G(c_i) \partial F(y_i) = \int \frac{1}{h^2} K\left(\frac{y_i - t}{h}\right)^2 \frac{1}{\bar{G}(y_i)} \partial F(y_i)$$

$$= \int \frac{1}{h} K(z)^2 \frac{f(t + zh)}{\bar{G}(t + zh)} dz \simeq \frac{1}{h} \frac{f(t)}{\bar{G}(t)} \int K(z)^2 dz.$$

Therefore, analogously to the usual kernel density-estimator asymptotic-distribution, we get

$$\sqrt{Nh}\{f_{KM}(t) - f(t) - \frac{h^2}{2} f''(t) \int z^2 K(z) dz\} \rightsquigarrow N\{0, \frac{f(t)}{\bar{G}(t)} \int K(z)^2 dz\}.$$

This is essentially shown in Mielniczuk (1986, p. 772) and Marron and Padgett (1987, pp. 1524–1525). The bias term may be ignored in practice, assuming under-smoothing.

5.3.2   Kernel Hazard Estimator

Having seen $f_{KM}(t)$, a natural estimator for hazard is

$$\lambda_{KM}(t) = \frac{f_{KM}(t)}{S_{KM}(t)}.$$

Since the denominator is $\sqrt{N}$-consistent while the numerator is only $\sqrt{Nh}$-consistent, the asymptotic distribution of $\sqrt{Nh}\{\lambda_{KM}(t) - \lambda(t)\}$ is the same as that of $\sqrt{Nh}\{\tilde{\lambda}_{KM}(t) - \lambda(t)\}$ where

$$\tilde{\lambda}_{KM}(t) = \frac{\tilde{f}_{KM}(t)}{S(t)}.$$

It then follows, ignoring the bias term with under-smoothing,

$$\sqrt{Nh}\{\lambda_{KM}(t) - \lambda(t)\} \rightsquigarrow N\{0, \ \frac{f(t)}{S(t)} \frac{1}{S(t)\bar{G}(t)} \int K(z)^2 dz\}$$

$$= N\{0, \ \frac{\lambda(t)}{H(t)} \int K(z)^2 dz\}$$

because $S(t)\bar{G}(t) = H(t)$; recall that $H(t)$ is the survival function of y and can be estimated with $N^{-1}\sum_i 1[y_i > t]$.

In the literature, other estimators for $\lambda(t)$ have appeared; see Hess et al. (1999) for the references. One of them, which is examined by Tanner and Wong (1983) and Ramlau-Hansen (1983), "smears out" the hazards at event times. Let R_i denote the rank of y_i (e.g., if $N = 5$ and $y_3 < y_4 < y_1 < y_2$, then $R_1 = 3$, $R_2 = 4$, $R_3 = 1$, $R_4 = 2$). The estimator is

$$\lambda_{NA}(t) \equiv \int \frac{1}{h}K\left(\frac{y-t}{h}\right) \partial \Lambda_N(y) = \sum_i \frac{1}{h}K\left(\frac{y_i-t}{h}\right)\frac{d_i}{N-R_i+1}$$

$$\left(= \sum_i \frac{1}{h}K\left(\frac{y_i-t}{h}\right)\frac{d_i}{N-i+1}, \quad \text{if } y_1 <, ..., < y_N\right)$$

where $N-R_i+1$ is the size of the risk set for duration i. Compared with $f_N(t)$, $\lambda_{NA}(t)$ replaces N^{-1} in $f_N(t)$ with $d_i/(N - R_i + 1)$. $\sqrt{Nh}\{\lambda_{NA}(t) - \lambda(t)\}$ follows the same asymptotic distribution as $\sqrt{Nh}\{\lambda_{KM}(t) - \lambda(t)\}$ follows. In the following, we apply $\lambda_{NA}(t)$ to the Korean women data, which is easier to compute than $\lambda_{KM}(t)$.

In practice, more often than not, duration is discrete even if it takes many different values. One way to go about this problem is doing grouped duration analysis as examined in the next chapter. Another way is making ad hoc adjustments in the above derivations for continuous time. Recall the Korean women data. With $N = 9312$ and the maximum duration about 450, there are many tied observations. First we sorted the durations and then used the formula in the last display, which is a kernel density estimator with N^{-1} replaced by $d_i/(N - i + 1)$. This weighting means that the tied observations are given different weights, depending on the order they are listed in the data; a better alternative than this ad hoc modification of $\lambda_{NA}(t)$ could be using the same weight based on the average rank. Figure 10 is the result with 95% pointwise CI's, for which we used $H_N \equiv N^{-1}\sum_i 1[y_i > t]$ (using $N^{-1}\sum_i 1[y_i \geq t]$ made no visible difference). The hazard increases in the early period up to about day 50, and then declines. After about 1 year, the hazard declines faster. These features in the scale of 0.0001 could not be seen in the earlier integrated hazard figure in the scale of 0.1, as integration blurs fine details in the derivative.

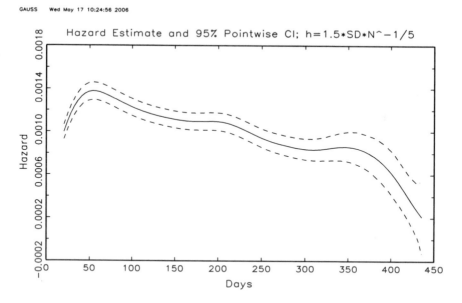

Figure 10: Kernel Hazard Estimate

6   Kernel Nonparametric Regression

6.1   Overview

In this section, we introduce kernel nonparametric estimation of the regression function $E(y|x = x_o) = E(y|x_o) \equiv r(x_o)$ where x is a $k \times 1$ regressor vector, $y = r(x) + u$, and $E(u|x) = 0$. More generally, we may consider functions of the conditional distribution, such as $\partial r(x)/\partial x$ and $V(y|x)$. But, as will be seen in detail later, estimation methods for such functions can be inferred from those for $E(y|x)$.

If many y_i's are available for a given x_o, we can estimate $E(y|x_o)$ by the sample mean of those observations. In a non-experimental setting with a continuous rv x, however, this is impossible, for we get all different x_i's by random sampling. Assuming that $E(y|x) = r(x)$ is continuous in x, y_i's with its x_i close to x_o may be treated as (y_i, x_o). This gives (pseudo) multiple observations for a given x_o so that $E(y|x_o)$ can be estimated by the sample mean. More generally, $E(y|x_o)$ can be estimated by a *local weighted average* $\sum_i w_i y_i$ subject to $\sum_i w_i = 1$, where w_i is large if x_i is close to x_o and small otherwise.

Typically, popular choices of w_i's in estimating $E(y|x_o)$ are various densities (or "kernels"). With K denoting a kernel, one potential estimator for $r(x_o)$ is $\sum_i K((x_i - x_o)/h)y_i$, where the role of the bandwidth h is the same as the group interval length in a histogram. To ensure for the weights to add up to one, normalize $\sum_i K((x_i - x_o)/h)y_i$ to get a "*kernel (nonparametric) regression estimator*"

$$r_N(x_o) \equiv \frac{\sum_i K((x_i - x_o)/h)y_i}{\sum_i K((x_i - x_o)/h)} = \sum_i \frac{K((x_i - x_o)/h)}{\sum_i K((x_i - x_o)/h)} y_i$$

$$= \sum_i w_{oi} y_i \quad \text{where } w_{oi} \equiv \frac{K((x_i - x_o)/h)}{\sum_i K((x_i - x_o)/h)};$$

w_{oi} is the weight given to the ith observation for the local averaging around x_o. The estimator can be written also as

$$r_N(x_o) = \frac{(Nh^k)^{-1} \sum_i K((x_i - x_o)/h)y_i}{(Nh^k)^{-1} \sum_i K((x_i - x_o)/h)} = \frac{g_N(x_o)}{f_N(x_o)}, \quad \text{where}$$

$$g_N(x_o) \equiv \frac{1}{Nh^k} \sum_i K\left(\frac{x_i - x_o}{h}\right) y_i \quad \text{and}$$

$$f_N(x_o) \equiv \frac{1}{Nh^k} \sum_i K\left(\frac{x_i - x_o}{h}\right).$$

This estimator was suggested by Nadaraya (1964) and Watson (1964), and is called a "Nadaraya–Watson kernel estimator." There are also other kernel estimators suitable for experimental data with nonrandom x.

Since $f_N(x_o) \to^p f(x_o)$, for $r_N(x_o) \to^p r(x_o)$ to hold, we should have

$$g_N(x_o) \to^p g(x_o) \equiv r(x_o) \cdot f(x_o).$$

This may be better understood in the following discrete case. Suppose x is discrete. Then we can estimate $E(y|x_o)$ by the sample average of y_i's with $x_i = x_o$:

$$\frac{\sum_i 1[x_i = x_o]y_i}{\sum_i 1[x_i = x_o]} = \frac{N^{-1} \sum_i 1[x_i = x_o]y_i}{N^{-1} \sum_i 1[x_i = x_o]} \to^p \frac{E(1[x = x_o] \cdot y)}{P(x = x_o)}$$

$$= E(y|x = x_o).$$

Thus it must be that

$$\frac{1}{N} \sum_i 1[x_i = x_o]y_i \to^p E(y|x = x_o)P(x = x_o) = r(x_o)P(x = x_o)$$

which is a discrete analog for $r(x_o)f(x_o)$.

Since the norming factor in g_N and f_N is Nh^k, not N, the asymptotic distribution of $r_N(x_o)$ is obtained multiplying $r_N(x_o) - r(x_o)$ by $\sqrt{Nh^k}$, not by $\sqrt{N}$. Since $r_N(x_o)$ is a ratio of two nonparametric estimators $g_N(x_o)$ and $f_N(x_o)$, each term's asymptotic variance and the two terms' asymptotic covariance contribute to the asymptotic variance of $(Nh^k)^{0.5}\{r_N(x_o) - r(x_o)\}$, which will be shown to be asymptotically normal with variance

$$\frac{V(y|x_o)}{f(x_o)} \int K(z)^2 dz.$$

This is different from the asymptotic variance $f(x_o) \int K(z)^2 dz$ of $\sqrt{Nh^k}$ $\{f_N(x_o) - f(x_o)\}$.

While the asymptotic variance of $r_N(x_o)$ is inversely related to $f(x_o)$, that of $f_N(x_o)$ is positively related to $f(x_o)$. This somewhat strange result may be understood as follows. For $r_N(x_o)$, a high $f(x_o)$ means more observations locally around x_o, which thus reduces the asymptotic variance. For $f_N(x_o)$, $f(x_o)$ itself is the target of the estimation. For instance, if $f(x_o) = 0$, then $K((x_i - x_o)/h) \simeq 0$ as all x_i's are away from x_o; here, having no observation around x_o is better in estimating $f(x_o)$. Essentially, $f(x_o)$ works as a "scale factor" for $f_N(x_o)$; the higher the scale, the higher the variance.

The kernel nonparametric method can be used for other functional estimations. For instance, suppose we want to estimate $V(y|x)$. Since $V(y|x) = E(y^2|x) - E^2(y|x)$, having already estimated $E(y|x)$, we only need to estimate $E(y^2|x)$ in addition. This is done by replacing y by y^2 in $r_N(x_o)$. Also $\partial E(y|x_o)/\partial x = \partial r(x_o)/\partial x$ can be estimated by numerically differentiating $r_N(x_o)$ wrt x_o. If $r(x_o) = x_o'\beta$, then $\partial r(x_o)/\partial x = \beta$. Hence a nonparametric estimator for $\partial r(x_o)/\partial x$ or $E\{\partial r(x_o)/\partial x\}$ reflects the marginal effect of x on $r(\cdot)$ at $x = x_o$.

One problem in the kernel regression estimator $r_N(x)$ (in fact, in most nonparametric methods based on "local weighted averaging" idea) is that $r_N(x)$ has a large bias near the boundary of the range of x. Suppose $k = 1$ and that $r(x)$ is increasing. If we estimate $r(x)$ at the right end of the boundary (i.e., at $\max_{1 \le i \le N} x_i$), then r_N has a downward bias, because $r(x)$ is increasing but the data come locally only from the left-hand side of $\max_{1 \le i \le N} x_i$. The opposite upward bias occurs near $\min_{1 \le i \le N} x_i$. There are ways to correct for the bias, but so long as $r(x)$ at the boundary is not of interest, we just have to restrict the evaluation points x_o well within the boundary or exercise caution in interpreting the estimation results near the boundaries.

As an illustration, recall the interest rate (US treasury rate) data used in Chapter 1. There we saw that y_{i-1} (and y_{i-2}) is highly influential on y_i and that there seems to be a good deal of heteroskedasticity. Applying kernel nonparametric regression with the ϕ kernel, we get Figure 11 which shows $E(y_i|y_{i-1})$ and 95% pointwise CI's; also shown at the bottom is $SD(y_i|y_{i-1})$. The bandwidth was chosen by a cross-validation method to be explained later. The figure reveals an interesting feature: there seems a trough at $y_{i-1} = 12\%$ and $SD(y_i|y_{i-1})$ is also the highest around 12%. One interpretation is that, as y_{i-1} approaches 12% that is quite high, some people expect y_i to fall as the rate might have reached its ceiling, while some other people expect y_i to go up further. This may lead to speculative activities in the economy. With

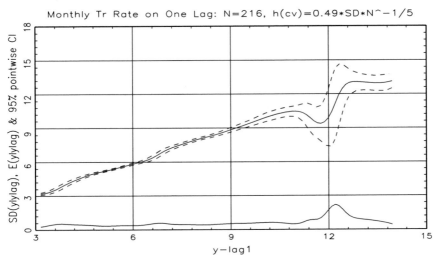

Figure 11: Nonparametric Regression for Interest Rate

LSE, this interesting feature was never detected, and the high level of linear model fitness—$R^2 = 0.98$—was "satisfying" enough to discourage further probing.

6.2   Consistency

Since we already know $f_N(x_o) \to^p f(x_o)$, if we show $g_N(x_o) \to^p g(x_o)$, then $r_N(x_o) \to^p r(x_o)$ follows. The only difference between $f_N(x_o)$ and $g_N(x_o)$ is the presence of the multiplicative y_i in $g_N(x_o)$. Other than this point, the steps to prove $g_N(x_o) \to^p g(x_o)$ are almost the same as those to prove $f_N(x_o) \to^p f(x_o)$. We will show $Eg_N(x_o) \to g(x_o)$ and $Vg_N(x_o) \to 0$ as $N \to \infty$, which are enough for $g_N(x_o) \to^p g(x_o)$.

Examine

$$Eg_N(x_o) = E\left\{\frac{1}{h^k} K\left(\frac{x - x_o}{h}\right) y\right\} = E\left\{\frac{1}{h^k} K\left(\frac{x - x_o}{h}\right) E(y|x)\right\}$$

$$= \int \frac{1}{h^k} K\left(\frac{x - x_o}{h}\right) E(y|x)f(x)dx = \int K(z)E(y|x_o + hz)f(x_o + hz)dz$$

$$= E(y|x_o)f(x_o) + O(h^2) \quad \text{applying Taylor's expansion of order 2}$$

assuming that $r(x)$ is twice continuously differentiable with bounded derivatives. Hence the bias is of order $O(h^2)$, and $Eg_N(x_o) \to g(x_o)$ under $h \to 0$ as $N \to \infty$.

Turning to the variance of $g_N(x_o)$, analogously to the density-estimation variance case,

$$V g_N(x_o) = \frac{\text{each term's variance}}{N} = \frac{\text{each term's 2nd moment}}{N}$$

$$+ o\left(\frac{1}{Nh^k}\right) = \frac{1}{N} E\left\{\frac{1}{h^{2k}} K\left(\frac{x-x_o}{h}\right)^2 y^2\right\} + o\left(\frac{1}{Nh^k}\right)$$

$$= \frac{1}{Nh^{2k}} \int K\left(\frac{x-x_o}{h}\right)^2 E(y^2|x)f(x)dx + o\left(\frac{1}{Nh^k}\right)$$

$$= \frac{1}{Nh^k} E(y^2|x_o)f(x_o) \int K^2(z)dz + o\left(\frac{1}{Nh^k}\right).$$

Hence, if $Nh^k \to \infty$, then we get $V g_N(x_o) \to 0$ as $N \to \infty$. Adding the condition $h \to 0$ as $N \to \infty$ for the bias of $g_N(x_N)$ to disappear, we get $g_N(x_o) \to^p g(x_o)$. Therefore, we have the desired result:

$$r_N(x_o) \to^p r(x_o) \quad \text{if } h \to 0 \text{ and } Nh^k \to \infty \text{ as } N \to \infty.$$

We just showed the pointwise consistency of $r_N(x_o)$ for $r(x_o)$. The remarks made for the uniform consistency of $f_N(x)$ hold as well for the uniform consistency $\sup_x |r_N(x) - r(x)| = o_p(1)$ under the convergence rate $Nh^k/\ln N \to \infty$. The h in $Nh^k/\ln(N) \to \infty$ should be greater than the h in $Nh^k \to \infty$.

As in kernel density estimation, the bias is $O(h^2)$ and the variance is $O\{(Nh^k)^{-1}\}$. The optimal MSE-minimizing h is then $O(N^{-1/(k+4)})$ and the resulting optimal MSE is $O(N^{-4/(k+4)})$. More generally, Stone (1982) showed that the "optimal" convergence rate for MISE-type criteria in nonparametric estimation of the νth derivative of $r(x)$ is $N^{-2(p-\nu)/(2p+k)}$ when $r(x)$ has its pth derivative that is bounded. Setting $\nu = 0$ and $p = 2$, the optimal rate is $N^{-4/(4+k)}$, which is the same as the MSE rate of $r_N(x)$. In the rest of this chapter, whenever kernel method asymptotic distribution is discussed, we will do away with the bias term, assuming under-smoothing.

6.3   Asymptotic Distribution

Observe

$$\sqrt{Nh^k}\{r_N(x_o) - r(x_o)\}$$

$$= \sqrt{Nh^k}\frac{g_N(x_o)}{f_N(x_o)} - \sqrt{Nh^k}\frac{g_N(x_o)}{f(x_o)} + \sqrt{Nh^k}\frac{g_N(x_o)}{f(x_o)} - \sqrt{Nh^k}\frac{g(x_o)}{f(x_o)}$$

$$= \sqrt{Nh^k}g_N(x_o)\left\{\frac{1}{f_N(x_o)} - \frac{1}{f(x_o)}\right\} + \frac{1}{f(x_o)}\sqrt{Nh^k}\{g_N(x_o) - g(x_o)\}$$

$$= \frac{-g_N(x_o)}{f_N(x_o)f(x_o)}\sqrt{Nh^k}\{f_N(x_o) - f(x_o)\} + \frac{1}{f(x_o)}\sqrt{Nh^k}\{g_N(x_o) - g(x_o)\}.$$

It can be shown that

- The first term's asymptotic variance is $r(x_o)^2 f(x_o)^{-1} \int K(z)^2 dz$.

- The second term's asymptotic variance is $E(y^2|x_o)f(x_o)^{-1} \int K(z)^2 dz$.

- The asymptotic covariance between the two is $-r(x_o)^2 f(x_o)^{-1} \int K(z)^2 dz$.

Putting together the three expressions gives the desired asymptotic variance

$$\frac{r(x_o)^2 + E(y^2|x) - 2r(x_o)^2}{f(x_o)} \int K(z)^2 dz = \frac{V(y|x_o)}{f(x_o)} \int K(z)^2 dz$$

$$= \frac{\sigma^2}{f(x_o)} \int K(z)^2 dz \quad \text{if homoskedasticity } (V(y|x) = \sigma^2 \ \forall x) \text{ holds.}$$

If $f(x_o)$ is higher, then the variance becomes lower, for the more observations are available for $r_N(x_o)$.

As for bivariate asymptotic normality for two different evaluation points x_a and x_b,

$$\sqrt{Nh^k}\{r_N(x_a) - r(x_a)\} \quad \text{and} \quad \sqrt{Nh^k}\{r_N(x_b) - r(x_b)\}$$

are asymptotically bivariate normal with the diagonal variance matrix, because the two estimators have zero asymptotic covariance. This is because each estimator uses only the observations local to each evaluation point. That is, there is no observation used for two different evaluation points under $h \to 0$ as $N \to \infty$. This bivariate result can be generalized to a multivariate asymptotic normality with a diagonal covariance matrix.

If we want a (pointwise) CI, we need an estimator for $V(y|x_o)$ (or for σ^2 in the homoskedasticity case). Estimating σ^2 is easy: take the squared residuals $\hat{u}_i^2 \equiv \{y_i - r_N(x_i)\}^2$, $i = 1, ..., N$, and get their sample average. Note that, to see the shape of the function $r_N(x)$ over x, we can pick the evaluation points of x, say $x^{(1)}, ..., x^{(m)}$, and then connect $r_N(x^{(1)}), ..., r_N(x^{(m)})$; that is, we need only m-many estimates. But getting $\hat{u}_i$ requires N-many estimates $r_N(x_i)$, $i = 1, ..., N$. Under heteroskedasticity, estimating $V(y|x_o)$ requires another kernel estimator:

$$E_N(u^2|x_o) \equiv \frac{\sum_i K((x_i - x_o)/h_u) \cdot \hat{u}_i^2}{\sum_i K((x_i - x_o)/h_u)}$$

where the bandwidth h_u should be chosen. A simpler estimator for $V(y|x_o)$ would be

$$V_N(y|x_o) \equiv E_N(y^2|x_o) - r_N(x_o)^2 \quad \text{where}$$

$$E_N(y^2|x_o) \equiv \frac{\sum_i K((x_i - x_o)/h) \cdot y_i^2}{\sum_i K((x_i - x_o)/h)}.$$

As mentioned in the kernel density estimation, collecting pointwise 95% CI's over $x^{(1)}, ..., x^{(m)}$ does not render joint 95% CI's. If we want joint 95% coverage probability at all m points, then we can use pointwise CI's with $1 - \alpha/m$ coverage probability to get $1 - \alpha$ coverage probability for all points jointly; here we are using the Bonferroni's inequality. For instance, $m = 20$ (40) gives $\alpha/m = 0.0025$ (0.00125), and the (0.0025/2)-quantile (0.00125/2-quantile) from $N(0,1)$ is -3.023 (-3.227).

Going further, we may get an "artificial" confidence band connecting the m-many CI's, which may paint too optimistic a picture, for m has to be ∞ in principle for the confidence band while only a finite m is used in reality. As noted in the density estimation section, one may use an uniform confidence band as a "limit" of $m \to \infty$. Härdle and Linton (1994, p. 2317) present an uniform confidence band when $k = 1$:

$$\lim_{N \to \infty} P[r(x) \in \{r_N(x)\pm$$

$$\frac{\{V_N(y|x)f_N(x)^{-1}\int K^2(t)dt\}^{0.5}}{\sqrt{Nh}}\left(d_N + \frac{\lambda}{\sqrt{2\ln h^{-1}}}\right)\} \ \forall x]$$

$$= \exp(-2e^{-\lambda}), \quad \text{where } d_N \equiv \sqrt{2\ln h^{-1}} + \frac{\ln\{(\hat{K}/2)^{1/2}\pi^{-1}\}}{\sqrt{2\ln h^{-1}}},$$

$$\hat{K} \equiv \frac{\int\{K'(t)\}^2dt}{2\int K^2(t)dt}.$$

This is the same as the density uniform band except that the multiplicative factor $V_N(y|x)f_N(x)^{-1}$ in the formula replaces $f_N(x)$ in the density uniform band in Bickel and Rosenblatt (1973). Härdle and Linton (1994, p. 2317) do not provide the proof for the display; instead they just refer to Härdle (1990) who only deals with density uniform confidence band. An alternative to uniform confidence band is using a bootstrap "simultaneous confidence bars" on many chosen points as in Härdle and Marron (1991), but this requires choosing two bandwidths—a substantial disadvantage. See Neumann and Polzehl (1998) for more on confidence bands for nonparametric regression.

6.4   Choosing Smoothing Parameter and Kernel

For kernel density estimation, we discussed choosing h by minimizing MISE or by the least squares cross validation. For $r_N(x)$, however, MSE is difficult to obtain. If $k = 1$ or 2, then the usual "eyeballing" (i.e., the graphical trial and error method) seems the best; if $k > 2$, then data-driven automatic CV-based methods shown below are recommended.

Define the "leave-one-out" kernel estimator for $r(x_j)$:

$$r_{Nj}(x_j) \equiv \frac{\sum_{i \neq j} K((x_i - x_j)/h)y_i}{\sum_{i \neq j} K((x_i - x_j)/h)}.$$

We can choose h minimizing the CV criterion

$$\frac{1}{N}\sum_{j=1}^{N}\{y_j - r_{Nj}(x_j)\}^2 \cdot w(x_j)$$

where $w(x_j)$ is a weighting function to downgrade the "prediction errors" when x_j falls near the boundary of its range. Choice of $w(x_j)$ is up to the researcher; obviously the simplest is $w(x_j) = 1$ for all j. The following shows that the CV minimand can be derived from ISE as in density estimation.

Introduce distances between r_N and r: with SE standing for "Squared Error,"

$$\text{Average SE: } d_A(r_N, r) \equiv N^{-1} \sum_j \{r_N(x_j) - r(x_j)\}^2 w(x_j),$$
$$\text{Integrated SE: } d_I(r_N, r) \equiv \int \{r_N(x) - r(x)\}^2 w(x) f(x) dx,$$
$$\text{Conditional Mean SE: } d_C(r_N, r) \equiv E\{d_I(r_N, r)|x_1, ..., x_N\}.$$

For notational simplicity, let $w(x) = 1$ from now on. Consider $d_I(r_N, r)$ which is analogous to ISE in density estimation:

$$d_I(r_N, r) = \int r_N(x)^2 f(x) dx - 2 \int r_N(x) r(x) f(x) dx + \int r(x)^2 f(x) dx.$$

Ignore the last term not depending on h. Approximate the first term with $N^{-1} \sum_j r_{Nj}(x_j)^2$. As for the middle term, observe, using $E(u|x) = 0$,

$$\int r_{Nj}(x) r(x) f(x) dx = E\{r_{Nj}(x) r(x)\} \simeq E\{r_N(x) y\} \simeq \frac{1}{N} \sum_j r_{Nj}(x_j) y_j.$$

Hence, $d_I(r_N, r)$ can be approximated by $N^{-1} \sum_j r_{Nj}(x_j)^2 - (2/N) \sum_j r_{Nj}(x_j) y_j$. Adding $N^{-1} \sum_j y_j^2$ to this yields the CV minimand.

Regard a bandwidth selection rule $\hat{h}$ as a function from $\{(x_i, y_i)\}_{i=1}^N$ to H_N (say, $H_N = [N^{-a}, N^{-b}]$). Then $\hat{h}$ is said to be "optimal wrt distance d," if

$$\lim_{N \to \infty} \frac{d\{r_N(x; \hat{h}), r(x)\}}{\inf_{h \in H_N} d\{r_N(x; h), r(x)\}} \to^p 1.$$

Härdle and Marron (1985) showed that the CV-minimizing $\hat{h}$ is optimal wrt d_A, d_I, and d_C under some conditions.

Consider a general nonparametric estimator $m_N(x; h)$ with $x_1, ..., x_N$ as the evaluation points:

$$\begin{bmatrix} m_N(x_1; h) \\ \vdots \\ m_N(x_N; h) \end{bmatrix} = \begin{bmatrix} \sum_{i=1}^N w_{N1}(x_1, ..., x_N; h) y_i \\ \vdots \\ \sum_{i=1}^N w_{NN}(x_1, ..., x_N; h) y_i \end{bmatrix} \equiv W_N(h) \cdot Y$$

where $Y \equiv (y_1, ..., y_N)'$ and $W_N(h)$ is a $N \times N$ matrix. For $(r_N(x_1), ..., r_N(x_N))'$, $W_N(h)$ is

$$\begin{bmatrix} K((x_1 - x_1)/h)/\sum_i K((x_i - x_1)/h), & \cdots, & K((x_N - x_1)/h)/\sum_i K((x_i - x_1)/h) \\ \vdots & & \vdots \\ K((x_1 - x_N)/h)/\sum_i K((x_i - x_N)/h), & \cdots, & K((x_N - x_N)/h)/\sum_i K((x_i - x_N)/h) \end{bmatrix}.$$

The *generalized cross validation* is another well-known bandwidth selection rule minimizing

$$\frac{1}{N}\sum_{j=1}^{N}\{y_j - m_N(x_j;h)\}^2 \cdot [1 - \frac{tr\{W_N(h)\}}{N}]^{-2}$$

$$= \frac{1}{N}\sum_{j=1}^{N}\{y_j - r_N(x_j;h)\}^2[1 - \frac{K(0)}{N}\sum_j \frac{1}{\sum_i K((x_i - x_j)/h)}]^{-2} \text{ for } r_N.$$

Compared with CV, generalized CV is slightly more convenient due to no need for leave-one-out. See, however, Andrews (1991) who showed that CV is better under heteroskedasticity; see also Härdle et al. (1988)

So far we used only one h even when $k > 1$. In practice, one may use k different bandwidths, say $h_1, ..., h_k$, because the regressors have different scales. Then we would have $\prod_{j=1}^{k} h_j$ instead of h^k. Although using different bandwidths should be more advantageous in principle, this can make choosing bandwidths too involved. A recommendable alternative is to standardize all regressors and use one bandwidth h_0. That is, use $h_j = h_0 SD(x_j)$, $j = 1, ..., k$, for non-standardized data; as in density estimation, one rule of thumb for h_0 is $N^{-1/(k+4)}$.

Choosing h may sound too difficult, but this is not necessarily the case as there exists a reasonable bound on h. Suppose that the kernel has the support $[-1, 1]$, x is a rv approximately normally distributed, and $h = 4 \cdot SD(x)$. Then $K\{(x_i - x_o)/(4 \cdot SD(x))\} \neq 0$ even when x_i is almost "4-$SD(x)$" away from x_o. For a normally distributed x, the distance $4 \cdot SD(x)$ is from one extreme end to the other extreme end; imagine -2 to 2 in $N(0, 1)$. This shows that h had better be kept below, say $1 \cdot SD(x)$, which is an upper bound on h; i.e., $|x_i - x_o| > SD(x)$ means that x_i is "too big" to be regarded as a neighboring point of x_o. This implies that $cN^{-1/(k+4)}$ in $h = cN^{-1/(k+4)} SD(x)$ should be less than 1. In $cN^{-1/(k+4)}$, since $100^{-1/5} \simeq 0.4$ and $1000^{-1/5} \simeq 0.25$ for $k = 1$, and $100^{-1/6} \simeq 0.46$ and $1000^{-1/6} \simeq 0.32$ for $k = 2$, the multiplicative constant c should be less than about 4.

As x gets standardized in kernel estimation, one may wonder how the original regression function is related to the regression function using the standardized regressors. For this, let $w = Sx \iff x = S^{-1}w$ for a $k \times k$ invertible matrix S. This includes

$$S = diag\left\{\frac{1}{SD(x_1)}, ..., \frac{1}{SD(x_k)}\right\}$$

to standardize x. It should be noted that $E(y|x = x_o) = E(y|w = w_o)$ where $w_o = Sx_o$. Intuitively, this holds as the amount of "information" is the same in "$|x$" and "$|w$". Formally, considering the transformations Sx of x and (y, Sx) of (y, x), the densities $g(w)$ and $g(y, w)$ are, respectively

$$g(w) = f(S^{-1}w)|S^{-1}|, \quad g(y, w) = f(y, S^{-1}w)|J| \quad \text{where}$$

$$|J| \equiv \det \begin{bmatrix} 1 & 0 \\ 0 & S^{-1} \end{bmatrix} = |S^{-1}|$$

where $f(x)$ and $f(y,x)$ are the densities for x and (y,x). Hence

$$
r(x_o) \equiv E(y|x = x_o) = \int y \frac{f(y, x_o)}{f(x_o)} dy
$$

$$
= \int y \frac{f(y, S^{-1}w_o)|S^{-1}|}{f(S^{-1}w_o)|S^{-1}|} dy = \int y \frac{g(y, w_o)}{g(w_o)} dy
$$

$$
= E(y|w = w_o) \equiv q(w_o).
$$

From the first and last expressions, we get $r(x_o) = q(w_o) = q(Sx_o)$. This, however, should not be construed as $r(x_o) = q(x_o)$. For instance, if $r(x) = \beta' x$, then

$$
r(x) = \beta' x = \beta' S^{-1} S x = \beta' S^{-1} w = q(w).
$$

Choosing a kernel in nonparametric regression is similar to that in density estimation; namely, one can set up an optimum criterion and choose an optimal kernel (see, e.g., Müller, 1988). There, however, seems to be a consensus that choosing a kernel does not matter much, being of a secondary importance compared with choosing a bandwidth. Usually in practice, there is little difference in using different kernels when a CV method is employed to choose h. Sometimes kernels satisfying certain properties are specifically needed—the issue of kernel choice among the kernels satisfying those properties still arises—and two such cases will be seen in the following section.

7   Topics in Kernel Nonparametric Regression

In this section, we discuss some topics related to kernel nonparametric regression. First, nonparametrics with "mixed" (discrete and continuous) regressors is examined, which has a practical importance as almost always there are discrete regressors in data; along with this topic, "structural breaks" in $r(x)$ will be briefly discussed. Second, estimating derivatives of $r(x)$, which has been mentioned once before, will be given a closer look. Third, combining MLE-based models with nonparametrics is studied, which is relevant when the regression functional form is unknown whereas the error term distributional form is. Fourth, kernel local linear regression is introduced, which is often used in practice as an improved version of kernel estimator.

7.1   Mixed Regressors and Structural Breaks

So far we dealt only with continuously distributed x with density $f(x)$. Suppose we have mixed regressors: $x = (x_c', x_d')'$, x_c is a $k_c \times 1$ continuously distributed regressor vector, and x_d is a $k_d \times 1$ discretely distributed regressor vector. We can form cells based on the values that x_d can take. For instance, if $x_d = (d_1, d_2)'$ where d_1 takes $0, 1, 2, 3$ and d_2 takes $1, ..., J$, then there will be $4 \times J$ cells. Within a given cell, we can apply the kernel estimator in the preceding subsections. The only change needed is to replace k by k_c and $f(x)$ by $f(x_c|x_d)$.

As an example, suppose $k_c = 3$ and $x_d = (d_1, d_2)'$. Let $d_1 = 0$ and $d_2 = 1$ for the first cell. Within the cell (i.e., conditioned on $d_1 = 0$ and $d_2 = 1$), we can estimate

$$r(x_o, 0, 1) = E(y|x_c = x_o, d_1 = 0, d_2 = 1)$$

with

$$r_N(x_o, 0, 1) \equiv \frac{\sum_i K((x_{ci} - x_o)/h) \, 1[d_{i1} = 0, d_{i2} = 1] \cdot y_i}{\sum_i K((x_{ci} - x_o)/h) \, 1[d_{i1} = 0, d_{i2} = 1]}$$

where $K(\cdot)$ is a three $(= k_c)$ dimensional kernel. The kernel estimator $r_N(x_o, 0, 1)$ is consistent for $r(x_o, 0, 1)$. As for the asymptotic distribution, ignoring the asymptotic bias,

$$\sqrt{N_{01} h^3}\{r_N(x_o, 0, 1) - r(x_o, 0, 1)\} \rightsquigarrow N\{0, \frac{V(y|x_o, 0, 1) \int K(z)^2 dz}{f(x_o|d_1 = 0, d_2 = 1)}\}$$

where N_{01} is the number of observations with $d_0 = 0$ and $d_1 = 1$. Everything is the same as before, except that we are operating on the subpopulation $d_1 = 0$ and $d_2 = 1$.

The convergence in law can be written also as

$$\sqrt{N h^3}\{r_N(x_o, 0, 1) - r(x_o, 0, 1)\}$$
$$\rightsquigarrow N\left\{0, \frac{V(y|x_o, 0, 1) \int K(z)^2 dz}{f(x_o|d_1 = 0, d_2 = 1)P(d_1 = 0, d_2 = 1)}\right\}$$

because

$$\sqrt{N_{01} h^3} = \sqrt{N h^3} \left(\frac{N_{01}}{N}\right)^{1/2} \simeq \sqrt{N h^3} \cdot P(d_1 = 0, d_2 = 1)^{1/2}.$$

It might be helpful to regard $f(x_o|d_1 = 0, d_2 = 1)P(d_1 = 0, d_2 = 1)$ as $f(x_0, 0, 1)$ as if all of x, d_0, and d_1 are continuously distributed. Then the preceding display with $\sqrt{N h^3}$ would look more in line with those in the previous subsections.

In the above asymptotic variance of $\sqrt{N_{01} h^3}\{r_N(x_o, 0, 1) - r(x_o, 0, 1)\}$, $V(y|x_o, 0, 1)$ and $f(x_o|d_1 = 0, d_2 = 1)$ can be estimated, respectively, by

$$E_N(u^2|x_o, 0, 1) \equiv \frac{\sum_i K((x_{ci} - x_o)/h_u) \, 1[d_{i1} = 0, d_{i2} = 1] \cdot \hat{u}_i^2}{\sum_i K((x_{ci} - x_o)/h_u) \, 1[d_{i1} = 0, d_{i2} = 1]},$$

$$f_N(x_o|d_1 = 0, d_2 = 1) \equiv \frac{(Nh^3)^{-1} \sum_i K((x_{ci} - x_o)/h) \, 1[d_{i1} = 0, d_{i2} = 1]}{N^{-1} \sum_i 1[d_{i1} = 0, d_{i2} = 1]}$$

where h_u is a bandwidth and $\hat{u}_i$ is the residual $y_i - r_N(x_o, 0, 1)$. The numerator in $f_N(x_o | d_1 = 0, d_2 = 1)$ estimates

$$f(x_o | d_1 = 0, d_2 = 1) \cdot P(d_1 = 0, d_2 = 1)$$

and we need to divide the numerator by the estimator for $P(d_1 = 0, d_2 = 1)$ in the denominator to get the conditional density.

The above method of applying nonparametrics on each cell can be cumbersome if there are too many cells. When all regressors are discrete, Bierens (1987, pp. 115–117) showed that applying the kernel method yields $r_N(x_o) \to^p r(x_o)$, and

$$\sqrt{N}\{r_N(x_o) - r(x_o)\} \rightsquigarrow N(0, \frac{V(y|x_o)}{P(x_o)});$$

the convergence rate is $\sqrt{N}$ and no kernel shows up in the asymptotic distribution. If we take the sample average of the observations with $x = x_o$, then we get the same asymptotic distribution result. Based on this, Bierens (1987, pp. 117–118) stated that, when the regressors are mixed (continuous and discrete), if we apply the kernel method as if all regressors are continuously distributed, then we still get consistent estimators and, for the above example,

$$\sqrt{Nh^3}\{r_N(x_o, 0, 1) - r(x_o, 0, 1)\}$$
$$\rightsquigarrow N\left\{0, \frac{V(y|x_o, 0, 1) \int K(z_1, 0)^2 dz_1}{f(x_o | d_1 = 0, d_2 = 1)P(d_1 = 0, d_2 = 1)}\right\}$$

where $K(z_1, z_2)$ is a kernel such that $\int K(z_1, 0)dz_1 = 1$ with z_1 and z_2 corresponding to continuous and discrete regressors, respectively.

For instance, if we use a product kernel with its marginal kernel being the bi-weight kernel $(15/16)(1 - t^2)^2 1[|t| \leq 1]$, then for the above case with $k_c = 3$ and $k_d = 2$, $\int K(z_1, 0)dz_1 = 1$ does not hold, because $\int K(z_1, 0)dz_1$ equals (with $z_1 = (z_{11}, z_{12}, z_{13})'$)

$$\int \int \int \left\{ \frac{15}{16}(1 - z_{11}^2)^2 1[|z_{11}| \leq 1] \frac{15}{16}(1 - z_{12}^2)^2 1[|z_{12}| \leq 1] \right.$$
$$\left. \cdot \frac{15}{16}(1 - z_{13}^2)^2 1[|z_{13}| \leq 1] \frac{15}{16} \frac{15}{16} \right\}^2 dz_{11} dz_{12} dz_{13} = \left(\frac{15}{16}\right)^2.$$

Instead, if we use $0.88^{-2}(0.88 - t^2)^2 1[|t| \leq 0.88]$ for the marginal kernel, then $\int K(z_1, 0)dz_1 = 1$ holds, because the marginal kernel takes one when its argument is zero. Note that 0.88 is the normalizing constant, because $\int (0.88 - t^2)^2 1[|t| \leq 0.88]dt \simeq 0.88$. If a ϕ-based product kernel is to be used, then since $\phi(0) = 0.400$, $\phi(\cdot/0.4)/0.4$ can be used instead of ϕ to satisfy $\int K(z_1, 0)dz_1 = 1$.

In short, whether the regressors are continuous or discrete, we can apply the kernel method indiscriminately. The only thing we should be careful about is the convergence rate and the asymptotic distribution. Still yet, we may

learn more by forming cells and doing nonparametric estimation on each
cell in mixed cases unless there are too many cells. Racine and Li (2004)
suggested a different way of smoothing: use a function taking on 1 and λ
($\to 0^+$ as $N \to \infty$) for discrete regressors. This is one of many different
ways of "smearing" discrete probability masses, and there are many studies
on nonparametric estimation and test with mixed regressors; see, e.g., Li and
Racine (2007) and the references therein.

In mixed regressor cases, the kernel $K(z_1, z_2)$ should satisfy
$\int K(z_1, 0)dz_1 = 1$. This is an example where a specific property is required
for a kernel. Another such example can be seen when $r(x)$ may jump at some
points—"structural break." Suppose that x is a rv and $r(x)$ consists of a
smooth component $r_c(x)$ and a "jumping" component with jump magnitude
m_j at jump point γ_j, $j = 1, ..., J$:

$$r(x) = r_c(x) + \sum_{j=1}^{J} m_j \cdot 1[x \geq \gamma_j].$$

One way to detect jumps is to use "one-sided" kernels to estimate $r(x_o)$ from
the lhs and rhs and then look at the difference. If there is no jump at x_o, the
difference from the two-sides becomes 0. Otherwise, the difference estimates
m_j.

Specifically, the difference estimator is

$$\frac{\sum_i K((x_i - x_o)/h)1[x_i \geq x_o]y_i}{\sum_i K((x_i - x_o)/h)1[x_i \geq x_o]} - \frac{\sum_i K((x_i - x_o)/h)1[x_i < x_o]y_i}{\sum_i K((x_i - x_o)/h)1[x_i < x_o]}.$$

Noting $K((x_i - x_o)/h)1[x_i \geq x_o] = K((x_i - x_o)/h)1[(x_i - x_o)/h \geq 0]$, the
kernel $K(\cdot)1[\cdot \geq 0]$ gives positive weights only to the right-neighboring observations of x_o, whereas $K(\cdot)1[\cdot < 0]$ gives positive weights only to the
left-neighboring observations of x_o. Once γ_j and m_j are estimated with $\hat{\gamma}_j$
and $\hat{m}_j$, respectively, $r_c(x)$ can be estimated by the usual kernel regression
with its y_i replaced by $y_j - \sum_{j=1}^{J} \hat{m}_j \cdot 1[x \geq \hat{\gamma}_j]$. See Qiu (2005) for more.

7.2   Estimating Derivatives

Define the gradient of $r(x)$ at x_o:

$$\nabla r(x_o) \equiv \frac{\partial r(x_o)}{\partial x} \equiv \left(\frac{\partial r(x_o)}{\partial x_1}, ..., \frac{\partial r(x_o)}{\partial x_k} \right)'.$$

The gradient is often of interest. For instance, if x is the factor of a production
function, then the gradient reflects the marginal contributions of x to the
average output $E(y|x_o) = r(x_o)$. Naturally, $\nabla r(x_o)$ can be estimated by the
gradient of the kernel estimator $r_N(x_o)$:

$$\nabla r_N(x_o) = \nabla \frac{g_N(x_o)}{f_N(x_o)} = \frac{1}{f_N(x_o)^2}\{f_N(x_o)\nabla g_N(x_o) - g_N(x_o)\nabla f_N(x_o)\}$$

$$= \frac{\nabla g_N(x_o)}{f_N(x_o)} - r_N(x_o)\frac{\nabla f_N(x_o)}{f_N(x_o)}$$

where, with $\nabla K(t) \equiv \partial K(t)/\partial t$,

$$\nabla g_N(x_o) = \frac{-1}{Nh^{k+1}} \sum_{i=1}^{N} \nabla K\left(\frac{x_i - x_o}{h}\right) y_i \quad \text{and}$$

$$\nabla f_N(x_o) = \frac{-1}{Nh^{k+1}} \sum_{i=1}^{N} \nabla K\left(\frac{x_i - x_o}{h}\right).$$

Call $\nabla r_N(x_o)$ "kernel regression-derivative estimator." Second-order derivatives are estimated by differentiating $\nabla r_N(x_o)$ again, but the number of terms goes up by twice due to the denominator of $r_N(x_o)$. The problem gets worse for higher-order derivatives.

Following essentially the same line of proof for $r_N(x_o) \to^p r(x_o)$, it can be proven (see, e.g., Vinod and Ullah, 1988) that $\nabla r_N(x_o) \to^p \nabla r(x_o)$, and denoting the jth component of $\nabla r_N(x_o)$ as $\nabla r_N^{(j)}(x_o)$ and the jth component of $\nabla r(x_o)$ as $\nabla r^{(j)}(x_o)$, it holds that

$$\sqrt{Nh^{k+2}}\{\nabla r_N^{(j)}(x_o) - \nabla r^{(j)}(x)\} \rightsquigarrow N\left(0, \frac{V(y|x_o)}{f(x_o)} \int \left\{\frac{\partial K(z)}{\partial z_j}\right\}^2 dz\right).$$

Compared to the asymptotic variance of $\sqrt{Nh^k}\{r_N(x_o) - r(x)\}$, there are two differences. One is the convergence rate that is slower by the factor $\sqrt{h^2} = h$, and the other is $\int\{\partial K(z)/\partial z_j\}^2 dz$ instead of $\int K(z)^2 dz$; $\partial K(z)/\partial z_j$ is due to the differentiation. If one uses h_j for x_j, $j = 1, ..., k$, Nh^{k+2} should be replaced by $N \cdot h_j^2 \prod_{j=1}^{k} h_j$. When $k = 1$ and $K = \phi$,

$$\int \{\partial K(z)/\partial z_j\}^2 dz = \int \{-z\phi(z)\}^2 dz = \int z^2 \phi(z)^2 dz$$

$$= E\{z^2 \phi(z)\} \text{ for } z \sim N(0,1).$$

CV bandwidth-choice methods use the prediction error $y_i - r_N(x_i)$ as its logical basis. But there is no analogous expression in choosing h for $\nabla r_N(x_o)$, because there is no "target" for $\nabla r_N(x_i)$ (in contrast, y_i is the target for $r_N(x_i)$). Although Müller (1988) made some suggestions on how to choose h for $\nabla r_N(x_o)$, there seems to be no particularly good way to choose h if $k > 2$; when $k \leq 2$, the trial and error method (i.e., "eyeballing") seems the best. An optimal h for $r_N(x_o)$ is not optimal for $\nabla r_N(x_o)$: the bandwidth for $\nabla r_N(x_o)$ should be larger than the bandwidth for $r_N(x_o)$, because $\nabla r_N(x_o)$

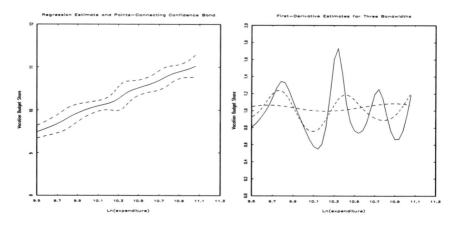

Figure 12: Regression and First-Derivative Estimates

will be a poor estimator of $\nabla r(x_o)$ if $r_N(x_o)$ is too wiggly as shown in the following.

The left panel of Figure 12 shows a kernel estimate (the middle solid line) for $E(y|x)$ where y is vacation budget share % and x is ln(total expenditure). The bandwidth h is $0.1 \times SD(x) \times N^{-1/5}$. This is 10 times smaller than the rule-of-thumb bandwidth $SD(x) \times N^{-1/5}$ which, however, gives almost the same (straight) line without local "wiggles." Also shown is a point-connecting confidence band (the dotted lines) from 40 evaluation points with the joint coverage probability 0.95. The expenditure data was already used once to present confidence intervals/bands in kernel density estimation. The vacation budget share falls around 10% and rises steadily as ln(total expenditure) increases. The left panel shows $\nabla r(x) \simeq 1$ for most x.

The right panel presents three estimates for $\nabla r(x)$. The solid line is for the same bandwidth $0.1 \times SD(x) \times N^{-1/5}$, whereas the two dotted lines are for the bandwidths twice and four times greater, respectively. When the same bandwidth as for $r_N(x)$ is used for $\nabla r(x)$ estimation, clearly the estimate is unstable. It is only when the four-times greater bandwidth is used that a reasonable estimate for $\nabla r(x)$ is obtained.

Mistakes are easily made in practice when the above analytic derivative of $\nabla r_N(x)$ is used to estimate $\nabla r(x_o)$. It is hence preferable to use numerical gradients, rather than the analytic ones. That is, for $\nabla r_N(x_o)$, it is safer to use the numerical derivative

$$\frac{r_N(x_o + \varepsilon) - r_N(x_o - \varepsilon)}{2\varepsilon} \quad \text{where } \varepsilon \text{ is a small constant}$$

rather than the analytic derivative $f_N(x_o)^{-1}\{\nabla g_N(x_o) - r_N(x_o)\nabla f_N(x_o)\}$ provided above. The figure was also obtained with the numerical derivative.

7.3   Nonparameric MLE and Quantile Regression

The kernel estimator $r_N(x_o)$ can be obtained by minimizing

$$\sum_i K\left(\frac{x_i - x_o}{h}\right)\{y_i - r_N(x_o)\}^2$$

wrt $r_N(x_o)$ because the first-order condition

$$\sum_i K\left(\frac{x_i - x_o}{h}\right)\{y_i - r_N(x_o)\} = 0$$

$$\Longleftrightarrow \quad \sum_i K\left(\frac{x_i - x_o}{h}\right) y_i = r_N(x_o) \sum_i K\left(\frac{x_i - x_o}{h}\right)$$

yields $r_N(x_o)$. As well known, if we minimize $\sum_i (y_i - \mu)^2$ wrt μ, we get $\mu = \bar{y}$. In view of this, $r_N(x_o)$ may be called a "local mean (averaging) estimator"— local around x_o. Using this idea of the minimizer representation of $r_N(x_o)$, it is possible to combine nonparametric regression with MLE specifying the error term distribution.

For instance, consider an Exponential duration model:

$$f(y|x) = \theta \exp(-\theta y) \quad \text{and} \quad \theta = e^{\mu(x)}$$

where $f(y|x)$ is the density of a duration y given x, and $\mu(x)$ is an unknown function. The parameter to estimate is $\mu(x)$, which is infinite-dimensional. Rewriting $f(y|x)$ as $f(y|x, \mu(x))$ to make it explicit that $\mu(x)$ is a parameter for the conditional distribution, we can maximize

$$\sum_i K\left(\frac{x_i - x_o}{h}\right) \cdot \ln f\{y_i|x_o, \mu(x_o)\}$$

wrt $\mu(x_o)$ to get a "kernel nonparametric MLE." Repeating this for all x_o will reveal the function $\mu(x)$.

Denoting the estimator for $\mu(x_o)$ as $\mu_N(x_o)$, Staniswalis (1989) showed $\mu_N(x_o) \to^p \mu(x_o)$ and

$$\sqrt{Nh^k}\{\mu_N(x_o) - \mu(x_o)\} \rightsquigarrow N(0, \frac{\int K(z)^2 dz}{E[\{\partial \ln f(y|x_o, \mu(x_o))/\partial \mu(x_o)\}^2 |x_o]}).$$

Compare this to the usual parametric MLE with a scalar parameter $\mu_o \equiv \mu(x_o)$ maximizing $\sum_{i \,:\, x_i = x_o} \ln f\{y_i|x_o, \mu(x_o)\}$ when multiple observations are available with $x = x_o$. The parametric MLE is $\sqrt{N}$-consistent with the asymptotic variance $E^{-1}[\{\partial \ln f(y|x_o, \mu_o)/\partial \mu_o\}^2]$—no conditioning on $x = x_o$ and nor $\int K(z)^2 dz$.

A further generalization of the above "nonparametric MLE" can be done. For instance, instead of going fully nonparametric, one may adopt a model

$$f(y|x) = \theta \exp(-\theta y) \quad \text{and} \quad \theta = \exp\{x_1'\beta_1 + \mu(x_2)\}$$

where $\mu(x)$ is replaced by a "semi-linear model" $x_1'\beta_1 + \mu(x_2)$; this type of models will be examined in the chapter for semi-nonparametrics in detail. There are two parameters: finite-dimensional β_1 and infinite-dimensional $\mu(\cdot)$. Severini and Staniswalis (1994) showed how to estimate iteratively both parameters in this kind of models using "profile likelihood," a more general and theoretical treatment of which appeared in Severini and Wong (1992). Note that a likelihood function $L\{\beta_1, \mu(\cdot)\}$ can be maximized wrt $\mu(\cdot)$ for each fixed β_1; denote the maximizer as $\hat{\mu}_{\beta_1}(\cdot)$. Then $L\{\beta_1, \hat{\mu}_{\beta_1}(\cdot)\}$ is called the *profile likelihood* for β_1 (with μ_1 "concentrated out"), which can be maximized for β_1.

Estimating a mean locally or doing MLE locally using the local weighting function $K((x_i - x_o)/h)$ is widely applicable. For instance, minimizing

$$\sum_i K\left(\frac{x_i - x_o}{h}\right) \{y_i - q_N(x_o)\}\{\alpha - 1[y_i - q_N(x_o) < 0]\}$$

wrt $q_N(x_o)$, we can estimate the αth quantile of $y|x = x_o$ nonparametrically; i.e., we get *nonparametric quantile regression*. An empirical example can be seen in Nahm (2001).

Another way to estimate the αth quantile nonparametrically can be seen by noting

$$P(y \le y_o|x_o) = E(1[y \le y_o]|x) \equiv F(y_o|x_o).$$

This leads to inverting a kernel nonparametric estimator of the conditional df $F(y_o|x_o)$ for its quantiles:

$$F_N(y_o|x_o) \equiv \frac{\sum_i K\{(x_i - x_o)/h\}1[y_i \le y_o]}{\sum_i K\{(x_i - x_o)/h\}}.$$

See Dette and Volgushev (2008) and the references therein for nonparametric quantile regression in general. The following subsection shows another estimator based on this line of local regression approach; see Loader (1999a) for more.

7.4   Local Linear Regression

Generalizing the local mean estimation, one can think of using, not just the "intercept," but also a slope (centered at x_o). *Local linear regression* (LLR) minimizes

$$\sum_i \{y_i - a - b'(x_i - x_o)\}^2 \cdot K\left(\frac{x_i - x_o}{h}\right)$$

wrt a and b. The estimator $a_N(x_o)$ for a is the LLR estimator for $r(x_o)$, whereas the estimator for b is the LLR estimator for $\nabla r(x_o)$. Compared with LLR, the kernel estimator $r_N(x_o)$ can be called a "local constant regression," because $r_N(x_o)$ fits a flat line (i.e., an intercept) locally around x_o while LLR fits an intercept and a slope.

The LLR estimator for $r(x_o)$ is

$$a_N(x_o) = (1, 0, ..., 0) \cdot \{X(x_o)'W(x_o)X(x_o)\}^{-1} \cdot \{X(x_o)'W(x_o)Y\}$$

where $Y \equiv (y_1, ..., y_N)'$, $W(x_o) \equiv diag\{K((x_1 - x_o)/h), ..., K((x_N - x_o)/h)\}$, and

$$\underset{N \times (1+k)}{X(x_o)} \equiv \begin{bmatrix} 1, & (x_1 - x_o)' \\ & \vdots \\ 1, & (x_N - x_o)' \end{bmatrix}.$$

The vector $(1, 0, ..., 0)$ is to pick up the first element; the remaining components are estimators for $\nabla r(x_o)$.

The details of $a_N(x_o)$ may be seen better by setting $k = 1$ and observing

$X(x_o)'W(x_o)$

$$= \begin{bmatrix} 1 & \cdots & 1 \\ (x_1 - x_o) & \cdots & (x_N - x_o) \end{bmatrix} \begin{bmatrix} K((x_1 - x_o)/h) & 0 & 0 \\ 0 & \ddots & 0 \\ 0 & 0 & K((x_N - x_o)/h) \end{bmatrix}$$

$$= \begin{bmatrix} K((x_1 - x_o)/h) & K((x_2 - x_o)/h) & \cdots & K((x_N - x_o)/h) \\ K((x_1 - x_o)/h)(x_1 - x_o) & K((x_2 - x_o)/h)(x_2 - x_o) & \cdots & K((x_N - x_o)/h)(x_N - x_o) \end{bmatrix}$$

$X(x_o)'W(x_o)X(x_o)$

$$= \begin{bmatrix} \sum_i K((x_i - x_o)/h) & \sum_i K((x_i - x_o)/h)(x_i - x_o) \\ \sum_i K((x_i - x_o)/h)(x_i - x_o) & \sum_i K((x_i - x_o)/h)(x_i - x_o)^2 \end{bmatrix},$$

$$X(x_o)'W(x_o)Y = \begin{bmatrix} \sum_i K((x_i - x_o)/h)y_i \\ \sum_i K((x_i - x_o)/h)(x_i - x_o)y_i \end{bmatrix}.$$

Recall that, when a product kernel $K(t) = \Pi_{j=1}^k L(t_j)$ is used,

$$Bias\{f_N(x_o)\} \simeq h^2 \kappa \frac{1}{2} \sum_{j=1}^k \frac{\partial^2 f(x_o)}{\partial x_{oj}^2} \quad \text{where } \kappa \equiv \int z^2 L(z)dz$$

$$V\{f_N(x_o)\} \simeq \frac{f(x_o)}{Nh^k} \int K(t)^2 dt.$$

Differently from $f_N(x_o)$, however, deriving the biases and variances of $r_N(x_o)$ and $a_N(x_o)$ for $E(y|x_o) = r(x_o)$ is not easy due to the random denominator $f_N(x)$ in $a_N(x_o)$ and $r_N(x_o)$. Nevertheless, Ruppert and Wand (1994) showed, for an interior point x_o in the support of x,

$$Bias\{a_N(x_o)|x_1, ..., x_N\} \simeq h^2 \kappa \frac{1}{2} \sum_{j=1}^k \frac{\partial^2 r(x_o)}{\partial x_{oj}^2}$$

$$V\{a_N(x_o)|x_1,...,x_N\} \simeq \frac{V(y|x_o)}{Nh^k f(x_o)} \int K(t)^2 dt.$$

The difference between this and the preceding display is that $r(x_o)$ replaces $f(x_o)$ in bias and $V(y|x_o)/f(x_o)$ replaces $f(x_o)$ in variance. In contrast to these, for the univariate case $k = 1$, Fan (1992) showed that

$$Bias\{r_N(x_o)|x_1,...,x_N\} \simeq h^2\kappa \left\{ \frac{1}{2}r''(x_o) + \frac{r'(x_o)f'(x_o)}{f(x_o)} \right\}.$$

Compared with $Bias\{a_N(x_o)|x_1,...,x_N\}$ with $k = 1$, there is one extra term $r'(x_o)f'(x_o)/f(x_o)$: even if $r(x)$ is linear such that $r(x) = \beta_1 + \beta_x x$, still there will be a bias in $r_N(x_o)$ due to the extra term with $r'(x_o)$.

LLR has some advantages over the kernel method. First, as just shown, whereas the kernel method has a bias even when $r(x) = x'\beta$, LLR has no bias in this case. Second, as a by-product, LLR provides an estimator for $\nabla r(x_o)$ as well. Third, LLR works better at the boundaries due to the local linear fitting. To appreciate this point better, suppose x has support $[0,1]$ and $r(x)$ is increasing around $x = 1$ with $r(1) = 3$. In this case, $r_N(x)$ is biased because the local weighted averaging is done with observations all less than 3 on the left-hand side of $x = 1$—$r(1)$ gets under-estimated. In contrast, LLR extrapolates to the right-hand side using a linear function; using only the left-hand side observations poses no problem, because the fitted up-sloping line at $x = 1$ continues to the right-hand side of $x = 1$.

LLR can be generalized to local quadratic (and higher polynomial) regression, although hardly ever one goes over local quadratic regression in practice; local quadratic regression is examined in detail by Rupper and Wand (1994). Local polynomial regression and its uniform convergence rates are examined in Masry (1996). When $k > 1$, local polynomial regression can be much simplified if we do the polynomial regression only for a selected regressor ("direction") of interest. For instance, suppose that the interest is only on the first regressor x_1. Then we may set

$$\underset{N\times 4}{X(x_o)} \equiv \begin{bmatrix} 1, & (x_{11}-x_{o1}), & (x_{11}-x_{o1})^2, & (x_{11}-x_{o1})^3 \\ & & \vdots & \\ 1, & (x_{N1}-x_{o1}), & (x_{N1}-x_{o1})^2, & (x_{N1}-x_{o1})^3 \end{bmatrix}.$$

This means a local cubic regression for x_1 and local constant regressions for the other regressors. This idea can be seen in Fan et al. (1998) and Severance-Lossin and Sperlich (1999).

For more on LLR and local regression, see Fan (1992), Hastie and Loader (1993), Ruppert and Wand (1994), Fan and Gijbels (1996), and Loader (1999a). Also there are some other (and earlier) versions of LLR using different weighting schemes; see Cleveland et al. (1988).

CHAPTER 8
BANDWIDTH-FREE SEMIPARAMETRIC METHODS

In parametric regression, the regression function and the error term conditional distribution $F_{u|x}$ of $u|x$ are specified. In nonparametrics, neither is specified. Thus, semantically speaking, anything between the two extremes can be called "*semiparametric*" or "*semi-nonparametric*." Under this classification, LSE and many method-of-moment estimators are semiparametric. Since these estimators have been studied already, semiparametric methods for LDV models are examined in this and the following chapters. This chapter examines semiparametric estimators that do not require any bandwidth although their asymptotic variance estimators may do. Typically such estimators specify the regression function, but not $F_{u|x}$. Those estimators make use of various semiparametric assumptions such as symmetry, unimodality, independence etc. The following chapter studies semiparametric estimators that do require a bandwidth. See Powell (1994), Horowitz (1998), Pagan and Ullah (1999), Ruppert et al. (2003), Yatchew (2003), Härdle et al. (2004), and Li and Racine (2007) for reviews on the semiparametric and nonparametric literature.

1 Quantile Regression for LDV models

Quantile regression takes a prominent position in semiparametric econometrics. This is because LDV models result from various transformations of the latent continuous responses, and quantile is "equivariant" to increasing (i.e., non-decreasing) transformations. This point is shown in the next paragraph, and the rest of this section introduces quantile-based semiparametric methods for binary, multinomial, ODR, count, and censored responses. Another prominent impetus—symmetry—for semiparametric econometrics will be reviewed in the next section.

Suppose a latent variable y^* is continuously distributed, and

$$y = \tau(y^*), \quad \text{where } \tau(\cdot) \text{ is increasing: } \tau(m_1) \le \tau(m_2) \text{ iff } m_1 \le m_2.$$

For instance, in the binary response $y = 1[y^* \ge 0]$, $\tau(\cdot) = 1[\cdot \ge 0]$. Recall the quantile function $F^{-1}(\cdot)$ of a df F:

$$\alpha\text{th quantile of } F \equiv \min\{\mu : F(\mu) \ge \alpha\} \quad \text{where } 0 < \alpha < 1.$$

Myoung-jae Lee, *Micro-Econometrics*, DOI 10.1007/b60971_8,

Denote the αth quantile of a rv z as $Q_\alpha(z)$; i.e., $Q_\alpha(z) = F_z^{-1}(\alpha)$ where F_z denotes the df of z.

An important fact is the *"quantile equivariance" to increasing transformations*:

$$Q_\alpha(z) = m \implies Q_\alpha\{\tau(z)\} = \tau(m).$$

For instance, with $\tau(\cdot) = \ln(\cdot)$ and $\alpha = 0.5$, the last equality becomes $Med\{\ln(z)\} = \ln\{Med(z)\}$ where "*Med*" stands for median. To understand this point which appeared in relation to nonlinear regression already, observe that, because m is the minimum for

$$P(z \le m) = P\{\tau(z) \le \tau(m)\} \ge \alpha,$$

if we replace m with any $m' < m$ in this display, then the display does not hold. The probability accumulated up to $\tau(m)$ is at least α whereas the probability accumulated up to any point strictly smaller than $\tau(m)$ is strictly smaller than α.

1.1 Binary and Multinomial Responses

Suppose $y^* = x'\beta + u$ but only its binary transformation $y = 1[y^* \ge 0]$ is observed along with x. To put it more formally,

$$y = \tau(y^*) \quad \text{where } \tau(\cdot) \equiv 1[\cdot \ge 0].$$

Binary response model has appeared already, and probit and logit were reviewed there. The main problem with parametric estimators is that the distribution of $u|x$ should be specified fully, and thus any misspecification can cause inconsistency of the estimators. For instance, suppose u is heteroskedastic in probit, say, $u = \sigma(x'\delta) \cdot v$ where $\sigma(x'\delta)$ is a function of $x'\delta$, δ is an unknown parameter vector, and $v \sim N(0,1)$ independently of x. In this case, ignoring $\sigma(x'\delta)$ means that the regression function becomes $x'\beta/\sigma(x'\delta)$, and probit will estimate a linear approximation to this nonlinear function. Of course, one can do probit with $\Phi\{x'\beta/\sigma(x'\delta)\}$ estimating both β and δ. But this requires both $\Phi(\cdot)$ and $\sigma(\cdot)$ to be the correct df and heteroskedasticity factor, respectively. This subsection introduces an estimator that does not require specifying either of them.

Applying the quantile equivariance to $\tau(y^*) = 1[y^* \ge 0]$ and $\alpha = 0.5$ yields

$$Med(y^*|x) = x'\beta \implies Med(y|x) = 1[x'\beta \ge 0].$$

Since the least absolute deviation loss function is minimized at the median, Manski's (1975, 1985) *maximum score estimator* (MSC) b_{msc} is obtained by minimizing

$$\frac{1}{N}\sum_i |y_i - 1[x_i'b \ge 0]|$$

wrt b. This minimand may be viewed as a nonlinear regression.

Although the estimator is consistent, since the derivative of $1[x'b \geq 0]$ wrt b is zero (except at $x'b = 0$ where the derivative does not exist), the usual method of deriving asymptotic distributions is not applicable. One critical assumption for MSC to overcome the lack of smoothness in the minimand is that there is a regressor, say x_k, with a non-zero coefficient such that $x_k|(x_1,...,x_{k-1})$ has a non-zero density on all of R^1 for all $(x_1,...,x_{k-1})$. Also the support of the distribution of $u|x$ should be the entire real line R for all x.

As other binary-response estimators, MSC cannot identify β fully, because if b_{msc} minimizes the minimand, λb_{msc} with $\lambda > 0$ also minimizes the same minimand. Also, if the threshold for y^* is not zero but unknown, say γ, it will be absorbed into the intercept estimate. Denote the regressors other than unity as $\tilde{x}$ so that $x = (1, \tilde{x}')'$; denote β accordingly as $\beta = (\beta_1, \tilde{\beta}')'$. Then

$$\beta_1 + \tilde{x}'\tilde{\beta} + u \geq \gamma \iff \beta_1 - \gamma + \tilde{x}'\tilde{\beta} + u \geq 0$$
$$\iff \beta_1^* + \tilde{x}'\tilde{\beta} + u \geq 0 \quad \text{where} \quad \beta_1^* \equiv \beta_1 - \gamma$$
$$\iff \lambda(\beta_1^* + \tilde{x}'\tilde{\beta} + u) \geq 0 \quad \text{for any } \lambda > 0 :$$

MSC can identify $(\beta_1^*, \tilde{\beta}')$ only up to a positive scale constant. That is, in the binary model, only the ratios of the estimates make sense and the ratios involving the intercept cannot be interpreted—a fact applying to all binary response estimators, not just to MSC.

In probit, we set $\lambda = 1/SD(u)$ to use $N(0,1)$, but $\lambda = 1/|\beta_k|$ is better for MSC where x_k is the continuous regressor with unbounded support. Then MSC estimates, as $\beta_k/|\beta_k|$ equals the sign of β_k,

$$\frac{\beta_1 - \gamma}{|\beta_k|}, \frac{\beta_2}{|\beta_k|}, \quad ..., \quad sgn(\beta_k) \quad \text{where } sgn(A) = 1 \text{ if } A \geq 0, \text{ and } -1 \text{ if } A < 0.$$

To implement this normalization in practice, first set $sgn(\beta_k) = 1$ and estimate the other parameters. Then set $sgn(\beta_k) = -1$ and again estimate the other parameters. Comparing the two minimands, choose one set of estimates between the two. If we have a prior information $\beta_k > 0$, then the step with $sgn(\beta_k) = -1$ is unnecessary.

More remarks are in order in comparing MSC with MLE such as probit. First, algorithm-wise, MSC needs one that uses no gradient. Second, whereas MLE does not produce a direct predictor for y, *MSC gives the natural predictor* $1[x'b_{msc} \geq 0]$. Third, whereas MLE does not allow heteroskedasticity of an unknown form, MSC does. Fourth, the main drawback of MSC is that its asymptotic distribution is not practical with $N^{1/3}$-consistency (Kim and Pollard, 1990). Finally, suppose $E(y|x) = G(x'\beta)$ but MLE misspecifies $G(x'\beta)$ as $F(x'\beta)$. Then the performance of the MLE depends on the difference between $G(x'\beta)$ and $F(x'\beta)$. If there is only one regressor x (other than 1) and if x is a dummy variable, then MLE requires G to agree with F only at two points; if x takes more values, it will become harder to have $G(x'\beta) = F(x'\beta)$.

On the contrary, MSC requires at least one continuous regressor as mentioned ahead. Hence the continuity of x works for MSC but against MLE when the likelihood function is misspecified. Since continuity is "realized" when N is large, the MLE may perform better in small samples where $G(x'\beta) = F(x'\beta)$ has better chance to hold, while MSC may work better in large samples.

Since $|y_i - 1[x_i'b \geq 0]|$ is either 0 or 1, square it to get

$$y_i - 2y_i \cdot 1[x_i'b \geq 0] + 1[x_i'b \geq 0] = y_i - (2y_i - 1) \cdot 1[x_i'b \geq 0].$$

Dropping y_i, the minimization problem is equivalent to maximizing

$$\frac{1}{N} \sum_i (2y_i - 1) \cdot 1\left[x_i'b \geq 0\right].$$

Further observing $sgn(x_i'b) = 2 \cdot 1[x_i'b \geq 0] - 1$, the maximization problem is equivalent to maximizing $N^{-1} \sum_i (2y_i - 1) \cdot sgn(x_i'b)$.

Horowitz (1992) improved on MSC by replacing the indicator function in MSC with a smooth function $J(\cdot)$ taking a value between 0 and 1. Specifically, Horowitz (1992) maximized

$$\frac{1}{N} \sum_i (2y_i - 1) \cdot J\left(\frac{x_i'b}{h}\right)$$

wrt b where h is a smoothing parameter, $J(-\infty) = 0$ and $J(\infty) = 1$. A df or the integral of a kernel can be used for $J(\cdot)$. Thus as $h \to 0$, J takes either 0 or 1 because $x_i'b/h$ goes to $\pm\infty$ depending on $sgn(x_i'b)$. The estimator is called "smoothed maximum score estimator," which has a faster convergence rate that still falls short of the usual $\sqrt{N}$-rate. See also Horowitz (1993, p. 66) for a quick overview. Horowitz (2002) suggested a bootstrap-based inference.

Mayer and Dorsey (1998) applied MSC to a disequilibrium model, and Moon (2004) extended MSC to non-stationary time-series data. Kordas (2006) extended the smoothed maximum score estimator to quantiles, and Lee and Seo (2007) considered a threshold-crossing "regime change" model. Abrevaya and Huang (2005) showed that bootstrap does not work for MSC, despite that some studies used bootstrap. Failure of bootstrap for $N^{1/3}$-consistent estimators is also examined in Léger and MacGibbon (2006). Although MSC kindled the semiparametric econometric literature for LDV models, the practicality of MSC and its smoothed version remains elusive despite some theoretical advances and applied studies, some of which seem ill-advised.

For a multinomial choice model $y_{ij}^* = x_{ij}'\beta + u_{ij}$ where j indexes the choice set $\{1, ..., J\}$, and $y_{ij} = 1$ if i chooses j and 0 otherwise, MSC for multinomial choice in Manski (1975) maximizes

$$\frac{1}{N} \sum_{i=1}^{N} \sum_{j=1}^{J} y_{ij} 1\left[x_{ij}'b \geq x_{i1}'b, ..., x_{ij}'b \geq x_{iJ}'b\right].$$

This works because $\sum_{j=1}^{J} y_{ij} 1[\cdot] = 1$ if the prediction is right and 0 otherwise. The population version is

$$E \left\{ \sum_{j=1}^{J} y_j 1 \left[x_j' b \geq x_1' b, ..., x_j' b \geq x_J' b \right] \right\}$$

$$= E_x \left\{ \sum_{j=1}^{J} P(y_j = 1|x) \cdot 1 \left[x_j' b \geq x_1' b, ..., x_j' b \geq x_J' b \right] \right\}$$

which is maximized by matching $1[\cdot]$ with the highest choice probability. Suppose $P(y_1 = 1|x)$ is the highest. Then, in order to identify β, it is necessary to have

$$P(y_1 = 1|x) \geq P(y_2 = 1|x), ..., P(y_1 = 1|x) \geq P(y_J = 1|x)$$
$$\text{iff} \quad x_1' \beta \geq x_2' \beta, ..., x_1' \beta \geq x_J' \beta.$$

Manski (1975) imposed two assumptions for this: all regressors are continuous, and $u_{i1}, ..., u_{iJ}$ are independent of one another. But these assumptions make the estimator too restrictive, and no applied study of multinomial MSC seems to exist.

1.2   Ordered Discrete Responses

Generalizing MSC "horizontally" (MSC with a multinomial choice model may be called a "vertical" generalization of MSC), suppose

$$y_i^* = x_i' \beta + u_i, \quad y_i = \sum_{r=1}^{R-1} 1[y_i^* \geq \gamma_r], \quad Med(y^*|x) = x'\beta;$$

i.e., omitting i,

$$y = r - 1 \quad \text{if } \gamma_{r-1} \leq y^* < \gamma_r, \quad \gamma_0 = -\infty, \quad \gamma_R = \infty, \quad r = 1, ..., R.$$

One example is income surveys: instead of the actual income y^*, a group representative value $y = r$ is reported if y^* belongs to the interval $[\gamma_r, \gamma_{r+1})$. Here, y takes an integer between 0 and $R - 1$. Another example is y^* being a continuous demand for a durable good. Since only positive integers can be realized, the observed y is an (integer) transformation of y^*.

Because of $Med(y^*|x) = x'\beta$ and the increasing transformation $\tau(\cdot) = \sum_{r=1}^{R-1} 1[\cdot \geq \gamma_r]$, it holds that

$$Med(y|x) = \sum_{r=1}^{R-1} 1[x'\beta \geq \gamma_r] :$$

the median of $y|x$ is the representative value of the group where $x'\beta$ belongs. Based on this idea, Lee (1992a) suggested minimizing the following wrt b and

$c \equiv (c_1, ..., c_{R-1})'$:

$$\frac{1}{N} \sum_i \left| y_i - \sum_{r=1}^{R-1} 1[x_i'b \geq c_r] \right|.$$

This is an ordered discrete response (ODR) version of MSC, and the motivations of this estimator are the same as those for MSC.

More generally, as was seen already when quantile regression was introduced, if the αth quantile $Q_\alpha(y^*|x)$ of $y^*|x$ is $x'\beta$, then one should minimize

$$\frac{1}{N} \sum_i \left\{ y_i - \sum_{r=1}^{R-1} 1[x_i'b \geq c_r] \right\} \cdot \left\{ \alpha - 1\left[y_i - \sum_{r=1}^{R-1} 1[x_i'b \geq c_r] < 0 \right] \right\}$$

to estimate the αth quantile parameter β, where the "predictor" $\sum_{r=1}^{R-1} 1$ $[x_i'b \geq c_r]$ is plugged into q in the "quantile loss function" $(y - q)(\alpha - 1$ $[y - q < 0])$.

The several remarks made for MSC in comparison to MLE also apply to the ODR version of MSC. A computation algorithm has been suggested in Pinkse (1993, appendix). Melenberg and Van Soest (1996b) smoothed the maximand following the idea of Horowitz (1992), and estimated household "equivalence scales" using ordered data. As noted for MSC, this ODR version of MSC is not quite practical. But if R is high (i.e., the number of categories is large) or if γ_r's are known, then the ODR version of MSC is likely to work much better.

As in MSC, β is identified up to a positive scale factor λ. But differently from MSC, there is a scope of finding an informative bound on λ, depending on assumptions on the thresholds. In the income survey example, $\gamma \equiv (\gamma_1, ..., \gamma_{R-1})'$ is known, whereas only some bounds on γ may be known in the durable good demand example. Also there exist cases where γ is completely unknown, in which case it seems difficult to bound λ. Suppose that y^* is a continuous demand for cars and

$$y = r - 1 \quad \text{if } \gamma_{r-1} \leq y^* < \gamma_r$$
$$\text{where} \quad r - 1 < \gamma_r \leq r, \quad r_0 = -\infty, \quad \gamma_R = \infty, \quad r = 1, ..., R;$$

i.e., if the realized demand y is 2, then y^* must be between 1 and 3. Using the bounds on γ_r's, Lee (1992a) derived the following bounds on λ:

$$\frac{R-3}{R-1} < \lambda \leq \frac{R-2}{R-3} \quad \text{for } R > 3.$$

For instance, $1/2 < \lambda \leq 3/2$ with $R = 5$. As $R \to \infty$, λ gets pinned down to one. Lee and Kimhi (2005) showed that bounds on thresholds helps identification in simultaneous equations with ODR's.

The above ODR version of MSC is also applicable to count responses, but a further simplification is possible when the following condition is imposed:

$$\gamma_r - \gamma_{r-1} = 1, \quad r = 2, ..., R - 1.$$

Add $-\gamma_1 + 1$ to $\gamma_{r-1} \le y^* < \gamma_r$, $r = 1, ..., R$, to get

$$y = r - 1 \quad \text{if} \quad \gamma_{r-1} - \gamma_1 + 1 \le y^* - \gamma_1 + 1 < \gamma_r - \gamma_1 + 1, \quad r = 1, ..., R.$$

Absorb $-\gamma_1 + 1$ into the intercept β_1 in $x'\beta$ (i.e., redefine y^* as $y^* - \gamma_1 + 1$)) and substitute $\gamma_{r-1} - \gamma_1 = r - 2$ and $\gamma_r - \gamma_1 = r - 1$ when $r = 2, ..., R - 1$ to get an ODR with rather simple thresholds:

$$\begin{aligned} y &= 0 \quad \text{if } y^* < 1 \\ &= r - 1 \quad \text{if} \quad r - 1 \le y^* < r, \quad r = 2, ..., R - 1 \\ &= R - 1 \quad \text{if} \quad R - 1 \le y^*. \end{aligned}$$

Rewrite the display succinctly as

$$y = \max\{0, \min(R - 1, \lfloor y^* \rfloor)\} \quad \text{where } \lfloor y^* \rfloor \text{ is the integer part of } y^*.$$

Then the ODR minimand becomes

$$\frac{1}{N} \sum_i |y_i - \max\{0, \min(R - 1, \lfloor x_i'b \rfloor)\}|.$$

Furthermore, if $\min(R-1, \lfloor x_i'b \rfloor) = \lfloor x_i'b \rfloor$, then this further becomes $N^{-1} \sum_i | y_i - \max(0, \lfloor x_i'b \rfloor)|$. An analogous minimand will appear in the censored response subsection shortly.

1.3   Count Responses

Although the preceding ordered response estimator is applicable to count responses, the quantile regression assumption was imposed on the latent response y^*. Is there any way to conduct quantile regression directly on the observed count y? The main obstacle to this is that $Q_\alpha(y|x)$ *is an integer and one cannot simply posit* $Q_\alpha(y|x) = x'\beta$ as x may be continuously distributed. Machado and Santos-Silva (2005, MSS in this section), however, found a way of "smoothing" $Q_\alpha(y|x)$ so that it can be linked to $x'\beta$, which is the topic of this subsection. Other than the empirical illustration on health care demand in MSS, applications of MSS can be seen in Winkelmann (2006) and Miranda (2008). STATA has the command qcount to implement the MSS estimator.

1.3.1   Main Idea

The key idea is using a smoothed (or "randomized") version of y_i:

$$z_i = y_i + u_i \quad \text{where } u_i \sim U[0, 1) :$$

z_i is generated by adding an independent, artificial uniform random number to y_i. Since $Q_\alpha(\cdot)$ is not linear,

$$Q_\alpha(z|x) \ne Q_\alpha(y|x) + \alpha \ \{= Q_\alpha(y|x) + Q_\alpha(u|x)\}.$$

But $Q_\alpha(z|x)$ should be greater than α as z is a non-negative y plus u: i.e.,

$$Q_\alpha(z|x) > \alpha \quad \text{so long as } P(y = 0|x) < 1.$$

This intuitive fact can be shown in the following way, which can be skipped. Suppose $Q_\alpha(z|x) = \alpha \iff \alpha \le P(z \le \alpha|x)$. Then

$$\alpha \le P(z \le \alpha|x) = P(y + u \le \alpha|x) = \int P(u \le \alpha - y \,|y, x)f(y|x)dy$$

$$= \int (\alpha - y)1[0 < \alpha - y < 1]f(y|x)dy$$

$$\text{as } P(u \le \alpha - y) = \alpha - y \quad \text{when } 0 < \alpha - y < 1$$

$$= \int (\alpha - y)1[y < \alpha < y + 1]f(y|x)dy = \int (\alpha - y)1[y = 0]f(y|x)dy$$

$$= \alpha 2 P(y = 0|x) < \alpha$$

which is a contradiction. Thus it must be that $Q_\alpha(z|x) > \alpha$ under $P(y = 0|x) < 1$.

For a given α, define a transformation of z_i: for a chosen constant ζ with $0 < \zeta < \min_i |z_i - \alpha|$,

$$t_{\alpha i} \equiv t_\alpha(z_i) = \ln\{(z_i - \alpha)1[z_i > \alpha] + \zeta 1[z_i \le \alpha]\}.$$

Then $t_\alpha(z)$ is increasing in z because, if $z_j \ge z_i$, then $t_\alpha(z_j) \ge t_\alpha(z_i)$:

$$(i) : z_j, z_i > \alpha \implies t_\alpha(z_j) = \ln(z_j - \alpha) \ge \ln(z_i - \alpha) = t_\alpha(z_i)$$

$$(ii) : z_j > \alpha > z_i \implies t_\alpha(z_j) = \ln(z_j - \alpha) > \ln\zeta = t_\alpha(z_i)$$

$$(iii) : \alpha > z_j, z_i \implies t_\alpha(z_j) = \ln\zeta = t_\alpha(z_i).$$

Quantile equivariance for increasing transformations yields

$$Q_\alpha\{t_\alpha(z)|x\} = \ln[\ \{Q_\alpha(z|x) - \alpha\}1[Q_\alpha(z|x) > \alpha] + \zeta 1[Q_\alpha(z|x) \le \alpha]\ ]$$

$$= \ln\{Q_\alpha(z|x) - \alpha\} \quad \text{as } Q_\alpha(z|x) > \alpha \quad \text{under} \quad P(y = 0|x) < 1.$$

Since $\ln\{Q_\alpha(z|x) - \alpha\}$ can take any value in $(-\infty, \infty)$, we may specify now

$$Q_\alpha(t_\alpha|x) = x'\beta_\alpha$$

$$\iff Q_\alpha(z|x) = \alpha + \exp(x'\beta_\alpha) \quad \text{from the preceding display.}$$

This would hold *if* $Q_\alpha(z|x) = Q_\alpha(y + u|x)$ were linear and $Q_\alpha(y|x) = \exp(x'\beta_\alpha)$ so that $Q_\alpha(z|x) = Q_\alpha(y|x) + Q_\alpha(u|x) = \exp(x'\beta_\alpha) + \alpha$. That is, despite that quantile is not a linear functional and $Q_\alpha(y|x) = \exp(x'\beta_\alpha)$ is not plausible because $Q_\alpha(y|x)$ is an integer, we still obtained the equation $Q_\alpha(z|x) = \alpha + \exp(x'\beta_\alpha)$. Note that "smoothing" can be done in different ways, and it is done to y by adding a random number to the integer y.

1.3.2   Quantile Regression of a Transformed Variable

Using $Q_\alpha(t_\alpha|x) = x'\beta_\alpha$, the αth quantile parameter β_α for z (not for y) in $Q_\alpha(z|x) = \alpha + \exp(x'\beta_\alpha)$ can be obtained by minimizing

$$\frac{1}{N}\sum_i (t_{\alpha i} - x_i'b_\alpha) \cdot (\alpha - 1[t_{\alpha i} - x_i'b_\alpha < 0])$$

wrt b_α, and it holds that

$$\sqrt{N}(b_{\alpha N} - \beta_\alpha) \rightsquigarrow N(0, D^{-1}AD^{-1}) \quad \text{where}$$
$$D \equiv E\{f_{t_\alpha|x}(x'\beta_\alpha) \cdot xx'\}, A \equiv \alpha(1-\alpha)E(xx')$$

and $f_{t_\alpha|x}$ is the density of $t_\alpha|x$. MSS propose a complicated estimator for D, but the following simple estimator should work just as well: for a kernel K and a bandwidth h,

$$D_N \equiv \frac{1}{N}\sum_i \frac{1}{h}K\left(\frac{t_{\alpha i} - x_i'b_{\alpha N}}{h}\right)x_i x_i'.$$

Clearly, the kernel part is for $f_{t_\alpha|x}(x'\beta_\alpha)$.

Theorem 2 in MSS shows

$$Q_\alpha(y|x) \quad = \quad \lceil Q_\alpha(z|x) - 1 \rceil \quad (= \lceil \alpha + \exp(x'\beta_\alpha) - 1 \rceil)$$
$$\text{where } \lceil c \rceil \text{ is the "ceiling function"}$$

giving the least upper integer bound for c. When c is not an integer, $\lceil c-1 \rceil$ is the integer part of c: $\lceil c-1 \rceil = \lfloor c \rfloor$. When c is an integer, however, $\lceil c-1 \rceil \neq \lfloor c \rfloor$ as can be seen in $\lceil 3-1 \rceil = 2 \neq 3 = \lfloor 3 \rfloor$. But, since the probability of $Q_\alpha(z|x)$ being an integer is zero, we can safely state

$$Q_\alpha(y|x) = \lfloor Q_\alpha(z|x) \rfloor = \lfloor \alpha + \exp(x'\beta_\alpha) \rfloor :$$

$Q_\alpha(y|x)$ *is the integer part of* $Q_\alpha(z|x)$ *despite* $Q_\alpha(z|x) \neq Q_\alpha(y|x) + \alpha$.

The last display also shows that a zero component of β_α implies zero influence of the corresponding regressor on $Q_\alpha(y|x)$. But does a non-zero component of β_α imply necessarily non-zero influence of the corresponding regressor on $Q_\alpha(y|x)$? This question is relevant because $Q_\alpha(z|x)$ may not take any integer value—recall that $Q_\alpha(y|x)$ is only the integer part of $Q_\alpha(z|x)$. The answer is positive when there is at least one continuously distributed regressor and $Q_\alpha(z|x)$ takes on an integer value for some value of the regressor, which MSS assume.

372   Ch. 8   Bandwidth-Free Semiparametric Methods

1.3.3   Further Remarks

The above estimator depends on the simulated u_i's used for smoothing (or "jittering"). To avoid the dependence, one may obtain estimates for many different sets of u_i's. Suppose this is done J-many times to yield $b_{\alpha N}^{(1)}, ..., b_{\alpha N}^{(J)}$. Then the averaged estimator is

$$\bar{b}_{\alpha N} \equiv \frac{1}{J} \sum_{j=1}^{J} b_{\alpha N}^{(j)}.$$

MSS show that $\bar{b}_{\alpha N}$ is more efficient than the above estimator, and analogously to the asymptotics seen for method of simulated moment,

$$\sqrt{N}(\bar{b}_{\alpha N} - \beta_\alpha) \rightsquigarrow N\left\{0, \frac{1}{J}D^{-1}AD^{-1} + \left(1 - \frac{1}{J}\right)D^{-1}BD^{-1}\right\}$$

where $B \equiv E[\{\alpha(1-\alpha) - \eta(x)\} \cdot xx']$,
$\eta(x) \equiv f_{y|x}\{Q_\alpha(y|x)\} \cdot \{\alpha + \exp(x'\beta_\alpha) - Q_\alpha(y|x)\} \cdot \{Q_\alpha(y|x) + 1 - \alpha - \exp(x'\beta_\alpha)\}$.

With the simulation number J large enough, the asymptotic variance becomes $D^{-1}BD^{-1}$.

As for estimating B, MSS use in their simulation study

$$B_N \equiv \frac{1}{N} \sum_i [\, \alpha^2 + (1 - 2\alpha)1\,[y_i \leq \alpha + \exp(x_i'b_{\alpha N}) - 1]$$

$$+ \{\alpha + \exp(x_i'b_{\alpha N}) - y_i\} \cdot 1[\alpha + \exp(x_i'b_{\alpha N}) - 1 < y_i \leq \alpha + \exp(x_i'b_{\alpha N})]$$
$$\cdot \{\alpha + \exp(x_i'b_{\alpha N}) - y_i - 2\alpha\}\,]\, x_i x_i'.$$

Rewriting B as $E\{\alpha(1-\alpha)xx'\} - E\{\eta(x)xx'\}$, the following estimator may be used for $E\{\eta(x)xx'\}$:

$$\frac{1}{N}\sum_i 1[y_i = Q_\alpha(y|x_i)]\{\alpha + \exp(x_i'b_{\alpha N}) - Q_\alpha(y|x_i)\}$$

$$\{Q_\alpha(y|x_i) + 1 - \alpha - \exp(x_i'b_{\alpha N})\}x_i x_i'$$

where $Q_\alpha(y|x_i)$ is to be replaced with $\lfloor \alpha + \exp(x'b_{\alpha N})\rfloor$. Using $\alpha(1-\alpha) \cdot N^{-1}\sum_i x_i x_i'$ minus this display would be simpler than using B_N.

The count-response quantile estimator by MSS is a novel idea, but it has a problem. Suppose $E(y|x) = \exp(x'\beta)$ as in Poisson or NB distributions with the mean parameter $\lambda(x) = \exp(x'\beta)$. Then we have

$$E(y|x) = \exp(x'\beta) \quad \text{and} \quad E(z|x) = \alpha + \exp(x'\beta).$$

But what should hold for MSS is

$$Q_\alpha(y|x) = \lfloor \alpha + \exp(x'\beta)\rfloor \quad \text{and} \quad Q_\alpha(z|x) = \alpha + \exp(x'\beta).$$

This would not hold for the Poisson or NB-distributed count response with $\lambda(x) = \exp(x'\beta)$. That is, these key equations would not be compatible with the popular parametric models for count responses. The source of this trouble is using z, not y, as the "latent" variable. If one is to do a Monte Carlo study, it is not clear how to generate y (with x and β given) subject to $Q_\alpha(y|x) = \lfloor \alpha + \exp(x'\beta) \rfloor$.

1.4   Censored Responses

1.4.1   Censored Quantile Estimators

Suppose that y^* is observed as censored from below at a constant c and that the median regression assumption holds:

$$y_i = \max(y_i^*, c) = \max(x_i'\beta + u_i, c) \quad \text{and} \quad Med(y^*|x) = x'\beta.$$

Since the transformation $\tau(\cdot) = \max(\cdot, c)$ is increasing, we get

$$Med(y|x) = \max(x'\beta, c).$$

Based on this idea, Powell (1984) proposed the censored least absolute deviation (CLAD) estimator minimizing

$$\frac{1}{N} \sum_i |y_i - \max(x_i'b, c)|.$$

STATA has the command `clad` to implement CLAD.

Recall that the asymptotic distribution of an M-estimator b_N minimizing $N^{-1} \sum_i q(b)$ is

$$\sqrt{N}(b_N - \beta) \rightsquigarrow N\left\{0, \ E^{-1}\left(\frac{\partial^2 q}{\partial b \partial b'}\right) \cdot E\left(\frac{\partial q}{\partial b}\frac{\partial q}{\partial b'}\right) \cdot E^{-1}\left(\frac{\partial^2 q}{\partial b \partial b'}\right)\right\}.$$

Although the CLAD minimand is not differentiable, as was seen for LAD before, still the asymptotic distribution of CLAD can be obtained analogously to yield

$$\sqrt{N}(b_{clad} - \beta) \rightsquigarrow N\left(0, H^{-1}\frac{1}{4}E\{xx' \ 1[x'\beta > c]\}H^{-1}\right),$$

where $H \equiv E\{f_{u|x}(0) \ xx' 1[x'\beta > c]\}$

$$= N\left(0, \ \frac{1}{4f_u(0)^2}E^{-1}\{xx'1[x'\beta > c]\}\right) \quad \text{under } u \amalg x.$$

For a bandwidth h, H can be estimated with

$$\frac{1}{N} \sum_i \frac{1[|y_i - x_i'b_{clad}| < h]}{2h} x_i x_i' 1[x_i'b_{clad} > c],$$

$$h \simeq N^{-1/5} SD(y - x'b_{clad}|x'b_{clad} > c).$$

The kernel part $(2h)^{-1}1[|y_i - x_i'b_{clad}| < h]$ can be replaced with a smooth version $h^{-1}K\{(y_i - x_i'b_{clad})/h\}$.

Due to the max function, censoring takes place on the left part of the *marginal* distribution of y. Even if 50% of y are censored, $Q_{0.25}(y^*|x)$ in the conditional distribution of $y|x$ may be still identified, because $y^*|x$ may be hardly censored depending on the value of x. For instance, let $y^* = \beta_1 + \beta_2 x + u$ where x is a scalar and $\beta_2 > 0$. If $x = x_o$ is large enough so that the corresponding y^* is always greater than c, then $y^*|(x = x_o)$ is not censored at all. This fact notwithstanding, in general, if the lower censoring percentage is high, then an αth quantile with $\alpha > 0.5$ may be a better location measure to estimate than the median.

In view of this consideration, to generalize CLAD for quantiles, suppose

$$Q_\alpha(y^*|x) = x'\beta \implies Q_\alpha(y|x) = \max(x'\beta, c);$$

although it is better to index β with α (i.e., β_α instead of β), we omit the index for brevity. Under this, the *"Censored Quantile Regression (CQR) Estimator"* (Powell, 1986b) minimizes

$$\frac{1}{N}\sum_i \{y_i - \max(x_i'b, c)\} \cdot \{\alpha - 1[y_i - \max(x_i'b, c) < 0]\}.$$

The asymptotic variance matrix is, with $y^* = x'\beta + u$ and $Q_\alpha(u|x) = 0$,

$$E^{-1}\{f_{u|x}(0)\, xx'1[x'\beta > c]\}\, \alpha(1-\alpha)E\{xx'1[x'\beta > c]\}$$

$$E^{-1}\{f_{u|x}(0)\, xx'1[x'\beta > c]\}$$

$$= \alpha(1-\alpha)\,\{f_u(0)\}^{-2}\, E^{-1}\{xx'1[x'\beta > c]\} \quad \text{under } u \perp\!\!\!\perp x.$$

Clearly, CQR includes CLAD as a special case when $\alpha = 0.5$.

In duration data, almost always censoring is from above: $y_i = \min(y_i^*, c)$. In this case, $Q(y|x) = \min(x'\beta, c)$, and the CQR minimand should be

$$\frac{1}{N}\sum_i \{y_i - \min(x_i'b, c)\} \cdot \{\alpha - 1[y_i - \min(x_i'b, c) < 0]\}.$$

Its asymptotic variance is

$$E^{-1}\{f_{u|x}(0)xx'1[x'\beta < c]\} \cdot \alpha(1-\alpha)E\{xx'1[x'\beta < c]\} \cdot$$

$$E^{-1}\{f_{u|x}(0)xx'1[x'\beta < c]\}.$$

CQR (and thus CLAD) is applicable to type I censoring with the censoring point c varying across the subjects so long as c is observed for all i. CQR is not applicable to random censoring, but there is a scope to overcome this problem as will be shown later.

It is possible to have both lower (left) and upper (right) censorings together as in fractional responses; e.g., a proportion variable $y^* \in [0, 1]$ with
$P(y^* = 0) > 0$ and $P(y^* = 1) > 0$. This may be modeled as

$$y = \max\{0, \ \min(1, y^*)\} \implies Q_\alpha(y|x) = \max\{0, \ \min(1, x'\beta)\}.$$

Plugging this into the quantile minimand, we can estimate β, and its asymptotic variance would be the same as the preceding display with $x'\beta < c$
replaced with $0 < x'\beta < 1$; alternatively, bootstrap may be applied as will be
explained below.

With censoring, one main advantage of using a quantile relative to the
mean or other location measures is in prediction. For instance, a predictor
is $Q_\alpha(y|x_o) = \max\{0, \min(1, x_o'\hat{\beta})\}$ in the proportion case, which can have
many 0's and 1's depending on α and x_o. If we use mean prediction, however,
$E(y|x_o)$ will fall always in $(0, 1)$ and never exactly equal 0 or 1, despite that
there always will be some i with $y_i = 0$ or 1.

A more efficient two stage version of CLAD was proposed by Newey
and Powell (1990). Hall and Horowitz (1990) discussed how to choose h in
CLAD and other related estimators. Yin et al. (2008) examined CQR for
Box-Cox transformed responses. CLAD has been applied by Horowitz and
Neumann (1987,1989) to duration data, by Lee (1995) to female labor sup-
ply data, and by Melenberg and Van Soest (1996a) to vacation expenditure
data, just to name a few. More applications of CQR can be found in the special edited volume of *Empirical Economics* (2001) and Hochguertel (2003).
Fitzenberger (1997) and Buchinsky (1998) reviewed CQR. See Honoré et al.
(2002) and Chernozhukov and Hong (2002) for further references regarding
CQR and the statistical literature for censored regression models including
Yang (1999) who estimated β by solving a weighted Kaplan-Meier estimator
for the medians.

Typically in the statistical literature, $u \amalg x$ is assumed (no heteroskedasticity of unknown form is allowed), and variation of c is required (a constant
c will not do); the usual assumption is random right censoring by c (i.e.,
$y_i = \min(y_i^*, c_i)$) with $c \amalg y^*|x$. The estimator of Honoré et al. (2002) is rare
in econometrics in that it allows random censoring, but it requires $c \amalg (u, x)$
which is rather restrictive, although the estimator does not require $u \amalg x$.
Under $c \amalg (u, x)$, the df of c is estimated nonparametrically (e.g., the Kaplan-
Meier product limit estimator), and then c is integrated out when it is not
observed (i.e., when $y^* < c$).

1.4.2   Two-Stage Procedures and Unobserved Censoring Point

CQR has two shortcomings: one is the non-convex minimand due to the
max (or min) function, and the other is the requirement of the censoring
points to be observed—c may vary across i but should be observed. This
subsection reviews two-stage procedures to overcome these two shortcomings. The two-stage procedures allow c_i to be unobserved so long as c_i fully

determined by x_i, i.e., so long as c_i gets fixed once x_i is conditioned on. This generalization, however, does not allow c_i to be determined by some unobserved variables (and x_i).

Lee et al. (1996) proposed a number of two-stage procedures for semiparametric estimators for LDV models. Among them, the ones for the truncated response and binary response appear in Lee and Kim (1998, pp. 216–217) and Lee (2002, pp. 107–108), respectively. The procedure for censored models is similar to the truncated model procedure and is presented in the following, drawing on Lee et al. (2007). Closely related ideas appeared in the literature as will be seen shortly. The main idea is that the observations with $Q_\alpha(y|x_i) > c$ are selected nonparametrically in the first stage, and then CQR is obtained only with those observations in the second stage.

For the first stage, observe

$$
\begin{aligned}
P(y \;&>\; c|x) = P(d = 1|x) > 1 - \alpha \\
[\;&\Longleftrightarrow\; \{P(y > c|x) - (1-\alpha)\}f(x) > 0 \quad \text{under } f(x) > 0] \\
&\Longleftrightarrow\; P(y \le c|x) < \alpha \quad \text{(multiplying by } -1 \text{ and adding 1)} \\
&\Longleftrightarrow\; Q_\alpha(y|x) > c
\end{aligned}
$$

where $f(x)$ is density/probability for x. Let k denote the dimension of $\tilde{x}$ that is the regressor vector other than unity (i.e., $x = (1, \tilde{x}')'$), and define

$$
P_N(y > c|x_i) \equiv \frac{\{(N-1)h^k\}^{-1} \sum_{j,j\neq i} K((\tilde{x}_j - \tilde{x}_i)/h)\, 1[y_j > c]}{\{(N-1)h^k\}^{-1} \sum_{j,j\neq i} K((\tilde{x}_j - \tilde{x}_i)/h)}.
$$

Subtract $1 - \alpha$ and multiply by $\{(N-1)h^k\}^{-1} \sum_{j,j\neq i} K((\tilde{x}_j - \tilde{x}_i)/h)$ to get a kernel estimator for $\{P(y > c|x_i) - (1-\alpha)\}f(x_i)$:

$$
g_N(x_i) \equiv \frac{1}{(N-1)h^k} \sum_{j,j\neq i} K\left(\frac{\tilde{x}_j - \tilde{x}_i}{h}\right) \{1[y_j > c] - (1-\alpha)\}.
$$

Getting this (and selecting the observations with positive $g_N(x_i)$) is the first stage. In the second stage, minimize wrt b

$$
\frac{1}{N} \sum_i (y_i - x_i'b)\{\alpha - 1[y_i - x_i'b < 0]\}\, 1[g_N(x_i) > \varepsilon]
$$

which is a convex minimization problem; a constant $\varepsilon > 0$ is used instead of 0 "to be on the safe side" and thus to possibly improve the finite-sample performance.

In practice, $g_N(x_i)$ requires selecting the bandwidth h. This may be done by minimizing the cross-validation (CV) criterion. Defining $w_j \equiv 1[y_j > c] - (1-\alpha)$, the CV criterion is

$$
\sum_i \{w_i - r_{-i}(\tilde{x}_i, h)\}^2 \quad \text{where} \quad r_{-i}(\tilde{x}_i, h) \equiv \frac{\sum_{j,j\neq i} K((\tilde{x}_j - \tilde{x}_i)/h)w_j}{\sum_{j,j\neq i} K((\tilde{x}_j - \tilde{x}_i)/h)}.
$$

That is, for a range of chosen values of h, this minimand is computed, and the bandwidth yielding the smallest value of the minimand is selected. Another thing to choose is the small constant $\varepsilon > 0$. As we do not have a good idea on the scale of $g_N(x_i)$, it may be in fact better to use $1[P_N(y > c|x_i) - (1-\alpha) > \varepsilon]$ because $P_N(y > c|x_i)$ is a probability whose scale is known. While choosing h is necessary, using ε is not because the improvement with $\varepsilon > 0$ over $\varepsilon = 0$ seems small in our experience.

Another computational issue for CQR is its asymptotic variance estimator, because the second-order matrix H includes the density component $f_{u|x}(0)$. Denoting the residuals as $r_i = y_i - x_i'b_{cqr}$, an estimator for H is

$$\frac{1}{N}\sum_i \frac{1[-h < r_i < h]}{2h} x_i x_i' 1[x_i'b_{cqr} > c];$$

$f_{u|x}(0)$ is estimated nonparametrically with $(2h)^{-1}1[-h < r_i < h]$. Note that $1[x_i'b_{cqr} > c]$ can be replaced with $1[g_N(x_i) > 0]$. The bandwidth h should be chosen, which is a disadvantage of using the CQR asymptotic variance. Instead, nonparametric bootstrap can be applied. The bootstrap consistency follows from the general bootstrap consistency result in Arcone and Giné's (1992) for M-estimators, which was made explicit for CQR in Hahn (1995).

One problem with bootstrap is that the repeated computation of CQR is computationally burdensome. This can be avoided as in Bilias et al. (2000) by minimizing

$$\frac{1}{N}\sum_i (\hat{y}_i - \hat{x}_i'b)\{\alpha - 1[\hat{y}_i - \hat{x}_i'b < 0]\} 1[x_i'b_{cqr} > c]$$

wrt b, where b_{cqr} is the original CQR estimator and $(\hat{y}_i, \hat{x}_i')'$, $i = 1, ..., N$, are pseudo bootstrap data. This idea is close to the above two-stage procedure of avoiding the max function. In $x_i'b_{cqr} > c$, to allow for the estimation error in b_{cqr}, one may add a small positive constant ε to use $x_i'b_{cqr} > c + \varepsilon$ as done in $g_N(x_i) > \varepsilon$.

A number of remarks are in order for the above two-stage procedure. First, in $g_N(x_i)$, $1[y_j > c]$ can be replaced with d_j which is free of c. This means that c does not have to be observed, and may change across i. Second, despite this generalization, in order to maintain "$P(y > c|x_i) > 1 - \alpha$ iff $Q_\alpha(y|x_i) > c$", c should be fixed once x is so. That is, c is allowed to be unobserved to the extent c is an unobserved function of only x; if c_i varies across i but observed, rewrite the model such that c_i becomes a regressor with its known slope 1. Honoré et al. (2002) proposed a version of CQR allowing for random censoring where c should be independent of x and y^*; they use the Kaplan–Meier estimator for c's survival function. Third, Lee et al. (1996) originally suggested the two-stage idea as a computational device to overcome the non-convexity of the minimand. Fourth, once the estimate is obtained through the two-stage method, one can find CQR using the estimate as an initial value, but our experience has been that searching further for CQR does not lead to any improvement. Fifth, if one takes the two-stage estimator as another estimator, then in order to take the first-stage estimation error into account, the indicator function may be replaced with a smooth function, say $\tilde{K}(t)$, such that $\tilde{K}(t) \to 0$ (1) as $t \to -\infty$ (∞). This replacement, however, makes hardly any difference in practice. Sixth, for the right-censoring model $y = \min(x'\beta + u, c)$, observe

$$Q_\alpha(y^*|x) \leq c \iff \alpha \leq P(y^* \leq c|x) = P(d = 1|x).$$

In this case, the first-stage of the above two-stage procedure selects the observations with $Q_\alpha(y^*|x) \leq c$, which can be done by

$$\frac{1}{(N-1)h^k} \sum_{j, j \neq i} K\left(\frac{\tilde{x}_j - \tilde{x}_i}{h}\right)(d_j - \alpha)\,\varepsilon.$$

Buchinsky and Hahn (1998) presented an idea similar to the above two-stage method. Observe that $P(y^* \leq x'\beta|x) = \alpha$ before censoring at c. When $x'\beta > c$ for lower censoring, they looked at

$$P(c < y^* \leq x'\beta \,|y^* > c,\ x) = \frac{\int_c^{x'\beta} f(y^*|x)dx}{P(y^* > c|x)} = \frac{\alpha - P(y^* \leq c|x)}{P(d = 1|x)}$$

$$= \frac{\alpha - P(d = 0|x)}{P(d = 1|x)} = \frac{P(d = 1|x) - (1 - \alpha)}{P(d = 1|x)} \equiv \pi(x).$$

Hence, $x'\beta$ is the $\pi(x)$th quantile of the truncated distribution of $y^*|(y^* > c, x)$ (i.e., $y^*|(d = 1, x)$). Let $\pi_N(x_i)$ denote the estimator for $\pi(x_i)$, obtained by replacing $P(d = 1|x)$ with $P_N(d = 1|x)$. This idea leads to minimizing

$$\frac{1}{N} \sum_i (y_i - x_i'b)\{\pi_N(x_i) - 1[y_i - x_i'b < 0]\} \cdot d_i 1[\pi_N(x_i) > 0].$$

Instead of $\pi_N(x_i) > 0$, $\pi_N(x_i) > \varepsilon$ may be used.

The first nonparametric stage in the two-stage method may not work well if the dimension of x is large. A practical alternative might be using probit (or logit) of d_i on z_i to select the observations with $P(d = 1|z) = \Phi(z_i'b_N) > 1 - \alpha$ in the first stage where the notation z is used to allow different regressors in the probit. This is essentially what Chernozhukov and Hong (2002) proposed——in fact using $\Phi(z_i'b_N) > 1 - \alpha + \varepsilon$ for a small $\varepsilon > 0$ as above. They showed conditions that can justify this practice. Khan and Powell (2001) examined using MSC, $P(d = 1|x)$, or nonparametric quantile regression for the first-stage to show that the first-stage estimation error does not influence the second-stage. Using nonparametric quantile regression for the first-stage was also considered by Lee et al. (1996), but $P(d = 1|x)$ was used there because it is computationally easier to obtain.

1.4.3   An Empirical Example

Different hospitals treat their patients differently and the survival rate of patients vary across hospitals. For a given disease, it is important to identify the best treatment protocol. In Finland, the occurrence rate of heart-related problems is high and hospitals use different treatments. In the following empirical example taken from Lee et al. (2007), we look at the survival duration at hospital after an acute myocardial infarction (AMI). It is desired to find the best treatment protocol for the survival duration by comparing 21 hospital districts in Finland (each district consists of multiple hospitals).

A large data set of size 5972 on males for 1996 is available that is, however, limited on covariates: only age and the distance (DIS) to the hospital are available. Let

z_i be the survival duration in days for person i at hospital

and $y_i^* \equiv \ln z_i$.

The duration is subject to the usual right censoring problem that, for some patients, the recorded value of z_i is not the actual survival duration, but only the censoring time, because the follow-up ended before death. In our data, the censoring time is fixed at one year for all i and only 27% are not right-censored. The particularly high censoring percentage (73%) in the marginal distribution of y^* (i.e., z) poses difficulties to estimators for the center(s) of a distribution (e.g., estimators under the assumption that the conditional mean of the error term given x is zero), because they are more or less for hard-to-identify "middle quantiles" of the $y^*|x$ distribution. Under a heavy right censoring, it makes sense to look at low quantiles.

Table 1 shows the mean and SD of the duration in days (z), DIS, and age. The table also lists the proportions for 12 age groups and 21 hospital districts. The mean duration is 282 days, but among the non-censored observations, the mean duration is only 56 days.

Table 2 provides three sets of CQR estimates for y: 10%, 17.5%, and 25% quantiles. In the marginal distribution of z, these quantiles correspond to 5, 24, and 253 days, respectively. The three sets of CQR estimates show how x affects the "short-term (days)," "mid-term (weeks)," and "long-term (months)" survivals differently. For CQR(25), CV was used for bandwidth choice, and the "selected number" of observations (i.e., the observations selected in the first nonparametric stage) was 2961. For CQR(17.5) and CQR(10), the same bandwidth was used and the selected number of observations was progressively higher than 2961 as the censoring is from above, not below. The base case d1 (Helsinki) is omitted.

Looking at the CQR columns, DIS is significantly negative in CQR(10) and the significance level drops in CQR (17.5) and CQR(25). This suggests that DIS affects at least the short-term survival: the farther away from the hospital is the residence, the less likely to survive at hospital. In terms of signs, there is no significant reversal of district effects across the three

Table 1: Mean (SD) of Variables

duration in days	282 (146)	non-censored duration		56.0 (91.0)	
		DIS in			
age	69.5 (12.4)	100km	0.29 (0.30)		
age 0 (-40)	0.013				
age 1 $(40^+ - 45)$	0.022	$d1$	0.075	d11	0.020
age 2 $(45^+ - 50)$	0.044	$d2$	0.114	d12	0.046
age 3 $(50^+ - 55)$	0.061	$d3$	0.082	d13	0.065
age 4 $(55^+ - 60)$	0.088	$d4$	0.048	d14	0.053
age 5 $(60^+ - 65)$	0.108	$d5$	0.035	d15	0.043
age 6 $(65^+ - 70)$	0.156	$d6$	0.080	d16	0.028
age 7 $(70^+ - 75)$	0.160	$d7$	0.042	d17	0.020
age 8 $(75^+ - 80)$	0.143	$d8$	0.045	d18	0.071
age 9 $(80^+ - 85)$	0.115	$d9$	0.035	d19	0.022
age 10 $(85^+ - 90)$	0.068	$d10$	0.032	d20	0.015
age 11 $(90^+ -)$	0.023			d21	0.024

Table 2: CQR for ln(duration)

x	CQR(10) (tv)	CQR(17.5) (tv)	CQR(25) (tv)
1	-5.040 (-12.8)	-6.158 (-10.2)	-6.212 (-7.85)
age	0.168 (13.2)	0.187 (9.69)	0.183 (7.29)
age^2/10	-0.015 (-14.5)	-0.016 (-10.2)	-0.015 (-7.66)
DIS (100km)	-0.163 (-2.63)	-0.135 (-1.67)	-0.181 (-1.78)
d2	-0.113 (-1.69)	-0.151 (-1.77)	-0.224 (-2.26)
d3	-0.057 (-0.76)	-0.088 (-0.91)	-0.163 (-1.45)
d4	0.053 (0.65)	0.029 (0.28)	-0.014 (-0.11)
d5	0.158 (1.86)	0.075 (0.69)	0.057 (0.45)
d6	0.022 (0.29)	-0.032 (-0.34)	-0.053 (-0.48)
d7	0.204 (2.40)	0.214 (1.99)	0.114 (0.92)
d8	-0.055 (-0.63)	-0.112 (-1.01)	-0.211 (-1.63)
d9	-0.088 (-0.87)	-0.109 (-0.85)	-0.119 (-0.78)
d10	0.157 (1.73)	0.164 (1.42)	0.210 (1.58)
d11	0.130 (1.30)	0.217 (1.69)	0.573 (3.89)
d12	0.210 (2.52)	0.228 (2.16)	0.124 (1.01)
d13	0.365 (4.69)	0.286 (3.16)	0.492 (4.73)
d14	0.094 (1.14)	0.037 (0.35)	-0.043 (-0.35)
d15	-0.215 (-2.39)	-0.244 (-2.12)	-0.287 (-2.07)
d16	0.046 (0.45)	-0.013 (-0.10)	-0.027 (-0.17)
d17	0.020 (0.19)	-0.050 (-0.37)	-0.084 (-0.53)
d18	0.066 (0.86)	0.024 (0.25)	0.083 (0.73)
d19	-0.119 (-1.07)	-0.149 (-1.03)	-0.083 (-0.50)
d20	0.160 (1.38)	0.071 (0.49)	-0.014 (-0.08)
d21	-0.167 (-1.31)	-0.213 (-1.30)	-0.204 (-1.06)

quantiles. That is, there is no district doing better in the short run than d1, but then worse for the longer terms. Comparing the district coefficients, the best performing district is d13 followed by d12 and d7, and the worst performing district is d15, followed by d2 or d21. Therefore, the treatment protocol to disseminate is that of d13.

1.4.4   Median Rational Expectation*

Rational expectation has played an important role in economics. For a variable y_t, often its expected future value at time $t-1$ is taken as $E(y_t|I_{t-1})$ where I_{t-1} is the information available at $t-1$. Given the linearity of $E(\cdot)$, it is understandable why the conditional mean has been used always to describe how people form their expectations. When a lower or upper ceiling (i.e., a censoring point) is present for y_t, however, using the conditional median $Med(y_t|I_{t-1})$ is more convenient than $E(y_t|I_{t-1})$ for the reason that should be obvious by now. Lee (1997) introduced "median rational expectation," which is reviewed in the following. Which one of $E(y_t|I_{t-1})$ and $Med(y_t|I_{t-1})$ is really used by economic agents—i.e., whether the squared or absolute loss in misprediction is used—would be an interesting empirical question.

Consider an agricultural product with the supply and demand equations

$$s_t = \alpha_1 p_t^e + x_{1t}'\beta_1 + u_{1t}, \quad \alpha_1 > 0 \quad \text{and} \quad d_t = \alpha_2 p_t + x_{2t}'\beta_2 + u_{2t}, \quad \alpha_2 < 0$$

where s_t is supply, d_t is demand, p_t is price, p_t^e is the expected price formed with I_{t-1}, x_{1t} is a vector of "demand shifters", x_{2t} is a vector of "supply shifters", and u_{1t} and u_{2t} are error terms. Assume that x_{1t} and x_{2t} are generated by "reduced forms"

$$x_{1t} = w_{1t}'\gamma_1 + v_{1t} \quad \text{and} \quad x_{2t} = w_{2t}'\gamma_2 + v_{2t}$$

where w_{1t} and w_{2t} are matrices of rv's (but fixed given I_{t-1}; e.g., lagged x_{1t} and x_{2t}), and v_{1t} and v_{2t} are error vectors conformably defined.

Suppose the government sets a lower bound p_{dt} ("d" from "down") for p_t. That is, if the equilibrium price is higher than p_{dt}, then the equilibrium price will prevail; otherwise, the government will intervene to buy the product at p_{dt}, which is a "price support." This means that the underlying market price gets censored from below at p_{dt}. The government policy works not just directly through the bound, but also indirectly through the policy announcement effect on price expectation formation. Assume that

$$p_{dt}, \; \max(p_t, p_{dt}), \; x_{1t}, x_{2t}, \; w_{1t}, w_{2t}, \quad t = 1, ..., T, \quad \text{are observed.}$$

It may be possible to extend the following analysis with a lower bound to lower and upper bounds, which are relevant to foreign exchange rate determination when there is an announced target zone for the exchange rate. The "target-zone exchange-rate analysis" was initiated by Krugman (1991); if further interested, see Iannizzotto and Taylor (1999), Kempa and Nelles (1999), and the references therein for the theoretical development and empirical evidences. Pesaran and Ruge-Murcia (1999) presented a LDV approach to the target zone model and showed more econometric references.

To see the difficulty in mean rational expectation, suppose

$$p_t^e = E(p_t | I_{t-1}).$$

To find the equilibrium price, solve the supply and demand equation for p_t using $s_t = d_t$ and substitute the x_{1t} and x_{2t} equations:

$$p_t = \frac{\alpha_1}{\alpha_2} p_t^e + \frac{1}{\alpha_2}(x_{1t}'\beta_1 - x_{2t}'\beta_2 + u_{1t} - u_{2t}) = \alpha p_t^e + c_t + \varepsilon_t \qquad \text{where}$$

$$\alpha \equiv \frac{\alpha_1}{\alpha_2}, \quad c_t \equiv \frac{1}{\alpha_2}\{(w_{1t}'\gamma_1)'\beta_1 - (w_{2t}'\gamma_2)'\beta_2\}$$

$$\text{and} \quad \varepsilon_t \equiv \frac{1}{\alpha_2}(v_{1t}'\beta_1 - v_{2t}'\beta_2 + u_{1t} - u_{2t}).$$

Without the price support, p_t^e can be found by taking $E(\cdot | I_{t-1})$ on this p_t equation under $E(\varepsilon_t | I_{t-1}) = 0$:

$$p_t^e = \alpha p_t^e + c_t \implies p_t^e = \frac{c_t}{1 - \alpha} \equiv p_{mt}.$$

With the price support, the p_t equation should be modified as

$$p_t = \max(p_{dt}, \ \alpha p_t^e + c_t + \varepsilon_t).$$

We can take $E(\cdot | I_{t-1})$ on this equation and then try to solve for p_t^e. But a closed-form solution for p_t^e is hard to obtain due to the max function even if the distribution of $\varepsilon_t | I_{t-1}$ is specified.

Contrary to this difficulty, suppose now "median rational expectation" holds:

$$p_t^e = Med(p_t | I_{t-1}) \quad \text{and} \quad Med(\varepsilon_t | I_{t-1}) = 0.$$

With the price support, take $Med(\cdot | I_{t-1})$ on $p_t = \max(p_{dt}, \alpha p_t^e + c_t + \varepsilon_t)$ to get

$$p_t^e = \max(p_{dt}, \ \alpha p_t^e + c_t).$$

In $y = \tau(y^*)$ with $\tau(.)$ increasing, we have $Med(y|x) = \tau\{Med(y^*|x)\} = \tau(x'\beta)$, and we just plugged $\tau(x'\beta)$ into the LAD minimand to estimate β in $Med(y^*|x)$. Here, p_t^e appears on both sides, and we need to find $p_t^e = Med(p_t | I_{t-1})$ first before we think about estimating the parameters in $Med(p_t | I_{t-1})$.

The right-hand side of the last display is a Lipschitz-continuous function of p_t^e with the Lipschitz constant $|\alpha|$. If $|\alpha| < 1$, then the mapping is a contraction, and there is a unique fixed point. In the next paragraph, this fixed point of p_t^e will be shown to be

$$\max(p_{dt}, p_{mt}).$$

Hence, *under median rational expectation and $|\alpha_1/\alpha_2| < 1$, the expected price under price support takes a simple closed form: the maximum of the supported price and the expected price without the price support.* This result does not require specifying the distribution of $\varepsilon_t|I_{t-1}$. Observe

$$p_{mt} = \frac{c_t}{1 - \alpha_1/\alpha_2} = \frac{\alpha_2}{\alpha_2 - \alpha_1} c_t = \frac{1}{\alpha_2 - \alpha_1} \{(w'_{1t}\gamma_1)'\beta_1 - (w'_{2t}\gamma_2)'\beta_2\}.$$

With $\hat{\gamma}_1$ and $\hat{\gamma}_2$ denoting the LSE for γ_1 and γ_2, $\delta_1 \equiv \beta_1/(\alpha_2 - \alpha_1)$ and $\delta_2 \equiv -\beta_2/(\alpha_2 - \alpha_1)$ can be estimated by minimizing

$$\frac{1}{T} \sum_t [p_t - \max\{p_{dt}, (w'_{1t}\hat{\gamma}_1)'\delta_1 + (w'_{2t}\hat{\gamma}_2)'\delta_2\}].$$

To prove that $\max(p_{dt}, p_{mt})$ is the fixed point, it is sufficient to show (recall $\alpha \equiv \alpha_1/\alpha_2 < 0$)

$$\max(p_{dt}, p_{mt}) = \max\{p_{dt}, \ \alpha \max(p_{dt}, p_{mt}) + c_t\}.$$

Suppose

$$p_{dt} > p_{mt} \iff p_{dt} > \frac{c_t}{1 - \alpha} \iff (1 - \alpha)p_{dt} > c_t \iff p_{dt} > \alpha p_{dt} + c_t.$$

Then the lhs of the equation becomes p_{dt}, and the rhs becomes $\max(p_{dt}, \alpha p_{dt} + c_t) = p_{dt}$ as well: the equation holds. Now suppose the opposite $p_{dt} \leq p_{mt}$. Then the lhs of the equation is p_{mt}, and the rhs becomes

$$\max(p_{dt}, \alpha p_{mt} + c_t) = \max(p_{dt}, \alpha \frac{c_t}{1 - \alpha} + c_t) = \max(p_{dt}, \frac{c_t}{1 - \alpha})$$

$$= \max(p_{dt}, p_{mt}) = p_{mt};$$

the equation holds.

2   Methods Based on Modality and Symmetry

As demonstrated amply so far, quantile-based ideas have been fruitful to semiparametric econometrics for LDV models. Another strain of semiparametric approaches based on modality and symmetry have been developed in parallel with the quantile-based ones. This section reviews such estimators for censored/truncated models and "censored-selection models."

Section 2.1 introduces "mode regression" under $Mode(y|x) = x'\beta$; the resulting estimator is robust to outliers and applicable to truncated regression models. Section 2.2 examines estimators based on the symmetry of $y|x$ distribution around $x'\beta$, and the estimators are applicable to censored/truncated models. Section 2.3 reviews partial-symmetry-based estimators, which look like combinations of the estimators in the first two sections. Finally, Section 2.4 studies a bivariate-symmetry-based estimator for censored-selection models. Although censored/truncated models may not look so prominent in economics, they are essential in (bio-) statistics where duration/survival models take the central position. As duration/survival models are getting more popular these days in econometrics, it seems fitting to discuss censored/truncated models extensively.

2.1   Mode Regression for Truncated Model and Robustness

Mean and median are not the only measures of location. Another well-known location measure is mode. Suppose we have a truncated model

$$y_i^* = x_i'\beta + u_i, \quad (x_i', y_i^*)' \text{ is observed only when } y_i^* = x_i'\beta + u_i > c,$$
$$i = 1, ..., N.$$

This is in contrast to the censored model $y_i = \max(y_i^*, c)$ where x is observed always. Denoting the truncated response by y, $E(y|x) \neq x'\beta$ in general when $E(y^*|x) = x'\beta$. As $y|x$ follows the truncated distribution—the distribution of $y^*|x$ divided by the conditional truncation probability—it is proper to use the different notation y, instead of y^*.

Consider maximizing

$$E\{1[|y^* - q(x)| < w]\} = E_x[F\{q(x) + w|x\} - F\{q(x) - w|x\}]$$

wrt $q(x)$, where $w > 0$ is a "tuning constant" to be chosen by the researcher and $F(y^*|x)$ is the df of $y^*|x$. The optimal choice $q^*(x)$ is the location measure of $y^*|x$ such that the interval $[q^*(x) - w, q^*(x) + w]$ captures the most probability mass under the density $f(y^*|x)$. *Assuming that f is strictly unimodal and symmetric around the mode up to $\pm w$, the mode maximizes* $E\{1[|y^* - q(x)| < w]\}$. That is, the optimal predictor $q^*(x)$ is $Mode(y^*|x)$.

Suppose $Mode(y^*|x) = x'\beta$ holds. The maximizer $\hat{q}(x)$ of $E\{1[|y - q(x)| < w]\}$ is not necessarily $x'\beta$ because y, not y^*, is in the objective function now. Consider Figure 1. The truncation at c is done in the interval $(-\infty, x'\beta - w)$, i.e. $c < x'\beta - w \Leftrightarrow x'\beta > c + w$, and we still capture the most probability mass under $f(y|x)$ with the interval $x'\beta \pm w$; $\hat{q}(x)$ is

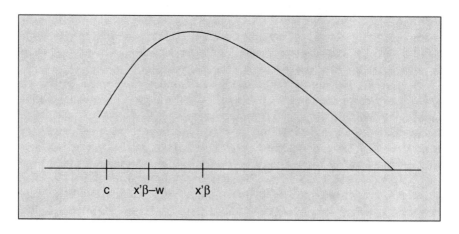

Figure 1: Modal Interval for Maximum Probability

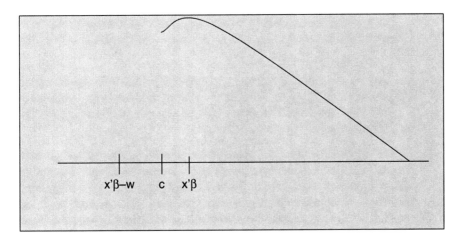

Figure 2: Boundary Interval for Maximum Probability

$x'\beta = Mode(y|x) = Mode(y^*|x)$. In contrast, in Figure 2, the truncation takes place somewhere in $(x'\beta - w, \infty)$, i.e. $c > x'\beta - w \Leftrightarrow x'\beta < c + w$, and thus $\hat{q}(x) = c + w$; the most probability mass under $f(y|x)$ is captured with the interval $[c, c + 2w]$—the interval width is still $2w$. Combining the two cases, the optimal predictor $\hat{q}(x)$ in the truncated case is

$$\max\{Mode(y^*|x), c + w\} = \max(x'\beta, c + w).$$

Based on the above fact, Lee (1989) proposed the *mode regression* estimator for the truncated regression model by maximizing

$$\frac{1}{N}\sum_i [|y_i - \max(x_i'b, c + w)| < w].$$

Letting $c \to -\infty$, we get no truncation where $N^{-1}\sum_i 1[|y_i - x_i'b| < w]$ is maximized. The mode is arguably the most attractive location parameter when the distribution is asymmetric. This estimator does not require specifying the distribution of $u|x$; the only essential requirement for $u|x$ is the strict unimodality (i.e., unimodal with an unique mode instead of a "plateau") as in the figure.

Several remarks are in order:

1. Owing to the nondifferentiability of the indicator function, following the approach in Kim and Pollard (1990), the asymptotic distribution of the mode estimator can be shown to be $N^{1/3}$-consistent with a nonpractical asymptotic distribution. Also the computation requires an algorithm not using gradients such as "downhill simplex" (as explained, e.g., in Himmelblau, 1972); see also Pinkse (1993, appendix).

2. For a random variable z, the optimal predictor maximizing $E\{1 [|z - q| < w]\}$ is not exactly the same as $Mode(z)$, which is clear if

we let f_z to be asymmetric. The optimal predictor is the middle value of the optimal interval of size $2w$ which captures the most probability under f_z.

3. Although we assumed the symmetry up to $\pm w$, it is not necessary under independence of u from x: without the symmetry, the slope coefficients are still identified in this case, which is essentially the same as the parallel shift assumption in quantiles.

4. In $N^{-1} \sum_i 1[|y_i - x_i'b| < w]$, each datum is given the equal weight of 1 or 0, no matter how large the error $y - x'b$ may be. Hence an outlier cannot influence the estimator by more than its share $1/N$. Therefore, *mode regression estimator is robust to outliers in x as well as in y*, resisting almost up to 50% data contamination—the best one can hope for.

5. Analogously to kernel density estimation with the uniform kernel, a small w may reduce the bias of the estimator and a large w may make the estimator more efficient. Although how to choose w is an open question, w may be chosen according to the researcher's tolerance level for misprediction—more on this in the next subsection.

6. A related problem is estimating the mode of a nonlinear regression function $r(x) = E(y|x)$ as examined in Shoung and Zhang (2001), who also lists the statistical literature on estimating the mode of a density.

2.2   Symmetrized LSE for Truncated and Censored Models

2.2.1   Symmetrically Trimmed LSE

In the linear model $y^* = x'\beta + u$, LSE satisfies the orthogonality condition $E(ux) = 0$, which is implied by $E(u|x) = 0$. Consider a truncation in $u|x$, which ruins the orthogonality condition. Assuming that $f_{u|x}$ is symmetric, Powell (1986a) suggested one way to restore the orthogonality condition by "trimming" the error density. Suppose $x_i'\beta > c$ and $y_i^*|x_i$ is truncated from below at c, which is equivalent to $u_i|x_i$ truncated from below at $-x_i'\beta + c$. Let y denote the truncated response versions of y^*. If we artificially trim $u|x$ from above at $x'\beta - c$, then $u \cdot 1[|u| < x'\beta - c]$ will be symmetric (it would be easier to understand this point with $c = 0$). Therefore

$$E_x\{x \ E_{u|x}(u \ 1[|u| < x'\beta - c])\} = E\{xu \cdot 1[|u| < x'\beta - c]\} = 0;$$

note that $|u| < x'\beta - c$ includes the condition $x'\beta > c$. Trimming $u|x$ at $\pm(x'\beta - c)$ is equivalent to trimming $y|x$ at c and $2x'\beta - c$ because y is just $x'\beta$-shifted version of u given x.

One minimand yielding an unique minimizer satisfying the moment condition as the asymptotic first-order condition is

$$\frac{1}{N}\sum_i \{y_i - \max(0.5y_i + 0.5c,\ x_i'b)\}^2.$$

The summand becomes the LSE summand $(y_i - x_ib)^2$ if $0.5y_i + 0.5c < x_i'b$ $\forall i$. Powell (1986b) named this estimator *symmetrically trimmed least squares estimator* (STLS). The main assumption for STLS is the symmetry and strict unimodality of $u|x$ (no need to specify the distribution of $u|x$). The reason for strict unimodality will be seen in the asymptotic variance.

The asymptotic distribution of STLS is

$$\sqrt{N}(b_{stls} - \beta) \rightsquigarrow N\{0,\ (W - V)^{-1}Z(W - V)^{-1}\} \quad \text{where}$$
$$W = E\{1[|u| < x'\beta - c]\ xx'\}, \quad Z = E\{1[|u| < x'\beta - c]\ u^2xx'\}$$
$$V = E\left[1[x'\beta > c]\ 2(x'\beta - c)\frac{f_{u|x}(x'\beta - c)}{F_{u|x}(x'\beta - c)}xx'\right].$$

Examination of $W - V$ reveals that $W - V$ is

$$E\left[1[x'\beta > c]\ xx'\ \{\text{area under } f_{u|x} \text{ between } \pm (x'\beta - c) \text{ above}\right.$$

$$f_{u|x}(x'\beta - c)\}]$$

where $\{\cdots\}$ is the crossed area in Figure 3; $F_{u|x}(x'\beta - c)$ is the normalizing constant for the truncation. Estimating V requires a nonparametric technique due to $f_{u|x}$. The figure shows that strict unimodality is necessary to assure the

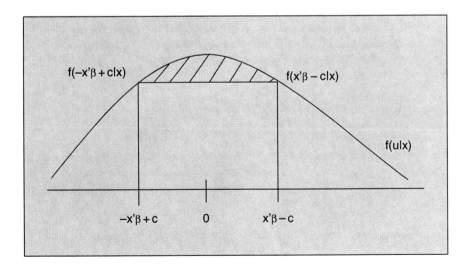

Figure 3: Second-Order Condition for STLS

p.d. of $W - V$. For the non-truncated case, let $c = -\infty$. Then the asymptotic variance becomes that of LSE.

Up to this point, we used the constant c as the censoring or truncation point, but c can be in fact set to zero without loss of generality so long as c is an unknown constant or an observed constant c_i that possibly varies across i, because we can transform the data such that $y = \max(x'\beta + u, 0)$ holds. For instance, when $y_i = \max(x'_i\beta + u_i, c_i)$, subtract c_i from both sides to get

$$y_i - c_i = \max(x'_i\beta - c_i + u_i, 0).$$

Transform y_i into $y_i - c_i$, and absorb $-c_i$ into the intercept of $x'_i\beta$ if $c_i = c \; \forall i$, or if c_i varies across i, redefine x_i as $(x'_i, c_i)'$ and β as $(\beta', -1)'$ to get $y_i = \max(x'_i\beta + u_i, 0)$.

As another example, suppose that y^* is censored from above at γ: $y = \min(x'\beta + u, \gamma)$. Then subtract γ from both sides and multiply by -1 to get

$$\gamma - y = -\min(x'\beta + u - \gamma, 0) = \max(-x'\beta - u + \gamma, 0).$$

Absorb γ into $x'\beta$ and define $v \equiv -u$ to get

$$\gamma - y = \max\{(-x)'\beta + v, 0\}.$$

Transforming y into $\gamma - y \equiv \tilde{y}$ and x into $-x \equiv \tilde{x}$, $\tilde{y} = \max(\tilde{x}'\beta + v, 0)$ holds. The case of γ varying across i can be handled doing analogously to what was done in the preceding paragraph. Hence, from now onward, set $c = 0$.

2.2.2   Symmetrically Censored LSE

Following STLS, Powell (1986a) proposed another estimator for y^* censored from below at 0. With $x'\beta > 0$, censoring $f_{u|x}$ at $-x'\beta$ in the lower tail results in a probability mass piling at $-x'\beta$, while truncation simply cuts off (and normalize) $f_{u|x}$ at $-x'\beta$. The idea is to censor the upper tail of $f_{u|x}$ at $x'\beta$ to restore the symmetry. Then under the symmetry of $u|x$ about 0, the following orthogonality condition holds:

$$E_x[\, x \cdot 1[x'\beta > 0] E_{u|x}\{u\, 1[|u| < x'\beta]$$
$$+\, x'\beta(1[u > x'\beta] - 1[u < -x'\beta])\} \,] = 0,$$

where u in $E(xu)$ is replaced by $x'\beta \cdot sgn(u)$ if $|u| > x'\beta$, which is absent in STLS. The symmetry of $f_{u|x}$ is sufficient for this moment condition.

One minimand yielding an unique minimum with the last display as its asymptotic first order condition is

$$\frac{1}{N}\sum_i [\, \{y_i - \max(0.5y_i, x'_ib)\}^2 + 1[y_i > 2x_ib] \cdot \{(0.5y_i)^2 - (\max(0, x'_ib))^2\} \,].$$

Minimizing this gives the *symmetrically censored least squares estimator* (SCLS). Observe that the first term is the minimand for STLS with $c = 0$.

For the censored response case, there is more information than the truncated case, and the second term makes use of the extra information.

The asymptotic distribution of SCLS does not involve any density component as STLS and CLAD do:

$$\sqrt{N}(b_{scls} - \beta) \rightsquigarrow N(0,\ H^{-1}\ E\{1[x'\beta > 0]\min(u^2, (x'\beta)^2)\ xx'\}\ H^{-1})$$
$$\text{where}\quad H \equiv E\{1[|u| < x'\beta] \cdot xx'\}.$$

The middle term in the variance indicates that u^2 is replaced by $(x'\beta)^2$ whenever $|u| > x'\beta$. Differently from STLS, SCLS does not require the strict unimodality of $u|x$.

Earlier we discussed a duration model under Weibull specification where the duration y follows the Weibull distribution: $P(y \leq t) = F(t) = 1 - \exp(-\theta t^\alpha)$. When $(x', y)'$ is fully observed, assuming $\theta = \exp(x'\beta)$, we showed $\ln y = -x'\beta/\alpha + u$ to which LSE is applicable. When y is censored from above, the LSE is not valid but the Weibull MLE is. But the log-linear model does not necessarily have to be motivated by Weibull, as it can be motivated as a way to achieve symmetry because typically a duration has a density with a long right tail. If $\ln y$ given x is symmetric, then we can use SCLS for the log-linear model under censoring.

Powell (1986a) suggested an iterative scheme to get b_{scls}. Start with an initial estimate b_0, say LSE, and then iterate the following until convergence:

$$b_1 = \left(\sum_i 1[x_i'b_0 > 0] \cdot x_i x_i'\right)^{-1} \sum_i \{1[x_i'b_0 > 0] \cdot \min(y_i, 2x_i'b_0) \cdot x_i\}.$$

This does not guarantee global convergence. Also the matrix to be inverted may not be invertible. If this problem occurs, then removing $1[x_i'b_0 > 0]$ in the inverted matrix may help. From our experience, however, this algorithm seems to work fairly well.

Honoré (1992) extended STLS and SCLS to panel data with truncated or censored dependent variables, and an application of the panel censored model estimator can be seen in Kang and Lee (2003). Newey (1991) proposed semiparametrically efficient versions of STLS and SCLS by taking the full advantage of the symmetry assumption. Lee (1995) applied SCLS to female labor supply data. Santos-Silva (2001a) proposed an improved algorithm and shows further related studies.

2.3   Partial-Symmetry-Based Estimators

2.3.1   Quadratic Mode Regression Estimator (QME)

Generalizing the mode regression further, Lee (1993) suggested to smooth the maximand of the mode regression estimator to obtain a $\sqrt{N}$-consistent estimator good for $y^* = x'\beta + u$ truncated from below at 0. Assume

$$(x', y^*) \text{ is observed only when } y^* = x'\beta + u > 0, \quad Mode(y^*|x) = x'\beta,$$

$\ln f_{u|x}$ is strictly concave, and $f_{u|x}$ is symmetric over $\pm w$;

the concavity of $\ln f_{u|x}$ is called the "strong unimodality" of $f_{u|x}$ (see Dharmadhikari and Joag-dev, 1988, and the references therein). As in the other estimators in this chapter, the distribution of $u|x$ is not specified.

Denoting the truncated response versions of y^* as y, the estimator named *quadratic mode regression estimator* (QME) maximizes

$$\frac{1}{N}\sum_i [w^2 - \{y_i - \max(x_i'b, w)\}^2] \cdot 1[|y_i - \max(x_i'b, w)| < w]$$

$$= \frac{1}{N}\sum_i \max[w^2 - \{y_i - \max(x_i'b, w)\}^2, \; 0]$$

where $w > 0$ is to be chosen by the researcher. The idea of QME is similar to that of STLS: QME trims $f_{u|x}$ at $\pm w$ when $x'\beta > w$, whereas STLS trims $f_{u|x}$ at $\pm x'\beta$ when $x'\beta > 0$. This QME trimming is possible, because $y^* > 0 \iff u > -x'\beta$, which implies $u > -w$ if $x'\beta > w$. The trimming points $\pm w$ in QME are constants, whereas the trimming points $\pm x'\beta$ in STLS vary as x varies.

Recall that, if the truncation point is $c \neq 0$, then $-c$ gets absorbed into $x'\beta$. In this case letting $c = -\infty$ for no truncation, $\max(x'\beta, w)$ becomes $x'\beta$ and the QME maximand becomes

$$\frac{1}{N}\sum_i \{w^2 - (y_i - x_i'b)^2\} \cdot 1[|y_i - x_i'b| < w].$$

This demonstrates that QME is similar to the mode regression but it imposes the quadratic weight $w^2 - (y_i - x_i'b)^2$ on the data with $|y_i - x_i'b| < w$. Instead of capturing the most probability mass under $f_{u|x}$ with the uniform interval $x'b \pm w$, the inverted U-shaped weighting interval is used. When the center of the interval matches the mode (0) of $f_{u|x}$, the maximand is maximized.

It is interesting to see that the maximizing problem is equivalent to minimizing

$$\frac{1}{N}\sum_i \{ (y_i - x_i'b)^2 1[|y_i - x_i'b| < w] \; + \; w^2 1[|y_i - x_i'b| \geq w] \}.$$

This minimizer is the regression version of Huber's "*trimmed (or skipped) mean*" (Huber, 1981), which was suggested to robustify LSE, limiting the effect of outliers by trimming the residuals at $\pm w$: any absolute error greater than w is replaced by w. Hence, QME is robust to outliers.

The asymptotic distribution of QME is analogous to that of STLS:

$$\sqrt{N}(b_{qme} - \beta) \rightsquigarrow N\{0, \; (W - V)^{-1}Z(W - V)^{-1}\}, \quad \text{where}$$

$$Z = E\{1[x'\beta > w, |u| < w] \; xx'u^2\}, \quad W = E\{1[x'\beta > w, |u| < w] \; xx'\}$$

$$V = E\left\{1[x'\beta > w] \cdot \left(2w\frac{f_{u|x}(w)}{1 - F_{u|x}(-x'\beta)}\right) \cdot xx'\right\}.$$

$W - V$ is analogous to the STLS second-order matrix (recall Figure 3). Consistent estimators for Z, W, and V are, with $\hat{u}_i \equiv y_i - x_i'b_{qme}$ and a bandwidth $h \to 0$,

$$Z_N \equiv \frac{1}{N} \sum_i 1[x_i'b_{qme} > w, \ |\hat{u}_i| < w]x_i x_i' \hat{u}_i^2$$

$$W_N \equiv \frac{1}{N} \sum_i 1[x_i'b_{qme} > w, \ |\hat{u}_i| < w]x_i x_i'$$

$$V_N \equiv \frac{1}{N} \sum_i 1[x_i'b_{qme} > w]2w$$

$$\cdot \frac{1[-w < \hat{u}_i < -w + h] + 1[w < \hat{u}_i < w + h]}{2h} x_i x_i';$$

the ratio with h in V_N is to estimate the truncated density component in V.

Whereas STLS becomes LSE with no truncation, QME becomes trimmed LSE with no truncation. Hence QME has two advantages over STLS. One is the weaker symmetry assumption on $f_{u|x}$ and the other is the robustness inherited from the trimmed LSE. On the other hand, STLS has two advantages over QME. One is no need to choose w as in QME, and the other is that STLS seems to work better computationally, possibly due to the random trimming points $\pm x'\beta$ where the randomness tends to smooth out the "edges" in the trimming operation.

Newey (2001) reviewed conditional moment restrictions in censored and truncated regression models, and derived the semiparametric efficiency bounds. Laitilla (2001) showed that QME still works for the slope coefficients even if u is asymmetric about 0 so long as u is independent of x; this point is also noted in Newey (2001). Karlsson (2004) conducted a Monte Carlo study to show that QME performs better than other estimators under asymmetric distributions, and applies bootstrap to avoid density estimation for the QME asymptotic variance.

2.3.2   Remarks for QME

One issue in QME (and mode regression) is how to choose w. Clearly, choosing w is not the same as choosing a bandwidth in nonparametric smoothing where the bandwidth should converge to zero. Our discussion so far points out a number of factors that can play roles in choosing w. The first factor is partial symmetry assumption: a smaller w is the less restrictive. The second is misprediction tolerance: a smaller w means the lower tolerance. The third is robustness concern: a smaller w mean the higher robustness to outliers.

Instead of choosing w, as a way of smoothing the trimming operation in QME, we may randomize (or use multiple fixed values of) w to obtain a "smoothed" QME. This can be done in three steps. First, choose a range for w, say $(0, \bar{w})$. As w is related to the error term scale, $\bar{w}$ may be chosen as follows: an arbitrary positive number $\bar{w}_o$ is used initially for w to get b_{qme} and the residual $\hat{u}_i \equiv y_i - \max(x_i'b_{qme}, w_o)$; then set $\bar{w} = SD(\hat{u})$, or $2 \cdot SD(\hat{u})$ (alternatively, set $\bar{w} = SD(\mathring{u}|x'b_{qme} > w_o)$ where $\mathring{u}_i \equiv y_i - x_i'b_{qme}$). If one wants to minimize the partial symmetry restriction, a small $\bar{w}$ should be chosen; otherwise if the full symmetry is to be imposed, a large value such as $2 \cdot SD(\hat{u})$ may be used. Second, draw multiple random numbers on $U(0, \bar{w})$ (or choose a number of fixed grid points), say w_j, $j = 1, ..., J$. Third, maximize the QME maximand $Q_N(0, \bar{w})$ that is a (weighted) sum across the individual QME maximands with the w_j's. The asymptotic inference may be done with the sandwich form asymptotic variance estimated with the first- and second-order numerical derivatives of $Q_N(0, \bar{w})$.

The fact that QME can be consistent for different values of w can be exploited further. For instance, if $u|x$ is symmetric, then we can use any w, and this implies that different w's can be used to test for the symmetry and heteroskedasticity of $u|x$ as shown in Lee (1996a). Also, instead of maximizing the smoothed QME maximand $Q_N(0, \bar{w})$ as above, we may obtain individual QME's separately to combine them later with MDE. This should be more efficient, but more involved as well. These statements regarding QME and w also apply more or less to another estimator "WME" to appear shortly.

Lee and Kim (1998) proposed a two-stage algorithm to avoid the max function in the QME maximand. In the first stage, minimize

$$\frac{1}{N}\sum_{j=1}^{N} K\left(\frac{x_j - x_i}{h}\right) \left\{ (y_j - q_i)^2 1[|y_j - q_i| < w] + w^2 1[|y_j - q_i| \geq w] \right\}$$

wrt q_i where K is a kernel and h is a bandwidth that may be selected by cross-validation with kernel regression of y on x. This is a one-dimensional minimization, and a grid search over q_i may be employed. Although this minimand is not convex in q_i, it has an U-shape trimmed at w^2, which is not ill-behaving in practice. Denoting the minimizer as $\hat{q}_i$, we do not in fact need to know the exact value of $\hat{q}_i$, because the second step is minimizing

$$\frac{1}{N}\sum_{i}\left\{ (y_i - x_i'b)^2 1[|y_i - x_i'b| < w] + w^2 1[|y_i - x_i'b| \geq w] \right\} \cdot 1[\hat{q}_i > w]$$

wrt b where it is sufficient to know only whether $\hat{q}_i > w$ or not in the first stage.

Lee and Kim (1998) reviewed econometric estimators for the truncated model. Other than the review, they also propose an improved estimator for the truncated response model, which removes the density component in the asymptotic variance of QME by replacing the inverted U-shaped weighting interval with one that approaches the horizontal axis smoothly. One such estimator is *cosine estimator* maximizing

$$\frac{1}{N}\sum_i \left[1 + \cos\left\{\frac{y_i - \max(x_i'b, w\pi)}{w}\right\}\right] \cdot 1\left[\frac{|y_i - \max(x_i'b, w\pi)|}{w} < \pi\right].$$

Since the argument of the function $1 + \cos(\cdot)$ lies between $\pm\pi$, $1 + \cos(\cdot)$ stays above zero while smoothly approaching zero (the horizontal axis), differently from the quadratic weights in QME and STLS. The asymptotic distribution is

$$\sqrt{N}(b_{\cos} - \beta) \rightsquigarrow N(0, \ G^{-1}LG^{-1}) \quad \text{where}$$

$$G \equiv E\left\{1[x'\beta > w\pi] \ xx' \frac{1[|u| < w\pi]}{w^2} \cos\left(\frac{u}{w}\right)\right\},$$

$$L \equiv E\left\{1[x'\beta > w\pi]xx' \frac{1[|u| < w\pi]}{w^2} \left(\sin(\frac{u}{w})\right)^2\right\}.$$

The sample analogs for G and L are easy to see.

Instead of the cosine "kernel," one may use the biweight kernel $(15/16)(1 - t^2)^2 1[|t| < 1]$ which also approaches zero smoothly, where t is to be replaced with $\{y - \max(x'b, w)\}/w$. Baldauf and Santos-Silva (2008, on the use of iteratively reweighted least squares M-estimators in econometrics, unpublished paper, University of essex.) show that a version of this estimator with "triweight" kernel $(35/32)(1-t^2)^3 1[|t| < 1]$ is implemented in STATA as a robust regression with command *rreg*. It is not clear in STATA which location measure in $y|x$ gets estimated, but our discussion reveals that it is $Mode(y|x)$ if the "tuning constant" w is small and the partial symmetry up to $\pm w$ holds.

QME may be implemented with the Newton–Raphson-type iteration:

$$b_1 = b_0 + \left\{\sum_i 1[x_i'b_0 > w] \cdot 1[|y_i - x_i'b_0| < w]x_i x_i'\right\}^{-1}$$

$$\sum_i 1[x_i'b_{qme} > w] \cdot 1[|y_i - x_i'b_0| < w]x_i(y_i - x_i'b_0).$$

If the inverted matrix is not p.d., then one of the indicator functions there may be dropped. The inverted matrix uses only W, not $W - V$, in the asymptotic variance. Unless there occurs the problem of non p.d., using $W - V$ would be better in the iteration.

The Newton–Raphson-type iteration can be applied also to the cosine estimator. But QME is likely to converge better than the cosine estimator, because the QME's second-order derivatives are non-negative whereas those of the cosine estimator can take either sign. This happens because the cosine function has to change its second derivative signs to touch the horizontal axis smoothly.

2.3.3   Winsorized Mean Estimator (WME)

In trimmed mean, we saw a way of robustifying LSE. One trade-off of doing it is the loss of efficiency by not using the quantitative information in the data with $|y - x'b| \geq w$. Another robust estimator "*winsorized mean*" is obtained by combining the mean and median regressions. Consider minimizing

$$\frac{1}{N} \sum_i \left\{ \frac{1}{2}(y_i - x_i'b)^2 \cdot 1[|y_i - x_i'b| < w] \right.$$
$$\left. + \left(w|y_i - x_i'b| - \frac{w}{2} \right) \cdot 1[|y_i - x_i'b| \geq w] \right\}.$$

The information in $|y - x'b| \geq w$ is not thrown away; instead its impact is downgraded by taking the absolute value, not the squared value. As it turns out, this version is applicable to censored models as follows.

For the censored model $y = \max(x'\beta + u, 0)$ where $f_{u|x}$ is symmetric around 0 up to $\pm w$, Lee (1992b) suggested to minimize

$$\frac{1}{N} \sum_i \left[\frac{1}{2}\{y_i - \max(x_i'b, w)\}^2 \cdot 1[|y_i - \max(x_i'b, w)| < w] \right.$$
$$\left. +\{w|y_i - \max(x_i'b, w)| - \frac{w}{2}\} \cdot 1[|y_i - \max(x_i'b, w)| \geq w] \right];$$

the first term was seen in QME. This estimator is called the *winsorized mean estimator* (WME). WME does not need the strict unimodality of $u|x$ as QME does.

Other than for the max function, the minimand is convex. A two-stage algorithm analogous to the one for QME goes as follows. In the first-stage, minimize wrt q_i

$$\frac{1}{N} \sum_{j=1}^N K\left(\frac{x_j - x_i}{h}\right) \left\{ \frac{1}{2}(y_j - q_i)^2 \cdot 1[|y_j - q_i| < w] \right.$$
$$\left. + \left(w|y_j - q_i| - \frac{w}{2} \right) \cdot 1[|y_j - q_i| \geq w] \right\}.$$

As in the QME two-stage procedure, the exact value of the minimizer $\hat{q}_i$ is not required: we only have to find out whether $\hat{q}_i > w$ or not. In the second stage, minimize wrt b

$$\frac{1}{N} \sum_i \left\{ \frac{1}{2}(y_i - x_i'b)^2 1[|y_i - x_i'b| < w] \right.$$
$$\left. + \left(w|y_i - x_i'b| - \frac{w}{2} \right) 1[|y_i - x_i'b| \geq w] \right\} \cdot 1[\hat{q}_i > w].$$

The asymptotic first-order condition for WME is

$$E_x[1[x'\beta > w] \, x \, E_{u|x}\{1[|u| < w]u + (1[u > w] - 1[u < -w])w\}] = 0;$$

given $x'\beta > w$, we replace u by $-w$ if $u < -w$ and by w if $u > w$. This is analogous to SCLS where u is replaced by $-x'\beta$ if $u < -x'\beta$ and by $x'\beta$ if

$u > x'\beta$ when $x'\beta > 0$. While SCLS needs the symmetry of $f_{u|x}$ up to $\pm x'\beta$ ($\pm\infty$ if $x'\beta$ has unbounded support), WME needs it only up to $\pm w$. WME is consistent for β and

$$\sqrt{N}(b_{wme} - \beta) \rightsquigarrow N[0,\ C^{-1} \cdot E\{1[x'\beta > w]\ \min(u^2, w^2)\ xx'\} \cdot C^{-1}]$$
$$\text{where}\quad C \equiv E\{1[x'\beta > w, |u| < w]\ xx'\}.$$

Lee (1995a) applied WME to female labor supply data.

Recall that the αth quantile is obtained when the asymmetric loss function $(y - q)(\alpha - 1[y - q < 0])$ is used where q is the predictor. As Newey and Powell (1987) examined an analogous asymmetric squared loss function to come up with "expectiles," it is possible to use asymmetric loss function for QME and WME—"modile" might be the right name here. Karlsson (2006) indeed explored this generalization, although the generalization did not go far enough to find any interesting characterization of the "modiles." Karlsson and Laitila (2008) combined QME and WME to deal with LTRC (left-truncated and right-censored) data.

The motivation for trimmed mean and winsorized mean in the statistics literature is their robustness to outliers (while not losing much in efficiency). Both truncation and censoring can be viewed as ways of data contamination, and thus it is only natural that QME and WME are applicable to truncated and censored models.

Initially, trimmed mean and winsorized mean were proposed for one sample problem (i.e., finding a location measure for y without x), and it has been less than straightforward to generalize the location trimmed or winsorized mean to linear regression models—no censoring/truncation here. Koenker and Bassett (1978) proposed a quantile-based "trimmed regression estimator" for the linear model $y = x'\beta + u$: do the LSE after removing the observations whose residual from the αth regression quantile is negative or whose residual from the $(1 - \alpha)$th regression quantile is positive, where α is a small positive number, say 0.05. The following trimmed regression estimator in Welsh (1987) is similar to this and may be preferred on some ground (see e.g., Chen et al., 2001).

Given a preliminary estimator for β and the residuals $\hat{u}_1, ..., \hat{u}_N$, let ξ_α be the αth quantile in the residuals. The Welsh (1987) estimator is

$$\left(\sum_i x_i x_i' 1[\xi_\alpha < \hat{u}_i < \xi_{1-\alpha}]\right)^{-1} \sum_i x_i \tilde{y}_i, \quad \text{where}$$
$$\tilde{y}_i \equiv \xi_\alpha(1[\hat{u}_i < \xi_\alpha] - \alpha)\ + y_i 1[\xi_\alpha < \hat{u}_i < \xi_{1-\alpha}]$$
$$+ \xi_{1-\alpha}\left(1\left[\xi_{1-\alpha} < \hat{u}_i\right] - \alpha\right).$$

The form of $\tilde{y}$ is reminiscent of winsorized mean, and surprisingly, the preliminary estimator does not affect this estimator's asymptotic variance. Although this kind of trimmed estimators are robust to outliers in y, they are not robust to outliers in x.

2.4   Estimators for Censored-Selection Models

Consider a *censored-selection* model:

$$\delta_i^* = x_i'\alpha + \varepsilon_i > 0, \qquad y_i = x_i'\beta + u_i$$
$$(1[\delta_i^* > 0]\delta_i^*, \ 1[\delta_i^* > 0]y_i, x_i'), \ i = 1, ..., N, \text{ are observed.}$$

In this model, δ^* as well as y are observed when $\delta^* > 0$. If only $(1[\delta^* > 0], 1[\delta^* > 0]y, x')$ is observed, then we get the usual *binary-selection model*. For instance, y is weakly wage and δ^* is work hours: when $\delta^* > 0$, both wage and work hours are observed. This model contains much more information than the corresponding binary-selection model, and exclusion restrictions are not required for this model. Since we have

$$\delta_i \equiv 1[\delta_i^* > 0]\delta_i^* = \max(\delta_i^*, 0) = \max(x_i'\alpha + \varepsilon_i, 0)$$

which is a familiar censored model, α can be estimated with any of the censored model estimators reviewed earlier. In the following, we introduce two LSE-based estimators for β in Honoré et al. (1997).

For the first estimator of Honoré et al. (1997), suppose (ε, u) is "*centrally symmetric*": $(\varepsilon, u)|x$ follows the same distribution as $(-\varepsilon, -u)|x$. This implies that the marginal distributions of $\varepsilon|x$ and $u|x$ are symmetric about zero:

$$f(u|x) = \int f(\varepsilon, u|x)d\varepsilon = \int f(-\varepsilon, -u|x)d\varepsilon$$

$$= \int f(v, -u|x)dv = f(-u|x).$$

When $\delta = 0 \iff -x'\alpha < \varepsilon$, the central symmetry breaks down, and $u|(x, -x'\alpha < \varepsilon)$ is not symmetric about zero. The idea is to artificially trim ε to restore the marginal symmetry:

$$u|(x, -x'\alpha < \varepsilon < x'\alpha) = u|(x, 0 < \delta^* < 2x'\alpha) \text{ is symmetric about zero.}$$

This can be seen by looking at the density of $u|(x, -x'\alpha < \varepsilon < x'\alpha)$:

$$\frac{\int_{-x'\alpha}^{x'\alpha} f(\varepsilon, u|x)d\varepsilon}{\int \int_{-x'\alpha}^{x'\alpha} f(\varepsilon, u|x)d\varepsilon du} = \frac{\int_{-x'\alpha}^{x'\alpha} f(-\varepsilon, -u|x)d\varepsilon}{\int \int_{-x'\alpha}^{x'\alpha} f(-\varepsilon, -u|x)d\varepsilon du}$$

$$= \frac{\int_{-x'\alpha}^{x'\alpha} f(v, -u|x)dv}{\int \int_{-x'\alpha}^{x'\alpha} f(v, -u|x)dv du};$$

i.e., the density with u replaced by $-u$ remains the same.

Denoting a $\sqrt{N}$-consistent estimator for α as a_N, a $\sqrt{N}$-consistent estimator for β can be obtained by minimizing wrt b

$$\frac{1}{N}\sum_i 1[0 < \delta_i^* < 2x_i'a_N] \cdot (y_i - x_i'b)^2$$

because $u|x$ is symmetric given $0 < \delta^* < 2x'\alpha$. Let $\gamma \equiv (\alpha', \beta')'$ and $g_N \equiv (a'_N, b'_N)'$. Honoré et al. (1997) presented the asymptotic distribution and consistent estimators for the asymptotic variance, which were, however, fairly complicated. Instead, they suggested to use bootstrap.

The second estimator of Honoré et al. (1997) requires (ε, u) to be independent of x and uses the pairwise differencing estimator in Honoré and Powell (1994) for a_N—this pairwise differencing idea will be examined in detail in the following section. The idea is that, given

$$\varepsilon_i > \max(-x'_i\alpha, -x'_j\alpha), \quad \varepsilon_j > \max(-x'_j\alpha, -x'_i\alpha) \iff$$
$$\delta_i > \max\{0, (x_i - x_j)'\alpha\}, \quad \delta_j > \max\{0, (x_j - x_i)'\alpha\}$$
$$\text{(adding } x'_i\alpha \text{ and } x'_j\alpha, \text{ respectively)}$$

$u_i - u_j$ is symmetric about zero because ε_i an ε_j are trimmed by the same threshold. The second estimator is obtained by minimizing

$$\sum_{i<j} 1[\delta_i > \max\{0, (x_i - x_j)'a_N\}, \ \delta_j > \max\{0, (x_j - x_i)'a_N\}] \cdot$$
$$\{y_i - y_j - (x_i - x_j)'b\}^2$$

wrt b. The asymptotic inference may be done with bootstrap.

Differently from the parametric approaches, it seems difficult to deal with binary-selection models in a semiparametric bandwidth-free fashion, which is why no semiparametric estimator has been introduced for binary-selection in this chapter. The next chapter will present some bandwidth-dependent semiparametric estimators for binary-selection models; for censored-selection models, see Lee and Vella (2006) and the references therein. An alternative is giving up on estimating β, but instead "bounding" the parameters of interest such as $Q_\alpha(y|x)$ or $E(y|x)$. See Lee and Melenberg (1998) for bounding quantiles and Lee (2005b) for bounding mean in binary-selection models.

3   Rank-Based Methods

In addition to quantiles, mode and symmetry-based ideas, rank and differencing have been adopted to deal with LDV models within semiparametric framework, which are reviewed in this and the following sections. Differently from the preceding section, however, rank and difference-based estimators are applicable not just to LDV models, but also to "single index models," transformation-of-variable models, and partially linear models. Section 3.1 introduces single index models along with one conceptually straightforward estimator; multiple index models are briefly discussed as well. Sections 3.2–3.4 review rank-based estimators: Section 3.2 uses pairs in a given data, Section 3.3 uses triples, and Section 3.4 uses quadruples. Section 3.5 introduces an estimator for $\Lambda(\cdot)$ where $\Lambda(y) = x'\beta + u$ and $\Lambda(\cdot)$ is an unknown increasing transformation.

3.1　Single Index Models (SIM)

3.1.1　Single Index and Transformation of Variables

In a wide sense, a "*single-index model*" *(SIM)* is a model where x affects y only through $x'\beta$. For example, some regression function in the conditional distribution of $y|x$ may depend on x, but only through $x'\beta$. In a narrow sense, the definition is restricted to models with $E(y|x) = G(x'\beta)$ for a function G whose form may be unknown. This may be called "*mean index sufficiency*" because knowing the scalar $x'\beta$ is as good as knowing the vector x, as far as the mean is concerned. We will adopt the narrow definition $E(y|x) = G(x'\beta)$ unless otherwise necessary as we did earlier for nonlinear regression models. In this definition, there is no restriction on $V(y|x)$; "variance index sufficiency" or variance independence (i.e., homoskedasticity) may not hold. Note that, even under the wide definition, β may still be estimable using $E(y|x) = G(x'\beta)$. If desired, one may further restrict the definition of SIM to models with a (strictly) monotonic $G(\cdot)$, or even to a (strictly) increasing $G(\cdot)$ if the relation between y and x is known to be positive.

Consider a *transformation-of-variable model*:

$$\Lambda(y_i) = x_i'\beta + u_i \quad \text{where } \Lambda(\cdot) \text{ is an unknown, continuous,}$$
$$\text{and strictly increasing function.}$$

Add a constant β_o and multiply both sides by a positive constant γ to get

$$\{\Lambda(y_i) + \beta_o\}\gamma = (x_i'\beta + \beta_o)\gamma + \gamma u_i$$

which can be rewritten as $\Lambda^*(y_i) = x_i'\beta^* + u_i^*$ for the appropriately defined Λ^*, β^*, and u_i^*—β_o is absorbed into the intercept in $x_i'\beta$; as Λ, β, and u_i are unknown, they can be rewritten "freely."

The last display illustrates that β is identified only up to a positive scale (and an additive constant). By inverting Λ, we get $y_i = \Lambda^{-1}(x_i'\beta + u_i)$ and $E(y|x) = \int \Lambda^{-1}(x'\beta + u)f_{u|x}(u)du$ that is not necessarily a SIM as x can appear in $f_{u|x}$ as well as in $x'\beta$. But if $u \amalg x$, then we get a SIM

$$E(y|x) = \int \Lambda^{-1}(x'\beta + u_o)f_u(u_o)du_o$$

that is a strictly increasing function of $x'\beta$.

If $\Lambda(\cdot)$ is not known to be increasing but only monotone, then the above scale factor γ can take on a negative as well as positive value, and consequently β is identified only up a scale (and an additive constant). If u depends on x only through $x'\beta$—i.e., if $f_{u|x'\beta}$ appears—then the last display is still a single index model but it is no longer clear whether $E(y|x)$ is monotone or not as a function of $x'\beta$.

To motivate the transformation-of-variable model, note "accelerated failure time" model:

$$\ln y = -x'\beta + u, \quad u \amalg x, \quad e^u \text{ has integrated hazard function } \Lambda(\cdot)$$

$$\Longrightarrow P(y>t|x) = P(e^{-x'\beta}e^u > t|x) = P(e^u > te^{x'\beta}|x) = \exp\{-\Lambda(te^{x'\beta})\}$$

where the role of $e^{x'\beta}$ in $\Lambda(te^{x'\beta})$ is to "accelerate (or decelerate)" the time t. Here, we are assuming that log transformation yields the linear model, which may be, however, too restrictive. A generalization is $\Gamma(y,\theta) = -x'\beta + u$ with θ being a parameter as in the Box-Cox transformation that appeared already for nonlinear regression. A further generalization of this is to allow an unknown transformation $\Lambda(y)$ as above.

3.1.2   Simple Single-Index Model Estimator

Consider a semiparametric single index model

$$y = G(x'\beta) + u, \quad E(u|x) = 0,$$

$G(\cdot)$ is an unknown continuously differentiable function.

Several examples for this appeared already in this and preceding chapters. Since any re-scaling of β can be absorbed into the unknown G, we can identify β only up to a scale. For an analogous reason, intercept is not identified either. If G is known to be increasing or decreasing, then β is identified up to a positive scale (i.e., up to a scale whose sign is identified).

Suppose that x has a density $f(x)$ with continuous gradient $\nabla f(x)$, and $f(x)$ is zero on the boundary $\partial\Omega$ of the support Ω of x. Then under $E\{\partial G(x'\beta)/\partial x\} < \infty$, using integration by parts,

$$E\left\{\frac{\partial G(x'\beta)}{\partial x}\right\} = \int_\Omega \frac{\partial G(x'\beta)}{\partial x} f(x)dx$$

$$= G(x'\beta)f(x)\mid_{\partial\Omega} - \int_\Omega G(x'\beta)\nabla f(x)dx.$$

As partial differentiation is involved, view $G(x'\beta)f(x)\mid_{\partial\Omega}$ also as "partially." For example, with $k = 2$ and $\Omega = [-1,1] \times [-2,2]$, $G(x'\beta)f(x)\mid_{\partial\Omega}$ for ∂x_1 is $G(x'\beta)f(x)\mid_{x_1=1} -G(x'\beta)f(x)\mid_{x_1=-1}$.

The first term on the right-hand side disappears, and we get

$$E\left\{\frac{\partial G(x'\beta)}{\partial x}\right\} = -\int_\Omega G(x'\beta)\nabla f(x)dx = -\int G(x'\beta)\nabla f(x)1[f(x)>0]dx$$

$$= -\int\{G(x'\beta) + u\}\frac{\nabla f(x)}{f(x)}1[f(x) > 0]f(x)dx \quad \text{(for } E(u|x) = 0\text{)}$$

$$= -\int y\frac{\nabla f(x)}{f(x)}1[f(x) > 0]f(x)dx = E\left\{y\frac{-\partial \ln f(x)}{\partial x}1[f(x) > 0]\right\}.$$

Since

$$E\left\{\frac{\partial G(x'\beta)}{\partial x}\right\} = E\left\{\frac{dG(x'\beta)}{d(x'\beta)}\right\} \cdot \beta \equiv \gamma\beta \text{ where}$$

$$\gamma \equiv E\left\{\frac{dG(x'\beta)}{d(x'\beta)}\right\} \text{ is a scalar}$$

we get the key equation

$$\gamma\beta = \gamma(\beta_1, ..., \beta_k)' = -E\left\{y\frac{\partial \ln f(x)}{\partial x}1[f(x) > 0]\right\}.$$

If $f(x)$ is known, then $\gamma\beta$ can be estimated by (Stoker, 1986)

$$\widehat{\gamma\beta} \equiv -\frac{1}{N}\sum_i y_i\frac{\partial \ln f(x_i)}{\partial x}1[f(x_i) > 0].$$

This equation links β to the marginal distribution of x, which may look somewhat strange at the first sight. In terms of the restrictiveness of assumptions,
this estimator is not any better than estimators specifying G, because f has
to be specified. The estimator, however, opened the way to other semiparametric methods for SIM. The obvious thing to do is replacing $-\partial \ln f(x_i)/\partial x$
with a nonparametric estimator. This idea and others for SIM will be examined in the next chapter.

3.1.3   Double or Multiple Indices

SIM provide a sensible compromise between parametric and purely non-
parametric models. The SIM $E(y|x) = G(x'\beta)$ limits nonlinearity into the
indexing function $G(\cdot)$, while preserving the linearity in $x'\beta$. Consequently it
is more manageable than the purely nonparametric model. As noted already,
however, one major drawback is the up-to-scale identification which more or
less limits its application to cases where the scale of the response variable is
not observed. Another major drawback is, as its name indicates, the effect
of x on y can be only through the sole index. In the following, we explore
double or multiple index models to relax the latter.

Generalizing SIM is *double index model* where the conditional mean of
interest depends on x and z through two indices, say $x'\beta$ and $z'\gamma$; $x = z$ is
allowed in principle. For instance, recall the sample selection model where
$y^* = x'\beta + u$ is observed only when $z'\gamma + v > 0$. Denoting the joint density
of u and v as $f(u,v)$ where $(u,v) \amalg (x,z)$, it holds that

$$
\begin{aligned}
E(y^*|x, z, v > -z'\gamma) &= x'\beta + E(u|x, z, v > -z'\gamma) \\
&= x'\beta + \frac{\int_{-\infty}^{\infty}\int_{-z'\gamma}^{\infty} uf(u,v)dvdu}{\int_{-\infty}^{\infty}\int_{-z'\gamma}^{\infty} f(u,v)dvdu}
\end{aligned}
$$

the right-hand side of which depends on x and z only through $x'\beta$ and $z'\gamma$. As
well known, if $f(u,v) = f(u)f(v)$ and $E(u) = 0$, then $E(y|x, z, v > -z'\gamma) =
x'\beta$: a SIM holds despite the selection process.

Another example of double index model is the transformation-of-variable model $\Lambda(y) = x'\beta + u \iff y = \Lambda^{-1}(x'\beta + u)$ where u may depend on regressors but only through a linear index $z'\gamma$:

$$E(y|x,z) = \int \Lambda^{-1}(x'\beta + u_o) f_{u|(x,z)}(u_o) du_o = \int \Lambda^{-1}(x'\beta + u_o) f_{u|z'\gamma}(u_o) du_o.$$

Yet another example is a bivariate-binary model:

$$y_j = 1[x'\beta_j + u_j > 0], \quad j = 1, 2, \qquad (u_1, u_2) \amalg x,$$
$$\implies E(y_1 y_2 | x) = P(u_1 > -x'\beta_1, u_2 > -x'\beta_2 | x) = H(x'\beta_1, x'\beta_2)$$

where $H(\cdot, \cdot)$ is the "joint survival" function of (u_1, u_2): $H(w_1, w_2) \equiv P(u_1 > w_1, u_2 > w_2)$. This example with two indices sharing the same regressor x, however, poses a problem as in the following.

A difficulty with double- or multiple-index models is that, unless some restrictions are imposed, the parameters are identified only under exclusion restrictions. To see this point, suppose $E(y|x) = H(x'\beta, x'\gamma)$, which is a special case of $H(x'\beta, z'\gamma)$ with $x = z$. Then

$$\frac{\partial E(y|x)}{\partial x} = H_1(x)\beta + H_2(x)\gamma \qquad \text{where } H_j(x) \equiv \frac{\partial H(w_1, w_2)}{\partial w_j}, \ j = 1, 2,$$

$$\implies E\left\{ \frac{\partial E(y|x)}{\partial x} \right\} = E\{H_1(x)\}\beta + E\{H_2(x)\}\gamma;$$

here β and γ are not separated, i.e., not identified. If the $(k-1)$th and kth elements of x are in $x'\beta$ but not in $z'\gamma$ (exclusion restrictions), then $E\{\partial E(y|x)/\partial x_j\} = E\{H_1(x)\}\beta_j, \ j = k-1, k$. In this case, β_k/β_{k-1} is identified (under $\beta_{k-1} \neq 0$) because

$$\frac{\beta_k}{\beta_{k-1}} = \frac{E\{\partial E(y|x)/\partial x_k\}}{E\{\partial E(y|x)/\partial x_{k-1}\}}.$$

Triple or higher index models can be easily thought of. We already saw that multiple choice models can have this structure where the linear indices come from the systematic parts of the alternative utilities. Another example is a sample selection model with multiple selection criteria: $y^* = x'\beta + u$ is observed only when $z_j'\gamma_j + v_j > 0, \ j = 1, ..., J$. Yet another example is a multivariate-binary model $y_j = 1[x'\beta_j + u_j > 0], \ j = 1, ..., J$.

3.2   Kendall Rank Correlation Estimator (KRE)

So far we introduced two consistent semiparametric estimators for the binary response model $y = 1[x'\beta + u \geq 0]$: maximum score estimator (MSC) and the single index estimator in the preceding subsection. In this subsection, we examine a $\sqrt{N}$-consistent estimator that is more practical than the two estimators. More estimators will be seen later.

3.2.1   Estimator and Identification

Han (1987a) proposed *"Kendall Rank Correlation Estimator (KRE)"* maximizing

$$Q_N(b) \equiv \frac{1}{N(N-1)} \sum_{i \neq j} 1[x_i'b > x_j'b,\ y_i > y_j]$$

$$= \frac{1}{N(N-1)} \sum_{i<j} \{1[x_i'b > x_j'b,\ y_i > y_j] + 1[x_j'b > x_i'b,\ y_j > y_i]\}$$

$$= \frac{2}{N(N-1)} \sum_{i<j} \frac{1}{2} \{1[x_i'b > x_j'b,\ y_i > y_j] + 1[x_j'b > x_i'b,\ y_j > y_i]\}.$$

The motivation comes from the fact that maximizing $Q_N(b)$ is equivalent to maximizing *Kendall's rank correlation* between y and $x'b$:

$$4Q_N(b) - 1 = \frac{2}{N(N-1)} \sum_{i<j} 2\{1[x_i'b > x_j'b,\ y_i > y_j]$$
$$+ 1[x_j'b > x_i'b,\ y_j > y_i]\} - 1.$$

Since multiplying $x'b$ by a positive constant does not change the ordering $x_i'b > x_j'b$, the parameters for KRE are identified only up to a positive scale. For this reason, we will discuss KRE only for binary responses, although KRE is also applicable to other LDV's. Note that x should not include unity, because only the difference $x_i - x_j$ matters for KRE in $x_i'b - x_j'b = (x_i - x_j)'b$ where the intercept disappears. In probit and MSC, x includes unity although the intercept in β is not identified—the intercept plus something is estimated with unity. In difference-based methods as KRE, x should not include unity and the intercept is not identified.

The main assumptions for KRE are $x \amalg u$ and the existence of at least one continuous regressor, say x_k, with unbounded support (conditional on all the other regressors). The latter also appeared for MSC, which makes excluding the ties $x_i'b = x_j'b$ in $Q_N(b)$ harmless. Assume that u has a continuous and strictly increasing distribution function $F_u(\cdot)$ and $S_u(\cdot) \equiv 1 - F_u(\cdot)$. The idea of KRE is similar to MSC: if $x_i'\beta > x_j'\beta$, then $y_i > y_j$ is more likely than $y_i < y_j$. That is,

$$P(y_i > y_j | x_i, x_j) > P(y_i < y_j | x_i, x_j) \quad \text{whenever } x_i'\beta > x_j'\beta.$$

Hence we can estimate β by maximizing the "pairwise prediction score" $Q_N(b)$.

To understand the idea of KRE better, observe that the population version of the summand in $Q_N(b)$ (with $\sum_{i<j}$) conditional on x_i and x_j is (the pairs with $y_i = y_j$ drop out)

$$P(y_i = 1, y_j = 0 | x_i, x_j) \; 1[x_i'b > x_j'b]$$

$$+ P(y_j = 1, y_i = 0 | x_i, x_j) \; 1[x_i'b > x_j'b]$$

$$= P(y_i = 1 | x_i) P(y_j = 0 | x_j) \; 1[x_i'b > x_j'b]$$
$$+ P(y_i = 0 | x_i) P(y_j = 1 | x_j) 1[x_i'b > x_j'b]$$

$$= P(u_i \geq -x'\beta | x_i) P(u_j < -x_j'\beta | x_j) 1[x_i'b > x_j'b]$$

$$+ P(u_i < -x_i'\beta | x_i) P(u_j \geq -x_j'\beta | x_j) 1[x_i'b > x_j'b]$$

$$= S_u(-x_i'\beta) F_u(-x_j'\beta) \; 1[x_i'b > x_j'b] + F_u(-x_i'\beta) \; S_u(-x_j'\beta) \; 1[x_j'b > x_i'b].$$

Suppose $x_i'\beta > x_j'\beta$. Since $F_u(-t)$ is strictly decreasing in t and $S_u(-t)$ is strictly increasing in t,

$$S_u(-x_i'\beta) \cdot F_u(-x_j'\beta) > S_u(-x_j'\beta) \cdot F_u(-x_i'\beta).$$

It is at this step that "$u \amalg x$" is invoked; otherwise, $x_i'\beta > x_j'\beta$ does not necessarily imply this display. Thus, when $x_i'\beta > x_j'\beta$, the conditional maximand is maximized by choosing $1[x_i'b > x_j'b]$, not $1[x_i'b < x_j'b]$; the opposite holds when $x_i'\beta < x_j'\beta$. In essence, maximizing $Q_N(b)$ is the same as matching the sign of $x_i'b - x_j'b$ with that of $x_i'\beta - x_j'\beta$ whereas LSE matches $x'b$ with $x'\beta$ as can be seen in

$$E(y - x'b)^2 = E\{(x'\beta - x'b) + u\}^2 = E(x'\beta - x'b)^2 + E(u^2).$$

KRE is better than probit and logit, as KRE does not specify the error-term distribution. KRE is better than MSC because KRE is $\sqrt{N}$-consistent with an useful asymptotic distribution, but worse because $x \amalg u$ is assumed. KRE is a SIM estimator, because the dependence of y on x takes place only through the index $x'\beta$ thanks to $x \amalg u$. KRE is better than the SIM estimators to appear in the next chapter, because KRE is bandwidth-free while those SIM estimators are not—this is why KRE is reviewed in this chapter. Also KRE is better than the other SIM estimators in general because KRE estimates the parameters up to a *positive* scale, but worse because KRE assumes $x \amalg u$. But the following example demonstrates that the difference between KRE and the other SIM estimators is actually small.

Consider a binary response model with heteroskedasticity:

$$y_i^* = x_i'\beta + u_i, \quad u_i = \sigma(x_i'\beta)v_i, \quad v \amalg x, \quad \text{and} \quad y_i = 1[y_i^* \geq 0]$$

$$\implies E(y|x) = 1 - F_v \left\{ \frac{-x'\beta}{\sigma(x'\beta)} \right\} = G(x'\beta) \quad \text{where } F_v \text{ is the df of } v$$

and $F_v(\cdot)$ and $\sigma(\cdot)$ are unknown functions. This is a SIM with a restricted form of heteroskedasticity: the heteroskedasticity factor is a function of the same index. KRE still works if we assume $x'\beta/\sigma(x'\beta)$ is increasing in $x'\beta$; $x \amalg u$ is not necessary, and thus the aforementioned advantage of SIM over KRE disappears. But in this case, the advantage of KRE also disappears because $G(\cdot)$ is increasing and the sign of the unknown scale factor is identified in the other SIM estimators. Therefore, KRE and the other SIM estimators become close. Justifying that $x'\beta/\sigma(x'\beta)$ is increasing is not easy because $\sigma(x'\beta)$ may depend on $x'\beta$ only through $|x'\beta|$, which is why $x \amalg u$ was assumed in KRE.

3.2.2   Asymptotic Distribution

Let $z_i \equiv (x_i', y_i)'$. With $Q(b) \equiv E\{1[y_i > y_j, x_i'b > x_j'b]\}$, Sherman (1993) rewrote $Q_N(b)$ with $\sum_{i \neq j}$ as

$$Q_N(b) = Q(b) \; + \frac{1}{N} \sum_i [E\{1[y_i > y_j, x_i'b > x_j'b]|z_i\} - Q(b)]$$

$$+\frac{1}{N} \sum_j [E\{1[y_i > y_j, x_i'b > x_j'b]|z_j\} - Q(b)] + \frac{1}{N(N-1)} \sum_{i \neq j} \lambda(z_i, z_j, b)$$

where

$$\lambda(z_i, z_j, b) \equiv 1[y_i > y_j, x_i'b > x_j'b] - [E\{1[y_i > y_j, x_i'b > x_j'b]|z_i\} - Q(b)]$$

$$- [E\{1[y_i > y_j, x_i'b > x_j'b]|z_j\} - Q(b)] - Q(b).$$

The first term $Q(b)$ yields the second-order matrix in the usual M-estimator, the middle two terms yield the asymptotic distribution for KRE, and that the last term with $\lambda(\cdot)$ is a negligible $o_p(N^{-1})$ term. This procedure of rewriting a double sum—a "U-statistic" (or "U-process" indexed by b; see Nolan and Pollard (1987, 1988) and Sherman (1994))—into single sums is called the *projection of U-statistic*; see, e.g., Serfling (1980) and Lehman (1999) for more on U-statistics. $Q_N(b)$ has an indicator function which is not differentiable. But $Q_N(b)$ has double sums, and one sum smooths the indicator function to yield $\sqrt{N}Q_N(b) = N^{-1/2} \sum_i q(b) + o_p(1)$, and the other provides a CLT to $N^{-1/2} \sum_i q(b)$. This is the key point for the $\sqrt{N}$-consistency and the asymptotic distribution of KRE.

Set $\beta_k = 1$ without loss of generality, and define

$$x_{kc} \equiv (x_1, ..., x_{k-1})', \quad \beta_{kc} \equiv (\beta_1, ..., \beta_{k-1})'.$$

Note that $sgn(\beta_k)$ can be treated as known, because any consistent estimator τ_N for $sgn(\beta_k)$ should satisfy $P\{\tau_N = sgn(\beta_k)\} \to 1$, which implies $P\{N^\nu|\tau_N - sgn(\beta_k)| \leq \varepsilon\} \to 1$ for any constant $\nu, \varepsilon > 0$. Hence, the estimation error $\tau_N - sgn(\beta_k)$ does not affect the asymptotic distribution of $\sqrt{N}(b_{kc} - \beta_{kc})$. Sherman (1993) proved

$$\sqrt{N}(b_{kc} - \beta_{kc}) \rightsquigarrow N(0, \; V_1^{-1}\Delta_1 V_1^{-1}) \quad \text{where}$$

$$V_1 \equiv E[\{x_{kc} - E(x_{kc}|x'\beta)\}\{x_{kc} - E(x_{kc}|x'\beta)\}'\ \psi(x'\beta)\ f_u(-x'\beta)\ ]$$

$$\Delta_1 \equiv E[\{x_{kc} - E(x_{kc}|x'\beta)\}\{x_{kc} - E(x_{kc}|x'\beta)\}'\ \psi(x'\beta)^2\ F_u(-x'\beta)$$

$$\{1 - F_u(-x'\beta)\}\ ]$$

$\psi(x'\beta)$ is the density of $x'\beta$ and $f_u = F'_u$.

Let $b_N \equiv (b'_{kc}, sgn(\beta_k))'$. In estimating the asymptotic variance, the main difficulty is in getting $f_u(-x'\beta)$, for u is not identified. But in binary response models, owing to $E(y|x'_i\beta) = 1 - F(-x'_i\beta)$, we can estimate $1 - F(-x'_i\beta)$ with

$$1 - F_N(-x'_i b_N) \equiv \frac{\sum_{j,j\neq i} K((x'_j b_N - x'_i b_N)/s) \cdot y_j}{\sum_{j,j\neq i} K((x'_j b_N - x'_i b_N)/s)}$$

where s is a smoothing parameter. Differentiating this wrt $x'_i b_N$, $f_u(-x'\beta)$ can be estimated by

$$f_N(-x'_i b_N) \;=\; -\frac{1}{s}\frac{\sum_{j,j\neq i} K'((x'_j b_N - x'_i b_N)/s)y_j}{\sum_{j,j\neq i} K((x'_j b_N - x'_i b_N)/s)}$$

$$+\frac{1 - F_N(-x'_i b_N)}{s}\frac{\sum_{j,j\neq i} K'((x'_j b_N - x'_i b_N)/s)}{\sum_{j,j\neq i} K((x'_j b_N - x'_i b_N)/s)}$$

where $K'(t) \equiv dK(t)/dt$. Alternatively, it may be simpler to use a numerical derivative: for a small positive constant ε,

$$\hat{f}_N(-x'_i b_N) = \frac{F_N(-x'_i b_N + \varepsilon) - F_N(-x'_i b_N - \varepsilon)}{2\varepsilon}.$$

The bandwidth for $f_N(-x'_i b_N)$ and $\hat{f}_N(-x'_i b_N)$ should be greater than the bandwidth for $1 - F_N(-x'_i b_N)$. In summary, V_1 may be estimated by (estimating Δ_1 is easier and so omitted)

$$\frac{1}{N}\sum_i \{x_{ikc} - E_N(x_{kc}|x'_i b_N)\}\{x_{ikc} - E_N(x_{kc}|x'_i b_N)\}'\ \psi_N(x'_i b_N)$$

$$\hat{f}_N(-x'_i b_N)\ \text{where}$$

$$E_N(x_{kc}|x'_i b_N) \equiv \frac{\sum_{j,j\neq i} K_{ji} x_{jkc}}{\sum_{j,j\neq i} K_{ji}}, \quad \psi_N(x'_i b_N) \equiv \frac{1}{(N-1)h}\sum_{j\neq i} K_{ji},$$

$$K_{ji} \equiv K\left(\frac{x'_j b_N - x'_i b_N}{h}\right).$$

Computationally, KRE needs an algorithm that does not require gradients (e.g., "downhill simplex"). If we want to use a derivative-based algorithm, we may smooth the maximand by replacing $1[x'_i b > x'_j b]$ with a smooth $J((x'_i b - x'_j b)/h)$ with $J(-\infty) = 0$ and $J(\infty) = 1$ and $h \to 0$; e.g.,

$J(\cdot) = \Phi(\cdot)$. This idea of smoothing already appeared in the "smoothed" MSC. Without smoothing, the KRE's maximand will return zero gradients as it consists of indicator functions. With $J((x_i'b - x_j'b)/h)$ used, the gradient is useful, but not the Hessian as it can be non n.d. That is, an algorithm using only gradients may work fine, but not the Newton–Raphson algorithm.

Although KRE is bandwidth-free, its asymptotic variance depends on the density functions of u and $x'\beta$, which require nonparametric estimation methods. In fact, if we use the above smoothing idea, a bandwidth will creep in even for the estimator itself. If an algorithm that does not require gradients is used and then followed by bootstrap inference, the bandwidth choice issue will be completely disposed of. Although not proven, this might be the best way to proceed because the above asymptotic variance estimator tends to over-estimate the variance in practice; the source of this problem seems to be $f_N(-x_i'b_N)$ in V_1 being too small relative to $F_N(-x_i'b_N)$ in Δ_1.

Although we focused on binary response models for KRE, KRE can be useful also for transformation-of-variable models where the scale is not identified; this will be shown next. In general, KRE works whenever the "sign-matching" idea works as evident in $Q_N(b)$.

3.2.3   Randomly Censored Duration with Unknown Transformation

Khan and Tamer (2007) proposed *"partial rank estimator (PRE)"* for randomly right-censored duration with an unknown transformation of the duration being linear in x and u. Specifically, for an unknown strictly increasing transformation $T(\cdot)$, suppose

$$T(y_i^*) = x_i'\beta + u_i, \quad T(c_i^*) = c_i, \quad y_i \equiv \min(y_i^*, c_i^*)$$
$$\Longrightarrow d_i \equiv 1[y_i^* \le c_i^*] = 1[T(y_i^*) \le T(c_i^*)] = 1[x_i'\beta + u \le c_i],$$
$$T(y_i) = T\{\min(y_i^*, c_i^*)\} = \min\{T(y_i^*), T(c_i^*)\} = \min(x_i'\beta + u_i, c_i)$$

where u is independent of x but its distribution unknown, and $y^* \amalg c^* | x$.

In terms of the "non-starred" variables, we have just

$$T(y_i) = \min(x_i'\beta + u_i, c_i) \quad \text{and} \quad d_i = 1[x_i'\beta + u_i \le c_i].$$

This model could have come from

$$y_i^* = x_i'\beta + u_i, \quad c_i^* = c_i, \quad T(y_i) = \min(y_i^*, c_i^*) \quad \text{and} \quad d_i \equiv 1[y_i^* \le c_i^*].$$

But thinking in this way would not make sense, because $T(y_i) = \min(y_i^*, c_i^*)$ is non-sensical: there is no reason to transform y_i which records merely the short end of y_i^* and c_i.

PRE is obtained by maximizing

$$\frac{1}{N(N-1)} \sum_{i \neq j} d_i 1[y_i < y_j] 1[x_i'b < x_j'b].$$

The adjective "partial" is used in PRE because censored observations are used "partially" as in Cox (1972) partial MLE. The appearance of d_i (but not d_j) can be understood from "$y_i < y_j$ iff $y_i^* < y_j^*$" under $d_i = 1$—right-censoring for y_j^* is irrelevant when $y_i < y_j$. This is the idea in Gehan (1965) that uses censored observations as much as possible, and this idea has been also adopted in Lee (2009).

An inefficient estimator that discards all censored observations can be thought of when $d_i d_j$ replaces d_i in the maximand. Abrevaya (1999) proposed an estimator reminiscent of this inefficient estimator for a truncated regression model: for unknown strictly increasing transformations $T(\cdot)$ and $G(\cdot)$,

$$T(y_i) = G(x_i'\beta) + u_i, \quad u_i \amalg x_i, \quad (x_i', y_i, t_i)' \text{ observed only when } y_i > t_i.$$

The estimator maximizes

$$\frac{1}{N(N-1)} \sum_{i \neq j} \delta_{ij} 1[y_i < y_j] 1[x_i'b < x_j'b] \quad \text{where}$$

$$\delta_{ij} = 1[y_i > t_j, \ y_j > t_i] = 1[y_i > \max(t_i, t_j), \ y_j > \max(t_i, t_j)].$$

Abrevaya (1999) proposed another version using the estimator in Cavanagh and Sherman (1998) which is explained in the following subsection.

Khan and Tamer (2007) also extended the estimator to interval censoring. Consider two random censoring durations c_{1i}^* and c_{2i}^* with $c_{1i}^* < c_{2i}^*$ such that the observed duration is $\max\{c_{1i}^*, \min(y_i^*, c_{2i}^*)\}$. Define

$$d_i = 1[c_{1i}^* \leq y_i^*] + 1[c_{2i}^* \leq y_i^*]$$
$$= 0 \cdot 1[y_i^* < c_{1i}^*] + 1 \cdot 1[c_{1i}^* \leq y_i^* < c_{2i}^*] + 2 \cdot 1[c_{2i}^* \leq y_i^*].$$

That is, d_i takes $0, 1, 2$ depending on which interval y_i^* falls on the real line marked by c_{1i}^* and c_{2i}^*. The question is when "$y_i < y_j$" implies "$y_i^* < y_j^*$," which is examined in the following.

Consider ("yes" means $y_i < y_j \implies y_i^* < y_j^*$)

$$d_i = 0, d_j = 1 : \text{yes} \quad \text{as } y_i^* < (c_{1i}^* =) \ y_i < y_j = y_j^*$$
$$d_i = 0, d_j = 2 : \text{yes} \quad \text{as } y_i^* < (c_{1i}^* =) \ y_i < y_j \ (= c_{2j}^*) \leq y_j^*$$
$$d_i = 1, d_j = 2 : \text{yes} \quad \text{as } y_i^* = y_i < y_j \ (= c_{2j}^*) \leq y_j^*$$
$$d_i = d_j : \text{yes only when } d_i = d_j = 1 \text{ as } y_i^* = y_i < y_j = y_j^*$$
$$d_i = 2, d_j = 0 : \text{no} \quad \text{as } y_i^* > y_i \ (= c_{2i}^*) < (c_{1j}^* =) \ y_j > y_j^*$$

$$d_i = 2, d_j = 1 : \text{no} \quad \text{as } y_i^* > y_i \; (= c_{2i}^*) < y_j = y_j^*$$
$$d_i = 1, d_j = 0 : \text{no} \quad \text{as } y_i^* = y_i < y_j \; (= c_{1j}^*) > y_j^*.$$

That is, only the cases with $d_i = 0, 1$ and $d_j = 1, 2$ are useful, and this yields the maximand

$$\frac{1}{N(N-1)} \sum_{i \neq j} \{1 - 1[d_i = 2]\}\{1 - 1[d_j = 0]\} \cdot 1[y_i < y_j] 1[x_i'b < x_j'b].$$

3.3   Spearman Rank Correlation Estimator (SRE)

Cavanagh and Sherman (1998) proposed a generalized version of KRE. Let $M(\cdot)$ be a *known increasing function* and define

$$R_N(a_i) \equiv \sum_{j=1}^{N} 1[a_j \leq a_i]$$

which is the rank of a_i in $\{a_1, ..., a_N\}$ under no ties. Cavanagh and Sherman (1998) maximized

$$\sum_i M(y_i) R_N(x_i'b) = \sum_i \sum_j M(y_i) 1[x_j'b \leq x_i'b].$$

Call this "*Spearman Rank Correlation Estimator (SRE)*" for a reason to be seen below. As in KRE, β is identified up to a positive constant and the intercept is not identified; x does not include 1.

The main condition for the consistency of this estimator is the *single index condition plus positive monotonicity*:

$$E\{M(y)|x\} = G(x'\beta) \text{ which is a non-constant, increasing}$$
$$\text{function of } x'\beta;$$

i.e., SRE works whenever there exists $M(\cdot)$ satisfying this display. In addition, as in MSC and KRE, there should be at least one regressor with unbounded support conditional on the other regressors. Because of this, "$\leq$" in the maximand can be replaced with "$<$." If desired, the increasing monotonicity of the single index can be verified with a nonparametric regression of $M(y)$ on $x'\hat{b}$ where $\hat{b}$ is the SRE.

Consider $y = C\{A(x'\beta, u)\}$ where C is increasing and $A(\cdot, \cdot)$ is strictly increasing in each of its arguments. The independence between x and u is sufficient (but not necessary) for the single-index positive-monotonicity condition for $E(y|x)$ as can be seen in $E(y|x) = \int C\{A(x'\beta, u)\} f_u(u) du$. For instance, the single-index positive monotonicity condition holds under

$$y_i = x_i'\beta + u_i, \quad u_i = \sigma(x_i)v_i, \quad \text{and} \quad v_i \amalg x_i$$

where $\sigma(x)$ is a heteroskedasticity factor and v's df is F_v. Although this linear model example makes SRE look non-restrictive, if we consider the binary version of this ($y^* = x'\beta + u$ and $y = 1[y^* \geq 0]$), then we get $E(y|x) = 1 - F_v\{-x'\beta/\sigma(x_i)\}$. This is not a single index model. Even if we replace $\sigma(x)$ with $\sigma(x'\beta)$ to get a single index model with $G(x'\beta) = 1 - F_v\{-x'\beta/\sigma(x_i'\beta)\}$, as mentioned in the preceding subsection, it is hard to justify why this should be increasing in $x'\beta$. Thus $x \amalg u$ will be maintained in the following.

To see the main idea of SRE, examine the population maximand

$$E[\ E\{M(y_i)1[x_j'b \leq x_i'b]\ |x_j\}\ ] = E[\ E\{G(x_i'\beta)1[x_j'b \leq x_i'b]\ |x_j\}\ ].$$

In the inner expected value conditional on x_j, the threshold $x_j'b$ is fixed (for any b). Call $x_i'b$ "large" if greater than $x_j'b$ and "small" otherwise. Since $G(x_i'\beta)$ increases as $x_i'\beta$ increases, one can maximize the maximand by equating $b = \beta$ so that $1[\cdot \leq x_i'\beta] = 1$ if $x_i'\beta$ is large and 0 otherwise. For any $b \neq \beta$, this matching of $1[\cdot \leq x_i'b]$ to a large $G(x_i'\beta)$ will fail for some x, making the maximand smaller.

Many choices are possible for $M(y_i)$. One choice is $M(y_i) = y_i$, and another is $M(y_i) = R_N(y_i) = \sum_m 1[y_m \leq y_i]$ if y_i's are continuously distributed, under which the maximand becomes

$$\sum_i R_N(y_i)R_N(x_i'b).$$

This maximand is reminiscent of the *Spearman rank correlation* between y and $x'b$, which is just the correlation coefficient between two variables" ranks. The main motivation for $M(y_i) = R_N(y_i)$ is to robustify the estimator to outliers in y. Robustification can be also done with a "winsorized version" of y: for some chosen constants $w_1 < w_2$,

$$M(y) = w_1 1[y < w_1] + y1[w_1 \leq y \leq w_2] + w_2 1[y > w_2].$$

When $M(y_i) = R_N(y_i)$, the maximand of SRE involves "triples" because

$$\sum_i R_N(y_i)R_N(x_i'b) = \sum_i \sum_j \sum_m 1[y_m \leq y_i]1[x_j'b \leq x_i'b].$$

The appearance of triples can be motivated also by

$$x_i'\beta > x_j'\beta \iff P(y_i > y_m|x_i, x_j, x_m) > P(y_j > y_m|x_i, x_j, x_m).$$

Compare this to the condition involving only a pair for KRE:

$$x_i'\beta > x_j'\beta \iff P(y_i > y_j|x_i, x_j) > P(y_j > y_i|x_i, x_j).$$

When $M(y_i)$ is either a deterministic function (such as the identity function) or $M(y_i) = R_N(y_i)$, the asymptotic distribution for b_{kc} in

$\hat{b} = (b'_{kc}, sgn(b_k))'$ is

$$\sqrt{N}(b_{kc} - \beta_{kc}) \rightsquigarrow N(0, V_2^{-1}\Delta_2 V_2^{-1}) \quad \text{where}$$

$$V_2 \equiv E\left[\{x_{kc} - E(x_{kc}|x'\beta)\}\{x_{kc} - E(x_{kc}|x'\beta)\}' \ \psi(x'\beta) \ \frac{\partial E(M(y)|x'\beta)}{\partial(x'\beta)}\right]$$

$$\Delta_2 \equiv E[\ \{x_{kc} - E(x_{kc}|x'\beta)\}\{x_{kc} - E(x_{kc}|x'\beta)\}' \ \psi(x'\beta)^2$$
$$\{M(y) - E(M(y)|x'\beta)\}^2 \ ]$$

and $\psi(x'\beta)$ is the density of $x'\beta$. An application of SRE and KRE as well as CLAD and SCLS to black–white earnings inequality appeared in Chay and Honoré (1998).

Cavanagh and Sherman (1998) showed that SRE becomes KRE for the binary response model with $M(y) = y$. Set $M(y) = y$ and $E(y|x'\beta) = 1 - F_u(-x'\beta)$ for binary response to get

$$\frac{\partial E(M(y)|x'\beta)}{\partial(x'\beta)} = \frac{\partial\{1 - F_u(-x'\beta)\}}{\partial(x'\beta)} = f_u(-x'\beta) \quad \text{and}$$
$$\{M(y) - E(M(y)|x'\beta)\}^2 = \{y - E(y|x'\beta)\}^2$$
$$\implies E[\{y - E(y|x'\beta)\}^2|x] = F_u(-x'\beta)\{1 - F_u(-x'\beta)\}.$$

With these plugged in, the asymptotic variance becomes that of KRE.

3.4   Pairwise-Difference Rank for Response Transformations

Recall that KRE is based on pairwise comparisons. We noted that the pairwise comparison was extended to a comparison with triples by SRE. Motivated by this, Abrevaya (2003) considered quadruple comparisons based on index *differences*, which is the topic of this subsection.

3.4.1   Main Idea and Estimator

For the transformation-of-variable model $\Lambda(y_i) = x_i'\beta + u_i$ with an unknown strictly increasing $\Lambda(\cdot)$, it holds that

$$\Lambda(y_i) - \Lambda(y_j) = (x_i - x_j)'\beta + u_i - u_j.$$

Assuming that $u_1, ..., u_N$ are iid with a continuous and strictly increasing df and $x \amalg u$, we get, since Λ is strictly increasing,

$$y_i > y_j|(x_i, x_j) = u_i - u_j > -(x_i - x_j)'\beta|(x_i, x_j)$$
$$\implies P(y_i > y_j|x_i, x_j) = T\{(x_i - x_j)'\beta\} \quad \text{where} \ T(t) \equiv P(u_i - u_j \leq t);$$

T is symmetric about 0.

From the last display,

$$(x_i - x_j)'\beta > (x_m - x_n)'\beta$$
$$\Longleftrightarrow \quad P(y_i > y_j | x_i, x_j, x_m, x_n) > P(y_m > y_n | x_i, x_j, x_m, x_n).$$

Using this idea, Abrevaya's (2003) *pairwise-difference rank estimator (PDE)* maximizes

$$\frac{1}{N(N-1)(N-2)(N-3)} \sum_{i \neq j \neq m \neq n} 1[(x_i - x_j)'b > (x_m - x_n)'b]$$

$$\cdot (1[y_i > y_j] - 1[y_m > y_n]) \equiv Q_{PDE}$$

where $\sum_{i \neq j \neq m \neq n}$ means the sum over the distinct ordered quadruples. As in KRE and SRE, x does not include 1.

As for the asymptotic variance of PDE, let

$$z_i \equiv (x_i', y_i)', \quad q_i \equiv x_i + x_j - x_m, \quad \psi(\cdot) \text{ be the density for } q'\beta$$
$$S(\nu, y) \equiv E_{z_i, z_j, z_m} \{sgn(y - y_i) - sgn(y_j - y_m) | q'\beta = \nu\},$$
$$S_1(\nu, y) \equiv \frac{\partial S(\nu, y)}{\partial \nu}.$$

With x_{kc} and q_{kc} denoting the elements of x and q other than x_k and q_k, respectively,

$$\sqrt{N}(b_{kc} - \beta_{kc}) \rightsquigarrow N(0, V_3^{-1} \Delta_3 V_3^{-1}) \quad \text{where}$$
$$V_3 \equiv E[\ E\{(x_{kc} - q_{kc})(x_{kc} - q_{kc})' | q'\beta = x'\beta\} \ \psi(x'\beta) \ S_1(x'\beta, y)\ ],$$
$$\Delta_3 \equiv E[\ \{x_{kc} - E(q_{kc}|q'\beta = x'\beta)\}\{x_{kc} - E(q_{kc}|q'\beta = x'\beta)\}'$$
$$\psi(x'\beta)^2 \ S(x'\beta, y)^2\ ].$$

In the outer-expected values of V_3 and Δ_3, the integrals are wrt $(x'\beta, y)$ and $(x_{kc}', x'\beta, y)$, respectively. Instead of the complicated asymptotic variance, bootstrap may be used, but bootstrap would be too costly in computation time since four sums are involved.

3.4.2   Remarks

In comparison to PDE, Abrevaya (2003) noted that SRE's maximand with $M(y) = R_N(y)$ can be written as

$$\frac{1}{N(N-1)(N-2)} \sum_{i \neq j \neq m} 1[x_i'b > x_j'b] \ (1[y_i > y_m] - 1[y_j > y_m]) \equiv Q_{SRE}.$$

Comparing Q_{KRE}, Q_{SRE}, and Q_{PDE}, one can see how the pairwise comparison "evolved." KRE, SRE, and PDE are all applicable to the transformation-
of-variable model and these estimators are likely to perform similarly, although
going from pairs to triples and then to quadruples may improve efficiency because the data get used more intensively at the cost of increasing computation
time.

Abrevaya (2003), in fact, also considered a triple-comparing estimator
of his own, that maximizes

$$\frac{1}{N(N-1)(N-2)} \sum_{i \neq j \neq m} 1[(x_i - x_j)'b > (x_j - x_m)'b]$$
$$(1[y_i > y_j] - 1[y_j > y_m]) \equiv Q_{PDE3}.$$

But differently from Q_{SRE} where i and j are compared symmetrically relative
to m, there is an "asymmetry" in Q_{PDE3}. This asymmetry seems to translate
to a rather complicated asymptotic variance, although the estimator is $\sqrt{N}$-
consistent and asymptotically normal as PDE is.

In introducing PDE, we motivated PDE with a transformation model.
A question that might arise is if PDE is applicable to binary responses. The
answer seems no. In Q_{PDE} for binary responses, it holds that

$$1[y_i > y_j] - 1[y_m > y_n] = 1 \iff y_i = 1, y_j = 0, y_m = 0, y_n = 0.$$

This would be compatible with $x_i'\beta > \max(x_j'\beta, x_m'\beta, x_n'\beta)$ which does not
imply $(x_i - x_j)'\beta > (x_m - x_n)'\beta$ in Q_{PDE}. In contrast, in Q_{SRE} for binary
responses, it holds that

$$1[y_i > y_m] - 1[y_j > y_m] = 1 \iff y_i = 1, y_j = 0, y_m = 0$$

which is compatible with $x_i'\beta > x_j'\beta$.

3.4.3   An Empirical Example

Abrevaya (2003) applied KRE, SRE, and PDE to baseball player duration data ($N = 702$) where y is the number of total games played (i.e., the
duration of career) and x consists of SLG (slugging proportion), OBP (on-
base proportion), AGE (age at which career began), MIDINF (one if middle
infielder), and CATCH (one if catcher). The last two dummy variables should
have a positive impact on y because middle infielders and catchers are valued
for their defensive as well as offensive skills. The descriptive statistics are in
Table 3

Part of his Table 6 is presented in Table 4 where the LSE is for $\ln y$.
The effect of SLG should be positive and its coefficient was normalized to
one, which is why no estimate for SLG appears in the table. Although not
presented here, LSE has its intercept estimate 3.56 and SLG estimate 8.19.
But for the sake of comparison, all LSE slope estimates were divided by 8.19
for the table.

Table 3: Baseball Data Mean and SD

	y	SLG	OBP	AGE	MIDINF	CATCH
Mean	769	0.349	0.308	23.1	0.228	0.147
SD	678	0.061	0.036	2.0	0.420	0.354

Table 4: Baseball Data Estimates (SD in $(\cdot)$)

Variables	KRE	PDE	LSE
OBP	1.04 (0.25)	1.06 (0.21)	1.08 (0.21)
AGE	−0.0144 (0.0029)	−0.0162 (0.0023)	−0.0168 (0.0025)
MIDINF	0.0737 (0.012)	0.0732 (0.0097)	0.0742 (0.0094)
CATCH	0.0322 (0.013)	0.0343 (0.011)	0.0412 (0.011)

A few remarks are in order. First, the effects of SLG and OBP are almost the same. Second, AGE has a negative impact, and MIDINF and CATCH have positive impacts; these findings match intuition. Third, PDE using quadruples is slightly more efficient than KRE, and PDE is as efficient as LSE. This observation matters much, as it dissipates the impression that semiparametric estimators tend to be inefficient compared with parametric ones and LSE.

3.5   Rank-Based Estimation of Transformation Function

Consider a "transformation of (response) variable" model:

$$\Lambda(y_i) = x_i'\beta + u_i, \quad \text{where } \Lambda(\cdot) \text{ is an increasing transformation}$$
$$\Lambda(y_c) = 0 \text{ for some } y_c \text{ (location normalization)},$$

u_i has a continuous and strictly increasing df F_u and $u_i \amalg x_i$.

Given a $\sqrt{N}$-consistent estimator b_N for β as those in the previous sections, Chen (2002) proposed a $\sqrt{N}$-consistent estimator for $\Lambda(\cdot)$ that does not require any bandwidth. The estimator is easier to implement than other estimators in the literature (e.g., Horowitz, 1996), and it allows $\Lambda(\cdot)$ to be nondifferentiable and y to be discrete. Chen (2002) extended the estimator to censored responses where c is the censoring variable, but this extension requires $c \amalg (x, y)$ which is too restrictive, rather than the usual $c \amalg y | x$. In this subsection, we review the Chen's estimator under no censoring.

Observe

$$1[y_i \geq y_o] = 1[\Lambda(y_i) \geq \Lambda(y_o)] = 1[x_i'\beta + u_i \geq \Lambda(y_o)]$$
$$\implies E\{1[y_i \geq y_o]|x_i\} = E\{1[u_i \geq \Lambda(y_o) - x_i'\beta]|x_i\}$$
$$= 1 - F_u\{\Lambda(y_o) - x_i'\beta\}.$$

From this, we get

$$E\{1[y_i \geq y_o] - 1[y_j \geq y_c]|x_i, x_j\} \quad \geq \quad 0 \text{ iff } \Lambda(y_c) - x_j'\beta \geq \Lambda(y_o) - x_i'\beta$$
$$\Longleftrightarrow \quad x_i'\beta - x_j'\beta \geq \Lambda(y_o).$$

Hence an estimator $\Lambda_N(y_o)$ for $\Lambda(y_o)$ is obtained by maximizing wrt Λ

$$\frac{1}{N(N-1)} \sum_{i \neq j} (1[y_i \geq y_o] - 1[y_j \geq y_c]) \cdot 1[x_i'b_N - x_j'b_N \geq \Lambda]$$

where one choice for y_c is $Med(y)$ as done in Chen (2002).

To better understand the normalization "$\Lambda(y_c) = 0$ for some y_c," suppose that $b_N = (b_{-k}', 1)'$ is KRE; i.e., $\beta_k > 0$ is used for the scale normalization of KRE. Start with $\Lambda^*(y_i) = x_i'\beta^* + u_i^*$ where x_i includes 1 as its first element to observe

$$\frac{\Lambda^*(y_i)}{\beta_k^*} = x_i'\frac{\beta^*}{\beta_k^*} + \frac{u_i^*}{\beta_k^*} \implies \frac{\Lambda^*(y_i)}{\beta_k^*} - \frac{\Lambda^*(y_c)}{\beta_k^*} = x_i'\frac{\beta^*}{\beta_k^*} - \frac{\Lambda^*(y_c)}{\beta_k^*} + \frac{u_i^*}{\beta_k^*}$$

$$\implies \Lambda(y_i) = x_i'\beta + u_i, \quad \text{where } \Lambda(t) \equiv \frac{\Lambda^*(y_i)}{\beta_k^*} - \frac{\Lambda^*(y_c)}{\beta_k^*},$$

$$\beta \equiv \frac{\beta^*}{\beta_k^*} \text{ and } u_i \equiv \frac{u_i^*}{\beta_k^*}.$$

Hence $\Lambda(y_c) = 0$. The intercept shifts by $-\Lambda^*(y_c)/\beta_k^*$, but the intercept does not matter as it drops out of $x_i'b_N - x_j'b_N$ in the above maximand. The identified transformation $\Lambda(\cdot)$ changes as β_k does.

With $z \equiv (x', y)'$, define

$$H(z_i, z_j, y_o, \Lambda, b) \equiv (1[y_i \geq y_o] - 1[y_j \geq y_c]) \cdot 1[x_i'b - x_j'b \geq \Lambda]$$

$$\tau(z, y_o, \Lambda, b) \equiv E_{z_j}\{H(z, z_j, y_o, \Lambda, b)\} + E_{z_i}\{H(z_i, z, y_o, \Lambda, b)\}$$

$$S(y_o) \equiv \frac{1}{2}E_z \left\{ \frac{\partial^2 \tau(z, y_o, \Lambda(y_o), \beta)}{\partial \Lambda^2} \right\}$$

where $E_z\{\cdot\}$ indicates that the expectation is taken only wrt z. Then,

$$\sqrt{N}\{\Lambda_N(y_o) - \Lambda(y_o)\} \rightsquigarrow N\left(0, \frac{E_z[\{\partial \tau(z, y_o, \Lambda(y_o), \beta)/\partial \Lambda\}^2]}{S(y_o)^2}\right).$$

The first-stage estimator b_N has no effect on this asymptotic distribution.

Instead of estimating a nonparametric $\Lambda(\cdot)$ at chosen points, a more parametric approach is specifying $\Lambda(y) \equiv \Gamma(y, \theta)$ for some parametric function Γ with a parameter θ as in the Box-Cox transformation. This has been in fact explored by Han (1987b) maximizing wrt θ

$$\frac{1}{N(N-1)(N-2)(N-3)} \sum_{i \neq j \neq m \neq n} 1[(x_i - x_j)'b_{KRE} > (x_m - x_n)'b_{KRE}]$$

$$\cdot \, 1[\Gamma(y_i, \theta) - \Gamma(y_j, \theta) > \Gamma(y_m, \theta) - \Gamma(y_n, \theta)].$$

The asymptotic distribution for this estimator was later derived by Asparouhova et al. (2002), who also suggested a different estimator: maximize wrt θ

$$\sum_{i \neq j} R_N \{\Gamma(y_i, \theta) - \Gamma(y_j, \theta)\} \cdot M\{(x_i - x_j)'b_N\}$$

where $M(\cdot)$ is an increasing function, b_N is SRE, and $R_N(a_{ij}) \equiv \sum_{s \neq t} 1[a_{st} \leq a_{ij}]$.

4   Differencing-Based Estimators

This section reviews two differencing estimators. Although the rank-based estimators in the preceding section can be viewed also as (pairwise) differencing estimators, the characteristics of the estimators in this section differ from those in the preceding section. Section 4.1 examines a pairwise differencing estimator for censored/truncated models where the symmetry resulting from differencing plays the key role. Section 4.2 introduces a differencing estimator for a partially linear model $y = \rho(z) + x'\beta + u$ where $\rho(\cdot)$ is unknown and z is a rv. The estimator orders the observations using z and first-difference the model (as in time-series) to get rid of ρ and then estimate β. Partially linear models will be discussed further in the next chapter.

4.1   Pairwise-Difference for Censored and Truncated Models

4.1.1   Differencing Idea

Recall the regression model censored at zero from below:

$$y_i^* = x_i'\beta + u_i, \quad y_i = \max(y_i^*, 0), \quad (x_i', y_i), \; i = 1, ..., N, \; \text{observed.}$$

Note that

$$y - x'\beta = \max(x'\beta + u, 0) - x'\beta = \max(u, -x'\beta).$$

Whereas the "structural form" error is $u \, (= y^* - x'\beta)$, the "reduced-form" error is $y - x'\beta = \max(u, -x'\beta)$, the use of which forms the basis for the following pairwise differencing idea.

Suppose that $u_1, ..., u_N$ are iid as usual, and $u \amalg x$. Although u_i and u_j are iid given x_i and x_j thanks to $u \amalg x$, the conditional RF errors $\max(u_i, -x_i'\beta)$ and $\max(u_j, -x_j'\beta)$ are not, because their trimming points $-x_i'\beta$ and $-x_j'\beta$ differ. Consider instead their artificially trimmed versions

$$e_{ij}(\beta) \equiv \max(u_i, -x_i'\beta, -x_j'\beta) \quad \text{and} \quad e_{ji}(\beta) \equiv \max(u_j, -x_j'\beta, -x_i'\beta)$$

which are iid given x_i and x_j. Then, because $e_{ij}(\beta)$ and $e_{ji}(\beta)$ are iid, $\{e_{ij}(\beta) - e_{ji}(\beta)\}|(x_i, x_j)$ is symmetric about 0 as can be seen in (omitting "$|x_i, x_j$")

$$P\{e_{ij}(\beta) - e_{ji}(\beta) \le -a\} = P\{e_{ji}(\beta) - e_{ij}(\beta) \le -a\}$$

(as $e_{ij}(\beta)$ and $e_{ji}(\beta)$ are iid)

$$= P\{e_{ij}(\beta) - e_{ji}(\beta) \ge a\} \quad \text{(multiplying both sides by } -1).$$

Consider an *odd function* $\zeta(\cdot)$ (i.e., $-\zeta(-a) = \zeta(a)$) and observe (omitting "$|x_i, x_j$" again)

$$P[\zeta\{e_{ij}(\beta) - e_{ji}(\beta)\} \le -a] = P[\zeta\{e_{ji}(\beta) - e_{ij}(\beta)\} \le -a]$$

(as $e_{ij}(\beta), e_{ji}(\beta)$ are iid)

$$= P[-\zeta\{e_{ij}(\beta) - e_{ji}(\beta)\} \le -a]$$

$$= P[\zeta\{e_{ij}(\beta) - e_{ji}(\beta)\} \ge a] \quad \text{(multiplying both sides by } -1):$$

$\zeta\{e_{ij}(\beta) - e_{ji}(\beta)\}|(x_i, x_j)$ is symmetric about 0 as well. Hence we get a moment condition

$$E[\, \zeta\{e_{ij}(\beta) - e_{ji}(\beta)\}|x_i, x_j \,] = 0.$$

If u were dependent on x, e.g., $u = \sigma(x'\beta)\varepsilon$ where $\varepsilon \amalg x$, then the artificially trimmed errors would not be iid, which shows why $u \amalg x$ is needed. Honoré and Powell (1994) used this moment condition as a basis to design estimators for censored and truncated regression models as follows. The assumption $u\amalg x$ is stronger than those for CLAD and WME (and QME) with a small tuning constant, but neither stronger nor weaker than the full symmetry for SCLS and STLS.

4.1.2   Censored Regression

For the censored regression model, one minimand that has the above moment condition as its asymptotic first-order condition is

$$S_N(b) \equiv \frac{2}{N(N-1)} \sum_{i<j} s\{y_i, y_j, (x_i - x_j)'b\} \quad \text{where}$$

$$s\{y_i, y_j, (x_i - x_j)'b\} = y_i^2 - \{y_j + (x_i - x_j)'b\}2y_i \text{ if } (x_i - x_j)'b \le -y_j$$

$$= \{y_i - y_j - (x_i - x_j)'b\}^2 \text{ if } -y_j < (x_i - x_j)'b < y_i$$

$$= (-y_j)^2 + \{(x_i - x_j)'b - y_i\}2y_j \text{ if } y_i \le (x_i - x_j)'b.$$

Examine the term for $-y_j < (x_i - x_j)'b < y_i \Longleftrightarrow y_j - x_j'b > -x_i'b$ and $-x_j'b < y_i - x_i'b$ under $b = \beta$:

$$\{y_i - y_j - (x_i - x_j)'\beta\}^2 = \{e_{ij}(\beta) - e_{ji}(\beta)\}^2.$$

From this, the derivative would yield $E\{e_{ij}(\beta) - e_{ji}(\beta)|x_i, x_j\} = 0$ as intended. To understand the first term ("left-tail) for $(x_i - x_j)'\beta \le -y_j$, substitute the "boundary equality" $(x_i - x_j)'\beta = -y_j$ of this inequality into $\{y_i - y_j - (x_i - x_j)'\beta\}^2$ to get y_i^2 which is the first part of $y_i^2 - \{y_j + (x_i - x_j)'\beta\}2y_i$; the second part is a linear function β so that the quadratic function in the middle portion branches out to the left-tail smoothly. If we recall QME and WME, the analogy is that QME uses only a constant for the left-tail whereas WME uses the constant plus a linear function of b. The analogy with QME and WME is more fitting than it looks, because using only the part y_i^2 in the left-tail works for the truncated regression model examined below. The third term in $S_N(b)$ can be understood analogously. Since this estimator is a differencing estimator, x should not include 1.

As for the asymptotic distribution, Theorem 3.2 in Honoré and Powell (1994) shows that

$$\sqrt{N}(b_N - \beta) \rightsquigarrow N(0, H^{-1}VH^{-1})$$

where H and V can be consistently estimated with, respectively,

$$V_N \equiv \frac{4}{N}\sum_i \hat{r}_i \hat{r}_i', \quad \text{where}$$

$$\hat{r}_i \equiv \frac{2}{(N-1)}\sum_{j=1, j\neq i}^{N} \{e_{ij}(b_N) - e_{ji}(b_N)\}(x_i - x_j)$$

$$H_N \equiv \frac{2}{N(N-1)}\sum_{i<j} 2(x_i - x_j)(x_i - x_j)'.$$

The "pesky" constants 4 and 2's all disappear in $H_N^{-1}V_N H_N^{-1}$ as an analogous cancelation happens in LSE. In fact, this sort of cancelation happened to the other estimators based on multiple sums. The above estimator is the LSE version with $\zeta(a) = 2a$, which is the derivative of a^2 for the squared loss function in LSE. Honoré and Powell (1994) also explore the LAD version using $|a|$ instead of a^2.

4.1.3   Truncated Regression

For the truncated regression model where $(y_i^*, x_i')'$ is observed only when $y_i^* > 0 \iff u_i > -x_i'\beta$, a modified version of $S_N(b)$ is

$$T_N(b) \equiv \frac{2}{N(N-1)}\sum_{i<j} t\{y_i, y_j, (x_i - x_j)'b\} \quad \text{where}$$

$$t\{y_i, y_j, (x_i - x_j)'b\} = y_i^2 \text{ if } (x_i - x_j)'b \leq -y_j$$
$$= \{y_i - y_j - (x_i - x_j)'b\}^2 \text{ if } -y_j < (x_i - x_j)'b < y_i$$
$$= (-y_j)^2 \text{ if } y_i \leq (x_i - x_j)'b.$$

This estimator requires the $\ln f_u$ to be strictly concave. As noted above, the tail portions of $t\{y_i, y_j, (x_i - x_j)'b\}$ are y_i^2 or y_j^2; no linear function of b appears there.

As for the asymptotic distribution, it is asymptotically normal with the variance taking the same form $\bar{H}^{-1}\bar{V}\bar{H}^{-1}$ as in the censored regression estimator. To estimate $\bar{V}$ and $\bar{H}$, use $\bar{V}_N \equiv 4N^{-1}\sum_i \hat{r}_i \hat{r}_i'$ for V where

$$\hat{r}_i \equiv \frac{2}{(N-1)} \sum_{j=1,j\neq i}^{N} -1[-y_j < (x_i - x_j)'b_N < y_i]$$

$$\{e_{ij}(b_N) - e_{ji}(b_N)\}(x_i - x_j).$$

Regarding $\bar{H}$, numerical derivatives can be used. For instance, the ath column of the estimate for $\bar{H}$ is

$$\frac{2}{N(N-1)}\frac{1}{2h}\sum_{i<j}\{-1[-y_j < (x_i - x_j)'(b_N + h\lambda_a) < y_i]$$

$$\cdot 2(e_{ij}(b_N + h\lambda_a) - e_{ji}(b_N + h\lambda_a))(x_i - x_j)$$

$$+ 1[-y_j < (x_i - x_j)'(b_N - h\lambda_a) < y_i]$$

$$\cdot 2(e_{ij}(b_N - h\lambda_a) - e_{ji}(b_N - h\lambda_a))(x_i - x_j)\}$$

where λ_a is the basis vector with 1 in its ath position and 0's elsewhere.

The assumption $x \amalg u$ and the relatively complicated asymptotic variance puts the pairwise-differencing estimators at disadvantage compared with the other semiparametric estimators for censored/ truncated models. But the idea of pairwise-differencing and symmetric trimming has been fruitfully applied to censored-selection models as shown already, and also to "fixed-effect" panel-data censored/truncated models as can be seen in Hóno2re and Kyriazidou (2000).

4.2   Differencing Estimator for Semi-linear Models

The "extreme" way to relax the usual linear model assumption is not specifying $E(y|x)$ at all to go fully nonparametric. A less extreme way is using a *semi-linear (or partially linear) model*, e.g., $y_i = \rho(z_i)+x_i'\beta+u_i$ where $\rho(z)$ is an unknown function of a regressor (vector) z. In a model like this, the goal is then estimating β with $\rho(\cdot)$ and the error term distribution unspecified. While semi-linear models will be examined more closely in the following chapter, if z is a continuously distributed rv, then there is a bandwidth-free estimator for β, which is introduced in this subsection.

Suppose

$$y_i = \rho(z_i) + x_i'\beta + u_i, \quad x_i = \mu(z_i) + v_i, \quad E(u|x,z) = 0, \ E(v|z) = 0,$$

where $\rho(\cdot)$ and $\mu(z)$ are unknown smooth functions ($\mu(z)$ is a vector), and z_i is a continuously distributed rv, ordered such that $z_1 \leq, ..., \leq z_N$.

There is no restriction in the model $x = \mu(z) + v$ with $E(v|x) = 0$, because x can be always written this way:

$$x = E(x|z) + x - E(x|z) = \mu(z) + v, \text{ where } E(x|z) \equiv \mu(z) \text{ and}$$
$$v \equiv x - E(x|z).$$

Note that $0 = E(u|x, z) = E(u|z, v)$, which implies $COR(u, v) = 0$.

Cross-section data have no natural ordering but z provides one. Using the ordering, difference the y equation:

$$y_i - y_{i-1} = \rho(z_i) - \rho(z_{i-1}) + (x_i - x_{i-1})'\beta + u_i - u_{i-1}, \quad i = 2, ..., N.$$

The LSE of $x_i - x_{i-1}$ on $y_i - y_{i-1}$ provides a $\sqrt{N}$-consistent estimator for β despite $\rho(z_i) - \rho(z_{i-1})$; see Yatchew (1997, 2003) and the references therein for more on this type of difference-based semiparametric methods; an empirical application can be seen in Yatchew and No (2001). The LSE works because $\rho(z_i) \simeq \rho(z_{i-1})$ as $z_i \simeq z_{i-1}$ due to the ordering when $N \to \infty$, whereas $x_i \neq x_{i-1}$ as x_i's are not ordered.

Let $\Delta y_i \equiv y_i - y_{i-1}$ and define Δx_i, Δu_i, and Δv_i analogously. Without loss of generality, suppose $z \sim U[0, 1]$; if not, rewrite $\rho(z) = \rho\{F_z^{-1}(F_z(z))\} \equiv \tilde{\rho}(F_z(z))$ and examine $F_z(z)$ instead of z where F_z is the df of z. Since z is not used explicitly in estimation, not knowing F_z does not pose any problem. The LSE is

$$b_N \equiv \left(\frac{1}{N-1} \sum_i \Delta x_i \Delta x_i' \right)^{-1} \frac{1}{N-1} \sum_i \Delta x_i \Delta y_i$$

$$= \left(\frac{1}{N-1} \sum_i \Delta x_i \Delta x_i' \right)^{-1}$$

$$\cdot \frac{1}{N-1} \sum_i \Delta x_i \{\rho(z_i) - \rho(z_{i-1}) + \Delta x_i'\beta + \Delta u_i\}$$

$$= \beta + \left(\frac{1}{N-1} \sum_i \Delta x_i \Delta x_i' \right)^{-1}$$

$$\cdot \frac{1}{N-1} \sum_i [\Delta x_i \{\rho(z_i) - \rho(z_{i-1})\} + \Delta x_i \Delta u_i].$$

The term involving Δu_i converges to zero in probability. Examine thus only

$$\frac{1}{N-1} \sum_i \Delta x_i \{\rho(z_i) - \rho(z_{i-1})\}$$

$$= \frac{1}{N-1} \sum_i \{\mu(z_i) - \mu(z_{i-1})\}\{\rho(z_i) - \rho(z_{i-1})\} + o_p(1)$$

because $COR(v, z) = 0$. If the probability limit of this term is zero, then $b_N \to^p \beta$.

With $\mu(z) = (\mu_1(z), ..., \mu_k(z))'$, the kth row of the last display is smaller than

$$\frac{1}{N-1} \sum_i \max_{2 \le i \le N} |\{\mu_k(z_i) - \mu_k(z_{i-1})\}\{\rho(z_i) - \rho(z_{i-1})\}|$$

$$= \max_{2 \le i \le N} |\{\mu_k(z_i) - \mu_k(z_{i-1})\}\{\rho(z_i) - \rho(z_{i-1})\}|$$

$$\le L_k L_\rho \max_{2 \le i \le N} |z_i - z_{i-1}|^2$$

assuming that both $\mu_k(z)$ and $\rho(z)$ are Lipschitz-continuous for some constants L_k and L_ρ:

$$|\mu_k(z_a) - \mu_k(z_b)| \le L_k|z_a - z_b| \text{ and } |\rho(z_a) - \rho(z_b)| \le L_\rho|z_a - z_b| \quad \forall z_a, z_b.$$

Although there is no restriction in $x = \mu(z) + v$, now a restriction is put on $\mu(z)$ due to the Lipschitz continuity.

When $z_1, ..., z_N$ fall on the unit interval, splitting $[0, 1]$ into intervals of size ε to get about ε^{-1}-many such intervals, we get

$\max_{2 \le i \le N} |z_i - z_{i-1}| > 2\varepsilon$ implies there is at least one interval of size

ε with no datum.

$\Longrightarrow P(\max_{2 \le i \le N} |z_i - z_{i-1}| > 2\varepsilon) \le P(\text{there is at least one interval of size}$

ε with no datum)

$\le P(\text{1st } \varepsilon \text{ interval with no datum})$

$+ P(\text{2nd } \varepsilon \text{ interval with no datum}) +, ... = (1 - \varepsilon)^N \frac{1}{\varepsilon} \to 0$ as $N \to \infty$.

Thus

$$P\left(\frac{1}{N-1} \sum_i \max_{2 \le i \le N} |\{\mu_k(z_i) - \mu_k(z_{i-1})\}\{\rho(z_i) - \rho(z_{i-1})\}| > 2\varepsilon\right) \to 0,$$

proving $b_N \to^p \beta$.

Turning to the asymptotic distribution, observe

$$\sqrt{N}(b_N - \beta) = \left(\frac{1}{N-1} \sum_i \Delta x_i \Delta x_i'\right)^{-1}$$

$$\frac{1}{\sqrt{N}} \sum_i [\Delta x_i \{\rho(z_i) - \rho(z_{i-1})\} + \Delta x_i \Delta u_i] + o_p(1).$$

Substituting $\Delta x_i = \Delta \mu(z_i) + \Delta v_i$, any term involving $\Delta \mu(z)$ or $\Delta \rho(z)$ is negligible, for $P(\max_{2 \le i \le N} |z_i - z_{i-1}| > \varepsilon) \to 0$ exponentially fast as shown above. This leaves only

$$\sqrt{N}(b_N - \beta) = E^{-1}(\Delta v \Delta v') \frac{1}{\sqrt{N}} \sum_i (\Delta v_i \Delta u_i) + o_p(1).$$

Using $COR(u,v) = 0$, we get $E(\Delta v \Delta u) = 0$. But across i, $\Delta v_2, ..., \Delta v_N$ are dependent (e.g., $v_2 - v_1$ and $v_3 - v_2$ are dependent), and the same holds for Δu_i. CLT's are known to hold for this type of "weak dependence" that applies only to adjacent terms. Hence,

$$\sqrt{N}(b_N - \beta) \rightsquigarrow N[0, \ E^{-1}(\Delta v \Delta v') \cdot V \cdot E^{-1}(\Delta v \Delta v')] \qquad \text{where}$$
$$V \equiv E\{\Delta v_i \Delta v_i' (\Delta u_i)^2\} + E\{\Delta v_i \Delta v_{i+1}' \Delta u_i \Delta u_{i+1}\}$$
$$+ \ E\{\Delta v_{i+1} \Delta v_i' \Delta u_{i+1} \Delta u_i\}.$$

A consistent estimator for the inverted Hessian is $\{(N-1)^{-1} \sum_{i=2}^{N} \Delta x_i \Delta x_i'\}^{-1}$ and a consistent estimator for V is

$$\frac{1}{N-1} \sum_{i=2}^{N} \Delta x_i \Delta x_i' (\Delta y_i - \Delta x_i' b_N)^2$$

$$+ \frac{1}{N-2} \sum_{i=3}^{N} \Delta x_i \Delta x_{i+1}' (\Delta y_i - \Delta x_i' b_N)(\Delta y_{i+1} - \Delta x_{i+1}' b_N)$$

$$+ \frac{1}{N-2} \sum_{i=3}^{N} \Delta x_{i+1} \Delta x_i' (\Delta y_{i+1} - \Delta x_{i+1}' b_N)(\Delta y_i - \Delta x_i' b_N).$$

5   Estimators for Duration Models

Some parametric estimators have been introduced for durations. The most controversial issue there has been the baseline hazard specification. This section introduces two relatively simple semiparametric estimators. One estimator specifies the baseline hazard as a discrete step-shaped function but otherwise unknown. The other leaves the baseline hazard completely arbitrary. Although the former can be applied to continuous time, it fits better discrete time. For this, we review discrete-time duration framework first.

5.1   Discrete Durations

Discrete duration can occur for two reasons. First, duration ends indeed at discrete times only, say, $t_1, ..., t_J$; e.g., job contracts may be on yearly basis such that employees may quit (or get fired) only at the end of each year. Second, duration occurs continuously but observed only at discrete times; for instance, machines can fail any time, but they are checked only periodically. Unless otherwise mentioned, we will assume the latter, as it relates better to the earlier LDV models.

Imagine a latent continuously distributed duration y^* with its survival function $S^*(t) \equiv P(y^* > t)$. Suppose that the observed duration y is discrete, taking on $0 \equiv t_0 = t_1 <, ..., < t_J$ and accumulating the y^* density over each interval such that

$$p_j \equiv P(y = t_j) = P(t_j \le y^* < t_{j+1}), \quad j = 1, ..., J, \quad \text{and } p_0 \equiv 0.$$

Further define

$$S(t_j^-) \equiv P(y \ge t_j) = p_j+, ..., +p_J = P(y > t_{j-1}) = S(t_{j-1}),$$
$$j = 2, ..., J; S(t_0^-) \equiv 1, \ S(t_0) \equiv 1.$$

As will become clear shortly, having t_0 and $p_0 = 0$ is notationally convenient, and for this, we deliberately set $0 = t_0 = t_1$ and $S(t_0^-) = 1 = S(t_0)$. If $t_1, ..., t_J$ are equally spaced, then instead of $t_1, ..., t_J$, we can simply use $1, ..., J$ for the values that y takes on. In this case, "latent discrete" duration is $t_1, ..., t_J$ whereas the "observed discrete" cardinal duration is $1, ..., J$. Even if $t_1, ..., t_J$ are not equally spaced, we can still denote the observed duration as $1, ..., J$, which is then a (grouped) ordinal duration.

Note $p_j = S(t_{j-1}) - S(t_j)$, and let, for $j = 1, ..., J$,

$$\lambda_j \equiv P(y = t_j | y \ge t_j) = \frac{P(y = t_j, y \ge t_j)}{P(y \ge t_j)} = \frac{P(y = t_j)}{P(y \ge t_j)} = \frac{p_j}{S(t_{j-1})}$$
$$\Longleftrightarrow \ p_j = \lambda_j S(t_{j-1});$$
$$S(t_{j-1}) = \frac{S(t_{j-1})}{S(t_0)} = \frac{S(t_{j-1})}{S(t_{j-2})} \frac{S(t_{j-2})}{S(t_{t-3})} \cdots \frac{S(t_1)}{S(t_0)} = \prod_{a=1}^{j-1} \frac{S(t_a)}{S(t_{a-1})}$$
$$= \prod_{a=1}^{j-1} \frac{S(t_{a-1}) - p_a}{S(t_{a-1})} = \prod_{a=1}^{j-1} \{1 - \frac{p_a}{S(t_{a-1})}\} = \prod_{a=1}^{j-1} (1 - \lambda_a)$$

which is the discrete analog for $S(t) = \exp\{-\Lambda(t)\}$ in continuous time case. Parametrizing p_j with $p_j = p_j(x_i, \theta)$, x_i can be taken into account and the parameter θ can be estimated. Alternatively, we can parametrize λ_j with $\lambda_j = \lambda_j(x_i, \theta)$. As seen in the parametric duration analysis, it is more convenient to parametrize λ_j, because the likelihood function can be written in terms of λ_j's only.

Suppose (x_i, y_i, d_i) is observed, $i = 1, ..., N$, and y_i is an event duration when $d_i = 1$ and an censoring duration when $d_i = 0$. Defining $y_{ij} = 1$ if $y_i = t_j$ and 0 otherwise (as in ODR), the log-likelihood function is

$$\sum_{i=1}^{N} \left\{ d_i \sum_{j=1}^{J} y_{ij} \ln p_j(x_i, \theta) + (1 - d_i) \sum_{j=1}^{J} y_{ij} \ln S(t_{j-1}, x_i, \theta) \right\}.$$

This log-likelihood function is similar to that of ODR, with one visible difference being y taking $1, ..., J$ instead of $0, 1, ...J - 1$ as $p_0 = P(y = 0) = 0$. Substituting $p_j = \lambda_j S(t_{j-1})$, the log-likelihood function becomes

$$
L_N(\theta) \equiv \sum_{i=1}^{N} \left\{ d_i \sum_{j=1}^{J} y_{ij} \ln \lambda_j(x_i, \theta) + \sum_{j=1}^{J} y_{ij} \ln S(t_{j-1}, x_i, \theta) \right\}
$$

$$
= \sum_{i=1}^{N} \left(d_i \sum_{j=1}^{J} y_{ij} \ln \lambda_j(x_i, \theta) + \sum_{j=1}^{J} y_{ij} \sum_{a=1}^{j-1} \ln\{1 - \lambda_a(x_i, \theta)\} \right).
$$

As already noted when parametric durations were studied, sometime left-truncation occurs as well: only durations greater than a certain threshold are observed. Differently from right-censoring, however, there is no information on x for those left-truncated; recall the distinction between truncation and censoring. For instance, suppose we observe the lifespan of old persons in a nursing home where an entry requirement to the nursing home is $age \geq 60$. In this case, those who died before 60 will not be in the data set. The two problems combined is often called "LTRC" (left-truncated and right-censored). Under left-truncation, all probabilities should be conditioned on the truncation event, say $y \geq t_c$, where t_c ($\geq t_1$) is the truncation point.

As in the no-truncation case, we still have

$$
\lambda_j \equiv P(y = t_j | y \geq t_j) = \frac{P(y = t_j, \, y \geq t_j)}{P(y \geq t_j)} = \frac{p_j}{S(t_{j-1})}.
$$

Define

$$
\pi_j \equiv P(y = t_j | y \geq t_c), \, j = c, ..., J
$$

$$
\implies S(t_{j-1} | y \geq t_c) = P(y \geq t_j | y \geq t_c) = \frac{P(y \geq t_j, \, y \geq t_c)}{P(y \geq t_c)}
$$

$$
= \frac{S(t_{j-1})}{S(t_{c-1})} = \frac{\prod_{a=1}^{j-1}(1 - \lambda_a)}{\prod_{a=1}^{c-1}(1 - \lambda_a)} = \prod_{a=c}^{j-1}(1 - \lambda_a).
$$

By the construction of π_j, for $j = c, ..., J$,

$$
\pi_j S(t_{c-1}) = P(y = t_j | y \geq t_c) S(t_{c-1}) = P(y = t_j, y \geq t_c) = P(y = t_j) = p_j
$$

$$
\implies \pi_j = \frac{p_j}{S(t_{c-1})} = \frac{\lambda_j S(t_{j-1})}{S(t_{c-1})}.
$$

The log-likelihood function is

$$
\sum_{i=1}^{N} \left\{ d_i \sum_{j=c}^{J} y_{ij} \ln \pi_j + (1 - d_i) \sum_{j=c}^{J} y_{ij} \ln S(t_{j-1} | y \geq t_c) \right\}
$$

$$= \sum_{i=1}^{N} \left\{ d_i \sum_{j=c}^{J} y_{ij} \ln \frac{\lambda_j S(t_{j-1})}{S(t_{c-1})} + (1 - d_i) \sum_{j=c}^{J} y_{ij} \ln \frac{S(t_{j-1})}{S(t_{c-1})} \right\}$$

$$= \sum_{i=1}^{N} \left\{ d_i \sum_{j=c}^{J} y_{ij} \ln \lambda_j + \sum_{j=c}^{J} y_{ij} \ln \frac{S(t_{j-1})}{S(t_{c-1})} \right\}$$

$$= \sum_{i=1}^{N} \left[d_i \sum_{j=c}^{J} y_{ij} \ln \lambda_j(x_i, \theta) + \sum_{j=c}^{J} y_{ij} \sum_{a=c}^{j-1} \ln\{1 - \lambda_a(x_i, \theta)\} \right].$$

The hazard-based log-likelihood function is the same as that without truncation; the only change needed is starting from $j = c$, not $j = 1$. The truncation point can be allowed to vary across $i = 1, ..., N$ so long as it is observed: replace c with c_i in the log-likelihood.

5.2   Piecewise Constant Hazard

5.2.1   Discrete-Time-Varying Regressors

Let y^* denote the latent duration with a continuous survival function S^* and y the observed discrete duration taking on $t_1, ..., t_J$. With $\lambda^*(t)$ denoting the hazard rate for y^*, recall

$$S^*(t) = \exp\left\{ - \int_0^t \lambda^*(v)dv \right\} = \exp\{-\Lambda^*(t)\}.$$

Since y^* is continuous, $S^*(t) = S^*(t^-)$. The jth discrete (piecewise constant) hazard, i.e., the failure probability over $[t_j, t_{j+1})$ given survival up to t_j is, for $j = 1, ..., J - 1$,

$$\lambda_j = \frac{P(y = t_j)}{P(y^* \geq t_j)} = \frac{P(t_j \leq y^* < t_{j+1})}{P(y^* \geq t_j)} = \frac{P(y^* \geq t_j) - P(y^* \geq t_{j+1})}{P(y^* \geq t_j)}$$

$$= \frac{S^*(t_j^-) - S^*(t_{j+1}^-)}{S^*(t_j^-)}$$

$$= 1 - \frac{S^*(t_{j+1}^-)}{S^*(t_j^-)} = 1 - \exp\left\{ - \int_0^{t_{j+1}^-} \lambda^*(v)dv + \int_0^{t_j^-} \lambda^*(v)dv \right\}$$

$$= 1 - \exp\left\{ - \int_{t_j}^{t_{j+1}^-} \lambda^*(v)dv \right\};$$

$$\lambda_J = 1 - \exp\left\{ - \int_{t_J}^{\infty} \lambda^*(v)dv \right\} \text{ with } t_{J+1} \equiv \infty.$$

Suppose that x is time-constant, or at most, changes only at discrete times $t_2, t_3, ...$ Also assume a proportional hazard for λ^*:

$$\lambda^*\{t, x(t)\} = \lambda_o^*(t) \cdot e^{x(t)'\beta}.$$

Then, for $j = 1, ..., J$,

$$\lambda_j = 1 - \exp\left\{ -\int_{t_j}^{t_{j+1}^-} \lambda_o^*(v) e^{x(v)'\beta} dv \right\}$$

$$= 1 - \exp\left\{ -e^{x(t_j)'\beta} \cdot \int_{t_j}^{t_{j+1}^-} \lambda_o^*(v) dv \right\}$$

$$= 1 - \exp\left\{ -e^{x(t_j)'\beta + \alpha_j} \right\}, \quad \text{where}$$

$$\alpha_j \equiv \ln \int_{t_j}^{t_{j+1}^-} \lambda_o^*(v) dv = \ln \int_{t_j}^{t_{j+1}} \lambda_o^*(v) dv.$$

Recalling $L_N(\theta)$ in the preceding subsection, the log-likelihood function with β and $\alpha_1, ..., \alpha_J$ is

$$\sum_{i=1}^{N} \left(d_i \sum_{j=1}^{J} y_{ij} \ln[1 - \exp\{-e^{x_i(t_j)'\beta + \alpha_j}\}] - \sum_{j=1}^{J} y_{ij} \sum_{\tau=1}^{j-1} e^{x_i(t_\tau)'\beta + \alpha_\tau} \right).$$

Analogous to the threshold identification problem in ODR is that one of α_j's is not identified, which requires modifying the log-likelihood function. To see this, observe

$$x_i(t_j)'\beta + \alpha_j = \beta_1 + \tilde{x}_i(t_j)'\tilde{\beta} + \alpha_j$$

$$= (\beta_1 + \alpha_1) + \tilde{x}_i(t_j)'\tilde{\beta} + (\alpha_j - \alpha_1),$$

$$j = 1, ..., J$$

where $x(t) = \{1, \tilde{x}(t)'\}'$; i.e., β_1 is the intercept and $\tilde{\beta}$ is the slope. In this display, α_1 is absorbed into β_1 and only $\alpha_j - \alpha_1$, $j = 2, ..., J$, are identified. Redefining $\alpha_j - \alpha_1$ as α_j and $\beta_1 + \alpha_1$ as β_1, the log-likelihood is maximized wrt β and $\alpha_2, ..., \alpha_J$ (with $\alpha_1 = 0$). This approach is semiparametric because it specifies $\exp(x'\beta)$ in the hazard, but not the baseline hazard.

5.2.2   Ordered Discrete Response Model for Time-Constant Regressors

While the above approach is valid for time-constant regressors or regressors that vary only at discrete times, when x_i is time-constant, it is possible to rewrite the log-likelihood such that the discrete duration model becomes an ODR model as done in Han and Hausman (1990). Suppose

$$y = 1 \quad \text{when } (0 =) t_1 \leq y^* < t_2$$

$$\vdots$$

$$y = J - 1 \quad \text{when } t_{J-1} \leq y^* < t_J$$

$$= J \quad \text{when } t_J \leq y^*.$$

That is,

$$y = j \quad \text{when } t_j \le y^* < t_{j+1}, \quad j = 1, ..., J \text{ with } t_1 \equiv 0 \text{ and } t_{J+1} \equiv \infty.$$

Recall the "invariance" fact that, for any continuous duration y^*, $-\ln \Lambda^*(y^*)$ follows the type-I extreme distribution where Λ^* denotes the cumulative hazard of y^*: if u' follows this distribution, then $P(u' \le a) = \exp(-e^{-a})$. Letting $u = \ln \Lambda^*(y^*)$,

$$
\begin{aligned}
P(u > a) &= P(-u < -a) = P(u' < -a) = \exp(-e^a) \\
&\text{and} \quad P(u \le a) = 1 - \exp(-e^a).
\end{aligned}
$$

Applying $\ln \Lambda^*(\cdot)$ to $t_j \le y^* < t_{j+1}$, we get

$$
\begin{aligned}
y &= j \quad \text{if } \ln \Lambda^*(t_j) \le \ln \Lambda(y^*) < \ln \Lambda^*(t_{j+1}), \ j = 1, ..., J, \\
\ln \Lambda^*(t_1) &\equiv -\infty, \ \ln \Lambda^*(t_{J+1}) = \infty.
\end{aligned}
$$

Assume the proportional hazard with exponential function:

$$\lambda^*(t) = \lambda_o^*(t) \exp(-x'\beta) \implies \Lambda^*(t) = \exp(-x'\beta) \int_0^t \lambda_o^*(s)$$

$$\implies \ln \Lambda^*(t) = -x'\beta + \ln \int_0^t \lambda_o^*(s)ds.$$

We use $-x'\beta$ instead of $x'\beta$ to make our derivation more analogous to the derivation for ODR. Substitute this display and $u = \ln \Lambda^*(y^*)$ into the preceding display to get

$$y = j \quad \text{when } -x'\beta + \tau_{j-1} \le u < -x'\beta + \tau_j, \quad j = 1, ..., J$$

where τ_j is the logged cumulative hazard up to t_{j+1}:

$$\tau_j \equiv \ln \int_0^{t_{j+1}} \lambda_o^*(s)ds, \quad j = 1, ..., J \quad (\tau_0 \equiv -\infty, \ \tau_J \equiv \infty).$$

Note that τ_j, which plays the role of ODR thresholds as can be seen in $\tau_{j-1} \le x'\beta + u < \tau_j$, is defined for the interval $[0, t_{j+1})$ to be coherent with the definition of λ_j for $[t_j, t_{j+1})$. Rewrite $x'\beta + \tau_j$ as $(\beta_1 + \tau_1) + \tilde{x}'\tilde{\beta} + \tau_j - \tau_1$ to see that only $\tau_j - \tau_1$ are identified, $j = 2, ..., J$. Compared with the ODR derivation, there are two visible differences: one is y taking on $1, ..., J$, instead of $0, ..., J-1$, and the other is that y takes the value of the upper threshold subscript, not the lower threshold subscript. These two differences will disappear once we rename $1, ..., J$ as $0, ..., J-1$; we will, however, stick to the differences and maintain $P(y = 0) = 0$.

Defining $\tau_2 - \tau_1, ..., \tau_{J-1} - \tau_1$ as $\tilde{\tau}_2, ..., \tilde{\tau}_{J-1}$, respectively, we get ($\tilde{\tau}_0 = -\infty, \tilde{\tau}_1 = 0, \tilde{\tau}_J = \infty$)

$$y = j \quad \text{when } -x'\beta + \tilde{\tau}_{j-1} \le u < -x'\beta + \tilde{\tau}_j, \quad j = 1, ..., J.$$

Recall $F(a) = 1 - \exp(-e^a)$ and $S(a) = \exp(-e^a)$ where F denotes the df of u and $S = 1 - F$. The log-likelihood contribution of individual i is

$$\sum_{j=1}^{J} y_{ij}[d_i \ln\{F(\tilde{\tau}_j - x_i'\beta) - F(\tilde{\tau}_{j-1} - x_i'\beta)\}$$

$$+ (1 - d_i)\ln\{1 - F(\tilde{\tau}_{j-1} - x'\beta)\}]$$

$$= \sum_{j=1}^{J} y_{ij}[d_i \ln\{S(\tilde{\tau}_{j-1} - x_i'\beta) - S(\tilde{\tau}_j - x_i'\beta)\}$$

$$+ (1 - d_i)\ln\{S(\tilde{\tau}_{j-1} - x'\beta)\}]$$

with which β and $\tilde{\tau}_2$, ..., $\tilde{\tau}_{J-1}$ can be estimated. Specifically, the summand is

for $y_i = 1 : d_i \ln F(-x_i'\beta)$ $[= d_i\{1 - \exp(-e^{-x_i'\beta})\}]$ as $\tilde{\tau}_1 = 0$;

for $y_i = 2 : d_i \ln\{S(-x_i'\beta) - S(\tilde{\tau}_2 - x_i'\beta)\}$ $+ (1 - d_i)\ln S(-x_i'\beta)$;

$$\vdots$$

for $y_i = J : \ln S(\tilde{\tau}_{J-1} - x_i'\beta)$ $[= \exp(-e^{\tilde{\tau}_{J-1} - x_i'\beta})]$.

Although x is assumed to be time-constant, it is possible to allow time-variants to some extent. Suppose x_k is a time-varying wage. A simple option is to use the temporally averaged wage, which is a time-constant variable for the given data. But wages for different intervals can be used as separate regressors as well; e.g., month-1 wage, month-2 wage, etc. may be used separately. Compared with using the average wage, the coefficients of these variables will show the time-varying effects of wage.

Once $\tilde{\tau}_2, ..., \tilde{\tau}_{J-1}$ are estimated, if $t_1, ..., t_{J-1}$ are equally spaced with small intervals, then we can identify the discrete hazard ratios $\lambda_{o2}/\lambda_{o1}$, ..., $\lambda_{o,J-1}/\lambda_{o1}$ where $\lambda_{oj} \equiv (t_{j+1} - t_j)\lambda_o^*(t_j)$. To see this, observe

$$\exp(\tilde{\tau}_2) = \exp(\tau_2 - \tau_1) = \frac{\Lambda_o^*(t_3)}{\Lambda_o^*(t_2)} = \frac{\int_0^{t_3} \lambda_o^*(s)ds}{\int_0^{t_2} \lambda_o^*(s)ds} = \frac{\int_{t_1}^{t_3} \lambda_o^*(s)ds}{\int_{t_1}^{t_2} \lambda_o^*(s)ds}$$

$$\simeq \frac{\lambda_{o1} + \lambda_{o2}}{\lambda_{o1}}$$

$$\exp(\tilde{\tau}_3) = \exp(\tau_3 - \tau_1) = \frac{\Lambda_o^*(t_4)}{\Lambda_o^*(t_2)} \simeq \frac{\lambda_{o1} + \lambda_{o2} + \lambda_{o3}}{\lambda_{o1}}, ...$$

$$\exp(\tilde{\tau}_{J-1}) = \exp(\tau_{J-1} - \tau_1) = \frac{\Lambda_o^*(t_J)}{\Lambda_o^*(t_2)} \simeq \frac{\lambda_{o1} + \lambda_{o2} +, ..., +\lambda_{o,J-1}}{\lambda_{o1}}.$$

Hence

$$\exp(\tilde{\tau}_2) - 1 \simeq \frac{\lambda_{o2}}{\lambda_{o1}}, \quad \exp(\tilde{\tau}_3) - \exp(\tilde{\tau}_2) \simeq \frac{\lambda_{o3}}{\lambda_{o1}}, ...,$$

$$\exp(\tilde{\tau}_{J-1}) - \exp(\tilde{\tau}_{J-2}) \simeq \frac{\lambda_{o,J-1}}{\lambda_{o1}}.$$

Therefore, taking exp $(\cdot)$ and then first-differencing the threshold estimates, we can estimate the baseline discrete hazards up to the multiplicative factor λ_{o1}. Plotting those versus the time $2, 3, ..., J-1$ will show the form of the baseline hazard up to the scale λ_{o1}.

5.2.3   An Empirical Example

Recall the Korean women data with $N = 9312$ whose two variables (age and ex-firm employment years) were used for density estimation. Here we use a few more variables to illustrate piecewise constant hazard estimation and compare it to Weibull MLE. The dependent variable is unemployment duration in days that is right-censored and

$$x = (1, \text{ age}, \text{ age}^2/100, \text{ ex-firm employment years}, \text{ education})'.$$

Education takes five completion levels 1,2,3,4,5 for primary, middle, high, college, and graduate school, respectively. But we will use it as a single cardinal variable to simplify illustration. The mean and SD of the variables are provided in the following table.

	Mean	SD
d (non-censoring dummy)	0.29	
y (unemployment days)	236	105
age in years	34.0	10.8
ex-firm employment years	2.39	1.53
education level (1 to 5)	3.19	1.03

For Weibull, y was used as such, but for piecewise hazard, y was grouped with 15 day intervals starting from 30:

$$t_1 = 0, \ t_2 = 30, \ t_3 = 45, \ t_4 = 60, \ ..., \ t_{27} = 420;$$

there are some observations with y greater than 420. The reason for starting from 30 not from 15 is that the number of observations over $[0, 15)$ is too small. This is inconsequential, because only the hazards at t_2 and onward

Table 5: Weibull MLE and Piecewise Hazard

	Weibull MLE (tv)	Piecewise (tv)
1	$-3.62(-12.6)$	$-0.316(-1.16)$
age	$-0.175(-11.6)$	$-0.172(-11.2)$
age^2/100	$0.189\ (9.27)$	$0.186\ (8.96)$
ex-firm	$-0.095(-6.23)$	$-0.092(-5.94)$
education	$0.124\ (5.55)$	$0.122\ (5.40)$
α	$1.027\ (43.5)$	

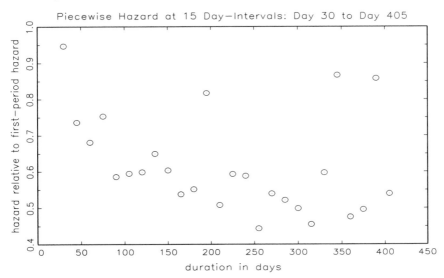

Figure 4: Piecewise Hazards

will be compared. Table 5 shows the estimation results. The estimates are almost the same except the intercept. Since Weibull includes exponential as a special case when $\alpha = 1$, it is interesting to test for $H_0 : \alpha = 1$: the test statistic value is $1.125 = 1.027/0.024$, failing to reject the H_0. That is, Weibull MLE concludes that the baseline hazard is a constant. Despite this, Figure 4 plotting $\lambda_{o2}/\lambda_{o1},...,\lambda_{o,J-1}/\lambda_{o1}$ shows either a monotonically declining or possibly quadratic (U-shaped) baseline hazard. Narendranathan and Stewart (1993) recorded a similar finding that Weibull MLE estimates β in $\exp(x'\beta)$ for the proportional hazard well, although its implied baseline hazard is rejected.

5.3   Partial Likelihood Estimator (PLE)

This subsection introduces the Partial Likelihood Estimator (PLE) in Cox (1972) which allows an arbitrary baseline hazard and time-varying regressors. In principle, PLE is for continuous time; we will thus drop the superscript * in notations. When PLE is applied to discrete time, there are ties—i.e., multiple failures at t. There are different ways to modify PLE to allow for ties, two of which will be examined below.

Distinguishing "event times" of interest from "censoring times," define

(j) as the "label" for the subject with the j-th event time, and

$R(t_{(j)})$ as the "risk set" at the j-th event time.

The risk set $R(t)$ includes subjects who either experienced the event of interest or censored at or after t. For a data set $\{2.7, 3.4^+, 5.6, 5.8^+, 8.9\}$, as there are three event times that occur to subjects 1, 3, and 5,

$$(1) = 1, \ (2) = 3, \text{ and } (3) = 5; \text{ also accordingly,}$$
$$R(t_{(1)}) = R(2.7) = \{1,2,3,4,5\}, \ R(t_{(2)}) = R(5.6) = \{3,4,5\}, \ R(t_{(3)})$$
$$= R(8.9) = \{5\}.$$

Recall the proportional hazard with exponential function: $\lambda(t, x_i(t))$ $= \lambda_o(t)\exp(x_i(t)'\beta)$, where x_i is allowed to depend on t. The main idea of PLE is that the "likelihood" of subject i experiencing the event at t given the risk set $R(t)$ is

$$\frac{\lambda_o(t)\exp(x_i(t)'\beta)}{\sum_{j \in R(t)}\{\lambda_o(t)\exp(x_j(t)'\beta)\}} = \frac{\exp(x_i(t)'\beta)}{\sum_{j \in R(t)}\exp(x_j(t)'\beta)}$$

which is free of $\lambda_o(t)$. One may regard this likelihood as a multinomial choice with a multinomial logit-form probability.

Let J be the number of the event times (the censoring times not counted). Since event occurs only at $t_{(j)}$, the "partial-likelihood" function to maximize for b is

$$\prod_{j=1}^{J} \frac{\exp\{x_{(j)}(t_{(j)})'b\}}{\sum_{m \in R(t_{(j)})}\exp\{x_m(t_{(j)})'b)\}}.$$

The partial likelihood may be taken as a "regular" likelihood in the sense that the estimator is $\sqrt{N}$-consistent with the asymptotic variance being -1 times the inverse of the second-order matrix (Andersen and Gill, 1982). Recalling the M-estimator asymptotic distribution theory, this can happen only if the second-order matrix equals the outer-product of the gradient times -1. That is, the asymptotic variance can be estimated in the usual "MLE-way."

Although $\lambda_o(t)$ drops out in PLE, if desired, $\lambda_o(t)$ can be estimated with

$$\lambda_{oN}(t_{(j)}) \equiv \frac{1}{\sum_{m \in R(t_{(j)})}\exp\{x_m(t_{(j)})'b_N\}}.$$

A rationale for $\lambda_{oN}(t)$ can be seen with "profile likelihood": specify the full likelihood function and maximize it wrt $\lambda_o(t)$ treating β as known. Then the solution is $\lambda_{oN}(t)$; see Klein and Moeschberger (2003, p. 258 and pp. 283–284). Summing up $\lambda_{oN}(t)$ yields a cumulative baseline hazard estimator $\Lambda_{oN}(t)$. Then

$$S_N\{t, x(t)\} = \exp[-\Lambda_{oN}(t) \cdot \exp\{x(t)'b_N\}]$$
$$= \{S_{oN}(t)\}^{\exp\{x(t)'b_N\}}, \quad \text{where } S_{oN}(t) \equiv \exp\{-\Lambda_{oN}(t)\}.$$

With this, the survival probability of a subject with a covariate path $x(t)$ can be predicted.

When there are ties in discrete event times, a modification in PLE is called for. For instance, suppose that subjects $7, 8, 9$ experience the event at time t. Breslow's (1974) modification is to use

$$\frac{\exp\{x_7(t)'b\} \cdot \exp\{x_8(t)'b\} \cdot \exp\{x_9(t)'b\}}{[\sum_{m \in R(t)} \exp\{x_m(t)'b\}]^3}$$

$$= \frac{\exp[\{x_7(t) + x_8(t) + x_9(t)\}'b]}{[\sum_{m \in R(t)} \exp\{x_m(t)'b\}]^3}$$

for the time-t partial likelihood. For a small number of ties, this modification seems adequate. But for many ties, this procedure may make the denominator too big, as the hazards of subjects $7, 8, 9$ appear three times additively.

If the latent duration of subject 7 is the smallest followed by those of subjects 8 and 9, then the hazard of subjects $7, 8, 9$ will appear, respectively, only once, twice, and three times in the partial likelihood denominators. On this account, Efron's (1977) modification seems better when there are many ties:

$$\frac{\exp\{x_7(t)'b\}}{\sum_{m \in R(t)} \exp\{x_m(t)'b\} - (0/3) \exp[\{x_7(t) + x_8(t) + x_9(t)\}'b]}$$

$$\cdot \frac{\exp\{x_8(t)'b\}}{\sum_{m \in R(t)} \exp\{x_m(t)'b\} - (1/3) \exp[\{x_7(t) + x_8(t) + x_9(t)\}'b]}$$

$$\cdot \frac{\exp\{x_9(t)'b\}}{\sum_{m \in R(t)} \exp\{x_m(t)'b\} - (2/3) \exp[\{x_7(t) + x_8(t) + x_9(t)\}'b]}.$$

Compared with the Breslow modification, the numerator is the same, but the denominator is smaller. It does not matter which subject gets listed first in the numerator. As for the baseline hazard estimator $\lambda_{oN}(t)$, the only adjustment needed is replacing the numerator 1 with 3.

6   Integrated-Moment Specification Tests*

When method-of-moment tests (MMT) were discussed, it was seen that a conditional moment $E(u|x) = 0$ yields infinitely many unconditional moments such as $E\{u \cdot g(x)\} = 0$. The unsettled question of which unconditional moments to use is answered fairly satisfactorily in this section. First, find a family of functions $\{g_t(x), \, t \in T\}$—each function $g_t(x)$ is indexed by t ranging over some set T—such that $E\{ug_t(x)\} = 0$ $\forall t$ *is equivalent* to $E(u|x) = 0$. Plotting $E\{ug_t(x)\}$ versus t will indicate in which direction $E(u|x)$ is violated—$g_t(x)$ serves as a "direction" or "axis" particularly when $g_t(x)$'s do not "overlap" (e.g., $g_t(x)$'s are "orthonormal"). Second, use, for some weighting function $\mu(t)$,

$$Kolmogorov\text{-}Smirnov \text{ type test} : \sup_t |E\{ug_t(x)\}|$$

$$Cram\acute{e}r\text{-}von\text{-}Mises \text{ type test} : \int E^2\{ug_t(x)\}d\mu(t).$$

χ^2 goodness-of-fit tests can be thought of as special cases of the latter test where u is the difference between an observed variable and its expected value. With the tests in the display, $E\{ug_t(x)\} \neq 0$ for any t can be detected, and the asymptotic inference can be done with bootstrap. These (and more) are discussed in this section.

6.1   Integrated Moment Tests (IMT)

Bierens (1990) noted that, when $x \in R^k$ is bounded and $E(u|x) \neq 0$ for some x, then $E\{u \cdot \exp(t'x)\} \neq 0$ for almost every t in R^k; the reverse also holds easily because $E\{u \cdot \exp(t'x)\} \neq 0$ for a.e. t in R^k cannot imply $E(u|x) = 0$ a.e. x. Intuitively, $E\{\exp(t'x)\}$ is the moment generating function (mgf) of x which exists when x is bounded, and as well known, if the mgf's of two rv's are the same $\forall t$ on a neighborhood of 0, then their distributions are the same. This suggests that the class of functions $\{\exp(t'x),\ t \in R^k\}$ is large enough and fits the role of the above $\{g_t(x),\ t \in T\}$ turning a conditional moment into infinite unconditional moments. Indeed, as can be seen in Bierens and Ploberger (1997), $E(u|x) \neq 0$ for some x iff $\int E^2\{u \cdot \exp(t'x)\}d\mu(t) > 0$ for any weighting function $\mu(t)$ (subject to some regularity conditions). If x is not bounded, then replace x in $\exp(t'x)$ with an one-to-one bounded function of x, e.g., $\tan^{-1}(x)$.

Formally, let $u(\beta) \equiv y - m(x, \beta)$, $\beta \in \Theta$, and consider

$$H_o : P[E\{u(\beta)|x\} = 0] = 1 \quad \text{for } some\ \beta \in \Theta \text{ where}$$
$$H_a : P[E\{u(b)|x\} = 0] < 1 \quad \text{for } any\ b \in \Theta.$$

A test statistic for a given t is

$$\sqrt{N}M_N(t) \equiv \frac{1}{\sqrt{N}} \sum_i \{u_i(b_N) \exp(t'x_i)\} \rightsquigarrow N\{0, s(t)^2\}$$

where $s(t)$ and estimators for $s(t)$ are in Bierens (1990, (13) to (16)). An application of this test to an Engel curve estimation can be found in the appendix of Bierens and Pott-Buter (1990); see also Bierens and Carvalho (2007).

With a consistent estimator $s_N(t)$ for $s(t)$ available, the test with $\sqrt{N}M_N$ $(t)/s_N(t)$ may be implemented with a chosen t, or this test statistic may be plotted against t to show at which t H_o gets violated. Better yet, we may use the Cramér-von-Mises-type test statistic $\int\{\sqrt{N}M_N(t)/s_N(t)\}^2d\mu(t)$ or its non-standardized version $\int\{\sqrt{N}M_N(t)\}^2d\mu(t)$. Bierens and Ploberger (1997) derived the asymptotic distribution of $\int\{\sqrt{N}M_N(t)\}^2d\mu(t)$ under local alternatives (i.e., the alternative model converging to the null model as $N \to \infty$) as well as under H_o. We may also use the Kolmogorov–Smirnov type test $\sup_t |\sqrt{N}M_N(t)/s_N(t)|$ or its non-standardized version $\sup_t |\sqrt{N}M_N(t)|$. The standardized versions are likely to perform better because they downweight regions of t with high variances. All of these tests can be called "*integrated moment tests (IMT)*." We will, however, consider mainly $\int\{\sqrt{N}M_N(t)\}^2d\mu(t)$

or its standardized version in the remainder of this section for the following
two reasons.

One reason is that integrating out t is likely to be more powerful than
using the maximum deviation in the Kolmogorov–Smirnov test when there
are small violations of H_o over many directions. The other is that Cramér-von-
Mises-type tests with quadratic functions can often be expressed as an infinite
sum of orthogonal components, called "principal component decomposition."
This is helpful in finding the asymptotic distribution and in exploring how
the null model is violated—i.e., in which direction it goes wrong—although
we will not discuss this aspect any further in the following.

Not just $\exp(t'x)$, there are functions $w(t,x)$ of t and x that appeared
in the literature for IMT, e.g., $w(t,x) = \exp(\mathbf{i}t'x)$ where $\mathbf{i}^2 = -1$ in Bierens
(1982) and $1[x \leq t]$ in Stute (1997); the vector inequality $x \leq t$ holds if all
element-by-element inequalities hold. The former corresponds to the characteristic function, while the latter corresponds to the distribution function.
Also, observing that

$$E(u|x) = 0 \text{ a.s.} \iff E\{u(\beta)|\alpha'x\} = 0 \text{ a.s.} \,\forall \alpha \text{ with } |\alpha| = 1$$

Escanciano (2006) proposed to use $1[\alpha'x \leq t]$, which reduces the dimension
of x into the scalar $\alpha'x$ (an advantage), but both α and t should be integrated
out (a disadvantage).

In implementing these tests, the weighting function $\mu(t)$ should be chosen. Although the choice is up to the researcher, typically the empirical distribution $F_{xN}(t) \equiv N^{-1} \sum_i 1[x_i \leq t]$ of x is used. For instance, in the case
of $w(t,x_i) = 1[x \leq t]$ in Stute (1997), the "evaluation point" t is set at the
observed values of x, which leads to $1[x_i \leq x_j]$ and the integral $\int(\cdot)d\mu(t)$
becomes $N^{-1} \sum_j(\cdot)$ as can be seen in

$$\int \left\{ \frac{1}{\sqrt{N}} \sum_i u_i(b_N) 1[x_i \leq t] \right\}^2 d\mu(t)$$

$$= \frac{1}{N} \sum_{j=1}^{N} \left\{ \frac{1}{\sqrt{N}} \sum_i u_i(b_N) 1[x_i \leq x_j] \right\}^2.$$

With $w(t,x) = \exp(\mathbf{i}t'x)$ and a normal-density weighting, a version of Bierens
(1982) test in Escanciano (2006) is

$$\frac{1}{N} \sum_{i=1}^{N} \sum_{j=1}^{N} u_i(b_N) u_j(b_N) \exp\left(-\frac{1}{2}|x_i - x_j|^2 \right)$$

which looks particularly easy to implement. Other versions of Bierens test
and more discussions can be found in Fan and Li (2000).

A common drawback of these IMT's is that their asymptotic
distributions are difficult to characterize and they depend on the null model.
Hence, no table of critical values for general use is available. One solution to

overcome this problem is following the Khmaladze (1981) approach to transform the test statistic such that its limiting distribution is known and free of the null-model ("asymptotically distribution free"); the limiting distribution becomes the distribution of a functional of Brownian motion. This approach reviewed in Koul (2006)—see also Khmaladze and Koul (2004) and the references therein—is, however, quite involved. Instead, bootstrap can be applied to IMT's which is shown in the next section where we will examine the Stute (1997) test in detail.

6.2   Integrated Regression Function Specification Test

6.2.1   Main Idea

Consider testing whether a parametric regression model $E(y|x) = m(x, \beta)$, $\beta \in \Theta$, is correct or not. The null and alternative hypotheses are, respectively,

$$H_o : P\{E(y|x) = m(x, \beta)\} = 1 \quad \text{for } some \ \beta \in \Theta$$
$$H_a : P\{E(y|x) = m(x, b)\} < 1 \quad \text{for } any \ b \in \Theta.$$

As H_a does not fix the alternative model—H_a just states that the null model is false—tests for H_a are usually called "*goodness-of-fit (GOF) tests.*" The well-known parametric GOF χ^2 tests compare the sample proportions and the probabilities under H_o in some specified cells, where the probabilities are the expected proportions computed using the parametric null model. The main weakness of those χ^2 tests is that choosing the cells is arbitrary, and the tests have no power if the true model has the same probability as the null model does on the cells, yet differs from the null model when finer cells are used. For instance, on a cell $(2, 3)$, both models may have the same probability 0.5, but different probabilities on finer cells $(2, 2.7)$ and $(2.7, 3)$.

One way to do the test nonparametrically is using a distance between $E_N(y|x)$ and $m(x, b_N)$ where $E_N(y|x)$ is a nonparametric estimator for $E(y|x)$ and b_N is a $\sqrt{N}$-consistent estimator for β. Once the difference for a given x_o is obtained, we can then use, e.g., $\int \{E_N(y|x_o) - m(x_o, b_N)\}^2 dx_o$ or $\sup_{x_o} |E_N (y|x_o) - m(x_o, b_N)|$ to "get rid of" x_o. While this procedure is relatively straightforward, a drawback is the dependence on the bandwidth in $E_N(y|x)$.

Another way to test the H_o while avoiding bandwidth dependence is comparing "integrated versions" of $E_N(y|x)$ and $m(x, b_N)$. Intuitively speaking, the bandwidth-dependent approach is analogous to comparing nonparametric and parametric density estimates, whereas the integrating approach is analogous to comparing nonparametric and parametric distribution function estimates. As a df can be estimated by the empirical df which does not require any bandwidth, the integrating approach is bandwidth-free.

Stute (1997) used the integrated regression function

$$\int_{-\infty}^{x_o} E(y|t) dF_x(t) = E\{1[x \le x_o] E(y|x)\} = E\{1[x \le x_o] y\}$$

where $F_x(t) \equiv P(x \leq t)$ is the df of x. The difference between the last two expressions is

$$E[1[x \leq x_o] \cdot \{y - E(y|x)\}] = 0$$
$$\implies E[1[x \leq x_o] \cdot \{y - m(x,\beta)\}] = 0 \qquad \text{under the } H_o.$$

A "marked empirical process" $R_N(x_o)$—marked by the residual—based on the sample analog for this difference (times $\sqrt{N}$) is

$$R_N(x_o) \equiv \frac{1}{\sqrt{N}} \sum_i 1[x_i \leq x_o]\{y_i - m(x_i, b_N)\}.$$

Integrating $R_N^2(x_o)$ over x_o using F_{xN} as the weighting function, we obtain a Cramér-von-Mises-type test statistic.

$F_{xN}(x_o) \equiv N^{-1}\sum_{j=1}^{N} 1[x_j \leq x_o]$ goes up by N^{-1} only at $x_o = x_j$, $j = 1, ..., N$, and $F_{xN}(x_o)$ does not change at the other points. We can thus write

$$F_{xN}(x_o) = \int_{-\infty}^{x_o} dF_{xN}(t) = \int_{-\infty}^{\infty} 1[t \leq x_o]dF_{xN}(t)$$
$$\left(\text{sum of } \frac{1[x_j \leq x_o]}{N} \text{ at } x_j, j = 1, ..., N \right).$$

That is, instead of assigning the weight N^{-1} to all $x_1, ..., x_N$, N^{-1} or 0 is assigned depending on $x_j \leq x_o$ or not. Replacing $1[t \leq x_o]$ in $\int_{-\infty}^{\infty} 1[t \leq x_o]dF_{xN}(t)$ with $R_N(t)^2$, we get the Cramér-von-Mises-type test statistic in Stute (1997):

$$W_N \equiv \int R_N(t)^2 dF_{xN}(t) \left(\text{sum of } \frac{R_N(x_j)^2}{N} \text{ at } x_j, j = 1, ..., N \right)$$
$$= \frac{1}{N} \sum_j R_N(x_j)^2 = \frac{1}{N} \sum_j \left[\frac{1}{\sqrt{N}} \sum_i 1[x_i \leq x_j]\{y_i - m(x_i, b_N)\} \right]^2.$$

If desired, we may also use the Kolmogorov–Smirnov-type test statistic $\sup_{1 \leq j \leq N} |R_N(x_j)|$.

6.2.2   Bootstrap Inference

$R_N(x_o)$ converges in law to a "centered Gaussian process" under the H_o. But using the asymptotic distribution of W_N resulting from this convergence-in-law makes the statistical inference depend on the H_o model. That is, the test is not asymptotically distribution-free and its limiting distribution should be characterized for each H_o. This is in contrast to the typical case, say,

$$\sqrt{N}(b_N - \beta) \rightsquigarrow N(0, C) \iff \sqrt{N}C^{-1/2}(b_N - \beta) \rightsquigarrow N(0, I_k)$$

where the limit distribution $N(0, I_k)$ is known and tabulated. Hence, instead of deriving and using the asymptotic distribution of W_N, Stute, et al. (1998) proved that the asymptotic distribution of W_N can be approximated by the following "*wild bootstrap.*"

First, obtain the residual $\hat{u}_i \equiv y_i - m(x_i, b_N)$ and then W_N. Second, draw a random sample v_i^*, $i = 1, \ldots, N$, from the two point distribution

$$P\left(v^* = \frac{1 - \sqrt{5}}{2}\right) = \frac{5 + \sqrt{5}}{10} \quad \text{and} \quad P\left(v^* = \frac{1 + \sqrt{5}}{2}\right) = \frac{5 - \sqrt{5}}{10}.$$

Third, construct a pseudo-sample (y_i^*, x_i), $i = 1, \ldots, N$ where $y_i^* = m(x_i, b_N) + v_i^* \hat{u}_i$. Fourth, obtain W_N^* using the pseudo-sample where b_N^* is based on (x_i', y_i^*). Fifth, denoting the number of bootstrap replications as B, construct the empirical distribution using the B-many W_N^*'s. Finally, see whether W_N falls in the upper tail of the empirical distribution to reject the H_o if yes. This bootstrap works also for the tests in Bierens (1982, 1990) and Escanciano (2006). Instead of simulating v^* as above, it is simpler to generate v^* such that v^* takes on ± 1 with probability 0.5.

A minor drawback of this bootstrap is that its power property for alternative models is difficult to analyze. By adopting the approach in Khmaladze (1981), the test can be transformed into an asymptotically distribution-free form, and then the transformed test statistic can be expressed as an infinite sum of orthogonal components, which is more informative for its power property analysis. This approach was pursued in Stute et al. (1998). Koul and Sakhanenko (2005) reported that finite sample behavior of Khmaladze-transformed test is better than that of bootstrap inference. This is a surprise, as bootstrap is often "sold" on its good finite sample behaviors. While the above Stute test uses the empirical x-process marked by the residuals, Stute et al. (2008) presented a related test using the empirical residual process marked by $g(x_1), \ldots, g(x_N)$ for a function g. This test was motivated by a better nonparametric "principal component" analysis, but the test raises the issue of choosing g.

Whang (2001) generalized the Stute test by considering $E\{\rho(y, x; \beta) | x\} = 0$ which includes the Stute regression function specification test as a special case when $\rho(y, x; \beta) = y - x'\beta$. The proposed test can thus be implemented almost the same way as the Stute test was implemented. One difference, however, is how bootstrap is done. Since $\rho(y, x; \beta)$ is not necessarily solvable for y, the wild bootstrap cannot be used to generate y^*. Whang (2001) does nonparametric bootstrap (i.e., drawing randomly with replacement from (x_i, y_i), $i = 1, \ldots, N$). To make sure that the bootstrap sample satisfies the conditional moment condition, Whang (2001) recentered ρ as Hall and Horowitz (1996) did for GMM tests. He and Zhu (2003) proposed a GOF test for quantile regression, which uses the $k \times 1$ asymptotic first-order condition for quantile regression in place of $y - m(x, \beta)$ in the Stute test.

6.2.3   Further Remarks

Kondo and Lee (2003) applied the Stute test to a "hedonic price index" estimation problem. For a repeated cross-section data over 1985–1991 with the total sample size $N = 1808$, the response variable is ln(monthly Tokyo office rent per m^2) and the regressors are the time dummies x and the office characteristics z. Kondo and Lee set up a quadratic model (linear in x however, as x is dummy variables):

$$\ln p_i = x'_i\beta + \sum_j \alpha_j z_j + \sum_{j,j'} \alpha_{jj'} z_j z_{j'} + u_i.$$

Three transformations of z_j" were tried: z_j, $z_j^{1/2}$, and $\ln z_j$, and the Stute test was applied with $B = 2000$. The resulting p-values were 0.005, 0.037, and 0.038, respectively. Thus, despite as many as 44 parameters in each model, the quadratic models were all rejected.

Kondo and Lee then set up a semi-linear model $y_i = x'_i\beta + \rho(z_i) + \varepsilon_i$ where $\rho(\cdot)$ is an unknown function of z, and went on to estimate only β using a kernel-based technique for $\rho(z)$ that will be shown in the next chapter. Taking the quadratic terms as a series approximation to $\rho(z)$, this empirical example illustrates that series approximation may require too many terms. Compared with this, kernel estimator seems more "parsimonious."

Often a test statistic can be "rigged" to produce estimators. For instance, suppose $T_N(\hat\theta)$ is a test statistic that depends on an estimator $\hat\theta$ obtained without H_o imposed, and we accept H_o if $T_N(\hat\theta) \simeq 0$. In this case, if we believe H_o, then we can estimate θ by minimizing $\{T_N(\theta)\}^2$ wrt θ. For instance, a MMT is based on $N^{-1/2}\sum_i m(z_i,\hat\theta) = 0$ where $\hat\theta$ is an estimator derived from a moment condition other than $E\{m(z,\theta)\} = 0$ but θ satisfies $E\{m(z,\theta)\} = 0$ under H_o. Instead of testing $E\{m(z,\theta)\} = 0$, if we believe the H_o, then we can estimate θ by imposing the H_o: do GMM for θ under $E\{m(z,\theta)\} = 0$.

In this regard, related to the Stute's (1997) test is the GMM by Dominguez and Lobato (2004) for a moment condition

$$E\{\psi(z,\beta)|x\} = 0 \iff E\{1[x \le x_o]\psi(z,\beta)\} = 0 \; \forall x_o$$

$$\iff \int E^2\{1[x \le x_o]\psi(z,\beta)\}dF_x(x_o) = 0.$$

They propose to estimate β by minimizing a sample analog for the last expression that is analogous to the above W_N:

$$\frac{1}{N}\sum_{j=1}^{N}\left\{\frac{1}{\sqrt{N}}\sum_{i=1}^{N}1[x_i \le x_j]\psi(z_i,b)\right\}^2.$$

6.3   Conditional Kolmogorov Test

Let $z_i = (x'_i, y'_i)'$ where x is $k_x \times 1$ and y is $k_y \times 1$; y is allowed to be multi-dimensional. Let $F(y_o|x)$ and $F_x(x)$ denote the df of $y|x$ and x, respectively:

$$F(y_o|x) = P(y \le y_o|x) \quad \text{and} \quad F_x(x_o) \equiv P(x \le x_o).$$

The null hypothesis is

$$H_0 : F(\cdot|x) = F_0(\cdot|x, \theta) \; \forall x$$

which is a parametric df indexed by an unknown θ while F_x is unrestricted. Andrews (1997) proposed "*conditional Kolmogorov (CK) test*" which is discussed in this subsection.

     With $z_o = (x'_o, y'_o)'$ being a fixed value, let

$$F_N(z_o) \equiv \frac{1}{N}\sum_i 1[z_i \le z_o] \quad \text{and} \quad F_{xN}(x_o) \equiv \frac{1}{N}\sum_i 1[x_i \le x_o].$$

and let $\hat{\theta}$ be a $\sqrt{N}$-consistent estimator for θ. Noting $1[z_i \le z_o] = 1[y_i \le y_o] \cdot 1[x_i \le x_o]$, take $E(\cdot|x_1, ..., x_N)$ on $F_N(z_o)$ to get

$$E\{F_N(z_o)|x_1, ...x_N\} = \frac{1}{N}\sum_i E(1[z_i \le z_o]|x_i)$$

$$= \frac{1}{N}\sum_i E\{1[y_i \le y_o] \; 1[x_i \le x_o]|x_i\}$$

$$= \frac{1}{N}\sum_i E\{1[y_i \le y_o]|x_i\}1[x_i \le x_o] = \frac{1}{N}\sum_i F_0(y_o|x_i, \theta) \; 1[x_i \le x_o].$$

As the Stute (1997) test is based on the empirical process with $1[x_i \le x_o]$ marked by $y_i - E(y|x_i)$, CK test is based on the empirical process with $1[x_i \le x_o]$ marked by $1[y_i \le y_o] - F_0(y_o|x_i, \theta)$:

$$\frac{1}{\sqrt{N}}\sum_i \{1[y_i \le y_o] - F_0(y_o|x_i, \theta)\} \; 1[x_i \le x_o].$$

     In the empirical process, replacing θ with $\hat{\theta}$ and then using the observed points $z_1, ..., z_N$ for the evaluation point z_o as in the Stute test, we get the CK test statistic:

$$CK_N \equiv \max_{1 \le j \le N} |\frac{1}{\sqrt{N}}\sum_i \{1[y_i \le y_j] - F_0(y_j|x_i, \hat{\theta})\}1[x_i \le x_j]|.$$

The difference between the Stute test and CK test is only in the "marks": the former uses $y_i - m(x_i, \beta)$ while the latter uses $1[y_i \le y_j] - F_0(y_j|x_i, \theta)$. As Andrews (1997) noted, instead of this Kolmogorov–Smirnov version, the Cramér-von-Mises version can be used as well:

$$CC_N \equiv \frac{1}{N}\sum_{j=1}^{N}\left\{\frac{1}{\sqrt{N}}\sum_i \{1\,[y_i \le y_j] - F_0(y_j|x_i, \hat{\theta})\}1[x_i \le x_j]\right\}^2.$$

For asymptotic inference, Andrews (1997) recommended *"fixed-x para-metric bootstrap."* First, draw y_i^* from the parametric distribution $F_0(\cdot|x_i, \hat{\theta})$, $i = 1, ..., N$—more on this below. Second, set $z_i^* = (x_i, y_i^{*\prime})'$, $i = 1, ..., N$, which is one pseudo-sample, with which we can get a pseudo CK_N value, say $CK_N^{(b)}$ where $\hat{\theta}^*$ is based on z_i^*, $i = 1, ..., N$,. Third, repeat these steps, say B times, to get $CK_N^{(b)}$, $b = 1, ..., B$. Then an asymptotic size-α test is rejecting the H_0 if

$$\frac{1}{B}\sum_{b=1}^{B} 1\left[CK_N \geq CK_N^{(b)}\right] < 0.05.$$

Regarding the first step above, we can draw $y_i^*|x_i$ from $F_0(\cdot|x_i, \hat{\theta})$ with "acceptance/rejection" method. For instance, if $F_0(\cdot|x_i, \hat{\theta})$ is $N(x_i'b_N, s_N^2)$, then keep drawing two uniform numbers (v_1, v_2) until they satisfy the condition

$$v_2 \leq \frac{1}{s_N\sqrt{2\pi}}\exp\left\{-\frac{1}{2}(\frac{v_1 - x_i'b_N}{s_N})^2\right\};$$

then $y_i^* = v_2$. Put it differently, imagine drawing the normal density on the two-dimensional plane. If the point (v_1, v_2) falls under the density, then we take v_2 as y_i^*; otherwise (v_1, v_2) gets discarded.

CHAPTER 9
BANDWIDTH-DEPENDENT
SEMIPARAMETRIC METHODS

Semiparametric estimators/tests that require nonparametric smoothing in getting the estimators/tests may be called "*semi-nonparametric*," while those that require smoothing at most for the asymptotic variance estimation may be called "semiparametric"; "semiparametrics" in its wide sense encompasses both. There are at least *two big themes in semi-nonparametrics: infinite-dimensional nuisance parameters and dimension reduction*. The former requires nonparametric estimators in the first stage, and as the consequence, the second-stage estimator becomes semi-nonparameric; the main difficulty is to find out how the first-stage nonparametric estimation error affects the second stage estimator/test. The latter appears in efforts to avoid the well-known dimension problem in pure nonparametric methods. The efforts took various forms: single- (or multiple-) index models, additive models, semi-linear models, transformation of response variables, etc. The econometric and statistical literature for semi-nonparametric estimation is immense. To keep our discussion at a reasonable length, only $\sqrt{N}$-consistent or easy-to-use estimators/tests will be studied, although there will be a few exceptions. Also semi-nonparametric methods for LDV models will be examined so long as they are relevant for the two themes, because most practitioners would prefer bandwidth-free methods.

1 Two-Stage Estimator with Nonparametric First-Stage

In the chapter for M-estimator, we examined estimators for a finite-dimensional parameter β with a finite-dimensional nuisance parameter α. There, we saw how to account for the effect of the first-stage estimation error $a_N - \alpha$ on the second-stage. This chapter deals with a more difficult case: infinite-dimensional nuisance parameters, and thus nonparametric first-stage estimators and their effects on the second-stage. Much of this section constitutes background materials for the other sections in this chapter for "dimension reduction," and the reader may refer back to this section when necessary, reading other sections first.

Before we proceed, it is helpful to classify four cases in two-stage estimators (TSE), depending on the dimension of the first-stage parameter α and the second-stage parameter β. Assuming that finite-dimensional parameters are estimable $\sqrt{N}$-consistently whereas infinite-dimensional parameters are estimable at slower rates, the asymptotic distribution of the TSE for β can be categorized as follows:

	Finite α	Infinite α
Finite β	covered in M-estimator chapter	covered in this section
Infinite β	α is as good as known	depends on convergence speeds

In the table, "α is as good as known" means that $a_N - \alpha$ has no effect, and thus replacing α with a_N is as good as knowing α. Also "depends on convergence speeds" has sub-cases. If the convergence speeds are the same in the two-stages, then we have a case similar to both α and β being $\sqrt{N}$-consistent; otherwise, the estimation error for the slower convergence would dominate. This case of both-stage estimators being nonparametric is hard to deal with and would not be practical, as at least two bandwidths are required. This section does not apply to this "both-nonparametric" case.

1.1 Density or Conditional Mean for First Stage

Recall our discussion on a two-stage estimator b_N with a finite dimensional nuisance parameter α with $a_N \to^p \alpha$. If b_N satisfies

$$\frac{1}{\sqrt{N}} \sum_i m(z_i, b_N, a_N) = o_p(1)$$

then applying Taylor's expansion around $b_N = \beta$ and $a_N = \alpha$, we get

$$o_p(1) = \frac{1}{\sqrt{N}} \sum_i m(z_i, \beta, \alpha) + \frac{1}{N} \sum_i m_{b'}(z_i, \beta, \alpha)\sqrt{N}(b_N - \beta)$$

$$+ \frac{1}{N} \sum_i m_{a'}(z_i, \beta, \alpha)\sqrt{N}(a_N - \alpha) \quad \text{where } m_{b'} \equiv \frac{\partial m(z, \beta, \alpha)}{\partial b'}$$
$$\text{and } m_{a'} \equiv \frac{\partial m(z, \beta, \alpha)}{\partial a'}.$$

Solve this for $\sqrt{N}(b_N - \beta)$ and replace $N^{-1} \sum_i m_{b'}(z_i, \beta, \alpha)$ and $N^{-1} \sum_i m_{a'}(z_i, \beta, \alpha)$ with $E\{m_{b'}(z_i, \beta, \alpha)\}$ and $E\{m_{a'}(z_i, \beta, \alpha)\}$ to get

$$\sqrt{N}(b_N - \beta) = -E^{-1}(m_b)\left\{ \frac{1}{\sqrt{N}} \sum_i m(z_i, \beta, \alpha) \right.$$

$$\left. +E(m_{a'})\sqrt{N}(a_N - \alpha) \right\} + o_p(1)$$

$$= -E^{-1}(m_b)\frac{1}{\sqrt{N}} \sum_i \{m(z_i, \beta, \alpha)$$

$$+E(m_{a'})\eta_i(\alpha)\} + o_p(1), \quad \text{where}$$

$$E(m_b) \;=\; E\{m_b(z_i,\beta,\alpha)\}, \;\; E(m_{a'}) = E\{m_{a'}(z_i,\beta,\alpha)\} \quad \text{and}$$

$$\sqrt{N}(a_N - \alpha) \;=\; \frac{1}{\sqrt{N}}\sum_i \eta_i(\alpha) + o_p(1).$$

The term η with $E(\eta\eta') < \infty$ is an influence function for α_N, and $E(m_a)$ is a "link matrix"—linking the first-stage estimation error to the second stage. In the following, we present an analogous correction term when α is not finite-dimensional—i.e., nonparametric.

For a $k \times 1$ random vector q with density $f(q)$, suppose that the nuisance parameter and its estimator are, respectively,

$$\alpha_i = E(w|q_i)f(q_i) \quad \text{and} \quad a_i = \frac{1}{Nh^k}\sum_{j,j\neq i} K\left(\frac{q_j - q_i}{h}\right)w_j$$

where K is a kernel, h is a bandwidth, and a_i is a leave-one-out estimator. There are in fact N-many nuisance parameters $\alpha_1, ..., \alpha_N$, and b_N now satisfies

$$\frac{1}{\sqrt{N}}\sum_i m(z_i, b_N, a_i) = o_p(1).$$

Denoting $m(z_i, \beta, \alpha_i)$ as m_i, it is known to hold that

$$\sqrt{N}(b_N - \beta) \;=\; -E^{-1}(m_b)\frac{1}{\sqrt{N}}\sum_i [m_i \;+\; E(m_{a'}|q_i)f(q_i)w_i$$

$$-E\{E(m_{a'}|q)f(q)w\}] + o_p(1) \qquad\qquad (\mathrm{CT}_{fE})$$

where the correction term is $E(m_{a'}|q_i)f(q_i)w_i - E\{E(m_{a'}|q)f(q)w\}$; CT stands for "Correction Term", and fE "density f times conditional mean E." The second term $E\{E(m_{a'}|q)f(q)w\}$ is to make the correction part to have mean zero. Here the link matrix is $E(m_{a'}|q_i)$.

As a special case, set $w = 1$ to get

$$\alpha_i = f(q_i) \quad \text{and} \quad a_i = \frac{1}{Nh^k}\sum_{j,j\neq i} K\left(\frac{q_j - q_i}{h}\right).$$

Then

$$\sqrt{N}(b_N - \beta) = -E^{-1}(m_b)\frac{1}{\sqrt{N}}\sum_i [m_i \;+\; E(m_{a'}|q_i)f(q_i)$$

$$-E\{E(m_{a'}|q)f(q)\}] + o_p(1). \qquad\qquad (\mathrm{CT}_f)$$

Suppose now that the nuisance parameter and its estimator are

$$\alpha_i = E(w|q_i) \quad \text{and} \quad a_i = \frac{(Nh^k)^{-1}\sum_{j,j\neq i} K((q_j - q_i)/h)w_j}{(Nh^k)^{-1}\sum_{j,j\neq i} K((q_j - q_i)/h)}.$$

It will be shown shortly that, with w being of dimension $p \times 1$,

$$\sqrt{N}(b_N - \beta) = -E^{-1}(m_b)\frac{1}{\sqrt{N}}\sum_i \left[m_i + E\left(\underset{1\times p}{m_{a'}}|q_i\right) \cdot \left\{ \underset{p\times 1}{w_i} - E(w|q_i) \right\} \right] + o_p(1) \qquad (\mathrm{CT}_E)$$

This shows that, when w is not a scalar, the correction term is just a sum of the individual correction terms. The link matrix is still $E(m_{a'}|q_i)$. CT_f and CT_E will be our "building blocks" for many parts in this chapter.

To get CT_E from CT_{fE} and CT_f, observe $E(w|q_i) = \{E(w|q_i) \, f(q_i)\}$ $f(q_i)^{-1}$. This means that product differentiation type of rules can be applied. That is, the correction term will be a sum of two terms: "the correction term for $E(w|q_i)f(q_i)$ with $m_{a'}$ multiplied by $f(q)^{-1}$" and "the correction term for $f(q_i)^{-1}$ with $m_{a'}$ multiplied by $E(w|q)f(q)$," because the effect of one term's estimation error gets channeled to m through the other term times $m_{a'}$. Specifically, the correction term is

$$E\left(\frac{m_{a'}}{f(q)}\Big|q_i\right) f(q_i)w_i - E\left\{ E\left(\frac{m_{a'}}{f(q)}\Big|q\right) f(q)w \right\}$$

$$- E\left(\frac{m_{a'}}{f(q)^2}E(w|q)f(q)\Big|q_i\right) f(q_i) + E\left\{ E\left(\frac{m_{a'}}{f(q)^2}E(w|q)f(q)\Big|q\right) f(q) \right\}$$

$$= E(m_{a'}|q_i)w_i - E\{E(m_{a'}|q)w\} - E(m_{a'}E(w|q)|q_i) + E\{m_{a'}E(w|q)\}.$$

The second and fourth terms cancel each other because

$$E\{E(m_{a'}|q)w\} = E[E\{E(m_{a'}|q)w|q\}] = E[E(m_{a'}|q)E(w|q)]$$
$$E\{m_{a'}E(w|q)\} = E[E\{m_{a'}E(w|q)|q\}] = E[E(m_{a'}|q)E(w|q)].$$

The first and third terms are thus

$$E(m_{a'}|q_i)w_i - E(m_{a'}|q_i)E(w|q_i) = E(m_{a'}|q_i)\{w_i - E(w|q_i)\}.$$

The above results hold when a high-order kernel is used in the first stage, and the bandwidth h is smaller than the optimal bandwidth minimizing the asymptotic mean squared error; such a small h reduces asymptotic bias faster than the optimal bandwidth. Robinson (1988) seems to be the first to use high-order kernels to facilitate similar derivations. Newey (1994) derived the above results using series-approximations for α_j. This along with the fact the correction terms do not depend on the kernel suggest that the choice of the nonparametric method in the first-stage would not matter.

1.2 Other Nonparametric Nuisance Parameters*

A slightly different case from all of the above occurs when $\alpha_i = f(w_i|q_i) = f(w_i, q_i)/f(q_i)$. The correction term can be found using again the product differentiation rule. First, the correction term for $f(w_i, q_i)$ with $m_{a'}$ multiplied by $f(q_i)^{-1}$ is

$$E\left(\frac{m_{a'}}{f(q)}\middle|w_i, q_i\right) f(w_i, q_i) - E\left\{E\left(\frac{m_{a'}}{f(q)}\middle|w, q\right) f(w, q)\right\}$$

$$= \quad E(m_{a'}|w_i, q_i) f(w_i|q_i) - E\{E(m_{a'}|w, q) f(w|q)\}.$$

Second, the correction term for $f(q_i)^{-1}$ with $m_{a'}$ multiplied by $f(w_i, q_i)$ is

$$E\left(\frac{-f(w,q)m_{a'}}{f(q)^2}\middle|q_i\right) f(q_i) - E\left\{E\left(\frac{-f(w,q)m_{a'}}{f(q)^2}\middle|q_i\right) f(q_i)\right\}$$

$$= \quad E\left\{-f(w|q)m_{a'}|q_i\right\} - E\{E(-f(w|q)m_{a'}|q)\}.$$

There are overall four terms, and putting them together, only two of them remain:

$$E(m_{a'}|w_i, q_i) f(w_i|q_i) - E\{E(m_{a'}|w, q) f(w|q)|q_i\} \qquad (\mathrm{CT}_{fc})$$

because the other two terms cancel each other, which can be seen by rewriting the fourth term as

$$E\{E(f(w|q)m_{a'}|q)\} \quad = \quad E[\; E\{E(f(w|q)m_{a'}|w, q)|q\}\;]$$

$$= \quad E\{f(w|q)E(m_{a'}|w, q)\}.$$

If $\alpha_i = \nabla f(q_i) = \partial f(q_i)/\partial q$, then the correction term is known to be

$$-\nabla_q\{E(m_{a'}|q_i) f(q_i)\} + E[\nabla_q\{E(m_{a'}|q) f(q)\}] \qquad (\mathrm{CT}_{\nabla_f})$$

where $\nabla_q\{E(m_{a'}|q_i) f(q_i)\} = \nabla_q E(m_{a'}|q_i) \cdot f(q_i) + E(m_{a'}|q_i) \cdot \nabla f(q_i).$

If $\alpha_i = \nabla_q E(w|q_i) = \partial E(w|q_i)/\partial q$, then the correction term is

$$\frac{-1}{f(q_i)} \nabla_q\{E(m_a|q_i) f(q_i)\}\{w_i - E(w|q_i)\}$$

$$= \quad -\left\{\nabla_q E(m_a|q_i) + E(m_a|q_i)\frac{\nabla_q f(q_i)}{f(q_i)}\right\}\{w_i - E(w|q_i)\}.$$

For the score function $\alpha_i = f(q_i)^{-1}\nabla f(q_i)$, using the product differentiation rule again, the correction term is

$$-\nabla_q\left\{E\left(\frac{m_{a'}}{f(q)}\middle|q_i\right) f(q_i)\right\} + E\left[\nabla_q\left\{E\left(\frac{m_{a'}}{f(q)}\middle|q\right) f(q)\right\}\right]$$

$$-E\left(m_{a'}\frac{\nabla f(q)}{f(q)}\middle|q_i\right) + E\left\{E\left(m_{a'}\frac{\nabla f(q)}{f(q)}\middle|q\right)\right\}$$

$$= \quad -\nabla_q E(m_{a'}|q_i) + E[\nabla_q E(m_{a'}|q)] - E(m_{a'}|q_i)\frac{\nabla f(q_i)}{f(q_i)}$$

$$+E\left\{E(m_{a'}|q)\frac{\nabla f(q)}{f(q)}\right\}$$

$$= \quad \frac{-1}{f(q_i)}\nabla_q\{E(m_{a'}|q_i) f(q_i)\} + E\left[\frac{1}{f(q)}\nabla_q\{E(m_{a'}|q) f(q)\}\right]. \quad (\mathrm{CT}_s)$$

This correction term is the $\nabla f(q_i)$-correction term divided by $f(q_i)$.

1.3   Examples

1.3.1   Moments with Nonparametric Nuisance Parameters

Suppose that there is no b_N in $m(z, b_N, a_i)$. Then the above derivations yield $o_p(1)$-equivalent expressions for $N^{-1/2} \sum_i m(z_i, a_i)$ which are useful for method-of-moment type tests. As an example, consider

$$\frac{1}{\sqrt{N}} \sum_i m(z_i, a_i) = \frac{1}{\sqrt{N}} \sum_i g(x_i)\{y_i - E_N(y|x_i)\}$$

where $\alpha_i \equiv E(y|x_i)$, $m(z_i, \alpha_i) = g(x_i)\{y_i - E(y|x_i)\}$ and $a_i = E_N(y|x_i)$. Observe

$$E(m_a|x) = E\left[\frac{\partial[g(x)\{y - E_N(y|x)\}]}{\partial E_N(y|x)|x}\right] = -g(x).$$

Hence CT_E yields

$$\frac{1}{\sqrt{N}} \sum_i g(x_i)\{y_i - E_N(y|x_i)\}$$

$$= \frac{1}{\sqrt{N}} \sum_i [g(x_i)\{y_i - E(y|x_i)\} - g(x_i)\{y_i - E(y|x_i)\}] + o_p(1) = o_p(1).$$

A similar but different example is

$$\frac{1}{\sqrt{N}} \sum_i g(x_i)\{E_N(y|x_i) - E(y|x)\}$$

$$\implies E(m_a|x) = E\left[\frac{\partial[g(x)\{E_N(y|x_i) - E(y|x)\}]}{\partial E_N(y|x)}|x\right] = g(x).$$

Thus

$$\frac{1}{\sqrt{N}} \sum_i g(x_i)\{E_N(y|x_i) - E(y|x)\}$$

$$= 0 + \frac{1}{\sqrt{N}} \sum_i g(x_i)\{y_i - E(y|x)\} \neq o_p(1) \qquad (\text{CT}_{2S})$$

where the first term (0) on the right-hand side is obtained because $m_i = E(y|x_i) - E(y|x_i) = 0$ and the second term comes by applying CT_E. It will become clear later why this equation is dubbed "CT_{2S}."

1.3.2   Nonparametric WLS

We have seen that the variance matrix of an efficient estimator under a $s \times 1$ conditional moment condition $E(\psi(\beta)|x) = 0$ for a $k \times 1$ vector β is

$$E_x^{-1}\{E(\psi_b(\beta)|x)\ E^{-1}(\psi(\beta)\psi(\beta)'|x)\ E(\psi_{b'}(\beta)|x)\} \quad \text{where } \underset{k \times s}{\psi_b} \equiv \frac{\partial \psi}{\partial b}.$$

Using this, with a given condition $E(u|x) = 0$ in the linear model $y = x'\beta + u$ (here $\psi(\beta) = y - x'\beta$), the efficiency bound for β is

$$E_x^{-1}\{x \cdot (V(u|x))^{-1} \cdot x'\} = E_x^{-1}\left\{\frac{xx'}{V(u|x)}\right\}.$$

Specifying a parametric form of heteroskedasticity, weighted LSE (WLS) attains this bound, when the specified form is correct. But it is possible to attain this bound *without* specifying the form of heteroskedasticity.

Robinson (1987) showed how to implement efficient estimation using a nonparametric estimator for $V(u|x)$ with LSE residuals: estimate $V(u|x)$ nonparametrically with the LSE residuals and then apply WLS. Robinson suggests a nearest-neighbor method for estimating $V(u|x)$. But we use a kernel estimator with $\hat{u}_j \equiv y_j - x_j' b_{lse}$:

$$V_N(u|x_i) = E_N(u^2|x_i) = \frac{\sum_{j,j \neq i} K((x_j - x_i)/h)\hat{u}_j^2}{\sum_{j,j \neq i} K((x_j - x_i)/h)}.$$

Another, perhaps better, estimator for $V(u|x) = V(y|x)$ that Robinson (1987) suggested was

$$\begin{aligned} V_N(y|x_i) &= E_N(y^2|x_i) - \{E_N(y|x_i)\}^2 \\ &= \frac{\sum_{j,j \neq i} K((x_j - x_i)/h)y_j^2}{\sum_{j,j \neq i} K((x_j - x_i)/h)} - \{E_N(y|x_i)\}^2 \end{aligned}$$

which does not depend on the LSE and the linearity assumption $E(y|x) = x'\beta$.

To see the effect of the nonparametric estimation on WLS, note that the moment condition for WLS b_{wls} is $E\{(y - x'b)/V(y|x)\} = 0$ and b_{wls} satisfies

$$\frac{1}{\sqrt{N}}\sum_i \frac{y_i - x_i' b_{wls}}{V_N(y|x_i)} = 0.$$

Here, $m(z_i, b, a_i) = (y_i - x_i'b)/V_N(y|x_i)$, $\alpha_i = V(y|x_i)$ and

$$m_a(z_i, \beta, \alpha_i) = -\frac{u_i}{\{V(y|x_i)\}^2} \implies E(m_a|x_i) = 0.$$

The link matrix is zero, and there is no effect of estimating $V(u|x)$ on WLS. That is, we can do as well as if the functional form of $V(y|x)$ were known.

1.3.3   Nonparametric Heteroskedasticity Test

In linear models, heteroskedasticity can be tested by the LSE of the residual squared on some functions of regressors. Consider a linear model with heteroskedasticity:

$$y_i = x_i'\beta + u_i \quad \text{and} \quad u_i^2 = z_i'\gamma + \varepsilon_i$$

where x is a $k \times 1$ vector, z is a $k_z \times 1$ vector consisting of functions of x, $E(u^2|x) = z'\gamma$, and $\varepsilon \equiv u^2 - z'\gamma$. Here we assume $z_i'\gamma > 0$ for all i, although it may be better to use a nonnegative function of z and γ such as $\exp(z'\gamma)$. With the first component of z being 1, we can test for homoskedasticity with $H_0 : \gamma_2 = \cdots = \gamma_{k_z} = 0$ by the LSE of $(y_i - x_i'b_{lse})^2$ on z_i. But this test can be misleading when $E(y|x) \neq x'\beta$. B.J. Lee (1992) suggested a nonparametric heteroskedasticity test regressing $\hat{u}_i^2$ where $\hat{u}_i = y_i - E_N(y|x_i)$ on z_i without specifying the regression function. B.J. Lee (1992) in fact suggested the LSE of $\hat{u}_i^2 f_N(x_i)$ on $z_i f_N(x_i)$ to remove the nonparametric denominator $f_N(x_i)$.

Specifically, suppose

$$y_i = E(y|x_i) + u_i \text{ with } E(u|x) = 0 \text{ and } u_i^2 = z_i'\gamma + \varepsilon_i \text{ with } E(\varepsilon|x) = 0.$$

The LSE of $\hat{u}_i^2 f_N(x_i)$ on $z_i f_N(x_i)$ is

$$g_N = \left\{ \frac{1}{N} \sum_i f_N(x_i)^2 z_i z_i' \right\}^{-1} \frac{1}{N} \sum_i f_N(x_i)^2 z_i \hat{u}_i^2.$$

Observe

$$\begin{aligned} \hat{u}^2 &= \{y - E_N(y|x)\}^2 = \{u + E(y|x) - E_N(y|x)\}^2 \\ &= z'\gamma + \varepsilon + 2u\{E(y|x) - E_N(y|x)\} + \{E(y|x) - E_N(y|x)\}^2 \\ &\quad \text{as } u_i^2 = z_i'\gamma + \varepsilon_i. \end{aligned}$$

Substitute this into g_N and solve for $\sqrt{N}(g_N - \gamma)$ to get

$$\sqrt{N}(g_N - \gamma) = \left\{ \frac{1}{N} \sum_i f_N(x_i)^2 z_i z_i' \right\}^{-1} \frac{1}{\sqrt{N}} \sum_i f_N(x_i)^2 z_i$$

$$\cdot \left[\varepsilon_i + 2u_i\{E(y|x_i) - E_N(y|x_i)\} + \{E(y|x_i) - E_N(y|x_i)\}^2 \right].$$

To find the correction term for the two nuisance parameters $\alpha_i = (f(x_i), E(y|x_i))'$, denote the summand in $N^{-1/2} \sum_i(\cdot)$ as m to differentiate m wrt a and then replace a with α:

$$\frac{\partial m}{\partial f_N(x)} = 2f(x)z\varepsilon \implies E\left\{ \frac{\partial m}{\partial f_N(x)} \Big| x \right\} = 0$$

$$\frac{\partial m}{\partial E_N(y|x)} = f(x)^2 z(-2u) \implies E\left\{ \frac{\partial m}{\partial E_N(y|x)} \Big| x \right\} = 0.$$

Hence the link function is zero and there is no first-stage estimation error effect. We thus have

$$\sqrt{N}(g_N - \gamma) = \left\{ \frac{1}{N} \sum_i f(x_i)^2 z_i z_i' \right\}^{-1}$$

$$\frac{1}{\sqrt{N}} \sum_i f(x_i)^2 z_i \varepsilon_i + o_p(1) \rightsquigarrow N(0, C)$$

where $C \equiv E^{-1}\{f(x)^2 zz'\} \cdot E\{f(x)^4 zz' \varepsilon^2\} \cdot E^{-1}\{f(x)^2 zz'\}$.

Clearly, $\varepsilon_i = u_i^2 - z_i'\gamma$ can be estimated with $\hat{u}_i^2 - z_i' g_N$ and we have $H_0 : \gamma_2 = \cdots = \gamma_{k_z} = 0$ under homoskedasticity. Under this H_0, ε_i can be estimated simply by $\hat{u}_i^2 - g_1$ where g_1 is the first element of g_N.

2   Nonparametric TSE for Endogenous Regressors

   In the preceding section, we examined two-stage estimators (TSE) with the first-stage nonparametric and the second-stage finite- dimensional, and then introduced some examples. One prominent case fitting this mold is replacing an endogenous regressor, say y_2, with $E(y_2|x)$, which is a nonparametric generalization of two-stage LSE (2SLSE) which replaces y_2 with its linear projection $L(y_2|x)$. Calling the approach "nonparametric substitution (SUB)" method, this section studies the approach in detail where the regression function can be linear/nonlinear or smooth/non-smooth. Other than nonparametric SUB, the so-called "control function (CF)" approach which adds a term—"control function"—into the model to account for the endogeneity of y_2 will be examined in the following section. Except in the last optional subsection, the second-stage parameter of interest is finite-dimensional in this section.

2.1   Linear Model and Nonparametric 2SLSE

   Consider a linear model:

$$y_1 = \alpha y_2 + x_1'\beta + u, \quad E(u|x) = 0$$

which is the first equation of a system consisting of two equations, where y_2 is the endogenous regressor and x is the exogenous variables of the system that includes at least one variable not in x_1. Suppose $E(y_2|x)$ is known for a while. Then we can estimate α and β applying IVE with $(E(y_2|x), x_1')'$ as the instrument. Defining

$$z \equiv \{E(y_2|x), x_1'\}', \quad w \equiv (y_2, x_1')', \quad \gamma \equiv (\alpha, \beta)', \quad g_N \equiv (a_N, b_N')'$$

it holds that

$$\sqrt{N}(g_N - \gamma) \rightsquigarrow N\{0, \ E^{-1}(wz') \ E(u^2 zz') \ E^{-1}(zw')\}.$$

Since $E(wz') = E_x[E(w|x) \cdot z'] = E(zz')$, the asymptotic variance becomes

$$E^{-1}(zz') \ E(u^2 zz') \ E^{-1}(zz') \quad \{= E(u^2) \cdot E^{-1}(zz')$$

under homoskedasticity}.

In reality, however, we do not know the form of $E(y_2|x)$, which means that the above IVE is infeasible. If we know that the other part of the simultaneous system (i.e., the y_2 equation) is linear, then $E(y_2|x)$ is a linear function of x. But even if $E(y_2|x)$ is not linear, or even if the y_2 equation is not specified, still we can have a feasible version replacing $E(y_2|x)$ with a nonparametric estimator $E_N(y_2|x)$. Let $z_{Ni} \equiv (E_N(y_2|x_i), x_i')'$. Then the feasible IVE with a nonparametric instrument, still denoted as g_N, satisfies

$$\frac{1}{\sqrt{N}} \sum_i (y_{1i} - w_i' g_N) z_{Ni} = 0.$$

With $m(y_i, w_i, g, z_i) = (y_i - w_i' g) z_i$ and $a_i = E_N(y_2|x_i)$ that is the first component of z_i, we get

$$E\{m_a(y_i, w_i, \gamma, z_i)|x\} = E(u|x) = 0 :$$

there is no effect of using an estimated instrument. Hence the feasible IVE has the same asymptotic distribution as the infeasible IVE.

It is interesting to note that the feasible IVE is equivalent to the nonparametric 2SLSE for $y_1 = \alpha y_2 + x_1' \beta + u$; in the nonparametric 2SLSE, first $E_N(y_2|x)$ is obtained, and then the LSE of y_1 on $E_N(y_2|x)$ and x_1 is done. Recall that this kind of equivalence also holds with the usual 2SLSE where y_2 is replaced by the linear projection $L(y_2|x)$. To see the equivalence, rewrite $y_1 = \alpha y_2 + x_1' \beta + u$ as

$$\begin{aligned} y_1 &= \alpha E(y_2|x) + x_1' \beta + u + \alpha \{y_2 - E(y_2|x)\} \\ &= \alpha E_N(y_2|x) + x_1' \beta + u + \alpha \{y_2 - E(y_2|x)\} - \alpha \{E_N(y_2|x) \\ &\quad - E(y_2|x)\} \end{aligned}$$

where the error term has three components. When LSE is applied in the second stage, the last two error terms cancel each other due to CT_{2S}, and only u is left as the error term. The asymptotic variance of the 2SLSE is then driven by, using CT_E,

$$\frac{1}{\sqrt{N}} \sum_i u_i E_N(y_2|x_i) = \frac{1}{\sqrt{N}} \sum_i u_i E(y_2|x_i) + o_p(1)$$

($E_N(y_2|x_i)$ is a nonparametric IV)

Hence the asymptotic variance becomes $E^{-1}(wz') E(u^2 zz') E^{-1}(zw')$ as appeared above.

Deriving the efficient variance matrix under the moment condition $E\{(y - \alpha_1 y_2 - x_1' \beta)|x\} = E(u|x) = 0$, we get

$$E_x^{-1}\{E(w|x) \cdot E(w'|x) \cdot (V(u|x))^{-1}\} = E_x^{-1}\left\{\frac{zz'}{V(u|x)}\right\}$$

$$= \ E(u^2) \ E^{-1}(zz') \quad \text{under homoskedasticity.}$$

This efficiency bound agrees with the asymptotic variance of the IVE/2SLSE under homoskedasticity. Thus the IVE/2SLSE, feasible or infeasible, is efficient under homoskedasticity. In this sense, $E(y_2|x)$ is the *best instrument* for y_2.

2.2   Smooth Nonlinear Models and Nonparametric SUB

Suppose we have a nonlinear model $\rho(m,\gamma) = u$ where ρ is a $s \times 1$ vector of functions known up to a $k \times 1$ parameter vector γ with

$$E\{\rho(m,\gamma)|x\} = E(u|x) = 0$$

and m includes a response variable y_1, an endogenous regressor y_2 and exogenous regressors x. Assume homoskedasticity:

$$E\{\rho(m,\gamma) \cdot \rho(m,\gamma)'|x\} \equiv \Omega \quad \text{(a constant matrix).}$$

For instance, let $\gamma = (\alpha, \beta')'$, $m_i \equiv (y_{1i}, y_{2i}, x_i')'$ and

$$\rho(m_i,\gamma) = y_{1i} - \alpha y_{2i} - r(x_{1i}, \beta) = u_i$$

where $r(x_1, \beta)$ is a nonlinear function of x_1 known up to β.

Recall the efficiency bound under $E\{\rho(m,\gamma)|x\} = 0$:

$$E_x^{-1}[E\{\rho_g(m,\gamma)|x\} \ E^{-1}\{\rho(m,\gamma)\rho(m,\gamma)'|x\} \ E\{\rho_g(m,\gamma)'|x\}]$$

$$\text{where } \rho_g(m,\gamma) \equiv \frac{\partial \rho}{\partial g}$$

$$= \ E_x^{-1}[E\{\rho_g(m,\gamma)|x\} \ \Omega^{-1} \ E\{\rho_g(m,\gamma)'|x\}]$$

$$\text{under the homoskedasticity.}$$

With an initial consistent estimator $\hat{\gamma}$ for γ, $\Omega_N \equiv N^{-1}\sum_i \rho(m_i,\hat{\gamma})\rho(m_i,\hat{\gamma})' \to^p \Omega$. The problem is in getting $E(\rho_g(m,\gamma)|x)$. In the example $\rho(m,\gamma) = y_1 - \alpha y_2 - r(x_1,\beta)$,

$$E\{\rho_g(m,\gamma)|x\} = E\{-(y_2, r_{b'}(x_1,\beta))'|x\} = \{-E(y_2|x), r_{b'}(x_1,\beta)\}'$$

$$\text{where } r_b \equiv \frac{\partial r}{\partial b}.$$

$E(y_2|x_i)$ can be replaced with a kernel estimator

$$E_N(y_2|x_i) \equiv \frac{\sum_{j,j\neq i}^N K((x_j - x_i)/h)y_{2j}}{\sum_{j\neq i}^N K((x_j - x_i)/h)}.$$

If both y_2 and x_1 reside in $r(\cdot)$ such that y_2 is not additively separable from x_1 as in $\rho(m,\gamma) = y_1 - r(y_2, x_1, \beta)$, then $E(\rho_g(m,\gamma)|x)$ can be estimated with $-E_N\{\partial r(y_2, x_1, g_0)/\partial g|x\}$ where g_0 is a $\sqrt{N}$-consistent estimator for γ. Using this kind of $E\{\rho_g(m, \gamma)|x\}$ does not lead to an efficient estimator in general. But it does attain the efficiency bound under homoskedasticity (Newey, 1990b) as illustrated in the preceding subsection.

Specifically, the following is a procedure for the efficient IVE of nonlinear (simultaneous) equations under homoskedasticity:

1. Use inefficient instruments to estimate γ by $\hat{\gamma}$ $\sqrt{N}$-consistently; e.g., $\hat{\gamma}$ may be the GMM under the moment conditions $E\{\rho(m, \gamma)$ $g_j(x)\} = 0$, $j = 1, ..., k$, where $g_1(x), ..., g_k(x)$ are some chosen functions of x.

2. Estimate Ω with Ω_N, and estimate $E\{\rho_g(m, \gamma)|x_i\}$ using a nonparametric estimator such as $E_N(y_2|x)$ if necessary. Denote the $k \times s$ estimator for $E\{\rho_g(m, \gamma)|x_i\}$ as $D_i \equiv D_i(x_i, \hat{\gamma})$.

3. Take one step from $\hat{\gamma}$ to get the efficient estimator g_N:

$$g_N = \hat{\gamma} - \left(\sum_i D_i \Omega_N^{-1} D_i' \right)^{-1} \cdot \sum_i D_i \Omega_N^{-1} \rho(m_i, \hat{\gamma}).$$

2.3   Non-smooth Models and Nonparametric SUB*

So far in this section, either linear or linearized versions of nonlinear models with an endogenous regressor have been examined. Here we study nonlinear but non-differentiable models with an endogenous regressor. Under non-differentiability, IVE is no longer applicable, but there are still various methods to handle endogeneity as reviewed in Lee (2008, semiparametric estimates for limited dependent variable (LDV) models with endogenous regressors, unpublished paper). One of them is the nonparametric 2SLSE as in Lee (1995b), which is reviewed in the following.

Consider a censored SF equation

$$y_1 = \max(\alpha y_2 + x_1'\beta + \varepsilon, 0)$$

where $y_2 = \tau_2(y_2^*)$ is a transformation of the latent continuous variable y_2^* with y_2^* related to ε, and x_1 is a $k_1 \times 1$ vector of exogenous variables. As an example, y_1 may be female labor supply, while y_2 is a dummy variable for a labor union membership (then τ_2 is an indicator function). More generally, we may consider

$$y_1 = \tau_1(y_1^*) = \tau_1\{\alpha \tau_2(y_2^*) + x_1'\beta + \varepsilon\} = \tau_1(\alpha y_2 + x_1'\beta + \varepsilon)$$

where τ_1 is a function whose form is known; e.g., $\tau_1(\cdot) = \max(\cdot, 0)$. The regressor y_2 may have its own SF with the regressors y_1 and x_2, although this is not explicit here. Define a $k \times 1$ exogenous regressor vector x as the collection of the variables in x_1 and x_2.

Rewrite the censored endogenous-regressor model as

$$
\begin{aligned}
y_1 &= \max[\alpha E(y_2|x) + x_1'\beta + \varepsilon + \alpha\{y_2 - E(y_2|x)\}, \ 0] \\
&= \max[\alpha \cdot E_N(y_2|x) + x_1'\beta + \varepsilon + \alpha v + \alpha\{E(y_2|x) - E_N(y_2|x)\}, 0], \\
&\quad v \equiv y_2 - E(y_2|x).
\end{aligned}
$$

The parameters α and β can be estimated in two stages. The first step is estimating $E(y_2|x)$ with a kernel nonparametric estimator $E_N(y_2|x)$. The second step is estimating the last equation with a semiparametric method which requires weak assumptions on $\varepsilon + \alpha v$. This strategy is applicable not only to the censored model, but also to $y = \tau_1(\alpha y_2 + x_1'\beta + \varepsilon)$ more generally. The first nonparametric step is uniform regardless of the form of τ_2 in $y_2 = \tau_2(y_2^*)$, for we need only $E_N(y_2|x)$. The second step, however, is τ_1-specific. If $\tau_1(\cdot) = \max(\cdot, 0)$, the second step needs a censored model estimator, and if τ_1 is an indicator function, the second step needs a binary model estimator. For the censored model, we will use Powell's (1986a) SCLS in the second stage as an illustration, although almost all $\sqrt{N}$-consistent semiparametric censored model estimators can be used. Lee (1996b) also showed a simpler version under the assumption $E(y_2|x) = x'\eta$, as well as extensions to the general model $y = \tau_1(\alpha y_2 + x_1'\beta + \varepsilon)$ and to cases with more than one endogenous regressors.

Turning to the details of the two-stage procedure for the censored model with SCLS as the second stage, define

$$
\begin{aligned}
\gamma &\equiv (\alpha, \beta')', \quad u \equiv \varepsilon + \alpha v, \\
z &\equiv (E(y_2|x), x_1')' \quad \text{and} \quad z_N \equiv (E_N(y_2|x), x_1')'
\end{aligned}
$$

where x and z are, respectively, $k \times 1$ and $(k_1 + 1) \times 1$ vectors. The censored model can be written as

$$
y_1 = \max(z'\gamma + u, 0).
$$

Minimize the SCLS minimand wrt g:

$$
Q_N(g) \equiv \frac{1}{N}\sum_i \left[\left\{ y_{1i} - \max\left(\frac{y_{1i}}{2}, z_{Ni}'g\right)\right\}^2 + 1[y_{1i} > 2z_{Ni}'g]\left\{\left(\frac{y_{1i}}{2}\right)^2 \right.\right.
$$
$$
\left.\left. -(\max(z_{Ni}'g, 0))^2 \right\} \right]
$$

to get the nonparametric two-stage SCLS g_N. Then

$$
\sqrt{N}(g_N - \gamma) \rightsquigarrow N(0, \ E^{-1}\{1[|u| < z'\gamma]\ zz'\}\ D\ E^{-1}\{1[|u| < z'\gamma]\ zz'\})
$$
$$
\text{where } D \equiv E[1[z'\gamma > 0]\min\{u^2, (z'\gamma)^2\}zz']
$$
$$
+ \alpha^2 E[1[z'\gamma > 0]v^2\{F_{u|x}(z'\gamma) - F_{u|x}(-z'\gamma)\}^2 zz']
$$

$$-2\alpha E[\{1[|u| < z'\gamma]u + (1[u > z'\gamma] - 1[u < -z'\gamma])z'\gamma\}$$
$$\cdot 1[z'\gamma > 0]v\{F_{u|x}(z'\gamma) - F_{u|x}(-z'\gamma)\}zz']$$

where $f_{u|x}$ is the density of $u|x$ and $F_{u|x}$ is the df.

As for estimating D, the second and third terms can be estimated by

$$a_N^2 \frac{1}{N} \sum_i 1[z'_{Ni}g_N > 0]\tilde{v}_i^2\{F_N(z'_{Ni}g_N)\}^2 z_{Ni}z'_{Ni}$$

$$-2a_N \frac{1}{N} \sum_i \{1[|\tilde{u}_i| < z'_{Ni}g_N]\tilde{u}_i + (1[\tilde{u}_i > z'_{Ni}g_N]$$

$$-1[\tilde{u}_i < -z'_{Ni}g_N])z'_{Ni}g_N\}$$

$$\cdot 1[z'_{Ni}g_N > 0]\tilde{v}_i F_N(z'_{Ni}g_N)z_{Ni}z'_{Ni} \quad \text{where}$$

$$\tilde{v}_i \equiv y_{2i} - E_N(y_2|x_i), \quad \tilde{u}_i \equiv y_{1i} - \tilde{z}'_i g_N,$$

$$F_N(z'_{Ni}g_N) \equiv \frac{\sum_{j,j\neq i} K\{(x_j - x_i)/s\}1[|\tilde{u}_j| < z'_{Ni}g_N]}{\sum_{j,j\neq i} K\{(x_j - x_i)/s\}}$$

and s is a smoothing parameter. Under the independence between x and u, $F_N(z'_{Ni}g_N)$ gets simplified to $(N-1)^{-1}\sum_{j,j\neq i} 1[|\tilde{u}_j| < z'_{Ni}g_N]$. If there is no censoring, then the indicator functions and $F_{u|x}$ in the asymptotic variance drop out to make the nonparametric SCLS equal to the nonparametric 2SLSE that appeared already with the asymptotic variance $E^{-1}(zz')E(\varepsilon^2 zz')E^{-1}(zz')$.

2.4   Nonparametric Second-Stage and Integral Equation*

Newey and Powell (2003) considered

$$y_1 = \rho(y_2, x_1) + u, \quad E(u|x) = 0$$

where y_2 is endogenous although no equation for y_2 appears explicitly. Taking $E(\cdot|x)$ on $y_1 = \rho(y_2, x_1) + u$ yields

$$\pi(x) \equiv E(y_1|x) = \int \rho(y_2, x_1)dF(y_2|x)$$

where $F(y_2|x)$ is the df of $y_2|x$. This equation may be regarded as a mapping from a space for $\rho(y_2, x_1)$ to a space for $\pi(x)$, say,

$$\pi = G_F(\rho).$$

Since $\pi(x)$ and $F(y_2|x)$ are identified, at least in principle, we can replace those with estimates $\pi_N(x)$ and $F_N(y_2|x)$ to get

$$\pi_N(x) = \int \rho_N(y_2, x_1)dF_N(y_2|x)$$

which defines implicitly the estimator ρ_N of ρ as a solution to this "Fredholm integral equation of the first kind"—"the second kind" would have ρ_N (times a constant) subtracted from the right-hand side. From this equation, we may get the inverse function $\rho = G_F^{-1}(\pi)$ and then replace F and π with F_N and π_N, which would then give an estimator $G_{F_N}^{-1}(\pi_N)$ for ρ.

Although the direction sounds right, the inverse $G_F^{-1}(\pi)$ may not exist for all possible π_N (i.e., G_F may not be "onto"); and even if it does under certain conditions, it may not be unique (i.e., G_F may not be "one-to-one"); even if it is unique, it may not be continuous in $\pi(x)$ (and $F(y_2|x)$). This means that the convergence of $\pi_N(x)$ (and $F_N(y_2|x)$) to the true function does not necessarily imply the convergence of ρ_N to ρ. Newey and Powell (2003) imposed assumptions such that G_F is bijective (one-to-one and onto) and ρ belongs to a compact set. These, combined with the continuity of $G_F(\rho)$ in ρ ($G_F(\rho)$ is linear in ρ and thus continuous), implies that $G_F^{-1}(\pi)$ is continuous in π.

Applying a series-approximation $\sum_{j=1}^{\infty} \beta_j p_j(y_2, x_1)$ to $\rho(y_2, x_1)$ where $p_1(y_2, x_1)$, $p_2(y_2, x_1)$, ... are known basis functions, we get (with the infinite series trimmed at J)

$$\pi(x) = E\{\rho(y_2, x_1)|x\} \simeq E\left\{\sum_{j=1}^{J} \beta_j p_j(y_2, x_1)|x\right\} = \sum_{j=1}^{J} \beta_j E\{p_j(y_2, x_1)|x\}.$$

The first-stage is replacing $\pi(x)$ and $E\{p_j(y_2, x_1)|x\}$ with nonparametric estimators $\pi_N(x)$ and $E_N\{p_j(y_2, x_1)|x\}$, and the second-stage is the LSE of $\pi_N(x)$ on $E_N\{p_j(y_2, x_1)|x\}$, which is in principle nonparametric when $J \to \infty$. This TSE is a SUB approach, because the endogenous part $\rho(y_2, x_1)$ is replaced by its projection on x. The actual implementation is, however, more complicated than this because of the aforementioned difficulties with G_F^{-1}. Another notable point for the implementation is that y_1 can be used instead of $\pi_N(x)$, because $E\{g(x)y_1\} = E\{g(x) \cdot E(y_1|x)\}$; i.e., we may do away with $\pi_N(x)$.

Hall and Horowitz (2005) proposed kernel-based estimators for the same model $y_1 = \rho(y_2, x_1) + u$ and $E(u|x) = 0$ and Horowitz (2007) derived its asymptotic distribution. Horowitz and Lee (2007) dealt with nonparameric quantile regression with endogenous regressors and instruments. These estimators, however, do not look any simpler than the above series-based idea. See also Florens (2003) and the references therein for more estimators. For the estimators on a nonparametric regression function with endogenous regressors, there is a lingering doubt on how feasible/workable those estimators could be. Indeed, Severini and Tripathi (2006) showed that the identification is "fragile," and various conditions used in the studies are either unverifiable or little better than restating the definition of identification. Severini and Tripathi illustrated that, even in a very special case where the endogenous regressor and an instrument are known to be jointly normal, still the problem is "ill-posed." They did show, however, that some linear functionals of the regression function can be identified even when the regression function itself is not—a positive message despite the grim picture painted on the problem in general.

3   Control-Function (CF) Approaches

"Control-function (CF) approach" turns an endogenous regressor into an exogenous one by adding an extra control function which is in essence the part of the error term related to the endogenous regressor. Viewed differently, the *control function is (a function of) the RF error of the endogenous regressor*. As noted in relation to parametric LDV estimators and as can be seen in Lee (2008, semiparametric estimates for limited dependent variable (LDV) models with endogenous regressors, unpublished paper), there are more ways to handle endogenous regressors other than SUB and CF (and IVE): model projection, system reduced form, and artificial instrumental regressor. But it seems that SUB and CF are the most popular ways of dealing with endogenous regressors. With SUB studied in the previous section, this section examines CF closely in linear and nonlinear models. In the last subsection, we also consider the case of an infinite-dimensional parameter of interest (i.e., nonparametric second stage) when there are nonparametric nuisance parameters.

3.1   Linear Models

Consider a linear SF for y_1 and a nonparametric RF for an endogenous regressor y_2:

$$y_1 = \alpha y_2 + x_1'\beta + u \text{ with } E(u) = 0 \quad \text{and} \quad y_2 = E(y_2|x) + v_2$$
$$\text{with } E(v_2|x) = 0.$$

Assume

(i) $E(u|x, v_2) = E(u|v_2)$: mean-independence of u from x given v_2

(ii) $E(u|v_2) = \zeta v_2$: linear projection where $\zeta \equiv E(v_2 u)/E(v_2^2)$.

As well known, the linear projection holds if $(u, v_2) \sim N(0, \Omega)$ for some Ω.
   CF for this model is based on

$$E(y_1|x, v_2) = \alpha y_2 + x_1'\beta + E(u|x, v_2) = \alpha y_2 + x_1'\beta + E(u|v_2)$$
$$= \alpha y_2 + x_1'\beta + \zeta v_2$$

where the first equality holds because y_2 is *determined by x and v_2*—this aspect is critical. In the second equality, if (i) does not hold, then a function of x coming from $E(u|x, v_2)$ will hamper the identification of β. Rewrite the y_1 SF as

$$y_1 = \alpha y_2 + x_1'\beta + \zeta v_2 + (u - \zeta v_2).$$

The error term $u - \zeta v_2$ satisfies

$$E(u - \zeta v_2 | x, v_2) = E(u|x, v_2) - \zeta v_2 = E(u|v_2) - \zeta v_2 = 0$$

which then implies $E(u - \zeta v_2 | x, y_2) = 0$.

The CF-based nonparametric TSE proceeds as follows. In the first stage, $E(y_2|x)$ is estimated nonparametrically by $E_N(y_2|x)$, and in the second stage, $(\alpha, \beta', \zeta)'$ is estimated by the LSE of y_1 on $(y_2, x_1', \hat{v}_2)$ where $\hat{v}_2 \equiv y_2 - E_N(y|x)$. The linear projection assumption can be relaxed, say $E(u|v_2) = \mu(v_2)$ where $\mu(\cdot)$ is an unknown function. In this case, we get

$$y_1 = \alpha y_2 + x_1'\beta + \mu(v_2) + \{u - \mu(v_2)\}$$

which is a semi-linear model to be studied in a later section.

Estimating the extra parameter ζ with $\hat{v}_2$ is a distinguishing characteristics of CF approach. By controlling the endogeneity source v_2 of u, the endogeneity problem gets avoided, whereas the endogenous regressor is replaced with its "predicted version" in SUB. One advantage of CF over SUB is that the endogeneity can be tested with $H_0: \zeta = 0$. Another advantage of CF is the ability to remove the endogeneity of multiple endogenous regressors: e.g., y_2 and y_2^2 may appear together in the y_1 equation, and both variables are turned exogenous by controlling the single variable v_2. Two disadvantages of CF can be seen in the above assumptions (i) and (ii): (i) is close to assuming $u \amalg x$ although somewhat weaker, and (ii) is a linearity assumption. Later it will be seen that (ii) may be avoided with "pairwise differencing," but assumptions like (i) are indispensable for CF.

For the LSE of y_1 on $(y_2, x_1', \hat{v}_2)$, define $\varepsilon \equiv u - \zeta v_2$ and rewrite $y_1 = \alpha y_2 + x_1'\beta + \zeta v + (u - \zeta v_2)$ as

$$y_1 = \alpha y_2 + x_1'\beta + \zeta \hat{v}_2 + \varepsilon - \zeta(\hat{v}_2 - v_2).$$

Define

$$\hat{q} \equiv (y_2', x_1', \hat{v}_2)', \quad q \equiv (y_2, x_1', v_2) \text{ and } \delta \equiv (\alpha, \beta', \zeta)'$$

to get $y_1 = \hat{q}'\delta + \varepsilon - \zeta(\hat{v}_2 - v_2)$ and

$$\sqrt{N}(\hat{\delta} - \delta) = \left(\frac{1}{N}\sum_i \hat{q}_i \hat{q}_i'\right)^{-1} \frac{1}{\sqrt{N}}\sum_i \hat{q}_i\{\varepsilon_i - \zeta(\hat{v}_{2i} - v_{2i})\}$$

$$= E^{-1}(qq')\frac{1}{\sqrt{N}}\sum_i \hat{q}_i[\varepsilon_i - \zeta\{E(y_2|x_i) - E_N(y_2|x_i)\}] + o_p(1).$$

The nuisance parameter is $E(y_2|x_i)$ in $v_{2i} = y_{2i} - E(y_2|x_i)$, and the estimator $E_N(y|x_i)$ appears in the last component of $\hat{q}_i$ as well as in $\varepsilon_i - \zeta\{E(y_2|x_i) - E_N(y_2|x_i)\}$. Use CT_E and the product-differentiation rule to get

$$\frac{1}{\sqrt{N}}\sum_i \hat{q}_i[\varepsilon_i - \zeta\{E(y_2|x_i) - E_N(y_2|x_i)\}]$$

$$= \frac{1}{\sqrt{N}} \sum_i [q_i \varepsilon_i + E\{(0, 0_{1 \times k_1}, -1)' \varepsilon_i + q_i(\varepsilon_i + \zeta)|x_i\}$$

$$\cdot \{y_{2i} - E(y_2|x_i)\}] + o_p(1)$$

$$= \frac{1}{\sqrt{N}} \sum_i [q_i \varepsilon_i + \zeta E(q|x_i)\{y_{2i} - E(y_2|x_i)\}] + o_p(1)$$

where $E(q|x_i) = \{E(y_2|x_i), x_1', 0\}'$.

Therefore, defining $\xi_i \equiv q_i \varepsilon_i + \zeta E(q|x_i)\{y_{2i} - E(y_2|x_i)\}$, it holds that

$$\sqrt{N}(\hat{\delta} - \delta) \rightsquigarrow N\{0, \ E^{-1}(qq')E(\xi\xi')E^{-1}(qq')\}.$$

As usual, the expected values can be estimated by their sample analogs with q and $E(y_2|x)$ replaced by $\hat{q}$ and $E_N(y_2|x)$. Under "H_0: no endogeneity $\iff \zeta = 0$," the correction term disappears and the usual LSE asymptotic variance works: as far as endogeneity testing goes, the correction term can be ignored.

3.2   Nonlinear Models

Defining $w \equiv (y_2, x_1')'$ and $\gamma \equiv (\alpha, \beta')'$, suppose

$$y_1 = T(w'\gamma + u) \quad \text{and} \quad y_2 = E(y_2|x) + v_2 \quad \text{where } u = \lambda v_2 + \varepsilon, \ v_2 \amalg \varepsilon$$

$T(\cdot)$ is a known increasing nonlinear function and x includes x_1. For example $T(\cdot)$ may be smooth and one-to-one as $\exp(\cdot)$, or non-smooth and many-to-one as $\max(\cdot, 0)$. Substitute $u = \lambda v_2 + \varepsilon$ into the y_1 equation to get

$$y_1 = T(w'\gamma + \lambda v_2 + \varepsilon) = T\{w'\gamma + \lambda \hat{v}_2 + \varepsilon - \lambda(\hat{v}_2 - v_2)\}.$$

Imposing some semiparametric assumptions on ε, (γ', λ) can be estimated using $(w', \hat{v}_2)$ as the regressors.

For instance, suppose that the αth conditional quantile $Q_\alpha(\varepsilon|x, v_2) = Q_\alpha(\varepsilon|x)$ is zero. Then the quantile equivariance renders

$$Q_\alpha(y_1|x, v_2) = T\{w'\gamma + \lambda v_2 + Q_\alpha(\varepsilon|x, v_2)\} = T(w'\gamma + \lambda v_2)$$

and we can estimate $Q_\alpha(y_1|x, v_2)$ by minimizing wrt (γ', λ)

$$\frac{1}{N} \sum_i \{y_{1i} - T(w_i'\gamma + \lambda \hat{v}_{2i})\}\{\alpha - 1[y_{1i} - T(w_i'\gamma + \lambda \hat{v}_{2i}) < 0]\}.$$

Note that the assumptions $u = \lambda v_2 + \varepsilon$, $v_2 \amalg \varepsilon$ and $Q_\alpha(\varepsilon|x) = 0$ replace the assumptions $E(u|x, v_2) = E(u|v_2)$ and $E(u|v_2) = \zeta v_2$ in the preceding subsection for linear models that would give only

$$u = \zeta v_2 + \omega \text{ and } E(\omega|x, v_2) = E(\omega|v_2) = 0.$$

Now suppose that $y_2 = \tau(x_2'\beta_2 + u_2)$ is a many-to-one mapping. Note that $y_2 = \tau(x_2'\beta_2 + u_2)$ is a SF whereas $y_2 = E(y_2|x) + v_2$ is a RF. The SF2 error u_2 is either not identified at all as in binary response models or identified only partly as in censored response models. Specifically, if y_2 is binary with $y_2^* = x_2'\beta_2 + u_2$ and $u_2 \sim N(0, \sigma^2)$ independently of x, then the identified RF2 error is only $w_2 \equiv y_2 - \Phi(x_2'\beta_2/\sigma)$ where Φ is the $N(0, 1)$ df. In this case, if $T(\cdot) = \max(\cdot, 0)$, then we may assume

$$ u = \lambda w_2 + \varepsilon \quad \text{and} \quad Q_\alpha(\varepsilon|x, w_2) = Q_\alpha(\varepsilon|w_2) = 0 $$

and use w_2 as a control variable in SF1 to remove the endogeneity of y_2. If $y_2 = \max(x_2'\beta_2 + u_2, 0)$, then the identified RF2 error is $w_2 \equiv y_2 - x_2'\beta_2 = \max(u_2, -x_2'\beta_2)$. That is, $y_2 - x_2'\beta_2$ may be used as a control variable in the y_1 equation. This kind of CF approach, however, could be controversial because of the assumption that u depends on the identified RF2 error w_2, not on the SF2 error u_2—more on this in the next subsection.

Sometimes we get a mixture of SUB and CF approaches. To see this, consider

$$ y_1 = \max\{0, \ \alpha y_2 + x_1'\beta + u\} $$
$$ = \max\{0, \ \alpha E(y_2|x) + x_1'\beta + \alpha(y_2 - E(y_2|x)) + u\} $$
$$ \implies \quad y_1 = \max\{0, \ \alpha E(y_2|x) + x_1'\beta + \alpha v_2 + u\} $$
$$ = \quad \max\{0, \ \alpha E_N(y_2|x) + x_1'\beta + \alpha v_2 + u + \alpha\{E(y_2|x) $$
$$ -E_N(y_2|x)\} \quad \text{for SUB} $$

Now suppose $u = \lambda v_2 + \varepsilon$. This gives two more equations

$$ (a): y_1 = \max\{0, \ \alpha y_2 + x_1'\beta + \lambda v_2 + \varepsilon\} $$
$$ = \max\{0, \ \alpha y_2 + x_1'\beta + \lambda \hat{v}_2 + \varepsilon + \lambda(v_2 - \hat{v}_2)\} $$
$$ \text{as } y_2 = E(y_2|x) + v_2 \quad \text{for CF}; $$

$$ (b) \quad : \quad y_1 = \max[0, \ \alpha\{E_N(y_2|x) + \hat{v}_2\} + x_1'\beta + \lambda \hat{v}_2 + \varepsilon $$
$$ +\lambda(v_2 - \hat{v}_2)] \quad \text{as } y_2 = E_N(y_2|x) + \hat{v}_2 $$
$$ = \max[0, \ \alpha E_N(y_2|x) + x_1'\beta + (\alpha + \lambda)\hat{v}_2 $$
$$ +\varepsilon + \lambda(v_2 - \hat{v}_2)\}] \quad \text{for SUB/CF.} $$

It is helpful to put together the three cases:

	Parameters	Regressors	Error Term
SUB	α, β	$E_N(y_2\|x), x_1$	$\alpha v_2 + u + \alpha\{E(y_2\|x) - E_N(y_2\|x)\}$
CF	α, β, λ	$y_2, x_1, \hat{v}_2$	$\varepsilon + \lambda(v_2 - \hat{v}_2)$
SUB/CF	$\alpha, \beta, \alpha + \lambda$	$E_N(y_2\|x), x_1, \hat{v}_2$	$\varepsilon + \lambda(v_2 - \hat{v}_2)$

SUB/CF has the SUB "flavor" because $E_N(y_2|x)$ replaces y_2 as a regressor, and also has the CF flavor because $\hat{v}_2$ is controlled. But this mixed case is equivalent to CF, because (b) is just a reparametrized version of (a).

3.3   Average Structural Function (ASF)

With $w = (y_2, x_1')'$, consider a nonlinear model

$$y_1 = T_1(w, u) \ \{= T_1(y_2, x_1, u)\}$$

which is a generalization of $y_1 = \alpha y_2 + x_1'\beta + u$. Also, T_1 generalizes T in the preceding subsection because T_1 allows w and u to enter separately whereas w and u get combined linearly and enter T as a single entity. Suppose there exists a variable v_2 such that

(i) $w = (y_2, x_1')'$ is a function of (x, v_2)

(ii) $u \amalg x | v_2$.

The former holds if $y_2 = E(y_2|x) + v_2$, as this implies that (y_2, x_1) is a function of (x, v_2).

Taking $E(\cdot|x, v_2)$ on $y_1 = T_1(w, u)$ yields

$$E(y_1|x, v_2) = E\{T_1(w, u)|x, v_2\} = \int T_1(w, u) f(u|x, v_2) du$$

$$= \int T_1(w, u) f(u|v_2) du.$$

Integrate v_2 out from the first and last terms to get

$$\int E(y_1|x, v_2) f(v_2) dv_2 = \int \int T_1(w, u) f(u|v_2) du f(v_2) dv_2$$

$$= \int T_1(w, u) f(u) du \equiv A(w).$$

This is the *average structural function (ASF)* in Blundell and Powell (2003). The unobserved u in $y_1 = T_1(w, u)$ is integrated out using its *marginal* density $f(u)$.

The finding $\int E(y_1|x, v_2) f(v_2) dv_2 = A(w)$ shows how to estimate $A(w)$. As the last two displays hold with '$|w, v_2$' instead of '$|x, v_2$', using "marginal integration" or "partial mean" as in Newey (1994b), ASF can be estimated with

$$\frac{1}{N} \sum_i E_N(y_1|w, \hat{v}_{2i}) \to^p \int E(y_1|w, v_2) f(v_2) dv_2 = A(w).$$

Note that only $\hat{v}_2$ is summed up while w is fixed. This sort of "marginal integration estimator" will be examined in detail later.

Strictly speaking, estimating ASF this way is applicable only to continuous endogenous regressors with unbounded support. To see this, observe that v_2 in the first stage should be estimable over the entire support of v_2 to integrate out v_2 using its marginal distribution in the second stage. But in the first stage, since v_2 is estimated using y_2, this may not be possible if $y_2 = \tau(x, v_2)$ is a LDV with v_2 residing inside the range-restricted $\tau(\cdot)$; e.g., $\tau(x, v_2) = \max(x'\eta_2 + v_2, 0)$.

It may be helpful to make clear that there can be in fact *three types of error terms*. For the censored median example, the three types of errors are: the SF error u_2 in $y_2 = \max(x_2'\beta_2 + u_2, 0)$, the RF error v_2 in $y_2 = \max(x'\eta_2 + v_2, 0)$ where $y_2^* = x'\eta_2 + v_2$, and the "truly" RF error

$$\omega_2 \equiv y_2 - \max(x'\eta_2, 0) = \max(x'\eta_2 + v_2, 0) - \max(x'\eta_2, 0).$$

We already mentioned the potential controversy in using ω_2, not v_2, in some CF approaches. The problem noted in the preceding paragraph occurs even if v_2 is used, as v_2 is not fully identified in the censored case. To avoid this criticism, one may suggest using ω_2 in the above assumptions (i) and (ii). Although (i) holds by construction $y_2 = \max(x'\eta_2, 0) + \omega_2$, (ii) is unlikely to hold with ω_2, because ω_2 depends on both x and v_2 while we need $u \amalg x|\omega_2$. For instance, $\omega_2 = 0 \iff (x'\eta_2 > -v_2, x'\eta_2 > 0)$ or $(x'\eta_2 < -v_2, x'\eta_2 < 0)$, which means that $x'\eta_2$ and v_2 both tend to be big or small together; with $COR(u, v_2) \neq 0$, this implies $COR(x, u) \neq 0 \,|(\omega_2 = 0)$. Hence ASF is feasible only for continuous endogenous regressors.

Instead of $A(w)$, Altonji and Matzkin (2005) examined

$$\int \frac{\partial T_1(w_o, u)}{\partial w} f(u|w_o) du$$

which is the effect of w averaged across u given $w = w_o$. Estimating this is, however, far more involved than estimating $A(w_o)$. In identifying this and ASF, the endogeneity source is "blocked" by conditioning on v_2 as well as on x (but x drops out due to $u \amalg x|v_2$), which is why this effect estimator and ASF are examined in relation to CF approach.

3.4   Pairwise Differencing for Nonparametric CF

As mentioned once, if we allow $E(u|v_2)$ to be a nonparametric function of v_2, then CF will end up with a regression function that is semi-linear. Semi-linear models will be examined in detail later, but in the following we provide an illustration of how to handle semi-linear models resulting from CF using the idea of pairwise differencing.

Assume quantile independence: for an unknown function $\lambda_\alpha(\cdot)$,

$$Q_\alpha(u|x, v_2) = Q_\alpha(u|v_2) \equiv \lambda_\alpha(v_2).$$

For a censored model with $y_1 = \max(w'\gamma + u, 0)$ where $w \equiv (y_2, x_1')'$ and $\gamma \equiv (\alpha, \beta')'$, this assumption yields

$$Q_\alpha(y_1|x, v_2) = \max\{w'\gamma + Q_\alpha(u|x, v_2),\ 0\} = \max\{w'\gamma + \lambda_\alpha(v_2), 0\}.$$

Define

$$q_{\alpha i} \equiv Q_\alpha(y_1|x_i, v_{2i}).$$

For an observation pair i and j with $q_{\alpha i}, q_{\alpha j} > 0$ (so that the max function drops out) and $v_{2i} = v_{2j}$, we get $q_{\alpha i} = w_i'\gamma + \lambda_\alpha(v_{2i})$ and thus

$$q_{\alpha i} - q_{\alpha j} = (w_i - w_j)'\gamma.$$

Based on this pairwise differencing idea, Blundell and Powell (2007) proposed a $\sqrt{N}$-consistent estimator for γ:

$$\left\{ \sum_{i<j} K\left(\frac{\hat{v}_{2i} - \hat{v}_{2j}}{h}\right) (w_i - w_j)(w_i - w_j)' \right\}^{-1}$$

$$\cdot \sum_{i<j} K\left(\frac{\hat{v}_{2i} - \hat{v}_{2j}}{h}\right) (w_i - w_j)(\hat{q}_{\alpha i} - \hat{q}_{\alpha j}).$$

In practice, however, it seems adequate to use a polynomial function of v_2 as a control function, say $\lambda_1 v_2 + \lambda_2 v_2^2 + \lambda_3 v_2^3$, and estimate $(\lambda_1, \lambda_2, \lambda_3)$ along with γ.

3.5   Nonparametric Second-Stage*

Newey et al. (1999) considered

$$y_1 = \rho(y_2, x_1) + u$$

$$y_2 = E(y_2|x) + v_2, \quad v_2 \equiv y_2 - E(y_2|x) \text{ which might be related to } u$$

where $\rho(\cdot)$ is an unknown function. While the y_1-equation is a SF, the y_2-equation is of RF type because y_1 does not appear in the y_2-equation—thus a "triangular" system. The endogeneity of y_2 comes from the relation between u and v_2. Of course, we may imagine a y_2-SF without y_1 as a regressor, in which case the above displayed y_2-equation is a SF. Estimating $\rho(y_2, x_1)$ non-parametrically despite the endogeneity of y_2 is the main goal, which is much more difficult than when the form of ρ is specified up to some parameters.

Newey et al. (1999) imposed the mean independence of u from x given v_2:

$$E(u|x, v_2) = E(u|v_2) \equiv \lambda(v_2) \; (\neq 0);$$

if $E(u|v_2) = 0$, then there would be no endogeneity of y_2 in the y_1-SF. Take $E(\cdot|x, v_2)$ on the y_1-SF to get

$$E(y_1|x, v_2) = \rho(y_2, x_1) + E(u|x, v_2) = \rho(y_2, x_1) + E(u|v_2)$$

$$= \rho(y_2, x_1) + \lambda(v_2) \equiv R(y_2, x_1, v_2).$$

From this, a TSE can be thought of. First, estimate the y_2-RF to get the residual $\hat{v}_2$. Then do a nonparametric regression of y_1 on $y_2, x_1, \hat{v}_2$ to estimate $R(y_2, x_1, v_2)$.

Since the goal is estimating $\rho(\cdot)$, $\rho(\cdot)$ has to be separated from $R(\cdot)$. An easy way to do this is using a series approximation for ρ and λ separately. That is, no interaction term between (y_2, x_1) and v_2 is included, which means imposing the separability of ρ from λ in $R(\cdot)$ on the estimation procedure. This convenience in imposing separability is an advantage of series-approximation. Newey et al. (1999) indeed used two types of series-approximations and then derived the asymptotic normal distributions of functionals of ρ—e.g, an weighted average of $\rho(y_2, x_1)$ or ρ at a chosen value of the arguments. Although they do not derive the convergence rates for the functionals in general, they show when the functional estimators are $\sqrt{N}$-consistent. Differently from the model $y_1 = \rho(y_2, x_1) + u$ and the methods as in Newey and Powell (2003) and Hall and Horowitz (2005), the procedure in Newey et al. (1999) makes use of more assumptions such as $E(u|x, v_2) = E(u|v_2)$ and is likely to work better in practice.

If we use a kernel estimator for the second stage instead, then it is difficult to impose the separability. In this case, one way to get ρ is integration wrt a weighting function $\omega(v_2)$ where $\int \omega(v_2)dv_2 = 1$:

$$\int R(y_2, x_1, v_2)\omega(v_2)dv_2 = \rho(y_2, x_1) + \int \lambda(v_2)\omega(v_2)dv_2$$

$$= \rho(y_2, x_1) + (\text{a constant}).$$

This idea of separating a component of a nonparametric regression function using "marginal integration" appears in Newey (1994b) and Linton and Nielsen (1995) and will be examined closely in a later section. Although the difficulty of imposing separability is a disadvantage of kernel estimator, an advantage is that it is relatively easy to find convergence rates in kernel estimation. For instance, Ahn (1995) dealt with a two-stage estimator with both stages being nonparametric kernel estimation. He proves that the convergence rate is the minimum of the first-stage rate and the 'second-stage rate with the first-stage parameter known', which is intuitively plausible.

4   Single Index Models

This section introduces a number of semiparametric estimators for single-index models. The first is the "density-weighted" average derivative estimator of Powell et al. (1989), which is analogous to the earlier single index model estimator of Stoker (1986); unweighted versions will be examined as well. The second is Ichimura's (1993) single index estimator, which is a nonparametric generalization of nonlinear LSE. The third is the Klein and Spady's (1993) estimator that is, however, good only for binary response models. In essence, the nonparametric dimension gets reduced to the number of indices in single and multiple index models.

4.1   Density-Weighted Average Derivative (WADE)

Consider a single-index model

$$y = r(x) + u \quad \text{with } E(y|x) = r(x) = G(x'\beta), \quad G(\cdot) \text{ unknown.}$$

From this, we get

$$\nabla r(x) \left(= \frac{\partial r(x)}{\partial x} \right) = \frac{dG(x'\beta)}{d(x'\beta)} \cdot \beta$$

$$\implies E_x\{\omega(x)\nabla r(x)\} = E_x \left\{ \omega(x) \frac{dG(x'\beta)}{d(x'\beta)} \right\} \beta \equiv \gamma_\omega \beta$$

for a weighting function $\omega(x)$

which is proportional to β provided $\gamma_\omega \neq 0$; i.e., β is identified up to the unknown scale factor γ_ω.

Setting $\omega(x) = f(x)$, the density of x, we get the *density weighted average derivative* δ:

$$\delta \equiv E\{f(x)\nabla r(x)\} = E \left\{ f(x) \frac{dG(x'\beta)}{d(x'\beta)} \right\} \cdot \beta \equiv \gamma_f \beta.$$

Suppose x has a continuous density function $f(x)$ which is zero at the boundary "∂X" of its support. With $r(x)$ bounded, integration by parts yields

$$\delta = \gamma_f \beta = E\{f(x)\nabla r(x)\} = \int \nabla r(x) \cdot f(x)^2 dx = r(x)f(x)^2|_{\partial X}$$

$$- 2 \int \{r(x)\nabla f(x) \cdot f(x)\} dx$$

$$= -2 \cdot E[\{r(x) + u\} \cdot \nabla f(x)] = -2 \cdot E\{y\nabla f(x)\};$$

note that x in $E\{f(x)\nabla r(x)\}$ is a rv whereas x in $\int \nabla r(x)f(x)^2 dx$ is just an integration dummy.

The elements of x that are functionally dependent are not identified because $\gamma_w \beta$ is identified through partial differentiation of $\nabla r(x)$. For example, suppose there are two regressors w and z, and $r(w,z) = G(\beta_w w + \beta_{w2} w^2 + \beta_z z)$. Then

$$\nabla r(w,z) = \begin{bmatrix} \partial r(x)/\partial w \\ \partial r(x)/\partial z \end{bmatrix} = \frac{dG(\beta_w w + \beta_{w2} w^2 + \beta_z z)}{d(\beta_w w + \beta_{w2} w^2 + \beta_z z)} \begin{bmatrix} \beta_w + \beta_{w2} 2w \\ \beta_z \end{bmatrix}:$$

β_w and β_{w2} are not identified although β_z is so. To simplify notations, we will assume no functionally dependent elements in x from now on, unless otherwise noted.

We can estimate δ by plugging a kernel estimator into $\nabla f(x)$ and using a sample moment for $E\{y\nabla f(x)\}$ as will be shown shortly. The reason why we use $f(x)$ as the weight is to prevent $f_N(x)$ from appearing in the denominator of the kernel estimator, because having $f_N(x)^{-1}$ is troublesome when $f_N(x) \simeq 0$. Although this problem may be overcome by "trimming" (using observations with $f_N(x_i) > \varepsilon$ for some $\varepsilon > 0$), trimming introduces yet another "tuning constant" ε to choose.

An estimator for $\nabla f(x_i)$ is

$$\nabla f_N(x_i) = \frac{1}{(N-1)h^{k+1}} \sum_{j,j\neq i} \nabla K\left(\frac{x_i - x_j}{h}\right) \quad \text{where}$$

$$f_N(x_i) \equiv \frac{1}{(N-1)h^k} \sum_{j,j\neq i} K\left(\frac{x_j - x_i}{h}\right).$$

as $-\nabla K(z) = \nabla K(-z)$ from the symmetry of K. Then the aforementioned density-weighted average derivative estimator (WADE) for δ is

$$\delta_N \equiv -\frac{2}{N} \sum_i y_i \nabla f_N(x_i) = \frac{2}{N(N-1)} \sum_i \sum_{j,j\neq i} \frac{1}{h^{k+1}} \nabla K\left(\frac{x_j - x_i}{h}\right) y_i$$

$$= \frac{2}{N(N-1)} \sum_{i<j} \frac{1}{h^{k+1}} \nabla K\left(\frac{x_j - x_i}{h}\right) (y_i - y_j)$$

again using $-\nabla K(z) = \nabla K(-z)$.

For an "U-statistic" U_N and its parameter θ, it is known that

$$U_N \equiv \frac{2}{N(N-1)} \sum_{i<j} g(z_i, z_j) \quad \text{and} \quad \theta \equiv E\{g(z_i, z_j)\}$$

$$\implies \sqrt{N}(U_N - \theta) = \frac{2}{\sqrt{N}} \sum_i [E\{g(z_i, z_j|z_i)\} - \theta] + o_p(1)$$

$$\rightsquigarrow N(0,\ 4 \cdot E[E\{g(z_i, z_j|z_i)\} - \theta]^2) \quad \text{unless}$$

$$E[E\{g(z_i, z_j|z_i)\} - \theta]^2 = 0.$$

Apply this to δ_N to obtain

$$\sqrt{N}(\delta_N - \delta) \rightsquigarrow N(0, 4\Omega) \quad \text{and it holds that}$$

$$\Omega_N = \frac{1}{N}\sum_i [E_N\{g(z_i, z_j)|z_i\}][E_N\{g(z_i, z_j)|z_i\}]' - \delta_N \delta_N' \to^P \Omega$$

$$E_N\{g(z_i, z_j)|z_i\} \equiv \frac{1}{N-1}\sum_{j,j\neq i} \frac{1}{h^{k+1}} \nabla K\left(\frac{x_j - x_i}{h}\right)(y_i - y_j).$$

When $\sqrt{N}(b_N - \beta) = N^{-1/2}\sum_i \psi_i + o_p(1)$, ψ_i is an influence function and b_N is said to be *asymptotically linear*. The influence function shows the contribution of each observation to the total estimation error $\sqrt{N}(b_N - \beta)$. For the U-statistic U_N, the influence function is $2 \cdot [E\{g(z_i, z_j|z_i)\} - \theta]$. But this does not show well the sources of the estimation error. To see this better, observe

$$\sqrt{N}(\delta_N - \delta) = \frac{-2}{\sqrt{N}}\sum_i [y_i \nabla f_N(x_i) - E\{y\nabla f(x)\}].$$

Invoking $CT_{\nabla f}$ with $m = y\nabla f(x)$ and $m_a = y$, we get

$$\sqrt{N}(\delta_N - \delta) = \frac{-2}{\sqrt{N}}\sum_i [y_i \nabla f(x_i) - \nabla\{E(y|x_i)f(x_i)\}$$
$$+ E(\nabla\{E(y|x)f(x)\}) - E\{y\nabla f(x)\}]$$
$$= \frac{-2}{\sqrt{N}}\sum_i [y_i \nabla f(x_i) - r(x_i)\nabla f(x_i) - f(x_i)\nabla r(x_i)$$
$$+ E\{r(x)\nabla f(x) + f(x)\nabla r(x)\} - E\{y\nabla f(x)\}]$$
$$= \frac{-2}{\sqrt{N}}\sum_i [u_i \nabla f(x_i) - f(x_i)\nabla r(x_i) + E\{r(x)\nabla f(x)$$
$$+ f(x)\nabla r(x)\} - E\{r(x)\nabla f(x)\}]$$
$$= \frac{-2}{\sqrt{N}}\sum_i [\, u_i \nabla f(x_i) - (f(x_i)\nabla r(x_i) - E\{f(x)\nabla r(x)\})\,].$$

This breaks down the influence function of WADE into two sources of error: u and $\nabla r(x)$. Since the covariance between the two terms is zero due to $E(u|x) = 0$, the asymptotic variance is the sum of the two variances.

In the usual linear model with $G(x'\beta) = x'\beta$, the density-weighted average derivative δ becomes $Ef(x) \cdot \beta$ for $dG(x'\beta)/d(x'\beta) = 1$. This motivates estimating $\delta^* \equiv \delta/Ef(x)$ instead of δ, and Powell et al. (1989) suggested an estimator for δ^*:

$$\delta_N^* \equiv \left\{\sum_i \nabla f_N(x_i) \cdot x_i'\right\}^{-1} \cdot \sum_i \nabla f_N(x_i)y_i.$$

The reason why this is a legitimate estimator for δ^* can be seen in the next subsection; δ_N^* has the form of an IVE with $\nabla f_N(x)$ as an instrument in the regression of y on x. In the next subsection, we will examine more of this kind of estimators. In the Monte Carlo study of Powell et al. (1989), δ_N^* performed better than δ_N. Härdle and Tsybakov (1993) and Powell and Stoker (1996) discussed how to choose a smoothing parameter for WADE. Nishiyama and Robinson (2000) provide an 'Edgeworth expansion' for WADE.

4.2   Average Derivative Estimators (ADE)

4.2.1   Motivations

In the preceding subsection, we examined the density-weighted average derivative estimator (WADE) for single-index models (SIM). Consider the "unweighted" average derivative estimator (ADE) for $E\{\nabla r(x)\}$ instead:

$$E\{\nabla r(x)\} = E\left\{\frac{dG(x'\beta)}{d(x'\beta)}\right\} \cdot \beta;$$

$E\{\nabla r(x)\} = \beta$ if $G(x'\beta) = x'\beta$. ADE becomes β if the model is indeed linear. In this sense, ADE is more natural than WADE.

Recall the following in the preceding chapter:

$$E\left\{\frac{\partial G(x'\beta)}{\partial x}\right\} = E\left\{\frac{dG(x'\beta)}{d(x'\beta)}\right\} \cdot \beta \equiv \gamma\beta \quad \text{where}$$

$$\gamma \equiv E\left\{\frac{dG(x'\beta)}{d(x'\beta)}\right\} \text{ is a scalar;}$$

$$E\left\{\frac{\partial G(x'\beta)}{\partial x}\right\} = E\{y\lambda(x)\}, \quad \text{where}$$

$$\lambda(x) \equiv \frac{-\nabla f(x)}{f(x)}1[f(x) > 0] = -\nabla \ln f(x) \cdot 1[f(x) > 0].$$

Observe $E\lambda(x) = 0$ because

$$
\begin{aligned}
E\lambda(x) &= E\left\{\frac{-\nabla f(x)}{f(x)}1[f(x) > 0]\right\} \\
&= E\left\{\frac{-\nabla f(x)}{f(x)}1[f(x) > 0] + 1[f(x) = 0]\right\} \\
&= \int \nabla f(x)dx = f(x)|_{\partial X} = 0.
\end{aligned}
$$

Thus,

$$\gamma\beta = E\{\nabla r(x)\} = E\{y\lambda(x)\} = COV\{\lambda(x), y\}.$$

Going further, observe

$$
\begin{aligned}
I_k &= E\left(\frac{\partial x}{\partial x}\right) = \int \frac{\partial x}{\partial x}f(x)dx = xf(x)|_{\partial X} - \int x\nabla f(x)dx \\
&= \int \{\lambda(x)x\}f(x)dx = COV\{\lambda(x), x\}.
\end{aligned}
$$

Combining this with the preceding display renders the key equations

$$\gamma\beta E\{\nabla_r(x)\} = COV\{\lambda(x), y\} = [COV\{\lambda(x), x)\}]^{-1} \cdot COV\{\lambda(x), y\}.$$

The presence of $1[f(x) > 0]$ entails a cumbersome "trimming" issue because $1[f_N(x) > \varepsilon_N]$ should be used in principle for some chosen value of $\varepsilon_N > 0$. In the following, however, we will ignore $1[f(x) > 0]$ and the trimming issue as if $\min_i f_N(x_i) > \varepsilon_N$. This is because throwing away a few observations with near zero (or negative) $f_N(x)$ would not bias the estimator much, and thus trimming does not seem to matter much in practice.

4.2.2   Estimators

Let $\lambda_N(x_i) \equiv -\nabla f_N(x_i)/f_N(x_i)$. Also let $\nabla r_N(x_i)$ be given by

$$\nabla r_N(x_i) = \frac{1}{h}\sum_{j,j\neq i}\frac{\nabla K((x_i - x_j)/h)y_j}{f_N(x_i)} - r_N(x_i)\frac{\nabla f_N(x_i)}{f_N(x_i)},$$

which is the derivative of the leave-one-out kernel estimator $r_N(x_i)$ with $K((x_j - x_i)/h)$. The key equations yield the following sample analogs (i.e., estimators) for $\gamma\beta$:

$$\text{ADE} \quad : \quad \frac{1}{N}\sum_i \nabla r_N(x_i)$$

$$\text{COV-ADE} \quad : \quad \frac{1}{N}\sum_i \lambda_N(x_i)(y_i - \bar{y})$$

$$\text{IVE-ADE} \quad : \quad \left\{\frac{1}{N}\sum_i \lambda_N(x_i)(x_i - \bar{x})'\right\}^{-1} \cdot \frac{1}{N}\sum_i \lambda_N(x_i)(y_i - \bar{y}).$$

where $\bar{x}$ and $\bar{y}$ may be omitted because $E\lambda(x) = 0$. IVE-ADE is similar to δ_N^* in the previous subsection; δ_N^* has $\nabla f_N(x_i)$ in place of $\lambda_N(x_i)$ due to the weighting by $f(x_i)$. All three estimators are asymptotically equivalent and $\sqrt{N}$-consistent (Stoker, 1991). In small samples, however, IVE-ADE seems to behave better than ADE and COV-ADE.

Denoting the COV-ADE as m_N, Härdle and Stoker (1989) showed that $\sqrt{N}(m_N - \mu)$ is asymptotically normal with the asymptotic variance

$$V[u\lambda(x) + \{\nabla r(x) - E\nabla r(x)\}];$$

compare this to the asymptotic variance of WADE which differs only by the multiplicative factor $f(x)$. To understand this display, consider instead ADE and invoke CT_{AE} with $m(x_i, a_i) = a_i = \nabla r_N(x_i)$ and $u_i = y_i - r(x_i)$ to get, as $m_a = 1$

$$\frac{1}{\sqrt{N}}\sum_i \nabla r_N(x_i) = \frac{1}{\sqrt{N}}\sum_i \left\{\nabla r(x_i) - \frac{\nabla f(x_i)}{f(x_i)}u_i\right\} + o_p(1)$$

$$\implies \quad \frac{1}{\sqrt{N}} \sum_i \{\nabla r_N(x_i) - E\nabla r(x)\}$$

$$= \quad \frac{1}{\sqrt{N}} \sum_i \left\{ \nabla r(x_i) - E\nabla r(x) - \frac{\nabla f(x_i)}{f(x_i)} u_i \right\} + o_p(1)$$

which gives the above variance.

The asymptotic variance can be consistently estimated with

$$\frac{1}{N} \sum_i \zeta_{Ni} \zeta_{Ni}' - \left(\frac{1}{N} \sum_i \zeta_{Ni} \right) \left(\frac{1}{N} \sum_i \zeta_{Ni} \right)' \quad \text{where}$$

$$\zeta_{Ni} \equiv \{y_i - r_N(x_i)\} \lambda_N(x_i) + \nabla r_N(x_i) - \frac{1}{N} \sum_i \nabla r_N(x_i).$$

A tempting but invalid estimator for the asymptotic variance is the sample analog

$$\frac{1}{N} \sum_i \nabla r_N(x_i) \{\nabla r_N(x_i)\}' - \left\{ \frac{1}{N} \nabla r_N(x_i) \right\} \left\{ \frac{1}{N} \nabla r_N(x_i) \right\}'$$

which misses $u\lambda(x)$. Obviously, the first-stage error of estimating $\nabla r(x)$ with $\nabla r_N(x)$ operates through $u\lambda(x)$. But this display may be useful in gauging the nonlinearity in $r(x)$ as x varies, because it becomes zero if either $r(x) = 0$ or $r(x) = x'\beta$.

4.2.3   Remarks

Newey and Stoker (1993, (3.6)) showed that the semiparametric efficiency bound for WADE $E\{\omega(x)\nabla r(x)\}$ is the expected value of the outer-product of

$$u \left\{ -\nabla\omega(x) - \omega(x)\frac{\nabla f(x)}{f(x)} \right\} + \{\omega(x)\nabla r(x) - E(\omega(x)\nabla r(x))\}.$$

That is, the expected value of the outer-product of this display is the smallest possible variance under the given semiparametric assumptions. With $\omega(x) = 1$, we get the influence function for m_N, which means that m_N is an efficient estimator for $E\{\nabla r(x)\}$.

ADE has been applied to a demand analysis by Härdle et al. (1991). If desired, $G(\cdot)$ can be estimated nonparametrically by regressing y_i on $x_i'b_N$, as Härdle and Stoker (1989) did. ADE $N^{-1}\sum_i \nabla r_N(x_i)$, which is also examined in Rilstone (1991), is interesting even if the model is not a SIM. In a non-parametric model $y = r(x) + u$, $\nabla r(x_o)$ measures the marginal effect of x on y at $x = x_o$; $\nabla r(x) = \beta$ for all x if $r(x) = x'\beta$. But since we have N different marginal effects $\nabla r(x_i)$, $i = 1, ..., N$, we may want to use $N^{-1}\sum_i \nabla r_N(x_i)$ to represent the N-many marginal effects; this is analogous to using a location measure to represent N observations. Both ADE and WADE have the advantage of no numerical maximization, but the disadvantage is that estimating f or ∇f brings back the very dimension problem in nonparametrics that SIM tries to avoid.

Banerjee (2007) divided the support of x into many cells to apply LSE in each cell, the size of which should go to 0 as $N \to \infty$. An ADE is obtained as a weighted average of the slopes across the cells. This estimator, which is reminiscent of local linear regression, works with bounded-support x and is likely to work better than other ADE's when $f(x)$ has discontinuities. But dividing the support is practical only when the dimension of x is low. Under $x \amalg u$, Banerjee's (2007) ADE is asymptotically normal with variance

$$E[\{\nabla r(x) - E\nabla r(x)\}\{\nabla r(x) - E\nabla r(x)\}'] + \sigma_u^2\{E(xx') - E(x)E(x')\}^{-1}$$
$$\leq E[\{\nabla r(x) - E\nabla r(x)\}\{\nabla r(x) - E\nabla r(x)\}'] + \sigma_u^2 E\{\lambda(x)\lambda(x)'\}.$$

The inequality follows from squaring the first and last terms in

$$\begin{aligned} I_k &= COV\{x, \lambda(x)\} = E[\{x - E(x)\}\lambda(x)'] \\ &\leq \{E(xx') - E(x)E(x')\}^{1/2} \cdot E^{1/2}\{\lambda(x)\lambda(x)'\}. \end{aligned}$$

Curiously, the asymptotic variance of the Banerjee's estimator is smaller than the above semiparametric efficiency bound. This might be due to different requisite conditions; e.g., WADE requires a smooth $f(x)$ that is zero on the x-support boundaries, while the Banerjee's estimator does not. But, given the "folklore" that regularity conditions do not alter efficiency bounds, this aspect may deserve a further look.

Gørgens (2004) proposed a single-index competing risk model related to ADE. Recall "cause-j sub-distribution function" $F(j,t)$, "cause-j sub-density function" $f(j,t)$, and cause-j sub-hazard function' $\lambda(j,t)$ in competing risks: for a continuous latent duration y^* with causes $j = 1, ..., J$,

$$F(j,t) \equiv P(r = j, y^* \leq t), \quad f(j,t) \equiv \frac{dF(j,t)}{dt} \quad \text{and} \quad \lambda(j,t) \equiv \frac{f(j,t)}{S^m(t)}$$

where $S^m(t) \equiv P(y^* \geq t)$. Suppose that SIM given x holds for cause-j sub-hazard:

$$\lambda(j,t|x) = \lambda(j,t|x'\beta).$$

This includes both proportional hazard $\lambda(j,t) = \lambda_j(t)\exp(x'\beta)$ and 'accelerated failure-time hazard' $\lambda_j\{t\exp(x'\beta)\}\exp(x'\beta)$ as special cases where $\lambda_j(\cdot)$ is the "cause-j basic hazard." The "x-conditional cause-j cumulative hazard function" is then

$$\Lambda(j,t|x) = \int_0^t \lambda(j,s|x)ds = \int_0^t \frac{f(j,s|x)}{S^m(s|x)}ds = \int_0^t \frac{f(j,s|x'\beta)}{S^m(s|x'\beta)}ds.$$

The main idea of Gørgens (2004) comes in observing that

(i) $\dfrac{\partial \lambda(j,t|x)}{\partial x} = \dfrac{\partial \lambda(j,t|x'\beta)}{\partial x} = \dfrac{\partial \lambda(j,t|x'\beta)}{\partial(x'\beta)} \cdot \beta$

(ii) $\dfrac{\partial \lambda(j,t|x)}{\partial x} = \dfrac{\partial\{f(j,t)/S^m(t)\}}{\partial x} = \dfrac{\partial f(j,t|x)/\partial x}{S^m(t|x)} - \dfrac{f(j,t|x)\cdot\partial S^m(t|x)/\partial x}{S^m(t|x)^2}$

Equating the last expressions of (i) and (ii), we can see that replacing the right-hand side of (ii) with nonparametric estimators yields an estimator for β up to the scale factor $\partial \lambda(j,t|x'\beta)/\partial(x'\beta)$. To speed up the convergence rate, instead of (i), Gørgens (2004) used its integrated (i.e., averaged) version

$$\frac{\partial \Lambda(j,t|x)}{\partial x} = \int_0^t \frac{\partial \lambda(j,s|x)}{\partial x} ds = \int_0^t \frac{\partial \lambda(j,s|t'\beta)}{\partial(x'\beta)} ds \cdot \beta.$$

The final estimator consists of double sums and it is $\sqrt{N}$-consistent.

4.3   Nonparametric LSE

One shortcoming of ADE is requiring x to have a continuous density and its components to be functionally independent of one another. Relaxing these, Ichimura (1993) proposed a semiparametric (nonlinear) LSE (SLS) for SIM under $u \amalg x$ by minimizing

$$Q_N(b) \equiv \frac{1}{N} \sum_i \{y_i - E_N(y|x_i'b)\}^2$$

wrt b where b is normalized to satisfy $||b|| = 1$ ($|| \cdot ||$ is the Euclidean norm) and

$$E_N(y|x_i'b) \equiv \frac{\sum_{j,j\neq i} K((x_j'b - x_i'b)/h)y_j}{\sum_{j,j\neq i} K((x_j'b - x_i'b)/h)}.$$

In $Q_N(b)$, $E_N(y|x_i'b)$ is a nonparametric estimator for $E(y|x'\beta = x_i'b)$: the expected value of y given that $x'\beta$ takes on the value $x_i'b$. As β is not known, we cannot estimate $E(y|x'\beta = t)$ for an arbitrary t. But the crucial point is that we can estimate $E(y|x'\beta = x_i'b)$ with $E_N(y|x_i'b)$. This way, the unknown functional form of $G(t) = E(y|x'\beta = t)$ has been "concentrated out" in $Q_N(b)$. The SLS b_{SLS} is not a two-stage estimator because $E_N(y|x_i'b)$ depends on b, but the fact that $E(y|x'\beta = x_i'b)$ is estimated makes SLS a two-stage estimator in a sense. The normalization $||b|| = 1$ indicates that $E(y|x'\beta)$ is not known to be increasing or decreasing in $x'\beta$, because we should be able to identify the sign of β if the direction of change were known.

SLS rests on the fundamental fact that the squared loss function $E\{y - q(x)\}^2$ is minimized when $q(x) = E(y|x)$. As in the other SIM estimators that have appeared so far, SLS requires at least one continuously distributed regressor with non-zero coefficient, and β is identified only up to a location and a scale. This identification feature also shows up in $E_N(y|x_i'b)$: the intercept is lost in $x_j'b - x_i'b$, and the scale factor is not identified with the scale normalization $||b|| = 1$.

In implementing SLS, imposing $||b|| = 1$ is cumbersome. Instead, assume that $G(t)$ is monotonic. Denoting the continuous regressor as x_k, divide $\beta = (\beta_1, ..., \beta_k)'$ by $|\beta_k|$ to identify $\{\beta_1/|\beta_k|, \cdots, \beta_{k-1}/|\beta_k|, sgn(\beta_k)\}'$. In estimation, both $+1$ and -1 should be tried for $sgn(\beta_k)$, but for the asymptotic distribution, $sgn(\beta_k)$ can be treated as known since it is estimated at a rate infinitely faster than $\sqrt{N}$. This was the case for a number of semiparametric estimators in the preceding chapter. Let $x = (x'_{-k}, x_k)'$ and $b = (b'_{-k}, b_k)'$ where $x_{-k} \equiv (x_1, ..., x_{k-1})'$ and $b_{-k} \equiv (b_1, ..., b_{k-1})'$. SLS for β_{-k} can be implemented with a Gauss–Newton type iteration: with an initial estimate $b^{(0)} = (b^{(0)}_{-k}, 1)$ for some $b^{(0)}_{-k}$,

$$b^{(1)}_{-k} = b^{(0)}_{-k} + \left\{ \sum_i \nabla E_N(y|x'_i b^{(0)}) \nabla E_N(y|x'_i b^{(0)})' \right\}^{-1}$$

$$\sum_i \{y_i - E_N(y|x'_i b^{(0)})\} \nabla E_N(y|x'_i b^{(0)})$$

$$\text{where } \nabla E_N(y|x'_i b^{(0)}) = \frac{\partial E_N(y|x'_i b)}{\partial b_{-k}}\Big|_{b=b^{(0)}}$$

(obtained with numerical derivatives).

Iterate this until convergence to denote the final estimate as $\hat{b}(1) = \{\hat{b}'_{-k}(1), 1\}'$. Also try $b^{(0)} = (b^{(0)}_{-k}, -1)$ for some $b^{(0)}_{-k}$ and iterate until convergence to denote the final estimate as $\hat{b}(-1) = \{\hat{b}'_{-k}(-1), -1\}'$. Finally, compare the minimand at $\hat{b}(1)$ and $\hat{b}(-1)$ to choose one of the two; denote the final choice as b_{SLS}.

SLS for β_{-k} is $\sqrt{N}$-consistent and asymptotically normal with the variance

$$E^{-1}\{\nabla E(y|x'\beta)\nabla E(y|x'\beta)'\} \; E\{V(u|x) \; \nabla E(y|x'\beta)\nabla E(y|x'\beta)'\}$$

$$E^{-1}\{\nabla E(y|x'\beta)\nabla E(y|x'\beta)'\}$$

$$\text{where } \nabla E(y|x'\beta) = \frac{\partial E(y|x'b)}{\partial b_{-k}}\Big|_{b=\beta}.$$

The middle matrix can be consistently estimated with

$$\frac{1}{N} \sum_i [\; \{y_i - E_N(y|x'_i b_{SLS})\}^2 \; \nabla E_N(y|x'_i b_{SLS}) \; \nabla E_N(y|x'_i b_{SLS})' \;].$$

In practice, it appears that the SD's tend to be under-estimated; i.e., the t-values tend to be too high. Also SLS does not seem to work well with functionally dependent regressors.

It can be shown that

$$\nabla E(y|x'\beta) = \{x_{-k} - E(x_{-k}|x'\beta)\} \frac{dG(x'\beta)}{d(x'\beta)}.$$

If the form of $G(\cdot)$ were known, we could do NLS whose Hessian matrix would contain $x \cdot dG(x'\beta)/d(x'\beta)$. Hence the difference in the asymptotic variance of SLS and that of the NLS is only in the regressor part: $x_{-k} - E(x_{-k}|x'\beta)$ vs. x. Technically, this is due to the b in the kernel for $E_N(y|x'b)$: when $E_N(y|x'b)$ is differentiated wrt b, $x_j - x$ comes out and x_j yields $E(x|x'\beta)$. Other than this, the fact that $G(\cdot)$ is estimated by a nonparametric method does not show up in the asymptotic variance. Newey (1994a, (3.11) and (3.12)) stated that this was essentially because $G(\cdot)$ has been concentrated out as noted above. Concentrating out happens also in 'profile likelihood' approaches.

Härdle et al. (1993) proposed to minimize the objective function jointly wrt b and h, and they showed that the bandwidth obtained this way is optimal. Denoting the minimand now as $Q_N(b,h)$, they also showed that

$$Q_N(b,h) \simeq \frac{1}{N}\sum_i \{y_i - E(y|x_i'b)\}^2 + \frac{1}{N}\sum_i \{E(y|x_i'\beta) - E_N(y|x_i'\beta)\}^2$$

$$= (\text{Minimand for } b) + (\text{Minimand for } h).$$

That is, although minimizing $Q_N(b,h)$ jointly wrt b and h looks difficult, this is equivalent to two separate minimizations: one for b with the infinite-dimensional single-index functional form known, and the other for h with the finite-dimensional β known. Although SIM estimators are not much used in practice, its theoretical development appears to proceed fast. See, e.g., Delecroix et al. (2006) for a further generalization of SLS and references on SIM.

4.4   Quasi-MLE for Binary Response

Consider a binary-response SIM

$$
\begin{aligned}
y &= 1[y^* \geq 0] = 1[x'\beta + u \geq 0] = 1[u \geq -x'\beta] \quad \text{where} \\
f(u|x) &= f(u|x'\beta) \text{ and } E(u|x) = 0 \\
&\Longrightarrow E(u|x'\beta) = 0 \text{ and } V(u|x) = V(u|x'\beta).
\end{aligned}
$$

This allows heteroskedastic u. For instance, if $u|x'\beta \sim N\{0, \exp(2x'\beta)\}$, then

$$P(y=1|x_i) = P(y=1|x_i'\beta) = \frac{1}{\exp(x_i'\beta)} \Phi\left\{\frac{x_i'\beta}{\exp(x_i'\beta)}\right\}.$$

Note that, thinking of a function $H(u,x) \equiv f(u|x)$, "$f(u|x) = f(u|x'\beta)$" should not be construed as $H(u,x'\beta)$; rather, "$f(u|x) = f(u|x'\beta)$" just means that conditionning on x yields the same information as conditioning on only $x'\beta$.

As in SLS in the preceding subsection, define

$$F_N(x_i'b) \equiv \frac{\sum_{j,j\neq i} K((x_j'b - x_i'b)/h)y_j}{\sum_{j,j\neq i} K((x_j'b - x_i'b)/h)} \to^p P(y=1|x'\beta = x_i'b);$$

$G(t) = E(y|x'\beta = t)$ gets "concentrated out." Klein and Spady (1993) proposed a quasi-MLE (QMLE) maximizing

$$Q_N(b) \equiv \frac{1}{N} \sum_{i=1}^{N} \{y_i \ln F_N(x_i'b) + (1 - y_i) \ln\{1 - F_N(x_i'b)\}$$

wrt b. QMLE is a two-stage estimator of a sort as SLS is. Klein and Spady (1993) allowed also nonlinear indices.

Define $F(x'\beta)$:

$$P(y = 1|x) = P(y = 1|x'\beta) = P(u \geq -x'\beta|x'\beta) \equiv F(x'\beta)$$

$$\implies \quad P(y = 1|x'\beta = x'b) = F(x'b).$$

Then the population version of $Q_N(b)$ can be written as

$$E_x[F(x'\beta) \ln F(x'b) + \{1 - F(x'\beta)\} \ln\{1 - F(x'b)\}].$$

$Q(s) \equiv t \ln s + (1 - t) \ln(1 - s)$ is maximized at $s = t$, which can be verified by differentiating $Q(s)$ wrt s twice. Using this fact, the population version of $Q_N(b)$ is maximized at $\beta = b$. This is the basis for identifying β in the Klein–Spady estimator, analogously to that SLS is based on the fact that the squared loss function is minimized at $E(y|x)$.

Assume that x_k is continuous with $\beta_k \neq 0$ as in SLS. Divide $\beta = (\beta_1, ..., \beta_k)'$ by $|\beta_k|$ to identify $\{\beta_1/|\beta_k|, \cdots, \beta_{k-1}/|\beta_k|, sgn(\beta_k)\}'$. The rest of the QMLE identification aspect is analogous to that for SLS with a monotonic $G(\cdot)$. With an initial estimate $b^{(0)} = (b_{-k}^{(0)}, \pm 1)$ for some $b_{-k}^{(0)}$, the iteration scheme is

$$b_{-k}^{(1)} = b_{-k}^{(0)} + \left\{ \sum_i S_N(y|x_i'b^{(0)}) S_N(y|x_i'b^{(0)})' \right\}^{-1} \sum_i S_N(y|x_i'b^{(0)})$$

where $\quad S_N(y|x_i'b^{(0)}) \equiv \dfrac{\partial\{y_i \ln F_N(x_i'b) + (1 - y_i) \ln\{1 - F_N(x_i'b)\}}{\partial b_{-k}}\Big|_{b=b^{(0)}}.$

Defining the population version of S_N as S, QMLE for β_{-k} is $\sqrt{N}$-consistent and asymptotically normal with the variance

$$E^{-1}\left[S(y|x'\beta) S(y|x'\beta)' \frac{1}{F(x'\beta)\{1 - F(x'\beta)\}} \right].$$

It can be shown that $S(y|x'\beta) = \{x_{-k} - E(x_{-k}|x'\beta)\} f(x'\beta)$ where $f(x'\beta) \equiv dF(x'\beta)/d(x'\beta)$, with which the variance can be written as

$$E^{-1}\left[\{x_{-k} - E(x_{-k}|x'\beta)\}\{x_{-k} - E(x_{-k}|x'\beta)\}' \frac{f(x'\beta)^2}{F(x'\beta)\{1 - F(x'\beta)\}} \right].$$

This is reminiscent of the probit variance except that the regressor part is $x_{-k} - E(x_{-k}|x'\beta)$ not x.

Cosslett (1987) and Chamberlain (1986) showed that *QMLE attains the semiparametric efficiency bound* under $u \amalg x$ that is stronger than the SIM assumption. This may make QMLE "the" semiparametric estimator to use for binary responses. But in practice, the SD's tend to be under-estimated. Also QMLE does not seem to work with functionally dependent regressors.

4.5   Discrete Regressors*

A shortcoming of the ADE-type semiparametric estimators for SIM is that they allow only continuously distributed regressors whereas a disadvantage of the nonparmetric LSE and the quasi-MLE for binary responses is that the estimators require iterations. Horowitz and Härdle (1996) introduced a direct estimator to overcome this shortcomings, which is examined in this subsection; they still require, however, at least one continuously distributed regressor. Although the estimator is not quite practical, the main idea—pulling out the index from a nonlinear function—led to other interesting semiparametric estimators in Lewbel (1998, 2000), Honoré and Lewbel (2002), and Khan and Lewbel (2007). Also Gørgens (2006) applied the idea to single-index hazard estimation where a hazard function depends on the regressors only through a single index.

Let a $k_w \times 1$ random vector w and a $k_x \times 1$ random vector x denote discrete and continuous regressors, respectively, with coefficient vectors α and β:

$$E(y|w, x) = G(w'\alpha + x'\beta).$$

Suppose β is estimated $\sqrt{N}$-consistently by b_N fixing w, say at w_o, and using one of the SIM estimators that have appeared already. The main remaining issue is then estimating α. Let $x'\beta = v$ and suppose that, for some finite constants $G_0 < G_1$ and $v_0 < v_1$,

$$G(\max_w w'\alpha + v_0) < G_0 < G_1 < G(\min_w w'\alpha + v_1) \quad \text{where}$$

$$w'\alpha + v_0 \leq t_{0,\max} \ (\equiv \max_w w'\alpha + v_0) < t_{1,\min} \ (\equiv \min_w w'\alpha + v_1)$$

$$\leq w'\alpha + v_1.$$

That is, v_0 should be chosen small enough so that adding the maximum $w'\alpha$ to v_0 still keeps the sum small enough such that $G(t_{0,\max}) < G_0$ for some G_0. Analogously, v_1 should be chosen large enough so that adding the minimum $w'\alpha$ to v_1 still keeps the sum large enough such that $G_1 < G(t_{1,\min})$ for some $G_1 > G_0$.

Let $x'\beta = v$ and define the integral of a "winsorzied" $G(w'\alpha + v)$:

$$\begin{aligned}
J(w) \equiv \int_{v_0}^{v_1} &\{G_0 1[G(w'\alpha + v) < G_0] \\
&+ G(w'\alpha + v)1[G_0 \leq G(w'\alpha + v) \leq G_1] \\
&+ G_1 1[G(w'\alpha + v) > G_1]\} dv
\end{aligned}$$

to get, setting $t = w'\alpha + v$,

$$J(w) = G_0 \int_{w'\alpha+v_0}^{t_{0,\max}} 1[G(t) < G_0]dt + G_0 \int_{t_{0,\max}}^{t_{1,\min}} 1[G(t) < G_0]dt$$

$$+ \int_{t_{0,\max}}^{t_{1,\min}} G(t)1[G_0 \le G(t) \le G_1]dt$$

$$+ G_1 \int_{t_{0,\max}}^{t_{1,\min}} 1[G(t) > G_1]dt + G_1 \int_{t_{1,\min}}^{w'\alpha+v_1} 1[G(t) > G_1]dt.$$

Carry out the first and last integrals to get, for some $\tilde{t}$ and $\bar{t}$,

$$J(w) = G_0(\tilde{t} - w'\alpha - v_0) + G_0 \int_{t_{0,\max}}^{t_{1,\min}} 1[G(t) < G_0]dt$$

$$+ \int_{t_{0,\max}}^{t_{1,\min}} G(t)1[G_0 \le G(t) \le G_1]dt$$

$$+ G_1 \int_{t_{0,\max}}^{t_{1,\min}} 1[G(t) > G_1]dt + G_1(w'\alpha + v_1 - \bar{t})$$

$$= (G_1 - G_0)w'\alpha + \text{ (terms not depending on } w).$$

Suppose w takes M-many different values $w^{(1)}, ..., w^{(m)}$. Then,

$$J\{w^{(m)}\} - J\{w^{(1)}\} = (G_1 - G_0)\{w^{(m)} - w^{(1)}\}'\alpha, \quad m = 2, ..., M.$$

Define

$$\underset{(M-1)\times 1}{\Delta J} \equiv \begin{bmatrix} J\{w^{(2)}\} - J\{w^{(1)}\} \\ \cdots \\ J\{w^{(M)}\} - J\{w^{(1)}\} \end{bmatrix} \quad \text{and} \quad \underset{(M-1)\times k_w}{W} \equiv \begin{bmatrix} w^{(2)\prime} - w^{(1)\prime} \\ \cdots \\ w^{(M)\prime} - w^{(1)\prime} \end{bmatrix}$$

to get

$$\alpha = \frac{1}{G_1 - G_0}(W'W)^{-1}W'\Delta J.$$

This is the key identifying equation for α.

To implement this equation, the four numbers in $G_0 < G_1$ and $v_0 < v_1$ should be chosen, which can be done by estimating $G(\cdot)$ nonparametrically. Then $J(\cdot)$ can be found by using v_0, v_1, G_0, and G_1 and plugging in the nonparametric estimator for $G(\cdot)$. Horowitz and Härdle (1996) showed that the resulting estimator is $\sqrt{N}$-consistent.

4.6   Extensions to Multiple Index Models*

Single-index models can be extended to double- or multiple-index models. As an example of triple-index model, consider a switching regression model with "unknown sample separation":

$$y = x_0'\beta_0 + u_0 \quad \text{if } d = 0 \iff z'\gamma + v < 0$$

$$= x_1'\beta_1 + u_1 \quad \text{if } d = 1 \iff z'\gamma + v \ge 0;$$

y is a mixture of two populations depending on $d = 1[z'\gamma + v \ge 0]$. Denoting the joint density of (u_0, u_1, v) as $f(u_0, u_1, v)$ that is independent of x,

$$
\begin{aligned}
E(y|x, z) &= E(y|x, z, d = 0)P(d = 0|x, z) \\
&\quad + E(y|x, z, d = 1)P(d = 1|x, z) \\
&= E(y|x, z, v < -z'\gamma) \cdot P(v < -z'\gamma|z) \\
&\quad + E(y|x, z, v \ge -z'\gamma) \cdot P(v \ge -z'\gamma|z) \\
&= \int \int \int^{-z'\gamma} (x_0'\beta_0 + u_0)f(u_0, u_1, v)dv du_0 du_1 \\
&\quad + \int \int \int_{-z'\gamma} (x_1'\beta_1 + u_1)f(u_0, u_1, v)dv du_0 du_1 = x_0'\beta_0 \\
&\quad \cdot P(v < -z'\gamma|z) + x_1'\beta_1 \cdot P(v \ge -z'\gamma|z) \\
&\quad + \int \int \int^{-z'\gamma} u_0 f(u_0, u_1, v)dv du_0 du_1 \\
&\quad + \int \int \int_{-z'\gamma} u_1 f(u_0, u_1, v)dv du_0 du_1
\end{aligned}
$$

where ∞ and $-\infty$ are omitted in $\int$. This is a triple index model:

$$E(y|x, z) = G(x_0'\beta_0, x_1'\beta_1, z'\gamma).$$

If $f(u_0, u_1, v) = f(u_0)f(u_1)f(v)$ and $Eu_0 = Eu_1 = 0$, then the integrals involving u_0 and u_1 disappear. Another example of multiple-index model can be seen in multinomial choice models where the regression functions in the choice utility equations become the indices.

A multiple-index model can also arise as a simplification of nonparametric models. One such method is "projection pursuit regression (PPR)" in Friedman and Stuetzle (1981) with

$$E(y|x) = \sum_{j=1}^{p} G_j(x'\beta_j)$$

where p is a integer to be chosen by the researcher, β_j is a $k \times 1$ parameter vector, and $G_j(\cdot)$ is an unknown univariate function. PPR is implemented sequentially. First, find b_1 and $G_j(\cdot)$ minimizing $N^{-1}\sum_i\{y_i - G_{N1}(x_i'b_1)\}^2$. Then find b_2 and $G_{N2}(\cdot)$ minimizing $N^{-1}\sum_i\{y_i - G_{N1}(x_i'b_1) - G_{N2}(x_i'b_2)\}^2$. This continues until the reduction in the sum of squared errors becomes small, which then determines p. The main idea of PPR is replacing one high-dimensional nonparametric problem with many one-dimensional nonparametric subproblems. In practice, one can try various combinations for β_j and estimate $G_j(\cdot)$ with a kernel method for a given $x'\beta_j$ at the jth step.

Alternatively, we may approximate $G_j(s)$ with a polynomial function, say $\alpha_{j1}s + \alpha_{j2}s^2 + \alpha_{j3}s^3$, and estimate α_{j1}, α_{j2}, α_{j3}, and β_j jointly. Either way, estimating β_j and $G_j(\cdot)$ is a difficult computational problem due to the high nonlinearity; almost countless local minima may be found.

A special case of PPR (at least from our econometric view point) is an artificial neural network model of "a single hidden layer with p neurons" (see, e.g., Kuan and White, 1994) in which

$$E(y|x) = \alpha_0 + \sum_{j=1}^{p} \alpha_j G(x'\beta_j),$$

where $G(\cdot)$ is a known nonlinear function, for instance, $G(s) = e^s/(1 + e^s)$. Here α_0, α_j, β_j, $j = 1, ..., p$, are the parameters to be estimated. But, computationally this seems as hard as PPR due to too many local minima. As in PPR, p should go to ∞ as $N \to \infty$; in fact, β_j's is not identified in this case (Bierens, 1994, p. 94), which explains the computational problem.

Xi et al. (2002) considered a multiple index model

$$E(y|x) = G(x'\beta_1, ..., x'\beta_p) \qquad \text{for some } p < k$$

$$\implies E(y|x) \simeq \alpha_0 + \alpha_1(x - x_o)'\beta_1 +, ..., +\alpha_p(x - x_o)'\beta_p$$

$$\text{where } \beta_p'\beta_q = 1[p = q].$$

The idea is reducing the k-dimensional nonparametric problem to a p-dimensional nonparametric problem such that $x'\beta_p$ stands for a direction of interest and the condition $\beta_p'\beta_q = 1[p = q]$ is to make each direction orthogonal to the other directions. Replacing the evaluation point with observation points, the parameters α and β can be estimated by minimizing

$$\frac{1}{N}\sum_j\sum_i \{y_i - \alpha_0 - \alpha_1(x_i - x_j)'\beta_1 -, ..., -\alpha_p(x_i - x_j)'\beta_p\}^2 w_{ij}$$

where w_{ij} is a kernel weighting function; e.g. $w_{ij} = K\{(x_i - x_j)/h\}/\sum_i K\{(x_i - x_j)/h\}$. The idea is reminiscent of local linear regression.

Using w_{ij}, however, seems to defeat the very motivation of dimension reduction. One way to avoid this problem is starting with an initial value for β's, say $\beta^{(0)}$, to use

$$\hat{w}_{ij} \equiv \frac{K\{\beta^{(0)'}(x_i - x_j)/h\}}{\sum_i K\{\beta^{(0)'}(x_i - x_j)/h\}}$$

where $\beta^{(0)'}x = (\beta_1^{(0)'}x, ..., \beta_p^{(0)'}x)'$. With this weight plugged in, we can estimate the α and β parameters by minimizing the above minimand, and $\hat{w}_{ij}$ is to get updated with the new β estimates to repeat the whole process until convergence.

Carroll et al. (1997) examined a model where

$$E(y|x, z) = G\{\eta(x'\beta) + z'\gamma\} \text{ where } G(\cdot) \text{ is known and } \eta(\cdot) \text{ is unknown.}$$

This is a double-index model and includes single index as a special case when either index drops out. One example of this model is logistic binary response with

$$P(y = 1|x, z) = \frac{\exp\{\eta(x'\beta) + z'\gamma\}}{1 + \exp\{\eta(x'\beta) + z'\gamma\}}.$$

Ichimura and Lee (1991) extended Ichimura (1993) to multiple-index models, and L.F. Lee (1995) extended Klein and Spady (1993) to multinomial choice models. Picone and Butler (2000) proposed also a semiparametric estimator for multiple index models. All of these studies rely on Ichimura and Lee (1991) for the identification proof of the multiple index parameters. But the proof seems inadequate, amounting more or less to a re-statement of the definition of identification. Lee and Kim (2007) proposed a two-stage multiple index estimator for multinomial choice models using ADE in the first-stage. The estimator does not involve iterations, and thus there is no identification problem that the other estimators faced.

While all of these approaches assume a known number of indices, Donkers and Schafgans (2008) proposed to test for the number of indices. For $G(x) \equiv E(y|x) = G(x'\beta_1, ..., x'\beta_p)$ with $\nabla G(x) \equiv \sum_{j=1}^{p}\{\partial G/\partial(x'\beta_j)\}\beta_j$, they estimated $E\{\nabla G(x)\nabla G(x)'\}$, the average outer-product matrix of the gradient to test for the number of indices using the rank of the matrix. With p determined this way, they went on to estimate parameters of interest with GMM using average derivative functionals.

5 Semi-linear Models

Single index model $E(y|x) = G(x'\beta)$ with G unknown is one way of reducing the nonparametric dimension problem. Another popular way is semi-linear or partially linear model $E(y|x, z) = x'\beta + \theta(z)$ with $\theta(\cdot)$ unknown. When the single index model is generalized to a multiple-index model $E(y|x) = G(x'\beta, z'\gamma)$ and when the semi-linear model is generalized for an "unknown link function" G so that $E(y|x, z) = G\{x'\beta + \theta(z)\}$, we can see how the two themes may overlap. Nevertheless, semi-linear model in its basic form provides a distinctive way of relaxing the linear model assumption while keeping the nonparametric dimension low, which is the topic of this subsection. Härdle et al. (2000) provided a review on semi-linear model estimation.

5.1 Two-Stage Procedure

Consider a *semi-linear (or partially linear) model*:

$$y = x'\beta + \theta(z) + u, \qquad E(u|x, z) = 0$$

where θ is an unknown function of z, x is a $k_x \times 1$ vector, and z is a $k_z \times 1$ vector; *x and z do not overlap*. This is a mixture of a parametric component $x'\beta$ and a nonparametric one $\theta(z)$. Typically β is a parameter of interest whereas $\theta(\cdot)$ is an infinite-dimensional nuisance parameter. There are various approaches to estimate β (and $\theta(\cdot)$); here we review the kernel-based one in Robinson (1988). The intercept in β is not identified as it can be absorbed into $\theta(z)$; i.e., x does not include 1 in semi-linear models.

Take $E(\cdot|z)$ on the y-equation to get

$$E(y|z) = E(x|z)'\beta + \theta(z).$$

Subtract this from the y-equation to remove $\theta(z)$:

$$y - E(y|z) = \{x - E(x|z)\}'\beta + u \iff y = E(y|z) + \{x - E(x|z)\}'\beta + u.$$

The systematic part of y is decomposed into two: one is $E(y|z)$—the effect of z on y, and the other is $\{x - E(x|z)\}'\beta$—the effect on y of x net of z.

In order to estimate β, first use kernel estimators

$$E_N(y|z_i) \equiv \frac{\sum_{j,j\neq i}^{N} K((z_j - z_i)/h)y_j}{\sum_{j,j\neq i}^{N} K((z_j - z_i)/h)} \quad \text{and}$$

$$E_N(x|z_i) \equiv \frac{\sum_{j,j\neq i}^{N} K((z_j - z_i)/h)x_j}{\sum_{j,j\neq i}^{N} K((z_j - z_i)/h)}$$

for $E(y|z)$ and $E(x|z)$, respectively. Replace $E(y|z)$ and $E(x|z)$ in the $y - E(y|z)$ equation with these estimators to get a new model $y - E_N(y|z) \simeq \{x - E_N(x|z)\}'\beta + u$ and its LSE

$$b_N = \left[\sum_i \{x_i - E_N(x|z_i)\}\{x_i - E_N(x|z_i)\}'\right]^{-1}$$

$$\sum_i \{x_i - E_N(x|z_i)\}\{y_i - E_N(y|z_i)\}.$$

The non-identifiability of the intercept is obvious as the LSE intercept term gets removed by the mean subtraction. Also if we allow variables to appear in both x and z, their coefficients in β are not identified for a similar reason.

The LSE b_N has the following asymptotic distribution:

$$\sqrt{N}(b_N - \beta) \rightsquigarrow N(0, A^{-1}BA^{-1}),$$

$$A_N \equiv \frac{1}{N}\sum_i \{x_i - E_N(x|z_i)\}\{x_i - E_N(x|z_i)\}' \to^p A,$$

$$B_N \equiv \frac{1}{N}\sum_i \{x_i - E_N(x|z_i)\}\{x_i - E_N(x|z_i)\}'\hat{u}_i^2 \to^p B,$$

$$\hat{u}_i \equiv \{y_i - E_N(y|z_i)\} - \{x_i - E_N(x|z_i)\}'b_N.$$

Despite the nonparametric first stage, the estimation errors do not affect the asymptotic distribution of b_N. This can be shown using CT_E. Chamberlain (1992) showed that this asymptotic variance is the semiparametric efficiency bound.

If the bandwidth is large enough so that $E_N(y|z) = \bar{y}$ (sample mean of y_i's) and $E_N(x|z) = \bar{x}$, then $y - E_N(y|z) \simeq \{x - E_N(x|z)\}'\beta + u$ becomes

$$y - \bar{y} \simeq (x - \bar{x})'\beta + u \iff y \simeq \bar{y} - \bar{x}'\beta + x'\beta + u$$

which is the linear model with $\theta(z)$ being the intercept term $\bar{y} - \bar{x}'\beta$. In this regard, the usual linear model $y = x'\beta + u$ (with $\bar{y} - \bar{x}'\beta$ absorbed into the intercept) is a special case of the semi-linear model, and it is a misspecified one if $\theta(z)$ indeed varies with z. If one wants to estimate $\theta(z)$, then this can be done by a nonparametric regression of $y - x'b_N$ on z, because of $E(y - x'\beta|z) = \theta(z)$.

The semi-linear model is applicable to labor supply models with z being age, because the labor supply profile of age is likely to be nonlinear (increasing and then leveling off after a certain age). The usual practice of including age and age^2 to capture this pattern may not be adequate, for it means declining labor supply after a peak age. Another case for semi-linear model is sample selection models. For instance, consider the selection equation $d = 1[w'\alpha + \varepsilon > 0]$ and outcome equation $y = x'\beta + u$ where $w = (x', c')'$ and $(\varepsilon, u) \amalg w$. Then, for the sub-population $d = 1$,

$$
\begin{aligned}
E(y|d &= 1, w) = x'\beta + E(u|d = 1, w) = x'\beta + E(u|\varepsilon > -w'\alpha, w) \\
&= x'\beta + \lambda(w'\alpha) \quad \text{where } \lambda(\cdot) \text{ is unknown}
\end{aligned}
$$

because, due to $(\varepsilon, u) \amalg w$, w can influence $E(u|\varepsilon > -w'\alpha, w)$ only through $w'\alpha$. This is a semi-linear model with $w'\alpha = z$ (as well as being a double index model). As mentioned already, the exclusion restriction is necessary; otherwise if $w = x$, then

$$\{x - E(x|x'\alpha)\}'\alpha = x'\alpha - x'\alpha = 0: \text{the columns of } x$$
$$-E(x|x'\alpha) \text{ are linearly dependent}$$
$$\implies E[\{x - E(x|x'\alpha)\}\{x - E(x|x'\alpha)\}'] \text{ is singular.}$$

If there are endogenous regressors in x, then the TSE can be modified as follows. Let w denote an instrument vector for x. Multiplying the conditional-mean-subtracted equation by w and then taking $E(\cdot)$, we get

$$E[w\{y - E(y|z)\}] = E[w\{x - E(x|z)\}']\beta.$$

This can be solved for β in "just-identified" cases $(\dim(w) = \dim(x))$; otherwise $(\dim(w) > \dim(x))$, we can proceed as in over-identified IVE cases.

Consider a nonlinear semi-linear model $E(y|x, z) = g\{x'\beta + \theta(z)\}$ where $g(\cdot)$ is a known function. For instance, $y = 1[x'\beta + \theta(z) + u \geq 0]$ with $u \sim$ *Logistic*, which implies

$$E(y|x, z) = \frac{\exp\{x'\beta + \theta(z)\}}{1 + \exp\{x'\beta + \theta(z)\}}.$$

Instead of specifying the distribution of u fully, Severini and Staniswalis (1994) set up a pseudo likelihood function which depends on $E(y|x, z)$ and $V(y|x, z)$ where $V(y|x, z)$ is posited to be determined solely by $E(y|x, z)$. Then a kernel $K((z - z_o)/h)$ is attached to the pseudo score function to do nonparametric pseudo MLE for $\theta(z_o; b)$ given $\beta = b$. Once $\theta(\cdot; b)$ is traced out by varying z_o, $\theta(\cdot; b)$ is plugged into the pseudo likelihood function (no kernel weighting at this stage), which is then maximized wrt b. This "profile likelihood approach" yields b_N and $\theta(\cdot; b_N)$. Note the difficulty in the second stage: $\theta(\cdot; b)$ changes as b changes.

Instead of the nonparametric (pseudo) MLE, Ai and McFadden (1997) solved the last display for $x'\beta + \theta(z)$:

$$\ln \left\{ \frac{E(y|x, z)}{1 - E(y|x, z)} \right\} = x'\beta + \theta(z).$$

Replacing $E(y|x, z)$ with $E_N(y|x, z)$, Robinson's (1988) procedure is applied to estimate β and $\theta(\cdot)$. But using $E_N(y|x, z)$ would ruin the very motivation to consider semi-linear models: reduce the nonparametric dimension problem.

Cai et al. (2007) examined semi-linear hazard model for multivariate hazards; e.g., survival durations of members in the same family. Although the multivariate durations are related to one another, they set up a pseudo-partial likelihood function simply adding up each member's partial likelihood function ignoring the relationship. Then they do nonparametric MLE for $\theta(z_o; b)$ given $\beta = b$ by doing analogously to what was done in Severini and Staniswalis (1994)—in fact Cai et al. (2007) do local polynomial estimation, not local constant as in Severini and Staniswalis (1994). Plugging $\theta(z_o; b)$ into the pseudo-partial likelihood to obtain the profile pseudo-partial likelihood, they estimate β.

The above approaches specify more or less the distribution of the error term u. If the distribution of u is unknown, semi-linear regression functions in LDV models are difficult to handle in general. Later, however, we will introduce pairwise differencing approach for this case.

Many applications of semi-linear models have appeared in the literature. For instance, Hausman and Newey (1995) estimated a demand function where the own price and income are the two main variables of interest. In their semi-linear model, the nonparametric component is a function of the two variables while the other variables enter the model linearly. Using the demand function, they estimated an "equivalent variation" ($\simeq$ consumer surplus) and deadweight loss from a price change. Anglin and Gencay (1996) used a semi-linear model for a "hedonic price function" where y is ln(house price), x is binary regressors, and z is continuous or discrete non-binary regressors; hedonic price functional forms are controversial in the literature. In the following, an empirical example in Kondo and Lee (2003) is provided, which shows some details on how to implement the above two-stage estimator. The example also demonstrates that there is a good reason to check out the independence between x and z in semi-linear models.

5.2   Empirical Application: Hedonic Price Indices

Hedonic price indices (HPI) show price changes controlling for quality changes, and are typically estimated by the LSE of $\ln(\text{price})$ on time dummies and quality-characteristics. This natural divide of the regressors into two sets (time dummies and quality characteristics) motivates semi-linear models: the linear part with time dummies and the other nonparametric part with quality-characteristics. Kondo and Lee (2003) applied the semi-linear HPI model and the above two-stage procedure to repeated cross-section data for rental offices in Tokyo during 1985–1991 ($T = 7$ years); assume that each office can be sampled at most only once during the 7-year period, as the number of offices is far greater than the sample size. Anglin and Gençay (1996) also used a semi-linear HPI model.

Consider

$$y_i = (x_i^*)'\beta^* + h(z_i) + u_i, \quad i = 1, \ldots, N,$$
$$(y_i, x_i', z_i') \text{ is observed, iid across } i,$$

where $y_i = \ln(p_i)$, p_i is the monthly rent for office i in the sampled year, $x_i^* = (1, x_i')'$ is the $T \times 1$ vector of year-dummies indicating from which year office i is sampled, z_i is a $q \times 1$ vector of office characteristics (among the q components, the first q_1 are continuous, and the other q_2 are discrete with $q_1 + q_2 = q$), $h(z_i)$ is an unknown function of z_i, $\beta^* = (\beta_1, \beta')' = (\beta_1, \ldots, \beta_T)'$ is a vector of unknown parameters, and u_i is an error term with $E(u|x, z) = 0$. For instance, with $T = 4$, x_i^* is

$$\begin{bmatrix} 1 \\ 0 \\ 0 \\ 0 \end{bmatrix} \text{ for year 1,} \quad \begin{bmatrix} 1 \\ 1 \\ 0 \\ 0 \end{bmatrix} \text{ for year 2,} \quad \begin{bmatrix} 1 \\ 0 \\ 1 \\ 0 \end{bmatrix} \text{ for year 3, and}$$

$$\begin{bmatrix} 1 \\ 0 \\ 0 \\ 1 \end{bmatrix} \text{ for year 4.}$$

Since the intercept is not identified in the semi-linear model, set $\beta_1 = 0$ (i.e., the year 1 time dummy is not used) so that $(x_i^*)'\beta^* = x_i'\beta$. Let $P_1, ..., P_T$ be the price level for year t. In view of the last display, the remaining time dummies' coefficients are, respectively,

$$\beta_2 = P_2 - P_1, \; \beta_3 = P_3 - P_1, \; ..., \; \beta_T = P_T - P_1.$$

Hence the annual price increase rate $P_{t,t+1}$ from year t to year $t + 1$ is

$$P_{t,t+1} = P_{t+1} - P_t = \beta_{t+1} - \beta_t, \quad t = 2, \ldots, T - 1.$$

Since x_i consists of 0's and 1's, the linearity assumption for $x_i'\beta$ is not restrictive. The only restriction for the semi-linear model is that x_i and z_i are separable with no interaction; i.e., the functional form of $h(\cdot)$ does not change across the years, while the fully nonparametric HPI model would allow such changes. Lack of interaction terms between x and z is required, not just for this HPI example, but also for all semi-linear models in general.

In choosing the bandwidth h in the semi-linear model, the following cross-validation criterion has been the most popular (Stock, 1991; Anglin and Gençay, 1996; Hausman and Newey, 1995): with $E_N(\cdot|z_i)$ denoting a leave-one-out kernel estimator for $E(\cdot|z_i)$,

$$\frac{1}{N}\sum_i(\tilde{y}_i - \tilde{x}_i'\hat{\beta})^2 \quad \text{where } \tilde{y}_i \equiv y_i - E_N(y|z_i),\ \tilde{x}_i \equiv x_i - E_N(x|z_i)$$

and $\hat{\beta}$ is the resulting two-stage estimator that consequently also depends on h. This was also used in Kondo and Lee (2003).

As for the kernel choice, although theoretically necessary for the TSE, high-order kernels are cumbersome if the regressor dimension is high. Robinson (1988) and Hausman and Newey (1995) used a high-order kernel, while Anglin and Gençay (1996) used a second-order kernel. Stock (1991) compared the performance of second-and high-order kernels, but his main results were based on second-order kernels. Kondo and Lee (2003, p. 39) found that a product kernel made up of second-order bounded-support kernels worked well in terms of the stability of the cross-validation criterion and the tolerance to changing bandwidth values.

For our data, x turns out to be (almost) mean-independent of z:

$$E(x|z) = E(x) \implies y - E(y|z) = \{x - E(x)\}'\beta + u$$

and $E(x)$ can be replaced by $\bar{x}$ in the second stage. Under the mean-independence, models misspecifying $\theta(z)$ still yield consistent LSE for β: the misspecified $\theta(z)$ makes a function of z enter the error term to become an omitted variable, but the LSE omitted variable bias for β is zero because the omitted variable is mean-independent of the included variable x. In this case, the differences among estimators misspecifying $\theta(z)$ are in efficiency.

To see this efficiency issue, recall that the asymptotic variance with $E(x|z) = E(x)$ is

$$E^{-1}(dd')\ E[dd'u^2]\ E^{-1}(dd') \quad \text{where } d = x - E(x|z) = x - E(x).$$

Define $w^* \equiv (1, w')'$ where w is a vector consisting of known functions of elements of z, and also define the linear projection and its orthogonal complement:

$$L(x|w^*) \equiv E\{x(w^*)'\}\,E^{-1}\{w^*(w^*)'\}w^* \quad \text{and} \quad L^\perp(x|w^*) \equiv x - L(x|w^*).$$

Suppose that $\theta(z)$ is misspecified as $(w^*)'\alpha^*$. That is, the model for LSE is

$$y_i = x_i'\beta + (w_i^*)'\alpha^* + v_i \quad \text{where } v_i \equiv u_i + \theta(z_i) - (w_i^*)'\alpha^* \text{ is the error term;}$$

a constant should be subtracted from v_i to assure $E(v) = 0$, but this is ignored to simplify presentation. The asymptotic variance becomes the above asymptotic variance with u replaced by

$$v - L(v|w^*) = u + \theta(z) - (w^*)'\alpha^* - L\{u + \theta(z) - (w^*)'\alpha^*|w^*\}$$
$$= u + \theta(z) - L\{\theta(z)|w^*\} = u + L^\perp\{\theta(z)|w^*\}.$$

This follows from the partial-regression LSE asymptotic variance: the equation is linearly projected on w_i^*, and then $y_i - L(y|w_i^*)$ is regressed on $x_i - L(x|w_i^*)$ where the error term is $v_i - L(v|w_i^*) = u_i + L^\perp\{\theta(z)|w_i^*\}$.

The efficiency of the LSE misspecifying $\theta(z)$ as linear depends on the extent $\theta(z)$ is explained by w_i^*: any part of $\theta(z)$ not accounted for becomes part of the error term, which makes the error variance greater because

$$V[u + L^\perp\{\theta(z)|w^*\}] = V(u) + V[L^\perp\{\theta(z)|w^*\}] \quad \text{as } E(u|x,z) = 0.$$

Therefore, when $E(x|z) = E(x)$, the two-stage procedure is as efficient as any LSE specifying $\theta(z)$ as linear, regardless of whether the linear specification is right or not.

Kondo and Lee employed linear models and their second-order linear models. That is, the latter uses all first- and second-order terms of the former; the second-order linear model had as many as 44 parameters. Despite this, both linear models were all rejected by the specification test in Stute (1997) and Stute et al. (1998), and they did worse than the above simple TSE using a kernel estimator. Table 1 presents four sets of HPI's and their SD's: the time-dummy-only model with no $\theta(z)$ ("DUM index"), the linear model with z ("LIN index"), the linear model with the first- and second-order terms of z ("QUAD index"), and the semi-linear model ("SLR index") using the biweight-kernel-based product kernel. The estimates indicate that the rent has increased at a rate higher than 9% during 1985–1991. From 1986 to 1987, the increase was particularly high, being at least 18%. The LIN indices are more variable across the years than the SLR indices, while the QUAD indices are less variable and greater than the SLR indices. With large standard errors, the DUM indices should not be used, for it does not take much to get at least the LIN indices. The SLR indices are not difficult to get either, and have parsimony as can be seen in the R^2's for the SLR model (around 0.72) and the QUAD indices model (around 0.76) with as many as 44 parameters.

5.3   Pair-wise Differencing for Semi-linear LDV Models

Honoré and Powell (2005) applied the pairwise differencing idea in Honoré (1992) to LDV models with semi-linear regression functions. Recall the semi-linear model

$$y_i = x_i'\beta + \theta(z_i) + u_i, \quad E(u|x,z) = 0, \quad \theta(\cdot) \text{ is an unknown function}$$

Table 1: HIP (SD) for Tokyo Offices

	DUM	LIN	QUAD	SLR
85-86	13.39(3.66)	10.62(2.32)	10.30(2.03)	9.37(2.18)
86-87	22.12(4.27)	20.43(2.59)	19.00(2.27)	18.11(2.46)
87-88	9.65(4.70)	9.07(2.80)	11.17(2.43)	9.56(2.65)
88-89	11.21(4.55)	15.60(2.81)	13.40(2.39)	11.20(2.62)
89-90	11.55(4.59)	9.30(2.99)	12.33(2.66)	9.61(2.78)
90-91	9.71(4.54)	13.33(2.91)	12.03(2.69)	10.83(2.91)

$$\implies \quad y_i - y_j = (x_i - x_j)'\beta + u_i - u_j \quad \text{if } z_i = z_j.$$

Imposing $z_i = z_j$ with a kernel function $K((z_i - z_j)/h)$, estimate β by minimizing

$$\frac{2}{N(N-1)} \sum_{i<j} K\left(\frac{z_i - z_j}{h}\right) \{y_i - y_j - (x_i - x_j)'\beta\}^2.$$

The resulting estimator is an alternative to the Robinson's (1988) TSE:

$$b_N \equiv \left\{ \sum_{i<j} K\left(\frac{z_i - z_j}{h}\right) (x_i - x_j)(x_i - x_j)' \right\}^{-1}$$

$$\sum_{i<j} K\left(\frac{z_i - z_j}{h}\right) (x_i - x_j)(y_i - y_j).$$

To see a slight variation on the theme, consider a sample selection model: with w strictly including x,

$$d = 1[w'\alpha + \varepsilon > 0], \quad y = x'\beta + u \quad \text{and} \quad y$$
$$\text{observed only when } d = 1$$
$$\implies \quad E(y|w, d = 1) = x'\beta + E(u|w, d = 1) = x'\beta$$
$$+E(u|w, \varepsilon > -w'\alpha)$$
$$\implies \quad y = x'\beta + \theta(w'\alpha) + v, \quad \text{where } v \equiv y - E(y|w, d = 1)$$
$$\text{under } (\varepsilon, u) \text{ II } w.$$

By construction, $E(v|w, d = 1) = 0$, and this is a semi-linear model where $w'\alpha$ is a rv (while z can be a random vector) but $w'\alpha$ has an unknown parameter α that needs to be replaced with an estimator a_N. That is, given a $\sqrt{N}$-consistent estimator a_N for α, we can consistently estimate β with

$$\left\{ \sum_{i<j} K\left(\frac{w_i'a_N - w_j'a_N}{h}\right) (x_i - x_j)(x_i - x_j)' \right\}^{-1}$$

$$\sum_{i<j} K\left(\frac{w_i'a_N - w_j'a_N}{h}\right) (x_i - x_j)(y_i - y_j).$$

The idea of removing the nonparametric component $\theta(z)$ or $\theta(w'\alpha)$ with pairwise differencing with $K((z_i - z_j)/h)$ or $K((w_i'a_N - w_j'a_N)/h)$ is applicable to many LDV models with $\theta(\cdot)$, and a couple of such LDV models will be examined shortly. Aradillas-Lopez and et al. (2007) further extended the idea when the argument in $\theta(\cdot)$ is a nonparametric estimator of a conditional mean. The asymptotic distribution of the TSE is model-specific and depends on whether the argument of $\theta(\cdot)$ is observed (as in $\theta(z)$) or unknown up to a finite-dimensional nuisance parameter $(\theta(w'\alpha))$ or an infinite-dimensional nuisance parameter (e.g., $\theta(E(d|w))$. In practice, using CI's based on nonparametric bootstrap or using the variance matrix of "jackknifed" pseudo estimates will be simpler, where jackknifed pseudo estimates are obtained by repeating the estimation procedure deleting one observation (or multiple observations) at a time. See Shao and Tu (1995) for bootstrap and jackknife in general.

Consider a fixed-effect panel logit model

$$y_{it} = 1[x_{it}'\beta + \delta_i + u_{it} > 0], \qquad u_{it},\ i = 1, ..., N,\ t = 1, 2 \text{ are iid logistic.}$$

The well-known "conditional logit" maximizes, with $\tau_i \equiv 1[y_{i1} \neq y_{i2}] = 1[y_{i1} + y_{i2} = 1]$,

$$\sum_i \tau_i \left[y_{i1} \ln \left\{ \frac{1}{1 + \exp((x_{i2} - x_{i1})'b)} \right\} + y_{i2} \ln \left\{ \frac{\exp((x_{i2} - x_{i1})'b)}{1 + \exp((x_{i2} - x_{i1})'b)} \right\} \right].$$

Analogously to this, consider a cross-section semi-linear logit model

$$y_i = 1[[x_i'\beta + \theta(z_i) + u_i > 0], \qquad u_i \text{'s are iid logistic, and } u \amalg x.$$

Then β can be estimated by maximizing $(\tau_{ij} \equiv 1[y_i \neq y_j])$

$$\sum_{i<j} \tau_{ij} K \left(\frac{z_i - z_j}{h} \right) \left[y_i \ln \left\{ \frac{1}{1 + \exp((x_j - x_i)'b)} \right\} \right.$$
$$+ y_j \ln \left\{ \frac{\exp((x_j - x_i)'b)}{1 + \exp((x_j - x_i)'b)} \right\} \right].$$

Consider a semi-linear censored model

$$y = \max(0, x'\beta + \theta(z) + u).$$

Recalling the discussion on the Honoré's (1992) pairwise differencing estimator for censored models, β can be estimated by minimizing the same minimand with a kernel function attached:

$$\frac{2}{N(N-1)} \sum_{i<j} K \left(\frac{z_i - z_j}{h} \right) s\{y_i, y_j, (x_i - x_j)'b\}$$

where

$$s\{y_i, y_j, (x_i - x_j)'b\} = y_i^2 - \{y_j + (x_i - x_j)'b\}2y_i \text{ if } (x_i - x_j)'b \leq -y_j$$
$$= \{y_i - y_j - (x_i - x_j)'b\}^2 \text{ if } -y_j < (x_i - x_j)'b < y_i$$
$$= (-y_j)^2 + \{(x_i - x_j)'b - y_i\}2y_j \text{ if } y_i \leq (x_i - x_j)'b.$$

6   Additive Models

For a nonparametric regression model $y = m(x) + u$ with a $k \times 1$ vector x and $E(u|x) = 0$, Stone (1982) showed that the "best attainable convergence rate" of nonparametric estimation is $N^{-s/(2s+k)}$ where s is a degree of smoothness of $m(x)$. The rate is slower than the usual parametric rate $N^{-1/2}$ although it approaches $N^{-1/2}$ as $s \to \infty$. Stone (1985) also showed that, for nonparametric additive models of the form $m(x) = \sum_{j=1}^{k} m_j(x_j)$, the best attainable convergence rate is $N^{-s/(2s+1)}$ as if there is only one regressor regardless of k. This section reviews nonparametric estimation for additive (or additively separable) models, which is yet another way of reducing nonparametric dimension. For ease of exposition, often we will use the special cases $k = 2$ or 3, as the general case can be inferred from these.

6.1   Backfitting

Buja et al. (1989) proposed "*(classical) backfitting*" method. For simplification, set $k = 3$. Take $E(\cdot|x_j)$, $j = 1, 2, 3$, on the additive model

$$y = m_0 + \sum_{j=1}^{3} m_j(x_j) + u, \quad E\{m_j(x_j)\} = 0, \ E(u|x) = 0$$

to get

$$\begin{aligned} E(y|x_1) &= m_0 + m_1(x_1) + E\{m_2(x_2)|x_1\} + E\{m_3(x_3)|x_1\}, \\ E(y|x_2) &= m_0 + E\{m_1(x_1)|x_2\} + m_2(x_2) + E\{m_3(x_3)|x_2\}, \\ E(y|x_3) &= m_0 + E\{m_1(x_1)|x_3\} + E\{m_2(x_2)|x_3\} + m_3(x_3). \end{aligned}$$

The condition $E\{m_j(x_j)\} = 0$ is a normalization, because $\sum_{j=1}^{3} E\{m_j(x_j)\}$ can be absorbed into the intercept m_0 which can be estimated with $\bar{y}$ because $E(y) = m_0$.

Rewrite the last display as

$$\begin{aligned} m_1(x_1) &= E(y|x_1) - m_0 - E\{m_2(x_2)|x_1\} - E\{m_3(x_3)|x_1\}, \\ m_2(x_2) &= E(y|x_2) - m_0 - E\{m_1(x_1)|x_2\} - E\{m_3(x_3)|x_2\}, \\ m_3(x_3) &= E(y|x_3) - m_0 - E\{m_1(x_1)|x_3\} - E\{m_2(x_2)|x_3\}. \end{aligned}$$

Imagine solving these equations for the three unknown functions $m_j(x_j)$, $j = 1, 2, 3$. As $E(\cdot|x_j)$ can be estimated by a kernel estimator $E_N(\cdot|x_j)$ and m_0 by $\bar{y}$, we get an iteration formula:

$$\begin{aligned} \hat{m}_1^{(1)}(x_1) &= E_N(y|x_1) - \bar{y} - E_N\{\hat{m}_2^{(0)}(x_2)|x_1\} - E\{\hat{m}_3^{(0)}(x_3)|x_1\}, \\ \hat{m}_2^{(1)}(x_2) &= E_N(y|x_2) - \bar{y} - E_N\{\hat{m}_1^{(1)}(x_1)|x_2\} - E\{\hat{m}_3^{(0)}(x_3)|x_2\}, \\ \hat{m}_3^{(1)}(x_3) &= E_N(y|x_3) - \bar{y} - E_N\{\hat{m}_1^{(1)}(x_1)|x_3\} - E\{\hat{m}_2^{(1)}(x_2)|x_3\} \end{aligned}$$

where, for a kernel L and bandwidth h,

$$E_N\{\hat{m}_j^{(0)}(x_j)|x_k\} = \frac{\sum_i L((x_{ik} - x_k)/h) \cdot \hat{m}_j^{(0)}(x_{ij})}{\sum_i L((x_{ik} - x_k)/h)},$$

$E_N\{\hat{m}_j^{(1)}(x_j)|x_k\}$ is analogously defined, and the superscript (0) denotes an early stage estimator whereas (1) denotes an updated estimator.

The rhs of the $\hat{m}_1^{(1)}(x_1)$ equation becomes

$$\frac{\sum_i L((x_{i1} - x_1)/h) \cdot \{y_i - \bar{y} - \hat{m}_2^{(0)}(x_{i2}) - \hat{m}_3^{(0)}(x_{i3})\}}{\sum_i L((x_{i1} - x_1)/h)}$$

where the "residual" $y_i - \bar{y} - \hat{m}_2^{(0)}(x_{i2}) - \hat{m}_3^{(0)}(x_{i3})$ is used to estimate $\hat{m}_1^{(1)}(x_1)$, which gives the name "*backfitting*." An alternative to the above iteration formula is using only the superscript (0) versions on the rhs. Then the $\hat{m}_2^{(1)}(x_2)$ line becomes

$$\frac{\sum_i L((x_{i2} - x_2)/h) \cdot \{y_i - \bar{y} - \hat{m}_1^{(0)}(x_{i1}) - \hat{m}_3^{(0)}(x_{i3})\}}{\sum_i L((x_{i2} - x_2)/h)}.$$

In contrast to this "Jacobi algorithm," using the updated estimates as much as possible on the rhs as in the above iteration formula is called "Gauss–Seidel algorithm." In the algorithms, $E_N(y|x_j)$ and $\bar{y}$ need to be computed only once.

We can start with an initial guess, say $m_j^{(0)}(x_j) = x_j$, $j = 2, 3$, and iterate until convergence, i.e., until changes in $m_j(x_j)$'s become small. For instance, with ε being a small positive constant, use a criterion function such as

$$\frac{\sup_{x_j} |\hat{m}_j^{(1)}(x_j) - \hat{m}_j^{(0)}(x_j)|}{\sup_{x_j} |\hat{m}_j^{(0)}(x_j)| + \varepsilon} < \varepsilon \quad \text{or}$$

$$\frac{N^{-1} \sum_i \{\hat{m}_j^{(1)}(x_{ij}) - \hat{m}_j^{(0)}(x_{ij})\}^2}{N^{-1} \sum_i \{\hat{m}_j^{(0)}(x_{ij})\}^2 + \varepsilon} < \varepsilon \quad \forall j$$

and then sum up these across $j = 1, ..., k$ to get a single stopping criterion. We can do the iteration at some chosen fixed evaluation points for (x_1, x_2, x_3). But programming-wise, it is simpler (but more time-consuming) to do the iteration at the observation points unless N is prohibitively large. Also at the end of the iteration, it would be a good idea to re-center each $\hat{m}_j(x_j)$ so that it has mean zero.

6.2   Smoothed Backfitting

One disadvantage of backfitting is that its asymptotic distribution is unknown. Mammen, et al. (1999) proposed "*smoothed backfitting*" and then derived its asymptotic distribution. But the paper is complicated, and here we introduce smoothed backfitting algorithm as explained in Nielsen and Sperlich (2005). We will use, as has been done most times, a product kernel $K(t) = \Pi_{l=1}^{k} L(t_k)$ with $t = (t_1, ..., t_k)'$.

Defining x_{-j} as x with its jth element x_j removed, observe

$$\int \hat{m}_k(x_k) \frac{f_N(x_j, x_k)}{f_N(x_j)} dx_k \to^p \int m_k(x_k) \frac{f(x_j, x_k)}{f(x_j)} dx_k$$

$$= \int m_k(x_k) f(x_k | x_j) dx_k = E\{m_k(x_k) | x_j\}.$$

Then a "Gauss–Seidel-type" iteration algorithm for *smoothed backfitting* is

$$\hat{m}_1^{(1)}(x_1) = E_N(y|x_1) - \bar{y} - \int \hat{m}_2^{(0)}(x_2) \frac{f_N(x_1, x_2)}{f_N(x_1)} dx_2$$

$$- \int \hat{m}_3^{(0)}(x_3) \frac{f_N(x_1, x_3)}{f_N(x_1)} dx_3$$

$$\hat{m}_2^{(1)}(x_2) = E_N(y|x_2) - \bar{y} - \int \hat{m}_1^{(1)}(x_1) \frac{f_N(x_2, x_1)}{f_N(x_2)} dx_1$$

$$- \int \hat{m}_3^{(0)}(x_3) \frac{f_N(x_2, x_3)}{f_N(x_2)} dx_3$$

$$\hat{m}_3^{(1)}(x_3) = E_N(y|x_3) - \bar{y} - \int \hat{m}_1^{(1)}(x_1) \frac{f_N(x_3, x_1)}{f_N(x_3)} dx_1$$

$$- \int \hat{m}_2^{(1)}(x_2) \frac{f_N(x_3, x_2)}{f_N(x_3)} dx_2.$$

Compared with backfitting, the difference falls in how expressions such as $E_N\{\hat{m}_j^{(0)}(x_j)|x_k\}$ gets computed: in backfitting and smoothed backfitting, respectively,

$$E_N\{\hat{m}_j^{(0)}(x_j)|x_k\} = \frac{(Nh)^{-1} \sum_i L((x_{ik} - x_k)/h) \hat{m}_j^{(0)}(x_{ij})}{(Nh)^{-1} \sum_i L((x_{ik} - x_k)/h)},$$

$$\int \hat{m}_j^{(0)}(x_j) \frac{f_N(x_k, x_j)}{f_N(x_k)} dx_j = \frac{\int \hat{m}_j^{(0)}(x_j)(Nh^2)^{-1} \sum_i L((x_{ik} - x_k)/h) L((x_{ij} - x_j)/h) dx_j}{(Nh)^{-1} \sum_i L((x_{ik} - x_k)/h)}$$

$$= \frac{(Nh^2)^{-1} \sum_i L((x_{ik} - x_k)/h) \int \hat{m}_j^{(0)}(x_j) L((x_{ij} - x_j)/h) dx_j}{(Nh)^{-1} \sum_i L((x_{ik} - x_k)/h)}$$

$$= \frac{(Nh)^{-1} \sum_i L((x_{ik} - x_k)/h) \int \hat{m}_j^{(0)}(x_{ij} - ht) L(t) dt}{(Nh)^{-1} \sum_i L((x_{ik} - x_k)/h)}.$$

Hence the difference between the two procedures is whether $\hat{m}_j^{(0)}(x_{ij})$ or its smoothed version $\int \hat{m}_j^{(0)}(x_{ij} - ht) L(t) dt$ is used.

In implementing the algorithm, $E_N(y|x_j)$, $f_N(x_j)$ and $f_N(x_j, x_k)$ need to be computed only once. Although $\hat{m}_j(x_j)$ may be estimated at some chosen fixed evaluation points of x_j, it is simpler (but more time-consuming) to use the observation points for x_j in programming the algorithm. In this case, a further simplification may be achieved by using

$$
\int \hat{m}_j^{(\cdot)}(x_j) \frac{f_N(x_k, x_j)}{f_N(x_k)} dx_j \;=\; \int \hat{m}_j^{(\cdot)}(x_j) \frac{f_N(x_k, x_j)}{f_N(x_k) f_N(x_j)} f_N(x_j) dx_j
$$

$$
\simeq \frac{1}{N} \sum_i \hat{m}_j^{(\cdot)}(x_{ij}) \frac{f_N(x_{ik}, x_{ij})}{f_N(x_{ik}) f_N(x_{ij})}.
$$

That is, the analytical integrations get replaced by averages.

Denoting the final smoothed-backfitting estimators for $m_j(x_j)$ as $\hat{m}_j(x_j)$, the asymptotic distribution is, ignoring the bias terms with under-smoothing,

$$
\begin{bmatrix} \sqrt{Nh}\{\hat{m}_1(x_1) - m_1(x_1)\} \\ \sqrt{Nh}\{\hat{m}_2(x_2) - m_2(x_2)\} \\ \sqrt{Nh}\{\hat{m}_3(x_3) - m_3(x_3)\} \end{bmatrix} \rightsquigarrow N(0, C)
$$

$$
\text{where } C \equiv diag \left\{ \frac{V(u|x_1)}{f(x_1)} \int L(t)^2 dt, \; \frac{V(u|x_2)}{f(x_2)} \int L(t)^2 dt, \right.
$$

$$
\left. \frac{V(u|x_3)}{f(x_3)} \in L(t)^2 dt \right\}.
$$

Despite the nonparametric model, each $m_j(x)$ is estimated $\sqrt{Nh}$-consistently with its variance being the usual kernel estimator asymptotic variance. Summing the component estimators yields then

$$
\sqrt{Nh}\{\hat{m}(x) - m(x)\} \rightsquigarrow N \left\{ 0, \; \int L(t)^2 dt \sum_{j=1}^{3} \frac{V(u|x_j)}{f(x_j)} \right\}
$$

$$
\text{where } \hat{m}(x) \equiv \sum_j \hat{m}_j(x_j).
$$

As for choosing h, Nielsen and Sperlich (2005) recommended a CV-type procedure:

$$
\min_h \sum_{i=1}^{N} \{y_i - \hat{m}_{-i}(x_i)\}^2
$$

where $\hat{m}_{-i}(x_i)$ is $\hat{m}(x_i)$ with the ith observation not used. Intuition suggests that this CV bandwidth is likely to be optimal for $m(x)$, but not necessarily optimal for the individual component $m_j(x_j)$.

6.3   Marginal Integration

Consider the simplest additive model with two regressors:

$$
m(x_1, x_2) = m_0 + m_1(x_1) + m_2(x_2) \quad \text{where } E\{m_j(x_j)\} = 0,
$$

$$j = 1, 2$$
$$\Longrightarrow E\{m(x_1, x_2)|x_j\} = m_0 + m_j(x_j), \quad j = 1, 2.$$

That is, by integrating out one regressor, we can estimate the nonparametric component for the other regressor. Let $\hat{m}(x_1, x_2)$ be a kernel estimator for $m(x_1, x_2)$ with a product kernel $K(t) = L(t_1)L(t_2)$ as usual. Linton and Nielsen (1995) proposed a marginal integration estimator for $m_1(x_1)$: for a weighting function $q(x_2)$,

$$\hat{m}_1(x_1) = \int \hat{m}(x_1, x_2)q(x_2)dx_2$$

$$\xrightarrow{p} \quad {}^p m_0 + m_1(x_1) + \int m_2(x_2)q(x_2)dx_2 = m_1(x_1) + \text{constant.}$$

Clearly, an analogous estimator can be thought of for $m_2(x_2)$. Newey (1994b) also considered marginal integration, calling it "partial mean."

Under the homoskedasticity assumption $E(u|x_1, x_2) = \sigma^2$, Linton and Nielsen (1995) stated

$$\sqrt{Nh}\{\hat{m}_1(x_1) - E(\hat{m}_1(x_1))\}$$

$$\rightsquigarrow \quad N\{0, \ \sigma^2 \int \frac{q(x_2)^2}{f(x_1, x_2)}dx_2 \int L(z)^2 dz\} \quad \text{where}$$

$$E(\hat{m}_1(x_1)) = m_1(x_1) + \frac{h_1^2}{2}\int z^2 L(z)dz \cdot m_1''(x_1)$$

$$+ \frac{h_2^2}{2}\int z^2 L(z)dz \cdot \int m_2''(x_2)q(x_2)dx_2$$

where h_1 and h_2 are the bandwidths for x_1 and x_2, respectively; as before, we may use $h_j = SD(x_j)h_0$ for a common h_0. As has been done so far, we will not bother with the asymptotic bias. The convergence rate is $\sqrt{Nh}$ as in one-dimensional nonparametric estimation despite $k = 2$. For heteroskedastic cases, the asymptotic variance becomes

$$\int \sigma(x_1, x_2)^2 \frac{q(x_2)^2}{f(x_1, x_2)}dx_2 \cdot \int L(z)^2 dz.$$

For $\hat{m}(x_1, x_2) = \hat{m}_1(x_1) + \hat{m}_2(x_2)$, its asymptotic variance (and bias) is just the sum of the individuals variances (and biases).

If we use the x_2-density $f_2(x_2)$ for $q(x_2)$, then the asymptotic variance becomes

$$\int \sigma(x_1, x_2)^2 \frac{f_2(x_2)}{f_{1|2}(x_1|x_2)}dx_2 \cdot \int L(z)^2 dz$$

where $f_{1|2}$ denotes the density for $x_1|x_2$. We will show this in the following for the case where the empirical distribution is used for $q(x_2)dx_2$ as in

Linton (1997) (this is equivalent to using $f(x_2)dx_2$), and thus the marginal integration estimator becomes

$$\hat{m}_1(x_1) = \frac{1}{N}\sum_i \hat{m}(x_1, x_{i2}) = \frac{1}{N}\sum_i \frac{\sum_{j\neq i} L\{(x_{j1}-x_1)/h\}L\{(x_{j2}-x_{i2})/h\}y_j}{\sum_{j\neq i} L\{(x_{j1}-x_1)/h\}L\{(x_{j2}-x_{i2})/h\}}.$$

For the estimator in the last display, it can be shown that

$$\sqrt{Nh}\{\hat{m}_1(x_1) - m_1(x_1)\}$$
$$-\frac{1}{\sqrt{N}}\sum_i\left[\frac{1}{\sqrt{h}}L\left(\frac{x_{i1}-x_1}{h}\right)\frac{1}{f_{1|2}(x_1|x_{i2})}\{y_i - m(x_1,x_{i2})\}\right] = o_p(1).$$

The asymptotic variance can be seen from the second moment of the summand:

$$E\left[\frac{1}{h}L\left(\frac{x_{i1}-x_1}{h}\right)^2 \frac{1}{f_{1|2}(x_1|x_{i2})^2}\{y_i - m(x_1,x_{i2})\}^2\right]$$

$$= \int \frac{1}{h}L\left(\frac{x_{i1}-x_1}{h}\right)^2 \frac{1}{f_{1|2}(x_1|x_{i2})^2}\sigma^2(x_1,x_{i2})f_{2|1}(x_{i2}|x_{i1})$$
$$f_1(x_{i1})dx_{i2}dx_{i1}$$

$$\rightarrow \int \sigma^2(x_1,x_2)\frac{f_{2|1}(x_2|x_1)f_1(x_1)}{f_{1|2}(x_1|x_2)^2}dx_2 \int L^2(t)dt$$

$$= \int \sigma^2(x_1,x_2)\frac{f_{1|2}(x_1|x_2)f_2(x_2)}{f_{1|2}(x_1|x_2)^2}dx_2 \int L^2(t)dt$$

$$= \int \sigma^2(x_1,x_2)\frac{f_2(x_2)}{f_{1|2}(x_1|x_2)}dx_2 \int L^2(t)dt$$

This is the same as the above asymptotic variance when $f_2(x_2)$ is used for $q(x_2)$.

The two-regressor case with the empirical distribution for $q(x_2)dx_2$ can be generalized to k-dimensional case as in

$$\hat{m}_1(x_1) = \frac{1}{N}\sum_i \hat{m}(x_1, x_{i,-1}) \quad \text{where } x_{-1} \equiv (x_2,...,x_k)'.$$

The only change that this generalization results in is that $\int\{f_2(x_2)/f_{1|2}(x_1|x_2)\}dx_2$ in the asymptotic variance gets replaced by $\int\{f_{-1}(x_{-1})/f_{1|-1}(x_1|x_{-1})\}dx_{-1}$.

Linton (1997) provided a better estimator which is "oracle-efficient" in the sense that the estimator's asymptotic variance is the same as that when the other component $m_2(x_2)$ is known. The idea is to add one backfitting step to the marginal integration. Suppose $m_2(x_2)$ were known. In this case, defining the residual $v_i \equiv y_i - m_0 - m_2(x_{i2})$, we would estimate $m_1(x_1)$ by a kernel regression of v on x_1 whose asymptotic distribution follows easily from that of the usual kernel estimator asymptotics. Linton (1997) showed that the above marginal integration estimator is inefficient than the infeasible "oracle" estimator under homoskedasticity. But if one does the backfitting—i.e., the nonparametric regression of $\hat{v}_i \equiv y_i - m_0 - \hat{m}_2(x_{i2})$ on x_{i1} where $\hat{m}_2(x_{i2})$ is the marginal integration estimator—then the resulting estimator is oracle-efficient. One caveat is that, for the two-stage idea to work, the first-stage marginal integration estimator should be under-smoothed.

Fan et al. (1998) propose another way to attain the oracle efficiency. Instead of adding the backfitting step as Linton (1997) did, they apply local linear regression only to the direction of x_1 in the first step while a local constant regression is maintained for x_{-1}; recall that this idea appeared in the nonparametrics chapter in relation to local linear regression. Then the second step is a marginal integration. One notable feature in this procedure is that, since x_{-1} is integrated out, this allows any functional form for the x_{-1} part, not necessarily the additive $m_2(x_2)+,...,+m_k(x_k)$.

Sperlich et al. (1999) compared finite sample properties of marginal integration and backfitting. They conclude that backfitting works better at boundary points and when there are high correlations among regressors, whereas the marginal integration works better in most other cases. But Martins-Filho and Yang (2007) compared marginal integration, backfitting, smoothed backfitting and an estimator in Kim et al. (1999), and recommend backfitting. In general, as integration tends to smooth out things, we would recommend marginal integration which is also simpler to implement without any iteration.Applications of marginal integration can be found in Rodriguez-Póo et al.(2005), Vollebergh et al. (2009) and Kan and Lee (2009).

6.4   Further Generalizations*

Linton and Härdle (1996) examined a 'generalized additive model: with $m(x) = E(y|x)$,

$$G\{m(x)\} = m_0 + \sum_{j=1}^{k} m_j(x_j) \iff m(x) = G^{-1}\{m_0 + \sum_{j=1}^{k} m_j(x_j)\}$$

where $G(\cdot)$ is a known link function, to propose a marginal integration-based estimator. The idea can be seen in

$$\int G\{m(x)\}f_{-1}(x_{-1})dx_{-1} = \int \left\{ m_0 + \sum_{j=1}^{k} m_j(x_j) \right\} f_{-1}(x_{-1})dx_{-1}$$

$$= m_1(x_1) + \text{constant}$$

and $m(x)$ and $f_{-1}(x_{-1})$ can be replaced with kernel estimators where x_{-1} is the elements of x other than x_1. Horowitz (2001b) dealt with a considerably more difficult model with an unknown G function.

Horowitz and Mammen (2004) proposed a two-stage oracle-efficient estimator for the additive model with a known link function G with $H \equiv G^{-1}$. The first-stage is a series approximation of $m(x)$ by series-approximating each $m_j(x_j)$, and the second-stage estimator for $m_1(x_{o1})$ is obtained by taking one Newton–Raphson minimization step from $(a = \tilde{m}_1(x_{o1}), b = 0)$ with the first-stage estimators $\tilde{m}_j(x_{ij})$, $j \neq 1$, and $\tilde{m}_0$ plugged in; i.e., minimize wrt a and b

$$\frac{1}{N} \sum_i [y_i - H\{\tilde{m}_0 + a + b(x_{i1} - x_{o1}) + \tilde{m}_2(x_{i2}) +, ..., +\tilde{m}_k(x_{ik})\}]^2$$

$$K\left(\frac{x_{i1} - x_{o1}}{h}\right).$$

Their estimator requires $m_j(x_j)$ to be continuously differentiable only twice whereas the estimator of Linton and Härdle (1996) requires m_j to get smoother as k goes up. Horowitz and Lee (2005) extended this idea to estimate additive quantile regression models.

Linton et al. (1997) considered a parametric transformation of y:

$$\theta(y; \lambda) \quad = \quad m_0 + \sum_{j=1}^{k} m_j(x_j) \quad \text{where } \theta(\cdot; \lambda) \text{ is parametrized by an}$$

unknown λ.

In the first step, the additive model is estimated for a given value of λ. In the second step, λ is estimated, e.g., using nonlinear IVE with z as an instrument and a moment condition such as

$$\frac{1}{N} \sum_i \left\{ \theta(y_i; \lambda) - m_0 + \sum_{j=1}^{k} \hat{m}_j(x_j; \lambda) \right\} z_i = 0$$

where $\hat{m}_j(x_j; \lambda)$ is the first-stage estimator using the given value of λ. Also, Linton et al. (2008) proposed a number of estimators for $\theta(y, \lambda) = G\{m_1 (x_1), ..., m_k(x_k)\}$ where G is known and m_j unknown.

Manzan and Zerom (2005) considered an additive semi-linear model

$$y = m(x, z) + u = x'\beta + m_1(z_1) +, ..., +m_q(z_q) + u \text{ where } E\{m_j(z_j)\} = 0.$$

Denote the densities for z, z_1, and z_{-1} as f_z, f_1, and f_{-1}, respectively. Observe

$$E\left\{ y\frac{f_1(z_1)f_{-1}(z_{-1})}{f_z(z_1, z_{-1})}|z_1 \right\} = E\left[E\left\{ y\frac{f_1(z_1)f_{-1}(z_{-1})}{f_z(z_1, z_{-1})}|z \right\}|z_1 \right]$$

$$= E\left\{ E(y|z)\frac{f_1(z_1)f_{-1}(z_{-1})}{f_z(z_1, z_{-1})}|z_1 \right\}$$

$$= \int \{\beta' E(x|z) + m_1(z_1) +, ..., +m_q(z_q)\}$$

$$\cdot \frac{f_1(z_1)f_{-1}(z_{-1})}{f_z(z_1,z_{-1})} \frac{f_z(z_1,z_{-1})}{f_1(z_1)} dz_{-1}$$

$$= \int \{\beta' E(x|z) + m_1(z_1)+, ..., +m_q(z_q)\} f_{-1}(z_{-1}) dz_{-1}$$

$$= \beta' \int E(x|z_1, z_{-1}) f_{-1}(z_{-1}) dz_{-1} + m_1(z_1)$$

and

$$E\left\{x \frac{f_1(z_1)f_{-1}(z_{-1})}{f_z(z_1,z_{-1})}|z_1\right\} = E\left\{E(x|z) \frac{f_1(z_1)f_{-1}(z_{-1})}{f_z(z_1,z_{-1})}|z_1\right\}$$

$$= \int E(x|z) \frac{f_1(z_1)f_{-1}(z_{-1})}{f_z(z_1,z_{-1})} \frac{f_z(z_1,z_{-1})}{f_1(z_1)} dz_{-1}$$

$$= \int E(x|z_1, z_{-1}) f_{-1}(z_{-1}) dz_{-1}.$$

Hence,

$$E\left\{y \frac{f_1(z_1)f_{-1}(z_{-1})}{f_z(z_1,z_{-1})}|z_1\right\} = \beta' E\left\{x \frac{f_1(z_1)f_{-1}(z_{-1})}{f_z(z_1,z_{-1})}|z_1\right\} + m_1(z_1).$$

Repeat this for each z_j and then sum up across j to get

$$\sum_{j=1}^{q} E\left\{y \frac{f_j(z_j)f_{-j}(z_{-j})}{f_z(z_j,z_{-j})}|z_j\right\}$$

$$= \sum_{j=1}^{q} \beta' E\left\{x \frac{f_j(z_j)f_{-j}(z_{-j})}{f_z(z_j,z_{-j})}|z_j\right\} + m_1(z_1)+, ..., +m_q(z_q).$$

Subtract this from $y = x'\beta + m_1(z_1)+, ..., +m_q(z_q) + u$ to get

$$y - \sum_{j=1}^{q} E\left\{y \frac{f_j(z_j)f_{-j}(z_{-j})}{f_z(z_j,z_{-j})}|z_j\right\}$$

$$= \left[x - \sum_{j=1}^{q} E\left\{x \frac{f_j(z_j)f_{-j}(z_{-j})}{f_z(z_j,z_{-j})}|z_j\right\}\right]' \beta + u.$$

Replacing the densities and conditional means with kernel estimators, we get a linear equation for β to which LSE can be applied. The resulting estimator for β is $\sqrt{N}$-consistent and asymptotically normal with its asymptotic variance analogous to that of the Robinson's (1988) two-stage estimator for semi-linear model. The idea of using the factor $f_j(z_j)f_{-j}(z_{-j})/f_z(z_j,z_{-j})$ is owed to Kim et al. (1999).

7   Transformation of Response Variables

For LDV's, the regression functions are nonlinear even if the latent response regression functions are linear. It would be nice if the linearity can be restored somehow, as this would enable trying familiar linear-model techniques. For some LDV models, transformed response models sometimes do admit linear models, which is the topic of this section. The form of the transformations may be unknown or known (up to some parameters, finite- or infinite-dimensional).

7.1   Density-Weighted Response Approach

Lewbel (2000) presented innovative two-stage estimators that allow an unknown error term distribution (including an unknown form of heteroskedasticity) for binary response models. The estimators require one special continuous regressor w that is independent of the error term conditional on the other regressors; the support of w should be large as well. The estimators improve on similar, but less practical, estimators in Lewbel (1998). Although the main motivation for the estimators is allowing for endogenous regressors as will be seen in the next subsection (the estimators take familiar IVE forms in this case), we examine binary response models with only exogenous regressors in this subsection as a preparation for the endogeneity case.

7.1.1   Main Idea

Consider a binary response model with a "special" regressors w and the other regressors x:
$$y_i = 1[\beta_w w_i + x_i'\beta^o + u_i^o > 0].$$
Suppose that the sign of β_w is known to be positive—the sign can be easily estimated as will be shown below. Normalizing the scale by dividing through with β_w yields

$$y_i = 1[w_i + x_i'\beta + u_i > 0] \quad \text{where } \beta \equiv \frac{\beta^o}{\beta_w} \text{ and } u_i^o \equiv \frac{u_i}{\beta_w}.$$

Denoting the conditional density and df of $w|x$ as $f_{w|x}(w)$ and $F_{w|x}(w)$, respectively, assume

(i) $E(xu) = 0$ and $u \amalg w|x$ (i.e., $F_{u|w,x}(u) = F_{u|x}(u)$)

(ii) the support of $f_{w|x}(w)$ is $[W_l, W_h]$ that includes the support of $-x'\beta - u$ where $-\infty \le W_l < 0 < W_h \le \infty$;

to assure $W_l < 0 < W_h$, we may "de-mean" w so that $E(w) = 0$. Define

$$\tilde{y} \equiv \frac{y - 1[w > 0]}{f_{w|x}(w)} \quad \text{and} \quad s(x, u) \equiv -x'\beta - u.$$

In the following , we prove $E(\tilde{y}|x) = x'\beta$: $\tilde{y}$ has a linear regression function, although y does not. A remark on notations is needed before we proceed: x in notations such as $x'\beta$ and $E(xx')$ includes unity (i.e., the intercept is estimated), but x in notations such as $w|x$ does not.

Observe

$$E(\tilde{y}|x) = E[\ E\{\frac{y - 1[w > 0]}{f_{w|x}(w)}|w, x\}\ |x]$$

$$= \int_{W_l}^{W_h} \frac{E\{y - 1[w > 0]|w, x\}}{f_{w|x}(w)} f_{w|x}(w) dw$$

$$= \int_{W_l}^{W_h} E\{1[w + x'\beta + u > 0] - 1[w > 0]|w, x\} dw, \text{ canceling } f_{w|x}(w)$$

$$= \int_{W_l}^{W_h} \int (1[w + x'\beta + u > 0] - 1[w > 0]) dF_{u|x} dw,$$

under $F_{u|w,x}(u) = F_{u|x}(u)$

$$= \int \int_{W_l}^{W_h} (1[w > s(x, u)] - 1[w > 0]) dw \cdot dF_{u|x}.$$

The inner integrand depends on $s(x, u)$: it is zero when $s(x, u) = 0$, and it is also zero when $s(x, u) \neq 0$ except

1. if $s(x, u) < 0$, then the inner integrand is 1 when $s(x, u) < w < 0$

2. if $s(x, u) > 0$, then the inner integrand is -1 when $0 < w < s(x, u)$.

Thus

$$E(\tilde{y}|x)$$

$$= \int \left\{ 1[s(x, u) < 0] \int_{s(x,u)}^{0} 1 dw - 1[s(x, u) > 0] \int_{0}^{s(x,u)} dw \right\} dF_{u|x}$$

$$= -\int s(x, u) dF_{u|x} = -\int (-x'\beta - u) dF_{u|x} = x'\beta + E(u|x).$$

Hence, using integral, the regression function is pulled out of the indicator function. This idea has its predecessor in Horowitz and Härdle (1996).

Multiply $E(\tilde{y}|x) = x'\beta + E(u|x)$ by x and take $E(\cdot)$ to get

$$E\{E(x\tilde{y}|x)\} = E(xx')\beta + E\{E(xu|x)\}$$

$$\implies E(x\tilde{y}) = E(xx')'\beta \quad \text{because } E\{E(xu|x)\} = E(xu) = 0$$

$$\implies \beta = E^{-1}(xx') \cdot E\left\{ x\frac{y - 1[w > 0]}{f_{w|x}(w)} \right\}.$$

An estimator for β is its sample analog

$$b_N = \left(\frac{1}{N} \sum_i x_i x_i' \right)^{-1} \cdot \frac{1}{N} \sum_i x_i \frac{y_i - 1[w_i > 0]}{\hat{f}_{w|x_i}(w_i)}$$

$$\text{where } \hat{f}_{w|x_i}(w_i) = \frac{\hat{f}_{w,x}(w_i, x_i)}{\hat{f}_x(x_i)}$$

and $\hat{f}_{w,x}(w_i, x_i)$ and $\hat{f}_x(x_i)$ are kernel estimators.

As for $sgn(\beta_w)$, recall $y_i = 1[\beta_w w_i + x_i'\beta^o + u_i^o > 0]$. For this model, we get

$$E(y|w, x) = 1 - F_{u^o|x}(-\beta_w w - x'\beta^o)$$
$$\implies \frac{\partial E(y|w, x)}{\partial w} = \beta_w f_{u^o|x}(-\beta_w w - x'\beta^o).$$

Estimate $\partial E(y|w, x)/\partial w$ to find $sgn(\beta_w)$; better yet, estimate $E\{\partial E(y|w, x)/\partial w\}$ with an average derivative estimator. The sign of the estimator is consistent for $sgn(\beta_w)$. The convergence rate should be faster than $\sqrt{N}$ because any consistent estimator τ_N for $sgn(\beta_w)$ should satisfy $P\{\tau_N = sgn(\beta_w)\} \to 1$, which implies $P\{N^\nu|\tau_N - sgn(\beta_w)| \le \varepsilon\} \to 1$ for any constant $\nu, \varepsilon > 0$. Hence, $sgn(\beta_w)$ is as good as known as in some other semiparametric estimators for binary responses.

7.1.2   Asymptotic Distribution

To find the asymptotic distribution of $\sqrt{N}(b_N - \beta)$, define $\tilde{y}_{Ni}$ as $\tilde{y}_i$ with its $f_{w|x}$ replaced by $\hat{f}_{w|x}$, and observe

$$
\begin{aligned}
b_N &= \left(\frac{1}{N}\sum_i x_i x_i'\right)^{-1}\frac{1}{N}\sum_i x_i\{\tilde{y}_{Ni} - (x_i'\beta - x_i'\beta)\} \\
&= \beta + \left(\frac{1}{N}\sum_i x_i x_i'\right)^{-1}\frac{1}{N}\sum_i x_i(\tilde{y}_{Ni} - x_i'\beta).
\end{aligned}
$$

Hence

$$\sqrt{N}(b_N - \beta) = \left(\frac{1}{N}\sum_i x_i x_i'\right)^{-1}\frac{1}{\sqrt{N}}\sum_i x_i(\tilde{y}_{Ni} - x_i'\beta).$$

We just have to account for the first-stage estimation error $\hat{f}_{w|x} - f_{w|x}$ in $\tilde{y}_{Ni}$.

Recall the CT_{fc} correction term modified for this case:

$$E(m_{a'}|w_i, x_i)f_{w|x_i}(w_i) - E\{E(m_{a'}|w, x)f_{w|x}(w)|x_i\}$$
$$\text{where } m\{\hat{f}_{w|x_i}(w_i)\} = x_i(\tilde{y}_{Ni} - x_i'\beta).$$

Use this to get

$$\frac{1}{\sqrt{N}}\sum_i x_i(\tilde{y}_{Ni} - x_i'\beta) = \frac{1}{\sqrt{N}}\sum_i \left[\, x_i(\tilde{y} - x_i'\beta)\right.$$

$$- x_i E \left\{ \frac{y - 1[w > 0]}{f_{w|x}(w)^2} \middle| w_i, x_i \right\} \cdot f_{w|x_i}(w_i)$$

$$+ x_i E \left(E \left\{ \frac{y - 1[w > 0]}{f_{w|x}(w)^2} \middle| w, x \right\} f_{w|x}(w) \middle| x_i \right) \right]$$

$$= \frac{1}{\sqrt{N}} \sum_i \left[x_i (\tilde{y} - x_i'\beta) - x_i E \left\{ \frac{y - 1[w > 0]}{f_{w|x}(w)} \middle| w_i, x_i \right\} \right.$$

$$+ x_i E \left\{ \frac{y - 1[w > 0]}{f_{w|x}(w)} \middle| x_i \right\} \right]$$

$$= \frac{1}{\sqrt{N}} \sum_i x_i [(\tilde{y} - x_i'\beta) - \{E(\tilde{y}|w_i, x_i) - E(\tilde{y}|x_i)\}] \equiv \frac{1}{\sqrt{N}} \sum_i \eta_i.$$

In η, the first term $\tilde{y} - x_i'\beta$ is for $f_{w|x}$ known. The correction term due to $\hat{f}_{w|x}$ is $E(\tilde{y}|w_i, x_i) - E(\tilde{y}|x_i)$ that appears with a negative sign because $\hat{f}_{w|x}$ appears in the denominator of $\tilde{y}$. Hence $\sqrt{N}(b_N - \beta)$ is asymptotically normal with the variance

$$E^{-1}(xx') \cdot E(\eta\eta') \cdot E^{-1}(xx')$$
$$\text{where } \eta = x[(\tilde{y} - x'\beta) - \{E(\tilde{y}|w, x) - E(\tilde{y}|x)\}].$$

Recalling $E(\tilde{y}|x) = x'\beta + E(u|x)$, since $E(u|x)$ eventually drops out when x is multiplied to u (or we may simply adopt the stronger condition $E(u|x) = 0$ instead of $E(xu) = 0$), we can take $E(\tilde{y}|x) = x'\beta$, and this $x'\beta$ gets canceled by $x'\beta$ in $\tilde{y} - x'\beta$. That is, we can take η as $x\{\tilde{y} - E(\tilde{y}|w, x)\}$. A nonparametric estimator is called for $E(y|w, x)$, because

$$E(\tilde{y}|w, x) = \frac{E(y|w, x) - 1[w > 0]}{f_{w|x}(w)}.$$

7.1.3   Further Remarks

Suppose

$$w_i = z_i'\gamma + \varepsilon_i, \text{ where } \varepsilon \text{ is a continuous rv with bounded support}$$
$$\text{and } \varepsilon \amalg z.$$

Regard z as x; here we use a more general "instrument" notation z so that this remark can be applied to the endogenous x case in the next subsection. Let $\hat{\varepsilon}$ be the LSE residual. Also let $\hat{\varepsilon}_i^+$ denote the smallest element of $\{\hat{\varepsilon}_1, ... \hat{\varepsilon}_N\}$ that is greater than $\hat{\varepsilon}_i$, and let $\hat{\varepsilon}_i^-$ denote the largest element of $\{\hat{\varepsilon}_1, ... \hat{\varepsilon}_N\}$ that is smaller than $\hat{\varepsilon}_i$. Define

$$\tilde{y}_i^* \equiv \frac{(y_i - 1[w_i > 0])}{2/\{(\hat{\varepsilon}_i^+ - \hat{\varepsilon}_i^-)N\}}, \quad \text{where } \frac{2}{(\hat{\varepsilon}_i^+ - \hat{\varepsilon}_i^-)N} \text{ is for } f_\varepsilon(\varepsilon_i) \simeq f_{w|z_i}(w_i).$$

Lewbel (2000, p. 158) stated that

$$b_N = \left(\frac{1}{N} \sum_i x_i x_i' \right)^{-1} \frac{1}{N} \sum_i z_i \tilde{y}_i^*$$

is consistent for β, without providing the asymptotic distribution of this simpler alternative under the extra linear model assumption for w.

Maurin (2002) applied the Lewbel's estimator to find the effect of parental income on repeating the same grade in elementary school. Maurin used the date of birth as the special regressor w. Khan and Lewbel (2007) extended the binary response estimator to truncated or censored models, but the estimators are rather complicated. Also, the truncated estimator requires a non-truncated auxiliary sample for (w, x) to estimate $f_{w|x}$ despite the truncation.

The above binary-response estimator can be extended also to ordered discrete response (ODR) models, as will be reviewed in the next subsection. Anton et al. (2001) applied an inefficient ODR extension to a grouped unemployment duration data where w is age. According to Stewart (2005) who compared a couple of semiparametric estimators for ODR through a simulation study (and two empirical analyses), the Lewbel's estimator did not perform well. The most obvious reason for this is $\hat{f}_{w|x}$ in the denominator, which can be close to zero to make $\tilde{y}$ too large.

7.2 Extensions of Density-Weighted Response Approach

The preceding subsection introduced Lewbel's (2000) estimator for binary responses with exogenous regressors. As there are many other better-performing semiparametric estimators, it is unlikely that the estimator gets used for ordinary binary response models. In fact, the real motivation for the estimator was dealing with endogenous regressors, which is examined in this subsection. Also, extensions for ODR and panel binary-response models will be studied in this subsection which draws partly on Lee, Huang and Kim (2008).

7.2.1 Endogenous Regressors

For endogenous regressors with an instrument z, conditions (i) and (ii) above should be replaced by

$(i)'$ $E(zu) = 0$ and $(u, x) \amalg w|z$ (i.e., $F_{u,x|w,z} = F_{u,x|z}$)

$(ii)'$ the support of $f_{w|z}(w)$ is $[W_l, W_h]$ that includes the support of $-x'\beta - u$ where $-\infty \leq W_l < 0 < W_h \leq \infty$.

The same notations W_l and W_h as used in (ii) of the preceding subsection are used here to avoid too many notations, although they refer to different things. The condition $x \amalg w|z$ in $(i)'$ is a restrictive feature, which was not present in the preceding no endogeneity case. Magnac and Maurin (2007) find that the support condition for $w|z$ requires $P(y = 1|w, x, z)$ to attain both 0 and 1 as w varies for a.e. (x, z); they show that the restrictive support condition can be weakened at the cost of a tail symmetry-type assumption that is more plausible. A remark on notations is that z in $E(zz')$ and $E(xz')$ includes unity while z in notations such as $w|z$ does not.

Let

$$\mathring{y}_i \equiv \frac{y_i - 1[w_i > 0]}{f_{w|z_i}(w_i)} \quad \text{and} \quad \mathring{y}_{Ni} \equiv \frac{y_i - 1[w_i > 0]}{\hat{f}_{w|z_i}(w_i)}$$

and observe

$$E(\mathring{y}|z) = E\left[E\left\{ \frac{y - 1[w > 0]}{f_{w|z}(w)} |w, z \right\} |z \right]$$

$$= \int_{W_l}^{W_h} \frac{E\{y - 1[w > 0]|w, z\}}{f_{w|z}(w)} f_{w|z}(w)dw$$

$$= \int_{W_l}^{W_h} E\{1[w + x'\beta + u > 0] - 1[w > 0]|w, z\}dw, \text{ canceling } f_{w|z}(w)$$

$$= \int_{W_l}^{W_h} \int (1[w + x'\beta + u > 0] - 1[w > 0])dF_{u,x|z}dw,$$

under $_{u,x|w,z} = F_{u,x|z}$

$$= \int \int_{W_l}^{W_h} (1[w > -x'\beta - u] - 1[w > 0])dw \cdot dF_{u,x|z}$$

$$= -\int (-x'\beta - u)dF_{u,x|z} = E(x'|z)\beta + E(u|z).$$

Multiply $E(\mathring{y}|z) = E(x'|z)\beta + E(u|z)$ by $E(xz')E^{-1}(zz')z$ and take $E(\cdot)$ to get

$$E(xz')E^{-1}(zz')E(z\mathring{y}) = E(xz')E^{-1}(zz')E(zx')\beta$$

$$\implies \beta = M \cdot E\left\{ z\frac{y - 1[w > 0]}{f_{w|z}(w)} \right\} \text{ where}$$

$$M \equiv \{E(xz')E^{-1}(zz')E(zx')\}^{-1}E(xz')E^{-1}(zz').$$

Instead of $E(xz')E^{-1}(zz')z$, the GMM version would use $E(xz')E^{-1}$ $\{zz'(\mathring{y} - x'\beta)^2\}z$.

Let b_{ive} denote the sample version. The main term for the asymptotic variance is

$$\frac{1}{\sqrt{N}} \sum_i z_i(\mathring{y}_{Ni} - x_i'\beta).$$

Recall the CT_{fc} correction term modified for this case:

$$E(m_{a'}|w_i, z_i)f_{w|z_i}(w_i) - E\{E(m_{a'}|w, z)f_{w|z}(w)|z_i\}$$

$$\text{where } m\{\hat{f}_{w|z_i}(w_i)\} = z_i(\mathring{y}_{Ni} - x_i'\beta).$$

Use this to get

$$\frac{1}{\sqrt{N}} \sum_i z_i(\mathring{y}_{Ni} - x_i'\beta) = \frac{1}{\sqrt{N}} \sum_i \Bigg[z_i(\mathring{y} - x_i'\beta)$$

$$-z_i E \left\{ \frac{y - 1[w > 0]}{f_{w|z}(w)^2} \Big| w_i, z_i \right\} f_{w|z_i}(w_i)$$

$$+z_i E \left(E \left\{ \frac{y - 1[w > 0]}{f_{w|z}(w)^2} \Big| w, z \right\} f_{w|z}(w) \Big| z_i \right) \Bigg]$$

$$= \frac{1}{\sqrt{N}} \sum_i \Bigg[z_i(\mathring{y} - x_i'\beta) - z_i E \left\{ \frac{y - 1[w > 0]}{f_{w|z}(w)} \Big| w_i, z_i \right\}$$

$$+z_i E \left\{ \frac{y - 1[w > 0]}{f_{w|z}(w)} \Big| z_i \right\} \Bigg]$$

$$= \frac{1}{\sqrt{N}} \sum_i z_i[(\mathring{y} - x_i'\beta) - \{E(\mathring{y}|w_i, z_i) - E(\mathring{y}|z_i)\}] \equiv \frac{1}{\sqrt{N}} \sum_i \zeta_i.$$

Differently from no endogeneity case, $\mathring{y} - \{E(\mathring{y}|w_i, z_i)$ cannot be taken as ζ because $E(\mathring{y}|z) \neq x'\beta$.

With ζ derived, $\sqrt{N}(b_{ive} - \beta)$ is asymptotically normal with variance

$$M \cdot E(\zeta\zeta') \cdot M, \quad \text{where } \zeta = z[(\mathring{y} - x'\beta) - \{E(\mathring{y}|w, z) - E(\mathring{y}|z)\}]$$
and $E(\zeta) = 0$.

Observe

$$E(\mathring{y}|w, z) = \frac{E(y|w, z) - 1[w > 0]}{f_{w|z}(w)} \quad \text{and} \quad E(\mathring{y}|z) = E(x'|z)\beta + E(u|z).$$

Hence, estimating $E(\mathring{y}|w, x)$ needs a nonparametric estimator for $E(y|w, z)$, and estimating $E(\mathring{y}|z)$ needs a nonparametric estimator for $E(x|z)$—here x includes unity.

The two-stage IVE-type estimator for binary response with endogenous regressors includes the two-stage LSE type estimator as a special case when $z = x$. Lewbel (2000, p. 156) defined, in essence, $q = z\mathring{y} + zE(\mathring{y}|z) - zE(\mathring{y}|w, z)$. Then Lewbel (2000, p. 157) introduced $q - zx'\beta$ which is the same as ζ. Our derivation shows better that $-E(\mathring{y}|w, z) + E(\mathring{y}|z)$ is the correction term, whereas $\mathring{y} - x'\beta$ is the error term with the nuisance parameter $f_{w|z}$ known.

As mentioned above, (u, x) II $w|z$ is restrictive because this requires w to be independent of x, whereas the no-endogeneity case condition u II $w|x$

has no such restriction. There is a way to avoid $(u, x) \amalg w|z$ at the cost of "aggravating" the dimension problem: assume instead

$$u \amalg w|(x, z)$$

which is a more natural extension of $u \amalg w|x$. As both x and z are in the conditioning set, the dimension problem gets worsened.

Define

$$\check{y} \equiv \frac{y - 1[w > 0]}{f_{w|x,z}(w)}$$

and observe

$$
\begin{aligned}
E(\check{y}|x, z) &= E\left[E\left\{ \frac{y - 1[w > 0]}{f_{w|x,z}(w)} \bigg| w, x, z \right\} \bigg| x, z \right] \\[2mm]
&= \int_{W_l}^{W_h} \frac{E\{y - 1[w > 0]|w, x, z\}}{f_{w|x,z}(w)} f_{w|x,z}(w) dw \\[2mm]
&= \int_{W_l}^{W_h} E\{1[w + x'\beta + u > 0] - 1[w > 0]|w, x, z\} dw, \\
&\quad \text{canceling } f_{w|x,z}(w) \\[2mm]
&= \int_{W_l}^{W_h} \int (1[w + x'\beta + u > 0] - 1[w > 0]) dF_{u|x,z} dw, \\
&\quad \text{as } F_{u|w,x,z} = F_{u|x,z} \\[2mm]
&= \int \int_{W_l}^{W_h} (1[w > -x'\beta - u] - 1[w > 0]) dw \cdot dF_{u|x,z} \\[2mm]
&= -\int (-x'\beta - u) dF_{u|x,z} = x'\beta + E(u|x, z).
\end{aligned}
$$

Multiply $E(\check{y}|x, z) = x'\beta + E(u|x, z)$ by $E(xz')E^{-1}(zz')z$ and take $E(\cdot)$ to get

$$E(xz')E^{-1}(zz')E(z\check{y}) = E(xz')E^{-1}(zz')E(zx')\beta$$

$$\implies \quad \beta = M \cdot E\left\{ z \frac{y - 1[w > 0]}{f_{w|x,z}(w)} \right\}.$$

Again, the GMM version may be used instead.

7.2.2   Ordered Discrete Response

Consider an ODR model

$$y_i = \sum_{r=1}^{R-1} 1[w_i + x_i'\beta + u_i \geq \gamma_r].$$

Let $x_i = (1, \tilde{x}_i')'$ and $\beta = (\beta_1, \tilde{\beta}')'$, and rewrite the ODR as a sum of binary response models:

$$y_i = \sum_{r=1}^{R-1} y_{ri} = 1[y_i \geq 1] + 1[y_i \geq 2] +, ..., + 1[y_i \geq R-1] \quad \text{where}$$

$$y_{ri} \equiv 1[y_i \geq r] = 1[w_i + x_i'\beta + u_i \geq \gamma_r]$$
$$= 1[\beta_1 - \gamma_r + w_i + \tilde{x}_i'\tilde{\beta} + u_i \geq 0].$$

Recall $M \equiv \{E(xz')E^{-1}(zz')E(zx')\}^{-1}E(xz')E^{-1}(zz')$ in the preceding subsection. With $M = (M_1', \tilde{M}')'$ where M_1 is the first row of M and $\tilde{M}$ is the remaining rows, it holds under $(i)'$ and $(ii)'$ that, for each binary y_r,

$$\beta_1 - \gamma_r = M_1 E\left(z\frac{y_r - 1[w > 0]}{f_{w|z}}\right), \quad r = 1, ..., R-1,$$

$$\tilde{\beta} = \tilde{M} \cdot E\left(z\frac{y_r - 1[w > 0]}{f_{w|z}}\right), \quad r = 1, ..., R-1.$$

Since the same $\tilde{\beta}$ satisfies the second condition for all r, Lewbel (2000, p. 161) suggested to use the unweighted average across r:

$$\tilde{\beta} = \tilde{M} \cdot E\left(z\frac{(R-1)^{-1}\sum_{r=1}^{R-1} y_r - 1[w > 0]}{f_{w|z}}\right)$$

for ODR. But, not just $(R-1)^{-1}\sum_{r=1}^{R-1} y_r$, but any combination $\sum_{r=1}^{R-1} w_r y_r$ with $\sum_{r=1}^{R-1} w_r = 1$ can be used. The efficient combination can be found using the minimum distance estimator (MDE) as follows.

For instance, with $R = 3$ for three-category ODR and $\tilde{k} = \dim(\tilde{\beta})$, Lee (2002, p. 94) showed (with modified notations)

$$\begin{bmatrix} \psi_{11} \\ \psi_{1s} \\ \psi_{21} \\ \psi_{2s} \end{bmatrix} = \begin{bmatrix} 1 & 0 & 0 \\ 0 & 0 & I_{\tilde{k}} \\ 0 & 1 & 0 \\ 0 & 0 & I_{\tilde{k}} \end{bmatrix} \begin{bmatrix} \beta_1 - \gamma_1 \\ \beta_1 - \gamma_2 \\ \tilde{\beta} \end{bmatrix}$$

where the left-hand side is the "RF" parameters: $\psi_1 \equiv (\psi_{11}, \psi_{1s}')'$ for y_{1i} and $\psi_2 \equiv (\psi_{21}, \psi_{2s}')'$ for y_{2i} such that ψ_{11} and ψ_{21} are the intercepts and ψ_{1s} and ψ_{2s} are the slopes. Defining the middle matrix as C, the efficient MDE for the right-hand side "SF" parameter $\Gamma \equiv (\beta_1 - \gamma_1, \beta_1 - \gamma_2, \tilde{\beta}')'$ is obtained by minimizing

$$\{(\hat{\psi}_1', \hat{\psi}_2')' - C\Gamma\}'\Omega_N^{-1}\{(\hat{\psi}_1', \hat{\psi}_2')' - C\Gamma\}$$

wrt Γ where $\hat{\psi}_1$ and $\hat{\psi}_2$ are the RF estimates and Ω_N is their asymptotic variance estimator. The resulting efficient MDE is

$$\hat{\Gamma} = (C'\Omega_N^{-1}C)^{-1}C'\Omega_N^{-1}(\hat{\psi}_1', \hat{\psi}_2')' \quad \text{with}$$

$$\sqrt{N}(\hat{\Gamma} - \Gamma) \quad \rightsquigarrow \quad N\{0, \ (C\Omega^{-1}C)^{-1}\}, \quad \text{where } \Omega_N \rightarrow^p \Omega.$$

To obtain Ω_N, we need influence functions for $(\hat{\psi}_1', \hat{\psi}_2')'$. Define $\mathring{y}_{ji}$ as the transformed response for y_{ji}, $j = 1, 2$, and define

$$\zeta_{ji} = z_i[\mathring{y}_{ji} - x_i'\beta - \{E(\mathring{y}_j|w_i, z_i) - E(\mathring{y}_j|z_i)\}], \quad j = 1, 2.$$

Let $\hat{\zeta}_{ji}$ be an estimator for ζ_{ji}. Then an estimator of the influence function for $\hat{\psi}_j$ is

$$\hat{\xi}_{ji} = M_N\hat{\zeta}, \, j = 1, 2 \quad \text{where } M_N \text{ is a sample analog for } M \text{ so that}$$

$$\hat{\psi}_j = \frac{1}{\sqrt{N}} \sum_i \hat{\xi}_{ji}.$$

Then $\Omega_N = N^{-1}\sum_i(\xi_{1i}', \xi_{2i}')'(\xi_{1i}', \xi_{2i}')$. The Lewbel's unweighted average amounts to replacing Ω_N in the MDE with I_N, which yields an inefficient MDE using the SF-RF restrictions

$$\beta_1 - \gamma_1 = \psi_{11}, \quad \beta_1 - \gamma_2 = \psi_{21}, \quad \text{and} \quad \tilde{\beta} = \frac{\psi_{1s} + \psi_{2s}}{2}.$$

The last equation is the unweighted simple average of ψ_{1s} and ψ_{2s}.

7.2.3   Panel Binary Response*

Honoré and Lewbel (2002) extended the binary response estimator to panel binary response models:

$$y_{it} = 1[w_{it} + x_{it}'\beta + v_{it} > 0], \quad v_{it} = \delta_i + u_{it}, \quad t = 0, 1$$

where δ_i is a time-constant error that may be related to x_{it}. Let z_i be an instrument vector; z_i would consist of "predetermined" regressor up to period 0, where a "predetermined" regressor ζ_{is} is orthogonal to v_{it} $\forall s \leq t$. Assume

(i)″ $E(u_{it}z_i) = 0$, $t = 0, 1$ and $v_{it} \perp\!\!\!\perp w_{it}|(x_{it}, z_i)$ (i.e., $F_{v_{it}|w_{it},x_{it},z_i} = F_{v_{it}|x_{it},z_i}$)

(ii)″ the support of $f_{w_{it}|x_{it},z_i}(w)$ is $[W_{lt}, W_{ht}]$ that includes the support of $-x_{it}'\beta - v_{it}$ where $-\infty \leq W_{lt} < 0 < W_{ht} \leq \infty$;

These assumptions do not require x_{it} to be "strictly exogenous" where a "strictly exogenous" regressor ζ_{is} is orthogonal to v_{it} $\forall s, t$. Although we use time-constant z_i, time-variant z_{it} can be accommodated as can be seen shortly. The assumption $v_{it} \perp\!\!\!\perp w_{it}|(x_{it}, z_i)$ is of "dimension-aggravating but more intuitive" type; instead of this, $(v_{it}, x_{it}) \perp\!\!\!\perp w_{it}|z_i$ may be adopted as in the cross-section case, which is easier on the dimension problem but less plausible.

Define

$$\tilde{y}_{it} \equiv \frac{y_{it} - 1[w_{it} > 0]}{\hat{f}_{w_{it}|x_{it},z_i}(w_{it})} \quad \text{and} \quad \tilde{y}_{Nit} \equiv \frac{y_{it} - 1[w_{it} > 0]}{f_{w_{it}|x_{it},z_i}(w_{it})}.$$

Doing analogously to what was done for the cross-section case, it can be shown that

$$E(\tilde{y}_{it}|x_{it}, z_i) = x_{it}'\beta + E(\delta_i + u_{it}|x_{it}, z_i).$$

Multiply this by z_i and take $E(\cdot)$ to get

$$\begin{aligned} E\{E(z_i\tilde{y}_{it}|x_{it}, z_i)\} &= E(z_ix_{it}')\beta + E\{E(z_i\delta_i + z_iu_{it}|x_{it}, z_i)\} \\ \implies E(z_i\tilde{y}_{it}) &= E(z_ix_{it}')\beta + E(z_i\delta_i), \quad \text{for } E(z_iu_{it}) = 0. \end{aligned}$$

Define $\Delta\tilde{y}_i \equiv \tilde{y}_{i1} - \tilde{y}_{i0}$ and $\Delta x_i \equiv x_{i1} - x_{i0}$. Take the first-difference to get

$$E(z_i\Delta\tilde{y}_i) = E(z_i\Delta x_i')\beta.$$

Multiply this by $E(\Delta x_i'z_i)E^{-1}(z_iz_i')$ to obtain

$$\beta \equiv [E(\Delta x_i'z_i)E^{-1}(z_iz_i')E(z_i\Delta x_i')]^{-1} \cdot E(\Delta x_i'z_i)E^{-1}(z_iz_i')E(z_i\Delta\tilde{y}_i).$$

Since $E(z_i\delta_i)$ is removed in the first differencing, this reveals that $E(z_iu_{it})$ does not have to be zero: what is needed is only a moment stationarity type assumption

$$E(z_iu_{i1}) = E(z_iu_{i0}).$$

Components of endogenous regressors can be used so long as they meet this condition.

As for the asymptotic variance, it again takes the form $QE(\nu\nu')Q$ where Q is the matrix before $E(z_i\Delta\tilde{y}_i)$ in the above β-equation and ν is the summand of the last expression in the following display:

$$\frac{1}{\sqrt{N}}\sum_i(z_i\Delta\tilde{y}_{Ni} - z_i\Delta x_i'\beta) = \frac{1}{\sqrt{N}}\sum_i\left[z_i\left(\frac{y_{i1} - 1[w_{i1} > 0]}{f_{w_{i1}|x_{i1},z_i}(w_{i1})}\right.\right.$$

$$\left.- \frac{y_{i0} - 1[w_{i0} > 0]}{f_{w_{i0}|x_{i0},z_i}(w_{i0})} - \Delta x_i'\beta\right)$$

$$- z_iE\left\{\frac{y_{i1} - 1[w_{i1} > 0]}{f_{w_{i1}|x_{i1},z_i}(w_{i1})}\middle|w_{i1}, x_{i1}, z_i\right\} + z_iE\left(E\left\{\frac{y_{i1} - 1[w_{i1} > 0]}{f_{w_{i1}|x_{i1},z_i}(w_{i1})}\right.\right.$$

$$\left.\left.\middle|w_{i1}, x_{i1}, z_i\right\} \middle|x_{i1}, z_i\right)$$

$$+ z_iE\left\{\frac{y_{i0} - 1[w_{i0} > 0]}{f_{w_{i0}|x_{i0},z_i}(w_{i0})}\middle|w_{i0}, x_{i0}, z_i\right\} + z_iE\left(E\left\{\frac{y_{i0} - 1[w_{i0} > 0]}{f_{w_{i0}|x_{i0},z_i}(w_{i0})}\right.\right.$$

$$\left.\left.\middle|w_{i0}, x_{i0}, z_i\right\} \middle|x_{i0}, z_i\right)\right]$$

$$= \frac{1}{\sqrt{N}}\sum_i z_i[(\Delta\tilde{y}_i - \Delta x_i'\beta) - \{E(\tilde{y}_{i1}|w_{i1}, x_{i1}, z_i) - E(\tilde{y}_{i1}|x_{i1}, z_i)\}$$

$$+ \{E(\tilde{y}_{i0}|w_{i0}, x_{i0}, z_i) - E(\tilde{y}_{i0}|x_{i0}, z_i)\}].$$

Suppose z is time-variant. This generalization is helpful as many instruments in panel data are time-specific; i.e., we have z_{it} instead of z_i. Then we would get

$$\tilde{y}_{it} \equiv \frac{y_{it} - 1[w_{it} > 0]}{f_{w_{it}|x_{it}, z_{it}}(w_{it})} \quad \text{and} \quad E(z_{it}\tilde{y}_{it}) = E(z_{it}x_{it}')\beta + E(z_{it}\delta_i + z_{it}u_{it}).$$

First-differencing yields

$$E\{\Delta(z_i\tilde{y}_i)\} = E\{\Delta(z_i x_i')\}\beta \quad \text{under } E(\delta_i\Delta z_i) = 0 \text{ and } E\{\Delta(z_i u_i)\} = 0.$$

The first equation can be inverted after being multiplied by a conformable matrix. What is interesting is the sufficient moment-stationarity-type conditions

$$E(\delta_i z_{i1}) = E(\delta_i z_{i0}) \quad \text{and} \quad E(u_{i1} z_{i1}) = E(u_{i0} z_{i0}).$$

Essentially, the first equation is the extra condition needed by allowing for time-variant instruments.

7.3   Unknown Transformation of Response*

Horowitz (1996) considered an unknown transformation-of-response model:

$$\Lambda(y) \;=\; x'\beta + u, \quad \Lambda(t) \text{ is unknown, strictly increasing with its}$$
$$\text{inverse } \Lambda^{-1}$$
$$\Lambda(y_0) \;=\; 0 \text{ for some } y_0, \quad u \amalg x, \quad u \sim F, \quad y|x'\beta \sim G(\cdot|x'\beta)$$

where F and G are df's with densities f and g, $x_k|x_{-k}$ has a density function for all $x_{-k} \equiv (x_1, ..., x_{k-1})'$, and β_k is either -1 or 1, and there is no intercept in $x'\beta$ as the intercept is not separately identified from the unknown Λ. Let $\lambda(t) \equiv d\Lambda(t)/dt$.

To understand these and the condition $\Lambda(y_0) = 0$, suppose that the initial model is

$$\Lambda^o(y) = \beta_0 + x_{-k}'\beta_{-k} + x_k\beta_k + u$$
$$\Longrightarrow \Lambda^o(y) - \beta_0 = x_{-k}'\beta_{-k} + x_k\beta_k + u$$
$$\Longrightarrow \Lambda^1(y) = x_{-k}'\beta_{-k} + x_k\beta_k + u$$
$$\text{defining } \Lambda^1(y) \equiv \Lambda^o(y) - \beta_0 \; (\Lambda^1(y_0) = 0 \text{ with } y_0 \equiv \Lambda^{o-1}(\beta_0))$$
$$\Longrightarrow \frac{\Lambda^1(y)}{|\beta_k|} = x_{-k}'\frac{\beta_{-k}}{|\beta_k|} + x_k sgn(\beta_k) + \frac{u}{|\beta_k|}$$
$$\text{dividing by } |\beta_k| \text{ assumed to be non-zero;}$$

let $\Lambda(y) \equiv \Lambda^1(y)/|\beta_k|$. The step of removing β_0 is a location normalization and the step dividing by $|\beta_k|$ is a scale normalization. Since $sgn(\beta_k)$ can be estimated at a rate faster than $\sqrt{N}$, we can assume that $sgn(\beta_k)$ is known as far as the asymptotic distribution of β_{-k} is concerned.

For any y_o and z_o,

$$
\begin{aligned}
G(y_o|z_o) &= P(y \le y_o|x'\beta = z_o) = P\{\Lambda(y) \le \Lambda(y_o)|x'\beta = z_o\} \\
&= P\{u \le \Lambda(y_o) - z_o|x'\beta = z_o\} = F\{\Lambda(y_o) - z_o\}.
\end{aligned}
$$

Differentiate the first and last expressions wrt y_o and z_o to get, respectively,

$$
\begin{aligned}
g(y_o|z_o) &= \lambda(y_o) \cdot f\{\Lambda(y_o) - z_o\} \\
\frac{\partial G(y_o|z_o)}{\partial z_o} &= -f\{\Lambda(y_o) - z_o\} \\
\Longrightarrow \lambda(y_o) &= -\frac{g(y_o|z_o)}{\partial G(y_o|z_o)/\partial z_o} \\
\Longrightarrow \Lambda(t) &= -\int_{y_0}^{t} \left\{ \frac{g(y_o|z_o)}{\partial G(y_o|z_o)/\partial z_o} \right\} dy_o
\end{aligned}
$$

(integrating both sides over $[y_0, t]$)

for any (y_o, z_o) with $\partial G(y_o|z_o)/\partial z_o \neq 0$. To simplify exposition, assume that $\partial G(y_o|z_o)/\partial z_o \neq 0$ holds $\forall (z_o, y_o)$.

Given a $\sqrt{N}$-consistent estimator b_N for $(\beta'_{-k}, sgn(\beta_k))'$—e.g., Han's (1987a) estimator—and a weighting function $w(z_o)$ such that $\int w(z_o)dz_o = 1$, a nonparametric estimator for $\Lambda(t)$ is

$$
\Lambda_N(t) \equiv -\int_{y_0}^{t} \int w(z_o) \left\{ \frac{g_N(y_o|z_o)}{\partial G_N(y_o|z_o)/\partial z_o} \right\} dz_o dy_o, \quad \text{where}
$$

$$
g_N(y_o|z_o) \equiv \frac{(N \cdot h_y h_z)^{-1} \sum_i K_z\{(x'_i b_N - z_o)/h_z\} K_y\{(y_i - y_o)/h_y\}}{(N \cdot h_z)^{-1} \sum_i K_z\{(x'_i b_N - z_o)/h_z\}},
$$

$$
G_N(y_o|z_o) \equiv \frac{(Nh_z)^{-1} \sum_i K_z\{(x'_i b_N - z_o)/h_z\} 1[y_i \le y_o]}{(Nh_z)^{-1} \sum_i K_z\{(x'_i b_N - z_o)/h_z\}},
$$

$\partial G_N(y_o|z_o)/\partial z_o$ is obtained by a numerical derivative of $G_N(y_o|z_o)$ wrt z_o, K_z and K_y are univariate kernels, and h_y and h_z are bandwidths. Clearly, $g_N(y_o|z_o)$ estimates $g(y_o|z_o)$ by estimating the joint density with the numerator and the marginal density with the denominator—a product kernel is used for the numerator. Also $G_N(y_o|z_o)$ is a kernel estimator for $E(1[y \le y_o]|x'\beta = z_o)$.

Although the nonparametric estimators' convergence rates are slower than $\sqrt{N}$, the averaging across z_o makes $\Lambda_N(t)$ $\sqrt{N}$-consistent. But $\Lambda(t)$ may not be bounded as $y \to \infty$ and the denominator in $\Lambda_N(t)$ should be kept away from zero, which makes the estimation possible only over a compact proper subset, say $[\underline{y}, \bar{y}]$, of the support of y. It is also possible to estimate F $\sqrt{N}$-consistently using the residual $\hat{u}_i \equiv \Lambda_N(y_i) - x'_i b_N$. But this statement should be also qualified, because $\Lambda(t)$ is estimable only over $[\underline{y}, \bar{y}]$, which then restricts the estimability of F. See Horowitz (1996) for the asymptotic distributions of $\sqrt{N}\{\Lambda_N(t) - \Lambda(t)\}$ and $\sqrt{N}\{F_N(t) - F(t)\}$ and more. Gørgens and Horowitz (1999) extended Horowitz (1996) to censored models.

8 Nonparametric Specification and Significance Tests

Consider nonparametric regression equation $y = E(y|x) + u$ with $E(u|x) = 0$. For a parametric function $r(x, \beta)$, $\beta \in B$ with B being a parameter space, it is desired to test

$$
\begin{array}{rl}
H_0 & : \quad P\{E(y|x) = r(x, \beta)\} = 1 \quad \text{for some } \beta \in B \\
H_1 & : \quad P\{E(y|x) = r(x, b)\} \neq 1 \quad \text{for any } b \neq \beta,\ b \in B.
\end{array}
$$

There are many tests for H_0, all of which use at least $r(x, b_N)$ where b_N is a $\sqrt{N}$-consistent estimator for β. Some bandwidth-free tests have been already reviewed in the previous chapter, and here bandwidth-dependent tests are reviewed. Test statistics in this section typically have nonparametric nuisance parameters replaced by nonparametric estimators.

8.1 Omitted-Variable-Based LM-Type Tests

One simple way to test for H_0 is examining $COR\{y - r(x, \beta), E(y|x)\}$, because $y - r(x, \beta) = u + E(y|x) - r(x'\beta)$ includes $E(y|x)$ if H_1 holds. That is, $E(y|x)$ is an omitted variable under H_1. Replacing u_i and $E(y|x_i)$, respectively, with $\hat{u}_i \equiv y_i - r(x_i, b_N)$ and $E_N(y|x_i)$, a LM-type test statistic with two nuisance parameters β and $E(y|x)$ is

$$
\frac{1}{\sqrt{N}} \sum_i \hat{u}_i E_N(y|x_i) = \frac{1}{\sqrt{N}} \sum_i \{y_i - r(x_i, b_N)\} E_N(y|x_i)
$$

$$
= \frac{1}{\sqrt{N}} \sum_i u_i E_N(y|x_i) + \frac{1}{\sqrt{N}} \sum_i E_N(y|x_i)\{E(y|x_i) - r(x_i, b_N)\}
$$

as $y_i = E(y|x_i) + u_i$.

Applying the mean value theorem to b_N in $r(x_i, b_N)$, this becomes

$$
\frac{1}{\sqrt{N}} \sum_i u_i E_N(y|x_i) + \frac{1}{\sqrt{N}} \sum_i E_N(y|x_i)\{E(y|x_i) - r(x_i, \beta)
$$

$$
- r_{b'}(x_i, b_N^*)(b_N - \beta)\}
$$

$$
= \frac{1}{\sqrt{N}} \sum_i u_i E(y|x_i) + \frac{1}{\sqrt{N}} \sum_i E_N(y|x_i)\{E(y|x_i) - r(x_i, \beta)\}
$$

$$
- \frac{1}{N} \sum_i E(y|x_i) r_{b'}(x_i, \beta)\ \sqrt{N}(b_N - \beta) + o_p(1),
$$

where $b_N^* \in (b_N, \beta)$, $r_b \equiv \partial r / \partial b$ and

$$\frac{1}{\sqrt{N}} \sum_i u_i E_N(y|x_i) = \frac{1}{\sqrt{N}} \sum_i u_i E(y|x_i) + o_p(1)$$

using CT_E.

Let λ_i be an influence function for $\sqrt{N}(b_N - \beta)$: $\sqrt{N}(b_N - \beta) = N^{-1/2} \sum_i \lambda_i + o_p(1)$. Hence $N^{-12} \sum_i \hat{u}_i E_N(y|x_i)$ is $o_p(1)$ equal to

$$\frac{1}{\sqrt{N}} \sum_i [u_i E(y|x_i) - E\{E(y|x)r_{b'}(x, \beta)\}\lambda_i]$$

$$+ \frac{1}{\sqrt{N}} \sum_i E_N(y|x_i)\{E(y|x_i) - r(x_i, \beta)\}.$$

Under H_0, the second term is zero, and thus the test can be easily implemented as the first term has mean zero and is asymptotically normal with the variance

$$C_1 \equiv E^2[uE(y|x) - E\{E(y|x)r_{b'}(x, \beta)\}\lambda].$$

Denoting the sample version for this as C_{1N}, a valid test statistic is

$$C_{1N}^{-1/2} \frac{1}{\sqrt{N}} \sum_i \hat{u}_i E_N(y|x_i) \rightsquigarrow N(0, 1).$$

Under H_1, the power of this test comes from $N^{-1/2} \sum_i E_N(y|x_i)$ $\{E(y|x_i) - r(x_i, \beta)\}$. There is, however, no guarantee that this term is nonzero despite $E(y|x) - r(x, \beta) \neq 0$ for some x. Shortly we will introduce a number of ways to avoid this problem.

A test statistic analogous to $N^{-1/2} \sum_i \hat{u}_i E_N(y|x_i)$ was in fact proposed by Wooldridge (1992b, (3.4)) although its asymptotic distribution was derived under homoskedasticity. Wooldridge (1992b, (2.8)) also looked at (recall $\hat{u}_i \equiv y_i - r(x_i, b_N)$)

$$\sqrt{N}\hat{m}_N = \frac{1}{\sqrt{N}} \sum_i \hat{u}_i \{E_N(y|x_i) - r(x_i, b_N)\}.$$

It will be seen later that most nonparametric specification tests are based on the differences

$$y_i - r(x_i, b_N) \quad \text{or} \quad E_N(y|x_i) - r(x_i, b_N).$$

Using $E_N(y|x_i) - r(x_i, b_N)$ instead of $E_N(y|x_i)$ looks like a fairly minor modification, but it will be shown that this is one of the ways to make the test to have power in all directions.

Hong and White (1995) also proposed a test using a version of $\hat{m}_N$. Let $\ddot{m}_N$ be $\hat{m}_N$ where series approximations is used for $E_N(y|x)$ with ν_N-many basis functions. Then

$$\frac{\hat{\sigma}_N^{-2} N \cdot \ddot{m}_N - \nu_N}{\sqrt{2\nu_N}} \rightsquigarrow N(0,1) \quad \text{where} \quad \hat{\sigma}_N^2 \equiv \frac{1}{N}\sum_i \hat{u}_i^2.$$

Another test (of LR-type) taking a similar form will appear later.

As for the asymptotic distribution of $\sqrt{N}\hat{m}_N$, substitute

$$\hat{u}_i \equiv y_i - r(x_i, b_N) = E(y|x_i) + u_i - r(x_i, b_N)$$

to rewrite $\sqrt{N}\hat{m}_N$ as

$$\frac{1}{\sqrt{N}}\sum_i \{u_i + E(y|x_i) - r(x_i, b_N)\}\{E_N(y|x_i) - r(x_i, b_N)\}$$

$$= \frac{1}{\sqrt{N}}\sum_i u_i\{E_N(y|x_i) - r(x_i, b_N)\} + \frac{1}{\sqrt{N}}\sum_i E_N(y|x_i)\{E(y|x_i) - r(x_i, b_N)\}.$$

Apply the mean value theorem to b_N in $r(x_i, b_N)$ to get

$$\sqrt{N}\hat{m}_N = \frac{1}{\sqrt{N}}\sum_i u_i\{E_N(y|x_i) - r(x_i, \beta) - r_{b'}(x_i, b_N^*)(b_N - \beta)\}$$

$$+ \frac{1}{\sqrt{N}}\sum_i E_N(y|x_i)\{E(y|x_i) - r(x_i, \beta) - r_{b'}(x_i, b_N^*)(b_N - \beta)\} + o_p(1).$$

The second term appeared in the preceding test, and it yields

$$\frac{1}{\sqrt{N}}\sum_i E_N(y|x_i)\{E(y|x_i) - r(x_i, \beta)\}$$

$$- \frac{1}{N}\sum_i E(y|x_i)r_{b'}(x_i, \beta)\sqrt{N}(b_N - \beta) + o_p(1).$$

As for the first term, it can be written as

$$\frac{1}{\sqrt{N}}\sum_i u_i\{E_N(y|x_i) - r(x_i, \beta)\} - \frac{1}{N}\sum_i u_i r_{b'}(x_i, b_N^*) \cdot \sqrt{N}(b_N - \beta)$$

in which the latter term is $o_p(1)$ whereas the former term is (CT_E is invoked)

$$\frac{1}{\sqrt{N}}\sum_i u_i\{E_N(y|x_i) - r(x_i, \beta)\} = \frac{1}{\sqrt{N}}\sum_i u_i\{E(y|x_i) - r(x_i, \beta)\} + o_p(1).$$

Hence

$$\sqrt{N}\hat{m}_N = \frac{1}{\sqrt{N}}\sum_i [u_i\{E(y|x_i) - r(x_i, \beta)\} - E\{E(y|x)r_{b'}(x, \beta)\}\lambda_i]$$

$$+\frac{1}{\sqrt{N}}\sum_i E_N(y|x_i)\{E(y|x_i) - r(x_i,\beta)\} + o_p(1).$$

The second term is zero under H_o, and the first term is asymptotically normal with variance

$$C_2 \equiv E^2[u\{E(y|x) - r(x,\beta)\} - E\{E(y|x)^{r_{b'}}(x,\beta)\}\lambda]$$

and a sample version C_{2N} of this can be used to implement the test with

$$C_{2N}^{-1/2}\frac{1}{\sqrt{N}}\sum_i \hat{u}_i\{E_N(y|x_i) - r_N(x_i,b_N)\} \rightsquigarrow N(0,1).$$

We may also drop the part $E(y|x) - r(x,\beta)$ in C_2 that is 0 under H_0, but this may make the test over-rejecting.

8.2   Wald-Type Tests with Parametric and Nonparametric Fits

Gozalo (1993) proposed a Wald-type test comparing the parametric regression function with a nonparametric one at a number of selected points of x. Let $E(y|x) = \rho(x)$, $\rho_N(x) = E_N(y|x)$, and

$$\sigma_N(x)^2 \equiv AV[\sqrt{Nh^k}\{\rho_N(x) - \rho(x)\}]$$

where "AV" stands for asymptotic variance. Choose a number of points in the range of x, say $(x^{(1)},...,x^{(J)})$ to define

$$T_N \equiv (T_N^{(1)},...,T_N^{(J)})' \quad \text{where} \quad T_N^{(j)} \equiv \frac{\sqrt{Nh^k}\{\rho_N(x^{(j)}) - r(x^{(j)},b_N)\}}{\sigma_N(x^{(j)})}.$$

Since $r(x^{(j)},b_N) \to^p r(x^{(j)},\beta)$ under H_0 at the rate $\sqrt{N}$ while $\rho_N(x^{(j)}) \to^p \rho(x^{(j)})$ at the rate $\sqrt{Nh^k}$, estimating β with b_N is innocuous. Hence, under H_0,

$$T_N^{(j)} = \frac{\sqrt{Nh^k}\{\rho_N(x^{(j)}) - r(x^{(j)},\beta)\}}{\sigma_N(x^{(j)})} + o_p(1) \rightsquigarrow N(0,1).$$

Using the asymptotic independence among $\rho_N(x^{(j)})$, $j = 1,...,J$, we get under H_0

$$\sum_{j=1}^J \{T_N^{(j)}\}^2 \rightsquigarrow \chi_J^2.$$

This test is also applicable to omitted variable tests. Suppose that z may be omitted in a null model $y = E(y|x) + u$. Then $E_N(y|x,z) - E_N(y|x)$ can be used to test the possible omission. The convergence rate of $E_N(y|x) \to^p E(y|x)$ is faster than that of $E_N(y|x,z) \to^p E(y|x,z)$, because the latter has more regressors. That is, for the test, $E(y|z)$ is as good as known. In this and above tests, a disadvantage is selecting the J evaluation points. Since the dof is J, selecting points where H_0 holds decreases the power of the test. While this may be overcome one way or another, the better way is integrating over the entire range of x as below. Of course, rather than selecting some evaluation points, one can always plot the test statistic vs. evaluation points to visually detect the set of x on which H_0 is violated.

Instead of selecting evaluation points, Härdle and Mammen (1993) proposed to use a weighted integrated quadratic difference between the parametric and nonparametric regression functions:

$$H_N \equiv N h^{k/2} \int [\rho_N(x) - \tau\{r(x, b_N)\}]^2 w(x) dx$$

where $w(x)$ is a weighting function and

$$\tau\{g(x)\} \equiv \frac{\sum_i K((x_i - x)/h) g(x_i)}{\sum_i K((x_i - x)/h)}.$$

The purpose of $\tau(\cdot)$ is to center $\rho_N(x)$ properly under H_0; this also appears in Staniswalis and Severini (1991).

Härdle and Mammen (1993) showed

$$N h^{k/2} \int [\rho_N(x) - \tau\{r(x, b_N)\}]^2 w(x) dx$$

$$- \frac{K^{(2)}(0)}{h^{k/2}} \int \frac{\sigma(x)^2 w(x)}{f(x)} dx$$

$$\rightsquigarrow N\{0, \ 2K^{(4)}(0) \int \frac{\sigma(x)^4 w(x)^2}{f(x)^2} dx\} \quad \text{under } H_0.$$

where $K^{(j)}$ denotes the j-time convolution product of K and $\sigma^2(x) = V(u|x)$. Härdle and Mammen (1993), however, do not recommend using this asymptotic distribution which depends on a stochastic expansion with $O_p(N^{-1/10})$ error terms; rather, they recommend wild bootstrap. Horowitz and Spokoiny (2001) applied this test with multiple bandwidths, and then proposed to use the maximum of the resulting test statistics. The critical value for this test is to be found with a bootstrap as well.

The wild bootstrap for Härdle and Mammen (1993) is done essentially in the same way as the wild bootstrap described in the preceding chapter for Stute test. Consider H_N using the observation points as the evaluation points:

$$H_N \equiv N h^{k/2} \frac{1}{N} \sum_i [\rho_N(x_i) - \tau\{r(x_i, b_N)\}]^2.$$

First, obtain the residual $\hat{u}_i \equiv y_i - r(x_i, b_N)$ and then H_N. Second, draw a random sample v_i^*, $i = 1, \ldots, N$, such that v^* takes ± 1 with probability 0.5. Third, construct a pseudo-sample (y_i^*, x_i), $i = 1, \ldots, N$ where

$y_i^* = r(x_i, b_N) + v_i^* \hat{u}_i$. Fourth, obtain H_N^* using the pseudo-sample and b_N^* that is based on (x_i', y_i^*), $i = 1, \ldots, N$. Fifth, denoting the number of bootstrap replications as B, construct the empirical distribution using the B-many H_N^*'s. Finally, see whether H_N falls in the upper tail of the empirical distribution to reject the H_0 if yes. Clearly, the multiplicative factor $Nh^{k/2}$ can be omitted for this bootstrap

Recall $\hat{u}_i = y_i - r(x_i, b_N)$. Zheng (1996) proposed a test with

$$
\begin{aligned}
T_N &\equiv \frac{1}{N(N-1)} \sum_{i=1}^{N} \sum_{j \neq i} \frac{1}{h^k} K\left(\frac{x_j - x_i}{h}\right) \hat{u}_i \hat{u}_j \\
&= \frac{1}{N} \sum_{i=1}^{N} \hat{u}_i \left\{ \frac{1}{N-1} \sum_{j \neq i} \frac{1}{h^k} K\left(\frac{x_j - x_i}{h}\right) \hat{u}_j \right\}.
\end{aligned}
$$

Defining $u \equiv y - r(x, \beta)$, T_N may be taken as a sample analog for the first term in

$$
E\{uE(u|x)\} = E[\ E\{uE(u|x)|x\}\ ] = E\{E^2(u|x)\}
$$

because the average with $\sum_{j \neq i}$ is for $E(u|x_i)$, and the outer average with $\sum_i$ is for the outer expected value for $uE(u|x)$.

Another way to look at the test comes from $E\{E^2(u|x)\}$ in the last display. Smoothing (i.e., taking $E_N(\cdot|x_i)$ on) $\hat{u}_i = y_i - r(x_i, b_N)$ gives $E_N(y|x_i) - E_N\{r(x, b_N)|x_i\}$. Squaring this renders the above test of Härdle and Mammen (1993) where the integral wrt $\omega(x)dx$ gets replaced by the integral wrt the empirical x-distribution. This is why the Zheng (1996) test is presented here although it is not exactly a Wald-type test. Zheng (1998) proposed an analogous test for quantile function specification.

It holds that

$$
Nh^{k/2} T_N \rightsquigarrow N(0, V), \quad V \equiv 2 \int K(t)^2 dt \int \sigma^4(x) f(x)^2 dx,
$$

$$
\frac{2}{N(N-1)} \sum_{i=1}^{N} \sum_{j \neq i} \frac{1}{h^k} K\left(\frac{x_i - x_j}{h}\right)^2 \hat{u}_i^2 \hat{u}_j^2 \equiv V_N \to^p V.
$$

Hence, dividing $Nh^{k/2} T_N$ by $V_N^{1/2}$, we get

$$
\frac{Nh^{k/2} T_N}{V_N^{1/2}} \rightsquigarrow N(0, 1).
$$

Li and Wang (1998) proposed to use wild bootstrap for Zheng (1996) test instead of the asymptotic distribution. First, let $u_i^* = v_i \hat{u}_i$, $i = 1, \ldots, N$, where v_i's are iid with $E(v) = 0$ and $SD(v) = 1$ independently of (x, y). Second, obtain $y_i^* \equiv r(x_i, b_N) + u_i^*$ and the bootstrap residual $\hat{u}_i^* \equiv y_i^* - r(x_i, b_N^*)$ where b_N^* is based on the bootstrap sample (x_i, y_i^*), $i = 1, \ldots, N$. Third, compute $T_N^{(m)}$ which is T_N with $\hat{u}_i$ replaced with $\hat{u}_i^*$. Repeat these steps M times to get $T_N^{(1)}, ..., T_N^{(M)}$ and their empirical distribution. Finally, locate where T_N stands in the distribution of $T_N^{(1)}, ..., T_N^{(M)}$; i.e., find $M^{-1}\{\#|T_N^{(m)}|$ greater than $|T_N|\}$ which is the p-value of T_N to reject H_0 if the p-value is smaller than, say, 5%. In our experience, this bootstrap test is much more powerful than the above test using the asymptotic distribution. Miles and Mora (2003) did a comparison study of various nonparametric specification tests, and found that a modified version of the Zheng's (1996) test in Ellison and Ellison (2000) performed well.

8.3   LR-Type Tests

Yatchew (1992) proposed to use

$$\tilde{m}_N \equiv \frac{1}{N}\sum_i \{y_i - r(x_i, b_N)\}^2 - \frac{1}{N}\sum_i \{y_i - E_N(y|x_i)\}^2$$

which is a LR-type test in that the objective function difference is involved. Yatchew (1992) required sample splitting to derive the asymptotic variance. Instead of looking at the difference $E_N(y|x) - r(x, b_N)$, this test measures the difference between the two regression function estimators using the objective functions, which is the idea behind LR test. Similar LR-type tests also appeared in Whang and Andrews (1993). Dette (1999) does away with sample splitting, but assumes fixed regressors. One may think that $\tilde{m}_N$ may never take a negative value because $E_N(y|x_i)$ would be a better predictor than $r(x_i, b_N)$ for all y_i. But under H_0, $r(x_i, b_N)$ with H_0 built in is a better predictor, which means that $\tilde{m}_N$ can be negative as well as positive under H_0.

Hong and White (1995) also used a LR-type test. Under homoskedasticity with no sample splitting, they showed that, with ν_N being the number of basis functions in their series approximation for $E(y|x)$,

$$\frac{\hat{\sigma}_N^{-2}N \cdot \tilde{m}_N - \nu_N}{\sqrt{2\nu_N}} \rightsquigarrow N(0,1) \quad \text{where} \quad \hat{\sigma}_N^2 \equiv \frac{1}{N}\sum_i \hat{u}_i^2.$$

This is reminiscent of the LM-type test of Hong and White (1995) that appeared already.

Azzalini et al. (1989) proposed LR-type test statistics. Suppose that H_0 specifies a parametric regression function $r(x, \beta)$, and there may be some other parameters θ in the model. Let $\hat{\beta}$ and $\hat{\theta}$ be estimators for β and θ, and let $E_N(y|x)$ be a nonparametric estimator for $E(y|x)$. Then one can construct a "generalized" LR-type test statistic

$$\sum_i \ln f\{E_N(y|x_i); \hat{\theta}\} - \sum_i \ln f\{r(x_i, \hat{\beta}); \hat{\theta}\}.$$

A natural question is if the asymptotic distribution is χ^2, and if yes, what would be its dof. This is a difficult question because a nonparametric $E(y|x)$

has infinite dimension. Azzalini et al. (1989) could not answer this question, but suggested to do parametric bootstrap from $f\{r(x_i, \hat{\beta}); \hat{\theta}\}$. Staniswalis and Severini (1991) also proposed two tests comparing the likelihood functions with the regression function parametrically estimated under H_0 and non-parametrically estimated under H_1 using the method in Staniswalis (1989). The comparisons were made at fixed m-many points. They also considered letting $m \to \infty$ as $N \to \infty$.

In parametric cases, the asymptotic distribution of LR test statistics are not affected by the fact that the parameters are replaced by estimators. Fan et al. (2001) called this "Wilks phenomenon" and looked at the generalized LR dof question. In a number of models they examined, it was found that, using local linear regression, the asymptotic null distribution of the generalized LR-type test statistic is nearly χ^2 with a large dof in the sense that

$$rLR \rightsquigarrow \chi^2_{c_N} \quad \text{where } r \simeq 2 \text{ and } c_N \to \infty; \text{ i.e.,} \frac{rLR - c_N}{\sqrt{2c_N}} \rightsquigarrow N(0,1).$$

See Fan and Jiang (2007) for a review on generalized LR tests. Note the similarity of this display to the above test of Hong and White (1995).

As shown in Zheng (1996), Yatchew's (1992) test is closely linked to a test in Wooldridge (1992b). Yatchew's (1992) test is based on the difference

$$\frac{1}{N} \sum_i [\{y_i - r(x_i, b_N)\}^2 - \{y_i - E_N(y|x_i)\}^2]$$

$$= \frac{1}{N} \sum_i [r(x_i, b_N)^2 - E_N(y|x_i)^2 - 2y_i\{r(x_i, b_N) - E_N(y|x_i)\}]$$

$$= \frac{1}{N} \sum_i \{r(x_i, b_N) - E_N(y|x_i)\}\{r(x_i, b_N) + E_N(y|x_i) - 2y_i\}$$

$$= \frac{1}{N} \sum_i \{r(x_i, b_N) - E_N(y|x_i)\}\{r(x_i, b_N) - y_i + E_N(y|x_i) - y_i\}$$

$$= \frac{1}{N} \sum_i \{E_N(y|x_i) - r(x_i, b_N)\}\{y_i - r(x_i, b_N)\}$$

$$+ \frac{1}{N} \sum_i \{E_N(y|x_i) - r(x_i, b_N)\}\{y_i - E_N(y|x_i)\}.$$

Intuitively, the second term in the last expression is $o_p(1)$ under both H_0 and H_1, and the first term is proportional to the test statistic $N^{-1/2} \sum_i \hat{u}_i \{E_N(y|x_i) - r(x_i, b_N)\}$ in Wooldridge (1992b) as shown already. It is likely that other LR-type tests have equivalent LM-type versions. This display shows that the Wooldridge test would have power in all directions because the Yatchew test statistic is asymptotically non-negative under H_a as $E_N(y|x_i)$ does better than $r(x_i, b_N)$ in predicting y_i.

When x is a rv (i.e., $k = 1$), to test the equality of two regression functions in two samples, Yatchew (1999) proposed to use the difference estimator

in Yatchew (1997) explained in the preceding chapter. The idea is similar to
the ANOVA approach where the error term variance is estimated separately
using the difference estimator in Yatchew (1997) for each sample and then for
the pooled sample. The former gives an unrestricted estimator whereas the
latter gives a restricted estimator. Define (the following exposition is drawn
from Yatchew et al. (2003))

$$s_{ur}^2 \equiv \frac{N_0}{N}s_0^2 + \frac{N_1}{N}s_1^2, \quad \text{where } s_j^2 \equiv \frac{1}{2N_j}\sum_{i=2}^{N_j}(y_i^j - y_{i-1}^j)^2, \quad j = 0,1$$

and y_i^j denote the ith observation in sample j; the observations within each
sample are ordered in terms of x. Define s_r^2 analogously to s_j^2 for the pooled
(and thus reordered) sample. Then, as stated in Yatchew et al. (2003), $\sqrt{N}(s_r^2 -
s_{ur}^2)/s_{ur}^2 \rightsquigarrow N(0,1)$.

8.4   Model-Selection-Based Tests*

In *lack-of-fit* or goodness-of-fit (GOF) tests—the former typically used
for regression function and the latter for distribution—there is a null model,
and we try to see if the null model is not rejected by the data at hand.
This is done sometimes with an alternative in mind and sometimes without.
Either way, the null model is given a favor, as the test procedure is designed
such that rejecting the null model is difficult. In *"model selection,"* there is no
favored model. Instead, all models under considerations are evaluated using a
model selection criterion and the model with the best score in the criterion is
selected. Also, differently from hypothesis testing, no probability of rejecting
the right/wrong model is provided.

Two best known model selection criteria are

1. *AIC*(Akaike Information Criterion): $2L_N(\theta) - 2\cdot\dim(\theta)$

2. *BIC*(Bayesian Information Criterion): $2L_N(\theta) - \ln(N)\cdot\dim(\theta)$

where $L_N(\theta)$ is a log-likelihood function for the sample and θ is the model
parameters. So long as $\ln(N) > 2$, BIC penalizes "over-fitting" more severely
than AIC does. Although model selection differs from goodness-of-fit tests
as just noted, Aerts et al. (1999) proposed a goodness-of-fit test using AIC.
This is explained in the following, drawing on Claeskens and Hjort (2008)
who review the related literature and model selection.

Let x be a rv and $m_\beta(x)$ a parametric function of x indexed by a $k \times 1$
vector β. Consider a sequence of *nested* regression function models:

$$m(x;J) = m_\beta(x) + \sum_{j=1}^{J}\gamma_j\psi_j(x), \quad J = 1,2,...$$

where $\psi_j(x)$, $j = 1,2,...,J$, are (orthonormal) functions that exclude $m_\beta(x)$.
In $m(x;J)$, there are $k + J$ parameters, and define

$$AIC(J) \equiv 2L_N(\hat\beta,\hat\gamma_1,...,\hat\gamma_J) - 2(k+J)$$

$$A(J) \equiv AIC(J) - AIC(0)$$
$$= 2\{L_N(\hat{\beta}, \hat{\gamma}_1, ..., \hat{\gamma}_J) - L_N(\hat{\beta}, 0, ..., 0)\} - 2J.$$

Observe $A(0) = 0$. $A(J)$ is a centered version of $AIC(J)$ and includes the LR test statistic for $H_0 : \gamma_1 =, ... = \gamma_J = 0$. Using this fact yields a specification test for the null model as shown next.

The model order J is selected by maximizing $AIC(J)$ wrt J, equivalently, maximizing $A(J)$ wrt J. The null model is rejected when

$$\max_{J \geq 1} A(J) > 0 \iff \max_{J \geq 1}[2\{L_N(\hat{\beta}, \hat{\gamma}_1, ..., \hat{\gamma}_J) - L_N(\hat{\beta}, 0, ..., 0)\}$$
$$-2J] > 0$$

$$\iff M_N > 2 \text{ where } M_N \equiv \max_{J \geq 1} \frac{L_N(\hat{\beta}, \hat{\gamma}_1, ..., \hat{\gamma}_J) - L_N(\hat{\beta}, 0, ..., 0)}{J}.$$

As one may conjecture, because the LR test follows χ_J^2 asymptotically, it holds that

$$M_N \rightsquigarrow M \equiv \max_{J \geq 1} \frac{\sum_{j=1}^{J} z_j^2}{J}, \quad \text{where } z_j^2\text{'s are iid } \chi_1^2.$$

The distribution of M can be simulated, but it holds that, with χ_j^2 following Chi-Square distribution with dof j,

$$P(M \leq m) \equiv \exp\left\{-\sum_{j=1}^{\infty} \frac{P(\chi_j^2 > jm)}{j}\right\};$$

this is a remarkable finding because M_N has its parameters replaced by estimators. Using this, if we follow the above test of rejecting H_0 when $M_N > 2$, we will get

$$P(M_N > 2) \to P(M > 2) = 0.288 \quad \text{under } H_0.$$

As this is unacceptably large, if 5% level is desired, change the critical value to 4.179 to get $P(M_N > 4.179) \to P(M > 4.179) = 0.05$.

One drawback of this kind of tests is that, when x is multi-dimensional, nesting the alternative sequence is rather restrictive. For instance, consider $m_\beta(\beta) = \beta_0 + \beta_1 x_1 + \beta_2 x_2$ and more general models

(i) $\beta_0 + \beta_1 x_1 + \beta_2 x_2 + \beta_3 x_1^2$, and (i)'$\beta_0 + \beta_1 x_1 + \beta_2 x_2 + \beta_4 x_2^2$

(ii) $\beta_0 + \beta_1 x_1 + \beta_2 x_2 + \beta_3 x_1^2 + \beta_4 x_2^2$.

The nesting restriction holds either with $\{m_\beta(\beta), (i), (ii)\}$ or $\{m_\beta(\beta), (i)', (ii)\}$, but both (i) and (ii) cannot be accommodated together.

The original idea of adding an extra term such as $\sum_j \gamma_j \psi_j(x)$ to the null model to test for GOF goes back to *Neyman's smooth test*; see e.g., Bera and

Ghosh (2001) for the original reference and the ensuing literature. For an iid rv's $z_1, ..., z_N$ with its specified df F, the test checks if $v_1 \equiv F(z_1), ..., v_N \equiv F(z_N)$ are iid $U[0,1]$. The test uses a generalized density on $[0,1]$:

$$\frac{\exp\left\{\sum_{j=1}^J \gamma_j \phi_j(v)\right\}}{C(\gamma_1, ..., \gamma_J)} \qquad \text{where } \phi_j\text{'s are orthornormal on } [0,1]$$

$$\text{with } \int_0^1 \phi_j(t)dt = 0$$

and $C(\gamma_1, ..., \gamma_J)$ is the normalizing constant. $\int_0^1 \phi_j(t)dt = 0$ implies that ϕ_j is orthogonal to the $U[0,1]$ density. Observe

$$C(\gamma_1, ..., \gamma_J) = \int_0^1 \exp\left\{\sum_{j=1}^J \gamma_j \phi_j(t)\right\} dt \quad (= 1 \text{ under } H_0)$$

$$\Longrightarrow \frac{\partial C(0)}{\partial \gamma_j} = \int_0^1 \phi_j(t)dt = 0.$$

The log-likelihood is $\sum_i \{\sum_{j=1}^J \gamma_j \phi_j(v_i) - \ln C(\gamma_1, ..., \gamma_J)\}$ and the score function for γ_j under H_0 is $\sum_i \phi_j(v_i)$. Observe

$$E\{\phi_j(v)\phi_{j'}(v)\} = \int_0^1 \phi_j(t)\phi_{j'}(t)dt = 1 \text{ if } j = j' \text{ and 0 otherwise.}$$

Using this, the score test statistic is

$$\sum_{j=1}^J \left\{\frac{1}{\sqrt{N}}\sum_{i=1}^N \phi_j(v_i)\right\}^2 \rightsquigarrow \chi_J^2.$$

The above model selection criterion-based test estimates γ_j's and uses the LR test idea. Instead of LR, score-test versions exist that do not require estimating γ_j's; see Aerts et al. (1999, 2000).

8.5   Single-Index Model Fitness Tests

Stute and Zhu (2005) proposed a GOF test for single-index models (SIM). Suppose

$$y = G(x'\beta) + u \quad \text{with } E(u|x) = E(u|x'\beta) = 0, \ \beta \text{ and } G(\cdot)$$
$$\text{unknown, } x'\beta \text{ has a continuous and strictly increasing df } F.$$

Let F^{-1} denote the inverse, which is also the quantile function of F: $F^{-1}(\alpha) \equiv \min\{t : F(t) \geq \alpha\}$. As F is continuous, $v_i \equiv F(x_i'\beta) \sim U(0,1)$. Define $\psi(\cdot) \equiv G\{F^{-1}(\cdot)\}$ so that

$$y = G[F^{-1}\{F(x'\beta)\}] + u = \psi(v) + u$$

$$\Longrightarrow E(y|x'\beta = v_o) = \psi(v_o) = \hat{E}(y|v = v_o).$$

An "infeasible" kernel estimator for $\psi(v_o)$ is

$$\frac{1}{N}\sum_j \frac{1}{h}K\left(\frac{v_j - v_o}{h}\right)y_i, \quad 0 < v_o < 1.$$

There is no denominator because v's density is 1 over $0 < v_o < 1$.
For some $\sqrt{N}$-consistent estimator b_N for β, define

$$\hat{v}_j \equiv F_N(x'_j b_N) = \frac{1}{N}\sum_{j'=1}^{N} 1[x'_{j'} b_N \leq x'_j b_N].$$

A "feasible" kernel estimator for $\psi(\hat{v}_i)$ is then

$$\psi_N(\hat{v}_i) \equiv \frac{1}{N}\sum_{j\neq i} \frac{1}{h}K\left(\frac{\hat{v}_j - \hat{v}_i}{h}\right)y_i.$$

For a weighting function $w_\gamma(x_i)$ indexed by γ, consider a test statistic

$$T_N(\gamma) \equiv \frac{1}{\sqrt{N}}\sum_i \hat{u}_i w_\gamma(x_i) \quad \text{where} \quad \hat{u}_i \equiv y_i - \psi_N(\hat{v}_i).$$

As was seen in relation to the single-index "nonparametric LSE" of Ichimura (1993), there is no effect of estimating ψ because ψ has been "concentrated out." There, the only consequence of estimating β in $x'\beta$ was that the regressor $x_{-k} = (x_1, ..., x_{k-1})'$ gets replaced by $x_{-k} - E(x_{-k}|x'\beta)$ in the asymptotic variance. Analogously, since $w_\gamma(x_i)$ plays the role of x_{-k} in $T_N(\gamma)$, we get

$$\begin{aligned}
T_N(\gamma) &= \frac{1}{\sqrt{N}}\sum_i u_i\{w_\gamma(x_i) - E(w_\gamma(x_i)|v_i)\} + o_p(1) \quad \text{under } H_0 \\
&\rightsquigarrow N(0, \ E[u^2\{w_\gamma(x) - E(w_\gamma(x)|v)\}^2]).
\end{aligned}$$

The asymptotic variance can be estimated with

$$\sigma_N^2 \equiv \frac{1}{N}\sum_i \hat{u}_i^2\{w_\gamma(x_i) - E_N(w_\gamma(x_i)|\hat{v}_i)\}^2 \quad \text{where}$$

$$E_N(w_\gamma(x_i)|\hat{v}_i) \equiv \frac{1}{N}\sum_{j\neq i} \frac{1}{h}K\left(\frac{\hat{v}_j - \hat{v}_i}{h}\right)w_\gamma(x_i).$$

A "directional" test with $T_N(\gamma)$ for the direction γ can be implemented with this asymptotic distribution.

The index γ can be chosen for a high power if we know the direction to which the H_0 of SIM may be violated. But knowing such a direction a priori is unlikely. In this case, we "cover" all directions to get an "omnibus test." Stute and Zhu (2005) used

$$w_\gamma(x_i) = \exp(\mathbf{i}\gamma'x_i) = \cos(\gamma'x_i) + \mathbf{i}\sin(\gamma'x_i)$$

$$\implies T_N(\gamma) \equiv \frac{1}{\sqrt{N}} \sum_i \hat{u}_i \cos(\gamma' x_i) + \mathbf{i}\frac{1}{\sqrt{N}} \sum_i \hat{u}_i \sin(\gamma' x_i).$$

Recall $|t| \equiv (a^2 + b^2)^{1/2}$ for a complex number $t = a + \mathbf{i}b$. From $T_N(\gamma)$, the Kolmogorov-Smirnov (KS)-type test statistic is

$$T_N \equiv \sup_\gamma |T_N(\gamma)| = \sup_\gamma \left[\left\{ \frac{1}{\sqrt{N}} \sum_i \hat{u}_i \cos(\gamma' x_i) \right\}^2 \right.$$

$$\left. + \left\{ \frac{1}{\sqrt{N}} \sum_i \hat{u}_i \sin(\gamma' x_i) \right\}^2 \right]^{1/2}.$$

Since the asymptotic distribution of T_N is difficult to use, Stute and Zhu (2005) recommended the following "simulation-based" procedure. Generate iid e_i's with $E(e) = 0$ and $SD(e) = 1$ and compute

$$T_N^{(r)} \equiv \sup_\gamma \left| \frac{1}{\sqrt{N}} \sum_i e_i \hat{u}_i \{\exp(\mathbf{i}\gamma' x_i) - E_N(\exp(\mathbf{i}\gamma' x_i)|\hat{v}_i)\} \right|.$$

Repeat this for $r = 1, ..., R$, and the upper 0.05th percentile can be used as a critical value for T_N.

In practice, because finding the supremum over γ can be cumbersome, instead of the KS version, one may use the following Cramér-von-Mises-type test with γ replaced by x_j:

$$T_N \equiv \frac{1}{N} \sum_j T_N(x_j)^2 = \frac{1}{N} \sum_j \left[\left\{ \frac{1}{\sqrt{N}} \sum_{i \neq j} \hat{u}_i \cos(x_j' x_i) \right\}^2 \right.$$

$$\left. + \left\{ \frac{1}{\sqrt{N}} \sum_i \hat{u}_i \sin(x_j' x_i) \right\}^2 \right].$$

For the simulation-based procedure, replace $\hat{u}_i$ with $e_i \hat{u}_i$ and do centering with $E_N(\exp(\mathbf{i}\gamma' x_i)|\hat{v}_i)$ as above.

In dealing with complex numbers, use $\exp(\mathbf{i}\gamma' x_i) = \cos(\gamma' x_i) + \mathbf{i}\sin(\gamma' x_i)$ and compute the real and imaginary parts separately; at the end, $\mathbf{i}$ disappears in the absolute value. Specifically,

$$E_N(\exp(\mathbf{i}\gamma' x_i)|\hat{v}_i) = \frac{1}{N} \sum_{j \neq i} \frac{1}{h} K\left(\frac{\hat{v}_j - \hat{v}_i}{h} \right) \cos(\gamma' x_j)$$

$$+ \mathbf{i}\frac{1}{N} \sum_{j \neq i} \frac{1}{h} K\left(\frac{\hat{v}_j - \hat{v}_i}{h} \right) \sin(\gamma' x_j).$$

Thus

$$\exp(\mathbf{i}\gamma' x_i) - E_N(\exp(\mathbf{i}\gamma' x_i)|\hat{v}_i) =$$

$$\left\{\cos(\gamma' x_i) - \frac{1}{N}\sum_{j\neq i}\frac{1}{h}K\left(\frac{\hat{v}_j - \hat{v}_i}{h}\right)\cos(\gamma' x_j)\right\}$$

$$+ \mathbf{i}\left\{\sin(\gamma' x_i) - \frac{1}{N}\sum_{j\neq i}\frac{1}{h}K\left(\frac{\hat{v}_j - \hat{v}_i}{h}\right)\sin(\gamma' x_j)\right\}$$

and

$$\left|\frac{1}{\sqrt{N}}\sum_i e_i \hat{u}_i\{\exp(\mathbf{i}\gamma' x_i) - E_N(\exp(\mathbf{i}\gamma' x_i)|\hat{v}_i)\}\right|$$

$$= \left(\left[\frac{1}{\sqrt{N}}\sum_i e_i \hat{u}_i\{\cos(\gamma' x_i) - \frac{1}{N}\sum_{j\neq i}\frac{1}{h}K(\frac{\hat{v}_j - \hat{v}_i}{h})\cos(\gamma' x_j)\}\right]^2\right.$$

$$\left.+ \left[\frac{1}{\sqrt{N}}\sum_i e_i \hat{u}_i\left\{\sin(\gamma' x_i) - \frac{1}{N}\sum_{j\neq i}\frac{1}{h}K\left(\frac{\hat{v}_j - \hat{v}_i}{h}\right)\sin(\gamma' x_j)\right\}\right]^2\right)^{1/2}.$$

Xia et al. (2004) also extended the Stute (1997) test to SIM using the moment condition

$$E\{(y - G(x'\beta))1[x \leq \gamma]\} = 0 \ \forall\gamma.$$

The resulting test statistic takes the almost the same form as $\hat{T}_N$ except that the residual is computed with $y_i - \hat{G}(x_i'b_N)$ where $\hat{G}$ is a kernel estimator of y on $x'b_N$. Not using the transformation $\psi(t) \equiv G\{F^{-1}(t)\}$, however, leads to a bias problem that cannot be easily dealt with, and the resulting test is far more complicated than the above Stute and Zhu (2005) test.

While SIM is the null model in the above test, Horowitz and Härdle (1994) considered SIM as an alternative model while $G_0(x'\beta)$ with a known $G_0(\cdot)$ is the null model. Their proposal is to compare $G_0(x'\hat{\beta})$ and $\hat{G}(x'\hat{\beta})$ where $\hat{\beta}$ is a $\sqrt{N}$-consistent estimator under H_0 and $\hat{G}(x'\hat{\beta})$ is a kernel estimator for the unknown $G(\cdot)$. The test statistic is

$$\frac{1}{\sqrt{N}}\sum_i \{y_i - G_0(x_i'\hat{\beta})\}\{\hat{G}_{-i}(x_i'\hat{\beta}) - G_0(x_i'\hat{\beta})\}$$

where $\hat{G}_{-i}(x_i'\hat{\beta})$ is the kernel estimator for $x_i'\hat{\beta}$ leaving out the ith observation. To see the idea, observe

$$E[\{y - G_0(x'\beta)\}\{G(x'\beta) - G_0(x'\beta)\}] = E(\ E[\{y - G_0(x'\beta)\}$$

$$\{G(x'\beta) - G_0(x'\beta)\}|x]\ )$$

$$= E(\ \{G(x'\beta) - G_0(x'\beta)\}\ E[\{y - G_0(x'\beta)\}|x]\ ) = 0 \text{ under}$$

$$H_0 : E(y|x) = G_0(x'\beta).$$

Other functions of x and y may be used instead of $y - G_0(x'\beta)$ so long as the test statistic is centered at 0. But using $y - G_0(x'\beta)$ gives the test power, because under $H_1 : P\{x : E(y|x) \neq G_0(x'\beta)\} > 0$

$$E[\{y - G_0(x'\beta)\}\{G(x'\beta) - G_0(x'\beta)\}] = E[\{G(x'\beta) - G_0(x'\beta)\}^2] > 0.$$

8.6   Nonparametric Significance Tests

So far we have discussed how to test a null model using nonparametric techniques—nonparametric "GOF tests." Another group of nonparametric tests have appeared for regressor selection, i.e., to see whether some variables are relevant or not among candidate regressors. In linear models, the test can be done easily, but this may be misleading if the linear model is misspecified. Tests examined in this subsection avoid this pitfall.

8.6.1   Two-bandwidth Tests

Fan and Li (1996) proposed a nonparametric test for

$$H_0 : E(y|w, z) = E(y|w);$$

i.e., z is irrelevant under H_0. The testing idea is similar to the one in Zheng's (1996) test. Let $x \equiv (w', z')$ be a $k \times 1$ vector ($k = k_w + k_z$), and let the "null" error term be

$$\varepsilon \equiv y - E(y|w).$$

Then

$$E\{\varepsilon E(\varepsilon|x)\} = E[\ E\{\varepsilon E(\varepsilon|x)|x\}\ ] = E\{E^2(\varepsilon|x)\} = 0 \quad \text{iff } H_0 \text{ holds}$$

because $E(\varepsilon|x) = E(y|w, z) - E(y|w) \neq 0$ for some x when H_0 does not hold . To avoid random denominators, Fan and Li (1996) used a density-weighted moment condition

$$E[\varepsilon f_w(w) \cdot E\{\varepsilon f_w(w)|x\} \cdot f_x(x)] = 0 \quad \text{iff} \quad H_0 \text{ holds.}$$

A sample version for the last display that is analogous to the Zheng's (1996) test is, defining $\hat{\varepsilon}_i \equiv y_i - E_N(y|w_i)$,

$$L_N \equiv \frac{1}{N} \sum_{i=1}^{N} \left\{ \hat{\varepsilon}_i \hat{f}_w(w_i) \frac{1}{(N-1)h^k} \sum_{j \neq i} K\left(\frac{x_j - x_i}{h}\right) \hat{\varepsilon}_j \hat{f}_w(w_j) \right\}.$$

The part other than $N^{-1} \sum_{i=1}^{N} \hat{\varepsilon}_i \hat{f}_w(w_i)$ is for $E\{\varepsilon f_w(w)|x\} \cdot f_x(x)$. With $\sigma^2(x) \equiv V(y|x)$,

$$Nh^{k/2} L_N \rightsquigarrow N(0, 2\sigma^2) \quad \text{where } \sigma^2 \equiv E\{f_x(x)\sigma^4(x)f_w(w)^4\}.$$

$$\int K^2(t)dt;$$

$$\sigma_N^2 \equiv \frac{1}{N}\sum_{i=1}^{N}\left\{\hat{\varepsilon}_i^2\hat{f}_w(w_i)^2\frac{1}{(N-1)h^k}\sum_{j\neq i}K\left(\frac{x_j-x_i}{h}\right)\hat{\varepsilon}_j^2\hat{f}_w(w_j)^2\right\} \to^p \sigma^2.$$

In σ_N^2, the part $\{(N-1)h^k\}^{-1}\sum_j K((x_i-x_j)/h)\hat{\varepsilon}_j^2\hat{f}_w(w_j)^2$ is for $E\{\varepsilon^2 f_w(w)^2|x\}f_x(x)$.

Fan and Li (1996) also proposed a test when $E(y|x)$ is semi-linear, say $E(y|x) = z'\gamma + \theta(w)$, and when $E(y|x)$ is of single index. All tests in Fan and Li (1996) need at least two bandwidths, which may make the tests "unstable" in practice; see, e.g., Kondo and Lee (2003). Recall that, in Zheng's test (1996), the null model is parametric, and thus only one bandwidth is needed.

Lavergne and Vuong (2000) proposed another test based on the same moment condition as used in Fan and Li (1996). Their test statistic requires quadruple sums, and thus the test is computationally very demanding:

$$\frac{(N-4)!}{N!}\sum_{4-\text{distinct}}\frac{1}{h_w^{k_w}}L\left(\frac{w_i-w_{i'}}{h_w}\right)\frac{1}{h_w^{k_w}}L\left(\frac{w_j-w_{j'}}{h_w}\right)\frac{1}{h_x^k}$$

$$K\left(\frac{x_i-x_j}{h_x}\right)(y_i-y_{i'})(y_j-y_{j'})$$

where the sum is over all four ordered distinct arrangements in $\{1,...,N\}$. Lavergne and Vuong (2000, p. 578) provided a decomposition of their test statistic that shows how their test is related to Fan and Li's (1996). The decomposition also shows that their test statistic can be computed with triple sums, instead of quadruple sums. The asymptotic distribution under the null is the same as that of Fan and Li's (1996) test. This means that the asymptotic variance can be computed as in Fan and Li (1996), but they propose different estimators for the asymptotic variance, one of which is

$$\frac{2}{N(N-1)h^k}\sum_{i=1}^{N}\sum_{j\neq i}K^2\left(\frac{x_i-x_j}{h}\right)\cdot\hat{\varepsilon}_i^2\hat{f}_w(w_i)^2\cdot\hat{\varepsilon}_j^2\hat{f}_w(w_j)^2.$$

This estimator is likely to behave better than the above σ_N^2.

Ait-Sahalia et al. (2001) examined integrated (i.e. averaged) versions of $\{E_N(y|w,z) - E_N(y|w)\}^2$, which are analogous to Härdle and Mammen's (1993) test. As in most other papers, instead of using the "raw" difference $E_N(y|w,z) - E_N(y|w)$, they weighted the difference with an weighting function $\omega(w,z)$, and the empirical measure is used for the integration. That is, the test statistic takes a form of

$$\frac{1}{N}\sum_i\{E_N(y|w_i,z_i) - E_N(y|w_i)\}^2\omega(w_i,z_i).$$

8.6.2   One-Bandwidth Test

One main disadvantage of the above tests is requiring two bandwidths, which could be one too many. Delgado and Manteiga (2001) proposed a test that needs only one bandwidth. The idea is very close to that in the Stute (1997) test: in comparing a parametric regression function with a non-parametric regression, use $y - r(x, b_N)$ instead of $r_N(x) - r(x, b_N)$ where $r_N(x) = E_N(y|x)$ because essentially y in $y - r(x, b_N)$ becomes $E(y|x)$ when $y - r(x, b_N)$ is conditioned on x. Applying this idea to the above $H_0 : E(y|w, z) = E(y|w)$, we can use $y - r_N(w)$ instead of $r_N(w, z) - r_N(w)$, and only one smoothing for the lower-dimensional w is required. Based on this idea, Delgado and Manteiga (2001) proposed the following tests, which are analogous to the Stute (1997) test.

With $x = (w', z')'$, consider a moment condition,

$$E\{f(w)\,(y - r(w))\,1[x \le x_o]\} = 0 \quad \text{which is 0 under } H_0.$$

Define its sample analog

$$
\begin{aligned}
T_N(x_o) &\equiv \frac{1}{N} \sum_i \hat{f}_w(w_i)\{y_i - r_N(w_i)\}1[x_i \le x_o] \\
&= \frac{1}{N} \sum_i \left\{ \sum_{j \ne i} \frac{1}{h^{k_w}} K\left(\frac{w_j - w_i}{h}\right) y_i \right. \\
&\qquad \left. - \sum_{j \ne i} \frac{1}{h^{k_w}} K\left(\frac{w_j - w_i}{h}\right) y_j \right\} 1[x_i \le x_o] \\
&= \frac{1}{Nh^{k_w}} \sum_i \sum_{j \ne i} K\left(\frac{w_j - w_i}{h}\right)(y_i - y_j)1[x_i \le x_o]
\end{aligned}
$$

which leads to Cramér-von-Mises-type (CM) and Kolmogorov-Smirnov-type (KS) test statistics:

$$CM_N \equiv \frac{1}{N} \sum_{q=1}^{N} \left\{\sqrt{N} T_N(x_q)\right\}^2 = \sum_q T_N(x_q)^2 \quad \text{and}$$

$$KS_N \equiv \sup_{x_o} \left| \sqrt{N} T_N(x_o) \right|.$$

$T_N(x_o)$ is, however, not asymptotically distribution-free. Delgado and Manteiga (2001) proposed a simulation-based test statistic, which is

$$
\begin{aligned}
T_N^{(a)}(x_o) &\equiv \frac{1}{N} \sum_i v_i^{(a)} \hat{f}_w(w_i)\{y_i - r_N(w_i)\} \left\{ 1[x_i \le x_o] \right. \\
&\qquad \left. - \frac{\sum_j K((w_j - w_i)/h)1[x_j \le x_o]}{\sum_j K((w_j - w_i)/h)} \right\}
\end{aligned}
$$

where $v_i^{(a)}$ are iid rv's independent of (y_i, x_i) such that $E(v^{(a)}) = 0$ and $SD(v^{(a)}) = 1$. Note the smoothed centering of $1[x_i \le x_o]$ in $T_N^{(a)}(x_o)$. With $T_N^{(a)}(x_o)$, the corresponding versions $CM_N^{(a)}$ and $KS_N^{(a)}$ of CM_N and KS_N can be obtained, and tests can be done with the p-value of CM_N and KS_N obtained from the empirical distribution of $CM_N^{(a)}$, $a = 1, ..., A$, and $KS_N^{(a)}$, $a = 1, ..., A$, respectively.

8.6.3   Cross-Validation Approach for Mixed Regressors

Hall et al. (2007) considered how to select relevant regressors when the regressors are mixed (i.e., discrete and continuous). Let x_i^c and x_i^d denote the continuous and discrete regressors in x_i; i.e., $x_i = (x_i^{c\prime}, x_i^{d\prime})'$. For each continuous component x_{is}^c in x_i^c, use a bandwidth h_s. For each component x_{is}^d in x_i^d, apply smoothing of the form

$$\begin{aligned} L(x_{is}^d - x_{js}^d) &= 1 \quad \text{if } x_{is}^d = x_{js}^d \\ &= \lambda_s \quad \text{if } x_{is}^d \ne x_{js}^d \quad \text{where } 0 \le \lambda_s \le 1; \end{aligned}$$

call λ_s a bandwidth. Then using a product kernel for x_i, a leave-one-out kernel estimator $r_{N,-i}(x_i)$ for $r(x_i)$ can be obtained.

The key idea is selecting h_s's and λ_s's by minimizing the CV criterion

$$\sum_i \{y_i - r_{N,-i}(x_i)\}^2$$

wrt all bandwidths λ_s's and h_s's. For irrelevant discrete regressors, their λ_s should be one as $N \to \infty$; for irrelevant continuous regressors, their h_s should diverge to ∞. In a given data set, $\lambda_s \simeq 1$ may be easy to see, but how big is big for h_s may not be so clear-cut. Also, minimizing the CV minimand wrt multiple bandwidths might be difficult.

8.7   Non-nested Model Tests and Multi-sample Tests*

8.7.1   LM-Type Tests for Non-nested Models

Delgado and Stengos (1994) considered, with x and z possibly overlapping,

$$H_0 : E(y|x, z) = r(x, \beta) \quad \text{vs.} \quad H_a : E(y|x, z) = E(y|z)$$

where the two hypotheses are non-nested and the null model is parametric. Following Davidson and MacKinnon (1981), they set up an artificial regression model

$$y = (1 - \delta) \cdot r(x, \beta) + \delta \cdot E(y|z) + u$$

and proposed to test the H_0 by extending the "J test" of Davidson and MacKinnon: replace $E(y|z)$ with $E_N(y|z)$, and estimate δ and β jointly (J is from the word "jointly") treating x and $E_N(y|z)$ as regressors. The H_0 is then equivalent to $\delta = 0$. This J test is of LM-type, because using an extra regressor to see its significance is analogous to omitted variable tests.

If estimating β with b_N is simple as in LSE, then instead of the J test, one can adopt a LR-type test using both b_N and $E_N(y|z)$. To simplify exposition, suppose $r(x'\beta) = x'\beta$ and b_N is LSE. Observe

$$y_i \simeq (1 - \delta)x_i'b_N + \delta E_N(y|z_i) + u_i \implies y_i - x_i'b_N \simeq \delta\{E_N(y|z_i) - x_i'b_N\} + u_i.$$

The simplest test would go as follows. Set up an artificial model where the LSE residual is the response variable and the regressors are unity and $E_N(y|x_i) - x_i'b_N$. Do the LSE for the artificial model and test whether the slope is zero or not using the usual LSE asymptotic variance formula. This simple test seems to work well in practice.

Li and Stengos (2007) further generalized the above non-nested test by considering nonparametric models under both H_0 and H_a:

$$H_0 : E(y|x, z) = E(y|x) \quad \text{vs.} \quad H_a : E(y|x, z) = E(y|z).$$

The artificial regression model for this is

$$\begin{aligned} y &= (1 - \delta)E(y|x) + \delta E(y|z) + u \\ &= \theta(x) + \delta E_N(y|z) + v, \quad \text{where } \theta(x) \equiv (1 - \delta)E(y|x) \end{aligned}$$

and $v \equiv u + \delta\{E(y|z) - E_N(y|z)\}$. Taking $E_N(y|x)$ as a regressor, this is a semi-linear model with $\delta E_N(y|z)$ being the linear part and $\theta(x)$ being the unknown part. Applying the Robinson's (1988) TSE, $\theta(x)$ can removed and $H_0 : \delta = 0$ can be tested. Li and Stengos (2007) used a density-weighted version of the following numerator in the second step for δ:

$$T_N \equiv \frac{1}{N} \sum_i [E_N(y|z_i) - E_N\{E_N(y|z_i)|x_i\}] \cdot [y_i - E_N(y|x_i)].$$

8.7.2   LR-Type Test for Non-nested Models

Lavergne and Vuong (1996) proposed a nonparametric regressor-selection test for non-nested models. Consider models 1 and 2 with regression functions $r_1(x)$ and $r_2(x)$, and

$$\begin{aligned} H_0 \quad &: \quad E\{y - r_1(x_1)\}^2 - E\{y - r_2(x_2)\}^2 = 0 : \text{model 1 and 2} \\ &\quad \text{are equally good} \\ H_1 \quad &: \quad E\{y - r_1(x_1)\}^2 - E\{y - r_2(x_2)\}^2 < 0 : \text{model 1 is better} \\ H_2 \quad &: \quad E\{y - r_1(x_1)\}^2 - E\{y - r_2(x_2)\}^2 > 0 : \text{model 2 is better.} \end{aligned}$$

The test uses the sample analog for the mean squared error difference in the two competing models:

$$T_N \equiv \frac{1}{\sqrt{N}} \sum_i [\{y_i - r_{1N}(x_{1i})\}^2 - \{y_i - r_{2N}(x_{2i})\}^2].$$

This is asymptotically normal with variance σ^2 unless $r_1(x_1) = r_2(x_2)$; if $r_1(x_1) = r_2(x_2)$, then $\sigma^2 = 0$—the degenerate case.

Define the "fourth moment" minus "squared second moment":

$$\sigma_N^2 \equiv \frac{1}{N} \sum_i [\{y_i - r_{1N}(x_{1i})\}^2 - \{y_i - r_{2N}(x_{2i})\}^2]^2$$

$$- \left(\frac{1}{N} \sum_i [\{y_i - r_{1N}(x_{1i})\}^2 - \{y_i - r_{2N}(x_{2i})\}^2] \right)^2 .$$

Then

$$\frac{T_N}{\sigma_N} \rightsquigarrow N(0,1) \text{ under } H_0; \quad \frac{T_N}{\sigma_N} \to^p -\infty \text{ under } H_1; \quad \frac{T_N}{\sigma_N} \to^p \infty \text{ under } H_2.$$

This test is a nonparametric analog for the parametric test in Vuong (1989). One example for "model 1 and 2 are equally good" is $y = x_1 + x_2 + u$ where x_1 and x_2 follow the same distribution and share the same relation with u.

8.7.3 Multi-sample Tests for Multiple Treatments

Although our focus has been on one-sample nonparametric regression-function specification tests, there is a substantial literature on "multi-sample nonparametric regression-function equality tests," which is relevant to the topic of multiple treatment effect analysis, where each sample comes from each treatment regime. The tests in the literature differ largely in terms of (i) stochastic or fixed x, (ii) the error term independent of x or not (dependence meaning heteroskedasticity), and (iii) the sample sizes equal or not. For instance, when x is fixed, the error term is homoskedastic and the samples sizes are equal, Hall and Hart (1990) removed the same regression function under the null with

$$d_i \equiv y_{1i} - y_{2i} \quad \text{where } y_j \text{ is from sample } j = 1, 2$$

and proposed a test statistic based on sums of $d_i{}^2$.

Delgado (1993) proposed a test adopting the idea in Hall and Hart (1990). Allowing for stochastic x in the setting of Hall and Hart (1990), let x be a continuously distributed rv. With $w(t)$ denoting a standard Brownian motion, the test statistic is

$$T_N \equiv \frac{\sup_t |\sum_{i=1}^N d_i 1[x_i \le t]|}{\{0.5 \sum_{i=1}^{N-1} (d_{i+1} - d_i)^2\}^{1/2}} \rightsquigarrow T \equiv \sup_{0 \le t \le 1} |w(t)| \quad \text{as } N \to \infty.$$

The quantiles of T can be seen in Shorack and Wellner (1986). For instance, the upper quantiles for 0.1, 0.05, and 0.01 are 1.96, 2.24, and 2.81, respectively. Note a similarity between T_N and the marked empirical process test of Stute (1997).

For econometric applications where x is stochastic, Lavergne (2001) adapted the test in Lavergne and Vuong (2000) because multiple treatments can be taken as extra (discrete) regressors, and Gørgens (2002) proposed to estimate each sample regression function separately with local linear regression to construct a Wald test for "$H_0 : r_j(x)$ are the same $\forall j = 1, ..., J$" when there are J sub-populations (treatments). Sun (2006) proposed a nonparametric test for the equality of quantile functions across J-many sub-populations.

Munk et al. (2007) proposed a nonparametric analysis of variance comparing the overall error term variance estimator with all samples pooled under the H_0 (thus all y_j's are equally centered) and not pooled (thus each y_j is differently centered). Their test requires fixed x, but allows heteroskedastic errors and different sample sizes. See Munk et al. (2007) and the references therein for more on the literature of nonparametric regression-function equality tests.

APPENDIX I: MATHEMATICAL BACKGROUNDS AND CHAPTER APPENDICES

This appendix collects supporting materials for the main text chapters, except the first section that reviews mathematical and statistical backgrounds. Some of them are put here for their technicality, some for additional or historical interest, and some for being digressive or tentative. Appendix II contains further supporting materials on various topics that are not specific to any particular chapter.

1 Mathematical and Statistical Backgrounds

This section provides a more or less self-contained review on the mathematical and statistical backgrounds for the main text; matrix algebra is not covered because it is readily available in many other sources. This section is not meant to be a systematic account of the topics covered below; rather it should be taken as an informal review. There are more discussions in the following than what is necessary to read the main text; the reader may refer to this section should the need arises, rather than trying to read everything in this section. If interested in further details, then there are many well known books to refer to for real analysis, probability theory, and asymptotic statistics. This section draws on Bartle (1976), Royden (1988), Luenberger (1969), Billingsley (1995), Dudley (1989), Van der Vaart (1998), and Pollard (1984, 2002) among others.

1.1 Bounds, Limits, and Functions

For a subset H of the Euclidean space R, the least upper bound of H is called the *supremum* of H; i.e.,

$$s = \sup H \iff h \leq s \; \forall h \in H, \text{ and if } h \leq s' \; \forall h \in H, \text{ then } s \leq s'.$$

The *infimum* of H is the greatest lower bound; i.e.

$$q = \inf H \iff h \geq q \; \forall h \in H, \text{ and if } h \geq q' \; \forall h \in H, \text{ then } q \geq q'.$$

Supremum and infimum are unique if they exist. For instance, $H = [0, 1)$ has no maximum, but its supremum is 1; whereas a maximum should be in H if it exists, the supremum does not have to be. Each non-empty set bounded above (below) has a supremum (infimum). The *extended real line* $[-\infty, \infty]$, denoted usually as $\overline{R}$, is convenient, because every non-empty set in $\overline{R}$ has a supremum and an infimum, as every non-empty set is bounded by $\pm\infty$.

A *sequence* in R^k is a function mapping the set $\mathcal{N}$ of natural numbers to R^k. For a sequence $\{z_n, n \in \mathcal{N}\}$, z is the *limit* of the sequence, if for any $\varepsilon > 0$ there exists $n(\varepsilon)$ such that

$$|z_n - z| < \varepsilon \ \forall n \geq n(\varepsilon).$$

When a limit exists, $\{z_n\}$ is said to be *convergent*. For a function $f(n)$ mapping $\mathcal{N}$ to $\mathcal{N}$ and a sequence $\{z_n\}$, $\{z_{f(n)}\}$ is a *subsequence* of $\{z_n\}$ if

$$n \leq f(n), \text{ and } f(n) < f(n') \text{ for } n < n'.$$

For instance, $f(n) = n^2$ and thus $f(1) = 1$, $f(2) = 4$, $f(3) = 9, ...$; the subsequence is $\{z_1, z_4, z_9, ...\}$. A sequence is *bounded* if there exists a constant C such that $|z_n| < C \ \forall n$. A convergent sequence is bounded, and a bounded sequence has a convergent subsequence.

An *open ball* of radius ε around $x \in R^k$ is the set $\{|x - y| < \varepsilon$, $y \in R^k\}$ where $|\cdot|$ is the Euclidean norm on R^k; $|x - y| \leq \varepsilon$ yields a "closed ball." A set A in R^k is an *open set*, if for each $a \in A$, there is an open ball around a contained in A; a set B is closed if B^c is open. For $x \in R^k$, any set that contains an open set containing x is a *neighborhood* of x; a neighborhood can be a closed set. A point z' is a *cluster point* of a set Z if every neighborhood of z' contains at least one point in Z that is not z'. If z' is a cluster point of $\{z_n\}$, then there is a subsequence of $\{z_n\}$ convergent to z'; a cluster point of $\{z_n\}$ is not necessarily the limit point, because $\{z_n\}$ can move in and out of a proximity of the cluster point. A set A is *compact* if every open cover has a finite subcover; i.e., if $A \subset \cup_j B_j$ for open sets B_j's, then there are finitely many B_j's, say $B_1, ..., B_H$, that still collectively cover A (i.e., $A \subset (B_1 \cup B_2 \cup ... \cup B_H)$). A set in R^k is compact iff it is closed and bounded.

A sequence $\{y_n\}$ is a *Cauchy sequence* if, for any $\varepsilon > 0$, there exists $n(\varepsilon)$ such that

$$|x_n - x_{n'}| < \varepsilon \ \forall n, n' \geq n(\varepsilon).$$

A sequence in R^k is convergent iff it is a Cauchy sequence. Since the limit of a sequence and its subsequences should be the same, if a subsequence converges to a limit, then the sequence must converge to the same limit if the sequence is convergent at all.

A function f with *domain* X and *range* Y assigns a unique value in Y for each element $x \in X$; equivalently, f can be viewed as a set of ordered pairs in $X \times Y$. This definition does not require Y to be "exhausted"; i.e., some values of Y may not be "reached" by f. If there exists at least some x such that $f(x) = y$ for each $y \in Y$, f maps *onto* Y; $f(x)$ is *one to one* if $f(x) = f(x')$ implies $x = x'$. If f is one-to-one and onto, then f has an *inverse* f^{-1} whose domain (range) is the range (domain) of f. If f is one-to-one but not onto, then f^{-1} exists whose domain is the proper subset of the range of f that makes f onto. Even if inverse does not exist, *inverse image* does: $\{x : f(x) \in A\}$ is the inverse image of $A \subset Y$. A function $f(x)$ is *bounded* if there is a constant C such that $|f(x)| < C \; \forall x \in X$, which implies that $\sup_{x \in X} f(x)$ and $\inf_{x \in X} f(x)$ exist.

The view regarding f as an ordered pair is convenient in understanding compositions of functions, say, $F(f)$ where F has domain Y and range Z: a *composition* $F(f)$ of F and f is an ordered pair (x, z) such that there exists some y with $(x, y) \in f$ and $(y, z) \in F$; the domain of $F(f)$ consists of the elements x in the domain of f such that $f(x)$ falls in the domain of F. A function g with domain $X_g \subset X$ is a *restriction* of f to X_g if $g(x) = f(x) \; \forall x \in X_g$. A function h with domain $X_h \supset X$ is an *extension* of f to X_h if $h(x) = f(x)$ $\forall x \in X$.

Combining sequences and functions, consider a *sequence of functions* $\{f_n(x)\}$; $f_n(x)$ should be viewed as a function of two arguments n and x. With n fixed, $f_n(x)$ is just a function of x, whereas $f_1(x), f_2(x), \ldots$ is just a sequence on R with x fixed. Depending on the value of x, $\{f_n(x)\}$ may be convergent or not; when convergent, denote the limit as $f(x)$ and denote the set of x for which $\{f_n(x)\}$ is convergent as X_c. Then we write

$$f(x) = \lim_{n \to \infty} f_n(x) \text{ for each } x \in X_c.$$

For instance, let $f_n(x) = x^n$ for x on $[0, \infty)$. Then $f_n(x)$ is convergent on $X_c = [0, 1]$ with $f(x) = 0$ for $0 \le x < 1$ and $f(x) = 1$ for $x = 1$. Interestingly, $f(x)$ is not continuous at 1 while $f_n(x)$ is.

Formally, $\{f_n\}$ on a domain X converges to f on $X_c \subset X$, if, for any $\varepsilon > 0$ and $x \in X_c$, there is $n(\varepsilon, x)$ such that

$$|f_n(x) - f(x)| < \varepsilon, \quad \forall n \ge n(\varepsilon, x).$$

If there is $n(\varepsilon)$ not depending on $x \in X_c$ such that

$$|f_n(x) - f(x)| < \varepsilon, \quad \forall n \ge n(\varepsilon),$$

then $\{f_n\}$ *converges uniformly* to $f(x)$ on X_c, which is denoted also as

$$\|f_n - f\|_{X_c} \equiv \sup_{x \in X_c} |f_n(x) - f(x)| \to 0 \text{ as } n \to \infty;$$

$\|f_n\|_{X_c}$ is called the *uniform norm* or *supremum norm* (on X_c).

If $\{z_n\}$ is a bounded sequence, then its *limit superior* is

$$\limsup_{n \to \infty} z_n \equiv \lim_{n \to \infty} (\sup_{q \ge n} z_q);$$

since $\{y_n\} \equiv \sup_{q \ge n} z_q = \sup\{z_n, z_{n+1}, \ldots\}$ is a bounded and monotonically decreasing sequence, it has a limit. Every bounded sequence has a limsup while it may not have a limit; e.g. $(-1)^n$ does not have any limit, but

$\limsup(-1)^n = 1$ because $\sup_{q \geq n}(-1)^q = 1$. Analogously to limsup, *limit inferior* is

$$\lim_{n \to \infty} \inf\, z_n \equiv \lim_{n \to \infty} (\inf_{q \geq n} z_q);$$

since $\{y_n\} \equiv \inf_{q \geq n} z_q = \inf\{z_n, z_{n+1}, ...\}$ is a bounded and monotonically increasing sequence, it has a limit. Observe $\liminf(-1)^n = -1$ because $\inf_{q \geq n}(-1)^q = -1$. In extended real space $\bar{R}$, we can drop the qualifier "bounded sequence" for limsup and liminf, because unbounded sequences in R are bounded by $\pm\infty$ in $\bar{R}$; in $\bar{R}$, limsup can be ∞ and liminf can be $-\infty$.

A *double sequence* $\{y_{mn}\}$ in R^k is a function mapping $\mathcal{N} \times \mathcal{N}$ to R^k. If there is a point y such that

$$|y_{mn} - y| < \varepsilon \;\; \forall m, n \geq n(\varepsilon),$$

then y is the limit of $\{y_{mn}\}$. A double sequence is convergent iff it is a Cauchy sequence: that is, iff

$$|y_{mn} - x_{pq}| < \varepsilon, \;\; \forall m, n, p, q \geq n(\varepsilon).$$

Define two *iterated limits* as, if they exists,

$$\lim_m \lim_n y_{mn} = \lim_m Y_m, \quad \text{where } Y_m \equiv \lim_n y_{mn},$$
$$\lim_n \lim_m y_{mn} = \lim_n Z_n, \quad \text{where } Z_n \equiv \lim_m y_{mn}.$$

For instance, consider $(-1)^m(m^{-1}+n^{-1})$. Because $m^{-1}+n^{-1} \to 0$ as $m, n \to \infty$, the double limit is 0, $Y_m = (-1)^m m^{-1}$, and $\lim_m Y_m = 0$, but Z_n does not exist because Z_n oscillates around $\pm n^{-1}$ for each n. *If the limit y for a double sequence exists and if Y_m exists for each m, then $y = \lim_m \lim_n y_{mn}$.* Analogously, if y exists and if Z_n exists for each n, then $y = \lim_n \lim_m y_{mn}$.

1.2   Continuity and Differentiability of Functions

A function $f(x)$ is *bounded on* A if there is a constant C such that $|f(x)| \leq C \; \forall x \in A$; a set $\mathcal{F}$ of functions f is *uniformly bounded on* A if there is a constant C such that $|f(x)| < C \; \forall x \in A$ and $\forall f \in \mathcal{F}$. A function $f(x)$ is *continuous at* x_o in its domain if, for any $\varepsilon > 0$, there exists $\delta(\varepsilon, x_o)$ such that

$$|f(x) - f(x_o)| < \varepsilon \;\; \forall x \text{ in the domain of } f(x) \text{ with } |x - x_o| < \delta(\varepsilon, x_o)$$
$$\Longleftrightarrow \;\; f(x_n) \to f(x_o) \; \forall x_n \to x_o \text{ where } \{x_n\} \text{ is in the domain of } f.$$

If there is no x with $|x - x_o| < \delta(\varepsilon, x_o)$ (or no sequence $x_n \to x_o$) other than x_o itself in the domain of f, then f is continuous at x_o by definition as x_n in any "$x_n \to x_o$" should be x_o—but continuity in this case is not really useful. If $f(x)$ is continuous at each point of its domain, then $f(x)$ is said to be *continuous*. One may think that the continuity at x_o implies continuity at points nearby, but this does not hold necessarily. There is a function that is continuous at all irrational points (points not of the form

$\pm m/n$, for $m, n \in \mathcal{N}$) while discontinuous at all rational points, despite that every neighborhood of an irrational point has a rational point.

If $f(x)$ is continuous on a subset X_u of its domain and if $\delta(\varepsilon, x_o)$ can be chosen independently of x_o on X_u, then $f(x)$ is *uniformly continuous* on X_u. For instance, $f(x) = 4x$ is uniformly continuous, because $|f(x) - f(z)| = 4|x - z|$: if we set $|x - z| < \varepsilon/4 = \delta(\varepsilon)$, then $|f(x) - f(z)| < \varepsilon$; here $\delta(\varepsilon)$ does not depend on x. On the other hand, for $f(x) = e^x$,

$$|f(x) - f(z)| = |e^x - e^z| = e^x|1 - e^{z-x}|.$$

Although we can make $|1 - e^{z-x}| < \varepsilon$ with $|z - x| < \delta(\varepsilon)$ regardless of x, since e^x is attached, $\delta(\varepsilon)$ cannot be chosen independently of x to make $|f(x) - f(z)| < \varepsilon$; e^x is not uniformly continuous. In e^x, the neighborhood should be chosen smaller as x gets greater to ensure $|e^x - e^z| < \varepsilon$, while this is not the case for $4x$. But *any continuous function on a compact interval is also uniformly continuous* on the interval; a bounded open interval will not do, as can be seen in $f(x) = 1/x$ on $(0, 1]$.

If a function $s(x)$ takes a finite number of values, then $s(x)$ is a *simple function*. If $g(x)$ is a continuous function on a compact set C, then there is a simple function $s(x)$ that is uniformly close to $g(x)$ on C in the sense that

$$\sup_{x \in C} |g(x) - s(x)| < \varepsilon, \text{ for any } \varepsilon > 0.$$

Polynomials in x or piecewise linear functions can be also used for this kind of uniform approximation of $g(x)$.

Let $f(x)$ be a function with domain $A \subset R^p$ and range R^q, which is often denoted simply as $f\colon A \to R^q$. The *directional differential* vector $f_a(\tau)$ at a given interior point $a \in A$ wrt a direction $\tau \in R^p$ exists, if

$$f_a(\tau) = \lim_{t \to 0} \frac{f(a + t\tau) - f(a)}{t} \iff \left| \frac{f(a + t\tau) - f(a)}{t} - f_a(\tau) \right| < \varepsilon$$
$$\text{if } |t| < \delta(\varepsilon, a, \tau) \text{ for any } \varepsilon > 0.$$

If $\tau = (1, 0, ..., 0)'$, then $f_a(\tau) = \partial f(a)/\partial x_1$, and if $\tau = (0, 0, ..., 1)'$, then $f_a(\tau) = \partial f(a)/\partial x_p$; these are *partial derivatives*. When $f_a(\tau)$ is viewed as a function of τ mapping R^p to R^q, the function f_a is the *directional derivative* vector, which is also called *Gateau derivative*.

A function $f(x)$ with an open domain $A \subset R^p$ and range R^q is *differentiable at* a if there exists a linear (and thus continuous) function $f'_a(\cdot)$ from R^p to R^q such that

$$\lim_{|\tau| \to 0} \frac{|f(a + \tau) - f(a) - f'_a\tau|}{|\tau|} = 0, \quad \text{or equivalently (Bickel et al.,}$$

$$\text{1993, p. 454),}$$

$$\left| \frac{f(a + t\tau) - f(a)}{t} - f'_a\tau \right| < \varepsilon \text{ if } |t| < \delta(\varepsilon, a) \text{ for any } \varepsilon > 0$$

$$\text{and } \tau \text{ in any bounded subset of } R^p.$$

The linear function f'_a is called the *(Fréchet) derivative* and $f'_a\tau$ is the *differential* at a with increment τ; if $q = 1$, then f'_a is also called the *gradient* and denoted as ∇f_a. The requirement for derivative is more stringent than that for directional derivative for two reasons. First, $f_a'\tau$ should be linear in τ while the Gateau derivative $f_a(\tau)$ does not have to. Second, the derivative remainder term $\{f(a + t\tau) - f(a)\}/t - f_a'\tau$ should converge uniformly over τ, whereas such uniformity is not required for the directional derivative remainder term.

Since any q-dimensional linear function on R^p can be expressed as a $q \times p$ matrix, if f_a' exists, it can be viewed as a $q \times p$ matrix (depending on the evaluation point a), whereas $f_a'\tau$ is the matrix times the vector τ. If f_a' exists, then the directional derivative $f_a(\tau)$ for any τ exists, and $f_a'\tau = f_a(\tau)$; as $f_a(\tau)$ for any τ exists as just noted, $\partial f(a)/\partial x_j \ \forall j$ exists and

$$f_a'\tau = \frac{\partial f(a)}{\partial x_1}\tau_1 +, ..., + \frac{\partial f(a)}{\partial x_p}\tau_p \quad \text{where } \tau = (\tau_1, ..., \tau_p)'.$$

Since finding a partial derivative vector $(\partial f(a)/\partial x_1, ..., \partial f(a)/\partial x_p)$ is easier than finding the derivative f_a', the following is helpful to see whether $f(x)$ is differentiable or not: *if $\partial f(a)/\partial x_j$, $j = 1, ..., p$, exist in a neighborhood of a and are continuous at a, then the derivative f_a' exists.*

Suppose that $f(x)$ is a real-valued function on $\Omega \subset R^p$ with continuous partial derivatives of order m in a neighborhood of every point on a line segment S joining two points a and $a + u$ in Ω. Then there exists a point c in S such that (Bartle, 1976, p. 371)

$$f(a + u) = f(a) + f_a'u + \frac{f_a^{(2)}(u)^2}{2!} + \frac{f_a^{(3)}(u)^3}{3!}, ..., + \frac{f_a^{(m-1)}(u)^{m-1}}{(m-1)!}$$

$$+ \frac{f_c^{(m)}(u)^m}{m!} \quad \text{where}$$

$$f_a^{(2)}(u)^2 \equiv \sum_{j,k} \frac{\partial^2 f(a)}{\partial x_j \partial x_k} u_j u_k, \quad f_a^{(3)}(u)^3 \equiv \sum_{j,k,l} \frac{\partial^3 f(a)}{\partial x_j \partial x_k \partial x_l} u_j u_k u_l,$$

$$u = (u_1, ..., u_p)'$$

and $f_a^{(m)}(u)^m$, $m = 4, ...,$ are defined analogously. This is a mean value theorem or a "Taylor expansion".

1.3   Probability Space and Random Variables

Imagine a statistical experiment. A *sample space* Ω is the collection of all possible outcomes of the experiment. A point $\omega \in \Omega$ is a *sample point*, and a subset $A \subset \Omega$ is an *event*. A collection (or "class") $\mathcal{A}$ of events is a *σ-algebra* or *σ-field* if

(*i*) $\Omega \in \mathcal{A}$,

(*ii*) $A \in \mathcal{A}$ implies $A^c \in \mathcal{A}$,

(*iii*) $A_1, A_2, ... \in \mathcal{A}$ implies $\cup_i A_i \in \mathcal{A}$;

(*iii*) is equivalent to "$\cap_i A_i \in \mathcal{A}$" due to (*ii*).

A set function $\mu : \mathcal{A} \longmapsto [0, \infty]$ is a *measure* if $\mu(\varnothing) = 0$ and

$$\mu(\cup_i B_i) = \sum_\iota \mu(B_i), \ \forall \text{ disjoint sequence } B_1, B_2, ... \in \mathcal{A};$$

this property is called *countable additivity*. A measure P is a *probability (measure)* if $P(\Omega) = 1$. Countable additivity implies the "continuity of P in set sequences" (see, e.g., Billingsley, 1995, p. 25): if $A_n \uparrow A$ (i.e., $A_1 \subset A_2 \subset \cdots$ and $A = \cup_n A_n$) or $A_n \downarrow A$ (i.e., $A_1 \supset A_2 \supset \cdots$ and $A = \cap_n A_n$) where $A_n, A \in \mathcal{A}$, then $P(A_n) \uparrow P(A)$ or $P(A_n) \downarrow P(A)$, respectively.

The pair $(\Omega, \mathcal{A})$ is a "*measurable space*"; the triplet $(\Omega, \mathcal{A}, \mu)$ is a "*measure space*"; $(\Omega, \mathcal{A}, P)$ is a *probability space*. The simplest probability space is $\mathcal{A} = \{\Omega, \varnothing\}$, $P(\Omega) = 1$, and $P(\varnothing) = 0$. The next simplest is $\mathcal{A} = \{\Omega, \varnothing, A, A^c\}$ where A is an event, $P(\Omega) = 1$, $P(\varnothing) = 0$, and $P(A)$ is a fraction. If $P(A) = 1$ for some $A \in \mathcal{A}$, then A is a *support* of the probability measure P; the smallest closed set $B \in \Omega$ such that $P(B) = 1$ is called "the" support.

The σ-field $\mathcal{B}$ "generated" by the open sets in R^k is the *Borel σ-field*, a member of which is a "*Borel set*"; "being generated" means that the Borel σ-field is the smallest collection of events containing all open sets in R^k while satisfying the three conditions for σ-field. That is, if $\mathcal{F}_j$ is a σ-field containing all open sets in R^k and j indexes all such σ-fields, then the Borel σ-field is $\cap_j \mathcal{F}_j$. The Borel σ-field exists, because there is at least one σ-field including all open sets in R^k (e.g., the collection of all subsets of R^k). On R, *Lebesque measure* $\mu(B)$ is the total length of $B \in \mathcal{B}$; Lebesque measure μ on R^k assigns the k-fold product of the one-dimensional lengths. If $v(A)$ is the number of points ("*cardinality*") in A, then v is *counting measure*; if A has infinitely many elements, then $v(A) = \infty$.

Given a probability space $(\Omega, \mathcal{A}, P)$, a *random variable* (rv) x is a "*measurable*" function $x : \Omega \longmapsto R$; here "*measurable*" means

$$x^{-1}(B) \equiv \{\omega : x(\omega) \in B\} \in \mathcal{A} \text{ for any Borel set } B \in \mathcal{B}.$$

A rv induces a probability measure P_x on $\mathcal{B}$:

$$P_x(B) \equiv P(x^{-1}(B)), \ \forall B \in \mathcal{B};$$

P_x for $(R, \mathcal{B}, P_x)$ is on R while P is on Ω. If we want to make it clear which σ-field is being used, we say that x is $\mathcal{A}$-*measurable* or $\mathcal{A}/\mathcal{B}$-*measurable*. A random vector is a vector consisting of rv's.

The *distribution function (df)* $F_x(t)$ of x is defined as $F_x(t) \equiv P(x \leq t)$ which is $P(x^{-1}(B))$ with $B = (-\infty, t]$. When $t_n \downarrow t$, we get $P(x \leq t_n) \downarrow P(x \leq t)$, because $(-\infty, t_n] \downarrow (-\infty, t]$; i.e., $F_x(t)$ is right-continuous in t. Note that $P(x < t_n) \neq P(x < t)$ when $P(x = t) > 0$, because $P(x < t_n)$ includes $P(x = t)$ while $P(x < t)$ does not. Whereas P_x is the probability measure (probability distribution or probability *law*) of x, F_x is just a function mapping R to $[0, 1]$. For a random vector $z \equiv (z_1, ..., z_k)'$, its joint distribution function $F_z(t)$ is defined as $F_z(t) \equiv P(z_1 \le t_1, ..., z_k \le t_k)$ where $t = (t_1, ..., t_k)'$.

The df F_x of x is discrete if $P(x = x_j) = p_j > 0$ for all j and $\sum_j p_j = 1$, where x_j is a support point of x; F_x is *continuous* if $P(x = r) = 0$ for any point $r \in R^k$; F_x is *absolutely continuous* if

$$F_x(t) = \int_{-\infty}^{t} f_x(s)ds$$

for a measurable function f_x. There exist rv's with both discrete and continuous components; e.g., $F_x(t) = \lambda F_1(t) + (1 - \lambda)F_2(t)$ where $0 < \lambda < 1$, F_1 is discrete and F_2 is continuous. In this case, $F_x(t)$ is continuous and increasing (i.e., non-decreasing) other than for occasional jumps at the support points of F_1.

1.4   Integrals

For a measure space $(\Omega, \mathcal{A}, \mu)$, a simple function f on Ω is defined as

$$f(\omega) = \sum_{j=1}^{J} a_j 1_{A_j}(\omega) \qquad \text{where } a_j\text{'s are constants, } A_j \in \mathcal{A} \ \forall j, \text{ and}$$

$1_{A_j}(\omega) = 1$ if $\omega \in A_j$ and 0 otherwise.

When $J = 2$ with two overlapping sets A_1 and A_2, define $B_1 \equiv A_1 \backslash A_2$ $(\equiv \{\omega : \omega \in A_1, \omega \notin A_2\})$, $B_2 \equiv A_1 \cap A_2$, and $B_3 \equiv A_2 \backslash A_1$ to get three disjoint sets and

$$f(\omega) = \sum_j b_j 1_{B_j}(\omega), \quad b_1 \equiv a_1, \ b_2 \equiv a_1 + a_2, \ b_3 \equiv a_2.$$

This shows that a simple function can be always rewritten as a linear combination of indicator functions for disjoint sets; thus we will use only simple functions with disjoint sets.

For any non-negative simple function $f(\omega) = \sum_j b_j 1_{B_j}(\omega)$ on a measure space $(\Omega, \mathcal{A}, \mu)$, its *(Lebesque) integral* $\int_C f(\omega)d\mu(\omega) = \int_C f d\mu$ over a set $C \in \mathcal{A}$ is defined as

$$\int f 1_C d\mu = \sum_j b_j \mu(B_j \cap C) \quad (\text{as } f(\omega)1_C(\omega) = \sum_j b_j 1_{B_j \cap C}(\omega))$$

$$\Longrightarrow \int f d\mu = \sum_j b_j \mu(B_j) \quad \text{when } C = \Omega.$$

If there is $0 \cdot \infty$ in the sum (e.g., $b_j = 0$ and $\mu(B_j) = \infty$ for some j), then define $0 \cdot \infty$ as 0.

The integration dummy w in $\int_C f(w)d\mu(w)$ serves the role of showing on which space the integration takes place; if this is clear, then w is omitted as in $\int_C f d\mu$. Instead of $d\mu(w)$, $\mu(dw)$ is also used as in $\int_C f(w)\mu(dw)$. If which measure (μ) is used is clear, then $\int f d\mu$ is denoted simply as $E(f)$. In fact, it is simpler and more informative to write $\mu(f)$ instead of $E(f)$ ($P(f)$ if $\mu = P$), because $\mu(f)$ shows better that integral $\int f d\mu$ is a mapping from the space of non-negative simple functions to R by the measure μ—but there may be a risk that $\mu(f)$ may be confused with $\mu(w)$. In the following, it is shown that the domain of the integral $\mu(\cdot)$ can be extended from the set of simple functions.

For any non-negative measurable function f (i.e., a non-negative rv f), there exists a sequence of non-negative simple functions $f_n(w) \uparrow f(w)$ $\forall w$ as $n \to \infty$, which means that $f_n(w) \le f(w)$ and $f_n(w) \to f(w)$ for each $w \in \Omega$; e.g., from Pollard (2002),

$$f_n(w) = \frac{1}{2^n} \sum_{j=1}^{4^n} 1\left[f(w) \ge \frac{j}{2^n}\right] = \frac{1}{2^n}\left\{1\left[f(w) \ge \frac{1}{2^n}\right]\right.$$

$$\left. + 1\left[f(w) \ge \frac{2}{2^n}\right] + , ..., +1\left[f(w) \ge \frac{4^n}{2^n}\right]\right\}.$$

To see that this function satisfies the requirement, observe that

$$f_n(w) = 0 \text{ if } f(w) < \frac{1}{2^n}, \quad f_n(w) = 2^n \text{ if } f(w) \ge \frac{4^n}{2^n} = 2^n$$

$$f_n(w) = \frac{j}{2^n} \text{ if } \frac{j}{2^n} \le f(w) < \frac{j+1}{2^n}, \quad j = 1, 2, ..., 4^n - 1;$$

$f_n(w) \le f(w)$ clearly holds. To see $f_n(w) \to f(w)$ as $n \to \infty$ for each w, first, observe that $f(w)$ is approximated by 0 when $f(w) \le 2^{-n}$ and thus $|f_n(w) - f(w)| \le 2^{-n}$. Second, $f(w)$ is approximated by $j2^{-n}$ when $j2^{-n} \le f(w) < (j+1)2^{-n}$; $|f_n(w) - f(w)| \le 2^{-n}$ holds here as well. Third, the approximation is not good when $f(w) \ge 2^n$, but w such that $f(w) \ge 2^n$ disappears as $n \to \infty$; i.e., for any w, $f(w)$ eventually falls in $[j2^{-n}, (j+1)2^{-n})$ for some j. Hence $f_n(w) \uparrow f(w)$ as $n \to \infty$ for each w.

Not just for the particular sequence in the last display, but for any sequence such that $f_n(w) \uparrow f(w)$ for each $w \in \Omega$, it holds that $\int f_n d\mu \uparrow \int f d\mu$. That is, for any non-negative measurable function f, its (Lebesque) *integral can be defined as the limit of the integral sequence of any approximating (from below) simple function sequence.*

If f takes both positive and negative values, then define two non-negative functions

$$f^+ \equiv \max(f, 0), \quad f^- = -\min(f, 0) \implies f = f^+ - f^-.$$

In this case, the integral is defined as

$$\int f d\mu \equiv \int f^+ d\mu - \int f^- d\mu;$$

if both terms are infinite, the integral is not defined; if $\int f^+ d\mu$ is infinite and $\int f^- d\mu$ is finite, then the integral does not exist but is defined as ∞; if $\int f^+ d\mu$ is finite and $\int f^- d\mu$ is infinite, then the integral does not exist but is defined as $-\infty$. When $\int f d\mu$ is finite, which is equivalent to $\int |f| d\mu \equiv \int f^+ d\mu + \int f^- d\mu$ being finite, f is said to be *integrable*. The domain of the integral $\mu(\cdot)$ is now all measurable functions, which has been extended from the set of simple functions.

Lebesque integral which uses a "horizontal approximation" as above is more general than Rieman integral which uses a "vertical approximation" to $f(\omega)$: whenever a Rieman integral exists, the Lebesque integral exists as well and the two are equal, but there are cases where Rieman integral does not exist while the Lebesque integral does. A well-known example is the function that takes 1 for rational numbers and 0 for irrational numbers.

Consider the original measure space $(\Omega, \mathcal{A}, \mu)$ and $(R, \mathcal{B})$ for a rv f: $\Omega \to R$, which induces a measure on $R : v \equiv \mu(f^{-1})$. For a $\mathcal{B}/\mathcal{B}$-measurable function g: $R \to R$, $g(f)$ is also a rv mapping $\Omega \to R$. The integral of g over $B \in \mathcal{B}$ can be obtained in two equivalent ways. One is using g and $v(B) \equiv \mu(f^{-1}(B))$, and the other is using $g(f(\omega))$ and the original measure μ:

$$\left\{ \int_B g dv = \right\} \int_B g(r) d\mu f^{-1}(r)$$
$$= \int_{f^{-1}(B)} g(f) d\mu \left\{ = \int_{f^{-1}(B)} g(f(\omega)) d\mu(\omega) \right\}.$$

This is a *change of variables* from f to the original ω.

A measure μ on $(\Omega, \mathcal{A})$ is *finite* if $\mu(\Omega) < \infty$, *infinite* if $\mu(\Omega) = \infty$, and $\sigma - finite$ if

$$\Omega = \cup_i A_i \text{ for some sequence of sets } A_i \text{'s with } \mu(A_i) < \infty \ \forall i.$$

Let $(X, \mathcal{X}, \mu)$ and $(Y, \mathcal{Y}, v)$ be σ-finite measure spaces. On the Cartesian product $X \times Y$, $A \times B$ with $A \in \mathcal{X}$ and $B \in \mathcal{Y}$ is a *measurable rectangle*. There exists the smallest σ-field (the *product σ-field*) generated by the measurable rectangles, i.e., the smallest collection of the measurable rectangles while satisfying the three requirements for σ-field.

On the product σ-field, there exists *product measure* π such that (Billingsley 1995)

$$\pi(A \times B) = \mu(A) \cdot v(B) \ \forall A \in \mathcal{X}, \ B \in \mathcal{Y}.$$

An integral wrt the product measure $\int_{X \times Y} f(x, y) d\pi(x, y)$, if it exists, can be calculated using the iterated integrals

$$\int_X \left\{ \int_Y f(x, y) dv(y) \right\} d\mu(x) \quad \text{or} \quad \int_Y \left\{ \int_X f(x, y) d\mu(x) \right\} dv(y).$$

To see whether the double integral exists or not, an iterated integral is computed with $|f|$ in place of f; if this is finite, then the double integral exists (Billingsley 1995, p. 234).

For a measure space $(\Omega, \mathcal{A}, \mu)$, the set of integrable functions are usually denoted as $L^1(\Omega, \mathcal{A}, \mu)$; if it is obvious which measure space is being involved, then this is abbreviated as L^1. More generally, if f^p is integrable where $0 < p < \infty$, then this is denoted as $f \in L^p(\Omega, \mathcal{A}, \mu)$. Among L^p, L^2 is probably the most often used; sometimes $L^2(x)$ is used to denote the set of square integrable functions of a rv (or a random vector) x. If $1 \le p < \infty$, then

$$||f||_p \equiv \left(\int |f|^p d\mu \right)^{1/p}$$

is the L^p *norm of* f. *Markov inequality* for a L^p integrable function f is

$$P(|f| > \varepsilon) \le \frac{E|f|^p}{\varepsilon^p} \quad \text{for any positive constant } \varepsilon.$$

Hölder inequality is: for $f \in L^p$ and $g \in L^q$ where $1 < p, q < \infty$, $p^{-1} + q^{-1} = 1$,

$$\left| \int fg d\mu \right| \le ||f||_p ||g||_q \quad (\Longrightarrow fg \in L^1).$$

Minkowski inequality is: for $f, g \in L^p$ where $1 \le p \le \infty$,

$$||f + g||_p \le ||f||_p + ||g||_p$$

where $||f||_\infty \equiv \inf\{B : |f| \le B \text{ a.e.}\}$ and "a.e." stands for "almost everywhere" as will be explained later.

1.5 Density and Conditional Mean

For a measure space $(\Omega, \mathcal{A}, \mu)$, let f be a nonnegative measurable function on Ω. Then a new measure v can defined on $(\Omega, \mathcal{A})$ using (μ and) f as a new "weight":

$$v(A) \equiv \int_A f(\omega) d\mu(\omega), \quad \forall A \in \mathcal{A};$$

v has *density* f wrt μ. Clearly, if $\mu(A) = 0$, then $v(A) = 0$; v is said to be *dominated* by μ or *absolutely continuous* wrt μ. For instance, if $\Omega = R$, $A = (-\infty, t]$, f is the $N(0,1)$ density, and μ is Lebesque measure on R, then $v(A)$ is the $N(0,1)$ density integrated up to t. As another example, if $\Omega = \{1, 2, 3, 4, 5, 6\}$, $A = \{1, 2\}$, $f(j) = p_j > 0$, and μ is the counting measure, then $f(\omega) = \sum_{j=1}^6 p_j 1_j(\omega)$, and

$$v(\{1, 2\}) = \int_{\{1,2\}} f(\omega) d\mu(\omega) = \sum_j p_j \mu(j \cap \{1, 2\}) = p_1 + p_2.$$

When v has density f wrt μ, an integral wrt v can be found using μ and f instead:

$$\int_C g dv = \int_C g f d\mu.$$

Radon–Nikodym theorem yields a converse to v being dominated by μ when v is defined by $v(A) \equiv \int_A f d\mu$: if μ and v are σ-finite measures on $(\Omega, \mathcal{A})$ and if v is dominated by μ, then there exists a nonnegative $\mathcal{A}$-measurable function f on Ω such that

$$v(A) = \int_A f d\mu \quad \forall A \in \mathcal{A}.$$

The density f is "μ-a.e." unique: if there is another density, say g, then $\mu(\{\omega : f(\omega) \neq g(\omega)\}) = 0$. The density is the *Radon–Nikodym derivative* of v wrt μ, denoted "$dv/d\mu$." But this derivative-like expression may cause a confusion, for which the following paragraph drawn from Van der Vaart (1998, pp. 86–87) helps.

For a measure space $(\Omega, \mathcal{A}, \mu)$, suppose F and G are probability measures with densities f and g wrt a dominating measure μ. The indicator function $1[f > 0]$ is also denoted as $\{f > 0\}$ in the following. Define

$$G_F(A) \equiv G(A \cap \{f > 0\}) \quad \text{and} \quad G_{F^c}(A) \equiv G(A \cap \{f = 0\}), \quad A \in \mathcal{A}$$
$$\Longrightarrow G(A) = G_F(A) + G_{F^c}(A) : \text{"Lebesque decomposition" of } G \text{ wrt } F.$$

Observe

$$G(A) = \int_A g d\mu = \int_A g(\{f > 0\} + \{f = 0\}) d\mu$$
$$\geq \int_A g\{f > 0\} d\mu = \int_A \frac{g}{f}\{f > 0\} f d\mu = \int_A \frac{g}{f}\{f > 0\} dF$$
$$= \int_A \frac{g}{f} dF \quad \text{(because } F(\{f = 0\}) = 0\text{).}$$

From the second and last expressions, we can see $\int_A g d\mu \geq \int_A (g/f) dF$, and it is wrong to state $\int_A g d\mu = \int_A g(d\mu/dF) dF = \int_A (g/f) dF$. Note that the last expression equals $G_F(A)$.

Let $\mathcal{B}$ be a sub σ-field of $\mathcal{A}$: $\mathcal{B}$ is a σ-field and $\mathcal{B} \subset \mathcal{A}$. For a non-negative integrable x, define a measure v on $\mathcal{B}$ with

$$v(B) = \int_B x dP \quad \forall B \in \mathcal{B}.$$

Since P and v are finite, they are σ-finite as well. Consider the restriction $P_{\mathcal{B}}$ of P to $\mathcal{B}$: $P_{\mathcal{B}}(B) = P(B) \forall B \in \mathcal{B}$. Because v is dominated by P, v is dominated by $P_{\mathcal{B}}$ on $\mathcal{B}$. Thus, there exists a Radon–Nikodym derivative, say $E(x|\mathcal{B})$, satisfying $v(B) = \int_B E(x|\mathcal{B}) P_{\mathcal{B}}$ and hence

$$\int_B E(x|\mathcal{B}) dP_{\mathcal{B}} = \int_B x dP_{\mathcal{B}} \left\{ = \int x(\omega) 1_B(\omega) dP_{\mathcal{B}}(\omega) \right\} \quad \forall B \in \mathcal{B}.$$

$E(x|\mathcal{B})$ is the *conditional mean* of x given $\mathcal{B}$; $E(x|\mathcal{B})$ *is B-measurable and integrable on* Ω; $E(x|\mathcal{B})$ *is unique a.e.* $P_{\mathcal{B}}$, which means that, if there is another rv $g(\omega)$ satisfying the equation, then $P_{\mathcal{B}}\{E(x|\mathcal{B}) = g\} = 1$. For x taking both positive and negative values, the Radon–Nikodym theorem can be applied to x^+ and x^- separately. In $\int_B E(x|\mathcal{B})dP_{\mathcal{B}} = \int_B xdP_{\mathcal{B}}$, $\mathcal{B}$ in $dP_{\mathcal{B}}$ may be dropped so long as it is understood that the equation is only for $B \in \mathcal{B}$, because $\int_B E(x|\mathcal{B})dP = \int_B xdP$ $\forall B \in \mathcal{B}$. Certainly, if $\mathcal{B} = \mathcal{A}$, then $E(x|\mathcal{B}) = x$.

For example, if $\mathcal{B} = \{\varnothing, \Omega\}$, then $E(x|\mathcal{B}) = E(x)$, because $E(x)$ satisfies the last display. Suppose $\mathcal{B} = \{\varnothing, \Omega, B, B^c\}$ with $0 < P(B) < 1$. In this case, the $\mathcal{B}$-measurable $E(x|\mathcal{B})$ should take the form

$$E(x|\mathcal{B})(\omega) = q_B 1_B(\omega) + q_{B^c} 1_{B^c}(\omega) \quad \text{for some constants } q_B, q_{B^c} > 0$$

as this form allows four different values for the four sets in $\mathcal{B}$. Recalling the definition of integral for simple functions,

$$\left(\int_B E(x|\mathcal{B})dP_{\mathcal{B}} = \right) q_B P(B) = \int_B xdP_{\mathcal{B}}$$

$$\Longrightarrow q_B = \frac{\int_B xdP_{\mathcal{B}}}{P(B)} \equiv E(x|B);$$

$$q_{B^c} = \frac{\int_{B^c} xdP_{\mathcal{B}}}{P(B^c)} \equiv E(x|B^c), \quad \text{doing analogously.}$$

Thus

$$E(x|\mathcal{B})(\omega) = E(x|B)1_B(\omega) + E(x|B_c)1_{B^c}(\omega).$$

The expression $E(x|z_1, ..., z_n)$ refers to $E\{x|\sigma(z_1, ..., z_n)\}$, where $\sigma(z_1, ..., z_n)$ is the smallest σ-field that makes $z_1, ..., z_n$ measurable. Dudley (1989, p. 340) showed that $E(x|z_1, ..., z_n)$ is indeed a (measurable) function of $z_1, ..., z_n$.

Conditional probability $P(A|\mathcal{B})$ for $A \in \mathcal{A}$ is a special case of $E(x|\mathcal{B})$ when $x(\omega) = 1_A(\omega)$. In this case, the requirement for $P(x|\mathcal{B})$ becomes

$$\int_B P(x|\mathcal{B})dP = P(A \cap B) = \left(\int_B 1_A dP\right), \quad \forall B \in \mathcal{B}.$$

For example, when $\mathcal{B} = \{\varnothing, \Omega, B, B^c\}$ with $0 < P(B) < 1$,

$$P(A|\mathcal{B}) = p_B 1_B(\omega) + p_{B^c} 1_{B^c}(\omega) \text{ for some } p_B, p_{B^c} > 0$$

$$\Longrightarrow \left(\int_B P(A|\mathcal{B})dP_{\mathcal{B}} = \right) p_B P(B) = P(A \cap B)$$

$$\Longrightarrow p_B = \frac{P(A \cap B)}{P(B)} \equiv P(A|B);$$

$$p_{B^c} = \frac{P(A \cap B^c)}{P(B^c)} \equiv P(A|B^c).$$

Hence $P(A|\mathcal{B})(\omega) = P(A|B)1_B(\omega) + P(A|B)1_{B^c}(\omega).$

1.6   Dominated and Monotone Convergences

Consider $(\Omega, \mathcal{A}, \mu)$ and measurable functions (i.e., rv's) defined on Ω. For a sequence of measurable functions f_n, there are three well-known convergence theorems useful for finding $\int f_n d\mu$ as $n \to \infty$ when we have some idea on f_n.

First, *dominated convergence theorem* is

$$\text{If } f_n, g \in L^1, \ |f_n(\omega)| \le g(\omega), \text{ and } f_n(\omega) \to f(\omega) \ \forall \omega, \text{ then}$$

$$\int f_n d\mu \to \int f d\mu < \infty.$$

Although we write "$\forall \omega$," the conditions $|f_n(\omega)| \le g(\omega)$ and $f_n(\omega) \to f(\omega)$ need to hold only for ω in a set A such that $\mu(A) = \mu(\Omega)$.

Second, *monotone convergence theorem* is

$$\text{If } f_n(\omega) \uparrow f(\omega) \ \forall \omega \text{ and } f_n \ge 0, \text{ then}$$

$$\int f_n d\mu \uparrow \int f d\mu \quad (\text{which can be } \infty);$$

we already saw an example of this "below-approximating" sequence $f_n(\omega)$ when integral was defined.

Third, *Fatou's Lemma* is ($f_n \to f$ is not assumed here)

$$\text{If } f_n \ge 0, \text{ then}$$

$$\int (\liminf f_n) d\mu \ \le \ \liminf \int f_n d\mu \quad (\text{which can be } \infty).$$

The following corollary of Fatou's Lemma is useful in showing $\int f d\mu < \infty$ when $\int f_n d\mu < \infty \ \forall n$ but there is no dominating function as $g(\omega)$ to invoke the dominated convergence theorem:

$$\text{If } f_n \ \ge \ 0 \text{ and } f_n(\omega) \to f(\omega) \ \forall \omega, \text{ then}$$

$$\int f d\mu \ \le \ \sup_n \int f_n d\mu \quad (\text{which can be } \infty).$$

One useful application of the dominated convergence is in establishing continuity and differentiability of integrals of a function $f(\omega, b)$. Suppose, for $\omega \in A$ where $\mu(A) = \mu(\Omega)$,

(i) $f(\omega, b)$ is continuous at $b = \beta \in R \ \forall \omega$ and

(ii) $\sup_{b \in N_\beta} |f(\omega, b)| \le g(\omega)$ in a neighborhood N_β of β
   with $\int g(\omega) d\mu(\omega) < \infty$.

Then, for any sequence $f(\omega, b_n)$ with $b_n \to \beta$, we get $f(\omega, b_n) \to f(\omega, \beta)$ due to (i), and $|f(\omega, b_n)| \le g(\omega)$ holds due to (ii)—$b_n \in N_\beta$ eventually as N_β is not a function of n. Due to $\int g(\omega) d\mu(\omega) < \infty$, we get $\int f(\omega, b_n) d\mu(\omega) \to \int f(\omega, \beta) d\mu(\omega)$ as $n \to \infty$, establishing the *continuity of* $\int f(\omega, b) d\mu$ at $b = \beta$ *when* $f(\omega, b)$ *is continuous at* $b = \beta$.

Going further, suppose

> (iii) $f(\omega, b)$ has a derivative $f'(\omega, b)$ wrt b on a neighborhood N_β $\forall \omega$ and

> (iv) $\sup_{b \in N_\beta} |f'(\omega, b)| \leq G(\omega)$ and $\int G(\omega) d\mu(\omega) < \infty$.

Then, for any sequence $h_n \to 0$,

$$r_n(\omega, \beta) \equiv \frac{f(\omega, \beta + h_n) - f(\omega, \beta)}{h_n} = f'(\omega, b_n^*)$$

for some b_n^* on the line joining β and $\beta + h_n$

$$\Longrightarrow \left| \frac{f(\omega, \beta + h_n) - f(\omega, \beta)}{h_n} \right| = |f'(\omega, b_n^*)| \leq G(\omega).$$

Since $r_n(\omega, \beta) \to f'(\omega, \beta)$ due to (iii) and $|r_n(\omega, \beta)| \leq G(\omega)$ due to (iv), we get

$$\frac{\int f(\omega, \beta + h_n) d\mu(\omega) - \int f(\omega, \beta) d\mu(\omega)}{h_n} = \int r_n(\omega, \beta) d\mu(\omega) \to$$

$$\int f'(\omega, \beta) d\mu(\omega)$$

which justifies *interchanging differentiation (at β) and integration*, drawing on Pollard (2002, p.33).

1.7 Convergence of Random Variables and Laws

Consider $(\Omega, \mathcal{A}, P)$ and rv's on Ω. A sequence of rv's $\{x_n\}$ *converge in probability* to x if

$$\lim_{n \to \infty} P(|x_n - x| < \varepsilon) = 1 \text{ for any constant } \varepsilon > 0$$

which is denoted as $x_n \to^P x$ or $x_n - x = o_p(1)$; x_n is said to be *consistent* for x. Notation $x_n = o_p(a_n)$ for a sequence a_n ($\neq 0$ $\forall n$) is used for $x_n/a_n \to^P 0$.

A sequence of rv's x_n *converge almost surely (a.s.), almost everywhere (a.e.) or with probability 1* if there is a set $A \in \mathcal{A}$ with $P(A) = 1$ such that

$$x_n(\omega) \to x(\omega) \text{ for each } \omega \in A$$

which is denoted as $x_n \to^{as} x$ or $x_n \to^{ae} x$. Almost sure convergence implies convergence in probability; the former states that the probability of x_n differing from x is 0 when n is large enough, whereas the latter states that the differing probability tends to 0. This may be understood better by writing $x_n \to^{as} x$ as (Serfling, 1980)

$$\lim_{n \to \infty} P(|x_{n'} - x| < \varepsilon \, \forall n' \geq n) = 1 \quad \text{for any constant } \varepsilon > 0;$$

compare this to the above display for convergence in probability.

Convergence in the rth mean is defined as

$$E|x_n - x|^r \to 0 \text{ as } n \to \infty.$$

This is not the same as $E|x_n|^r \to E|x|^r$, nor as $Ex_n^r \to Ex^r$, which are the "convergence of the rth absolute moment" and the "convergence of the rth moment," respectively.

Using Markov inequality, convergence in the rth mean implies convergence in probability. Convergence in the rth mean is neither stronger nor weaker than a.s. convergence, but if $\sum_n E|x_n - x|^r \to 0$, then $x_n \to^{as} x$. For a sort of the reverse implication, we need the following: a sequence $\{x_n\}$ is *uniformly integrable* if

$$\sup_n E|x_n| < \infty \quad \text{and} \quad \sup_n E(|x_n| \cdot 1[|x_n| > \lambda]) \to 0 \text{ as } \lambda \to \infty.$$

If $x_n \to^p x$ and $\{x_n\}$ *is uniformly integrable, then* $E|x_n - x| \to 0$. A sufficient condition for the uniform integrability of $\{x_n\}$ is dominance: $|x_n| \le y$ with $E(y) < \infty$. Another sufficient condition is $E|x_n|^{1+\delta} < \infty$ for some $\delta > 0$. The rv sequence $x_n(\omega) = n1_{[0,1/n]}(\omega)$ has $E|x_n| = 1$ $\forall n$ for Lebesque measure, but is not uniformly integrable because $\lim_{\lambda \to \infty} \sup_n E(|x_n| \cdot 1[|x_n| > \lambda]) = 1$.

If

$$\sup_n P(|x_n| > \lambda) \to 0 \text{ as } \lambda \to \infty,$$

then $\{x_n\}$ is *uniformly tight* (or *bounded in probability*), denoted $x_n = O_p(1)$. Notation $x_n = O_p(a_n)$ for a sequence a_n ($\ne 0$ $\forall n$) is used for

$$\sup_n P(|\frac{x_n}{a_n}| > \lambda) \to 0 \text{ as } \lambda \to \infty.$$

A rv sequence $\{x_n\}$ *converge in distribution* to x if

$$\lim_{n \to \infty} P(x_n \le t) = P(x \le t) \text{ for all } t \text{ such that } P(x = t) = 0.$$

Since the probabilities on both sides are obtained separately, x does not have to be on the same probability space as x_n is on. The convergence in distribution is equivalent to *convergence in law*:

$$Ef(x_n) \to Ef(x) \text{ for any bounded continuous function } f \text{ as } n \to \infty.$$

The convergence in law of $\{x_n\}$ to x does not imply $Ex_n \to Ex$, because the identity mapping $f(x) = x$ is not bounded on R although continuous.

A *relation* T on a set A is *partially ordered* if T is transitive ($a_1 T a_2$ and $a_2 T a_3$ implies $a_1 T a_3$ $\forall a_1, a_2, a_3 \in A$) and antisymmetric ($a_1 T a_2$ and $a_2 T a_1$ implies $a_1 = a_2$ $\forall a_1, a_2 \in A$). One example of T is "$\le$" on R. For a "random element" not necessarily on R, convergence in law is more general than convergence in distribution, because the latter is not useful when "$x \le t$" does not make sense, i.e., when the range of x is not *totally (or linearly)*

ordered, which means partially ordered and "total" ($a_1 T a_2$ or $a_2 T a_1$ $\forall a_1, a_2 \in$
A). Convergence in law (or in distribution) is denoted $x_n \to^L x$, $x_n \to^d x$, or
$x_n \rightsquigarrow x$. If $x_n \rightsquigarrow x$, then $\{x_n\}$ is $O_p(1)$. For every $O_p(1)$ sequence, there is a
subsequence converging in law to some x (*Prohorov theorem*).

With the original probability space being $(\Omega, \mathcal{A}, P)$, denote the probability on R induced by x_n as P_n (i.e., $P_n(B) = P(x_n^{-1}(B))$ for any Borel set $B \in$
$\mathcal{B}$ on R) and the probability induced by x as P_o (i.e., $P_o(B) = P(x^{-1}(B)))$,
where $\mathcal{B}$ is the Borel σ-algebra. That is, we consider $(R, \mathcal{B}, P_n)$ for each n.
The convergence in law is equivalently expressed as: P_n is said to *converge
weakly* to P_o ($P_n \rightsquigarrow P_o$) if for any bounded continuous function f on R,

$$\{Ef(x_n) =\} \int f dP_n \to \int f dP_o \{= Ef(x)\} \quad \text{as } n \to \infty.$$

The notation $P_n \rightsquigarrow P_o$ may be preferred to $x_n \rightsquigarrow x$, because there is a lesser
danger of thinking of x_n and x being "related" on the same space. Note
that *a relation between two rv's on the same space depends on their joint
distribution, whereas $x_n \rightsquigarrow x$ depends only on their marginal distributions
P_n and P_o.* In $P_n \rightsquigarrow P_o$, the underlying x_n and x lose their presence, because
P_n could have been induced by another rv sequence y_n living on another
probability space $(\Omega', \mathcal{A}', P')$.

Let f be a function from R^p to R^q, continuous on X_c where $P(x \in X_c) =$
1 for some rv x inducing P_o on $(R, \mathcal{B})$. If $x_n \to^{as} x$, $x_n \to^P x$, or $x_n \rightsquigarrow x$,
then $f(x_n) \to^{as} f(x)$, $f(x_n) \to^P f(x)$, or $f(x_n) \rightsquigarrow f(x)$, respectively. This
is known as the *continuous mapping theorem*, which is implied by a stronger
version that f is continuous on all of R^p.

If a real-valued function f is continuous on C and $x_n \rightsquigarrow x$ where x takes
its values in C, then $Ef(x_n) \to Ef(x)$ iff $\{f(x_n)\}$ is uniformly integrable
(see, e.g., Van der Vaart, 1998, p.17). This implies that, when $f(x) = x^k$,
convergence in law along with the uniform integrability of the kth moment
implies the convergence in the kth moment.

If $x_n \to^P x$, then $x_n \rightsquigarrow x$; the converse holds if x is a constant. If $x_n \to^P x$,
and $y_n \rightsquigarrow c$ that is a constant, then $(x_n, y_n) \rightsquigarrow (x, c)$. This, combined with
the continuous mapping theorem, implies the following *Slutsky Lemma*:

$$x_n + y_n \rightsquigarrow x + c, \quad x_n y_n \rightsquigarrow xc \text{ and } x_n/y_n \rightsquigarrow x/c \text{ so long as } c \neq 0.$$

1.8   LLN and CLT

Let $\{x_i\}$ be a sequence of iid rv. Then a necessary and sufficient condition
that $\bar{x}_N \equiv N^{-1}\sum_i x_i$ converges to $E(x)$ a.e. is that $E(x)$ exists. If $\{x_i\}$ is
inid (independent but non-identically distributed), then

$$\sum_{i=1}^{\infty} \frac{E|x_i - E(x_i)|^{p_i}}{i^{p_i}} < \infty \quad \text{for } 1 \leq p_i \leq 2$$

is sufficient for $N^{-1}\sum_i(x_i-Ex_i)$ to converge to 0 a.e. "$N^{-1}\sum_i(x_i-Ex_i) \to^p$ 0 (or $\to^{as}$ 0)" is called a *law of large numbers (LLN)*. The first LLN for iid sequences is the *Kolmogorov LLN*, and the second for inid sequences is from Chow and Teicher (1988, p.124). An a.e. LLN is called a "strong LLN," whereas a LLN in probability is called a "weak LLN."

An inid sequence can occur when $y_i|x_i$ is heteroskedastic and *exogenous sampling* is done (i.e., x_i is fixed first and then y_i gets drawn from the distribution of $y|x_i$); because $V(y|x_i)$ varies across i, an inid sequence is obtained. If (x_i', y_i) is drawn together from the joint distribution of (x', y), which may be called a random sampling, then heteroskedasticity does not imply inid. To better understand the above display for the inid LLN, suppose $p_i = 2$ $\forall i$ to have $\sum_{i=1}^{\infty} V(x_i)/i^2 < \infty$. Now imagine $V(x_i) = i$, growing with i. Then the condition becomes $\sum_{i=1}^{\infty} i^{-1}$ which is divergent. However, since $\sum_{i=1}^{\infty} i^{-(1+\varepsilon)} < \infty$ for any $\varepsilon > 0$, $\sum_{i=1}^{\infty} V(x_i)/i^2 < \infty$ so long as $V(x_i)$ increases at rate $i^{1-\varepsilon}$. Hence we can allow different (and growing) $V(x_i)$ across i. For iid cases, the condition is trivially satisfied.

Let $\{x_{N,n_N}\}$ be a "*triangular array*" of zero mean rv's. The word "triangular" may be understood from the following display "shape":

$$N = 2 \text{ and } n_2 = 1 : x_{2,1};$$
$$N = 3 \text{ and } n_3 = 2 : x_{3,1}, \ x_{3,2};$$
$$N = 4 \text{ and } n_4 = 4 : x_{4,1}, \ x_{4,2}, \ x_{4,3}, \ x_{4,4};$$

note that we could have $n_3 = 3$ and $x_{3,1}, x_{3,2}, x_{3,3}$. The elements in each row should be independent of one another, but elements across different rows may be dependent.

Define $s_N^2 \equiv \sum_{i=1}^{n_N} \sigma_{Ni}^2$ where $\sigma_{Ni}^2 \equiv V(x_{Ni}) = E(x_{Ni}^2) < \infty$ under $E(x_{Ni}) = 0$. If

$$\frac{1}{s_N^2} \sum_{i=1}^{n_N} E\left(x_{Ni}^2 1\left[|x_{Ni}| > \varepsilon s_N\right]\right) = \sum_{i=1}^{n_N} E\left(\frac{x_{Ni}^2}{s_N^2} 1\left[\frac{x_{Ni}^2}{s_N^2} > \varepsilon^2\right]\right) \to 0$$
$$\forall \text{ constant } \varepsilon > 0 \quad \text{as } N \longrightarrow \infty$$

then

$$\frac{1}{s_N} \sum_{i=1}^{n_N} x_{Ni} \rightsquigarrow N(0,1).$$

This is the *Lindeberg central limit theorem (CLT)*, and the condition is the "Lindeberg condition." If $E(x_{Ni}) \neq 0$, redefine x_{Ni} as $x_{Ni} - E(x_{Ni})$ for the CLT.

For a sum $\sum_{i=1}^{N} x_i$ of an iid zero-mean rv sequence $x_1, ..., x_N$, we get $s_N^2 = N\sigma^2$ where $\sigma^2 \equiv V(x)$. The Lindeberg condition easily holds, because

$$\frac{1}{N\sigma^2} \sum_{i=1}^{N} E\left(x_i^2 1\left[x_i > \varepsilon\sigma\sqrt{N}\right]\right) = \frac{E\left(x^2 1\left[|x| > \varepsilon\sigma\sqrt{N}\right]\right)}{E(x^2)} \to 0 \quad \text{as } N \to \infty$$

due to the dominated convergence theorem as x is square-integrable. The Lindeberg CLT is then

$$\frac{1}{\sqrt{N}} \sum_{i=1}^{N} \frac{x_i}{\sigma} \rightsquigarrow N(0,1):$$

the standardized sum of N-many rv's divided by $\sqrt{N}$ converges to $N(0,1)$ in law.

For an iid random vector sequence $z_1, ..., z_N$ with finite second moments, the CLT is stated in the following simple form:

$$\frac{1}{\sqrt{N}} \sum_{i=1}^{N} \{z_i - E(z)\} \rightsquigarrow N\left(0, E\left[\{z - E(z)\}\{z - E(z)\}'\right]\right) \quad \text{as } N \to \infty.$$

2 Appendix for Chapter 2

2.1 Seemingly Unrelated Regression (SUR)

2.1.1 Two-Equation SUR

In the main text, we allowed for heteroskedasticity of unknown form with occasional drift to homoskedasticity. In this subsection, we assume homoskedasticity $E(uu'|x) = E(uu')$ and review *seemingly unrelated regression (SUR)*, which looks somewhat "old" from the modern GMM perspective. Nevertheless, under homoskedasticity, the roles of regressors and error terms get separated, and the analysis can be done more neatly to provide helpful insights. Because endogenous regressors are not considered in this subsection, notations here are somewhat different from those in the main text. A survey on SUR can be found in Fiebig (2001).

Consider two equations under homoskedasticity:

$$y_1 = x_1'\beta_1 + u_1, \quad y_2 = x_2'\beta_2 + u_2, \quad E(x_j u_{j'}) = 0 \ \forall j, j' = 1, 2$$
$$E(u_1 u_2 | x_1, x_2) = \sigma_{12}, \quad E(u_1^2 | x_1, x_2) = \sigma_1^2, \quad E(u_2^2 | x_1, x_2) = \sigma_2^2,$$

where x_1 and x_2 are respectively $k_1 \times 1$ and $k_2 \times 1$ vectors, and σ_1^2, σ_2^2, and σ_{12} are constants. We observe $(y_{1i}, y_{2i}, x_{1i}', x_{2i}')'$, $i = 1, \cdots, N$, which are iid. We can apply LSE to each equation separately. Combining the two equations, however, it is possible to obtain a more efficient estimator than the LSE.

Although the two equations look unrelated, they can be still related through $COR(u_1, u_2)$. Define

$$y \equiv (y_1, y_2)', \quad \beta \equiv (\beta_1', \beta_2')', \quad u \equiv (u_1, u_2)'$$

to rewrite the two equations as

$$y = x'\beta + u, \quad E\{(I_2 \otimes w) \cdot u\} = 0, \quad E(uu') = C, \quad \text{where}$$

$$x' \equiv \begin{bmatrix} x_1' & 0 \\ 0 & x_2' \end{bmatrix} \quad \beta \equiv \begin{bmatrix} \beta_1 \\ \beta_2 \end{bmatrix} \quad w \equiv \begin{bmatrix} x_1 \\ x_2 \end{bmatrix} \quad C \equiv \begin{bmatrix} \sigma_1^2 & \sigma_{12} \\ \sigma_{12} & \sigma_2^2 \end{bmatrix}.$$

In matrix notations, $y = x'\beta + u$ and $E(uu') = C$ are

$$Y = X\beta + U,$$

$$E(UU') = \begin{bmatrix} E(U_1U_1') & E(U_1U_2') \\ E(U_2U_1') & E(U_2U_2') \end{bmatrix} = \begin{bmatrix} \sigma_1^2 I_N & \sigma_{12} I_N \\ \sigma_{12} I_N & \sigma_2^2 I_N \end{bmatrix}_{2N \times 2N} = C \otimes I_N,$$

$$\text{where} \quad \underset{2N \times 1}{Y} = \begin{bmatrix} Y_1 \\ Y_2 \end{bmatrix} \quad \underset{2N \times (k_1+k_2)}{X} = \begin{bmatrix} X_1 & 0 \\ 0 & X_2 \end{bmatrix} \quad \underset{2N \times 1}{U} = \begin{bmatrix} U_1 \\ U_2 \end{bmatrix}.$$

Since the variance matrix of U is not diagonal, we can think of applying GLS to $Y = X\beta + U$, where the (m,n)th component of C can be estimated consistently with

$$\frac{1}{N} \sum_i (y_{mi} - x_{mi}'\hat{b}_m)(y_{ni} - x_{ni}'\hat{b}_n), \quad m, n = 1, 2$$

and $\hat{b}_m$ is the LSE for the mth equation. The GLS is the SUR estimator: denoting the estimator for C as C_N,

$$\begin{aligned} b_{sur} &= \{X'(C_N \otimes I_N)^{-1}X\}^{-1} \{X'(C_N \otimes I_N)^{-1}Y\} \\ &= \{X'(C_N^{-1} \otimes I_N)X\}^{-1} \{X'(C_N^{-1} \otimes I_N)Y\}. \end{aligned}$$

The consistency of b_{sur} is easy to show.

Although we dealt with only two equations, it is straightforward to extend the above derivation to more than two equations; all we have to do is stacking the equations. For instance, if we have three equations, we will need

$$Y = \begin{bmatrix} Y_1 \\ Y_2 \\ Y_3 \end{bmatrix} \quad X = \begin{bmatrix} X_1 & 0 & 0 \\ 0 & X_2 & 0 \\ 0 & 0 & X_3 \end{bmatrix} \quad U = \begin{bmatrix} U_1 \\ U_2 \\ U_3 \end{bmatrix}.$$

2.1.2 Asymptotic Distribution

As for the asymptotic distribution, suppose that x_i is a fixed constant, not a rv; also replace C_N with C as this replacement does not alter the asymptotic distribution. Substitute $Y = X\beta + U$ into the b_{sur} formula to get

$$\begin{aligned} b_{sur} &= \beta + \{X'(C^{-1} \otimes I_N)X\}^{-1} \{X'(C^{-1} \otimes I_N)U\} \\ &\implies E\{(b_{sur} - \beta)(b_{sur} - \beta)'\} \\ &= \{X'(C^{-1} \otimes I_N)X\}^{-1} \{X'(C^{-1} \otimes I_N)(C \otimes I_N)(C^{-1} \otimes I_N)X\} \\ &\quad \{X'(C^{-1} \otimes I_N)X\}^{-1} = \{X'(C^{-1} \otimes I_N)X\}^{-1}. \end{aligned}$$

This suggests that $\{X'(C^{-1} \otimes I_N)X\}^{-1}$ might be a consistent estimator for the asymptotic variance of $b_{sur} - \beta$ even when x is random. Indeed this is the case as shown next.

To get the asymptotic distribution when x is random, observe

$$\sqrt{2N}(b_{sur} - \beta) = \left\{ \frac{1}{2N} X'(C^{-1} \otimes I_N)X \right\}^{-1} \cdot \left\{ \frac{1}{\sqrt{2N}} X'(C^{-1} \otimes I_N)U \right\}.$$

Define

$$X^* \equiv \underset{2N \times 2N}{(C^{-1/2} \otimes I_N)} \underset{2N \times 2k}{X} \quad \text{where } C^{-1} = C^{-1/2} C^{-1/2}.$$

Also define $x_i^{*\prime}$ as the ith row of X^* and u_i^* as the ith row of $(C^{-1/2} \otimes I_N)U$, $i = 1, \cdots, 2N$, to get

$$\sqrt{2N}(b_{sur} - \beta) = \left(\frac{1}{2N} X^{*\prime} X^* \right)^{-1} \cdot \left\{ \frac{1}{\sqrt{2N}} X^{*\prime}(C^{-1/2} \otimes I_N)U \right\}$$

$$= \left(\frac{1}{2N} \sum_i x_i^* x_i^{*\prime} \right)^{-1} \cdot \frac{1}{\sqrt{2N}} \sum_i x_i^* u_i^*.$$

The error terms u_i^* in $(C^{-1/2} \otimes I_N)U$ are iid with variance 1, because

$$E[(C^{-1/2} \otimes I_N)UU'(C^{-1/2} \otimes I_N)]$$
$$= E[(C^{-1/2} \otimes I_N)(C \otimes I_N)(C^{-1/2} \otimes I_N)] = E(I_2 \otimes I_N) = I_{2N}.$$

Hence, we get
$$\sqrt{2N}(b_{sur} - \beta) \rightsquigarrow N\{0, E^{-1}(x^* x^{*\prime})\}.$$

The asymptotic variance can be consistently estimated with $\{X^{*\prime}X^*/(2N)\}^{-1}$. Since we use the asymptotic variance for b_{sur}, not for $(2N)^{1/2}(b_{sur} - \beta)$ in practice, the confusing normalizing factor $2N$ disappears in the often used practical statement

$$b_{sur} \sim N\{\beta, (X^{*\prime}X^*)^{-1}\} = N[\beta, \{X'(C^{-1} \otimes I_N)X\}^{-1}].$$

2.1.3 Efficiency Gain

As a version of GLS, SUR is more efficient than the separate LSE. However, there are two cases where SUR is equivalent to the separate LSE. One is $\sigma_{12} = 0$ and the other is $x_1 = x_2$. If $\sigma_{12} = 0$, then the two equations are truly—not just seemingly—unrelated, and there is nothing to be gained by pooling the equations. If $x_1 = x_2$, define $X_C \equiv X_1 = X_2$. Then $X = I_2 \otimes X_C$. Substitute this into b_{sur} to get

$$b_{sur} = \{(I_2 \otimes X_C')(C^{-1} \otimes I_N)(I_2 \otimes X_C)\}^{-1} \{(I_2 \otimes X_C')(C^{-1} \otimes I_N)Y\}$$
$$= (C^{-1} \otimes X_C'X_C)^{-1} \cdot (C^{-1} \otimes X_C')Y = \{C \otimes (X_C'X_C)^{-1}\}.$$

$$(C^{-1} \otimes X_C')Y = (I_2 \otimes (X_C'X_C)^{-1}X_C')Y = (X'X)^{-1}X'Y,$$
owing to $X = I_2 \otimes X_C$.

Since X is block-diagonal, this is the same as applying LSE to each equation separately. Thus SUR = LSE when $X_1 = X_2$.

To see better the efficiency gain of SUR over the separate LSE, define $\rho \equiv COR(u_1, u_2)$ and observe that the inverse of C is

$$C^{-1} = \frac{1}{|C|} \begin{bmatrix} \sigma_2^2 & -\sigma_{12} \\ -\sigma_{21} & \sigma_1^2 \end{bmatrix} \equiv \begin{bmatrix} \sigma^{11} & \sigma^{12} \\ \sigma^{21} & \sigma^{22} \end{bmatrix} \quad \text{where}$$

$$|C| = \sigma_1^2\sigma_2^2 - \sigma_{12}^2 = \sigma_1^2\sigma_2^2 \cdot (1 - \rho^2).$$

With $b_{sur} \equiv (b_1', b_2')'$, the asymptotic variance of b_1 is the upper left submatrix of $\{X'(C^{-1} \otimes I_N)X\}^{-1}$, i.e.

$$\left(\begin{bmatrix} X_1' & 0 \\ 0 & X_2' \end{bmatrix} \begin{bmatrix} \sigma^{11}I_N & \sigma^{12}I_N \\ \sigma^{21}I_N & \sigma^{22}I_N \end{bmatrix} \begin{bmatrix} X_1 & 0 \\ 0 & X_2 \end{bmatrix} \right)^{-1}$$

$$= \begin{bmatrix} \sigma^{11}X_1'X_1 & \sigma^{12}X_1'X_2 \\ \sigma^{21}X_2'X_1 & \sigma^{22}X_2'X_2 \end{bmatrix}^{-1}.$$

Using the formula for the inverse of a partitioned matrix, the upper left submatrix is

$$\left[\sigma^{11}X_1'X_1 - \left(\frac{\sigma^{12}\sigma^{21}}{\sigma^{22}} \right) X_1'X_2(X_2'X_2)^{-1}X_2'X_1 \right]^{-1}$$

$$= \left[\frac{\sigma_2^2}{|C|}X_1'X_1 - \frac{\sigma_{12}^2}{|C|\,\sigma_1^2}X_1'X_2\,(X_2'X_2)^{-1}\,X_2'X_1 \right]^{-1}$$

$$\text{for } \sigma^{11} = \frac{\sigma_2^2}{|C|}, \quad \frac{\sigma^{12}\sigma^{21}}{\sigma^{22}} = \frac{\sigma_{12}^2}{|C|\,\sigma_1^2} \right)$$

$$= \left[\frac{\sigma_2^2}{|C|}X_1'X_1 - \frac{\sigma_2^2}{|C|}\frac{\sigma_{12}^2}{\sigma_1^2\sigma_2^2}X_1'X_2\,(X_2'X_2)^{-1}\,X_2'X_1 \right]^{-1}$$

$$= \left(\frac{\sigma_2^2}{|C|} \right)^{-1} \left[X_1'X_1 - \left(\frac{\sigma_{12}^2}{\sigma_1^2\sigma_2^2} \right) X_1'X_2\,(X_2'X_2)^{-1}\,X_2'X_1 \right]^{-1}$$

$$= \sigma_1^2\,(1 - \rho^2) \cdot \left[X_1'X_1 - \rho^2 X_1'X_2\,(X_2'X_2)^{-1}\,X_2'X_1 \right]^{-1}$$

$$\text{for } |C| = \sigma_1^2\sigma_2^2 \cdot (1 - \rho^2).$$

This shows that the efficiency gain depends on two factors. One is $\rho \neq 0$, and the other is the relation between X_1 and X_2—further below on this. Observe

$$\sigma_1^2\,(1 - \rho^2) = \sigma_1^2 - \frac{\sigma_{12}^2}{\sigma_2^2} = \sigma_1^2 - 2\frac{\sigma_{12}^2}{\sigma_2^2} + \frac{\sigma_{12}^2}{\sigma_2^2} = V\left(u_1 - \frac{\sigma_{12}}{\sigma_2^2}u_2 \right)$$

which is the variance of the part of u_1 not explained by u_2.

To see the efficiency gain better, bring $1-\rho^2$ into $[\cdot]^{-1}$ and further rewrite the asymptotic variance as (Binkley and Nelson, 1988)

$$\sigma_1^2 \left\{ \frac{1}{1-\rho^2} X_1' X_1 - \frac{\rho^2}{1-\rho^2} X_1' X_2 (X_2' X_2)^{-1} X_2' X_1 \right\}^{-1}$$

$$= \sigma_1^2 \left\{ \frac{1}{1-\rho^2} X_1' X_1 - \frac{\rho^2}{1-\rho^2} X_1' X_1 + \frac{\rho^2}{1-\rho^2} X_1' X_1 \right.$$

$$\left. - \frac{\rho^2}{1-\rho^2} X_1' X_2 (X_2' X_2)^{-1} X_2' X_1 \right\}^{-1}$$

$$= \sigma_1^2 \left\{ X_1' X_1 + \frac{\rho^2}{1-\rho^2} X_1' Q_{X_2} X_1 \right\}^{-1} \quad \text{as } Q_{X_2} = I_N - X_2 (X_2' X_2)^{-1} X_2'.$$

$X_1' Q_{X_2} X_1$ is the variation of X_1 unexplained by X_2 ($X_1' Q_{X_2} X_1$ is p.s.d.) and $\rho^2/(1-\rho^2)$ is non-negative. Thus SUR is more efficient than the separate LSE. This clearly shows that the *source of the efficiency gain is $\rho \neq 0$ and $Q_{X_2} X_1 \neq 0$*. If $\rho = 0$ or $X_1 \subset X_2$, then there is no efficiency gain. This raises an interesting possibility: if the LSE for a single equation $y_i = x_i' \beta + u_i$ has a large asymptotic variance, then we can do better with SUR by looking for an equation with its error term highly correlated with u_1 and the regressors low correlated with x_1.

2.2   On System GMM Efficiency Gain

Recall the subsection in Chapter 2 for multiple equation GMM. Suppose $H = 2$ and the second equation has one endogenous regressor, for which only one instrument is available; i.e., the second equation is just identified. In this case, the system GMM asymptotic variance is

$$\left[\begin{bmatrix} E(x_1 x') & 0 \\ 0 & E(x_2 x') \end{bmatrix} \begin{bmatrix} E(xx' u_1^2) & E(xx' u_1 u_2) \\ E(xx' u_1 u_2) & E(xx' u_2^2) \end{bmatrix}^{-1} \begin{bmatrix} E(xx_1') & 0 \\ 0 & E(xx_2') \end{bmatrix} \right]^{-1}$$

$$= \begin{bmatrix} E(x_1 x') D_1 E(xx_1') & E(x_1 x') C E(xx_2') \\ E(x_2 x') C' E(xx_1') & E(x_2 x') D_2 E(xx_2') \end{bmatrix}^{-1}, \quad \text{where}$$

$$\begin{bmatrix} D_1 & C \\ C' & D_2 \end{bmatrix} \equiv \begin{bmatrix} E(xx' u_1^2) & E(xx' u_1 u_2) \\ E(xx' u_1 u_2) & E(xx' u_2^2) \end{bmatrix}^{-1}$$

$$\Longleftrightarrow \begin{bmatrix} D_1 & C \\ C' & D_2 \end{bmatrix}^{-1} \equiv \begin{bmatrix} E(xx' u_1^2) & E(xx' u_1 u_2) \\ E(xx' u_1 u_2) & E(xx' u_2^2) \end{bmatrix}.$$

Using the formula for the inverse of a partitioned matrix that

the upper-left submatrix of $\begin{bmatrix} A & B \\ C & D \end{bmatrix}^{-1}$ is $(A - BD^{-1}C)^{-1}$

when D^{-1} exists,

the upper left submatrix of the system GMM asymptotic variance for $\sqrt{N}(b_1 - \beta_1)$ is

$$\left[E\left(x_1 x'\right) D_1 E\left(xx'_1\right) - E\left(x_1 x'\right) C E\left(xx'_2\right) \left\{ E\left(x_2 x'\right) D_2 E\left(xx'_2\right) \right\}^{-1} \right.$$

$$\left. E\left(x_2 x'\right) C' E\left(xx'_1\right) \right]^{-1}$$

$$= \left[E(x_1 x') D_1 E(xx'_1) - E(x_1 x') C E(xx'_2) E^{-1}(xx'_2) D_2^{-1} E^{-1}(x_2 x') \right.$$

$$\left. E(x_2 x') C' E(xx'_1) \right]^{-1}$$

$$= \left\{ E(x_1 x') D_1 E(xx'_1) - E(x_1 x') C D_2^{-1} C' E(xx'_1) \right\}^{-1} = \left\{ E(x_1 x') \right.$$

$$\left. (D_1 - C D_2^{-1} C') E(xx'_1) \right\}^{-1}$$

$$= \left\{ E(x_1 x') E^{-1}(xx' u_1^2) E(xx'_1) \right\}^{-1} \text{ because the upper-left submatrix of}$$

$\begin{bmatrix} D_1 & C \\ C' & D_2 \end{bmatrix}^{-1}$ is $(D_1 - C D_2^{-1} C')^{-1}$ which is also $E(xx' u_1^2)$

as shown above.

Hence in a two-equation system, if one equation is just identified, then the other equation has no efficiency gain with the system GMM.

2.3   Classical Simultaneous Equation Estimators

In the main text, we examined simultaneous equation estimation from the modern MOM framework. Here we present some "classical" estimators for simultaneous equations which the reader may encounter from time to time.

2.3.1   Full-Information MLE (FIML)

Consider H-many simultaneous equations

$$\underset{H \times H}{\Gamma} \underset{H \times 1}{y_i} - \underset{H \times K}{B} \underset{K \times 1}{x_i} = \underset{H \times 1}{u_i}, \quad u_i \sim N(0, C_u), \quad u_i \amalg x_i$$

$$f_u(u) = \frac{1}{(2\pi)^{H/2} |C_u|^{1/2}} \exp\left(-\frac{u' C_u^{-1} u}{2} \right)$$

$$\implies y\text{-density given } x \text{ is } ||\Gamma|| f_u(\Gamma y - Bx)$$

where $|C_u|$ and $|\Gamma|$ denote the determinants and $||\Gamma||$ is the absolute value of $|\Gamma|$. Thus the joint density function L_N of $y_1, ..., y_N$ given $x_1, ..., x_N$ is

$$L_N(\Gamma, B, C_u) = \frac{||\Gamma||^N}{(2\pi)^{HN/2}|C_u|^{N/2}}$$

$$\cdot \prod_{i=1}^{N} \exp\left\{-\frac{(\Gamma y_i - Bx_i)'C_u^{-1}(\Gamma y_i - Bx_i)}{2}\right\}.$$

From this,

$$\ln L_N(\Gamma, B, C_u) = N\ln||\Gamma|| - \frac{HN}{2}\ln(2\pi) - \frac{N}{2}\ln|C_u|$$

$$- \frac{1}{2}\sum_i (\Gamma y_i - Bx_i)'C_u^{-1}(\Gamma y_i - Bx_i)$$

$$= N\ln||\Gamma|| - \frac{HN}{2}\ln(2\pi) - \frac{N}{2}\ln|C_u|$$

$$- \frac{1}{2}\sum_i tr\left\{C_u^{-1}(\Gamma y_i - Bx_i)(\Gamma y_i - Bx_i)'\right\}$$

$$= N\ln||\Gamma|| - \frac{HN}{2}\ln(2\pi) - \frac{N}{2}\ln|C_u|$$

$$- \frac{1}{2}tr\left\{C_u^{-1}\sum_i (\Gamma y_i - Bx_i)(\Gamma y_i - Bx_i)'\right\}.$$

The "full information MLE (FIML)" for the parameters Γ, B, C_u is obtained maximizing this log-likelihood.

Note the following matrix differentiation rules:

$$\frac{\partial \ln|A|}{\partial A} = (A^{-1})' \quad \text{and} \quad \frac{\partial tr(AB)}{\partial A} = B' \text{ where } A \text{ is p.d. and } B \text{ is square.}$$

For example, suppose

$$A = \begin{bmatrix} a_{11} & a_{12} \\ a_{21} & a_{22} \end{bmatrix} \quad \text{and} \quad B = \begin{bmatrix} b_{11} & b_{12} \\ b_{21} & b_{22} \end{bmatrix}$$

$$\Longrightarrow \ln|A| = \ln(a_{11}a_{22} - a_{12}a_{21}) \quad \text{and} \quad tr(AB) = a_{11}b_{11}$$
$$+ a_{12}b_{21} + a_{21}b_{12} + a_{22}b_{22}$$

$$\Longrightarrow \frac{\partial \ln|A|}{\partial A} = \frac{1}{(a_{11}a_{22} - a_{12}a_{21})}\begin{bmatrix} a_{22} & -a_{21} \\ -a_{12} & a_{11} \end{bmatrix} \quad \text{and}$$

$$\frac{\partial tr(AB)}{\partial A} = \begin{bmatrix} b_{11} & b_{21} \\ b_{12} & b_{22} \end{bmatrix}.$$

Instead of C_u, it is easier to maximize the log-likelihood function wrt $\Omega_u \equiv C_u^{-1}$ using $|C_u| = |\Omega_u|^{-1}$. Maximize

$$N \ln ||\Gamma|| - \frac{HN}{2} \ln(2\pi) + \frac{N}{2} \ln |\Omega_u|$$

$$- \frac{1}{2} tr \left\{ \Omega_u \sum_i (\Gamma y_i - B x_i)(\Gamma y_i - B x_i)' \right\}$$

wrt Ω_u to get the first-order condition:

$$0 = \frac{N}{2} C_u - \frac{1}{2} \sum_i (\Gamma y_i - B x_i)(\Gamma y_i - B x_i)'$$

$$\iff C_u = \frac{1}{N} \sum_i \left(\underset{H \times H}{\Gamma} \underset{H \times 1}{y_i} - \underset{H \times K}{B} \underset{K \times 1}{x_i} \right)(\Gamma y_i - B x_i)'$$

$$\left\{ = \frac{1}{N} \left(\underset{N \times H}{Y} \underset{H \times H}{\Gamma}' - \underset{N \times K}{X} \underset{K \times H}{B}' \right)' (Y\Gamma' - XB') \quad \text{in matrices} \right\}$$

where Y is the $N \times H$ endogenous variable matrix and X is the $N \times K$ exogenous variable matrix.

Substitute the C_u into the log-likelihood function to get the "*concentrated log-likelihood function*" for Γ and B:

$$N \ln ||\Gamma|| - \frac{HN}{2} \ln(2\pi) - \frac{N}{2} \ln | \frac{1}{N} \sum_i (\Gamma y_i - B x_i)(\Gamma y_i - B x_i)' | - \frac{H}{2}$$

$$= N \ln ||\Gamma|| - \frac{HN}{2} \ln(2\pi)$$

$$- \frac{N}{2} \ln |\Gamma \left\{ \frac{1}{N} \sum_i (y_i - \Gamma^{-1} B x_i)(y_i - \Gamma^{-1} B x_i)' \right\} \Gamma'| - \frac{H}{2}$$

$$\implies - \frac{N}{2} \ln | \frac{1}{N} \sum_i (y_i - \Gamma^{-1} B x_i)(y_i - \Gamma^{-1} B x_i)' |,$$

$$\text{dropping} \; - \frac{HN}{2} \ln(2\pi) - \frac{H}{2}$$

and using $|ABA'| = |A| \cdot |B| \cdot |A'| = |B| \cdot |A|^2$ when A and B are square (here $A = \Gamma$). Although this resembles the simple LSE minimand, $N^{-1} \sum_i (y_i - \Gamma^{-1} B x_i)(y_i - \Gamma^{-1} B x_i)'$ is a $H \times H$ matrix, and one has to minimize its determinant wrt Γ and B. The consistency of FIML and its asymptotic distribution follow from those of MLE.

Although FIML may be difficult to implement, as far as estimating the RF parameters $\Pi \equiv \Gamma^{-1} B$ goes, LSE to each equation is the same as the MLE for Π. To see this, define the columns of Y as $Y_1, ... Y_H$ and denote the RF error matrix as V to get

$$\underset{N \times H}{Y} = (\underset{N \times 1}{Y_1}, ..., \underset{N \times 1}{Y_H}) = \underset{N \times K}{X} \cdot \underset{K \times H}{\Pi} + \underset{N \times H}{V}.$$

Consider the LSE $\hat{\Pi}_h$ of Y_h on X $\forall h$; $\hat{\Pi}_h$ is $K \times 1$ as all system regressors are included in each RF. The residual vector is $\hat{V}_h \equiv Y_h - X\hat{\Pi}_h$ with the first-order condition $\hat{V}_h' X = 0_{1\times K}$. Defining

$$\underset{K\times H}{\hat{\Pi}} \equiv (\hat{\Pi}_1, ..., \hat{\Pi}_H) \quad \text{and} \quad \underset{N\times H}{\hat{V}} \equiv (\hat{V}_1, ..., \hat{V}_H)$$

we get $Y = X\hat{\Pi} + \hat{V}$; $\hat{V}$ is consistent for $V = \Gamma^{-1}U$. The above MLE minimand can be written as

$$|(Y - X\Pi)'(Y - X\Pi)|$$
$$= |(Y - X\hat{\Pi} + X\hat{\Pi} - X\Pi)'(Y - X\hat{\Pi} + X\hat{\Pi} - X\Pi)|$$
$$= |\{\hat{V} + (X\hat{\Pi} - X\Pi)\}'\{\hat{V} + (X\hat{\Pi} - X\Pi)\}|$$
$$= |(\hat{V}'\hat{V} + (X\hat{\Pi} - X\Pi)'(X\hat{\Pi} - X\Pi)|, \quad \text{as} \quad \underset{H\times N}{\hat{V}}' \cdot \underset{N\times K}{X} = 0_{H\times K}$$
$$= |(\hat{V}'\hat{V} + (\hat{\Pi} - \Pi)'X'X(\hat{\Pi} - \Pi)| \geq |\hat{V}'\hat{V}| \quad \text{as} \; \hat{V}'\hat{V} \text{ and } X'X \text{ are p.d.}$$

Therefore, the LSE to each equation is the MLE for Π.

2.3.2 Limited-Information MLE (LIML)

Suppose now that we aim to estimate only one SF, say the first SF, that has a 2×1 endogenous regressor vector w_i:

$$\underset{1\times1}{y_i} = \underset{1\times2}{\alpha'} \underset{2\times1}{w_i} + \underset{1\times k_1}{\beta'} \underset{k_1\times1}{x_{1i}} + \underset{1\times1}{u_i} \quad \text{and} \quad \underset{2\times1}{w_i} = \underset{2\times k_1}{\eta_1'} \underset{k_1\times1}{x_{1i}} + \underset{2\times k_2}{\eta_2'} \underset{k_2\times1}{x_{2i}} + \underset{2\times1}{\varepsilon_i}$$

where x_1 is a vector of exogenous regressors in SF1, x_2 is not the exogenous regressors in SF2 but the other exogenous regressors in the system so that $x_i = (x_{1i}', x_{2i}')'$ is a $K \times 1$ vector, the equations for w are RF's, and the error term vector $(u, \varepsilon')'$ follows $N(0, \Omega)$ independently of x. In this case, the MLE is called the *"limited information MLE (LIML)"*—"limited" in the sense that only the first SF for y_i appears. Note that the number "2" for the endogenous variables is just to simplify the exposition.

Rewrite the three equations (one for y_i and two for w_i) as

$$\begin{bmatrix} 1 & -\alpha' \\ 0_{2\times1} & I_2 \end{bmatrix} \begin{bmatrix} y_i \\ w_i \end{bmatrix} - \begin{bmatrix} \beta' & 0_{1\times k_2} \\ \eta_1' & \eta_2' \end{bmatrix} \begin{bmatrix} x_{1i} \\ x_{2i} \end{bmatrix} = \begin{bmatrix} u_i \\ \varepsilon_i \end{bmatrix}.$$

This may be seen as a special case of $\Gamma y_i - B x_i = u_i$ where Γ and B are block-diagonal. Since the w-equations are RF's, there is no restriction on η_1 and η_2. In contrast, the SF has a critical restriction that x_2 is excluded.

Let the matrix versions for y, w, x_1, x_2, x be Y, W, X_1, X_2, X, respectively. The exclusion restriction means that removing only the X_1-component from $Y - W\alpha = X_1\beta + U$ is the same as removing the X-component. Hence, one can think of estimating α with the exclusion restriction by minimizing, with $Q_X \equiv I_N - X(X'X)^{-1}X'$,

$$\kappa(\alpha) \equiv \frac{(Y - W\alpha)'Q_{X_1}(Y - W\alpha)}{(Y - W\alpha)'Q_X(Y - W\alpha)};$$

note $\kappa(\alpha) \geq 1$, because the variation of $Y - W\alpha$ devoid of the X_1-component is as large as that devoid of the X-component. Maximizing the log-likelihood function, the LIML α-estimator is the same as the one minimizing $\kappa(\alpha)$ as can be seen in, e.g., Davidson and MacKinnon (1993) (the original references for LIML are Anderson and Rubin, 1949 and 1950). Defining

$$\underset{N \times (2+k_1)}{M} \equiv (W, X_1) \quad \text{and} \quad \delta = (\alpha', \beta')'$$

the LIML $\hat{\delta}$ for y_1-SF SF1 turns out to be

$$\hat{\delta} = \{M'(I_N - \hat{\kappa}Q_X)M\}^{-1} \cdot M'(I_N - \hat{\kappa}Q_X)Y.$$

The LIML is a special case of the "*k-class estimators*" where "k" refers to $\hat{\kappa}$. If $\hat{\kappa} = 0$, then LSE is obtained. If $\hat{\kappa} = 1$, the k-class estimator becomes

$$(M'P_X M)^{-1} \cdot M'P_X Y \quad \text{where} \quad P_X \equiv X(X'X)^{-1}X'$$

which is the IVE for SF1 with X as the instruments for M. This IVE is also the two-stage LSE (2SLSE). It is known that $\sqrt{N}(\hat{\kappa} - 1) = o_p(1)$, and thus the asymptotic distribution of the LIML is the same as that of the IVE. The difference is, however, that the IVE does not require homoskedasticity nor the independence of the error terms from x whereas the LIML does.

2.3.3   Three-Stage LSE (3SLSE)

There is a simultaneous system estimator that falls in between FIML and 2SLSE in terms of its strength of the requisite assumptions. The estimator requires homoskedasticity and it improves on 2SLSE in terms of efficiency by incorporating the information in the other SF's. The estimator is called the *three stage LSE (3SLSE)*, which is a full information variety, for it uses all SF's. The remainder of this subsection reviews 3SLSE.

Rewrite SF h as

$$y_h = m_h' \delta_h + u_h, \quad h = 1, ..., H$$

where m_h is a $k_h \times 1$ regressor vector, and u_h is homoskedastic; m_h includes endogenous as well as exogenous regressors. Denote the corresponding matrix for y_h, m_h, and u_h as Y_h, M_h, and U_h respectively. Stack the equations to get

$$Y = diag(M_1, \cdots, M_H) \cdot \delta + U \equiv M \cdot \delta + \underset{HN \times 1}{U},$$

where $Y \equiv (Y_1', \cdots, Y_H')'$, $\delta \equiv (\delta_1', \cdots, \delta_H')'$, $U \equiv (U_1', \cdots, U_H')'$ and $M = diag(M_1, \cdots, M_H)$; note that the definition of Y (of dimension $NH \times 1$) here differs from the Y (of dimension $N \times H$) that appeared for FIML and LIML.

Denoting the system exogenous variable vector as x (of dimension $K \times 1$) and its matrix version as X, the instrument matrix for each SF is X, and thus for the system, it is $I_H \otimes X$ of dimension $HN \times K$. Also, the system error term U does not have a diagonal covariance matrix; rather,

$$E(UU') \equiv \Omega = C \otimes I_N, \quad \text{where } \underset{H \times H}{C} = \begin{bmatrix} E(u_1 u_1) & \cdots & E(u_1 u_H) \\ \vdots & & \vdots \\ E(u_H u_1) & \cdots & E(u_H u_H) \end{bmatrix}.$$

Hence, combining the ideas of GLS and IVE, we can think of multiplying $Y = M\delta + U$ with $\Omega^{-1/2} = C^{-1/2} \otimes I_N$, and then applying IVE. This is essentially the 3SLSE. As in GLS, replacing C with a consistent estimator, say $\hat{C}$, does not affect the asymptotic distribution; the $(h,j)th$ component of $\hat{C}$ can be consistently estimated with

$$\frac{1}{N} \sum_i (y_{hi} - m_{hi} d_{h,2SLSE})(y_{ji} - m_{ji} d_{j,2SLSE}), \quad h, j = 1, ..., H$$

where $d_{h,2SLSE}$ is the 2SLSE for SF h.

Firstly, multiply $Y = M\delta + U$ with $C^{-1/2} \otimes I_N$ to get (we use C instead of $\hat{C}$ to ease notations as this makes no difference for the asymptotic distribution)

$$\left(C^{-1/2} \otimes I_N\right) \cdot Y = \left(C^{-1/2} \otimes I_N\right) M \cdot \delta + \left(C^{-1/2} \otimes I_N\right) \cdot U.$$

Secondly, in applying IVE to this equation that has the standardized error terms, recall the IVE with more than enough instruments: we need to (linearly) project $(C^{-1/2} \otimes I_N)M$ on $I_H \otimes X$ to get the effective instrument matrix:

$$(I_H \otimes X) \cdot \left\{(I_H \otimes X)' (I_H \otimes X)\right\}^{-1} \cdot (I_H \otimes X)' \cdot \left(C^{-1/2} \otimes I_N\right) M$$

$$= (I_H \otimes X) \cdot \left\{I_H \otimes (X'X)^{-1}\right\} \cdot (I_H \otimes X)' \cdot \left(C^{-1/2} \otimes I_N\right) M$$

$$= \left\{C^{-1/2} \otimes X (X'X)^{-1} X'\right\} \cdot M.$$

Thirdly, the IVE with this instrument matrix is the 3SLSE:

$$d_{3SLSE} = \left(\left[\left\{C^{-1/2} \otimes X(X'X)^{-1}X'\right\} M\right]' \cdot \left(C^{-1/2} \otimes I_N\right) M\right)^{-1}$$

$$\cdot \left[\left\{C^{-1/2} \otimes X(X'X)^{-1}X'\right\} M\right]' \cdot \left(C^{-1/2} \otimes I_N\right) Y$$

$$= \left[M' \left\{C^{-1} \otimes X(X'X)^{-1}X'\right\} M\right]^{-1}$$

$$\cdot M' \left\{C^{-1} \otimes X (X'X)^{-1} X'\right\} Y.$$

Denoting the element of C^{-1} as σ^{ij}, $M' \left\{C^{-1} \otimes X(X'X)^{-1}X'\right\}$ equals

$$diag(M'_1, ..., M'_H) \cdot \begin{bmatrix} \sigma^{11} X(X'X)^{-1}X' & \cdots & \sigma^{1H} X(X'X)^{-1}X' \\ \vdots & & \vdots \\ \sigma^{H1} X(X'X)^{-1}X' & \cdots & \sigma^{HH} X(X'X)^{-1}X' \end{bmatrix}$$

$$= \begin{bmatrix} \sigma^{11}M_1'X(X'X)^{-1}X' & \cdots & \sigma^{1H}M_H'X(X'X)^{-1}X' \\ \vdots & & \vdots \\ \sigma^{H1}M_1'X(X'X)^{-1}X' & \cdots & \sigma^{HH}M_H'X(X'X)^{-1}X' \end{bmatrix}$$

$$= diag\left\{X(X'X)^{-1}X'M_1, ..., X(X'X)^{-1}X'M_H\right\}' \cdot (C^{-1} \otimes I_N).$$

Each block in the diagonal matrix is the LSE fitted value of M_h on X. Hence defining

$$\hat{M}_h \equiv X(X'X)^{-1}X'M_h \quad \text{and} \quad \hat{M} \equiv diag(\hat{M}_1, ..., \hat{M}_H),$$

3SLSE can be written as its usual textbook form

$$d_{3slse} = \left[\hat{M}'\left\{C^{-1} \otimes I_N\right\}M\right]^{-1} \cdot \hat{M}'\left\{C^{-1} \otimes I_N\right\}Y$$

which is reminiscent of $b_{sur} = \left\{X'(C_N^{-1} \otimes I_N)X\right\}^{-1}\left\{X'(C_N^{-1} \otimes I_N)Y\right\}$.

As for the asymptotic distribution, substitute $Y = M\delta + U$ into the δ_{3slse} formula to get

$$\sqrt{NH}(d_{3slse} - \delta) = \left\{M'\left\{C^{-1} \otimes X(X'X)^{-1}X'\right\} \cdot M\right\}^{-1} \cdot M'$$
$$\left\{C^{-1} \otimes X(X'X)^{-1}X'\right\}U.$$

Applying CLT and using $E(UU') = C \otimes I_N$, we can obtain the asymptotic variance, the steps for which are similar to those for SUR. The asymptotic distributional result to be used in practice is

$$d_{3slse} \sim N\left[\delta, \left\{M'(C^{-1} \otimes X(X'X)^{-1}X')M\right\}^{-1}\right]$$

which is reminiscent of $b_{sur} \sim N\left[\beta, \left\{X'(C^{-1} \otimes I_N)X\right\}^{-1}\right]$.

3   Appendix for Chapter 3

3.1   Details on Four Issues for M-Estimator

First, regarding identification, the M-estimator b_N is for the parameter β defined by

$$\beta \equiv argmax_{b \in B}Q(b) = argmax_{b \in B}Eq(z, b).$$

If such β is unique, then β is said to be *"identified"*; the uniqueness is established by restricting q, z, or the parameter space B. There is no way to establish identification (ID) in a general term; it should be done case by case. For instance, in LSE with $Eq(z, b) = -E\left\{(y - x'b)^2\right\}$ and $E(xu) = 0$, assuming a regularity condition

$$\sup_{b \in B}|x(y - x'b)| \le g(z), \quad \text{where } Eg(z) < \infty,$$

the order of differentiation and integration can be interchanged to give

$$-\frac{\partial E\left\{(y - x'b)^2\right\}}{\partial b} = -E\left[\frac{\partial \left\{(y - x'b)^2\right\}}{\partial b}\right] = 2E\left\{x(y - x'b)\right\},$$

which is 0 at $b = \beta$.

The second derivative matrix is $-2E(xx')$. Since this is n.d., the maximizer β is unique. More examples will appear later.

Second, with ID holding, the following three conditions together imply $b_N \to^p \beta$: the compactness of B, the continuity of $Eq(b)$ in b, and the uniform LLN

$$\sup_{b \in B}\left|\frac{1}{N}\sum_i q(z_i, b) - Eq(z, b)\right| \to^p 0.$$

To see $b_N \to^p \beta$ under the three conditions, observe

$$\frac{1}{N}\sum_i q(b_N) \geq \frac{1}{N}\sum_i q(\beta) \quad \text{by the construction of } b_N$$

$$\Longrightarrow \frac{1}{N}\sum_i q(b_N) - Eq(b_N) \geq \frac{1}{N}\sum_i q(\beta) - Eq(b_N),$$

subtracting $Eq(b_N)$

$$\Longrightarrow \frac{1}{N}\sum_i q(b_N) - Eq(b_N) \geq \frac{1}{N}\sum_i q(\beta) - Eq(\beta) + Eq(\beta) - Eq(b_N).$$

The left-hand side is less than or equal to $\sup_{b \in B}|N^{-1}\sum_i q(b) - Eq(b)|$ which is $o_p(1)$ by the uniform LLN, and on the right-hand side, $N^{-1}\sum_i q(\beta) - Eq(\beta) = o_p(1)$ as $N \to \infty$. Hence,

$$o_p(1) \geq Eq(\beta) - Eq(b_N) \implies Eq(b_N) \geq Eq(\beta) + o_p(1) \text{ as } N \to \infty.$$

Combine this with $Eq(\beta) \geq Eq(b_N)$ (by the construction of β) to get

$$Eq(\beta) - Eq(b_N) = o_p(1) \quad \text{as } N \to \infty.$$

For any open neighborhood U_β of β, $Eq(b)$ attains a maximum on U_β^c at a point β^* due to the continuity of $Eq(b)$ and the compactness of U_β^c. But since ID implies $Eq(\beta^*) < Eq(\beta)$, b_N cannot stay out of U_β as $N \to \infty$ while satisfying $Eq(b_N) - Eq(\beta) = o_p(1)$ however small U_β may be. This implies $b_N = \beta + o_p(1)$.

An usual (pointwise) LLN under iid is not sufficient for $N^{-1}\sum_i q(b_N) - Eq(b_N) = o_p(1)$, because $q(z_i, b_N)$, $i = 1...N$, are dependent on one another through b_N. But when a pointwise LLN holds, the uniform LLN holds as well in most cases; see, e.g., Andrews (1987a) and Pötcher and Prucha (1989).

Among the aforementioned three conditions, the compactness of B is usually assumed, and the continuity of $Eq(b)$ holds in most cases. With these, an uniform a.e. LLN implies $b_N \to^{as} \beta$.

Third, the asymptotic distribution $\sqrt{N}(b_N - \beta)$ is almost always $N(0, C)$ for some p.d. symmetric matrix C. This is derived from the first-order condition for b_N:

$$\frac{1}{\sqrt{N}} \sum_i q_b(b_N) = 0, \quad \text{where } q_b(b_N) \equiv \frac{\partial q(b)}{\partial b} \bigg|_{b=b_N}.$$

Apply the mean-value theorem to b_N around β to get

$$0 = \frac{1}{\sqrt{N}} \sum_i q_b(\beta) + \frac{1}{N} \sum_i q_{bb'}(b_N^*) \sqrt{N}(b_N - \beta)$$

where $q_{bb'} = \partial^2 q / \partial b \partial b'$ and $b_N^* \in (b_N, \beta)$. The mean value theorem applies to each component of q_b separately, which means that each component of q_b may need a different b_N^* although this is not explicit in the display. Invert the second-order matrix to solve for $\sqrt{N}(b_N - \beta)$:

$$\sqrt{N}(b_N - \beta) = \{-\frac{1}{N} \sum_i q_{bb'}(b_N^*)\}^{-1} \cdot \frac{1}{\sqrt{N}} \sum_i q_b(\beta).$$

Applying the uniform LLN again, we get

$$\frac{1}{N} \sum_i q_{bb'}(b_N^*) - E q_{bb'}(b_N^*) = o_p(1).$$

Let $|A| = \{tr(A'A)\}^{1/2}$ for a matrix A. Assuming a regularity condition $|q_{bb'}(z, b)| \leq g(z)$ for all b in a neighborhood of β with $Eg(z) < \infty$, the dominated convergence theorem yields

$$E q_{bb'}(b_N^*) = E q_{bb'}(\beta).$$

Since $q_{bb'}\{z(\omega), b_N^*(\omega)\} - q_{bb'}\{z(\omega), \beta\} \to^{as} 0$ and since $|q_{bb'}|$ is dominated by the integrable g, the integral $\int q_{bb'}\{z(\omega), b_N^*(\omega)\} d\mu(\omega)$ converges to $\int q_{bb'}\{z(\omega), \beta\} d\mu(\omega)$.
Hence

$$\sqrt{N}(b_N - \beta) = - \left[E^{-1}\{q_{bb'}(\beta)\} + o_p(1) \right] \frac{1}{\sqrt{N}} \sum_i q_b(\beta)$$

$$= - E^{-1}\{q_{bb'}(\beta)\} \frac{1}{\sqrt{N}} \sum_i q_b(\beta) + o_p(1).$$

Due to the CLT, the first term on the rhs is asymptotically normal under the assumption $E\{q_b(\beta) q_{b'}(\beta)\} < \infty$, and

$$\sqrt{N}(b_N - \beta) \rightsquigarrow N\left[0, \, E^{-1}\{q_{bb'}(\beta)\} \cdot E\{q_b(\beta) q_{b'}(\beta)\} \cdot E^{-1}\{q_{bb'}(\beta)\}\right].$$

It is straightforward to derive the asymptotic distribution of b_{lse} as a special case with $q(y - x'b) = -(y - x'b)^2$.

Fourth, the variance matrix can be consistently estimated with

$$\left\{ \frac{1}{N} \sum_i q_{bb'}(b_N) \right\}^{-1} \cdot \frac{1}{N} \sum_i q_b(b_N) q_{b'}(b_N) \cdot \left\{ \frac{1}{N} \sum_i q_{bb'}(b_N) \right\}^{-1};$$

the proof goes analogously to the proof for $N^{-1} \sum_i q_{bb}(b_N^*) = E(q_{bb}(\beta)) + o_p(1)$. Indeed, in almost all cases we will encounter, we will have

$$\frac{1}{N} \sum_i m(z_i, b_N) = \frac{1}{N} \sum_i m(z_i, \beta) + o_p(1) = E\{m(z, \beta)\} + o_p(1).$$

3.2 MLE with LSE First-Stage and Control Function

Often in practice, we have MLE with nuisance parameters that can be estimated by an estimator other than MLE, say LSE. This leads to a two-stage M-estimator. Here we examine this case and give a specific example (MLE with endogenous regressors) for two reasons. One is to provide a detailed illustration of M-estimator with nuisance parameters, and the other is to show how endogenous regressors can be dealt with in MLE easily using "control function (CF)" approach.

Let α be a first stage LSE parameter of dimension $k_1 \times 1$ and a_N be the LSE. Let β be a likelihood function parameter of dimension $k_2 \times 1$ for the second stage, and b_N be the MLE. Denote the second stage score function as $s(z_i, \alpha, \beta)$; omit z_i for simplification. Define

$$\nabla_{\alpha'} s(\alpha, \beta) \equiv \frac{\partial s(\alpha, \beta)}{\partial a'} \quad \text{and} \quad \nabla_{\beta'} s(\alpha, \beta) \equiv \frac{\partial s(\alpha, \beta)}{\partial b'}.$$
$$\phantom{\nabla_{\alpha'} s(\alpha, \beta)}_{k_2 \times k_1} \phantom{\nabla_{\beta'} s(\alpha, \beta)}_{k_2 \times k_2}$$

By the construction of b_N, it holds that, using Taylor's expansion,

$$0 = \frac{1}{\sqrt{N}} \sum_i s(a_N, b_N) \implies$$

$$0 = \frac{1}{\sqrt{N}} \sum_i s(a_N, \beta) + \left\{ \frac{1}{N} \sum_i \nabla_{\beta'} s(\alpha, \beta) \right\} \sqrt{N}(b_N - \beta) + o_p(1)$$

$$\implies \sqrt{N}(b_N - \beta)$$

$$= -\left\{ \frac{1}{N} \sum_i \nabla_{\beta'} s(\alpha, \beta) \right\}^{-1} \frac{1}{\sqrt{N}} \sum_i s(a_N, \beta) + o_p(1)$$

$$\implies \sqrt{N}(b_N - \beta)$$

$$= \left\{ \frac{1}{N} \sum_i s(\alpha, \beta) s(\alpha, \beta)' \right\}^{-1} \frac{1}{\sqrt{N}} \sum_i s(a_N, \beta) + o_p(1).$$

To account for the first-stage estimation error $a_N - \alpha$, define

$$H_N \equiv \frac{1}{N} \sum_i s(\alpha, \beta) s(\alpha, \beta)' \quad \text{and}$$

$$L_N \equiv \frac{1}{N} \sum_i \nabla_{\alpha'} s(\alpha, \beta).$$

Apply Taylor's expansion to $s(a_N, \beta)$ in the above expression to get

$$\sqrt{N}(b_N - \beta) = H_N^{-1} \cdot \left\{ \frac{1}{\sqrt{N}} \sum_i s(\alpha, \beta) + L_N \sqrt{N}(a_N - \alpha) \right\} + o_p(1).$$

With r_i denoting the first-stage LSE residual with Z_i as the regressor vector, up to $o_p(1)$,

$$\sqrt{N}(a_N - \alpha) = \frac{1}{\sqrt{N}} \sum_i \left(\frac{1}{N} \sum_i Z_i Z_i' \right)^{-1} Z_i r_i \equiv \frac{1}{\sqrt{N}} \sum_i \eta_i,$$

$$\eta_i \equiv \left(\frac{1}{N} \sum_i Z_i Z_i' \right)^{-1} Z_i r_i.$$

Hence

$$\sqrt{N}(b_N - \beta) = H_N^{-1} \left\{ \frac{1}{\sqrt{N}} \sum_i s(\alpha, \beta) + L_N \frac{1}{\sqrt{N}} \sum_i \eta_i \right\} + o_p(1)$$

$$= H_N^{-1} \frac{1}{\sqrt{N}} \sum_i q_i + o_p(1) \quad (\text{where } q_i \equiv s(\alpha, \beta) + L_N \eta_i)$$

$$\rightsquigarrow N(0, H^{-1} E(qq') H^{-1}) \quad \text{where} \quad H \equiv E\{s(\alpha, \beta) s(\alpha, \beta)'\}.$$

Consistent estimators for H and $E(qq')$ are

$$\hat{H}_N \equiv \frac{1}{N} \sum_i s(a_N, b_N) s(a_N, b_N)',$$

$$Q_N \equiv \frac{1}{N} \sum_i \left\{ s(a_N, b_N) + \hat{L}_N \eta_i \right\} \left\{ s(a_N, b_N) + \hat{L}_N \eta_i \right\}' \quad \text{where}$$

$$\hat{L}_N \equiv \frac{1}{N} \sum_i \nabla_{\alpha'} s(a_N, b_N).$$

In practice, s can be obtained by numerical derivatives, and $\nabla_{\alpha'} s$ can be obtained using numerical derivatives once more.

If the first-stage LSE involves two equations, each with regressor Z_{ji}, residual r_{ji}, influence function η_{ji}, and the parameter α_j, $j = 1, 2$,

then set
$$\alpha \equiv (\alpha_1', \alpha_2')', \quad a_N \equiv (a_{1N}', a_{2N}')', \quad \eta_i \equiv (\eta_{1i}', \eta_{2i}')'$$
and proceed as above.

The two-stage approach can be particularly useful when the first-stage LSE residuals are plugged into an equation of interest to deal with endogeneity of some regressors. For instance, suppose

$$y_{1i} = 1 \left[\beta_y y_{2i} + x_{1i}'\beta_x + u_i > 0 \right] \quad \text{and}$$
$$y_{2i} = x_i'\alpha + v_i = x_{1i}'\alpha_1 + x_{2i}'\alpha_2 + v_i \quad \text{where}$$
$$x_i = (x_{1i}', x_{2i}')', \quad u_i = \rho v_i + \varepsilon_i, \quad (v, x) \amalg \varepsilon;$$

v can be heteroskedastic whereas ε cannot. The y_1 equation is a SF and the y_2 equation is a RF; y_2 is an endogenous regressor in the y_1 SF because u is related to v. This is a "triangular system."

One way to deal with the endogeneity of y_2 is to plug $u = \rho v + \varepsilon = \rho(y_2 - x'\alpha) + \varepsilon$ into the y_1 SF:

$$y_{1i} = 1 \left[\beta_y y_{2i} + x_{1i}'\beta_x + \rho(y_{2i} - x_i'\alpha) + \varepsilon_i > 0 \right]$$

where α is the nuisance parameter to be estimated with LSE. Suppose $\varepsilon \sim N(0, \sigma_\varepsilon^2)$. Then we can do MLE (probit) for y_1 using y_2, x_1 and $\hat{v} \equiv y_2 - x'a_{lse}$ as the regressors, which is a special case of the above TSE. This way of handling an endogenous regressor is *"control function" (CF) approach*, as the endogeneity source v is controlled; LDV models with endogenous regressors and CF and other approaches are examined in detail in Chapter 6. Kang and Lee (2009) presented the details on how to estimate the asymptotic variance in various TSE's (including CF) to handle endogenous regressors. The model they dealt with is not a binary response model, but the procedures described there can be easily adapted to binary responses.

The essence of CF approach is that y_2 is related to u only through v so that the endogeneity of y_2 is removed once v is controlled. The assumption $u = \rho v + \varepsilon$ can be relaxed to, e.g.,

$$u = \rho_0 + \rho_1 v + \rho_2 v^2 + \rho_3 v^3 + \varepsilon;$$

$\varepsilon \sim N(0, \sigma_\varepsilon^2)$ is not essential as other distributions can be used instead. Nevertheless, it is often assumed in practice that

$$(u, v) \sim N(0, \Omega) \text{ independently of } x, \text{ which is sufficient for}$$
$$u = \rho v + \varepsilon \text{ and } \varepsilon \sim N(0, \sigma_\varepsilon^2).$$

This assumption unnecessarily rules out heteroskedastic v, and the joint normality is also not necessary.

4   Appendix for Chapter 4

4.1   LR and LM tests in NLS

Recall H_0: $R'\beta = c$ where R is $k \times g$ with $k \geq g$. The restricted estimator b_{Nr} is obtained maximizing wrt b

$$\frac{1}{2} \sum_i \{y_i - r(x_i, b)\}^2 + N\lambda'(R'b - c)$$

where λ is a $g \times 1$ Lagrangian multiplier and b_{Nr} satisfies the first-order condition

$$0 = \sum_i - \{y_i - r(x_i, b_{Nr})\} r_b(b_{Nr}) + NR\lambda.$$

Expand this display for b_{Nr} around b_N: for some $b_{Nr}^* \in (b_{Nr}, b_N)$, because $\sum_i - \{y_i - r(x_i, b_N)\} r_b(b_N) = 0$ (the first-order condition for b_N),

$$0 = \sum_i [r_b(b_N^*) r_{b'}(b_N^*) - \{y_i - r(x_i, b_N^*)\} r_{bb'}(b_N^*)]$$

$$\cdot (b_{Nr} - b_N) + NR\lambda \implies$$

$$0 = \frac{1}{N} \sum_i [r_b(b_N^*) r_{b'}(b_N^*) - \{y_i - r(x_i, b_N^*)\} r_{bb'}(b_N^*)]$$

$$\cdot \sqrt{N}(b_{Nr} - b_N) + \sqrt{N}R\lambda.$$

Observe

$$\frac{1}{N} \sum_i r_b(b_N^*) r_{b'}(b_N^*) = \frac{1}{N} \sum_i r_b(\beta) r_{b'}(\beta) + o_p(1) \quad \text{and}$$

$$\frac{1}{N} \sum_i \{y_i - r(x_i, b_N^*)\} r_{bb'}(b_N^*) = E\{u \cdot r_{bb'}(\beta)\} + o_p(1) = o_p(1).$$

Plugging this into the preceding display gives

$$0 = H\sqrt{N}(b_{Nr} - b_N) + R\sqrt{N}\lambda + o_p(1), \quad \text{where } H \equiv E\{r_b(\beta) r_{b'}(\beta)\}.$$

Multiply both sides by $R'H^{-1}$ to get

$$0 = R'\sqrt{N}(b_{Nr} - b_N) + R'H^{-1}R \cdot \sqrt{N}\lambda + o_p(1)$$

$$\implies \sqrt{N}\lambda = (R'H^{-1}R)^{-1}\sqrt{N}(R'b_N - c) + o_p(1), \quad \text{for } R'b_{Nr} = c.$$

Substitute this back into $0 = H\sqrt{N}(b_{Nr} - b_N) + R\sqrt{N}\lambda + o_p(1)$ to obtain

$$0 = H\sqrt{N}(b_{Nr} - b_N) + R(R'H^{-1}R)^{-1}\sqrt{N}(R'b_N - c) + o_p(1)$$

$$\implies \sqrt{N}(b_N - b_{Nr}) = H^{-1}R(R'H^{-1}R)^{-1}\sqrt{N}(R'b_N - c)$$

$$\implies b_{Nr} = b_N - H^{-1}R(R'H^{-1}R)^{-1}(R'b_N - c)$$

which shows that b_{Nr} can be obtained from b_N instead of maximizing the above constrained maximand.

Apply Taylor expansion of second order to $\sum_i \{y_i - r(x_i, b_{Nr})\}^2$ around b_N to get, for some $b_N^* \in (b_{Nr}, b_N)$,

$$\sum_i \{y_i - r(x_i, b_{Nr})\}^2 - \sum_i \{y_i - r(x_i, b_N)\}^2$$
$$= \sqrt{N}(b_{Nr} - b_N)' H \sqrt{N}(b_{Nr} - b_N) + o_p(1)$$
$$= N(R'b_N - c)'(R'H^{-1}R)^{-1}(R'b_N - c) + o_p(1)$$

as $\sqrt{N}(b_N - b_{Nr}) = H^{-1}R(R'H^{-1}R)^{-1}\sqrt{N}(R'b_N - c)$. Note that $\sqrt{N}(R' b_N - c) = \sqrt{N}\{R'(b_N - \beta)\}$ under H_0 and $\sigma^2 H^{-1}$ is the asymptotic variance of $\sqrt{N}(b_N - \beta)$ under homoskedasticity. Hence, under H_0 and homoskedasticity, the LR-type test statistic for NLS is

$$LR_{nls} = \frac{\sum_i \{y_i - r(x_i, b_{Nr})\}^2 - \sum_i \{y_i - r(x_i, b_N)\}^2}{N^{-1}\sum_i \hat{u}_i^2} \rightsquigarrow \chi_g^2.$$

Turning to the LM-type test, we saw $\sqrt{N}\lambda = (R'H^{-1}R)^{-1}\sqrt{N}$ $(R'b_N - c) + o_p(1)$ for the Lagrangian multiplier, and we can turn this to a χ^2 test statistic:

$$\sqrt{N}R\lambda = R(R'H^{-1}R)^{-1}\sqrt{N}(R'b_N - c) + o_p(1) \implies$$
$$N\lambda'R'H^{-1}R\lambda = \sqrt{N}(R'b_N - c)'(R'H^{-1}R)^{-1}\sqrt{N}(R'b_N - c) + o_p(1)$$
$$\implies N\frac{\lambda'R'H^{-1}R\lambda}{\sigma^2}$$
$$= \sqrt{N}(R'b_N - c)'(R'\sigma^2 H^{-1}R)^{-1}\sqrt{N}(R'b_N - c) + o_p(1) \rightsquigarrow \chi_g^2$$

under homoskedasticity.

Define $\tilde{u}_i \equiv y - r(b_{Nr})$ and observe

$$R\lambda = \frac{1}{N}\sum_i \tilde{u}_i r_b(b_{Nr}) \quad \text{from the above } b_{Nr} \text{ first-order condition;}$$

$$\frac{1}{N}\sum_i \tilde{u}_i^2 r_b(b_{Nr}) r_{b'}(b_{Nr}) \to^p E\{u^2 r_b(\beta) r_{b'}(\beta)\}$$

$$= \sigma^2 H \quad \text{under homoskedasticity.}$$

Hence, $N\lambda'R'H^{-1}R\lambda/\sigma^2$ can be written as $o_p(1)$ plus

$$N \cdot \frac{1}{N}\sum_i \tilde{u}_i r_{b'}(b_{Nr}) \cdot \left\{\frac{1}{N}\sum_i \tilde{u}_i^2 r_b(b_{Nr}) r_{b'}(b_{Nr})\right\}^{-1} \frac{1}{N}\sum_i \tilde{u}_i r_b(b_{Nr})$$

$$= \sum_i \tilde{u}_i r_{b'}(b_{Nr}) \left\{\sum_i \tilde{u}_i^2 r_b(b_{Nr}) r_{b'}(b_{Nr})\right\}^{-1} \sum_i \tilde{u}_i r_b(b_{Nr}) = LM_{nls}.$$

Only b_{Nr} is needed for LM_{nls}.

4.2   Topics for GMM

4.2.1   LR and LM tests

Recall the GMM for $E\psi(\beta) = 0$ and $H_0 : h(\beta) = 0$ where $h(\beta)$ is a R^g-valued function and

$$Wald_{gmm} = N \cdot h(b_N)' \left\{ \frac{\partial h(b_N)}{\partial b'} \, \Omega_N^{-1}(b_N) \, \frac{\partial h(b_N)}{\partial b} \right\}^{-1} h(b_N),$$

$$LR_{gmm} = N \left\{ \frac{1}{N} \sum_i \psi(b_{Nr}) W_N^{-1} \frac{1}{N} \sum_i \psi(b_{Nr}) - \frac{1}{N} \sum_i \psi(b_N) \right.$$

$$\left. W_N^{-1} \frac{1}{N} \sum_i \psi(b_N) \right\},$$

$$LM_{gmm} = \left\{ \frac{1}{N} \sum_i \psi_b(b_{Nr}) W_N^{-1} \frac{1}{\sqrt{N}} \sum_i \psi(b_{Nr}) \right\}' \Omega_N(b_{Nr})^{-1}$$

$$\cdot \left\{ \frac{1}{\sqrt{N}} \sum_i \psi_b(b_{Nr}) W_N^{-1} \frac{1}{N} \sum_i \psi(b_{Nr}) \right\}.$$

The restricted estimator b_{Nr} is obtained by minimizing wrt b

$$Q_N(b) + \lambda' h(b), \quad \text{where } Q_N(b) \equiv 0.5 \frac{1}{N} \sum_i \psi(b)' W_N^{-1} \frac{1}{N} \sum_i \psi(b)$$

and λ is a $g \times 1$ Lagrangian multiplier. Hence b_{Nr} satisfies the first-order condition

$$\frac{1}{N} \sum_i \psi_b(b_{Nr}) W_N^{-1} \frac{1}{\sqrt{N}} \sum_i \psi(b_{Nr}) + \frac{\partial h(b_{Nr})}{\partial b} \sqrt{N} \lambda = 0.$$

Expand this display for b_{Nr} around b_N. Using b_N's first-order condition, the expansion yields, for some $b_{Nr}^* \in (b_{Nr}, b_N)$ and up to $o_p(1)$,

$$0 = \frac{1}{N} \sum_i \psi_b(b_{Nr}^*) W_N^{-1} \frac{1}{N} \sum_i \psi_{b'}(b_{Nr}^*) \sqrt{N}(b_{Nr} - b_N) + \frac{\partial h(b_N)}{\partial b} \sqrt{N} \lambda;$$

the term that would have the derivatives of ψ_b wrt b is negligible because it includes

$$\frac{1}{N} \sum_i \psi(b_{Nr}^*) \sqrt{N}(b_{Nr} - b_N) = o_p(1) \quad \text{as } \frac{1}{N} \sum_i \psi(b_{Nr}^*) \to^p E\psi(\beta) = 0.$$

Also the term that would have the second derivatives of $h(\cdot)$ is negligible as it includes $(b_{Nr_j} - b_{Nj})\sqrt{N}\lambda$ which is of smaller order than the other terms. Hence, since $N^{-1} \sum_i \psi_b(b_{Nr}^*) \to^p E\psi_b(\beta)$ and $\partial h(b_N)/\partial b \to^p \partial h(\beta)/\partial b \equiv R$,

$$o_p(1) = H\sqrt{N}(b_{Nr} - b_N) + R\sqrt{N}\lambda \quad \text{where } H \equiv E\psi_b(\beta)W^{-1}E\psi_{b'}(\beta);$$

H is the probability limit of $\Omega_N(b_N)$ (and $\Omega_N(b_{Nr})$ under the H_0).
Multiply both sides by $R'H^{-1}$ to get

$$o_p(1) = R'\sqrt{N}(b_{Nr} - b_N) + R'H^{-1}R\sqrt{N}\lambda \implies$$

$$\sqrt{N}\lambda = (R'H^{-1}R)^{-1}R'\sqrt{N}(b_N - b_{Nr})$$
$$= (R'H^{-1}R)^{-1}R'\sqrt{N}\{b_N - \beta - (b_{Nr} - \beta)\}$$
$$= (R'H^{-1}R)^{-1}R'\sqrt{N}(b_N - \beta), \quad \text{as } R'\sqrt{N}(b_{Nr} - \beta) = o_p(1)$$

$$\text{under } H_0$$

which follows from

$$\sqrt{N}h(b_{Nr}) = 0 \implies \sqrt{N}h(\beta) + \frac{\partial h(b_{Nr}^*)}{\partial b'}\sqrt{N}(b_{Nr} - \beta) = 0$$

$$\implies \{R' + o_p(1)\}\sqrt{N}(b_{Nr} - \beta) = 0 \quad \text{as } h(\beta) = 0 \text{ under } H_0.$$

Substitute the $\sqrt{N}\lambda$ equation into $o_p(1) = H\sqrt{N}(b_{Nr} - b_N) + R\sqrt{N}\lambda$ to get

$$o_p(1) = H\sqrt{N}(b_{Nr} - b_N) + R(R'H^{-1}R)^{-1}R'\sqrt{N}(b_N - \beta)$$

$$\implies \sqrt{N}(b_{Nr} - b_N) = -H^{-1}R(R'H^{-1}R)^{-1}R'\sqrt{N}(b_N - \beta).$$

The LR-type test statistic for GMM is obtained by applying Taylor expansion of second order to $2Q_N(b_{Nr})$ around b_N: for some $b_N^* \in (b_{Nr}, b_N)$,

$$LR_{gmm} = 2N\{Q_N(b_{Nr}) - Q_N(b_N)\}$$
$$= \sqrt{N}(b_{Nr} - b_N)'H\sqrt{N}(b_{Nr} - b_N) + o_p(1)$$
$$= N\{R'(b_N - \beta)\}'(R'H^{-1}R)^{-1}\{R'(b_N - \beta)\}$$
$$+ o_p(1) \rightsquigarrow \chi_g^2.$$

Turning to the LM-type test, use $\sqrt{N}\lambda = (R'H^{-1}R)^{-1}\sqrt{N}R'(b_N - \beta) + o_p(1)$ under H_0 to get

$$\sqrt{N}R\lambda = R(R'H^{-1}R)^{-1}\sqrt{N}R'(b_N - \beta) + o_p(1)$$
$$\implies N\lambda'R'H^{-1}R\lambda$$
$$= \sqrt{N}\{R'(b_N - \beta)\}'(R'H^{-1}R)^{-1}\sqrt{N}R'(b_N - \beta) + o_p(1) \rightsquigarrow \chi_g^2.$$

From the b_{Nr} first-order condition, we get

$$\sqrt{N}R\lambda = -\frac{1}{N}\sum_i \psi_b(b_{Nr})W_N^{-1}\frac{1}{\sqrt{N}}\sum_i \psi(b_{Nr}).$$

Substitute this into $N\lambda'R'H^{-1}R\lambda$ to obtain LM_{gmm}.

4.2.2 Optimal Weighting Matrix

Define A and C such that $W^{-1} = AA'$ and $E(\psi\psi') = CC'$ with C^{-1} existing. Also define D

$$D \equiv (E\psi_b AA' E\psi_{b'})^{-1} E\psi_b AA'C - (E\psi_b E^{-1}(\psi\psi')E\psi_{b'})^{-1} E\psi_b C^{-1'}.$$

Observe

$$D \cdot C^{-1} E\psi_{b'} = \left\{ (E\psi_b AA' E\psi_{b'})^{-1} E\psi_b AA'C - (E\psi_b E^{-1}(\psi\psi')E\psi_{b'})^{-1} \right.$$

$$\left. E\psi_b C^{-1'} \right\} \cdot C^{-1} E\psi_{b'} = 0$$

and

$$DD' = \left\{ (E\psi_b AA' E\psi_{b'})^{-1} E\psi_b AA'C - (E\psi_b E^{-1}(\psi\psi')E\psi_{b'})^{-1} E\psi_b C^{-1'} \right\}$$

$$\cdot \left\{ (E\psi_b AA' E\psi_{b'})^{-1} E\psi_b AA'C - (E\psi_b E^{-1}(\psi\psi')E\psi_{b'})^{-1} E\psi_b C^{-1'} \right\}'$$

$$= (E\psi_b AA' E\psi_{b'})^{-1} E\psi_b AA'CC'AA' E\psi_{b'}(E\psi_b AA' E\psi_{b'})^{-1}$$

$$- (E\psi_b AA' E\psi_{b'})^{-1} E\psi_b AA'CC^{-1} E\psi_{b'}(E\psi_b E^{-1}(\psi\psi')E\psi_{b'})^{-1}$$

$$- (E\psi_b E^{-1}(\psi\psi')E\psi_{b'})^{-1} E\psi_b C^{-1'}C'AA' E\psi_{b'}(E\psi_b AA' E\psi_{b'})^{-1}$$

$$+ (E\psi_b E^{-1}(\psi\psi')E\psi_{b'})^{-1} E\psi_b C^{-1'}C^{-1} E\psi_{b'}(E\psi_b E^{-1}(\psi\psi')E\psi_{b'})^{-1}.$$

Among the four terms, the first term is the GMM variance indexed by W, and each of the second and third terms is $-(E\psi_b E^{-1}(\psi\psi')E\psi_{b'})^{-1}$, which cancels the fourth term. Hence

$$\left\{ E\psi_b W^{-1} E\psi_{b'} \right\}^{-1} E\psi_b W^{-1} E\psi\psi' W^{-1} E\psi_{b'} \left\{ E\psi_b W^{-1} E\psi_{b'} \right\}^{-1}$$

$$= DD' + \left\{ E\psi_b E^{-1}(\psi\psi')E\psi_{b'} \right\}^{-1}.$$

This is therefore minimized when $D = 0$ with the minimum $\{E\psi_b E^{-1}(\psi\psi')E\psi_{b'}\}^{-1}$.

4.2.3 Over-Identification Test

To see why the GMM over-id test statistic follows χ^2_{s-k}, observe first

$$\frac{1}{\sqrt{N}} \sum_i \psi(\beta)' W_N^{-1} \frac{1}{\sqrt{N}} \sum_i \psi(\beta) \rightsquigarrow \chi^2_s.$$

Taylor-expand $N^{-1/2} \sum_i \psi(\beta)$ around $\beta = b_N \, (= b_{gmm})$

$$\frac{1}{\sqrt{N}} \sum_i \psi(\beta) = \frac{1}{\sqrt{N}} \sum_i \psi(b_N) - \frac{1}{N} \sum_i \psi_{b'}(b_N^*) \sqrt{N}(b_N - \beta).$$

Substitute this into the preceding display to get $o_p(1)$ plus

$$\left\{\frac{1}{\sqrt{N}}\sum_i \psi(b_N) - \frac{1}{N}\sum_i \psi_{b'}(\beta)\sqrt{N}(b_N-\beta)\right\}' W_N^{-1}$$

$$\cdot\left\{\frac{1}{\sqrt{N}}\sum_i \psi(b_N) - \frac{1}{N}\sum_i \psi_{b'}(\beta)\sqrt{N}(b_N-\beta)\right\}.$$

Because $N^{-1}\sum_i \psi_b(b_N)W_N^{-1}N^{-1/2}\sum_i \psi(b_N) = 0$ (the first-order condition for b_N), the cross-products disappear to leave

$$\frac{1}{\sqrt{N}}\sum_i \psi(b_N)' W_N^{-1}\frac{1}{\sqrt{N}}\sum_i \psi(b_N)$$

$$+ \sqrt{N}(b_N-\beta)'\frac{1}{N}\sum_i \psi_b(\beta)W_N^{-1}\frac{1}{N}\sum_i \psi_{b'}(\beta)\sqrt{N}(b_N-\beta).$$

The whole term is χ_s^2 whereas the second term is χ_k^2 because $N^{-1}\sum_i \psi_b$ $(\beta)W_N^{-1}N^{-1}\sum_i \psi_{b'}(\beta)$ is the inverse of the asymptotic variance of $\sqrt{N}(b_N-\beta)$. Also the two terms in the display are asymptotically independent, which is shown in the following. Hence, the first term which is the GMM over-id test statistic follows χ_{s-k}^2.

Observe

$$\sqrt{N}(b_N-\beta) = -C\frac{1}{\sqrt{N}}\sum_i \psi(\beta) + o_p(1) \quad\text{where}$$

$$C \equiv \{E\psi_b E^{-1}(\psi\psi')E\psi_{b'}\}^{-1}E\psi_b E^{-1}(\psi\psi').$$

$$\frac{1}{\sqrt{N}}\sum_i \psi(b_N) = \frac{1}{\sqrt{N}}\sum_i \psi(\beta) + \frac{1}{N}\sum_i \psi_{b'}(\beta)\sqrt{N}(b_N-\beta) + o_p(1)$$

(Taylor expansion)

$$= (I_s - E\psi_{b'}C)\frac{1}{\sqrt{N}}\sum_i \psi(\beta) + o_p(1) \quad (\text{using}\sqrt{N}(b_N-\beta)$$

$$= -C\frac{1}{\sqrt{N}}\sum_i \psi(\beta) + o_p(1)).$$

From these

$$\left[\begin{array}{c}\sqrt{N}(b_N-\beta)\\ N^{-1/2}\sum_i \psi(b_N)\end{array}\right] = \left[\begin{array}{c}-C\\ I_s - E\psi_{b'}C\end{array}\right]\frac{1}{\sqrt{N}}\sum_i \psi(\beta) + o_p(1)$$

which is asymptotically normal with its variance

$$\left[\begin{array}{c}-C\\ I_s - E\psi_{b'}C\end{array}\right]\cdot E(\psi\psi')\cdot[-C',\ I_s - C'E\psi_b].$$

The off-diagonal term (i.e., the asymptotic covariance) is

$$-C\,E\left(\psi\psi'\right)\left(I_s - C'E\psi_b\right)$$
$$= -\left\{E\psi_b E^{-1}\left(\psi\psi'\right)E\psi_{b'}\right\}^{-1}E\psi_b E^{-1}\left(\psi\psi'\right)\cdot E\left(\psi\psi'\right)\left(I_s - C'E\psi_b\right)$$
$$= -\left\{E\psi_b E^{-1}\left(\psi\psi'\right)E\psi_{b'}\right\}^{-1}E\psi_b\left(I_s - C'E\psi_b\right)$$
$$= -\left\{E\psi_b E^{-1}\left(\psi\psi'\right)E\psi_{b'}\right\}^{-1}E\psi_b + \left\{E\psi_b E^{-1}\left(\psi\psi'\right)E\psi_{b'}\right\}^{-1}E\psi_b$$
$$\cdot E^{-1}\left(\psi\psi'\right)E\psi_{b'}\left\{E\psi_b E^{-1}\left(\psi\psi'\right)E\psi_{b'}\right\}^{-1}\cdot E\psi_b = 0.$$

5 Appendix for Chapter 5

5.1 Proportional Hazard and Accelerated Failure Time

5.1.1 Proportional Hazard

Proportional hazard includes Weibull hazard $\lambda(t,x) = \alpha t^{\alpha-1}\cdot e^{x'\beta}$ and exponential power hazard $\lambda(t,x) = \alpha t^{\alpha-1}\exp(\theta t^\alpha)\cdot e^{x'\beta}$. The latter hazard, however, is not a proportional hazard if θ depends on x. Of course, the more general form of hazard is the ones not separable in t and x; one example is the log-logistic hazard with $\theta(x) = \exp(x'\beta)$.

To see the motivation to adopt proportional hazard, observe that, when person i ends duration at y_i without being censored, everybody who survived up to y_i was at risk of ending the duration along with person i. The likelihood of person i to be "chosen" out of all those people at risk is, with $\lambda(t,x) = \lambda_0(t)\phi(x_i,\beta)$ and $\phi(x_i,\beta) = \exp(x_i'\beta)$,

$$\frac{\lambda_0(y_i)\cdot\phi(x_i,\beta)}{\sum_{j\in\{\text{those surviving at } y_i\}}\lambda_0(y_i)\cdot\phi(x_j,\beta)}$$
$$= \frac{\exp(x_i'\beta)}{\sum_{j\in\{\text{those surviving at } y_i\}}\exp(x_j'\beta)}.$$

Constructing the likelihood function that is the product of these ratios across the time points at which the "event duration" (not censored duration) is observed, we can estimate β as shown by Cox (1972)—this is further examined in the semiparametrics chapters.

One advantage of this method relative to the other estimation methods that appeared in the main text is that $\lambda_0(t)$ does not have to be specified as it drops out of the likelihood function. Another advantage is allowing for time-varying regressors $x(t)$:

$$\frac{\exp\{x_i(y_i)'\beta\}}{\sum_{j\in\{\text{those surviving at } y_i\}}\exp\{x_j(y_i)'\beta\}};$$

$x_j(y_i)$ is the value of the person-j regressors when person-i duration ends.

Lancaster (1979) introduced a positive error term (i.e., unobserved heterogeneity) v with its df F_v independent of x in the proportional hazard model:

$$\lambda(t, x; v) = \lambda_0(t)\phi(x, \beta) \cdot v$$

$$\implies \Lambda(t, x; v) = \Lambda_0(t)\phi(x, \beta)v \quad \text{where } \Lambda_0(t) \equiv \int_0^t \lambda_0(s)ds.$$

$\Lambda(t, x; v)$ is the cumulative hazard conditional on x and v. This model is called "*mixed proportional hazard (MPH)*." The survival function conditional on x and v is $\exp\{-\Lambda_0(t)\phi(x, \beta)v\}$. Integrating out v and subtracting the outcome from 1 yields the duration df free of v:

$$1 - \int_0^\infty \exp\{-\Lambda_0(t)\phi(x, \beta)v\}dF_v(v).$$

By specifying parametric functional forms of $\Lambda_0(t)$, $\phi(x, \beta)$ and F_v, MLE can be applied to estimate β as done in Lancaster (1979). But Elbers and Ridder (1982) showed that, if there is at least one non-constant regressor and $E(v) < \infty$, then all three functions are nonparametrically identified. Heckman and Singer (1984) further examined this nonparametric identification issue and provided other identification conditions trading off assumptions from those in Elbers and Ridder (1982).

5.1.2   Accelerated Failure Time (AFT)

Suppose

$$\ln y = (-x)'\beta + u \ (\Longleftrightarrow y = e^{-x'\beta}e^u), \quad u \amalg x$$

e^u has the survival and integrated hazard functions $S_e(\cdot)$ and $\Lambda_e(\cdot)$, respectively.

Then

$$S(t|x) = P(y > t|x) = P(e^u > te^{x'\beta}|x) = S_e(te^{x'\beta}) = \exp\{-\Lambda_e(te^{x'\beta})\}.$$

The role of x is only to multiply t in $S_e(te^{x'\beta})$—i.e., accelerating or decelerating t. This model is called *accelerated failure-time (AFT)* model. From $-\partial \ln S(t|x)/\partial t = \lambda(t|x)$, taking ln on the last display and differentiating it, we get

$$\lambda(t|x) = \lambda_e(te^{x'\beta})e^{x'\beta}.$$

Although AFT accords a nice interpretation for the role of regressors, there should be no good reason to be "obsessed" with AFT and insist on $u \amalg x$. We may just take the log linear model allowing for, for instance, an unknown form of heteroskedasticity to use econometric techniques available for log-linear models. In this case, what is lost is the AFT interpretation that x affects y only through the accelerating/decelerating route, because x can affect y through more than one channel if $u|x$ is a non-constant function of x.

Rewrite the AFT equation as $e^u = ye^{x'\beta} = y/e^{-x'\beta}$. This can be viewed as discounting y with its systematic component $e^{-x'\beta}$ to get an error term e^u. Alternatively, e^u may be viewed as the duration when $x = 0$; then $x \neq 0$ can accelerate or deccelerate the "baseline duration" e^u.

If x is time-varying, then the discounting factor should be replaced by an integral of the time-varying discounting factor, which motivates (Cox and Oakes, 1984)

$$e^u = \int_0^y \exp\{x(\tau)'\beta\}d\tau \quad \{= ye^{x'\beta} \text{ if } x \text{ is time-constant}\}.$$

Then

$$S(t|x) = P(y > t|x) = P\left(e^u > \int_0^t \exp\{x(\tau)'\beta\}d\tau \,|x\right)$$

$$= S_e\left[\int_0^t \exp\{x(\tau)'\beta\}d\tau\right]$$

$$\implies \Lambda(t|x) = -\ln S(t|x) = -\ln S_e\left[\int_0^t \exp\{x(\tau)'\beta\}d\tau\right]$$

$$= \Lambda_e\left[\int_0^t \exp\{x(\tau)'\beta\}d\tau\right]$$

$$\implies \lambda(t|x) = \Lambda_e\left[\int_0^t \exp\{x(\tau)'\beta\}d\tau\right] \cdot \exp\{x(t)'\beta\}.$$

Specifying Λ_e (i.e., S_e), MLE for AFT can be applied allowing for time-varying regressors. Robins and Tsiatis (1992) proposed a semiparametric estimator using the relation $e^u = \int_0^y \exp\{x(\tau)'\beta\}d\tau$ without specifying S_e. See Zeng and Lin (2007) and the references therein for more.

5.1.3 Further Remarks

In Weibull distribution, both proportional hazard and AFT hold. To see that AFT holds for Weibull, first recall the "invariance" property that, regardless of the distribution of y,

$\Lambda(y) \sim Expo(1)$ and $-\ln \Lambda(y) \sim$ "standard" type-1 extreme value distribution.

Now recalling $\Lambda(t) = \theta t^\alpha$ for Weibull, when $y|x$ follows Weibull with $\theta(x) = \exp(x'\beta)$ and α, we get $-\ln \Lambda(y) = -\alpha \ln y - \ln \theta(x)$. Hence

$-\alpha \ln y - x'\beta$ follows the "standard" type-I extreme value distribution.

That is, with $\varepsilon \equiv -\alpha \ln y - \ln \theta(x)$ following the standard type-I extreme value distribution, we get $\ln y = -x'\beta/\alpha + (-\varepsilon/\alpha)$ as in the AFT model where $-\varepsilon/\alpha$ is independent of x and its distributional form is known up to α.

In practice, Weibull MLE is widely used for censored durations. Also the $\ln y$ model ($\ln y = x'\beta + u$) assuming $u \sim N(0, \sigma^2)$ to deal with censoring is often used. Note that when $u \sim N(\mu, \sigma^2)$ with $\mu = 0$, the error term e^u in the AFT model $y = e^{-x'\beta} e^u$ follows *lognormal distribution* with

$$E(e^u) = \exp\left(\mu + \frac{\sigma^2}{2}\right) \quad \text{and} \quad V(e^u) = \exp(2\mu + 2\sigma^2) - \exp(2\mu + \sigma^2).$$

An advantage of log-linear models is that they are easier to interpret than hazard-based models; e.g., β_k in $\beta_k \ln x_k$ is the elasticity of the duration wrt x_k. Also we can apply many familiar estimators/tests to log-linear models. A disadvantage is, however, that log-linear models are inconvenient if there are time-varying regressors, although they may be accommodated as follows. Suppose that x_k is time-varying at some time points $t_1, ..., t_J$, and its values at those times $x_k^{(t_1)}, ..., x_k^{(t_J)}$ are either observed or predicted easily. Then $x_k^{(t_1)}, ..., x_k^{(t_J)}$ can be used as separate regressors in log-linear models.

A generalization of AFT is a transformation-of-variable model

$$T(y; \alpha) = (-x)'\beta + u$$

where $T(y; \alpha)$ is a function parametrized by α that should be estimated along with β. For instance, $T(y; \alpha)$ can be the Box-Cox transformation of y. In the Weibull case, recall that we had $\alpha \ln y = (-x)'\beta + u$; i.e., $T(y; \alpha) = \alpha \ln y$.

Estimating α jointly with β can be troublesome as both α and β have to do with a single "entity"—the effect of x on y. In the semiparametrics chapters, it will be seen that β can be estimated first without estimating α, and then if desired, α can be estimated with β replaced by $\hat{\beta}$. Ridder (1990) called the model with a nonparametric transformation $T(y)$ "*Generalized AFT (GAFT)*," and showed that $T(y)$, β and the error term distribution that is assumed to be independent of x are nonparametrically identified up to a normalization.

6   Appendix for Chapter 6

6.1   Type-I Extreme Errors to Multinomial Logit

Here we will show that, for each i, if $u_{i1}, ..., u_{iJ}$ are iid with type-I extreme value distribution where alternative-j utility is

$$s_{ij} = q'_{ij}\alpha_j + u_{ij} \equiv R_{ij} + u_{ij}, \quad \text{where } R_{ij} \equiv q'_{ij}\alpha_j,$$

then the choice probabilities take the multinomial logit form

$$P(y_{ij} = 1 | q_i) = \frac{\exp(R_{ij})}{\sum_{j=1}^{J} \exp(R_{ij})}$$

where q_i consists of the elements in $q_{i1}, ..., q_{iJ}$. In the following, we will omit the subscript i. We will also often omit "$|q_i$," under which R_{ij} is a constant.

Recall the df $\exp(-e^{-u})$ and density $f(u) = \exp(-u - e^{-u})$. Observe now

$$P(y_j = 1|q) = P(R_j + u_j > R_{j'} + u_{j'}, \forall j' \neq j)$$
$$= P(u_{j'} < u_j + R_j - R_{j'}, \forall j' \neq j)$$
$$= \int \prod_{j' \neq j} P(u_{j'} < u_j + R_j - R_{j'}, \forall j' \neq j \,|u_j)f(u_j)du_j$$

as $u_{j'}$'s are independent

$$= \int \prod_{j' \neq j} \exp(-e^{-u_j - R_j + R_{j'}}) \cdot \exp(-u_j - e^{-u_j})du_j.$$

It is helpful to rewrite the $\prod_{j' \neq j} \exp(\cdot)$ part as

$$\prod_{j' \neq j} \exp(-e^{-u_j - R_j + R_{j'}}) = \exp\left[\ln\left\{\prod_{j' \neq j} \exp(-e^{-u_j - R_j + R_{j'}})\right\}\right].$$

Now observe that the logarithm part is

$$\ln\left\{\prod_{j' \neq j} \exp\left(-e^{-u_j - R_j + R_{j'}}\right)\right\} = \sum_{j' \neq j} -e^{-u_j - R_j + R_{j'}}$$
$$= \sum_{j' \neq j} -e^{-u_j - R_j} \cdot e^{R_{j'}} = -e^{-u_j} \cdot e^{-R_j} \sum_{j \neq j} e^{R_{j'}}$$
$$= -e^{-u_j} \cdot e^{-R_j} \left(\sum_{j'} e^{R_{j'}} - e^{R_j}\right) = -e^{-u_j}\left(\sum_{j'} \frac{e^{R_{j'}}}{e^{R_j}} - 1\right)$$
$$= -e^{-u_j} \sum_{j'} \left(e^{R_j}/e^{R_{j'}}\right) + e^{-u_j}.$$

Thus the $\prod_{j' \neq j} \exp(\cdot)$ part is $\exp\{-e^{-u_j} \sum_{j'}(e^{R_j}/e^{R_{j'}}) + e^{-u_j}\}$, and hence

$$P(y_j = 1|q) = \int \exp\left\{-e^{-u_j} \sum_{j'}(e^{R_j}/e^{R_{j'}}) + e^{-u_j}\right\}$$
$$\cdot \exp(-u_j - e^{-u_j})du_j$$
$$= \int \exp\left\{-e^{-u_j} \sum_{j'}\left(e^{R_j}/e^{R_{j'}}\right)\right\} \cdot \exp(-u_j)du_j$$
$$= \int \exp(-u_j) \cdot \exp\left\{-e^{-u_j} e^{\ln(\sum_{j'} e^{R_{j'}}/e^{R_j})}\right\} du_j$$

$$= \int \exp(-u_j) \cdot \exp\left[-e^{-\{u_j - \ln(\sum_{j'} e^{R_{j'}}/e^{R_j})\}}\right] du_j.$$

Multiply and then divide by $\exp\{-\ln(\sum_{j'} e^{R_{j'}}/e^{R_j})\}$ to get

$$\exp\left\{-\ln\left(\sum_{j'} e^{R_{j'}}/e^{R_j}\right)\right\}\int \exp\left\{-\left(u_j - \ln\left(\sum_{j'} e^{R_{j'}}/e^{R_j}\right)\right)\right\}$$

$$\exp\left[-e^{-\{u_j - \ln\left(\sum_{j'} e^{R_{j'}}/e^{R_j}\right)\}}\right] du_j$$

$$= \exp\left\{-\ln\left(\sum_{j'} e^{R_{j'}}/e^{R_j}\right)\right\}\int \exp(-u_j^*)\exp\left(-e^{-u_j^*}\right) du_j^*$$

$$(\text{with } u_j^* \equiv u_j - \ln\left(\sum_{j'} e^{R_{j'}}/e^{R_j}\right))$$

$$= \exp\left\{-\ln\left(\sum_{j'} e^{R_{j'}}/e^{R_j}\right)\right\} \quad (\text{as the integral is } \int f(u)du = 1)$$

$$= \frac{e^{R_j}}{\sum_{j'} e^{R_{j'}}}.$$

6.2 Two-Level Nested Logit

6.2.1 Lower-Level MNL

For alternative j in the total J-many alternatives, define

$$w_{ij} \equiv (-x'_{i1},\ 0'_{k_x}, \dots, 0'_{k_x},\ x'_{ij},\ 0'_{k_x}, \dots, 0'_{k_x},\ 0'_{k_z}, \dots, 0'_{k_z},\ z'_i,\ 0'_{k_z}, \dots, 0'_{k_z})'$$

so that

$$-x_{i1}\delta_1 + x_{ij}\delta_j + z'_i(\eta_j - \eta_1) + u_{ij} - u_{i1} = w'_{ij}\beta + u_{ij} - u_{i1};$$

this generalizes w_{i2} and w_{i3} in three alternative cases. Recall that w_{ij} have many zero vectors whereas q_{ij} is the zero-stripped version of w_{ij} and that the regression function for alternative j is $w'_{ij}\beta = q'_{ij}\alpha_j$; $\alpha_j = (-\delta'_1, \delta'_j, \eta'_j - \eta'_1)'$ is the alternative-j-specific parameters. We will use w_{ij} or q_{ij} at our convenience.

Let

$$P_{i,j_k} = P(i \text{ chooses house } j \text{ in town } k| \ w_i)$$
$$k = 1, \dots, K,\ j_k = 1, \dots, J_k \ (J = \Sigma_{k=1}^K J_k).$$

For example, there are four houses and two towns, each with two houses. Then, $K = 2$ (number of towns), $J_1 = 2$ (number of houses in town 1),

$J_2 = 2$ (number of houses in town 2), $J = 4$ (the total number of houses in all towns), with the correspondence

house $1 : 1_1$ (first house in town 1),   house2 $: 2_1$ (second house in town1),

house $3 : 1_2$ (first house in town 2),   house4 $: 2_2$ (second house in town2).

Let
$$P_{i,j_k|k} \text{ denote } P(i \text{ chooses } j_k | \text{ town } k \text{ is chosen}).$$

For the first (lower) stage of nested logit (NES), we have, for a given k (town)

$$P_{i,j_k|k} = \frac{\exp(q'_{ij_k}\alpha_{j_k}/\sigma_k)}{\Sigma_{j_k=1}^{J_k} \exp(q'_{ij_k}\alpha_{j_k}/\sigma_k)}, \quad j_k = 1, ..., J_k$$

which is 1 if $J_k = 1$; $q'_{ij_k}\alpha_{j_k}/\sigma_k$ is the regression function before location normalization within town (or "nest") k. When $J_k \geq 2$ (at least two houses), normalize with the first house within town k to get

$$P_{i,1_k|k} = \frac{1}{1 + \Sigma_{j_k=2}^{J_k} \exp(w'_{ij_k}\beta^k/\sigma_k)},$$

$$P_{i,j_k|k} = \frac{\exp(w'_{ij_k}\beta^k/\sigma_k)}{1 + \Sigma_{j_k=2}^{J_k} \exp(w'_{ij_k}\beta^k/\sigma_k)}, \quad j_k = 2, ..., J_k$$

where w_{ij_k} and β^k are defined relative to the first house 1_k in town k analogously to w_{ij} and β:

$$\beta_k \equiv (\delta'_{1_k}, ..., \delta'_{J_k}, (\eta_{2_k} - \eta_{1_k})', ..., (\eta_{J_k} - \eta_{1_k})')',$$
$$w_{ij_k} \equiv (-x'_{i1_k}, 0'_{k_x}, ..., 0'_{k_x}, x'_{ij_k}, 0'_{k_x}, ..., 0'_{k_x}, 0'_{k_z}, ..., 0'_{k_z}, z'_i,$$
$$0'_{k_z}, ..., 0'_{k_z})', j_k = 2, ..., J_k.$$

MNL is applied to each town to estimate β^k/σ_k if $J_k \geq 3$; binary logit is sufficient if $J_k = 2$; no estimation is necessary if $J_k = 1$. In β^k, δ_j's for the houses only in the same town are included, and each η_{j_k} is centered at η_{1_k}, the first houses in town k, not at the very first house among the total J houses. Note that w_{i1_k} is not defined yet.

For the 2×2 town example, $J_1 = J_2 = 2$, and thus binary logit can be applied to each town with

$$w_{i2_k} \equiv (-x'_{i1_k}, x'_{i2_k}, z'_i)' \text{ and } \beta^k \equiv (-\delta'_{1_k}, \delta'_{2_k}, (\eta_{2_k} - \eta_{1_k})')'.$$

Consequently, *what is identified in the first (lower) stage of NES for $k = 1$ and $k = 2$ is*

$$\frac{\delta_{1_1}}{\sigma_1} \left(= \frac{\delta_1}{\sigma_1}\right), \quad \frac{\delta_{2_1}}{\sigma_1} \left(= \frac{\delta_2}{\sigma_1}\right), \quad \frac{\delta_{1_2}}{\sigma_2} \left(= \frac{\delta_3}{\sigma_2}\right), \quad \frac{\delta_{2_2}}{\sigma_2} \left(= \frac{\delta_4}{\sigma_2}\right),$$

$$\frac{\eta_{2_1} - \eta_{1_1}}{\sigma_1} \left(= \frac{\eta_2 - \eta_1}{\sigma_1}\right), \quad \frac{\eta_{2_2} - \eta_{1_2}}{\sigma_2} \left(= \frac{\eta_4 - \eta_3}{\sigma_2}\right).$$

Note that there should be three "η-vectors" for the total 4 ($= J$) alternatives, although there are only two here. This is because two alternatives (one house from each town) are left out for normalization.

6.2.2   Upper-Level MNL

For the second (higher) stage of NES, define the *"inclusive value"* m_{ik}:

$$m_{ik} \equiv \ln\{\Sigma_{j_k=1}^{J_k} \exp(q'_{ij_k}\alpha_{j_k}/\sigma_k)\} \quad \text{to get}$$

$$P_{ik} \equiv P(i \text{ chooses town } k \mid w_i) = \frac{\exp(\sigma_k m_{ik})}{\sum_{k=1}^{K} \exp(\sigma_k m_{ik})};$$

if $J_k = 1$, let $\sigma_k = 1$.

But m_{ik} is not identified; define thus the *"normalized inclusive value"* v_{ik} that is identified from the first-stage:

$$v_{ik} \equiv \ln\{1 + \Sigma_{j_k=2}^{J_k} \exp(w'_{ij_k}\beta^k/\sigma_k)\};$$

let $v_{ik} = 0$ for k with $J_k = 1$. Multiplying and dividing $\exp(m_{ik})$ by $(q'_{i1_k}\alpha_{1_k}/\sigma_k)$, we can see that

$$m_{ik} = \ln[\exp(q'_{i1_k}\alpha_{1_k}/\sigma_k) \cdot \{1 + \Sigma_{j=2}^{J_k} \exp(w'_{ij_k}\beta^k/\sigma_k)\}] = (q'_{i1_k}\alpha_{1_k}/\sigma_k) + v_{ik};$$

for those k with $J_k = 1$, we get $m_{ik} = q'_{i1_k}\alpha_{1_k}$, because $\sigma_k = 1$ and $v_{ik} = 0$. Hence,

$$P_{ik} = \frac{\exp(q'_{i1_k}\alpha_{1_k} + \sigma_k v_{ik})}{\sum_{k=1}^{K} \exp(q'_{i1_k}\alpha_{1_k} + \sigma_k v_{ik})}.$$

Normalizing with the first town,

$$P_{i1} = \frac{1}{1 + \sum_{k=2}^{K} \exp(q'_{i1_k}\alpha_{1_k} - q'_{i1_1}\alpha_{1_1} + \sigma_k v_{ik} - \sigma_1 v_{i1})},$$

$$P_{ik} = \frac{\exp(q'_{i1_k}\alpha_{1_k} - q'_{i1_1}\alpha_{1_1} + \sigma_k v_{ik} - \sigma_1 v_{i1})}{1 + \sum_{k=2}^{K} \exp(q'_{i1_k}\alpha_{1_k} - q'_{i1_1}\alpha_{1_1} + \sigma_k v_{ik} - \sigma_1 v_{i1})} \quad \text{for } k \neq 1.$$

Observe

$$q'_{i1_k}\alpha_{1_k} - q'_{i1_1}\alpha_{1_1} = -x'_{i1_1}\delta_{1_1} + x'_{i1_k}\delta_{1_k} + z'_i(\eta_{1_k} - \eta_{1_1}) = w'_{i1_k}\beta^0 \quad \text{where}$$

$$w_{i1_k} \equiv (-x'_{i1_1}, 0'_{k_x}, \dots, 0'_{k_x}, x'_{i1_k}, 0'_{k_x}, \dots, 0'_{k_x},$$

$$0'_{k_z}, \dots, 0'_{k_z}, z'_i, 0'_{k_z}, \dots, 0'_{k_z})',$$

$$\beta^0 \equiv (\delta'_{1_1}, \dots, \delta'_{1_K}, (\eta_{1_k} - \eta_{1_1})', \dots, (\eta_{1_K} - \eta_{1_1})')';$$

β^0 consists of the parameters for the first houses in all towns. Hence

$$P_{i1} = \frac{1}{1 + \sum_{k=2}^{K} \exp(w'_{i1_k}\beta^0 + \sigma_k v_{ik} - \sigma_1 v_{i1})},$$

$$P_{ik} = \frac{\exp\{w'_{i1_k}\beta^0 + \sigma_k v_{ik} - \sigma_1 v_{i1}\}}{1 + \sum_{k=2}^{K} \exp(w'_{i1_k}\beta^0 + \sigma_k v_{ik} - \sigma_1 v_{i1})}.$$

Using these, we can estimate

$$\beta^0 \quad \text{and} \quad \sigma_1, \dots, \sigma_K$$

with MNL (or binary logit if $K = 2$).

6.2.3　Final-Stage MLE and Remarks

After the second stage, multiply the identified first-stage parameters (β^k/σ_k) by σ_k to identify β^k for each town k. This identifies all δ_j parameters. As for "η-parameters,"

$$\eta_{j_1} - \eta_{1_1},\ \eta_{j_2} - \eta_{1_2}, ..., \eta_{j_K} - \eta_{1_K}$$

were identified in the first stage where each η_{j_k} in town k was centered around η_{1_k}, the first house within the town. If we want η_{j_k} centered around the very first town among all the houses, we can do that after the second stage with

$$(\eta_{j_k} - \eta_{1_k}) + (\eta_{1_k} - \eta_{1_1}) = \eta_{j_k} - \eta_{1_1}$$

where the first term is from the first-stage and the second term is from the second-stage; note $\eta_{1_1} = \eta_1$.

We may stop after the second-stage, but to enhance efficiency and to get standard errors, we can do MLE using the two-stage estimator as initial values; alternatively, MNL with $\sigma_k = 1$ may be used as initial values. Asymptotically, taking just one Newton–Raphson step from the two-stage estimator is equivalent to the fully iterated version. For either version, MLE requires setting up the likelihood function for which we need

$$P_{ij_k} = P(i \text{ chooses } j_k | w_i) = P_{i,j_k|k} \cdot P_{ik}.$$

Using the normalized probabilities (recall $\alpha_j = (-\delta'_1, \delta'_j, \eta'_j - \eta'_1)'$)

$$P_{i1_1} = \frac{1}{1 + \Sigma_{j_k=2}^{J_1} \exp(w'_{ij_1}\beta^1/\sigma_1)}$$
$$\cdot \frac{1}{1 + \Sigma_{k=2}^{K} \exp(q'_{i1_k}\alpha_{1_k} - q'_{i1_1}\alpha_{1_1} + \sigma_k v_{ik} - \sigma_1 v_{i1})},$$

$$P_{ij_1} = \frac{\exp(w'_{ij_1}\beta^1/\sigma_1)}{1 + \Sigma_{j_k=2}^{J_1} \exp(w'_{ij_1}\beta^1/\sigma_1)}$$
$$\cdot \frac{1}{1 + \Sigma_{k=2}^{K} \exp(q'_{i1_k}\alpha_{1_k} - q'_{i1_1}\alpha_{1_1} + \sigma_k v_{ik} - \sigma_1 v_{i1})},$$

$$P_{i1_k} = \frac{1}{1 + \Sigma_{j_k=2}^{J_k} \exp(w'_{ij_k}\beta^k/\sigma_k)}$$
$$\cdot \frac{\exp(q'_{i1_k}\alpha_{1_k} - q'_{i1_1}\alpha_{1_1} + \sigma_k v_{ik} - \sigma_1 v_{i1})}{1 + \Sigma_{k=2}^{K} \exp(q'_{i1_k}\alpha_{1_k} - q'_{i1_1}\alpha_{1_1} + \sigma_k v_{ik} - \sigma_1 v_{i1})},$$

$$P_{ij_k} = \frac{\exp(w'_{ij_k}\beta^k/\sigma_k)}{1 + \Sigma_{j_k=2}^{J_k} \exp(w'_{ij_k}\beta^k/\sigma_k)}$$
$$\cdot \frac{\exp(q'_{i1_k}\alpha_{1_k} - q'_{i1_1}\alpha_{1_1} + \sigma_k v_{ik} - \sigma_1 v_{i1})}{1 + \Sigma_{k=2}^{K} \exp(q'_{i1_k}\alpha_{1_k} - q'_{i1_1}\alpha_{1_1} + \sigma_k v_{ik} - \sigma_1 v_{i1})}.$$

In summary, NES proceeds as follows:

1. Do (multinomial) logit across houses for each town to estimate β^k/σ_k.

2. Obtain the normalized inclusive value v_{ik}, and do (multinomial) logit across towns to estimate β^0 (for w_{i1_k}) and σ_1 (for $-v_{i1}$),..., σ_K (for v_{iK}).

3. Find δ_{j_k} by multiplying the estimates at step 1 by σ_k estimated at step 2; find $\eta_{j_k} - \eta_{1_1}$ by adding $\eta_{j_k} - \eta_{1_k}$ from step 1 to $\eta_{1_k} - \eta_{1_1}$ from step 2.

4. To do the MLE instead of the two-stage estimation for NES, obtain $L(\beta,\sigma_1,...,\sigma_K) = \Sigma_i\Sigma_{k=1}^K\Sigma_{j_k=1}^{J_k}y_{ij_k}\ln P_{ij_k}$, where P_{ij_k}'s are functions of $\beta,\sigma_1,...,\sigma_K$, y_{ij_k} is the choice dummy that takes 1 if i chooses j_k and 0 otherwise, and $\beta = (\delta_1',...,\delta_J', \eta_2'-\eta_1',..,\eta_J'-\eta_1')'$.

To see that NES includes MNL as a special case when $\sigma_k = 1\ \forall k$, write P_{ij_k} using the non-normalized probabilities:

$$P_{ij_k} = \frac{\exp(q_{ij_k}'\alpha_{j_k}/\sigma_k)}{\Sigma_{j_k=1}^{J_k}\exp(q_{ij_k}'\alpha_{j_k}/\sigma_k)} \cdot \frac{\exp(\sigma_k m_{ik})}{\Sigma_{k=1}^K\exp(\sigma_k m_{ik})}.$$

As $m_{ik} \equiv \ln\{\Sigma_{j_k=1}^{J_k}\exp(q_{ij_k}'\alpha_{j_k}/\sigma_k)\}$, we get

$$\exp(\sigma_k m_{ik}) = \Sigma_{j_k=1}^{J_k}\exp(q_{ij_k}'\alpha_{j_k}/\sigma_k)\quad\text{and}$$

$$\Sigma_{k=1}^K\exp(\sigma_k m_{ik}) = \Sigma_{k=1}^K\Sigma_{j=1}^{J_k}\exp(q_{ij_k}'\alpha_{j_k}/\sigma_k).$$

Hence $P_{ij_k} = \exp(q_{ij_k}'\alpha_{j_k}/\sigma_k)/\Sigma_{k=1}^K\Sigma_{j=1}^{J_k}\exp(q_{ij_k}'\alpha_{j_k}/\sigma_k)$. With $\sigma_k = 1\ \forall k$, this is nothing but the MNL probability when all J houses are taken together.

NES relaxes the IIA assumption of MNL. Suppose $J = 3$ and $K = 2$ with car ($j = 1$ with $J_1 = 1$), red bus ($j = 2$), and blue bus ($j = 3$) with $J_2 = 2$; the two buses are nested. Observe, with $v_{i1} = 0$ due to $J_1 = 1$,

$$P_{i1_1} = \frac{1}{1 + \exp(q_{i1_2}'\alpha_{1_2} - q_{i1_1}'\alpha_{1_1} + \sigma_2 v_{i2})},$$

$$P_{i1_2} = \frac{1}{1 + \exp(w_{i2_2}'\beta^2/\sigma_2)} \cdot \frac{\exp(q_{i1_2}'\alpha_{1_2} - q_{i1_1}'\alpha_{1_1} + \sigma_2 v_{i2})}{1 + \exp(q_{i1_2}'\alpha_{1_2} - q_{i1_1}'\alpha_{1_1} + \sigma_2 v_{i2})}.$$

The ratio of the choice probabilities for car and red bus is, as $\exp(v_{i2}) = \{1 + \exp(w_{ij_2}'\beta^2/\sigma_2)\}$,

$$\frac{P_{i1_1}}{P_{i1_2}} = \frac{1}{1 + \exp\left(w_{i2_2}'\beta^2/\sigma_2\right)}\exp\left(q_{i1_2}'\alpha_{1_2} - q_{i1_1}'\alpha_{1_1}\right) \cdot \left\{1 + \exp\left(w_{i2_2}'\beta^2/\sigma_2\right)\right\}^{\sigma_2}.$$

$$= \left\{1 + \exp\left(w'_{i2_2}\beta^2/\sigma_2\right)\right\}^{\sigma_2-1} \cdot \exp\left(q'_{i1_2}\alpha_{1_2} - q'_{i1_1}\alpha_{1_1}\right)$$

$$= \left\{1 + \exp\left(x'_{i3}\frac{\delta_3}{\sigma_2} - x'_{i2}\frac{\delta_2}{\sigma_2} + z'_i\frac{\eta_3 - \eta_2}{\sigma_2}\right)\right\}^{\sigma_2-1} \cdot \exp\left(x'_{i2}\delta_2 + z'_i\eta'_2\right)$$

$$-x'_{i1}\delta_1 - z'_i\eta_1).$$

Unless $\sigma_2 = 1$ or $\delta_3 = 0$, x_{i3} matters for this ratio.

6.3　Asymptotic Distribution of MSM estimators

Recall the Multinomial probit(MNP) first-order condition

$$\sum_{i=1}^{N}\sum_{j=1}^{J}\{y_{ij} - P_{ij}(g)\}z_{ij}(g) = 0, \quad z_{ij}(g) \equiv \frac{\partial \ln P_{ij}(g)}{\partial g} \quad \Longrightarrow$$

$$\sum_{i=1}^{N}\underset{k \times J}{z_i}(g)\{y_i - \underset{J \times 1}{P_i(g)}\} = 0$$

where $P_i(g)$ and $z_i(g)$ are the stacked versions of $P_{ij}(g)$ and $z_{ij}(g)$ across $j = 1, ..., J$, respectively. Viewed as MOM, the MOM estimator satisfying this condition is MNP with $\sqrt{N}(g_{mom} - \gamma) \rightsquigarrow N(0, I_f^{-1})$. In the following, the MSM g_{msm} that simulates $P_i(g)$ and $z_{ij}(g)$ will be shown to have an asymptotic variance slightly larger than I_f^{-1} due to the simulation error.

As is the case usually, the error in estimating the instrument $z_i(g)$ does not affect the asymptotic variance of g_{msm}, so we will denote $z_i(g)$ just as z_i as if it does not depend on g. Denoting the simulated estimator for $P_i(g)$ as $f_i(g)$, the asymptotic distribution of g_{msm}, denoted also as g_N, will be derived using the decomposition

$$\frac{1}{\sqrt{N}}\sum_i z_i\{y_i - f_i(g_N)\}$$

$$= \frac{1}{\sqrt{N}}\sum_i z_i[\{y_i - f_i(\gamma)\} - \{P_i(g_N) - P_i(\gamma)\} + \{f_i(\gamma) - P_i(\gamma)\}$$

$$- \{f_i(g_N) - P_i(g_N)\}] = 0.$$

Denote all regressors determining the choice probabilities as w_i; both z_i and $P_i(g)$ depend on w_i, whereas $f_i(g)$ depends on w_i and the simulated error vector ε_i. Note $E_\varepsilon\{f_i(g)|w_i\} = P_i(g)$ where $E_\varepsilon\{\cdot|w_i\}$ is taken wrt the simulated error; this unbiasedness of $f_i(g)$ for $P_i(g)$ holds due to a LLN and the fact that $f_i(g)$ is an average of the simulated samples. Using "stochastic equicontinuity," it holds that

$$\frac{1}{\sqrt{N}}\sum_i z_i[\{f_i(\gamma) - P_i(\gamma)\} - \{f_i(g_N) - P_i(g_N)\}] = o_p(1).$$

This display substituted into the preceding one yields

$$\frac{1}{\sqrt{N}}\sum_i z_i\{y_i - f_i(\gamma)\} - \frac{1}{\sqrt{N}}\sum_i z_i\{P_i(g_N) - P_i(\gamma)\} = o_p(1).$$

Apply the mean value theorem to $P_i(g_N)$ around γ to get, for some $g_N^* \in (g_N, \gamma)$,

$$P_i(g_N) - P_i(\gamma) = \frac{\partial P_i(g_N^*)}{\partial g'}(g_N - \gamma).$$

Substitute this into the preceding display to obtain

$$\frac{1}{\sqrt{N}}\sum_i z_i\{y_i - f_i(\gamma)\} - \frac{1}{N}\sum_i z_i\frac{\partial P_i(g_N^*)}{\partial g'}\cdot\sqrt{N}(g_N - \gamma) = o_p(1)$$

$$\implies \sqrt{N}(g_N - \gamma) = E^{-1}\left\{z_i\frac{\partial P_i(\gamma)}{\partial g'}\right\}\cdot\frac{1}{\sqrt{N}}\sum_i z_i\{y_i - f_i(\gamma)\} + o_p(1).$$

From $E[z_i\{y_i - P_i(\gamma)\}] = 0$, which is the MLE population first-order condition written as a moment condition, we get the information equality in the form

$$(I_f =) E[z_i\{y_i - P_i(\gamma)\}\{y_i - P_i(\gamma)\}'z_i] = -E\left\{z_i\frac{\partial P_i(\gamma)}{\partial g}\right\}.$$

This leads to

$$\sqrt{N}(g_N - \gamma) = -I_f^{-1}\cdot\frac{1}{\sqrt{N}}\sum_i z_i\{y_i - f_i(\gamma)\} + o_p(1)$$

$$= -I_f^{-1}\left[\frac{1}{\sqrt{N}}\sum_i z_i(y_i - P_i) + \frac{1}{\sqrt{N}}\sum_i z_i\{P_i - f_i(\gamma)\}\right] + o_p(1).$$

The first term $-I_f^{-1}N^{-1/2}\sum_i z_i(y_i - P_i)$ yields the MNP asymptotic distribution. As can be conjectured from the next paragraph, the first and second terms are uncorrelated. Thus we just have to obtain the second term variance. Observe

$$E[z_i\{P_i - f_i(\gamma)\}] = E[z_i E_\varepsilon\{P_i - f_i(\gamma)|w_i\}] = 0 \quad \text{as } E_\varepsilon\{P_i - f_i(\gamma)|w_i\} = 0.$$

For the variance, denote f_i as $n^{-1}\sum_{t=1}^n g(w_i, \varepsilon_t)$ where $g(w_i, \varepsilon_t)$ is an indicator function that differs from y_i only in that y_i is generated with w_i and the true errors while $g(w_i, \varepsilon_t)$ is generated with w_i and the simulated errors following the same distribution as the true errors do. Then

$$E\{(P_i - f_i)(P_i - f_i)'|w_i\} = E\left[\left\{\frac{1}{n}\sum_{t=1}^n(g(w_i, \varepsilon_t) - P_i)\right\}\right.$$

$$\left\{ \left(\frac{1}{n} \sum_{t=1}^{n} (g(w_i, \varepsilon_t) - P_i) \right)' |w_i \right]$$

$$= \frac{1}{n} E \left[\{ g(w_i, \varepsilon_t) - P_i \} \{ g(w_i, \varepsilon_t) - P_i \}' |w_i \right]$$

$$= \frac{1}{n} E \{ g(w_i, \varepsilon_t) g(w_i, \varepsilon_t)' - P_i g(w_i, \varepsilon_t)' - g(w_i, \varepsilon_t) P_i' + P_i P_i' |w_i \}$$

$$= \frac{1}{n} E \{ g(w_i, \varepsilon_t) g(w_i, \varepsilon_t)' - P_i P_i' |w_i \} \quad \{ \text{as } E \{ P_i g(w_i, \varepsilon_t)' |w_i \}$$

$$= P_i E \{ g(w_i, \varepsilon_t)' |w_i \} = P_i P_i' \}$$

$$= \frac{1}{n} E \{ y_i y_i' - P_i P_i' |w_i \} = \frac{1}{n} E \{ (y_i - P_i)(y_i - P_i)' |w_i \} = \frac{1}{n} I_f.$$

Hence we get

$$\sqrt{N}(g_N - \gamma) \rightsquigarrow N \left\{ 0, I_f^{-1} \left(I_f + \frac{I_f}{n} \right) I_f^{-1} \right\} = N \left\{ 0, \left(1 + \frac{1}{n} \right) I_f^{-1} \right\}.$$

7   Appendix for Chapter 7

7.1   Other Density Estimation Ideas

In the following, we introduce three more nonparametric density estimation ideas with $k = 1$. More nonparametric methods will appear later for nonparametric regression.

7.1.1   Nearest-Neighbor Method

On average, $N \cdot 2h f(x_o)$ observations in a data set will fall in the interval $[x_o - h, x_o + h]$, because $2h f(x_o) \simeq P(x_o - h \leq x_i \leq x_o + h)$. Rearrange the data in an increasing order of distance from x_o and denote the distance of the ith nearest datum by $d_i(x_o)$. Then in the interval of $[x_o - d_s(x_o), x_o + d_s(x_o)]$, there are s observations. Thus, s should be approximately equal to $N \cdot 2d_s(x_o) f_N(x_o)$, and solving this for $f_N(x_o)$ gives

$$f_N(x_o) = \frac{s}{N 2 d_s(x_o)}$$

which is called the sth *nearest-neighbor estimator*. Here, s is the smoothing parameter as h is in the kernel method, and s should be chosen such that $s/N \to 0$ as $N \to \infty$; e.g., $s = \sqrt{N}$.

The advantage of nearest neighbor (NN) method over kernel method is that the same number of observations are used for estimating $f(x_a)$ as well as $f(x_b)$. This is achieved by letting the size of neighborhood flexible depending on the availability of neighboring observations. In kernel method, the size of neighborhood is fixed by h and the number of observations used for $f(x_a)$ and $f(x_b)$ differ. The disadvantage of nearest neighbor method is that the curve estimate is not smooth and it is not a density in general—i.e., $\int f_N(x_o)dx_o \neq 1$.

7.1.2 Maximum Penalized Likelihood Estimator

Extending the idea of MLE, we may consider maximizing the likelihood function over the space of all possible probability density functions of x. There is a major difference between this idea and the usual MLE: in the latter we have only a finite number of parameters, while in the former we have an infinite number of parameters. One modification to make this idea operational is penalizing the "overfit." The *maximum penalized likelihood estimator* is defined by

$$\max_{g \in G} \sum_i \left\{ \ln g(x_i) - \alpha \int g''(x)^2 dx \right\}$$

where G is a set of probability densities satisfying certain properties.

In maximum penalized likelihood estimator, α determines the degree of smoothing. With $\alpha = 0$, the resulting estimate will be too jagged, because g can be chosen to have a peak (mode) at each $x_i = 1, ..., N$ subject to the unit integral constraint. With a high α, the total variation in g is highly penalized, and the density with a peak at each x_i will be no longer optimal. Note that we use g'', not g', to measure the variation of g; a straight line with a slope has $g' \neq 0$ but it is just a transposition of a flat line. This idea is related to "spline smoothing" to be seen later in nonparametric regression.

7.1.3 Series Approximation

Suppose we suspect that $f(x)$ is a normal density, but not quite sure of this. In this case, we may specify the density as

$$f(x; \mu, \sigma, \beta) = \frac{1}{\sigma} \phi \left(\frac{x - \mu}{\sigma} \right) + \beta \cdot \psi(x)$$

where $\phi(\cdot)$ is the $N(0,1)$ density, (μ, σ, β) are unknown parameters, and $\int \psi(x)dx = 0$ so that $\int f(x; \mu, \sigma, \beta)dx = 1$. An estimate of β will shed light on the validity of the normality assumption. One problem is, however, that $f(x; \mu, \sigma, \beta)$ may not be a density because $\psi(x)$ can take negative values. Ideas to avoid this problem will appear shortly.

Generalizing the idea of writing f as a sum of terms, suppose that $f(x)$ admits the following series expansion:

$$f(x) = \sum_{j=1}^{\infty} \beta_j \psi_j(x) \text{ in the sense } \lim_{J \to \infty} \int \left\{ f(x) - \sum_{j=1}^{J} \beta_j \psi_j(x) \right\}^2 dx = 0$$

where $\{\psi_j(x)\}$ are the "basis" of the series expansion; when this kind of expansion is valid will be discussed below when series expansion appears again for nonparametric regression. Usually $\{\psi_j(x)\}$ are chosen to be "orthonormal": $\int \psi_j(x)\psi_m(x)dx = \delta_{jm}$ (Kronecker delta). Then

$$\int f(x)\psi_j(x)dx = \beta_j : \beta_j \text{ shows the contribution}$$

of the ("directional") component $\psi_j(x)$.

Orthogonality makes estimating β_j's "less wasteful," but the overall explanatory power of the series remains the same, orthogonal or not. In practice, only a finite number of terms, say s, can be used, and s becomes a smoothing parameter (a small s means over-smoothing).

As a variant of $\sum_{j=1}^{s}\beta_j\psi_j(x)$, Gallant and Nychka (1987) proposed to use

$$H(x) = \frac{1}{C(\beta)}\left(\sum_{|\alpha|=0}^{s}\beta_\alpha x^\alpha\right)^2 \cdot \phi(x)$$

where $\alpha \equiv (\alpha_1, ..., \alpha_k)'$, α_j's are nonnegative integers, and

$$|\alpha| \equiv \sum_{j=1}^{k}\alpha_j, \quad x^\alpha \equiv \prod_{j=1}^{k}x_j^{\alpha_j}, \quad C(\beta) \equiv \int \left\{\sum_{|\alpha|=0}^{s}\beta_\alpha x^\alpha\right\}^2 \cdot \phi(x)dx;$$

$C(\beta)$ is a normalizing constant. Estimating the β parameters, we get to estimate the density nonparametrically. For instance, for $k = 3$ (trivariate density), we get $|\alpha| = 1$ when $\alpha = (1,0,0), (0,1,0),$ or $(0,0,1)$ which yield $x^\alpha = x_1, x_2,$ or x_3, respectively. Also $|\alpha| = 2$ when

$$\begin{array}{ccccccc}
\alpha: & (1,1,0), & (1,0,1), & (0,1,1), & (2,0,0), & (0,2,0), & (0,0,2), \\
x^\alpha: & x_1 x_2, & x_1 x_3, & x_2 x_3, & x_1^2, & x_2^2, & x_3^2,
\end{array}$$

More specifically for k-variate density, if $s = 2$, then

$$H(x) = \frac{1}{C(\beta)}\left(\beta_0 + \sum_{j=1}^{k}\beta_j x_j + \sum_{j=1}^{k}\sum_{\ell=1}^{k}\beta_{j\ell}x_j x_\ell\right)^2 \cdot \phi(x).$$

This becomes $\phi(x)$ when all β_j's and β_{jm}'s but β_0 are zero. If $s = 3$, $H(x)$ becomes

$$\frac{1}{C(\beta)}\left(\beta_0 + \sum_{j=1}^{k}\beta_j x_j + \sum_{j=1}^{k}\sum_{\ell=1}^{k}\beta_{j\ell}x_j x_\ell + \sum_{j=1}^{k}\sum_{\ell=1}^{k}\sum_{m=1}^{k}\beta_{j\ell m}x_j x_\ell x_m\right)^2 \cdot \phi(x).$$

The idea of embedding a parametric density function in a general parametric family of densities in a smooth fashion is called the "Neyman's smooth test for goodness of fit." The original reference for the test going back to 1937 is hard to find; see, e.g., Rayner and Best (1990) for the reference and the literature after that. The Neyman's smooth test will be examined when non-parametric goodness-of-fit tests are studied later.

7.2 Asymptotic Distribution for Kernel Regression Estimator

Recall the expression in the main text

$$\sqrt{Nh^k}\{r_N(x_o) - r(x_o)\} = \frac{-g_N(x_o)}{f_N(x_o)f(x_o)}\sqrt{Nh^k}\{f_N(x_o) - f(x_o)\}$$

$$+ \frac{1}{f(x_o)}\sqrt{Nh^k}\{g_N(x_o) - g(x_o)\}$$

$$= \begin{bmatrix} -r(x_o)f(x_o)^{-1} & f(x_o)^{-1} \end{bmatrix} \begin{bmatrix} \sqrt{Nh^k}\{f_N(x_o) - f(x_o)\} \\ \sqrt{Nh^k}\{g_N(x_o) - g(x_o)\} \end{bmatrix} + o_p(1)$$

as $\dfrac{-g_N(x_o)}{f_N(x_o)f(x_o)} \to^p \dfrac{-r(x_o)}{f(x_o)}$.

We will show, for any constants a_1 and a_2,

$$a_1\sqrt{Nh^k}\{f_N(x_o) - f(x_o)\} + a_2\sqrt{Nh^k}\{g_N(x_o) - g(x_o)\}$$
$$\rightsquigarrow N\{0, (a_1, a_2)\Omega(a_1, a_2)'\} \text{ for some } \Omega$$

as this is equivalent to

$$\begin{bmatrix} \sqrt{Nh^k}\{f_N(x_o) - f(x_o)\} \\ \sqrt{Nh^k}\{g_N(x_o) - g(x_o)\} \end{bmatrix} \rightsquigarrow N(0, \Omega).$$

From this, the asymptotic normality of $\sqrt{Nh^k}\{r_N(x_o) - r(x_o)\}$ easily follows.
Observe

$$a_1\sqrt{Nh^k}\{f_N(x_o) - f(x_o)\} + a_2\sqrt{Nh^k}\{g_N(x_o) - g(x_o)\}$$

$$= \frac{1}{N}\sum_i \sqrt{Nh^k}\left[a_1\left\{\frac{1}{h^k}K\left(\frac{x_i - x_o}{h}\right) - f(x_o)\right\}\right.$$

$$\left. + a_2\left\{\frac{1}{h^k}K\left(\frac{x_i - x_o}{h}\right)y_i - g(x_o)\right\}\right].$$

The expected value of this average is the individual expected value

$$\sqrt{Nh^k}E\left[a_1\left\{\frac{1}{h^k}K\left(\frac{x_i - x_o}{h}\right) - f(x_o)\right\}\right.$$

$$+ a_2 \left\{ \frac{1}{h^k} K \left(\frac{x_i - x_o}{h} \right) y_i - g(x_o) \right\} \right] = \sqrt{N h^k} O(h^2)$$

$$= O(\sqrt{N h^{k+4}}) = o(1) \quad \text{with under-smoothing } N h^{k+4} \to 0.$$

As for the variance, it is the individual variance divided by N, which is smaller than the individual second moment divided by N:

$$h^k E \left[a_1 \left\{ \frac{1}{h^k} K \left(\frac{x_i - x_o}{h} \right) - f(x_o) \right\} + a_2 \left\{ \frac{1}{h^k} K \left(\frac{x_i - x_o}{h} \right) y_i - g(x_o) \right\} \right]^2$$

$$= h^k a_1^2 E \left\{ \frac{1}{h^k} K \left(\frac{x_i - x_o}{h} \right) - f(x_o) \right\}^2 + h^k a_2^2 E \left\{ \frac{1}{h^k} K \left(\frac{x_i - x_o}{h} \right) y_i - g(x_o) \right\}^2$$

$$+ h^k 2 a_1 a_2 E \left[\left\{ \frac{1}{h^k} K \left(\frac{x_i - x_o}{h} \right) - f(x_o) \right\} \left\{ \frac{1}{h^k} K \left(\frac{x_i - x_o}{h} \right) y_i - g(x_o) \right\} \right].$$

The dominant terms are those with $1/h^k$, and putting together only those terms, we get

$$h^k a_1^2 E \left\{ \frac{1}{h^{2k}} K \left(\frac{x_i - x_o}{h} \right)^2 \right\} + h^k a_2^2 E \left\{ \frac{1}{h^{2k}} K \left(\frac{x_i - x_o}{h} \right)^2 y_i^2 \right\} a_2$$

$$+ h^k 2 a_1 E \left\{ \frac{1}{h^{2k}} K \left(\frac{x_i - x_o}{h} \right)^2 y_i \right\}.$$

With the change of variables, this is $o(1)$ equal to

$$a_1^2 f(x_o) \int K(z)^2 \, dz + a_2^2 E(y^2 | x_o) f(x_o) \int K(z)^2 \, dz$$

$$+ 2 a_1 a_2 r(x_o) f(x_o) \int K(z)^2 \, dz.$$

Hence, with the Lindeberg condition holding easily, we get

$$\Omega = \begin{bmatrix} 1 & r(x_o) \\ r(x_o) & E(y^2 | x_o) \end{bmatrix} f(x_o) \int K(z)^2 dz.$$

Finally, observe

$$\begin{bmatrix} \dfrac{-r(x_o)}{f(x_o)}, & \dfrac{1}{f(x_o)} \end{bmatrix} \begin{bmatrix} 1 & r(x_o) \\ r(x_o) & E(y^2 | x_o) \end{bmatrix} f(x_o) \int K(z)^2 dz \begin{bmatrix} -r(x_o) f(x_o)^{-1} \\ f(x_o)^{-1} \end{bmatrix}$$

$$= [-r(x_o), 1] \begin{bmatrix} 1 & r(x_o) \\ r(x_o) & E(y^2|x_o) \end{bmatrix} \int K(z)^2 dz \begin{bmatrix} -r(x_o)f(x_o)^{-1} \\ f(x_o)^{-1} \end{bmatrix}$$

$$= \int K(z)^2 dz \cdot [\ 0 \quad -r(x_o)^2 + E(y^2|x_o)\ ] \begin{bmatrix} -r(x_o)f(x_o)^{-1} \\ f(x_o)^{-1} \end{bmatrix}$$

$$= \int K(z)^2 dz \cdot [\ 0 \quad V(y|x_o)\ ] \begin{bmatrix} -r(x_o)f(x_o)^{-1} \\ f(x_o)^{-1} \end{bmatrix} = \int K(z)^2 dz \cdot \frac{V(y|x_o)}{f(x_o)}.$$

7.3 Other Nonparametric Regression Methods

7.3.1 Nearest-Neighbor Estimator

Generalizing the nearest-neighbor method for density estimation, we can estimate $r(x_o)$ with an (weighted) average of y_i's whose x_i's fall within the sth nearest neighbor (NN) of x_o. One such NN estimator is

$$\hat{r}_N(x_o) \equiv \frac{1}{s} \sum_i 1[x_i \in s\text{th NN of } x_o] \cdot y_i$$

$$= \sum_i w_i(x_o, s) \cdot y_i, \quad \text{where } w_i(x_o, s) \equiv \frac{1[x_i \in s\text{th NN of } x_o]}{s}.$$

Note that $\sum_i w_i(x_o, s) = 1$ for all x_o and s, so long as we include x_o in the sth NN observations when $x_o = x_i$ for some i. One advantage of NN over kernel methods is that the same number of observations are used for each $\hat{r}_N(x_o)$ as x_o varies, which makes estimates in data-scarce areas more reliable. One disadvantage is the non-smoothness in $\hat{r}_N(x_o)$, which makes the asymptotic analysis somewhat difficult; it is possible to make the weighting function smooth (still subject to $\sum_i w_i(x_o, s) = 1$), which then yields a smoother estimator.

For the case $k = 1$, with $s/N \to 0$ as $N \to \infty$, Härdle (1990, p. 43) shows

$$E\{\hat{r}_N(x_o)\} - r(x_o) \simeq \frac{1}{24f(x_o)^2} \{r''(x_o) + \frac{2r'(x_o)f'(x_o)}{f(x_o)}\}(\frac{s}{N})^2 \quad \text{and}$$

$$V\{\hat{r}_N(x_o)\} \simeq \frac{V(y|x_o)}{s}.$$

Regarding s/N as h in the kernel method, the bias is of order h^2 while the variance is of order $(Nh)^{-1}$, which is the same as in the kernel method. Minimizing the MSE, we get $s = O(N^{4/5})$, which makes the MSE converge to 0 at $O(N^{-4/5})$—the same rate as in the kernel method. The variance does not have $f(x_o)^{-1}$ which the kernel method has; i.e., even if $f(x_o)$ is small, the variance is not affected. In this case, however, the bias gets larger than that of the kernel method due the leading factor $f(x_o)^{-2}$.

When $k = 1$, Stute (1984) showed a "hybrid" between the kernel estimator and NN estimator:

$$\tilde{r}_N(x_o) \equiv \frac{1}{Nh} \sum_i K(\frac{F_N(x_o) - F_N(x_i)}{h}) y_i$$

$$\text{where} \quad F_N(x_o) = \frac{1}{N} \sum_{j=1}^{N} 1[x_j \leq x_o] \text{ is the empirical df.}$$

Here the distance between x_o and x_i is gauged using $F_N(\cdot)$. Alternatively, Stute (1984) proposed the normalized version $\tilde{r}_N(x_o)/[(Nh)^{-1} \sum_i K\{(F_N(x_o) - F_N(x_i))/h\}]$. Both are consistent for $E(y|x_o)$ and follow the same asymptotic normal distribution

$$\sqrt{Nh}\{\tilde{r}_N(x_o) - r(x_o)\} \rightsquigarrow N(0, \; V(y|x_o) \int K(z)^2 dz).$$

No $f(x)$ appears in the asymptotic variance.

7.3.2   Spline Smoothing

Suppose that x is a rv taking values in $[0, 1]$, and $r(x_o)$ and the derivatives $r'(x_o), ..., r^{(m)}(x_o)$ are continuous on $[0, 1]$. Imagine (x_i, y_i) scattered over the x–y plane and we want to fit a line that has a good fit as well as smoothness. These two contradicting goals can be achieved by minimizing the following wrt q for a $\lambda > 0$ over a function space to which $r(x_o)$ belongs:

$$Q_N(q) \equiv \frac{1}{N} \sum_i \{y_i - q(x_i)\}^2 + \lambda \int_0^1 \{q^{(m)}(x)\}^2 dx.$$

Here λ penalizes overfit, and thus it is a smoothing parameter. The solution to this minimization problem is a piecewise polynomial. Choosing the optimal $q(x)$ this way is a *spline smoothing* (Wahba, 1990). If we replace the first term by $-N^{-1} \sum_i \ln f(y_i, x_i)$ and the second term by a measure of likelihood variation, then we get "likelihood spline smoothing." The smoothing spline is attractive not so much for its practicality (at least when $k > 1$) but rather for its relation to prior information on $r(x_o)$. We will show this point below, drawing on Eubank (1999).

Apply the Taylor expansion with integral remainder to $r(x)$ around $x = 0$:

$$r(x) = \sum_{j=0}^{m-1} \beta_j x^j + \frac{1}{(m-1)!} \int_0^1 r^{(m)}(z) \cdot (1 - z)^{m-1} dz.$$

If we want to approximate $r(x)$ with a polynomial in x with $m-1$ degree, the result depends on the extent that the remainder term is negligible. Using the Cauchy–Schwartz inequality,

$$\left[\int_0^1 r^{(m)}(z) \cdot (1-z)^{m-1} dz\right]^2 \leq \left[\int_0^1 \{r^{(m)}(z)\}^2 dz\right] \cdot \int_0^1 (1-z)^{2m-2} dz$$

$$= \left[\int_0^1 \{r^{(m)}(z)\}^2 dz\right] \frac{1}{1-2m}(1-z)^{2m-1}\big|_0^1 = \left[\int_0^1 \{r^{(m)}(z)\}^2 dz\right] \frac{1}{2m-1}.$$

With this, we get for the remainder term

$$\left|\int_0^1 r^{(m)}(z)(1-z)^{m-1} dz\right| \leq \frac{1}{(2m-1)^{0.5}}\left[\int_0^1 \{r^{(m)}(z)\}^2 dz\right]^{0.5}$$

$$\equiv \frac{1}{(2m-1)^{0.5}} J_m(r)^{0.5}.$$

An assumption such as $J_m(r) \leq \rho$ reflects our prior belief on how much the model deviates from the polynomial regression. It is known that, if we minimize $N^{-1}\sum_i \{y_i - q(x_i)\}^2$ over q subject to the condition that q has $(m-1)$ continuous derivatives and $J_m(q) \leq \rho$, then there exists a $\lambda > 0$ such that the same q minimizes $Q_N(q)$. Choosing $\lambda = 0$ implies a polynomial regression where y is regressed on $1, x, x^2, ..., x^{m-1}$. Hence smoothing spline is an extension of polynomial regressions guarding against departures from the assumption that the regression function is polynomial.

Implementing smoothing spline in practice is far more involved than implementing kernel methods. Instead we mention a result in Silverman (1984): for a smoothing spline, there exists an equivalent adaptive kernel estimator with its local bandwidth proportional to $f(x)^{-1/4}$. Therefore, we will not lose much by using adaptive kernel-type methods. Also Jennen-Steinmetz and Gasser (1988) presented a further generalization where various nonparametric methods are shown to be equivalent to a kernel method with the local bandwidth proportional to $f^{-\alpha}$, $0 \leq \alpha \leq 1$. They also suggest selecting α adaptively instead of fixing it (at $1/4$) in advance. These findings suggest that we will not miss much in practice by focusing on kernel estimator or its variations.

7.3.3 Series Approximation

If x is in R^k, then we can write x as $x = \sum_{j=1}^k x_j e_j$ where e_j's are the orthonormal basis vectors for R^k; $e_j' e_{j'} = \delta_{jj'}$ (Kronecker delta). A regression function $r(x)$ at m different x points, $r(x_{(1)}), ..., r(x_{(m)})$, may be viewed as a point in R^m. More generally, $r(x)$ with $x \in X$ where X has infinite elements can be regarded as a "point" in an infinite-dimensional space. Suppose $r(x)$ admits the following series expansion:

$$r(x) = \sum_{j=-\infty}^{\infty} \beta_j \psi_j(x) \; \forall x \in X \quad \text{in the sense that}$$

$$\lim_{J \to \infty} \int_X |r(x) - \sum_{j=-J}^{J} \beta_j \psi_j(x)|^2 dx = 0$$

where $\int_X \psi_j(x)\psi_{j'}(x)dx = \delta_{jj'}$. Here $r(x)$ is decomposed into orthogonal components, and $\psi_j(x)$ may be regarded as an "axis" or "direction"; $\sum_{j=-\infty}^{\infty} \beta_j \psi_j(x)$ is an "orthogonal series."

The best-known example of orthogonal series expansion is the following Fourier series expansion. For a scalar x with $X = [-\pi, \pi]$, suppose that $r(x)$ is square-integrable on $[-\pi, \pi]$: $\int_{-\pi}^{\pi} r(x)^2 dx < \infty$; i.e., $r(x)$ belongs to $L^2([-\pi, \pi])$. Because $(2\pi)^{-1/2} \exp(\mathbf{i}jx)$, $j = 0, \pm 1, \pm 2, ...$, is a "complete orthonormal basis" of $L^2([-\pi, \pi])$ where $\mathbf{i}^2 = -1$ (see the next paragraph), any element in $L^2([-\pi, \pi])$ can be written as an infinite linear combination of the basis; an orthonormal basis is complete if any element orthogonal to the orthonormal basis is 0. Thus, for some β_j's, we get

$$r(x) = \sum_{j=-\infty}^{\infty} \beta_j \frac{1}{\sqrt{2\pi}} \exp(\mathbf{i}jx) = \sum_{j=-\infty}^{\infty} \beta_j \frac{1}{\sqrt{2\pi}} \{\cos(jx) + \mathbf{i}\sin(jx)\}$$

$$= \frac{\beta_0}{\sqrt{2\pi}} + \sum_{j=1}^{\infty} \left\{ \frac{\beta_j + \beta_{-j}}{\sqrt{2\pi}} \cos(jx) + \mathbf{i}\frac{\beta_j - \beta_{-j}}{\sqrt{2\pi}} \sin(jx) \right\}$$

using $\cos(jx) = \cos(-jx)$ and $\sin(-jx) = -\sin(jx)$.

The orthonormality of the basis can be seen in the inner product between $(2\pi)^{-1/2} \exp(\mathbf{i}jx)$ and $(2\pi)^{-1/2} \exp(\mathbf{i}j'x)$:

$$\int_{-\pi}^{\pi} \frac{1}{\sqrt{2\pi}} \exp(\mathbf{i}jx)\frac{1}{\sqrt{2\pi}} \exp(-\mathbf{i}j'x)dx \quad (\frac{1}{\sqrt{2\pi}} \exp(-\mathbf{i}j'x)$$

is the complex conjugate)

$$= \frac{1}{2\pi} \int_{-\pi}^{\pi} \{\cos(jx) + \mathbf{i}\sin(jx)\}\{\cos(-j'x) + \mathbf{i}\sin(-j'x)\}dx$$

$$= \frac{1}{2\pi} \int_{-\pi}^{\pi} \{\cos(jx) + \mathbf{i}\sin(jx)\}\{\cos(j'x) - \mathbf{i}\sin(j'x)\}dx$$

$$= \frac{1}{2\pi} \int_{-\pi}^{\pi} [\{\cos(jx)\cos(j'x) + \sin(jx)\sin(j'x)\} + \mathbf{i}\{\cos(j'x)\sin(jx)$$

$$- \cos(jx)\sin(j'x)\}]dx$$

$$= \frac{1}{2\pi} (\pi\delta_{jj'} + \pi\delta_{jj'}) = \delta_{jj'} \quad \text{using the fact}$$

$$\int_{-\pi}^{\pi} \cos(jx)\cos(j'x)dx = \pi\delta_{jj'}, \quad \int_{-\pi}^{\pi} \sin(jx)\sin(j'x)dx = \pi\delta_{jj'},$$

$$\int_{-\pi}^{\pi} \cos(jx) \sin(j'x)dx = 0.$$

Suppose now that, with X being the support of x,

$$y_i = r(x_i) + u_i = \sum_{j=0}^{\infty} \beta_j \psi_j(x_i) + u_i.$$

With a given data set, we can only estimate a finite number of β_j's. So we need to trim the series at a number, say h, to get

$$y_i = r_h(x_i) + \left\{ u_i - \sum_{j=h+1}^{\infty} \beta_j \psi_j(x_i) \right\}, \quad \text{where } r_h(x_i) \equiv \sum_{j=0}^{h} \beta_j \psi_j(x_i).$$

Here, h plays the role of a reversed smoothing parameter—the higher h the less smoothing. This equation can be estimated with LSE where $\{\cdot\}$ is the error term. Although $\int_X \psi_j(x_o)\psi_{j'}(x_o)dx_o = 0$ if $j \neq j'$, $\int_X \psi_j(x_o)\psi_{j'}(x_o)f(x_o)$ $dx_o = E\{\psi_j(x)\psi_{j'}(x)1[x \in X]\} \neq 0$ in general. Thus, we will incur an omitted variable bias—$\sum_{j=h+1}^{\infty} \beta_j \psi_j(x_i)$ is omitted and related to $\psi_1(x), ..., \psi_h(x)$. This is natural, as other nonparametric estimators have biases. If $f(x)$ is known as in experimental data, then we may choose $\psi_j(x)$ such that $\int_X \psi_j(x)$ $\psi_{j'}(x)f(x)dx = 0$, in which case the bias would be zero. But $f(x)$ is not known in observational data; also $\int_X \psi_j(x_o)\psi_{j'}(x_o)dx_o = \delta_{jj'}$ will not hold in general anymore if $\int_X \psi_j(x_o)\psi_{j'}(x_o)f(x_o)dx_o = 0$.

While the Fourier series expansion does not require $r(x)$ to be continuous, it has trigonometric functions which are not often used in econometrics. It is known that the set of continuous functions on $[a_0, a_1]$ where $-\infty < a_0 < a_1 < \infty$ is "dense" for $L^2[a_0, a_1]$ in the sense, for any $r(x)$ with $\int_{a_0}^{a_1} r(x)^2 dx < \infty$, there exists a continuous function $c(x)$ on $[a_0, a_1]$ such that $\{\int_{a_0}^{a_1} |r(x) - c(x)|^2 dx\}^{1/2} < \varepsilon$ for any constant $\varepsilon > 0$. Furthermore, according to the *Weierstrass approximation theorem*, any continuous function $r(x)$ on $[a_0, a_1]$ can be uniformly approximated by some polynomial function $p(x)$:

$$\sup_{x \in [a_0, a_1]} |r(x) - p(x)| < \varepsilon, \quad \text{for any } \varepsilon > 0.$$

This motivates using orthonormal polynomial functions for $\{\psi_j\}$.

There are many sets of orthonormal functions for $\{\psi_j\}$. One example when $X = [-1, 1]$ is the *Legendre polynomials*, which are obtained by applying the so-called "Gram-Schmidt procedure" to $1, x, x^2, x^3, ...$ The orthonormal *Legendre polynomials for $L^2([-1, 1])$* are

$$\left(\frac{2n+1}{2}\right)^{1/2} \cdot \frac{1}{2^n n!} \cdot \frac{\partial^n \{(x^2 - 1)^n\}}{\partial x^n}, \quad n = 0, 1, 2, ...;$$

without the first term $\{(2n+1)/2\}^{1/2}$, we get the Legendre polynomials which are orthogonal, but not orthonormal. Specifically, substituting $n = 0, 1, 2, 3, ...$, we get

$$\frac{1}{\sqrt{2}}, \quad \frac{x}{(2/3)^{0.5}}, \quad \frac{0.5\left(3x^2 - 1\right)}{(2/5)^{0.5}}, \quad \frac{0.5\left(5x^3 - 3x\right)}{(2/7)^{0.5}}, \quad ...$$

While a kernel method uses a local approximation idea, a series estimator $\bar{r}_N(x_o)$ for $r(x_o)$ uses a *global approximation*, because the same basis functions are used for all x and all observations contribute to $\bar{r}_N(x_o)$. Compared with local approximation estimators, series estimators have a couple of advantages. First, they can be computationally convenient; e.g., using polynomial series for $\bar{r}_N(x)$, the series estimator is nothing but a LSE. Second, they are convenient for imposing certain restrictions such as additive separability. For example, with $k = 3$, if $E(y|x_1, x_2, x_3) = \rho(x_1, x_2) + \alpha x_3$—a semi-linear model—where $\rho(\cdot)$ is an unknown function, then we may use

$$r(x) = \beta_0 + \beta_1 x_1 + \beta_2 x_2 + \beta_{11} x_1^2 + \beta_{12} x_1 x_2 + \beta_{22} x_2^2 + \alpha x_3$$

where no interaction term between (x_1, x_2) and x_3 appears. This lack of interaction terms is an easy way of imposing the additive separability, which is, however, difficult in kernel estimators. The regression function can be then estimated easily with LSE. Imposing an extra restriction such as additive separability on nonparametric estimation should enhance the efficiency.

Series estimators also have disadvantages. First, a high degree of local nonlinearity at a single point may force many terms in series expansion. For instance, when $r(x)$ is linear around most points but quadratic around a single point, series approximation needs quadratic terms to account for the nonlinearity around the single point, which is due to the global nature of series approximation. Second, although the degree of polynomials serves as a (reversed) smoothing parameter, ambiguity arises when there is a "fine" room for choice between degrees, say, 2 and 3; e.g., when a polynomial of degree 3 is used with two regressors x_1 and x_2, we may use all of x_1^3, x_2^3, $x_1 x_2^2$, $x_1^2 x_1$, or only some of them. Third, convergence rates and asymptotic normal distribution are not easily characterized. Newey (1997) presented, however, an asymptotic normality result. The result is something of an "anticlimax," because applying nonparametric power-series estimation turns out to be the same as doing LSE with polynomial regressors for all practical purposes.

7.4   Asymptotic Normality of Series Estimators

In this subsection, we show the asymptotic normality of power series estimators for regression function, drawing on Newey (1997) who dealt also with series estimators using series other than power series. For more generality, let the parameter of interest be a function $a\{r(x_o)\}$. We will show the asymptotic normality for an estimator of $a\{r(x_o)\}$, and then specialize the result for $r(x_o)$.

With K denoting the degree of series approximation, consider a K-vector $p^K(x)$ of approximation functions, and its matrix version P:

$$\underset{K \times 1}{p^K(x)} = \begin{bmatrix} p_{1K}(x) \\ \vdots \\ p_{KK}(x) \end{bmatrix} \quad \text{and} \quad \underset{N \times K}{P} = \begin{bmatrix} p^K(x_1)' \\ \vdots \\ p^K(x_N)' \end{bmatrix}$$

$$= \begin{bmatrix} p_{1K}(x_1) & \cdots & p_{KK}(x_1) \\ \vdots & & \vdots \\ p_{1K}(x_N) & \cdots & p_{KK}(x_N) \end{bmatrix}.$$

For instance, with $k = 2$ and $K = 6$,

$$p_{1K}(x_i) = 1, \; p_{2K}(x_i) = x_{i1}, \; p_{3K}(x_i) = x_{i2}, \; p_{4K}(x_i) = x_{i1}^2,$$
$$p_{5K}(x_i) = x_{i2}^2, \; p_{6K}(x_i) = x_{i1}x_{i2};$$

P is nothing but the regressor matrix in LSE using the polynomial regressors. A series estimator for $r(x_o) = E(y|x = x_o)$ is

$$\hat{r}(x_o) \equiv p^K(x_o)'b_N, \quad \text{where} \quad \underset{K \times 1}{b_N} = (P'P)^{-1}P'Y$$

For $a\{r(x_o)\}$, define

$$\underset{K \times 1}{\hat{A}} \equiv \frac{\partial a\{p^K(x_o)'b_N\}}{\partial b_N}, \quad \underset{K \times K}{\hat{Q}} \equiv \frac{P'P}{N},$$

$$\underset{K \times K}{\hat{\Omega}} \equiv \frac{1}{N}\sum_i p^K(x_i)p^K(x_i)'\{y_i - \hat{r}(x_i)\}^2 \quad \text{and} \quad \underset{1 \times 1}{\hat{V}} \equiv \hat{A}'\hat{Q}^{-1}\hat{\Omega}\hat{Q}^{-1}\hat{A}.$$

In the last matrix, $\hat{Q}^{-1}\hat{\Omega}\hat{Q}^{-1}$ is nothing but the LSE variance estimator in estimating $r(x_o)$, and $\hat{A}$ is the Jacobian (i.e., the derivative of the function $a\{\hat{r}(x_o)\} = a\{p^K(x_o)'b_N\}$ wrt b_N). For a matrix B, let $\|B\| \equiv \{trace(B'B)\}^{1/2}$, and let X be the support of x that is compact.

Assume, for a sequence of constants $\zeta_o(K)$ (note that K is an increasing function of N),

$$\sup_{x \in X} \|p^K(x)\| \le \zeta_o(K) \quad \text{and} \quad \frac{\zeta_o(K)^2 K}{N} \to 0 \text{ as } N \to \infty;$$

$r(x)$ is continuously differentiable up to order s on X;

$$\frac{\sqrt{N}}{K^{s/k}} \to 0; \text{ either } \frac{K^6}{N} \to 0, \text{ or } a(\cdot) \text{ is linear and } \frac{K^3}{N} \to 0,$$

Then

$$a\{\hat{r}(x_o)\} = a\{r(x_o)\} + O_p\left(\frac{K}{\sqrt{N}}\right) \quad \left(\implies \text{convergence rate is } \frac{K}{\sqrt{N}}\right)$$

$$\sqrt{N}\hat{V}^{-1/2}\left[a\{\hat{r}(x_o)\} - a\{r(x_o)\}\right] \rightsquigarrow N(0,1).$$

When $a(\cdot)$ is the identity function, we get $\hat{A} = p^K(x_o)$, and hence

$$\sqrt{N} \cdot \left\{p^K(x_o)'\hat{Q}^{-1}\hat{\Omega}\hat{Q}^{-1}p^K(x_o)\right\}^{-1/2} \cdot \{\hat{r}(x_o) - r(x_o)\} \rightsquigarrow N(0,1).$$

Here the asymptotic variance estimator for $\hat{r}(x_o)$ is

$$p^K(x_o)' \cdot (P'P)^{-1}\sum_i p^K(x_i)p^K(x_i)' \{y_i - \hat{r}(x_i)\}^2 (P'P)^{-1} \cdot p^K(x_o)$$

which is nothing but the usual LSE asymptotic variance estimator. That is, if one applies nonparametric power-series estimation to $r(x)$, the estimates and their t-values will be the same as those in doing LSE with polynomial regressors. The only theoretical caveat is that the convergence rate is slower than $\sqrt{N}$, which is, however, not noticeable in practice.

If $a(\cdot)$ depends on the entire function $r(x)$, not just on a single point $r(x_o)$, and if some type of averaging takes place in $a(\cdot)$, then the convergence rate can be $N^{-1/2}$, which is faster than $K/\sqrt{N}$. Newey (1997) showed when this rate of convergence takes place.

8 Appendix for Chapter 8

8.1 U-Statistics

8.1.1 Motivations

Consider an iid sample x_i, $i = 1, ..., N$, with $E(x) = \mu$ and $V(x) = \sigma^2$. In estimating "dispersion" in the population distribution, we can use $s_x^2 \equiv (N-1)^{-1}\sum_i(x_i - \bar{x})^2$ for σ^2 where the "deviation" $(x_i - \bar{x})^2$ of x_i is relative to $\bar{x}$. Another way to estimate deviation of x_i is $(x_i - x_j)^2$ where the deviation of x_i is relative to x_j. Doing this with all pairs leads to the average of $(x_i - x_j)^2$ across all pairs—this in fact equals s_x, as will be shown below. Going further, we may pick a triple (x_i, x_j, x_k) and use $x_i - (x_j + x_k)/2$ as a deviation of x_i where $(x_j + x_k)/2$ plays the role of $\bar{x}$. But this is asymmetric in x_i, x_j, x_k. "Symmetrizing" this gives the average across all possible triples of

$$\frac{1}{3}\left\{\left(x_i - \frac{x_j + x_k}{2}\right) + \left(x_j - \frac{x_i + x_k}{2}\right) + \left(x_k - \frac{x_i + x_j}{2}\right)\right\}$$

which may be consistent for some dispersion measure other than σ^2. Going for the extreme, we can think of the average of $\{x_i - (N-1)^{-1}\sum_{j \neq i}x_j\}^2$ across i. Using singletons, pairs, triples, etc. gives "U-statistics" of order $1, 2, 3 \ldots$, which is examined in this subsection.

Suppose there is a parameter of interest θ such that, for some function $h(x_1, ..., x_m)$ with $1 \leq m < N$,

$$E\{h(x_1, ..., x_m)\} = \theta.$$

Then $h(x_1, ..., x_m)$ per se is an unbiased estimator for θ. But we can also use all possible subsets of the data with m ordered elements (there are $N!/(N-m)!$-many of them) by averaging them:

$$T_N \equiv \frac{(N-m)!}{N!} \sum_{m-ordered} h(x_{i_1}, ..., x_{i_m})$$

where $\sum$ ranges over all $N!/(N-m)!$ permutations with $(i_1, ..., i_m)$ being a subset of $(1, ..., N)$. T_N is called an U-statistic of order m.

Intuition suggests $V(T_N) \leq V\{h(x_1, ..., x_m)\}$ because averaging would reduce the variance. The inequality indeed holds, which is nothing but Rao–Blackwell theorem because the order statistics $x_{(1)}, ..., x_{(N)}$ such that $x_{(1)} \leq , ..., \leq x_{(N)}$ are sufficient statistics for $x_1, ..., x_N$, and T_N is the mean of $h(\cdot)$ conditional on the order statistics.

The simplest U-statistic is U-statistic of order 1, and the best-known example is sample mean. Because $E(x_i) = \mu \; \forall i$, each x_i is an unbiased estimator for μ, and its U-statistic version is nothing but $N^{-1} \sum_i x_i$. Also sample moments are U-statistics of order 1 as can be seen in $E(N^{-1} \sum_i x_i^k) = E(x^k)$. It will be shown shortly that sample variance is an U-statistic of order 2. In U-statistic, $h(\cdot)$ is called the *kernel*, which is not to be confused with "kernel" in kernel nonparametric estimator. To ease exposition, we will review mainly U-statistics of order 2 in the following, while pointing out how the findings there can be generalized for $m > 2$. See, e.g., Serfling (1980) and Lehman (1999) for more on U-statistics.

8.1.2   Symmetrization

U-statistic of order m can be further rewritten so that the kernel becomes symmetric in its arguments:

$$\frac{(N-m)!m!}{N!} \frac{1}{m!} \sum_{m-ordered} h(x_{i_1}, ..., x_{i_m})$$

$$= \frac{1}{\binom{N}{m}} \sum_{m-unordered} g(x_{i_1}, ..., x_{i_m}) \quad \text{where}$$

$$g(x_{i_1}, ..., x_{i_m}) \equiv \frac{1}{m!} \sum h(x_{i_1}, ..., x_{i_m});$$

$g(x_{i_1}, ..., x_{i_m})$ is the *symmetrized* version of $h(x_{i_1}, ..., x_{i_m})$ by averaging across all possible rearrangements of the index $(i_1, ..., i_m)$.

For instance, with $m = 2$ and $N = 3$, we get $(3-2)!/3! = 1/6$ and

$$\frac{1}{6} \{ h(x_1, x_2) + h(x_2, x_1) + h(x_1, x_3) + h(x_3, x_1) + h(x_2, x_3) + h(x_3, x_2) \}$$

$$= \frac{1}{3} \left[\frac{1}{2} \{ h(x_1, x_2) + h(x_2, x_1) \} + \frac{1}{2} \{ h(x_1, x_3) + h(x_3, x_1) \} \right.$$

$$\left.+\frac{1}{2}\left\{h(x_2,x_3)+h(x_3,x_2)\right\}\right]$$

$$=\frac{1}{3}\left[g(x_1,x_2)+g(x_1,x_3)+g(x_2,x_3)\right].$$

Here $g(x_1,x_2)$ is "(*permutation-*) *symmetric*" because

$$g(x_1,x_2)=\frac{1}{2}\left\{h(x_1,x_2)+h(x_2,x_1)\right\}=\frac{1}{2}\left\{h(x_2,x_1)+h(x_1,x_2)\right\}$$

$$=g(x_2,x_1);$$

$g(x_1,x_2)$ is the average over all possible rearrangements of the index 1 and 2. With $m = 2$ but a general N, for $E\{h(x_1,x_2)\} = \theta$, its U-statistic of order 2 is

$$U_N \equiv \frac{1}{N(N-1)}\sum_{i\neq j}h(x_i,x_j)=\frac{2}{N(N-1)}\sum_{i<j}g(x_i,x_j),$$

$$g(x_i,x_j)\equiv\frac{h(x_i,x_j)+h(x_j,x_i)}{2}.$$

From now on, we will assume that the kernel is symmetric without loss of generality.

    Observe

$$V(U_N)=E\left\{(U_N-\theta)^2\right\}$$

$$=\frac{4}{N^2(N-1)^2}\sum_{i<j}\sum_{i'<j'}E\left[\{g(x_i,x_j)-\theta\}\{g(x_{i'},x_{j'})-\theta\}\right].$$

All terms with $i \neq i'$ and $j \neq j'$ which are of order N^4 are zero. The remaining terms are of order N^3 at most. Thus $V(U_N) \to 0$ as $N \to \infty$, from which $U_N \to^p \theta$ follows.

    For the asymptotic distribution, the well-known *U-statistic projection* result holds:

$$\sqrt{N}(U_N-\theta)=\frac{2}{\sqrt{N}}\sum_i\left[E\{h(x_i,x_j)|x_i\}-\theta\right]+o_p(1)$$

$$\rightsquigarrow N\left(0,\ 4E\left[\{E(h(x_i,x_j)|x_i)-\theta\}^2\right]\right)$$

where the number 2 is due to the kernel order. The conditional mean $E\{h(x_i,x_j)|x_i\}$ is called the *projection* of $h(x_i,x_j)$ on x_i. More generally for U-statistic of order m, if $E\{h(x_1,x_2,...,x_m)^2\} < \infty$, then

$$\sqrt{N}(U_N-\theta)\rightsquigarrow N(0,\ m^2 E[\{E(h(x_1,x_2,...,x_m)|x_i)-\theta\}^2]).$$

8.1.3 Examples

As an example, consider estimating $E^2(x) \equiv \mu^2$. One estimator for μ^2 is $\bar{x}^2$, and with δ-method,

$$\sqrt{N}(\bar{x}^2 - \mu^2) \rightsquigarrow N\{0,\ (2\mu)^2\sigma^2\}.$$

Another estimator for μ^2 is the U-statistic

$$Z_N \equiv \frac{2}{N(N-1)} \sum_{i<j} x_i x_j;$$

the kernel $x_i x_j$ is symmetric. Since $E(Z_N) = E(x_i)E(x_j) = \mu^2$, Z_N is unbiased for μ^2. Because $V(Z_N) \to 0$ as $N \to \infty$ as shown in the next paragraph, Z_N is consistent for its expected value μ^2. For the asymptotic distribution, observe that

$$E(x_i x_j | x_i) = x_i E(x_j) = x_i \mu$$
$$\Longrightarrow E[\{E(x_i x_j | x_i) - \mu^2\}^2] = E\{(x_i\mu - \mu^2)^2\} = \mu^2\sigma^2$$
$$\Longrightarrow \sqrt{N}(Z_N - \mu^2) \rightsquigarrow N(0, 4\mu^2\sigma^2)$$

which is the same as the asymptotic distribution of $\sqrt{N}(\bar{x}^2 - \mu^2)$, although the small sample behaviors of $\bar{x}^2$ and Z_N should differ. This example also raises the possibility of "degenerate" U-statistic when $\mu = 0$, although we do not discuss this case any further.

To see $V(Z_N) \to 0$ as $N \to \infty$ in the above example, observe

$$Z_N - E(Z_N) \equiv \frac{2}{N(N-1)} \sum_{i<j}(x_i x_j - \mu^2),$$

$$E\{Z_N - E(Z_N)\}^2 \equiv \frac{4}{N^2(N-1)^2} \sum_{i<j}\sum_{i'<j'} E\left\{(x_i x_j - \mu^2)(x_{i'} x_{j'} - \mu^2)\right\}.$$

Firstly, if $i \neq i'$ and $j \neq j'$, then $E\{(x_i x_j - \mu^2)(x_{i'} x_{j'} - \mu^2)\} = 0$. Second, if $i = i'$ and $j \neq j'$, then

$$E\{(x_i x_j - \mu^2)(x_i x_{j'} - \mu^2)\} = E(x_i^2 x_j x_{j'} - \mu^2 x_i x_{j'} - \mu^2 x_i x_j + \mu^4)$$
$$= E(x_i^2)E(x_j)E(x_{j'}) - \mu^2 E(x_i)E(x_{j'}) - \mu^2 E(x_i)E(x_j) + \mu^4$$
$$= E(x^2)\mu^2 - \mu^4.$$

The number of the terms with $i = i'$ and $j \neq j'$ is less than $N(N-1)(N-2)$, which is negligible as it is of smaller order than $N^2(N-1)^2$ in the denominator of the variance. Third, there are terms with $i = i'$ and $j = j'$, but these terms are also of order smaller than $N^2(N-1)^2$. Therefore $V(Z_N) \to 0$ as $N \to \infty$.

As another example of U-statistic of order 2, because $\sigma^2 = E(x^2) - E^2(x)$, an unbiased estimator for σ^2 is $x_i^2 - x_i x_j$: $E(x_i^2 - x_i x_j) = \sigma^2$. But $x_i^2 - x_i x_j$ is asymmetric. Its U-statistic version is the pair average after symmetrization:

$$\frac{1}{\binom{N}{m}}\sum_{i<j}\frac{(x_i-x_j)^2}{2} \quad \text{as} \quad \frac{(x_i-x_j)^2}{2}=\frac{1}{2}\left\{(x_i^2-x_ix_j)+(x_j^2-x_jx_i)\right\}.$$

To see that this is in fact the sample variance s_{N-1}^2, observe

$$\sum_i(x_i-\bar{x})^2=\sum_i x_i^2-N\bar{x}^2=\sum_i x_i^2-\frac{1}{N}\left(\sum_i x_i\right)^2$$

$$=\sum_i x_i^2-\frac{1}{N}\left(\sum_i x_i^2+\sum_{i\neq j}x_ix_j\right)=\frac{N-1}{N}\sum_i x_i^2-\frac{1}{N}\sum_{i\neq j}x_ix_j.$$

Divide the first and last expressions by $N-1$ to get

$$(s_{N-1}^2=)\;\frac{1}{N-1}\sum_i(x_i-\bar{x})^2=\frac{1}{N}\sum_i x_i^2-\frac{1}{N(N-1)}\sum_{i\neq j}x_ix_j$$

$$=\frac{1}{N(N-1)}\sum_{i\neq j}(x_i^2-x_ix_j)$$

because $\sum_{i\neq j}x_i^2=\sum_i\sum_{j,j\neq i}x_i^2=(N-1)\sum_i x_i^2$.

We can also find the asymptotic distribution of s_{N-1}^2 using its U-statistic representation. Observe

$$E\left\{\frac{(x_i-x_j)^2}{2}\Big|x_i\right\}=\frac{E\left(x_i^2-2x_ix_j+x_j^2|x_i\right)}{2}=\frac{x_i^2-2x_iE(x_j)+E(x_j^2)}{2}$$

$$=\frac{x_i^2-2x_i\mu+\sigma^2+\mu^2}{2}$$

$$\Longrightarrow E\left(\left[E\{\frac{(x_i-x_j)^2}{2}|x_i\}-\sigma^2\right]^2\right)=E\left[\left(\frac{x_i^2-2x_i\mu+\mu^2-\sigma^2}{2}\right)^2\right]$$

$$=\frac{1}{4}E\left[\{(x-\mu)^2-\sigma^2\}^2\right]=\frac{1}{4}\left[E\{(x-\mu)^4\}-\sigma^4\right].$$

Hence, with $1/4$ canceled by $m^2=4$ in the asymptotic variance, we obtain

$$\sqrt{N}(s_{N-1}^2-\sigma^2)\rightsquigarrow N\left(0,\;E\{(x-\mu)^4\}-\sigma^4\right).$$

8.2   GMM with Integrated Squared Moments

Given a conditional moment condition $E\{\psi(z,\beta)|x\}=0$, it holds that

$$E\{\psi(z,\beta)|x\}=0\;\Longrightarrow\;E\{\psi(z,\beta)1[x\le a]\}=0\;\forall a$$

$$\Longrightarrow\int[E\{\psi(z,\beta)1[x\le a]\}]^2\,dF_x(a)=0$$

Dominguez and Lobato (2004) proposed to estimate β by minimizing a sample analog for the last expression:

$$\frac{1}{\sqrt{N}} \sum_{j=1}^{N} \left\{ \frac{1}{N} \sum_{i=1}^{N} \psi(z_i, b) 1[x_i \le x_j] \right\}^2.$$

Although the estimator is not efficient under $E\{\psi(z,\beta)|x\} = 0$, integration tends to "smooth rough edges," and thus the estimator may behave better than other nonlinear GMM minimands, leading to fewer convergence problems in practice.

Define

$$\nabla \Psi(a) \equiv E\left\{\psi_b(\beta) 1[x \le a]\right\}, \quad H \equiv \int \nabla \Psi(a) \nabla \Psi(a)' dF_x(a) \quad \text{and}$$

$$\Gamma(a_1, a_2) \equiv E\left\{\psi(\beta)^2 1\left[x \le \min(a_1, a_2)\right]\right\},$$

$$\Omega \equiv \int \int \nabla \Psi(a_1) \nabla \Psi(a_2)' \Gamma(a_1, a_2) dF_x(a_1) dF_x(a_2).$$

The asymptotic distribution is

$$\sqrt{N}(b_N - \beta) \rightsquigarrow N(0, \ H^{-1}\Omega H^{-1}).$$

H and Ω can be estimated consistently with

$$H_N \equiv \frac{1}{N} \sum_{j=1}^{N} \left[\left\{ \frac{1}{N} \sum_{i=1}^{N} \psi_b(\beta) 1[x_i \le x_j] \right\} \left\{ \frac{1}{N} \sum_{i=1}^{N} \psi_{b'}(\beta) 1[x_i \le x_j] \right\} \right]$$

$$\Omega_N \equiv \frac{1}{N^2} \sum_{j=1, j'=1}^{N} \left[\left\{ \frac{1}{N} \sum_{i=1}^{N} \psi_b(\beta) 1[x_i \le x_j] \right\} \left\{ \frac{1}{N} \sum_{i=1}^{N} \psi_{b'}(\beta) 1[x_i \le x_{j'}] \right\} \right.$$

$$\left. \cdot \left\{ \frac{1}{N} \sum_{i=1}^{N} \psi(z_i, b_N)^2 1\left[x_i \le \min(x_j, x_{j'})\right] \right\} \right].$$

To understand the asymptotic distribution, examine the first-order condition (divided by 2):

$$\frac{1}{N} \sum_{j=1}^{N} \left\{ \frac{1}{N} \sum_{i=1}^{N} \psi_b(z_i, b_N) 1[x_i \le x_j] \right\} \left\{ \frac{1}{\sqrt{N}} \sum_{i=1}^{N} \psi(z_i, b_N) 1[x_i \le x_j] \right\} = 0.$$

Apply Taylor's expansion around $b_N = \beta$ to get (the term with $\psi_{bb'}$ is negligible because the term includes $N^{-1} \sum_{i=1}^{N} \psi(z_i, \beta) 1[x_i \le x_j] = o_p(1)$)

$$\frac{1}{N} \sum_{j=1}^{N} \left\{ \frac{1}{N} \sum_{i=1}^{N} \psi_b(z_i, \beta) 1[x_i \le x_j] \right\} \left\{ \frac{1}{\sqrt{N}} \sum_{i=1}^{N} \psi(z_i, \beta) 1[x_i \le x_j] \right\}$$

$$+ H_N \sqrt{N}(b_N - \beta) = o_p(1)$$

$$\implies \sqrt{N}(b_N - \beta) = -H_N^{-1} \frac{1}{N} \sum_{j=1}^{N} \left\{ \frac{1}{N} \sum_{i=1}^{N} \psi_b(z_i, \beta) 1[x_i \leq x_j] \right\}$$

$$\left\{ \frac{1}{\sqrt{N}} \sum_{i=1}^{N} \psi(z_i, \beta) 1[x_i \leq x_j] \right\} + o_p(1).$$

The term next to H_N^{-1} determines the asymptotic distribution as follows.

First, the last term $N^{-1/2} \sum_{i=1}^{N} \psi(z_i, \beta) 1[x_i \leq x_j]$ converges in law to a Gaussian process indexed by x_j. Second, the term $N^{-1} \sum_{i=1}^{N} \psi_b$ $(z_i, \beta) 1[x_i \leq x_j]$ converges to a constant matrix indexed by x_j. Third, the outer sum $N^{-1} \sum_j$ yields the convergence in law to a weighted integral of the Gaussian process—integral due to x_j. Fourth, it is known that a weighted integral of a Gaussian process is also Gaussian (i.e., normal). This means that $\sqrt{N}(b_N - \beta)$ is asymptotically normal. The expected value of the term next to H_N^{-1} goes to 0 as $N \to \infty$, because it is

$$E \left[\left\{ \frac{1}{N} \sum_{i=1}^{N} \psi_b(z_i, \beta) 1 [x_i \leq x_j] \right\} \left\{ \frac{1}{\sqrt{N}} \sum_{i'=1}^{N} \psi(z_{i'}, \beta) 1 [x_{i'} \leq x_j] \right\} \right]$$

$$= \frac{1}{N^{3/2}} E\{\psi_b(z_i, \beta) \psi(z_i, \beta) 1[x_i \leq x_j]\} \quad \text{(the terms with } i \neq i'$$

have mean zero).

In the following, we derive the asymptotic variance which will then explain the form of Ω_N.

As for the variance, replace $N^{-1} \sum_{i=1}^{N} \psi_b(z_i, \beta) 1[x_i \leq x_j]\}$ with $E\{\psi_b$ $(z, \beta) 1[x \leq x_j] | x_j\}$ to examine instead

$$\frac{1}{N} \sum_{j=1}^{N} \left[E \{\psi_b(z, \beta) 1[x \leq x_j] | x_j\} \left\{ \frac{1}{\sqrt{N}} \sum_{i=1}^{N} \psi(z_i, \beta) 1[x_i \leq x_j] \right\} \right].$$

The variance of this display is its second moment since the mean term is negligible:

$$\frac{1}{N^3} \sum_{j=1}^{N} \sum_{j'=1}^{N} \sum_{i=1}^{N} \sum_{i'=1}^{N} E[\ E\{\psi_b(z, \beta) 1[x \leq x_j] | x_j\}$$

$$E\{\psi_b(z, \beta) 1[x \leq x_{j'}] | x_{j'}\} \cdot \psi(z_i, \beta) 1[x_i \leq x_j] \ \psi(z_{i'}, \beta) 1[x_{i'} \leq x_{j'}] \ ].$$

All terms with $i \neq i'$ disappear, and thus this becomes

$$\frac{1}{N^3} \sum_{j=1}^{N} \sum_{j'=1}^{N} \sum_{i=1}^{N} E[\ E\{\psi_b(z, \beta) 1[x \leq x_j] | x_j\} \ E\{\psi_b(z, \beta) 1[x \leq x_{j'}] | x_{j'}\}$$

$$\cdot \psi(z_i, \beta)^2 1[x_i \le \min(x_j, x_{j'})]\;]$$

$$= \frac{1}{N^2} \sum_{j=1}^{N} \sum_{j'=1}^{N} E[\; E\{\psi_b(z, \beta) 1[x \le x_j] | x_j\} \; E\{\psi_b(z, \beta) 1[x \le x_{j'}] | x_{j'}\}$$

$$\cdot \frac{1}{N} \sum_{i=1}^{N} \psi(z_i, \beta)^2 1[x_i \le \min(x_j, x_{j'})]\;].$$

Ω_N is a sample analog for this, i.e., a version without the outermost expected value.

In practice, the estimator may be implemented by the usual Gauss–Newton type algorithm. That is, first, take the minimand as $N^{-1} \sum_j q_j$ (b) where $q_j(b) \equiv \{N^{-1} \sum_{i=1}^{N} \psi(z_i, b) 1[x_i \le x_j]\}^2$. Second, obtain the numerical derivative $\nabla q_j(b)$ of $q_j(b)$ wrt b. Third, iterate until convergence using

$$b_1 = b_0 - \left\{ \sum_j \nabla q_j(b_0) \nabla q_j(b_0)' \right\}^{-1} \sum_j \nabla q_j(b_0).$$

The asymptotic variance may be estimated in the usual "sandwich form"— the outer-product of the gradient flanked by the Hessians. The Hessian can also be estimated with numerical derivatives. Some simulation studies suggest, however, that neither the sandwich form nor the above estimator for the asymptotic variance works reliably; bootstrap may be a better alternative for the asymptotic variance.

8.3 Goodness-of-Fit Tests for Distribution Functions

8.3.1 Brownian Motion and Brownian Bridge

Suppose we have iid data $y_1, ..., y_N$ from a df F and desire to test

$$H_0 : F = F_0 \quad \text{for some specified continuous df } F_0.$$

With the empirical df $F_N(t) \equiv N^{-1} \sum_i 1[y_i \le t]$, the *Kolmogorov–Smirnov test* statistic is

$$KS_N \equiv \sup_t \sqrt{N} |F_N(t) - F_0(t)| = \sup_t |\frac{1}{\sqrt{N}} \sum_i \{1[y_i \le t] - F_0(t)\}|$$

$$= \sup_t |\frac{1}{\sqrt{N}} \sum_i \{1[F_0(y_i) \le F_0(t)] - F_0(t)\}|$$

$$= \sup_{\tau \in [0,1]} |U_N(\tau)|, \quad \text{where } U_N(\tau) \equiv \frac{1}{\sqrt{N}} \sum_i \{1[F_0(y_i) \le \tau] - \tau\}$$

with $\tau = F_0(t)$.

Under H_0, $U_N(\tau)$ is an empirical df of the uniform rv's $F_0(y_i) \sim U[0,1]$, $i = 1, ..., N$. $\sqrt{N}\{F_N(t) - F_0(t)\}$ is called an *empirical process*, and $U_N(\tau)$ is an uniform empirical process. A natural question to arise is how to find the asymptotic distribution of KS_N, for which "Brownian bridge" is needed.

"*Brownian motion*" or "*Wiener process*" $W(t)$, $0 \le t \le T$, has continuous sample path that starts from 0 (i.e., $W(0) = 0$) and, for any finite number of points $t_1, ..., t_m$,

$$\{W(t_1), ..., W(t_m)\} \text{ is Gaussian with 0-mean and}$$
$$E\{W(t_j)W(t_k)\} = \min(t_j, t_k).$$

This implies $E\{W(t)^2\} = t$ when $t_j = t_k = t$ and

$$W(t) - W(s) \sim N(0, t - s) \ \forall s < t \ (\Longrightarrow W(t) \sim N(0,t)) \text{ because}$$
$$E\{W(t) - W(s)\}^2 = E\{W(t)^2\} - 2E\{W(t)W(s)\} + E\{W(s)^2\}$$
$$= t - 2s + s = t - s.$$

Also $W(t)$ has independent increments:

$$W(t_4) - W(t_3) \text{ is independent of } W(t_2) - W(t_1), \quad \forall \ t_1 < t_2 \le t_3 < t_4$$
$$\text{because } E[\{W(t_4) - W(t_3)\}\{W(t_2) - W(t_1)\}] = t_2 - t_2 - t_1 + t_1 = 0.$$

As well known as Brownian motion is a *Brownian bridge* $B(\tau)$ with $0 \le \tau \le 1$: $B(0) = B(1) = 0$ with continuous sample path, and for any finite number of points $\tau_1, ..., \tau_m$,

$$B(\tau_1), ..., B(\tau_m) \text{ is Gaussian with 0-mean, and}$$
$$E\{B(\tau_j)B(\tau_k)\} = \min(\tau_j, \tau_k) - \tau_j\tau_k$$
$$\Longrightarrow V\{B(\tau)\} = \tau - \tau^2 = \tau(1 - \tau).$$

Note the difference between the covariance functions of $W(t)$ and $B(\tau)$. While $W(t)$ has independent increments, $B(\tau)$ does not because $B(\tau)$ has to satisfy $B(1) = 0$. Because of this, $B(\tau)$ is also called "tied-down" Brownian motion— down to 0 at $t = 0, 1$.

Going back to $U_N(\tau)$, observe, for any two points $\tau_1, \tau_2 \in [0, 1]$,

$$E[U_N(\tau_1)U_N(\tau_2)] = E\left[\left\{ \frac{1}{\sqrt{N}} \sum_i (1[u_i \le \tau_1] - \tau_1) \right\} \cdot \right.$$
$$\left. \frac{1}{\sqrt{N}} \sum_j (1[u_j \le \tau_2] - \tau_2)\} \right]$$
$$= E\{1[u \le \tau_1] \, 1[u \le \tau_2]\} - E1[u \le \tau_1] \, E1[u \le \tau_2]$$
$$= \min(\tau_1, \tau_2) - \tau_1\tau_2$$

which implies, setting $\tau_1 = \tau_2 = \tau$,

$$V\{U_N(\tau)\} = \tau(1 - \tau) \Longrightarrow$$

$$V\left[\sqrt{N}\{F_N(t) - F_0(t)\}\right] = F_0(t)\{1 - F_0(t)\}, \text{recalling } \tau = F_0(t).$$

The covariance of $U_N(\tau)$ at any finite number of points is the same as that of $B(\tau)$. This suggests that the distribution of $U_N(\tau)$ may be found from that of $B(\tau)$, and this convergence in law indeed holds:

$$U_N(\tau) \rightsquigarrow B(\tau) \text{ as } N \to \infty, \quad \text{i.e.,}$$

$$\int f\{U_N(\omega, \tau)\} dP(\omega) \to \int f\{B(\omega', \tau)\} dP(\omega')$$

for each bounded and continuous function $f(\cdot)$. In essence, this gives the distribution not just over a finite number of points, but over the entire range $[0, 1]$ of τ, which enables invoking "continuous mapping theorem" as follows.

8.3.2 Kolmogorov–Smirnov (KS) test

Since $\sup_{\tau \in [0,1]} |\cdot|$ is a continuous function, applying the continuous mapping theorem to $U_N(\tau) \rightsquigarrow B(\tau)$ yields

$$KS_N = \sup_{\tau \in [0,1]} |U_N(\tau)| \rightsquigarrow \sup_{\tau \in [0,1]} B(\tau).$$

We thus obtain, for any constant $\varepsilon > 0$,

$$P(KS_N \le \varepsilon) \to P\left\{ \sup_{\tau \in [0,1]} |B(\tau)| \le \varepsilon \right\} = 1 + 2\sum_{j=1}^{\infty} (-1)^j \exp(-2j^2\varepsilon^2);$$

this equality is well known, as can be seen in Serfling (1980).

As a specific example, suppose $\{y_1, ..., y_N\}$ is from $N(0, 1)$. Using the observation points as discrete evaluation points, we get

$$KS_N \simeq \sup_{i=1,...,N} \sqrt{N}|F_N(y_i) - \Phi(y_i)|.$$

Suppose this yields $KS_N = 0.65$. Then the p-value is

$$P(KS_N > 0.65) = -2\sum_{j=1}^{\infty} (-1)^j \exp(-2j^2 0.65^2)$$

$$\simeq -2\sum_{j=1}^{500} (-1)^j \exp(-2j^2 0.65^2).$$

This test is asymptotically distribution-free, as this display does not depend on any aspect of $F_0 = \Phi$. If we do not want to compute this sum, then we can just use the 90, 95, and 99% quantiles of KS_N which are, respectively, 1.23, 1.36, and 1.63; see e.g., Shorack and Wellner (1986).

In contrast to $U_N(\tau) \rightsquigarrow B(\tau)$, when we have v_i's which are iid $(0, \sigma_v^2)$, with $[N\tau]$ denoting the integer part of $N\tau$ and $0 \leq \tau \leq 1$, another well-known convergence in law for a "partial sum process" is

$$S_N(\tau) \equiv \frac{1}{\sqrt{N}} \sum_{i=1}^{[N\tau]} \frac{v_i}{\sigma_v} \rightsquigarrow W(\tau), \ 0 \leq \tau \leq 1.$$

To see this at a finite number of points, observe $E\{S_N(\tau)\} = 0$, and with $\tau_1 \leq \tau_2$,

$$E\left\{S_N(\tau_1)S_N(\tau_2)\right\} = E\left\{\frac{1}{\sigma_v\sqrt{N}} \sum_{i=1}^{[N\tau_1]} v_i \cdot \frac{1}{\sigma_v\sqrt{N}} \sum_{j=1}^{[N\tau_2]} v_j\right\}$$

$$= \frac{1}{\sigma_v^2 N} \sum_{i=1}^{[N\tau_1]} E(v_i^2) = \frac{[N\tau_1]}{N} = \tau_1;$$

$$\frac{[N\tau]}{N} = \tau\frac{[N\tau]}{N\tau} = \tau\left(1 - \frac{N\tau - [N\tau]}{N\tau}\right) \to \tau \text{ as } N \to \infty$$

$$\text{because } \frac{N\tau - [N\tau]}{N\tau} \to 0.$$

Hence, for any finite number of points, the covariance of $S_N(\tau)$ is the same as that of $W(\tau)$. This suggests $S_N(\tau) \rightsquigarrow W(\tau)$, which indeed holds.

8.3.3   Cramer–von-Mises (CM) and Anderson–Darling (AD) tests

Another test statistic for $H_0 : F = F_0$ is the *Cramer–von-Mises test* (CM) statistic:

$$CM_N \equiv \int_{-\infty}^{\infty} [\sqrt{N}\{F_N(t) - F_0(t)\}]^2 dF_0(t).$$

While KS_N is determined by the largest deviation, all deviations (small or large) contribute to CM_N. That is, if F_0 is violated by small magnitudes over many points, then CM_N is likely to be more powerful than KS_N. Using the convergence in law again, we get, setting $\tau = F_0(t) \Longrightarrow d\tau = dF_0(t)$,

$$CM_N = \int_0^1 U_N(\tau)^2 d\tau \rightsquigarrow \int_0^1 B(\tau)^2 d\tau \sim \sum_{j=1}^{\infty} \frac{z_j^2}{j^2\pi^2} \text{ where } z_j\text{'s are iid } N(0,1).$$

The last expression that the "quadratic" test statistic CM_N (and "AD_N" below) follows the same distribution as an infinite "weighted" sum of independent χ_1^2 rv's can be shown using "principal component decomposition" (Shorack and Wellner, 1986, Chapter 5).

The 90, 95, and 99% quantiles of CM_N are, respectively, 0.347, 0.461, and 0.743 (see Shorack and Wellner, 1986); e.g., $P\left(\sum_{j=1}^{\infty} z_j^2/(j^2\pi^2) \leq 0.347\right)$

$= 0.9$. As a specific example, suppose $\{y_1, ..., y_N\}$ is from $N(0,1)$. Choosing grid points $t_1, ..., t_J$ over $[-5.5]$ in increment of 0.01 (i.e., $t_1 = -5$, $t_2 = -5 + 0.01$, $t_3 = -5 + 0.02, ...$), we can compute

$$CM_N \simeq \sum_{j=1}^{J} \left[\sqrt{N}\{F_N(t_j) - \Phi(t_j)\} \right]^2 \phi(t_j) \cdot 0.01.$$

Various weighted versions of KS_N and CM_N are available as well. One well-known version of CM_N is "*Anderson–Darling test statistic*" (AD):

$$AD_N \equiv \int_{-\infty}^{\infty} \frac{\left[\sqrt{N}\{F_N(t) - F_0(t)\} \right]^2}{F_0(t)\{1 - F_0(t)\}} dF_0(t)$$

where the weighting is done by the asymptotic variance at each t. Analogously to the above convergence in law, it holds that

$$AD_N \rightsquigarrow \int_0^1 \frac{B(\tau)^2}{\tau(1-\tau)} d\tau \sim \sum_{j=1}^{\infty} \frac{z_j^2}{j(j+1)}.$$

The 90, 95, and 99% quantiles of AD_N are, respectively, 1.93, 2.49, and 3.85. See Del Barrio (2007) for a review on goodness-of-fit (GOF) tests.

8.4   Joint Test for All Quantiles

For a linear model $y = x'\beta_\alpha + u$ with $Q_\alpha(u|x) = 0$, recall the αth quantile minimand:

$$V_N(\alpha) \equiv \frac{1}{\sqrt{N}} \sum_i (y_i - x_i'b_\alpha) \cdot (\alpha - 1[y_i - x_i'b_\alpha < 0])$$

and the "asymptotic first-order condition"

$$\frac{1}{\sqrt{N}} \sum_i -(\alpha - 1[y_i - x_i'b_\alpha < 0]) x_i = o_p(1).$$

Also recall the asymptotic variance for $\sqrt{N}(\hat{b}_\alpha - \beta_\alpha)$:

$$E^{-1}\{f_{u|x}(0)xx'\} \cdot \alpha(1-\alpha)E(xx') \cdot E^{-1}\{f_{u|x}(0)xx'\}$$

and a kernel estimator for $E\{f_{u|x}(0)xx'\}$:

$$\frac{1}{N} \sum_i \frac{1}{h} K\left(\frac{r_i}{h}\right) x_i x_i' \quad \text{where } h \to 0 \text{ as } N \to \infty \text{ and } r_i \equiv y_i - x_i'\hat{b}_\alpha.$$

With these, one can conduct tests for β_α.

Differently from the mean regression, however, there are many quantiles as β_α is indexed by α. Thus, if we want to know whether a regressor x_k has any influence at all on any quantile, then

$$H_o : \beta_\alpha = 0 \; \forall \alpha$$

should be tested. We explore this topic in this subsection, drawing on Koenker and Machado (1999). This topic could have been discussed in the chapter for nonlinear models, but given its difficulty, the discussion has been postponed and done here now as quantiles are heavily used in semiparametrics.

Assume that u has a continuous df F with density f and $u \amalg x$, although the independence assumption is weakened somewhat in Koenker and Machado (1999). Denote the αth quantile of u as $F^{-1}(\alpha)$ that is zero for some α where $F^{-1}(\alpha) \equiv \min\{a : F^{-1}(a) \geq \alpha\}$. Under this and $u \amalg x$, the above asymptotic variance becomes

$$\frac{\alpha(1-\alpha)}{[f\{F^{-1}(\alpha)\}]^2} E^{-1}(xx') = \left(\frac{[f\{F^{-1}(\alpha)\}]^2}{\alpha(1-\alpha)} E(xx') \right)^{-1}.$$

With x being a $k \times 1$ random vector, consider

$$H_o : \text{the last } q \text{ components of } \beta_\alpha \text{ are zero } \forall \alpha \in (0,1).$$

Denoting the first $k - q$ components of x and β_α as z and γ_α, respectively, imposing this H_o means minimizing

$$V_N^r(\alpha) \equiv \sum_i (y_i - z_i' g_\alpha) \cdot (\alpha - 1[y_i - z_i' g_\alpha < 0]).$$

A LR-type test statistic at a fixed α is

$$\begin{aligned} L_N(\alpha) &\equiv \frac{f\{F^{-1}(\alpha)\}}{\alpha(1-\alpha)} 2\{V_N^r(\alpha) - V_N(\alpha)\} \rightsquigarrow Q_q(\alpha)^2 \\ &\equiv \frac{B_1(\alpha)^2 +, ..., + B_q(\alpha)^2}{\alpha(1-\alpha)} \end{aligned}$$

where $B_j(\alpha)$, $j = 1, ..., q$, are independent Brownian bridges. The process $Q_q(\alpha)$, the square-root of $Q_q(\alpha)^2$, is called a "standardized" *Bessel process* of order q. We have $Q_q(\alpha)^2 \sim \chi_q^2 \; \forall \alpha$ because of the standardization $B_j(\alpha)/\sqrt{\alpha(1-\alpha)} \sim N(0,1)$, $j = 1, ..., q$, and because $Q_q(\alpha)^2$ is a sum of q-many squared independent $N(0,1)$ variables. The intuition for $L_N(\alpha) \rightsquigarrow \chi_q^2$ is provided in the following.

Denote the unrestricted and restricted estimators as $b_N(\alpha)$ and $g_N(\alpha)$, respectively; the last q components of $g_N(\alpha)$ are zero. Although $V_N^r(\alpha)$ is not differentiable everywhere, a second-order asymptotic expansion around

$b_N(\alpha)$ holds that is analogous to the second-order expansion of log-likelihood functions for LR tests. That is,

$$V_N^r(\alpha) = V_N(\alpha) + \frac{1}{2}\sqrt{N}\left\{g_N(\alpha) - b_N(\alpha)\right\}' \; f\left\{F^{-1}(\alpha)\right\} E(xx')$$
$$\sqrt{N}\{g_N(\alpha) - b_N(\alpha)\} + o_p(1)$$

where the first-order term disappears because it is the asymptotic first-order condition for b_N. Hence

$$\frac{f\{F^{-1}(\alpha)\}}{\alpha(1-\alpha)}2\{V_N^r(\alpha) - V_N(\alpha)\} \; \{= L_N(\alpha)\}$$
$$= \sqrt{N}\{g_N(\alpha) - b_N(\alpha)\}' \cdot \frac{[f\{F^{-1}(\alpha)\}]^2}{\alpha(1-\alpha)}E(xx') \cdot \sqrt{N}\{g_N(\alpha) - b_N(\alpha)\}$$
$$+ o_p(1) \rightsquigarrow \chi_q^2$$

because the middle matrix is the inverse of the asymptotic variance of the adjacent vector.

The appearance of Brownian bridges can be understood in view of the asymptotic first order condition: with 0 in $1[u < 0]$ replaced by $F^{-1}(\alpha)$,

$$o_p(1) = \frac{1}{\sqrt{N}}\sum_i -(\alpha - 1[u < F^{-1}(\alpha)])x_i$$
$$= \frac{1}{\sqrt{N}}\sum_i -(\alpha - 1[F(u) < \alpha])x_i \quad \text{and}$$
$$E\{(\alpha - 1[F(u) < \alpha]) \cdot (\alpha' - 1[F(u) < \alpha'])\}$$
$$= \min(\alpha, \alpha') - \alpha\alpha', \quad \text{as } F(u) \sim U[0,1].$$

The above convergence in law is for a given α. For the H_o, owing to the continuous mapping theorem, we can use the KS-type test statistic that jointly takes all $\alpha \in (0,1)$ into account:

$$\sup_\alpha L_N(\alpha) \rightsquigarrow \sup_\alpha Q_q(a)^2.$$

Since the critical values for the upper tail probabilities 1, 5, and 10% of $Q_q(t)^2$ are tabulated in Andrews (1993), we can conduct the test comparing $\sup_\alpha L_N(\alpha)$ to a critical value. Koenker and Machado (1999) also explored LM and Wald-type tests.

Koenker and Xiao (2002) examined more general tests allowing nuisance parameters in the null hypothesis. Specifically, Koenker and Xiao (2002) looked at the null hypothesis that x influences only the location or scale in the $y|x$ distribution versus the more general alternative that x can influence the conditional distributional shape as well. Since the location and scale have to be estimated under the null, this causes a nuisance-parameter problem that the nuisance-parameter estimator affects the asymptotic distribution of

the ensuing test statistic that is no longer asymptotically distribution-free
(ADF). Koenker and Xiao (2002) overcame this problem by transforming
the test statistic so that it becomes ADF (see also Koenker, 2005); we will
call such a tranformtion "ADF transformation." ADF transformation is also
called the "Khamaladze" transformation," following the name of the inven-
tor in Khamaladze (1981, 1993). Bai (2003) provided a relatively easier-to-
understand account of ADF transformation; see also Koul (2006) for a review.
ADF transformation will be examined later.

9 Appendix for Chapter 9

9.1 Asymptotic Variance of Marginal Integration

In the main text, we used the following representation of the marginal
integration estimator $\hat{m}_1(x_1)$ to derive the asymptotic variance of $\sqrt{Nh}\{\hat{m}_1$
$(x_1) - m_1(x_1)\}$:

$$\sqrt{Nh}\{\hat{m}_1(x_1) - m_1(x_1)\} - \frac{1}{\sqrt{N}}\sum_i \left[\frac{1}{\sqrt{h}}L\left(\frac{x_{i1} - x_1}{h}\right)\frac{1}{f_{1|2}(x_1|x_{i2})}\right.$$

$$\left.\{y_i - m(x_1, x_{i2})\}\right] = o_p(1).$$

In this subsection, we sketch the main steps to obtain this display.
Recall the marginal integration estimator $\hat{m}_1(x_1) = N^{-1}\sum_i$
$\hat{m}(x_1, x_{i2})$. For its asymptotic distribution, define $\hat{g}$ and $\hat{f}$ such that

$$\hat{m}(x_1, x_{i2}) = \frac{\hat{g}(x_1, x_{i2})}{\hat{f}(x_1, x_{i2})} = \frac{(Nh^2)^{-1}\sum_{j\neq i}L((x_{j1} - x_1)/h))L((x_{j2} - x_{i2})/h)y_j}{(Nh^2)^{-1}\sum_{j\neq i}L((x_{j1} - x_1)/h))L((x_{j2} - x_{i2})/h)}$$

to see that

$$\sqrt{Nh^2}\{\hat{m}(x_1, x_{i2}) - m(x_1, x_{i2})\}$$

$$= \frac{-g(x_1, x_{i2})}{f(x_1, x_{i2})^2}\sqrt{Nh^2}\left\{\hat{f}(x_1, x_{i2}) - f(x_1, x_{i2})\right\}$$

$$+ \frac{1}{f(x_1, x_{i2})}\sqrt{Nh^2}\{\hat{g}(x_1, x_{i2} - g(x_1, x_{i2})\} + o_p(1).$$

Observe

$$\sqrt{Nh}\{\hat{m}_1(x_1) - m_1(x_1)\} = \sqrt{Nh}\left\{\frac{1}{N}\sum_i\{\hat{m}(x_1, x_{i2}) - m(x_1, x_{i2})\} + o_p(1)\right.$$

$$= \frac{1}{N\sqrt{h}}\sum_i\sqrt{Nh^2}\left\{\hat{m}(x_1, x_{i2}) - m(x_1, x_{i2})\right\}\right\}$$

and substitute $\sqrt{Nh^2}\{\hat{m}(x_1, x_{i2}) - m(x_1, x_{i2})\}$ to get

$$\sqrt{Nh}\{\hat{m}_1(x_1) - m_1(x_1)\}$$

$$= \frac{1}{N\sqrt{h}}\sum_i \frac{-g(x_1, x_{i2})}{f(x_1, x_{i2})^2}\sqrt{Nh^2}\{\hat{f}(x_1, x_{i2}) - f(x_1, x_{i2})\}$$

$$+ \frac{1}{N\sqrt{h}}\sum_i \frac{1}{f(x_1, x_{i2})}\sqrt{Nh^2}\{\hat{g}(x_1, x_{i2}) - g(x_1, x_{i2})\} + o_p(1).$$

The two terms are examined in detail in the following one by one.
Note

$$\hat{f}(x_1, x_{i2}) - f(x_1, x_{i2})$$

$$= \frac{1}{N-1}\sum_{j \neq i}\left\{\frac{1}{h^2}L\left(\frac{x_{j1} - x_1}{h}\right)L\left(\frac{x_{j2} - x_{i2}}{h}\right) - f(x_1, x_{i2})\right\}$$

and examine the first term of $\sqrt{Nh}\{\hat{m}_1(x_1) - m_1(x_1)\}$:

$$\frac{1}{N(N-1)}\sum_i\sum_{j \neq i}\frac{\sqrt{Nh^2}}{\sqrt{h}}\frac{-g(x_1, x_{i2})}{f(x_1, x_{i2})^2}$$

$$\cdot\left\{\frac{1}{h^2}L\left(\frac{x_{j1} - x_1}{h}\right)L\left(\frac{x_{j2} - x_{i2}}{h}\right) - f(x_1, x_{i2})\right\}$$

$$= \frac{1}{\sqrt{N}(N-1)}\sum_i\sum_{j \neq i}\frac{-g(x_1, x_{i2})}{f(x_1, x_{i2})^2}$$

$$\cdot\left\{\frac{1}{h^{3/2}}L(\frac{x_{j1} - x_1}{h})L(\frac{x_{j2} - x_{i2}}{h}) - \sqrt{h}f(x_1, x_{i2})\right\}.$$

Rewrite the summand to get

$$U_N \equiv \frac{1}{\sqrt{N}(N-1)}\sum_i\sum_{j > i}\left[-\frac{g(x_1, x_{i2})}{f(x_1, x_{i2})^2}\right.$$

$$\cdot\left\{\frac{1}{h^{3/2}}L\left(\frac{x_{j1} - x_1}{h}\right)L\left(\frac{x_{j2} - x_{i2}}{h}\right) - \sqrt{h}f(x_1, x_{i2})\right\}$$

$$\left. - \frac{g(x_1, x_{j2})}{f(x_1, x_{j2})^2}\left\{\frac{1}{h^{3/2}}L\left(\frac{x_{i1} - x_1}{h}\right)L\left(\frac{x_{i2} - x_{j2}}{h}\right) - \sqrt{h}f(x_1, x_{j2})\right\}\right].$$

Denote the summand as $S(z_i, z_j)$ and use the U-statistic projection theorem to get

$$U_N = \frac{1}{\sqrt{N}}\sum_i E\{S(z_i, z_j)|z_i\} + o_p(1)$$

where

$$E\{S(z_i, z_j)|z_i\} = -\frac{g(x_1, x_{i2})}{f(x_1, x_{i2})^2}\left\{\frac{1}{h^{3/2}}\right.$$

$$\cdot \int L\left(\frac{x_{j1}-x_1}{h}\right) L\left(\frac{x_{j2}-x_{i2}}{h}\right) f(x_{j1},x_{j2}) dx_{j1} dx_{j2} - \sqrt{h} f(x_1,x_{i2}) \bigg\}$$

$$- \bigg\{ \frac{1}{h^{3/2}} L\left(\frac{x_{i1}-x_1}{h}\right) \int \frac{g(x_1,x_{j2})}{f(x_1,x_{j2})^2} L\left(\frac{x_{i2}-x_{j2}}{h}\right) f(x_{j2}) dx_{j2}$$

$$- \sqrt{h} \int \frac{g(x_1,x_{j2})}{f(x_1,x_{j2})} f(x_{j2}) dx_{j2} \bigg\}$$

$$= -\frac{1}{\sqrt{h}} L\left(\frac{x_{i1}-x_1}{h}\right) \frac{m(x_1,x_{i2})}{f(x_1|x_{i2})} + o_p(1)$$

because the first term in $E\{S(z_i,z_j)|z_i\}$ is negligible and the second term uses

$$\frac{g(x_1,x_{i2})}{f(x_1,x_{i2})^2} f(x_{i2}) = \frac{g(x_1,x_{i2})}{f(x_1,x_{i2})} \frac{f(x_{i2})}{f(x_1,x_{i2})} = \frac{m(x_1,x_{i2})}{f(x_1|x_{i2})}.$$

As for the second term of $\sqrt{Nh}\{\hat{m}_1(x_1) - m_1(x_1)\}$, use

$$\hat{g}(x_1,x_{i2}) - g(x_1,x_{i2})$$

$$= \frac{1}{N-1} \sum_{j\neq i} \bigg\{ \frac{1}{h^2} L\left(\frac{x_{j1}-x_1}{h}\right) L\left(\frac{x_{j2}-x_{i2}}{h}\right) y_j - g(x_1,x_{i2}) \bigg\}$$

to rewrite it as

$$\frac{1}{\sqrt{N}(N-1)} \sum_i \sum_{j\neq i} \frac{1}{f(x_1,x_{i2})}$$

$$\cdot \bigg\{ \frac{1}{h^{3/2}} L\left(\frac{x_{j1}-x_1}{h}\right) L\left(\frac{x_{j2}-x_{i2}}{h}\right) y_j - \sqrt{h} g(x_1,x_{i2}) \bigg\}.$$

Rewrite this further as

$$V_N \equiv \frac{1}{\sqrt{N}(N-1)} \sum_i \sum_{j>i} \bigg[\frac{1}{f(x_1,x_{i2})}$$

$$\cdot \bigg\{ \frac{1}{h^{3/2}} L\left(\frac{x_{j1}-x_1}{h}\right) \quad L\left(\frac{x_{j2}-x_{i2}}{h}\right) y_j - \sqrt{h} g(x_1,x_{i2}) \bigg\}$$

$$+ \frac{1}{f(x_1,x_{j2})} \bigg\{ \frac{1}{h^{3/2}} L\left(\frac{x_{i1}-x_1}{h}\right) L\left(\frac{x_{i2}-x_{j2}}{h}\right) y_i - \sqrt{h} g(x_1,x_{j2}) \bigg\} \bigg]$$

to get

$$U_N = \frac{1}{\sqrt{N}} \sum_i E\{T(z_i,z_j)|z_i\} + o_p(1)$$

where

$$E\{T(z_i,z_j)|z_i\}$$

$$
= \frac{1}{f(x_1, x_{i2})} \left\{ \frac{1}{h^{3/2}} \int L\left(\frac{x_{j1} - x_1}{h}\right) L\left(\frac{x_{j2} - x_{i2}}{h}\right) E(y|x_{j1}, x_{j2}) \right.
$$

$$
\left. f(x_{j1}, x_{j2}) dx_{j1} dx_{j2} - \sqrt{h} g(x_1, x_{i2}) \right\}
$$

$$
+ \left\{ \frac{1}{h^{3/2}} L\left(\frac{x_{i1} - x_1}{h}\right) y_i \int \frac{1}{f(x_1, x_{j2})} L\left(\frac{x_{i2} - x_{j2}}{h}\right) f(x_{j2}) dx_{j2} \right.
$$

$$
\left. - \sqrt{h} \int \frac{g(x_1, x_{j2})}{f(x_1, x_{j2})} f(x_{j2}) dx_{j2} \right\}
$$

$$
= \frac{1}{\sqrt{h}} L\left(\frac{x_{i1} - x_1}{h}\right) \frac{y_i}{f(x_1|x_{i2})} + o_p(1).
$$

Put the two terms together to obtain the desired expression:

$$
\sqrt{Nh} \{\hat{m}_1(x_1) - m_1(x_1)\}
$$

$$
= \frac{1}{\sqrt{N}} \sum_i \left[\frac{1}{\sqrt{h}} L\left(\frac{x_{i1} - x_1}{h}\right) \frac{1}{f(x_1|x_{i2})} \{y_i - m(x_1, x_{i2})\} \right] + o_p(1).
$$

9.2 CLT for Degenerate U-Statistics

Since degenerate U-statistics are used to derive the asymptotic distributions for some test statistics in the main text, here we present a CLT in De Jong (1987).

Consider a degenerate U-statistic

$$
W_N \equiv \sum_{i<j} w_{ijN}(z_i, z_j) \quad \text{where}
$$

$$
w_{ijN}(z_i, z_j) = w_{jiN}(z_j, z_i), \quad E\{w_{ijN}(z_i, z_j)|z_i\} = 0
$$

z_i's are independent, and w_{ijN} is square-integrable. Defining

$$
\sigma_N^2 \equiv V(W_N) \quad \text{and} \quad \sigma_{ij}^2 \equiv V(w_{ijN}),
$$

Theorem 2.1 of De Jong (1987) is that

$$
\frac{W_N}{\sigma_N} \rightsquigarrow N(0, 1) \quad \text{if} \quad \frac{1}{\sigma_N^2} \max_{1 \le i \le N} \sum_{j=1}^{N} \sigma_{ij}^2 \to 0 \quad \text{and} \quad \frac{E(W_N^4)}{\sigma_N^4} \to 3 \quad \text{as } N \to \infty.
$$

De Jong (1987) presented another CLT (Theorem 5.3) without the fourth moment condition. The theorem is more general than a CLT in Hall (1984) for degenerate U-statistics.

Let μ_{iN}, $i = 1...N$, be the eigenvalues of a symmetric matrix $[a_{ijN}]$, and let

$$
W_N \equiv \sum_{i<j} a_{ijN} \cdot w_N(z_i, z_j) \quad \text{where}
$$

$$w_N(z_i, z_j) = w_N(z_j, z_i), \quad E\{w_N(z_i, z_j)|z_i\} = 0$$

z_i's are iid and $w_N(z_i, z_j)$ is square-integrable with

$$E\{w_N(z_i, z_j)\}^2 = 1 \quad \text{and} \quad \sigma_N^2 \equiv V(W_N).$$

Then $W_N/\sigma_N \rightsquigarrow N(0,1)$, if there exists a sequence of real numbers $\{A_N\}$ such that the following (i) and (ii) hold:

$$(i): \frac{A_N^2}{\sigma_N^2} \max_{1 \le i \le N} \sum_{j=1}^{N} a_{ij}^2 \to 0, \quad E\{w_N^2(z_i, z_j) \cdot 1[|w_N(z_i, z_j)| > A_N]\} \to 0$$

$$(ii): \frac{1}{\sigma_N^2} \max_{1 \le i \le N} \mu_{iN}^2 \to 0 \quad \text{or}$$

$$E\{w_N(z_1, z_2)w_N(z_1, z_3)w_N(z_4, z_2) \quad w_N(z_4, z_3)\} \to 0.$$

APPENDIX II: SUPPLEMENTARY TOPICS

1   Appendix for Hypothesis Test

This section collects a number of topics on hypothesis test: comparison of tests and local alternatives, non-nested hypotheses, and χ^2 goodness-of-fit tests. As hypothesis testing is a huge topic, the review here is brief and relatively informal. Textbook discussion of the topics can be found in mathematical statistics books such as Shao (2003) and Hogg et al. (2005), and a detailed review is provided by Lehmann and Romano (2005).

1.1   Basics

Consider a parameter of interest $\beta \in B$, and two hypotheses

$$H_0 : \beta \in B_0 \quad \text{versus} \quad H_1 : \beta \in B_1 \quad \text{where } B_0 \cap B_1 = \varnothing.$$

If a hypothesis contains only a single point (e.g., $B_0 = \beta_0$) then the hypothesis is *simple*; otherwise, *composite*. Consider a test statistic $T_N \equiv T(x_1, ..., x_N)$ (e.g. $\bar{x} = N^{-1} \sum_i x_i$), a critical region C, and a test such that

$$H_0 \text{ is rejected, if } T_N \in C.$$

Then $1[T_N \in C]$ is called the *test function*, and the *power function* mapping β to $[0, 1]$ is defined as

$$\pi_N(\beta) \equiv E_\beta\{1[T(x_1, ..., x_N) \in C]\} = P(H_0 \text{ rejected} \mid \beta \text{ is true}), \ \beta \in B.$$

A test is said to be of *(significance) level* α if the supremum (i.e., the least upper bound) of the *Type I error probability* is α or smaller:

$$\alpha_N(B_0) \equiv \sup_{\beta \in B_0} \pi_N(\beta) \leq \alpha;$$

the supremum $\alpha_N(B_0)$ is called the *size*. If a test is of level α, it is also of level α' whenever $\alpha < \alpha'$. Thus, using only the smallest possible level (i.e., size) makes sense, and often level and size are used interchangeably. For $\beta \in B_1$, $\pi_N(\beta)$ is *power*, and $1 - \pi_N(\beta)$ is the *Type II error probability*.

We desire a test that minimizes $\pi_N(\beta)$ when $\beta \in B_0$ and maximizes $\pi_N(\beta)$ when $\beta \in B_1$. But no sensible test can achieve this feat, because the deterministic test that rejects always has power 1 and the deterministic test that accepts always has size 0. Instead, we look for a test with the maximum power among a set of tests sharing the same size. Such a test is hard to find, but when both H_0 and H_1 are simple, the LR test is optimal achieving this goal. This is known as the *Neyman–Pearson Theorem* (or Lemma), from which it follows that LR test is optimal also for one-sided alternative (H_0: $\beta = \beta_0$ and H_1: $\beta > \beta_0$) for a scalar parameter β because LR test has the largest power (optimal) for each point in B_1.

The test is *unbiased* if its power is at least as large as its size: $\pi_N(\beta) \geq \alpha_N(B_0)$ for all $\beta \in B_1$. Unbiasedness is a minimum requirement for any sensible test. For example, consider an "irrelevant" test of flipping a biased coin with $P(heads) = \alpha$ and rejecting the H_0 if heads come up. As this test does not depend on the data $x_1, ..., x_N$, its size and power are $P(heads) = \alpha$ always. An optimal test with size α should have a power greater than α, the power of the irrelevant test.

Given a test, if $\pi_N(\beta)$ and $\alpha_N(B_0)$ are difficult to assess with a finite N, then we let $N \to \infty$ and see what happens to $\pi_N(\beta)$ and $\alpha_N(B_0)$. A sequence of tests $T_1, T_2, ...$ is said to be of "*asymptotically level* α if

$$\lim_{N \to \infty} \sup \alpha_N(B_0) \leq \alpha;$$

$\limsup_{N \to \infty} \alpha_N(B_0)$ is the *limiting size* of the test sequence. A sequence of tests is "*consistent*" if $\pi_N(\beta) \to 1$ as $N \to \infty$ for all $\beta \in B_1$.

As a digression, concepts analogous to level and size in tests are used for confidence intervals (CI): if

$$\inf_{\beta \in B} P(T_{N1} \leq \beta \leq T_{N2}) \geq 1 - \alpha$$

then (T_{N1}, T_{N2}) is a CI of *(confidence) level* $1 - \alpha$ where T_{N1} and T_{N2} are some statistics with $T_{N1} \leq T_{N2}$. The infimum (i.e., the largest lower bound) is called the *confidence coefficient*. Typically $P(T_{N1} \leq \beta \leq T_{N2})$ does not depend on β, and in this case, the confidence level equals the confidence coefficient. If a CI is a level $1 - \alpha$ CI, then it is also a level $1 - \alpha'$ CI whenever $\alpha < \alpha'$. Thus, using only the largest possible level makes sense, and often confidence level and confidence coefficient are used interchangeably. In the CI, if $T_{N1} \to -\infty$, we get an one-sided confidence bound $(-\infty, T_{N2})$, and if $T_{N2} \to \infty$, we get an one-sided confidence bound (T_{N2}, ∞). More general than CI's is a *confidence set* R_N where R_N is a set depending on $x_1, ..., x_N$; R_N may take a shape other than an interval.

For instance, from a sample $x_1, ..., x_N$ with $E(x) = \beta$ with $SD(x) = \sigma$, we obtain its sample mean $\bar{x}_N$ and sample SD s_N. The usual test statistic for H_0: $\beta \leq 0$ (i.e., $B_0 = (-\infty, 0]$ and $B_1 = (0, \infty)$) is the t-value $T_N = \sqrt{N}\bar{x}_N/s_N$, and suppose that the critical region is $C = (1.645, \infty)$. In this case,

$$\pi_N(\beta) = P\left\{ \frac{\sqrt{N}\bar{x}_N}{s_N} \in (1.645, \infty) \text{ under } \beta \right\}$$

$$= P\left\{ \frac{\sqrt{N}(\bar{x}_N - \beta)}{s_N} \in \left(1.645 - \frac{\sqrt{N}\beta}{s_N}, \infty\right) \text{ under } \beta \right\}$$

$$\alpha_N\{(-\infty, 0]\}$$

$$= \sup_{\beta \in (-\infty, 0]} P\left\{ \frac{\sqrt{N}(\bar{x}_N - \beta)}{s_N} \in \left(1.645 - \frac{\sqrt{N}\beta}{s_N}, \infty\right) \text{ under } \beta \right\}.$$

In finite samples, these are difficult to assess. In large samples, letting $N \to \infty$ in $1.645 - \sqrt{N}\beta/s_N$ (note $s_N \to^p \sigma$), $\pi_N(\beta)$ converges to

$$0 = P(\, N(0,1) \in (\infty, \infty) \text{ under } \beta < 0\,)$$
$$0.05 = P(\, N(0,1) \in (1.645, \infty) \text{ under } \beta = 0\,)$$
$$1 = P(\, N(0,1) \in (-\infty, \infty) \text{ under } \beta > 0\,).$$

The last line with probability 1 shows that the test is consistent. As for $\alpha_N\{(-\infty, 0]\}$, observe

$$\sup_{\beta \in (-\infty, 0]} P\left\{ \frac{\sqrt{N}(\bar{x}_N - \beta)}{s_N} \in \left(1.645 - \frac{\sqrt{N}\beta}{s_N}, \infty\right) \text{ under } \beta \right\}$$
$$= P\{\, N(0,1) \in (1.645, \infty) \text{ under } \beta = 0\,\} = 0.05,$$
for the sup is attained at $\beta = 0$.

The test sequence (or simply the test) is of asymptotically level (size) 0.05.

1.2   Comparison of Tests and Local Alternatives

Almost all tests are consistent if the alternative model is at a fixed distance from the null model. Thus if we are to compare a group of tests, the alternative model should converge to the null model. Such an alternative model is called a *"local alternative"* or a *Pitman drift*. As most estimators are $\sqrt{N}$-consistent, interesting comparison can be made when the local alternative converges to the null model at rate $N^{-1/2}$.

1.2.1   Efficacy and Relative Efficiency

Consider two simple hypotheses

$$H_0 : \beta = \beta_0 \quad \text{and} \quad H_1 : \beta = \beta_0 + \frac{\delta}{\sqrt{N}} \equiv \beta_N \quad \text{where } \delta > 0.$$

Consider a test T_N that rejects H_0 with a large value (because $\delta > 0$) and satisfies, for some smooth functions $\mu(\cdot)$ and a constant σ,

$$\sqrt{N}\{T_N - \mu(\beta)\} \rightsquigarrow N(0, \sigma^2) \quad \text{as } N \to \infty \quad \text{where } \beta \text{ is the true}$$
parameter value;

note that the limiting distribution $N(0, \sigma^2)$ is free of δ. The asymptotic distribution of T_N under H_0 is a special case of this with $\beta = \beta_0$, and the test statistic for H_0 is $\sqrt{N}\{T_N - \mu(\beta_0)\}/\sigma$. Thus the power function is

$$P(\text{test statistic} > \text{critical value}) = P\left\{\frac{\sqrt{N}\{T_N - \mu(\beta_0)\}}{\sigma} > z_{1-\alpha}\right\}.$$

To assess the power, we need to derive the asymptotic distribution of the test statistic under H_1. Let $\nabla\mu$ denote the gradient of μ. Rewrite the power function as

$$P\left[\frac{\sqrt{N}\{T_N - \mu(\beta_N)\}}{\sigma} + \frac{\sqrt{N}\{\mu(\beta_N) - \mu(\beta_0)\}}{\sigma} > z_{1-\alpha}\right]$$

$$= P\left[\frac{\sqrt{N}\{T_N - \mu(\beta_N)\}}{\sigma} > z_{1-\alpha} - \frac{\sqrt{N}\{\mu(\beta_N) - \mu(\beta_0)\}}{\sigma}\right]$$

$$= P\left[\frac{\sqrt{N}\{T_N - \mu(\beta_N)\}}{\sigma} > z_{1-\alpha} - \frac{\sqrt{N}\{\delta'\nabla\mu(\beta_0)/\sqrt{N} + O(N^{-1})\}}{\sigma}\right]$$

$$= P\left[\frac{\sqrt{N}\{T_N - \mu(\beta_N)\}}{\sigma} > z_{1-\alpha} - \delta'\frac{\nabla\mu(\beta_0)}{\sigma} + O(N^{-1/2})\right\}$$

$$\to \Phi\left\{\delta'\frac{\nabla\mu(\beta_0)}{\sigma} - z_{1-\alpha}\right\} \text{ as } N \to \infty.$$

The term $\nabla\mu(\beta_0)/\sigma$ next to δ' is called *efficacy*. The greater $\nabla\mu(\beta_0)/\sigma$ is, the greater the power. That is, given multiple tests of the form $\sqrt{N}\{T_N - \mu(\beta_0)\} \rightsquigarrow N(0, \sigma^2)$ that have the same asymptotic size, the one with the highest efficacy should be used, as it is the most powerful.

For example, suppose $x_1, ..., x_N$ are iid (β, σ^2) and $T_N = \bar{x}$. It holds that $\sqrt{N}\{T_N - \mu(\beta)\} \rightsquigarrow N(0, \sigma^2)$ where $\mu(\beta) = \beta$ with $\nabla\mu(\beta) = 1$. Observe, as $E(x) = \beta_0 + \delta/\sqrt{N}$ under the alternative,

$$\sqrt{N}\{T_N - \mu(\beta_N)\} = \frac{1}{\sqrt{N}}\sum_i\left(x_i - \beta_0 - \frac{\delta}{\sqrt{N}}\right) \rightsquigarrow N(0, \sigma^2);$$

the limiting distribution is free of δ. The efficacy is $\nabla\mu(\beta)/\sigma = 1/\sigma$.

Recall the above display with $\Phi\{\delta'\nabla\mu(\beta_0)/\sigma - z_{1-\alpha}\}$ and consider another test statistic W_N such that

$$P\left[\frac{\sqrt{N}\{W_N - \mu_W(\beta_N)\}}{\sigma_W} > z_{1-\alpha} - \delta'\frac{\nabla\mu_W(\beta_0)}{\sigma_W} + 0(N^{-1/2})\right\} \to$$

$$\Phi\left\{\delta'\frac{\mu_W(\beta_0)}{\sigma_W} - z_{1-\alpha}\right\}.$$

If H_1: $\beta = \beta_0 + \delta$ while H_0: $\beta = \beta_0$, then the powers of T_N with N_T and W_N with N_W are determined by, respectively $\sqrt{N_T}\nabla\mu(\beta_0)/\sigma$ and $\sqrt{N_W}\nabla\mu_W(\beta_0)/\sigma_w$ ($\sqrt{N_T}$ and $\sqrt{N_W}$ appear next to efficacy now because $\delta/\sqrt{N}$ in H_1 is replaced with δ).

Suppose that β is a scalar and equate these two terms to get

$$\frac{N_T}{N_W} = \left(\frac{\nabla\mu_W(\beta_0)/\sigma_W}{\nabla\mu(\beta_0)/\sigma}\right)^2.$$

This ratio is called the "*(Pitman) asymptotic relative efficiency*" of W_N relative to T_N, which shows in essence how many more observations W_N needs to reach the same power as T_N attains with the observation number N_T.

Suppose the x-distribution df F is differentiable at the αth quantile β_α with $f(\beta_\alpha) > 0$. Denoting the sample αth quantile as $Q_{\alpha N}$, it holds that

$$\sqrt{N}(Q_{\alpha N} - \beta_\alpha) = \frac{1}{f(\beta_\alpha)}\frac{-1}{\sqrt{N}}\sum_i(1[x_i \leq \beta_\alpha] - \alpha) + o_p(1)$$

$$\rightsquigarrow N\left\{0, \frac{\alpha(1-\alpha)}{f(\beta_\alpha)^2}\right\}.$$

Suppose $x_1, ..., x_N$ are iid (β, σ^2) and the distribution is symmetric about β. Due to the symmetry, β can be estimated with the sample median $W_N \equiv Q_{0.5N}$ as well as with the sample mean T_N. It holds that

$$\sqrt{N}\left(W_N - \beta_{0.5} - \frac{\delta}{\sqrt{N}}\right) \rightsquigarrow N\left[0, \left\{\frac{0.5}{f(\beta_{0.5})}\right\}^2\right] \quad \text{under } H_1$$

$$\implies P\left\{\frac{\sqrt{N}(W_N - \beta_{0.5})}{0.5/f(\beta_{0.5})} > z_{1-\alpha}\right\} = P\left\{\frac{\sqrt{N}(W_N - \beta_{0.5} - \delta/\sqrt{N})}{0.5/f(\beta_{0.5})}\right.$$

$$\left. > z_{1-\alpha} - \frac{\delta}{0.5/f(\beta_{0.5})}\right\} = \Phi\left(\delta\frac{f(\beta_{0.5})}{0.5} - z_{1-\alpha}\right).$$

The efficacy of this median test is $f(\beta_{0.5})/0.5$.

Further suppose that $x \sim N(\beta, \sigma^2)$. Then

$$\frac{f(\beta_{0.5})}{0.5} = \frac{\sigma^{-1}\phi(0)}{0.5} \simeq \frac{0.4}{\sigma 0.5} = \frac{0.8}{\sigma}.$$

Recalling that the efficacy of the mean test is σ^{-1}, for the sample mean and median tests under normality, the asymptotic relative efficiency is $\{(0.8/\sigma)/\sigma^{-1}\}^2 = 0.64$: the median test is only 64% efficient compared with the mean test. Namely, the mean test attains the same power as the median test does using only 64% of the sample.

Pitman asymptotic relative efficiency is not the only way to compare tests. In practice, often "p-values" are presented as an evidence against H_0 in a given sample. Suppose that T_N follows a distribution under β with its df $G(\cdot; \beta)$. Then the *p-value* for T_N is

$$PV(T_N) \equiv \sup_{\beta \in B_0} 1 - G(T_N; \beta).$$

For any consistent test, $PV(T_N)$ approaches 0 as $N \to \infty$ under H_1 because T_N generated from H_1 goes to ∞. Typically, the zero-approaching rate is exponential, and the rate is called the "*Bahadur slope*." For two such tests, the test with the higher rate is preferred, and the ratio, say R_1/R_0, is the Bahadur asymptotic relative efficiency of test 1 relative to test 0. See Serfling (1980, Chapter 10) and Nikitin (1995) for more on asymptotic comparisons of tests.

1.2.2   Finding Distribution Under Alternatives

A difficulty in finding asymptotic relative efficiency is in verifying the condition $\sqrt{N}\{T_N - \mu(\beta_N)\} \rightsquigarrow N(0, \sigma^2)$. Typically, we know the (asymptotic) distribution of a test statistic under H_0, but not under H_1. In finding the distribution under H_1, a convenient tool is the so-called *Le Cam's third lemma*: defining the log-likelihood ratio $\ln L_N \equiv \sum_i \ln\{g(x_i; \beta_N)/g(x_i; \beta_0)\}$ where $g(x, \beta)$ is the likelihood for x under β, if for some statistic Z_N

$$\begin{bmatrix} Z_N \\ \ln L_N \end{bmatrix} \rightsquigarrow N\left(\begin{bmatrix} \mu \\ -0.5\Omega_{ll} \end{bmatrix}, \begin{bmatrix} \Omega_{zz} & \Omega_{zl} \\ \Omega'_{zl} & \Omega_{ll} \end{bmatrix} \right) \quad \text{under } H_0$$

then

$$Z_N \rightsquigarrow N(\mu + \Omega_{zl}, \Omega_{zz}) \quad \text{under } H_1.$$

For example, suppose that x is a rv, the x-density is $g(x - \beta)$, H_0: $\beta = 0$ and H_1: $\beta = \delta/\sqrt{N} \equiv \beta_N$. We get

$$\ln L_N = \sum_i \ln g(x_i - \delta/\sqrt{N}) - \sum_i \ln g(x_i).$$

Taylor-expand around $\delta = 0$ and denote the score function and its derivative at $\beta = \beta_0 = 0$ as $s(x; \beta_0)$ and $s'(x; \beta_0)$ to get

$$\ln L_N = \frac{\delta}{\sqrt{N}} \sum_i s(x_i; \beta_0) + \frac{\delta^2}{2} \frac{1}{N} \sum_i s'(x_i; \beta_0) + o_p(N^{-1})$$

$$\rightsquigarrow N\{-0.5\delta^2 I_f, \ \delta^2 I_f\} \text{ where } I_f \equiv E\{s(x; \beta_0)^2\} \text{ and } E\{s'(x; \beta_0)\} = -I_f$$
under H_0;

note that the expansion yields

$$\frac{g'(x_i)}{g(x_i)} \frac{-\delta}{\sqrt{N}} = s(x_i) \frac{\delta}{\sqrt{N}} \quad \text{because } s(x_i) = \frac{d \ln g(x_i - \beta)}{d\beta}\Big|_{\beta=0} = \frac{-g'(x_i)}{g(x_i)}.$$

The mean is -0.5 times the asymptotic variance as necessary for Le Cam's third lemma.

Further suppose

$$\sqrt{N}\{T_N - \mu(\beta_0)\} = \frac{1}{\sqrt{N}} \sum_i \eta(x_i) + o_p(1), \quad E\{\eta(x)\} = 0 \quad \text{under } H_0.$$

Then we can apply Le Cam's third lemma with $Z_N = \sqrt{N}\{T_N - \mu(\beta_0)\}$, because

$$\left[\begin{array}{c} N^{-0.5} \sum_i \eta(x_i) \\ \delta N^{-0.5} \sum_i s(x_i; \beta_0) + 0.5\delta^2 N^{-1} \sum_i s'(x_i; \beta_0) \end{array} \right]$$

$$\rightsquigarrow N\left(\left[\begin{array}{c} 0 \\ -0.5\delta^2 I_f \end{array} \right], \left[\begin{array}{cc} \Omega_{zz} & \Omega_{zl} \\ \Omega'_{zl} & \delta^2 I_f \end{array} \right] \right)$$

under H_0 where $\Omega_{zz} = E_0\{\eta(x)^2\}$ and $\Omega_{zl} = \delta E_0\{\eta(x)s(x; \beta_0)\}$ and $E_0(\cdot)$ denotes the expected value under H_0. Hence

$$\sqrt{N}\{T_N - \mu(\beta_0)\} \rightsquigarrow N[\delta E_0\{\eta(x)s(x; \beta_0)\}, \Omega_{zz}] \quad \text{under } H_1:$$

the asymptotic distribution of $\sqrt{N}\{T_N - \mu(\beta_0)\}$ under H_1 differs from that under H_0 only in the mean shift by $\delta E_0\{\eta(x)s(x; \beta_0)\}$.

For a specific but simple example, suppose $x \sim N(\beta, 1)$ and thus

$$g(x - \beta) = \frac{1}{\sqrt{2\pi}} \exp\left\{ -\frac{1}{2}(x - \beta)^2 \right\}.$$

Estimating the mean, set $T_N = \bar{x}$ and $\mu(\beta) = \beta$. Observe $s(x; \beta_0) = x$ under $H_0: \beta = 0$, and the test statistic for H_0 is

$$\sqrt{N}\{T_N - \mu(\beta_0)\} = \frac{1}{\sqrt{N}} \sum_i x_i \rightsquigarrow N(0, 1) \quad \text{as } \beta_0 = 0 \text{ under } H_0.$$

What is desired is the asymptotic distribution of this test statistic under H_1. It is obvious in this simple case that

$$\frac{1}{\sqrt{N}} \sum_i x_i \rightsquigarrow N(\delta, 1) \iff$$

$$\frac{1}{\sqrt{N}} \sum_i \left(x_i - \frac{\delta}{\sqrt{N}} \right) \rightsquigarrow N(0, 1) \quad \text{under } H_1 : \beta = \frac{\delta}{\sqrt{N}}$$

but this result can be obtained also from Le Cam's third lemma as follows, which comes handy when T_N takes a complicated form.

Because $\eta(x) = x$ for $\bar{x}$ and $s(x; \beta_0) = x$, we get

$$E_0\{\eta(x)s(x; \beta_0)\} = E_0(x^2) = 1.$$

Hence under H_1,

$$\sqrt{N}\{T_N - \mu(\beta_0)\} \rightsquigarrow N[\delta E_0\{\eta(x)s(x; \beta_0)\}, 1] = N(\delta, 1).$$

1.2.3 Wald Test Under Local Alternatives to Linear Hypotheses

When H_0 is g-dimensional, we get multiple test statistics (one for each restriction in H_0). In this case, the issue of which test statistic vector to use appears. Suppose this issue has been settled and we have g-many asymptotically normally distributed test statistics. Typically then we convert those to a quadratic form and do a χ^2 test, although this is not the only way of converting the chosen test statistic vector into a scalar test statistic. In the following, we examine the power of Wald test under a local alternative for linear hypotheses, and then compare two tests that share the same $\mu(\cdot)$ but are different in their asymptotic variances so that the relative efficiency depends only on the asymptotic variance ratio (or difference).

Consider

$$H_0 : R'\beta = c \quad \text{and} \quad H_1 : R'\beta = c + \frac{\delta}{\sqrt{N}}$$

where R is a known $k \times g$ matrix with $rank(R) = g$ and δ is a $g \times 1$ vector. Suppose $\sqrt{N}(b_N - \beta) \rightsquigarrow N(0, V)$ and we have a matrix S such that $S(R'VR)S' = I_g$ and $S'S = (R'VR)^{-1}$. Substitute H_1: $R'\beta = c + \delta/\sqrt{N}$ into $\sqrt{N}S(R'b_N - R'\beta) \rightsquigarrow N(0, I_g)$ to get

$$\sqrt{N}S(R'b_N - c) \rightsquigarrow N(S\delta, I_g).$$

Hence, under the H_1, the Wald test statistic

$$N(R'b_N - c)'(R'VR)^{-1}(R'b_N - c) = \left\{ \sqrt{N}S(R'b_N - c) \right\}' \left\{ \sqrt{N}S(R'b_N - c) \right\}$$

asymptotically follows a noncentral χ_g^2 with the noncentrality parameter (NCP) $\delta'S'S\delta = \delta'(R'VR)^{-1}\delta$, which is positive because $(R'VR)^{-1}$ is p.d. The larger the NCP, the higher the power is, for it becomes easier to tell H_0 from H_1; imagine H_0 and H_1 "centered" at 0 and the NCP, respectively.

In Wald test, we turn the vector $R'b_N - c$ into a scalar using a quadratic form. This guarantees that each non-zero element of $R'b_N - c$ contributes to the test statistic as it gets "squared." But the disadvantage is the resulting higher dof, which means that the critical value of the test becomes the higher. For instance, suppose $R'\beta - c = (5, 0, ..., 0)'$: only one part of H_0 is false. Because only the first component of $R'b_N - c$ contributes to the test statistic while the other components just increase the dof (and thus the critical value) of the Wald test, the test may fail to reject H_0 despite it is false. Put it differently, while the dimension of $R'\beta - c$ goes up by adding more true hypotheses, the increase in NCP is small relative to the dof increase, which lowers the power of the Wald test.

Consider two asymptotically normal estimators with the same probability limit but different asymptotic variance matrices V_0 and V_1, respectively, where $V_0 \leq V_1$ in the matrix sense (i.e., $V_1 - V_0$ is p.s.d). The Wald tests with the two estimators $\hat{V}_0$ and $\hat{V}_1$, respectively, have the same asymptotic size, for both follows χ_g^2 under H_0: $R'\beta = c$. But they have different powers under the local alternative. To see this, observe $R'V_1R \geq R'V_0R \iff (R'V_0R)^{-1} \geq (R'V_1R)^{-1}$. Hence,

$$\delta'\{(R'V_0R)^{-1} - (R'V_1R)^{-1}\}\delta \geq 0 \quad \forall \delta.$$

So, although intuitively obvious, the more efficient estimator which is the one with V_0 has the higher power for the local alternative.

1.3   Non-nested Hypothesis Testing

In the main text, we dealt mostly with "nested hypotheses" where the set of parameters $\{\beta : R'\beta = c\}$ under the H_0 is a subset of the parameter space B and β can be estimated without the restriction. There are, however, "non-nested hypotheses" as well. To get an idea, consider two alternative linear models

$$H_\beta : y = x'\beta + u \quad \text{and} \quad H_\gamma : y = z'\gamma + v$$

where x is a $k_x \times 1$ vector and z is a $k_z \times 1$ vector. As far as the LSE goes, the three models are

- *nested* if $x \subset z$ or $z \subset x$

- *non-nested and overlapping* if $x \cap z \neq \varnothing, x \not\subset z, z \not\subset x$

- *non-nested and non-overlapping (or strictly non-nested)* if $x \cap z = \varnothing$.

In the nested case, if $x \subset z$, then some zero restrictions on γ yield $z'\gamma = x'\beta$, whereas no restriction on β and γ can give $x'\beta = z'\gamma$ in the other two cases.

In the following, we examine non-nested hypothesis test. Firstly, we introduce some terminologies. Secondly, Vuong's (1989) approach and a related one are examined. Thirdly, the other approaches are studied. Surveys on non-nested hypotheses testing can be found in Gourieroux and Monfort (1994) and Pesaran and Weeks (2001).

1.3.1   Terminologies

Consider two alternative hypotheses (or models) for the likelihood function of a single observation z in an iid sample:

$$H_f : f(z,\alpha) \text{ for some } \alpha \in A \quad \text{and} \quad H_g : g(z,\beta) \text{ for some } \beta \in B$$

where A and B are parameter spaces. Let a_N and b_N be the MLE for the two models:

$$a_N = \text{argmax}_{a \in A} \sum_{i=1}^{N} \ln f(z_i, a) \quad \text{and} \quad b_N = \text{argmax}_{b \in B} \sum_{i=1}^{N} \ln g(z_i, b).$$

Define α^* and β^* as

$$\alpha^* = \text{argmax}_{a \in A} E \ln f(z, a) \quad \text{and} \quad \beta^* = \text{argmax}_{b \in B} E \ln g(z, b)$$
$$\Longrightarrow a_N \to^p \alpha^* \quad \text{and} \quad b_N \to^p \beta^*.$$

If H_f holds, then $\alpha^* = \alpha$, but β^* depends on α because $E \ln g(z, b) = \int \ln g(z, b) f(z, \alpha) dz$; denote the β^* as $\beta^*(\alpha)$ and call the $\beta^*(\alpha)$ a "pseudo true value." If H_g holds, then $\beta^* = \beta$ and $\alpha^*(\beta)$ is the pseudo true value. If neither holds while the true model is $h(z, \gamma)$, then both $\alpha^*(\gamma)$ and $\beta^*(\gamma)$ are the pseudo true values.

For instance, consider probit and logit for binary response with $z = (x', y)'$:

$$H_f : P(y = 1|x) = \Phi(x'\alpha) \quad \text{and} \quad H_g : P(y = 1|x) = \frac{\exp(x'\beta)}{1 + \exp(x'\beta)}$$

$$\Longrightarrow a_N = \text{argmax}_{a \in A} \sum_{i=1}^{N} [y_i \ln \Phi(x_i'a) + (1 - y_i) \ln\{1 - \Phi(x_i'a)\}],$$

$$b_N = \text{argmax}_{b \in B} \sum_{i=1}^{N} \left[y_i \ln \left\{ \frac{\exp(x'b)}{1 + \exp(x'b)} \right\} + (1 - y_i) \ln \left\{ \frac{1}{1 + \exp(x'b)} \right\} \right].$$

If neither model holds and the true model is $P(y = 1|x) = H(x'\gamma)$, then

$$\alpha^*(\gamma) = \text{argmax}_{a \in A} E[y \ln \Phi(x'a) + (1 - y) \ln\{1 - \Phi(x'a)\}]$$
$$= \text{argmax}_{a \in A} E[H(x'\gamma) \ln \Phi(x'a) + \{1 - H(x'\gamma)\} \ln\{1 - \Phi(x'a)\}].$$

Here, $\alpha^*(\gamma)$ is not the true value as the data were not drawn from $f(z, \alpha^*(\gamma))$; $\alpha^*(\gamma)$ is simply defined by this display.

Recall the Kullback–Leibler information criterion (KLIC) with $f(z, a)$ where α is the true value:

$$\int \left\{ \ln \frac{f(z, \alpha)}{f(z, a)} \right\} f(z, \alpha) dz.$$

Using the inequality $\ln w \leq 2(w^{1/2} - 1)$, KLIC times -1 is

$$\int \left\{ \ln \frac{f(z, a)}{f(z, \alpha)} \right\} f(z, \alpha) dz \leq 2 \int \left(\frac{f(z, a)}{f(z, \alpha)} \right)^{1/2} f(z, \alpha) dz - 2$$

$$= 2 \int f(z, a)^{1/2} f(z, \alpha)^{1/2} dz - \int f(z, \alpha) dz - \int f(z, a) dz$$

$$= - \int \left\{ f(z, a)^{1/2} - f(z, \alpha)^{1/2} \right\}^2 dz \leq 0.$$

That is, KLIC is non-negative always and positive iff $f(z, \alpha) \neq f(z, a)$ (Van der Vaart, 1998). This fact can be used to define "proximity" between two models—KLIC is not a distance as it is not symmetric. Let

$$K_{fg}(\alpha) \equiv \int \left\{ \ln \frac{f(z, \alpha)}{g(z, \beta^*(\alpha))} \right\} f(z, \alpha) dz \quad \text{and}$$

$$K_{gf}(\beta) \equiv \int \left\{ \ln \frac{g(z, \beta)}{f\{z, \alpha^*(\beta)\}} \right\} g(z, \beta) dz.$$

$K_{fg}(\alpha)$ shows how far H_g is away from the true model H_f, and $K_{gf}(\beta)$ shows how far H_f is away from the true model H_g. We will often omit z to simplify notations in the following.

Drawing on Pesaran and Weeks (2001), first, if

$$(i) \ K_{fg}(a) = 0 \ \forall a \in A \quad \text{and} \quad K_{gf}(b) = 0 \ \forall b \in B,$$

then H_f and H_g are *observationally equivalent*; the data simply cannot tell one model from the other. Second, if

$$(ii) \ K_{fg}(a) = 0 \ \forall a \in A \quad \text{but} \quad K_{gf}(b) \neq 0 \text{ for some } b \in B,$$

then $g(b)$ can "mimic" $f(a)$ regardless of a in the sense $g\{z, \beta^*(\alpha)\} = f(z, \alpha)$ for any true value α, whereas $f(a)$ cannot "mimic" $g(b)$ for some b; in this case, H_f is *nested* in H_g. Analogously, we can define the case H_g being nested in H_f. Third, if

$$(iii) \ K_{fg}(a) \neq 0 \text{ for some } a \in A \quad \text{and} \quad K_{gf}(b) \neq 0 \text{ for some } \in B,$$

then H_f and H_g are *non-nested* (but maybe overlapping). Fourth, if

$$(iv) \ K_{fg}(a) \neq 0 \ \forall a \in A \quad \text{and} \quad K_{gf}(b) \neq 0 \ \forall b \in B,$$

then H_f and H_g are *strictly non-nested*. The above example with probit and logit is a strictly non-nested case.

1.3.2 LR Test for Strictly Non-nested Hypotheses

In reality, strictly speaking, any specified model would be false and we should choose the one that is less wrong. Suppose we consider two strictly non-nested models $f(a)$ and $g(b)$ and wonder what the maximized log-likelihood difference reflects:

$$\tau_N \equiv \sum_i \ln f(z, a_N) - \sum_i \ln g(z, b_N)$$

where a_N is the MLE for $f(z, a)$ and b_N is the MLE for $g(z, b_N)$. Suppose the true model is $h(\gamma)$. Then the population version of this difference is

$$E \ln f\{z, \alpha^*(\gamma)\} - E \ln g\{z, \beta^*(\gamma)\}$$

$$= E \ln h(z,\gamma) - E \ln g\{z,\beta^*(\gamma)\} - [E \ln h(z,\gamma) - E \ln f\{z,\alpha^*(\gamma)\}]$$
$$= K_{hg}(\gamma) - K_{hf}(\gamma).$$

Thus the maximized log-likelihood difference τ_N shows which one between f and g is further away from the truth h. If positive, g is further away (i.e., f is closer to the truth); if negative, f is further away; if zero, both f and g are equally away. Vuong (1989, Theorem 5.1) provided the following LR test to tell which is the case.

Define

$$T_N \equiv \frac{1}{\omega_N \sqrt{N}} \sum_i \ln \frac{f(z_i, a_N)}{g(z_i, b_N)},$$

$$\omega_N^2 \equiv \frac{1}{N} \sum_i \left\{ \ln \frac{f(z_i, a_N)}{g(z_i, b_N)} \right\}^2 - \left\{ \frac{1}{N} \sum_i \ln \frac{f(z_i, a_N)}{g(z_i, b_N)} \right\}^2.$$

Then

$$(i)\ f, g\ \text{are equally away} : T_N \rightsquigarrow N(0,1),$$
$$(ii)\ g\ \text{is further away} : T_N \to \infty,$$
$$(iii)\ f\ \text{is further away} : T_N \to -\infty.$$

For example, if $|T_N| < 1.96$, then the null hypothesis (i) is not rejected at 5% level. If $T_N > 1.96$, then we conclude (ii) that f is closer to the truth. If $T_N < -1.96$, then we conclude (iii). To account for the possible dimension difference between α and β, Vuong (1989) suggested two modifications of T_N whose asymptotic distributions are still the same as T_N:

$$T'_N \equiv \frac{1}{\omega_N \sqrt{N}} \left[\sum_i \ln \frac{f(z_i, a_N)}{g(z_i, b_N)} - \{\dim(\alpha) - \dim(\beta)\} \right],$$

$$T''_N \equiv \frac{1}{\omega_N \sqrt{N}} \left[\sum_i \ln \frac{f(z_i, a_N)}{g(z_i, b_N)} - \left\{ \frac{\dim(\alpha)}{2} \ln N - \frac{\dim(\beta)}{2} \ln N \right\} \right].$$

While T_N looks at the nullity of the mean of the LR, Clarke (2007) proposed a median version: under (i), the median of the individual LR difference would be zero. Specifically, the test statistic is

$$\sum_{i=1}^N 1[[f(z_i, a_N) > g(z_i, b_N)] \sim B(N, 0.5) \quad \text{or}$$

$$\frac{\sum_i 1[f(z_i, a_N) > g(z_i, b_N)] - 0.5N}{\sqrt{0.25N}} \sim N(0,1).$$

Clarke (2007) also suggested a modification for the dimension difference as in T''_N. Since a_N and b_N depend on all observations, strictly speaking, the terms in the sum are not independent, invalidating the distributional result. Furthermore, it is not clear how the nuisance parameter problem (i.e., how the estimation errors $a_N - \alpha$ and $b_N - \beta$ affect the asymptotic distribution) is handled for the test. But some simulation studies suggest that the test behaves fairly well, possibly better than the Vuong's test.

1.3.3 Centered LR Test and Encompassing

Instead of saying which model is further away or both are equally away, we may desire to say that one of them is true. But in non-nested tests, this runs into a difficulty. Recall that, for the nested H_0: $R'\beta = c$, the LR test asymptotics was driven by a quadratic form of $b_N - b_{Nr} = b_N - \beta - (b_{Nr} - \beta)$ where b_N is the unrestricted MLE and b_{Nr} is the restricted one. Under the H_0, both b_N and b_{Nr} are consistent for β, and the LR test is thus centered at zero. For two non-nested hypotheses, however, the LR test statistic τ_N is not centered at zero.

Cox (1962) proposed to subtract the centering constant and use the "*centered LR*" test: when H_f is taken as the truth, the test statistic is

$$\sum_i \ln f(z_i, a_N) - \sum_i \ln g(z_i, b_N) - E_\alpha \left\{ \sum_i \ln f(z_i, a_N) - \sum_i \ln g(z_i, b_N) \right\}$$

where $E_\alpha(\cdot)$ is the expected value wrt $f(z, \alpha)dz$. The Cox centered-LR test should be done twice: first with H_f as the truth, and then H_g as the truth in reverse. Depending on the outcomes of the two tests, there are four combinations of H_f accepted/rejected and H_g accepted/rejected.

The main difficulty in the Cox LR test is finding the form of the centering constant, and "parametric bootstrap" may provide an easy answer to this problem. Suppose that H_f is taken as the true model. First, obtain a_N, b_N, and τ_N. Second, draw a pseudo sample of size N from the distribution $f(z, a_N)$ to get a pseudo estimate $\tau_N^{(j)}$; repeat this for J-many times to get $\tau_N^{(j)}$, $j = 1, ..., J$. Third, obtain the bootstrap p-value $J^{-1} \sum_{j=1}^{J} 1[\tau_N^{(j)} \leq \tau_N]$ as the null model gets rejected when the test statistic is too small. But the validity of this bootstrap has been questioned by Godfrey and Santos-Silva (2004) and Godfrey (2007).

Other than this centered-LR test idea, the *encompassing* idea (Mizon and Richard, 1986) looks at the difference between

$$\sqrt{N}\{a_N - \alpha^*(b_N)\} \quad \text{or} \quad \sqrt{N}\{b_N - \beta^*(a_N)\}.$$

If the former is almost zero, then $\alpha^*(b_N)$ can "match" a_N closely; i.e., H_g is more general, encompassing (or "explaining") H_f. Analogously, if the latter is almost zero, H_f encompasses H_g. If H_f encompasses H_g while H_g cannot encompasses H_f, then H_f is a better model. The encompassing approach, however, seems difficult to implement in general, not least because the functions $\alpha^*(\cdot)$ and $\beta^*(\cdot)$ should be found.

1.3.4 J-Test and Score Test Under Artificial Nesting

One may try *artificial nesting* of H_f and H_g in a comprehensive model. For example, Cox (1962, p. 407) considered, but did not elaborate on, a multiplicative form

$$\frac{f(z, \alpha)^\lambda g(z, \beta)^{1-\lambda}}{\int f(z, \alpha)^\lambda g(z, \beta)^{1-\lambda} dz}, \quad 0 \leq \lambda \leq 1$$

as an artificial nesting model. Here, testing for $\lambda = 0$ or 1 will tell which
model holds. If λ is not 0 nor 1 but takes a value in-between, say 0.3, then
the truth may be the mixture of 30% $f(z,\alpha)$ and 70% $g(z,\beta)$. As good as this
may sound, a straightforward application of this idea runs into difficulties:
$\lambda = 0, 1$ are the boundary values for λ on which the usual asymptotics does
not hold, and α (β) is not identified if $\lambda = 0$ ($\lambda = 1$). The latter problem can
be better seen in the following linear model regressor selection example.

Suppose

$$H_f : y = x'\alpha + u \quad \text{and} \quad H_g : y = z'\beta + v$$
$$\implies y = \lambda x'\alpha + (1-\lambda)z'\beta + \varepsilon \quad \text{(additive form of artificial nesting)}.$$

In this comprehensive model, λ is not identified. Davidson and MacKinnon's
(1981) *J-test* for H_f as the null model estimates α and μ in

$$y = x'\alpha + \mu \cdot z'b_N + \varepsilon, \quad \text{where} \quad b_N = (Z'Z)^{-1}Z'Y$$

and tests for $\mu = 0$. If $\mu = 0$ is rejected, then H_f is rejected to the direction
of H_g. Analogously, the J-test for H_g as the null model can be done as well.
J-test is an artificial regressor test.

Santos-Silva (2001b) examined yet another form of artificial nesting:

$$R(\zeta,\lambda) \equiv \frac{\{(1-\zeta)f(z,\alpha)^\lambda + \zeta g(z,\beta)^\lambda\}^{1/\lambda}}{\int\{(1-\zeta)f(z,\alpha)^\lambda + \zeta g(z,\beta)^\lambda\}^{1/\lambda}dz}, \quad 0 \le \zeta \le 1,\ 0 < \lambda \le 1.$$

The multiplicative and additive forms can be obtained from this:

$$\lim_{\lambda\to 0+} R(\zeta,\lambda) = \frac{f(z,\alpha)^\lambda g(z,\beta)^{1-\lambda}}{\int f(z,\alpha)^\lambda g(z,\beta)^{1-\lambda}dz}$$
$$\text{applying the L'Hospital's rule to } \ln R(\zeta,\lambda)$$
$$R(\zeta,1) = (1-\zeta)f(z,\alpha) + \zeta g(z,\beta).$$

That is, the multiplicative form is a limiting case when $\lambda \downarrow 0$, and the additive
form is a special case when $\lambda = 1$.

Suppose $f(z,\alpha)$ is the null model and thus we want to test for $H_0: \zeta = 0$
against the alternative model $R(\zeta,\lambda)$. Since $\zeta = 0$ is a boundary value for ζ,
plugging in an estimator for ζ is troublesome. Instead, use the score test: the
score function for ζ is obtained by differentiating $R(\zeta,\lambda)$ wrt ζ and then set
$\zeta = 0$ in the resulting score. This score test still needs α and β to be replaced
by some value. Santos-Silva (2001b) replaced α and β with the MLE's $\hat\alpha$ and
$\hat\beta$ using $f(z,a)$ and $g(z,\beta)$ as the true likelihood, respectively.

1.4   Pearson Chi-Square Goodness-of-Fit Test

Suppose we have data $y_1,...,y_N$ from a discrete distribution with J different support points (or "J cells") $v_1,...,v_J$ and the corresponding probabilities $p_1,...,p_J$. Let N_j denote the number of observations falling in cell j.

One well-known test for the distributional assumption is the *"Pearson χ^2 goodness of fit (GOF)"* test:

$$M_N \equiv \sum_{j=1}^{J} \frac{(N_j - Np_j)^2}{Np_j} = \sum_{j=1}^{J} \frac{\{(N_j - Np_j)/\sqrt{N}\}^2}{p_j}$$

$$= \sum_{j=1}^{J} \frac{\left\{\sqrt{N}(N_j/N - p_j)\right\}^2}{p_j} \rightsquigarrow \chi^2_{J-1}$$

which is proven below. The qualifier "GOF" refers to the fact that there is no specific alternative distributional hypothesis under consideration: the test is simply to assess how well the assumed distribution fits the data at hand. There are a number of interesting features in this test.

First, the dof in χ^2 is $J - 1$, not J, despite that there are J many terms in the sum. This is because one probability (i.e., one cell) is redundant owing to the restriction $\sum_j p_j = 1$. Recall that, in a score test, the dof is k (the number of unrestricted parameters) minus the number of restrictions, despite that the dimension of the score function is k. If we are to drop one cell, then we will have to decide which cell to drop. But the test statistic shows that dropping any cell is unnecessary: just use all cells with dof $J - 1$.

Second, since $N_j = \sum_i 1[y_i = v_j]$ and $p_j = E(1[y = v_j])$, the numerator of the summand in M_N is

$$\sqrt{N}\left(\frac{N_j}{N} - p_j\right) = \frac{1}{\sqrt{N}} \sum_i \{1[y_i = v_j] - E(1[y = v_j])\}$$

$$\rightsquigarrow N\{0, \ p_j(1 - p_j)\}$$

which compares the sample moment with the population moment, reminiscent of MOM. M_N is a sum of J-many squared asymptotically normal terms, which are related with one another because being in one cell means not being in any other cell. In M_N, weighting is going on, but somewhat strangely, the weight is p_j—the denominator of the summand in M_N—not the inverse of the asymptotic variance $p_j(1 - p_j)$. Again, this has to do with one redundant cell.

Third, without changing the asymptotic distribution, the denominator Np_j in M_N can be replaced with N_j to yield

$$M_N \equiv \sum_{j=1}^{J} \frac{(N_j - Np_j)^2}{N_j} = \sum_{j=1}^{J} \frac{\{\sqrt{N}(N^{-1}N_j - p_j)\}^2}{N_j/N} \rightsquigarrow \chi^2_{J-1}.$$

This is called *"Neyman modified χ^2."* Clearly, the parameter p_j is replaced by the estimator N_j/N. This may look like a bad idea, because we are replacing a sure number p_j with an estimator N_j/N, but it comes handy as can be seen in the following.

Fourth, suppose that p_j is a parametric function of a SF parameter β whose dimension is smaller than $J-1$: $p_j = p_j(\beta)$. Then we can estimate β by minimizing wrt b

$$\sum_{j=1}^{J} \frac{\{N_j - Np_j(b)\}^2}{Np_j(b)} = \sum_{j=1}^{J} \frac{N\{N^{-1}N_j - p_j(b)\}^2}{p_j(b)}.$$

As usual, having something to estimate in the denominator can be involved. Hence, it is convenient to use the Neyman's modified χ^2

$$\sum_{j=1}^{J} \frac{\{N_j - Np_j(b)\}^2}{N_j} = \sum_{j=1}^{J} \frac{N\{N^{-1}N_j - p_j(b)\}^2}{N_j/N}.$$

Here we can see a connection to MDE, because N_j/N is an estimator for the RF parameter p_j and we want to estimate the SF parameter β using the relation $p_j = p_j(\beta)$ between the two parameters. Interestingly, minimizing either minimand in the two preceding displays renders an estimator asymptotically equivalent to the MLE maximizing the likelihood function $(N!/\prod_j N_j!) \prod_j p_j(b)^{N_j}$; see, e.g., Sen and Singer (1993, Chapter 6).

Fifth, not just the MDE interpretation, a MOM interpretation is possible as well. Observe

$$\sqrt{N}\left\{\frac{N_j}{N} - p_j(\beta)\right\} = \frac{1}{\sqrt{N}} \sum_i \{1[y_i = v_j] - p_j(\beta)\}.$$

Setting $\zeta_j(y_i,\beta) = 1[y_i = v_j] - p_j(\beta)$, we get J many (or $J-1$ many, because one is redundant) moment conditions $E\{\zeta_j(y_i,\beta)\} = 0$, and M_N is a way to combine the moments. The rest of this subsection derives the asymptotic distribution of M_N.

To derive the asymptotic distribution of M_N, recall the first display of this subsection

$$M_N = N \sum_{j=1}^{J} \frac{\{(N_j/N - p_j)\}^2}{p_j}$$

and define

$$m_j \equiv \frac{N_j/N - p_j}{\sqrt{p_j}}, \quad m \equiv \begin{bmatrix} m_1 \\ \vdots \\ m_J \end{bmatrix}, \quad p \equiv \begin{bmatrix} p_1 \\ \vdots \\ p_J \end{bmatrix}, \quad p_{1/2} \equiv \begin{bmatrix} p_1^{1/2} \\ \vdots \\ p_J^{1/2} \end{bmatrix}$$

$$\implies M_N = N \cdot m'm \quad \text{and}$$

$$m'p_{1/2} = \sum_j \left(\frac{N_j}{N} - p_j\right) = \frac{\sum_j N_j}{N} - \sum_j p_j = 1 - 1 = 0$$

which shows that m_j's are linear dependent. Using $m'p_{1/2} = 0$, further rewrite M_N as

$$M_N = N \cdot m' I_J m - N \cdot m' p_{1/2}(p'_{1/2}m) = N \cdot m'(I_J - p_{1/2}p'_{1/2})m.$$

The middle matrix $I_J - p_{1/2}p'_{1/2}$ is idempotent (and symmetric):

$$(I_J - p_{1/2}p'_{1/2})(I_J - p_{1/2}p'_{1/2}) = I_J - 2p_{1/2}p'_{1/2} + p_{1/2}p'_{1/2}p_{1/2}p'_{1/2}$$
$$= I_J - 2p_{1/2}p'_{1/2} + p_{1/2}(1)p'_{1/2} = I_J - p_{1/2}p'_{1/2}.$$

Also $rank(I_J - p_{1/2}p'_{1/2}) = J - 1$, because the rank of an idempotent matrix is its trace: $tr(I_J - p_{1/2}p'_{1/2}) = J - 1$. We will show in the following that $I_J - p_{1/2}p'_{1/2}$ is the asymptotic variance of $\sqrt{N}m$.

Rewrite the jth component $\sqrt{N}m_j$ of $\sqrt{N}m$ as

$$\sqrt{N}m_j = \sqrt{N}\left(\frac{N_j}{N} - p_j\right)/\sqrt{p_j}$$
$$= \frac{1}{\sqrt{N}}\sum_i \{1[y_i = v_j] - E(1[y = v_j])\}/\sqrt{p_j} \leadsto N\{0, (1 - p_j)\}.$$

For the cross terms of $\sqrt{N}m_j$ and $\sqrt{N}m_q$ with $j \neq q$, it holds that

$$\frac{E\{(1[y = v_j] - p_j)(1[y = v_q] - p_q)\}}{\sqrt{p_j}\sqrt{p_q}} = \frac{-2p_jp_q + p_jp_q}{\sqrt{p_jp_q}} = -\sqrt{p_jp_q}.$$

Hence,

$$\sqrt{N}m \quad \leadsto \quad N\left(0_J, \begin{bmatrix} 1 - p_1 & -\sqrt{p_1p_2} & \cdots & -\sqrt{p_1p_J} \\ -\sqrt{p_2p_1} & 1 - p_2 & \cdots & \cdots \\ \cdots & \cdots & \cdots & -\sqrt{p_{J-1}p_J} \\ -\sqrt{p_Jp_1} & \cdots & -\sqrt{p_Jp_{J-1}} & 1 - p_J \end{bmatrix}\right)$$
$$= N\left(0, \ I_J - p_{1/2}p'_{1/2}\right).$$

For a $k \times 1$ random vector $z \sim N(0, V)$, as well known, V can be written as

$$V = H\Lambda H' \quad \text{where } H'H = I_k \text{ and } \Lambda \text{ is the diagonal}$$
$$\text{matrix of the eigenvalues of } V.$$

For the quadratic form $z'Vz$,

$$z'Vz = z'H\Lambda H'z = w'\Lambda w \quad \text{where } w \equiv H'z \sim N(0, \Lambda) \text{ because}$$
$$H'VH = H'H\Lambda H'H = \Lambda.$$

Observe that, with $w = (w_1, ..., w_k)'$,

$$w'\Lambda w = [w_1, ..., w_k] \begin{bmatrix} \lambda_1 & 0 & 0 \\ 0 & \ddots & 0 \\ 0 & 0 & \lambda_k \end{bmatrix} \begin{bmatrix} w_1 \\ \vdots \\ w_k \end{bmatrix} = \sum_{j=1}^{k} \lambda_j w_j^2.$$

But the eigenvalues of an idempotent matrix is either 0 or 1 and the number of one's equals the rank of the matrix. Thus, when V is idempotent, $w'\Lambda w$ is a sum of "$rank(V)$-many" independent $N(0,1)$ rv's, and hence follows $\chi^2_{rank(V)}$. Using this fact, since $rank(I_J - p_{1/2}p'_{1/2}) = J - 1$, we get the desired result

$$M_N = \sqrt{N}m' \left(I_J - p_{1/2}p'_{1/2} \right) \sqrt{N}m \rightsquigarrow \chi^2_{J-1}.$$

Usually when we look at quadratic forms such as $z'V^{-1}z$ where $z \sim N(0, V)$, the role of V^{-1} is "weighting (i.e., standardizing)" z. But in M_N, weighting has been partly built into m by the denominator $\sqrt{p_j}$ and the middle matrix is the adjacent vector's asymptotic variance, not its inverse. In $\sqrt{N}m'(I_J - p_{1/2}p'_{1/2})\sqrt{N}m$, the role of the middle matrix is thus not so much weighting as it is to pick up $J - 1$ terms in the adjacent $J \times 1$ vector $\sqrt{N}m$.

Many economic data have a continuously distributed y along with regressors, and the Pearson χ^2 test needs to be modified for those data. For example, y has to be grouped and the number of groups and grouping intervals have to be chosen. This seems to be an impediment to the Pearson χ^2 test. Also there exist χ^2 GOF tests for regression function specification as reviewed in Lee (1996). But the tests are subject to the same grouping problems, and they are "dominated" (in terms of power and ease in use) by other specification tests examined in the semiparametrics chapters. See Lee (1996, pp. 63–67) and the references therein if interested.

2   Stratified Sampling and Weighted M-Estimator

Suppose we have a study population of interest with a parameter β; as usual, let β be a parameter governing the relationship between a response variable y and regressors x. We want to find β using a sample of size N drawn from the population. The question is how exactly N-many observations are to be sampled. In the main text, we assumed that the observations (x_i, y_i) are iid across $i = 1, ..., N$, i.e., the data is obtained by random sampling from the population. We can then identify the joint density/probability of (x, y)— omit the part "/probability" from now on. Random sampling is the yardstick again which other sampling schemes are compared.

Random sampling can be taken as synonymous as having an iid sample from an "infinite population" because drawing a finite number of observations hardly alters the population. But in a finite population with M-many units, random sampling does not yield an iid sample; rather (simple) random sampling of m units is a sampling such that each set with m units is equally likely to be selected. This then implies that each unit has the equal probability m/M to be selected: as there are $\binom{M}{m}$-many combinations choosing m out of M-many units,

$$P(\text{1st unit included in the selected } m \text{ units})$$

$$= \frac{\#\text{combinations choosing } m - 1 \text{ units out of } (M - 1)\text{-many}}{\#\text{combinations choosing } m \text{ units out of } M\text{-many}}$$

$$= \frac{\binom{M-1}{m-1}}{\binom{M}{m}} = \frac{(M-1)!/\{(m-1)!(M-m)!\}}{M!/\{m!(M-m)!\}} = \frac{m}{M}$$

where the reason for the numerator in the first equality is that we have to choose $m - 1$ out of $N - 1$ excluding the first unit because the first unit has been already chosen. As our framework of analysis is asymptotic, unless otherwise noted, we will assume a population to be large enough so that random sampling yields an (almost) iid sample.

But random sampling is not necessarily easy (and might be too costly) to implement. Whereas a general discussion on sampling can be found in textbooks such as Lohr (1999) and Thompson (2002), "stratified sampling" which is popular in practice is examined in this section. Under stratified sampling, estimators need modifications, and we will focus on M-estimator; in econometrics, Wooldridge (1999, 2001) has dealt with M-estimator in stratified sampling.

2.1 Three Stratified Sampling Methods

Suppose we have a S-many mutually exclusive and exhaustive cells (i.e., strata) drawn in the population using the values of (x, y), say $C_1, ..., C_S$; S is a fixed number. Following the terminologies adopted as in Imbens and Lancaster (1996), stratified sampling can be further classified into three categories: standard stratified sampling, variable probability sampling, and multinomial sampling. All three sampling methods can yield the cell proportions different from the corresponding population proportions, but the way to arrive at those proportions differs across the three methods. For instance, we may want to have 2/3 of the observations in the data from the poor people, and 1/3 from the rest, and this can be done in different ways.

2.1.1 Standard Stratified Sampling (SSS)

In *standard stratified sampling (SSS)*, N_s observations are drawn randomly from C_s, $s = 1, ..., S$, where N_s is fixed and $N = \sum_{s=1}^{S} N_s$; $N_1, ..., N_S$ as well as N are all fixed. The observations from SSS are independent but non-identically distributed, although the observations belonging to the same cell are iid: defining the cell variable c_i taking $s = 1, ..., S$,

$$(y_i, x_i)|(c_i = s) \text{ are iid } P(y, x|c = s), \text{ where } P(y, x|c = s)$$
may vary across s.

The sampling proportion N_s/N for cell C_s may differ from the *population cell proportion* p_s, as N_s/N is chosen by the researcher—to be more precise, N would be chosen by the data collection budget constraint first and then N_s by the researcher. SSS is useful when it is easy to see who belongs to which cell before data collection—e.g., geographical stratification. For the poor-oversampling example, however, this requires knowing who is poor before data collection—not necessarily an easy requirement.

2.1.2 Variable Probability Sampling (VPS)

In *variable probability sampling (VPS)*, an observation is drawn randomly from the population and gets retained with probability r_s ("r" from "retained") when it falls in cell s. In VPS, $N_1, ..., N_S$ are random as well as their sum N. VPS is useful when it is difficult to see who belongs to which cell before data collection. For instance, in the poor-oversampling example, as it is typically difficult to know in advance who is poor or not, an observation is drawn randomly first and then if the person is poor, he/she is retained with probability $2/3$; if not, with probability $1/3$. As some observations get discarded, VPS makes an economic sense when screening people for their cell membership is cheap but the ensuing information collection on (x, y) is expensive.

For an observation in the data to be in C_s, it has to be drawn first with probability p_s and then retained with probability r_s subject to $\sum_{s=1}^{S} r_s = 1$. Differently from SSS, VPS gives iid data:

$$(y_i, x_i, c_i) \text{ are iid } P(y, x|c)P(c), \text{ where } P(c = s) = \frac{p_s r_s}{\sum_s p_s r_s}.$$

Note that $P(c = s)$ denotes the cell probability in the sample whereas the population probability is denoted as $P^*(c = s) = p_s$. Use these and analogous notations $E(\cdot)$ and $E^*(\cdot)$ from now on. In stratified sampling,

$$P(y, x|c) = P^*(y, x|c) \quad \text{whereas } P(c) \neq P^*(c) \quad \text{in general.}$$

Let $\bar{N}$ be the fixed number of total observations initially drawn including the discarded ones. Define

$d_{js} = 1$ if the jth draw falls in cell s and 0 otherwise;
$$P^*(d_{js} = 1) = p_s \; \forall j.$$
$k_j = 1$ if the jth draw is kept ("k" from "kept") and 0 otherwise

where

$$P^*(k_j = 1) = \sum_{s=1}^{S} P(k_j = 1|d_{js} = 1)P^*(d_{js} = 1) = \sum_{s=1}^{S} r_s p_s.$$

Observe that N_s follows $B(\bar{N}, r_s p_s)$, and

$$N_s = \sum_{j=1}^{\bar{N}} \delta_{js} k_j \implies$$

$$E^*(N_s) = \bar{N} \cdot E^*(\delta_{js} k_j) = \bar{N} \cdot E(k_j | \delta_{js} = 1) E^*(\delta_{js}) = \bar{N} r_s p_s;$$

$$N = \sum_{s=1}^{S} N_s \implies$$

$$E^*(N) = \sum_s E^*(N_s) = \sum_s \bar{N} r_s p_s = \bar{N} \sum_s r_s p_s$$

$$= \bar{N} \cdot P(\text{observation kept}).$$

This shows that VPS is somewhat complicated due to the extra randomness coming from the retain/discard feature. The real complication comes because N_s and N might have information on β through $p_s = p_s(\beta)$ which appears in $E^*(N_s)$ and $E^*(N)$. Weighted M-estimator for VPS is studied in detail in Wooldridge (1999).

2.1.3 Multinomial Sampling (MNS)

In *multinomial sampling (MNS)*, a stratum is chosen randomly with probabilities $r_1, ..., r_S$ subject to $\sum_s r_s = 1$, and then an observation is drawn randomly from the chosen cell; $N_1, ..., N_S$ are random but N is fixed as no draw gets discarded. A theoretically nice feature of multinomial sampling is that the observations in the data are iid

$$(y_i, x_i, c_i) \text{ are iid } P(y, x|c) P(c), \text{ where } P(c = s) = r_s.$$

The most often used stratified sampling in practice is SSS as the name suggests, followed by VPS; MNS is rarely used. Nevertheless, MNS is easier to deal with in asymptotic theory because of its "iidness" as will be seen below for "weighted M-estimator," and because MNS is essentially the same as standard stratified sampling once the cell selection probability r_s is replaced with the cell proportion N_s/N in the data. Also VPS has an equivalent form of MNS through a certain reparametrization as Imbens and Lancaster (1996) showed.

2.2 Infeasible MLE

In *SSS*, the sampling density of $(x, y)|s$—*the density of (x, y) given that (x, y) is in cell s*—is

$$\frac{P(y|x, \beta) P^*(x) 1[(x, y) \in C_s]}{p_s(\beta)} \quad \text{where}$$

$$p_s(\beta) \equiv P\{(x, y) \in C_s\} = \int_{(x,y) \in C_s} P(y|x, \beta) P^*(x) dy dx.$$

In the sampling density, the numerator is for the joint density and the denominator is for the marginal density for cell s so that the ratio becomes the conditional density. For those $(x, y) \notin C_s$, the conditional density is zero.

Define the person i cell indicator

$$\delta_{is} = 1[(x_i, y_i) \in C_s].$$

The likelihood function for an observation (x, y) in SSS can be written as

$$\prod_{s=1}^{S} g(x, y|s)^{\delta_s}, \quad \text{where} \quad g(x, y|s) \equiv \frac{P(y|x, \beta)P^*(x)}{p_s(\beta)};$$

$g(x, y|s)$ has no indicator function $1[(x, y) \in C_s]$ attached as δ_s appears as the exponent.

The sample log-likelihood function for β is

$$\sum_i \sum_{s=1}^{S} \delta_{is} \ln g(x_i, y_i|s) \implies \sum_i \sum_s \delta_{is} \ln \frac{P(y_i|x_i, \beta)}{p_s(\beta)}$$

$$\left(\text{dropping} \sum_i \sum_s \delta_{is} \ln P^*(x_i)\right)$$

$$= \sum_i \sum_s \delta_{is} \ln P(y_i|x_i, \beta) - \sum_i \sum_s \delta_{is} \ln p_s(\beta)$$

$$= \sum_i \sum_s \delta_{is} \ln P(y_i|x_i, \beta) - \sum_s \left\{ \sum_i \delta_{is} \cdot \ln p_s(\beta) \right\}$$

$$= \sum_i \sum_s \delta_{is} \ln P(y_i|x_i, \beta) - \sum_s N_s \ln p_s(\beta).$$

Note that $p_s(\beta)$ is not indexed by i. Random sampling is a special case of SSS with a single cell where $p_1(\beta) = 1$, $p_2(\beta) = 0$, ..., $p_S(\beta) = 0$, in which case the log-likelihood function becomes the usual one $\sum_i \ln P(y_i|x_i, \beta)$. It is impossible to get MLE for $\sum_i \sum_s \delta_{is} \ln g(x_i, y_i|s)$ unless $P^*(x)$ is either known or parametrized as $P^*(x)$ appears in $p_s(\beta)$. But neither option is appealing because x typically has different types of components (continuous, binary, ordered discrete, etc.) and knowing or parametrizing the x-distribution would be far-fetching. The MLE as such is infeasible.

Suppose that $p_s(\beta)$ is known from an auxiliary information source. But plugging in the known value of $p_s(\beta)$ into the above log-likelihood function does not yield a consistent estimator because the known value part is no longer a function of β, thus dropping out of the log-likelihood function in the maximization—this is equivalent to ignoring $\sum_s N_s \ln p_s(\beta)$ in the log-likelihood function. Although the plugging-in fails, using the known $p_s(\beta)$ value for weighting leads to a consistent estimator as will be shown later. Before we proceed to this weighting idea, however, we will examine two informative polar cases of SSS, exogenous sampling and endogenous sampling.

If we fix x first and then sample y ("x-stratified" sampling), then this is *exogenous sampling*. For instance, fix x (gender) at male and sample males to find out their y_i, $i = 1, ..., N_m$; fix x at female and sample females to find their y_i, $i = 1, ..., N_f$; pool the two samples to get a single sample where $P^*(y|x)$ is identified because $P(y|x) = P^*(y|x)$, but not $P^*(x)$, because the proportion N_m/N_f of males to females in the data is chosen arbitrarily. In this case, only y is random while x is fixed. The log-likelihood function is $\sum_{i=1}^{N} P(y_i|x_i, \beta)$, which is the same as that in random sampling. This suggests that the MLE for this is as efficient as the MLE for random sampling, and indeed this is the case as shown in the following.

The population version of $\sum_{i=1}^{N} P(y_i|x_i, \beta)$ divided by N is, under exogenous sampling with fixed x_i's,

$$\frac{1}{N} \sum_i E_i\{P(y|x, \beta)\} \quad \text{where } E_i\{P(y|x, \beta)\} \equiv \int P(y|x_i, \beta) dP(y|x_i).$$

The MLE b_N satisfies the usual M-estimator expansion

$$\sqrt{N}(b_N - \beta) = -\left\{ \frac{1}{N} \sum_{i=1}^{N} \frac{\partial s(y_i|x_i, \beta)}{\partial b} \right\}^{-1} \frac{1}{\sqrt{N}} \sum_{i=1}^{N} s(y_i|x_i, \beta) + o_p(1).$$

The observations (x_i, y_i), $i = 1, ..., N$, are independent, but not iid. LLN and CLT for independent but non-identically distributed sequences are needed, and there are many versions of them available in the literature. Using them and the fact that the information equality holds for each $E_i\{P(y|x, \beta)\}$ leads to the fact that the MLE under exogenous sampling is asymptotically equivalent to the MLE under random sampling. A formalization of this intuitive arguments can be seen in Wooldridge (2001).

If we fix y first and then sample x ("y-stratified sampling"), then this is an endogenous sampling, biased sampling, or *response-based sampling* where $P^*(x|y)$ is identified because $P^*(x|y) = P(x|y)$, but not $P^*(y)$; this is the opposite to exogenous sampling. Since (x, y)-stratified sampling also leads to endogeneity problems causing biases in general, we will call y-stratified sampling "response-based sampling," which is also called "*choice-based sampling*" although y is not always "chosen" by the subject. Differently from exogenous sampling, however, dealing with response-based sampling is not any easier than dealing with general SSS. Hence we will simply note that response-based sampling is a special case of SSS and will not further mention response-based sampling unless otherwise necessary.

2.3   Weighted M-Estimator

Although MLE is difficult to implement for endogenous samples, if the population strata probabilities $p_s \equiv p_s(\beta)$ are known from an extra source, then *weighted MLE* can be easily done. In the following, for more generality, we will examine weighted M-estimator that includes weighted MLE as a special case. Suppose we have MNS with the cell sampling probabilities r_s and

$$z_i \equiv (x_i', y_i)' \quad \text{and} \quad r_s = \frac{N_s}{N} \text{ for all } s \text{ and } N.$$

As already mentioned, the assumption $r_s = N_s/N \; \forall s, N$ makes MNS essentially equivalent to SSS, and makes deriving the asymptotic distribution much easier than otherwise.

2.3.1 Consistency

Recall the cell indicator δ_{is} for person i. Consider a *weighted M-estimator* maximizing

$$Q_N(b) \equiv \frac{1}{N} \sum_i \sum_s \frac{p_s}{r_s} \delta_{is} q(z_i, b) = \sum_s \frac{p_s}{r_s} \frac{1}{N} \sum_i \delta_{is} q(z_i, b)$$

$$\rightarrow^p \sum_s \frac{p_s}{r_s} E\{\delta_s q(z, b)\} \quad \text{as } N \rightarrow \infty$$

$$= \sum_s \frac{p_s}{r_s} E\{q(z, b)|z \in C_s\} r_s \quad \text{as "joint" } E\{\delta_s q(z, b)\}$$

is "conditional" times "marginal"

$$= \sum_s \frac{p_s}{r_s} E^*\{q(z, b)|z \in C_s\} r_s = \sum_s E^*\{q(z, b)|z \in C_s\} p_s$$

$$= E^*\{q(z, b)\}.$$

Recall that $E^*(\cdot)$ denotes an expected value using the population density, not the sampling density, whereas $E(\cdot)$ is the expected value using the sampling density. $E\{q(z, b)|z \in C_s\}$ equals $E^*\{q(z, b)|z \in C_s\}$ because random sampling is done within a given stratum although $p_s \neq r_s$.

The display shows that the sample maximand under MNS converges in probability to the population maximand (for random sampling), and thus the weighted M-estimator is consistent for β that maximizes $E^*\{q(z, b)\}$. Typically in M-estimator, β is the maximizer of $E^*\{q(z, b)|x\}$ for any given x, and thus β becomes also the maximizer of $E^*\{q(z, b)\}$. In the last display, however, the decomposition of $E^*\{q(z, b)\}$ is not wrt x, but wrt the stratum index s. Unless s is determined solely by x, $E^*\{q(z, b)|z \in C_s\}$ is not maximized by β which maximizes $E^*\{q(z, b)\}$. Setting $q(z, b) = \ln f(y|x, b)$ yields a weighted MLE as in Manski and Lerman (1977).

2.3.2 Asymptotic Distribution

Apply the Taylor expansion to the first-order condition to get

$$\sqrt{N}(b_N - \beta) = E^{-1}\left\{\sum_s \frac{p_s}{r_s} \delta_s q_{bb'}(z, \beta)\right\}$$

$$\cdot \frac{-1}{\sqrt{N}} \sum_i \sum_s \frac{p_s}{r_s} \delta_{is} q_b(z_i, \beta) + o_p(1)$$

where $q_{bb'}(z, \beta)$ and $q_b(z, \beta)$ denote, respectively, the gradient and Hessian wrt b evaluated at β. From this,

$$\sqrt{N}(b_N - \beta) \rightsquigarrow N(0, H^{-1}CH^{-1}) \quad \text{where}$$

$$H \equiv \sum_s \frac{p_s}{r_s} E\{\delta_s q_{bb'}(z, \beta)\} \quad \left[= \sum_s \frac{p_s}{r_s} E^*\{q_{bb'}(z, \beta)|\delta_s = 1\} r_s \right.$$

$$\left. = E^*\{q_{bb'}(z, \beta)\} \right]$$

$$C \equiv \sum_s \left(\frac{p_s}{r_s}\right)^2 E\{\delta_s q_b(z, \beta) q_{b'}(z, \beta)\} \quad \left[\neq E^*\{q_b(z, \beta) q_{b'}(z, \beta)\} \right.$$

$$\left. \text{due to 2 in } \left(\frac{p_s}{r_s}\right)^2 \right].$$

Consistent estimators for H and C are, respectively,

$$H_N \equiv \sum_s \frac{p_s}{r_s} \frac{1}{N} \sum_i \delta_{is} q_{bb'}(z_i, b_N) \quad \text{and}$$

$$C_N \equiv \sum_s \left(\frac{p_s}{r_s}\right)^2 \frac{1}{N} \sum_i \delta_{is} q_b(z_i, b_N) q_{b'}(z_i, b_N).$$

As a special case, for MLE with $q(z, b) = \ln f(y|x, b)$, we get

$$\frac{1}{N} \sum_i w_i w_i' = C_N \ (\neq -H_N) \quad \text{where} \quad w_i \equiv \sum_s \frac{p_s}{r_s} \delta_{is} q_b(z_i, b_N)$$

is the *weighted score function*,

$$\frac{1}{N} \sum_i \tilde{w}_i \tilde{w}_i' = -H_N \quad \text{where} \quad \tilde{w}_i \equiv \sum_s \left(\frac{p_s}{r_s}\right)^{1/2} \delta_{is} q_b(z_i, b_N).$$

The usual way of estimating the asymptotic variance of MLE needs this modification. That is, the sandwich form $H_N^{-1} C_N H_N^{-1}$ should be used, neither $-H_N^{-1}$ nor C_N^{-1}.

The asymptotic variance can be written in a more informative way: a sum of stratum variances. Use the population first-order condition

$$E\left\{\sum_s \frac{p_s}{r_s} \delta_s q_b(z_i, \beta)\right\} = \sum_s \frac{p_s}{r_s} E\{\delta_s q_b(z_i, \beta)\} = 0$$

to get

$$\frac{1}{\sqrt{N}} \sum_i \sum_s \frac{p_s}{r_s} \delta_{is} q_b(z_i, \beta) = \frac{1}{\sqrt{N}} \sum_i \sum_s \frac{p_s}{r_s} \{\delta_{is} q_b(z_i, \beta) - E(\delta_s q_b(z_i, \beta))\}$$

$$= \sum_s \frac{p_s}{r_s} \frac{1}{\sqrt{N}} \sum_i \{\delta_{is} q_b(z_i, \beta) - E(\delta_s q_b(z_i, \beta))\}.$$

This is asymptotically normal with the variance

$$\sum_s \left(\frac{p_s}{r_s}\right)^2 V_s r_s = \sum_s \frac{p_s^2}{r_s} V_s, \quad \text{where } V_s \text{ is the stratum-}s \text{ variance:}$$

$$V_s \equiv E[\{q_b(z, \beta) - E(q_b(z, \beta)|z \in C_s)\} \{q_b(z, \beta) - E(q_b(z, \beta)|z \in C_s)\}' \, |z \in C_s].$$

V_s is nothing but the variance of the vector $q_b(z, \beta)$ in C_s. Be aware that, although the above population first-order condition with $\sum_s$ holds, $E(q_b(z, \beta)| z \in C_s)$ for a given s is not necessarily 0—this is what endogenous sampling would do: only the sum of these over all strata is zero. Thus if we are to estimate V_s, then $E(q_b(z, \beta)|z \in C_s)$ should be estimated with the stratum-s sample average of $q_b(z_i, b_N)$'s.

There is a special case where weighting is not necessary although the sampling is endogenous: in logit (binary as well as multinomial logit), the inconsistency resulting from response-based sampling is known to be restricted to the intercept estimator; this will be shown shortly in the next subsection. Hence, unless the intercept is of interest, logit can be applied to endogenous samples without any modification. This is an advantage of logit compared with probit, whereas the advantage of probit is that a probit equation can be part of a multivariate model where joint normality is used fruitfully— multivariate logistic distribution exists only in a limited form.

2.3.3 An Example: Weighted M-Estimator for Mean

As a simple example of weighted M-estimator, ignore x and suppose we use $q(z, b) = -0.5(y - b)^2$. Define the stratum-s average

$$\bar{y}_s \equiv \frac{1}{N_s} \sum_{i=1}^N \delta_{is} y_i = \frac{1}{\sum_{i=1}^N \delta_{is}} \sum_{i=1}^N \delta_{is} y_i.$$

The weighted M-estimator first-order condition is

$$\frac{1}{N} \sum_i \sum_s \frac{p_s}{r_s} \delta_{is}(y_i - b_N) = 0 \iff \sum_s \frac{p_s}{r_s} \frac{1}{N} \sum_i \delta_{is}(y_i - b_N) = 0$$

$$\iff \sum_s \frac{p_s}{r_s} \frac{1}{N} \sum_i \delta_{is} y_i = \sum_s \frac{p_s}{r_s} \frac{1}{N} \sum_i \delta_{is} \cdot b_N.$$

The term in front of b_N is one because, recalling $r_s = N_s/N$ and $N_s = \sum_i \delta_{is}$,

$$\sum_s \frac{p_s}{r_s} \frac{1}{N} \sum_i \delta_{is} = \sum_s \frac{p_s}{r_s} \frac{N_s}{N} \frac{1}{N_s} \sum_i \delta_{is} = \sum_s p_s = 1.$$

Hence b_N is an weighted average of $\bar{y}_s$'s:

$$b_N = \sum_s \frac{p_s}{r_s} \frac{1}{N} \sum_i \delta_{is} y_i = \sum_s \frac{p_s}{r_s} \frac{N_s}{N} \frac{1}{N_s} \sum_i \delta_{is} y_i = \sum_s p_s \bar{y}_s.$$

It follows that $b_N \to^P \beta \equiv \sum_s p_s \beta_s$ as $\bar{y}_s \to^P \beta_s \equiv E(y | y \in C_s)$ $\forall s$. The estimator b_N includes the usual estimator $\bar{y}$ as a special case when $p_s = r_s\ (= N_s/N)$:

$$b_N = \sum_s \frac{P_s}{r_s} \frac{1}{N} \sum_i \delta_{is} y_i = \frac{1}{N} \sum_i y_i \sum_s \delta_{is} = \bar{y} \quad \text{because} \sum_s \delta_{is} = 1.$$

In the asymptotic variance of $\sqrt{N}(b_N - \beta)$, with $q(z,b) = -0.5$ $(y-b)^2$ and thus $q_{bb'} = -1$, we get $H = -\sum_s p_s r_s^{-1} E(\delta_s) = -1$ because $E(\delta_s) = r_s$. Also,

$$C = \sum_s \left(\frac{p_s}{r_s}\right)^2 E\{\delta_s (y - \beta)^2\} = \sum_s \left(\frac{p_s}{r_s}\right)^2 E^*\{(y-\beta)^2 | \delta_s = 1\} r_s = \sum_s \frac{p_s^2}{r_s} V_s$$

$$\left(= \sigma^2 \quad \text{if } V_s = \sigma^2 \ \forall s \text{ and } p_s = r_s \right).$$

Hence

$$\sqrt{N}(b_N - \beta) \rightsquigarrow N\left(0, \sum_s \frac{p_s^2}{r_s} V_s\right).$$

Consistent estimators for the asymptotic variance are

$$\sum_s \frac{p_s^2}{r_s^2} \frac{1}{N} \sum_i \delta_{is}(y_i - b_N)^2 \quad \text{or}$$

$$\sum_s \frac{p_s^2}{r_s} V_{Ns} \quad \text{where } V_{Ns} \equiv \frac{1}{N_s} \sum_{i \in C_s} \left\{ y_i - b_N - \frac{1}{N_s} \sum_{i \in C_s}(y_i - b_N) \right\}^2 \to^P V_s.$$

The first estimator is a sample analog for $\sum_s (p_s/r_s)^2 E\{\delta_s(y-\beta)^2\}$ in the above display involving C.

It is possible to minimize the asymptotic variance using r_s. For instance, suppose $S = 2$, $p_1 = p$ and $r_1 = r$, and thus the asymptotic variance is

$$\frac{p^2}{r} V_1 + \frac{(1-p)^2}{1-r} V_2.$$

The first-and the second-order conditions for r are, respectively,

$$\frac{-p^2}{r^2} V_1 + \frac{(1-p)^2}{(1-r)^2} V_2 = 0 \quad \text{and} \quad \frac{2p^2}{r^3} V_1 + \frac{2(1-p)^2}{(1-r)^3} V_2 > 0.$$

From the first-order condition,

$$\frac{p^2}{r^2} V_1 = \frac{(1-p)^2}{(1-r)^2} V_2 \iff \frac{(1-p)^2}{p^2} \frac{V_2}{V_1} = \frac{(1-r)^2}{r^2}$$

$$\implies \frac{1}{r} - 1 = \frac{1-p}{p}\left(\frac{V_2}{V_1}\right)^{1/2} \implies r = \left\{1 + \frac{1-p}{p}\left(\frac{V_2}{V_1}\right)^{1/2}\right\}^{-1}.$$

The optimal r is decreasing in V_2/V_1; $r = p$ if $V_2 = V_1$. Therefore, when the variance changes across the cells, do stratified sampling such that the cell with the larger (smaller) variance gets over-sampled (under-sampled). Substituting the optimal r into the asymptotic variance gives (in fact, substituting $(p^2/r^2)V_1(1-r) = \{(1-p)^2/(1-r)\}V_2)$, with $p_2 = 1 - p$,

$$p^2 V_1 + (1-p)V_2 + 2p(1-p)V_1^{1/2}V_2^{1/2} = \left(p_1 V_1^{1/2} + p_2 V_2^{1/2}\right)^2$$

which can be taken as an "average" of V_1 and V_2.

As was just seen, the optimal r is not necessarily p. This means that stratified sampling is better than random sampling. The improvement will be greater if the "within stratum variances" (V_s's) are small (and the "between stratum variances" are large), which is achieved if the sampling units are homogenous within the same stratum (and heterogenous across different strata). This is because *only the stratum variances appear in the asymptotic variance*. Recalling ANOVA $V(y) = E\{V(y|c)\} + V\{E(y|c)\}$, since $V(y)$ is fixed, we can reduce the within stratum variances $V(y|c)$'s (and thus $E\{V(y|c)\}$) only at the cost of increasing the second term $V\{E(y|c)\}$. Designing strata such that the strata averages $E(y|c)$'s differ much across the cells can deliver this feature. A natural question is then "why not always use stratified sampling?" This is because stratified sampling is complicated compared with random sampling, and the strata should be designed *before* data collection. It is not necessarily easy to see who are similar to whom ahead of data collection.

2.4   Logit Slope Consistency in Response-Based Samples

For more generality, we will deal with multinomial, instead of binary, logit. Suppose there are $j = 1, ..., J$ alternatives to choose from, with the choice probabilities (omitting "$|w_i$")

$$P(i \text{ chooses } 1) = \frac{1}{1 + \sum_{j=2}^{J} \exp\left(w'_{ij}\beta\right)},$$

$$P(i \text{ chooses } j) = \frac{\exp\left(w'_{ij}\beta\right)}{1 + \sum_{j=2}^{J} \exp\left(w'_{ij}\beta\right)}, \quad j \neq 1.$$

This includes binary logit as a special case when $J = 2$. Define $\tilde{w}_{ij}$ as the elements of w_{ij} other than unity and $\tilde{\beta}$ as the corresponding parameters for $\tilde{w}_{ij}$—w_{ij} includes the "alternative-constant" regressors z_i and the first element of z_i is unity. Denoting the alternative-j intercept as τ_j, we get

$$P(i \text{ chooses } 1) = \frac{1}{1 + \sum_{j=2}^{J} \exp\left(\tau_j + \tilde{w}'_{ij}\tilde{\beta}\right)},$$

$$P(i \text{ chooses } j) = \frac{\exp(\tau_j + \tilde{w}'_{ij}\tilde{\beta})}{1 + \sum_{j=2}^{J} \exp\left(\tau_j + \tilde{w}'_{ij}\tilde{\beta}\right)}, \quad j \neq 1.$$

In the following, we will show that the slope parameters are identified in response-based samples; part of the proof is borrowed from Prentice and Pyke (1979). That is, MNL inconsistency for response-based samples is restricted to the intercepts only. We will use f for the population density/probability and g for the sampling density/probability. The key step in the proof is that "odds ratios," which are identified with response-based samples, are exponential functions of the $w'_{ij}\beta$ differences.

Dropping the index i, let w be the collection of the elements in w_j, $j = 1, ..., J$ and $\tilde{w}$ be the part of w other than unity. Let $f(j|\tilde{w})$ denote $P(j \text{ chosen}|\tilde{w})$, and $f(j|0)$ denote $P(j \text{ chosen}|\tilde{w} = 0)$. The above MNL probabilities imply

$$\frac{f(j|\tilde{w})/f(1|\tilde{w})}{f(j|0)/f(1|0)} = \frac{\exp\left(\tau_j + \tilde{w}'_j\tilde{\beta}\right)}{\exp(\tau_j)} = \exp\left(\tilde{w}'_j\tilde{\beta}\right), \quad j \neq 1.$$

The reverse also holds: if this display is true, then the above MNL probability follows. To see this, take ln on the first and last expressions on this display to get

$$\ln\frac{f(j|\tilde{w})}{f(1|w)} - \tilde{w}'_j\tilde{\beta} = \ln\frac{f(j|0)}{f(1|0)} \equiv \tau_j \implies \frac{f(j|\tilde{w})}{f(1|\tilde{w})} = \exp\left(\tau_j + \tilde{w}'_j\tilde{\beta}\right)$$

$$\implies f(j|\tilde{w}) = \exp(w'_j\beta) \cdot f(1|\tilde{w}), \quad \text{with } w'_j\beta \equiv \tau_j + \tilde{w}'_j\tilde{\beta}$$

$$\implies 1 - f(1|\tilde{w}) = \sum_{j=2}^{J} f(j|\tilde{w}) = f(1|\tilde{w})\sum_{j=2}^{J}\exp\left(w'_j\beta\right).$$

From the first and last terms, we get $f(1|\tilde{w}) = \left\{1 + \sum_{j=2}^{J}\exp(w'_j\beta)\right\}^{-1}$, substitution of which into $f(j|w) = \exp(w'_j\beta) \cdot f(1|w)$ yields the desired MNL probabilities.

Using $f(j|\tilde{w}) = f(\tilde{w}|j)f(j)/f(\tilde{w})$, it holds that

$$\frac{f(j|\tilde{w})/f(1|\tilde{w})}{f(j|0)/f(1|0)} \left(= \frac{\{f(\tilde{w}|j)f(j)/f(\tilde{w})\} / \{f(\tilde{w}|1)f(1)/f(\tilde{w})\}}{\{f(0|j)f(j)/f(0)\} / \{f(0|1)f(1)/f(0)\}}\right)$$
$$= \frac{f(\tilde{w}|j)/f(0|j)}{f(\tilde{w}|1)/f(0|1)}.$$

This equation also holds for any probability/density including the sampling density g as well:

$$\frac{g(j|\tilde{w})/g(1|\tilde{w})}{g(j|0)/g(1|0)} = \frac{g(\tilde{w}|j)/g(0|j)}{g(\tilde{w}|1)/g(0|1)} = \frac{f(\tilde{w}|j)/f(0|j)}{f(\tilde{w}|1)/f(0|1)}$$

where the last equality holds because $g(\tilde{w}|j) = f(\tilde{w}|j)\ \forall j$ in response-based samples. Combining the last equality with $\{f(j|\tilde{w})/f(1|\tilde{w})\} / \{f(j|0)/f(1|0)\}$ $= \exp\left(\tilde{w}_j'\tilde{\beta}\right)\ \forall j \neq 1$ yields

$$\frac{g(j|\tilde{w})/g(1|\tilde{w})}{g(j|0)/g(1|0)} = \exp\left(\tilde{w}_j'\tilde{\beta}\right), \quad \forall j \neq 1$$

$$\implies \ln\frac{g(j|\tilde{w})}{g(1|w)} - \tilde{w}_j'\tilde{\beta} = \ln\frac{g(j|0)}{g(1|0)} \equiv \xi_j, \quad \text{taking } \ln.$$

Hence, when MNL holds but the sampling is response-based, only the intercepts differ:

$$P(i \text{ chooses } 1) = \frac{1}{1 + \sum_{j=2}^J \exp\left(\xi_j + \tilde{w}_{ij}'\tilde{\beta}\right)},$$

$$P(i \text{ chooses } j) = \frac{\exp\left(\xi_j + \tilde{w}_{ij}'\tilde{\beta}\right)}{1 + \sum_{j=2}^J \exp\left(\xi_j + \tilde{w}_{ij}'\tilde{\beta}\right)}, \quad j \neq 1.$$

The true intercept is $\ln\{f(j|0)/f(1|0)\}$ while the identified intercept is $\ln\{g(j|0)/g(1|0)\}$.

2.5   Truncated Samples with Zero Cell Probability

The weighted M-estimator examined above applies to cells with non-zero sampling probabilities. What if there is a cell with $r_s = 0$? For this case, we turn back to the infeasible MLE. Earlier we mentioned the difficulty of knowing or specifying the x distribution in maximizing

$$\sum_i \sum_s \delta_{is} \ln P(y_i|x_i, \beta) - \sum_s N_s \ln p_s(\beta) \quad \text{where}$$

$$p_s(\beta) \equiv \int_{C_s} P(y|x, \beta)P^*(x)dydx.$$

Suppose we regard the sample as an exogenous sample: instead of sampling (x, y) jointly from a given cell, suppose x is fixed arbitrarily first in the cell and then y is sampled. In this case, x is a fixed constant and $p_s(\beta)$ gets replaced by

$$p_s(x_i; \beta) \equiv \int_{(x_i, y)\in C_s} P(y|x_i, \beta)dy$$

and there is no need to bother with the x-density.

For instance, consider the "zero-truncated model" under normality:

$$y = x'\beta + u, \quad u \sim N(0, \sigma^2) \text{ independently of } x$$
$$(x, y) \text{ observed only when } y > 0 \iff u > -x'\beta.$$

There are two cells ($y \leq 0$ and $y > 0$) and the cell $y \leq 0$ has zero cell selection probability. The log-likelihood function for the sample becomes

$$\sum_i \ln \frac{1}{\sigma} \phi \left(\frac{y_i - x_i' b}{\sigma} \right) - \ln \int \int_{-x'b}^{\infty} \frac{1}{\sigma} \phi \left(\frac{u}{\sigma} \right) P^*(x) du dx.$$

But if we regard x fixed, then this becomes

$$\sum_i \ln \frac{1}{\sigma} \phi \left(\frac{y_i - x_i' b}{\sigma} \right) - \ln \Phi \left(\frac{x_i' b}{\sigma} \right)$$

which is the MLE proposed by Hausman and Wise (1977).

The MLE for the last-display avoids the problem of specifying $P^*(x)$, but it should be borne in mind that the MLE incurs information loss by treating "the sample with random x and zero cell probability for the cell $y \leq 0$" as a "sample with fixed x and zero cell probability for the cell $y \leq 0$."

2.6   Truncated Count Response Under On-Site Sampling

One well-known method to evaluate non-market (e.g. public) good demand is contingent valuation methods (CVM) as shown in the main text. But a problem of CVM is that CVM is based on intentions (e.g., willingness to pay), not on the actual behaviors. For public goods such as mountain trails, their demand depends on the cost involved in traveling to the trails (and the entrance fee, if there is any); the cost includes transportations, lodgings, etc. forth. Using the actual travel cost data avoids the "intention" criticism. But random sampling to find out the demand for mountain trails would not yield enough observations with $y > 0$ as most people have $y = 0$. One way to avoid this problem is "on-site sampling": sample (say, for one month) those who come to the trails; y is the number of visits per month, and x consists of trail and individual characteristics including travel cost, age, gender, income etc. This is a response-based sampling because only those with $y > 0$ are sampled.

With $P(j|x) \equiv P(y = j|x)$, the truncated probability is

$$\frac{P(y = j|x)}{P(y \neq 0|x)} = \frac{P(j|x)}{1 - P(0|x)}.$$

For instance, using the Poisson model,

$$\frac{P(y = j|x)}{P(y \neq 0|x)} = \frac{e^{-\lambda(x)} \lambda(x)^j / j!}{1 - e^{-\lambda(x)}} \text{ where } \lambda(x) = \exp(x'\beta) \text{ is the Poisson parameter.}$$

In this case, we can use the log-likelihood function

$$\sum_i [-\exp(x_i' b) + x_i' b \cdot y_i - \ln\{1 - \exp(-e^{x_i' b})\}].$$

Unfortunately, on-site sampling typically has an additional problem: those with a large y is more likely to be picked up than those with a small y. This problem is called "*size-biased sampling*" or "*length-biased sampling*," and should be dealt with as well. The conditional probability of visit in this case becomes (e.g., Shaw, 1988)

$$\frac{yP(y|x)}{\sum_{a=0}^{\infty} aP(a|x)} = \frac{yP(y|x)}{E(y|x)} \left\{ = \frac{y}{\lambda(x)} \frac{e^{-\lambda(x)}\lambda(x)^y}{y!} = \frac{e^{-\lambda(x)}\lambda(x)^{y-1}}{(y-1)!} \right.$$

$$\left. \text{for Poisson model} \right\}.$$

The reason for the appearance of y in $yP(y|x)$ is that those with $y \geq 1$ is y-times more likely to be picked up compared to those with $y = 1$; $\sum_{a=1}^{\infty} aP(a|x)$ is the normalizing constant. For the on-site sampling Poisson MLE, the only change required from the usual Poisson MLE is using $\tilde{y} \equiv y - 1$ that can take 0 now. Then $\lambda(x) \equiv E(\tilde{y}|x) = E(y - 1|x)$, from which we get $E(y|x) = \lambda(x) + 1$ while $\lambda(x) = V(\tilde{y}|x) = V(y|x)$. Note that the size-biased sampling density $yf(y|x)/E(y|x)$ also holds even if y were continuously distributed.

Still a less-than-satisfactory feature of the Poisson MLE is that it cannot handle the over-dispersion problem, for which negative binomial distribution (NB) can be used, as suggested by Englin and Shonkwiler (1995). Using $\{y_i/E(y)\}P(y_i|x_i)$, we get (define $\lambda_i \equiv \lambda(x_i)$ and recall $\psi_i = \lambda_i^\kappa/\alpha$) the probability of $(y = y_i)|(x = x_i)$:

$$\frac{y_i}{\lambda_i} \cdot \frac{\Gamma(y_i + \psi_i)}{\Gamma(\psi_i)\Gamma(y_i + 1)} \cdot \left(\frac{\psi_i}{\lambda_i + \psi_i}\right)^{\psi_i} \left(\frac{\lambda_i}{\lambda_i + \psi_i}\right)^{y_i}$$

and Englin and Shonkwiler (1995) show

$$E(y|x_i) = 1 + \lambda_i + \frac{\lambda_i}{\psi_i} \quad \text{and} \quad V(y|x_i) = \lambda_i \left(1 + \frac{1}{\psi_i} + \frac{\lambda_i}{\psi_i} + \frac{\lambda_i}{\psi_i^2}\right).$$

As in our earlier NB parametrization for count responses, we may further set $\psi_i = \alpha^{-1}$ (with $\kappa = 0$), which differs from the parametrization in Englin and Shonkwiler (1995) who used instead $\psi_i = \lambda_i/\alpha$ (with $\kappa = 1$).

Yet another alternative to the on-site-sampling Poisson is the GMM using the moment condition $E[\{y - 1 - \exp(x'\beta)\}x] = 0$, which is likely to be much simpler than the on-site-sampling NB. Also, going beyond visits to a single site, Moeltner and Shonkwiler (2005) examined joint visits to multiple sites under on-site sampling in a parametric context.

3   Empirical Likelihood Estimator

Although GMM is a useful framework to unify many apparently different estimation ideas, it is sometimes criticized for its poor small sample performance. Here we examine "empirical likelihood estimation" and related methods which are touted to be better than GMM in small sample performance. Since small sample performance is not of our main interest, we will just show that empirical likelihood-based methods are equivalent to GMM in their asymptotic distributions. See, e.g., Kitamura (2007) and the references therein. Despite much "fanfare," however, it is not clear whether empirical likelihood estimators are worth the extra efforts in implementing them.

3.1 Empirical Likelihood (EL) Method

Suppose we have a $s \times 1$ moment condition $E\psi(z_i, \beta) = 0$ and want to find the "*empirical likelihood (EL)*" π_i, $i = 1, ..., N$, satisfying the moment condition. In EL-based ideas (see Owen (2001) and the references therein), the probabilities π_i, $i = 1, ..., N$, are estimated along with β by maximizing

$$\max_{b, p_1, ..., p_N} \sum_{i=1}^{N} \ln p_i \quad \text{subject to } p_i \geq 0, \ \sum_i p_i = 1, \ \sum_i p_i \psi(z_i, b) = 0.$$

The last restriction is that the moment condition holds exactly in the "restricted empirical world"—restricted by the moment condition—with $P(z = z_i) = p_i$, whereas $P(z = z_i) = N^{-1}$ in the unrestricted empirical world because the likelihood of observing $z = z_i$ in a given sample $z_1, ..., z_N$ is N^{-1}. Surprisingly, the estimator for β obtained this way avoids the issue of the weighting matrix in GMM and yet follows the same asymptotic distribution. We show this in the following, drawing on Qin and Lawless (1994).

EL method sets up the maximand

$$Q_N(b, t) \equiv \sum_i \ln p_i + \lambda \left(1 - \sum_i p_i \right) - Nt' \sum_i p_i \psi(z_i, b)$$

where λ and t are the Lagrangian multipliers; we suppress $p_1, ..., p_N$ and λ in the arguments of Q_N. The first-order condition for p_{Ni} and b_N is

$$\frac{\partial Q_N(b_N, t_N)}{\partial p_i} = \frac{1}{p_{Ni}} - \lambda - Nt_N' \psi(z_i, b_N) = 0$$

$$\implies p_{Ni} = \frac{1}{\lambda + Nt_N' \psi(z_i, b_N)};$$

$$\frac{\partial Q_N(b_N, t_N)}{\partial b'} = \sum_i p_{Ni} \underset{1 \times s}{t_N'} \cdot \underset{s \times k}{\psi_{b'}(z_i, b_N)} = 0.$$

From $\partial Q_N / \partial p_i = 0$, using $\sum_i p_{Ni} \psi(z_i, b_N) = 0$,

$$0 = \sum_i p_{Ni} \frac{\partial Q_N}{\partial p_i} = \sum_i p_{Ni} \left\{ \frac{1}{p_{Ni}} - \lambda - N\psi(z_i, b_N)' t_N \right\} = N - \lambda$$

$$\implies \lambda = N \quad \text{and} \quad p_{Ni} = \frac{1}{N} \cdot \frac{1}{1 + t_N' \psi(z_i, b_N)}.$$

The term next to N^{-1} is the adjusting factor for the unrestricted empirical likelihood.

From $\partial Q_N(b_N, t_N)/\partial t = 0$ and $\partial Q_N(b_N, t_N)/\partial b' = 0$, it holds that

$$\sum_i p_{Ni}\psi(z_i, b_N) = E^*\psi(z_i, b_N) = 0,$$

$$\sum_i p_{Ni}\underset{k \times s}{\psi_b}(z_i, b_N)\,t_N = \underset{s \times 1}{E^*\psi_b(z_i, b_N) \cdot t_N} = 0$$

where $E^*(\cdot)$ is the expectation using $\{p_{N1}, ..., p_{NN}\}$. Substituting the above p_{Ni} into these two gives

$$\underset{s \times 1}{Q_{N1}(b_N, t_N)} \equiv \frac{1}{N}\sum_i \frac{\psi(z_i, b_N)}{1 + t_N'\psi(z_i, b_N)} = 0,$$

$$\underset{k \times 1}{Q_{N2}(b_N, t_N)} \equiv \frac{1}{N}\sum_i \frac{\psi_b(z_i, b_N)}{1 + t_N'\psi(z_i, b_N)} t_N = 0 \quad (\implies Q_{N2}(b_N, 0) = 0).$$

These two equations determine the asymptotic distribution of $(\sqrt{N}(b_N - \beta)', \sqrt{N}t_N')'$.

We have (note that $t = 0$ eliminates many terms in the following):

$$\underset{s \times k}{\frac{\partial Q_{N1}(b_N, 0)}{\partial b'}} = \frac{1}{N}\sum_i \psi_{b'}(z_i, b_N) \quad \text{and}$$

$$\underset{s \times s}{\frac{\partial Q_{N1}(b_N, 0)}{\partial t'}} = -\frac{1}{N}\sum_i \psi(z_i, b_N)\psi(z_i, b_N)',$$

$$\underset{k \times k}{\frac{\partial Q_{N2}(b_N, 0)}{\partial b'}} = 0 \quad \text{and} \quad \underset{k \times s}{\frac{\partial Q_{N2}(b_N, 0)}{\partial t'}} = \frac{1}{N}\sum_i \psi_b(z_i, b_N).$$

Using this, expand $0 = \{Q_{N2}(b_N, t_N)', Q_{N1}(b_N, t_N)'\}'$ around $(\beta', 0_s')'$ to get, for some $b_N^* \in (b_N, \beta)$,

$$0 = \begin{bmatrix} 0 \\ (1/\sqrt{N})\sum_i \psi(z_i, \beta) \end{bmatrix}$$
$$+ \begin{bmatrix} 0 & (1/N)\sum_i \psi_b(z_i, b_N^*) \\ (1/N)\sum_i \psi_{b'}(z_i, b_N^*) & -(1/N)\sum_i \psi(z_i, b_N^*)\psi(z_i, b_N^*)' \end{bmatrix}$$
$$\begin{bmatrix} \sqrt{N}(b_N - \beta) \\ \sqrt{N}t_N \end{bmatrix}.$$

Solve this for the last vector to get an $o_p(1)$ term plus

$$\begin{bmatrix} \sqrt{N}(b_N - \beta) \\ \sqrt{N}t_N \end{bmatrix} = \begin{bmatrix} 0 & E\psi_b(z, \beta) \\ E\psi_{b'}(z, \beta) & -E\psi(z_i, \beta)\psi(z_i, \beta)' \end{bmatrix}^{-1}$$
$$\begin{bmatrix} 0 \\ -(1/\sqrt{N})\sum_i \psi(z_i, \beta) \end{bmatrix}.$$

The rhs is asymptotically normal with the variance (omit (z, β) in $\psi(z, \beta)$)

$$\begin{bmatrix} 0 & E\psi_b \\ E\psi_{b'} & -E\psi\psi' \end{bmatrix}^{-1} \begin{bmatrix} 0 & 0 \\ 0 & E\psi\psi' \end{bmatrix} \begin{bmatrix} 0 & E\psi_b \\ E\psi_{b'} & -E\psi\psi' \end{bmatrix}^{-1}.$$

Define matrices A, C, D such that

$$\begin{bmatrix} A & C \\ C' & D \end{bmatrix} = \begin{bmatrix} 0 & E\psi_b \\ E\psi_{b'} & -E\psi\psi' \end{bmatrix}^{-1}.$$

Then the variance matrix is

$$\begin{bmatrix} A & C \\ C' & D \end{bmatrix} \begin{bmatrix} 0 & 0 \\ 0 & E\psi\psi' \end{bmatrix} \begin{bmatrix} A' & C \\ C' & D' \end{bmatrix}$$

$$= \begin{bmatrix} 0 & C \cdot E\psi\psi' \\ 0 & D \cdot E\psi\psi' \end{bmatrix} \begin{bmatrix} A' & C \\ C' & D' \end{bmatrix} = \begin{bmatrix} CE\psi\psi'C' & CE\psi\psi'D' \\ DE\psi\psi'C' & DE\psi\psi'D' \end{bmatrix}.$$

Use the formula for the inverse of a partitioned matrix to get

$$A = \{E\psi_b E^{-1}(\psi\psi')E\psi_{b'}\}^{-1}, \quad C = \{E\psi_b E^{-1}(\psi\psi')E\psi_{b'}\}^{-1} E\psi_b E^{-1}(\psi\psi'),$$
$$D = -E^{-1}(\psi\psi') + E^{-1}(\psi\psi')E\psi_{b'}\{E\psi_b E^{-1}(\psi\psi')E\psi_{b'}\}^{-1} E\psi_b E^{-1}(\psi\psi').$$

Hence, the asymptotic variance of $\sqrt{N}(b_N - \beta)$ is $CE\psi\psi'C' = \{E\psi_b E^{-1}(\psi\psi')E\psi_{b'}\}^{-1}$, which is the same as that of GMM.

Several remarks are in order. First, the asymptotic variance of $\sqrt{N}(b_N - \beta)$ can be estimated with

$$\left[\sum_i p_{Ni}\psi_b(z_i, b_N) \left\{ \sum_i p_{Ni}\psi(z_i, b_N)\psi(z_i, b_N)' \right\}^{-1} \sum_i p_{Ni}\psi_{b'}(z_i, b_N) \right]^{-1};$$

if we replace p_{Ni} with N^{-1}, then we get the GMM asymptotic variance estimator. Second, $\sqrt{N}t_N$ is asymptotically normal with variance $DE\psi\psi'D'$. This yields a LM over-id test statistic

$$Nt_N'(DE\psi\psi'D')^+ t_N \rightsquigarrow \chi^2_{s-k}$$

where $(DE\psi\psi'D')^+$ is the "Moore–Penrose generalized inverse": the Moore–Penrose generalized inverse "A^+" of a matrix A which may be singular or even non-square satisfies uniquely

$$AA^+A = A, \quad A^+AA^+ = A^+, \quad (AA^+)' = AA^+, \quad (A^+A)' = A^+A.$$

To understand the dof $s - k$, note that, for a $s \times k$ matrix W with $rank(W) = k$, $W(W'W)^{-1}W'$ has its rank k and $I_s - W(W'W)^{-1}W'$ has rank $s - k$. Defining A such that $-E^{-1}(\psi\psi') = A'A$ and a $s \times k$ matrix $W \equiv A \cdot E\psi_{b'}$, D can be written as

$$A'(I_s - W(W'W)^{-1}W')A \implies rank(D) = rank(I_s - W(W'W)^{-1}W') = s - k.$$

Third, there is an LR over-id test comparing p_{Ni} and N^{-1}:

$$2\left(\sum_i \ln\frac{1}{N} - \ln p_{Ni}\right) = 2\sum_i \ln\{1 + t'_N\psi(z_i, b_N)\} \rightsquigarrow \chi^2_{s-k}.$$

This test may be over-rejecting, and $2\sum_i t'_N\psi(z_i, b_N) \rightsquigarrow \chi^2_{s-k}$ may work better in practice using $\ln(1 + a) \simeq a$ when $a \simeq 0$.

To see how EL may be implemented, expand $Q_{N1}(b_N, t_N) = 0$ wrt $t_N = t_0$ and then replace t_N with t_1 to get an iteration formula:

$$0 \simeq \frac{1}{N}\sum_i \frac{\psi(z_i, b_N)}{1 + t'_0\psi(z_i, b_N)} - \frac{1}{N}\sum_i \frac{\psi(z_i, b_N)\psi(z_i, b_N)'}{\{1 + t'_0\psi(z_i, b_N)\}^2}(t_1 - t_0)$$

$$\Longrightarrow t_1 = t_0 + \left\{\sum_i \frac{\psi(z_i, b_N)\psi(z_i, b_N)'}{\{1 + t'_0\psi(z_i, b_N)\}^2}\right\}^{-1}\sum_i \frac{\psi(z_i, b_N)}{1 + t'_0\psi(z_i, b_N)}.$$

Once b_N is found, getting t_N that satisfies $Q_{N1}(b_N, t_N) = 0$ can be done using this formula; then, for a given b_N and t_N, p_{Ni} is $N^{-1}\{1 + t'_N\psi(z_i, b_N)\}^{-1}$. The only question left is how to get b_N. One may use the unweighted GMM as an initial b_N, obtain t_1 with $t_0 = 0$ in the last display, and then get p_{Ni}; then a new b_N may be found using $\sum_i p_{Ni}\psi(z_i, b_N) = 0$ or $\sum_i p_{Ni}\psi_b(z_i, b_N)t_N = 0$ to go for a new round of iteration. Shortly, we will show an estimator related to EL but easier to obtain computationally.

3.2   Exponential Tilting Estimator

Imbens et al. (1998) proposed "*exponential tilting (ET)*" estimator:

$$\min_{b, p_1, \ldots, p_N} \sum_i p_i \ln p_i \quad \text{subject to } p_i \geq 0, \quad \sum_i p_i = 1, \quad \sum_i p_i\psi(z_i, b) = 0.$$

Set up the minimand

$$Q_N(b, t) = \sum_i p_i \ln p_i + \lambda(1 - \sum_i p_i) - t'\sum_i p_i\psi(z_i, b);$$

following Imbens et al. (1998), we use just t instead of Nt, which results in t_N, not $\sqrt{N}t_N$, following an asymptotic normal distribution.

Minimizing this is in fact the same as minimizing the Kullback–Leibler information criterion (KLIC) between two distributions $\{p_1, \ldots, p_N\}$ and $\{N^{-1}, \ldots, N^{-1}\}$; this equivalence will become clearer when "power-divergence statistic" is discussed later. Kitamura and Stutzer (1997) also suggested ET independently of Imbens et al. (1998), minimizing the KLIC. ET estimation is also called the *maximum entropy estimation* where the *entropy* for a distribution $\{p_1, \ldots, p_N\}$ is

$$\sum_i p_i(-\ln p_i).$$

The entity $-\ln p_i$ is the "information contents" of the event with probability p_i in the following sense.

The less likely an event is, the more informative it is. For instance, "the president attended an official meeting" is highly likely and consequently not informative, whereas "the president danced tango" is unlikely and consequently highly informative. Thus, the information contents of an event should be a decreasing function of the probability, and $-\ln p_i$ is such a function. If two events are independent, then the information contents of the two events should be the sum over each information contents: this is also satisfied by $-\ln p_i$ because $-\ln p_i p_j = -\ln p_i - \ln p_j$ for two independent events with probabilities p_i and p_j, respectively. Adding requirements like this leads to $\{-\ln p_1, ..., -\ln p_N\}$ reflecting the information contents of the distribution $\{p_1, ..., p_N\}$, and $\sum_i p_i(-\ln p_i)$ is a weighted sum of $\{-\ln p_1, ..., -\ln p_N\}$.

Turning back to ET, the first-order condition for p_{Ni} and b_N is

$$\frac{\partial Q_N(p_{Ni}, b_N)}{\partial p_i} = 1 + \ln p_{Ni} - \lambda - t'_N \psi(z_i, b_N) = 0$$

$$\implies p_{Ni} = \exp\{\lambda - 1 + t'_N \psi(z_i, b_N)\}$$

$$\frac{\partial Q_N(p_{Ni}, b_N)}{\partial b} = \sum_i p_{Ni} \underset{k \times s}{\psi_b}(z_i, b_N) \cdot \underset{s \times 1}{t_N} = 0.$$

Substituting p_{Ni} into $\sum_i p_{Ni} = 1$ and solving it for $\exp(\lambda)$, we get

$$\exp(\lambda) = \frac{1}{\sum_i \exp\{-1 + t'_N \psi(z_i, b_N)\}} \implies p_{Ni} = \frac{\exp\{t'_N \psi(z_i, b_N)\}}{\sum_i \exp\{t'_N \psi(z_i, b_N)\}}.$$

Substitute this p_{Ni} into $\sum_i p_{Ni}\psi(z_i, b_N) = 0$ and $\sum_i p_{Ni}\psi_b(z_i, b_N)t_N = 0$ and multiply by $1/\sqrt{N}$ to get

$$\underset{s \times 1}{Q_{N1}(b_N, t_N)} \equiv \frac{1}{\sqrt{N}} \sum_i \psi(z_i, b_N) \exp\{t'_N \psi(z_i, b_N)\} = 0,$$

$$\underset{k \times 1}{Q_{N2}(b_N, t_N)} \equiv \frac{1}{\sqrt{N}} \sum_i \psi_b(z_i, b_N)t_N \exp\{t'_N \psi(z_i, b_N)\} = 0.$$

The rest of the steps are analogous to the EL method; the first-order asymptotics for $(\sqrt{N}(b'_N - \beta'), t'_N)'$ is the same as that of the EL method.

Newey and Smith (2004) proposed a *generalized empirical likelihood (GEL)* estimation where β is estimated in a "saddle point" problem:

$$\max_b \min_t \sum_i \rho\{t'\psi(z_i, b)\}$$

where $\rho(v)$ is concave in v which ranges over a domain including 0. Newey and Smith (2004) showed that GEL includes EL as a special case when $\rho(v) = \ln(1-v)$ and ET when $\rho(v) = -e^v$. Furthermore, if $\rho(v) = -(1+v)^2/2$, then

GEL becomes the "continuously updating GMM" in Hansen et al. (1996). In GEL, once b_N and t_N are obtained, the associated probabilities become

$$p_{Ni} = \frac{\rho'\{t_N'\psi(z_i, b_N)\}}{\sum_i \rho'\{t_N'\psi(z_i, b_N)\}}.$$

Newey and Smith (2004) showed that EL is least biased and go on to propose bias-corrected versions of GMM and GEL.

The joint maximization wrt b and minimization wrt t may not work well in practice. Some iteration ideas were shown above to implement EL. But the GEL with $\rho(v) = -(1+v)^2/2$ may be modified into a two-stage estimator to ease the computation. First, obtain a GMM estimator b_N to be plugged into the GEL:

$$\min_t \sum_i \frac{1}{2}\{1 + t'\psi(z_i, b_N)\}^2.$$

Second, minimizing this wrt t yields

$$0 = \sum_i \psi(z_i, b_N)\{1 + t_N'\psi(z_i, b_N)\}$$

$$\implies t_N = -\left\{\sum_i \psi(z_i, b_N)\psi(z_i, b_N)'\right\}^{-1}\sum_i \psi(z_i, b_N)$$

$$\text{and } p_{Ni} = \frac{1 + t_N'\psi(z_i, b_N)}{\sum_i\{1 + t_N'\psi(z_i, b_N)\}};$$

i.e., t_N and p_{Ni} can be obtained in closed forms, not requiring iteration. Third, obtain

$$\sum_i p_{Ni}\psi(z_i, b_N)\psi(z_i, b_N)' \quad \text{and} \quad \sum_i p_{Ni}\psi_b(z_i, b_N)$$

to do GMM with these matrices. The three steps may be iterated until convergence.

3.3   Minimum Discrepancy Estimator

Consider two probability distributions $A \equiv \{a_1, ..., a_N\}$ and $P \equiv \{p_1, ..., p_N\}$ for the same support points. For a given constant λ, a "*power-divergence statistic*" for the two distributions is (Cressie and Read, 1984)

$$D_\lambda(A, P) = \frac{2}{\lambda(1+\lambda)}\sum_i a_i\left\{\left(\frac{a_i}{p_i}\right)^\lambda - 1\right\}.$$

Cressie and Read (1984) in fact suggested this as a test statistic for the difference between two multinomial distributions where the sum is over the number of categories in the distributions.

The family $D_\lambda(A, P)$ indexed by λ includes several well-known test statistics, and Cressie and Read (1984) recommended using $\lambda = 2/3$. Since we are interested in connecting $D_\lambda(A, P)$ to GMM and EL-related methods, we will let $A = \{N^{-1}, ..., N^{-1}\}$ and the sum in $D_\lambda(A, P)$ will be over N distinct data points. When P depends on an unknown parameter vector θ, θ can be estimated consistently by minimizing $D_\lambda(A, P(\theta))$. But since $D_\lambda(A, P(\theta))$ is not a distance as $D_\lambda(A, P(\theta))$ is not symmetric in A and $P(\theta)$, Cressie and Read (1984) called the idea *"minimum discrepancy estimation,"* not minimum distance estimation.

To get the probabilities satisfying $\sum_i p_i \psi(z_i, \beta) = 0$ while being close to the unrestricted empirical likelihood N^{-1} for each observation, we can think of minimizing

$$\min_{b, p_1, ..., p_N} \sum_i D_\lambda(\{N^{-1}, ..., N^{-1}\}, P) \quad \text{subject to } p_i \geq 0, \ \sum_i p_i = 1,$$

$$\sum_i p_i \psi(z_i, b) = 0.$$

For any λ, the estimator b_N obtained this way has the same first-order asymptotic distribution as the GMM. But λ should matter for small sample performance. There are a number of well-known special cases of minimum discrepancy estimation as examined in the following.

First, letting $\lambda \to 0$, we get EL, because (recall the Box-Cox transformation and set $a_i = N^{-1}$)

$$\left\{ \left(\frac{a_i}{p_i} \right)^\lambda - 1 \right\} / \lambda \to \ln \frac{a_i}{p_i} = -\ln(Np_i) = -\ln N - \ln p_i \quad \text{as } \lambda \to 0;$$

$-\ln N$ is irrelevant for maximizing $\sum_i p_i$. Furthermore, as $\lambda \to 0$,

$$\frac{2}{1+\lambda} \sum_i a_i \left\{ \left(\frac{a_i}{p_i} \right)^\lambda - 1 \right\} / \lambda$$

$$\to -2 \sum_i \frac{1}{N} \ln(Np_i) = 2 \sum_i \frac{1}{N} \{ \ln \left(\frac{1}{N} \right) - \ln p_i \}.$$

This is a kind of LR test statistic for the restricted likelihood $\prod_i p_i$ and the unrestricted likelihood $\prod_i N^{-1}$: the difference $\ln(N^{-1}) - \ln p_i$ is weighted by the unrestricted empirical likelihood N^{-1}.

Second, as $\lambda \to -1$ (recall $d \ln g^\lambda / d\lambda = g^\lambda \ln g$),

$$\left\{ \left(\frac{a_i}{p_i} \right)^\lambda - 1 \right\} / (1+\lambda) \to \left(\frac{a_i}{p_i} \right)^{-1} \ln \frac{a_i}{p_i} \quad \text{due to L'Hospital's rule.}$$

Hence $D_\lambda(A, P)$ includes ET as a special case when $\lambda \to -1$. Furthermore, as $\lambda \to -1$, again with $a_i = N^{-1}$,

$$\frac{2}{\lambda} \sum_i a_i \left\{ \left(\frac{a_i}{p_i} \right)^\lambda - 1 \right\} / (1+\lambda)$$

$$\to -2 \sum_i p_i \ln\left(\frac{a_i}{p_i}\right) = -2 \sum_i p_i (\ln \frac{1}{N} - \ln p_i),$$

which is reminiscent of the LR test statistic: the difference $\ln(N^{-1}) - \ln p_i$ is weighted by the restricted p_i.

Third, with $\lambda = 1$, we get (using $\sum_i a_i = \sum_i p_i = 1$)

$$\sum_i a_i \left(\frac{a_i}{p_i} - 1\right) = \sum_i \frac{a_i^2}{p_i} - 1 = \sum_i \frac{a_i^2}{p_i} - 2 + 1 = \sum_i \frac{a_i^2}{p_i} - 2 \sum_i \frac{p_i a_i}{p_i}$$

$$+ \sum_i \frac{p_i^2}{p_i} = \sum_i \frac{a_i^2 - 2 p_i a_i + p_i^2}{p_i} = \sum_i \frac{(a_i - p_i)^2}{p_i}$$

which is the Pearson χ^2 goodness-of-fit test statistic. Cressie and Read (1984) showed that $D_\lambda(A, P) = D_1(A, P) + o_p(1)$, using Taylor expansion; i.e., the asymptotic distribution of the power-divergence test statistics is the same as that of the Pearson χ^2 test statistic. Recall that Cressie and Read (1984) recommended $\lambda = 2/3$, i.e., using something between EL and the Pearson minimum χ^2 if one intends to estimate θ in $p_i(\theta)$.

4 Stochastic-Process Convergence and Applications

4.1 Motivations

Recall the Kolmogorov-Smirnov (KS) goodness-of-fit (GOF) test: for iid data $y_1, ..., y_N$ from a continuous df F, we test H_0: $F = F_0$. With the empirical df $F_N(t) \equiv N^{-1} \sum_i 1[y_i \le t]$, the test statistic is

$$KS_N \equiv \sup_t \sqrt{N}|F_N(t) - F_0(t)| = \sup_t |\frac{1}{\sqrt{N}} \sum_i \{1[y_i \le t] - F_0(t)\}|$$

$$= \sup_{\tau \in [0,1]} |U_N(\tau)| \left\{ \text{ where } U_N(\tau) \equiv \frac{1}{\sqrt{N}} \sum_i \{1[F_0(y_i) \le \tau] - \tau\} \right.$$

$$\left. \text{with } \tau = F_0(t) \right\}$$

$$\rightsquigarrow \sup_{\tau \in [0,1]} |B(\tau)| \quad \text{under } H_0 \text{ where } B(\tau) \text{ is a Brownian bridge.}$$

The last expression follows from the continuous mapping theorem applied to the convergence in law "$U_N(\tau) \rightsquigarrow B(\tau)$ on $\tau \in [0,1]$"—more on this below. This test is ADF as its limit distribution can be found from the distribution of $\sup_{\tau \in [0,1]} |B(\tau)|$ that does not depend on F.

In many other GOF tests, however, the null df depends on a $k \times 1$ parameter θ which should be replaced by an estimator $\hat{\theta}_N$ to make the test operational. Although the test with θ known might be ADF as KS_N is, the test with $\hat{\theta}_N$ is no longer so in general. Consider

$$H_0 : F(t) = \Phi(t; \theta)$$

where Φ denotes the specified distribution in H_0 up to the unknown parameter θ. The natural question is then how to find the asymptotic distribution of the resulting test statistic; for KS_N, the test statistic is $\sup_t \sqrt{N}|F_N(t) - \Phi(t; \hat{\theta}_N)|$.

There are three approaches to this question. The first is finding the asymptotic distribution with a "correction term" due to $\hat{\theta}_N - \theta$, which is analogous to dealing with the usual TSE. The second is ADF-transformation by transforming the test statistic such that it becomes orthogonal to $\hat{\theta}_N - \theta$; this is also called "Khamaladze transformation." The third is bootstrap. Bootstrap in general will be reviewed in the next section, and this section examines the first two approaches. This task requires studying stochastic processes and convergence in law (or "weak convergence") as prerequisites. As "spin-offs," studying the prerequisites renders extra benefits such as showing consistency for asymptotic variance estimators and handling M-estimators in non-differentiable cases such as LAD.

Recall what the main use of convergence in law was in the case of usual multivariate CLT's. When $\sqrt{N}(b_N - \beta) \rightsquigarrow N(0, V)$, we use this to get

$$P\{\sqrt{N}(b_N - \beta) \leq a\} \to P\{N(0, V) \leq a\} \quad \text{for any } a.$$

This is in fact using the continuous mapping theorem as follows. Set $h\{\sqrt{N}(b_N - \beta)\} = 1[\sqrt{N}(b_N - \beta) \leq a]$, and observe that $h(\cdot)$ is continuous a.e. $N(0, V)$ because $h(\cdot)$ is continuous except at point a but $P\{N(0, V) = a\} = 0$. Hence

$$h\{\sqrt{N}(b_N - \beta)\} \rightsquigarrow h\{N(0, V)\} = P\{1[N(0, V) \leq a]\}.$$

This example differs from $U_N(\tau) \rightsquigarrow B(\tau)$ on $\tau \in [0, 1]$ in that $\sqrt{N}(b_N - \beta)$ is a k-dimensional vector and the convergence in law holds easily with a multivariate CLT, whereas there are uncountably many points in $[0, 1]$ for $U_N(\tau) \rightsquigarrow B(\tau)$.

A "direct" way a sequence of $k \times 1$ random vector z_N "converges" to a random vector z is that the difference $z_N - z$ converges to 0 in some sense. This needs a distance concept and both z_N and z should be defined on the same space so that we can look at $|z_N - z|$ and check, e.g., $P(|z - z_N| > \varepsilon) \to 0$ as $N \to \infty$ for any constant $\varepsilon > 0$. But for convergence in law, we are interested only in the *marginal* distributions of z_N and z. So long as the marginal distribution of z_N converges to that of z, the convergence in law holds; z_N and z do not have to be defined on the same probability space because $z_N - z$ is not needed. An "indirect" way of convergence is checking how close $a'z_N$ is to $a'z$ for any fixed $k \times 1$ vector a; here the difference between z_N and z

is indirectly looked at through a. But this idea does not yield anything new, because any a can be written as a linear combination of the basis vectors in R^k; i.e., as well known

$$z_N \rightsquigarrow z \iff a'z_N \rightsquigarrow a'z \text{ for all fixed vector } a.$$

For $U_N(\tau) \rightsquigarrow B(\tau)$, "$k$-variate distributional convergence" is not enough because $U_N(\tau)$ and $B(\tau)$ should be compared at uncountably many points; we need a "functional CLT" relative to the usual multivariate CLT's. Despite this difference, however, convergence in law such as $U_N(\tau) \rightsquigarrow B(\tau)$ can be defined "indirectly" by transformations analogous to $a'z_N$: a "bounded continuous function" of $U_N(\tau)$ and $B(\tau)$ will be used as an analog for $a'z_N$ and $a'z$. Compared with "strong convergence" concepts requiring $|z - z_N|$ or "$|U_N(\tau) - B(\tau)|$," the indirect convergences are "weak convergences". Also, whereas both z_N and z take their values in R^k, the appropriate space for $U_N(\tau)$ and $B(\tau)$ should be defined because $U_N(\tau)$ is not a continuous function of τ while $B(\tau)$ is.

4.2   Stochastic Process and Weak Convergence

4.2.1   Stochastic Process

A *metric space* M is a set with a metric m:

$$m(a, b) = m(b, a) \quad \text{(symmetry)},$$
$$m(a, c) \leq m(a, b) + m(b, c) \quad \text{(triangle inequality)},$$
$$m(a, b) \geq 0 \quad \text{(nonnegativity), and } 0 \text{ iff } a = b.$$

If $a \neq b$ is allowed when $m(a, b) = 0$, then m is a *semi-metric*. Since the basic concepts such as what is "open/closed," "close/distant" etc. in metric spaces are analogous to those for Euclidean spaces, in the following, we list only a few not mentioned in relation to Euclidean spaces but needed later.

The *closure* of a subset A of M is the smallest closed set that contains A. A subset of M is *dense* if its closure is M. M is *separable* if there is a countable dense subset. M is *totally bounded* (or *precompact*) if finitely many balls of radius ε can cover M for any $\varepsilon > 0$. A set A in M is *compact* if every open cover has a finite subcover. Although compactness is equivalent to A being closed and bounded when $M = R^k$, this does not hold in general; instead, a set A in M is *compact iff it is closed and totally bounded* ($\iff$ closed and there is a convergent subsequence for any sequence in M). For example, the extended real line $\bar{R} = [-\infty, \infty]$ is compact because it is closed and any sequence in $[-\infty, \infty]$ is bounded and thus has a convergent subsequence. Because of the theoretical convenience/necessity accorded by compactness, compact sets such as $[a, b]$ and $\bar{R}$ will appear below, but one may as well think of (a, b) and R in most cases, as typically functions/processes defined on (a, b) and R can be extended to $[a, b]$ and $\bar{R}$.

Define two spaces of functions on an interval $[a, b] \subset [-\infty, \infty]$:

$D[a, b]$: the set of "*cadlag*" functions on $[a, b]$
(right-continuous with left limit existing),
$C[a, b]$: the set of continuous functions on $[a, b]$.

It holds that $C[a, b] \subset D[a, b]$. Since each element of $D[a, b]$ is bounded, we can equip both spaces with

$$\text{uniform norm} \quad \sup_{x \in [a,b]} |f(x)| \quad \forall f \in D[a, b].$$

With this norm, $C[a, b]$ is known to be separable but $D[a, b]$ is not.

With a probability space $(\Omega, \mathcal{F}, P)$, a collection of rv's $\{v(\omega, t), a \leq t \leq b\}$ is a *stochastic process* "on $[a, b]$" or "indexed by $t \in [a, b]$." For each given t, $v(\omega, t)$ is a rv; for each given ω, $v(\omega, t)$ is a *sample path*. Depending on the context, we may write $v(\omega, t)$ as $v(\omega)$, $v(t)$ or just v. The *(coordinate) projection* of v on the coordinate t is the rv $v(\omega, t)$ with t fixed, and the projection is a functional of v denoted

$$\pi_t(v) \equiv v(t).$$

A stochastic process is also called a *random function* (or *random element*), because it is a function of t as well as ω. A well-known stochastic process is a Brownian motion or a Wiener process $W(t)$, $0 \leq t \leq T$, which has continuous sample path with $W(0) = 0$; $W(t) \in C[0, T]$.

A sequence of random functions v_n in $D[a, b]$ "*converge in law*" to a random function $v_o \in D[a, b]$ (denoted $v_n \rightsquigarrow v_o$) if

$$Ef(v_n) \to Ef(v_o) \quad \forall \text{ bounded continuous function } f : D[a, b] \mapsto R.$$

One example for $f(\cdot)$ is the projection $\pi_t(\cdot)$; $\pi_t(\cdot)$ is bounded as each element of $D[a, b]$ is so, and $\pi_t(\cdot)$ is continuous because $v_n \to v$ in the uniform metric implies $v_n \to v$ at t. When $f(\cdot) = \pi_t(\cdot)$,

$$Ef(v_n) = E\pi_t(v_n) = E\{v_n(t)\} = \int v_n(\omega, t)dP(\omega).$$

Although v_n can be any stochastic process, our main interest will be on empirical processes.

When $f(\cdot)$ is not as simple as $\pi_t(\cdot)$, $Ef(v_n)$ can be understood using the approximating function sequence $f_\nu \uparrow f$ that appeared in Lebesque integral construction:

$$f_\nu(\cdot) = 0 \text{ if } f(\cdot) < \frac{1}{2^\nu}, \quad f_\nu(\cdot) = 2^\nu \text{ if } f(\cdot) \geq 2^\nu,$$

$$f_\nu(\cdot) = \frac{j}{2^\nu} \text{ if } \frac{j}{2^\nu} \leq f(\cdot) < \frac{j+1}{2^\nu}, \quad j = 1, 2, ..., 4^\nu - 1.$$

For the events $f(\cdot) \geq 2^{\nu}$ and $2^{-\nu}j \leq f(\cdot) < 2^{-\nu+1}j$, consider, respectively,

$$P\{\omega : f(v_n(\omega)) \geq 2^{\nu}\} \quad \text{and} \quad P\{\omega : 2^{-\nu}j \leq f(v_n(\omega)) < 2^{-\nu+1}j\}.$$

$Ef_{\nu}(v_n)$ is the sum of many terms: 2^{ν} times the first probability, and $2^{-\nu}j$ times the second probability, $j = 1, 2, ..., 4^{\nu} - 1$; then $Ef(v_n)$ is the limit of $Ef_{\nu}(v_n)$ as $\nu \to \infty$.

Consider a sigma-field $\mathcal{G}$ on $D[a,b]$ and the probability measures P_n and P_o on $(D[a,b], \mathcal{G})$ induced by v_n and v_o, respectively:

$$P_n(G) \equiv P(v_n^{-1}(G)) \quad \text{and} \quad P_o \equiv P(v^{-1}(G)) \; \forall G \in \mathcal{G}.$$

Then convergence in law can be stated equivalently as: P_n is said to *converge weakly* to P_o ($P_n \rightsquigarrow P_o$) when

$$\{Ef(v_n) =\} \int f dP_n \to \int f dP_o \{= Ef(v_o)\}.$$

For $F_N(t) = N^{-1} \sum_{i=1}^{N} 1[y_i \leq t]$, we get $F_N(t) \to^{as} F(t)$ for each t owing to the LLN; i.e., there is $N_o = N_o(t, \varepsilon)$ such that $|F_N(t) - F(t)| < \varepsilon$ for all $N \geq N_o$ and any $\varepsilon > 0$. Not just this pointwise convergence, an uniform convergence holds in that the entire function $F_N(t)$, $t \in \bar{R}$, converges in the uniform metric:

$$\text{\textit{Glivenko Cantelli} Theorem: } \sup_{t \in \bar{R}} |F_N(t) - F(t)| \to^{as} 0.$$

That is, there is $N_o = N_o(\varepsilon)$ not depending on t such that $|F_N(t) - F(t)| < \varepsilon$ for all $N \geq N_o$. $F_1(t), F_2(t), ...$ is in fact a sequence of rv's, and this can be made explicit with $F_1(\omega, t), F_2(\omega, t), ...$ where $F_N(\omega, t) = N^{-1} \sum_i 1[y_i(\omega) \leq t]$ and each y_i is a rv mapping Ω to R. $F_N(\omega, t)$ is a stochastic process, and $F_N(\omega, \cdot)$ is cadlag.

Owing to CLT's, we have, for each given $t \in R$,

$$G_N(t) \equiv \sqrt{N}\{F_N(t) - F(t)\} = \frac{1}{\sqrt{N}} \sum_i \{1[y_i \leq t] - F(t)\}$$

$$\rightsquigarrow N\{0, F(t)(1 - F(t))\}.$$

The presence of $F(t)(1 - F(t))$ is natural because $N \cdot F_N(t)$ is binomial with N-many trials with "success" probability $F(t)$. $G_N(t)$ is an *empirical process* indexed by $t \in R$. As all indicator functions can take on 0 or 1, $G_N(t)$ is bounded by $-\sqrt{N}F(t)$ or $\sqrt{N}(1 - F(t))$ for a given N. Define $G_N(-\infty) \equiv 0$ and $G_N(\infty) \equiv 0$ when $t \in \bar{R}$; this gives a natural extension of $G_N(t)$, $t \in R$, to $G_N(t)$, $t \in \bar{R}$.

Recalling that F is continuous, replace $1[y_i \leq t]$ with $1[F(y_i) \leq F(t)]$ where $F(y_i) \sim U[0,1]$. Then $G_N(t)$ can be written as an *uniform empirical process*: with $u_i \equiv F(y_i)$ and $\tau = F(t)$,

$$U_N(\tau) \equiv \frac{1}{\sqrt{N}} \sum_i \{1[u_i \leq \tau] - \tau\}, \quad \tau \in [0,1];$$

$U_N(0) \equiv 0$ and $U_N(1) \equiv 1$.

There are empirical processes more general than the above $G_N(t)$, e.g.,

$$G_N\{t(y)\} \equiv \frac{1}{\sqrt{N}} \sum_i [t(y_i) - E\{t(y)\}]$$

which is an *empirical process indexed by a function* $t(y) \in \mathcal{T}$ for a class $\mathcal{T}$ of functions. This includes the preceding $G_N(t)$ as a special case when $t(y) = 1[y \leq t]$. This generalization is needed when functions such as $m(x, y, \theta)$ appear in the sum where $m(x, y, \theta)$ is a parametric function of (x, y) indexed by a parameter $\theta \in \Theta$.

4.2.2   Weak Convergence

To discuss weak convergence of empirical processes in the space of cadlag functions $D(R)$ (or $D(\bar{R})$), we need to have a σ-algebra on $D(R)$. Recall that, for a rv sequence $\{x_N\}$, we considered $(\Omega, \mathcal{F}, P)$ to $(R, \mathcal{B}, P_N)$ where $\mathcal{B}$ is the Borel σ-algebra on R and the induced probability measure is $P_N(B) \equiv P(x_N^{-1}(B))$ for $B \in \mathcal{B}$ on R. For an empirical process $G_N(\omega, t)$, we need to consider $(\Omega, \mathcal{F}, P)$ to $(D(R), \mathcal{G}, P_N)$ for a σ-algebra $\mathcal{G}$ on $D(R)$ where

$$P_N(C) \equiv P\{G_N^{-1}(C)\} = P\{\omega : G_N(\omega, \cdot) \in C\} \quad \text{for each } C \in \mathcal{G};$$

e.g., $C = \{c(t) : \sup_{t \in R} |c(t)| < \varepsilon, \text{ for a constant } \varepsilon > 0\}$. But using the Borel σ-algebra under the uniform metric on $D(R)$ is known to cause a measure-theoretic difficulty, as $D(R)$ is not separable under the uniform metric.

Instead, set $\mathcal{G}$ as the smallest σ-algebra containing all *open balls* in $D(R)$ generated by the uniform metric. For example, an ε-open ball in $D(R)$ around $c(t)$ under the uniform metric is $\{g \in D(R) : \sup_t |g(t) - c(t)| < \varepsilon\}$. This σ-algebra, which is smaller than the Borel σ-algebra containing all open sets in $D(R)$, is the same as the σ-algebra that makes all coordinate projections $\{\pi_t(\cdot), t \in R\}$ continuous. That is, considering $(D(R), \mathcal{G}, P_N)$ to $(R, \mathcal{B})$, it holds that

$$\pi_t^{-1}(B) \in \mathcal{G} \quad \text{for each Borel set } B \in \mathcal{B} \quad \text{for any } t \in R;$$

here we are using the "topological" definition of continuity that the inverse image of an open set is open, as σ-algebras's are topologies and a set in a σ-algebra is open. For instance, with $B = (b_1, b_2)$, $\pi_t^{-1}(B)$ is the elements in $D(R)$ with their values at t falling in (b_1, b_2). This choice of $\mathcal{G}$ makes $\pi_t(\cdot)$ a continuous function from $D(R)$ to R; as each element of $D(R)$ is bounded, $\pi_t(\cdot)$ is also bounded. Hence $\mathcal{G}$ assures at least that each marginal distribution at t of v_n converge to the marginal distribution at t of v for $v_n \rightsquigarrow v$ to hold. So long as the limit process, say $G(t)$ in $G_N(t) \rightsquigarrow G(t)$, belongs to $C(\bar{R})$, the measure-theoretic difficulty with the uniform metric does not matter for $G_N(t) \rightsquigarrow G(t)$; see, e.g., Pollard (1984, p. 92).

Recall

$$G_N(t) = \sqrt{N}\{F_N(t) - F(t)\} = \frac{1}{\sqrt{N}}\sum_i \{1[y_i \le t] - F(t)\}.$$

Although the weak convergence of G_N with its induced measure P_N to G_o with its induced measure P_o is defined in terms of $\int f dP_N \to \int f dP_o$ for each bounded and continuous function f, we do not evaluate the integrals for each continuous and bounded function f to verify the weak convergence. Instead, the weak convergence of $G_N(t)$ to a tight limit process $G_o(t)$ can be established in the following two steps, where $G_o(t)$ is *tight* if, for each $\varepsilon > 0$, there is a compact set $\bar{G}$ such that $P(G_o(t) \in \bar{G}) > 1 - \varepsilon$.

First, $G_N(t)$ at any finite number of points is shown to converge weakly to a tight process $G_o(t)$ evaluated at those points—this typically follows from a multivariate CLT as the finite point distribution of $G_o(t)$ is jointly normal in most cases. Second, $G_N(t)$ is shown to be "stochastically equicontinuous." These two facts are equivalent to $G_N(t) \rightsquigarrow G_o(t)$ that is tight. In essence, the role of stochastic equicontinuity is to make sure that $G_N(t)$ is not too variable as a function of t so that the finite-dimensional weak convergence leads to the convergence across all t. This way of establishing $P_N \rightsquigarrow P_o$ applies also when P_N is induced by the function-indexed empirical process $G_N\{t(y)\}$ with $t(y) \in \mathcal{T}$.

Recall a Brownian bridge $B(\tau)$ on $[0,1]$: $B(0) = B(1) = 0$ with continuous sample path, and for any given $\tau_1, ..., \tau_m$,

$$B(\tau_1), ..., B(\tau_m) \text{ is Gaussian with 0-mean and}$$
$$E\{B(\tau_j)B(\tau_k)\} = \min(\tau_j, \tau_k) - \tau_j\tau_k.$$

A generalized version with an increasing $F(\cdot)$ is *F-Brownian bridge* $B_F(t)$, $t \in [-\infty, \infty]$: $B_F(-\infty) = B_F(\infty) = 0$ with continuous sample path,

$$B_F(t_1), ..., B_F(t_m) \text{ is Gaussian with 0-mean and}$$
$$E\{B_F(t_j)B_F(t_k)\} = F\{\min(t_j, t_k)\} - F(t_j)F(t_k)$$
$$= \min\{F(t_j), F(t_k)\} - F(t_j)F(t_k).$$

This covariance function shows that $B_F(t)$ is the "F-transformed version" $B\{F(t)\}$ of t.

Now we will identify $B_F(t)$ as the limit process of $G_N(t)$ in comparison to the fact that Brownian bridge is the limit process of uniform empirical processes. That is, the finite dimensional distribution from $G_N(t)$ should equal the last display and $G_N(t)$ should be stochastically equicontinuous. Observe, for any two points $t_j, t_k \in R$,

$$E[G_N(t_j)G_N(t_k)] = E[\;\frac{1}{\sqrt{N}}\sum_i \{1[y_i \le t_j] - F(t_j)\}$$

$$\cdot \frac{1}{\sqrt{N}}\sum_i \{1[y_i \le t_k] - F(t_k)\}\;] = E\{1[y \le t_j] \cdot 1[y \le t_k]\}$$

$$- E1[y \le t_j] \cdot E1[x \le t_k] = F\{\min(\tau_j, \tau_k)\} - F(\tau_j)F(\tau_k).$$

That is, modulo stochastic equicontinuity, the limit process of $G_N(t)$ is $B_F(t)$.

An empirical process $G_N(t)$ indexed by a class $\{t(y)\epsilon T\}$ of functions that is totally bounded with a semi-metric ρ on T is *stochastically equicontinuous* if, for any $\varepsilon_1, \varepsilon_2 > 0$, there exists $\delta = \delta(\varepsilon_1, \varepsilon_2)$ such that

$$\limsup_{N \to \infty} P(\sup_{\rho(t_1, t_2) < \delta} |G_N(t_1) - G_N(t_2)| > \varepsilon_1) < \varepsilon_2.$$

We will examine when this holds in the following subsection. As $G_N\{t(y)\}$, $t \in T$, includes both $G_N\{t(y)\}$ with $t(y) = 1[y \le t]$ and $U_N\{\tau(u)\}$ with $\tau(u) = 1[u \le \tau]$ as special cases, stochastic equicontinuity along with the above finite dimensional convergence in law implies $G_N(t) \rightsquigarrow B_F(t)$ for $t \in [-\infty, \infty]$ and $U_N(\tau) \rightsquigarrow B(\tau)$ for $\tau \in [0, 1]$.

Weak convergence to $(F\text{-})$ Brownian bridge is called the *Donsker Theorem*. Donsker (1952) proved the weak convergence of the uniform empirical process to $B(\tau)$ and the continuous mapping theorem thereafter. See Pollard (1984), Van der Vaart and Wellner (1996), Van der Vaart (1998), and Billingsley (1999) for further details on this and the next subsections.

4.2.3   Stochastically Equicontinuous Empirical Processes

Consider an induced probability space $(X, \mathcal{X}, P_x)$, an iid sample $x_1, ..., x_N$ from P_x, and a measurable real-valued function f in a class $\mathcal{F}$ of functions mapping X to R. Define the *empirical measure*

$$P_N \equiv \frac{1}{N} \sum_i \delta_{x_i}$$

that assigns weight N^{-1} to each x_i; let $\delta_{x_i}(f) \equiv f(x_i)$. This leads to

$$P_N f \equiv \int f(x) dP_N(x) = \frac{1}{N} \sum_i f(x_i), \quad \text{whereas } Pf \equiv \int f(x) dP(x).$$

For instance, when f is the identity mapping I, it holds that $\delta_{x_i}(I) = x_i$ and $\int x dP_N = N^{-1} \sum_i x_i$ whereas $\int x dP = E(x)$. In the notation $P_N f$, P_N can be viewed as a stochastic process indexed by $F \in \mathcal{F}$; for a given index f, $P_N f$ is a rv (the sample average of $f(x_i)$'s); for a given ω (i.e., given $x_1, ..., x_N$), $P_N f$ is a sample path indexed by f as f varies over $\mathcal{F}$.

Analogously to the Glivenko-Cantelli Theorem, uniform consistency holds for a certain $\mathcal{F}$ that

$$\sup_{f \in \mathcal{F}} |P_N f - Pf| \to^{as} 0;$$

then $\mathcal{F}$ is *Glivenko-Cantelli* (or "$\mathcal{F}$ is a Glivenko-Cantelli class"). Also, analogously to the Donsker Theorem, for a certain $\mathcal{F}$ with $Pf^2 < \infty$, the empirical process indexed by f

$$G_N(f) \equiv \sqrt{N}(P_N f - Pf)$$

is stochastically equicontinuous and $G_N(f) \rightsquigarrow G_P(f)$ that is tight. For finite dimensional distributions, observe that $EG_P(f_1) = 0 = EG_P(f_2)$ for $f_1, f_2 \in \mathcal{F}$, and

$$E\{G_P(f_1)G_P(f_2)\} = P\{(f_1 - P(f_1)) \cdot (f_2 - P(f_2))\} = P(f_1 f_2) - P(f_1) \cdot P(f_2).$$

When the weak convergence holds, $\mathcal{F}$ is *Donsker* ("$\mathcal{F}$ is a Donsker class"). The question is then what classes are Glivenko-Cantelli or Donsker. In the following, we will show Donsker classes of functions drawing on Van der Vaart (1998, pp. 270–277) and Pollard (1984). If $\mathcal{F}$ is Donsker, it is also Glivenko-Cantelli.

Consider a class $\mathcal{F}$ of functions $f(x)$ with an *envelope function* $F(x)$: $|f(x)| \leq F(x)$ for every x and f. For an $\varepsilon > 0$, an *ε-bracket* is a pair of functions g and h in $L^2(P)$ such that $g \leq h$ and $\{\int |g - h|^p dP\}^{1/p} \leq \varepsilon$; g and h may not be in $\mathcal{F}$. The *bracketing number* $N(\varepsilon, \mathcal{F}, L^p(P))$ is the minimum number of ε-brackets to "cover" $\mathcal{F}$, where "cover" means $g_j \leq f \leq h_j$ for some bracket $[g_j, h_j] \ \forall f \in \mathcal{F}$. $N(\varepsilon, \mathcal{F}, L^p(P))$ is increasing as $\varepsilon \to 0$. Define the *bracketing integral* as

$$J(\delta, \mathcal{F}, L^2(P)) \equiv \int_0^\delta \{\ln N(\varepsilon, \mathcal{F}, L^2(P))\}^{1/2} d\varepsilon.$$

If this is finite for some $\delta > 0$, then $\mathcal{F}$ is Donsker.

Showing directly $J(\delta, \mathcal{F}, L^2(P)) < \infty$ for a given $\mathcal{F}$ is not easy. Fortunately, it is known that most classes of functions we encounter have finite bracketing integrals. One example is functions of *bounded variation*: these functions can be written as a difference of two increasing (i.e., non-decreasing) functions. Another example is $\mathcal{F} = \{f(x, \theta), \theta \in \Theta\}$ with a bounded indexing set Θ where there exists a function $m(x) \in L^2(P)$ satisfying a Lipschitz continuity:

$$|f(x, \theta_1) - f(x, \theta_2)| \leq m(x)|\theta_1 - \theta_2|, \ \forall \theta_1, \theta_2 \in \Theta.$$

To see more ways to check if $\mathcal{F}$ is Donsker, let a collection $\mathcal{C}$ of sets "pick out" a subset $A \subset W = \{\omega_1, ..., \omega_m\}$ if $A = W \cap C$ for some $C \in \mathcal{C}$. The collection $\mathcal{C}$ "shatters" W if $\mathcal{C}$ can pick out any of the 2^m subsets in W. If $\mathcal{C}$ cannot shatter W for any given m, then $\mathcal{C}$ is a *VC (Vapnik–Cervonenkis) class of sets* or a *class of polynomial discrimination*. For instance, consider $\mathcal{G}$ consisting of intervals $(-\infty, x]$ for $x \in R$ and $W = \{x_1, x_2\}$ with $x_1 < x_2$. The subset $\{x_1\}$ of W can be written as $W \cap (-\infty, x_1]$, and $\{x_1, x_2\}$ can be written as $W \cap (-\infty, x_2]$. But $\{x_2\}$ cannot be written as $W \cap (-\infty, x]$ for any x: $\mathcal{G}$ cannot shatter any set with 2 elements, and thus $\mathcal{G}$ is a VC class of sets.

For a $f \in \mathcal{F}$, define the *subgraph* of $f(x)$ as

$$\{(x, z) : f(x) < z\}.$$

If the collection of the subgraphs over $f \in \mathcal{F}$ is a VC class of sets in the (x, z) space, then call $\mathcal{F}$ a "VC class" (of functions). If $\mathcal{F}$ is a VC class and if $\int F^2 dP < \infty$, then $\mathcal{F}$ is Donsker. For example, the subgraph of $1[x \le t]$ is $\{(x, z) : 1[x \le t] < z\}$, and the collection of subgraphs indexed by t is

$$\{(x, z) : (x \le t, \ 1 < z) \ \text{ or } \ (x > t, \ 0 < z)\}.$$

This collection on R^2 is easily a VC class of sets with the constant envelope one; $\mathcal{F} = \{1[x \le t], \ t \in R\}$ is a VC class. Hence the empirical process $G_N(t)$ is stochastically equicontinuous.

Another VC class of functions is a finite-dimensional vector space of real functions, e.g., a three-dimensional vector space $\mathcal{G}$ of functions

$$g(x) = \zeta_1 f_1(x) + \zeta_2 f_2(x) + \zeta_3 f_3(x) \ \text{ indexed by } (\zeta_1, \zeta_2, \zeta_3) \in R^3.$$

With the envelope condition holding, $\mathcal{G}$ becomes Donsker. The class $\tilde{\mathcal{G}}$ of functions consisting of $1[g(x) > 0]$ is also a VC class, and is Donsker as well.

Given two Donsker classes $\mathcal{F}$ and $\mathcal{G}$ with envelope function F and G, respectively, the class of functions $f \pm g$, $\max(f, g)$ and $\min(f, g)$ are also Donsker. For example, with (x, y) playing the role of x now,

$$|y - x'b| = \max(y - x'b, 0) + \max(x'b - y, 0)$$

is Donsker. Also $\{sign(y - x'b), b \in B\} = \{1[y \ge x'b] - 1[y < x'b], \ b \in B\}$ is Donsker.

4.2.4 Applications

As an application of uniform consistency (i.e., Glivenko-Cantelli), we can prove the consistency of a LSE variance estimator. Recall

$$\sqrt{N}(b_{lse} - \beta) \rightsquigarrow N\{0, \ E^{-1}(xx')E(xx'u^2)E^{-1}(xx')\}$$

where the "natural" estimator for $E(xx'u^2)$ is $N^{-1}\sum_i x_i x_i r_i^2$ with $r_i \equiv y_i - x_i' b_{lse}$. Observe

$$|\frac{1}{N}\sum_i x_i x_i' r_i^2 - E(xx'u^2)| \ \le \ |\frac{1}{N}\sum_i x_i x_i' r_i^2 - E(xx'r^2)|$$

$$+ \ |E(xx'r^2) - E(xx'u^2)|.$$

For the first term on the rhs, consider $\mathcal{F}_{jk} = \{x_j x_k (y - x'b)^2, \ b \in B\}$ where $|b| < C \ \forall b \in B$ for some constant C, and x_j and x_k are the jth and kth components of x. Under some moment conditions, $\mathcal{F}_{jk}$ is Lipschitz-continuous and thus Glivenko-Cantelli. Then the uniform consistency over $\mathcal{F}_{jk}$ implies

$$|\frac{1}{N}\sum_i x_i x_i' (y_i - x_i' b_{lse})^2 - E\{xx'(y - x'b_{lse})^2\}| = o_p(1).$$

As for the second term, since $xx'(y - x'b_{lse}) \to^{ae} xx'(y - x'\beta)$, the dominated convergence yields $E(xx'r^2) \to E(xx'u^2)$ as $N \to \infty$.

Consider an empirical process indexed by a function $f \in \mathcal{F}$ parametrized by a parameter θ. Suppose that θ is replaced by $\hat{\theta}_N \to^p \theta_o$. If $\mathcal{F}$ is Donsker, then stochastic equicontinuity implies

$$\frac{1}{\sqrt{N}} \sum_i \{f(x_i, \hat{\theta}_N) - Ef(x, \theta_N)\} - \frac{1}{\sqrt{N}} \sum_i \{f(x_i, \theta_o) - Ef(x, \theta_o)\} = o_p(1).$$

It is remarkable that, although $\hat{\theta}_N - \theta_o$ may be of order $O_p(N^{-1/2})$ or slower, the error $\hat{\theta}_N - \theta_o$ does not affect the behavior of the empirical process around θ_o in the sense of this display.

In the last display, the centering by $Ef(x, \theta)$ is critical, without which the display does not hold in general. But if $\partial Ef(x, \theta_o)/\partial\theta = 0$, then the display holds without $Ef(x, \theta)$. To see how, observe that the lhs can be written as, for some $\theta_N^* \in (\theta_N, \theta_o)$,

$$\frac{1}{\sqrt{N}} \sum_i f(x_i, \hat{\theta}_N) - \frac{1}{\sqrt{N}} \sum_i f(x_i, \theta_o) - \sqrt{N}\{Ef(x, \theta_N) - Ef(x, \theta_o)\}$$

$$= \frac{1}{\sqrt{N}} \sum_i f(x_i, \hat{\theta}_N) - \frac{1}{\sqrt{N}} \sum_i f(x_i, \theta_o)$$

$$- \sqrt{N}\frac{\partial Ef(x, \theta_o)}{\partial\theta'}(\theta_N - \theta) + \frac{1}{2}N^{1/4}(\theta_N - \theta)'\frac{\partial^2 Ef(x, \theta_N^*)}{\partial\theta\partial\theta'}N^{1/4}(\theta_N - \theta).$$

If the Hessian matrix is bounded and if

$$\frac{\partial Ef(x, \theta_o)}{\partial\theta'} = 0 \text{ and } \theta_N - \theta = o_p\left(N^{-1/4}\right)$$

then the Taylor-expansion terms are $o_p(1)$, and we get

$$\frac{1}{\sqrt{N}} \sum_i f(x_i, \hat{\theta}_N) - \frac{1}{\sqrt{N}} \sum_i f(x_i, \theta_o) = o_p(1).$$

As another application of Donsker class, for an M-estimator b_N maximizing $N^{-1} \sum_i q(z_i, b)$ over $b \in B$ where the true value β is an interior point of a compact set B, let V be the second derivative matrix of $Eq(z, b)$ at β; assume that V is non-singular. Also define $r(z, b)$ such that

$$r(z, b) = \frac{q(z, b) - q(z, \beta) - (b - \beta)'\Delta(z, \beta)}{|b - \beta|} \text{ for } b \neq \beta,$$

$$= 0 \quad \text{when } b = \beta$$

for some $\Delta(z, \beta)$ with $E\{\Delta(z, \beta)\Delta(z, \beta)'\} < \infty$. If $N^{-1/2} \sum_i r(z_i, b)$ is stochastically equicontinuous, then Theorem 5 in Pollard (1984, p. 141) yields the asymptotic distribution for b_N:

$$\sqrt{N}(b_N - \beta) \rightsquigarrow N[0, \ V^{-1}E\{\Delta(z, \beta)\Delta(z, \beta)'\}V^{-1}].$$

In applying this to the LAD estimator maximizing $N^{-1}\sum_i -|y_i - x_i'b|$ where $u_i = y_i - x_i'\beta$ is continuously distributed with density $f_{u|x}$ and $Med(u|x) = 0$, let $z_i = (x_i', y_i)'$ and

$$q(z, b) = -|y - x'b| \quad \text{and} \quad \Delta(z, \beta) = \{1[y \geq x'b] - 1[y < x'b]\}x_i;$$

$\Delta(z, \beta)$ is an "asymptotic first derivative" of $q(z, b)$ at $b = \beta$. Since $V = E(2f_{u|x}(0)xx')$ (assume that V^{-1} exists) and $E\{\Delta(z, \beta)\Delta(z, \beta)'\} = E(xx')$, the asymptotic variance is

$$E^{-1}\{2f_{u|x}(0)xx'\} \cdot E(xx') \cdot E^{-1}\{2f_{u|x}(0)xx'\}.$$

What is left to be shown is the stochastic equicontinuity of $N^{-1/2}\sum_i r(z_i, b)$, for which it is sufficient to prove that $\{r(z, b), b \in B\}$ is Donsker where

$$r(z, b) = \frac{|y - x'b| - |y - x'\beta| - (b - \beta)'(1[y \geq x'b] - 1[y < x'b])x}{|b - \beta|}.$$

If we show that $\{r(z, b), b \in B\}$ has a square-integrable envelope, and that $\{r(z, b), b \in B\}$ is a VC class, then $\{r(z, b), b \in B\}$ is Donsker. Showing the VC class part can be done by doing analogously to the proof for "spatial median" in Pollard (1984, p. 153). Here, we will show only the square-integrable envelope part. Observe the well-known inequality:

$$|a + b| \leq |a| + |b| \implies |a + b| - |a| \leq |b| \implies |c| - |a| \leq |c - a| \text{ with } b = c - a.$$

For $|r(z, b)|$, we have

$$\frac{|y - x'b| - |y - x'\beta|}{|b - \beta|} \leq \frac{|x'\beta - x'b|}{|b - \beta|} = \frac{|x'(\beta - b)|}{|b - \beta|} \leq |x|;$$

$$\frac{|(b - \beta)'(1[y \geq x'b] - 1[y < x'b])x|}{|b - \beta|} \leq |x|.$$

Hence $|r(z, b)| \leq 2|x|$ and $2|x|$ is a square-integrable envelope under $E|x|^2 < \infty$.

4.3   Goodness-of-Fit Tests with Nuisance Parameters

4.3.1   Some Stochastic Integrals

As a preliminary to GOF tests with nuisance parameters replaced by estimators, we need to examine stochastic integrals, which is done in the following.

In a stochastic integral, say $\int g(\omega, t)F(\omega, dt)$, where both the integrand and integrator may be random and the integration is done wrt t, we can think of a number of cases: (i) only the integrand is random; (ii) only the integrator is random; (iii) both are random. We may try to think of the integral as an usual integral that is just a function of ω. But we will have a Brownian bridge $(B(t))$ integrator, and $B(t)$ is known to have "unbounded variation." Because the usual definition of integrals does not apply to an integrator with unbounded variation, a stochastic integral with $B(t)$ as an integrator has to be defined. In the following, we examine the case (i) first and explain "unbounded variation." Then (ii) will be examined informally, drawing on Del Barrio (2007). A formal treatment of stochastic integrals can be found, e.g., in Steele (2001) and Protter (2005). For our purpose, (ii) is enough, to which (i) serves as a "preparation." Whereas the usual integral is based on "the first-order variation", stochastic integral with an unbounded-variation integrator is based on "the second order (quadratic) variation", although this point may not be obvious in the following presentation.

In an integral such as $\int_0^1 X(\omega, t) f(t) dt$ where $E\{X(t)\} = 0$, the integrand $X(t)$ is random. Taking the integral as "conditional on ω," the integral is a rv, but its distribution is far from obvious. Consider a finite approximation to the integral using a partition $0 = t_0, t_1, ..., t_J = 1$ of $[0, 1]$: $\sum_{j=1}^{J} X(\omega, t_{j-1})\{f(t_j) - f(t_{j-1})\}$. Note that $X(\omega, \cdot)$ is evaluated at t_{j-1}, not at t_j, nor somewhere in-between. The variance of this finite approximation is

$$E[\sum_{j=1}^{J} X(\omega, t_{j-1})\{f(t_j) - f(t_{j-1})\} \cdot \sum_{s=1}^{J} X(\omega, t_{s-1})\{f(t_s) - f(t_{s-1})\}]$$

$$= \sum_{j,s} E\{X(t_{j-1})X(t_{s-1})\} \cdot \{f(t_j) - f(t_{j-1})\}\{f(t_s) - f(t_{s-1})\}$$

$$= \sum_{j,s} C(t_{j-1}, t_{s-1})\{f(t_j) - f(t_{j-1})\}\{f(t_s) - f(t_{s-1})\} \quad (C(t_{j-1}, t_{s-1})$$

$$\equiv E\{X(t_{j-1})X(t_{s-1})\})$$

$$\simeq \int_0^1 \int_0^1 C(s, t) f(s) f(t) ds dt.$$

Formally, Shorack and Wellner (1986, p. 42) showed that, if $H(t) = H_1(t) - H_2(t)$ with H_1 and H_2 increasing and right-(or left-) continuous,

$$\int_0^1 X(\omega, t) dH(t) \sim N(0, \sigma^2) \quad \text{if} \quad \sigma^2 \equiv \int_0^1 \int_0^1 C(s, t) dH(s) dH(t) \text{ exits.}$$

The condition on $H(t)$ is equivalent to $H(t)$ having "bounded variation," which is required for the integral (conditional on ω) to exist; $H(t)$, $t \in [0, 1]$, is of *bounded variation* if for each partition $0 = t_0, t_1, ..., t_J = 1$ of $[0, 1]$, $\sum_{j=1}^{J} |H(t_j) - H(t_{j-1})| < \infty$. The supremum over all possible partitions is the *total variation*. $H(t)$ is thus of bounded variation when its total variation is bounded. If H is increasing, then the total variation is simply $H(1) - H(0)$.

Now consider a stochastic integral $\int_0^1 g(t) dB(\omega, t)$ with $\int_0^1 g(t)^2 dt < \infty$ where the integrator is a Brownian bridge. Suppose $g(t)$ is a simple left-continuous function, say

$$g(t) = \sum_{j=1}^{J} \beta_j 1[\, t \in (t_{j-1}, t_j]\,].$$

Then the stochastic integral is defined as

$$\int_0^1 g(t)dB(t) \equiv \sum_{j=1}^{J} \beta_j \{B(t_j) - B(t_{j-1})\} = \sum_{j=1}^{J} \beta_j \Delta B_j \quad \text{where}$$

$$\Delta B_j \equiv B(t_j) - B(t_{j-1}).$$

Clearly this is Gaussian with mean zero as $E\Delta B_j = 0$; we just need to find its covariance function.

Recall $E\{B(s)B(t)\} = s - st$ where $s < t$. Defining $\Delta t_j \equiv t_j - t_{j-1} \iff t_j = \Delta t_j + t_{j-1}$, it holds that

$$E\{(\Delta B_j)^2\} = E\{B(t_j)^2 + B(t_{j-1})^2 - 2B(t_j)B(t_{j-1})\}$$
$$= (t_j - t_j^2) + (t_{j-1} - t_{j-1}^2) - 2(t_{j-1} - t_j t_{j-1})$$
$$= \{\Delta t_j + t_{j-1} - (\Delta t_j + t_{j-1})^2\} + (t_{j-1} - t_{j-1}^2) - 2t_{j-1} + 2(\Delta t_j + t_{j-1})t_{j-1}$$
$$= \Delta t_j + t_{j-1} - (\Delta t_j)^2 - 2(\Delta t_j)t_{j-1} - t_{j-1}^2 + t_{j-1} - t_{j-1}^2 - 2t_{j-1}$$
$$+ 2(\Delta t_j)t_{j-1} + 2t_{j-1}^2$$
$$= \Delta t_j - (\Delta t_j)^2 = \Delta t_j (1 - \Delta t_j).$$

Also, for $j < k$,

$$E\{\Delta B_j \Delta B_k\} = E[\{B(t_j) - B(t_{j-1})\}\{B(t_k) - B(t_{k-1})\}]$$
$$= E\{B(t_j)B(t_k)\} - E\{B(t_j)B(t_{k-1})\} - E\{B(t_{j-1})B(t_k)\} + E\{B(t_{j-1})B(t_{k-1})\}$$
$$= (t_j - t_j t_k) - (t_j - t_j t_{k-1}) - (t_{j-1} - t_{j-1} t_k) + (t_{j-1} - t_{j-1} t_{k-1})$$
$$= -t_j t_k + t_j t_{k-1} + t_{j-1} t_k - t_{j-1} t_{k-1} = -t_j \Delta t_k + t_{j-1} \Delta t_k = -\Delta t_j \cdot \Delta t_k.$$

Hence $\int_0^1 g(t)dB(t)$ is a zero-mean Gaussian process with variance

$$E\{(\sum_{j=1}^{J} \beta_j \Delta B_j)^2\} = \sum_{j=1}^{J}\sum_{k=1}^{J} \beta_j \beta_k \Delta B_j \Delta B_k = \sum_{j} \beta_j^2 \Delta t_j (1 - \Delta t_j)$$
$$- \sum_{j \neq k} \beta_j \beta_k \Delta t_j \Delta t_k$$
$$= \sum_{j} \beta_j^2 \Delta t_j - \sum_{j=1}^{J}\sum_{k=1}^{J} \beta_j \beta_k \Delta t_j \Delta t_k = \sum_{j} \beta_j^2 \Delta t_j - (\sum_{j} \beta_j \Delta t_j)^2$$
$$= \int_0^1 g(\tau)^2 d\tau - \{\int_0^1 g(\tau)d\tau\}^2.$$

Doing analogously, we can find the covariance between two stochastic integrals $\int_0^1 g(\tau)dB(\tau)$ and $\int_0^1 f(\tau)dB(\tau)$:

$$E\{\int_0^1 g(\tau)dB(\tau) \cdot \int_0^1 f(\tau)dB(\tau)\} = \int_0^1 g(\tau)f(\tau)d\tau - \int_0^1 g(\tau)d\tau \int_0^1 f(\tau)d\tau.$$

As a special case of this product expression, consider

$$B(\tau) = \int_0^\tau dB(t) = \int_0^1 1[t \leq \tau]dB(t)$$

and observe

$$E\{B(\tau) \cdot \int_0^1 f(t)dB(t)\} = \int_0^1 1[t \leq \tau]f(t)dt - \int_0^1 1[t \leq \tau]dt \int_0^1 f(t)dt$$

$$= \int_0^\tau f(t)dt - \tau \int_0^1 f(t)dt.$$

Although we assumed a simple function for $g(t)$, the variance and covariances hold more generally for smooth functions as well, so long as $\int_0^1 g(\tau)^2 d\tau < \infty$ and $\int_0^1 f(\tau)^2 d\tau < \infty$. Stochastic integral with a smooth function is defined as the limit of stochastic integrals of approximating simple-function sequences in the L^2 space. Replacing the upper integration number '1' with t, stochastic integral becomes a stochastic process.

4.3.2 Weak Limit of GOF tests with Nuisance Parameters

For $y_1, ..., y_N$ which are iid with a continuous df F, recall the empirical processes

$$U_N(\tau) = \frac{1}{\sqrt{N}} \sum_i \{1[u_i \leq \tau] - \tau\} \quad \text{where } u \sim U[0,1];$$

$$G_N(t) = \frac{1}{\sqrt{N}} \sum_i \{1[y_i \leq t] - E(1[y_i \leq t])\} = \frac{1}{\sqrt{N}} \sum_i \{1[F(y_i) \leq F(t)] - F(t)\}$$

$$= \frac{1}{\sqrt{N}} \sum_i \{1[u_i \leq F(t)] - F(t)\} = U_N\{F(t)\} \quad (\text{with } u_i \equiv F(y_i) \sim U[0,1]).$$

$$\Longrightarrow G_N(t) = U_N\{F(t)\}.$$

Consider H_0: $F(t) = \Phi(t; \theta)$ where Φ denotes the specified df in H_0 that depends on a parameter θ and is continuous and strictly increasing in t. Assume that $\Phi(t; \theta)$ has a continuous first derivative vector $\nabla_{\theta'}\Phi(t, \theta)$ wrt θ and $\sqrt{N}(\hat{\theta}_N - \theta)$ has an influence function $\eta(y)$. Observe

$$\sqrt{N}\{F_N(t) - \Phi(t; \hat{\theta}_N)\} = \frac{1}{\sqrt{N}} \sum_i \{1[y_i \leq t] - \Phi(t; \hat{\theta}_N)\}$$

$$= \frac{1}{\sqrt{N}} \sum_i \{1[y_i \le t] - \Phi(t, \theta) - \nabla_{\theta'} \Phi(t; \theta_N^*)(\hat{\theta}_N - \theta)\} \quad \text{(for some } \theta_N^*\text{)}$$

$$= \frac{1}{\sqrt{N}} \sum_i \{1[y_i \le t] - \Phi(t, \theta)\} - \frac{1}{N} \sum_i \nabla_{\theta'} \Phi(t; \theta_N^*) \sqrt{N}(\hat{\theta}_N - \theta)$$

$$= \frac{1}{\sqrt{N}} \sum_i \{1[F(y_i) \le F(t)] - \Phi(t, \theta)\} - \nabla_{\theta'} \Phi(t; \theta) \frac{1}{\sqrt{N}} \sum_i \eta(y_i) + o_p(1).$$

We get, up to an $o_p(1)$ term, as $F(t) = \Phi(t, \theta)$ under H_0,

$$\sqrt{N} \{F_N(t) - \Phi(t; \hat{\theta}_N)\} = G_N(t) - \nabla_{\theta'} \Phi(t; \theta) \frac{1}{\sqrt{N}} \sum_i \eta(y_i)$$

$$= G_N(t) - \nabla_{\theta'} \Phi(t; \theta) \int_{-\infty}^{\infty} \eta(t) G_N(dt)$$

$$= U_N\{F(t)\} - \nabla_{\theta'} \Phi(t; \theta) \int_{-\infty}^{\infty} \eta(t) U_N\{F(dt)\} \quad \text{using } G_N(t) = U_N\{F(t)\}$$

where $G_N(dt)$ gives the "charge" $N^{-1/2} \eta(y_i)$ at $t = y_i$, $i = 1, ..., N$, so that $\int_{-\infty}^{\infty} \eta(t) G_N(dt) = N^{-1/2} \sum_i \eta(y_i)$. Let F^{-1} be the quantile function (i.e., the generalized inverse) of F.

Setting $\tau = F(t)$, the last display can be written as (the next paragraph discusses this point further)

$$U_N(\tau) - \nabla_{\theta'} \Phi\{F^{-1}(\tau); \theta\} \int_0^1 \eta\{F^{-1}(\tau)\} U_N(d\tau);$$

$U_N(d\tau)$ gives the "charge" $N^{-1/2} \eta\{F^{-1}(u_i)\} = N^{-1/2} \eta(y_i)$ at $\tau = u_i$, $i = 1, ..., N$, so that both $\int_{-\infty}^{\infty} \eta(t) G_N(dt)$ and $\int_0^1 \eta\{F^{-1}(\tau)\} U_N(d\tau)$ are the sum of $N^{-1/2} \eta(y_i)$'s.

For F, we assumed only that $F(t)$ is continuous, not necessarily strictly increasing; i.e., $\tau = F(t)$ does not necessarily imply $t = F^{-1}(\tau)$ because the usual inverse exists iff F is continuous and strictly increasing. Recall the properties of the quantile function:

$$F^{-1}(\tau) \le t \iff \tau \le F(t)$$
$$F\{F^{-1}(\tau)\} = \tau \quad \forall\, 0 < \tau < 1 \quad \text{iff } F \text{ is continuous}$$
$$F^{-1}\{F(t)\} = t \quad \forall t \in R \quad \text{iff } F \text{ is strictly increasing.}$$

For $\Phi(t) = \Phi\{F^{-1}(\tau); \theta\}$ in the preceding display, observe that, under H_0,

$$\Phi(t) = P(y \le t) = P\{F(y) \le F(t)\} = P(u \le \tau) = \tau$$
$$\Phi\{F^{-1}(\tau)\} = P\{y \le F^{-1}(\tau)\} = P\{F(y) \le F(F^{-1}(\tau))\} = P(u \le \tau) = \tau.$$

As for $\eta\{F^{-1}(u_i)\} = \eta(y_i)$ for the integral, we do not necessarily have $y_i = F^{-1}(u_i)$ when $u_i \equiv F(y_i)$, because there may be flat portions in continuous

$F(t)$. But for an observation y_i, the probability that it equals a boundary value (e.g., one end of a flat portion) is zero; i.e., F must be strictly increasing at y_i, which yields $y_i = F^{-1}(u_i)$.

Under H_0, we get

$$\sup_t \sqrt{N}|F_N(t) - \Phi(t;\hat{\theta}_N)| = \sup_{\tau\in[0,1]} |\hat{B}(\tau)| + o_p(1)$$

$$\text{where} \quad \hat{B}(\tau) \equiv U_N(\tau) - \nabla_{\theta'}\Phi\{F^{-1}(\tau);\theta\}\int_0^1 \eta\{F^{-1}(\tau)\}dU_N(\tau).$$

Using $U_N(\tau) \rightsquigarrow B(\tau)$ and continuous mapping theorem, it further holds that

$$\hat{B}(\tau) \rightsquigarrow \int_0^\tau dB(t) - \nabla_{\theta'}\Phi\{F^{-1}(\tau);\theta\}\int_0^1 \eta\{F^{-1}(t)\}dB(t)$$

Although we drew on Del Barrio (2007) for this, the original reference goes back to Durbin (1973).

The weak limit of $\hat{B}(\tau)$ is a sum of two stochastic integrals $\int_0^\tau dB(t)$ and $\int_0^1 \eta\{F^{-1}(t)\}dB(t)$, which have zero expected values. Hence the weak limit of $\hat{B}(\tau)$ is a centered Gaussian process with its covariance, for $0 \le s,t \le 1$,

$$C(s,t) = \min(s,t) - st$$

$$- \nabla_{\theta'}\Phi\{F^{-1}(s);\theta\}\int_0^t \eta\{F^{-1}(\tau)\}d\tau - \nabla_{\theta'}\Phi\{F^{-1}(t);\theta\}\int_0^s \eta\{F^{-1}(\tau)\}d\tau$$

$$+ \nabla_{\theta'}\Phi\{F^{-1}(s);\theta\} \cdot \int_0^1 \eta\{F^{-1}(\tau)\}\eta\{F^{-1}(\tau)\}'d\tau \cdot \nabla_\theta\Phi\{F^{-1}(t);\theta\}.$$

As this depends on F, the GOF test with $\sqrt{N}|F_N(t) - \Phi(t;\hat{\theta}_N)|$ is no more ADF.

As an example, consider H_0: $y \sim N(\mu,1)$, i.e., $\theta = \mu$ with known variance 1 and we test for normality. In this case, $\hat{\theta} = \bar{y}$ is the MLE. Suppose F has density f. Then H_0: $F(t) = \Phi(t - \theta)$ where Φ is the N(0,1) df and

$$\nabla_\theta\Phi(t;\theta) = \frac{\partial\Phi(t-\theta)}{\partial\theta} = -\phi(t-\theta) \;(= -f(t)) = -\frac{\partial F(t)}{\partial t}).$$

With I_f denoting the information matrix, the fact that the influence function $\eta(y)$ is I_f^{-1} times $-f'(y)/f(y)$ implies

$$\int_0^1 \eta\{F^{-1}(\tau)\}\eta\{F^{-1}(\tau)\}'d\tau = \int_{-\infty}^\infty \eta(s)\eta(s)'f(s)ds = I_f^{-1}.$$

Also, with $\tau = F(s) \implies F^{-1}(\tau) = s$ and $d\tau = f(s)ds$,

$$\int_0^t \eta\{F^{-1}(\tau)\}d\tau = -I_f^{-1}\int_0^t \frac{f'\{F^{-1}(\tau)\}}{f\{F^{-1}(\tau)\}}d\tau = -I_f^{-1}$$

$$\int_{-\infty}^{F^{-1}(t)} f'(s)ds = -I_f^{-1}f\{F^{-1}(t)\}.$$

Substituting this and $\nabla_\theta\Phi(t,\theta) = -f(t)$ into the above $C(s,t)$ renders a much simpler expression

$$C(s,t) = \min(s,t) - st - f\{F^{-1}(s)\} \cdot I_f^{-1} \cdot f\{F^{-1}(t)\}$$
$$= \min(s,t) - st - \zeta(s)'\zeta(t), \quad \text{where } \zeta(t) \equiv I_f^{-1/2}f\{F^{-1}(t)\}.$$

4.3.3 Asymptotically Distribution-Free (ADF) Transformation

Here we show ADF transformation drawing on Koul (2006); the original idea is due to Khamaladze (1981, 1993). Then we revert to the GOF test with nuisance parameters.

For a function $g(\cdot)$ with its domain $[0,1]$, let $w(\cdot) = (1, g(\cdot))'$ and define

$$\underset{2\times 2}{A(s)} \equiv \int_0^1 w(t)w(t)'1[s \leq t]dt = \int_s^1 w(t)w(t)'dt, \quad 0 \leq s \leq 1$$
$$= E_u\{w(u)w(u)'1[s \leq u]\} \quad \text{for } u \sim U[0,1].$$

Assume that $A(s)$ is p.d. $\forall\, 0 \leq s < 1$; $A(s)$ is indexed by s, the lower bound in $s \leq t$.

The function g in w is the "source" for the correction term. When the nuisance parameter gets replaced by the MLE, g is the score function. This is understandable because ADF transformation restores the ADF property by transforming the original test statistic such that it becomes orthogonal to the nuisance parameter estimator; an example will be seen below. The constant 1 in w is to make the ADF transformation have mean zero.

Consider "explaining" a function $\psi(\cdot)$ with $w(\cdot)$, i.e.,

$$\underset{2\times 1}{Q_\psi(\tau)} \equiv \int_0^1 \psi(s)A^{-1}(s)w(s)1[s \leq \tau]ds = \int_0^\tau \psi(s)A^{-1}(s)w(s)ds$$
$$= E_v\{\psi(v)A^{-1}(v)w(v)1[v \leq \tau]\} \quad \text{for } v \sim U[0,1];$$
$$T_\psi(\tau) \equiv \psi(\tau) - Q_\psi(\tau)'w(\tau), \quad 0 \leq \tau < 1.$$

$Q_\psi(\tau)$ is reminiscent of the linear projection coefficient, and $T_\psi(\tau)$ the residual.

Define
$$H \equiv \{\psi \in L^2[0,1] : \int_0^1 \psi(\tau)w(\tau)d\tau = 0\}.$$

$\psi \in L^2[0,1]$ can be regarded as $\psi(u)$, $u \sim U[0,1]$, for a probability space $([0,1], \mathcal{B}, P)$ with the Lebesgue measure P; the norm is $\{E\psi(u)^2\}^{1/2} = \{\int_0^1 \psi(\tau)^2 d\tau\}^{1/2}$. Both $L^2[0,1]$ and H are Hilbert spaces with the same inner product $\int_0^1 \psi_1(\tau)\psi_2(\tau)d\tau$, and H is a subspace of $L^2[0,1]$ orthogonal to $w(\cdot)$—"*Hilbert space*" is a "complete" inner-product space with its norm defined as the square root of the "self inner product" where 'complete' means that each 'Cauchy sequence' has its limit in the same space, and a 'Cauchy sequence' $\{\psi_n\}$ satisfies that, for any $\varepsilon > 0$, there exists $N(\varepsilon)$ with $|\psi_n - \psi_m| < \varepsilon \ \forall n, m \geq N(\varepsilon)$.

T_ψ maps $L^2[0,1]$ to H, for which we need to verify $\int_0^1 T_\psi(\tau) w(\tau)' \, d\tau = 0$. Observe

$$\int_0^1 Q_\psi(\tau)' w(\tau) w(\tau)' d\tau = \int_0^1 \left\{ \int_{s \leq \tau} \psi(s) w(s)' A^{-1}(s) \right\} ds \ w(\tau) w(\tau)' d\tau$$

$$= \int_0^1 \psi(s) w(s)' A^{-1}(s) \left\{ \int_{s \leq \tau} w(\tau) w(\tau)' d\tau \right\} ds = \int_0^1 \psi(s) w(s)' ds$$

which cancels the $\psi(\tau)$ part of $T_\psi(\tau)$. Hence $\int_0^1 T_\psi(\tau) w(\tau) d\tau = 0$. As $w(\tau)$ includes 1, this also gives $\int_0^1 T_\psi(\tau) d\tau = E\{T_\psi(u)\} = 0$: $T_\psi(u)$ with $u \sim U[0,1]$ has zero mean.

Also T_ψ is *norm-preserving*:

$$\int_0^1 T_{\psi_1}(\tau) T_{\psi_2}(\tau)' d\tau = \int_0^1 \psi_1(\tau) \psi_2(\tau)' d\tau.$$

To see the norm preservation, observe

$$\int_0^1 T_{\psi_1}(\tau) T_{\psi_2}(\tau)' d\tau = \int_0^1 \{\psi_1(\tau) - Q_{\psi_1}(\tau)' w(\tau)\}\{\psi_2(\tau)'$$

$$- w(\tau)' Q_{\psi_2}(\tau)\} d\tau = \int_0^1 \psi_1(\tau) \psi_2(\tau)' d\tau$$

$$- \int_0^1 \psi_1(\tau) w(\tau)' Q_{\psi_2}(\tau) d\tau - \int_0^1 Q_{\psi_1}(\tau)' w(\tau) \psi_2(\tau)' d\tau$$

$$+ \int_0^1 Q_{\psi_1}(\tau)' w(\tau) w(\tau)' Q_{\psi_2}(\tau)' d\tau.$$

We need to verify that the last three terms on the rhs cancel one another.

Rewrite the second and third terms as

$$\int \psi_1(\tau) w(\tau)' Q_{\psi_2}(\tau) d\tau = \int \psi_1(\tau) w(\tau)' \left\{ \int_{s \leq \tau} \psi_2(s) A^{-1}(s) w(s) ds \right\} d\tau$$

$$= \int \int_{s \leq \tau} \psi_1(\tau) \psi_2(s) w(\tau)' A^{-1}(s) w(s) \ ds d\tau.$$

$$\int Q_{\psi_1}(\tau)' w(\tau) \psi_2(\tau) d\tau = \int \left\{ \int_{s \leq \tau} \psi_1(s) w(s)' A^{-1}(s) ds \right\} w(\tau) \psi_2(\tau) d\tau$$

$$= \int \int_{s \leq \tau} \psi_1(s) \psi_2(\tau) w(s)' A^{-1}(s) w(\tau) \ ds d\tau.$$

Also rewrite the fourth term as

$$\int \left\{ \int_{s\leq\tau} \psi_1(s)w(s)'A^{-1}(s)ds \cdot w(\tau)w(\tau)' \cdot \int_{q\leq\tau} \psi_2(q)A^{-1}(q)w(q)dq \right\} d\tau$$

$$= \int \int \psi_1(s)\psi_2(q)w(s)'A^{-1}(s)A(\max(s,q))A^{-1}(q)w(q)\, dsdq$$

$$= \int \int_{s\leq q} \psi_1(s)\psi_2(q)w(s)'A^{-1}(s)w(q)\, dsdq$$

$$+ \int \int_{q\leq s} \psi_1(s)\psi_2(q)w(s)'A^{-1}(q)w(q)\, dsdq.$$

Hence the fourth term cancels the second and third terms.

Recall that a Wiener process $W(\tau)$, $\tau \in [0,1]$ satisfies, for any finite number of points $\tau_1, ..., \tau_m$,

$$\{W(\tau_1), ..., W(\tau_m)\} \text{ is Gaussian with 0-mean and}$$

$$E\{W(\tau_j)W(\tau_k)\} = \min(\tau_j, \tau_k).$$

Viewing each τ as a function $1[u \leq \tau]$ where $u \sim U[0,1]$, we can regard the Wiener process as indexed by $1[u \leq \tau] \in L^2[0,1]$ such that

$$\{W(1[u \leq \tau_1]), ..., W(1[u \leq \tau_m])\} \text{ is Gaussian with 0-mean and}$$

$$E\{W(1[u \leq \tau_j])W(1[u \leq \tau_k])\} = \min(\tau_j, \tau_k) \ \{= \int_0^1 1[s \leq \tau_j]1[s \leq \tau_k]ds\}.$$

Then $W(1[u \leq \tau])$ maps the subspace of $L^2[0,1]$ consisting of $\{1[u \leq \tau], \tau \in [0,1]\}$ into a Hilbert space with the same inner product. With $\tau_j < \tau_k$,

$$\{\int |1[s \leq \tau_j] - 1[s \leq \tau_k]|^2 ds\}^{1/2} = \{\int 1[\tau_j \leq s \leq \tau_k]ds\}^{1/2} = |\tau_k - \tau_j|^{1/2}.$$

Hence, as $\tau_j \to \tau_k$, we get $W(1[u \leq \tau_j]) \to W(1[u \leq \tau_k])$ as the norms are the same in the two spaces: $W(1[u \leq \tau])$ has continuous sample path.

Recalling $T_\psi(\cdot)$, consider now the empirical process $G_N(\tau) = N^{-1/2} \sum_i T_\tau(u_i)$ for iid $u_i \sim U[0,1]$, $i = 1, ..., N$, and $\tau(u) = 1[u \leq \tau]$; note

$$T_\tau(u_i) = \tau(u_i) - Q_\tau(u_i)'w(u_i)$$

$$= 1[u_i \leq \tau] - \int_0^1 w(s)'A^{-1}(s)1[s \leq \tau]1[s \leq u_i]ds \cdot w(u_i).$$

Recall $E\{T_\tau(u)\} = 0$. Any finite dimensional distribution of this process equals that of $W(\tau)$ as can be seen in

$$E\{G_N(\tau_1)G_N(\tau_2)\} = E\{T_{\tau_1}(u)T_{\tau_1}(u)\} = \int_0^1 T_{\tau_1}(v)T_{\tau_2}(v)'d\tau$$

$$= \int_0^1 \tau_1(v)\tau_2(v)'dv = \min(\tau_1, \tau_2).$$

Therefore, under stochastic equicontinuity, we get $G_N(\tau) \rightsquigarrow W(\tau)$.

Recall the GOF test with the nuisance parameter being the mean and H_0: $F(t) = \Phi(t; \theta)$ where $\Phi(t; \theta) = \Phi(t - \theta)$. It was shown that

$$\sqrt{N}|F_N(t) - \Phi(t; \hat{\theta}_N)| = |\hat{B}(\tau)|, \text{ where}$$

$$\hat{B}(\tau) \equiv U_N(\tau) - \nabla_{\theta'}\Phi\{F^{-1}(\tau), \theta\}\int_0^1 \eta\{F^{-1}(\tau)\}dU_N(\tau);$$

$$\hat{B}(\tau) \rightsquigarrow \int_0^\tau dB(t) - \nabla_{\theta'}\Phi\{F^{-1}(\tau), \theta\}\int_0^1 \eta\{F^{-1}(t)\}dB(t)$$

$$C(s, t) = \min(s, t) - st - I_f^{-1}f\{F^{-1}(s)\}f\{F^{-1}(t)\}$$

where $C(s, t)$ is the covariance kernel for the weak limit. Note that the covariance kernel in a stochastic processe is informative as it yields the eigen-values and eigen-functions, with which often the stochastic process can be expressed as a sum of terms where each term depends on an eigen-function and the corresponding eigen-value (and some others).

Assume now that F is strictly increasing; with F being continuous, this means that the usual inverse exists for F. Let $q(t) \equiv F^{-1}(t) \Longleftrightarrow F\{q(t)\} = t$. Observe

$$1 = \frac{dF\{q(t)\}}{dt} = f\{q(t)\}\frac{dq(t)}{dt} \implies \frac{dq(t)}{dt} = \frac{1}{f\{q(t)\}}$$

$$\Longleftrightarrow \frac{dF^{-1}(t)}{dt} = \frac{1}{f\{F^{-1}(t)\}};$$

$$m(t) \equiv f\{F^{-1}(t)\} \implies m'(t) = \frac{f'\{F^{-1}(t)\}}{f\{F^{-1}(t)\}}$$

(score function at the tth quantile);

$$\int_0^1 m'(t)^2 dt = I_f \text{ because } \int_0^1 \left(\frac{f'\{F^{-1}(t)\}}{f\{F^{-1}(t)\}}\right)^2 dt = \int \left\{\frac{f'(s)}{f(s)}\right\}^2 f(s)ds = I_f.$$

Hence $C(s, t)$ can be written as

$$C(s, t) = \min(s, t) - st - \frac{m(s)m(t)}{\int_0^1 m'(a)^2 da}, \quad 0 \leq s, t \leq 1.$$

Now set $g(\tau) = m'(\tau)$ in $w(\tau)$; i.e., $w(\tau) = (1, m'(\tau))'$. With $\tau(u) = 1[u \leq \tau]$,

$$Q_\tau(u) = \int_{s \leq u} \tau(s)A^{-1}(s)w(s)ds = \int_{s \leq u} 1[s \leq \tau]A^{-1}(s)w(s)ds$$

$$= \int_0^{\min(\tau,u)} A^{-1}(s)w(s)ds;$$

$T_\tau(u)$ was presented already. As shown above, $G_N(\tau) = N^{-1/2}\sum_i T_\tau(u_i) \rightsquigarrow W(\tau)$. It further holds that replacing u with $\hat{u}_i \equiv \Phi(y_i - \bar{y})$ is innocuous, and thus

$$\frac{1}{\sqrt{N}}\sum_i T_\tau(\hat{u}_i) \rightsquigarrow W(\tau), \quad \tau \in [0,1].$$

The intuition for "$\hat{u}_i$ being as good as u_i" seems to be that any error term can be subjected to the linear projection whether it is estimated or not such that the resulting residual becomes a "martingale difference," which then yields the limit Wiener process.

5 Bootstrap

This section reviews bootstrap. For more details, refer to Hall (1992), Efron and Tibshirani (1993), Shao and Tu (1995), Davison and Hinkley (1997), Horowitz (2001a), Efron (2003) and the references therein. Other than these bootstrap-dedicated studies, Van der Vaart (1998), Lehmann and Romano (2005) and DasGupta (2008) also have substantial coverages of bootstrap, and further references can be found there.

5.1 Review on Asymptotic Statistical Inference

Before we introduce bootstrap, it is helpful to look back at how we typically conduct statistical inferences with large samples, as this contrasts the conventional large sample inferences with bootstrap.

Statistical inference (i.e., learning about β using a sample) is typically done in two ways: confidence intervals (CI) and hypothesis test (HT) with test statistics (TS). For a $k \times 1$ parameter β and an estimator $b_N \to^p \beta$, both CI and HT are done using the asymptotic distribution of a transformation of b_N. Usually, the transformation is a location- and scale-normalized version of b_N so that the transformation follows a known distribution; in most cases,

$$\sqrt{N}(b_N - \beta) \rightsquigarrow N(0,V) \implies \sqrt{N}V^{-1/2}(b_N - \beta) \rightsquigarrow N(0,I_k).$$

Here $\sqrt{N}V^{-1/2}(b_N - \beta)$ asymptotically follows a known distribution free of any (nuisance) parameters, and the TS $\sqrt{N}V^{-1/2}(b_N - \beta)$ is then said to be *asymptotically pivotal*. In practice, the unknown V should be replaced with an estimator $V_N \to^p V$, and due to the Slutsky Lemma, using $\sqrt{N}V_N^{-1/2}(b_N - \beta)$ is asymptotically as good as using $\sqrt{N}V^{-1/2}(b_N - \beta)$.

For a fixed $k \times 1$ known vector t, we have

$$\sqrt{N}(t'b_N - t'\beta) \rightsquigarrow N(0,t'Vt) \implies \frac{\sqrt{N}(t'b_N - t'\beta)}{\sqrt{t'V_N t}} \rightsquigarrow N(0,1).$$

As $N \to \infty$, with ζ_α denoting the αth quantile of $N(0,1)$,

$$P\left\{-\zeta_{1-\alpha/2} < \frac{\sqrt{N}(t'b_N - t'\beta)}{\sqrt{t'V_N t}} < \zeta_{1-\alpha/2}\right\}$$

$$\to^p P\{-\zeta_{1-\alpha/2} < N(0,1) < \zeta_{1-\alpha/2}) = 1 - \alpha$$

$$\Longrightarrow P\left\{t'b_N - \zeta_{1-\alpha/2}\frac{\sqrt{t'V_N t}}{\sqrt{N}} < t'\beta < t'b_N + \zeta_{1-\alpha/2}\frac{\sqrt{t'V_N t}}{\sqrt{N}}\right) \to 1 - \alpha;$$

the lhs is a CI for $t'\beta$. For instance, $t = (0,...,0,1)'$ and $\alpha = 0.05$ yields a symmetric asymptotic 95% CI for β_k. For a hypothesis H_0: $t'\beta = c$ for a specified value of c (typically $c = 0$), we reject the H_0 if c is not captured by the CI. Here the false rejection probability (i.e., the type I error) is α. This is doing a test using a CI.

Alternatively to using CI, we can use an asymptotically pivotal TS to conduct a HT. If the realized value of the TS is "extreme" (i.e., an "unlikely" value) for the known asymptotic distribution under H_0, then the H_0 gets rejected. For instance, under H_0: $t'\beta = c$, we can use

$$\frac{\sqrt{N}(t'b_N - c)}{\sqrt{t'V_N t}} \rightsquigarrow N(0,1) \quad \text{where the unknown } t'\beta \text{ is replaced by } c \text{ in } H_0.$$

For two-sided tests, we choose the critical region $(-\infty, -\zeta_{1-\alpha/2})$ and $(\zeta_{1-\alpha/2}, \infty)$, and H_0 is rejected if the realized value of the TS falls in the critical region with the false rejection probability α. A better way of doing inference might be looking at the *p-value*

$$2 \times P\left\{N(0,1) > \left|\text{realized value of } \frac{\sqrt{N}(t'b_N - c)}{\sqrt{t'V_N t}}\right|\right\}$$

showing how extreme the realized value of the TS is; TS being extreme leads to rejecting the H_0. For one-sided tests, this scenario requires minor modifications.

For multiple parameters, say $(t'_1\beta, ..., t'_g\beta)$, CI gets replaced by confidence regions. Stacking $t'_1\beta, ..., t'_g\beta$ to get $R'\beta$ (a $g \times 1$ vector) and

$$\sqrt{N}(R'b_N - R'\beta) \rightsquigarrow N(0, R'VR) \implies N(R'b_N - R'\beta)'(R'V_N R)^{-1}$$
$$(R'b_N - R'\beta) \rightsquigarrow \chi_g^2$$
$$\implies P\{N(R'b_N - R'\beta)'(R'V_N R)^{-1}(R'b_N - R'\beta) \le \psi_{g,1-\alpha}) \to 1 - \alpha$$

where $\psi_{g,1-\alpha}$ denotes the $(1-\alpha)$th quantile of χ_g^2. We can get an asymptotic confidence region (or volume) for $(t'_1\beta, ..., t'_g\beta)$. For example, with $g = 2$, this display leads to a circle or ellipse on the two-dimensional space for $(t'_1\beta, t'_2\beta)$. For HT with H_0: $(t'_1\beta, ..., t'_g\beta) = c$, we can check if c falls in the confidence region, or equivalently, reject the H_0

$$\text{if } N(R'b_N - c)'(R'V_N R)^{-1}(R'b_N - c) > \psi_{g,1-\alpha}, \quad \text{or}$$

if the p-value $P\left\{\chi_g^2 > \text{realized value of } N(R'b_N - c)'(R'V_N R)^{-1}\right.$
$\left.(R'b_N - c)\right\} < \alpha$.

Although using χ^2 distribution to turn the vector $\sqrt{N}(R'b_N - R'\beta)$ into a scalar is a widespread practice, this is not the only way. For instance, we may look at $\max\{V_N^{-1/2}\sqrt{N}(R'b_N - R'\beta)\}$ or $|V_N^{-1/2}\sqrt{N}(R'b_N - R'\beta)|$, although their asymptotic distributions are not as easy to obtain as in the above quadratic form. Other than for this aspect, CI and HT would proceed in the same manner. For instance, since $\max(\cdot)$ is a continuous function, its asymptotic distribution is that of the maximum of g independent $N(0,1)$ rv's, and thus an asymptotic 95% confidence region for $R\beta$ is

$$\{\tau : \max\{V_N^{-1/2}\sqrt{N}(R'b_N - \tau)\} \le \xi_{1-\alpha}\}$$

where $\xi_{1-\alpha}$ is the $(1-\alpha)$th quantile of $\max\{g$ independent $N(0,1)$ rv's$\}$, which can be tabulated by simulation. HT can be done by checking if the realized value of $\max\{V_N^{-1/2}\sqrt{N}(R'b_N - R'\beta)\}$ falls in the confidence region or not. Alternatively, the TS or its p-value can be used for HT.

Although CI and HT are equivalent to (i.e., "dual" to) each other in the case of using $\sqrt{N}(b_N - \beta) \rightsquigarrow N(0, V)$, there are many HT's whose corresponding CI's are hard to think of; e.g., H_0: the df of y is symmetric about 0, or H_0: $E(y^4) = 3E(y^2)$.

5.2   Bootstrap for Distribution Functions

5.2.1   Main Idea

Define the exact df for a statistic $T_N(F)$:

$$G_N(c; F) \equiv P[T_N(F) \le c] \quad \text{where}$$
$$T_N(F) \equiv V_N(F)^{-1/2}\sqrt{N}\{b_N(F) - \beta(F)\}$$

where F denotes the distribution for the original sample and V_N is a "scaling constant (matrix)." Regard β as a scalar for simplification. We want to find out $G_N(c; F)$: how $T_N(F)$ behaves with a given sample of size N when the sample was drawn from the true distribution F. This display makes it explicit that the exact, not asymptotic, distribution of $T_N(F)$ depends on the underlying distribution F. Large sample inference uses the approximation (the "asymptotic df" of $T_N(F)$) for $G_N(c, F)$:

$$G_\infty(c; F) \equiv \lim_{N\to\infty} G_N(c, F).$$

Often $T_N(F)$ is *asymptotically pivotal*: $G_\infty(c; F)$ does not depend on F; e.g., $G_\infty(c, F) = P\{N(0, I_k) \le c\}$. We may then write just $G_\infty(c)$ instead of $G_\infty(c; F)$. In this case, the large sample approximation $G_\infty(c; F)$ to $G_N(c; F)$

is done only through one route ("through the subscript"). "Two-route" approximation is shown next.

Suppose $T_N(F)$ is not asymptotically pivotal; e.g., $G_\infty(c, F) = \Phi\{c/\sigma(F)\}$ where the parameter of interest is the mean and σ is the SD. In this non-pivotal case, the nuisance parameter $\sigma(F)$ gets replaced by an estimator, say, $s_N \equiv \sigma(F_N)$. In general, when $T_N(F)$ is not asymptotically pivotal, $G_\infty(c, F_N)$ is used as a large sample approximation for $G_N(c; F)$: two routes of approximation are done between $G_N(c; F)$ and $G_\infty(c, F_N)$, through the subscript ∞ and F_N. It should be kept in mind that, *large sample inference uses $G_\infty(c)$ or $G_\infty(c, F_N)$ as an approximation to $G_N(c; F)$*, not the other way around.

Suppose that $G_N(c, F)$ is smooth in F in the sense

$$G_N(c; F_N) - G_N(c; F) \to^p 0 \text{ as } N \to \infty$$

where F_N is the empirical distribution for F;

recall that F_N gives probability N^{-1} to each z_i, $i = 1, ..., N$. *Bootstrap uses $G_N(c; F_N)$ as an approximation to $G_N(c; F)$ where the approximation is done only through F_N.* This is in contrast to the two-route large sample approximation $G_\infty(c, F_N)$ to $G_N(c, F)$. As Beran and Ducharme (1991) noted, this fact suggests that bootstrap may work better than the large sample approximation unless $T_N(F)$ is asymptotically pivotal.

Whether the last display holds or not depends on the smoothness of $G_N(c; F)$ as a functional of F. This also shows that consistent estimators for F other than F_N (e.g., a smoothed version of F_N) may be used in place of F_N. This is the *basic bootstrap idea: replace F with F_N and do the same thing as done with F.* Since the smoothness of $G_N(c, F)$ is the key ingredient for bootstrap, if the "source" $T_N(F)$ is not smooth in F, bootstrap either will not work as well (e.g., quantile regression is "one-degree"less smoother than LSE, and bootstrap works for quantile regression in a weaker sense than for LSE), or work not at all (in some M-estimators with b residing inside an indicator function). Bear in mind the different versions of G that appeared so far:

	Non-Operational	Operational
Finite-Sample	$G_N(c; F)$ for	$G_N(c; F_N)$ in bootstrap
Asymptotic	target $G_\infty(c; F)$	$G_\infty(c)$ (pivotal); $G_\infty(c; F_N)$
		(non-pivotal)

Using $G_N(c; F_N)$ means treating the original sample $(z_1, ..., z_N)$ as the population—i.e., the population distribution is multinomial with $P(z = z_i) = N^{-1}$. Specifically, with F replaced by F_N, we have

$$G_N(c; F_N) = P[T_N(F_N) \le c] = P[V_N(F_N)^{-1/2}\sqrt{N}\{b_N(F_N) - \beta(F_N)\} \le c]$$

and $\beta(F_N)$ is the parameter for the empirical distribution, i.e., $\beta(F_N) = b_N$. In M-estimator, β maximizes $E\{q(z,b)\} = \int q(z_o,b)dF(z_o)$ whereas b_N maximizes the empirical counterpart $N^{-1}\sum_i q(z_i,b) = \int q(z_o,b)dF_N(z_o)$, which shows that $\beta(F_N) = b_N$: b_N *plays the role of the parameter when F_N is taken as the population.*

For instance, suppose that $\beta(F)$ is the mean: $\beta(F) = \int z_o dF(z_o)$. Then considering a pseudo sample $z_1^*, ..., z_N^*$ drawn from F_N with replacement—some observations in the original sample get drawn multiple times while some get never drawn—we have

$$\beta(F_N) = \int z_o dF_N(z_o) = \bar{z} = b_N \text{ as } F_N \text{ assigns weight } \frac{1}{N} \text{ to each}$$

support point z_i in F_N

$$b_N(F_N) = \bar{z}^* \equiv \frac{1}{N}\sum_i z_i^*, \text{ pseudo sample mean estimator for the}$$

parameter $\beta(F_N)$ in F_N

$$V(F_N) = \frac{1}{N}\sum_i z_i^2 - \bar{z}^2 = \frac{1}{N}\sum_i (z_i - \bar{z})^2, \text{ which is also the sample}$$

variance "$V_N(F)$"

$$V_N(F_N) = \frac{1}{N}\sum_i z_i^{*2} - \bar{z}^{*2} = \frac{1}{N}\sum_i (z_i^* - \bar{z}^*)^2, \text{ pseudo sample variance}$$

to estimate $V(F_N)$.

This example illustrates that bootstrap approximates the distribution of (scaled) $\bar{z} - E(z)$ with that of (scaled) $\bar{z}^* - \bar{z}$. That is, the relationship of $\bar{z}$ to $E(z)$ is inferred from that of $\bar{z}^*$ to $\bar{z}$.

$G_N(c; F_N)$ may look hard to get, but it can be estimated as precisely as desired because F_N is known. One pseudo sample of size N gives one realization of $T_N(F_N)$. Repeating this N_B times yields N_B-many pseudo realizations, $b_N^{*(1)}, ... b_N^{*(N_B)}$. Then, thanks to the LLN applied with the "population distribution F_N for the pseudo sample,"

$$\frac{1}{N_B}\sum_{j=1}^{N_B} 1\left[V_N^{*(j)-1/2}\sqrt{N}(b_N^{*(j)} - b_N) \le c\right] \to G_N(c; F_N) \text{ as } N_B \to \infty$$

where $*$ means using a pseudo sample and the superscript j refers to the jth pseudo sample. This convergence is "in probability" or "a.e." conditional on the original sample $z_1, ..., z_N$. Hence there are two phases of approximation in bootstrap: the first is as $N_B \to \infty$ for a given N (as in the last display), and the second is as $N \to \infty$ for $G_N(c; F_N) - G_N(c; F) \to^p 0$. Since we can increase N_B as much as we want (so long as the computing power allows), the first phase of approximation can be ignored, and our focus will be on the second phase.

5.2.2   Percentile-t, Centered-Percentile, and Percentile

Suppose that T_N is asymptotically pivotal. Under $G_N(c; F_N) - G_N(c; F) \to^p 0$ as $N \to \infty$ (this is a "bootstrap consistency," to be examined later), we can use the bootstrap quantiles $\xi_{N,\alpha/2}$ and $\xi_{N,1-\alpha/2}$ found from the last display. A $(1-\alpha)100\%$ bootstrap CI for β is obtained from

$$\xi_{N,\alpha/2} < V_N^{-1/2}\sqrt{N}(b_N - \beta) < \xi_{N,1-\alpha/2}$$
$$\implies \left(b_N - \xi_{N,1-\alpha/2}\frac{V_N^{1/2}}{\sqrt{N}}, \ b_N - \xi_{N,\alpha/2}\frac{V_N^{1/2}}{\sqrt{N}} \right) \text{ for } \beta.$$

This way of constructing a CI with an asymptotically pivotal T_N is called *percentile-t method*—"percentile" because percentiles (i.e., quantiles) are used and "t" because T_N takes the form of the usual t-value that is asymptotically pivotal.

Suppose we use the non-scaled version $\sqrt{N}(b_N - \beta)$, in which case we would want to know the exact df of $\sqrt{N}\{b_N(F) - \beta(F)\}$:

$$H_N(c; F) \equiv P\left[\sqrt{N}\left\{ b_N(F) - \beta(F) \right\} \le c \right].$$

We will call bootstrap with this kind of a nonpivotal TS *"centered-percentile method"* for the lack of a better expression. The multiplicative factor $\sqrt{N}$ means that centered-percentile method in fact has a scale adjustment as well, although the scale adjustment is not "complete" so that the resulting TS is not asymptotically pivotal. Under $H_N(c; F_N) - H_N(c; F) \to^p 0$ as $N \to \infty$ (bootstrap consistency), the bootstrap estimator for $H_N(c, F)$ is $N_B^{-1}\sum_j 1\left[\sqrt{N}(b_N^{*(j)} - b_N) \le c \right]$. With $\psi_{N,\alpha/2}$ and $\psi_{N,1-\alpha/2}$ denoting the quantiles, a $(1-\alpha)100\%$ CI for β can be obtained from

$$\psi_{N,\alpha/2} < \sqrt{N}(b_N - \beta) < \psi_{N,1-\alpha/2}$$
$$\implies \left(b_N - \psi_{N,1-\alpha/2}\frac{1}{\sqrt{N}}, \ b_N - \psi_{N,\alpha/2}\frac{1}{\sqrt{N}} \right) \text{ for } \beta.$$

This method is easier to apply than bootstrap-t method because V_N is not needed. But the method is inferior to percentile-t method in finite sample performances as will be shown later.

There is another bootstrap method using b_N instead of $\sqrt{N}(b_N - \beta)$, which is called *percentile method*. Define the exact df for b_N as

$$J_N(c; F) \equiv P[b_N(F) \le c].$$

Under $J_N(c; F_N) - J_N(c; F) \to^p 0$ as $N \to \infty$ (bootstrap consistency), the bootstrap estimator for $J_N(c, F)$ is $N_B^{-1}\sum_{j=1}^{N_B} 1\left[b_N^{*(j)} \le c \right]$. Denoting the empirical df of $b_N^{*(1)}, ..., b_N^{*(N_B)}$ as K_N^*, a $(1-\alpha)100\%$ CI for β is

$$\left[K_N^{*-1}\left(\frac{\alpha}{2}\right), \ K_N^{*-1}\left(1 - \frac{\alpha}{2}\right) \right].$$

One disadvantage with this CI is that b_N may fall outside this CI (or near one end of the CI). To avoid this problem, sometimes a "bias-corrected CI" gets used as in the following paragraph. An advantage is, however, that a percentile method CI is invariant to monotonic transformations as will be shown shortly. Percentile method is closely related to centered-percentile method, because $\sqrt{N}(b_N^* - b_N)$ is a linear transformation of b_N^*, which implies that $\sqrt{N}\{K_N^{*-1}(\alpha) - b_N\}$ is the αth quantile in the bootstrap distribution in the centered-percentile method. In contrast, quantiles of asymptotically pivotal $V_N^{*-1/2}\sqrt{N}(b_N^* - b_N)$ cannot be obtained from those of $\sqrt{N}(b_N^* - b_N)$, nor from those of b_N^*, because of the random multiplicative factor $V_N^{*-1/2}$.

A *two-sided* $(1 - \alpha)100\%$ *bias-corrected CI* when the asymptotic distribution is normal is, with Φ being the df of $N(0,1)$(see, e.g., DiCiccio and Efron, 1996),

$$\left(\ K_N^{*-1}\left[\Phi\{\ \zeta_{\alpha/2} + 2\Phi^{-1}(K_N^*(b_N))\ \}\right], \right.$$
$$\left. K_N^{*-1}[\Phi\{\ \zeta_{1-\alpha/2} + 2\Phi^{-1}(K_N^*(b_N))\ \}]\ \right).$$

This is a special case of "bias-corrected and accelerated CI" when the "acceleration factor" is set to zero. If b_N is the exact median among the pseudo estimates so that $K_N^*(b_N) = 0.5$, then $\Phi^{-1}(K_N^*(b_N)) = 0$: the bias-corrected CI reduces to $\{K_N^{*-1}(\alpha/2), K_N^{*-1}(1-\alpha/2)\}$ that appeared above. If b_N is smaller than the median, however, then $K_N^*(b_N) < 0.5$, and $\Phi^{-1}(K_N^*(b_N)) < 0$: the bias-corrected CI shifts to the left so that b_N moves to the center of the interval.

A natural question at this stage is why bootstrap inference might be preferred. First, in terms of convenience, so long as the computing power allows, bootstrap is easier to use as it just repeats the same estimation procedure N_B times, which makes bootstrap a "no-brain" method. Second, estimating the asymptotic variance may be difficult; e.g. the asymptotic variance for the LAD estimator has $f_{u|x}(0)$ or $E(f_{u|x}(0)xx')$, whose estimation requires a bandwidth. Here bootstrap avoids choosing a bandwidth. Third, the bootstrap approximation error may be of smaller order than the asymptotic approximation error; e.g.,

$$G_\infty(c; F_N) - G_N(c; F) = O_p(N^{-1/2}) \quad \text{whereas}$$
$$G_N(c; F_N) - G_N(c; F) = O_p(N^{-1}).$$

Proving this requires "higher-order expansions" beyond the usual first term of order $O_p(N^{-1/2})$.

In bootstrap, there are many different versions—we saw already a bootstrap with an asymptotically pivotal statistic $(V_N^{-1/2}\sqrt{N}(b_N - \beta))$ and bootstraps with non-pivotal statistics $(\sqrt{N}(b_N - \beta)$ and $b_N)$. Showing why using a pivotal statistic is better requires again high-order expansions. In general, high-order *improvements of bootstrap over asymptotic inference hold only when asymptotically pivotal statistics are used* for the bootstrap.

5.2.3   Transformation and Percentile Method Invariance

Generalizing the transformation $T_N = V_N^{-1/2}\sqrt{N}(b_N - \beta)$ of b_N, suppose that there is a continuous and increasing transformation $\tau_N(b_N)$ such that, as $N \to \infty$,

$$P\{\tau_N(b_N) - \tau_N(\beta) \le c\} \to \Psi(c)$$

that is continuous, strictly increasing, and symmetric about 0;

$\tau_N(b_N) - \tau_N(\beta)$ is asymptotically pivotal, and Ψ is its known asymptotic df. With $\Psi(\xi_\alpha) \equiv \alpha$ and $\tau_N^{-1}(a) \equiv \min\{t : \tau_N(t) \ge a\}$ as in quantile functions, observe

$$P\left[\tau_N^{-1}\{\tau_N(b_N) + \xi_\alpha\} \le \beta\right] = P\left[\tau_N(b_N) + \xi_\alpha \le \tau_N(\beta)\right]$$

$$= P\left[\tau_N(b_N) - \tau_N(\beta) \le -\xi_\alpha\right] \to \Psi(-\xi_\alpha) = 1 - \Psi(\xi_\alpha) = 1 - \alpha:$$

$\tau_N^{-1}\{\tau_N(b_N) + \xi_\alpha\}$ is an asymptotic lower confidence bound for β with confidence coefficient $1 - \alpha$.

Intuitively, a first attempt to construct a lower confidence bound for $\tau_N(\beta)$ might be just $\tau_N(b_N)$ (then the CI would be $\{\tau_N(b_N), \infty\}$). To allow for the sampling error we want to lower the lower bound $\tau_N(b_N)$, which is done by adding a negative number ξ_α. After this, $\tau_N^{-1}\{\tau_N(b_N) + \xi_\alpha\}$ transforms the lower bound $\tau_N(b_N) + \xi_\alpha$ for $\tau_N(\beta)$ to that for β.

Recall the percentile method and the empirical df K_N^* of $b_N^{*(1)}, ...,$ $b_N^{*(N_B)}$. Suppose that the bootstrap consistency hold as well. Then we get

$$P^*\{\tau_N(b_N^*) - \tau_N(b_N) \le c\} \to^p \Psi(c)$$

because the lhs is $o_p(1)$-equal to $P\{\tau_N(b_N) - \tau_N(\beta) \le c\}$. Substitute $\tau_N(K_N^{*-1}(\alpha)) - \tau_N(b_N)$ into c in the display to get

$$\Psi\{\tau_N(K_N^{*-1}(\alpha)) - \tau_N(b_N)\} + o_p(1)$$

$$= P^*\{\tau_N(b_N^*) - \tau_N(b_N) \le \tau_N(K_N^{*-1}(\alpha)) - \tau_N(b_N)\}$$

$$= P^*\{\tau_N(b_N^*) \le \tau_N(K_N^{*-1}(\alpha))\} = P^*\{b_N^* \le K_N^{*-1}(\alpha)\} = \alpha.$$

Comparing the first and the last terms, we get, up to $o_p(1)$ terms

$$\tau_N(K_N^{*-1}(\alpha)) - \tau_N(b_N) = \xi_\alpha \iff K_N^{*-1}(\alpha) = \tau_N^{-1}\{\tau_N(b_N) + \xi_\alpha\};$$

recall that $\tau_N^{-1}\{\tau_N(b_N) + \xi_\alpha\}$ is the asymptotic lower confidence bound for β when the transformation $\tau_N(\cdot)$ is used to induce Ψ in the first display of this subsection. This shows that *bootstrap percentile method is invariant to continuous and increasing transformations*. That is, no matter which transformation of that type is used to induce the distribution Ψ ($= \Phi$ in most cases) asymptotically, percentile method can be used blindly.

As for two-sided intervals, observe, for some $\alpha' > \alpha$,

$$P\left[\tau_N^{-1}\{\tau_N(b_N) + \xi_\alpha\} \le \beta \le \tau_N^{-1}\{\tau_N(b_N) + \xi_{\alpha'}\}\right]$$
$$= P\left[\tau_N(b_N) + \xi_\alpha \le \tau_N(\beta) \le \tau_N(b_N) + \xi_{\alpha'}\right]$$
$$= P\left[\xi_\alpha \le \tau_N(\beta) - \tau_N(b_N) \le \xi_{\alpha'}\right] = P\left[-\xi_\alpha \ge \tau_N(b_N) - \tau_N(\beta) \ge -\xi_{\alpha'}\right]$$
$$\to \Psi(-\xi_\alpha) - \Psi(-\xi_{\alpha'}) = 1 - \Psi(\xi_\alpha) - \{1 - \Psi(\xi_{\alpha'})\} = \alpha' - \alpha.$$

Setting e.g., $\alpha' = 1 - \tilde\alpha/2$ and $\alpha = \tilde\alpha/2$ yields a $(1 - \tilde\alpha)100\%$ asymptotic CI for β:

$$[\tau_N^{-1}\{\tau_N(b_N) + \xi_\alpha\}, \ \tau_N^{-1}\{\tau_N(b_N) + \xi_{\alpha'}\}].$$

The percentile method CI is still just $[K_N^{*-1}(\alpha), \ K_N^{*-1}(\alpha')]$.

5.3 Bootstrap Consistency and Confidence Intervals

5.3.1 Defining Bootstrap Consistency

Let
$$C_N = \sqrt{N}(b_N - \beta) \quad \text{and} \quad C_N^* = \sqrt{N}(b_N^* - b_N).$$

To show that the bootstrap distribution is consistent, it should be proven that

$$P(C_N \le t) - P^*(C_N^* \le t | z_1, ..., z_N) \to^P 0$$

where P^* is for F_N which takes $z_1, ..., z_N$ as given whereas P is for F. The convergence in probability comes from the fact that P^* is conditional on the original sample $z_1, ..., z_N$. Here we state bootstrap consistency, using the centered version C_N, not the (asymptotically pivotal) $T_N = V_N^{-1/2}\sqrt{N}(b_N - \beta)$, because dealing with C_N is easier than T_N. As for b_N, the bootstrap consistency of b_N follows that of C_N; this will be also examined later.

Formally, for any constant $\varepsilon > 0$,

$$P[\ |P(C_N \le t) - P^*(C_N^* \le t | z_1, ..., z_N)| \ \le \varepsilon\ ] \to 1 \text{ as } N \to \infty$$

which is weak *bootstrap consistency*. That is, for any constants $\varepsilon, \varepsilon' > 0$, there is $N_o(\varepsilon, \varepsilon', t)$ such that, for all $N \ge N_o(\varepsilon, \varepsilon', t)$,

$$P[\ |P(C_N \le t) - P^*(C_N^* \le t | z_1, ..., z_N)| \ \le \varepsilon\ ] > 1 - \varepsilon'.$$

Replacing the convergence in probability with a.s. convergence yields "strong bootstrap consistency."

Instead of the consistency above, as we are estimating a function, we would desire the uniform consistency

$$\sup_t |P(C_N \le t) - P^*(C_N^* \le t | z_1, ..., z_N)| \to^P 0.$$

That is, for any positive constants $\varepsilon, \varepsilon' > 0$, there is $N_o(\varepsilon, \varepsilon')$ not depending on t such that, for all $N \ge N_o(\varepsilon, \varepsilon')$,

$$P[\ |P(C_N \le t) - P^*(C_N^* \le t | z_1, ..., z_N)| \ \le \varepsilon\ ] > 1 - \varepsilon'.$$

The uniform consistency is needed when t is replaced by a random sequence τ_N, because the uniformity implies

$$P(C_N \le \tau_N) - P^*(C_N^* \le \tau_N | z_1, ..., z_N) \to^p 0.$$

Subtracting and adding the limit df $\Psi(t; F)$, the uniform consistency holds if

$$(a): \sup_t |P(C_N \le t) - \Psi(t; F)| \to 0 \quad \text{and}$$

$$(b): \sup_t |P^*(C_N^* \le t | z_1, ..., z_N) - \Psi(t; F)| \to^p 0.$$

Recall convergence in distribution: a sequence of df's $\{J_N(t)\}$ converges to a df $J(t)$ at each continuity point t of $J(t)$. *Polya's theorem* is that, if $J(t)$ is continuous in t, then the convergence is uniform:

$$\lim_{N \to \infty} \sup_t |J_N(t) - J(t)| = 0.$$

With Φ being the $N(0, 1)$ df, in almost all cases we encounter, $\Psi = \Phi$ or at least Ψ is continuous, so that it is enough to establish pointwise convergence. Since the pointwise version of (a) holds by assumption (typically $C_N \rightsquigarrow N(0, V)$), we just have to show the pointwise version of (b):

$$P^*\{\sqrt{N}(b_N^* - b_N) \le t | z_1, ..., z_N\} \to^p \Psi(t; F) \ (= \Phi(t) \text{ typically}).$$

In short, we do not have to bother with the uniformity, as it will follow from the pointwise convergence.

5.3.2   Bootstrap Consistency with Empirical Processes

With bootstrap consistency defined, we want to know when it holds. For this query, empirical process approach gives answers, using empirical processes $G_N(f)$, $f \in \mathcal{F}$, in $l^\infty(\mathcal{F})$ where $l^\infty(\mathcal{F})$ denotes the space of bounded functions on $\mathcal{F}$ (i.e., $l^\infty(\mathcal{F})$ consists of functions G mapping $\mathcal{F} \to R$ such that $\sup_{f \in \mathcal{F}} |G(f)| < \infty$). The empirical process approach is shown here drawing on Van der Vaart (1998), Giné and Zinn (1990) and Arcone and Giné (1992).

Define the *bootstrap empirical measure* P_N^* and *bootstrap empirical process* G_N^* as

$$P_N^* \equiv \frac{1}{N} \sum_i \delta_{z_i^*} = \frac{1}{N} \sum_i m_{ni} \delta_{z_i}, \quad \text{and} \quad G_N^* \equiv \sqrt{N}(P_N^* - P_N) = \frac{1}{\sqrt{N}} \sum_i (m_{ni} - 1) \delta_{z_i}$$

where m_{ni} is the number of times z_i appears in the pseudo sample. Then for a function $f(z)$,

$$P_N^* f = \frac{1}{N} \sum_i f(z_i^*) = \frac{1}{N} \sum_i m_{ni} f(z_i) \quad \text{and} \quad G_N^* f = \frac{1}{\sqrt{N}} \sum_i (m_{ni} - 1) f(z_i).$$

Consider a Donsker class $\mathcal{F}$ of functions on z with $F(z) \equiv \sup_{f \in \mathcal{F}} |f(z)|$ being finite for each $z \in Z$. Then G_N^* converges weakly to G_P in $l^\infty(\mathcal{F})$

in probability where G_p is the weak convergence limit for the corresponding empirical process $\sqrt{N}(P_N - P)$ using the original data. That is, the pointwise version of (b) above holds:

$$P(G_p f \leq t) - P^*(G_N^* f \leq t | z_1, ..., z_N) \to^p 0$$

If $PF^2 < \infty$, then the weak convergence holds a.e., not just in probability. For instance, if $f(z) = z$, this display combined with (a) above gives

$$P\left(\frac{1}{\sqrt{N}} \sum_i (z_i - E(z)) \leq t\right) - P^*\left(\frac{1}{\sqrt{N}} \sum_i (z_i^* - \bar{z}) \leq t | z_1, ..., z_N\right) \to^p 0$$

justifying bootstrap for non-standardized sample means.

For parameters other than mean, the last two displays are not of direct help and one needs findings more involved than those. For this, Theorem 3.2 in Arcone and Giné (1992) proves a.e. bootstrap consistency for $\sqrt{N}(b_N - \beta)$ under some conditions where b_N is a M-estimator. Remark 3.3 and Example 4.10 in Arcone and Giné (1992) provide conditions for "in-probability" bootstrap consistency. The conditions, which are weaker than those for the a.s. consistency, are satisfied for almost all M-estimators (including the LAD estimator). In essence, if Theorem 2 of Pollard (1985) or Theorem 5 of Pollard (1984, p. 141) holds for the M-estimator maximizing $Eq(z, b)$, then *nonparametric bootstrap* (resampling from the original sample with replacement, repeating the same estimation procedure to get pseudo estimates along with their empirical df, and then using the quantiles thereof) is consistent. The stochastic equicontinuity of the "remainder process" $r(z, b)$ in the M-estimator plays the key role for the bootstrap consistency.

5.3.3   Confidence Intervals with Bootstrap Quantiles

Recall the quantile function $F^{-1}(\alpha) \equiv \min\{t : F(t) \geq \alpha\}$, $0 < \alpha < 1$. A sequence $\{F_N^{-1}\}$ is said to converge weakly to F^{-1} if $F_N^{-1}(\alpha) \to F^{-1}(\alpha)$ for every continuity point α of F^{-1} as $N \to \infty$. As shown in Van der Vaart (1998, p. 305),

$$F_N \rightsquigarrow F \iff F_N^{-1} \rightsquigarrow F^{-1}.$$

From bootstrap consistency and this display, *the αth quantile of the bootstrap empirical distribution of $C_N^* = \sqrt{N}(b_N^* - b_N)$ is consistent for the αth quantile of Ψ, assuming that $\Psi(t)$ is continuous and strictly increasing in t.* Observe

$$\sqrt{N}(b_N - \beta) \leq \Psi^{-1}(\alpha; F) \iff b_N - \frac{\Psi^{-1}(\alpha; F)}{\sqrt{N}} \leq \beta$$

$$\implies P\left\{\beta \in [b_N - \frac{\Psi^{-1}(\alpha; F)}{\sqrt{N}}, \infty)\right\} = P\left\{\sqrt{N}(b_N - \beta) \leq \Psi^{-1}(\alpha; F)\right\}$$

$$\to \Psi\{\Psi^{-1}(\alpha; F); F\} = \alpha \quad \text{as } \Psi \text{ is continuous.}$$

To make the confidence bound to be operational, $\Psi^{-1}(\alpha; F)$ can be replaced with the *bootstrap quantile* $\Psi^{-1}(\alpha; F_N)$, and we would be using

$$P\left\{\beta \in [b_N - \frac{\Psi^{-1}(\alpha; F_N)}{\sqrt{N}}, \infty)\right\} = P\left\{\sqrt{N}(b_N - \beta) \le \Psi^{-1}(\alpha; F_N)\right\}.$$

We show that this centered-quantile-based confidence interval yields the asymptotic coverage probability α.

Because $\Psi\{\Psi^{-1}(\alpha; F); F\} = \alpha$,

$$P\left\{\sqrt{N}(b_N - \beta) \le \Psi^{-1}(\alpha; F_N)\right\} - \alpha$$
$$= P\left\{\sqrt{N}(b_N - \beta) \le \Psi^{-1}(\alpha; F_N)\right\} - \Psi\{\Psi^{-1}(\alpha; F_N); F\}$$
$$+ \Psi\{\Psi^{-1}(\alpha; F_N); F\} - \Psi\{\Psi^{-1}(\alpha; F); F\}.$$

Due to the uniform convergence of $P\{\sqrt{N}(b_N - \beta) \le t\} \rightsquigarrow \Psi(t; F)$ over t, the first difference with $t = \Psi^{-1}(\alpha; F_N)$ is $o_p(1)$. As for the second difference, it is also $o_p(1)$ because $\Psi(t; F)$ is continuous in t and $\Psi^{-1}(t; F_N) \longrightarrow^p \Psi^{-1}(t; F)$ as mentioned above. For instance, if $\Psi(t; F) = \Phi\{t/\sigma(F)\}$, then

$$\Psi(t; F_N) = \Phi\{t/\sigma(F_N)\} = \Phi(t/s_N) \to^p \Phi\{t/\sigma(F)\} = \Psi(t; F)$$

because $s_N \to^p \sigma$ and Φ is continuous; note

$$\Psi^{-1}(\cdot; F_N) = s_N \Phi^{-1}(\cdot) \to^p \sigma \Phi^{-1}(\cdot) = \Psi^{-1}(\cdot; F).$$

Although we discussed bootstrap consistency only for $\sqrt{N}(b_N - \beta)$, suppose that bootstrap consistency holds for the asymptotically pivotal case $T_N = V_N^{-1/2}\sqrt{N}(b_N - \beta)$ as well. Observe

$$V_N^{-1/2}\sqrt{N}(b_N - \beta) \le \Psi^{-1}(\alpha) \iff b_N - \beta \le \Psi^{-1}(\alpha)\frac{V_N^{1/2}}{\sqrt{N}} \iff b_N - \Psi^{-1}(\alpha)\frac{V_N^{1/2}}{\sqrt{N}} \le \beta.$$

The coverage error is, as Ψ is continuous,

$$P\left\{\beta \in [b_N - \Psi^{-1}(\alpha)\frac{V_N^{1/2}}{\sqrt{N}}, \infty)\right\} = P\left\{V_N^{-1/2}\sqrt{N}(b_N - \beta) \le \Psi^{-1}(\alpha)\right\}$$
$$\to \Psi\{\Psi^{-1}(\alpha)\} = \alpha$$

as desired. The bootstrap consistency of $V_N^{-1/2}\sqrt{N}(b_N - \beta)$ holds at least for the standardized sample averages: recall the example $\Psi(t; F) = \Phi\{t/\sigma(F)\}$ and $\Psi^{-1}(\cdot; F_N) = s_N \Phi^{-1}(\cdot)$ and thus

$$P\left\{\sqrt{N}(b_N - \beta) \le \Psi^{-1}(\alpha; F_N)\right\} = P\left\{\frac{\sqrt{N}(b_N - \beta)}{s_N} \le \Phi^{-1}(\alpha)\right\} \to \alpha.$$

For percentile method, under the assumption that the limit distribution of the centered percentile method is symmetric about 0, we can obtain bootstrap consistency using the fact that the αth quantile of centered-percentile method is $\sqrt{N}\{K_N^{*-1}(\alpha) - b_N\}$. Observe

$$P\left\{K_N^{*-1}(\alpha) \leq \beta\right\} = P\left[\sqrt{N}\left\{K_N^{*-1}(\alpha) - b_N\right\} \leq \sqrt{N}(\beta - b_N)\right]$$

$$= P\left[\sqrt{N}(b_N - \beta) \leq \sqrt{N}\left\{b_N - K_N^{*-1}(\alpha)\right\}\right]$$

$$= \Psi\left\{\sqrt{N}(b_N - K_N^{*-1}(\alpha)); F\right\} + o_p(1)$$

$$= 1 - \Psi\left\{\sqrt{N}(K_N^{*-1}(\alpha) - b_N); F\right\} + o_p(1) \quad \text{(using the symmetry of } \Psi\text{)}$$

$$= 1 - \alpha + o_p(1).$$

Hence, we get the desired expression

$$P\{K_N^{*-1}(\alpha) \leq \beta\} \to^p 1 - \alpha :$$

the lower confidence bound with level $1 - \alpha$ is $K_N^{*-1}(\alpha)$.

5.4 High-Order Improvement for Asymptotic Normality

In most cases we encounter in econometrics, a consistent estimator is asymptotically normal, converging to the parameter at the rate $N^{-1/2}$, and the estimator can be written as a sum of terms with order $N^{-\nu/2}$, $\nu = 1, 2, 3, \ldots$ Under some regularity conditions, this also holds for the distribution function of $V_N^{-1/2}\sqrt{N}(b_N - \beta)$:

$$P\left\{V_N^{-1/2}\sqrt{N}(b_N - \beta) \leq c\right\} = \Phi(c) + \frac{q_1(c; F)}{N^{1/2}} + \frac{q_2(c; F)}{N} +, \ldots$$

for some functions $q_1(c; F)$, $q_2(c; F)$,.... This display shows that the *asymptotic approximation error using only* $\Phi(c)$ *for* $T_N \equiv V_N^{-1/2}\sqrt{N}(b_N - \beta)$ *is* $O(N^{-1/2})$.

As will be seen in "Edgeworth expansion," suppose that $q_1(c; F)$ is an even function and $q_2(c; F)$ is an odd function:

$$q_1(c; F) = q_1(-c; F) \quad \text{and} \quad q_2(c; F) = -q_2(-c; F).$$

Denoting the $N(0, 1)$ αth quantile as ζ_α, we get

$$P\{V_N^{-1/2}\sqrt{N}(b_N - \beta) \leq \zeta_{1-\alpha/2}\} = \Phi(\zeta_{1-\alpha/2}) + \frac{q_1(\zeta_{1-\alpha/2}; F)}{N^{1/2}} + \frac{q_2(\zeta_{1-\alpha/2}; F)}{N} +, \ldots$$

$$P\{V_N^{-1/2}\sqrt{N}(b_N - \beta) \leq \zeta_{\alpha/2}\} = \Phi(\zeta_{\alpha/2}) + \frac{q_1(\zeta_{\alpha/2}; F)}{N^{1/2}} + \frac{q_2(\zeta_{\alpha/2}; F)}{N} +, \ldots$$

Subtract the latter from the former to get,

$$P\{\zeta_{\alpha/2} \le V_N^{-1/2}\sqrt{N}(b_N - \beta) \le \zeta_{1-\alpha/2}\} = 1 - \alpha + \frac{2q_2(\zeta_{1-\alpha/2}; F)}{N} + O(\frac{1}{N^{3/2}})$$

as $q_1(\zeta_{\alpha/2}; F) = q_1(-\zeta_{1-\alpha/2}; F) = q_1(\zeta_{1-\alpha/2}; F)$ and $-q_2(\zeta_{\alpha/2}; F) = q_2(-\zeta_{\alpha/2}; F) = q_2(\zeta_{1-\alpha/2}; F)$.

Consider now $|T_N| = |V_N^{-1/2}\sqrt{N}(b_N - \beta)|$ with its exact df

$$P(|T_N| \le c) = G_N(c; F) - G_N(-c; F).$$

Setting $c = \zeta_{1-\alpha/2} \ (= -\zeta_{\alpha/2})$, this becomes, using the preceding display,

$$P(|T_N| \le \zeta_{\alpha/2}) = P\left\{\zeta_{\alpha/2} \le V_N^{-1/2}\sqrt{N}(b_N - \beta) \le \zeta_{1-\alpha/2}\right\} = 1 - \alpha$$
$$+ \frac{2q_2(\zeta_{1-\alpha/2}; F)}{N} + O\left(\frac{1}{N^{3/2}}\right).$$

The *asymptotic approximation error* for the df of $|V_N^{-1/2}\sqrt{N}(b_N - \beta)|$ is $O(N^{-1})$, which is said to be "second-order accurate."

Typically, in expansions including higher-order terms with functions $q_1(c; F)$, $q_2(c; F)$, $q_3(c; F)$, $q_4(c; F)$,..., every other functions from $q_1(c; F)$ and on are all even functions, and every other functions from $q_2(c; F)$ and on are all odd functions. This means that, when q_1 gets canceled, $q_3, q_5, \dots$ get all canceled as well. Hence,

$$P\left\{\zeta_{\alpha/2} \le V_N^{-1/2}\sqrt{N}(b_N - \beta) \le \zeta_{1-\alpha/2}\right\} = 1 - \alpha + \frac{2q_2(\zeta_{1-\alpha/2}; F)}{N}$$
$$+ \frac{2q_4(\zeta_{1-\alpha/2}; F)}{N^2} + O\left(\frac{1}{N^3}\right).$$

Suppose that, not just bootstrap consistency, but the approximation for $P\left\{V_N^{-1/2}\sqrt{N}(b_N - \beta) \le c\right\}$ holds with F in all terms replaced by F_N:

$$P\left\{V_N^{*-1/2}\sqrt{N}(b_N^* - b_N) \le c\right\} = \Phi(c) + \frac{q_1(c; F_N)}{N^{1/2}} + \frac{q_2(c; F_N)}{N} +, \dots$$

Under this, we get

$$P\left\{V_N^{-1/2}\sqrt{N}(b_N - \beta) \le c\right\} - P\left\{V_N^{*-1/2}\sqrt{N}(b_N^* - b_N) \le c\right\}$$
$$= \frac{q_1(c; F) - q_1(c; F_N)}{N^{1/2}} +, \dots$$

Typically, $q_1(c; F)$ is "smooth" in F, to result in

$$\frac{q_1(c; F) - q_1(c; F_N)}{N^{1/2}} \simeq \frac{O_p(N^{-1/2})}{N^{1/2}} = O_p\left(\frac{1}{N}\right):$$

the *bootstrap approximation error for the df of* T_N *is of order* O_p
(N^{-1})—second-order accurate.

As for symmetric CI's, observe

$$P\{\zeta_{\alpha/2} \le V_N^{*-1/2}\sqrt{N}(b_N^* - b_N) \le \zeta_{1-\alpha/2}\} = 1 - \alpha + \frac{2q_2(\zeta_{1-\alpha/2}; F_N)}{N} + O\left(\frac{1}{N^2}\right).$$

Subtract this from the above expansion for $P\{\zeta_{\alpha/2} \le V_N^{-1/2}\sqrt{N}(b_N - \beta) \le \zeta_{1-\alpha/2}\}$ to get

$$\frac{2q_2(\zeta_{1-\alpha/2}; F) - 2q_2(\zeta_{1-\alpha/2}; F_N)}{N} + O\left(\frac{1}{N^2}\right) = \frac{O_p(N^{-1/2})}{N} + O\left(\frac{1}{N^2}\right)$$

$$= O_p\left(\frac{1}{N^{3/2}}\right).$$

The *bootstrap approximation error for the df of* $|T_N|$ *is of order*
$O_p(N^{-3/2})$.

Now consider the nonpivotal versions and their expansions:

$$P\{\sqrt{N}(b_N - \beta) \le c\} = \Phi\left(\frac{c}{\sqrt{V}}\right) + \frac{r_1(c; F)}{N^{1/2}} + \frac{r_2(c; F)}{N} +, \ldots$$

$$P\{\sqrt{N}(b_N^* - b_N) \le c\} = \Phi\left(\frac{c}{\sqrt{V_N}}\right) + \frac{r_1(c; F_N)}{N^{1/2}} + \frac{r_2(c; F_N)}{N} +, \ldots$$

for some functions $r_1(c; F)$ and $r_2(c; F)$. Subtract the latter from the former
to get

$$P\left\{\sqrt{N}(b_N - \beta) \le c\right\} - P\left\{\sqrt{N}(b_N^* - b_N) \le c\right\} = \Phi\left(\frac{c}{\sqrt{V}}\right)$$

$$- \Phi\left(\frac{c}{\sqrt{V_N}}\right) + \frac{r_1(c; F) - r_1(c; F_N)}{N^{1/2}} +, \ldots$$

$$= O_p\left(\frac{1}{N^{1/2}}\right) \quad \text{as } V - V_N = O_p\left(\frac{1}{N^{1/2}}\right).$$

This shows that *non-pivotal bootstrap has the approximation error of order*
$O_p(N^{-1/2})$, which is the same as the asymptotic approximation error; high-
order improvement is possible only for pivotal statistics.

For $|\sqrt{N}(b_N - \beta)|$, observe for $c_2 > 0$ and $c_1 = -c_2$,

$$P\{\sqrt{N}(b_N - \beta) \le c_2\} = \Phi\left(\frac{c_2}{\sqrt{V}}\right) + \frac{r_1(c_2; F)}{N^{1/2}} + \frac{r_2(c_2; F)}{N} +, \ldots$$

$$P\{\sqrt{N}(b_N - \beta) \le c_1\} = \Phi\left(\frac{c_1}{\sqrt{V}}\right) + \frac{r_1(c_1; F)}{N^{1/2}} + \frac{r_2(c_1; F)}{N} +, \ldots$$

With $r_1(c; F)$ being an even function of c, subtracting the latter from the former yields, because of $r_1(c_1; F) = r_1(c_2; F)$,

$$P\{c_1 \le \sqrt{N}(b_N - \beta) \le c_2\} = \Phi\left(\frac{c_2}{\sqrt{V}}\right) - \Phi\left(\frac{c_1}{\sqrt{V}}\right) + \frac{r_2(c_2; F) - r_1(c_1; F)}{N}$$

$$+ O\left(\frac{1}{N^{3/2}}\right) = \Phi\left(\frac{c_2}{\sqrt{V}}\right) - \Phi\left(\frac{c_1}{\sqrt{V}}\right) + O\left(\frac{1}{N}\right):$$

asymptotic approximation error for the df of $|\sqrt{N}(b_N - \beta)|$ *is of order* $O(N^{-1})$ *which is the same as the one using a pivotal statistic. Thus the asymptotic approximation error order is the same regardless of pivotal or non-pivotal statistics.*

Subtracting the bootstrap version of this display from this, we get

$$P\{c_1 \le \sqrt{N}(b_N - \beta) \le c_2\} - P\{c_1 \le \sqrt{N}(b_N^* - b_N) \le c_2\}$$

$$= \frac{r_2(c_2; F) - r_2(c_1; F)}{N} - \frac{r_2(c_2; F_N) - r_2(c_1; F_N)}{N} + O\left(\frac{1}{N^{3/2}}\right)$$

$$= O_p\left(\frac{1}{N^{3/2}}\right).$$

nonpivotal bootstrap has the approximation error of order $O_p(N^{-3/2})$ *for symmetric CI's, which is the same as pivotal bootstrap approximation error order.*

In summary, the approximation orders of df's are as follows:

	Order of Approximation Error		
	Asymptotic	Pivotal bootstrap	Non-pivotal bootstrap
Asymmetric CI	$N^{-1/2}$	N^{-1}	$N^{-1/2}$
Symmetric CI	N^{-1}	$N^{-3/2}$	$N^{-3/2}$

These approximation orders get carried over to coverage probabilities in CI's and false rejection probabilities (i.e., type I error) in HT. But Horowitz (2001a) showed that, in some cases, the bootstrap approximation orders N^{-1} and $N^{-3/2}$ get even smaller to become, respectively, $N^{-3/2}$ and N^{-2}.

5.5   Edgeworth Expansion

Formally showing the expansions in the preceding subsection is hard, but the intuition can be gained from Edgeworth expansion, which is explained in this subsection.

5.5.1   Cumulant Generating Function

Let z is a rv with distribution F. On a neighborhood of $t = 0$, moment generating function (mgf) is defined as $E \exp(tz)$ if this exists. If the mgf is finite in a neighborhood of $t = 0$, then all moments of z are finite. The characteristic function is

$$E \exp(\mathrm{i} t z) = \int \exp(\mathrm{i} t z_o) dF(z_o) = E \cos(tz) + \mathrm{i} E \sin(tz).$$

The last expression shows clearly that characteristic function does not require any moment of z, as it needs only $E \cos(tz)$ and $E \sin(tz)$ which are certainly finite. For instance, $N(\mu, \sigma^2)$ has mgf $\exp(\mu t + \sigma^2 t^2/2)$ and characteristic function $\exp(\mu t - \sigma^2 t^2/2)$; -1 in front of σ^2 comes from $\mathrm{i}^2 = -1$.

The "*cumulant generating function (cgf)*" is the logarithm of the characteristic function:

$$\kappa(t) \equiv \ln\{E \exp(\mathrm{i} t z)\}.$$

For the sum $\sum_{i=1}^{N} z_i$ of independent $\{z_i\}$, observe

$$\ln\left[E\left\{ \exp(\mathrm{i} t \sum_{i=1}^{N} z_i) \right\} \right] = \ln\left[E\left\{ \prod_{i=1}^{N} \exp(\mathrm{i} t z_i) \right\} \right]$$

$$= \ln\left[\prod_{i=1}^{N} E \exp(\mathrm{i} t z_i) \right] = \sum_{i} \ln\{E \exp(\mathrm{i} t z_i)\} :$$

the *cgf of $\sum_{i=1}^{N} z_i$ is the sum of the individual cgf's.*

Suppose all moments of z exist and $E(z) = 0$ (or redefine z as $z - E(z)$). Then Taylor expansion gives

$$E \exp(\mathrm{i} t z) - 1 = \frac{\mathrm{i}^2 t^2 E(z^2)}{2} + \frac{\mathrm{i}^3 t^3 E(z^3)}{3!} + \frac{\mathrm{i}^4 t^4 E(z^4)}{4!} +, \dots \quad \text{as the first term} \quad \frac{\mathrm{i} t E(z)}{1} \text{ is } 0$$

Use the expansion

$$\ln(1 + x) = x - \frac{x^2}{2!} + \frac{x^3}{3!} - \frac{x^4}{4!} +, \dots$$

to get

$$\kappa(t) = \ln[1 + \{E \exp(\mathrm{i} t z) - 1\}] = \left\{ \frac{\mathrm{i}^2 t^2 E(z^2)}{2} + \frac{\mathrm{i}^3 t^3 E(z^3)}{3!} + \frac{\mathrm{i}^4 t^4 E(z^4)}{4!} +, \dots \right\}$$

$$- \frac{1}{2} \left\{ \frac{\mathrm{i}^2 t^2 E(z^2)}{2} +, \dots \right\}^2 + \frac{1}{3!} \left\{ \frac{\mathrm{i}^2 t^2 E(z^2)}{2} +, \dots \right\}^3 +, \dots$$

$$= E(z^2) \frac{\mathrm{i}^2 t^2}{2} + E(z^3) \frac{\mathrm{i}^3 t^3}{3!} + \{E(z^4) - 3E^2(z^2)\} \frac{\mathrm{i}^4 t^4}{4!} +, \dots$$

$$\equiv \sum_{j=1}^{\infty} \kappa_j \frac{(\mathrm{i} t)^j}{j!}, \quad \text{where } \kappa_1 = 0 \ (= E(z)), \ \kappa_2 = V(z),$$

$$\kappa_3 = E(z^3), \ \kappa_4 = E^4(z) - 3E^2(z^2), \dots;$$

the jth cumulant is κ_j

Consider now the standardized sum of iid z_i's with $E(z) = 0$ and $V(z) = \sigma^2$:

$$\frac{1}{\sqrt{N}} \sum_i^N \frac{z_i}{\sigma} = \frac{1}{\sigma\sqrt{N}} \sum_i^N z_i \implies \frac{t}{\sigma\sqrt{N}} \sum_i^N z_i = t' \sum_i^N z_i, \quad t' \equiv \frac{t}{\sigma\sqrt{N}}$$

$$\implies \text{ the cgf is } N\kappa\left(\frac{t}{\sigma\sqrt{N}}\right) = N \sum_{j=2}^{\infty} \kappa_j \frac{(it)^j}{(\sigma\sqrt{N})^j \cdot j!} = \sum_{j=2}^{\infty} \kappa_j' \frac{(it)^j}{j!} \frac{1}{N^{(j-2)/2}},$$

where $\kappa_j' \equiv \dfrac{\kappa_j}{\sigma^j}$;

$$\kappa_2' = \frac{V(z)}{\sigma^2} = 1, \quad \kappa_3' = \frac{E(z^3)}{\sigma^3} \text{ (skewness)},$$

$$\kappa_4' = \frac{E(z^4) - 3E^2(z^2)}{\sigma^4} = \frac{E(z^4)}{\sigma^4} - 3 \text{ (kurtosis)}.$$

5.5.2   Density of Normalized Sums

As $N(\mu, \sigma^2)$ has cgf $\mu t - \sigma^2 t^2/2$, when z_i's are iid $N(0, \sigma^2)$, the cgf of $N^{-1/2} \sum_i z_i/\sigma = (\sigma\sqrt{N})^{-1} \sum_i z_i$ is N times the individual cgf with t replaced by $t/(\sigma\sqrt{N})$:

$$N \cdot \left\{ -\frac{\sigma^2}{2} \left(\frac{t}{\sigma\sqrt{N}}\right)^2 \right\} = -\frac{t^2}{2}.$$

In comparison, when z_i's are iid $(0, \sigma^2)$—not necessarily normal—the cgf of $N^{-1/2} \sum_i^N z_i/\sigma$ is, as just derived,

$$\kappa(t) = -\frac{t^2}{2} + \kappa_3' \frac{(it)^3}{3!} \frac{1}{\sqrt{N}} + \kappa_4' \frac{(it)^4}{4!} \frac{1}{N} +, \dots :$$

the cgf of a standardized iid rv's sum is the cgf of $N(0, 1)$ plus $O(N^{-1/2})$.

The characteristic function of $N^{-1/2} \sum_i^N z_i/\sigma$ is then

$$\psi(t) \equiv \exp \kappa(t) = \exp \left\{ -\frac{t^2}{2} + \kappa_3' \frac{(it)^3}{3!} \frac{1}{\sqrt{N}} + \kappa_4' \frac{(it)^4}{4!} \frac{1}{N} +, \dots \right\}$$

$$= e^{-t^2/2} \exp \left\{ \kappa_3' \frac{(it)^3}{3!} \frac{1}{\sqrt{N}} + \kappa_4' \frac{(it)^4}{4!} \frac{1}{N} +, \dots \right\}.$$

Using

$$e^x = 1 + x + \frac{x^2}{2!} + \frac{x^3}{3!} + \frac{x^4}{4!} + \dots$$

$\psi(t)$ can be written as

$$\psi(t) = e^{-t^2/2} \left[1 + \left\{ \kappa_3' \frac{(it)^3}{3!} \frac{1}{\sqrt{N}} + \kappa_4' \frac{(it)^4}{4!} \frac{1}{N} +, \dots \right\} + \frac{1}{2!} \left\{ \kappa_3' \frac{(it)^3}{3!} \frac{1}{\sqrt{N}} \right. \right.$$

$$+ \kappa_4' \frac{(it)^4}{4!} \frac{1}{N} +, \dots \Big\}^2 +, \dots \Big]$$

$$= e^{-t^2/2} \Big[1 + \kappa_3' \frac{(it)^3}{3!} \frac{1}{\sqrt{N}} + \Big\{ \kappa_4' \frac{(it)^4}{4!} + \kappa_3'^2 \frac{(it)^6}{72} \Big\} \frac{1}{N} +, \dots \Big]$$

If a rv x has a characteristic function $\xi(t)$ that is integrable (i.e., $\int_{-\infty}^{\infty} |\xi(t)| \, dt < \infty$), then x has an uniformly continuous density f_x obtained by (e.g., Shorack, 2000, p. 347)

$$f_x(x_o) = \frac{1}{2\pi} \int_{-\infty}^{\infty} e^{-itx_o} \xi(t) dt.$$

Apply this to $\psi(t)$ to get the density $g_N(w)$ of $N^{-1/2} \sum_i^N z_i / \sigma$:

$$g_N(w) = \frac{1}{2\pi} \int_{-\infty}^{\infty} e^{-itw} e^{-t^2/2} \Big[1 + \kappa_3' \frac{(it)^3}{3!} \frac{1}{\sqrt{N}} + \Big\{ \kappa_4' \frac{(it)^4}{4!} + \kappa_3'^2 \frac{(it)^6}{72} \Big\} \frac{1}{N} +, \dots \Big] dt$$

$$= \frac{1}{2\pi} \int_{-\infty}^{\infty} e^{-itw} e^{-t^2/2} dt \; + \; \frac{1}{2\pi} \int_{-\infty}^{\infty} e^{-itw} e^{-t^2/2} (it)^3 dt \cdot \frac{\kappa_3'}{3!} \frac{1}{\sqrt{N}}$$

$$+ \; \frac{1}{2\pi} \int_{-\infty}^{\infty} e^{-itw} e^{-t^2/2} \Big\{ \frac{\kappa_4'}{4!} (it)^4 + \frac{\kappa_3'^2}{72} (it)^6 \Big\} dt \frac{1}{N} +, \dots$$

Observe, for $p = 0, 1, 2, \dots$,

$$\frac{1}{2\pi} \int_{-\infty}^{\infty} e^{-itw} e^{-t^2/2} (it)^p dt = (-1)^p \frac{1}{2\pi} \int_{-\infty}^{\infty} \frac{d^p(e^{-itw})}{dw^p} e^{-t^2/2} dt$$

$$= (-1)^p \frac{d^p}{dw^p} \Big\{ \frac{1}{2\pi} \int_{-\infty}^{\infty} e^{-itw} e^{-t^2/2} dt \Big\} = (-1)^p \frac{1}{\sqrt{2\pi}} \frac{d^p}{dw^p} \Big\{ \int_{-\infty}^{\infty} e^{-itw} \frac{1}{\sqrt{2\pi}} e^{-t^2/2} dt \Big\}$$

$$= (-1)^p \frac{1}{\sqrt{2\pi}} \frac{d^p}{dw^p} \{ \text{characteristic function of } t \sim N(0,1) \text{ with } w \text{ fixed} \}$$

$$= (-1)^p \frac{d^p}{dw^p} \Big\{ \frac{1}{\sqrt{2\pi}} e^{-w^2/2} \Big\} = (-1)^p \frac{d^p}{dw^p} \phi(w) \quad \text{where } \phi(\cdot) \text{ is the } N(0,1) \text{ density.}$$

Carrying out the differentiation, we can see that

$$\phi'(w) = -w\phi(w)$$
$$\phi''(w) = -\phi(w) - w\phi'(w) = -\phi(w) - w\{-w\phi(w)\} = \phi(w)(w^2 - 1)$$
$$\phi'''(w) = \phi(w)(-w^3 + 3w) \quad \text{doing analogously.}$$

The w-polynomials $H_p(w)$ next to $\phi(w)$ such that

$$(-1)^p \phi^{(p)}(w) = \phi(w) \cdot H_p(w)$$

are called *Hermite polynomials*:

$$H_0(w) = 1, \quad H_1(w) = w, \quad H_2(w) = w^2 - 1, \quad H_3(w) = w^3 - 3w;$$

for $p \geq 2$, they can be found from the recursive formula

$$H_p(w) = wH_{p-1}(w) - (p-1)H_{p-2}(w); \text{ for example}$$
$$H_3(w) = w(w^2 - 1) - 2w = w^3 - 3w,$$
$$H_4(w) = w(w^3 - 3w) - 3(w^2 - 1) = w^4 - 6w^2 + 3,$$
$$H_5(w) = w(w^4 - 6w^2 + 3) - 4(w^3 - 3w) = w^5 - 10w^3 + 15w.$$

$H_p(w)$ is an even (odd) function if p is even (odd).

Substitute the just derived finding

$$\frac{1}{2\pi} \int_{-\infty}^{\infty} e^{-itw} e^{-t^2/2} (\mathbf{i}t)^p dt = (-1)^p \frac{d^p}{dw^p} \phi(w) = \phi(w) \cdot H_p(w)$$

into $g_N(w)$ to get

$$g_N(w) = \phi(w) \left[1 + H_3(w) \frac{\kappa_3'}{3!} \frac{1}{\sqrt{N}} + \left\{ H_4(w) \frac{\kappa_4'}{4!} + H_6(w) \frac{\kappa_3'^2}{72} \right\} \frac{1}{N} +, ... \right]$$

which is the *Edgeworth approximation* of the density $g_N(w)$.

5.5.3 Distribution Function of Normalized Sums

Integrating $g_N(w)$ wrt w, we get the *Edgeworth approximation* of the df $G_N(w)$. But we need to integrate $H_p(w)\phi(w)$. Observe now (keeping in mind that p in $\phi^{(p)}$ denotes the pth order differentiation, not power)

$$\int_{-\infty}^{w} H_p(w_o)\phi(w_o)dw_o = \int_{-\infty}^{w} (-1)^p \phi^{(p)}(w_o)dw_o = (-1)^p \phi^{(p-1)}(w_o)|_{-\infty}^{w}$$
$$= -(-1)^{p-1} \phi^{(p-1)}(w_o)|_{-\infty}^{w} = -\phi(w_o)H_{p-1}(w_o)|_{-\infty}^{w} = -\phi(w)H_{p-1}(w)$$

because $\phi(-\infty)H_{p-1}(-\infty) = 0$—$\phi$ decreases to 0 exponentially fast whereas H_{p-1} has only polynomial functions. Using this, we get

$$G_N(w) = \Phi(w) - \phi(w)H_2(w) \frac{\kappa_3'}{3!} \frac{1}{\sqrt{N}} + \left\{ -\phi(w)H_3(w) \frac{\kappa_4'}{4!} - \phi(w)H_5(w) \frac{\kappa_3'^2}{72} \right\} \frac{1}{N} +, ...]$$
$$= \Phi(w) - \phi(w) \left[H_2(w) \frac{\kappa_3'}{3!} \frac{1}{\sqrt{N}} + \left\{ H_3(w) \frac{\kappa_4'}{4!} + H_5(w) \frac{\kappa_3'^2}{72} \right\} \frac{1}{N} +, ... \right]$$
$$= \Phi(w) - \phi(w)(w^2 - 1) \frac{\kappa_3'}{3!} \frac{1}{\sqrt{N}} +, ... \quad \text{recalling } H_2(w) = w^2 - 1.$$

As σ is unknown typically, we would be using s_N for σ to get $N^{-1/2} \sum_i^N z_i/s_N$. For this, Hall (1992) showed an analogous but different expansion:

$$\Phi(w) + \phi(w)(2w^2 + 1) \frac{\kappa_3'}{3!} \frac{1}{\sqrt{N}} +, ...$$

In this Edgeworth expansion of df, the term with $N^{-1/2}$ has an even function whereas the term with N^{-1} has an odd function. More generally, for a term with $N^{-(j-1)/2}$, $j = 2, ...$, the polynomial function in w is an even function if j is even and an odd function if j is odd. This explains why even and odd functions were used for $O(N^{-1/2})$ and $O(N^{-1})$ terms in explaining high-order improvements in bootstrap. One caution in Hall (1992) is that Edgeworth expansion hardly ever converges as an infinite series; rather, it should be understood as an "asymptotic series" in the sense that if the series is stopped after some term, the remainder is of smaller order than the last included term. This suggests that "Edgeworth approximation" seems to be a better name than "Edgeworth expansion."

5.5.4 Moments

Consider $t_N \equiv \sqrt{N}(b_N - \beta)/\sigma$; note that σ is not estimated in t_N. The Edgeworth approximation for the df of t_N leads to an approximation for the moment $E(b_N - \beta)^p$ through the formula

$$E(w^p) = \int_0^\infty w_o^p dP(w \le w_o) - \int_0^\infty (-w_o)^p dP(w \le -w_o).$$

Observe, drawing on Hall (1992, p. 50),

$$E(b_N - \beta)^p = \frac{\sigma^p}{N^{p/2}} E(t_N^p) = \frac{\sigma^p}{N^{p/2}} \int_0^\infty w_o^p dP(t_N \le w_o)$$

$$- \int_0^\infty (-w_o)^p dP(t_N \le -w_o)$$

$$= \frac{\sigma^p}{N^{p/2}} \left\{ \int_0^\infty w_o^p dQ_N(w_o) - (-1)^p \int_0^\infty w_o^p dQ_N(-w_o) \right\}$$

where $Q_N(w_o) \equiv P(t_N \le w_o)$.

Define the survival function $S_N(w_o) \equiv 1 - Q_N(w_o)$, which goes to 0 exponentially fast with the asymptotic normality holding for t_N. Then

$$0 = w_o^p S_N(w_o)|_0^\infty = -\int_0^\infty w_o^p dQ_N(w_o) + p \int_0^\infty w_o^{p-1} S_N(w_o) dw_o$$

$$\implies \int_0^\infty w_o^p dQ_N(w_o) = p \int_0^\infty w_o^{p-1} S_N(w_o) dw_o;$$

$$0 = w_o^p Q_N(-w_o)|_0^\infty = \int_0^\infty w_o^p dQ_N(-w_o) + p \int_0^\infty w_o^{p-1} Q_N(-w_o) dw_o$$

$$\implies \int_0^\infty w_o^p dQ_N(-w_o) = -p \int_0^\infty w_o^{p-1} Q_N(-w_o) dw_o.$$

Use these to obtain

$$E(b_N - \beta)^p = \frac{\sigma^p}{N^{p/2}} \left\{ p \int_0^\infty w_o^{p-1} S_N(w_o) dw_o + (-1)^p p \int_0^\infty w_o^{p-1} Q_N(-w_o) dw_o \right\}$$

$$= \frac{\sigma^p}{N^{p/2}} p \int_0^\infty w_o^{p-1} \{S_N(w_o) + (-1)^p Q_N(-w_o)\} dw_o.$$

Applying the Edgeworth approximation,

$$P(t_N \le w_o) = \Phi(w_o) - \phi(w_o) \left[H_2(w_o) \frac{\kappa_3'}{3!} \frac{1}{\sqrt{N}} + \left\{ H_3(w_o) \frac{\kappa_4'}{4!} \right. \right.$$
$$\left. \left. + H_5(w_o) \frac{\kappa_3'^2}{72} \right\} \frac{1}{N} +, \dots \right],$$

$$1 - P(t_N \le w_o) = \Phi(-w_o) + \phi(w_o) \left[H_2(w_o) \frac{\kappa_3'}{3!} \frac{1}{\sqrt{N}} + \left\{ H_3(w_o) \frac{\kappa_4'}{4!} \right. \right.$$
$$\left. \left. + H_5(w_o) \frac{\kappa_3'^2}{72} \right\} \frac{1}{N} +, \dots \right];$$

$$P(t_N \le -w_o) = \Phi(-w_o) - \phi(-w_o) \left[H_2(-w_o) \frac{\kappa_3'}{3!} \frac{1}{\sqrt{N}} \right.$$
$$\left. + \left\{ H_3(-w_o) \frac{\kappa_4'}{4!} + H_5(-w_o) \frac{\kappa_3'^2}{72} \right\} \frac{1}{N} +, \dots \right]$$
$$= \Phi(-w_o) - \phi(w_o) \left[H_2(w_o) \frac{\kappa_3'}{3!} \frac{1}{\sqrt{N}} - \left\{ H_3(w_o) \frac{\kappa_4'}{4!} \right. \right.$$
$$\left. \left. + H_5(w_o) \frac{\kappa_3'^2}{72} \right\} \frac{1}{N} +, \dots \right]$$

When p is even,

$$S_N(w_o) + (-1)^p Q_N(-w_o) = S_N(w_o) + Q_N(-w_o)$$
$$= 1 - P(t_N \le w_o) + P(t_N \le -w_o)$$
$$= 2\Phi(-w_o) + 2\phi(w_o) \left\{ H_3(w_o) \frac{\kappa_4'}{4!} + H_5(w_o) \frac{\kappa_3'^2}{72} \right\} \frac{1}{N} + O\left(\frac{1}{N^2} \right)$$

using the above display.

When p is odd,

$$S_N(w_o) - Q_N(-w_o) = 1 - P(t_N \le w_o) - P(t_N \le -w_o)$$
$$= 2\phi(w_o) H_2(w_o) \frac{\kappa_3'}{3!} \frac{1}{\sqrt{N}} + O\left(\frac{1}{N^{3/2}} \right)$$

Therefore, with $p = 1$,

$$E(b_N - \beta) = \frac{\sigma}{N^{1/2}} \int_0^\infty 2\phi(w_o) H_2(w_o) \frac{\kappa_3'}{3!} \frac{1}{\sqrt{N}} dw_o + O\left(\frac{1}{N^2} \right)$$
$$= \frac{2\sigma}{N} \frac{\kappa_3'}{3!} \int_0^\infty \phi(w_o) H_2(w_o) dw_o + O\left(\frac{1}{N^2} \right)$$

$$= \frac{2\sigma}{N} \frac{\kappa_3'}{3!} \int_0^\infty \phi(w_o)(w_o^2 - 1)dw_o + O\left(\frac{1}{N^2}\right)$$

$$= \frac{2\sigma}{N} \frac{\kappa_3'}{3!}(0.5 - 0.5) + O\left(\frac{1}{N^2}\right) = O\left(\frac{1}{N^2}\right).$$

Also, with $p = 2$,

$$E(b_N - \beta)^2 = \frac{\sigma^2}{N} 2 \int_0^\infty w_o \left[2\Phi(-w_o)\right.$$

$$\left. +2\phi(w_o)\left\{H_3(w_o)\frac{\kappa_4'}{4!} + H_5(w_o)\frac{\kappa_3'^2}{72}\right\}\right] dw_o \frac{1}{N} + O\left(\frac{1}{N^3}\right)$$

$$= \frac{4\sigma^2}{N^2} \int_0^\infty w_o \left[\Phi(-w_o)\right.$$

$$\left. +\phi(w_o)\left\{H_3(w_o)\frac{\kappa_4'}{4!} + H_5(w_o)\frac{\kappa_3'^2}{72}\right\}\right] dw_o + O\left(\frac{1}{N^3}\right).$$

5.6 Other Bootstrap Topics

5.6.1 Bootstrap Test

As noted once, hypothesis test (HT) can be done with bootstrap CI's (or confidence regions), but there are cases where confidence intervals are irrelevant concepts—e.g., various model GOF tests. In such cases, the issue of bootstrap test appears. The key issue in bootstrap test is *how to impose the null hypothesis in generating pseudo samples*. Although we only mentioned sampling from the original sample with replacement so far—this is called "*nonparametric bootstrap*" or "*empirical bootstrap*"—bootstrap test brings about a host of other ways to generate pseudo samples as will be seen later, depending on how the null hypothesis is imposed on the pseudo samples.

To appreciate the importance of imposing H_0 on pseudo samples, suppose "H_0: F is $N(0,1)$." Under the H_0, nonparametric bootstrap would yield a pseudo sample consisting of "nearly" $N(0,1)$ rv's, and the test with non-parametric bootstrap would work because the realized TS for the original sample will be "similar" to the pseudo sample TS's. Now suppose that H_0 is false because the true model is $N(5,1)$. In this case, we want to have the realized TS to be much different from the pseudo TS's so that the bootstrap test becomes powerful. If we do not impose the H_0 in generating the pseudo samples, then both the original data and pseudo samples will be similar because they all follow more or less $N(5,1)$, leading to a low power. But if we impose the H_0 on the pseudo samples, then the realized TS for the original sample observations (centered around 5) will differ much from the TS's from the pseudo sample observations (centered around 0), leading to a high power.

Suppose H_0: $f = f_o(\theta)$; i.e., the null model is parametric with an unknown parameter θ. In this case, θ may be estimated by the MLE $\hat{\theta}$, and the pseudo data can be generated from $f_o(\hat{\theta})$. This is called *parametric bootstrap* where imposing the H_0 on pseudo data is straightforward. But often we have the null model that is not fully parametric, which makes imposing the null on pseudo data less than straightforward. For instance, the null model may be just a linear model $y_i = x_i'\beta + u_i$ without the distribution of (x_i', y_i) specified. In this case, one way of imposing the null goes as follows. Step 1: sample x_i^* from the empirical distribution of $x_1, ..., x_N$. Step 2: sample a residual r_i^* from the empirical distribution of the residuals $r_i \equiv y_i - x_i'b_N$, $i = 1, ..., N$. Step 3: generate $y_i^* \equiv x_i^*b_N + r_i^*$. Repeat this N times to get a pseudo-sample of size N.

In the bootstrap scheme for the linear model, r_i is drawn independently of x_i, which is fine if u_i is independent of x_i. But if we want to allow for heteroskedasticity, then the above bootstrap does not work because r_i^* is generated independently of x_i; instead *wild bootstrap* is suitable: when $x_i^* = x_i$, generate $y_i^* = x_i^{*'}b_N + v_i^*r_i$ where v_i^* is drawn from the two point distribution:

$$P\left(v^* = \frac{1-\sqrt{5}}{2}\right) = \frac{5+\sqrt{5}}{10} \text{ and } P\left(v^* = \frac{1+\sqrt{5}}{2}\right) = \frac{5-\sqrt{5}}{10}.$$

This distribution has mean 0 and variance 1, which implies that

$$E\left(v_i^*r_i|x_i\right) = E\left(v_i^*|x_i\right)E(r_i|x_i) = 0 \text{ and}$$
$$E(v_i^{*2}r_i^2|x_i) = E(v_i^{*2}|x_i)E(r_i^2|x_i) \simeq E(u_i^2|x_i)$$

preserving the heteroskedasticity in the pseudo-sample. As noted in the main text, v_i^* that takes ± 1 with probability 0.5 works just as well.

It is often reported that GMM small sample distribution differs much from the asymptotic distribution. To avoid this problem, one may do bootstrap. Hall and Horowitz (1996) showed how to do GMM bootstrap properly for t-tests and the over-id test. Let β_k be the kth element of β and let b_{Nk} denote the GMM estimator. We want to test for H_0: $\beta_k = \beta_{ko}$ with the usual t-ratio $t_{Nk} = (b_{Nk} - \beta_{ko})/ASD(b_{Nk})$ where the asymptotic standard deviation $ASD(b_{Nk})$ is from the usual GMM asymptotic variance estimator. The main idea of the GMM bootstrap is on "centering" the moments, which appears in essence in Lahiri (1992). Brown and Newey (2002) showed how to do "empirical-likelihood" bootstrap, where pseudo samples are drawn from the empirical distribution constrained by the moment condition.

Hall and Horowitz (1996) GMM bootstrap goes as follows:

1. For an integer j, draw the jth pseudo sample $z_1^{(j)}, z_2^{(j)}, ..., z_N^{(j)}$ of size N from the original sample with replacement.

2. From the pseudo sample, calculate the pseudo estimator $b_{Nk}^{(j)}$ for β_k, the pseudo t-value $t_{Nk}^{(j)} \equiv (b_{Nk}^{(j)} - b_{Nk})/ASD(b_{Nk}^{(j)})$, and the pseudo

over-id test statistic $\tau_N^{(j)}$ using the GMM with the *"centered moment"* $N^{-1}\sum_{i=1}^{N}\{\psi(z_i^*, b) - N^{-1}\sum_{i=1}^{N}\psi(z_i^*, b_N)\}$.

3. Repeat Steps 1 and 2 for $j = 1, ..., J$ to get $t_{Nk}^{(1)}, ..., t_{Nk}^{(J)}$ and over-id test statistics $\tau_N^{(1)}, ..., \tau_N^{(J)}$.

4. Reject H_0: $\beta_k = \beta_{ko}$ if the bootstrap p-value $J^{-1}\sum_j 1[|t_{Nk}^{(j)}| \geq |t_{Nk}|]$ is smaller than, say, 5%.

5. As for the over-id test, reject the over-identifying moment condition if the bootstrap p-value $J^{-1}\sum_j 1[\tau_N^{(j)} \geq \tau_N]$ is smaller than, say, 5%.

5.6.2 Bootstrap Bias-Correction

Let $\theta(F)$ denote the parameter of interest. An estimator for $\theta(F)$ is $\theta(F_N)$, which may be biased where

$$bias \equiv E\{\theta(F_N)\} - \theta(F).$$

For instance,

$$\theta(F) = \int \left\{z_o - \int z_o dF(z_o)\right\}^2 dF(z_o) = \sigma^2$$

$$\theta(F_N) = \int \left\{z_o - \int z_o dF_N(z_o)\right\}^2 dF_N(z_o) = \frac{1}{N}\sum_i (z_i - \bar{z})^2 \equiv s_N^2.$$

As well known, $\theta(F_N) = s_N^2$ is a biased estimator for σ^2 whereas $(N-1)^{-1}\sum_i (z_i - \bar{z})^2$ is unbiased.

The bootstrap idea of correcting for bias is approximating the bias with

$$E^*\{\theta(F_N^*)\} - \theta(F_N) = E^*\left\{\frac{1}{N}\sum_i (z_i^* - \bar{z}^*)^2\right\} - \frac{1}{N}\sum_i (z_i - \bar{z})^2$$

$$\simeq \frac{1}{N_B}\sum_{j=1}^{N_B} s_N^{2*j} - s_N^2, \quad \text{where } s_N^{2*j} \text{ is the sample variance for the } j\text{th}$$

$$\text{pseudo sample}$$

and $E^*(\cdot)$ is the expected value obtained using the original sample's empirical distribution (this is the population distribution for a pseudo sample). Subtracting this bootstrap bias estimator from s_N^2 yields a bias-corrected estimator for $\theta(F)$:

$$s_N^2 - \left(\frac{1}{N_B}\sum_j s_N^{2*j} - s_N^2\right) = 2s_N^2 - \frac{1}{N_B}\sum_j s_N^{2*j}.$$

Why this works can be seen intuitively as follows, drawing partly on Horowitz (2001a).

Suppose b_N is an unbiased estimator for β, but we are interested in a nonlinear function $g(\beta)$. A natural estimator for $g(\beta)$ is $g(b_N)$, but it is biased in general, as $Eg(b_N) \neq g\{E(b_N)\} = g(\beta)$. Consider the second-order Taylor expansion of $g(b_N)$ around $b_N = \beta$:

$$g(b_N) - g(\beta) = \nabla g(\beta)'(b_N - \beta) + \frac{1}{2}(b_N - \beta)' \cdot \nabla^2 g(\beta) \cdot (b_N - \beta) + \frac{\lambda(F)}{N^{3/2}} + O_p\left(\frac{1}{N^2}\right).$$

Because $E\{\nabla g(\beta)'(b_N - \beta)\} = \nabla g(\beta)' E\{(b_N - \beta)\} = 0$, the quadratic term of order $O_p(N^{-1})$ is the leading bias term. Now consider the bootstrap version of the Taylor expansion:

$$g(b_N^*) - g(b_N) = \nabla g(\beta)'(b_N^* - b_N) + \frac{1}{2}(b_N^* - b_N)' \cdot \nabla^2 g(b_N) \cdot (b_N^* - b_N)$$

$$+ \frac{\lambda(F_N)}{N^{3/2}} + O_p\left(\frac{1}{N^2}\right)$$

$$\implies \frac{1}{N_B}\sum_{j=1}^{N_B} g(b_N^{*j}) - g(b_N) \simeq \frac{1}{N_B}\sum_{j=1}^{N_B}\left\{\frac{1}{2}(b_N^{*j} - b_N)'\nabla^2 g(b_N)(b_N^{*j} - b_N)\right\}$$

$$+ \frac{\lambda(F_N)}{N^{3/2}} + O_p\left(\frac{1}{N^2}\right)$$

where the averaging removes the first-order term on the rhs.

Suppose now

$$\frac{1}{2}(b_N - \beta)' \cdot \nabla^2 g(\beta) \cdot (b_N - \beta) = \frac{\mu_1}{N} + \frac{\mu_2(F)}{N^{3/2}} + O_p\left(\frac{1}{N^2}\right)$$

$$\frac{1}{N_B}\sum_{j=1}^{N_B}\frac{1}{2}(b_N^{*j} - b_N)'\nabla^2 g(b_N)(b_N^{*j} - b_N) \simeq \frac{\mu_1}{N} + \frac{\mu_2(F_N)}{N^{3/2}} + O_p\left(\frac{1}{N^2}\right)$$

where μ_1 is presumed to be the same for both the original quadratic form and its bootstrap version; this is analogous to the bootstrap consistency that the original statistic and its bootstrap version have the same first-order term.

Hence the bias-corrected estimator for $g(\beta)$ is

$$g(b_N) - \left\{\frac{1}{N_B}\sum_{j=1}^{N_B} g(b_N^{*j}) - g(b_N)\right\}$$

and we have

$$g(b_N) - \left\{\frac{1}{N_B}\sum_{j=1}^{N_B} g(b_N^{*j}) - g(b_N)\right\} - g(\beta)$$

$$=\{g(b_N) - g(\beta)\} - \left\{\frac{1}{N_B}\sum_{j=1}^{N_B} g(b_N^{*j}) - g(b_N)\right\}$$

$$=\nabla g(\beta)'(b_N - \beta) + \frac{\mu_2(F) - \mu_2(F_N)}{N^{3/2}} + \frac{\lambda(F) - \lambda(F_N)}{N^{3/2}} + O_p\left(\frac{1}{N^2}\right)$$

using the above approximation. Taking $E(\cdot)$ on the rhs yields a bias term of $O_p(N^{-2})$, because typically

$$\mu_2(F) - \mu_2(F_N) = O_p(N^{-1/2}) \quad\text{and}\quad \lambda(F) - \lambda(F_N) = O_p(N^{-1/2}).$$

As an illustration, take variance $\theta(F) = \sigma^2$. For the sample variance $s_N^2 = \theta(F_N)$, we can in fact compute $E^*\{N^{-1}\sum_i(z_i^* - \bar z^*)^2\}$ as follows instead of invoking the Monte Carlo estimator. Observe

$$E^*\left\{\frac{1}{N}\sum_i(z_i^* - \bar z^*)^2\right\} = E^*(z^{*2}) - E^*(\bar z^{*2}).$$

The first term $E^*(z^{*2})$ is just $N^{-1}\sum_i z_i^2$ as z^* takes z_i with $P(z^* = z_i) = N^{-1}$. As for the second term, it is

$$E^*\left\{\left(\frac{1}{N}\sum_i z_i^*\right)^2\right\} = \frac{1}{N^2}\sum_{i,j} E^*(z_i^* z_j^*) = \frac{1}{N}E^*(z^{*2}) + \frac{N(N-1)}{N^2}\{E^*(z^*)\}^2$$

$$= \frac{1}{N^2}\sum_i z_i^2 + \left(1 - \frac{1}{N}\right)\bar z^2 = \bar z^2 + \frac{1}{N}\left\{\frac{1}{N}\sum_i z_i^2 - \bar z^2\right\}.$$

Therefore,

$$E^*\left\{\frac{1}{N}\sum_i(z_i^* - \bar z^*)^2\right\} = \frac{1}{N}\sum_i z_i^2 - \bar z^2 - \frac{1}{N}\left\{\frac{1}{N}\sum_i z_i^2 - \bar z^2\right\} = s_N^2 - \frac{1}{N}s_N^2,$$

$$\text{Bias}: E^*\left\{\frac{1}{N}\sum_i(z_i^* - \bar z^*)^2\right\} - s_N^2 = -\frac{1}{N}s_N^2 = O_p\left(\frac{1}{N}\right),$$

$$\text{Bias-Corrected Estimator}: s_N + \frac{1}{N}s_N^2 = \frac{N+1}{N}s_N^2.$$

Comparing this to the unbiased estimator

$$\frac{1}{N-1}\sum_i(z_i - \bar z)^2 = \frac{N}{N-1}s_N^2$$

which blows up s_N^2 by the factor $N/(N-1)$, the bias-corrected estimator blows up s_N^2 by the factor $(N+1)/N$. The difference between the two factors is

$$\frac{N+1}{N} - \frac{N}{N-1} = \frac{(N+1)(N-1) - N^2}{N(N-1)} = \frac{1}{N(N-1)}.$$

That is, the bootstrap bias-adjusted estimator reduces the bias of order $O_p(N^{-1})$ to a bias of order $O_p(N^{-2})$.

5.6.3 Estimating Asymptotic Variance with Bootstrap Quantiles

In asymptotic inference, $G_\infty(t; F_N)$ is used for $G_N(t; F)$. Although the asymptotic distribution of $V_N^{-1/2}\sqrt{N}(b_N - \beta)$ is the same as that of $V^{-1/2}\sqrt{N}(b_N - \beta)$, sometimes estimating V is difficult; e.g., V includes a density component or V_N is nearly singular. In this case, one may wonder whether

$$V_{NB} \equiv \frac{1}{N_B} \sum_{j=1}^{N_B} \left(b_N^{*(j)} - b_N \right) \left(b_N^{*(j)} - b_N \right)'$$

can be used instead of V_N. This will be particularly handy for various two-stage estimators with complicated asymptotic variances.

Note that, justifying V_{NB} for V does not simply follow from the bootstrap consistency $G_N(t; F_N) - G_N(t; F) \to^p 0$ as $N \to \infty$, which is good only for finding quantiles using bootstrap approximation. For instance, if the asymptotic distribution does not have the second moment, then although we can still compute V_{NB}, V_{NB} would be a nonsensical entity. Despite that V_{NB} is often used for V in practice, justifying this practice is not easy other than for simple cases such as averages (see Bickel and Freedman, 1981). In the following, we present a simple way of estimating V using bootstrap quantiles.

An estimator for the asymptotic variance can be obtained with bootstrap quantiles (Shao, 2003, p. 382). Suppose $\sqrt{N}(b_N - \beta) \rightsquigarrow N(0, V)$. Let $\zeta_{N\alpha}$ denote the nonparametric bootstrap αth quantile. Then we get

$$P\left\{ \sqrt{N}(b_N^* - b_N) \le t | F_N \right\} - \Phi\left(\frac{t}{\sqrt{V}} \right) \to^p 0$$

$$\Longrightarrow \; P\left\{ \sqrt{N}(b_N^* - b_N) \le \zeta_\alpha \sqrt{V} | F_N \right\} - \Phi(\zeta_\alpha) \to^p 0 \; \left(\text{setting } \zeta_\alpha = \frac{t}{\sqrt{V}} \right)$$

$$\Longrightarrow \; \zeta_{N\alpha} \to^p \zeta_\alpha \sqrt{V} \quad \text{and} \quad \zeta_{N(1-\alpha)} \to^p \zeta_{1-\alpha} \sqrt{V}.$$

Hence

$$\left\{ \frac{\zeta_{N(1-\alpha)} - \zeta_{N\alpha}}{\zeta_{(1-\alpha)} - \zeta_\alpha} \right\}^2 \to^p V.$$

This raises the question "which α to use or whether we can do better by using ζ_α and $\zeta_{\alpha'}$ where α' is not necessarily $1 - \alpha$." Although no best answer is available yet, some answers can be found in Machado and Parente (2005).

5.6.4 Bootstrap Iteration and Pre-pivoting

Bootstrap may be iterated for further improvement. To see how bootstrap test can be iterated, consider a test with its H_0-rejecting interval on the lower tail. Bootstrap test with $\sqrt{N}(b_N - \beta)$ rejects H_0 if the bootstrap p-value is less than α:

$$H_N^* \left\{ \sqrt{N}(b_N - \beta) \right\} \leq \alpha \quad \text{where } H_N^* \text{ is the empirical df of } \sqrt{N}(b_N^* - b_N).$$

This is analogous to rejecting H_0 in an asymptotic test when $\Psi\{\sqrt{N}(b_N - \beta)\} \leq \alpha$ where Ψ is the asymptotic df of $\sqrt{N}(b_N - \beta)$. Recall that the justification for this asymptotic test comes from $\Psi(W; F) \sim U[0, 1]$ where W is a rv following the weak limit of $\sqrt{N}(b_N - \beta)$. But $H_N^*\{\sqrt{N}(b_N - \beta)\}$ is not exactly $U[0, 1]$. Consider the df H^* of $H_N^*\{\sqrt{N}(b_N - \beta)\}$. Then *pre-pivoting* (Beran 1988) the bootstrap test statistic $H_N^*\{\sqrt{N}(b_N - \beta)\}$ is inserting this into H^* to use the resulting transformation as a TS:

$$\text{reject } H_0 \quad \text{if} \quad H^* \left[H_N^* \left\{ \sqrt{N}(b_N - \beta) \right\} \right] \leq \alpha;$$

this new test statistic would be closer to $U[0, 1]$ than $H_N^*\{\sqrt{N}(b_N - \beta)\}$ is.

A problem in implementing this idea is that H^* would be difficult to obtain analytically. Instead, we can apply bootstrap again to estimate H^*. Recall

$$H_N^* \left\{ \sqrt{N}(b_N - \beta) \right\} = \frac{1}{N_B} \sum_{j=1}^{N_B} 1 \left[\sqrt{N} \left(b_N^{*(j)} - b_N \right) \leq \sqrt{N}(b_N - \beta) \right].$$

- Step 1. Draw a pseudo sample F_N^* from F_N to obtain b_N^*.

- Step 2. Draw a second-stage pseudo sample F_N^{**} from F_N^* to obtain b_N^{**}. Repeat this N_B'-many times to obtain second-stage pseudo estimates $b_N^{**(j)}$, $j = 1, ..., N_B'$. This yields

$$H_N^{**} \left\{ \sqrt{N}(b_N^* - b_N) \right\} \equiv \frac{1}{N_{B'}} \sum_{j=1}^{N_B''} 1 \left[\sqrt{N} \left(b_N^{**(j)} - b_N^* \right) \leq \sqrt{N}(b_N^* - b_N) \right]$$

  which is a bootstrap estimator for $H_N^* \left\{ \sqrt{N}(b_N - \beta) \right\}$.

- Step 3. Repeat steps 1 and 2 N_B-times to obtain N_B-many estimates for $H_N^* \left\{ \sqrt{N}(b_N - \beta) \right\}$. Then a *double-bootstrap (i.e., iterated bootstrap, or nested bootstrap)* estimator of $H^* \left[H_N^* \left\{ \sqrt{N}(b_N - \beta) \right\} \right]$ is

$$\frac{1}{N_B} \sum_{k=1}^{N_B} 1 \left[H_N^{**(j)} \left\{ \sqrt{N}(b_N^* - b_N) \right\} \leq H_N^* \left\{ \sqrt{N}(b_N - \beta) \right\} \right].$$

APPENDIX III: SELECT GAUSS PROGRAMS

This appendix provides some GAUSS programs to illustrate that applying many estimators and tests in the main text is not difficult at all. Most estimators and tests take about one page of programming to implement. More GAUSS programs can be found in Lee (1995, 1996a and 2002) to supplement the programs below. All programs are numerically stable and reliable; i.e., they converge well.

Most programs in this appendix are a little longer than necessary for pedagogical reasons, namely to explain how things work and to introduce different programming techniques gradually over many programs. Hence, if desired, they can be much shortened. Although the programs are simple using simulated data only, they can be applied to large data with some minor modifications. These programs are given for the reader's benefit; no further support on how to tailor the programs to meet the reader's need will be provided.

1 LSE, IVE, GMM and Wald Test

```
new; format /m1 /rd 6,2;
n=100; one=ones(n,1);
x2=rndu(n,1); x3=rndn(n,1); w1=rndu(n,1); w2=rndn(n,1);

u=rndn(n,1); x4=w2+u; x4=x4/stdc(x4); /* x4 is endogenous */
y=1+x2+x3+x4+u; x=one~x2~x3~x4; k=cols(x);
z=one~x2~x3~w1~w2; /* z is the IV vector */

proc tv(b,cov); /* tv procedure */
retp(b./sqrt(diag(cov))); endp;

proc (6) = ols(x,y);
  local n,k,invx,est,res,res2,s,rsq,covhe,covho;
 n=rows(x); k=cols(x); invx=invpd(x'*x); est=invx*(x'*y);
 res=y-x*est; res2=res^2; s=sqrt(sumc(res2)/(n-k));
 rsq=1-sumc(res2)/sumc((y-meanc(y))^2);
 covhe=invx*(x'*(x.*res2))*invx; covho=(s^2)*invx;
retp(est,covhe,covho,s,rsq,res); endp;

{est1,covhe1,covho1,s1,rsq1,res1}=ols(x,y);   /* LSE */
```

```
tvlsehe=tv(est1,covhe1); tvlseho=tv(est1,covho1);
"LSE, tvhe,tvho: " est1~tvlsehe~tvlseho; "s,rsq: " s1~rsq1;

invz=invpd(z'z); invzxz=invpd(x'z*invz*z'x);   /* IVE */
ive=invzxz*(x'z)*invz*(z'y); uive=y-x*ive; s2=sumc(uive^2)/
  (n-k);
covive=s2*invzxz; tvive=tv(ive,covive); /* covive under
  homo. */

zdz=z'*(z.*(uive^2)); izdz=invpd(zdz); covgmm=invpd
  (x'z*izdz*z'x);
gmm=covgmm*(x'z)*izdz*(z'y); tvgmm=tv(gmm,covgmm); /* GMM */
?; "ive~tvive~gmm~tvgmm: " ive~tvive~gmm~tvgmm;

ugmm=y-x*gmm; ugmm2=ugmm^2;
idts=(z'*ugmm)'*invpd(z'*(z.*ugmm2))*(z'*ugmm); pvidts=
  cdfchic(idts,1);
"over-id test: "; idts~pvidts; /* over-id test */

/* Wald test for beta(2)=beta(4)=0 */
eyek=eye(k); g=eyek[.,2]~eyek[.,4]; wald=(g'*gmm)'*invpd
  (g'*covgmm*g)*(g'*gmm);
pvwald=cdfchic(wald,2); ?; "beta2=beta4=0; "; wald~pvwald;
end;
```

2   System LSE

```
new; format /m1 /rd 6,2;
n=500; one=ones(n,1);

q1=rndn(n,1); q2=rndn(n,1); u1=rndn(n,1); u2=rndn(n,1);
x1=one~q1; x2=one~q2; k1=cols(x1); k2=cols(x2); k=k1+k2;
y1=1+q1+u1; y2=1+q2+u2; /* two equations */

ww=zeros(k,k); wy=zeros(k,1); i=1;
do until i>n;
 wi=(x1[i,.]'~zeros(k1,1))|
        (zeros(k2,1)~x2[i,.]'); /* wi: regressor matrix */
 yi=y1[i,1]|y2[i,1]; ww=ww+wi*wi'; wy=wy+wi*yi;
i=i+1; endo;

iww=invpd(ww); g=iww*wy; wuuw=zeros(k,k); i=1; /* g is
   system LSE */
do until i>n;
 wi=(x1[i,.]'~zeros(k1,1))|
        (zeros(k2,1)~x2[i,.]');
 yi=y1[i,1]|y2[i,1]; ui=yi-wi'*g;
 wuuwi=wi*ui*ui'*wi'; wuuw=wuuw+wuuwi;
i=i+1; endo;

cov=iww*wuuw*iww; sd=sqrt(diag(cov)); tv=g./sd;
r=eye(2)|(-eye(2)); wald=(r'*g)'*invpd(r'cov*r)*(r'*g);
?; "Wald,pv(equal para.):" wald~cdfchic(wald,2);

/* To get a nicer output with variable names */
let varlist[4,1] = int1 slope1 int2 slope2;
let mask[1,4] = 0 1 1 1; /* 0 for alphabet; # for 1 in
   4 columns */
let fmt[4,3] = "-*.*s" 7 7 "*.*lf," 7 2 "*.*lf" 6 3 "
   *.*lf" 7 2;
                        /* display formats for 4 columns */

?; printfm(varlist~g~sd~tv,mask,fmt);
end;
```

3   Method-of-Moment Test for Symmetry

```
new; format /m1 /rd 7,3;
n=200; one=ones(n,1);

x2=rndn(n,1);
u=rndn(n,1)^2+rndn(n,1)^2; u=(u-meanc(u))/stdc(u);
y=1+x2+u; x=one~x2; /* u is asymmetric with chi-2 */

proc tv(b,cov);
retp(b./sqrt(diag(cov))); endp;

proc (6) = ols(x,y);
 local k,invx,est,res,res2,s,rsq,covhe,covho;
  k=cols(x); invx=invpd(x'*x); est=invx*(x'*y); res=y-x*est;
  res2=res^2; s=sqrt(sumc(res2)/(n-k));
 rsq=1-sumc(res2)/sumc((y-meanc(y))^2);
  covhe=invx*(x'*(x.*res2))*invx; covho=(s^2)*invx;
retp(est,covhe,covho,s,rsq,res); endp;

/*
#include h:\procedures\ProcTvLse;
/* This command pastes the file "ProcTvLse" in
  h:\procedures\ right here;
     ProcTvLse is nothing but the above tv and ols
        procedures;
     This obviates repeating often used procedures in each
        program. */
*/

{lse,covlsehe,covlseho,slse,rsqlse,reslse}=ols(x,y);
r=reslse; del=(r^3)-3*(x.*r)*invpd(x'x)*(x'*(r^2));
tvgood=sumc(r^3)/sqrt(sumc(del^2)); tvbad=sumc(r^3)/sqrt
  (sumc(r^6));
                        /* tvbad ignores 1st stage error */

"correct and wrong tv's: " tvgood~tvbad;
end;
```

4 Quantile Regression

```
new; format /m1 /rd 7,3; cr=0.00001; /* cr is stopping
  criterion */
n=400; one=ones(n,1); iterlim=100; /* iteration limit */

alp=0.5; beta=1|1|1; /* alp for alp-quantile */
x2=rndn(n,1); x3=rndu(n,1); u=rndn(n,1);
x=one~x2~x3; k=cols(x); y=x*beta+u;

proc obj(g);
retp( -(y-x*g).*(alp-(y.<x*g)) ); endp; /* output is N
  by 1 */

b0=0.5*invpd(x'x)*(x'y); /* initial value */
niter=1; /* niter is iteration counter */
bestobj=sumc(obj(b0)); bestb=b0;

JOB:
  gradi=gradp(&obj,b0); /* numerical derivative */
 b1=b0+invpd(gradi'gradi)*sumc(gradi); newobj=sumc
  (obj(b1));

 if abs(newobj-bestobj)<cr; goto DONE; endif;
 if newobj>bestobj; bestobj=newobj; bestb=b1; endif;
  if niter>iterlim; goto DONE; endif;
  b0=b1; niter=niter+1; goto JOB;

DONE:
b=bestb; r=y-x*b; h=stdc(r)*(n^(-1/5)); /*h is bandwidth */
ker=pdfn(r/h)/h; hessi=x'*(x.*ker);
cov=alp*(1-alp)*invpd(hessi)*(x'x)*invpd(hessi);

tv=b./sqrt(diag(cov)); b~tv;
end;
```

5   Univariate Parametric LDV Models

5.1   Probit

```
new; format /m1 /rd 7,3;
n=300; one=ones(n,1);

x2=rndn(n,1).>0; x3=rndn(n,1); x=one~x2~x3; k=cols(x);
beta=1|1|1; u=rndn(n,1);
/* u=rndn(n,1)^2+rndn(n,1)^2; u=(u-meanc(u))/stdc(u); */
ys=x*beta+u; y=ys.>0;

proc like(a); local c; c=cdfn(x*a);
retp( -sumc(y.*ln(c)+(1-y).*ln(1-c)) ); endp; /* output
  is scalar */

proc first(a); local xa,d,c;
 xa=x*a; d=pdfn(xa); c=cdfn(xa);
retp( x.*( ((y-c).*d)./(c.*(1-c)) ) ); endp;

a0=invpd(x'x)*(x'y); {pro,obj,grad,ret}=qnewton(&like,a0);
/* GAUSS-provided minimization procedure qnewton is used;
 pro is the minimizer and obj is the minimum; */

hessi=hessp(&like,pro); covhes=invpd(hessi);
gradi=first(pro); covgra=invpd(gradi'gradi);
covrob=covhes*(gradi'gradi)*covhes; /* robust cov */

tvgra=pro./sqrt(diag(covgra));
tvhes=pro./sqrt(diag(covhes));
tvrob=pro./sqrt(diag(covrob)); /* 3 ways of getting tv */

pro~tvgra~tvhes~tvrob; "log-like: " -obj;
end;
```

5.2   Ordered Probit

```
new; format /m1 /rd 7,3; /* 4 category ODR */
n=500; one=ones(n,1); iterlim=100; cr=0.00001;
beta1=1; beta2=1; g1=-0.5; g2=0.5; g3=1; /* thresholds */

x2=rndn(n,1); u=rndn(n,1);
ys=beta1+beta2*x2+u; y=(ys.>g1)+(ys.>g2)+(ys.>g3);
y0=(y.==0); y1=(y.==1); y2=(y.==2); y3=(y.==3);
x=one~x2; k=cols(x)+2; /* k is # parameters */

proc like(b); local a,xa,tau1,tau2,p0,p1,p2,p3;
 a=b[1:k-2,1]; tau1=b[k-1,1]; tau2=b[k,1]; xa=x*a;
  p0=cdfn(-xa); p1=cdfn(tau1-xa)-cdfn(-xa);
 p2=cdfn(tau2-xa)-cdfn(tau1-xa); p3=1-cdfn(tau2-xa);
retp( y0.*ln(p0)+y1.*ln(p1)+y2.*ln(p2)+y3.*ln(p3) ); endp;

b0=inv(x'x)*(x'y)|0.5|1; bestobj=sumc(like(b0)); bestb=b0;
   niter=1;

JOB:
 gradi=gradp(&like,b0); b1=b0+invpd(gradi'gradi)*sumc
   gradi);
 newobj=sumc(like(b1)); niter=niter+1;

 if abs(newobj-bestobj)<cr; goto DONE; endif;
 if niter>iterlim; goto DONE; endif;
 if newobj>bestobj; bestobj=newobj; bestb=b1; endif;
 b1'~newobj~bestobj~niter; b0=b1; goto JOB;

DONE:
opro=bestb; loglike=sumc(like(opro)); gradi=gradp(&like,
   opro);
cov=invpd(gradi'gradi); tv=opro./sqrt(diag(cov)); opro~tv;
end;
```

5.3   Tobit

```
new; format /m1 /rd 7,3; n=400; one=ones(n,1); eps=0.00001;
x2=rndn(n,1); u=2*rndn(n,1);

beta1=1; beta2=1; ys=beta1+beta2*x2+u; y=maxc(ys'|
   zeros(1,n));
d=ys.>0; /* d is non-censoring indicator */
x=one~x2; kx=cols(x); k=kx+1; /* k is # parameters */

proc likesum(g); local b,s,xb,r,l1,l2;
 b=g[1:kx,1]; s=g[k,1]; xb=x*b; r=y-xb;
  l1=d.*ln(pdfn(r/s)/s); l2=(1-d).*ln(cdfn(-xb/s));
retp(-sumc(l1+l2)); endp; /* output is a scalar */

proc like(g); local b,s,xb,r,l1,l2;
 b=g[1:kx,1]; s=g[k,1]; xb=x*b; r=y-xb;
  l1=d.*ln(pdfn(r/s)/s); l2=(1-d).*ln(cdfn(-xb/s));
retp(l1+l2); endp; /* output is N*1 */

proc first(g); local j,bas,fder,ghi,glo;
 j=1; bas=eye(rows(g)); fder=zeros(n,k);
  do until j>k;
     ghi=g+eps*bas[.,j]; glo=g-eps*bas[.,j];
   fder[.,j]=(like(ghi)-like(glo))/(2*eps);
  j=j+1; endo;
retp(fder); endp;

/* "first" is a numerical derivative procedure; gradp can
   be used instead; but understanding this should be helpful when
      second order (cross) derivatives are needed */

b0=invpd(x'x)*(x'y); res=y-x*b0; s0=sqrt(meanc(res^2));
g0=b0|s0; {g,obj,grad,ret}=qnewton(&likesum,g0);
gradi=first(g); cov=invpd(gradi'gradi);
tv=g./sqrt(diag(cov)); "g,tv:" g~tv; "loglike: " -obj;
end;
```

5.4   Weibull MLE under Random Censoring

```
new; format/m1/rd 7,2; n=500; one=ones(n,1); cr=0.00001;
  iterlim=50;
x2=rndn(n,1); x=one~x2; kx=cols(x); k=kx+1; /* k is # para.   */

beta=1|1; alpha=1;
u=rndu(n,1); ys=( -exp(-x*beta).*ln(u) )^(1/alpha);

/* Weibull ys simulation using S(y) follows U[0,1];
   logic: (S(ys)=) exp(-theta*ys^alp)=u; ln(u)=-theta*ys^alp;
   ys=(-ln(u)/theta)^(1/alp); let theta=exp(x*beta) */

c=rndn(n,1)^2; /* random censoring point */
y=minc(ys'|c'); d=ys.<=c; /* d is non-censoring indicator */

proc like(g); local a,b,xb;
 b=g[1:kx,1]; a=g[k,1]; xb=x*b;
retp( d.*(ln(a)+(a-1)*ln(y)+xb) - (y^a).*exp(xb) ); endp;

lny=ln(y); lse=-invpd(x'x)*(x'lny); res2=(lny+x*lse)^2;
s2=meanc(res2); a0=sqrt(1.645/s2); b0=lse*a0; g0=b0|a0;
bestobj=-100000; bestg=g0; niter=1;
/* no guarantee for global max; try different values
   for g0*/

JOB:
gradi=gradp(&like,g0); g1=g0+invpd(gradi'gradi)*sumc(gradi);
newobj=sumc(like(g1)); niter=niter+1;

if abs(newobj-bestobj)<cr; goto DONE; endif;
if niter>iterlim; goto DONE; endif;
if newobj>bestobj; bestobj=newobj; bestg=g1; endif;
newobj~bestobj~niter; g0=g1; goto JOB;

DONE:
g=bestg; gradi=gradp(&like,g); covwei=invpd(gradi'gradi);
tvwei=g./sqrt(diag(covwei)); g~tvwei; "Non-censoring %:"
  meanc(d);
end;
```

6   Multivariate Parametric LDV Models

6.1   Multinomial Logit (MNL)

```
new; format/m1/rd 6,3; /* 3 alternatives */
n=1000; one=ones(n,1); cr=0.00001; iterlim=100;
x11=rndn(n,1); x21=rndn(n,1); x31=rndn(n,1);
x12=rndn(n,1); x22=rndn(n,1); x32=rndn(n,1);

/* x1,x2 are two alternative-variants */
x1=x11~x12; x2=x21~x22; x3=x31~x32; kx=cols(x1);
z=one~rndn(n,1); kz=cols(z); /* z is an alternative-
  constant */

d11=1; d12=1; d21=1; d22=1; d31=1; d32=1;
eta1=0|0; eta2=1|1; eta3=1|1; eta21=eta2-eta1;
  eta31=eta3-eta1;
d1=d11|d12; d2=d21|d22; d3=d31|d32;
true=d1|d2|d3|eta21|eta31; /* identified parameters */

w2=(-x1)~x2~zeros(n,kx)~z~zeros(n,kz); w2b=w2*true;
w3=(-x1)~zeros(n,kx)~x3~zeros(n,kz)~z; w3b=w3*true;
  k=cols(w2);

/* survival function for type-1 extreme is exp(-exp(-t)) */
u1=-ln(ln(1./rndu(n,1))); u2=-ln(ln(1./rndu(n,1)));
u3=-ln(ln(1./rndu(n,1)));
s1=x1*d1+z*eta1+u1; s2=x2*d2+z*eta2+u2; s3=x3*d3+z*eta3+u3;
y1=(s1.>s2).*(s1.>s3); y2=(s2.>s1).*(s2.>s3); y3=1-y1-y2;

proc mnllike(b); local e2,e3,norm,p1,p2,p3;
 e2=exp(w2*b); e3=exp(w3*b); norm=1+e2+e3;
  p1=1./norm; p2=e2./norm; p3=e3./norm;
retp( -sumc(y1.*ln(p1) +y2.*ln(p2) +y3.*ln(p3)) ); endp;

b0=0.5*ones(k,1); {b,obj,grad,ret}=qnewton(&mnllike,b0);
/* If qnewton does not work well, try Newton-Raphson with
  grap */
loglike=-obj; hessi=hessp(&mnllike,b); covmnl=invpd(hessi);
tv=b./sqrt(diag(covmnl)); true~b~tv; "log-like: "; loglike;
end;
```

6.2   Two-Stage Estimator for Sample Selection

```
new; format/m1/rd 7,3; n=300; one=ones(n,1);
x2=rndn(n,1); x3=rndn(n,1); e=rndn(n,1);

ds=1+x2+x3+e; d=ds.>0; w=one~x2~x3; kw=cols(w);
u=rndn(n,1)+e; u=u/stdc(u); /* x3 excluded from y eq. */
y=1+x2+u; yd=y.*d; x=one~x2; kx=cols(x);

proc prolike(a); local prob; prob=cdfn(w*a);
retp(-d'*ln(prob)-(1-d)'*ln(1-prob)); endp;

proc profir(a); local wa,den, prob,e,denom;
 wa=w*a; den=pdfn(wa); prob=cdfn(wa);
  denom=prob.*(1-prob); e=(d-prob).*den./denom;
retp(w.*e); endp;

a0=invpd(w'w)*(w'd); {pro,obj,grad,ret}=qnewton
  (&prolike,a0);
gradi=profir(pro); info=gradi'*gradi; covpro=invpd(info);
wa=w*pro; pd=pdfn(wa); cd=cdfn(wa);lam=pd./cd; lam2=lam^2;

zd=(x~lam).*d; b=invpd(zd'zd)*zd'yd; pyd=zd*b; vd=yd-pyd;
rhosig=b[cols(zd),1]; qbb=zd'zd; iqbb=invpd(qbb);
capa=rhosig*( zd'*( w.*(-wa.*lam-lam2) ) ); /* Link matrix  */

qb=zd.*vd - gradi*covpro*capa';/*gradi*covpro*n is eta */
qbqb=qb'*qb; cov=iqbb*qbqb*iqbb';
qb0=zd.*vd; qb0qb0=qb0'*qb0; cov0=iqbb*qb0qb0*iqbb';
"b,tvgood,tvbad:" b~(b./sqrt(diag(cov)))~(b./sqrt(diag
  (cov0)));

c=invpd(x'x)*(x'lam); v=lam-x*c; sse=sumc(v^2);
"R-sq for lambda:" 1-sse/sumc((lam-meanc(lam))^2);
 /* this checks the multicollinearity problem of pd./cd */

end;
```

7   Nonparametric Regression and Hazard

7.1   Univariate Density

```
new; format/m1/rd 7,3; library pgraph;
n=500; x=rndn(n,1)^2+rndn(n,1)^2;
h0=n^(-1/5); /* h0 is the "base" bandwidth for univariate
  density */

evalo=meanc(x)-2*stdc(x); /* minimum evaluation point */
evahi=meanc(x)+2*stdc(x); /* maximum evaluation point */
neva=100; /* # total evaluation points */
inc=(evahi-evalo)/neva;
eva=seqa(evalo,inc,neva); /* eva is the actual evaluation
  points*/

proc dens(z,mf,zo); /* z is n*1, mf is bandwidth multi.
  factor */
 local sd,ker,c; /* zo is a scalar evaluation point */
  sd=stdc(z); c=(z-zo)/(sd*mf*h0);
  ker=pdfn(c); /* N(0,1) kerel used */
  /* ker=(15/16)*((1-c^2)^2).*(abs(c).<1); /* biweigt
    kernel */ */
retp(meanc(ker)/(sd*mf*h0)); endp;

mf=1; j=1; est=zeros(neva,1);
do until j>neva;
 est[j,1]=dens(x,mf,eva[j,1]);
j=j+1; endo;

xlabel("Regressor"); title("Kernel Density Estimate");
xy(eva,est);
end;
```

7.2 Bivariate Regression Function

```
new; format /m1 /rd 7,3; library pgraph; n=400;
x1=rndn(n,1); x2=3*rndu(n,1); sd1=stdc(x1); sd2=stdc(x2);

u=rndn(n,1); y=x1+sin(x2)+0.5*x1.*x2+u; k=2;
h0=n^(-1/(k+4)); /* h0 is base bandwidth */

eva1lo=meanc(x1)-2*sd1; eva1hi=meanc(x1)+2*sd1;
neva1=50; inc1=(eva1hi-eva1lo)/neva1; eva1=seqa(eva1lo,
  inc1,neva1);
                                /* evaluation point for x1 */

eva2lo=meanc(x2)-2*sd2; eva2hi=meanc(x2)+2*sd2;
neva2=50; inc2=(eva2hi-eva2lo)/neva2; eva2=seqa(eva2lo,
  inc2,neva2);
                                /* evaluation point for x2 */

proc bireg(x,y,p,eva); /* x is n*2, y is n*1, p is mul.
  factor, eva is 1*2 */
 local c1,c2,k1,k2,ker;
  c1=(x[.,1]-eva[1,1])/(p*sd1*h0); k1=pdfn(c1);
  c2=(x[.,2]-eva[1,2])/(p*sd2*h0); k2=pdfn(c2); ker=k1.*k2;
retp(meanc(ker.*y)/meanc(ker)); endp;

p=1.5; j1=1; reg=zeros(neva2,neva1);
do until j1>neva1; j2=1;
 do until j2>neva2;
  reg[j2,j1]=bireg(x1~x2,y,p,eva1[j1,1]~eva2[j2,1]);
  j2=j2+1; endo;
j1=j1+1; endo;

xlabel("x1"); ylabel("x2"); title("Regression Function");
ztics(-3,5,2,5); /* min, max, major tick interval, # minor
 subdivisions */
surface(eva1',eva2,reg); /* 3 dimensional graph */
end;
```

7.3   Regression Derivative and Confidence Interval

```
new; format /m1 /rd 7,4; library pgraph; eps=0.00001;
intkerd2=0.14; /* Integral of (kernel derivative)^2;
   computed by
   Monte Carlo integration: v=rndn(5000,1); meanc(pdfn(v).
   *(v^2)); */

n=300; x=rndn(n,1); u=x.*rndn(n,1); u=u/stdc(u); /* u is
   heteroskedastic */
y=x^2+u; h0=n^(-1/5); /* base bandwidth for univariate npr
   regression */
evalo=meanc(x)-2*stdc(x); evahi=meanc(x)+2*stdc(x); /*
   evaluation points */
neva=100; inc=(evahi-evalo)/neva; eva=seqa(evalo,inc,neva);

proc (3) = regdeva(x,y,hm,eva);
 local hsd,neva,regd,den,reg1,reg2,j,c,clo,chi,ker,kerlo,
    kerhi,var,lb,ub;
    hsd=hm*h0*stdc(x); neva=rows(eva); regd=zeros(neva,1);
      den=zeros(neva,1);
    reg1=zeros(neva,1); reg2=zeros(neva,1); j=1;
    do until j>neva;
      c=(x-eva[j,1])/hsd; ker=pdfn(c); den[j,1]=sumc(ker)/
        (n*hsd); /* density */
      clo=(x-eva[j,1]+eps)/hsd; kerlo=pdfn(clo);
      chi=(x-eva[j,1]-eps)/hsd; kerhi=pdfn(chi); /* for
        regression derivative */
      regd[j,1]=(sumc(kerhi.*y)/sumc(kerhi)-sumc(kerlo.*y)/
        sumc(kerlo))/(2*eps);
      reg1[j,1]=sumc(ker.*y)/sumc(ker);
        reg2[j,1]=sumc(ker.*(y^2))/sumc(ker);
      j=j+1; endo;

 var=reg2-reg1^2; /* conditional variance function */
    lb=regd-1.96*sqrt((intkerd2/(n*(hsd^3)))*var./den); /*
      CI lower bound */
    ub=regd+1.96*sqrt((intkerd2/(n*(hsd^3)))*var./den); /*
      CI upper bound */
retp(lb,regd,ub); endp;

hm=2; {cilb,regd,ciub}=regdeva(x,y,hm,eva); title("dr/dx &
   95% pointwise CI");
xtics(-2,2,1,5); ytics(-4,4,2,5); xlabel("x"); ylabel("y");

_pgrid={3,2}; /* 3 for sold grid (0 for no); 2 for ticks at
   subdivisions */
```

```
_pltype = {3,6,3}; /* 1 dahsed, 2 dotted, 3 short dashes, 6
   solid */
xy(eva,cilb~regd~ciub); /* xy is for two dimensional graph */
end;
```

8   Bandwidth-Free Semiparametric Methods

8.1   Winsorized Mean Estimator (WME) for Censored Model

```
new; format/m1/rd 7,3; cr=0.00001; n=200; one=ones(n,1);
  iterlim=50;
x1=rndn(n,1); x2=rndn(n,1); x=x1~x2; sdx=stdc(x);
  k=cols(x); xx=one~x;
u=rndn(n,1); beta=1|1|1; ys=xx*beta+u; y=maxc(ys'|
  zeros(1,n));

lse=invpd(xx'xx)*(xx'y); w=stdc(y-xx*lse); /* 0.25w,
  0.5w,... can be used */
h0=n^(-1/(k+4)); /* base bandwidth */
qlo=minc(y); qhi=maxc(y); nq=101; /* larger nq gives
  better 1st stage est. */
qinc=(qhi-qlo)/nq; q=seqa(qlo,qinc,nq); /* q is 1st stage
  est. to be tried */

proc obj1(q,eva,m); /* eva is 1*k; m is bandwidth multi.
  factor */
 local c,ker,aymq; aymq=abs(y-q); /* produt normal kernel
  used */
  c=(x-one.*.eva)./(one.*.(h0*m*sdx')); ker=exp(sumc
    (ln(pdfn(c))'));
retp(meanc(ker.*(0.5*(aymq^2).*(aymq.<w)+(w*aymq-0.5*w).
 *(aymq.>=w)))); endp;

m=1; pass=zeros(n,1); i=1; /* 0.5m,2m,3m,... can be used
  as well */
do until i>n;
  comp=zeros(nq,2); j=1;
  do until j>nq;
  comp[j,.]=obj1(q[j,1],x[i,.],m)~q[j,1];
  j=j+1; endo; pass[i,1]=comp[minindc(comp[.,1]),2]>w;
i=i+1; endo; /* For pass, try also a # greater than w,
  e.g., 1.5w */

proc obj2(b); local ar;
 ar=abs(y-xx*b); /* only those with pass=1 are used in 2nd
  stage */
retp(pass.*(0.5*(ar^2).*(ar.<w)+(w*ar-0.5*w).*(ar.>=w) ));
 endp;

b0=invpd(xx'xx)*(xx'y); niter=1; bestobj=-100000; bestb=b0;

JOB:
```

```
gradi=gradp(&obj2,b0); xxb0=xx*b0; ar0=abs(y-xxb0);
 invh=invpd(xx'( xx.*(pass.*(xxb0.>w).*(ar0.<w)) ));
 /* if invh is not p.d., remove either (xxb0.>w) or (ar0.<w) */
b1=b0-invh*sumc(gradi); newobj=sumc(obj2(b1)); niter=
 niter+1;

if abs(newobj-bestobj)<cr; goto DONE; endif;
 if niter>iterlim; goto DONE; endif;
 if newobj>bestobj; bestobj=newobj; bestb=b1; endif;
 b0=b1; goto JOB;

DONE:
b=bestb; xxb=xx*b; ar=abs(y-xx*b); r2=ar^2;
invh=invpd(xx'(xx.*((xxb.>w).*(ar.<w))));

cov=invh*(xx'(xx.*(xxb.>w).*minc(r2'|(w*one'))))*invh;
tv=b./sqrt(diag(cov)); lse~b~tv; "non-censoring %:
 " meanc(y.>0);
end;
```

8.2   Differencing for Semi-Linear Model

```
new; format/m1/rd 7,3; n=200;
x1=rndn(n,1); x2=x1^2; x3=rndu(n,1).>0.3; z=rndn(n,1);
x=x1~x2~x3; k=cols(x);

beta=2|1|0; u=rndn(n,1); y=1+sin(z)+x*beta+u;
xyz=sortc(x~y~z,k+2); /* sorting on z */
x=xyz[.,1:k]; y=xyz[.,k+1]; z=xyz[.,k+2];

dx=x[2:n,.]-x[1:n-1,.]; dy=y[2:n,1]-y[1:n-1,1];
b=invpd(dx'dx)*(dx'dy); du=dy-dx*b; du2=du^2;

dx1=dx[2:rows(dx),.]; dx0=dx[1:rows(dx)-1,.];
du1=du[2:rows(du),1]; du0=du[1:rows(du)-1,1];
v1=dx'(dx.*du2)/(n-1); v2=dx0'(dx1.*du0.*du1)/(n-2);

hessi=dx'dx/(n-1);
cov=invpd(hessi)*(v1+v2+v2')*invpd(hessi)/n;
b~(b./sqrt(diag(cov)));
end;
```

9 Bandwidth-Dependent Semiparametric Methods

9.1 Two-Stage Estimator for Semi-Linear Model

```
new; format /m1 /rd 7,3; n=200;
hlo=1; hhi=4; hinc=0.1; hrep=int((hhi-hlo)/hinc)+1;  /* for
 CV search */

x1=2*rndu(n,1); x2=rndn(n,1).>0; x=x1~x2; k=cols(x);
z1=rndn(n,1); z2=rndn(n,1); sd1=stdc(z1); sd2=stdc(z2);
u=rndn(n,1); y=1+x1+x2+((z1+z2+z1.*z2)/2)+u; /* npr part
  consists of z1,z2 */
h0=n^(-1/6); /* h0 is base bandwidth with 2-dimensional
  smoothing */

proc npr(y,z1,z2,m);
 local reg,i,z1i,z2i,yi,c1i,c2i,k1,k2,k1ij,k2ij; reg=zeros
  (n,1); i=1;
  do until i>n;
  if i==1; z1i=z1[2:n,1]; z2i=z2[2:n,1]; yi=y[2:n,1];
  elseif i==n; z1i=z1[1:n-1,1]; z2i=z2[1:n-1,1]; yi=y
    [1:n-1,1];
  else;z1i=z1[1:i-1,1]|z1[i+1:n,1];z2i=z2[1:i-1,1]|z2[i+1:n,1];
  yi=y[1:i-1,1]|y[i+1:n,1]; endif;
  c1i=(z1[i,1]-z1i)/(m*h0*sd1); c2i=(z2[i,1]-z2i)/
    (m*h0*sd2);
  k1=pdfn(c1i); k1ij=1.5*k1-0.5*k1.*(c1i^2);
  k2=pdfn(c2i); k2ij=1.5*k2-0.5*k2.*(c2i^2); /* 3rd-order
   kernel */
       reg[i,1]=sumc((k1ij.*k2ij).*yi)/sumc(k1ij.*k2ij); i=i+1;
   endo;
retp(reg); endp;

j=1; comp=zeros(hrep,2); h=hlo;    /* CV step to find the
  optimal h for y */
do until j>hrep;  comp[j,1]=meanc((y-npr(y,z1,z2,h))^2);
  comp[j,2]=h;
j=j+1; h=h+hinc; endo; h=comp[minindc(comp[.,1]),2];
  my=npr(y,z1,z2,h);

j=1; mx=zeros(n,k);
do until j>k;         mx[.,j]=npr(x[.,j],z1,z2,h); /* same h used
 to save time */
j=j+1; endo;
```

```
mdx=x-mx; mdy=y-my; invmdx=invpd(mdx'mdx); lse=invmdx*
  (mdx'mdy);
res=mdy-mdx*lse; res2=res^2; cov=invmdx*(mdx'*(mdx.*res2))
  *invmdx;
tvlse=lse./sqrt(diag(cov)); "b,tv:" lse~tvlse; "multi.
  factor :" h; end;
```

9.2   Quasi-MLE for Single-Index Binary Response

```
new; format/m1/rd 7,3; cr=0.00001; iterlim=50; n=300;
  one=ones(n,1);
x1=rndn(n,1); x2=rndn(n,1); xk=rndn(n,1); u=rndn(n,1);
x=x1~x2; k=cols(x); xx=x~xk; beta=1|1|1; ys=1+xx*beta+u;
  y=ys.>0;

h0=n^(-1/5); /* h0 is the base bandwidth; try h0*(a multi.
  factor) */
bk=1; /* try bk=-1 also unless the sign of beta-k is known   */

proc reg(b); local xxb,hxxb,rg,i,yi,xxbi,ci,ker;
 xxb=x*b+bk*xk; hxxb=h0*stdc(xxb); rg=zeros(n,1); i=1;
  do until i>n;
    if i==1; yi=y[2:n,1]; xxbi=xxb[2:n,1];
    elseif i==n; yi=y[1:n-1,1]; xxbi=xxb[1:n-1,1];
     else; yi=y[1:i-1,1]|y[i+1:n,1]; xxbi=xxb[1:i-1,1]|
       xxb[i+1:n,1]; endif;
     ci=(xxbi-xxb[i,1])/hxxb; ker=pdfn(ci); rg[i,1]=
       sumc(ker.*yi)/sumc(ker);
   i=i+1; endo;       /* in theory, a high order kernel
     should be used */
retp(rg); endp;                 /* but, high order kernel can
  cause ln(0) */

proc obj(b); local p; p=reg(b);
retp(y.*ln(p)+(1-y).*ln(1-p)); endp;

a0=invpd(xx'xx)*(xx'y); b0=a0[1:k,1]/a0[k+1,1];
niter=1; bestobj=-100000; bestb=b0;

JOB:
gradi=gradp(&obj,b0); b1=b0+invpd(gradi'gradi)*sumc(gradi);
newobj=sumc(obj(b1)); niter=niter+1;
if abs(newobj-bestobj)<cr; goto DONE; endif;
if niter>iterlim; goto DONE; endif;
if newobj>bestobj; bestobj=newobj; bestb=b1; endif;
b1'~newobj~bestobj~niter; b0=b1; goto JOB;

DONE:         /* tv tends to be over-estimated */
b=bestb; p=reg(b); pd=gradp(&obj,b); cov=invpd(pd'*(pd./
  (p.*(1-p))));
tv=b./sqrt(diag(cov)); b~tv; sumc(obj(b)); end;
```

REFERENCES

AS: Annals of Statistics
ECA: Econometrica
ECMJ: Econometrics Journal
ER: Econometric Reviews
ET: Econometric Theory
IER: International Economic Review
JAE: Journal of Applied Econometrics
JASA: Journal of The American Statistical Association
JBES: Journal of Business and Economic Statistics
JOE: Journal of Econometrics
JRSS-B: Journal of The Royal Statistical Society, Series B.
REStat: Review of Economics and Statistics
REStud: Review of Economic Studies.

Abbring, J.H. and G.J. van den Berg, 2003, The identifiability of the mixed proportional hazards competing risks model, JRSS-B 65, 701–710.

Abrevaya, J.,1999, Rank estimation of a transformation model with observed truncation, ECMJ 2, 292–305.

Abrevaya, J., 2001, The effects of demographics and maternal behavior on the distribution of birth outcomes, Empirical Economics 26, 247–257.

Abrevaya, J., 2002, Computing marginal effects in the Box-Cox model, ER 21, 383–393.

Abrevaya, J., 2003, Pairwise-difference rank estimation of the transformation model, JBES 21, 437–447.

Abrevaya, J. and J. Huang, 2005, On the bootstrap of the maximum score estimator, ECA 73, 1175–1204.

Aerts, M., G. Claeskens, J.D. Hart, 1999, Testing the fit of a parametric function, JASA 94, 869–879.

Aerts, M., G. Claeskens, J.D. Hart, 2000, Testing lack of fit in multiple regression, Biometrika 87, 405–424.

Ahn, H.T., 1995, Nonparametric two-stage estimation of conditional choice probabilities in a binary choice model under uncertainty, JOE 67, 337–378.

Ai, C. and D. McFadden, 1997, Estimation of some partially specified nonlinear models, JOE 76, 1–37.

Ai, C. and E. Norton, 2008, A semiparametric derivative estimator in log transformation models, ECMJ 11, 538–553.

Ait-Sahalia, Y., P.J. Bickel and T.M. Stoker, 2001, Goodness-of-fit tests for kernel regression with an application to option implied volatilities, JOE 105, 363–412.

Altonji, J.G. and R.L. Matzkin, 2005, Cross-section and panel data estimators for nonseparable models with endogenous regressors, ECA 73, 1053–1102.

Amemiya, T., 1974, The nonlinear two-stage least squares estimator, JOE 2, 105–110.

Amemiya, T., 1985, Advanced econometrics, Harvard University Press.

Anderson, E.B., 1970, Asymptotic properties of conditional maximum likelihood estimators, JRSS-B 32, 283–301.

Andersen, P.K. and R.D. Gill, 1982, Cox's regression model for counting processess: a large sample study, AS 10, 1100–1120.

Anderson, T.W. and H. Rubin, 1949, Estimator of the parameters of a single equation in a complete system of stochastic equations, Annals of Mathematical Statistics 20, 46–63.

Anderson, T.W. and H. Rubin, 1950, The asymptotic properties of estimates of the parameters of a single equation in a complete system of stochastic equations, Annals of Mathematical Statistics 21, 570–582.

Andrews, D.W.K., 1987a, Consistency in nonlinear econometric models: A generic Uniform LLN, ECA 55(6), 1465–1471.

Andrews, D.W.K., 1987b, Asymptotic results for generalized Wald tests, ET 3, 348–358.

Andrews, D.W.K., 1991, Asymptotic optimality of generalized C_L, cross-validation and generalized cross-validation in regression with heteroskedastic errors, JOE 47, 359–377.

Andrews, D.W.K., 1993, Tests for parameter instability and structural change with unknown change point, ECA 61, 821–856.

Andrews, D.W.K., 1994, Asymptotics for semiparametric econometric models via stochastic equicontinuity, ECA 62, 43–72.

Andrews, D.W.K., 1997, A conditional Kolmogorov test, ECA 65, 1097–1128.

Andrews, D.W.K., 1999, Consistent moment selection procedures for generalized method of moment estimation, ECA 67, 543–564.

Andrews, D.W.K. and J.C. Monahan, 1992, An improved heteroskedasticity and autocorrelation consistent covariance matrix, ECA 60, 953–966.

Anglin, P.M. and R. Gençay, 1996, Semiparametric estimation of a hedonic price function, JAE 11, 633–648.

Angrist, J.D., V. Chernozhukov, and I. Fernández-Val, 2006, Quantile regression under misspecification, with an application to the U.S. wage structure, ECA 74, 539–563.

Angrist, J.D. and W.N. Evans, 1998, Children and their parent's labor supply: evidence from exogenous variation in family size, AER 88, 450–477.

Angrist, J. and A.B. Krueger, 2001, Instrumental variables and the search for identification: from suppy and demand to natural experiments, Journal of Economic Perspectives 15, 69–85.

Anton, A.A., A.F. Sainz, and J. Rodriguez-Poo, 2001, Semiparametric estimation of a duration model, Oxford Bulletin of Economics and Statistics 63, 517–533.

Aradillas-Lopez, B.E. Honoré and J.L. Powell, 2007, Pairwise difference estimation with nonparametric control variables, IER 48, 1119–1158.

Arcone, M. and E. Giné, 1992, On the bootstrap of M-estimators and other statistical functional, in Exploring the Limits of Bootstrap, edited by R. LePage and L. Billard, Wiley.

Asparouhova, E., R. Golanski, K. Kasprzyk, and R.P. Sherman, 2002, Rank estimators for a transformation model, ET 18, 1099–1120.

Bai, J., 2003, Testing parametric conditional distributions of dynamic models. REStat 85, 531–549.

Baltagi, B.H., 2005, Econometric analysis of panel data, 3rd ed., Wiley.

Banerjee, A.N., 2007, A method of estimating the average derivative, JOE 136, 65–88.

Bartle, R.G., 1976, The elements of real analysis, 2nd ed., Wiley.

Ben-Akiva, M.E., 1974, Structure of passenger travel demand models, Transportation Research Record 526, 26–42.

Bera, A.K. and Y. Bilias, 2001, Rao's score, Neyman's $C(\alpha)$ and Silvey's LM test: an essay on historical developments and some new results, Journal of Statistical Planning and Inference 97, 9–44.

Bera, A.K. and Y. Bilias, 2002, The MM, ME, ML, EL, EF and GMM approaches to estimation: a synthesis, JOE 107, 51–86.

Bera, A.K. and A. Ghosh, 2001, Neyman's smooth test and its applications in econometrics, in Handbook of Applied Econometrics and Statistical Inference, edited by A. Ullah, A. Wan and A. Chaturvedi, Chapter 10, 170–230, Marcel Dekker.

Bera, A.K. and G. Premaratne, 2001, General hypothesis testing, in A Companion to Theoretical Econometrics, edited by B.H. Baltagi, Blackwell.

Beran, R., 1988, Prepivoting test statistics: a bootstrap view of asymptotic refinements, JASA 83, 687–697.

Beran, R. and G.R. Ducharme, 1991, Asymptotic theory for bootstrap methods in statistics, Centre de Recherches Mathématiques, Université de Montréal.

Bianchi, M., 1997, Testing for convergence: evidence from nonparametric multimodality tests, JAE 12, 393–409.

Bickel, P.J. and K.A. Doksum, 1981, An analysis of transformations revisited, JASA 76, 296–311.

Bickel, P.J. and D.A. Freedman, 1982, Some asymptotic theory for the bootstrap, AS 9, 1196–1217.

Bickel, P.J. and M. Rosenblatt, 1973, On some global measures of the deviations of density function estimates, AS 1, 1071–1095.

Bickel, P.J., C.A.J. Klaassen, Y. Ritov and J.A. Wellner, 1993, Efficient and adaptive estimation for semiparametric models, Johns Hopkins University Press.

Bierens, H.J., 1982, Consistent model specification test, JOE 20, 105–134.

Bierens, H.J., 1987, Kernel estimators of regression function, in Advances in Econometrics, Cambridge University Press.

Bierens, H.J., 1990, A consistent conditional moment test of functional form, ECA 58, 1443–1458.

Bierens, H.J., 1994, Comments on "Artificial Neural Networks: An Econometric Perspective," ER 13, 93–97.

Bierens, H.J. and J.R. Carvalho, 2007, Semi-nonparametric competing risks analysis of recidivism, JAE 22, 971–993.

Bierens, H.J. and W. Ploberger, 1997, Asymptotic theory of integrated conditional moment tests, ECA 65, 1129–1151.

Bierens, H.J. and H.A. Pott-Buter, 1990, Specification of household Engel curves by nonparametric regression, ER 9, 123–184.

Bilias, Y., S. Chen, and Z. Ying, 2000, Simple resampling methods for censored regression quantiles, JOE 99, 373–386.

Billingsley, P., 1995, Probability and measure, 3rd ed., John Wiley & Sons.

Billingsley, P., 1999, Convergence of probability measures, 2nd ed., John Wiley & Sons.

Binkley, J.K. and C.H. Nelson, 1988, A note on the efficiency of seemingly unrelated regression, American Statistician 42, 137–139.

Blundell, R.W. and J.L. Powell, 2003, Endogeneity in nonparametric and semiparametric regression models, Advances in Economics and Econometrics: Theory and Applications, Eighth World Congress, Vol. II, edited by M. Dewatripont, L.P. Hansen, and S.J. Turnovsky, Cambridge University Press.

Blundell, R.W. and J.L. Powell, 2007, Censored regression quantiles with endogenous regressors, JOE 141, 65–83.

Blundell, R.W. and R.J. Smith, 1993, Simultaneous microeconometric models with censored or qualitative dependent variables, in Handbook of Statistics 11, edited by G.S. Maddala, C.R. Rao and H.D. Vinod, North-Holland.

Bourguignon, F., M. Fournier, and M. Gurgand, 2007, Selection bias corrections based on the multinomial logit model: Monte Carlo comparison, Journal of Economic Surveys 21, 174–205.

Bowden, R.J., 1973, The theory of parametric identification, ECA 41, 1068–1074.

Bowman, A.W., 1984, An alternative method of cross-validation for the smoothing of density estimates, Biometrika 71, 353–360.

Box, G.E.P. and D.R. Cox, 1964, An analysis of transformation, JRSS-B 26, 211–246.

Breiman, L. and J.H. Friedman, 1985, Estimating optimal transformations for multiple regression and correlation, JASA 80, 580–597.

Breslow, N.E., 1974, Covariance analysis of censored survival data, Biometrics 30, 89–99.

Brown, B.W. and W.K. Newey, 1998, Efficient semiparametric estimation of expectations, ECA 66, 453–464.

Brown, B.W. and W.K. Newey, 2002, Generalized method of momnets, efficient bootstrapping, and improved inference, JBES 20, 507–517.

Brownstone, D. and K.A. Small, 1989, Efficient estimation of nested logit models, JBES 7, 67–74.

Buchinsky, M., 1998, Recent advances in quantile regression models—a practical guideline for empirical research, Journal of Human Resources 33, 88–126.

Buchinsky, M. and J. Hahn, 1998, An alternative estimator for the censored quantile regression model, ECA 66, 653–671.

Buja, A., T. Hastie and R. Tibshirani, 1989, Linear smoothers and additive models, AS 17, 453–510.

Cai, J., J. Fan, J. Jiang, and H. Zhou, 2007, Partially linear hazard regression for multivariate survival data, JASA 102, 538–551.

Cameron, A.C. and P.K. Trivedi, 1998, Regression analysis of count data, Cambridge University Press.

Cameron, A.C. and P.K. Trivedi, 2005, Microeconometrics, Cambridge University Press.

Cameron, A.C. and P.K. Trivedi, 2009, Microeconometrics using Stata, STATA Press.

Carroll, R.J., J. Fan, I. Gijbels, M.P. Wand, 1997, Generalized partially linear single-index models, JASA 92, 477–489.

Carroll, R.J. and D. Ruppert, 1988, Transformation and Weighting in Regression, Chapman and Hall.

Carrasco, J.A. and J.D. Ortúzar, 2002, Review and assessment of the nested logit model, Transport Reviews 22, 197–218.

Carson, R.T., R.C. Mitchell, M. Hanemann, R.J. Kopp, S. Presser, and P.A. Ruud, 2003, Contingent valuation and lost passive use: damages from the Exxon Valdez oil spill, Environmental and Resource Economics 25, 257–286

Carson, R.T. and W.M. Hanemann, 2005, Contingent valuation, in Handbook of Environmental Economics 2. edited by K.G. Mäler and J.R. Vincent, North-Holland.

Cavanagh, C. and R.P. Sherman, 1998, Rank estimators for monotone index models, JOE 84, 351–381.

Chamberlain, G., 1980, Analysis of covariance with qualitative data, REStud 47, 225–238.

Chamberlain, G., 1982, Multivariate regression models for panel data, JOE 18, 5–46.

Chamberlain, G., 1986, Asymptotic efficiency in semiparametric models with censoring, JOE 32, 189–218.

Chamberlain, G., 1987, Asymptotic efficiency in estimation with conditional moment restrictions, JOE 34, 305–334.

Chamberlain, G., 1992, Efficiency bounds for semiparametric regression, ECA 60, 567–596.

Chay, K.Y. and B.E. Honoré, 1998, Estimation of semiparametric censored regression models: an application to changes in black-white earnings inequality during the 1960s, Journal of Human Resources 33, 4–38.

Chen, L.A., A.H. Welsh, and W. Chan, 2001, Estimators for the linear regression model based on winsorized observations, Statistica Sinica 11, 147–172.

Chen, S., 2002, Rank estimation of transformation models, ECA 70, 1683–1697.

Chernozhukov, V., 2005, Extremal quantile regression, AS 33, 806–839.

Chernozhukov, V. and C. Hansen, 2006, Instrumental quantile regression inference for structural and treatment effect models. JOE 132, 491–525.

Chernozhukov, V. and C. Hansen, 2008, Instrumental variable quantile regression: a robust inference approach, JOE 142, 379–398.

Chernozhukov, V. and H. Hong, 2002, Three-step censored quantile regression and extramarital affaris, JASA 97, 872–882.

Chow, Y.S., and H. Teicher, 1988, Probability theory, 2nd ed., Springer-Verlag.

Claeskens, G. and N.L. Hjort, 2008, Model selection and model averaging, Cambrige University Press.

Clarke, K.A., 2007, A simple distribution-free test for nonnested model selection, Political Analysis 15, 347–363.

Cleveland, W.S., S.J. Devlin, and E. Grosse, 1988, Regression by local fitting, JOE 37, 87–114.

Clinch, J.P. and A. Murphy, 2001, Modelling winners and losers in contingent valuation of public goods: appropriate wlefare measures and econometric analysis, Economic Journal 111, 420–443.

Cosslett, S.R., 1987, Efficiency bounds for distribution-free estimators of the binary choice and the censored regression models, ECA 55, 559–585.

Cox, D.R., 1962, Further results on tests of separate families of hypotheses, JRSS-B 24, 406–424.

Cox, D.R., 1972, Regression models and life tables, JRSS-B 34, 187–220.

Cox, D.R. and D. Oakes, 1984, Analysis of survival data, Chapman & Hall.

Cragg, J., 1971, Some statistical models for limited dependent variables with application to the demand for durable goods, ECA 39, 829–844.

Cressie, N. and T.R.C. Read, 1984, Multinomial goodness-of-fit tests, JRSS-B 46, 440–464.

Crowder, M., 2001, Classical competing risks, Chapman & Hall/ CRC.

Dagenais, D.G. and J.M. Dufour, 1991, Invariance, nonlinear models and asymptotic tests, ECA 59, 1601–1615.

Daly, A.J. and S. Zachary, 1978, Improved multiple choice models, in Hensher, D.A., Dalvi, M.Q. (eds.), Determinants of Travel Choice, Saxon House, Sussex.

DasGupta, A., 2008, Asymptotic theory of statistics and probability, Springer.

Davidson, R. and J.G. Mackinnon, 1981, Several model specification tests in the presence of alternative hypotheses, ECA 49, 781–793.

Davidson, R. and J.G. Mackinnon, 1992, A new form of the information matrix test, ECA 60, 145–157.

Davidson, R. and J.G. Mackinnon, 1993, Estimation and inference in econometrics, Oxford University Press.

Davison, A.C. and D.V. Hinkley, 1997, Bootstrap methods and their application, 1997, Cambridge University Press.

Deaton, A., 1995, Data and econometric tools for development analysis, in Handbook of Development Economics III, eds., J. Behrman and T.N. Srinivasan, North-Holland.

De Jong, P., 1987, A central limit theorem for generalized quadratic forms, Probability Theory and Related Fields 75, 261–277.

De Jong, R.M. and H.J. Bierens, 1994, On the limit behavior of a chi-square type test if the number of conditional moments tested approaches infinity, ET 9, 70–90.

Del Barrio, E., 2007, Empirical and qunatile processes in the asymptotic theory of goodness-of-fit tests, in Lecture on Empirical Processes, pp. 1–99, EMS Series of Lectures in Mathematics, European Mathematical Society.

Delecroix, M., M. Hristache, V. Patilea, 2006, On semiparametric M-estimation in single-index regression, Journal of Statistical Planning and Inference 136, 730–769.

Delgado, M.A., 1993, Testing the equality of nonparametric regression curves, Statistics and Probability Letters 17, 199–204.

Delgado, M.A. and T. Stengos, 1994, Semiparametric specification testing of non-nested econometric models, REStud 61, 291–303.

Delgado, M.A. and W.G. Manteiga, 2001, Significance testing in nonparametric regression based on the bootstrap, AS 29, 1469–1507.

Dette, H., 1999, A consistent test for the functional form of a regression based on a difference of variance estimators, AS 27, 1012–1050.

Dette, H. and S. Volgushev, 2008, Non-crossing non-parametric estimates of quantile curves, JRSS-B 70, 609–627.

Dharmadhikari, S. and K. Joag-dev, 1988, Unimodality, convexity, and applications, Academic Press.

DiCiccio, T.J., and B. Efron, 1996, Bootstrap confidence intervals, Statistical Science 11, 189–228.

Dominguez, M.A. and I.N. Lobato, 2004, Consistent estimation of models defined by conditional moment restrictions, ECA 72, 1601–1615.

Donald, S.G., G.W. Imbens, and W.K. Newey, 2003, Empirical likelihood estimation and consistent tests with conditional moment restrictions, JOE 117, 55–93.

Donald, S.G. and W.K. Newey, 2001, Choosing the number of instruments, ECA 69, 1161–1191.

Donker, B. and M. Schafgans, 2008, Specification and estimation of semiparametric multiple-index models, ET 24, 1584–1606.

Donsker, M.D., 1952, Justification and extension of Doob's heuristic approach to the Kolmogorov-Smirnov theorems, Annals of Mathematical Statistics 23, 277–281.

Dubin, J.A. and D. Rivers, 1989, Selection bias in linear regression, Sociological Methods and Research 18, 360–390.

Dubin, J.A. and D. Rivers, 1993, Experimental estimates of the impact of wage subsidies, JOE 56, 219–242.

Dudley, R.M., 1989, Real Analysis and Probability, Chapman and Hall.

Durbin, J., 1973, Weak convergence of the sample distribution function when parameters are estimated, AS 1, 279–290.

Efron, B., 1977, The efficiency of Cox's likelihood function for censored data, JASA 72, 557–565.

Efron, B., 1991, Regression percentiles using asymmetric squared error loss, Statistica Sinica 1, 93–125.

Efron, B., 2003, Second thoughts on the bootstrap, Statistical Science 18, 135–140.

Efron, B. and R.J. Tibshirani, 1993, An introduction to the bootstrap, Chapman and Hall.

Einmahl, U. and D.M. Mason, 2005, Uniform in bandwidth consistency of kernel-type function estimators, AS 33, 1380–1403.

Elbers, C. and G. Ridder, 1982, True and spurious duration dependence: The identifiability of the proportional hazard model, REStud 49, 402–411.

Ellison, G. and S.F. Ellison, 2000, A simple framework for nonparametric specification testing, JOE 96, 1–23.

Engle, R.F., 1984, Wald, likelihood ratio and Lagrange multiplier tests in econometrics, in Handbook of Econometrics II, North-Holland.

Englin, J. and J.S. Shonkwiler, 1995, Estimating social welfare using count data models: an application to long-run recreation demand under conditions of endogenous stratification and truncation, REStat 77, 104–112.

Escanciano, J.C., 2006, A consistent diagnostic test for regression models using projections, ET 22, 1030–1051.

Eubank, R., 1999, Nonparametric regression and spling smoothing, 2nd ed., Marcel Dekker.

Evans, W.N. and R.M. Schwab, 1995, Finishing high school and starting college: do Catholic schools make a difference?, QJE 110, 941–974.

Fan, J., 1992, Design-adaptive nonparametric regression, JASA 87, 998–1004.

Fan, J. and I. Gijbels, 1996, Local polynomial modelling and its applications, Chapman & Hall.

Fan, J., W. Härdle and E. Mammen, 1998, Direct estimation of low dimensional components in additive models, AS 26, 943–971.

Fan, J. and J. Jiang, 2007, Nonparametric inference with generalized likelihood ratio tests, Test 16, 409–444.

Fan, J., C. Zhang, and J. Zhang, 2001, Generalized likelihood ration statistics and Wilks phenomenon, AS 29, 153–193.

Fan, Y. and Q. Li, 1996, Consistent model specification tests: omitted variables and semiparametric functional forms, ECA 64, 865–890.

Fan, Y. and Q. Li, 2000, Consistent model specification tests, ET 2000, 1016–1041.

Fermanian, J.D., 2003, Nonparametric estimation of competing risks models with covariates, Journal of Multivariate Analysis 85, 156–191.

Fiebig, D.G., 2001, Seemingly unrelated regression, in A Companion to Theoretical Econometrics, edited by B.H. Baltagi, Blackwell.

Fitzenberger, B., 1997, A guide to censored quantile regressions, in Handbook of Statistics 15, edited by G.S. Maddala and C.R. Rao, North-Holland.

Florens, J.P., 2003, Inverse problems and structural econometrics, Advances in Economics and Econometrics: Theory and Applications, Eighth World Congress, Vol. II, edited by M. Dewatripont, L.P. Hansen, and S.J. Turnovsky, Cambridge University Press.

Freedman, D.A., 1985, The mean versus the median: A case study in 4-R Act litigation, JBES 3, 1–13.

Freeman, R.B., 1999, The economics of crime, in Handbook of Labor Economics 3C, edited by O.C. Ashenfelter and D. Card, North-Holland.

Friedman, J. and W. Stuetzle, 1981, Projection pursuit regression, JASA 76, 817–823.

Fry, T.R.L. and M.N. Harris, 1996, A Monte Carlo study of tests for the independence of irrelevant alternatives property, Transportation Research B 30, 19–30.

Gallant, A.R. and D. Nychka, 1987, Semi-nonparametric maximum likelihood estimation, ECA 55(2), 363–390.

Gehan, E., 1965, A generalized Wilcoxon test for comparing arbitrarily singly censored samples, Biometrika 52, 203–223.

Giné, E., V. Koltchinskii, and J. Zinn, 2004, Weighted uniform consistency of kernel density estimators, Annals of Probability 32, 2570–2605.

Giné, E. and J. Zinn, 1990, Bootstrapping general empirical measures, Annals of Probability 18, 851–869.

Godambe, V.P., 1960, An optimum property of regular maximum likelihood estimation, Annals of Mathematical Statistics 31, 1208–1212.

Godfrey, L.G., 2007, On the asymptotic validity of a bootstrap method for testing nonnested hypotheses, Economics Letters 94, 408–413.

Godfrey, L.G. and J.M.C. Santos-Silva, 2004, Bootstrap tests of nonnested hypotheses: some further results, ER 23, 325–340.

Gørgens, T., 2002, Nonparametric comparison of regression curves by local linear fitting, Statistics & Probability Letters 60, 81–89.

Gørgens, T., 2004, Average derivatives for hazard functions, ET 20, 2004, 437–463.

Gørgens, T., 2006, Semiparametric estimation of single-index hazard functions without proportional hazards, Econometrics Journal 9, pp. 1–22.

Gørgens, T. and J.L. Horowitz, 1999, Semiparametric estimation of a censored regression model with an unknown transformation of the dependent variable, JOE 90, 155–191.

Gould, E.D., B.A. Weinberg, D.B. Mustard, 2002, Crime rates and local labor market opportunities in the United States: 1979–1997, REStat 84, 45–61.

Gourieroux, C. and J. Jasiak, 2007, The econometrics of individual risk: credit, insurance, and marketing, Princeton University Press.

Gourieroux, C., J.J. Laffont, and A. Monfort, 1980, Coherency conditions in simultaneous linear equation models with endogenous switching regimes, ECA 48, 675–695.

Gourieroux, C. and A. Monfort, 1994, Testing non-nested hypotheses, in Handbook of Econometrics IV, edited by R.F. Engle and D.L. McFadden, North-Holland.

Gourieroux, C. and A. Monfort, 1996, Simulation-based econometric methods, Oxford University Press.

Gourieroux, C., A. Monfort and E. Renault, 1993, Indirect inference, JAE 8, S85–S118.

Gozalo, P.L., 1993, A consistent model specification tests for nonparametric estimation of regression function models, ET 9, 451–477.

Greene, W., 1999, Marginal effects in the censored regression model, Economics Letters 64, 43–49.

Greene, W., 2007, Econometric analysis, 6th ed., Prentice Hall.

Gregory, A.W. and M.R. Veall, 1985, Formulating Wald tests of nonlinear restrictions, ECA 53, 1465–1468.

Haab, T.C. and K.E. McConnell, 2002, Valuing environmental and natural resources: the econometrics of non-market valuation, Edward Elgar Publishing Ltd.

Hahn, J., 1995, Bootstrap quantile regression estimators, ET 11, 105–121.

Hajivassiliou, V.A. and D.L. McFadden, 1998, The method of simulated scores for the estimation of LDV models, ECA 66, 863–896.

Hajivassiliou, V.A., D. McFadden, P. Ruud, 1996, Simulation of multivariate normal rectangle probabilities and their derivatives: theoretical and computational results, JOE 72, 85–134.

Hajivassiliou, V.A. and P.A. Ruud, 1994, Classical estimation methods for LDV models using simulation, in Handbook of Econometrics IV, North-Holland.

Hall, A.R., 2000, Covariance matrix estimation and the power of the overidentifying restrictions test, ECA 68, 1517–1527.

Hall, A.R., 2005, Generalized method of moments, Oxford University Press.

Hall, A.R. and F.P.M. Peixe, 2003, A consistent method for the selection of relevant instruments, ER 22, 269–287.

Hall, P., 1983, Large sample optimality of least squares cross validation in density estimation, AS 11, 1156–1174.

Hall, P., 1984, Central limit theorem for integrated square error of multivariate nonparametric density estimators, Journal of Multivariate Analysis 14, 1–16.

Hall, P., 1992, The bootstrap and Edgeworth expansion, Springer.

Hall, P. and J.D. Hart, 1990, Bootstrap test for difference between means in nonparametric regression, JASA 85, 1039–1049.

Hall, P. and J.L. Horowitz, 1990, Bandwidth selection in semiparametric estimation of censored linear regression models, ET 6, 123–150.

Hall, P. and J.L. Horowitz, 1996, Bootstrap critical values for tests based on generalized-method-of-moments estimators, ECA 64, 891–916.

Hall, P. and J.L. Horowitz, 2005, Nonparametric methods for inference in the presence of instrumental variables, AS 33, 2904–2929.

Hall, P. and M. York, 2001, On the calibration of Silverman's test for multimodality, Statistica Sinica 11, 515–536.

Han, A., 1987a, Nonparametric analysis of a generalized regression model, JOE 35, 303–316.

Han, A., 1987b, A nonparametric analysis of transformations, JOE 35, 191–209.

Han, A. and J.A. Hausman, 1990, Flexible parametric estimation of duration and competing risk models, JAE 5, 1–28.

Hansen, L.P., 1982, Large sample properties of generalized method of moments estimators, ECA 50, 1029–1054.

Hansen, L.P., J. Heaton, and A. Yaron, 1996, Finite sample properties of some alternative GMM estimators, JBES 14, 262–280.

Härdle, W., 1990, Applied Nonparametric Regression, Cambridge University Press.

Härdle, W., P. Hall, H. Ichimura, 1993, Optimal smoothing in single-index models, AS 21, 157–178.

Härdle, W., P. Hall, and J.S. Marron, 1988, How far are automatically chosen regression smoothing parameters from their optimum? JASA 83, 86–101.

Härdle, W., W. Hildenbrand, and M. Jerison, 1991, Empirical evidence on the law of demand, ECA 59, 1525–1549.

Härdle, W., H. Liang, and J. Gao, 2000, Partially linear models, Springer.

Härdle, W. and O. Linton, 1994, Applied nonparametric methods, in Handbook of Econometrics IV, edited by R.F. Engle and D.L. McFadden, Elsevier Science.

Härdle, W. and E. Mammen, 1993, Comparing nonparametric versus parametric regression fits, AS 21, 1926–1947.

Härdle, W. and J.S. Marron, 1985, Optimal bandwidth selection in nonparametric regression function estimation, AS 13, 1465–1481.

Härdle, W. and J.S. Marron, 1991, Bootstrap simultaneous error bars for nonparametric regression, AS 19, 778–796.

Härdle, W., M. Müller, S. Sperlich, and A. Werwatz, 2004, Nonparametric and semiparametric models, Springer.

Härdle, W. and T.M. Stoker, 1989, Investigating smooth multiple regression by the method of average derivatives, JASA 84, 986–995.

Härdle, W. and A.B. Tsybakov, 1993, How sensitive are average derivatives? JOE 58, 31–48.

Hastie, T.J. and C. Loader, 1993, Local regression: Automatic kernel carpentry, Statistical Science 8, 120–143.

Hausman, J.A., 1978, Specification tests in econometrics, ECA 46, 1251–1272.

Hausman, J.A. and W.K. Newey, 1995, Nonparametric estimation of exact consumers surplus and deadweight loss, ECA 63, 1445–1476.

Hausman, J.A. and D.A. Wise, 1977, Social experimentation, truncated distributions, and efficient estimation, ECA 45, 919–938.

Hausman, J.A. and D.A. Wise, 1978, A conditional probit model for qualitative choice: discrete decisions recognizing interdependence and heterogenous preferences, ECA 46, 403–426.

He, X. and L.X. Zhu, 2003, A Lack-of-fit test for quantile regression, JASA 98, 1013–1022.

Heckman, J.J., 1978, Dummy endogenous variables in a simultaneous equation system, ECA 46, 931–959.

Heckman, J.J., 1979, Sample selection bias as a specification error, ECA 47, 153–161.

Heckman, J.J. and B. Honoré, 1989, The identifiability of the competing risks model, Biometrika 76, 325–330.

Heckman, J.J. and B. Singer, 1984, The Identifiability of the proportional hazard model, REStud 51, 231–241.

Hensher, D.A. , J.M. Rose, and W.H. Green, 2005, Applied choice analysis: a primer, Cambridge University Press.

Hensher, D.A. and W.H. Green, 2003, The mixed logit model: the state of practice, Transportation 30, 135–176.

Hess, K.R., D.M. Serachitopol, and B.W. Brown, 1999, Hazard function estimators: a simulation study, Statistics in Medicine 18, 3075–3088.

Himmelblau, D.M., 1972, Applied nonlinear programming, McGraw-Hill.

Hitomi, K., Y. Nishiyama and R. Okui, 2008, A puzzling phenomenon in semiparametric estimation problems with infinite-dimensional nuisance parameters, ET 24, 1717–1728.

Hochguertel, S., 2003, Precautionary motives and portfolio decisions, JAE 18, 61–77.

Hoffman, S.D. and G.J. Duncan, 1988, A comparison of choice-based multinomial and nested logit models: the family structure and welfare use decisions of divorced or separated women, Journal of Human Resources 23, 550–562.

Hong, Y. and H. White, 1995, Consistent specification testing via nonparametric series regression, ECA 63, 1133–1159.

Hogg, R.V., J.W. McKean, and A.T. Craig, 2005, Introduction to mathematical statistics, 6th ed., Pearson/Prentice-Hall.

Honoré, B., 1992, Trimmed LAD and LSE of truncated and censored regression models with fixed effects, ECA 60, 533–565.

Honoré, B., S. Khan, and J.L. Powell, 2002, Quantile regression under random censoring, JOE 109, 67–105.

Honoré, B., E. Kyriazidou, and C. Udry, 1997, Estimation of type 3 tobit models using symmetric trimming and pairwise comparisons, JOE 76, 107–128.

Honoré, B. and E. Kyriazidou, 2000, Estimation of tobit-type models with individual specific effects, ER 19, 341–366.

Honoré, B. and A. Lewbel, 2002, Semiparametric binary choice panel data models without strictly exogenous regressors, ECA 70, 2053–2063.

Honoré, B. and A. Lleras-Muney, 2006, Bounds in competing risks models and the war on cancer, ECA 74, 1675–1698.

Honoré, B. and J.L. Powell, 1994, Pairwise difference estimators of censored and truncated regression models, JOE 64, 241–278.

Honoré, B. and J.L. Powell, 2005, Pairwise difference estimators for nonlinear models, in Identification and Inference for Econometric Models, pp. 520–553, edited by D.W.K. Andrews and J.H. Stock, Cambridge University Press.

Horowitz, J.L., 1992, A smoothed maximum score estimator for the binary response model, ECA 60, 505–531.

Horowitz, J.L., 1993, Semiparametric and nonparametric estimation of quantal response models, in Handbook of Statistics 11, edited by G.S. Maddala, C.R. Rao, and H.D. Vinod, Elsevier Science Publishers.

Horowitz, J.L., 1994, Bootstrap-based critical values for the information matrix test, JOE 61, 395–411.

Horowitz, J.L., 1996, Semiparametric estimation of a regression model with an unknown transformation of the dependent variable, ECA 64, 103–137.

Horowitz, J.L., 1998, Semiparametric methods in econometrics, Springer-Verlag.

Horowitz, J.L., 2001a, The bootstrap, Chapter 52, in Handbook of Econometrics V, edited by J.J. Heckman and E. Leamer, North-Holland.

Horowtiz, J.L., 2001b, Nonparametric estimation of a generalized additive model with an unknown link function, ECA 69, 499–513.

Horowitz, J.L., 2002, Bootstrap critical values for tests based on the smoothed maximum score estimator, JOE 111, 141–167.

Horowitz, J.L., 2007, Asymptotic normality of a nonparametric instrumental variables estimator, IER 48, 1329–1349.

Horowitz, J.L. and W. Härdle, 1994, Testing a parametric model against a semiparametric alternative, ET 10, 821–848.

Horowitz, J.L. and W. Härdle, 1996, Direct semiparametric estimation of a single index model with discrete covariates, JASA 91, 1632–1640.

Horowitz, J.L. and S. Lee, 2005, Nonparametric estimation of an additive quantile regression model, JASA 100, 1238–1249.

Horowitz, J.L. and S. Lee, 2007, Nonparametric instrumental variables estimation of a quantile regression model, ECA 75, 1191–1208.

Horowitz, J.L. and C.F. Manski, 2000, Nonparametric analysis of randomized experiments with missing covariate and outcome data, JASA 95, 77–84.

Horowitz, J.L. and E. Mammen, 2004, Nonparametric estimation of an additive model with a link function, AS 32, 2412–2443.

Horowitz, J.L. and G.R. Neumann, 1987, Semiparametric estimation of employment duration models, ER 6, 5–40.

Horowitz, J.L. and G.R. Neumann, 1989, Specification testing in censored models: Parametric and semiparametric methods, JAE 4, S61–S86.

Horowitz, J.L. and V.G. Spokoiny, 2001, An adaptive, rate-optimal test of a parametric mean-regression model against a nonparametric alternative, ECA 69, 599–631.

Hougaard, P., 2000, Analysis of multivariate survival data, Springer

Hsiao, C., 2003, Analysis of panel data, 2nd ed., Cambridge University Press.

Huber, P.J., 1981, Robust Statistics, Wiley.

Iannizzotto, M. and M.P. Taylor, 1999, The target zone model, nonlinearity and mean reversion: is the honeymoon really over?, Economic Journal 109, C96–C110.

Ichimura, H., 1993, Semiparametric least squares (SLS) and weighted SLS estimation of single index models, JOE 58, 71–120.

Ichimura, H. and L.F. Lee, 1991, Semiparametric least squares estimation of multiple index models, in Nonparametric and Semiparametric Methods in Econometrics and Statistics, edited by W.A. Barnett, J. Powell, and G. Tauchen, Cambridge University Press.

Ida, T. and T. Kuroda, 2009, Discrete choice model analysis of mobile telephone service demand in Japan, Empirical Econ 36, 65–80.

Imbens, G.W., and T. Lancaster, 1996, Efficient estimation and stratified sampling, JOE 74, 289–318.

Imbens, G.W., R.H. Spady, and P. Johnson, 1998, Information theoretic approaches to inference in moment condition models, ECA 66, 333–357.

Izenman, A.J., 1991, Recent developments in nonparametric density estimation, JASA 86, 205–224.

Jagannathan, R., G. Skoulakis, and Z. Wang, 2002, Generalized method of moments: applications in finance, JBES 20, 470–481.

Jennen-Steinmetz, C. and T. Gasser, 1988, A unifying approach to nonparametric regression estimation, JASA 83, 1084–1089.

Johnson, N.L., S. Kotz and N. Balakrishnan, 1995, Continuous univariate distributions Volume 2, 2nd ed., Wiley.

Jones, M.C., J.S. Marron, and S.J. Sheather, 1996, A brief survey of bandwidth selection for density estimation, JASA 91, 401–407.

Kan, K. and M.J. Lee, 2009, Lose weight for money only if over-weight: marginal integration for semi-linear panel models, unpublished paper.

Kan, K. and W.D. Tsai, 2004, Obesity and risk knowledge, Journal of Health Economics 23, 907–934.

Kang, S.J. and M.J. Lee, 2003, Analysis of private transfers with panel fixed effect censored model estimator, Economics Letters 80, 233–237.

Kang, S.J. and M.J. Lee, 2005, Q-Convergence with interquartile ranges, Journal of Economic Dynamics and Control 29, 1785–1806

Kang, C.H. and M.J. Lee, 2009, Performance of various estimators for censored response models with endogenous regressors, Pacific Economic Review, forthcoming.

Kaplan, E.L. and P. Meier, 1958, Nonparametric estimation from incomplete observations, JASA 53, 457–481.

Karlsson, M., 2004, Finite sample properties of the QME. Communications in Statistics (Simulation and Computation) 33, 567–583.

Karlsson, M., 2006, Estimators of regression parameters for truncated and censored data, Metrika 63, 329–341.

Karlsson, M. and T. Laitila, 2008, A semiparametric regression estimator under left truncation and right censoring, Statistics and Probability Letters 78, 2567–2571.

Kauermann G. and R.J. Carroll, 2001, A note on the efficiency of sandwich covariance matrix estimation, JASA 96, 1387–1396.

Keane, M.P., 1992, A note on identification in the multinomial probit model, JBES 10, 193–200.

Keane, M.P. and R. Sauer, 2009, Classification error in dynamic discrete choice models: implications for female labor supply behavior, ECA 77, 975–991.

Kelly, M., 2000, Inequality and crime, REStat 82, 530–539.

Kempa, B. and M. Nelles, 1999, The theory of exchange rate target zones, Journal of Economic Surveys 13, 173–210.

Khan, S. and A. Lewbel, 2007, Weighted and two stage least squares estimation of semiparametric truncated regression models, ET 23, 309–347.

Khan, S. and J.L. Powell, 2001, Two-step estimation of semiparametic censored regression models, JOE 103, 73–110.

Khan, S. and E. Tamer, 2007, Partial rank estimation of duration models with general forms of censoring, JOE 136, 251–280.

Khmaladze, E.V., 1981, Martingale approach in the theory of goodness-of-fit test, Theory of Probability and Its Applications 26, 240–257.

Khmaladze, E.V., 1993, Goodness of fit problem and scanning innovation martingales, AS 21, 798–829.

Khmaladze, E.V. and H. Koul, 2004, Martingale transforms goodness-of-fit tests in regression models, AS 32, 995–1034.

Kim, J.K. and D. Pollard, 1990, Cube-root asymptotics, AS 18, 191–219.

Kim, T.H. and C. Muller, 2004, Two-stage quantile regression when the first stage is based on quantile regression, ECMJ 7, 218–231.

Kim, W., O. Linton, N.W. Hengartner, 1999, A computationally efficient oracle estimator for additive nonparametric regression with bootstrap confidence intervals, Journal of Computational and Graphical Statistics 8, 278–297.

Kimhi, Ayal and M.J. Lee, 1996, Joint farm and off-farm work decisions of farm couples: estimating structural simultaneous equations with ordered categorical dependent variables, American Journal of Agricultural Economics 78, 687–698.

Kitamura, Y., 2007, Empirical likelihood methods in econometrics: theory and practice, in Advances in Economics and Econometrics: Theory and Applications, 9th World Congress, vol. 3, edited by R. Blundell, W.K. Newey, and T. Persson, Cambridge University Press.

Kitamura, Y. and M. Stutzer, 1997, An information-theoretic alternative to generalized method of moment estimation, ECA 65, 861–874.

Kitamura, Y., G. Tripathi and H.T. Ahn, 2004, Empirical likelihood-based inference in conditional moment restriction models, ECA 72, 1667–1714.

Klein, J.P. and M.L. Moeschberger, 2003, Survival analysis, 2nd ed., Springer.

Klein, R.W. and R.H. Spady, 1993, An efficient semiparametric estimator for binary response models, ECA 61, 387–421.

Koenker, R., 2005, Quantile regression, Cambridge University Press.

Koenker, R. and G. Bassett, 1978, Regression quantiles, ECA 46, 33–50.

Koenker, R. and O. Geling, 2001, Reappraising medfly longevity: a quantile regression survival analysis, JASA 96, 458–468.

Koenker, R. and K.F. Hallock, 2001, Quantile regression, Journal of Economic Perspectives 15 (4), 143–156.

Koenker, R. and J.A.F. Machado, 1999, Goodness of fit and related inference processes for quantile regression, JASA 94, 1296–1310.

Koenker, R. and Z. Xiao, 2002, Inference on the quantile regression process, ECA 70, 1583–1612.

Kondo, Y. and M.J. Lee, 2003, Hedonic price index estimation under mean independence of time dummies from quality-characteristics, ECMJ 6, 28–45.

Kordas, G., 2006, Smoothed binary regression quantiles, JAE 21, 387–407.

Koul, H.L., 2006, Model diagnostics via martingale transforms: a briew review, in Frontiers in Statistics, pp. 183–206, edited by J. Fan and H.L. Koul, Imperial College Press.

Koul, H.L. and L. Sakhanenko, 2005, Goodness-of-fit testing in regression: a finite sample comparison of bootstrap methodology and Khmaladze tranformation, Statistics and Probability Letters 74, 290–302.

Krugman, P.R., 1991, Target zones and exchange rate dynamics, Quarterly Journal of Economics 106, 669–682.

Kuan, C.M. and H. White, 1994, Artificial neural networks: An econometric perspective, ER 13, 1–91.

Lafontaine, F. and K.J. White, 1986, Obtaining any Wald statistic you want, Economics Letters 21, 35–40.

Lahiri, S.N., 1992, Bootstrapping M-estimators of a multiple linear regression parameter, AS 20, 1548–1570.

Laitila, T., 1993, A pseudo-R^2 measure for limited and qualitative dependent variable models, JOE 56, 341–356.

Laitila, T., 2001, Properties of the QME under asymmetrically distributed disturbances, Statistics and Probability Letters 52, 347–352.

Lam, K.F., H. Xue, Y.B. Cheung, 2006, Semiparametric analysis of zero-inflated count data, Biometrics 62, 996–1003.

Lambert, D., 1992, Zero-inflated Poisson regression, with an application to defects in manufacturing. Technometrics 34, 1–14.

Lancaster, T., 1979, Econometric methods for the duration of unemployment, ECA 47, 939–956.

Lancaster, T., 1984, The covariance matrix of the information matrix test, ECA 52, 1051–1053.

Lancaster, T., 1992, The Econometric Analysis of Transition data, Cambridge University Press.

Lavergne, P., 2001, An equality test across nonparametric regressions, JOE 103, 307–344.

Lavergne, P. and Q. Vuong, 1996, Nonparametric selection of regressors: the nonnested case, ECA 64, 207–219.

Lavergne, P. and Q. Vuong, 2000, Nonparametric significance testing, ET 16, 576–601.

Lee, B.J., 1992, A heteroskedasticity test robust to conditional mean specification, ECA 60, 159–171.

Lee, L.F., 1992, Amemiya's generalized least squares and tests of overidentification in simultaneous equation models with qualitative or limited dependent variables, ER 11, 319–328.

Lee, L.F., 1993, Multivariate tobit models in econometrics, in Handbook of Statistics 11, edited by G.S. Maddala, C.R. Rao and H.D. Vinod, North-Holland.

Lee, L.F., 1995, Semiparametric maximum likelihood estimation of polychotomous and sequential choice models, JOE 65, 381–428.

Lee, L.F., 2001, Self-selection, in A Companion to Theoretical Econometrics, edited by B.H. Baltagi, Blackwell.

Lee, M.J., 1989, Mode regression, JOE 42, 337–349.

Lee, M.J., 1992a, Median regression for ordered discrete response, JOE 51, 59–77.

Lee, M.J., 1992b, Winsorized mean estimator for censored regression model, ET 8, 368–382.

Lee, M.J., 1993, Quadratic mode regression, JOE 57, 1–19.

Lee, M.J., 1995, A semiparametric estimation of simultaneous equations with limited dependent variables: A case study of female labor supply, JAE 10, 187–200.

Lee, M.J., 1996a, Methods-of-moments and semiparametric econometrics for limited dependent variable models, Springer-Verlag.

Lee, M.J., 1996b, Nonparametric two stage estimation of simultaneous equations with limited endogenous regressors, ET 12, 305–330.

Lee, M.J., 1997, A Limited dependent variable model under median rationality, Economics Letters 54, 221–225.

Lee, M.J., 1999, Nonparametric estimation and test for quadrant correlations in multivariate binary response models, ER 18, 387–415.

Lee, M.J., 2000, Median treatment effect in randomized trials, JRSS-B 62, 595–604.

Lee, M.J., 2002, Panel data econometrics: methods-of-moments and limited dependent variables, Academic Press.

Lee, M.J., 2003, Exclusion bias in sample-selection model estimators, Japanese Economic Review 54, 229–236.

Lee, M.J., 2004a, Efficiency Gain of System GMM and MDE over Individual Equation Estimation, Japanese Economic Review 55, 451–459.

Lee, M.J., 2004b, Selection correction and sensitivity analysis for ordered treatment effect on count response, JAE 19, 323–337.

Lee, M.J., 2005a, Micro-Econometrics for Policy, Program, and Treatment Effects, Advanced Text Series in Econometrics, Oxford University Press.

Lee, M.J., 2005b, Monotonicity conditions and inequality imputation for sample-selection and non-response problems, ER 24, 175–194.

Lee, M.J., 2008, Method-of-moment view of linear simultaneous equation systems, Statistica Neerlandica 62, 230–238.

Lee, M.J., 2009, Nonparametric tests for distributional treatment effects for censored responses, JRSS-B 71, 1–22.

Lee, M.J., and P.L. Chang, 2007, Avoiding arbitrary exclusion restrictions using ratios of reduced-form estimates, Empirical Economics 33, 339–357.

Lee, M.J., B. Donkers, and H.J. Kim, 1996, Nonparametric two-stage algorithms for some semiparametric estimators in censored and discrete response models, Presented at 1997 North American Winter Meeting of the Econometric Society (New Orleans, Louisiana).

Lee, M.J., U. Häkkinen, and G. Rosenqvist, 2007, Finding the best treatment under heavy censoring and hidden bias, JRSS-A 170, 133–147.

Lee, M.J., F. Huang, and Y.S. Kim, 2008, Asymptotic variance and extensions of a density-weighted-response semiparametric estimator, Journal of Economic Theory and Econometrics 19, 49–58, The Korean Econometric Society.

Lee, M.J. and S.J. Kang, 2009, Strategic voting and multinomial choice in US presidential elections, Journal of Economic Theory and Econometrics, forthcoming, The Korean Econometric Society.

Lee, M.J. and H.J. Kim, 1998, Semiparametric econometric estimators for a truncated regression model: a review with an extension, Statistica Neerlandica 52, 200–225.

Lee, M.J. and Y.S. Kim, 2007, Multinomial choice and nonparametric average derivatives, Transportation Research Part B: Methodological 41, 63–81.

Lee, M.J. and A. Kimhi, 2005, Simultaneous Equations in Ordered Discrete Responses with Regressor-Dependent Thresholds, ECMJ 8, 176–196.

Lee, M.J. and B. Melenberg, 1998, Bounding quantiles in sample selection models, Economics Letters 61, 29–35.

Lee, M.J. and S.J. Lee, 2005, Analysis of job-training effects on Korean women, JAE 20, 549–562.

Lee, M.J. and W.J. Li, 2005, Drift and diffusion function specification for short-term interest rates, Economics Letters 86, 339–346.

Lee, M.J. and Y.H. Tae, 2005, Analysis of labor-participation behavior of Korean women with dynamic probit and conditional logit, Oxford Bulletin of Economics and Statistics 67, 71–91.

Lee, M.J. and F. Vella, 2006, A semi-parametric estimator for censored selection models with endogeneity, JOE 130, 235–252.

Lee, S.B., 2006, Identification of a competing risks model with unknown transformations of latent failure times, Biometrika 93, 996–1002.

Lee, S.B. and M.H. Seo, 2007, Semiparametric estimation of a binary response model with a change-point due to a covariate threshold, JOE 144, 492–499.

Léger, C. and B. MacGibbon, 2006, On the boostrap in cube root asymptotics, Canadian Journal of Statistics 34, 29–44.

Lehmann, E.L., 1983, Theory of point estimation, Wiley.

Lehmann, E.L., 1999, Elements of large-sample theory, Springer.

Lehmann, E.L. and J.P. Romano, 2005, Testing statistical hypotheses, 3rd ed., Springer.

Levitt, S.D, 1997, Using electoral cycles in police hiring to estimate the effects of police on crime, AER 87, 270–290.

Levitt, S.D., 2002, Using electoral cycles in police hiring to estimate the effects of police on crime: reply, AER 92, 1244–1250.

Lewbel, A., 1998, Semiparametric latent variable model estimation with endogenous or mismeasured regressors, ECA 66, 105–121.

Lewbel, A., 2000, Semiparametric qualitative response model estimation with unknown heteroskedasticity or instrumental variables, JOE 97, 145–177.

Lewbel, A., 2007, Coherency and completeness of structural models containing a dummy endogenous variable, IER 48, 1379–1392.

Li, Q. and J.S. Racine, 2007, Nonparametric econometrics: theory and practice, Princeton University Press.

Li, Q. and T. Stengos, 2007, Non-parametric specification testing of nonnested econometric models, in The Refinement of Econometric Estimation and Test Procedures, pp. 205–219, edited by G.D.A. Phillips and E. Tzavalis, Cambridge University Presss.

Li, Q. and W. Wang, 1998, A simple consistent bootstrap test for a parametric regression function, JOE 87, 145–165.

Linton, O.B., 1997, Efficient estimation of additive nonparametric regression models, Biometrika 84, 469–473.

Linton, O.B., R. Cheng, N. Wang, W. Härdle, 1997, An analysis of transformations for additive nonparametric regression, JASA 92, 1512–1521.

Linton, O.B. and W. Härdle, 1996, Estimation of additive regression models with known links, Biometrika 83, 529–540.

Linton, O.Bl and J.B. Nielsen, 1995, A kernel method of estimating structural nonparametric regression based on marginal integration, Biometrika 82, 93–100.

Linton, O.B., S. Sperlich, and I. van Keilegom, 2008, Estimation of a semiparametric transformation model, AS 36, 686–718.

Loader, C., 1999a, Local regression and likelihood, Springer.

Loader, C., 1999b, Bandwidth selection: classical and plug-in, AS 27, 415–438.

Lohr, S.L., 1999, Sampling: design and analysis, Duxbury Press.

Luenberger, D.G., 1969, Optimization by vector space methods, Wiley.

Machado, J.A.F. and P. Parente, 2005, Bootstrap estimation of covariance matrices via the percentile method, ECMJ 8, 70–78.

Machado, J.A.F. and J.M.C. Santos Silva, 2005, Quantiles for Counts, JASA 100, 1226–1237.

MacKinnon, J.G. and L. Magee, 1990, Transforming the dependent variable in regression models, IER 31, 315–339.

MacKinnon, J.G. and H. White, 1985, Some heteroskedasticity-consistent covariance matrix estimators with improved finite sample properties, JOE 29, 305–325.

Maddala, G.S., 1983, Limited-Dependent and Qualitative Variables in Econometrics, Cambridge University Press.

Magnac, T. and E. Maurin, 2007, Identification and information in monotone binary models, JOE 139, 76–104.

Malinvaud, E., 1970, Statistical Methods of Econometrics, North Holland.

Mammen, E., O. Linton and J.P. Nielsen, 1999, The existence and asymptotic properties of a backfitting projection algorithm under weak conditions, AS 27, 1443–1490.

Manski, C.F., 1975, Maximum score estimation of the stochastic utility, model of choice, JOE 3, 205–228.

Manski, C.F., 1985, Semiparametric analysis of discrete response, JOE 27, 313–333.

Manski, C.F., 1988, Analog Estimation Methods in Econometrics, Chapman and Hall.

Manski, C.F., 1991, Regression, Journal of Economic Literature 29, 34–50.

Manski, C.F. and S.R. Lerman, 1977, The estimation of choice probabilities from choice-based samples, ECA 45, 1977–1988.

Manzan, S. and D. Zerom, 2005, Kernel estimation of a partially linear additive model, Statistics and Probability Letters 72, 313–327.

Marron, J.S., 1988, Automatic smoothing parameter selection, Empirical Economics 13, 187–208.

Marron, J.S. and W.J. Padgett, 1987, Asymptotically optimal bandwidth selection for kernel density estimators from randomly right-censored samples, AS 15, 1520–1535.

Martins-Filho, C. and K. Yang, 2007, Finite sample performance of kernel-based regression methods for non-parametric additive models under common bandwidth selection criterion, Journal of Nonparametric Statistics 19, 23–62.

Masry, E., 1996, Multivariate local polynomial regression for time series: uniform strong consistency and rates, Journal of Time Series Analysis 17, 571–599.

Maurin, E., 2002, The impact of parental income on early schooling transitions: a re-examination using data over three generations, Journal of Public Economics 85, 301–332.

Mayer, W.J. and R.E. Dorsey, 1998, Maximum score estimation of disequilibrium models and the role of anticipatory price-setting, JOE 87, 1–24.

McCrary, J., 2002, Do electoral cycles in police hiring really help us estimate the effect of police on crime? Comment, AER 92, 1236–1243.

McFadden, D., 1974, Conditional logit analysis of qualitative choice behavior, in Frontiers in Econometrics, edited by P. Zarembka, Academic Press.

McFadden, D., 1978, Modelling the choice of residential location, in Spatial Interaction Theory and Planning Models, edited by A. Karlqvist, L. Lundqvist, F. Snickars, J.W. Weibull, North-Holland.

McFadden, D., 1981, Econometric models of probabilistic choice, in Structural Analysis of Discrete Data, edited by C. Manski and D. McFadden, MIT Press.

McFadden, D., 1989, A method of simulated moments for estimation of discrete response models without numerical integration, ECA 57, 995–1026.

McFadden, D., 2001, Economic choices, American economic review 91, 351–378.

McFadden, D. and K. Train, 2000, Mixed MNL models for discrete response, JAE 5, 447–470.

McKelvey, R. and W. Zavoina, 1975, A statistical model for the analysis of ordinal level dependent variables, Journal of Mathematical Sociology 4, 103–120.

Melenberg, B. and A. van Soest, 1996a, Parametric and semiparametric modeling of vacation expenditures, JAE 11, 59–76.

Melenberg, B. and A. van Soest, 1996b, Measuring the cost of children: parametric and semiparametric estimators, Statistica Neerlandica 50, 171–192.

Meng, C.L. and P. Schmidt, 1985, On the cost of partial observability in the bivariate probit model, IER 26, 71–85.

Mielniczuk, J., 1986, Some asymptotic properties of kernel estimators of a density function in case of censored data, AS 14, 766–773.

Miles, D. and J. Mora, 2003, On the performance of nonparametric specification tests in regression models, Computational Statistics and Data Analysis 42, 477–490.

Miranda, A., 2008, Planned fertility and family background: a quantile regression for counts analysis, Journal of Population Economics 21, 67–81.

Mizon, G.E. and J. Richard, 1986, The encompassing principle and its application to testing non-nested hypotheses, ECA 54, 657–678.

Moeltner, K. and J.S. Schonkwiler, 2005, Correcting for on-site sampling in random utility models, American Journal of Agricultural Economics 87, 327–339.

Moon, H.R., 2004, Maximum score estimation of a nonstationary binary choice model, JOE 122, 385–403.

Mullahy, J., 1997, Instrumental-variable estimation of count data models: applications to models of cigarette smoking behavior, REStat 79, 586–593.

Müller, H.G., 1988, Nonparametric regression analysis of longitudinal data, Lecture Notes in Statistics 46, Springer-Verlag.

Munk, A., N. Neumeyer and A. Scholz, 2007, Non-parametric analysis of covariance–the case of inhomogeneous and heteroscedastic noise, Scandinavian Journal of Statistics 34, 511–534.

Nadaraya, E.A., 1964, On estimating regression, Theory of Probability and Its Applications 9, 141–142.

Nahm, J.W., 2001, Nonparametric quantile regression analysis of R&D-sales relationship for Korean firms, Empirical Economics 26, 259–270.

Narendranathan, W. and M. Stewart, 1993, Modelling the probability of leaving unemployment: competing risks models with flexible base-line hazards, Applied Statistics 42, 63–83.

Nawata, K. and N. Nagase, 1996, Estimation of sample selection bias models, ER 15, 387–400.

Neumann, M.H. and J. Polzehl, 1998, Simultaneous bootstrap confidence bands in nonparametric regression, Journal of Nonparametric Statistics 9, 307–333.

Newey, W., 1985, Maximum likelihood specification testing and conditional moment tests, ECA 53, 1047–1070.

Newey, W., 1988, Adaptive estimation of regression models via moment restrictions, JOE 38, 301–339.

Newey, W., 1990a, Semiparametric efficiency bounds, JAE 5, 99–135.

Newey, W., 1990b, Efficient instrumental variables estimation of nonlinear models, ECA 58, 809–837.

Newey, W., 1991, Efficient estimation of Tobit models under conditional symmetry, in W.A. Barnett, J. Powell and G. Tauchen, eds., Nonparametric and Semiparametric Methods in Econometrics and Statistics, Cambridge University Press.

Newey, W., 1993, Efficient estimation of models with conditional moment restrictions, in Handbook of Statistics 11, edited by G.S. Maddala, C.R. Rao, and H.D. Vinod, Elsevier Science Publishers.

Newey, W., 1994a, The asymptotic variance of semiparametric estimators, ECA 62, 1349–1382.

Newey, W., 1994b, Kernel estimation of partial means and a general variance estimator, ET 10, 233–253.

Newey, W., 1997, Convergence rates and asymptotic normality for series estimators, JOE 79, 147–168.

Newey, W., 2001, Conditional moment restrictions in censored and truncated regression models, ET 17, 863–888.

Newey, W. and D. McFadden, 1994, Large sample estimation and hypothesis testing, Handbook of Econometrics IV, North-Holland.

Newey, W. and J.L. Powell, 1987, Asymmetric least squares estimation and testing, ECA 55, 819–847.

Newey, W. and J.L. Powell, 1990, Efficient estimation of linear and type I censored regression models under conditional quantile restrictions, ET 6, 295–317.

Newey, W. and J.L. Powell, 2003, Instrumental variable estimation of nonparametric models, ECA 71, 1565–1578.

Newey, W. J.L. Powell, and F. Vella, 1999, Nonparametric estimation of triangular simultaneous equations models, ECA 67, 565–603.

Newey, W. and R.J. Smith, 2004, High-order properties of GMM and generalized empirical likelihood estimators, ECA 72, 219–255.

Newey, W. and T.M. Stoker, 1993, Efficiency of weighted average derivative estimators and index models, ECA 61, 1199–1223.

Newey, W. and K. West, 1987, A simple positive semidefinite, heteroskedasticity and autocorrelation consistent covariance matrix, ECA 55, 703–708.

Newey, W. and K. West, 1994, Automatic lag selection in covariance matrix estimation, REStud 61, 631–654.

Nielsen, J.P. and S. Sperlich, 2005, Smooth backfitting in practice, JRSS-B 67, 43–61.

Nikitin, Y., 1995, Asymptotic efficiency of nonparametric tests, Cambridge University Press.

Nishiyama, Y. and P.M. Robinson, 2000, Edgeworth expansions for semiparametric average derivatives, ECA 68, 931–979.

Nolan, D. and D. Pollard, 1987, U-Processes: Rates of convergence, AS 15, 780–799.

Nolan, D. and D. Pollard, 1988, Functional limit theorems for U-processes, Annals of Probability 16, 1291–1298.

Orme, C., 1990, The small-sample performance of the information-matrix test, JOE 46, 309–331.

Ortúzar, J.D., 2001, On the development of the of nested logit model, Transportation Research (Part B) 35, 213–216.

Owen, A.B., 2001, Empirical likelihood, CRC Press.

Pagan, A.R. and A. Ullah, 1988, The econometric analysis of models with risk terms, JAE 3, 87–105.

Pagan, A.R. and A. Ullah, 1999, Nonparametric econometrics, Cambridge University Press.

Pagan, A.R. and F. Vella, 1989, Diagnostic tests for models based on individual data, JAE 4, S29–S59.

Papke, L.E. and J.M. Wooldridge, 1996, Econometric methods for fractional response variables with an application to 401 (k) plan participation rates, JAE 11, 619–632.

Park, B.U. and J.S. Marron, 1990, Comparison of data-driven bandwidth selectors, JASA 85, 66–72.

Park, B.U. and B.A. Turlach, 1992, Practical performance of several data driven bandwidth selector, Computational Statistics 7, 251–270.

Pesaran, M.H. and F.J. Ruge-Murcia, 1999, Analysis of exchange-rate target zones using a limited-dependent rational-expectations model with jumps, JBES 17, 50–66.

Pesaran, M.H. and R.J. Smith, 1994, A generalized R^2 criterion for regression models estimated by the instrumental variables method, ECA 62, 705–710.

Pesaran, M.H. and M. Weeks, 2001, Nonnested hypothesis testing: an overview, in A Companion to Theoretical Econometrics, edited by B.H. Baltagi, Blackwell.

Phillips, P.C.B and J.Y. Park, 1988, On the formulation of Wald tests of nonlinear restrictions, ECA 56, 1065–1083.

Picone, G.A. and J.S. Butler, 2000, Semiparametric estimation of multiple equation models, ET 16, 551–575.

Pierce, D.A., 1982, The asymptotic effect of substituting estimators for parameters in certain types of statistics, AS 10, 475–478.

Pinkse, C.A.P., 1993, On the computation of semiparametric estimates in limited dependent variable models, JOE 58, 185–205.

Pollard, D., 1984, Convergence of Stochastic Processes, Springer-Verlag.

Pollard, D., 2002, A user's guide to measure theoretic probability, Cambridge University Press.

Pötcher, B.M. and I.R. Prucha, 1989, A uniform LLN for dependent and heterogeneous data processes, ECA 57(3), 675–683.

Powell, J.L., 1984, Least absolute deviations estimation for the censored regression model, JOE 25, 303–325.

Powell, J.L., 1986a, Symmetrically trimmed least squares estimation for Tobit models, ECA 54, 1435–1460.

Powell, J.L., 1986b, Censored regression quantiles, JOE 32, 143–155.

Powell, J.L., 1994, Estimation of semiparametric models, in Handbook of Econometrics IV, edited by R.F. Engle and D.L. McFadden, Elsevier Science.

Powell, J.L., J.H. Stock, and T.S. Stoker, 1989, Semiparametric estimation of index coefficients, ECA 57, 1403–1430.

Powel, J.L. and T.M. Stoker, 1996, Optimal bandwidth choice for density-weighted averages, JOE 75, 291–316.

Prakasa Rao, B.L.S., 1983, Nonparametric Functional Estimation, Academic Press.

Pratt, J.W., 1981, Concavity of the log likelihood, JASA 76, 103–106.

Press, W.H., B.P. Flannery, and S.A. Teukolsky, 1986, Numerical Recipes: The Art of Scientific Computing, Cambridge University Press.

Protter, P.E., 2005, Stochastic integration and differential equations, 2nd ed., Springer.

Pudney, S. and M. Shields, 2000, Gender, race, pay and promotion in the British nursing profession: estimation of a generalized ordered probit model, JAE 15, 367–399.

Qin, J. and J. Lawless, 1994, Empirical likelihood and general estimating equations, AS 22, 300–325.

Qiu, P., 2005, Image processing and jump regression analysis, Wiley.

Racine, J. and Q. Li, 2004, Nonparametric estimation of regression functinos with both categorical and continous data, JOE 119, 99–130.

Ramlau-Hanse, H., 1983, Smoothing counting process intensities by means of kernel functions, AS 11, 453–466.

Ramsey, J.B., 1969, Tests for specification errors in classical linear least squares regression analysis, JRSS-B 31, 350–371.

Rayner, J.C.W. and D.J. Best, 1990, Smooth tests of goodness of fit: an overview, International Statistical Review 58, 9–17.

Ridder, G., 1990, The non-parametric identification of generalized accelerated failure-time models, REStud 57, 167–181.

Rilstone, P., 1991, Nonparametric hypothesis testing with parametric rates of convergence, IER 32, 209–227.

Robins, J. and A.A. Tsiatis, 1992, Semiparametric estimation of an accelerated failure time model with time-dependent covariates, Biometrika 79, 311–319.

Robinson, P.M., 1987, Asymptotically efficient estimation in the presence of heteroskedasticity of unknown form, ECA 55, 875–891.

Robinson, P.M., 1988, Root-N consistent semiparametric regression, ECA 56, 931–954.

Robinson, P.M., 1991, Best nonlinear three-stage least squares estimation of certain econometric models, ECA 59, 755–786.

Rodríguez-Póo, J.M., S. Sperlich and A.I. Fernández, 2005, Semiparametric three-step estimation methods for simultaneous equation systems, JAE 20, 699–721.

Rosenblatt, M., 1956, Remarks on some nonparametric estimates of a density function, Annals of Mathematical Statistics 27, 832–837.

Rosenblatt, M., 1991, Stochastic curve estimation, NSF-CBMS Regional Conference Series in Probability and Statistics, Vol 3, Institute of Mathematical Statistics, Hayward, California.

Royden, H.L., 1988, Real analysis, 3rd ed., Prentice Hall.

Rudemo, M., 1982, Empirical choice of histograms and kernel density estimators, Scandinavian Journal of Statistics 9, 65–78.

Ruppert, D. and M.P. Wand, 1994, Multivariate locally weighted least squares regression, AS 22, 1346–1370.

Ruppert, D., M.P. Wand, and R.J. Carroll, 2003, Semiparametric regression, Cambridge University Press.

Saha, A. and L. Hilton, 1997, Expo-power: a flexible hazard function for duration data models, Economics Letters 54, 227–233.

Sakata, S., 2007, Instrumental variable estimation based on conditional median restriction, JOE 141, 350–382.

Santos-Silva, J.M.C., 2001a, Influence diagnostics and estimation algorithms for Powell's SCLS, JBES 19, 55–62.

Santos Silva, J.M.C., 2001b, A score test for non-nested hypotheses with applications to discrete data models, JAE 16, 577–597.

Schaefer, S., 1998, The dependence of pay-performance sensitivity on the size of the firm, REStat 80, 436–443.

Schellhorn, M., 2001, The effect of variable health insurance deductibles on the demand for physician visits, Health Economics 10, 441–456.

Schott, J.R., 2005, Matrix analysis for statistics, 2nd ed., Wiley.

Scott, D.W., 1992, Multivariate Density Estimation, Wiley.

Searle, S.R., 1982, Matrix algebra useful for statistics, Wiley.

Seetharaman, P.B. and P.K. Chintagunta, 2003, The proportional hazard model for purchase timing: a comparison of alternative specifications, JBES 21, 368–382.

Sen P.K. and J.M. Singer, 1993, Large sample methods in statistics, Chapman and Hall.

Serfling, R., 1980, Approximation Theorems of Mathematical Statistics, Wiley.

Severance-Lossin, E. and S. Sperlich, 1999, Estimation of derivatives for additive separable models, Statistics 33, 241–265.

Severini, T.A. and J.G. Staniswalis, 1994, Quasi-likelihood estimation in semiparametric models, JASA 89, 501–511.

Severini, T.A. and G. Tripathi, 2006, Some identification issues in nonparametric linear models with endogenous regressors, ET 22, 258–278.

Severini, T.A. and W.H. Wong, 1992, Generalized profile likelihood and conditionally parametric models, AS 20, 1768–1802.

Shao, J., 2003, Mathematical statistics, 2nd ed., Springer.

Shao, J. and D. Tu, 1995, The jackknife and bootstrap, Springer.

Shaw, D., 1988, On-site samples regression: problems of non-negative integers, truncation and endogenous stratification, JOE 37, 211–223.

Sheather, S.J. and M.C. Jones, 1991, A reliable data-based bandwidth selection method for kernel density estimation, JRSS-B 53, 683–690.

Sherman, R., 1993, The limiting distribution of the maximum rank correlation estimator, ECA 61, 123–137.

Sherman, R., 1994, Maximal inequalities for degenerate U-processes with applications to optimization estimators, AS 22, 439–459.

Shorack, G.R., 2000, Probability for statisticians, Springer.

Shorack, G.R. and J.A. Wellner, 1986, Empirical processes with applications to statistics, Wiley.

Shoung, J.M. and C.H. Zhang, 2001, Least squares estimators of the mode of a unimodal regression function, AS 29, 648–665.

Silverman, B.W., 1981, Using kernel density estimates to investigate multimodality, JRSS-B 43, 97–99.

Silverman, B.W., 1984, Spline smoothing: the equivalent variable kenel method, AS 12, 898–916.

Silverman, B.W., 1986, Density Estimation for Statistics and Data Analysis, Chapman and Hall.

Smith M.D., 2002, On specifying double hurdle models, in Handbook of Applied Econometrics and Statistical Inference, edited by A. Ullah, A. Wan and A. Chaturvedi, Marcel-Dekker.

Smith, R.J., 2005, Automatic positive semidefinite HAC covariance matrix and GMM estimation, ET 21, 158–170.

Sperlich, S., O.B. Linton, W. Härdle, 1999, Integration and backfitting methods in additive models—finite sample properties and comparison, Test 8, 419–458.

Sperlich, S., D. Tjostheim, and L. Yang, 2002, Nonparametric estimation and testing of interaction in additive models, ET 18, 197–251.

Staniswalis, J.G., 1989, The kernel estimate of a regression function in likelihood-based models, JASA 84, 276–283.

Staniswalis, J.G. and T.A. Severini, 1991, Diagnostics for assessing regression models, JASA 86, 684–692.

Steele, M.J., 2001, Stochastic calculus and financial applications, Springer.

Stern, S., 1997, Simulation-based estimation, Journal of Economic Literature 35, 2006–2039.

Stewart, M.B., 2005, A comparison of semiparametric estimators for the ordered response model, Computational Statistics and Data Analysis 49, 555–573.

Stock, J.H., 1991, Nonparametric policy analysis: an application to estimating hazardous waste cleanup benefits, in Nonparametric and Semiparametric Methods in Econometrics and Statistics, edited by W.A. Barnett, J. Powell and G.E. Tauchen, Cambridge University Press, 77–98.

Stock, J.H. and F. Trebbi, 2003, Retrospectives: who invented instrumental variable regression?, Journal of Economic Perspectives 17, 177–194.

Stock, J.H., J.H. Wright, M. Yogo, 2002, A survey of weak instruments and weak identification in generalized method of moments, JBES 20, 518–529.

Stoker, T.M., 1986, Consistent estimation of scaled coefficients, ECA 54, 1461–1481.

Stoker, T.M., 1991, Equivalence of direct, indirect and slope estimators of average derivatives, in W.A. Barnett, J. Powell, and G. Tauchen, eds., Nonparametric and Semiparametric Methods in Econometrics and Statistics, Cambridge University Press.

Stone, C.J., 1982, Optimal global rates of convergence for nonparametric regression, AS 10, 1040–1053.

Stone, C.J., 1984, An asymptotically optimal window selection rule for kernel density estimates, AS 12, 1285–1297.

Stone, C.J., 1985, Additive regression and other nonparametric models, AS 13, 689–705.

Stute, W., 1984, Asymptotic normality of nearest neighbor regression function estimates, AS 12, 917–926.

Stute, W., 1997, Nonparametric model checks for regression, AS 25, 613–641.

Stute, W., W. González Manteiga, and M.P. Quindimil, 1998, Bootstrap approximation in model checks for regression, JASA 93, 141–149.

Stute, W., S. Thies, and L.X. Zhu, 1998, Model checks for regression: an innovation process approach, AS 26, 1916–1934.

Stute, W., W.L. Xu and L.X. Zhu, 2008, Model diagnosis for parametric regression in high-dimensional spaces, Biometrika 95, 451–467.

Stute, W. L.X. Zhu, 2005, Nonparametric checks for single-index models, AS 33, 1048–1083.

Sun, J., 2006, A consistent nonparametric equality test of conditional quantile functions, ET 22, 614–632.

Susarla, V., W.Y. Tsai and J. Van Ryzin, 1984, A Buckley-James-type estimator for the mean with censored data, Biometrika 71, 624–625.

Tamer, E., 2003, Incomplete simultaneous discrete response model with multiple equilibria, REStud 70, 147–165.

Tanner, M.A. and W.H. Wong, 1983, The estimation of the hazard function from randomly censored data by the kernel method, AS 11, 989–993.

Tauchen, G., 1985, Diagnostic testing and evaluation of maximum likelihood models, JOE 30, 415–443.

Terza, J.V., 1985, Ordinal probit: a generalization, Communication in Statistics: Theory and Method 14, 1–11.

Terza, J.V., 1998, Estimating count data models with endogenous switching: sample selection and endogenous treatment effects, JOE 84, 129–154.

Therneau, T.M. and P.M. Grambsch, 2000, Modeling survival data: extending the Cox model, Springer.

Thompson, S.K., 2002, Sampling, 2nd ed., Wiley.

Tibshirani, R., 1988, Estimating transformations for regression via additivity and variance stabilization, JASA 83, 394–405.

Tobin, J., 1958, Estimation of relationships for limited dependent variables, ECA 26, 24–36.

Train, K.E., 2003, Discrete choice methods with simulation, Cambridge University Press.

Tsiatis, A., 1975, A nonidentifiability aspect of the problem of competing risks, Proceedings of the National Academy of Sciences 72 (1), 20–22.

Van den Berg, G.J., 2001, Duration models: specification, identification and multiple durations, in Handbook of Econometrics V, edited by J.J. Heckman and E. Leamer, North-Holland.

Van der Vaart, A.W., 1998, Asymptotic statistics, Cambridge University Press.

Van der Vaart, A.W. and J.A. Wellner, 1996, Weak convergence and empirical processes, Springer.

Van Soest, A. and, A. Kapteyn, and P. Kooreman, 1993, Coherency and regularity of demand systems with equality and inequality constraints, JOE 57, 161–188.

Veall, M.R. and K.F. Zimmermann, 1996, Pseudo R^2 measures for some common limited dependent variable models, Journal of Economic Surveys 10, 241–259.

Vella, F. and M. Verbeek, 1998, Whose wages do unions raise? A dynamic model of unionism and wage determination for young men, JAE 13, 163–183.

Vinod, H.D. and A. Ullah, 1988, Flexible production function estimation by nonparametric kernel estimators, Advances in Econometrics 7, 139–160, JAI Press.

Vollebergh, H.R.J., B. Melenberg and E. Dijkgraaf, 2009, Identifying reduced-form relations with panel data: the case of pollution and income, Journal of Environmental Economics and Management 58, 27–42.

Vuong, Q.H., 1989, Likelihood ratio tests for model selection and non-nested hypotheses, ECA 57, 307–333.

Wand, M.P. and M.C. Jones, 1995, Kernel smoothing, Chapman and Hall.

Wasserman, L., 2006, All of nonparametric statistics, Springer.

Watson, G.S., 1964, Smooth regression analysis, Sankhya A 26, 359–372.

Werner, M., 1999, Allowing for zeros in dichotomous-choice contingent valuation models, JBES 17, 479–486.

Welsh, A.H., 1987, The trimmed mean in the linear model, AS 15, 20–36.

West, K.D., 1997, Another heteroskedasticity- and autocorrelation-consistent covariance matrix estimator, JOE 76, 171–191.

Whaba, G., 1990, Spline models for observational data, CBMS-NSF #59, SIAM Society.

Whang, Y.J., 2001, Consistent specification testing for conditional moment restrictions, Economics Letters 71.

Whang, Y.J. and D.W.K. Andrews, 1993, Tests of specification for parametric and semiparametric models, JOE 57, 277–318.

White, H., 1980, A heteroskedasticity-consistent covariance matrix estimator and a direct test for heteroskedasticity, ECA 48, 817–838.

White, H., 1982, Maximum likelihood estimation of misspecified models, ECA 50, 1–25.

White, H. and I. Domowitz, 1984, Nonlinear regression with dependent observations, ECA 52, 143–161.

Williams, H.C.W.L., 1977, On the formation of travel demand models and economic evaluation measures of user benefit, Environment and Planning A 9, 285–344.

Windmeijer, F., 2005, A finite sample correction for the variance of linear efficient two-step GMM estimators, JOE 126, 25–51.

Windmeijer, F. and J.M. Santos-Silva, 1997, Endogeneity in count data models: an application to demand for health care. JAE 12, 281–294.

Winkelmann, R., 2003, Econometric analysis of count data, 4th ed., Springer.

Winkelmann, R., 2006, Reforming health care: evidence from quantile regressions for counts. Journal of Health Economics 25, 131–145.

Winter-Ebmer, R. and J. Zweimuller, 1997, Unequal assignment and unequal promotion in job ladders, Journal of Labor Economics 15, 43–71.

Wooldridge, J.M., 1992a, Some alternatives to the Box-Cox regression model, IER 33, 935–955.

Wooldridge, J.M., 1992b, A test for functional form against nonparametric alternatives, ET 8, 452–475.

Wooldridge, J.M., 1999, Asymptotic properties of weighted M-estimators for variable probability samples, ECA 67, 1385–1406.

Wooldridge, J.M., 2001, Asymptotic properties of weighted M-estimators for standard stratified samples, ET 17, 451–470.

Wooldridge, J.M., 2002, Econometric analysis of cross section and panel data, MIT Press.

Xia, Y., W.K. Li, H. Tong, and D. Zhang, 2004, A goodness-of-fit test for single-index models, Statistica Sinica 14, 1–39.

Xia, Y., H. Tong, W.K. Li, and L.X. Zhu, 2002, An adaptive estimation of dimension reduction space, JRSS-B 64, 363–410.

Yang, S., 1999, Censored median regression using weighted empirical survival and hazard functions, JASA 94, 137–145.

Yatchew, A., 1992, Nonparametric regression tests based on least squares, ET 8, 435–451.

Yatchew, A., 1997, An elementary estimator for the partial linear model, Economics Letters 57, 135–143.

Yatchew, A., 1999, An elementary nonparametric differencing test of equality of regression functions, Economics Letters 62, 271–278.

Yatchew, A., 2003, Semiparametric regression for the applied econometrician, Cambridge University Press.

Yatchew, A. and J.A. No, 2001, Household gasoline demand in Canada, ECA 69, 1697–1709.

Yatchew, A., Y. Sun and C. Deri, 2003, Efficient estimation of semiparametric equivalence scales with evidence from South Africa, JBES 21, 247–257.

Yin, G., D. Zeng and H. Li, 2008, Power-transformed linear quantile regression with censored data, JASA 103, 1214–1224.

Zeng, D. and D.Y. Lin, 2007, Efficient estimation for the accelerated failure time model, JASA 102, 1387–1396.

Zhang, J. and Y. Peng, 2007, A new estimation method for the semiparametric accelerated failure time mixture cure model, Statistics in Medicine 26, 3157–3171.

Zheng, J.X., 1996, A consistent test of functional form via nonparametric estimation techniques, JOE 75, 263–289.

Zheng, J.X., 1998, A Consistent nonparametric test of parametric regression models under conditional quantile restrictions, ET 14, 123–138.

Zhu, F., 2005, A nonparametric analysis of the shape dynamics of the US personal income distribution: 1962–2000, Bank for International Settlements, Working Paper 184.

Index